ENCYCLOPEDIA OF PHYSICS

EDITED BY

S. FLÜGGE

VOLUME IX

FLUID DYNAMICS III

CO-EDITOR

C. TRUESDELL

WITH 248 FIGURES

SPRINGER-VERLAG
BERLIN · GÖTTINGEN · HEIDELBERG
1960

HANDBUCH DER PHYSIK

HERAUSGEGEBEN VON

S. FLÜGGE

BAND IX

STRÖMUNGSMECHANIK III

MITHERAUSGEBER

C. TRUESDELL

MIT 248 FIGUREN

SPRINGER-VERLAG
BERLIN · GÖTTINGEN · HEIDELBERG
1960

ISBN-13: 978-3-642-45946-7 e-ISBN-13: 978-3-642-45944-3
DOI: 10.1007/978-3-642-45944-3

Softcover reprint of the hardcover 1st edition 1960

Contents.

Analytical Theory of Subsonic and Supersonic Flows.

By

M. Schiffer.

With 24 Figures.

Introduction.

This article deals with the mathematical theory of motion of a compressible fluid. On account of the non-linear character of the differential equations involved, an analytical treatment of the flow under most general circumstances has not yet been possible. Therefore, in order to facilitate treatment we have throughout made various idealizing assumptions. In particular we neglect viscosity, thermal conductivity and any external forces acting on the fluid and, moreover, we restrict ourselves to stationary irrotational flows. Even with these simplifications the deeper mathematical problems regarding the existence and uniqueness theory have been satisfactorily treated only under the additional requirement of plane flow. And even in this special theory we ran into some unsolved problems of the theory of partial differential equations of mixed elliptic-hyperbolic type which are connected with the transition from subsonic to supersonic flow regime. We have not touched upon the theory of shock discontinuities in the flow as this theory is developed systematically in the article by Cabannes.

The structure of the article is as follows: In the first chapter, we give the basic physical and mathematical concepts of the theory and the fundamental equations of motion. Some explicit solutions illustrate the general theory.

The second chapter deals with the theory of linearized flows. This is a theory of approximations applicable to thin or slender bodies in an otherwise uniform flow. In this case, the fundamental non-linear equations can be replaced by linear differential equations with constant coefficients. Since only a small number of explicit solutions for the correct fundamental equations is known, it is important to possess approximation methods which permit qualitative statements on the nature of the flow considered. The theory of linearization is applicable to flows in the plane and in three-dimensional space and is extensively used in airfoil theory and other branches of applied aerodynamics.

The third chapter is devoted to the hodograph method. This more refined method achieves a rigorous linearization of the differential equations of motion of a compressible fluid. It works, however, only in the case of plane flow and is, moreover, an indirect procedure. Indeed, it yields exact solutions of the correct equations, but not necessarily those solutions which correspond to the desired boundary conditions. The adjustment to the correct side conditions implies approximation procedures which are quite involved. However, at present the hodograph method seems the most promising analytical tool in the theory of plane flows.

The fourth chapter deals with the uniqueness and existence theory of plane subsonic flows. We describe the standard analytical methods of this theory and sketch some fundamental proofs in detail sufficient to allow an understanding of the underlying principles and methods. Existence proofs may appear as rather esoteric to the applied aerodynamist, but a good existence proof contains the germs of a constructive procedure, and the role of the Janzen-Rayleigh iteration is considered in this context. Finally, we describe a method of obtaining inequalities and estimates for the velocity and pressure fields of a flow; it stands in analogy to the distortion theorems in conformal mapping which allow similar estimates in the case of an incompressible flow. When exact solutions are not easily available such estimates may be quite valuable.

The fifth chapter covers the theory of two-dimensional transonic flows. The methods and problems in this theory are described and, in particular, the Taylor problem of the existence of continuous transonic flows is discussed. While the existence of such flows for arbitrary boundaries seems dubious, explicit solutions can be given for special problems. We consider the method of idealized fluids which permits the construction of some transonic flow patterns of practical importance, such as flow through a nozzle.

We have tried to make the different chapters to a large extent independent of each other. In view of the extensive literature in the field, we have aimed less at completeness of results and methods than at a clear description of the basic ideas; for more detail we refer to the general references and textbooks listed at the end of the paper as well as to the bibliography, organized by chapters.

I. Physical and mathematical foundations.

1. Basic assumptions and fundamental equations. The motion of a fluid is described mathematically by the vector field of velocity $\boldsymbol{q}$ and the three scalar fields of pressure p, density ϱ and temperature T. These six quantities are considered as functions of the three space variables x, y, z and the time t.

We shall consider in this article only non-viscous fluids and neglect the influence of external forces like gravity. Then the equations of motion are

$$\varrho \frac{D}{Dt} \boldsymbol{q} = - \nabla p. \tag{1.1}$$

The symbol D/Dt is the Eulerian or hydrodynamical derivative: it denotes differentiation of a field quantity relative to an observer moving with the flow. It takes into account, therefore, the spatial as well as the temporal change and is defined by the well-known operator formula:

$$\frac{D}{Dt} = \frac{\partial}{\partial t} + (\boldsymbol{q} \cdot \nabla). \tag{1.2}$$

We can thus give to (1.1) the alternative form

$$\frac{\partial}{\partial t} \boldsymbol{q} + (\boldsymbol{q} \cdot \nabla)\, \boldsymbol{q} = - \frac{1}{\varrho} \nabla p. \tag{1.1'}$$

The conservation of matter in the flow leads to another fundamental differential relation between the velocity and the density field, namely, the equation of continuity

$$\frac{\partial}{\partial t} \varrho + \nabla \cdot (\varrho\, \boldsymbol{q}) = 0. \tag{1.3}$$

The physical nature of the fluid considered enters the theory through the equation of state which connects p, ϱ and T. There are various other thermodynamical quantities like the internal energy per unit of mass U or the entropy per unit mass S. Any two of these variables can be used in order to express all the others in terms of them. In theoretical fluid dynamics it is particularly useful to take the entropy S as one basic variable. This is evident from the definition of the entropy change ΔS with a heat increase ΔQ

$$\Delta S = \frac{\Delta Q}{T} \tag{1.4}$$

and from the first fundamental law of thermodynamics

$$\Delta Q = \Delta U + p\,\Delta\left(\frac{1}{\varrho}\right). \tag{1.5}$$

It is often permissible to neglect heat conduction in the fluid and we shall do so consistently. We can then state that for a fixed unit of mass

$$\Delta Q = 0 \tag{1.6}$$

throughout the entire flow motion. This guarantees by Eq. (1.4) that the specific entropy of the moving matter remains constant:

$$\frac{DS}{Dt} = \frac{\partial}{\partial t} S + (\boldsymbol{q}\cdot\nabla)\, S = 0, \tag{1.7}$$

while we see from Eq. (1.5) that T will in general vary during the motion.

A flow which has a constant specific entropy for each moving particle is called an isentropic flow. The theory of most continuous flow phenomena can be carried out under the assumption of isentropy: only in the case of discontinuity surfaces and shock phenomena does the change of entropy play an important role and need to be taken into account.

The six Eqs. (1.1′), (1.3), (1.7) and the equation of state

$$p = f(\varrho, S) \tag{1.8}$$

are a system for the six unknown quantities $\boldsymbol{q}$, ϱ, p, S describing the motion of the fluid. They have to be complemented by proper initial and boundary conditions in order to make the problem of integration a well determined one.

In order to bring the basic equations of motion (1.1′) into a simpler form we make use of the identity of vector analysis

$$(\boldsymbol{q}\cdot\nabla)\,\boldsymbol{q} = \tfrac{1}{2}\nabla(q^2) - \boldsymbol{q}\times(\nabla\times\boldsymbol{q}). \tag{1.9}$$

This identity is useful since it introduces the vector

$$\boldsymbol{\zeta} = \nabla\times\boldsymbol{q} \tag{1.10}$$

which represents the vorticity of the flow and is significant for the flow pattern.

The meaning of the term $\frac{1}{\varrho}\nabla p$ is best understood from the thermodynamical relations (1.4) and (1.5). Since S, U, p and ϱ are well determined functions of the physical state variables in the fluid, we obtain after elimination of ΔQ the generally valid identity:

$$\nabla U = T\,\nabla S - p\,\nabla\left(\frac{1}{\varrho}\right). \tag{1.11}$$

We introduce now the specific enthalpy [1]

$$H = U + \frac{1}{\varrho}\,p \tag{1.12}$$

and bring Eq. (1.11) into the form

$$\nabla H = T \nabla S + \frac{1}{\varrho} \nabla p. \tag{1.11'}$$

Combining Eqs. (1.1'), (1.9) and (1.11'), we then obtain finally

$$\frac{\partial}{\partial t} \boldsymbol{q} + \nabla\left(H + \frac{1}{2} q^2\right) - \boldsymbol{q} \times \boldsymbol{\zeta} = T \nabla S. \tag{1.12'}$$

This formulation of the equations of motion is due to CROCCO; it is advantageous since it exhibits clearly the various quantities which have an immediate physical significance [2].

As an illustration of the usefulness of CROCCO's formula we shall derive from it the relation between entropy and vorticity in the flow. Let $C(t)$ be a closed curve moving with the flow, and let it be represented in the parametric form $\boldsymbol{r} = \boldsymbol{r}(s, t)$. The quantity

$$Z(t) = \oint_{C(t)} \boldsymbol{q}\, d\boldsymbol{r} = \int \boldsymbol{q} \frac{\partial}{\partial s} \boldsymbol{r}(s, t)\, ds \tag{1.13}$$

is called the circulation around the curve $C(t)$. From Eqs. (1.2), (1.9) and (1.12') we compute

$$\frac{D}{Dt} Z(t) = \int_s \frac{D}{Dt} \boldsymbol{q} \cdot \frac{\partial}{\partial s} \boldsymbol{r}(s, t)\, ds + \int_s \boldsymbol{q} \frac{\partial}{\partial s} \boldsymbol{q}\, ds = \oint_{C(t)} T \nabla S \cdot d\boldsymbol{r}, \tag{1.14}$$

since $\frac{D}{Dt}\left(\frac{\partial}{\partial s} \boldsymbol{r}\right) = \frac{\partial}{\partial s} \boldsymbol{q}$ on the curve $C(t)$ which moves with the flow and since q^2 and H are single-valued functions in space. The change of circulation is thus closely related to the specific entropy along the curve considered. On the other hand, we have by STOKES' theorem

$$Z = \oint_C \boldsymbol{q} \cdot d\boldsymbol{r} = \iint_\Sigma \boldsymbol{\zeta} \cdot \boldsymbol{n}\, d\sigma, \tag{1.15}$$

where Σ is an arbitrary surface spanned through the curve C and $\boldsymbol{n}$ is the local normal on Σ. This identity and Eq. (1.14) show the close relation between circulation, entropy and vorticity.

A flow in which the fluid has the same constant entropy S is called homentropic. Since in this case $\nabla S \equiv 0$, we have in every homentropic flow the Helmholtz-Kelvin circulation theorem

$$\frac{D}{Dt} Z = 0 \tag{1.16}$$

for every closed curve C.

If a flow is made up of particles which come all from a region with constant entropy it must be homentropic by virtue of the isentropy condition (1.7). If all particles pass through a region where, moreover, the vorticity vector vanishes identically we can conclude from Eqs. (1.16) and (1.15) that the flow has everywhere vorticity zero, that is:

$$\nabla \times \boldsymbol{q} = 0. \tag{1.17}$$

Such a flow is called irrotational.

We shall deal from now on with steady (that is, time independent), irrotational and homentropic flows. Our preceding considerations show that such flows will arise under very general assumptions and are of great importance; on the other hand, the assumptions made will lead to a simple mathematical theory which can be handled without too great complications.

Because of the simplifying assumptions, we are now dealing with the five functions $\boldsymbol{q}$, p and ϱ which depend only on the three space variables x, y, z. Between p and ϱ there exists the (adiabatic) equation of state

$$p = f(\varrho). \tag{1.18}$$

CROCCO's equation (1.12) reduces to

$$H + \tfrac{1}{2} q^2 = \text{const} \tag{1.19}$$

and the continuity equation has now the form

$$\nabla(\varrho \cdot \boldsymbol{q}) = 0. \tag{1.20}$$

In the homentropic case, Eq. (1.11′) simplifies to

$$H = H(\varrho) = \int \frac{dp}{\varrho} \tag{1.11''}$$

and (1.19) becomes the classical Bernoulli equation

$$\frac{1}{2} q^2 + \int \frac{dp}{\varrho} = \text{const}, \tag{1.19'}$$

which establishes the speed-density relation.

From the condition of irrotationality (1.17) follows the existence of a function $\varphi(x, y, z)$ such that

$$\boldsymbol{q} = \nabla\varphi. \tag{1.21}$$

φ is called the velocity potential of the flow. If we insert Eq. (1.21) into the continuity equation (1.20), we obtain the second order partial differential equation for the velocity potential

$$\nabla(\varrho \nabla\varphi) = 0. \tag{1.22}$$

The mathematical theory of a steady, homentropic and irrotational flow reduces, therefore, to the following procedure: (a) From the adiabatic equation of state (1.18) we compute the enthalpy function (1.11″). (b) By means of BERNOULLI's equation, we express the density as a function of the local speed

$$\varrho = P\big((\nabla\varphi)^2\big) = P(q^2). \tag{1.23}$$

(c) We insert Eq. (1.23) into Eq. (1.22) and obtain the partial differential equation

$$\nabla\big(P((\nabla\varphi)^2) \cdot \nabla\varphi\big) = 0 \tag{1.24}$$

which is to be integrated in accordance with specified boundary conditions.

The entire analytic theory of steady homentropic and irrotational flow is the theory of the non-linear partial differential equation (1.24) and its boundary value problems.

Because of the great importance of Eq. (1.24) for the whole theory we write it out in detail:

$$\varrho \nabla^2 \varphi + \frac{\partial\varrho}{\partial q^2} \cdot \nabla\big((\nabla\varphi)^2\big) \cdot \nabla\varphi = 0. \tag{1.25}$$

From Eq. (1.19′) we deduce

$$\frac{\partial\varrho}{\partial q^2} = -\frac{1}{2}\frac{\varrho}{c^2}, \qquad c^2 = \frac{dp}{d\varrho}. \tag{1.26}$$

The quantity $c = \sqrt{p'(\varrho)}$ is called the local speed of sound since it is the speed at which small perturbations travel, as is apparent from the linearized theory discussed in Chap. II. By means of Eq. (1.26), we obtain for Eq. (1.25) the form:

$$\left.\begin{aligned}\varphi_{xx}\left(1-\frac{\varphi_x^2}{c^2}\right)+\varphi_{yy}\left(1-\frac{\varphi_y^2}{c^2}\right)+\varphi_{zz}\left(1-\frac{\varphi_z^2}{c^2}\right)-\\ -2\varphi_{xy}\frac{\varphi_x\varphi_y}{c^2}-2\varphi_{xz}\frac{\varphi_x\varphi_z}{c^2}-2\varphi_{yz}\frac{\varphi_y\varphi_z}{c^2}=0.\end{aligned}\right\}\tag{1.27}$$

The Eq. (1.27) is non-linear; it is, however, quasi-linear since the highest order derivatives (2nd order) occur linearly. It reduces to LAPLACE's equation in the limit case $c=\infty$ which can, therefore, be considered as the case of an incompressible fluid. The deviation from LAPLACE's equation will obviously be small if the ratio

$$M^2=\frac{q^2}{c^2}=\frac{\varphi_x^2+\varphi_y^2+\varphi_z^2}{c^2}\tag{1.28}$$

is small.

The quantity $M=q/c$ is called the local Mach number of the flow at the point x, y, z considered. It is a well determined function of the local speed q since c depends on q only. Its importance for the general theory is already obvious in the form (1.27) of the fundamental partial differential equation and will become more so in the further developments.

2. The case of the ideal gas. The most important equation of state considered in gas dynamics is that of an ideal adiabatic gas. With a proper choice of units we have POISSON's pressure density relation:

$$p=\varrho^\gamma,\tag{2.1}$$

where γ is the ratio of the specific heats of the gas for constant pressure and for constant volume:

$$\gamma=\frac{c_p}{c_v}.\tag{2.1'}$$

In the case of air the value $\gamma=1.4$ is theoretically and experimentally satisfactory.

The enthalpy of a fluid with the pressure density relation (2.1) is by Eq. (1.11'')

$$H(\varrho)=\frac{\gamma}{\gamma-1}\varrho^{\gamma-1},\tag{2.2}$$

and the Bernoulli equation takes the form

$$q^2+\frac{2\gamma}{\gamma-1}\varrho^{\gamma-1}=\text{const}.\tag{2.3}$$

The significance of the right hand constant is obvious; it represents the maximum velocity of the fluid corresponding to the value $\varrho=0$ and is attained when the fluid flows into vacuum. This constant is called the escape velocity and is denoted by $q_{\max}$. We can derive from Eqs. (2.3) and (2.1) the formulas

$$\frac{\gamma-1}{2\gamma}[q_{\max}^2-q^2]=\varrho^{\gamma-1}\tag{2.4}$$

and

$$q^2=q_{\max}^2-\frac{2\gamma}{\gamma-1}p^{\frac{\gamma-1}{\gamma}}.\tag{2.4'}$$

We normalize the units in such a way that to the state of rest, i.e. to $q=0$, correspond the values $\varrho=1$, $p=1$. We then derive from Eq. (2.4')

$$q_{\max}=\sqrt{\frac{2\gamma}{\gamma-1}}\tag{2.5}$$

and can rewrite Eq. (2.4′) in the form

$$q^2 = q_{\max}^2\left(1 - p^{\frac{\gamma-1}{\gamma}}\right). \tag{2.4''}$$

This is the formula of de ST. VENANT and WANTZEL [3].

We obtain from Eq. (2.1) the formulas for the local speed of sound

$$c^2 = p'(\varrho) = \gamma\,\varrho^{\gamma-1} = \gamma\frac{p}{\varrho} = \gamma\,p^{\frac{\gamma-1}{\gamma}}. \tag{2.6}$$

In view of Eqs. (2.5) and (2.4″) we have the alternative forms

$$c^2 = \frac{\gamma-1}{2}(q_{\max}^2 - q^2) = \gamma - \frac{\gamma-1}{2}q^2. \tag{2.7}$$

c is thus expressed explicitly in terms of the local speed q. We observe that $q = q_{\max}$ implies $c = 0$ and that the maximum value of c

$$c_0 = \sqrt{\gamma} \tag{2.8}$$

is attained for the state of rest.

The speed q which coincides with its corresponding velocity of sound is called the critical speed q_{crit}. From (2.7) we compute

$$q_{\text{crit}} = \sqrt{\frac{2\gamma}{\gamma+1}} = \sqrt{\frac{2}{\gamma+1}}\,c_0. \tag{2.9}$$

To q_{crit} corresponds the Mach number $M = 1$; for $q < q_{\text{crit}}$ we have $M(q) < 1$ and for $q > q_{\text{crit}}$ holds $M(q) > 1$. A flow is called subsonic at a place where its local Mach number is less than 1 and supersonic where it is larger than 1. We have therefore the criterion

$$q < \sqrt{\frac{2\gamma}{\gamma+1}}, \quad \text{subsonic;} \qquad q > \sqrt{\frac{2\gamma}{\gamma+1}}, \quad \text{supersonic flow.} \tag{2.10}$$

It is of interest to point out a mathematical accident which is of considerable importance in gas dynamics. We can apply the binomial theorem and derive from Eq. (2.4″) the series development for the pressure in terms of the speed variable q:

$$p = \left[1 - \frac{\gamma-1}{2\gamma}q^2\right]^{\frac{\gamma}{\gamma-1}} = 1 - \frac{1}{2}q^2 + \frac{1}{8\gamma}q^4 + \cdots. \tag{2.11}$$

In the case of an incompressible fluid of constant density 1 and the pressure 1 at rest, we have the Bernoulli formula

$$p = 1 - \frac{1}{2}q^2. \tag{2.12}$$

It happens that the two series developments coincide in their first two terms. If we treat an ideal gas as incompressible, we commit an error in all pressure effects which is only of the order $c_0^{-2}q^4$ and relatively small for smaller Mach numbers. This is the reason why the theory of incompressible fluid flows is quite satisfactory for gas dynamics even up to the Mach number 0.5.

From Eqs. (2.7) and (2.5), we get the formula for the Mach number

$$M^2 = \frac{1}{\gamma}\,\frac{q^2}{1 - \frac{q^2}{q_{\max}^2}}, \qquad q^2 = \frac{\gamma M^2}{1 + \frac{\gamma-1}{2}M^2} \tag{2.13}$$

and from Eq. (2.4) we derive

$$\varrho^{\gamma-1} = \left[1 + \frac{\gamma-1}{2} M^2\right]^{-1}. \tag{2.13'}$$

Let us consider a steady flow of a compressible medium. All stream lines through a fixed small circle form a narrow tube, called a stream tube. In the flow regime, no fluid passes through the walls of a stream tube. Consider next the stream line through the center of the original circle and denote by $P(s)$ a variable point on it; the parameter s may be chosen as the arc length along this curve. Finally, let $\pi(s)$ be the plane through $P(s)$ which is orthogonal to the stream line.

The stream tube will cut off from $\pi(s)$ a cross section of area $A(s)$. We assume the tube so narrow that the variation of density and speed over each cross section may be neglected and we denote the corresponding values of density and speed by $\varrho(s)$ and $q(s)$. We may then formulate the law of conservation of matter as

$$A(s)\,\varrho(s)\,q(s) = \text{const}. \tag{2.14}$$

By logarithmic differentiation of Eq. (2.14) we obtain

$$\frac{dA}{A} + \frac{d\varrho}{\varrho} + \frac{dq}{q} = 0. \tag{2.15}$$

On the other hand, we may express $d\varrho$ and dq in terms of dp by use of Eqs. (2.1) and (2.3); we have

$$\frac{dp}{p} = \gamma \frac{d\varrho}{\varrho}, \qquad q\,dq + \frac{1}{\varrho}\,dp = 0 \tag{2.16}$$

and, hence, using Eq. (2.6), we may bring Eq. (2.15) into the form

$$\frac{dA}{A} = \frac{dp}{\varrho q^2}\left(1 - \frac{\varrho q^2}{\gamma p}\right) = \frac{dp}{\varrho q^2}\left(1 - \frac{q^2}{c^2}\right). \tag{2.17}$$

It is easily seen that in the case of an incompressible fluid we have instead of Eq. (2.17)

$$\frac{dA}{A} = \frac{dp}{\varrho q^2}. \tag{2.17'}$$

The significance of the Mach number $M = q/c$ in the comparison between compressible and incompressible fluid flow is obvious.

From Eqs. (2.16) and (2.17) we find

$$\frac{dA}{A} = -\frac{dq}{q}\left(1 - \frac{q^2}{c^2}\right). \tag{2.18}$$

The cross section of a stream tube decreases with increasing speed until the sonic velocity is attained. At this moment, A attains its minimum and increases thereafter with increasing supersonic speed. This shows the basic difference in flow geometry and dynamics between the subsonic and supersonic flow regimes. The preceding considerations play an important role in turbine theory and in the theory of Laval nozzles. One purpose of such a nozzle is to create of gas flow of high supersonic speed. In order to surpass the sonic velocity in a flow through a Laval nozzle, the latter must narrow down until sonic velocity is attained and widen again after this point in order to make possible a further increase in the speed of flow.

3. Some explicit solutions of the fundamental equations. In finding particular solutions for the differential equations (1.27) of the velocity potential, it is often useful to bear in mind its genesis from the continuity equation. Let us ask, for example, for a solution $\varphi(r)$ which depends only on the distance r from a fixed point O, say the origin. Since $\boldsymbol{q} = \nabla\varphi$, the corresponding flow will be radial; that is, its stream lines will be radii from O and its speed $q(r)$ will depend on r only.

The conservation of matter will be expressed by the formula

$$4\pi r^2 \varrho q = \text{const} \tag{3.1}$$

which guarantees that the same amount of matter enters per unit of time through the inner wall of every concentric spherical shell as leaves through the outer wall. This result could also have been obtained by integration of the equation of continuity.

From Eqs. (3.1), (2.13) and (2.13′) we compute easily

$$r^2 = C \cdot \frac{1}{M}\left[1 + \frac{\gamma - 1}{2} M^2\right]^{\frac{\gamma+1}{2(\gamma-1)}}. \tag{3.2}$$

We observe now that the right hand term in Eq. (3.2) cannot decrease indefinitely. It has a minimum for $M = 1$, namely

$$r_{\min} = C \cdot \left(\frac{\gamma + 1}{2}\right)^{\frac{\gamma+1}{2(\gamma-1)}}. \tag{3.2′}$$

Thus, a radially symmetric solution of Eq. (1.27) is only possible outside of some critical sphere. While in the theory of an incompressible fluid we have point sources and sinks giving rise to radial flows, in the theory of compressible fluids we find radial sources and sinks possessing a spherical nucleus inside of which the mathematical solution breaks down and where the physical idealization become inapplicable. Here for the first time we encounter the phenomenon of a limit surface beyond which the mathematical theory of the physical assumptions cannot be continued; we shall meet this situation later in a more general context; see Sect. 33.

We observe that the acceleration of a radial flow is given by

$$a = q \cdot \frac{dq}{dr} = q \cdot \frac{dq}{dM} \cdot \frac{dM}{dr}. \tag{3.3}$$

As long as $q < q_{\max}$ we have clearly $dq/dM > 0$. On the other hand, it follows from Eq. (3.2) that $dr/dM = 0$ for $M = 1$. Hence, if we approach the critical radius $r_{\min}$ the acceleration in the flow approaches infinity.

Flows of radial nature occur in conical pipes and are well-described by formula (3.2) over the range of validity of this solution.

In precisely the same way we could solve the problem of a plane radial flow. Using the law of conservation of matter, we would find

$$r = r_{\min} \frac{1}{M}\left[\frac{2}{\gamma + 1} + \frac{\gamma - 1}{\gamma + 1} M^2\right]^{\frac{\gamma+1}{2(\gamma-1)}}. \tag{3.4}$$

We would encounter a limit circle for the flow of radius $r_{\min}$ and corresponding to the Mach number $M = 1$.

Another explicit solution of the problem of a steady irrotational and homentropic flow is obtained when we consider a plane flow whose stream lines are concentric circles around a point O, say the origin, and where the speed $q(r)$

depends only on the distance r from O. We call such a flow a twodimensional vortex flow around O.

By the geometric nature of the flow the conservation of matter is automatically fulfilled and we have now only to guarantee the irrotationality of the motion. We observe that by Eq. (1.13) the circulation over a circle of radius r is given by

$$Z = 2\pi r q(r). \tag{3.5}$$

If the vorticity ζ is identically zero in every concentric ring around 0, the application of Eq. (1.15) to the ring domain shows immediately that Z is a constant. Thus Eq. (3.5) represents $q(r)$ as a function of r; by use of Eq. (2.13) we have

$$r^2 = C\,\frac{2+(\gamma-1)\,M^2}{M^2}. \tag{3.6}$$

We see again that the radius r cannot decrease indefinitely but has the minimum

$$r_{\min}^2 = (\gamma-1)\,C \tag{3.7}$$

which corresponds to the Mach number infinity, that is, to the escape velocity $q_{\max}$.

If we write

$$r^2 = r_{\min}^2\left[1+\frac{2}{\gamma-1}\cdot\frac{1}{M^2}\right] \tag{3.6'}$$

we see clearly that the two-dimensional vortex flow is subsonic for

$$r > r_{\min}\sqrt{\frac{\gamma+1}{\gamma-1}} \tag{3.8}$$

and supersonic for

$$r < r_{\min}\sqrt{\frac{\gamma+1}{\gamma-1}}. \tag{3.8}$$

Since by Eq. (2.4) $q=q_{\max}$ implies the value $\varrho=0$, we see that the two-dimensional vortex flow has a vacuum core of radius $r_{\min}$.

The two plane flows considered so far are both radially symmetric; that is, in both the velocity vector $\boldsymbol{q}(r)$ depends only upon the distance from the origin. It is equally easy to determine the most general radially symmetric flow [*4*] to [*6*] by applying at the same time the conditions of irrotationality and of conservation of matter. For this purpose decompose the velocity $\boldsymbol{q}$ into the radial component $a(r)$ and the angular component $b(r)$. The law of conservation of matter requires

$$2\pi r\varrho a(r) = C \tag{3.9}$$

while the condition of irrotationality or constant circulation affects only $b(r)$:

$$2\pi r b(r) = Z. \tag{3.10}$$

Since $q^2 = a^2 + b^2$, we derive from Eqs. (2.13), (2.13'), (3.9) and (3.10):

$$r^2 = \frac{C^2}{4\pi^2\gamma}\,\frac{1}{M^2}\left[1+\frac{\gamma-1}{2}M^2\right]^{\frac{\gamma+1}{(\gamma-1)}} + \frac{Z^2}{4\pi^2\gamma}\cdot\frac{1}{M^2}\left[1+\frac{\gamma-1}{2}M^2\right]. \tag{3.11}$$

It is again apparent that r has a minimum value $r_{\min}$, which can be derived from Eq. (3.11) by differentiation. We obtain a more intuitive characterization by using the identity

$$4\pi^2 r^2 q^2 = \frac{C^2}{\varrho^2} + Z^2 \tag{3.12}$$

and differentiating it with respect to q^2; using Eq. (1.26) we find

$$4\pi^2 q^2 \frac{dr^2}{dq^2} + 4\pi^2 r^2 = + \frac{C^2}{\varrho^2 c^2}. \qquad (3.12')$$

For the minimum value of r, the first term in Eq. (3.12') must vanish and we find

$$4\pi^2 r_{\min}^2 \varrho^2 c^2 = C^2. \qquad (3.12'')$$

Comparing Eqs. (3.9) with (3.12'') we see that the minimum radius is attained when the radial speed equals the speed of sound, that is, for sonic radial speed.

There are two values of M^2 possible for a given value of r^2; the one M-value corresponds to subsonic radial speed and the other to supersonic values of $a(r)$. Two different flow patterns, therefore, are possible according to the branch $M(r)$ chosen in the domain $r > r_{\min}$.

In order to calculate the streamlines of the flow obtained, we start with the differential equation

$$\frac{d\vartheta}{dr} = \frac{b(r)}{r\,a(r)} = \frac{Z}{C} \cdot \frac{\varrho}{r}. \qquad (3.13)$$

Since ϱ is a simple function of M and the relation between r and M is given by Eq. (3.11) we can always find the function $\vartheta(r)$ describing the streamlines in polar coordinates. The calculation becomes quite elementary when we choose $\gamma = 1.4$, that is, $\frac{7}{5}$ exactly, and introduce the variable

$$l = (1 + \tfrac{1}{5} M^2)^{-\frac{1}{2}}. \qquad (3.14)$$

We find from Eqs. (2.13') and (3.11)

$$\varrho = l^5, \qquad r^2 = \frac{\alpha^2 l^{-10} + \beta^2}{1 - l^2} \qquad (3.15)$$

and from Eq. (3.13) we can compute $\vartheta(l)$ in terms of elementary functions. We find

$$\vartheta = \frac{\beta}{\alpha}\left[\log\sqrt{\frac{1+l}{1-l}} - l - \frac{1}{3} l^3 - \frac{1}{5} l^5\right] - \arctan\frac{\beta}{\alpha} l^5 + \text{const}. \qquad (3.15')$$

All streamlines are obtained from a representative one by turning by a fixed angle since the general equation is $\vartheta = \vartheta_0(r) + \text{const}$.

The flow described by Eqs. (3.15), (3.15') has streamlines in forms of spirals starting from the limit circle $r = r_{\min}$. For a detailed description, see Ringleb [7]. See also [8], [9].

4. Plane and axially symmetric flows. A useful simplification of the partial differential equation for the velocity potential φ is possible in the case of a two-dimensional flow. Here φ depends only on the two variables, x, y, and the Eq. (1.22) takes the form

$$\frac{\partial}{\partial x}(\varrho\,\varphi_x) + \frac{\partial}{\partial y}(\varrho\,\varphi_y) = 0. \qquad (4.1)$$

This can be interpreted as the integrability condition for a new function $\psi(x, y)$ such that

$$\frac{\partial\psi}{\partial x} = -\varrho\frac{\partial\varphi}{\partial y} = -\varrho\,v, \qquad \frac{\partial\psi}{\partial y} = \varrho\frac{\partial\varphi}{\partial x} = \varrho\,u. \qquad (4.2)$$

The hydrodynamical derivative of ψ is

$$\frac{D\psi}{Dt} = \psi_x u + \psi_y v = \varrho(-\varphi_x\varphi_y + \varphi_x\varphi_y) = 0; \qquad (4.3)$$

that is, $\psi(x, y)$ is constant along each streamline of the flow. $\psi(x, y)$ is called the *stream function* of the flow.

If C is an arbitrary curve in the (x, y)-plane connecting the points x_0, y_0 and x_1, y_1 the integral

$$Q = \int_{x_0, y_0}^{x_1, y_1} \varrho(u\,dy - v\,dx) \tag{4.4}$$

extended over C represents the amount of matter carried per unit of time across this curve. By virtue of Eq. (4.2) we can write

$$Q = \psi(x_1, y_1) - \psi(x_0, y_0). \tag{4.4'}$$

Thus, the difference of the stream function values at two points represents the flux of matter through any curve connecting them. This is another intuitive interpretation of the stream function which reveals its close connection with the continuity equation.

If we eliminate ψ from the system of first order differential equations (4.2) we obtain again Eq. (4.1), and by virtue of Eq. (1.26) follows

$$\left(1 - \frac{\varphi_x^2}{c^2}\right)\varphi_{xx} + \left(1 - \frac{\varphi_y^2}{c^2}\right)\varphi_{yy} - 2\frac{\varphi_x \varphi_y}{c^2}\varphi_{xy} = 0, \tag{4.5}$$

which is a particular case of Eq. (1.27). In order to eliminate φ from the system (4.2) we observe the identity

$$(\psi_x^2 + \psi_y^2) = \varrho^2 q^2. \tag{4.6}$$

We differentiate this with respect to x and y and use the differential relation (1.26). We obtain

$$\psi_x \psi_{xx} + \psi_y \psi_{xy} = \varrho_x \varrho (q^2 - c^2), \qquad \psi_x \psi_{xy} + \psi_y \psi_{yy} = \varrho_y \varrho (q^2 - c^2). \tag{4.7}$$

On the other hand, we obtain from Eq. (4.2) directly

$$\varrho(\psi_{xx} + \psi_{yy}) - (\varrho_x \psi_x + \varrho_y \psi_y) = 0. \tag{4.8}$$

Combining Eqs. (4.7) with (4.8) we are led finally to the differential equation for the stream function

$$\psi_{xx}\left(1 - \frac{\psi_y^2}{c^2\varrho^2}\right) + \psi_{yy}\left(1 - \frac{\psi_x^2}{c^2\varrho^2}\right) + 2\psi_{xy}\frac{\psi_x \psi_y}{c^2\varrho^2} = 0. \tag{4.9}$$

In this equation $c\varrho$ is to be considered as a function of $\psi_x^2 + \psi_y^2$ computed by means of Eq. (4.6) and the known relations between c, ϱ and q.

The stream function is particularly useful in the study of flows with prescribed fixed boundaries. The presence of rigid walls subjects the velocity potential to the boundary condition $\partial\varphi/\partial n = 0$, which guarantees that the flow does not cross the boundary; the stream function, on the other hand, satisfies the boundary condition $\psi = \text{const}$, which means that the boundary curve must be a streamline. In general, it is much easier to find the solution of a partial differential equation with prescribed boundary values (DIRICHLET's problem) than to find a solution with specified normal derivative on the boundary (NEUMANN's problem). This fact explains the importance of the stream function in the theory of two-dimensional flows around given profiles.

The significance of the system (4.2) is well illustrated by the limit case of an incompressible fluid with density 1. In this case, the system reduces to the classical Cauchy-Riemann equations

$$\varphi_x = \psi_y, \qquad \varphi_y = -\psi_x \tag{4.10}$$

and the Eqs. (4.5) and (4.9) coincide with the Laplace equation. The system (4.2) will have solutions which are qualitatively similar to those of Eqs. (4.10) as long as the flow is subsonic. This similarity will appear clearly in the sequel.

Let us consider next an axially symmetric flow; that is, a flow whose velocity potential depends only on the two variables x and $r=\sqrt{y^2+z^2}$. The entire flow is symmetric with respect to the x-axis and is completely described by its pattern in the meridian plane x, r. The differential equation (1.22) can be easily computed under the assumption that everything depends on x and r only; we fid

$$\frac{\partial}{\partial x}\left(\varrho\frac{\partial\varphi}{\partial x}\right)+\frac{\partial}{\partial r}\left(\varrho\frac{\partial\varphi}{\partial r}\right)+\frac{\varrho}{r}\frac{\partial\varphi}{\partial r}=0 \tag{4.11}$$

which can be simplified to

$$\frac{\partial}{\partial x}\left(\varrho\, r\frac{\partial\varphi}{\partial x}\right)+\frac{\partial}{\partial r}\left(\varrho\, r\frac{\partial\varphi}{\partial r}\right)=0. \tag{4.11'}$$

This equation may again be understood as an integrability condition. There must exist a function $\psi(x, r)$ which satisfies the differential equations

$$\frac{\partial\psi}{\partial r}=r\varrho\frac{\partial\varphi}{\partial x}, \qquad \frac{\partial\psi}{\partial x}=-r\varrho\frac{\partial\varphi}{\partial r}. \tag{4.12}$$

We observe that in the meridian plane x, r we have the velocity field

$$\frac{\partial\varphi}{\partial x}=u, \qquad \frac{\partial\varphi}{\partial r}=v, \tag{4.13}$$

and we find for the hydrodynamical derivative of ψ

$$\frac{D\psi}{Dt}=\psi_x u+\psi_r v=-r\varrho\left[\varphi_x\varphi_r-\varphi_x\varphi_r\right]=0. \tag{4.14}$$

Thus, $\psi(x, r)=\text{const}$ along the streamlines in the meridian plane; ψ may be called again the stream function for the axially symmetric flow.

Observe that now the integral

$$Q=\int_{x_0, r_0}^{x_1, r_1} 2\pi r\varrho\,(u\,dr-v\,dx), \tag{4.15}$$

extended over an arbitrary curve C connecting the points x_0, r_0 and x_1, r_1 in the meridian plane, represents the flux of matter through the surface of rotation with the meridian curve C. We have, by Eq. (4.12)

$$Q=2\pi\left[\psi(x_1, r_1)-\psi(x_0, r_0)\right] \tag{4.16}$$

which stands in complete analogy to the relation (4.4') in the two-dimensional case.

When we eliminate ψ from Eq. (4.12) and make use of Eq. (1.26), we obtain for the velocity potential the differential equation

$$\left(1-\frac{\varphi_x^2}{c^2}\right)\varphi_{xx}+\left(1-\frac{\varphi_r^2}{c^2}\right)\varphi_{rr}-2\frac{\varphi_x\varphi_r}{c^2}\varphi_{xr}+\frac{\varphi_r}{r}=0. \tag{4.17}$$

This equation could also have been obtained from Eq. (1.27) by the use of cylindrical coordinates. It is very similar to the differential equation (4.5) for the velocity potential of a plane flow; the only difference comes from the term $\frac{1}{r}\varphi_r$, which thus makes the entire difference between the theory of plane and of axially symmetric flows. The role of the axis as a singular line of the differential equation deserves particular attention.

When we eliminate φ from the system (4.12) and make use of the equation

$$\psi_x^2 + \psi_r^2 = r^2 \varrho^2 q^2, \tag{4.18}$$

after some calculation we obtain the partial differential equation for the stream function in an axially symmetric flow:

$$\left(1 - \frac{\psi_r^2}{r^2 \varrho^2 c^2}\right) \psi_{xx} + \left(1 - \frac{\psi_x^2}{r^2 \varrho^2 c^2}\right) \psi_{rr} + 2 \frac{\psi_x \psi_r}{r^2 \varrho^2 c^2} \psi_{xr} - \frac{\psi_r}{r} = 0. \tag{4.19}$$

The analogy to the plane case (4.9) is obvious.

The fluid dynamical theory of plane and axially symmetric flows is by far the best developed part of the general theory of steady, irrotational and homentropic flows. This is due partially to the simpler form of the differential equation for the velocity potential and partially to the existence of a stream function. No single stream function exists for a general three-dimensional flow.

5. Elliptic and hyperbolic equations; the concept of characteristics [*10*] to [*13*]. We want to discuss in this section the differential equation for the two-dimensional velocity potential,

$$A \varphi_{xx} + 2B \varphi_{xy} + C \varphi_{yy} = 0 \tag{5.1}$$

with

$$A = 1 - \frac{\varphi_x^2}{c^2}, \qquad B = -\frac{\varphi_x \varphi_y}{c^2}, \qquad C = 1 - \frac{\varphi_y^2}{c^2}, \tag{5.1'}$$

from a formal mathematical point of view and to obtain various general properties of solutions. We shall be led to the concept of characteristics, fundamental for the qualitative as well as the numerical aspects of the solution. The concepts obtained in the particularly simple case of two dimensions generalize to a considerable extent to axially symmetric and arbitrary three-dimensional problems.

Let us consider an arbitrary quasi-linear partial differential equation of the type (5.1); that is, we allow the coefficients A, B and C to depend on the variables x, y, the solution φ and its first derivatives $\varphi_x = u$, $\varphi_y = v$. Let Γ be a curve in the (x, y)-plane and $x(s)$, $y(s)$ be its parametric representation in terms of its arc length s. Suppose that we specify along Γ the values of φ and its first derivatives, say as functions $\varphi(s)$, $u(s)$ and $v(s)$. Observe that we are not entirely free in this prescription. Indeed, we have the identity

$$\varphi(s) = \varphi\big(x(s), y(s)\big); \tag{5.2}$$

we obtain therefore from Eq. (5.2) by differentiation with respect to s the linear relation

$$\varphi'(s) = u(s)\, x'(s) + v(s)\, y'(s) \tag{5.2'}$$

between the data $\varphi(s)$, $u(s)$ and $v(s)$. If the solution $\varphi(x, y)$ is represented geometrically as a surface over the (x, y)-plane the data $x(s)$, $y(s)$, $\varphi(s)$, $u(s)$, $v(s)$ represent a curve on the surface together with the normal vector. We may conceive of this geometric entity as an infinitesimal surface element which is shifted along the curve (x, y, φ) of the surface. The five functions of s determine an infinitesimal strip along the solution surface and the condition (5.2′) is called the strip condition for the data [*10*].

We arrive now at the question whether for a given strip (x, y, φ, u, v) there exists always a unique solution $\varphi(x, y)$ of Eq. (5.1) which as a surface in the (x, y, φ)-space contains this strip. In order to discuss this problem, we start with the identities

$$u(s) = \varphi_x\big(x(s), y(s)\big), \qquad v(s) = \varphi_y\big(x(s), y(s)\big) \tag{5.3}$$

and obtain from them by differentiation

$$u'(s) = \varphi_{xx}\, x'(s) + \varphi_{xy}\, y'(s), \qquad v'(s) = \varphi_{xy}\, x'(s) + \varphi_{yy}\, y'(s). \tag{5.4}$$

The original differential equation (5.1) and the two Eqs. (5.4) represent a system of three linear equations for the three unknown functions φ_{xx}, φ_{xy} and φ_{yy}. The determinant of this system is

$$D = \begin{vmatrix} A & 2B & C \\ x' & y' & 0 \\ 0 & x' & y' \end{vmatrix} = A\, y'^2 - 2B\, x'\, y' + C\, x'^2. \tag{5.5}$$

If this determinant does nowhere vanish on Γ we can compute all the second order derivatives of φ along Γ, and it is easily seen that under the same assumption all higher derivatives of φ can also be computed along Γ by continued differentiation; we have only to assume that $x(s)$ and $y(s)$ can be differentiated indefinitely with respect to their arguments. Hence, two solutions of Eq. (5.1) which coincide on Γ and have the same first derivatives there must coincide in all their derivatives. Two solution surfaces $\varphi_1(x, y)$ and $\varphi_2(x, y)$ which contain such a strip must have a contact of infinite order. If a solution were discontinuous across Γ it must be discontinuous at least in its first derivatives; no weaker discontinuity is possible. The values of φ, u, v along Γ determine the solution locally in a unique way.

If D vanished at isolated points of Γ, we could decompose Γ into arcs along which $D \neq 0$ and repeat the preceding argument. An entirely different situation arises, however, if we consider an arc Γ along which

$$D = A\, y'^2 - 2B\, x'\, y' + C\, x'^2 \equiv 0. \tag{5.6}$$

Since A, B, C depend on φ, u, v and since Γ is the projection of the strip x, y, φ, u, v into the (x, y)-plane, the condition (5.6) refers indeed to the strip. We call a set of five functions $x(s)$, $y(s)$, $\varphi(s)$, $u(s)$ and $v(s)$ a characteristic strip for the differential equation if the condition (5.6) is satisfied. For a characteristic strip we cannot solve the linear system at all, except when $u'(s)$, $v'(s)$ satisfy certain compatibility conditions [see Eq. (6.4)]; and when we can solve the linear system the solution is again not unique. Thus a characteristic strip is a strip along which two distinct solution surfaces $\varphi(x, y)$ can have a first order contact. In other words, weak discontinuities of a solution of Eq. (6.1) (that is discontinuities, only in the second or higher derivatives) can take place only on a characteristic strip.

When we are dealing with a fixed solution $\varphi = \varphi(x, y)$, then the prescription of a curve $\Gamma \equiv x(s)$, $y(s)$ determines already the entire strip on this fixed surface. If Γ leads to a characteristic strip we call it a characteristic curve of the differential equation (6.1) with respect to the particular solution $\varphi(x, y)$.

For the particular case of coefficients (5.1') in the quasi-linear differential equation (5.1) the equation (5.6) for characteristic strips simplifies to

$$(\varphi_x\, y' - \varphi_y\, x')^2 = c^2. \tag{5.7}$$

For a given solution $\varphi(x, y)$ the characteristic curves Γ are determined therefore by the requirement that they form with the local velocity vector (u, v) an angle α which satisfies the condition

$$\sin^2\alpha = \frac{c^2}{q^2}. \tag{5.8}$$

This direction field for the characteristics Γ defines a first order differential equation for these curves. We obtain two sets of characteristic curves according to our choice of sign in

$$\sin\alpha = \pm \frac{c}{q}. \tag{5.8'}$$

The characteristic curves play an important role in the theory of two-dimensional supersonic flow; they are called the Mach lines of the flow.

Since $\sin^2\alpha \leq 1$, it is clear that real characteristics in the (x, y)-plane can only occur wherever the flow is supersonic and that the character of the solution must be entirely different in the subsonic and the supersonic case. Indeed, in the subsonic case there are no characteristics and no discontinuities in the higher derivatives are possible; a strong inner coherence of the solution becomes necessary. In the supersonic region different solutions may be fused together continuously and with continuous derivatives along the characteristic curves.

If we return to the general quasi-linear equation (5.1) we recognize from Eq. (5.6) that the necessary and sufficient condition for the existence of real characteristics is the discriminant condition

$$\Delta = B^2 - AC \geq 0. \tag{5.9}$$

A differential equation (5.1) with $\Delta > 0$ is called a hyperbolic differential equation; it has two real characteristics which satisfy the differential equations

$$\frac{dy}{dx} = \frac{B}{A} \pm \frac{1}{A}\sqrt{B^2 - AC}. \tag{5.10}$$

Observe that the fact of being hyperbolic depends on the solution $\varphi(x, y)$ considered and the point x, y itself.

A differential equation with $\Delta < 0$ is called elliptic; it has no real characteristics. The prototype of all elliptic differential equations is LAPLACE's equation with $A = C = 1$, $B = 0$. If φ is a solution of an elliptic differential equation with analytic coefficients in a domain of the (x, y)-plane, it is infinitely often differentiable there and determined inside of the entire domain by its values on the boundary of the domain. This is well-known for the harmonic functions which solve the Laplace equation and holds for all solutions of elliptic equations because of the strong coherence due to the absence of characteristics [*10*].

The prototype of the hyperbolic equations is the wave equation:

$$\varphi_{xx} - \varphi_{yy} = 0. \tag{5.11}$$

By Eq. (5.10) it has the characteristics $dy/dx = \pm 1$, and the general solution of Eq. (5.11) due to D'ALEMBERT,

$$\varphi = f(x + y) + g(x - y), \tag{5.12}$$

shows the central role of the characteristics in the actual solution of the equation. Observe that f and g may have discontinuous derivatives of order three and yet remain solutions of the Eq. (5.11). This behavior is typical for hyperbolic differential equations.

6. Characteristics and the initial value problem for hyperbolic equations. We wish to show how the concept of characteristics can be used to calculate the solution of a quasi-linear differential equation (5.1) from the given initial values of φ, u and v along a curve Γ which is not a characteristic [*10*], [*14*]. If we had a solution $\varphi(x, y)$ satisfying on Γ the prescribed specifications we could compute near Γ the field of characteristics from the differential equations

$$\frac{dy}{dx} = K_1(x, y, \varphi, u, v), \qquad \frac{dy}{dx} = K_2(x, y, \varphi, u, v), \tag{6.1}$$

where K_1 and K_2 are the two determinations of the right hand side in Eq. (5.10).

We introduce a curvilinear net of coordinates made up from the two sets of characteristics and denote them by λ and μ. That is: along the characteristics of the first kind μ is constant and λ can be used as parameter on the curve, while on the characteristics of the second kind λ is constant and μ is the running curve parameter. Near Γ we may then consider the coordinates $x(\lambda, \mu)$ and $y(\lambda, \mu)$ as functions of the characteristic variables.

In order to introduce the characteristic variables λ, μ we had to assume that we have a solution $\varphi(x, y)$ of the differential equation. We shall now analyze the dependence of x, y, φ, u and v on these new variables and find such a simple differential relation that it is possible conversely to start with the initial data on Γ and to construct all five quantities explicitly in terms of λ and μ.

We have the following differential relations:

$$y_\lambda = K_1 x_\lambda, \qquad y_\mu = K_2 x_\mu. \tag{6.2}$$

Next, we remember that the values u and v cannot be prescribed arbitrarily along a characteristic but must satisfy certain compatibility conditions. The linear system considered was

$$\left.\begin{aligned} A\varphi_{xx} + 2B\varphi_{xy} + C\varphi_{yy} &= 0, \\ x'\varphi_{xx} + y'\varphi_{xy} &= u', \\ x'\varphi_{xy} + y'\varphi_{yy} &= v', \end{aligned}\right\} \tag{6.3}$$

and the dash denoted differentiation with respect to the parameter of the curve considered. Along a characteristic the determinant of the system vanishes, and a solution can exist only if

$$\begin{vmatrix} A & C & 0 \\ x' & 0 & u' \\ 0 & y' & v' \end{vmatrix} = -(A y' u' + C x' v') = 0. \tag{6.4}$$

We can apply the compatibility condition (6.4) to either type of characteristics, that is, use λ or μ as curve parameter. By means of Eq. (6.2) we then obtain

$$K_1 A u_\lambda + C v_\lambda = 0, \qquad K_2 A u_\mu + C v_\mu = 0. \tag{6.5}$$

Finally, we note the strip relation (5.2') which has also to hold along each characteristic:

$$\varphi_\lambda = u x_\lambda + v y_\lambda, \qquad \varphi_\mu = u x_\mu + v y_\mu. \tag{6.6}$$

The three pairs of differential relations (6.2), (6.5) and (6.6) represent six partial differential equations for the five unknown functions x, y, φ, u, v of λ and μ.

In order to avoid an over-determined system, we restrict ourselves to the system of five partial differential equations

$$\left.\begin{aligned} y_\lambda - K_1 x_\lambda &= 0, & K_2 A u_\mu + C v_\mu &= 0, \\ y_\mu - K_2 x_\mu &= 0, & \varphi_\lambda - u x_\lambda - v y_\lambda &= 0; \\ K_1 A u_\lambda + C v_\lambda &= 0 \end{aligned}\right\} \tag{6.7}$$

we wish to show that the equation omitted is essentially a consequence of this system. Indeed, let

$$r(\lambda, \mu) = \varphi_\mu - u\, x_\mu - v\, y_\mu. \tag{6.8}$$

We have

$$r_\lambda = \varphi_{\lambda\mu} - u\, x_{\lambda\mu} - v\, y_{\lambda\mu} - u_\lambda\, x_\mu - v_\lambda\, y_\mu. \tag{6.9}$$

We subtract from Eq. (6.9) the last equation (6.7) differentiated with respect to μ and find

$$r_\lambda = (x_\lambda\, u_\mu - x_\mu\, u_\lambda) + (y_\lambda\, v_\mu - y_\mu\, v_\lambda). \tag{6.9'}$$

Using the first four Eq. (6.7) to eliminate y_λ, y_μ and v_λ, v_μ and by the identity $K_1 K_2 = C/A$ which follows from the definition, we finally arrive at

$$r_\lambda \equiv 0. \tag{6.10}$$

Thus, if we can only guarantee that $r=0$ along a single curve, say Γ, which is cut by all characteristics $\mu=\text{const}$, we can assert $r(\lambda, \mu)\equiv 0$; that is, the equation omitted will be satisfied in consequence of Eqs. (6.1).

We have not specified anything about the choice of parameters λ and μ except that each should be constant on one family of characteristics. This still leaves considerable freedom to satisfy additional requirements. We shall suppose that along the non-characteristic curve Γ along which we wish to prescribe the initial values $\varphi(s)$, $u(s)$ and $v(s)$ the relations

$$\lambda = s, \qquad \mu = -s \tag{6.11}$$

are satisfied. From this normalization of the characteristic parameters along Γ we can gauge the entire characteristic net.

Since the initial data φ, u an dv have to satisfy the strip condition (5.2′) on Γ and since we have the last Eq. (6.7) valid there, it is easily seen that $r=0$ is fulfilled identically along Γ and will hold everywhere when the system (6.7) has been successfully integrated.

We have thus reduced our differential equation problem to the following initial value problem in the characteristic (λ, μ)-plane:

To find solutions of the quasi-linear system of differential equations (6.7) which take specified values x, y, φ, u, and v along the line $\lambda+\mu=0$.

The mathematical approach to this initial value problem is standard; we differentiate those Eqs. (6.7) which contain λ-derivatives with respect to μ and those which contain μ-derivatives with respect to λ. We obtain

$$\left.\begin{aligned} y_{\lambda\mu} - K_1 x_{\lambda\mu} &= F_1, & y_{\lambda\mu} - K_2 x_{\lambda\mu} &= F_2, \\ K_1 A\, u_{\lambda\mu} + C\, v_{\lambda\mu} &= F_3, & K_2 A\, u_{\lambda\mu} + C\, v_{\lambda\mu} &= F_4, \\ \varphi_{\lambda\mu} - u\, x_{\lambda\mu} - v\, y_{\lambda\mu} &= F_5 \end{aligned}\right\} \tag{6.12}$$

where all F_i depend only on x, y, φ, u, v and their first partial derivatives with respect to λ and μ. This is a linear system of equations for the mixed second derivatives of the functions sought; its determinant is easily shown to be

$$(K_2 - K_1)^2 A\, C, \tag{6.13}$$

which is supposed different from zero on the line $\lambda+\mu=0$. Hence, we can derive expressions of the form

$$\frac{\partial^2 \varphi_n(\lambda, \mu)}{\partial\lambda\,\partial\mu} = F_n\left(\varphi_1, \ldots \varphi_5, \frac{\partial\varphi_1}{\partial\lambda}, \ldots \frac{\partial\varphi_5}{\partial\lambda}, \frac{\partial\varphi_1}{\partial\mu}, \ldots \frac{\partial\varphi_5}{\partial\mu}\right), \tag{6.14}$$

where φ_n stands for the five unknowns x, y, φ, u, v.

This system of equations has to be solved with given values of the φ_n and their first derivatives along the line $\lambda+\mu=0$; in fact, we can calculate from the original system (6.7) the values of all first derivatives on that line, since this system has the non-vanishing determinant (6.13).

We are now able to transform the actual solution of Eq. (6.14) to a system of convenient integral equations. Let $\Delta(\lambda_0, \mu_0)$ denote the triangle made up of the three straight line $\lambda+\mu=0$, $\lambda=\lambda_0$, $\mu=\mu_0$ (Fig. 1). The integral

$$J(\lambda,\mu) = \iint_{\Delta(\lambda,\mu)} f(\alpha,\beta)\, d\alpha\, d\beta \tag{6.15}$$

satisfies, as is well-known, the simple partial differential equation

$$\frac{\partial^2 J}{\partial\lambda\,\partial\mu} = f(\lambda,\mu) \tag{6.15'}$$

and vanishes on the line $\lambda+\mu=0$. Hence, we can transform (6.14) into the system

$$\left.\begin{aligned}\varphi_n(\lambda,\mu) = \iint_{\Delta(\lambda,\mu)} F_n\left(\varphi_\nu, \frac{\partial\varphi_\nu}{\partial\lambda}, \frac{\partial\varphi_\nu}{\partial\mu}\right) d\lambda\, d\mu\, + \\ + a_n(\lambda) + b_n(\mu).\end{aligned}\right\} \tag{6.16}$$

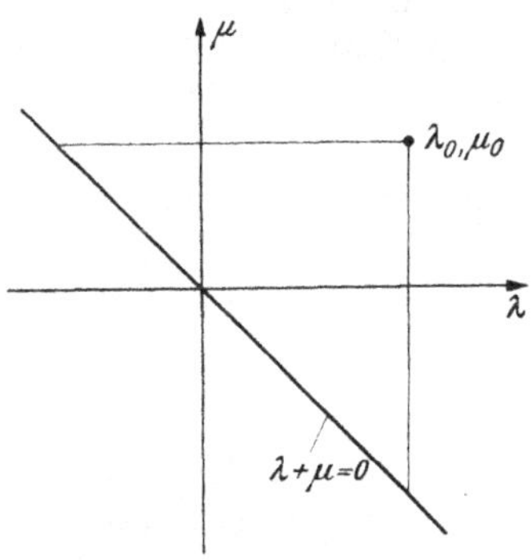

Fig. 1.

The functions $a_n(\lambda)$ and $b_n(\mu)$ can easily be computed from the initial values required for φ_n and its first derivatives on $\lambda+\mu=0$.

The actual determination of the solutions $\varphi_n(\lambda,\mu)$ can now be carried out by successive approximation. One starts with a set of trial functions $\varphi_n^{(0)}$ and by puting them into the right hand side of Eqs. 6.16) determines an improved set $\varphi_n^{(1)}$. Next one substitutes the $\varphi_n^{(1)}$ into the right hand side of Eqs. (6.16) and obtains $\varphi_n^{(2)}$. Continuing in this way one obtains a convergent sequence whose limits $\varphi_n(\lambda,\mu)$ can be shown to be solutions of the set (6.7) with the right initial values.

The procedure described yields the solution $\varphi(x, y)$ and its first partial derivatives u, v in parametric form. Eliminating the auxiliary variables λ and μ we obtain φ, u and v as explicit functions of x and y near the curve Γ which carries the initial data.

This procedure yields first of all an existence proof for the initial value problem. It also clearly indicates the region of influence of the various initial data on the line $\lambda+\mu=0$, i.e. on the curve Γ. Indeed, the integral equations (6.16) over the triangle $\Delta(\lambda,\mu)$ show that the values of φ, u, v at λ, μ depend only on the initial values on the segment cut out by the two parallels to the coordinate curves and not on the rest of the line. Hence, a change of these initial values outside of the segment would not affect the solution at λ, μ. A change in the solution occurs only when the characteristics through λ, μ reach the portion of the line where the initial values have been changed. This fact explains why for hyperbolic differential equations different solutions can be fused along a characteristic line.

If we stay near to the line $\lambda+\mu=0$, the area of the triangle of integration is small and the successive correction in Eq. (6.16) will decrease rapidly. Under these circumstances the theoretical procedure of successive approximation may yield a quickly convergent numerical procedure for solving the initial value problem.

For most numerical purposes it is more convenient and intuitive to operate directly with the linear system (6.7). Considering the differential equations as difference equations and the characteristic Mach lines as polygonal curves, one develops a step-by-step procedure for calculating the velocity field at not too distant points. This semi-graphical procedure was first considered by MASSAU [*15*] and extended to actual problems of fluid dynamics by PRANDTL-BUSEMAN [*16*] and GUDERLEY [*17*]. For an example of a difference equation approach along these lines see Sect. 34.

A very similar analysis is possible in the case of the differential equation (4.11) for the velocity potential of an axially symmetric flow. Since we are still dealing with a problem in two independent variables, the entire characteristic theory is still applicable and the final results and methods are completely analogous. For a more general treatment of the theory of hyperbolic differential equations see COURANT-HILBERT II [*10*]; for a more detailed description of the characteristics in application to fluid dynamics see MEYER [*19*].

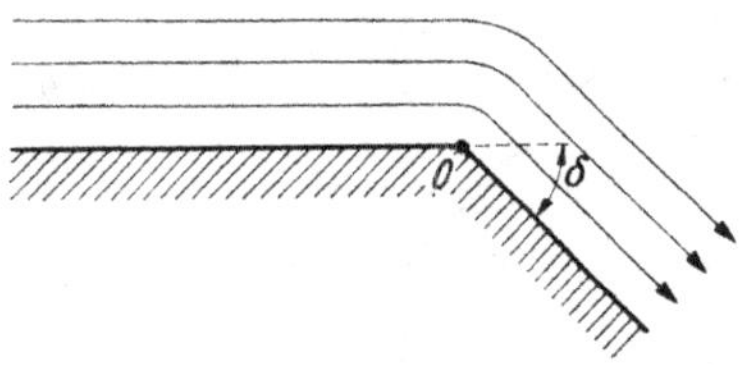

Fig. 2.

7. Supersonic flow around a corner. The significance of the concept of characteristics in the general theory and the Mach lines in supersonic fluid dynamics is clearly exhibited in the classical problem of a steady irrotational flow around a corner (PRANDTL-MEYER flow [*21*]).

We consider a supersonic flow in the (x, y)-plane which takes place in the angular region described in polar coordinates by the inequality $-\delta<\vartheta<\pi$ (Fig. 2). We suppose that the flow is uniform at infinity and parallel there to the boundary of the flow region, and we wish to study how it turns around the corner at the origin.

Since the geometric configuration is invariant under a similarity transformation it is natural to try a solution whose velocity field depends only on the angular variable ϑ and not on the radial coordinate r. The same will then hold also for the density, which is a known function of the local speed. We set up

$$\varphi(r, \vartheta) = r\,\Phi(\vartheta) \tag{7.1}$$

and verify easily that the velocity field

$$u = \frac{x}{r}\,\Phi(\vartheta) - \frac{y}{r}\,\Phi'(\vartheta), \qquad v = \frac{y}{r}\,\Phi(\vartheta) + \frac{x}{r}\,\Phi'(\vartheta) \tag{7.2}$$

depends on ϑ only.

The differential equation (1.22) for the velocity potential becomes

$$-\sin\vartheta\,\frac{d}{d\vartheta}\,(\varrho\,\Phi\cos\vartheta - \varrho\,\Phi'\sin\vartheta) + \cos\vartheta\,\frac{d}{d\vartheta}\,(\varrho\,\Phi\sin\vartheta + \varrho\,\Phi'\cos\vartheta) = 0. \tag{7.3}$$

Rearranging terms, we can reduce Eq. (7.3) to the following ordinary differential equation for Φ:

$$(\varrho\,\Phi')' + \varrho\,\Phi = 0. \tag{7.4}$$

We observe that ϱ is a function of

$$q^2 = u^2 + v^2 = \Phi^2 + \Phi'^2. \tag{7.5}$$

Hence, using Eq. (1.26), we can bring Eq. (7.4) into the form

$$[\Phi'' + \Phi]\left[1 - \frac{1}{c^2}\,\Phi'^2\right] = 0. \tag{7.6}$$

There are two possible ways of satisfying Eq. (7.6) by equating one of the factors to zero. One solution is

$$\Phi = a\sin(\vartheta + \beta) \tag{7.7}$$

while, in view of Eqs. (2.7) and (7.5), the other solution must satisfy

$$\Phi'^2 = c^2 = \gamma - \frac{\gamma-1}{2} q^2 = \gamma - \frac{\gamma-1}{2}[\Phi^2 + \Phi'^2]. \tag{7.8}$$

This differential equation has the solution

$$\Phi(\vartheta) = q_{\max} \sin\left(\sqrt{\frac{\gamma-1}{\gamma+1}}\,\vartheta + \beta\right), \qquad q_{\max} = \sqrt{\frac{2\gamma}{\gamma-1}}. \tag{7.9}$$

The local speed at every point is given by Eqs. (7.5) and (7.9):

$$\left.\begin{aligned} q^2 &= q_{\max}^2\left[\frac{\gamma-1}{\gamma+1} + \frac{2}{\gamma+1}\sin^2\left(\sqrt{\frac{\gamma-1}{\gamma+1}}\,\vartheta + \beta\right)\right] \\ &= q_{\text{crit}}^2\left[1 + \frac{2}{\gamma-1}\sin^2\left(\sqrt{\frac{\gamma-1}{\gamma+1}}\,\vartheta + \beta\right)\right]. \end{aligned}\right\} \tag{7.10}$$

We observe next that the velocity component in the radial direction is

$$\frac{\partial\varphi}{\partial r} = \Phi = \sqrt{q^2 - \Phi'^2} = \sqrt{q^2 - c^2} = q\sqrt{1 - \frac{c^2}{q^2}} \tag{7.11}$$

by virtue of the differential equation (7.8); on the other hand, we have on geometrical grounds

$$\frac{\partial\varphi}{\partial r} = q\cos(r, q). \tag{7.12}$$

Hence, we see that

$$\sin(r, q) = \pm\frac{c}{q}; \tag{7.13}$$

the radii are Mach lines of the flow determined by Eq. (7.9).

We can now utilize the fact that various solutions of the same differential equation can be fused together along a characteristic in order to satisfy the conditions of our problem in the large. We start with the velocity potential, prescribed at the horizontal wall far away from the corner 0:

$$\varphi_1 = u_0 x \tag{7.14}$$

which corresponds to the solution (7.7) of the differential equation (7.6). We have to connect it continuously to a solution $\varphi_2 = r\Phi(\vartheta)$ given by Eq. (7.9). Observe that

$$\varphi_r = \Phi(\vartheta), \qquad \frac{1}{r}\varphi_\vartheta = \Phi'(\vartheta) \tag{7.15}$$

represent the radial and angular velocity components. Along the radius where we wish to fuse the two solutions, the radius corresponding, say, to the angle ϑ_0, we have

$$q_r = u_0\cos\vartheta_0, \qquad q_\vartheta = -u_0\sin\vartheta_0. \tag{7.16}$$

The conditions of continuity

$$\varphi_r = q_r, \qquad \frac{1}{r}\varphi_\vartheta = q_\vartheta \tag{7.17}$$

yield two equations for the unknowns ϑ_0 and β. In view of Eq. (7.8), we find from Eqs. (7.15), (7.16) and (7.17):

$$\sin\vartheta_0 = \frac{1}{u_0}\Phi'(\vartheta) = \pm\frac{c(u_0)}{u_0}. \tag{7.18}$$

This result means that the Mach line of fusion must have the fixed direction of the Mach lines of the incoming uniform flow. From the remaining equation

$$u_0 \cos\vartheta_0 = q_{\max} \sin\left(\sqrt{\frac{\gamma-1}{\gamma+1}}\,\vartheta_0 + \beta\right) \tag{7.19}$$

we can then easily determine β.

Next, we wish to connect φ_2 continuously with the solution

$$\varphi_3 = U(x\cos\delta - y\sin\delta) \tag{7.20}$$

which represents the final downstream velocity potential describing a uniform flow parallel to the second wall. We have to combine φ_2 and φ_3 along a radial Mach line with angle ϑ. The conditions of continuity

$$U\cos\delta = \frac{\partial}{\partial x}\varphi_2, \qquad -U\sin\delta = \frac{\partial}{\partial y}\varphi_2 \tag{7.21}$$

serve now to determine the two unknown values U and ϑ_1. Thus, the entire flow can be pieced together along properly chosen Mach lines. This illustrates clearly the importance of characteristics as lines of weak discontinuities of solutions of hyperbolic differential equations; it allows in practice a piecing together of different solutions in order to fulfill given boundary conditions. No such method is available in the case of elliptic differential equations where local changes of boundary conditions affect the solution in its entire domain of existence. For this reason, the mathematical theory of purely supersonic flows is in many aspects simpler than that of subsonic ones.

We may extend the method of fusing solutions along Mach lines in order to construct supersonic flows in regions with very general convex polygonal boundaries. Indeed, our preceding method enables us to pass from one uniform flow in a fixed direction and in a given angular space to another uniform flow with a new direction and lying in another sector of the (x, y)-plane. Repeating this procedure, we can obtain general supersonic flow patterns. It is even possible to carry out a limit process and to obtain at least graphically flow patterns for a smooth convex boundary curve.

A supersonic flow pattern in which one family of Mach lines consists of straight lines is called a simple wave. The importance of simple waves as construction elements for more general flows is obvious in view of the foregoing remarks.

8. Supersonic conical flow. We can obtain in an analogous way a solution for the differential equation (4.17) of the velocity potential for an axially symmetric flow. We consider the flow past a right circular cone with the vertex O at the origin of our system of cylindrical coordinates x and r and with the x-axis as axis. We introduce in the meridian plane x, r the polar coordinates

$$R = \sqrt{x^2 + r^2}, \qquad \vartheta = \operatorname{arc\,tan}\frac{r}{x}. \tag{8.1}$$

Using again the similarity invariance of the flow pattern, we expect that the velocity vector q in the meridian plane depends only on the angle ϑ. Thus, we set up for the velocity potential

$$\varphi(R, \vartheta) = R\,\Phi(\vartheta). \tag{8.2}$$

It is then easily verified that the velocity field obtained depends only on ϑ and this implies also that the speed q and the density ϱ depend solely on ϑ.

If we insert now the function (8.2) into the differential equation (4.11′) for the axially-symmetric potential, we obtain after an easy and straightforward calculation

$$\frac{d}{d\vartheta}(\varrho \sin\vartheta\, \Phi') + 2\varrho \sin\vartheta\, \Phi = 0. \tag{8.3}$$

Finally, we make use of Eq. (1.26) and the fact that

$$q^2 = \Phi^2 + \Phi'^2. \tag{8.4}$$

We obtain then for $\Phi(\vartheta)$ the ordinary differential equation

$$[\Phi'' + \Phi]\left[1 - \frac{1}{c^2}\Phi'^2\right] + [\Phi' \cot\vartheta + \Phi] = 0. \tag{8.5}$$

Here, c^2 is a known function of $\Phi^2 + \Phi'^2$.

If it is required that $\varphi(R, \vartheta)$ be the velocity potential of a flow around a cone with vertex angle ϑ_0, we have to satisfy the condition $\partial\varphi/\partial n = 0$ on the boundary of the flow region, which means by Eq. (8.2)

$$\Phi'(\vartheta_0) = 0. \tag{8.6}$$

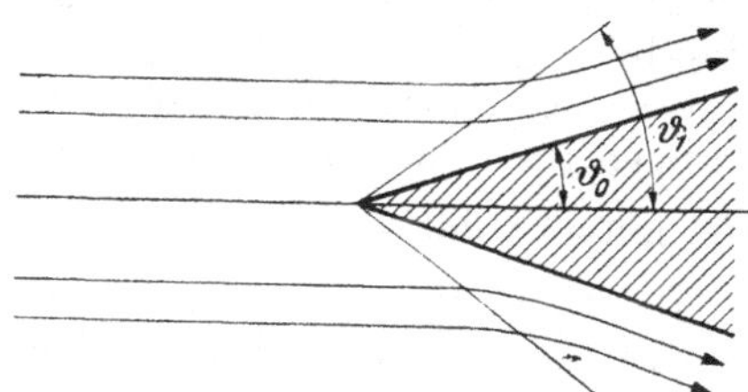

Fig. 3.

One can choose an arbitrary value of $\Phi(\vartheta_0)$ and $\Phi'(\vartheta_0) = 0$ and integrate the differential equation (8.5) for these initial values. One will obtain a solution $\Phi(\vartheta)$ at least in an angular interval $\vartheta_0 \leqq \vartheta \leqq \vartheta_1$. Obviously it cannot be expected that a solution of this form exists in the entire flow region, and we shall have to combine it somehow with the velocity potential of a uniform flow parallel to the x-axis. Observe that indeed

$$\Phi(\vartheta) = u_0 \cos\vartheta \tag{8.7}$$

is for every choice of u_0 a solution of (8.5).

But it is easily seen that there exists no Mach line along which this particular solution (8.7) can be fused continuously with another solution of the Eq. (8.5). One has to expect, therefore, strong discontinuities if a uniform flow hits the cone. In fact Taylor and Maccoll [22] have calculated flow patterns which yield a uniform flow in the sector $\vartheta_1 < \vartheta < 2\pi - \vartheta_1$, of the meridian plane, while in the sectors $\vartheta_0 < \vartheta < \vartheta_1$ and $2\pi - \vartheta_1 < \vartheta < 2\pi - \vartheta_0$ adjacent to the cone a non-uniform flow takes place. The radii with the angles ϑ_1, $2\pi - \vartheta_1$ represent lines of discontinuity along which the two flows are connected according to the shock conditions (Fig. 3).

The shock angle ϑ_1 depends on the uniform speed u_0; however, in actual calculations it is found more expedient to start integrating the Eq. (8.5) from $\vartheta = \vartheta_0$ with arbitrary initial value $\Phi(\vartheta_0)$ and to determine ϑ_1 and u_0 simultaneously from the shock conditions in terms of this parameter. For extensive numerical work see the M.I.T. tables of supersonic flow around cones [25].

The flow described by the assumption (8.2) becomes particularly interesting when we consider its image in the velocity plane (u, v). We find easily

$$u = \Phi\cos\vartheta - \Phi'\sin\vartheta, \qquad v = \Phi\sin\vartheta + \Phi'\cos\vartheta \tag{8.8}$$

and

$$u' = -\sin\vartheta\ [\Phi + \Phi''], \qquad v' = \cos\vartheta\ [\Phi + \Phi'']. \tag{8.8'}$$

All (u, v)-points lie on one single curve depending on the parameter ϑ. If s is the arc length of this curve and τ the angle of its tangent against the u-axis, we read off from Eq. (8.8′)

$$\frac{ds}{d\vartheta} = \Phi + \Phi'', \qquad \tau = \vartheta + \frac{\pi}{2}. \tag{8.9}$$

Since the curvature of the (u, v)-curve is obviously

$$K = \frac{d\tau}{ds} = [\Phi + \Phi'']^{-1}, \tag{8.10}$$

we have a simple geometric interpretation of the term $\Phi + \Phi''$ which occurs in the differential equation (8.5). If R is the radius of curvature at a given point (u, v) we may put Eq. (8.5) into the form

$$R = \frac{-v}{\sin\vartheta\left[1 - \frac{1}{c^2}(v\cos\vartheta - u\sin\vartheta)^2\right]}. \tag{8.11}$$

BUSEMANN [23], [24] made this relation the basis for a graphical construction for the image of the stream lines in the velocity plane. He uses the following step-by-step procedure: Suppose that we have initial values $u(\vartheta_1)$, $v(\vartheta_1)$ prescribed; we approximate a small arc of the sought curve by a small circular arc through the same point where radius R is given by Eq. (8.11) and which has at the initial point the tangent angle $\tau_1 = \vartheta_1 + \frac{2}{\pi}$. At the end of this small arc we determine the new values u_2, v_2 and the corresponding value ϑ_2 from the new slope angle τ_2 there. In this way, the entire curve can be approximated by circular arcs to any desired precision.

If we know the values u, v for all values ϑ in an interval $\vartheta_1 \leqq \vartheta \leqq \vartheta_2$ we can easily construct the streamlines in the meridian plane itself. On each ray through the origin with angle ϑ we know the direction of the streamlines; we have thus a direction field over the entire sector $\vartheta_1 \leqq \vartheta \leqq \vartheta_2$ and can find the streamlines by graphical integration.

9. Characteristics and bi-characteristics in three-dimensional problems. We shall give a brief sketch on the extension of the concept of characteristics to the case of three-dimensional problems [10] to [13]. We consider the quasi-linear differential equation

$$\sum_{i,k=1}^{3} a_{ik} \frac{\partial^2 \varphi}{\partial x_i \partial x_k} = 0, \tag{9.1}$$

where the coefficients a_{ik} are functions of the three space variables x_1, x_2, x_3, the solution function φ and its first partial derivatives. We ask for a surface Σ given by the equation

$$g(x_1, x_2, x_3) = 0 \tag{9.2}$$

such that a solution φ of (Eq. 9.1) exists which is continuous and has continuous first derivatives along and across Σ but whose second derivatives jump when we cross the surface Σ.

Clearly, the derivatives of every function $\varphi_i = \partial\varphi/\partial x_i$ in the direction of a tangent to Σ must be continuous across Σ. Since ∇g is the normal vector to Σ, we have for every choice of i, k and l:

$$g_k \varphi_{il} - g_l \varphi_{ik} \quad \text{continuous across } \Sigma, \qquad g_i = \frac{\partial g}{\partial x_i}. \tag{9.3}$$

Let us denote the jump of a quantity Q across Σ by the symbol $[Q]$. We can then put Eq. (9.3) in the form of an equation

$$g_k[\varphi_{il}] = g_l[\varphi_{ik}]. \tag{9.3'}$$

This leads at once to the condition

$$[\varphi_{il}] = \lambda g_i g_l \tag{9.4}$$

where λ is a factor of proportionality which will, in general, vary on Σ. It is not identically zero if we assume actual discontinuities of the second derivatives along Σ.

We observe now that Eq. (9.1) states the continuity of a certain combination of second derivatives of φ everywhere. It implies, therefore,

$$\sum_{i,k=1}^{3} a_{ik} g_i g_k = 0. \tag{9.5}$$

This is a first order partial differential equation for a surface $g=0$ along which the second derivatives of the solution φ can admit a discontinuity while the solution and its first derivatives pass through continuously.

The surface Σ is called a characteristic surface for the differential equation (9.1) with respect to the solution φ. It should be observed that because of the quasi-linearity of Eq. (9.1), the first order equation (9.5) for g depends on the solution φ.

We can analyze a characteristic surface Σ in greater detail by introducing the concept of a bi-characteristic. This is a curve $x_i(\sigma)$ defined by the ordinary differential system

$$\frac{d}{d\sigma} x_i = \sum_{k=1}^{3} a_{ik} g_k. \tag{9.6}$$

It is easily seen that this curve lies on the surface Σ and that it is closely related to the theory of characteristics for the first order partial differential equation (9.5). The entire surface Σ can be covered by such bi-characteristics and may be considered as generated by a one-parameter family of such curves.

We want to show now that the saltus terms $[\varphi_{il}]$ satisfy ordinary differential equations along each bi-characteristic and therefore cannot be prescribed quite arbitrarily on Σ.

For this purpose we introduce a new set of curvilinear coordinates $\xi_\alpha(x_i)$ $(\alpha=1, 2, 3)$ such that the surface $\xi_3=\text{const}$ coincide with the characteristic manifolds; we put

$$g(x_1, x_2, x_3) = \xi_3. \tag{9.7}$$

In the new coordinates, the differential equation (9.1) becomes obviously

$$\sum_{\alpha,\beta=1}^{3} A_{\alpha\beta} \frac{\partial^2\varphi}{\partial\xi_\alpha\,\partial\xi_\beta} + \sum_{\alpha=1}^{3} B_\alpha \frac{\partial\varphi}{\partial\xi_\alpha} = 0 \tag{9.8}$$

with

$$A_{\alpha\beta} = \sum_{i,k=1}^{3} a_{ik} \frac{\partial\xi_\alpha}{\partial x_i}\,\frac{\partial\xi_\beta}{\partial x_k}, \qquad B_\alpha = \sum_{i,k=1}^{3} a_{ik} \frac{\partial^2\xi_\alpha}{\partial x_i\,\partial x_k}. \tag{9.8'}$$

Observe that $\varphi_\alpha = \partial\varphi/\partial\xi_\alpha$ denotes a tangential derivative to the characteristic manifold when $\alpha=1, 2$ and a normal derivative to it when $\alpha=3$. The lowest order discontinuity for any derivative of φ occurs for $\partial^2\varphi/\partial\xi_3^2$; but by our choice of ξ_3, we have by Eqs. (9.5), (9.7) and (9.8')

$$A_{33} = 0, \tag{9.9}$$

so that this discontinuous term drops out of Eq. (9.8). However, we can differentiate Eq. (9.8) with respect to ξ_3 and take the saltus across Σ of this new identity. One finds easily:

$$\sum_{\alpha=1}^{2} A_{\alpha 3} \frac{\partial}{\partial\xi_\alpha}\left[\frac{\partial^2\varphi}{\partial\xi_3^2}\right] + B_3\left[\frac{\partial^2\varphi}{\partial\xi_3^2}\right] + C\left[\frac{\partial^2\varphi}{\partial\xi_3^2}\right] = 0. \tag{9.10}$$

Regarding the genesis of this expression, we remark that all terms which contain only derivatives in ξ_3 of order less than the second are continuous through Σ. On the other hand it should be observed that because of the quasi-linearity of Eq. (9.1) the $A_{\alpha\beta}$ and B_α may depend on $\partial\varphi/\partial\xi_3$ and that the differentiation of Eq. (9.8) with respect to ξ_3 may thus introduce new terms of discontinuity. The contribution of these terms to the saltus is obviously of the form $C[\partial^2\varphi/\partial\xi_3^2]$, where C is a complicated function on Σ.

By the definitions (9.6), (9.7) and (9.8′), we have

$$A_{\alpha 3} = \sum_{i=1}^{3} \frac{\partial \xi_\alpha}{\partial x_i} \frac{d x_i}{d\sigma} = \frac{d\xi_\alpha}{d x_\sigma}, \tag{9.11}$$

and Eq. (9.10) reduces, therefore, to

$$\frac{d}{d\sigma}\left[\frac{\partial^2\varphi}{\partial\xi_3^2}\right] + (B_3 + C)\left[\frac{\partial^2\varphi}{\partial\xi_3^2}\right] = 0. \tag{9.12}$$

This is a linear homogeneous differential equation for the saltus term and can be integrated along the bi-characteristic:

$$\left[\frac{\partial^2\varphi}{\partial\xi_3^2}\right] = s_0 e^{-\int (B_3 + C)\, d\sigma}. \tag{9.13}$$

We learn, in particular, that if we have a discontinuity at one given point of the characteristic surface Σ this discontinuity must actually extend along the entire bi-characteristic through this point. This result will be of importance when we try to fuse together different solutions of the same differential equation (9.1) along characteristic manifolds.

In the particular case of the differential equation (1.27) for the velocity potential of a general flow in three-dimensions we find

$$(\nabla g)^2 = \frac{1}{c^2}(\nabla g \cdot \nabla\varphi)^2 \tag{9.14}$$

as the differential equation for the characteristic manifold. We can, therefore, define a characteristic surface as a surface with respect to which the local flow velocity has a normal component of amount c. This is a direct generalization of an analogous property of the Mach lines in two-dimensional flow.

The differential equation for the bi-characteristics (9.6) reduces now to

$$\frac{d x_i}{d\sigma} = g_i - \frac{|\nabla g|}{c}\varphi_i. \tag{9.15}$$

Since $\nabla\varphi$ is the local velocity vector and ∇g the normal vector on Σ and since the vector $\frac{d x_i}{d\sigma}$ lies in the surface, Eq. (9.15) has a simple geometric interpretation. The bi-characteristics on Σ have at every point of Σ the direction of the tangential velocity vector. This defines a direction field on Σ and the bi-characteristics are its integral curves.

II. Linearized theory.

10. Linearization of the fundamental equation. The main source of difficulties in the theory of compressible fluid flow is the non-linear character of the basic differential equation (1.27) for the velocity potential. While various explicit solutions for this equation can be found, and some have been discussed in Chap. I, it has been possible to develop a systematic and constructive theory for general fluid dynamical problems only when the classical theory of linear partial differential equations was brought into play.

In the important but special case of two-dimensional flow the differential equations for the velocity potential and the stream function become linear when the velocity components are introduced as the independent variables. A great many particular solutions can be provided, and numerous boundary value problems can be solved by proper linear superposition of these solutions. This approach constitutes the hodograph method, which will be described in detail in Chap. III.

A more flexible procedure for linearization which has also a wider range of applicability is provided by the method of perturbations. We consider a solution $\varphi(x, y, z; t)$ of the non-linear differential equation

$$\nabla\cdot(\varrho\,\nabla\varphi) = 0 \tag{10.1}$$

which depends on a parameter t. This parameter may be connected with the geometry of the boundary of the flow region or with the speed of the flow at infinity or with some other physical quantity. In every case, we can establish a linear partial differential equation for the variation of the solution,

$$\omega(x, y, z; t) = \frac{\partial\varphi}{\partial t}. \tag{10.2}$$

In fact, differentiating the identity (10.1) with respect to t and using the relation (1.26) we obtain

$$\nabla\cdot\left[\varrho\,\nabla\omega - \frac{\varrho}{c^2}(\nabla\varphi\cdot\nabla\omega)\,\nabla\varphi\right] = 0. \tag{10.3}$$

The coefficients of this linear partial differential equation depend only on the solution φ but are independent of the particular meaning of the parameter t. The significance of t expresses itself in the singularity and boundary conditions for the solution ω. The Eq. (10.3) describes thus all solutions infinitesimally near to φ; it is called the variational equation belonging to the non-linear equation (10.1).

If a solution φ depends analytically on the parameter t near $t=0$, we can develop it there in a Taylor series

$$\varphi(x, y, z; t) = \varphi_0(x, y, z) + \sum_{\nu=1}^{\infty} t^{\nu}\,\omega_\nu(x, y, z). \tag{10.4}$$

By continued differentiation of Eq. (10.3) with respect to t, we find for $\omega_\nu(x, y, z)$ the non-homogeneous but linear differential equation

$$\nabla\cdot\left[\varrho_0\,\nabla\omega_\nu - \frac{\varrho_0}{c_0^2}(\nabla\varphi_0\cdot\nabla\omega_\nu)\,\nabla\varphi_0\right] = E_\nu, \tag{10.5}$$

where E_ν depends on $\varphi_0, \omega_1, \ldots, \omega_{\nu-1}$ only and ϱ_0, c_0 are computed from $\nabla\varphi_0$. In this way, we may use the variational equation of Eq. (10.1) in order to compute recursively the entire series development (10.4).

In general, it is not easy to carry out the development (10.4) for more than a very few terms; there arises also the question of convergence of the series obtained. Therefore one is usually satisfied by solving the Eq. (10.3) for a given known solution $\varphi_0(x, y, z)$ and then using

$$\varphi(x, y, z, t) = \varphi_0(x, y, z) + t\,\omega(x, y, z) \tag{10.6}$$

as an approximate new solution of Eq. (10.1). All we can assert is that if we introduce this φ into the left side of Eq. (10.1) we shall obtain a deviation from zero of the order t^2 only. In many cases, however, the first approximation leads to acceptable and useful results, and we shall discuss it now to some extent. For a discussion of higher order approximations see [5], [15], [16], [53]. We refer also to the Janzen-Rayleigh iteration procedure, described in Sect. 47, which may be considered as a particular instance of the general method of series development in terms of a small parameter.

The simplest choice of a solution $\varphi_0(x, y, z)$ for the fundamental equation (10.1) is probably

$$\varphi_0(x, y, z) = U_0\, x. \tag{10.7}$$

This is the velocity potential of a uniform flow in the direction of the x-axis with the speed U_0. The variational equation (10.3) reduces in this case to

$$(1 - M_0^2)\,\omega_{xx} + \omega_{yy} + \omega_{zz} = 0 \tag{10.8}$$

where in view of Eq. (2.7)

$$M_0^2 = \frac{U_0^2}{c_0^2} = \frac{U_0^2}{\gamma - \frac{\gamma - 1}{2} U_0^2}\,. \tag{10.8'}$$

The variational equation for a uniform flow is a linear differential equation with constant coefficients and is, therefore, particularly easy to handle. For this reason, the uniform flow is used most frequently as the starting point for a perturbation procedure. This particular approximation method is called the linearized theory of fluid dynamics.

The linearized theory can be applied to flows which are uniform at infinity and pass around obstacles whose boundaries make only small angles with the direction of the uniform flow. Observe that a developable surface in the (x, y, z)-space which is generated by segments parallel to the x-direction will not distort the uniform flow at all. If we inflate such a surface slightly we will obtain a thin body for which the linearized theory can be expected to hold. Another type of obstacle for which this approximation is applicable is a slender body; that is, a needle shaped body obtained by inflating slightly a linear segment oriented in the stream direction. If we turn a thin or slender body by a small angle against the direction of the uniform flow we may still expect the linearized theory to yield an acceptable approximation to the correct solution.

In the linearized theory, it is usual to set up the solution in the form

$$\varphi(x, y, z) = U_0\, x + \omega(x, y, z)\,. \tag{10.9}$$

This is equivalent to the approximation (10.6); but the factor t has now been absorbed into ω, so that this quantity is to be considered small. In particular we shall assume $|\nabla\omega| \ll U_0$ and neglect powers higher than the first of $\nabla\omega$ in our calculations.

For example, BERNOULLI'S equation (1.19) becomes in this approximation

$$\int \frac{dp}{\varrho} + \frac{1}{2}\,(U_0^2 + 2\,U_0\,\omega_x) = \text{const}. \tag{10.10}$$

If we put $p_0 = p(U_0^2)$ and

$$\Delta p = p - p_0 \tag{10.11}$$

we obtain from Eq. (10.10) the linearized Bernoulli equation

$$\frac{\Delta p}{\frac{1}{2}\,\varrho_0\, U_0^2} = -2\,\frac{\omega_x}{U_0}\,. \tag{10.12}$$

This formula expresses the pressure perturbation at every point in terms of the local speed perturbation and is decisive in all dynamical questions connected with the linearized flow.

We observe that the free stream velocity potential φ_0 is a solution of the differential equation (10.8) and that because of the linearity of this equation the solution φ itself is also a solution of Eq. (10.8).

The detailed theory of Eq. (10.8) depends decisively on the sign of the constant $1-M_0^2$ and is basically different in the elliptic case $M_0 < 1$ and the hyperbolic case $M_0 > 1$. We have thus to distinguish between a linearized theory of subsonic flow which is very similar to the problem of incompressible fluid flow and a linearized theory of supersonic flow. Near $M_0 = 1$, both theories of linearization break down and a continuous transition from the linearized subsonic theory to the linearized supersonic theory is, therefore, not possible.

11. PRANDTL's correspondence rule for affine deformation. In the subsonic case the linearized differential equation for the velocity potential has the form

$$\beta^2 \varphi_{xx} + \varphi_{yy} + \varphi_{zz} = 0, \qquad \beta = \sqrt{1-M_0^2}. \tag{11.1}$$

We can construct at once an infinity of solutions by referring to the highly developed theory of LAPLACE's equation. In fact, let $f(x, y, z)$ be a solution of this latter equation; then

$$\varphi(x, y, z) = f(x, \beta y, \beta z) \tag{11.2}$$

will obviously be a solution of Eq. (11.1) and, conversely, every solution of Eq. (11.1) can be obtained in this way.

The theory of harmonic functions $f(x, y, z)$ is closely related to the theory of incompressible fluid flow; indeed, each harmonic function may be interpreted as the velocity potential of such a flow. It is, therefore, natural to look for a correspondence principle which will connect flow patterns in the incompressible and the compressible cases. Such a principle has been developed by PRANDTL [*1*] to [*3*] and has been applied systematically in wing theory by GLAUERT [*4*]. See also for detailed discussion: [*5*], [*6*], [*7*], [*9*], [*10*].

We start with an incompressible flow around a thin body B' with boundary surface S' and write its velocity potential in the form

$$\varphi' = U_0 x + f(x, y, z). \tag{11.3}$$

Here $U_0 x$ is the potential of a uniform flow far from the body B' and f is a harmonic function expressing the distortion of the uniform flow due to the presence of B'. Since B' is supposed thin, we can expect f and its partial derivatives to be small. Since we have on S' the boundary conditions $\partial\varphi'/\partial n = 0$, f can be characterized by the boundary condition on S':

$$(U_0 + f_x)\nu_x + f_y \nu_y + f_z \nu_z = 0, \qquad \boldsymbol{\nu} = \text{normal of } S'. \tag{11.4}$$

Finally, the fact that B' is a thin body expresses itself in the inequality

$$\nu_x \ll 1; \tag{11.5}$$

that is, the normal to S' is at every point almost perpendicular to the direction of the uniform speed. Within the framework of our approximation theory we may neglect the product of two small quantities $f_x \nu_x$ and can put Eq. (11.4) into the simpler form

$$U_0 \nu_x + f_y \nu_y + f_z \nu_z = 0. \tag{11.6}$$

Consider now the function

$$\varphi(x, y, z) = U_0 x + \frac{1}{\beta^2} f(x, \beta y, \beta z). \tag{11.7}$$

It is a solution of the differential equation (11.1) and may be interpreted as the velocity potential of a linearized flow coming from infinity with a uniform velocity U_0 in the x-direction.

It will be convenient to plot the compressible fluid flow in the (x, y, z)-space while the incompressible flow shall be represented in a space with the coordinates x', y', z'. We put the two spaces into the affine relation

$$x' = x, \qquad y' = \beta y, \qquad z' = \beta z. \tag{11.8}$$

We observe that the velocity potential (11.3) leads at the point x', y', z' to the velocity

$$u' = U_0 + f_{x'}, \qquad v' = f_{y'}, \qquad w' = f_{z'}(x', y', z') \tag{11.9}$$

while the compressible flow described by Eq. (11.7) has at the corresponding point x, y, z the velocity

$$u = U_0 + \frac{1}{\beta^2} f_x, \qquad v = \frac{1}{\beta} f_y, \qquad w = \frac{1}{\beta} f_z(x, \beta y, \beta z). \tag{11.10}$$

If we neglect f_x in comparison to U_0 we see that the velocity vectors at corresponding points stand in the relation

$$u' = u, \qquad v' = \beta v, \qquad w' = \beta w. \tag{11.11}$$

In other words: *Under the affine transformation (11.8) the velocity vectors of Eq. (11.7) are carried covariantly into the velocity vectors of Eq. (11.3).*

Since the streamlines of each flow are obtained by integration of the direction field of velocities, we see that the transformation

$$\left.\begin{aligned} &x = x', \quad y = \beta^{-1} y', \quad z = \beta^{-1} z'; \\ &\varphi(x, y, z) = (1 - \beta^{-2})\, U_0 x' + \beta^{-2} \varphi'(x', y', z') \end{aligned}\right\} \tag{11.12}$$

carries the entire flow pattern of an incompressible fluid in the (x', y', z')-space into a flow pattern of a linearized compressible fluid in the (x, y, z)-space.

The surface S' in the (x', y', z')-space has been carried over into a surface S in the (x, y, z)-space and it is easily seen that $\partial\varphi/\partial n = 0$ on S is again satisfied. Therefore, if we can solve the problem of an incompressible fluid flow around a thin body B', we have solved at the same time the problem of a linearized compressible flow around a body B obtained from B' by an affine thickening of the body in the ratio $1:\beta$ in all directions perpendicular to the free stream velocity.

If we apply the approximation formula (10.12) to the incompressible fluid flow we find for it a pressure perturbation from the state of uniform flow:

$$\frac{\Delta p'}{\frac{1}{2}\varrho_0 U_0^2} = -2\,\frac{f_{x'}(x', y', z')}{U_0}. \tag{11.13}$$

Similarly, we find for the deviation from the uniform pressure p_0 in the case of the corresponding compressible fluid flow:

$$\frac{\Delta p}{\frac{1}{2}\varrho_0 U_0^2} = -2\,\frac{f_x(x, \beta y, \beta z)}{\beta^2 U_0}. \tag{11.14}$$

Hence, we have the correspondence

$$\Delta p = \beta^{-2} \Delta p'. \tag{11.15}$$

We summarize our results in the following rule, due to PRANDTL:

An incompressible fluid flow around a thin body is carried, streamline by streamline, into a linearized compressible fluid flow around another thin body by an affine stretching of the entire space in the ration 1:β perpendicular to the uniform flow. The speed at infinity is unchanged and the pressure perturbations in the corresponding points under the affinity stand in the ratio 1:β².

In the case of slender bodies the above reasoning can be applied without change. However, it can be shown that in this case the term f_x becomes small of higher than the first order in the perturbation parameters and the second order terms in the Bernoulli equation cannot be neglected relative to it [*10*]. It can, however, be seen easily that the relation (11.15) remains valid even when the second order terms in the Bernoulli equation are taken into consideration. Hence, PRANDTL's rule remains valid for slender bodies.

It is well known that the theory of incompressible irrotational fluids leads to the paradox of D'ALEMBERT; no force is exerted by a steady flow on any body immersed in it. PRANDTL's rule implies then the same paradox for a linearized compressible fluid flow. Indeed, it can be shown that even for the exact theory of a compressible fluid flow one is led again to D'ALEMBERT's paradox [*61*].

12. Further Prandtl rules for correspondence between flow of an incompressible fluid and linearized flow of a compressible fluid. We started in the preceding section with the potential (11.3) of an incompressible fluid flow and compared its flow pattern with that belonging to the solution (11.7) of the linearized potential equation (11.1). This correspondence is particularly interesting because of the simple transformation law for the streamlines. There are, however, many other possibilities of deriving a potential φ for the linearized theory from a harmonic potential φ' [*6*], [*9*], [*10*].

Given the harmonic potential (11.3), we consider the solution

$$\varphi(x, y, z) = U_0 x + k f(x, \beta y, \beta z) \tag{12.1}$$

of Eq. (11.1). We shall dispose of the coefficient k later. We have now to consider three families of streamlines:

a) The streamlines $y = y_0$, $z = z_0$ of the uniform flow.

b) The streamlines of the incompressible fluid flow in the (x', y', z')-space with the velocity potential (11.3). They satisfy the differential equations

$$dx' : dy' : dz' = U_0 : f_{y'}(x', y', z') : f_{z'}(x', y', z'). \tag{12.2}$$

Since we assume the obstacle to be a thin body the new streamlines will not deviate much from the straight lines $y' = y_0'$, $z' = z_0'$. We can integrate Eq. (12.2) and obtain

$$y' - y_0' = \frac{1}{U_0} \int_{-\infty}^{x'} f_{y'}\, dx', \qquad z' - z_0' = \frac{1}{U_0} \int_{-\infty}^{x'} f_{z'}\, dx'. \tag{12.3}$$

Since $f_{y'}$ and $f_{z'}$ are small we may, within the precision of our approximation, integrate over the original streamlines $y' = y_0'$, $z' = z_0'$. Formulas (12.3) give then the distortion of these straight streamlines due to the presence of the thin body.

c) The streamlines of the linearized compressible fluid flow in the (x, y, z)-space with the velocity potential (12.1). They satisfy the differential equations

$$dx : dy : dz = U_0 : k\beta f_y : k\beta f_z. \tag{12.4}$$

In analogy to the integration from Eq. (12.2) to (12.3) we find for the distortion of these streamlines

$$y - y_0 = \frac{k\beta}{U_0} \int_{-\infty}^{x} f_y\, dx, \qquad z - z_0 = \frac{k\beta}{U_0} \int_{-\infty}^{x} f_z\, dx. \tag{12.5}$$

Comparing Eqs. (12.3) and (12.5), we obtain

$$y - y_0 = k\beta(y' - y_0'), \qquad z - z_0 = k\beta(z' - z_0'); \tag{12.6}$$

these formulas connect the distortion of streamlines through corresponding points in the affinity transformation

$$x' = x, \qquad y' = \beta y, \qquad z' = \beta z. \tag{12.7}$$

It is natural to choose $k = \beta^{-1}$ since in this case the distortion of streamlines at corresponding points becomes equal.

In order to apply this result let us consider a thin airfoil which may be represented by its profile in the (x, y)-plane. We consider the ensuing flow as two-dimensional and disregard the z-dependence entirely; we assume that the profile deviates very little from the x-axis and that the streamline around the profile is due to a distortion of the unperturbed streamline $y = y_0$. If we know the solution for a flow past this profile in the incompressible case, we have immediately a solution for the linearized compressible case with respect to the same profile. Indeed, since the streamline $y = y_0$ undergoes the same deformation in the (x', y')-plane of the incompressible flow as in the plane (x, y) of the compressible one, it is clear that it deforms into the same profile.

Thus, the transformation

$$x = x', \qquad y = \beta^{-1} y', \qquad \varphi(x, y) = (1 - \beta^{-1})\, U_0 x' + \beta^{-1} \varphi'(x', y') \tag{12.8}$$

leads from an incompressible fluid flow around a profile to a linearized compressible fluid flow around the same profile.

From the linearized Bernoulli equation (10.12) we find for the pressure perturbation of the two flows the relation

$$\Delta p = \beta^{-1} \Delta p'. \tag{12.9}$$

The velocity fields at corresponding points stand in the relation

$$u - U_0 = \beta^{-1}(u' - U_0), \qquad v = v'. \tag{12.10}$$

We summarize our results in PRANDTL's rule:

If an incompressible fluid flow around a thin two-dimensional profile is known, one can determine the linearized compressible fluid flow around the same profile. The velocity perturbation of the second flow is obtained from the velocity perturbation of the first at the affinely corresponding point by multiplying the x-component by β^{-1} and leaving the y-component unchanged. The pressure perturbation of the linearized flow is β^{-1} times the pressure perturbation of the incompressible flow at the corresponding points.

This Prandtl rule can be generalized to a certain extent for thin airfoils of finite length. This is obvious from our derivation of this rule.

Another interesting choice of k is $k = 1$. In this case, the value of the velocity potentials φ' and φ is the same at corresponding points in the affine transformation (12.7). Since we can map simply the level surfaces $\varphi' = \text{const}$ into the surface $\varphi = \text{const}$ and determine the streamlines of the linearized flow as the orthogonal trajectories of these surfaces this correspondence is also very convenient for the construction of flow patterns. It is easy to determine the distortion of the boundary obtained under this transition from the incompressible to the linearized compressible theory.

This question may again be answered more easily in the case of a thin two-dimensional profile. We assume that the given profile for the incompressible fluid flow is formed by a streamline, which has been obtained by distorting the unperturbed streamline $y=0$. According to Eq. (12.6), we have to distort the streamline in the linearized case proportionally with the factor β. Thus, the profile in the compressible case will be thinner by this factor β.

We have thus PRANDTL's rule:

The transformation

$$x = x', \quad y = \beta^{-1} y', \quad \varphi(x, y) = \varphi'(x', y') \tag{12.11}$$

carries an incompressible fluid flow around a thin profile in the (x', y')-plane into a compressible fluid flow around a profile in the (x, y)-plane which is obtained from the former by shrinking in the direction perpendicular to the flow in the ratio $\beta:1$. The pressure perturbation at corresponding points is the same in both flows.

The Prandtl rules creating correspondence between incompressible fluid flow and linearized compressible fluid flow are very extensively used in applied fluid dynamics. The three transformations (11.12), (12.8) and (12.11) are distinguished by the properties of affinely transforming streamlines, of preserving profiles and of affinely transforming isobars, respectively. All three are good approximations in the case of thin obstacles but only the first can be justified in the case of slender bodies [*8*], [*10*] to [*14*]. The cause of difficulty in the two transformations (12.8) and (12.11) in the case of a slender body is the fact that near the obstacle the perturbation potential $f(x, y, z)$ will have, in general, unbounded higher derivatives, and the truncations made in establishing the distortion relations (12.6) cannot be justified.

13. Application to thin wings in linearized subsonic flow. By means of PRANDTL's rule of analogy between incompressible and linearized compressible fluid flow one can draw immediate conclusions for the linearized theory from known solutions of incompressible fluid flow problems. Consider, for example, a thin two-dimensional profile P in a uniform flow with speed U_0 in direction of the positive x-axis. Its velocity potential for the incompressible case is

$$\varphi'(x, y) = U_0 x + f(x, y), \tag{13.1}$$

where f is a harmonic function outside of the profile P. We may assume f and its derivatives to be small perturbation terms since P is supposed thin and oriented in the direction of the flow. Near infinity, we can represent f as the real part of an analytic function in the form

$$f(x, y) = \mathrm{Re}\left\{\frac{K' i}{2\pi} \log Z + \sum_{n=1}^{\infty} \frac{a_n}{Z^n}\right\}, \qquad Z = x + i y. \tag{13.2}$$

Here K' is a real number, the circulation created by the perturbation; K' and the coefficients a_n are small.

We consider next the velocity potential

$$\varphi(x, y) = U_0 x + \frac{1}{\beta} f(x, \beta y). \tag{13.3}$$

In view of the profile preserving property of the transformation (12.8) this function describes a linearized compressible flow around the same thin profile P with the same uniform speed U_0.

Now let Δp and $\Delta p'$ denote the pressure perturbations of the two flows at the same point x, y. We find from Eq. (10.12)

$$\frac{\Delta p}{\Delta p'} = \frac{f_x(x, \beta y)}{\beta f_x(x, y)}. \tag{13.4}$$

We want to compute this ratio for large values of y; that means, we shall compare the influence of the two perturbations at large distances from the profile in a direction perpendicular to the flow and the profile P. We find easily that in general

$$\lim_{y\to\infty} \frac{\Delta p}{\Delta p'} = \beta^{-2}. \tag{13.5}$$

In the case of zero circulation the ratio will even become larger in the limit.

We observe next that the correspondence of velocity potentials (13.1) and (13.3) leads to the corresponding values for the circulation around the same profile

$$K = \frac{1}{\beta} K'. \tag{13.6}$$

If we wish to compute the lift forces L and L' on the profile P we have to calculate the integrals

$$L = \oint_P \Delta p\, dx, \qquad L' = \int_P \Delta p'\, dx. \tag{13.7}$$

We have established the rule (12.9) for the perturbation pressures under a profile preserving transformation (12.8). However, this formula compares p and p' at affinely corresponding points, while we need in Eqs. (13.7) the values of p and p' at the same points on the profile P. But if we suppose the profile P to be small and to lie entirely in a very narrow neighborhood of the x-axis, the error committed by taking points on P in Eq. (12.9) is very small and we may assume within our degree of approximation

$$\Delta p = \beta^{-1} \Delta p' \tag{13.8}$$

which yields by Eq. (13.7)

$$L = \beta^{-1} L'. \tag{13.9}$$

The same relation holds by the same argument for the drag forces in the x-direction. Since an incompressible flow does not exert a force in the direction of its motion, we can conclude the same for a linearized compressible flow. There is no drag in this case either.

If M and M' are the moments of the fluid forces on the profile the above argument shows that

$$M = \beta^{-1} M'. \tag{13.10}$$

The preceding rules for transition from the theory of a wing profile in the incompressible case to that for the linearized compressible case is due to PRANDTL [*3*] and GLAUERT [*4*].

It should also be remarked that the Kutta-Joukowski formula

$$L' = \varrho_0 U_0 K' \tag{13.11}$$

relating lift and circulation in the incompressible case remains unaltered:

$$L = \varrho_0 U_0 K, \tag{13.12}$$

by virtue of the formulas (13.6) and (13.9).

It is evident that as $M_0 \to 1$ the factor β^{-1} will become indefinitely large and that the transformation rules given will become unreliable. The reason for this is the fact that the series development of the velocity potential φ in terms of the smallness parameter t (discussed in Sect. 1) is not uniformly convergent in M_0 and that the range of convergence becomes progressively smaller as M_0 approaches the sonic limit 1.

In the theory of the flow past a thin body, we assume that the velocities induced by the perturbation potential are small compared to the free stream velocity U_0. This is true in general but obviously cannot hold at the stagnation points of the flow, where the perturbation velocities have to compensate the free stream velocity. However, the region within the flow where our approximation assumptions are not valid is so small that the consequences for the total lift and momentum are negligible [*10*], [*60*].

14. Potential theory for thin bodies. By the Prandtl rules we have reduced the theory of linearized compressible fluid flows around thin obstacles to that of incompressible fluid flows. Since the latter theory is essentially a boundary value problem for solutions of LAPLACE's equation, we have connected the problem of linearized flow past an obstacle with one of the most highly developed theories of analysis. However, it is not even necessary to apply the full power of potential theory; the thinness of the obstacle can be used again in order to simplify the solution of the new boundary value problem.

We consider, for the sake of simplicity, the case of a thin two-dimensional profile P which lies near to the segment $-l \leqq x \leqq l$ of the x-axis. We suppose that the profile is cut by each parallel to the y-axis in two points, say in $y^+(x)$ and $y^-(x)$; we denote the normal to the profile at the upper point by $\boldsymbol{\nu}^+$ and at the lower point by $\boldsymbol{\nu}^-$. The velocity potential of the incompressible fluid flow around P which has at infinity the speed U_0 and the direction of the positive x-axis is

$$\varphi(x, y) = U_0 x + f(x, y). \tag{14.1}$$

The function f is determined by the conditions of being harmonic outside of P with the possible exception of the point at infinity where it may have a vortex-like singularity and by the boundary condition on P:

$$U_0 \nu_x + f_y \nu_y = 0. \tag{14.2}$$

Since the profile is thin and oriented in the x-direction, ν_x is small and $\nu_y = \sqrt{1-\nu_x^2}$ can be replaced by the value ± 1 in our approximation. Thus Eq. (14.2) leads to the boundary value problem

$$f_y = -U_0 \nu_x^+, \quad f_y = U_0 \nu_x^- \quad \text{on } P \tag{14.3}$$

for the harmonic function $f(x, y)$.

Now we replace the boundary value problem (14.3) with respect to the thin profile P by the following boundary value problem: To find a harmonic function $f(x, y)$ outside of the segment $-l \leqq x \leqq l$ of the x-axis which has the boundary values of the normal derivative

$$f_y = -U_0 \nu_x^+ \quad \text{or} \quad f_y = U_0 \nu_x^- \tag{14.4}$$

on the segment according as we approach the segment from above or from below, respectively.

In order to solve this problem of two-dimensional potential theory, we introduce the complex variable

$$Z = x + i y \tag{14.5}$$

and the analytic function

$$F(Z) = f_x - i f_y. \tag{14.6}$$

We set up the representation

$$F(Z) = \frac{1}{2\pi i} \int_{-l}^{l} \frac{\mu(\xi)}{\xi - Z} d\xi \tag{14.7}$$

where the complex weight function $\mu(\xi)$ has to be adjusted to the requirements of the boundary value problem.

It is well known from the elements of analytic function theory that $F(Z)$ represents an analytic function of Z outside of the segment $\langle -l, l\rangle$ of the x-axis and that

$$\lim_{z\to x^+} F(Z) = \frac{1}{2}\mu(x) + \frac{1}{2\pi i}\int_{-l}^{+l} \frac{\mu(\xi)}{\xi - x}\, d\xi, \tag{14.8}$$

$$\lim_{z\to x^-} F(Z) = \frac{-1}{2}\mu(x) + \frac{1}{2\pi i}\int_{-l}^{+l} \frac{\mu(\xi)}{\xi - x}\, d\xi \tag{14.8'}$$

where the two integrals are improper and have to be evaluated as Cauchy principal values.

Subtracting Eq. (14.8') from Eq. (14.8) and using the boundary condition (14.4), we find

$$\operatorname{Im}\{\mu(x)\} = U_0(v_x^+ + v_x^-). \tag{14.9}$$

On the other hand, addition of Eqs. (14.8) and (14.8') together with Eq. (14.4) yields the integral equation

$$\frac{1}{\pi}\int_{-l}^{l} \frac{\operatorname{Re}\{\mu(\xi)\}}{\xi - x}\, d\xi = U_0(v_x^+ - v_x^-) \tag{14.10}$$

for the real part of $\mu(\xi)$.

The integral equation

$$\frac{1}{\pi}\int_{-l}^{l} \frac{\gamma(\xi)}{\xi - x}\, d\xi = G(x) \tag{14.11}$$

can be solved by the Betz inversion formula [17] to [19]

$$\sqrt{l^2 - x^2}\,\gamma(x) = \Gamma - \frac{1}{\pi}\int_{-l}^{l} G(\xi)\sqrt{l^2 - \xi^2}\,\frac{d\xi}{\xi - x} \tag{14.12}$$

with an arbitrary constant Γ.

By using Eqs. (14.9), (14.10) and (14.12), we can therefore determine the velocity field of the perturbed motion in terms of definite integrals. The arbitrary constant Γ appears in the solution since we can indeed prescribe the circulation of the flow around the profile in an arbitrary way.

By Eqs. (14.8) and (14.8'), the following formula for the forward speed u on the profile P is obtained:

$$u - U_0 = f_x = \frac{U_0}{2\pi}\int_{-l}^{l} \frac{(v_\xi^+ + v_\xi^-)}{\xi - x}\, d\xi \pm \frac{1}{2\sqrt{l^2 - x^2}}\left[\Gamma - \frac{U_0}{\pi}\int_{-l}^{l} \sqrt{l^2 - \xi^2}\,\frac{(v_\xi^+ - v_\xi^-)}{\xi - x}\, d\xi\right]. \tag{14.13}$$

The upper or the lower sign has to be taken on the upper or lower side of the profile, respectively. The approximation used leads to infinite velocities at the two end-points $x = \pm l$ of the profile. Even worse, we obtain different asymptotic behavior when we approach these endpoints from above or below. The Kutta-Joukowski condition on the trailing edge of P is reasonably formulated as the requirement that at this endpoint at least the second antisymmetric term be zero. This leads to a uniform asymptotic behavior at the trailing edge; it determines the circulation constant Γ and enables us to calculate lifts and moments on the profile.

After having solved the approximation problem for a thin profile in the incompressible case, we can immediately apply the results to the linearized compressible fluid flow problem for the same profile by using the Prandtl rules of correspondence.

In the case of a thin obstacle in three dimensions it is sometimes possible also to express the perturbation on the uniform flow as a definite integral. In fact, we may suppose that the entire obstacle lies near the plane $z=0$ and let D be the two-dimensional domain in this plane obtained by orthogonal projection of the obstacle. As before, we denote by $\boldsymbol{\nu}^+$ and $\boldsymbol{\nu}^-$ the normal vectors at the upper and lower side of the obstacle at the point with the projection (x, y) in D. The velocity potential of the incompressible fluid flow around the obstacle is

$$\varphi(x, y, z) = U_0 x + f(x, y, z) \tag{14.14}$$

where f is a small harmonic perturbation function determined by the boundary conditions

$$U_0 \nu_x^+ + f_z = 0, \qquad U_0 \nu_x^- - f_z = 0. \tag{14.15}$$

We made here use of the fact that ν_x, ν_y are small because of the thinness of the obstacle and that $\nu_z = \pm 1$ up to a second order error in our approximation.

Consider now the integral

$$f(x, y, z) = -\frac{1}{2\pi} \iint\limits_D \frac{\mu(\xi, \eta)\, d\xi\, d\eta}{\sqrt{(x-\xi)^2 + (y-\eta)^2 + z^2}}. \tag{14.16}$$

It is an even function in z and is harmonic in the entire (x, y, z)-space outside of the plane domain D. Its z-derivative f_z is an odd function of z; it vanishes, therefore, at all points of the plane $z=0$ which do not lie in D. From the elements of potential theory, we know that if we approach $z=0$ at a point in D from above and from below, we obtain two different limit values f_z^+ and f_z^-, respectively such that at the point $(x, y) \in D$

$$f_z^+ - f_z^- = 2\mu(x, y). \tag{14.17}$$

Since f_z is odd we find

$$f_z^+ = \mu(x, y) \qquad f_z^- = -\mu(x, y). \tag{14.18}$$

In view of Eqs. (14.16) and (14.18) we can now easily solve the harmonic boundary value problem (14.15) for the perturbation potential $f(x, y, z)$ in Eq. (14.14) provided that the thin obstacle is symmetric with respect to the plane $z=0$. Indeed, the symmetry implies

$$\nu_x^+ = \nu_x^- = \nu_x(x, y) \tag{14.19}$$

and the integral

$$f(x, y, z) = \frac{U_0}{2\pi} \iint\limits_D \frac{\nu_x(\xi, \eta)\, d\xi\, d\eta}{\sqrt{(x-\xi)^2 + (y-\eta)^2 + z^2}} \tag{14.20}$$

is harmonic outside of D and satisfies on D the boundary conditions (14.15). It is true, that we ask in fact that Eq. (14.15) shall hold on the boundary of the thin obstacle and not on D; but we can again assert that the error committed by replacing the one boundary condition by the other is negligible in our approximation.

The case of non-symmetric thin bodies in three-space is much more complicated and will not be discussed here (see [*10*]). However, it should be remarked that the boundary value problem considered again belongs to classical potential theory and that the transition to the problem of a linearized compressible fluid around the same obstacle is, in general, quite simple because of the Prandtl rules.

15. Linearized supersonic flow. Mach cones. Domains of dependence and influence. In the supersonic case the linearized differential equation for the velocity potential, discussed in Sect. 10, has the form

$$B^2 \varphi_{xx} - \varphi_{yy} - \varphi_{zz} = 0, \qquad B = \sqrt{M_0^2 - 1}. \tag{15.1}$$

This is a linear hyperbolic differential equation with constant coefficients. It has real characteristic surfaces Σ defined by the equation

$$g(x, y, z) = 0 \tag{15.2}$$

where g satisfies the first order partial differential equation (9.5). It reduces in our particular case to the simple form

$$B^2 \left(\frac{\partial g}{\partial x}\right)^2 - \left(\frac{\partial g}{\partial y}\right)^2 - \left(\frac{\partial g}{\partial z}\right)^2 = 0. \tag{15.3}$$

We find easily a complete integral to the partial differential equation (15.3); namely

$$g(x, y, z) = (x - x_0)^2 - B^2[(y - y_0)^2 + (z - z_0)^2] = 0. \tag{15.4}$$

This is the general equation of a right circular cone whose axis is along the x-direction and whose vertex semi-angle is given by

$$\alpha = \operatorname{arc\,cot} \sqrt{M_0^2 - 1}, \qquad \sin\alpha = \frac{1}{M_0} = \frac{c(U_0)}{U_0}. \tag{15.5}$$

This solution admits a simple and illuminating physical interpretation. Suppose we have a uniform flow in the positive x-direction with the supersonic speed U_0. If we produce a permanent perturbation of the flow at one fixed point x_0, y_0, z_0 it will spread from that point along concentric spheres with the local speed of sound $c_0 = c(U_0)$. However, these spheres are washed downstream with the larger speed of flow U_0 and the influence of the perturbation at x_0, y_0, z_0 is confined permanently to the interior of the cone (15.4) which is the envelope of the moving and expanding spheres $(x - x_0 - U_0 t)^2 + (y - y_0)^2 + (z - z_0)^2 = c_0^2 t^2$ — more precisely, to the half-cone (15.4) which satisfies the additional inequality

$$x \geqq x_0. \tag{15.6}$$

The presence of a permanent obstacle at x_0, y_0, z_0 does not make itself felt at all at the points upstream $x < x_0$ and outside of the half-cone. In the outside, we have the velocity potential $\varphi = U_0 x$ of a free stream while inside of the cone appears a different flow pattern, depending on the perturbation at the vertex. The role of the cone as a characteristic surface along which two solutions of the wave equation (15.1) are fused continuously but non-analytically is quite obvious. Our interpretation of the cone (15.4) proves also that small perturbations in the flow spread with the local speed of sound, relative to a point moving with the flow.

From the elementary theory of partial differential equations it follows that every solution of Eq. (15.3) can be obtained as the envelope of a properly chosen one-parameter family of cones (15.4). This means physically that the general characteristic surface can be built up from elementary cones due to the perturbation of the flow at the individual points.

The cones (15.4) are called the Mach cones of the uniform flow with speed U_0. They can be visualized experimentally by the Toepler "Schlieren method" which brings out optically the fast changes of density across the characteristic surfaces. The angle α defined by Eq. (15.5) is called the Mach angle of the flow [20] to [22].

We have shown that a point x_0, y_0, z_0 in a uniform flow affects only the flow pattern in the down-stream half-cone (15.4). We wish to show, on the other hand, that the flow at x_0, y_0, z_0 depends only on the flow in the up-stream half of the same Mach cone. While this fact is geometrically rather obvious in view of our preceding arguments, we give an analytical proof which exhibits the power of GREEN's integral identity in considerations of this kind. Let S be an arbitrary closed surface and let $\boldsymbol{\nu} = (\nu_x, \nu_y, \nu_z)$ denote the inner normal to S at the point x, y, z. Let $\Phi(x, y, z)$ be twice continuously differentiable in the domain D interior to S and satisfy there the wave equation (15.1). GREEN's identity, valid for arbitrary vector fields $\boldsymbol{V}$ in D,

$$\iint_S \boldsymbol{V}\cdot\boldsymbol{\nu}\, d\sigma = -\iint_D (\nabla\cdot V)\, dx\, dy\, dz \tag{15.7}$$

if applied to

$$\boldsymbol{V} = (B^2\Phi_x^2 + \Phi_y^2 + \Phi_z^2, -2\Phi_x\Phi_y, -2\Phi_x\Phi_z)$$

implies obviously by virtue of Eq. (15.1)

$$\left.\begin{aligned} &\iint_S [2\Phi_x(\nu_y\Phi_y + \nu_z\Phi_z) - \\ &\qquad - \nu_x(B^2\Phi_x^2 + \Phi_y^2 + \Phi_z^2)]\, d\sigma = 0. \end{aligned}\right\} \tag{15.8}$$

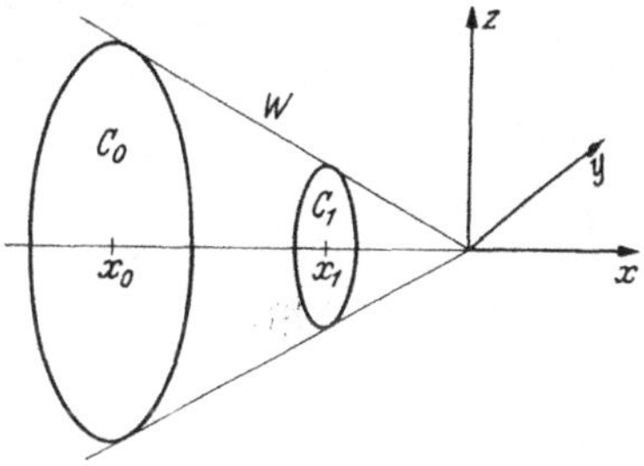

Fig. 4.

In order to utilize this identity we select an arbitrary point in the flow; we may assume without loss of generality that it is the origin. Consider the up-stream half of its Mach cone (Fig. 4).

$$x^2 = B^2(y^2 + z^2), \qquad x < 0. \tag{15.9}$$

We cut this cone with the planes $x = x_0$ and $x = x_1$, $x_0 < x_1 < 0$. Let D be the frustrum cut by these planes from the cone and let S denote the boundary of D. Thus, S consists of the two circles C_0 and C_1 in the planes $x = x_0$ and $x = x_1$, respectively, and of the conical wall W. We observe that the inner normal of S is $(1, 0, 0)$ on C_0, $(-1, 0, 0)$ on C_1 and $\frac{1}{M_0}\left(-1, B^2\frac{y}{x}, B^2\frac{z}{x}\right)$ on W. It is easy to see that for this particular choice of the surface of integration S the identity (15.8) takes the elegant form

$$\iint_{C_0} [B^2\Phi_x^2 + \Phi_y^2 + \Phi_z^2]\, dy\, dz = \frac{1}{M_0}\iint_W \left[\left(\Phi_y + \frac{y}{x}B^2\Phi_x\right)^2 + \left(\Phi_z + \frac{z}{x}B^2\Phi_x\right)^2\right] d\sigma + \\ + \iint_{C_1} [B^2\Phi_x^2 + \Phi_y^2 + \Phi_z^2]\, dy\, dz.$$

Suppose now that $\nabla\Phi$ vanishes identically on C_0; the sum of the two integrals on the right side must vanish, and, since they are non-negative by definition, each must vanish separately. In particular, we conclude

$$\nabla\Phi \equiv 0 \quad \text{on } C_1. \tag{15.10}$$

Now, the abscissa x_1 is entirely arbitrary except for the condition $x_0 < x_1 < 0$. Hence, we can conclude that

$$\nabla\Phi \equiv 0 \quad \text{for} \quad x^2 > B^2(y^2 + z^2), \qquad x_0 < x < 0. \tag{15.10'}$$

An arbitrary solution of Eq. (15.1) which has a vanishing gradient on C_0 is constant in the entire cone over C_0 until $x = 0$.

More generally: A solution of the wave equation (15.1) is determined in the above cone by its gradient on the basis C_0 up to an additive constant. In fact, if there were two solutions of Eq. (15.1), say Φ_1 and Φ_2, with the same values of the gradient on C_0, then the difference function $\Phi = \Phi_1 - \Phi_2$ would also be a solution of Eq. (15.1) but with vanishing gradient on C_0. Hence, Φ must be a constant in the cone by our preceding result.

If we cut the upstream Mach cone through the origin with an arbitrary surface S_1 instead of the plane $x = x_0$ we obtain a cone with a curved base. The preceding argument shows that the values of $\nabla\Phi$ on this base determine $\nabla\Phi$ uniquely in the cone up to the vertex. If there are rays from the vertex which intersect the base surface more than once we obviously cannot prescribe $\nabla\Phi$ arbitrarily on the entire base, since some pieces of it lie in the region of influence of other pieces. Thus, while the uniqueness of a solution can be asserted in every case, the existence problem may lead to new difficulties.

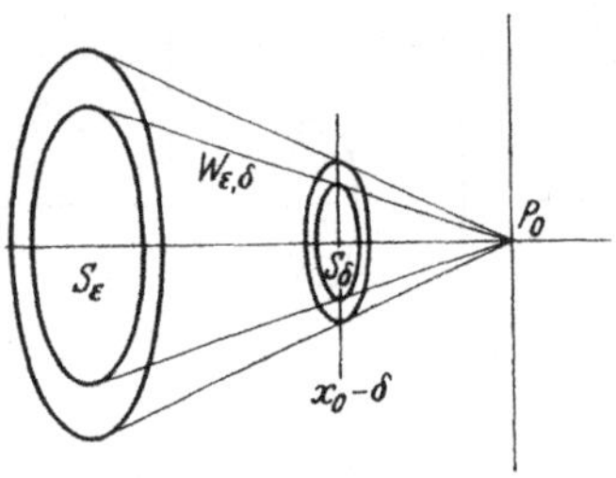

Fig. 5.

Given a point P in the flow, we draw the Mach cone through it. Its upstream half consists of all points which can influence the flow pattern at P; its down-stream half consists of all points which can be affected by perturbations of the flow at P. We may call the up stream half of the Mach cone the domain of dependence of P and the downstream half of the cone the domain of influence of the point P.

16. The initial value problem for the wave equation. Method of finite parts. The partial differential equation (15.1) has the particular solution

$$\Gamma(x, y, z; x_0, y_0, z_0) = [(x - x_0)^2 - B^2(y - y_0)^2 - B^2(z - z_0)^2]^{-\frac{1}{2}} \tag{16.1}$$

which plays in the theory of the wave equation a role analogous to that of r^{-1} in the theory of LAPLACE'S equation. Γ depends on the source point $P_0 \equiv (x_0, y_0, z_0)$; Γ is a real and continuous function inside the Mach cone of P_0, but becomes infinite on its surface and is imaginary outside of it. We can use this singularity function in order to give an explicit representation formula for any solution of Eq. (15.1) in terms of its data on some properly chosen surface S. The main tool will be GREEN'S integral identity just as in the harmonic case; but the fact that the fundamental singularity now becomes infinite on an entire cone instead of an isolated point will introduce a great difference in method [23] to [25].

We start with an arbitrary but fixed point P_0 and its upstream Mach half-cone. We cut from this half-cone a finite subdomain D by means of a surface S which is supposed to be intersected at most once by every ray from P_0. Let $\Phi(x, y, z)$ be twice continuously differentiable in D and be a solution of Eq. (15.1) there. We shall express Φ at P_0 in terms of its data on the part of the surface S in the Mach cone. This will be achieved by applying GREEN'S identity to Φ and Γ with respect to D; but we have to overcome the typical difficulty that Γ becomes infinite on the conical boundary of D.

We introduce for this purpose the conical sub-domain (Fig. 5)

$$(1 - \varepsilon)^2 (x - x_0)^2 - B^2(y - y_0)^2 - B^2(z - z_0)^2 > 0, \qquad x < x_0 \tag{16.2}$$

of the Mach half-cone. Let $D_{\varepsilon,\delta}$ be the domain cut from the subcone by the surface S, on the one side, and by the plane $x = x_0 - \delta$, $\delta > 0$, on the other. The boundary of $D_{\varepsilon,\delta}$ consists of a piece S_ε of S, a circle s_δ in the plane $x = x_0 - \delta$ and of a conical wall $W_{\varepsilon,\delta}$.

We denote the inner normal on the boundary of $D_{\varepsilon,\delta}$ by $\boldsymbol{\nu}$ and define there

$$\{\Gamma, \Phi\} = (B^2 \Phi_x \nu_x - \Phi_y \nu_y - \Phi_z \nu_z)\, \Gamma - \Phi (B^2 \Gamma_x \nu_x - \Gamma_y \nu_y - \Gamma_z \nu_z). \quad (16.3)$$

Since Γ and Φ satisfy in $D_{\varepsilon,\delta}$ the wave equation (15.1) we have in view of GREEN's identity

$$\iint\limits_{S_\varepsilon + s_\delta + W_{\varepsilon,\delta}} \{\Gamma, \Phi\}\, d\sigma = 0. \quad (16.4)$$

This identity holds for every choice of $\varepsilon > 0$, $\delta > 0$, and we wish to pass to the limits $\varepsilon = 0$, $\delta = 0$. For this purpose, the contributions of the three boundary pieces have to be considered separately.

We introduce cylindrical coordinates with respect to the point P_0:

$$\xi = x - x_0, \qquad r = \sqrt{(y - y_0)^2 + (z - z_0)^2}, \qquad \vartheta = \operatorname{arc\,tan} \frac{z - z_0}{y - y_0}. \quad (16.5)$$

The boundary piece $W_{\varepsilon,\delta}$ is characterized by the two equivalent equations

$$(1 - \varepsilon)^2 \xi^2 = B^2 r^2, \qquad \Gamma = \frac{1}{\xi} [\varepsilon (2 - \varepsilon)]^{-\frac{1}{2}}. \quad (16.6)$$

An easy calculation shows that on $W_{\varepsilon,\delta}$ holds:

$$B^2 \Gamma_x \nu_x - \Gamma_y \nu_y - \Gamma_z \nu_z = 0. \quad (16.7)$$

Therefore we can write

$$\iint\limits_{W_{\varepsilon,\delta}} \{\Gamma, \Phi\}\, d\sigma = \frac{1}{\sqrt{\varepsilon}}\, a(\varepsilon, \delta), \quad (16.8)$$

where, for fixed $\delta > 0$, $a(\varepsilon, \delta)$ is a continuously differentiable function of ε. If we use the mean value theorem of calculus, we may put

$$a(\varepsilon, \delta) = a(0, \delta) + O(\varepsilon) \quad (16.9)$$

and, hence, finally

$$\iint\limits_{W_{\varepsilon,\delta}} \{\Gamma, \Phi\}\, d\sigma = \frac{a(0, \delta)}{\sqrt{\varepsilon}} + o(1) \quad (16.10)$$

with $\lim o(1) = 0$ as $\varepsilon \to 0$.

We consider next the contribution of s_δ:

$$\iint\limits_{s_\delta} \{\Gamma, \Phi\}\, d\sigma = B^2 \iint\limits_{s_\delta} (\Gamma_x \Phi - \Gamma \Phi_x)\, r\, dr\, d\vartheta. \quad (16.11)$$

It can be shown by straightforward calculation that the integral (16.11) admits the asymptotic development in ε:

$$\iint\limits_{s_\delta} \{\Gamma, \Phi\}\, d\sigma = \frac{b(\delta)}{\sqrt{\varepsilon}} + \delta c - \frac{2 B^2}{\delta^2} \iint\limits_{r \leq \frac{\delta}{B}} \Phi\, d\sigma + o(1), \quad (16.12)$$

with a constant c which is independent of ε and δ.

Similarly, we have the asymptotic relation for the contribution of S_ε:

$$\iint\limits_{S_\varepsilon} \{\Gamma, \Phi\}\, d\sigma = \frac{E}{\sqrt{\varepsilon}} + L + o(1). \quad (16.13)$$

Here, L is a value which does not depend on ε but only on the values of Φ and $\nabla\Phi$ on the piece S_0 of S inside the Mach cone of P_0.

Using Eqs. (16.10), (16.12) and (16.13), we find instead of the identity (16.4):

$$\frac{1}{\sqrt{\varepsilon}}\left[a(0,\delta)+b(\delta)+E\right]+L+\delta c-\frac{2B^2}{\delta^2}\iint\limits_{r\leq\frac{\delta}{B}}\Phi\,d\sigma+o(1)=0. \quad (16.14)$$

This has to hold for every choice of ε and δ; holding δ fixed and sending $\varepsilon\to 0$, we find

$$a(0,\delta)+b(\delta)+E\equiv 0 \quad \text{for all } \delta\text{-values}. \quad (16.15)$$

Thus, Eq. (16.14) leads to

$$L+\delta c-\frac{2B^2}{\delta^2}\iint\limits_{r\leq\frac{\delta}{B}}\Phi\,d\sigma=0. \quad (16.16)$$

Now, we may let $\delta\to 0$; observe that the last term in Eq. (16.16) represents 2π times the average of Φ over a small circle near P_0. Hence, we obtain the limit

$$\Phi(x_0,y_0,z_0)=\frac{1}{2\pi}L. \quad (16.17)$$

This formula expresses $\Phi(x_0,y_0,z_0)$ in terms of Φ and $\nabla\Phi$ taken on the piece S_0 of S inside the Mach cone of P_0. The uniqueness result of the preceding section has thus been proved anew.

We observe the genesis of the decisive term L; it was obtained by considering the integral (16.13) which diverges as $\varepsilon\to 0$. It is possible, however, to split off in a unique way the term responsible for the divergence and to determine the remainder term without ambiguity. L is called the "finite part" of the divergent integral (16.13). The concept of the finite part of a divergent integral was introduced by HADAMARD [25] in connection with the initial value problem for hyperbolic differential equations. His definition may be summarized as follows [28]:

Let $g(x,y,z)$ be a positive analytic function of the three variables x, y, z in a domain D and let $g=0$ on the boundary Σ of D. Let

$$f(x,y,z)=\frac{h(x,y,z)}{[g(x,y,z)]^{n+\frac{1}{2}}} \quad (16.18)$$

where $h(x,y,z)$ is likewise analytic in D. If M_0 is a manifold in D (domain, surface or curve) which has parts of its boundary in common with D the integral of f over M_0 will, in general, diverge. Let $M_0(\varepsilon)$ be the part of M_0 which lies in the subdomain of D defined by the inequality

$$g(x,y,z)>\varepsilon. \quad (16.19)$$

Consider now the integral

$$J(\varepsilon)=\int\limits_{M_0(\varepsilon)} f\,dM_0. \quad (16.20)$$

It can then easily be shown that $J(\varepsilon)$ is finite and has a development

$$J(\varepsilon)=\sum_{\nu=0}^{n}a_\nu\,\varepsilon^{-(\nu+\frac{1}{2})}+{}^*J+o(1). \quad (16.21)$$

The term $^*J$ is uniquely defined by this rule and is called the finite part of the divergent integral; in notation:

$${}^*J={}^*\!\!\int\limits_{M_0} f\,dM_0. \quad (16.22)$$

Thus, we may now formulate the result (16.17) as follows [26] to [28]:

$$\Phi(x_0, y_0, z_0) = \frac{1}{2\pi} \overset{*}{\int\limits_{S_0}} \{\Gamma, \Phi\}\, d\sigma. \tag{16.23}$$

The following remark is useful in various applications. Suppose that the function $g_\tau(x, y, z)$ depends on a parameter τ; let D_τ be its domain of positiveness. If a fixed surface S is given in the (x, y, z)-space, let S_τ denote its part which lies in D_τ. Finally, let $h_\tau(x, y, z)$ depend likewise on τ and in analogy to Eq. (16.18) be $f_\tau = h_\tau g_\tau^{-(n+\frac{1}{2})}$. It can be shown from the definition (16.21) that

$$\frac{d}{d\tau} \overset{*}{\iint\limits_{S_\tau}} f_\tau\, d\sigma = \overset{*}{\iint\limits_{S_\tau}} \frac{\partial f_\tau}{\partial \tau}\, d\sigma. \tag{16.24}$$

Thus we may interchange the processes of taking finite parts and differentiation.

17. Application to thin supersonic wings. We illustrate the general initial value theory of the preceding section by considering an important special case. Let $\Phi(x, y, z)$ be a solution of the same Eq. (15.1) which is identically zero for $x < 0$ and has specified values on the half-plane $z = 0$, $x > 0$. The plane $z = 0$ will now play the role of the surface S considered in Sect. 16. Given a point $P_0 \equiv (x_0, y_0, z_0)$ with $x_0 > 0$, $z_0 > 0$, we intersect the plane S with the upstream Mach cone of P_0; in this way, we cut from the plane $z = 0$ a domain S_0 defined by the inequalities

$$(x - x_0)^2 - B^2 (y - y_0)^2 \geq B^2 z_0^2, \qquad x < x_0. \tag{17.1}$$

S_0 is one half of a hyperbola with the center x_0, y_0.

In view of Eq. (16.23) and since $\nu = (0, 0, 1)$, every solution Φ satisfies the equation

$$\Phi(x_0, y_0, z_0) = -\frac{1}{2\pi} \overset{*}{\iint\limits_{S_0}} (\Gamma \Phi_z - \Phi \Gamma_z)\, dx\, dy. \tag{17.2}$$

We observe that Γ satisfies the partial differential equation

$$\frac{\partial}{\partial z} \Gamma = -\frac{\partial}{\partial z_0} \Gamma. \tag{17.3}$$

Hence, using the law (16.24) on interchange of differentiation and integration, we find:

$$\Phi(x_0, y_0, z_0) = -\frac{1}{2\pi} \overset{*}{\iint\limits_{S_0}} \Gamma \Phi_z\, dx\, dy - \frac{1}{2\pi} \frac{\partial}{\partial z_0} \overset{*}{\iint\limits_{S_0}} \Gamma \Phi\, dx\, dy. \tag{17.4}$$

We can write

$$\overset{*}{\iint\limits_{S_0}} \Gamma \Phi_z\, dx\, dy = \iint\limits_{S_0} \frac{\Phi_z(x, y, 0)\, dx\, dy}{[(x - x_0)^2 - B^2 (y - y_0)^2 - B^2 z_0^2]^{\frac{1}{2}}} \tag{17.5}$$

and

$$\overset{*}{\iint\limits_{S_0}} \Gamma \Phi\, dx\, dy = \iint\limits_{S_0} \frac{\Phi(x, y, 0)\, dx\, dy}{[(x - x_0)^2 - B^2 (y - y_0)^2 - B^2 z_0^2]^{\frac{1}{2}}}. \tag{17.6}$$

The integrals on the right sides are improper but convergent and there is no need to consider finite parts. The second integral possesses a well determined derivative with respect to z_0 according to our general theory. It should also be observed that the integration over S_0 is, in fact, restricted to the finite section in the half-plane $x > 0$, $z = 0$ since Φ and $\nabla\Phi$ vanish identically in the rest of S_0.

We wish to prove next that the two integrals occurring in Eq. (17.4) are equal to each other. For this purpose, we consider the reflected image $P_0' \equiv (x_0, y_0, -z_0)$ of P_0 with respect to the plane $z=0$. Let Γ' be the singularity function with the source point at P_0'. Consider next the upstream Mach half-cone of P_0' and let Δ be its finite section in the quadrant $z>0$, $x>0$ (Fig. 6). The boundary of Δ consists of a piece of conical wall, of a plane piece with $x=0$ and of the part of the hyperbola S_0 in the half-plane $x>0$, $z=0$. Applying GREEN's identity to Φ and Γ' in Δ and taking finite parts leads easily to the result:

$$0 = {}^*\!\iint_{S_0} (\Gamma' \Phi_z - \Phi \Gamma_z'] \, dx \, dy, \tag{17.7}$$

since the vertex P_0' lies outside of Δ. But observe that if $z=0$ we have $\Gamma=\Gamma'$ and $\Gamma_z = -\Gamma_z'$. Hence, Eq. (17.7) implies

$$ {}^*\!\iint_{S_0} \Gamma \Phi_z \, dx \, dy = \frac{\partial}{\partial z_0} {}^*\!\iint_{S_0} \Gamma \Phi \, dx \, dy. \tag{17.8}$$

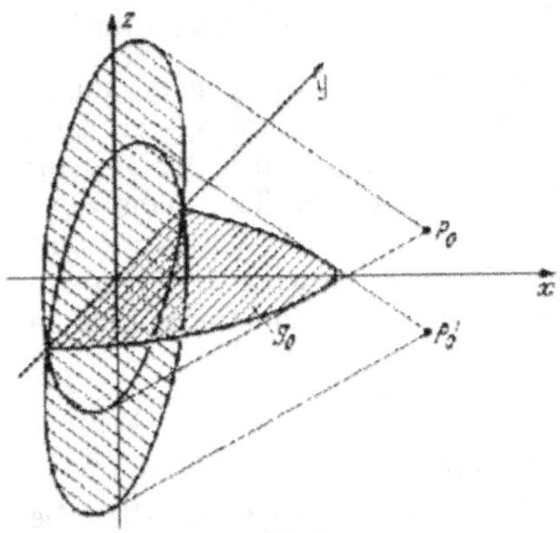

Fig. 6.

Thus, we may put Eq. (17.4) into the two equivalent simpler forms:

$$\left.\begin{aligned} \Phi(x_0, y_0, z_0) &= -\frac{1}{\pi} \iint_{S_0} \Gamma \Phi_z \, dx \, dy \\ &= -\frac{1}{\pi} \frac{\partial}{\partial z_0} \iint_{S_0} \Gamma \Phi \, dx \, dy. \end{aligned}\right\} \tag{17.9}$$

We can thus express Φ either in terms of its values in the half-plane $z=0$, $x>0$ only or in terms of is normal derivative Φ_z there. In general, a solution of the wave equation is determined only if the values of this solution and of its normal derivative are given on a surface and both data can be prescribed independently. In our present problem, however, Φ and Φ_z are tied together by the condition (17.8); this is due to the fact that we have the additional information that Φ and $\nabla\Phi$ vanish identically in the plane $x=0$.

We can apply Eq. (17.9) immediately to determine a linearized compressible supersonic flow past a thin three-dimensional wing. We suppose the obstacle near the plane $z=0$ and in the half-space $x>0$. It shall be symmetric with respect to the plane $z=0$ and its slope along curves $x=\text{const}$ shall be negligible compared to its slope along curves $y=\text{const}$. Thus, the upper normal $\boldsymbol{\nu}^+$ shall have the approximate form $(\nu_x^+, 0, 1)$. This assumption describes a long wing extending into the y-direction; $\nu_x \ll 1$ is the further condition necessary to guarantee a thin wing.

We consider a uniform flow in the x-direction with supersonic speed U_0. Due to the presence of the wing arises a total velocity potential

$$\varphi(x, y, z) = U_0 x + f(x, y, z), \tag{17.10}$$

where f is the potential of the perturbation and, consequently, small compared to $U_0 x$. The function f is a solution of the wave equation (15.1), and from the condition that φ have a vanishing normal derivative on the wing we find within the precision of our approximation

$$U_0 \nu_x^+ + f_z = 0, \qquad U_0 \nu_x^- - f_z = 0. \tag{17.11}$$

This condition holds on the surface of the wing; but we simplify the problem by projecting the entire wing into a plane domain D_0 in the plane $z=0$ and ask for a solution of the wave equation in the whole space outside D_0 with the boundary condition (17.11), taken now on D_0.

For reasons of symmetry, the potential φ is even in the variable z and, consequently, f is even and f_z is odd in z. This is indeed fulfilled on the boundary D_0 of the domain considered, in view of Eq. (17.11) and the fact that $v_x^+ = v_x^-$. At all points of the plane $z=0$ which do not lie in D_0 the oddness of f_z implies $f_z=0$. Thus, the values of f_z are known in the entire plane $z=0$, and from Eq. (17.9) we find the representation for the perturbation potential [29]:

$$f(x_0, y_0, z_0) = \frac{U_0}{\pi} \iint\limits_{S_0} \Gamma \cdot v_\lambda \, dx\, dy \tag{17.12}$$

where S_0 is the part of D_0 cut out by the upstream Mach half-cone through the point x_0, y_0, z_0.

We derived Eq. (17.12) under the assumption $z_0 > 0$; but since f is an even function of z_0 and since the same holds for the integral on the right, the formula is valid everywhere in the flow.

Formula (17.12) has been used to calculate the perturbation velocity and the perturbation pressure for various forms of wings. In the case of triangular plane wings the integration in Eq. (17.12) can be carried out in closed form and simple formulas for pressure distribution and drag can be obtained [30].

The above method can be extended to the case of non-symmetric wings. Because of the linearity of the boundary value problem one may split the required solution f into a symmetric and an antisymmetric function in z. For the first component the preceding method can be applied immediately. The determination of the antisymmetric component can be reduced to an integral equation, say for f_z, in the part of the plane $z=0$ outside of D_0. However, the amount of computational effort is very much larger than in the preceding symmetric case [31] to [33].

18. Two-dimensional linearized supersonic flow. The linearized theory of supersonic flow becomes particularly simple if we consider the two-dimensional case. The velocity potential φ is now a function of the two variables x, y and satisfies the two-dimensional wave equation:

$$B^2 \varphi_{xx} - \varphi_{yy} = 0, \qquad B = \sqrt{M_0^2 - 1}. \tag{18.1}$$

The equation has the general solution

$$\varphi(x, y) = a(x + By) + b(x - By) \tag{18.2}$$

where a and b are twice differentiable but otherwise arbitrary functions.

The lines

$$x + By = \text{const}, \qquad x - By = \text{const} \tag{18.3}$$

which play obviously a distinguished role in the theory of solutions of the wave equation form with the direction of the basic uniform flow the angles

$$\pm\alpha = \operatorname{arc\,cot} B \tag{18.4}$$

and are called the Mach lines of the flow. By Eqs. (18.1) and (18.4) we have

$$|\sin\alpha| = M_0^{-1} = \frac{c(U_0)}{U_0}. \tag{18.5}$$

The role of the Mach lines in the linearized theory will become evident from the following application. Let P be a thin profile near the x-axis lying in a uniform flow in the positive x-direction with supersonic speed U_0. The velocity potential of the perturbed flow will have the form

$$\varphi(x, y) = U_0 x + f(x, y), \tag{18.6}$$

and since $f(x, y)$ must satisfy the wave equation (18.1) we can set it up as

$$f(x, y) = a(x + B y) + b(x - B y). \tag{18.7}$$

The functions a and b have to be determined in such a way that $\partial\varphi/\partial n = 0$ on P. If we denote as before by $\boldsymbol{v}^+$ and $\boldsymbol{v}^-$ the normals on P at the upper and lower side and observe that because of the thinness of P: $v_y^+ = 1$, $v_y^- = -1$, we have the conditions on f:

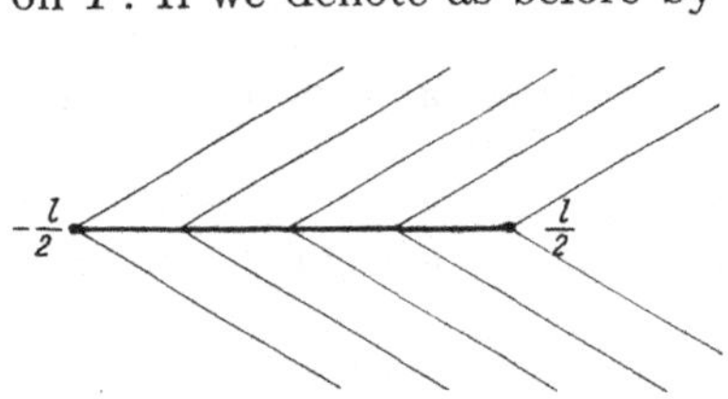

Fig. 7.

$$U_0 v_x^+ + f_y = 0 \quad \text{and} \quad U_0 v_x^- - f_y = 0 \tag{18.8}$$

on the upper and lower side of P, respectively. We project the profile onto the segment $\langle -\frac{l}{2}, \frac{l}{2} \rangle$ of the x-axis and ask for a solution f of the form (18.7) which satisfies Eq. (18.8) on the upper and lower edge of the segment. This slight change of the boundary value problem is negligible under our assumptions but simplifies the analysis.

We observe that for $x < -\frac{l}{2}$ no effect of the perturbation can exist and that a uniform flow regime must hold in this half-plane. Hence a and b must vanish identically in this region and on every Mach line which passes through this upstream half-plane and has the equation

$$x + B y = r \quad \text{or} \quad x - B y = r \tag{18.9}$$

must hold

$$a(r) = 0 \quad \text{or} \quad b(r) = 0, \tag{18.10}$$

respectively.

However, if a Mach line intersects the segment $-\frac{l}{2} \leqq x \leqq \frac{l}{2}$, the downstream half of it will obtain a non-zero value. Observe that in the upper half-plane $y > 0$ all Mach lines $x - B y = r$ with $-\frac{l}{2} \leqq r \leqq \frac{l}{2}$ intersect the segment and go downstream. In the lower half-plane $y < 0$ the Mach lines $x + B y = r$ with $-\frac{l}{2} \leqq r \leqq \frac{l}{2}$ intersect the segment and go downstream. Thus, the entire flow region consists of four different domains (Fig. 7): (a) The unperturbed upstream region bounded by two Mach lines through $x = -\frac{l}{2}$. (b) The unperturbed downstream region bounded by two Mach lines through $x = +\frac{l}{2}$. In these two regions we have $f = 0$. (c) The perturbed region in the upper half-plane. Here $a = 0$ and

$$f(x, y) = b(x - B y). \tag{18.11}$$

From Eq. (18.8) we find

$$b'(x) = \frac{U_0}{B} v_x^+, \tag{18.12}$$

and $b(x)$ can be found by integration and by use of the condition $b\left(-\frac{l}{2}\right)=0$.
(d) The perturbed region in the lower half plane. Here $b=0$ and

$$f(x, y) = a(x + By). \tag{18.13}$$

We obtain from Eq. (18.8) the differential condition on a:

$$a'(x) = \frac{U_0}{B} v_x^- \tag{18.14}$$

which determines $a(x)$ in a unique way because of $a\left(-\frac{l}{2}\right)=0$.

The velocity at a point x, y in the upper perturbation zone can be determined as follows. We draw the Mach line $x-By=r$ through this point; it will intersect the segment $\langle -\frac{l}{2}, \frac{l}{2}\rangle$ at the abscissa r. The velocity vector at x, y is then given by

$$u = U_0 + \frac{U_0}{B} v_x^+(r), \qquad v = -U_0 v_x^+(r). \tag{18.15}$$

This means geometrically that the velocity perturbation $u-U_0, v$ is perpendicular to the Mach line through the point considered.

From the linearized Bernoulli equation (10.12) we deduce

$$\Delta p = -\frac{2}{B} v_x^+(r) \cdot \frac{1}{2} \varrho_0 U_0^2. \tag{18.16}$$

If $y=y_+(x)$ is the equation of the upper boundary of P, we can relate v_x^+ to the slope angle ϑ_+ of this curve; in fact:

$$\vartheta_+ \sim \tan\vartheta_+ = y'_+(x) = -v_x^+. \tag{18.17}$$

We rewrite Eq. (18.16) in the form

$$\Delta p = \frac{2}{B} \vartheta_+ \cdot \frac{1}{2} \varrho_0 U_0^2 \tag{18.16'}$$

and see that we have an overpressure on all Mach lines which meet a rising arc of the upper profile and that we have an underpressure on all Mach lines which meet a descending arc of the upper side of P.

We obtain analogous results in the lower perturbation zone. The Mach line $x+By=r$ through a point in that region intersects the perturbing segment at the abscissa r. We find now:

$$u = U_0 + \frac{U_0}{B} v_x^-(r), \qquad v = U_0 v_x^-(r). \tag{18.18}$$

The pressure formula for the lower half-plane is

$$\Delta p = -\frac{2}{B} \vartheta_- \frac{1}{2} \varrho_0 U_0^2, \tag{18.19}$$

where ϑ^- is the tangent angle against the positive x-direction on the underside of P.

The lift exerted by the uniform flow on the thin profile P is obtained by integrating the pressure differences to both sides of the segment. We obtain an upward force

$$L = -\frac{2}{B} \cdot \frac{1}{2} \varrho_0 U_0^2 \int_{-\frac{l}{2}}^{l/2} (\vartheta_+ + \vartheta_-)\, dx. \tag{18.20}$$

By virtue of Eq. (18.17), we have

$$\int_{-\frac{l}{2}}^{l/2} \vartheta_+ \, dx = \int_{-\frac{l}{2}}^{l/2} y'_+(x)\, dx = y_+\left(\frac{l}{2}\right) - y_+\left(-\frac{l}{2}\right). \tag{18.21}$$

But at $\pm\frac{l}{2}$ the y-values of the upper and lower side coincide. Hence, the integral over ϑ_- gives the same value, and we have finally

$$L = \frac{4}{B}(y_1 - y_2) \cdot \frac{1}{2} \varrho_0 U_0^2, \tag{18.22}$$

where y_1 and y_2 are the ordinates of the leading and trailing edge of the profile.

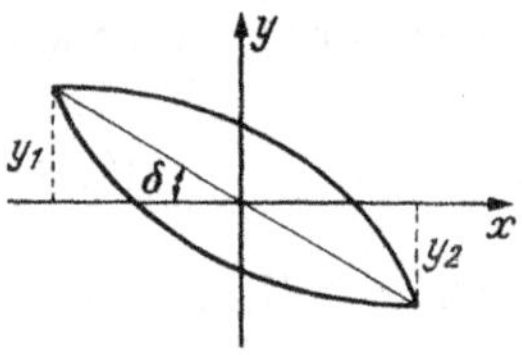

Fig. 8.

In general, a supersonic profile will have two sharp endpoints and a well defined chord connecting them. If Λ is the length of the chord and δ is the small angle of the chord against the x-axis (Fig. 8), within our limits of precision, we can write Eq. (18.22) in the form

$$L = \frac{4}{B} \Lambda \delta \cdot \frac{1}{2} \varrho_0 U_0^2. \tag{18.23}$$

The components of all pressure forces on the profile in the x-direction combine to produce the drag D of P. The x-component of Δp acting on an element ds of the boundary curve of P is (with the usual identification of ϑ with $\sin\vartheta$)

$$\frac{2}{B} \vartheta^2 \cdot \frac{1}{2} \varrho_0 U_0^2 \, ds, \tag{18.24}$$

and hence by integration

$$D = \frac{2}{B} \cdot \frac{1}{2} \varrho_0 U_0^2 \int_{-\frac{l}{2}}^{l/2} (\vartheta_+^2 + \vartheta_-^2)\, dx. \tag{18.24'}$$

The most remarkable features of the formulas (18.23) and (18.24') are the following:

(a) The lift of a supersonic profile in the linearized theory is independent of its particular shape and depends only on its chord and the angle of incidence.

(b) In contradistinction to the subsonic case, every profile has a non-zero drag except for a linear profile with zero angle of incidence. The drag is, however, of second order in the parameter of thickness; it depends on the geometry of the profile.

(c) The gliding angle D/L is independent of the speed U_0 and depends only on the profile and the angle of incidence.

(d) The number $B = \sqrt{M_0^2 - 1}$ appears in the denominator of L and D; if $M_0 \to 1$ both quantities increase indefinitely. For Mach numbers near one, the approximation of the linearized theory breaks down.

Finally, the discontinuous character of the above solution should be observed. In general, the slopes ϑ_+ and ϑ_- will be different from zero at the end points $x = \pm\frac{l}{2}$ of the segment. Hence, according to the formulas (18.15), (18.18) and (18.16'), (18.19), there will be a sudden change of normal velocity and pressure across the Mach lines issuing in downstream direction from these endpoints. These lines foreshadow the occurence of shock discontinuities in the exact theory of supersonic profiles.

Another discontinuity arises in the velocity potential itself; if we integrate Eqs. (18.12) and (18.14) from $-\frac{l}{2}$ to $\frac{l}{2}$ and use Eqs. (18.17), (18.21) and analogous formulas for v_x^- and ϑ_-, we obtain

$$a\left(\frac{l}{2}\right) = -\frac{U_0}{B}(y_1 - y_2), \qquad b\left(\frac{l}{2}\right) = \frac{U_0}{B}(y_1 - y_2), \tag{18.25}$$

while in the adjacent unperturbed zone $a = b = 0$. This discontinuity of the potential may, however, be conceived as due to the multivaluedness of the potential; this stands in analogy to the subsonic case and the period $K = b\left(\frac{l}{2}\right) - a\left(\frac{l}{2}\right)$ of φ plays the same role in the lift formula (18.22) as does the circulation in the Kutta-Joukovsky formula (13.12).

The linearized theory of a two-dimensional supersonic airfoil was first developed by ACKERET [*34*], [*35*], who found the formulas for lift, drag and moment of such a profile. For additional discussion see [*9*], [*10*], [*35*], [*37*], [*38*]. BUSEMANN considered higher order approximations for the pressure field on the airfoil [*39*] and LIGHTHILL determined the velocity potential up to the second approximation [*40*], [*41*]. It is doubtful, however, if the additional calculations are worth the great labor since the effects of other idealizations and omissions in the set-up of the theory are probably of the same order of magnitude as the terms involved.

19. Conical flows. In Sect. 8 we studied the supersonic flow against a cone and were able to reduce the number of independent variables in the differential equation since we could foresee on geometrical grounds that the velocity field obtained had to be constant on rays through the vertex of the cone.

In a similar manner the linearized supersonic theory can be simplified considerably when the obstacle in the flow remains unchanged if subjected to a similarity transformation. We may set up in such a case

$$\varphi(x, y, z) = x f(\eta, \zeta) \quad \text{with} \quad \eta = \frac{y}{x}, \quad \zeta = \frac{z}{x}. \tag{19.1}$$

In fact, the velocity field described under this assumption is

$$u = \varphi_x = f - \eta f_\eta - \zeta f_\zeta, \qquad v = \varphi_y = f_\eta, \qquad w = \varphi_z = f_\zeta; \tag{19.2}$$

it is the same on rays through the origin and of a form to be expected for an obstacle of the type considered. A flow with a velocity potential of the form (19.1) is called a conical flow [*43*] to [*48*].

The differential equation (15.1) satisfied by the velocity potential φ reduces to the differential equation in two variables:

$$(B^2\eta^2 - 1) f_{\eta\eta} + 2B^2\eta\zeta f_{\eta\zeta} + (B^2\zeta^2 - 1) f_{\zeta\zeta} = 0. \tag{19.3}$$

Each solution of Eq. (19.3) in the (η, ζ)-plane will provide us by Eq. (19.2) with a linearized supersonic velocity field of the required type.

Let $g(\eta, \zeta) = 0$ be the equation of a characteristic of Eq. (19.3). We have by Eq. (9.5) the differential equation

$$B^2(\eta g_\eta + \zeta g_\zeta)^2 = g_\eta^2 + g_\zeta^2, \tag{19.4}$$

which has the integral

$$\alpha\eta + \beta\zeta \pm \frac{1}{B}\sqrt{\alpha^2 + \beta^2} = 0. \tag{19.5}$$

This formula characterizes all straight lines which touch the circle $C: \eta^2 + \zeta^2 = B^{-2}$.

Through each point (η, ζ) on the outside of C pass two straight characteristics of Eq. (19.3), namely the tangents from the point to C. For no point interior to C can Eq. (19.4) have a real solution, since SCHWARZ' inequality applied to Eq. (19.4) requires

$$B^2(\eta^2+\zeta^2) \geqq 1. \tag{19.6}$$

We see, therefore, that the differential equation (19.3) is elliptic for $\eta^2+\zeta^2 < B^{-2}$ and hyperbolic for $\eta^2+\zeta^2 > B^{-2}$. It is easy to obtain closed forms for the general solution of Eq. (19.3) inside and outside of the critical circle C.

Outside of C, it is natural to introduce new variables which are constant along the characteristics; we define such characteristic variables r, s by

$$B(\eta\cos r+\zeta\sin r)=1, \qquad B(\eta\cos s+\zeta\sin s)=1. \tag{19.7}$$

An elementary and straightforward calculation shows that for an arbitrary function $F(\eta,\zeta)$ the following holds:

$$\left.\begin{aligned}\frac{\partial^2 F}{\partial r\,\partial s} &= -\frac{1}{4B^2\cos^2\frac{r-s}{2}}\times\\ &\times\{(B^2\eta^2-1)F_{\eta\eta}+2B^2\eta\zeta F_{\eta\zeta}+(B^2\zeta^2-1)F_{\zeta\zeta}+2B^2(\eta F_\eta+\zeta F_\zeta)\}.\end{aligned}\right\} \tag{19.8}$$

If we differentiate Eq. (19.3) with respect to η and ζ, we obtain two differential equations for f_η and f_ζ which by means of Eq. (19.8) can be written in the form:

$$\frac{\partial^2}{\partial r\,\partial s} f_\eta = 0, \qquad \frac{\partial^2}{\partial r\,\partial s} f_\zeta = 0. \tag{19.9}$$

Hence, we can conclude that f_η and f_ζ have the functional dependence:

$$f_\eta = A_1(r)+B_1(s), \qquad f_\zeta = A_2(r)+B_2(s). \tag{19.10}$$

The four functions are not independent in view of the integrability condition $f_{\eta\zeta}=f_{\zeta\eta}$, which leads easily to the relations

$$\frac{A_1'(r)}{\cos r} = \frac{A_2'(r)}{\sin r} = L(r); \qquad \frac{B_1'(s)}{\cos s} = \frac{B_2'(s)}{\sin s} = M(s). \tag{19.11}$$

We obtain, therefore, the following representation for the velocity components v and w:

$$v=\int L(r)\cos r\,dr+\int M(s)\cos s\,ds; \quad w=\int L(r)\sin r\,dr+\int M(s)\sin s\,ds \tag{19.12}$$

with arbitrary functions L and M. From Eqs. (19.2) and (19.10) we derive

$$\frac{\partial u}{\partial r} = -\eta A_1'-\zeta A_2', \qquad \frac{\partial u}{\partial s} = -\eta B_1'-\zeta B_2', \tag{19.13}$$

which, in view of Eqs. (19.11) and (19.7), we may simplify as follows:

$$\frac{\partial u}{\partial r} = -\frac{1}{B}L(r), \qquad \frac{\partial u}{\partial s} = -\frac{1}{B}M(s). \tag{19.14}$$

Thus, finally, the forward component of the velocity is given by

$$u=-\frac{1}{B}\left\{\int L(r)\,dr+\int M(s)\,ds\right\}. \tag{19.15}$$

Formulas (19.12) and (19.15) represent the general velocity field connected with a conical flow in the hyperbolic region $\eta^2+\zeta^2>B^{-2}$. We see that the general solution depends on two entirely arbitrary functions L and M which have to be adjusted to the various boundary conditions of the specific problem considered. In order to obtain single-valued velocity fields we have to choose the functions L and M periodic with the period 2π and with vanishing Fourier coefficients of order zero and one.

In the case of the elliptic region no real characteristic coordinates exist. However, the purely formal transformation (19.7) will lead to an analogous result if we admit complex values for the new variables. Writing instead of $\cos\nu$ and $\sin\nu$ now $\frac{1}{2}\left(t+\frac{1}{t}\right)$ and $\frac{1}{2i}\left(t-\frac{1}{t}\right)$ respectively, we may transform Eq. (19.7) into

$$\frac{1}{2} B\left[(\eta - i\zeta)\, t + (\eta + i\zeta)\frac{1}{t}\right] = 1, \tag{19.16}$$

an equation which connects a complex variable t with the given point η, ζ. We have

$$\eta + i\zeta = \frac{2}{B}\frac{t}{1+t\bar{t}}, \tag{19.17}$$

where $\bar{t}$ is the conjugate complex value of t, and it is easily seen that when t runs through all complex values $|t|<1$ the corresponding points η, ζ will fill the interior of the critical circle C exactly once. If $F(\eta, \zeta)$ is an arbitrary function of its two variables, we have the identity

$$\left.\begin{aligned}\frac{\partial^2 F}{\partial t\,\partial\bar{t}} &= -\frac{1}{B^2(1+t\bar{t})^2}\times\\ &\times\{(B^2\eta^2-1)F_{\eta\eta}+2B^2\eta\zeta F_{\eta\zeta}+(B^2\zeta^2-1)F_{\zeta\zeta}+2B^2(\eta F_\eta+\zeta F_\zeta)\}.\end{aligned}\right\} \tag{19.18}$$

This equation is analogous to Eq. (19.8) and we can again conclude that f_η and f_ζ split into a function of t and another function of $\bar{t}$. However, we must now require that the arbitrary functions be defined and formally differentiable with respect to their complex variables and that their combinations give real valued functions f_μ and f_ζ. Hence, we have necessarily

$$f_\mu = \mathrm{Re}\{A_1(t)\}, \qquad f_\zeta = \mathrm{Re}\{A_2(t)\}, \tag{19.19}$$

where A_1 and A_2 are analytic functions of their argument t. The integrability condition $f_{\mu\zeta} = f_{\zeta\mu}$ leads now to

$$A_1'(t)\left(t+\frac{1}{t}\right)^{-1} = A_2'(t)\, i\left(t-\frac{1}{t}\right)^{-1} = H(t). \tag{19.20}$$

Thus we find the following general representation for the velocity field of a conical flow in the elliptic domain:

$$v = \mathrm{Re}\left\{\int H(t)\left(t+\frac{1}{t}\right)dt\right\}, \qquad w = \mathrm{Re}\left\{\frac{1}{i}\int H(t)\left(t-\frac{1}{t}\right)dt\right\}. \tag{19.21}$$

Finally, differentiating Eq. (19.2) formally with respect to t and applying Eqs. (19.19), (19.20) and (19.16), one arrives at

$$\left.\begin{aligned}\frac{\partial u}{\partial t} &= -\eta\frac{\partial f_\eta}{\partial t} - \zeta\frac{\partial f_\zeta}{\partial t}\\ &= -\frac{1}{2}H(t)\left\{\eta\left(t+\frac{1}{t}\right) - i\zeta\left(t-\frac{1}{t}\right)\right\} = -\frac{1}{B}H(t).\end{aligned}\right\} \tag{19.22}$$

Hence we find for the forward velocity component

$$u = -\frac{1}{B}\mathrm{Re}\left\{\int H(t)\,dt\right\}. \tag{19.23}$$

Formulas (19.21) and (19.23) represent the velocity field of a conical flow in terms of an arbitrary analytic function $H(t)$. The variable t is connected with the point η, ζ by the relation (19.16).

Various generalizations of conical velocity fields have been considered. We mention the theory of homogeneous velocity fields where the velocity vectors on each ray through the vertex have the same direction but their magnitude is multiplied by $(x^2+y^2+z^2)^{n/2}$. It is possible in this generalized case, too, to give the general solution for a velocity field of this type [*49*] to [*51*].

20. Application to delta wings. The linearized theory of conical flow can be applied to thin profiles oriented in the direction of a uniform supersonic flow whose boundary is generated by rays from one distinguished point, called the vertex of the profile. Among the possible cones so obtained the limit case of the triangular wing, the delta wing, is of particular interest. We shall illustrate the method of Sect. 19 by considering this special case.

We consider a flat delta wing which lies near to the plane $z=0$ and is symmetric with respect to the plane $y=0$. Let the wing form an angle of incidence γ with the uniform flow and let γ be small. Let the angle at the vertex of the triangle be 2ω.

The image of the delta wing in the (η, ζ)-plane will be very nearly the segment $\langle -\tan\omega, \tan\omega\rangle$ of the η-axis. We suppose, for the sake of definiteness, that

$$\tan\omega < \frac{1}{B} = \tan\alpha, \qquad \alpha = \text{Mach angle}, \tag{20.1}$$

that is, that the entire wing lies within the Mach cone from the vertex. The flow outside of this Mach cone lies outside the domain of influence of the wing and has, therefore, a uniform flow regime. Inside of the Mach cone, that is, inside of the critical circle C in the (η, ζ)-plane, the velocity field is described by an analytic function $H(t)$ and the Eqs. (19.16), (19.21) and (19.23).

We have now to determine $H(t)$ from the boundary conditions on the flow. We require that the upward speed on the wing is determined by its slope, which implies

$$w = -U_0\gamma \quad \text{on} \quad \zeta = 0, \quad |\eta| < \tan\omega. \tag{20.2}$$

On the critical circle, on the other hand, we have for reasons of continuity $w=0$.

Let us pass now to the t-plane in which the function $H(t)$ is analytic. Observe that by Eq. (19.17) the circle C corresponds to the unit circle in the t-plane and that the diameters $\zeta=0$ and $\mathrm{Im}\{t\}=0$ correspond. The segment $|\eta|<\tan\omega$ on the first goes now into the segment $|\mathrm{Re}\{t\}|<t_0$ on the latter with

$$\frac{t_0}{1+t_0^2} = \frac{1}{2} B\tan\omega. \tag{20.3}$$

The analytic function

$$F(t) = \frac{1}{i}\int H(t)\left(t - \frac{1}{t}\right) dt \tag{20.4}$$

has a single-valued real part w in the doubly-connected domain obtained from the unit circle by removal of the segment $\langle -t_0, t_0\rangle$. This real part takes the boundary values zero and $-U_0\gamma$ on the unit circumference and on the segment, respectively. We find by differentiating $F(t)$ with respect to arc length on the boundary

$$t F'(t) = \begin{cases} \text{imaginary on segment} \\ \text{real on circumference.} \end{cases} \tag{20.5}$$

On the other hand, in view of Eq. (20.4)

$$H(t) = \frac{i}{t^2-1}\, t\, F'(t). \tag{20.6}$$

The points $t=\pm 1$ correspond to the points $\eta=\pm B^{-1}$, $\zeta=0$ on the critical circle and the velocity field is obviously continuous there; in particular, the velocity v must be bounded and, in view of Eq. (19.21), $H(t)$ must have a finite integral at these points. It is easily seen by means of the Schwarz reflection principle that $tF'(t)$ is a regular analytic function in the entire t-plane, except for the points $t=\pm t_0$ where it may become infinite and their image points $t=\pm 1/t_0$ with respect to the unit circumference. One finds easily that

$$t\,F'(t) = \frac{k_0\,(t^2-1)^2\,t}{(t^2-t_0^2)^{\frac{3}{2}}\,(1-t_0^2 t^2)^{\frac{3}{2}}}\,, \qquad k_0 \text{ real} \tag{20.7}$$

fulfills all our requirements. We could find other functions also satisfying our conditions but they would either not yield lifting forces or they would have higher order infinities at the points $\pm t_0$ and $\pm t_0^{-1}$ and would lead to divergent integrals for the pressure forces near the edges of the wing. Necessarily, therefore, we have to choose the form (20.7) for $tF'(t)$. This result leads to

$$H(t) = i\,k_0 \frac{(t^2-1)\,t}{(t^2-t_0^2)^{\frac{3}{2}}\,(1-t_0^2 t^2)^{\frac{3}{2}}} \tag{20.8}$$

and, in particular, to the forward velocity perturbation

$$u = \frac{k_0}{B}\,\mathrm{Re}\left\{\frac{1}{i}\int \frac{(t^2-1)\,t\,dt}{(t^2-t_0^2)^{\frac{3}{2}}\,(1-t_0^2 t^2)^{\frac{3}{2}}}\right\}. \tag{20.9}$$

This formula simplifies under the change of variable

$$\tau = \left[\frac{1}{2}\left(t+\frac{1}{t}\right)\right]^{-1}, \qquad \tau_0 = \left[\frac{1}{2}\left(t_0+\frac{1}{t_0}\right)\right]^{-1}, \tag{20.10}$$

which leads to

$$u = K\,\mathrm{Re}\left\{\frac{1}{\sqrt{\tau_0^2-\tau^2}}\right\}. \tag{20.11}$$

Here the real factor K is related in a simple manner to the above constant k_0, and the constant of integration has been chosen in such a way that $u=0$ for $\tau=1$, which corresponds to a point η, ζ on the critical circle C.

We observe that $|\tau|<\tau_0$ corresponds to points on the wing and that because of the branch points at $\tau=\pm\tau_0$ the function u takes opposite values on the upper and lower side of the wing surface. Thus, u will be an odd function of z and, by Bernoulli's law (10.12), the same will hold for the pressure distribution in the flow. An elementary calculation leads to the following determination of the constants:

$$u = \pm \frac{U_0\,\gamma\,\tau_0^2}{B\,E\sqrt{\tau_0^2-\tau^2}}\,. \tag{20.12}$$

The signs depend on the side of the wing surface, and E is the complete elliptic integral of the second kind with modulus $k=1-\tau_0^2$. The pressure perturbations at a point x, y, z in the flow is given by [45], [52]

$$\frac{\Delta p}{\frac{1}{2}\varrho_0 U_0^2} = \mp \frac{U_0\,\gamma\,x\tan^2\omega}{E\sqrt{x^2\tan^2\omega-y^2}}\,. \tag{20.13}$$

The case of a symmetric delta wing whose vertex semi-angle ω is larger than the Mach angle α of the uniform flow has to be treated differently. The upper and lower side of the wing lie now in different domains of influence and can be treated independently of each other. We can start in the half-plane $\zeta > 0$ and outside of the critical circle C; here we have a hyperbolic initial value problem which can be solved by characteristics as indicated by formulas (19.12) and (19.15). Having solved this problem, we know already the velocity field on the boundary of the critical circle; this information together with the known conditions on the diameter of the circle C allows us to solve an elliptic boundary value problem for the upper half of C and thus to determine the velocity field in the entire upper half-plane. In the same way, the problem can be solved for the lower half-plane.

The method can be extended to non-symmetric wings and various possibilities have to be considered; namely, that both edges, one edge or no edge of the wing lie outside of the Mach cone from the vertex. However, the general representation of the solution given in Sect. 19 permits us in each case to solve the problem completely.

Because of the linearity of the basis equation (15.1) we can superimpose solutions. Hence we can combine conical fields with different vertices in order to obtain flow patterns past more complicated wing forms. In particular, if we consider a wing whose projection on the plane $z=0$ is a polygon, we can express the flow around it by a proper combination of various conical fields whose vertices are the corners of the polygon. It is necessary, however, that the different corners of the polygon lie outside of each other's domains of influence.

21. Slender bodies of revolution. There exists a considerable literature on the theory of linearized flows past slender bodies (see [*9*], [*10*]). We shall restrict ourselves to the simplest case, a body with axial symmetry. Let the (x, r)-plane be its meridian plane and let

$$r = \varepsilon f(x), \qquad f(0) = 0, \qquad f(1) = 0 \tag{21.1}$$

be the boundary curve of its meridian cut. We suppose that $f(x)$ has three continuous derivatives in the interval $0 \leq x \leq 1$ and that ε is a small positive factor.

We suppose that a stream in the direction of the x-axis with supersonic free stream speed U_0 meets the obstacle and that the velocity potential of the ensuing flow is

$$\varphi(x, r) = U_0 x + f(x, Br), \qquad r = \sqrt{y^2 + z^2}. \tag{21.2}$$

By virtue of Eq. (15.1), the function $f(x, u)$ must satisfy the partial differential equation

$$\frac{\partial^2 f}{\partial x^2} - \left(\frac{\partial^2 f}{\partial u^2} + \frac{1}{u}\frac{\partial f}{\partial u}\right) = 0. \tag{21.3}$$

It is easily seen that

$$s\,(x, u;\, x_0) = [(x - x_0)^2 - u^2]^{-\frac{1}{2}} \tag{21.4}$$

is a solution of Eq. (21.3) with a singularity at the point x_0 on the axis of symmetry and with a singular surface

$$(x - x_0)^2 - u^2 = 0. \tag{21.5}$$

It is natural, therefore, to try to construct the solution (21.2) as a linear superposition of elementary singularities (21.4) whose singular points are located on the axis of symmetry. We arrive thus at the form

$$\varphi(x, r) = U_0 x + \varepsilon \int_0^{x - Br} \frac{\mu(s)\, ds}{\sqrt{(x-s)^2 - B^2 r^2}}, \qquad x \geq Br. \tag{21.6}$$

We have to use the variable upper boundary in order to ensure the reality of the integral. It is sufficient to define φ for $x \geq Br$; if $x \leq Br$, the point considered lies outside of the region of influence of the perturbing body and, consequently, in the uniform regime. At points with $x \leq Br$, therefore, we have obviously

$$\varphi = U_0 x.$$

In order to show that $\varphi(x, r)$ satisfies the differential equation (15.1) for the velocity potential in the linearized theory, it is convenient to eliminate the improper integral in Eq. (21.6). This is achieved by the change of variables

$$s = x - Br\cosh t \tag{21.7}$$

under which

$$\int_0^{x-Br} \frac{\mu(s)\,ds}{\sqrt{(x-s)^2 - B^2 r^2}} = \int_0^{\alpha(x,r)} \mu(x - Br\cosh t)\,dt \tag{21.8}$$

with

$$\cosh\alpha = \frac{x}{Br}. \tag{21.9}$$

It is now easy to verify that φ indeed satisfies the Eq. (15.1) for every choice of $\mu(s)$.

We have to dispose of the weight function $\mu(s)$ in such a way that the boundary condition

$$\frac{\partial\varphi}{\partial n} = 0 \tag{21.10}$$

is fulfilled along the boundary curve (21.1). Within the precision of the linearized theory this is equivalent to the integral equation

$$\left[\frac{\partial}{\partial r}\int_0^{\alpha} \mu(x - Br\cosh t)\,dt\right]_{r=\varepsilon f(x)} = U_0 f'(x). \tag{21.11}$$

A straightforward calculation leads to the following asymptotic formula for $\mu(x)$:

$$\mu(x) = -\varepsilon U_0 f(x) f'(x) + O(\varepsilon\log\varepsilon). \tag{21.12}$$

From Eq. (21.12) we can easily compute the velocity potential (21.6), the entire flow pattern and the drag and moments acting on the body. For a careful discussion of the asymptotics in the ε-development see [*10*], [*58*].

The technique of solving boundary value problems for axially symmetric bodies by superposition of singularities on the axis is due to v. Kármán. He developed it originally in the theory of airships in an incompressible fluid [*55*]; v. Kármán-Moore [*56*] and Tsien [*57*] extended it to the case of the subsonic and supersonic flow of a compressible fluid (see also [*54*]).

The above method can be generalized to the case of a body of revolution whose axis forms a small angle ϑ with the direction of the incident uniform flow. We introduce for this purpose cylindrical coordinates x, r, ϑ in space and express the Eq. (15.1) for the velocity potential in the form

$$B^2\varphi_{xx} - \varphi_{rr} - \frac{1}{r}\varphi_r - \frac{1}{r^2}\varphi_{\vartheta\vartheta} = 0. \tag{21.13}$$

If we set up

$$\varphi(x, r, \vartheta) = f^{(n)}(x, Br)\,\begin{matrix}\sin\\ \cos\end{matrix}\, n\vartheta \tag{21.14}$$

we find the differential equation

$$f^{(n)}_{xx} - \left(f^{(n)}_{uu} + \frac{1}{u} f^{(n)}_{u} - \frac{n^2}{u^2} f^{(n)}\right) = 0. \tag{21.15}$$

We observe that if $f^{(0)}(x, u)$ is a solution of Eq. (21.3), then

$$f^{(1)}(x, u) = \frac{\partial}{\partial u} f^{(0)}(x, u) \tag{21.16}$$

will be a solution of Eq. (21.15) for $n=1$. Thus

$$\varphi(x, r) = \begin{matrix}\cos\\ \sin\end{matrix}\vartheta \int\limits_0^{x-Br} \frac{\mu(s)\, r\, ds}{[(x-s)^2 - B^2 r^2]^{\frac{3}{2}}} \tag{21.17}$$

will also be a solution of Eq. (15.1). The new solution is no longer axially symmetric and can be used to enforce the proper boundary conditions on an inclined body of revolution.

It is easily verified that the functions

$$f^{(n)}(x, u) = \int\limits_0^\infty \mu(x - u\cosh t)\cosh n\, t\, dt \tag{21.18}$$

are solutions of the differential equation (21.15) for an arbitrary function $\mu(s)$, provided that μ is twice continuously differentiable and vanishes for negative argument. Thus a great number of solutions for the linearized potential equation (15.1) is obtained, each depending on a weight function $\mu(s)$. This set of solutions may be utilized in order to adjust the boundary conditions on a general slender body.

III. The hodograph method.

22. Legendre transformation for the velocity potential. The velocity potential $\varphi(x, y)$ of a steady irrotational fluid flow in the x, y-plane satisfies the partial differential equation

$$(c^2 - \varphi_x^2)\,\varphi_{xx} - 2\varphi_x\varphi_y\varphi_{xy} + (c^2 - \varphi_y^2)\,\varphi_{yy} = 0. \tag{22.1}$$

From this equation with the given boundary conditions and the prescribed sources and sinks of the flow, the unknown function $\varphi(x, y)$ has to be determined. Since Eq. (22.1) is a non-linear differential equation in $\varphi(x, y)$, we have to deal with a rather difficult boundary value problem.

We remark, however, that Eq. (22.1) is a quasi-linear, that is, linear in the highest (second order) derivatives. Furthermore, the coefficients of these highest derivatives depend only on φ_x and φ_y, but not directly on the independent variables x, y. In fact, from BERNOULLI's equation

$$\frac{q^2}{2} + \int \frac{dp}{\varrho} = \text{const} \tag{22.2}$$

we may express the density ϱ as a function of $q^2 = \varphi_x^2 + \varphi_y^2$ and, hence, consider $c^2 = dp/d\varrho$ as a function of this term.

The special structure of Eq. (22.1) suggests use of the well known Legendre transformation in order to bring it into linear form. For this purpose, in the usual manner we may introduce

$$\varphi_x = u \qquad \varphi_y = v \tag{22.3}$$

as new indepedent variables and the combination

$$\tilde{\varphi} = u\,x + v\,y - \varphi \tag{22.4}$$

as the new depedent variable. Provided that the Jacobian

$$\Delta = \begin{vmatrix} \varphi_{xx} & \varphi_{xy} \\ \varphi_{xy} & \varphi_{yy} \end{vmatrix} \tag{22.5}$$

is different from zero, we may use Eq. (22.3) in order to eliminate x and y as functions of u and v and therefore may consider $\tilde{\varphi}$ as a function of u and v. If we differentiate Eq. (22.4) with respect to u and v, respectively, and take the Eqs. (22.3) into consideration, we obtain immediately the identities

$$\tilde{\varphi}_u = x, \quad \tilde{\varphi}_v = y \tag{22.6}$$

which together with Eqs. (22.3) and (22.4) display clearly the great symmetry of the Legendre transformation.

Next we differentiate Eq. (22.3) with respect to u and v but use Eq. (22.6) in order to express $\partial x/\partial u$, $\partial x/\partial v$ and $\partial y/\partial u$, $\partial y/\partial v$. We find

$$\left.\begin{aligned} \varphi_{xx}\tilde{\varphi}_{uu} + \varphi_{xy}\tilde{\varphi}_{uv} &= 1, \qquad \varphi_{xy}\tilde{\varphi}_{uu} + \varphi_{yy}\tilde{\varphi}_{uv} = 0, \\ \varphi_{xx}\tilde{\varphi}_{uv} + \varphi_{xy}\tilde{\varphi}_{vv} &= 0, \qquad \varphi_{xy}\tilde{\varphi}_{uv} + \varphi_{yy}\tilde{\varphi}_{vv} = 1. \end{aligned}\right\} \tag{22.7}$$

The four relations (22.7) may be written as the matrix equation

$$\begin{pmatrix} \varphi_{xx} & \varphi_{xy} \\ \varphi_{xy} & \varphi_{yy} \end{pmatrix} \begin{pmatrix} \tilde{\varphi}_{uu} & \tilde{\varphi}_{uv} \\ \tilde{\varphi}_{uv} & \tilde{\varphi}_{vv} \end{pmatrix} = \begin{pmatrix} 1 & 0 \\ 0 & 1 \end{pmatrix}; \tag{22.7'}$$

they enable us to express the second derivatives of φ in terms of the second derivatives of $\tilde{\varphi}$. By means of these formulas, we may finally transform Eq. (22.1) into the linear partial differential equation

$$(c^2 - u^2)\,\tilde{\varphi}_{vv} + 2u\,v\,\tilde{\varphi}_{uv} + (c^2 - v^2)\,\tilde{\varphi}_{uu} = 0. \tag{22.8}$$

The value of the linear equivalent (22.8) to the given non-linear differential equation (22.1) consists in the possibility of obtaining from two known solutions a new solution by the principle of linear superposition [*1*]. Also, it is frequently easier to obtain singular solutions and to study their behavior by use of the equivalent linear equation.

We may obtain a useful alternative form of Eq. (22.8) by introducing the polar coordinates in the velocity plane

$$q = \sqrt{u^2 + v^2}\,. \qquad \vartheta = \arc\tan\frac{v}{u}\,. \tag{22.9}$$

A simple calculation transforms Eq. (22.8) into

$$c^2 q^2\,\tilde{\varphi}_{qq} + q\,(c^2 - q^2)\,\tilde{\varphi}_q + (c^2 - q^2)\,\tilde{\varphi}_{\vartheta\vartheta} = 0. \tag{22.10}$$

This particular form of the linear equation lends itself readily to separation of variables and leads to numerous particular solutions. These may, in turn, be superimposed with arbitrary coefficients and one obtains thus very many solutions of Eq. (22.1).

We set up

$$\tilde{\varphi}_n(q, \vartheta) = A_n(q)\,\begin{matrix}\cos n\,\vartheta \\ \sin n\,\vartheta\end{matrix} \tag{22.11}$$

and obtain for $A_n(q)$ the ordinary differential equation

$$c^2 q^2 A_n''(q) + q(c^2 - q^2) A_n'(q) - n^2(c^2 - q^2) A_n(q) = 0. \tag{22.12}$$

For sake of illustration we shall discuss the solution $\tilde{\varphi} = \tilde{\varphi}(q) = A_0(q)$ where

$$c^2 q^2 A_0'' + q(c^2 - q^2) A_0' = 0. \tag{22.12'}$$

By virtue of Eq. (22.6) and the special form of $\tilde{\varphi}$, we have

$$x = A_0'(q) \frac{u}{q}, \qquad y = A_0'(q) \frac{v}{q}, \tag{22.13}$$

which shows that the fluid moves radially and that its velocity depends on the distance r from the origin in the x, y-plane:

$$r^2 = A_0'(q)^2. \tag{22.14}$$

In view of the Bernoulli equation, we have

$$\frac{1}{2} q^2 + \int \frac{dp}{\varrho} = \frac{1}{2} q^2 + \int c^2 \frac{d\varrho}{\varrho} = \text{const} \tag{22.15}$$

and hence

$$\frac{q\,dq}{c^2} = -\frac{d\varrho}{\varrho}. \tag{22.15'}$$

Thus, we can bring Eq. (22.12') into the form

$$\frac{d}{dq}(q A_0') = \frac{q}{c^2} q A_0' = -\frac{\varrho'(q)}{\varrho(q)} \cdot (q A_0'), \tag{22.16}$$

which can be integrated to

$$\varrho q A_0' = \varrho q r = \text{const}. \tag{22.17}$$

This integral of the motion expresses the conservation of matter under the radial flow and could have been expected, once the radial nature of the motion had been established. Conversely, it can be verified easily that every radial motion in the (x, y)-plane is connected to a Legendre potential of the particular form $\tilde{\varphi} = \tilde{\varphi}(q)$.

We derive from Eqs. (22.12') and (22.14) that

$$\frac{d}{dq} \log r = q\left(\frac{1}{c^2} - \frac{1}{q^2}\right), \tag{22.18}$$

that is, that r decreases as a function of q as long as $q < c$ but is stationary for $q = c$ and starts to increase with q if $q > c$. Thus, there exists a minimum value r_0 of $r(q)$. The flow defined by the function $\tilde{\varphi}(q)$ is only defined outside of a circle of radius r_0 in the (x, y)-plane and cannot be continued into this circle.

23. The Chaplygin-Molenbroek transformation for the stream function. While the Legendre transformation of the differential equation (22.1) is of use in studying and constructing solutions, it does not help much if one wishes to satisfy boundary conditions for the solution $\varphi(x, y)$. The differential equation (22.1), however, is of a particular form permitting another transformation into a linear equation which is much more useful since it does not change the dependent variable and enables us, therefore, to carry over the boundary conditions. This transformation was considered first by MOLENBROEK [2] and developed systematically by CHAPLYGIN [3] (see also [4] to [7]).

We observe that by means of the stream function $\psi(x, y)$ and the velocity potential $\varphi(x, y)$ we can express the velocity components u, v as follows:

$$u = \frac{\partial \varphi}{\partial x} = \frac{1}{\varrho} \frac{\partial \psi}{\partial y}, \qquad v = \frac{\partial \varphi}{\partial y} = -\frac{1}{\varrho} \frac{\partial \psi}{\partial x}. \tag{23.1}$$

The differential equation (23.1) for $\varphi(x, y)$ is nothing but the corresponding integrability condition

$$\frac{\partial}{\partial x}\left(\varrho \frac{\partial \varphi}{\partial x}\right) + \frac{\partial}{\partial y}\left(\varrho \frac{\partial \varphi}{\partial y}\right) = 0. \tag{23.2}$$

In a similar way, we obtain the partial differential equation for the stream function

$$\frac{\partial}{\partial x}\left(\frac{1}{\varrho} \frac{\partial \psi}{\partial x}\right) + \frac{\partial}{\partial y}\left(\frac{1}{\varrho} \frac{\partial \psi}{\partial y}\right) = 0. \tag{23.3}$$

The close relation with integrability conditions enables us to perform a transformation of the Eq. (22.1) as follows: We consider u, v as the independent variables and express x and y as functions of them by solving the system (23.1). This is possible if the Jacobian Δ defined in Eq. (22.5) is different from zero. If we perform a change du, dv of the dependent variables, they will induce a change $dx, dy, d\varphi, d\psi$ of the dependent variables. We have by Eq. (23.1) the following linear relations between these differentials:

$$d\varphi = u\,dx + v\,dy, \qquad \frac{1}{\varrho}\,d\psi = -v\,dx + u\,dy \tag{23.4}$$

which can be solved for dx and dy:

$$dx = \frac{u}{q^2}\,d\varphi - \frac{v}{\varrho q^2}\,d\psi, \qquad dy = \frac{v}{q^2}\,d\varphi + \frac{u}{\varrho q^2}\,d\psi. \tag{23.5}$$

Since x and y are uniquely determined as functions of u and v (at least in a sufficiently small neighborhood of the point u, v considered) the differentials on the right hand sides have to satisfy integrability conditions which lead to the desired differential equations for φ and ψ, considered as functions of u and v.

Since by the Bernoulli equation (22.2) ϱ is a function of q alone, and since q plays a distinguished role in the Eqs. (23.5), it is useful to introduce the polar coordinates q, ϑ defined in Eq. (22.9) as the new independent variables. We find

$$\left.\begin{aligned} dx &= \left(\frac{\cos\vartheta}{q}\,\varphi_q - \frac{\sin\vartheta}{\varrho q}\,\psi_q\right) dq + \left(\frac{\cos\vartheta}{q}\,\varphi_\vartheta - \frac{\sin\vartheta}{\varrho q}\,\psi_\vartheta\right) d\vartheta, \\ dy &= \left(\frac{\sin\vartheta}{q}\,\varphi_q + \frac{\cos\vartheta}{\varrho q}\,\psi_q\right) dq + \left(\frac{\sin\vartheta}{q}\,\varphi_\vartheta + \frac{\cos\vartheta}{\varrho q}\,\psi_\vartheta\right) d\vartheta. \end{aligned}\right\} \tag{23.6}$$

Hence, we are led to the integrability conditions

$$\left.\begin{aligned} -\frac{\sin\vartheta}{q}\,\varphi_q + \frac{\cos\vartheta}{q^2}\,\varphi_\vartheta &= \frac{\cos\vartheta}{\varrho q}\,\psi_q + \frac{\sin\vartheta}{(\varrho q)^2}\,\frac{d(\varrho q)}{dq}\,\psi_\vartheta, \\ \frac{\cos\vartheta}{q}\,\varphi_q + \frac{\sin\vartheta}{q^2}\,\varphi_\vartheta &= \frac{\sin\vartheta}{\varrho q}\,\psi_q - \frac{\cos\vartheta}{(\varrho q)^2}\,\frac{d(\varrho q)}{dq}\,\psi_\vartheta. \end{aligned}\right\} \tag{23.7}$$

We can solve this system easily and obtain

$$\varphi_q = q\,\frac{d}{dq}\left(\frac{1}{\varrho q}\right)\psi_\vartheta, \qquad \varphi_\vartheta = \frac{q}{\varrho}\,\psi_q. \tag{23.8}$$

This is a simple system of linear differential equations between the stream function and the velocity potential considered as functions of the variables u, v in the velocity or hodograph plane.

Using the identity (22.15′), we can write

$$q\frac{d}{dq}\left(\frac{1}{\varrho q}\right)=\frac{q}{\varrho c^2}-\frac{1}{q\varrho}=-\frac{1}{q\varrho}(1-M^2) \tag{23.9}$$

where

$$M=\frac{q}{c} \tag{23.9′}$$

is the local Mach number of the flow. M depends on the variable q only; the precise form of the function $M(q)$ depends on the equation of state considered.

If we should use a different scale for the radial coordinate in the hodograph plane, say by introducing the new variable $\lambda=\lambda(q)$, we could write instead of Eq. (23.8)

$$\lambda'(q)\frac{\partial\varphi}{\partial\lambda}=-\frac{1-M^2}{q\varrho}\frac{\partial\psi}{\partial\vartheta},\qquad \frac{\partial\varphi}{\partial\vartheta}=\frac{q}{\varrho}\lambda'(q)\frac{\partial\psi}{\partial\lambda}. \tag{23.8′}$$

We wish to choose $\lambda(q)$ in such a manner that the system (23.8′) is particularly symmetric. This will be achieved if we require that the coefficients in both Eqs. (23.8′) be the same, except for the sign. This leads to the condition

$$\frac{1-M^2}{q^2}=\lambda'(q)^2,\qquad \lambda(q)=-\int\limits_q^c\frac{\sqrt{1-M^2}}{q}\,dq. \tag{23.10}$$

By this choice of the integral, we have attained the normalization of $\lambda(q)$, namely

$$\lambda(q)=0,\qquad \text{if } M=1. \tag{23.10′}$$

We introduce the function

$$z(\lambda)=\frac{\sqrt{1-M^2}}{\varrho} \tag{23.11}$$

and obtain then from Eq. (23.8′) the final system of differential equations

$$\frac{\partial\varphi}{\partial\lambda}=-z(\lambda)\frac{\partial\psi}{\partial\vartheta},\qquad \frac{\partial\varphi}{\partial\vartheta}=z(\lambda)\frac{\partial\psi}{\partial\lambda}. \tag{23.12}$$

We may now eliminate φ or ψ from the system (23.12) and obtain a second order partial differential equation for the remaining variable. It is preferable to deal with the stream function ψ, since the boundary condition $\psi=$const is unchanged if we pass from the physical to the hodograph plane, while the boundary condition $\partial\varphi/\partial n=0$ becomes more complicated. Therefore we study the equation

$$\frac{\partial^2\psi}{\partial\vartheta^2}+\frac{1}{z(\lambda)}\frac{\partial}{\partial\lambda}\left(z(\lambda)\frac{\partial\psi}{\partial\lambda}\right)=0. \tag{23.13}$$

We can put Eq. (23.13) into an alternative form by introducing the new dependent variable

$$\psi^*(\lambda,\vartheta)=z(\lambda)^{\frac{1}{2}}\psi(\lambda,\vartheta). \tag{23.14}$$

Observe that the change of ψ does not affect the boundary condition $\psi=0$ which now reads $\psi^*=0$. We compute easily

$$\frac{\partial^2\psi^*}{\partial\lambda^2}+\frac{\partial^2\psi^*}{\partial\vartheta^2}=f(\lambda)\,\psi^*,\qquad f(\lambda)=\frac{\frac{d^2}{d\lambda^2}(\sqrt{z})}{\sqrt{z}}. \tag{23.15}$$

We see that $\psi^*(\lambda,\vartheta)$ satisfies a very simple differential equation of a type which has been extensively studied (CHAPLYGIN's equation).

It should be remarked that the transformation (23.10) from q into $\lambda(q)$ is real only so long as $M<1$, i.e. in the subsonic case. A more detailed discussion of the transformation is possible only if we make specific assumptions with respect to the equation of state of the fluid in motion.

In the case of a purely supersonic motion, we may introduce a new variable $\lambda_1(q)$ by the definition

$$\frac{M^2-1}{q^2} = \lambda_1'(q)^2 \tag{23.16}$$

and the function of λ_1

$$z_1(\lambda_1) = \frac{\sqrt{M^2-1}}{\varrho}. \tag{23.17}$$

Let $\psi_1^*(\lambda_1, \vartheta)$ be defined analogously to $\psi^*(\lambda, \vartheta)$ by

$$\psi_1^*(\lambda_1, \vartheta) = z_1(\lambda_1)^{\frac{1}{2}}\,\psi(\lambda_1, \vartheta). \tag{23.18}$$

The same calculations which led to Eq. (23.15) will now lead us to

$$\frac{\partial^2\psi_1^*}{\partial\lambda_1^2} - \frac{\partial^2\psi_1^*}{\partial\vartheta^2} = f_1(\lambda_1)\,\psi_1^*, \quad f_1(\lambda_1) = \frac{1}{(\sqrt{z_1})}\,\frac{d^2}{d\lambda_1^2}\,(\sqrt{z_1}). \tag{23.19}$$

We now have a hyperbolic differential equation for the supersonic stream function. Its characteristic directions are given by the differential equations

$$\frac{d\lambda_1}{d\vartheta} = \pm 1 \tag{23.20}$$

which leads in the (q, ϑ)-plane to the curves defined by

$$\frac{dq}{d\vartheta} = \pm\frac{q}{\sqrt{M^2-1}}. \tag{23.21}$$

We shall show in Sect. 34 that these characteristic curves correspond, in general, to the Mach curves in the physical (x, y)-plane under the hodograph transformation. The branch lines discussed in Sect. 33 correspond also to solution curves of Eq. (23.21). Branch lines are singular curves for the hodograph transformation characterized by the vanishing of the Jacobian (22.5).

24. The ideal adiabatic gas. The functions $\lambda(q)$ and $f(\lambda)$ depend on the choice of the equation of state for the fluid considered. The most important case is that of an adiabatic gas, which shall now be discussed. We start with the pressure-density relation

$$p = a\varrho^\gamma, \quad \gamma > 1 \tag{24.1}$$

from which we derive the local velocity of sound

$$c^2 = \gamma a \varrho^{\gamma-1} \tag{24.2}$$

and the particular form of the Bernoulli equation

$$\frac{q^2}{2} + \frac{1}{\gamma-1}c^2 = \frac{q_1^2}{2}. \tag{24.3}$$

The constant q_1 on the right side represents obviously the maximum speed of the flow, which can only be attained when $c = \varrho = 0$, i.e. for flow into a vacuum.

We introduce for the sake of simplicity the variable $m = M^2$ and bring Eq. (24.2) into the form

$$q^2\left(\frac{2}{\gamma-1} + m\right) = q_1^2 m, \tag{24.2'}$$

which shows that q^{-2} is a linear function of m^{-1}. We have, in particular,

$$\frac{d q^{-2}}{d m} = -\frac{2}{(\gamma-1)\, q_1^2} \cdot \frac{1}{m^2}, \qquad \frac{d q}{d m} = \frac{1}{\gamma-1} \frac{q^3}{q_1^2 m^2}. \tag{24.4}$$

We can now compute from Eqs. (23.10) and (24.4)

$$\lambda(q) = -\int_m^1 \frac{\sqrt{1-m}\, d m}{m\,(2+(\gamma-1)\, m)} = \frac{1}{2k} \operatorname{arc\,tan}\left(k\sqrt{1-m}\right) - \frac{1}{2} \operatorname{arc\,tan}\sqrt{1-m} \tag{24.5}$$

with

$$k = \sqrt{\frac{\gamma-1}{\gamma+1}}. \tag{24.6}$$

Since λ is thus obtained as a function of m, it is convenient to express also the function $f(\lambda)$ in terms of this variable. A straightforward computation gives

$$f(\lambda) = \frac{\gamma+1}{16} m^2 (1-m)^{-3} \left[16 - 4(3-2\gamma)\, m - (3\gamma-1)\, m^2\right]. \tag{24.7}$$

The most important feature in this formula is the strong singularity of $f(\lambda)$ for the value $m = M = 1$.

It is also of interest to consider the original differential system (23.8) in the case of an adiabatic fluid. This system is independent of the subsonic character of the flow. In view of the Bernoulli equation (24.3) and of Eq. (24.2) we have

$$\varrho = \varrho_0 (1-\tau)^{\frac{1}{\gamma-1}}, \qquad \varrho_0 = \left(\frac{\gamma-1}{2\gamma a} q_1^2\right)^{\frac{1}{\gamma-1}}, \qquad \tau = \frac{q^2}{q_1^2}, \tag{24.8}$$

where ϱ_0 can be interpreted as the density of the fluid at rest. Therefore, it is advantageous to replace the speed variable q in the differential system by the dimensionless quantity τ. A simple calculation gives

$$\frac{\partial \varphi}{\partial \vartheta} = A(\tau) \frac{\partial \psi}{\partial \tau}, \qquad \frac{\partial \psi}{\partial \vartheta} = B(\tau) \frac{\partial \varphi}{\partial \tau} \tag{24.9}$$

with

$$A(\tau) = \frac{2\tau}{\varrho_0} (1-\tau)^{-\frac{1}{\gamma-1}}, \qquad B(\tau) = \frac{2\varrho_0 \tau (\gamma-1)(1-\tau)^{\frac{\gamma}{\gamma-1}}}{(\gamma+1)\tau - (\gamma-1)}. \tag{24.10}$$

In the case of an isothermal gas, $\gamma = 1$ and these formulas are not applicable. In view of Eq. (24.2) we have in this case $c^2 = a$ and it is convenient to use the dimensionless variable $\tau = q^2/a$. We can transform the system (23.8) into the normal form (24.9) with

$$A(\tau) = \frac{2\tau}{\varrho_0} e^{\frac{1}{2}\tau}, \qquad B(\tau) = \frac{2\tau \varrho_0}{\tau - 1} e^{-\frac{1}{2}\tau}. \tag{24.10'}$$

We may eliminate φ from the differential system (24.9) and obtain the following second order equation for the stream function:

$$\frac{\partial^2 \psi}{\partial \vartheta^2} = B(\tau) \frac{\partial}{\partial \tau} \left[A(\tau) \frac{\partial \psi}{\partial \tau}\right]. \tag{24.11}$$

This equation holds in the general case $\gamma \geq 1$ with the proper choice (24.10) or (24.10′) of the coefficient functions.

25. Solutions of CHAPLYGIN's equation by separation of variables. The special form of Eq. (24.11) suggests the construction of solutions by the method of separation of variables. We set up

$$\psi(\tau, \vartheta) = \tau^\nu F_\nu(\tau)\, e^{2 i \nu \vartheta} \tag{25.1}$$

and obtain for $F_\nu(\tau)$ the ordinary differential equation

$$\tau(1-\tau)F_\nu''(\tau)+\left[(2\nu+1)-\left(2\nu-\frac{1}{\gamma-1}+1\right)\tau\right]F_\nu'(\tau)+\frac{(2\nu+1)\nu}{\gamma-1}F_\nu(\tau)=0. \quad (25.2)$$

This is a special case of the general hypergeometric differential equation

$$\tau(1-\tau)F''+[c-(a+b+1)\tau]F'-abF=0 \quad (25.2')$$

which is solved by the hypergeometric function

$$F(a,b,c;\tau)=1+\sum_{k=1}^{\infty}\frac{a(a+1)\dots(a+k-1)b(b+1)\dots(b+k-1)}{1\cdot 2\cdot\dots k\,c(c+1)\dots(c+k-1)}\tau^k. \quad (25.3)$$

If we put

$$a_\nu+b_\nu=2\nu-\frac{1}{\gamma-1},\qquad a_\nu b_\nu=-\frac{(2\nu+1)\nu}{\gamma-1},\qquad c=2\nu+1 \quad (25.4)$$

we can express the set of particular solutions of Eq. (24.11), obtained by separation of variables, as follows:

$$\psi(\tau,\vartheta)=\tau^\nu F(a_\nu,b_\nu,2\nu+1;\tau)\,e^{2i\nu\vartheta}. \quad (25.5)$$

Here, the parameter ν can take arbitrary real or complex values. It is, however, to be observed that the hypergeometric function $F(a, b, c; \tau)$ becomes singular if c approaches any non-positive integer; thus, our solutions (25.5) are only defined for values of $2\nu+1$ which are not zero or negative integers.

The particular solutions (25.5) of the differential equation for the stream function in the hodograph plane were first considered by CHAPLYGIN [3]. He used them in order to solve boundary value problems by linear superposition of particular solutions. He had to overcome one of the principal difficulties inherent in the method of the hodograph plane, namely the determination of the hodograph domain for which the boundary value problem is to be solved. In fact, in general, the flow region in the physical plane is given and in order to find the hodograph image of this region one would first have to solve the boundary value problem in question. We shall discuss this characteristic difficulty in detail in Sect. 27. CHAPLYGIN overcame this problem by considering a situation where one can predict the shape of the hodograph image of the flow a priori.

We consider a flow in a region of the physical plane whose boundary consists of straight lines (walls) and of free stream lines. Along the linear walls, the velocity of the flow may change but its direction is constant, namely that of the bounding straight line. Thus, the images of the boundary walls in the hodograph plane are straight lines with the equation $\vartheta=$ const. The free stream lines are, by definition, lines along which the fluid pressure p is constant and hence, by BERNOULLI's equation, also $q=$ const. Thus, the image of a free stream line in the hodograph plane is a circular arc around the origin with the equation $q=$ const. Likewise, in the (τ, ϑ)-plane the image of a rigid linear wall is a straight line $\vartheta=$ const and the image of a free stream line is a segment $\tau=$ const. Thus, except for the finitely many parameters which determine the end points of the linear segments and circular arcs in the hodograph plane, the flow region image in the hodograph plane is entirely determined. Its general structure does not depend on the equation of state of the fluid, it is the same for a compressible or an incompressible fluid. It is no wonder that the hodograph method works particularly well in problems of this kind.

As an illustration of the method we consider two infinite linear segments which run from $x_0=-a\cos\alpha$, $y_0=a\sin\alpha$ and $x_1=-a\cos\alpha$, $y_1=-a\sin\alpha$

towards infinity under an angle $\pm(\pi-\alpha)$, respectively. To the left of these lines which are rigid walls, a compressible fluid is stored, and it streams through the gap between the walls into the empty exterior of the container. The flow forms a jet with free stream lines as boundaries (Fig. 9).

The image of the flow region in the hodograph plane is the circular sector (Fig. 10) $-\alpha\leqq\vartheta\leqq\alpha$, $0\leqq q\leqq q_0$ where q_0 is the speed of flow along the free stream lines.

If we had to deal with an incompressible fluid flow, the transition from the physical (x, y)-plane to the hodograph plane would be a conformal mapping. In this case, the stream function is harmonic in x, y and, therefore, also in the velocity variables u, v. It is easily seen that

$$\psi(u, v) = 2C\,\mathrm{Im}\left\{\log\frac{\zeta^{\frac{\pi}{2\alpha}}}{1-\zeta^{\frac{\pi}{\alpha}}}\right\}, \qquad \zeta=\frac{1}{q_0}(u-i\,v) \tag{25.6}$$

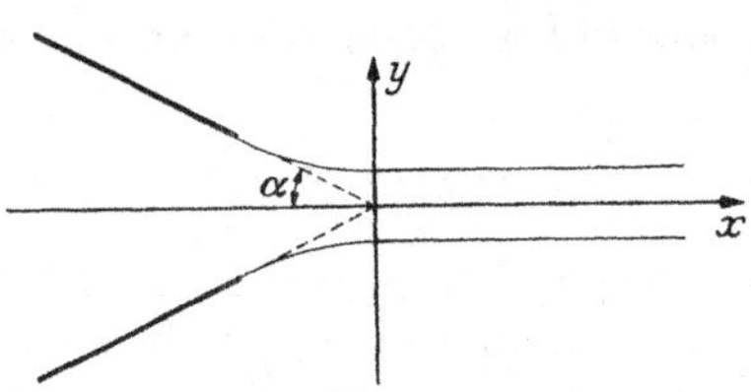

Fig. 9.

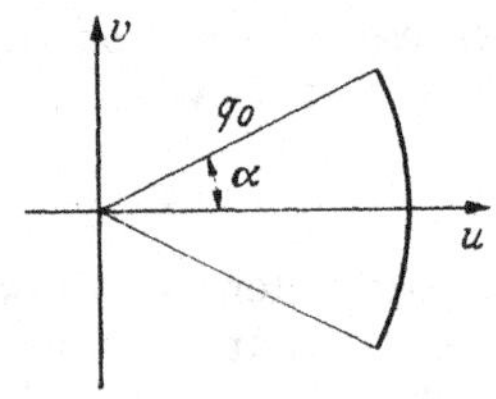

Fig. 10.

satisfies all our requirements. Indeed, ψ is harmonic in u, v, except for a logarithmic singularity near $\zeta=0$ which corresponds to the source of the flow at infinity and a logarithmic singularity near $\zeta=1$ which corresponds to the sink into which the jet is runing for $x\to+\infty$. For $v>0$ one has on the boundary of the hodograph domain $\psi=C\pi$, while for $v<0$ the boundary relation $\psi=-C\pi$ is fulfilled. Thus, the harmonic function ψ is piecewise constant on the boundary of the hodograph domain.

In the special case of an incompressible fluid the stream function (25.6) admits the series development

$$\psi=-C\left[\frac{\pi}{\alpha}\vartheta+2\sum_{n=1}^{\infty}\frac{1}{n}\left(\frac{q}{q_0}\right)^{\frac{\pi}{\alpha}n}\sin\frac{\pi}{\alpha}n\vartheta\right]. \tag{25.7}$$

CHAPLYGIN'S ingenious way of solving the corresponding flow problem for an arbitrary adiabatic fluid consists in replacing

$$q^{2\nu}\sin 2\nu\,\vartheta\to c_\nu\,\tau^\nu F(a_\nu, b_\nu, 2\nu+1;\tau)\sin 2\nu\vartheta. \tag{25.8}$$

Thus, we are led to the function

$$\psi(\tau,\vartheta)=-C\left[\frac{\pi}{\alpha}\vartheta+2\sum_{n=1}^{\infty}\frac{c_\nu}{n}\tau^\nu F(a_\nu, b_\nu, 2\nu+1;\tau)\sin 2\nu\vartheta\right] \tag{25.9}$$

with $\nu=\frac{\pi}{2\alpha}n$. Observe that ϑ is a solution of the differential equation (24.11) for every choice of γ; hence Eq. (25.9) is a solution of this differential equation whereever its convergence behavior admits the formal operations. For $\vartheta=\pm\alpha$ we have $\sin 2\nu\vartheta=0$ and hence

$$\psi(\tau, \pm\alpha)=\pm C\pi. \tag{25.9'}$$

For the particular choice of coefficients

$$c_\nu=[\tau_0^\nu F(a_\nu, b_\nu, 2\nu+1;\tau_0)]^{-1} \tag{25.10}$$

we have also

$$\psi(\tau_0, \vartheta) = -C\left[\frac{\pi}{\alpha}\vartheta + 2\sum_{n=1}^{\infty}\frac{1}{n}\sin n\left(\frac{\pi}{\alpha}\vartheta\right)\right] \tag{25.11}$$

and it can be easily seen that the correct boundary conditions are fulfilled in the entire circular sector $-\alpha \leqq \vartheta \leqq \alpha$, $0 \leqq \tau \leqq \tau_0$.

Thus, we can solve some of the classical free boundary problems as easily in the case of a compressible fluid as in the incompressible case. The constant C allows us to adjust the solution to the geometric constant a in the physical plane. It can be shown that the solution series converges for $\tau < \tau_0$ as long as τ_0 is subsonic.

The limitations of the Chaplygin method of separation of variables are rather obvious. We observe that the particular solutions (25.5) can have their singularities only for the values $\tau = 0$, 1 or ∞. In many practical problems singularities of different location and also different nature are required and these singularities cannot be obtained in the above way.

We consider, for example, the flow of an incompressible fluid which has a doublet at infinity and streams around a circle of radius r. Let us suppose that the fluid moves at infinity with the speed 1 in direction of the positive x-axis. If we introduce the complex variable $\zeta = u - iv$ in the hodograph plane and use $z = x + iy$ in the physical plane, we find

$$\zeta = 1 - \frac{r^2}{z^2}, \qquad \psi = \operatorname{Im}\left\{z + \frac{r^2}{z}\right\}. \tag{25.12}$$

We see that the image from the flow region $|z| > r$ onto the hodograph plane is not-univalent, and that the points $\pm z$ have the same image ζ. Therefore, we shall have to consider Riemann domains over the ζ-plane and the single-valued particular Chaplygin solutions will not be sufficient there. Moreover, we have

$$\psi = r \operatorname{Im}\left\{\sqrt{1-\zeta} + \frac{1}{\sqrt{1-\zeta}}\right\} \tag{25.13}$$

which shows that the singularities connected with simple doublets lead to algebraic singularities for the stream function even in the simplest case of an incompressible fluid. A general method for constructing solutions of the stream function equation in the hodograph plane must, therefore, be able to provide singularities which converge in the limit to the algebraic singularities of the incompressible fluid.

An alternative approach to the problem of singularities of the solution of the differential equation (24.11) consists in a systematic theory of analytic continuation of the series (25.9) around the singular points. This method has been developed by CHERRY [32], [33], [34]. For additional discussions and applications see [30], [31], [35].

26. Solutions of CHAPLYGIN's equation depending on an artitrary analytic function. We wish to discuss a general method for constructing solutions of the differential equations in the hodograph plane. It will be convenient to operate with the variables λ, ϑ defined in Sect. 23 and to consider the Eq. (23.15). We set up a solution of this equation in the form [8]

$$\psi^*(\lambda, \vartheta) = \sum_{n=0}^{\infty} g_n(\lambda, \vartheta)\, G_n(\lambda) \tag{26.1}$$

where the functions $g_n(\lambda, \vartheta)$ are harmonic in their two variables. This form may be considered as a variation of constants; indeed, in the limit case $f(\lambda) \equiv 0$ the general solution will be of the form (26.1) with constant factors G_n. We shall try to adjust the $G_n(\lambda)$ to the exact equation.

We proceed at first in a formal way and postpone questions of convergence until later. We start with the identity

$$\Delta(AB) = A\Delta B + B\Delta A + 2\left(\frac{\partial A}{\partial\lambda}\frac{\partial B}{\partial\lambda} + \frac{\partial A}{\partial\vartheta}\frac{\partial B}{\partial\vartheta}\right) \tag{26.2}$$

where $\Delta = \frac{\partial^2}{\partial\lambda^2} + \frac{\partial^2}{\partial\vartheta^2}$ and A, B are two arbitrary functions of λ and ϑ. We apply this identity to $A = g_n$, $B = G_n$ and use the fact that

$$\Delta g_n = 0, \qquad \frac{\partial}{\partial\vartheta} G_n(\lambda) = 0. \tag{26.3}$$

Hence

$$\Delta(g_n G_n) = G_n''(\lambda)\, g_n(\lambda,\vartheta) + 2G_n'(\lambda)\frac{\partial}{\partial\lambda} g_n(\lambda,\vartheta). \tag{26.4}$$

Thus, inserting Eq. (26.1) into the differential equation (23.15) and using Eq. (26.4) we find

$$\sum_{n=0}^{\infty}\left\{g_n(\lambda,\vartheta)\,[G_n''(\lambda) - f(\lambda)\, G_n(\lambda)] + 2G_n'(\lambda)\frac{\partial}{\partial\lambda} g_n(\lambda,\vartheta)\right\} = 0. \tag{26.5}$$

In order to elaborate the last equation we wish to make some assumption about $\frac{\partial}{\partial\lambda} g_n$. Various possibilities arise and we shall discuss two different procedures which have been stressed in the literature.

a) Let us assume at first the differential relation

$$\frac{\partial}{\partial\lambda} g_n(\lambda,\vartheta) = p_n g_n(\lambda,\vartheta) + h(\lambda,\vartheta) \tag{26.6}$$

where $h(\lambda,\vartheta)$ is an arbitrary but fixed harmonic function. We can satisfy Eq. (26.5) by choosing

$$G_n''(\lambda) + 2p_n G_n'(\lambda) - f(\lambda)\, G_n(\lambda) = 0 \tag{26.7}$$

and demanding

$$\sum_{n=0}^{\infty} G_n'(\lambda) = 0. \tag{26.8}$$

It will be shown in the next section how one can find sequences of real numbers p_n such that the corresponding solutions of Eq. (26.7) satisfy the identities (26.8); we postpone also till then the study of convergence of the series. We proceed here formally and first integrate Eq. (26.6). We introduce the complex variable $\zeta = \lambda - i\vartheta$ and the analytic function $v(\zeta)$ such that

$$\mathrm{Im}\{v'(\zeta)\} = h(\lambda,\vartheta). \tag{26.9}$$

Then we find easily

$$g_n(\lambda,\vartheta) = \mathrm{Im}\left\{A_n\, e^{p_n\zeta} + \int_{\zeta_0}^{\zeta} e^{p_n(\zeta-\eta)}\, dv(\eta)\right\}, \tag{26.10}$$

with the arbitrary constant of integration A_n. In particular, we find that

$$\psi^*(\lambda,\vartheta) = \mathrm{Im}\left\{\sum_{n=0}^{\infty} G_n(\lambda)\, e^{p_n\zeta}\cdot\int_{\zeta_0}^{\zeta} e^{-p_n\zeta}\, dv(\eta)\right\} \tag{26.11}$$

is a solution of the hodograph differential equation (23.15). This series development is due to LIGHTHILL [9] to [13]; it is characterized by the analytic function $v(\zeta)$ and the exponents p_n. It will be discussed at greater length in Sect. 27.

b) Instead of requiring the identity (26.6), we might have set up the recursion formula

$$\frac{\partial}{\partial\lambda} g_n(\lambda,\vartheta) = -\frac{1}{2} g_{n-1}(\lambda,\vartheta) \tag{26.12}$$

and satisfied Eq. (26.5) by the recursion differential equation

$$G'_{n+1}(\lambda) = G''_n(\lambda) - f(\lambda)\, G_n(\lambda), \qquad n = 1, 2, \ldots \tag{26.13}$$

with

$$G_0(\lambda) = 1. \tag{26.13'}$$

One may start with an arbitrary harmonic function $g_0(\lambda,\vartheta)$ and determine the rest by recursion. Let $v(\zeta)$ be analytic in $\zeta = \lambda - i\vartheta$ and such that

$$g_0(\lambda,\vartheta) = \mathrm{Im}\{v(\zeta)\}. \tag{26.14}$$

We have then clearly

$$g_n(\lambda,\vartheta) = \mathrm{Im}\left\{\frac{(-1)^n}{(n-1)!\,2^n}\int\limits_{\zeta_0}^{\zeta} v(\eta)\,(\zeta-\eta)^{n-1}\,d\eta\right\}. \tag{26.15}$$

Finally, let

$$U(t;\lambda,\vartheta) = \sum_{n=1}^{\infty} G_n(\lambda)\,\frac{(-1)^n(\zeta-t)^{n-1}}{2^n(n-1)!}; \tag{26.16}$$

it is then easily seen from Eqs. (26.1) and (26.15) that the solution $\psi^*(\lambda,\vartheta)$ can be written in the form

$$\psi^*(\lambda,\vartheta) = \mathrm{Im}\left\{v(\zeta) + \int\limits_{\zeta_0}^{\zeta} U(t;\lambda,\vartheta)\,v(t)\,dt\right\}. \tag{26.17}$$

The validity of the formula and the verification that $\psi^*(\lambda,\vartheta)$ is a solution of Eq. (23.15) depends, of course, upon the convergence behavior of the series (26.16) and will be considered in Sect. 29. The series (26.17) for the solution of the differential equation (23.15) was given by BERGMAN [*8*], [*14*] to [*17*].

The common feature of both integration methods is the arbitrary analytic function $v(\zeta)$ which occurs in them. It may be a multivalued function in the ζ-plane, it may have branch points and singularities. It serves, therefore, to create the rather complicated singularity functions needed in the actual problems of a two-dimensional irrotational flow of a compressible fluid. It allows also the adjustment to the boundary value problem which is, in general, posed in the physical x, y-plane.

The procedure of obtaining a very general solution of a given differential equation with enough parameters to approximate the conditions of an actual physical problem is typical of the inverse methods occurring frequently in applied mathematics [*18*]. The value of such a solution depends considerably upon the ease with which it can be adapted to these conditions. There exist alternative procedures for obtaining general solutions for the partial differential equation (24.11). We mention BERGMAN's general method of complex integral operators [*15*], [*16*] and the theory of BERS-GELBART of sigma-monogenic functions [*39*] which permits the construction of an infinity of particular solutions with various types of singularity behavior.

27. LIGHTHILL's method. In order to obtain a precise understanding of the Lighthill series solution (26.11) we have to consider the general second order differential equation

$$\frac{d^2 G(\lambda;\sigma)}{d\lambda^2} + 2\sigma \frac{dG(\lambda;\sigma)}{d\lambda} = f(\lambda)\, G(\lambda;\sigma). \tag{27.1}$$

We wish to normalize the solutions at the stagnation point of the flow; in view of the definition (24.5), this point corresponds to $\lambda = -\infty$. We observe that in view of Eqs. (24.5) and (24.7), we have near that point

$$m \sim e^{2\lambda}, \qquad f(\lambda) = \sum_{n=2}^{\infty} c_n e^{2n\lambda}. \tag{27.2}$$

Therefore, it is useful to introduce $s = e^{2\lambda}$ as a new variable and to consider the equivalent equation

$$s^2 \Gamma''(s;\sigma) + s(1+\sigma)\,\Gamma'(s;\sigma) = \tfrac{1}{4} \sum_{n=2}^{\infty} c_n s^n \cdot \Gamma(s;\sigma) \tag{27.3}$$

with $\Gamma(e^{2\lambda};\sigma) = G(\lambda;\sigma)$. The stagnation point corresponds now to $s=0$, and we observe that this point is a singular point of the differential equation. We normalize $\Gamma(s;\sigma)$ by the requirement of being analytic at this point and having the value $\Gamma(0;\sigma)=1$. This determines $\Gamma(s;\sigma)$ and consequently $G(\lambda;\sigma)$ in a unique way.

We may set up

$$\Gamma(s;\sigma) = \sum_{n=0}^{\infty} a_n(\sigma)\, s^n, \qquad a_0(\sigma) = 1 \tag{27.4}$$

and obtain from Eq. (27.3) the recursion formula

$$(n+1)(n+1+\sigma)\, a_{n+1}(\sigma) = \tfrac{1}{4} \sum_{\nu=0}^{n-1} c_{n-\nu+1} a_\nu(\sigma). \tag{27.5}$$

Thus we always obtain a solution to the problem provided that σ is not a negative integer, and from the general theory of differential equations it follows that $\Gamma(s;\sigma)$ is an analytic function of σ, except at these negative integer singularities, for all values s for which $f(\lambda)$ is regular in s.

Let k be a fixed integer; we write the recursion formula (27.5) in the form

$$\left.\begin{aligned} &(k+r+1)(k+r+1+\sigma)\, a_{k+r+1}(\sigma)\cdot(k+\sigma) \\ &\quad = \tfrac{1}{4} \sum_{\nu=0}^{k-1} c_{k+r-\nu+1} a_\nu(\sigma)\cdot(k+\sigma) + \tfrac{1}{4} \sum_{\nu=0}^{r-1} c_{r-\nu+1} a_{k+\nu}(\sigma)\cdot(k+\sigma). \end{aligned}\right\} \tag{27.6}$$

We observe that even for $\sigma = -k$ the recursion formula (27.5) yields finite values for $a_\nu(\sigma)$ as long as $\nu \leqq k-1$. For higher values of ν the a_ν will become infinite. However, we can calculate from Eq. (27.6) the limit values

$$b_r(k) = \lim_{\sigma\to -k} a_{k+r}(\sigma)\cdot(k+\sigma). \tag{27.7}$$

We obtain the recursion formula

$$(r+1)(r+1+k)\, b_{r+1}(k) = \tfrac{1}{4} \sum_{\nu=0}^{r-1} c_{r-\nu+1} b_\nu(k). \tag{27.8}$$

This formula leaves $b_0(k)$ undetermined; but it coincides exactly with the recursion formula (27.5) for $a_r(k)$. Hence, we have proved that

$$\lim_{\sigma\to -k} (k+\sigma)\,\Gamma(s;\sigma) = A_k \Gamma(s;k)\, s^k, \qquad k = 1, 2, \ldots, \tag{27.9}$$

where A_k is a constant factor.

It is easy to see that $A_1 = 0$. In fact, due to the particular circumstance that $c_0 = c_1 = 0$ in the development of $f(\lambda)$ the right side of Eq. (27.5) vanishes always for $n = 0$ and, hence, we do not obtain an infinity even for $\sigma = -1$. Thus, we have proved:

The function $G(\lambda; \sigma)$ is meromorphic in σ; at $\sigma = -k\,(k = 2, 3, \ldots)$ it has simple poles whose residues are proportional to

$$H(\lambda; k) = e^{2\lambda k}\, G(\lambda; k). \tag{27.10}$$

Observe that $H(\lambda; k)$ satisfies the differential equation

$$H'' - 2kH' = fH. \tag{27.11}$$

It is the solution of the Eq. (27.1) for $\sigma = -k$ which is regular for $\lambda = -\infty$.

Consider the complex σ-plane from which the set of circles of radius δ around the negative integers has been removed. It can be shown [13] that in the remaining domain we have

$$\lim_{|\sigma| \to \infty} G(\lambda; \sigma) = 1 \tag{27.12}$$

uniformly for all $\lambda \leqq \lambda_0 < 0$.

Because of the simple structure of the differential equation (27.1) and its close relation to the hypergeometric differential equation (25.2′), the function $G(\lambda; \sigma)$ can be expressed explicitely in terms of hypergeometric functions. We remember that $\psi^*(\lambda, \vartheta)$ satisfies the differential equation (23.15) and assume the special form

$$\psi^*(\lambda, \vartheta) = G(\lambda; \sigma)\, e^{\sigma\lambda}\, e^{i\sigma\vartheta}. \tag{27.13}$$

This leads immediately to the differential equation (27.1) for the factor $G(\lambda; \sigma)$. On the other hand by Eq. (23.14) the corresponding stream function is of the form

$$\psi(\lambda, \vartheta) = z(\lambda)^{-\frac{1}{2}}\, G(\lambda; \sigma)\, e^{\sigma\lambda}\, e^{i\sigma\vartheta}, \tag{27.14}$$

with variables separated. But in Sect. 25 we calculated the general form of a solution ψ from which the ϑ-variable is split. We found

$$\psi(\tau, \vartheta) = \tau^{\sigma/2} F(a_{\sigma/2}, b_{\sigma/2}, \sigma + 1; \tau)\, e^{i\sigma\vartheta}; \tag{27.15}$$

therefore, it is easily seen that

$$z(\lambda)^{-\frac{1}{2}}\, G(\lambda; \sigma)\, e^{\sigma\lambda} = \tau^{\sigma/2} F(a_{\sigma/2}, b_{\sigma/2}, \sigma + 1; \tau) = \psi_\sigma(\tau). \tag{27.16}$$

This result allows the application of the highly developed theory of the hypergeometric functions to the fluid dynamical problem.

We note, in particular, that the right side of Eq. (27.16) becomes identically 1 for $\sigma = 0$. Hence

$$G(\lambda; 0) = z(\lambda)^{\frac{1}{2}}. \tag{27.17}$$

We have now completed the mathematical preparations for the construction of solutions (26.11). We introduce an analytic function $\varphi(\sigma)$ with the following properties:

1. $\varphi(\sigma)$ is meromorphic in σ; its poles are simple poles with real residues; they lie on the real axis but not at zero or at negative integers.

2. There exists in the σ-plane a sequence of closed curves L_r surrounding the origin, having a distance $> \delta$ from all negative integers and minimum distances r from the origin increasing to infinity in such a way that

$$\max_{\sigma \in L_r} |\varphi(\sigma) - 1| \to 0 \quad \text{as} \quad r \to \infty. \tag{27.18}$$

Consider now the contour integral

$$\frac{1}{2\pi i} \oint_{L_r} \varphi(\sigma)\, G(\lambda;\sigma)\, \frac{d\sigma}{\sigma} = J_r(\lambda). \tag{27.19}$$

The residues of the integral come from the poles p_n of the integrand which lie at $\sigma=0$, at the (real) poles of $\varphi(\sigma)$ and at the negative integers $-k$ inside of L_r. By Eq. (27.11), the residues at the poles p_n are in each case functions $G_n(\lambda)$ satisfying the differential equation

$$G_n''(\lambda) + 2p_n G_n'(\lambda) - f(\lambda)\, G_n(\lambda) = 0. \tag{27.20}$$

On the other hand, the asymptotic behavior (27.12) of $G(\lambda;\sigma)$ and the assumption (27.18) on $\varphi(\sigma)$ imply

$$\lim_{r\to\infty} J_r(\lambda) = 1. \tag{27.21}$$

Hence, we find

$$\sum_{n=0}^{\infty} G_n(\lambda) = 1. \tag{27.22}$$

Thus, we have indeed constructed a set $G_n(\lambda)$ of solutions of Eq. (27.20) satisfying the decisive condition (26.8).

Each function $\varphi(\sigma)$ of the prescribed character provides a set $G_n(\lambda)$, and each such set leads to an infinity of solutions $\psi^*(\lambda,\vartheta)$ of the hodograph differential equation, provided that the series (26.11) converges. The great freedom in the choice of $\varphi(\sigma)$ can be used to enforce additional requirements with respect to the solution; the most important one is the single-valuedness of $\psi^*(\lambda,\vartheta)$ in the entire flow domain. However, it is often sufficient to make the simplest admissible choice of $\varphi(\sigma)$, namely $\varphi(\sigma)\equiv 1$. In this case, the residues of the integral in Eq. (27.19) lie at the points $\sigma=0, -2, -3, \ldots$ and we find

$$\psi^*(\lambda,\vartheta) = \operatorname{Im}\left\{G(\lambda,0)\int_{\zeta_0}^{\zeta} dv(\eta) + \sum_{k=2}^{\infty} A_k\, G(\lambda;k)\, e^{k\bar{\zeta}} \int_{\zeta_0}^{\zeta} e^{k\eta}\, dv(\eta)\right\} \tag{27.23}$$

where $\bar{\zeta} = 2\lambda - \zeta = \lambda + i\vartheta$. The coefficients A_k of this development can be determined from the general theory of the hypergeometric function. It can be shown that the series (27.23) converges in the entire subsonic region $\lambda<0$ [*13*].

Using Eqs. (23.14), (27.16) and (27.17), we can return to the stream function itself:

$$\psi(\lambda,\vartheta) = \operatorname{Im}\left\{v(\zeta) - v(\zeta_0) + \sum_{k=2}^{\infty} A_k\, \psi_\sigma(\tau)\, e^{-ik\vartheta} \int_{\zeta_0}^{\zeta} e^{k\eta}\, dv(\eta)\right\}. \tag{27.23'}$$

This formula clarifies the role of the parameter function $v(\zeta)$. If we had an incompressible fluid flow, it would have given rise to a harmonic stream function in the plane of the variable $\zeta = \log q - i\vartheta$, say $\psi(\log q,\vartheta) = \operatorname{Im}\{w(\zeta)\}$ where $w(\zeta)$ is an analytic function of ζ. On the other hand, it is easily seen from the definition of λ that for very large negative values of λ the two variables λ and $\log q$ are asymptotic to each other [see Eq. (23.10)]. In the same region the most important term in the development (27.23′) is precisely the harmonic term $\operatorname{Im}\{v(\zeta) - v(\zeta_0)\}$. Hence, we may conceive the remainder of this series to be correction terms to the first term which characterizes an incompressible fluid flow.

The actual solution of a flow problem by means of the hodograph method may then be described as follows. One first solves the problem under the assumption of incompressibility. This leads to an image of the flow region in the logarithmic hodograph plane $\zeta = \log q - i\vartheta$ which will, in general, cover a Riemann surface over this plane. The stream function referred to these new variables becomes an harmonic function $\psi = \mathrm{Im}\,\{v(\zeta)\}$. One then uses this analytic function $v(\zeta)$ as the generating function in a series development (27.23′). This yields a stream function $\psi(\lambda, \vartheta)$ which solves the correct differential equation and will, in general, have the same overall character as the stream function for the incompressible cases. In particular, we may find a closed curve in the hodograph plane on which $\psi(\lambda, \vartheta)$ will have a constant value. We may then map the region enclosed by this curve back into the physical (x, y)-plane by means of the relations (23.6) which can be expressed in terms of ψ only, by use of Eq. (23.8). The function $\psi(x, y)$ will then be the stream function in the physical plane of a flow which has the boundary as a stream line. The new boundary of the flow will be near to the original flow boundary, provided that the deformation of the stream function (27.23′) was not too great. This indirect procedure for obtaining flow patterns is characteristic for all hodograph methods.

28. Flows with circulation. Relation to CHAPLYGIN's solution. The stream function

$$\psi(x, y) = \mathrm{Im}\left\{q_0\left[z + \frac{\alpha}{2\pi i}\log z + \frac{1}{z}\right]\right\}, \qquad z = x + i y, \qquad \alpha \text{ real} \tag{28.1}$$

represents an incompressible fluid flow coming from infinity with speed q_0 in the direction of the x-axis and streaming past a circle of radius 1 around the origin. This example shows that the stream function needs not necessarily be single-valued. It has a constant period αq_0 if we decribe a closed loop around the boundary of the flow region; αq_0 is called the circulation of the flow.

If we pass to the logarithmic hodograph plane of the variable $\zeta = \log q - i\vartheta$ we derive from Eq. (28.1) the relation

$$\zeta = \log\left(1 + \frac{\alpha}{2\pi i}\cdot\frac{1}{z} - \frac{1}{z^2}\right) + \log q_0 \tag{28.2}$$

and hence, with $\lambda = \log q$,

$$\psi(\lambda, \vartheta) = \mathrm{Im}\left\{q_0\left[z(\zeta) + \frac{\alpha}{2\pi i}\log z(\zeta) + \frac{1}{z(\zeta)}\right]\right\} = \mathrm{Im}\,\{v(\zeta)\}, \tag{28.3}$$

a rather complicated function of ζ which is also multivalued. The situation becomes much worse if we insert this expression for $v(\zeta)$ as the generating function into a series (26.11) for a compressible fluid stream function.

We can write $\eta = \eta(z)$ in Eq. (26.11), and using Eqs. (28.2) and (28.3) we obtain

$$\int e^{-p_n\eta}\,dv(\eta) = q_0^{1-p_n}\int\left(1 + \frac{\alpha}{2\pi i}\cdot\frac{1}{z} - \frac{1}{z^2}\right)^{-p_n}\left(1 + \frac{\alpha}{2\pi i}\cdot\frac{1}{z} - \frac{1}{z^2}\right)dz. \tag{28.4}$$

If we compute the right side integral over a sufficiently large closed curve which surrounds the unit circle once, we obtain the value

$$\oint e^{-p_n\eta}\,dv(\eta) = \alpha(1 - p_n)\,q_0^{1-p_n}; \tag{28.5}$$

hence the series (26.11) is determined only up to a period

$$\mathrm{Im}\left\{\sum_{n=0}^{\infty} q_0^{1-p_n}G_n(\lambda)\,e^{p_n\zeta}\alpha(1 - p_n)\right\} = -\alpha\sum_{n=0}^{\infty} q_0^{1-p_n}G_n(\lambda)\,e^{\lambda p_n}\sin p_n\vartheta\cdot(1 - p_n). \tag{28.6}$$

There is no hope of obtaining a well determined velocity field in the physical plane if this periodicity modulus does not reduce to a constant. From the structure of the series (28.6), on the other hand, it is clear that this term can become indepentend of λ and ϑ only if the coefficients of each term $\sin p_n\vartheta$ vanish separately.

In our construction by means of the auxiliary analytic function $\varphi(\sigma)$, described in Sect. 27, the p_n were the poles of the integrand in Eq. (27.19). The factor $G(\lambda;\sigma)$ necessarily introduces the values $-2, -3, \ldots$ for p_n with residues

$$\frac{\varphi(-k)}{-k} A_k H(\lambda; k). \tag{28.7}$$

In order to cancel these poles, $\varphi(\sigma)$ must have simple poles at $2, 3, \ldots$ with the residues

$$R_k = -\frac{1+k}{1-k} \cdot \frac{\varphi(-k)}{G(\lambda; k)} A_k H(\lambda; k)\, e^{-2\lambda k} q_0^{2k}. \tag{28.8}$$

It is easily seen that under this assumption the terms in the series (28.6) will cancel each other pairwise. By use of Eq. (27.10), we have the simpler condition

$$R_k = -\frac{1+k}{1-k} A_k \varphi(-k)\, q_0^{2k} \tag{28.8'}$$

for the residues of the analytic function $\varphi(\sigma)$ at the points $\sigma = k$, $k = 2, 3, \ldots$.

It is easy to guess a function $\varphi(\sigma)$ with the required residue behavior. In fact, we conclude from Eq. (27.10) that

$$\varphi_0(\sigma) = \frac{G(\lambda_0; -\sigma)}{1-\sigma} \tag{28.9}$$

satisfies this condition provided we select $\lambda_0 = \log q_0$. This function, however, is not admissible in our construction since we have by Eq. (27.12) $\varphi(\sigma) = O(1/|\sigma|)$, which contradicts the requirement (27.18). It can be easily verified that the function

$$\varphi_1(\sigma) = \frac{1}{\sigma - 1} [G'(\lambda_0; \sigma) + \sigma G(\lambda_0; \sigma)] \tag{28.10}$$

still satisfies the residue condition (28.8′) and has, moreover, the limit behavior (27.18). (The prime denotes differentiation of G with respect to its first variable.) Thus, every combination $\varphi_1(\sigma) + \beta\varphi_0(\sigma) = \varphi(\sigma)$ represents an admissible function which yields a single-valued solution $\psi^*(\lambda, \vartheta)$ of the hodograph equation. We observe that $\varphi(\sigma)$ will, in general, have a simple pole at $p = 1$, but in view of the factor $(1-p)$ in Eq. (28.6) it will not give rise to multivaluedness of the solution.

The preceding analysis shows the flexibility of LIGHTHILL's method for constructing stream functions and the usefulness of the analytic function $\varphi(\sigma)$ at our disposal. It should be also observed that in the case of singularities of $v(\zeta)$, the integrability conditions (23.8) do not guarantee a single-valued mapping in the large from the hodograph plane into the physical plane. Lighthill shows, however, by a more elaborate analysis that even in the case of non-zero circulation the function $\varphi(\sigma)$ can be chosen in such a way that a single-valued mapping is obtained.

The close relation between the series (26.11) and the Chaplygin solutions (25.5) is seen if we specialize the generating analytic function to

$$v(\zeta) = e^{\alpha\zeta}. \tag{28.11}$$

The Lighthill series becomes

$$\left.\begin{aligned} \psi^*(\lambda, \vartheta) &= \operatorname{Im}\left\{\sum_{n=0}^{\infty} G_n(\lambda)\, e^{p_n\zeta} \frac{e^{(\alpha-p_n)\zeta} - e^{(\alpha-p_n)\zeta_0}}{\alpha - p_n}\right\} \\ &= \operatorname{Im}\left\{\sum_{n=0}^{\infty} e_n G_n(\lambda)\, e^{p_n\zeta}\right\} + \operatorname{Im}\left\{\sum_{n=0}^{\infty} \frac{G_n(\lambda)}{\alpha - p_n} e^{\alpha\zeta}\right\} \end{aligned}\right\} \tag{28.12}$$

with

$$e_n = \frac{e^{(\alpha-p_n)\zeta_0}}{p_n - \alpha}. \tag{28.13}$$

If we assume $\alpha - p_n > 0$ and put $\operatorname{Re}\{\zeta_0\} = -\infty$, we may obtain $e_n = 0$.

Observe next that in view of Eqs. (27.12) and (27.18)

$$\lim_{r\to\infty} \frac{1}{2\pi i} \int_{L_r} \varphi(\sigma)\, G(\lambda;\sigma)\, \frac{d\sigma}{\sigma(\sigma-\alpha)} = 0. \tag{28.14}$$

Hence, when we apply the residue theorem to Eq. (28.14) and remember the definition of the $G_n(\lambda)$ as residues of the integrand in Eq. (27.19), we find

$$\sum_{n=0}^{\infty} \frac{G_n(\lambda)}{\alpha - p_n} = \frac{1}{\alpha}\, \varphi(\alpha)\, G(\lambda;\alpha). \tag{28.15}$$

Thus the solution $\psi^*(\lambda, \vartheta)$ becomes

$$\psi^*(\lambda,\vartheta) = \operatorname{Im}\left\{\frac{1}{\alpha}\, \varphi(\alpha)\, G(\lambda;\alpha)\, e^{\alpha\zeta}\right\}. \tag{28.16}$$

Finally we can pass from $\psi^*(\lambda, \vartheta)$ to $\psi(\tau, \vartheta)$ by dividing by $G(\lambda; 0) = z(\lambda)^{\frac{1}{2}}$ and by use of the identity (27.16); we arrive at

$$\psi(\tau,\vartheta) = \operatorname{Im}\left\{\frac{1}{\alpha}\, \varphi(\alpha)\, \psi_\alpha(\tau)\, e^{-i\alpha\vartheta}\right\} \tag{28.17}$$

which, by Eq. (27.16), coincides with Eq. (25.5).

If the function $v(\zeta)$ can be developed into a Dirichlet series

$$v(\zeta) = \sum \beta_n e^{\gamma_n \zeta} \tag{28.18}$$

converging in a certain strip $a \leqq \operatorname{Re}\{\zeta\} \leqq b$ of the ζ-plane, we can compute the Lighthill series formally in an analogous way. However, we shall not be able to assume that all the e_n will vanish, and we shall get more terms than just the Chaplygin solutions. This procedure permits a controlled analytic continuation of a solution and is of particular interest when one is concerned with the extension of a solution into a transonic region. It is obvious, however, that the formulas will become very complicated and the value of the entire procedure becomes questionable.

For further applications of the Lighthill solution method see [*19*], [*20*]. Flows with circulation were also treated in [*33*], [*36*], [*37*].

29. BERGMAN's method of integration for CHAPLYGIN's equation. BERGMAN's integration method sketched in Sect. 26 provides an alternative approach to solutions of the differential equation (23.15) for the stream function in the hodograph plane. It is somewhat less dependent on the special equation of state of the fluid which enters into Eq. (23.15) through the coefficient $f(\lambda)$; it provides a particularly simple solution for an equation of state which represents a idealized fluid near enough to an adiabatic gas to be of considerable interest.

The main problem in the present approach is to obtain the basic functions $G_n(\lambda)$ which are defined by the recursion formula (26.13) and (26.13′) and the normalization

$$\lim_{\lambda\to-\infty} G_n(\lambda) = 0, \qquad n = 1, 2, \ldots. \tag{29.1}$$

A theoretical understanding of the $G_n(\lambda)$ is facilitated by the use of a set of elementary dominating functions $Q_n(\lambda)$. We say that a function $Q(\lambda)$ dominates the function $G(\lambda)$ in the interval $a \leqq \lambda \leqq b$ if for all λ in this interval we have

$$|G(\lambda)| \leqq Q(\lambda), \qquad |G^{(\nu)}(\lambda)| \leqq Q^{(\nu)}(\lambda), \qquad \nu = 1, 2, \ldots; \tag{29.2}$$

we express this relation by the symbol

$$G(\lambda) \ll Q(\lambda), \quad a \leqq \lambda \leqq b. \tag{29.3}$$

Let us now suppose that the coefficient function $f(\lambda)$ in (23.15) satisfies the relation

$$f(\lambda) \ll C(\varepsilon-\lambda)^{-2}, \quad \lambda \leqq \varepsilon < 0. \tag{29.3'}$$

We then define a sequence of functions $Q_n(\lambda)$ in the infinite interval $\lambda \leqq \varepsilon$ by the recursion formulas

$$Q'_{n+1}(\lambda) = Q''_n(\lambda) + C(\varepsilon-\lambda)^{-2} Q_n(\lambda), \qquad Q_0(\lambda) = 1, \tag{29.4}$$

and the normalization $Q_n(-\infty) = 0$ for $n \geqq 1$. The $Q_n(\lambda)$ satisfy the same recursion formulas as the $G_n(\lambda)$, but we have replaced $f(\lambda)$ by a dominating function. Since we calculate the $Q_n(\lambda)$ recursively by integration, differentiation and multiplication with the dominating function, it is easily seen that

$$G_n(\lambda) \ll Q_n(\lambda), \quad \lambda \leqq \varepsilon. \tag{29.5}$$

On the other hand, we find immediately

$$Q_n(\lambda) = \mu_n \cdot \frac{n!}{(\varepsilon-\lambda)^n}, \tag{29.6}$$

where the μ_n satisfy the recursion formula

$$(n+1)^2 \mu_{n+1} = \mu_n[n^2+n+C] = \mu_n(n+a)(n+b), \qquad \mu_0 = 1 \tag{29.7}$$

with

$$a = \tfrac{1}{2} - \sqrt{\tfrac{1}{4} - C}, \qquad b = \tfrac{1}{2} + \sqrt{\tfrac{1}{4} - C}. \tag{29.7'}$$

Thus, the μ_n satisfy the recursion formula of the coefficients in the hypergeometric series

$$F(a, b, 1; x) = \sum_{\nu=0}^{\infty} \mu_\nu x^\nu. \tag{29.8}$$

This series converges uniformly in the domain $|x| \leqq \alpha < 1$.

We are now able to determine the region of convergence of the series (26.16), which is the decisive tool in BERGMAN'S integration method. The general term in this series can be estimated in view of Eq. (29.6) as follows:

$$\left| G_n(\lambda) \frac{(-1)^n (\zeta-\tau)^{n-1}}{2^n (n-1)!} \right| \leqq \mu_n n \frac{|\zeta-\tau|^{n-1}}{(2(\varepsilon-\lambda))^n}. \tag{29.9}$$

Thus, the series (26.16) is majorized by the series with positive terms

$$\frac{1}{2(\varepsilon-\lambda)} \sum_{n=1}^{\infty} \mu_n n \left(\frac{|\zeta-\tau|}{2(\varepsilon-\lambda)} \right)^{n-1} = \frac{1}{2(\varepsilon-\lambda)} F'\left(a, b, 1; \frac{|\zeta-\tau|}{2(\varepsilon-\lambda)}\right) \tag{29.10}$$

which clearly converges if

$$|\zeta-\tau| < 2(\varepsilon-\lambda). \tag{29.11}$$

For a given point τ the function $U(\tau; \lambda, \vartheta)$ is defined for all values $\zeta = \lambda - i\vartheta$ within the above hyperbola with τ as one focus. If we want to use $U(\tau; \lambda, \vartheta)$ in the formula (26.17) and to integrate from a given initial point ζ_0, we have to restrict ourselves to values ζ in the hyperbola

$$|\zeta-\zeta_0| < 2(\varepsilon-\lambda). \tag{29.11'}$$

Conversely, if $\zeta_1 = \lambda_1 - i\vartheta_1$ lies in this hyperbola, we consider the circle

$$|\zeta_1-\zeta| < 2(\varepsilon-\lambda_1) \tag{29.12}$$

which contains, in particular, the point ζ_0 and which has a domain of intersection with the hyperbola (29.11′) inside of which we can connect the points ζ_0 and ζ_1 by a curve Γ. On this curve, $U(\tau;\lambda_1,\vartheta_1)$ is obviously an analytic function of τ and, hence, the integral (26.17) between ζ_0 and ζ_1 can be computed along Γ. Thus we have proved that for every given point ζ_0 with $\operatorname{Re}\{\zeta_0\}\leqq\varepsilon<0$ we can construct solutions of the stream function equation in the form (26.17) inside of the hyperbola (29.11′). The analytic function $v(\tau)$ in this expression is subjected only to the condition of being analytic in this hyperbola except for isolated singularities.

We make the special choice $\zeta_0=0$ and pass to the limit $\varepsilon=0$; we recognize that we can construct solutions of the problem within the angular space $|\zeta|<2|\lambda|$ which is bounded by the two straight lines through the origin which form with the negative λ-axis the angles $\pm 60°$.

In order to solve actual flow problems by the integration formula (26.17), we proceed in complete analogy to the method described in Sect. 27. We observe that if the velocity of the flow is small everywhere, we may choose $\operatorname{Re}\{\zeta_0\}$ as very large and negative, and we may restrict ourselves to large negative values of λ. In view of the estimates (29.5) and (29.6), we see that $U(\tau;\lambda,\vartheta)$ is very small and that the stream function (26.17) coincides essentially with $\operatorname{Im}\{v(\zeta)\}$. The effect of the U-integral may be considered as a small compressibility correction to the harmonic solution $\operatorname{Im}\{v(\zeta)\}$ in the incompressible case. This remark determines our choice of the generating function $v(\zeta)$ in Eq. (26.17). We solve at first the given flow problem with the required boundary conditions in the incompressible case; then we determine the image of the flow in the logarithmic hodograph plane and compute there the complex potential $v(\zeta)$. If the image domain lies in a hyperbola (29.11′) we can use $v(\zeta)$ as the generating function for the new stream function $\psi^*(\lambda,\vartheta)$. Returning, finally, to the physical plane, we obtain the flow of the compressible fluid past an obstacle which is not too distorted a copy of the original boundary curve.

There remains the question for what coefficient functions $f(\lambda)$ in Eq. (23.15) the domination relation (29.3′) can be asserted. It is easily seen that each elementary function

$$f(\lambda)=c(a-\lambda)^{-n},\qquad n=2,3,\ldots,\qquad a>0$$

admits such domination. Consequently, all finite sums

$$s(\lambda)=\sum_{n=2}^{N}c_n(1-\lambda)^{-n}\tag{29.13}$$

are admissible in our above consideration. The particular function (24.7) has for large negative values of λ the asymptotic character (27.2), hence, $f(\lambda)(1-\lambda)^2$ is continuous for $\lambda\leqq\beta<0$ and vanishes as $\lambda\to-\infty$. Thus, in view of the Weierstrass approximation theorem we can approximate this $f(\lambda)$ uniformly for all $\lambda\leqq\beta<0$ by a combination (29.13) to any desired degree of precision. Since the functions $G_n(\lambda)$ are calculated from Eqs. (26.13) by numerical integration only with a finite degree of precision with respect to $f(\lambda)$, this remark satisfies to justify the application of BERGMAN's method to the function (24.7).

30. A set particular solutions. We return to the recursion formula for the $G_n(\lambda)$:

$$G'_{n+1}(\lambda)=G''_n(\lambda)-f(\lambda)\,G_n(\lambda).\tag{30.1}$$

If we take

$$G_0(\lambda)=\sqrt{z(\lambda)},\tag{30.2}$$

in view of the definition (23.15) of $f(\lambda)$ we find that we can choose

$$G_1(\lambda) = 0, \quad G_n(\lambda) = 0, \quad n > 0. \tag{30.3}$$

On the other hand, we can start with this particular function and define recursively $G_{-n}(\lambda)$. In order to ensure the regularity of the $G_{-n}(\lambda)$ for all values $\lambda \leq \beta < 0$ we have to consider specially the character of $f(\lambda)$ for $\lambda = -\infty$. We suppose that

$$f(\lambda) = \sum_{n=2}^{\infty} c_n e^{n\lambda} \tag{30.4}$$

for large negative values of λ. Introducing the new variable $s = e^{2\lambda}$, we can write instead of Eq. (30.1) with $\Gamma_n(s) = G_n(\lambda)$

$$\Gamma'_{n+1}(s) = 2s\,\Gamma''_n(s) + 2\Gamma'_n(s) - \tfrac{1}{2} \sum_{m=2}^{\infty} c_m s^{m-1} \Gamma_n(s). \tag{30.1'}$$

We ask for solutions $\Gamma_{-n}(s)$ which have near $s = 0$ (i.e. $\lambda = -\infty$) the series development

$$\Gamma_{-n}(s) = \sum_{\nu=2}^{\infty} a_{-n,\nu} s^{\nu}, \quad n = 1, 2, \ldots. \tag{30.5}$$

It is easily seen that $\Gamma_0(s)$ has the development

$$\Gamma_0(s) = a_0 + a_2 s^2 + \cdots, \quad a_0 \neq 0, \tag{30.6}$$

and since the $a_{-n,\nu}$ are determined recursively by

$$2\nu^2 a_{-n,\nu} = \tfrac{1}{2} \sum_{\mu=2}^{\nu-2} c_{\nu-\mu} a_{-n,\mu} + \nu a_{1-n,\nu} \tag{30.7}$$

we see that such solution $\Gamma_{-n}(s)$ exists and is uniquely determined. Let $G_{-n}(\lambda)$ be the sequence of corresponding G-functions.

We define next the sequence of harmonic functions

$$g_\nu(\lambda, \vartheta) = \operatorname{Im}\left\{\frac{(a-\zeta)^\nu}{2^\nu \nu!}\right\}, \quad \nu = 0, 1, \ldots \tag{30.8}$$

with arbitrary a. Clearly

$$\frac{\partial}{\partial\lambda} g_\nu(\lambda, \vartheta) = -\frac{1}{2} g_{\nu-1}(\lambda, \vartheta). \tag{30.9}$$

We construct now the finite series

$$\psi_n^*(\lambda, \vartheta) = \sum_{\nu=0}^{n} G_{-\nu}(\lambda)\, g_{n-\nu}(\lambda, \vartheta) \tag{30.10}$$

and show just as we did in Sect. 26 that this function is a solution of Eq. (23.15). We can thus, construct an infinity of particular solutions ψ_n^* to the stream function equation which are regular in the entire sub-sonic region $\lambda < 0$ and behave near $\lambda = -\infty$ like $\operatorname{Im}\left\{\frac{\zeta^n}{2^n n!}\right\}$. Many further solutions can then be obtained by linear superposition.

We made the assumption (30.4) on the nature of $f(\lambda)$ in order to have regular solutions at $\lambda = -\infty$. Except for this requirement, we can use the recursion formula (30.1) to determine $G_{-n}(\lambda)$ even in the case that $f(\lambda)$ is only a continuous function of λ. Indeed, we are proceeding in our sequence only by processes of integration and not by differentiation as was done in the computation of G_n for $n > 0$. Thus, particular solutions to Eq. (23.15) can be obtained for every given $f(\lambda)$; in general, these solutions will be singular for $\lambda = -\infty$.

Instead of focusing our attention to the point $\lambda = -\infty$, which is a singular point of the differential operator, we might concentrate on the point $\lambda = 0$, which is, in general, a singularity of the coefficient $f(\lambda)$. If we choose the function $f(\lambda)$ defined in Eq. (24.7) and connected with the adiabatic equation of state, we can find an infinite sequence of $G_{-n}(\lambda)$ satisfying the recursion relation (30.1) and possessing at $\lambda = 0$ a series development

$$G_{-n}^{(\varkappa)}(\lambda) = (-\lambda)^{n+\varkappa} \sum_{\nu=0}^{\infty} a_{-n,\nu}^{(\varkappa)}(-\lambda)^{\frac{\nu}{3}}, \tag{30.11}$$

where we may choose $\varkappa = \frac{1}{6}$ or $\varkappa = \frac{5}{6}$. BERGMAN used these sequences in order to construct solutions of Eq. (23.15) in the angular domains bounded by the λ-axis and the straight lines which form with it an angle of 30° and go into the subsonic half-plane $\lambda < 0$ [38].

31. Exact solutions for idealized fluids. In order to discuss the convergence of the series for $U(t; \lambda, \vartheta)$ one has to compare it with the hypergeometric series. It is possible to find an equation of state and a corresponding function $f(\lambda)$ such that the function U becomes exactly a hypergeometric series. The mathematical theory of fluid dynamics becomes very easy for such a kind of fluid. It is hard to decide if approximate mathematical treatment of a more realistic fluid is preferable to exact mathematical treatment of an idealized fluid. Several methods in fluid dynamics are indeed based on the exact mathematical solution of an approximated physical situation; the first and best known approach is due to v. KÁRMÁN and TSIEN and will be discussed in Sect. 32. In the theory of transonic flow, TOMOTIKA and TAMADA have used various idealized fluids; see Chap. V. We want to show that BERGMAN'S integration method is particularly suited to an idealized gas for which the coefficient function $f(\lambda)$ has the special form [8]:

$$f(\lambda) = -\frac{C}{\lambda^2}, \qquad C > 0. \tag{31.1}$$

In this case the basic functions $G_n(\lambda)$ defined by Eq. (26.13) coincide with the elementary functions $Q_n(\lambda)$ defined by Eq. (29.4) with $\varepsilon = 0$. The function $U(t; \lambda, \vartheta)$ becomes now

$$U(t; \lambda, \vartheta) = \sum_{n=1}^{\infty} n\,\mu_n (-1)^n \frac{(\zeta - t)^{n-1}}{2^n(-\lambda)^n} = \frac{1}{2\lambda} F'\left(a, b, 1; \frac{\zeta - t}{2\lambda}\right) \tag{31.2}$$

with

$$a = \tfrac{1}{2} - \sqrt{\tfrac{1}{4} - C} \qquad b = \tfrac{1}{2} + \sqrt{\tfrac{1}{4} - C}. \tag{31.3}$$

The general solution of the stream function equation, given in the general form by Eq. (26.17), reduces now to the simple expression

$$\psi^*(\lambda, \vartheta) = \operatorname{Im}\left\{v(\zeta) - \int_{\zeta_0}^{\zeta} \frac{d}{dt} F\left(a, b, 1; \frac{\zeta - t}{2\lambda}\right) v(t)\, dt\right\}. \tag{31.4}$$

The complex potential $v(\zeta)$ is determined only up to an additive constant, and we may, therefore, assume without loss of generality that $v(\zeta_0) = 0$. Using the fact that $F(a, b, 1; 0) = 1$, we obtain by integration by parts

$$\psi^*(\lambda, \vartheta) = \operatorname{Im}\left\{\int_{\zeta_0}^{\zeta} F\left(a, b, 1; \frac{\zeta - t}{2\lambda}\right) v'(t)\, dt\right\} \tag{31.5}$$

as a general solution of Eq. (23.15) for the simplified fluid defined by Eq. (31.1).

The solution (31.5) represents an analytic continuation of the series solution given in Eq. (26.16). It is well known that the hypergeometric function $F(a, b, c; x)$ is regular analytic for all values of x except for $x=1$ and $x=\infty$; the general hypergeometric function can also have a singularity at $x=0$ but the branch with which we are concerned is regular even there. Except for these critical values of the argument $\frac{\zeta-t}{2\lambda}=x$, the integrand in Eq. (31.5) is well determined and regular analytic. The values $x=\infty$ and $x=1$ correspond to $\lambda=0$ and $t=-(\lambda-i\vartheta)$. In the case of subsonic flow, which is considered here, we may restrict the path of integration to the subsonic half-plane $\operatorname{Re}\{t\}<0$ and the integration in Eq. (31.5) will never lead us to singularities of the integrand.

It is easy to verify that $F\left(a, b, 1; \frac{\zeta-t}{2\lambda}\right)$ is itself a solution of the Eq. (23.15) for every fixed t and Eq. (31.5) is not much more than a linear superposition of these particular solutions. If instead of F we took the solution F^* of the hypergeometric equation which is singular for $x=0$, we should find in $F^*\left(a, b, 1; \frac{\zeta-t}{2\lambda}\right)$ another particular solution.

We have now to study the fluids characterized by the particular function (31.1) and to determine how well a real gas can be approximated by such an idealized medium. We use the definition (23.15) for $f(\lambda)$ and derive from Eq. (31.1) the relation

$$z(\lambda)=\frac{1}{\varrho}\sqrt{1-M^2}=(A_1|\lambda|^{\delta_1}+A_2|\lambda|^{\delta_2})^2 \tag{31.6}$$

where $\delta_{1,2}$ are the roots of the quadratic equation

$$\delta^2-\delta+C=0 \tag{31.6'}$$

and the A_i arbitrary constants at our disposal.

For the sake of simplicity, we restrict ourselves to the particular case

$$z(\lambda)=A^{-2\delta}|\lambda|^{2\delta}, \tag{31.7}$$

which we wish to study in detail. We shall express the flow variables M, q. and p as functions of the density ϱ and describe in this way the fluid obtained, We derive from Eq. (31.7)

$$\frac{d\lambda}{d\varrho}=-A(2\delta)^{-1}z^{\frac{1}{2\delta}-1}\frac{dz}{d\varrho}, \tag{31.8}$$

while in view of Eqs. (23.10) and (23.11) we have

$$\frac{d\lambda}{dq}=\frac{\varrho}{q}z. \tag{31.9}$$

The Bernoulli equation (22.2) leads, on the other hand, to the (ϱ, q)-relation

$$q\frac{dq}{d\varrho}+\frac{1}{\varrho}\frac{dp}{d\varrho}=0. \tag{31.10}$$

Thus, Eqs. (31.9) and (31.10) combined yield

$$\frac{d\lambda}{d\varrho}=\frac{d\lambda}{dq}\cdot\frac{dq}{d\varrho}=-zq^{-2}\frac{dp}{d\varrho}. \tag{31.11}$$

We have by definition

$$M^2=q^2\left(\frac{dp}{d\varrho}\right)^{-1}=1-z^2\varrho^2; \tag{31.12}$$

therefore we can put Eq. (31.11) into the form

$$\frac{d\lambda}{d\varrho} = -\frac{z}{M^2} = -\frac{z}{1 - z^2\varrho^2}. \tag{31.13}$$

Comparing finally Eqs. (31.8) and (31.13), we obtain the following differential equation for $z(\varrho)$:

$$\frac{dz}{d\varrho} = \frac{2\delta}{A} \cdot \frac{z^{2-\frac{1}{2\delta}}}{1 - z^2\varrho^2}. \tag{31.14}$$

From Eq. (31.14) we can calculate z as a function of ϱ; though in general one cannot express z as an elementary function of ϱ, one can carry out the integration numerically as long as $0 < z\varrho < 1$. Assuming the factor $\frac{2\delta}{A}$ positive, we see that $z(\varrho)$ is monotonically increasing. Observe that by Eq. (31.6) the limits 0 and 1 for $z\varrho$ correspond to the values 1 and 0 of the Mach number M. From the same formula follows also that

$$M(\varrho) = \sqrt{1 - \varrho^2 z^2} \tag{31.15}$$

is known, once $z(\varrho)$ has been determined. In the solution $z(\varrho)$ occurs an arbitrary constant of integration; we normalize it in such a way that $\varrho = 1$ implies $z = 1$ and $M = 0$. Thus, the choice of the arbitrary integration constant will fix the units for ϱ.

We introduce next the quantity

$$J(\varrho) = \int_{\varrho_0}^{\varrho} \frac{dp(x)}{x} \tag{31.16}$$

and observe that

$$\frac{dJ}{d\varrho} = \frac{1}{\varrho} p'(\varrho) = \frac{q^2}{M^2 \varrho}, \tag{31.17}$$

while BERNOULLI'S equation can be written as

$$\tfrac{1}{2} q^2 + J(\varrho) = c_1. \tag{31.16'}$$

Elimination of q^2 leads to the first order differential equation for $J(\varrho)$

$$\tfrac{1}{2} \varrho M(\varrho)^2 J'(\varrho) + J(\varrho) = c_1 \tag{31.18}$$

with the integral

$$J(\varrho) = c_1 - c_2 \exp\left[-2\int_{\varrho_0}^{\varrho} \frac{dx}{x M(x)^2}\right]. \tag{31.19}$$

From the important quantity $J(\varrho)$ all other variables can be easily computed. We have immediately by the Bernoulli equation

$$q^2 = 2[c_1 - J(\varrho)] = 2c_2 \exp\left[-2\int_{\varrho_0}^{\varrho} \frac{dx}{x M(x)^2}\right]. \tag{31.20}$$

On the other hand, we find from Eqs. (31.17) and (31.19)

$$p'(\varrho) = 2c_2 \frac{1}{M(\varrho)^2} \exp\left[-2\int_{\varrho_0}^{\varrho} \frac{dx}{x M(x)^2}\right], \tag{31.21}$$

which yields the equation of state $p = p(\varrho)$.

We normalized our solution by the correspondence $M(1)=0$, $z(1)=1$. Since M^{-2} occurs in various formulas, we have to study the convergence of the various integrals at the point $\varrho=1$. In view of Eq. (31.14), we have

$$\frac{d}{d\varrho}\left[z^{\frac{1}{2\delta}-1}\right]=\frac{1-2\delta}{A}M(\varrho)^{-2} \tag{31.22}$$

and this guarantees the convergence of

$$\int_{\varrho_0}^{1}\frac{dx}{M(x)^2}=\frac{A}{1-2\delta}\left[1-z(\varrho_0)^{\frac{1}{2\delta}-1}\right] \tag{31.23}$$

and of all the integrals occurring in the above formulas. A similar argument applies also in the case $\delta=\frac{1}{2}$.

From the Eq. (31.20) and the convergence of the integral for $J(\varrho)$, we conclude that $M=0$ corresponds to a positive value of q; in other words, that the state of rest cannot be obtained in the idealized fluid. A solution of a flow problem for such an idealized fluid will never admit points of stagnation. In the presence of an obstacle, it is necessary that at least one stream line splits up and this can happen only at a stagnation point or a sharp corner of the obstacle. In the case of a smooth obstacle the flow has to leave out cavities in front of and behind the obstacle; the free boundaries of the cavities must have sharp corners which will allow a separation of streamlines without zero velocities. This particular feature of the idealized fluid seems the most serious drawback in the method described. It may be overcome partially by a choice of the disposable constants in such a way that the minimum velocity of the fluid is small. For a numerical discussion of the problem see [*8*].

It is of interest to deal in particular with the limit case

$$z(\lambda)=1 \tag{31.24}$$

which corresponds to $C=0$ in Eq. (31.1). In this case, we have

$$M(\varrho)=\sqrt{1-\varrho^2},\qquad J(\varrho)=c_1-c_2\left(\frac{1}{\varrho^2}-1\right). \tag{31.25}$$

This leads to

$$q^2=2c_2\left(\frac{1}{\varrho^2}-1\right),\qquad p'(\varrho)=\frac{2c_2}{\varrho^2}. \tag{31.26}$$

We see that in this particular limit case the velocity $q=0$ is still attained if $\varrho=1$ and $M=0$. This case corresponds to the equation of state

$$p(\varrho)=\frac{\alpha}{\varrho}+\beta. \tag{31.27}$$

It was remarked by CHAPLYGIN himself that an idealized gas with this equation of state leads to a stream function which satisfies LAPLACE's equation. v. KÁRMÁN and TSIEN developed the theory of such a fluid systematically and applied it successfully to various flow problems [*22*], [*23*]. We shall discuss this particular theory in the next section. It will obviously be least satisfactory for high speed flow since the $f(\lambda)$ of the real gas has a singularity for $\lambda=0$ while the Kármán-Tsien approximation consists in replacing $f(\lambda)$ by zero. The more general form $f(\lambda)=-\frac{c}{\lambda^2}$ allows a better adjustment for high speeds to the real equation of state, though at the cost of a certain distortion near the velocity zero.

It is worth pointing out that the Eq. (23.15) in the special case

$$\frac{\partial^2 \psi^*}{\partial \vartheta^2} + \frac{\partial^2 \psi^*}{\partial \lambda^2} + \frac{c}{\lambda^2} \psi^* = 0 \tag{31.28}$$

can be brought into a particularly simple standard form. Indeed, let us introduce new independent variables, l, ϑ_1 by the definitions

$$-\lambda = l^{2\mu+1}, \quad \vartheta = (2\mu + 1)\,\vartheta_1 \tag{31.29}$$

and the new dependent variable $\chi(l, \vartheta_1)$ by

$$\psi^*(\lambda, \vartheta) = l^\mu \chi(l, \vartheta_1). \tag{31.29'}$$

We choose μ as one of the roots of the quadratic equation

$$\mu(\mu + 1) = C(1 - 4C)^{-1} \tag{31.30}$$

and find after easy calculation the following differential equation for $\chi(l, \vartheta_1)$:

$$l^{-4\mu} \frac{\partial^2 \chi}{\partial l^2} + \frac{\partial^2 \chi}{\partial \vartheta_1^2} = 0. \tag{31.31}$$

In the particular case $\mu = \frac{1}{4}$ (that is $C = \frac{5}{36}$), we obtain the very simple form

$$\frac{\partial^2 \chi}{\partial l^2} + l \frac{\partial^2 \chi}{\partial \vartheta_1^2} = 0. \tag{31.31'}$$

This is the Tricomi equation, which has been extensively studied in the general theory of second order partial differential equations [*21*]. It is obviously one of the simplest equations with non-constant coefficients and is of mixed type; namely, it is elliptic for $l > 0$, hyperbolic for $l < 0$ and parabolic for $l = 0$. The admissible boundary value problems, their uniqueness and existence theorems and the various numerical procedures for this type of differential equation have been investigated best for this simplest but quite representative form.

Another important cross-connection between the differential equations for the stream function of an idealized fluid and a classical problem of analysis exists in the case of the equation of state (31.27), the Kármán-Tsien limit case. In view of Eq. (31.26), we can choose our units in such a way that $c^2 = p'(\varrho)$ satisfies

$$q^2 = c^2 - 1, \tag{31.32}$$

and the original partial differential equation (22.1) for the velocity potential $\varphi(x, y)$ becomes

$$(1 + \varphi_y^2)\,\varphi_{xx} - 2\,\varphi_x \varphi_y \varphi_{xy} + (1 + \varphi_x^2)\,\varphi_{yy} = 0, \tag{31.33}$$

which is precisely the differential equation for minimal surfaces [*24*]. It was well known in differential geometry that this non-linear equation can be linearized and related to the theory of harmonic and analytic functions by a Legendre transformation. The hodograph transformations in the case of a general equation of state may be considered as generalizations of this theory.

32. The Kármán-Tsien fluid. The simplest application of the hodograph method is connected with the Kármán-Tsien approximation to the equation of state [*22*], [*23*]. In this section, therefore, we wish to collect the main formulas appropriate to this approach. We have by Eqs. (31.25) and (31.26)

$$M^2 = 1 - \varrho^2, \quad q^2 = A^2(\varrho^{-2} - 1). \tag{32.1}$$

We put $\tau = A^{-2} q^2$ and express ϱ and M by means of τ:

$$\varrho = \left(\sqrt{1+\tau}\right)^{-1}, \qquad M = \sqrt{\tau}/\sqrt{1+\tau}. \tag{32.2}$$

In this particular case it is advisable to introduce the variable

$$\lambda = -\frac{1}{2}\int_{\tau}^{\infty} \frac{d\tau}{\tau\sqrt{1+\tau}} = \frac{1}{2}\log\frac{\sqrt{1+\tau}-1}{\sqrt{1+\tau}+1}, \tag{32.3}$$

which stands in analogy to the definition (23.10).

We have conversely in terms of λ (observe that $\lambda < 0$):

$$q = -A(\operatorname{Sinh}\lambda)^{-1}, \qquad \varrho = -\operatorname{Tanh}\lambda \tag{32.4}$$

and

$$\frac{1}{\varrho q} = A^{-1}\operatorname{Cosh}\lambda. \tag{32.4'}$$

If we repeat the calculations at the end of Sect. 23 with the new variables λ, ϑ, we find that the velocity potential $\varphi(\lambda, \vartheta)$ and the stream function $\psi(\lambda, \vartheta)$ are conjugate harmonic functions and give rise to the analytic function

$$w = \varphi(\lambda, \vartheta) + i\,\psi(\lambda, \vartheta) \tag{32.5}$$

of the complex variable $\zeta = \lambda - i\vartheta$. If we introduce, moreover, the complex variable

$$z = x + i\,y \tag{32.6}$$

in the physical plane, we can simplify considerably the general formulas (23.5) which give the transition from the hodograph to the physical plane. We obtain

$$\left.\begin{aligned} dz &= -A^{-1}e^{i\vartheta}\left[\operatorname{Sinh}\lambda\, d\varphi - i\operatorname{Cosh}\lambda\, d\psi\right] \\ &= \frac{1}{2}A^{-1}e^{i\vartheta}\left[e^{-\lambda}dw - e^{\lambda}d\overline{w}\right] = \frac{1}{2A}\left[e^{-\zeta}dw - e^{\overline{\zeta}}d\overline{w}\right]. \end{aligned}\right\} \tag{32.7}$$

Formula (32.7) exhibits clearly the difference between the theory of an incompressible fluid flow and a Kármán-Tsien flow. In both cases, one has to start with an analytic function $w(\zeta)$ in the hodograph plane. In the first case, however, we have the simple relation

$$\frac{dw}{dz} = q\,e^{-i\vartheta} = e^{\zeta_1}, \qquad \zeta_1 = \log q - i\vartheta; \tag{32.8}$$

in the second case, we have the transition formula (32.7). We may use this analogy in order to give a concise solution formula for the Kármán-Tsien fluid flow problem which is expressed in the physical plane variables.

In order to determine the flow past a given obstacle in the z-plane, we first solve the problem under the assumption of incompressibility and determine the corresponding complex potential $w(z)$. We then effect the conformal mapping

$$\zeta_1 = \log\frac{dw}{dz} + \log 2A = \zeta + \log 2A; \tag{32.8'}$$

the function $w(z(\zeta))$ is analytic in the ζ-plane. We now interpret $\zeta = \lambda - i\vartheta$ as the (λ, ϑ)-plane for a Kármán-Tsien flow, and $w(z(\zeta))$ will give rise to solutions of the corresponding stream function equation. The flow will take place in a new physical plane, say with the coordinate Z. The transition from the

ζ-plane is given by Eq. (32.7); by our choice of the relation between ζ_1 and ζ, we have

$$dZ = e^{-\zeta_1} dw - \frac{1}{4A^2} e^{\bar{\zeta}_1} d\bar{w} = dz - \frac{1}{4A^2} \overline{\left(\frac{dw}{dz}\right)^2 dz} \tag{32.9}$$

since $e^{-\zeta_1} dw = dz$ by virtue of Eq. (32.8).

We have in Eq. (32.9) a continuous mapping from the z-plane of the incompressible fluid flow to the Z-plane of the Kármán-Tsien flow. It is characterized by the fact that at corresponding points the complex potential has the same value. Since the boundary of the obstacle in the z-plane is characterized by $\psi = \text{const}$, its image in the z_1-plane may be interpreted as the boundary of the flow region in the Kármán-Tsien case. The relation

$$Z = z - \frac{1}{4A^2} \left[\int \left(\frac{dw}{dz}\right)^2 dz\right]^- \tag{32.9'}$$

may be considered as a deformation of the z-plane, adjusting the compressibility conditions. If the velocities q are small relative to A the distortion will also be small and the new obstacle obtained will not differ too much from the original one.

It is, of course, important for our theory that the function $Z(z)$ be single-valued. We derived the entire hodograph transformation from integrability conditions; hence, any possible multivaluedness of the mapping must be due to singularities of the complex potential $w(z)$. Consider, for example, the representative case (28.1), which yields by Eqs. (28.2) and (32.8)

$$\frac{dw}{dz} = q_0 \left[1 + \frac{\alpha}{2\pi i} \frac{1}{z} - \frac{1}{z^2}\right]. \tag{32.10}$$

The single-valuedness of the mapping $Z(z)$ is equivalent to the condition $\alpha = 0$, i.e. to absence of circulation.

We may apply the above method to a flow with circulation by the following artifice. We apply formula (32.9′) to a mapping from a plane t to the Z-plane. We define

$$t = z + a \log z \tag{32.11}$$

and have then the mapping formula

$$Z = z + a \log z - \frac{1}{4A^2} \left[\int \left(\frac{dw}{dz}\right)^2 \left(1 + \frac{a}{z}\right)^{-1} dz\right]^- \tag{32.12}$$

from the z-plane to the Z-plane. If we insert now the function (32.10) for dw/dz we find

$$Z = z + \left[a\left(1 + \frac{q_0^2}{4A^2}\right) - \frac{\alpha q_0^2}{4\pi i A^2}\right] \log z + \cdots. \tag{32.13}$$

Hence, if we choose

$$a = \frac{\alpha q_0^2}{(4A^2 + q_0^2)\pi i}, \tag{32.13'}$$

we can cause the multivalued terms to cancel and obtain a $Z(z)$ that is single-valued in spite of a non-vanishing circulation term. The stream function ψ is not single-valued and in fact describes in the Z-plane a circulatory Kármán-Tsien flow. For small values of α and $q_0 \ll A$ the locus $\psi = 0$ will give a closed curve which is not very different from a circle of radius 1. The intermediate transformation (32.11) was possible because of the degree of indeterminateness in the correspondence between the complex potentials of the incompressible fluid flow and of the Kármán-Tsien flow considered.

We return to the transformation (32.9′) in the non-circulatory case and compute by means of this formula the velocity field U, V at the point Z which corresponds to the point z in the plane of the incompressible flow; let u, v be the velocity field in the z-plane.

Since φ has the same value at Z and z, we calculate (with $Z=X+iY$):

$$U=\frac{\partial\varphi}{\partial X}=u\left(1-\frac{q^2}{4A^2}\right)^{-1},\quad V=\frac{\partial\varphi}{\partial Y}=v\left(1-\frac{q^2}{4A^2}\right)^{-1} \tag{32.14}$$

with $q^2=u^2+v^2$. It was to be expected that the velocity vectors at corresponding points are parallel because of the common value of ϑ in the two hodograph plane interpretations. Differentiating ψ with respect to X and Y and using Eq. (23.1), we can compute ϱ:

$$\varrho=\frac{1-\dfrac{q^2}{4A^2}}{1+\dfrac{q^2}{4A^2}}, \tag{32.15}$$

a result which can also be obtained from the (q,ϱ)-relation (32.1) and the formulas (32.14).

The pressure correction follows from the formula

$$p(\varrho)=A^2\left(1-\frac{1}{\varrho}\right)+1 \tag{32.16}$$

which is equivalent to Eq. (31.26) and the normalization $p(1)=1$, and from the Bernoulli equation

$$p_i=1-\frac{q^2}{2} \tag{32.17}$$

for an incompressible fluid flow with density 1. We find from Eqs. (32.15) to (32.17)

$$p_i-p=\frac{q^4}{2(4A^2-q^2)} \tag{32.18}$$

for the pressure difference in corresponding points.

The remarkable success of the Kármán-Tsien approximation for flows with Mach numbers not too near to 1 can be understood from a consideration of the coefficient function $f(\lambda)$ in Eq. (23.15) for real gases. We see in formula (24.7) that $f(\lambda)$ has a factor $m^2=M^4$. Thus, the error committed in replacing $f(\lambda)$ by zero in Eq. (23.15) is only of the order M^4 and hence negligible for small Mach numbers.

33. Limit lines and branch lines. A solution of the stream function equation in the hodograph plane does not always lead to a significant flow pattern in the physical plane. We have to require that the transition from the region in the hodograph plane considered gives a non-self-overlapping domain in the physical plane. In general, it is not easy to check this condition; but it obviously implies the necessary condition that the mapping from the (q,ϑ)-plane into the (x,y)-plane be locally univalent. This means that the Jacobian $\dfrac{\partial(x,y)}{\partial(q,\vartheta)}$ must not vanish in the hodograph region. By virtue of Eq. (23.1), we have

$$\frac{\partial(\varphi,\psi)}{\partial(q,\vartheta)}=\frac{\partial(\varphi,\psi)}{\partial(x,y)}\cdot\frac{\partial(x,y)}{\partial(q,\vartheta)}=\varrho\, q^2\frac{\partial(x,y)}{\partial(q,\vartheta)}; \tag{33.1}$$

hence our requirement leads to the demand that this new Jacobian does not vanish in the hodograph domain, except at the stagnation point $q=0$.

On the other hand, we derive from Eq. (23.8) immediately

$$\frac{\partial(\varphi,\psi)}{\partial(q,\vartheta)}=q\frac{d}{dq}\left(\frac{1}{\varrho q}\right)\psi_\vartheta^2-\frac{q}{\varrho}\psi_q^2; \tag{33.2}$$

in view of (23.9) this can be simplified to

$$\frac{\partial(\varphi,\psi)}{\partial(q,\vartheta)}=\frac{1}{q\varrho}\left[(M^2-1)\psi_\vartheta^2-q^2\psi_q^2\right]. \tag{33.3}$$

This Jacobian can vanish for $M<1$ only if $\psi_q=\psi_\vartheta=0$. Since the curves $\psi_\vartheta=0$ and $\psi_q=0$ will intersect, in general, in a discrete set of points, the singularities of the hodograph mapping in the subsonic region will be represented by isolated points. The situation is different in the supersonic case $M>1$. Here one may have entire curves

$$\psi_\vartheta^2=\frac{q^2}{M^2-1}\psi_q^2 \tag{33.4}$$

along which the mapping breaks down. If we consider a regular solution of the stream function equation in the hodograph plane and consider its stream lines in the physical plane, we will find, in general, that these lines have sudden changes in direction and self-intersections if they cross the image of a locus (33.4) in the physical plane. The lines described by Eq. (33.4) and their image curves in the physical plane are called "limit lines" of the flow [*25*] to [*29*]. Only that part of the physical plane which is covered by stream lines not intersecting limit lines can be considered of physical significance.

As illustration, we consider the solution (25.1) of the stream function equation (24.11) in the special case $\nu=-\frac{1}{2}$. Since in this case by Eq. (25.4) we have $a_\nu b_\nu=0$, we recognize that $F_{-\frac{1}{2}}(\tau)=1$ and consequently, using the definition (24.8) of τ, we find

$$\psi(q,\vartheta)=C\frac{q_1}{q}\sin\vartheta \tag{33.5}$$

as a possible particular solution. The flow connected with Eq. (33.5) was discussed by RINGLEB [*27*]. The limit lines (33.4) of this particular flow have in the hodograph plane the equation

$$M^2=\frac{1}{\cos^2\vartheta}. \tag{33.6}$$

If we use again the variable $\tau=q^2/q_1^2$, we find from Eqs. (24.2′) and (33.6)

$$\sin\vartheta=\pm\sqrt{\frac{\gamma+1}{2}-\frac{\gamma-1}{2}\tau^{-1}}, \tag{33.7}$$

and, consequently, we have on the limit line

$$\psi(q,\vartheta)=C\tau^{-1}\sqrt{\frac{\gamma+1}{2}\tau-\frac{\gamma-1}{2}}. \tag{33.7′}$$

Since τ can vary only from zero to one, we see easily that the values of ψ on the limit line vary from zero to C. Streamlines $\psi=k$ with $k>C$ will not meet a limit line and thus represent possible particle motions, while stream lines with $k<C$ will be reverted or broken if they meet the limit line. RINGLEB showed that the solution (33.5) can be used to describe a flow around an edge [*27*]. It should be observed that on part of the admissible streamlines the motion becomes

supersonic on certain stretches and that the edge flow obtained will admit supersonic pockets.

Once one has determined a limit line in the hodograph plane one can map it back into the physical plane by means of the formulas (23.6). LIGHTHILL [13] analyzed the geometry of the limit lines in the physical plane. He showed that a stream line never touches a limit line but always meets it at a finite angle. Suppose now that we run along a limit line in the sense of increasing ψ; either ψ will increase indefinitely, or it will reach a maximum value. This cannot happen at a point where the limit line has a continuous tangent, since in this case a stationary value of ψ would imply a contact between the limit line and the stream line; but this has been excluded. Hence, the limit line must have a cusp if it has a maximum ψ-value. The same argument can be developed for points of minimum of ψ on the limit line. Hence, a limit line must either intersect all stream lines or it must necessarily have cusps.

The difficulty of interpreting a flow up to a limit line becomes evident when we observe that the acceleration of a particle in the flow has the components

$$a_x = u_x u + u_y v, \qquad a_y = v_x u + v_y v; \tag{33.8}$$

in the case of a vanishing Jacobian $\frac{\partial(x, y)}{\partial(q, \vartheta)}$ some terms $(u_x, \ldots, v_y)$ must become infinite and, in general, lead to infinite accelerations. The occurence of infinite accelerations certainly invalidates the idealizations of the non-viscous fluid theory, which is the basis of our mathematical argument.

Thus far we have discussed the breakdown of a regular stream function solution in the hodograph plane due to difficulties of transition to the physical plane. The question arises whether there might not be regular solutions of the hydrodynamical equations in the physical plane which conversely do not admit a regular mapping into the hodograph plane. Such solutions might be overlooked in the hodograph theory. In order to clarify this problem, we have to consider the Jacobian

$$\Delta = \frac{\partial(u, v)}{\partial(x, y)} = \begin{vmatrix} \varphi_{xx} & \varphi_{xy} \\ \varphi_{xy} & \varphi_{yy} \end{vmatrix} \tag{33.9}$$

and to ask for the points where it vanishes. At these points the preceding Jacobian would become infinite. We mentioned this kind of critical points briefly after Eq. (22.5).

We observe that $\Delta = 0$ implies $\varphi_{xx}\varphi_{yy} = \varphi_{xy}^2 \geqq 0$ and that we may assume without loss of generality $\varphi_{xx} \geqq 0$, $\varphi_{yy} \geqq 0$. The differential equation (22.1) for the velocity potential leads at the singular points to the relation

$$c^2(\varphi_{xx} + \varphi_{yy}) = \left(\varphi_x \sqrt{\varphi_{xx}} + \varphi_y \sqrt{\varphi_{yy}}\right)^2 \tag{33.10}$$

and, in view of SCHWARZ' inequality, we find

$$c^2(\varphi_{xx} + \varphi_{yy}) \leqq q^2(\varphi_{xx} + \varphi_{yy}). \tag{33.10'}$$

If $M < 1$, this inequality is possible only if $\varphi_{xx} = \varphi_{xy} = \varphi_{yy} = 0$, and then it will hold, in general, only at isolated points. However, in the supersonic region $M > 1$ the equation $\Delta = 0$ may hold along whole curves.

In order to study the images of the singular curves in the hodograph plane, we remark the identity

$$J = \frac{\partial(q, \vartheta)}{\partial(\varphi, \psi)} = \frac{\partial(q, \vartheta)}{\partial(u, v)} \cdot \frac{\partial(u, v)}{\partial(x, y)} \frac{\partial(x, y)}{\partial(\varphi, \psi)} = \frac{1}{\varrho q^3} \Delta. \tag{33.11}$$

Hence the zero lines of Δ and of J are the same. Let us proceed along the image of the zero line of J in the (φ, ψ)-plane. We have

$$dq = \frac{\partial q}{\partial \varphi} d\varphi + \frac{\partial q}{\partial \psi} d\psi, \qquad d\vartheta = \frac{\partial \vartheta}{\partial \varphi} d\varphi + \frac{\partial \vartheta}{\partial \psi} d\psi. \tag{33.12}$$

Since $J=0$ on that line, we have for every choice of $d\varphi$, $d\psi$

$$\frac{dq}{d\vartheta} = \frac{\partial q}{\partial \varphi} \Big/ \frac{\partial \vartheta}{\partial \varphi}. \tag{33.13}$$

By the law of differentiation of inverse functions, we have

$$\frac{\partial q}{\partial \varphi} \Big/ \frac{\partial \vartheta}{\partial \varphi} = -\frac{\partial \psi}{\partial \vartheta} \Big/ \frac{\partial \psi}{\partial q} = \frac{dq}{d\vartheta}. \tag{33.14}$$

On the other hand, the law of differentiation of inverse functions leads also to the equation

$$\frac{\varphi_q \psi_\vartheta}{\psi_q \varphi_\vartheta} = \frac{q_\varphi \vartheta_\psi}{q_\psi \vartheta_\varphi}. \tag{33.15}$$

The left hand side of Eq. (33.15) can be simplified by use of the differential relations (23.8) between φ and ψ and by the identity (23.9). We find

$$\frac{M^2-1}{q^2} \frac{\psi_\vartheta^2}{\psi_q^2} = \frac{q_\varphi \vartheta_\psi}{q_\psi \vartheta_\varphi}. \tag{33.15'}$$

Eqs. (33.14) and (33.15) are valid as long as all terms occurring are finite. On the singular line $\Delta=0$ the derivatives of φ and ψ with respect to q and ϑ become infinite. But the condition $J=0$ implies that the right hand side of Eq. (33.15') has the value 1 on the singular line. Hence, we must have

$$\lim \left(\frac{\psi_\vartheta}{\psi_q}\right)^2 = \frac{q^2}{M^2-1} \tag{33.16}$$

if this line is approached. We obtain again the condition $M \geqq 1$ and from Eqs. (33.14) and (33.16) we conclude

$$\frac{dq}{d\vartheta} = \pm \frac{q}{\sqrt{M^2-1}}. \tag{33.17}$$

This is the differential equation of the singular line in the hodograph plane. A comparison with Eq. (23.21) shows that it must coincide with a characteristic.

The singular lines in the hodograph plane due to the vanishing of the Jacobian Δ were discovered by LIGHTHILL [*9*], [*13*] and called "branch lines" of the flow. He was led to these lines by a discussion of transonic flow patterns which have a straight stream line. If we use φ and ψ as the independent variables, a stream line is characterized by the condition $\psi=$const, and on a straight stream line we must have in addition

$$\vartheta_\varphi = 0. \tag{33.18}$$

By the law of differentiation of inverse functions, Eq. (33.18) leads to $\psi_q=0$ and, therefore, by Eq. (23.8) and (23.9) to

$$J^{-1} = \frac{\partial(\varphi, \psi)}{\partial(q, \vartheta)} = \varphi_q \psi_\vartheta = \frac{M^2-1}{q\varrho} \psi_\vartheta^2. \tag{33.19}$$

It should be remarked that $\psi_\vartheta = J^{-1} q_\varphi$ and that q_φ has a simple physical significance. Indeed, the velocity gradient along the stream line

$$g = \frac{dq}{ds} = \frac{\partial q}{\partial \varphi} \cdot \frac{\partial \varphi}{\partial s} = q \frac{\partial q}{\partial \varphi} \tag{33.20}$$

is closely related to q_φ. Hence Eq. (33.19) leads to

$$J = \frac{M^2-1}{\varrho q^3} g^2. \tag{33.21}$$

If the velocity gradient remains finite on the straight stream line, we have necessarily $J=0$ at the sonic points. It can be shown that the points $J=0$ so obtained are not isolated but lie indeed on a branch line. On the other hand, it can also happen that J vanishes because of $g=0$; if this occurs at a subsonic speed the singularity can be shown to be isolated, while for $M>1$ the singular point lies on a branch lines [*13*]. We see that a transonic flow with a straight stream line must necessarily possess a branch line which intersects this stream line at its sonic point.

The occurrence of limit and branch lines prevents in many cases the application of the hodograph method to the theory of transonic flows. This fact is not surprising since we shall see in Chap. V that the existence of a transonic flow past a given obstacle is probably rather the exeption than the rule. A breakdown mechanism within the mathematical apparatus is, therefore, necessary; it is provided by these singular lines and by the shock discontinuities of the flow.

34. The principal net and its application. In the theory of supersonic flow the Mach directions play a central role. They are defined relative to the local velocity vector (u, v) and form with it an angle α such that

$$\sin\alpha = \pm\frac{c}{q}, \qquad q^2 = u^2 + v^2. \tag{34.1}$$

We have shown in Chap. I that the Mach directions are the characteristic directions of the system of partial differential equations for the fluid motion. At this stage, we wish to stress their distinguished role under the hodograph transformation.

The Mach directions determine two direction fields in the (x, y)-plane and the integral curves of these fields form a curvilinear net, the Mach net of the flow. Each curve of the net is called a Mach curve. While the streamlines depend to a large extend on the individual flow considered, the Mach lines after transformation into the hodograph plane become a set of easily characterized curves. This fact is consistently used in the numerical approach to supersonic flow problems.

Again let q and ϑ denote the local speed and the angle of the local velocity with the positive x-axis. If $x(s)$, $y(s)$ is the parametric representation of a Mach curve, say in terms of its arc length, we can write Eq. (34.1) in the form

$$(dy\cos\vartheta - dx\sin\vartheta)^2 = \frac{c^2}{q^2}(dx^2 + dy^2). \tag{34.2}$$

We use the relations (23.5) in order to replace (x, y) by (φ, ψ) and find easily that along a Mach curve

$$\frac{1}{\varrho^2}(M^2-1)\,d\psi^2 = d\varphi^2, \qquad M = \frac{q}{c}. \tag{34.3}$$

Thus along each Mach curve we have one of the equations

$$\frac{1}{\varrho}\sqrt{M^2-1}\,d\psi \pm d\varphi = 0. \tag{34.4}$$

By considering φ and ψ as functions of q and ϑ, we now pass easily into the hodograph plane; moreover, we can eliminate φ by means of the basic differential equations (23.8) and the identity (23.9). We find from Eq. (34.4):

$$\left(\frac{1}{q}\sqrt{M^2-1}\,dq \pm d\vartheta\right)\left(q\psi_q \pm \sqrt{M^2-1}\,\psi_\vartheta\right) = 0. \tag{34.5}$$

If we equate the first factor in Eq. (34.5) to zero, we obtain an ordinary differential equation for $q(\vartheta)$ which leads to the two systems of integral curves

$$\vartheta - \vartheta_0 = \pm \int_{q_0}^{q} \sqrt{M^2-1}\,\frac{dq}{q}. \tag{34.6}$$

This set of images of the Mach net is called the principal net, and each curve is called a principal curve in the hodograph plane. The various principal curves are obtained from any fixed one by rotating it around the origin by an angle ϑ_0. All principal curves therefore, are congruent; they depend on the equation of state of the fluid, which determines the function $M(q)$. But they are independent of the particular flow pattern considered. The principal curves were already considered in Sect. 23 as the characteristics of the hodograph differential equation (23.19).

In the particular case of an adiabatic ideal gas when $p=\varrho^\gamma$, we can write Eq. (34.6) in the form

$$d\vartheta = \sqrt{\frac{q^2-q_s^2}{1-q^2}}\,\frac{dq}{q_s q}, \qquad q_s = \sqrt{\frac{\gamma-1}{\gamma+1}} = \text{sonic speed}, \tag{34.7}$$

if we choose our units so that $q_{\max}=1$. Introducing a parameter t, we may put

$$q^2 = R_1^2 + R_2^2 + 2R_1R_2\cos\left(\frac{R_1}{R_2}-1\right)t, \qquad \tan\vartheta = \frac{R_1\sin t + R_2\sin\frac{R_1}{R_2}t}{R_1\cos t + R_2\cos\frac{R_1}{R_2}t} \tag{34.8}$$

with

$$R_1 = \frac{1+q_s}{2}, \qquad R_2 = \frac{1-q_s}{2}. \tag{34.8'}$$

If we calculate $q'(\vartheta)$ from Eq. (34.8) we obtain exactly the differential equation (34.7). We may replace Eq. (34.8) by the parametric representation in Cartesian coordinates

$$u = R_1\cos t + R_2\cos\frac{R_1}{R_2}t, \qquad v = R_1\sin t + R_2\sin\frac{R_1}{R_2}t. \tag{34.9}$$

This curve is obviously described by a point (u, v) located on the periphery of a circle of radius R_2 which rolls inside of a circle around the origin with radius R_1+R_2. The principal curves are thus shown to be epicycloids [*41*]. Observe that $R_1+R_2=1$ and $R_1-R_2=q_s$; each principal curve lies, therefore, in the supersonic ring $q_s \leq q \leq 1$, where $q_{\max}=1$ is the escape velocity of the fluid. The curve runs up to the circle $q=1$, where it experiences a cusp.

We can conclude from the geometry of the epicycloid that as it runs from the sonic circle $q=q_s$ to the circle $q=1$, q increases monotonically. Hence we have the theorem: *The local speed and, hence, the pressure and density, vary monotonically on a Mach line in the physical plane.*

The differential equation

$$\sqrt{M^2-1}\,\frac{dq}{d\vartheta} = \pm q \tag{34.5'}$$

of the principal curve reads in Cartesian coordinates

$$\frac{dv}{du} = \frac{\tan\vartheta \mp \sqrt{M^2-1}}{1 \pm \tan\vartheta\sqrt{M^2-1}}. \tag{34.10}$$

Observe that by the definition (34.1)

$$\tan\alpha = (M^2 - 1)^{-\frac{1}{2}}, \tag{34.11}$$

and, therefore, by the addition theorem for trigonometric functions

$$\frac{dv}{du} = \cot(\alpha \pm \vartheta). \tag{34.12}$$

On the other hand, the stream line in the physical plane at the point corresponding to u, v has the direction ϑ and the two Mach lines at the point have the slopes

$$\frac{dy}{dx} = \tan(\vartheta \pm \alpha). \tag{34.13}$$

Thus we have proved that at corresponding points (u, v) and (x, y) the lines of the principal system in the hodograph plane have tangents perpendicular to the

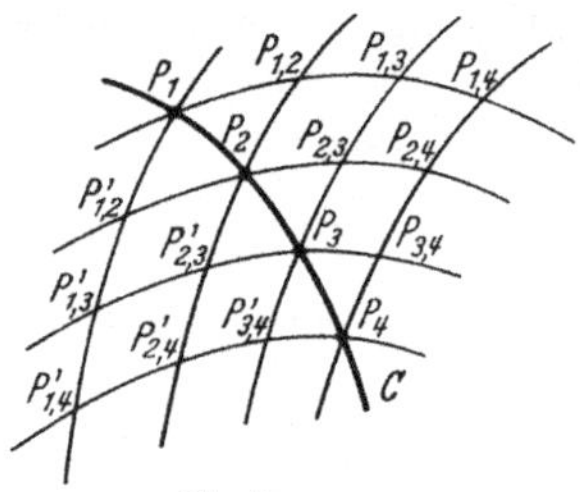

Fig. 11.

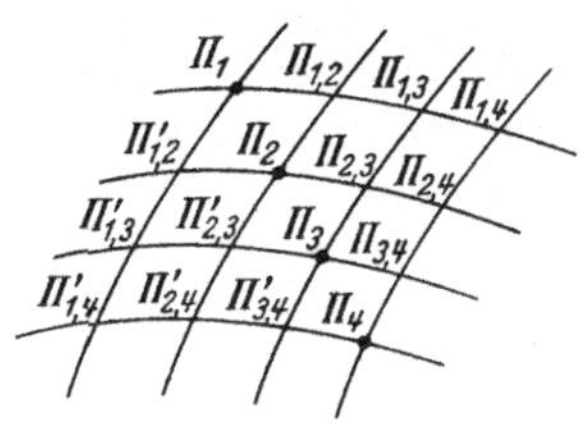

Fig. 12.

Mach lines in the physical plane. This is a second fundamental fact for the graphical and numerical solution methods in supersonic flow theory.

We sketch very briefly a graphical procedure due to PRANDTL and BUSEMANN [42] which is based on the principal net and the preceding orthogonality property. It uses the characteristic diagram, which is the supersonic ring $q_s \leq q \leq 1$ in the hodograph plane covered by a network of epicycloids obtained from one by rotation about consecutive angles of degree ε ($\varepsilon \sim 2°$), and by reflecting the whole set at the line $\vartheta = 0$. Suppose that along an arbitrary curve C in the (x, y)-plane the velocities of a supersonic flow are prescribed. The problem is to determine the velocity field in a neighborhood of C.

We select on C (Fig. 11) a set of sufficiently near points P_i $(i = 1, 2, \ldots, n)$; since we know the velocity at these points, we can determine their images π_i in the hodograph plane (Fig. 12). By assumption, all points π_i lie in the supersonic ring, and the characteristic diagram provides a discrete net of principal curves passing through these points. We denote the intersection points of the principal curves through π_i and π_k by $\pi_{i,k}$ and $\pi'_{i,k}$. The corresponding Mach lines through P_i and P_k, which we do not know, would intersect at corresponding points $P_{i,k}$ and $P'_{i,k}$. Our approximation consists now in replacing the unknown Mach curves by linear segments in the Mach directions. We know, indeed, the Mach directions at the points P_i; they are orthogonal to the epicycloids at the points π_i. We draw straight lines in the Mach directions through the points P_i; lines through consecutive points will intersect very near to the curve C and we may assume these intersection points to be $P_{i,i+1}$ and $P'_{i,i+1}$ without appreciable error. At the points $P_{i,i+1}$ and $P'_{i,i+1}$ we again know the Mach directions since these points correspond to $\pi_{i,i+1}$ and $\pi'_{i,i+1}$ where the Mach directions are known. We can draw tangents to the Mach lines at the newly obtained points and determine approximately new P_{ik} and P'_{ik}. If the original points P_i are near enough, this procedure will provide near the curve C a Mach net which is very close to the correct one. Once a sufficiently dense Mach net has been obtained, the velocity field of the flow can be easily constructed. One has only to observe that at each point the velocity vector bisects the angle between the intersecting Mach lines.

The Prandtl-Busemann graphical procedure has been used extensively. It has been applied, in particular, to construct the supersonic flow between given fixed walls, in nozzles and in jets.

There exists a simple numerical procedure based directly on the Eq. (34.6) for the Mach lines; it replaces a differential quotient by a difference quotient and is thus the numerical counterpart to the above graphical procedure where a curved arc was replaced by a tangent. We write

$$\int_{q_0}^{q} \sqrt{M^2-1}\,\frac{dq}{q} = \omega(q) \tag{34.14}$$

and in view of Eq. (34.6) we can assert that the two equations

$$\vartheta + \omega(q) = \lambda, \qquad \vartheta - \omega(q) = \mu \tag{34.15}$$

determine the net of principal lines in the hodograph plane; λ and μ may serve as a new system of coordinates there. We introduce a net of parameter curves $\lambda_n = n\lambda_0$, $\mu_m = m\mu_0$ and consider points x, y as functions of these coordinates. Thus:

$$x_{n,m} = x(n\lambda_0, m\mu_0), \qquad y_{n,m} = y(n\lambda_0, m\mu_0). \tag{34.16}$$

Observe further that $\vartheta = \frac{1}{2}(\lambda+\mu)$ and $\omega(q) = \frac{1}{2}(\lambda-\mu)$. From the last equation, q can be computed as a function of $\lambda-\mu$, and so can α by virtue of Eq. (34.1). Hence

$$\tan(\vartheta+\alpha) = f(\lambda,\mu), \qquad \tan(\vartheta-\alpha) = g(\lambda,\mu). \tag{34.17}$$

We can now replace the differential quotient in Eq. (34.13) by a difference quotient and obtain the following system of linear equations for the calculation of the $x_{n,m}$, $y_{n,m}$:

$$\frac{y_{n+1,m} - y_{n,m}}{x_{n+1,m} - x_{n,m}} = f_{n,m}, \qquad \frac{y_{n,m+1} - y_{n,m}}{x_{n,m+1} - x_{n,m}} = g_{n,m} \tag{34.18}$$

with

$$f_{n,m} = f(\lambda_n, \mu_m), \qquad g_{n,m} = g(\lambda_n, \mu_m). \tag{34.18'}$$

These equations must be conceived as determining $x_{n,m}$ and $y_{n,m}$ from two known neighboring points; therefore they permit determination of $x_{n,m}$, $y_{n,m}$ if the values $x_{n',m'}$, $y_{n',m'}$ with $n'+m' > n+m$ are known. Observe that the numbers $f_{n,m}$ and $g_{n,m}$ are dependent only upon the equation of state, not on the particular flow pattern considered. They can be tabulated once and for all for a given net of principal curves.

It can be shown that the procedure is considerably improved if one uses the midpoints on the principal curves instead of the mesh points in the determination of f and g [*43*], [*44*].

Observe that at every given point x_n, y_m the direction of the stream line is given explicitly by

$$\vartheta = \lambda_n + \mu_m. \tag{34.19}$$

Therefore, the computation described yields the determination of the velocity field in the (x, y)-plane by giving the velocity direction in a dense point lattice. For further details and generalizations see [*43*], [*44*], [*45*].

We have discussed so far the Eq. (34.5) for the image of a Mach curve in the hodograph plane under the assumption that the first factor vanishes. The vanishing of the second factor implies

$$(M^2 - 1)\,\psi_\vartheta^2 = q^2 \psi_q^2 \tag{34.20}$$

which was given in Eq. (33.4) as the condition for a limit line. Since Eq. (34.20) implies the differential relation (34.2) in the physical plane, we see that at each of its points the limit line touches a Mach line of the flow; it is an envelope of Mach lines.

IV. The analytical theory of two-dimensional subsonic flows.

35. The basic equation. The velocity potential $\varphi(x, y)$ and the stream function $\psi(x, y)$ which describe the steady plane irrotational motion of a fluid with a speed-density relation $\varrho = \varrho(q^2)$, $q^2 = u^2 + v^2$, are related by the differential system

$$u = \varphi_x = \frac{1}{\varrho}\,\psi_y, \qquad v = \varphi_y = -\frac{1}{\varrho}\,\psi_x. \tag{35.1}$$

If we eliminate one of the unknown functions from the system, we obtain one second order partial differential equation for the remaining dependent variable:

$$\frac{\partial}{\partial x}(\varrho\,\varphi_x) + \frac{\partial}{\partial y}(\varrho\,\varphi_y) = 0 \tag{35.2}$$

and

$$\frac{\partial}{\partial x}\left(\frac{1}{\varrho}\,\psi_x\right) + \frac{\partial}{\partial y}\left(\frac{1}{\varrho}\,\psi_y\right) = 0, \tag{35.2'}$$

respectively. Since ϱ depends on $q^2 = \varphi_x^2 + \varphi_y^2$, these differential equations are non-linear, and the mathematical problems of existence and uniqueness which arise for the usual boundary conditions are by no means simple nor are they yet completely solved.

The structure of the differential equation (35.2) becomes clearer when we introduce a function $F(u^2 + v^2) = F(q^2)$ such that

$$F'(q^2) = \tfrac{1}{2}\,\varrho(q^2), \qquad F(q^2) = \int \varrho\, q\, dq. \tag{35.3}$$

With this notation we can bring Eq. (35.2) into the form

$$\frac{\partial}{\partial x}\left(\frac{\partial F}{\partial u}\right) + \frac{\partial}{\partial y}\left(\frac{\partial F}{\partial v}\right) = 0 \tag{35.4}$$

and Eq. (35.2) may be interpreted as the Euler-Lagrange equation of the variational problem

$$\delta \iint F(\varphi_x^2 + \varphi_y^2)\, dx\, dy = 0. \tag{35.5}$$

When written out, Eq. (35.4) takes the form

$$F_{uu}\,\varphi_{xx} + 2F_{uv}\,\varphi_{xy} + F_{vv}\,\varphi_{yy} = 0. \tag{35.6}$$

This is a quasi-linear equation of the second order since the highest order derivatives occur linearly. In order to determine the type of the equation we have to consider the discriminant

$$\Delta(x, y) = F_{uu}\,F_{vv} - F_{uv}^2. \tag{35.7}$$

We compute easily from Eq. (35.3) that

$$\Delta(x, y) = \varrho^2 + 2\varrho\, q^2 \frac{d\varrho}{dq^2} = \frac{d}{dq^2}(\varrho^2 q^2) = \varrho \frac{d}{dq}(\varrho q). \tag{35.8}$$

On the other hand, the Bernoulli equation

$$\frac{1}{2} q^2 + \int \frac{dp}{\varrho} = \text{const} \tag{35.9}$$

connects the speed-density relation with the equation of state of the fluid; differentiating Eq. (35.9) with respect to q we find

$$\frac{d\varrho}{dq} = -\frac{q\varrho}{c^2}, \qquad c^2 = \frac{dp}{d\varrho} = \text{local speed of sound}. \tag{35.9'}$$

From Eqs. (35.8) and (35.9) we obtain finally

$$\Delta(x, y) = \varrho^2 \left(1 - \frac{q^2}{c^2}\right). \tag{35.10}$$

Thus the partial differential equation (35.6) is elliptic in domains where the speed is subsonic, hyperbolic where the speed is supersonic, and of mixed type in domains where both super- and subsonic speeds occur.

In the case of an adiabatic gas we have the pressure-density relation

$$p = a\varrho^{\gamma}. \tag{35.11}$$

We have for Eq. (35.9) the specific form

$$\frac{1}{2} q^2 + a \frac{\gamma}{\gamma - 1} \varrho^{\gamma - 1} = \frac{1}{2} q_{\max}^2 \tag{35.12}$$

where $q_{\max}$ is the largest possible speed in the flow considered. This leads to the speed-density relation

$$\varrho = \left(\frac{\gamma - 1}{2\gamma a} \left[q_{\max}^2 - q^2\right]\right)^{\frac{1}{\gamma - 1}}. \tag{35.12'}$$

The speed q_s which corresponds to the velocity of sound is easily computed to be

$$q_s = \sqrt{\frac{\gamma - 1}{\gamma + 1}}\, q_{\max}. \tag{35.13}$$

As long as $q < q_s$ the flow is subsonic; the range $q_s < q < q_{\max}$ represents the possible supersonic speeds.

The best known variational problem in two independent variables of the form (35.5) is undoubtedly the problem of the minimal surfaces, which corresponds to

$$F(q^2) = \sqrt{1 + q^2}. \tag{35.14}$$

The analytical tools developed for the theory of minimal surfaces play an essential role in the existence theory of fluid dynamics; see Sect. 44. We may inquire into the nature of a fluid for which $F(q^2)$ has this particular form. We find for it from Eq. (35.3) the speed-density relation

$$\frac{1}{\varrho^2} = 1 + q^2 \tag{35.15}$$

and from Eq. (35.9) the pressure-density relation

$$p = -\frac{1}{\varrho} + b. \tag{35.16}$$

These are the formulas for the idealized fluid discussed by KÁRMÁN-TSIEN; see Sect. 32. For this fluid the classical results of minimal surface theory can be applied immediately. For a more general fluid, the methods if not the immediate results can be extended, as will be seen later.

36. The maximum and minimum principle for linear elliptic differential equations. The theory of the non-linear differential equation (35.6) can be reduced in many aspects to the much simpler theory of linear elliptic differential equations [*1*], [*2*]. We assume that the function $F(q^2)$ is analytic and has, therefore, derivatives of arbitrarily high order. If we consider now an arbitrary solution $\Phi(x,y)$ of Eq. (35.6), we are sure that $\Phi_x^2+\Phi_y^2$ is a differentiable function and that

$$a(x,y)=F_{uu}(\Phi_x^2+\Phi_y^2),\quad b(x,y)=F_{uv}(\Phi_x^2+\Phi_y^2),\quad c(x,y)=F_{vv}(\Phi_x^2+\Phi_y^2) \tag{36.1}$$

are differentiable functions of x and y. Thus $\Phi(x,y)$ can be considered as a solution of the linear differential equation

$$L[\Phi]=a\,\Phi_{xx}+2b\,\Phi_{xy}+c\,\Phi_{yy}=0. \tag{36.2}$$

This equation is of no value for an actual calculation of $\Phi(x,y)$ since its coefficients depend on Φ_x and Φ_y. But if $\Phi(x,y)$ is a subsonic solution in some domain D of the (x,y)-plane, that is, if it gives rise to a positive discriminant

$$\Delta=ac-b^2>0 \tag{36.3}$$

in D, then Eq. (36.2) is an elliptic linear equation and the solutions of such equations have various characteristic common properties independently of the specific nature of the coefficients. This reduction to linear equations is a major tool in the general theory of non-linear differential equations; it is particularly simple in the special case (35.6).

If $\Phi(x,y)$ is a solution of an Eq. (36.2) in a domain D where the equation is elliptic, it will take its maximum and minimum value on the boundary of the domain and experience an extremum value at an interior point of D only if it is a constant. Since this result is of great importance in the theory of linear and non-linear elliptic differential equations, we give a sketch of the proof which is due to E. HOPF [*3*].

We consider first an arbitrary twice continuously differentiable function $\chi(x,y)$ which takes its maximum value at an interior point x_0, y_0 of the domain D.

By the well known characterization of maximum points, χ must have vanishing first partial derivatives at this point, and the quadratic form

$$\chi_{xx}(x_0,y_0)\,\alpha^2+2\chi_{xy}(x_0,y_0)\,\alpha\beta+\chi_{yy}(x_0,y_0)\,\beta^2 \tag{36.4}$$

must be non-positive for every choice of α,β. The elliptic character of the differential equation (36.2) at x_0, y_0 implies that the quadratic form

$$a(x_0,y_0)\,\alpha^2+2b(x_0,y_0)\,\alpha\beta+c(x_0,y_0)\,\beta^2 \tag{36.4'}$$

be definite; we may assume without loss of generality that it is positive-definite. From the algebra of definite quadratic forms, we may then conclude the inequality

$$a(x_0,y_0)\,\chi_{xx}(x_0,y_0)+2b(x_0,y_0)\,\chi_{xy}(x_0,y_0)+c(x_0,y_0)\,\chi_{yy}(x_0,y_0)\le 0. \tag{36.5}$$

This leads us to the important intermediate result: If $\chi(x,y)$ satisfies the differential inequality

$$L[\chi]>0 \tag{36.6}$$

in D, it cannot take its maximum in an interior point of D.

Suppose next that D is a bounded domain and that a solution $\Phi(x, y)$ of Eq. (36.2) takes its maximum value M at an interior point x_0, y_0 of D while on the boundary of D it is always less than or equal to $M-c$, $c>0$. We introduce then the function

$$h(x, y) = e^{\alpha[(x-x_0)^2+(y-y_0)^2]} - 1, \qquad \alpha > 0. \tag{36.7}$$

We find by simple calculation

$$L[h] > 0. \tag{36.7'}$$

Consider now the combination

$$\chi(x, y) = \Phi(x, y) + \varepsilon h(x, y), \qquad \varepsilon > 0.$$

This function satisfies the differential inequality (36.6) since $L[\Phi]=0$. At x_0, y_0 it assumes the value M, and if we choose ε small enough we can insure that $\chi(x, y) < M - \frac{c}{2}$ on the boundary of D. But this means that χ would take its maximum inside of D in contradiction to our intermediate theorem. Thus we have shown that $\Phi(x, y)$ cannot achieve its maximum at an interior point of D while retaining a lesser value everywhere on the boundary.

By a simple geometric argument we can reduce the general case of an interior maximum to the preceding one. It is easy to prove that Φ cannot have an interior maximum except if $\Phi(x, y)$ is a constant. In a similar way we can proceed in the case of an interior minimum. It is also obvious that our reasoning would have been valid for the more general partial differential equation

$$L[\Phi] = a\,\Phi_{xx} + 2b\,\Phi_{xy} + c\,\Phi_{yy} + d\,\Phi_x + e\,\Phi_y = 0 \tag{36.2'}$$

provided again that $ac > b^2$ in D.

We can conclude from the maximum principle that the velocity potential φ takes its maximum on the boundary of any domain D in the subsonic flow region inside of which no singularities of the solution occur. If we differentiate the Eq. (35.6) for $\varphi(x, y)$ with respect to x or to y, obviously we obtain partial differential equations of the form (36.2') for the velocity components u and v of the flow. Thus the minimum-maximum statements can be applied to these components, too. This implies, in particular, that the speed cannot have its maximum in the interior of a regular subsonic flow region. For we can always choose our coordinates in such a way that at the maximum point the flow direction points in the x-axis and that the maximum value of q becomes the maximum value of u, which is impossible.

37. FINN-GILBARG's extension of the maximum and minimum principle. The maximum-minimum principle of the preceding section becomes inapplicable when the solution $\Phi(x, y)$ of the partial differential equation (35.6) has a singularity in the domain considered. For flows in infinite domains the point at infinity will, in general, appear as a singular point of the velocity potential and prevent the direct application of our preceding results. There exists an interesting extension of the minimum-maximum principle to such situations. It was formulated by FINN and GILBARG [4] and used for a detailed study of the possible singularities of the velocity potential at infinity. Their main result is the following theorem:

Let $\Phi(x, y)$ be a subsonic solution of the differential equation (35.6) in the region Γ_{r_0} outside of the circle $x^2+y^2=r_0^2$. We suppose that $u=\Phi_x$ is not constant. Then, for all points in Γ_{r_0},

$$\mu_1 = \min_{r=r_0} u < u(x, y) < \max_{r=r_0} u = \mu_2. \tag{37.1}$$

The theorem asserts that the maximum and minimum of $u(x, y)$ are taken on the boundary $x^2+y^2=r_0^2$ of Γ_{r_0}. We do not know anything about the regularity of $u(x, y)$ at infinity. But the subsonic character of $\Phi(x, y)$ guarantees the boundedness of u and this suffices to establish the maximum-minimum principle. It is not even necessary that the solution $\Phi(x, y)$ be single-valued in Γ_{r_0}; it may have a logarithmic branch point at infinity. We assume, however, that the partial derivatives $u=\Phi_x$ and $v=\Phi_y$ are single-valued in Γ_{r_0}.

In order to prove the theorem we proceed by contradiction. Suppose that there are points in Γ_{r_0} where $u(x, y)>\mu_2$. Let D be one component domain of Γ_{r_0} consisting of such points. If we put

$$\bar{u}(x, y) = u(x, y) - \mu_2, \tag{37.2}$$

we have $\bar{u}(x, y)>0$ at interior points of D and $\bar{u}(x, y)=0$ at all finite boundary points of D. If D were bounded, all its boundary points would be finite and $u(x, y)$ would have its maximum at an interior point of D; this contradicts the maximum principle of Sect. 36 and is impossible. Hence, D must extend to infinity.

We denote by D_r that part of D which lies inside of the circle $x^2+y^2=r^2$ and by C_r the set of boundary points of D_r with $x^2+y^2=r^2$. For large enough values of r, neither D_r nor C_r will be empty. Consider now the identity

$$\left.\begin{aligned} &\iint_{D_r}\left[\frac{\partial \bar{u}}{\partial x}\frac{\partial}{\partial x}(F_u)+\frac{\partial \bar{u}}{\partial y}\frac{\partial}{\partial x}(F_v)\right]dx\,dy \\ &\qquad = \int_{C_r}\bar{u}\left[\frac{\partial}{\partial x}(F_u)\cos(n, x)+\frac{\partial}{\partial x}(F_v)\cos(n, y)\right]ds. \end{aligned}\right\} \tag{37.3}$$

Indeed, if we transform the left side integral in Eq. (37.3) by integration by parts and observe that $\bar{u}=0$ on the boundary of D_r with $x^2+y^2<r^2$ and that $F(x, y)$ satisfies the differential equation (35.4), we arrive precisely at the Eq. (37.3).

Let us introduce the notations

$$F_{uu}u_x^2+2F_{uv}u_x u_y+F_{vv}u_y^2 = J(x, y) \tag{37.4}$$

and

$$\iint_{D_r} J(x, y)\,dx\,dy = D(r). \tag{37.4'}$$

We can then write Eq. (37.3) in the form

$$D(r) = \int_{C_r}\bar{u}[(F_{uu}u_x+F_{uv}u_y)\cos(n, x)+(F_{uv}u_x+F_{vv}u_y)\cos(n, y)]\,ds. \tag{37.5}$$

In order to connect the right hand side of Eq. (37.5) with the function $J(x, y)$ we consider the quadratic form

$$Q[\alpha, \beta] = F_{uu}\alpha^2+2F_{uv}\alpha\beta+F_{vv}\beta^2. \tag{37.6}$$

Because of the assumed elliptic character of the differential equation the form (37.6) is definite, and if we suppose (without loss of generality) that $F_{uu}>0$, it will be positive definite. We have $|F_{uv}|<F_{uu}F_{vv}$ and hence the inequality

$$Q(\alpha, \beta) \leq (\sqrt{F_{uu}}\,\alpha+\sqrt{F_{vv}}\,\beta)^2 \leq (\alpha^2+\beta^2)(F_{uu}+F_{vv}). \tag{37.7}$$

The last estimate is due to SCHWARZ' inequality. On the other hand, we can apply SCHWARZ' inequality to the positive-definite quadratic form Q:

$$[F_{uu}\alpha\alpha'+F_{uv}(\alpha\beta'+\alpha'\beta)+F_{vv}\beta\beta']^2 \leq Q[\alpha, \beta]\,Q[\alpha', \beta']. \tag{37.8}$$

We apply this estimate with $\alpha = u_x$, $\beta = u_y$; $\alpha' = \cos(n, x)$, $\beta' = \cos(n, y)$ and, using the definition (37.4) and the inequality (37.7), we obtain

$$\left.\begin{aligned}&[(F_{uu} u_x + F_{uv} u_y)\cos(n, x) + (F_{uv} u_x + F_{vv} u_y)\cos(n, y)]^2\\ &\qquad \leq J(x, y)\,(F_{uu} + F_{vv}).\end{aligned}\right\} \tag{37.9}$$

In view of the definition (35.3) of F, we have by Eq. (35.9')

$$0 \leq F_{uu} + F_{vv} = 2\left[\varrho + q^2 \frac{\partial\varrho}{\partial q^2}\right] = 2\varrho\left(1 - \frac{q^2}{2c^2}\right) \leq A^2 \tag{37.10}$$

because of the assumed subsonic character of the flow (A denotes a fixed constant). Hence, from Eq. (37.5) and from the inequalities (37.9) and (37.10), we derive the inequality

$$D(r) \leq A \int_{C_r} \bar{u}\, J(x, y)^{\frac{1}{2}}\, r\, d\vartheta. \tag{37.11}$$

By a final application of SCHWARZ' inequality we obtain

$$D(r)^2 \leq B r \int_{C_r} J(x, y)\, r\, d\vartheta, \tag{37.12}$$

where B is a properly chosen constant.

After these formal preparations, we observe that from Eq. (37.4') follows

$$\frac{d}{dr} D(r) = \int_{C_r} J\, r\, d\vartheta. \tag{37.13}$$

Thus, inequality (37.12) can be interpreted as a differential inequality for the function $D(r)$:

$$r^{-1} \leq B\, D(r)^{-2} \cdot \frac{d}{dr} D(r). \tag{37.14}$$

We integrate this inequality between the limits r_1 and r_2 and find that

$$\log r_2 - \log r_1 \leq B\,[D(r_1)^{-1} - D(r_2)^{-1}]. \tag{37.15}$$

By its definition, $D(r)$ is a non-negative and increasing function of r. Hence the right hand side of inequality (37.15), for fixed r_1 and increasing r_2 is bounded by $B\,D(r_1)^{-1}$. The left-hand side, however, will grow beyond any limit if $r_2 \to \infty$. Thus we have obtained a contradiction which is a consequence of our initial assumption that $u(x, y) > \mu_2$ holds somewhere in Γ_{r_0}. Hence we have proved that $u(x, y) \leq \mu_2$ in Γ_{r_0}.

By means of the maximum principle of Sect. 36 we can now exclude the possibility $u = \mu_2$ at some finite point of Γ_{r_0}, except for the case $u \equiv \mu_2$. But we assumed a non-constant velocity component $u(x, y)$. We proceed in complete analogy with respect to the minimum problem and thus complete the proof of the inequality (37.1).

The above proof is a typical example of the method of differential inequalities for Dirichlet integrals as now frequently applied in the theory of partial differential equations. The general idea is to effect integration by parts over a variable domain D_r with boundary C_r. Using proper boundary conditions and the given partial differential equation one obtains an identity between a domain integral over D_r and a boundary integral over C_r. If the latter integral can be compared with the r-derivative of the domain integral one obtains a differential inequality for the domain integral. This leads frequently to very valuable insight into the growth properties of the solutions considered [5].

38. Application to subsonic flows. We can draw numerous conclusions from the generalized minimum-maximum principle of Sect. 37. Let μ_1 and μ_2 again denote the minimum and maximum of the velocity component $u(x, y)$ on the circle $x^2+y^2=r_0^2$. If $r_1>r_0$ and if u is not constant, we have by the preceding theorem,

$$\mu_1<\min_{r=r_1} u=\mu_1^*, \qquad \mu_2>\max_{r=r_1} u=\mu_2^*. \tag{38.1}$$

Hence in Γ_{r_1} we have the sharpened inequality $\mu_1^*<u<\mu_2^*$. Thus if we approach the point at infinity we have the inequalities

$$\mu_1<\underline{\lim}\, u(x, y)\leq\overline{\lim}\, u(x, y)<\mu_2, \tag{38.2}$$

provided $u(x, y)$ is not constant.

The same conclusions hold, of course, for the v-component of the velocity. We can further assert that everywhere in Γ_{r_0}

$$q=\sqrt{u^2+v^2}\leq\max_{r=r_0} q. \tag{38.3}$$

Indeed, suppose that at x_0, y_0 in Γ_{r_0} we had $q\geq\max\limits_{r=r_0} q$. Since we can choose our coordinate axes arbitrarily, we may assume that the velocity at x_0, y_0 has the direction of the x-axis and that at this particular point $u=q$. But clearly, $\mu_2\leq\max\limits_{r=r_0} q$ and hence we should have necessarily $\mu_2=\max\limits_{r=r_0} q$ and $u=\mu_2=\max\limits_{r=r_0} q$ by virtue of Sect. 37. Moreover, we should have $v\equiv 0$ in Γ_{r_0}, and the flow obtained would be uniform. Thus we have proved the inequality (38.3) and shown that equality can hold only for uniform flow.

From the inequality (38.3) we can draw the important corollary:

If a flow is subsonic outside of and on the circle $x^2+y^2=r_0^2$ then it is uniformly subsonic in the exterior Γ_{r_0} of this circle.

In fact, nowhere in Γ_{r_0} can the speed come nearer to sonic speed than on the circle, and therefore it is uniformly bounded away from this speed.

Our considerations give also a mathematical proof for the following physically evident statement:

If a subsonic flow is regular in the entire (x, y)-plane it must necessarily be a uniform flow.

Indeed, if we draw an arbitrary circle $x^2+y^2=r_0^2$, our result in Sect. 37 shows that each velocity component must take its maximum value inside of or on the circle. Thus, the functions $u(x, y)$ and $v(x, y)$ take their absolute maximum at a finite point and must be constant by the maximum principle.

39. The asymptotic behavior of the velocity potential at infinity. The velocity components $u(x, y)$ and $v(x, y)$ of a flow satisfy the system of differential equations

$$-v_y=A\,u_x+B\,v_y, \qquad v_x=u_y, \tag{39.1}$$

with $A=F_{vv}^{-1}F_{uu}$, $B=2F_{vv}^{-1}F_{uv}$. We shall obtain considerable information about their behavior from two basic theorems from the general theory of linear partial differential equations.

Theorem I: Let u, v be solutions of the linear system

$$-v_y=a\,u_x+b_1\,u_y, \qquad v_x=b_2\,u_x+c\,u_y, \tag{39.2}$$

where the coefficients are bounded functions of x, y in a domain D and satisfy there the inequality

$$4a\,c-(b_1+b_2)^2\geq k>0, \qquad k=\text{const}. \tag{39.3}$$

Then $u(x, y)$ and $v(x, y)$ possess Hölder-continuous derivatives at all points of D.

A function $f(x, y)$ is called Hölder-continuous in D if there exists a constant K and an exponent $0<\alpha<1$ such that for any two points (x, y) and (x_0, y_0) in D holds

$$|f(x, y) - f(x_0, y_0)| \leq K[(x-x_0)^2 + (y-y_0)^2]^{\alpha/2}. \tag{39.4}$$

The concept of Hölder-continuity is as important in the theory of partial differential equations as the Lipschitz continuity is for ordinary differential equations.

Theorem II: If the coefficients of the partial differential equation (39.2) satisfy the inequality (39.3) and are Hölder-continuous in D, then every Hölder-continuous solution u, v which is twice continuously differentiable in D except for an isolated interior point has Hölder-continuous derivatives in D, even at the exceptional point.

The preceding theorems have been adapted for our needs from much deeper results due to KORN, LICHTENSTEIN and MORREY [*6*], [*7*], [*9*], [*10*], [*11*]. For a convenient arrangement and discussion of the results on elliptic equations from the point of view needed here and for an extensive bibliography, see [*7*].

In order to apply the theorems I and II to subsonic flows, we observe first that for the specific differential equation (39.1) the expression on the left side of (39.3) has the particular form

$$4\frac{F_{uu}F_{vv} - F_{uv}^2}{F_{vv}^2} = 4\frac{1-\dfrac{q^2}{c^2}}{\left(1-\dfrac{v^2}{c^2}\right)^2}, \tag{39.5}$$

as can be easily seen from Eqs. (35.3) and (35.9′). This expression is surely bounded from below by a positive bound if the flow is uniformly subsonic. A certain complication arises, however, from the fact that we are operating in the neighborhood of infinity; this difficulty can be removed easily by the transformation

$$\zeta = \frac{1}{z}, \qquad z = x + iy, \qquad \zeta = \xi + i\eta, \tag{39.6}$$

which brings the point at infinity of the (x, y)-plane into the origin of the (ξ, η)-plane.

The transformation of the differential system (39.2) under such a change of variables is best expressed if one introduces the complex differentiations

$$\left.\begin{aligned} \frac{\partial}{\partial z} &= \frac{1}{2}\left(\frac{\partial}{\partial x} - i\frac{\partial}{\partial y}\right), & \frac{\partial}{\partial \bar{z}} &= \frac{1}{2}\left(\frac{\partial}{\partial x} + i\frac{\partial}{\partial y}\right), \\ \frac{\partial}{\partial \zeta} &= \frac{1}{2}\left(\frac{\partial}{\partial \xi} - i\frac{\partial}{\partial \eta}\right), & \frac{\partial}{\partial \bar{\zeta}} &= \frac{1}{2}\left(\frac{\partial}{\partial \xi} + i\frac{\partial}{\partial \eta}\right). \end{aligned}\right\} \tag{39.7}$$

Because of the relation (39.6), we have obviously

$$\frac{\partial}{\partial z} = -\frac{1}{\zeta^2}\frac{\partial}{\partial \zeta}, \qquad \frac{\partial}{\partial \bar{z}} = -\frac{1}{\bar{\zeta}^2}\frac{\partial}{\partial \bar{\zeta}}. \tag{39.8}$$

The system (39.2) takes the complex form

$$\frac{\partial v}{\partial z} = a^*\frac{\partial u}{\partial z} + b^*\frac{\partial u}{\partial \bar{z}} \tag{39.9}$$

with

$$a^* = \tfrac{1}{2}[(b_2 - b_1) + i(a+c)], \qquad b^* = \tfrac{1}{2}[b_2 + b_1 + i(a-c)], \tag{39.10}$$

and the inequality (39.3) appears as

$$\operatorname{Im}\{a^*\}^2 - |b^*|^2 \geq \tfrac{1}{4}k > 0. \tag{39.11}$$

In view of Eqs. (39.8) the Eq. (39.9) becomes a complex differential equation in ζ:

$$\frac{\partial v}{\partial \zeta} = a^* \frac{\partial u}{\partial \zeta} + b^* \frac{\zeta^2}{\bar{\zeta}^2} \frac{\partial u}{\partial \bar{\zeta}}, \tag{39.12}$$

which may again be interpreted as a system of real differential equations.

It is evident that the coefficients of the new differential system are bounded near the origin and that an inequality analogous to Eq. (39.3) can be assured for the corresponding real system. In fact, the equivalent inequality (39.11) holds since $|b^*|$ does not change under the transformation.

Applying Theorem I, we recognize that $u(x, y)$, $v(x, y)$ tend continuously to limit values u_0, v_0 if $\zeta \to 0$ and correspondingly $z \to \infty$. The velocity field of a uniformly subsonic flow can be defined continuously at infinity.

We observe next that the coefficients A, B in Eq. (39.1) are analytic functions of u and v. Hence, the corresponding coefficients a^* and b^* will be Hölder-continuous near $\zeta = 0$, and we might try to apply Theorem II in order to obtain additional information about u and v near infinity. Unfortunately, the term $\zeta^2 \bar{\zeta}^{-2}$ is not continuous for $\zeta = 0$, and Theorem II does not apply to the differential equation (39.12). We have to use a simple artifice in order to remove this difficulty.

We may assume without loss of generality that the velocity vector has at infinity the limit values $u_0 = q_0$, $v_0 = 0$. We show then from Eqs. (35.3), (35.6) and (35.9′) that the coefficients A and B take at infinity the limits

$$A_0 = 1 - M_0^2, \quad B_0 = 0, \quad M_0 = \frac{q_0}{c_0} = \text{Mach number at infinity}. \tag{39.13}$$

By the change of variables

$$x_1 = x, \qquad y_1 = \beta y, \qquad \beta = \sqrt{1 - M_0^2}, \tag{39.14}$$

we bring Eq. (39.1) into the form

$$-v_{y_1} = \frac{1}{\beta} A u_{x_1} + B u_{y_1}, \qquad v_{x_1} = \beta u_{y_1}. \tag{39.1′}$$

Using the transformation

$$z_1 = \zeta_1^{-1}, \qquad z_1 = x_1 + i y_1, \qquad \zeta_1 = \xi_1 + i \eta_1, \tag{39.15}$$

we now obtain in the ζ_1-plane a differential system

$$\frac{\partial v}{\partial \zeta_1} = a_1^* \frac{\partial u}{\partial \zeta_1} + b_1^* \frac{\zeta_1^2}{\bar{\zeta}_1^2} \frac{\partial u}{\partial \bar{\zeta}_1} \tag{39.16}$$

with

$$a_1^* = \frac{1}{2}\left[-B + i \frac{1}{\beta}(A + \beta^2)\right], \quad b_1^* = \frac{1}{2}\left[-B + i \frac{1}{\beta}(A - \beta^2)\right]. \tag{39.16′}$$

But in view of Eqs. (39.13) and (39.14), we have $\lim b_1^* = 0$ as $\zeta_1 \to 0$. The differential system (39.16) is of the form (39.2) and (39.3) near $\zeta_1 = 0$ and has Hölder-continuous coefficients. Hence we can apply the Theorem II and find:

Near $\zeta_1 = 0$ the partial derivatives of u and v with respect to ξ_1 and η_1 exist and are Hölder-continuous. Thus, we have the asymptotic equations:

$$\left.\begin{aligned} u_{\xi_1} &= \mu + O(|\zeta|^\alpha), & u_{\eta_1} &= \nu + O(|\zeta|^\alpha), \\ v_{\xi_1} &= \varkappa + O(|\zeta|^\alpha), & v_{\eta_1} &= \lambda + O(|\zeta|^\alpha) \end{aligned}\right\} \tag{39.17}$$

where μ, ν, $\varkappa$, λ are properly chosen constants. If we insert Eqs. (39.17) into Eqs. (39.16) and pass to the limit $\zeta = 0$, we find the relation

$$\varkappa - i\lambda = i\beta(\mu - i\nu); \tag{39.18}$$

that is,

$$\varkappa = \sqrt{1 - M_0^2}\,\nu, \qquad \lambda = -\sqrt{1 - M_0^2}\,\mu. \tag{39.18'}$$

By integrating Eqs. (39.17) and then returning to the original variables x, y, we obtain finally

$$\left.\begin{aligned} u &= u_0 + \frac{\mu x}{x^2 + \beta^2 y^2} - \frac{\varkappa y}{x^2 + \beta^2 y^2} + O(r^{-\alpha-1}),\\ v &= \frac{\varkappa x}{x^2 + \beta^2 y^2} + \frac{\beta^2 \mu y}{x^2 + \beta^2 y^2} + O(r^{-\alpha-1}). \end{aligned}\right\} \tag{39.19}$$

Since $u = \varphi_x$ and $v = \varphi_y$, by integration of Eqs. (39.19) we are led to the velocity potential

$$\varphi(x, y) = u_0 x + \frac{1}{2}\mu \log(x^2 + \beta^2 y^2) + \frac{\varkappa}{\beta} \operatorname{arc\,tan}\left(\beta \frac{y}{x}\right) + O(r^{-\alpha}). \tag{39.20}$$

Thus, we have the important result [4]:

Every solution of the differential equation (35.6) for the velocity potential which is subsonic outside of a circle has (for proper choice of the coordinate system) the asymptotic form (39.20) at infinity.

The physical significance of the three disposable constants u_0, $\varkappa$ and μ is obvious. u_0 is the uniform speed at infinity (the free stream velocity) and determines in turn the parameter β. The coefficient $\varkappa$ is obviously connected with the circulation of the flow. Indeed, let C be a closed curve outside of which the flow is regular and subsonic, except for the point at infinity itself. We have then

$$\oint_C (\varphi_x\,dx + \varphi_y\,dy) = \frac{2\pi\varkappa}{\beta} = \Gamma. \tag{39.21}$$

Finally, by Eq. (35.1) we have

$$\oint_C (\psi_x\,dx + \psi_y\,dy) = \oint_C \varrho(\varphi_x\,dy - \varphi_y\,dx) = \Sigma \tag{39.22}$$

as the total outflow of fluid from the source at infinity. From Eqs. (39.20) and (39.22), we obtain after easy calculations

$$\Sigma = 2\pi\mu\varrho_0. \tag{39.23}$$

Finally, we express $\varphi(x, y)$ in terms of the circulation Γ and the source strength Σ:

$$\varphi(x, y) = u_0 x + \frac{1}{\varrho_0} \cdot \frac{\Sigma}{4\pi} \log(x^2 + \beta^2 y^2) + \frac{\Gamma}{2\pi} \operatorname{arc\,tan}\left(\beta \frac{y}{x}\right) + O(r^{-\alpha}). \tag{39.24}$$

The fluid coming from the source of strength Σ at infinity has to be absorbed by singularities in the finite part of the (x, y)-plane. Suppose, on the other hand, that a flow under consideration is subsonic and regular outside of a closed profile P, except at infinity. Since the integral (39.22) is independent of the particular curve C we may deform the path of integration in such a way that it coincides with P. But along P the stream function ψ is constant; hence the integral vanishes and $\Sigma = 0$. We have proved:

A subsonic flow past a profile P has the source strength $\Sigma = 0$ at infinity.

It should be observed that the asymptotic development for the velocity potential of a compressible fluid flow near infinity goes over into the corresponding development in the incompressible case if we let $\beta \to 1$. The quantities Σ and Γ have the same physical interpretation in both cases.

40. A uniqueness theorem for subsonic flows past a profile. We are now able to prove the following uniqueness theorem for a steady, irrotational, two-dimensional flow [4], [7]:

A subsonic flow past a given profile P is uniquely determined by its free stream velocity u_0 and its circulation Γ.

Let us assume, indeed, that there exist two velocity potentials φ and φ_1 which describe flows past the same profile and which have at infinity the same constants u_0 and Γ in their asymptotic development (39.24). Since by virtue of the theorem at the end of Sect. 39 $\Sigma = 0$, we can, therefore, assert that their difference satisfies

$$\tilde{\varphi}(x, y) = \varphi_1(x, y) - \varphi(x, y) = O(r^{-\alpha}). \tag{40.1}$$

We introduce the corresponding stream functions $\psi(x, y)$ and $\psi_1(x, y)$ and assume that both vanish on the profile P; hence the same will be true for the difference function

$$\tilde{\psi}(x, y) = \psi_1(x, y) - \psi(x, y). \tag{40.2}$$

We denote the velocity vectors belonging to φ and φ_1 by u, v and u_1, v_1, respectively; we introduce the interpolated velocity field:

$$u_t = u + t(u_1 - u), \qquad v_t = v + t(v_1 - v) \tag{40.3}$$

and denote by q_t the corresponding speed, by $\varrho(q_t^2) = \varrho_t$ the corresponding density field.

Because of the differential equations (35.1), we have the identites:

$$\left.\begin{aligned} \tilde{\psi}_y &= \varrho_1 u_1 - \varrho u = \int_0^1 \frac{d}{dt}[\varrho_t u_t]\,dt, \\ -\tilde{\psi}_x &= \varrho_1 v_1 - \varrho v = \int_0^1 \frac{d}{dt}[\varrho_t v_t]\,dt. \end{aligned}\right\} \tag{40.4}$$

We perform the differentiations in the right side integrals and find:

$$\tilde{\psi}_y = a\tilde{\varphi}_x + b\tilde{\varphi}_y, \qquad -\tilde{\psi}_x = b\tilde{\varphi}_x + c\tilde{\varphi}_y \tag{40.5}$$

with

$$a = \int_0^1 [\varrho_t + \varrho'(q_t^2)\,2u_t^2]\,dt, \quad b = 2\int_0^1 \varrho'(q_t^2)\,u_t v_t\,dt, \quad c = \int_0^1 [\varrho_t + \varrho'(q_t^2)\,2v_t^2]\,dt. \tag{40.6}$$

The difference functions $\tilde{\varphi}$ and $\tilde{\psi}$ satisfy the first order differential system (40.5). It has the important property that the quadratic form

$$a(x, y)\,\alpha^2 + 2b(x, y)\,\alpha\beta + c(x, y)\,\beta^2 = Q[\alpha, \beta] \tag{40.7}$$

based on its coefficients is positive-definite, implying, in particular:

$$ac - b^2 > 0. \tag{40.7'}$$

In fact, we derive from Eqs. (40.6), (40.7) and (35.9′)

$$\left.\begin{aligned} Q[\alpha,\beta] &= \int_0^1 [(\alpha^2+\beta^2)\,\varrho_t + 2\varrho'(q_t^2)\,(\alpha u_t+\beta v_t)^2]\,dt \\ &= \int_0^1 \varrho_t\left[\alpha^2+\beta^2-\frac{1}{c_t^2}(\alpha u_t+\beta v_t)^2\right]dt. \end{aligned}\right\} \tag{40.8}$$

By SCHWARZ' inequality we have

$$(\alpha u_t+\beta v_t)^2 \leqq (\alpha^2+\beta^2)\,q_t^2, \tag{40.9}$$

which shows that $Q(\alpha,\beta)$ is positive-definite if $q_t < c_t$. But if we remember that by assumption q and q_1 are both subsonic, this inequality is assured. Thus, $Q[\alpha,\beta]$ has been proved to be positive-definite.

Next, we make use of the identity

$$\iint_D [a\tilde{\varphi}_x^2 + 2b\tilde{\varphi}_x\tilde{\varphi}_y + c\tilde{\varphi}_y^2]\,dx\,dy = \iint_D (\tilde{\varphi}_x\tilde{\psi}_y - \tilde{\psi}_x\tilde{\varphi}_y)\,dx\,dy = -\int_C \tilde{\psi}\,d\tilde{\varphi} \tag{40.10}$$

which holds for every domain D with boundary C in which the system (40.5) is valid. Now let C_R be a circle having sufficiently large radius R and enclosing the given profile P; let D_R be the ring domain bounded by C_R and P. Since $\tilde{\psi}=0$ on P, application of Eq. (40.10) to D_R yields:

$$\iint_{D_R} Q[\tilde{\varphi}_x,\tilde{\varphi}_y]\,dx\,dy = -\int_{C_R} \tilde{\psi}\,d\tilde{\varphi}\,. \tag{40.11}$$

We conclude from Eq. (40.1) that on C_R

$$\frac{d\tilde{\varphi}}{ds} = O(R^{-1-\alpha}), \tag{40.12}$$

while Eq. (40.1) combined with Eq. (40.5) yields clearly on C_R

$$\tilde{\psi} = O(R^{-\alpha}). \tag{40.12′}$$

Thus, when we let $R\to\infty$, we find

$$\lim_{R\to\infty} \iint_{D_R} Q[\tilde{\varphi}_x,\tilde{\varphi}_y]\,dx\,dy = 0. \tag{40.13}$$

But the integrand is non-negative, and the integral over D_R cannot decrease with increasing R. Hence we have proved:

$$Q[\tilde{\varphi}_x,\tilde{\varphi}_y] = 0, \tag{40.14}$$

which by the definite character of $Q[\alpha,\beta]$ implies

$$\nabla\tilde{\varphi} = 0, \qquad \tilde{\varphi} = \text{const}. \tag{40.15}$$

Finally, from Eq. (40.1) we can deduce that the value of the constant is zero; hence

$$\varphi = \varphi_1, \qquad \psi = \psi_1 \tag{40.16}$$

and the uniqueness theorem is proved.

By refining the preceding reasoning Finn and Gilbarg prove the stronger theorem [4]:

Given a uniformly subsonic flow past a profile P with prescribed free stream velocity u_0 and circulation Γ, there is no other flow past P with the same asymptotic behavior near infinity.

This means that a uniformly subsonic flow is characterized uniquely by u_0 and Γ even among all mixed flows past the same profile P.

A different approach to the uniqueness theory of subsonic flows past given profiles has been developed by BERS [7]. He makes use of the concept of pseudo-analytic functions. A complex-valued function $w(x, y) = u + iv$ is called pseudo-analytic if its real and imaginary parts satisfy a system of differential equations

$$u_x = \tau v_x + \sigma v_y, \qquad u_y = -\sigma v_x + \tau v_y \tag{40.17}$$

where τ and σ are real Hölder-continuous functions of x, y and where $\sigma > 0$. The case $\tau = 0, \sigma = 1$ leads to the classical Cauchy-Riemann equations and to the theory of analytic functions. A systematic extension of numerous results of function theory to pseudo-analytic functions is possible; it has been formulated by BERS in a concise similarity principle. The theory of pseudo-analytic functions applies in subsonic flows since the system (39.1) can be easily brought into the normal form (40.17). The method of pseudo-analytic functions may also be used in the existence theory of fluid dynamics, which will be considered below.

41. Series development for the velocity potential at infinity. In deriving the asymptotic formulas (39.19) for the velocity components we used an iterative procedure which is characteristic for the theory of elliptic non-linear differential equations. We started with relatively little information on the coefficients of the differential system (39.1) for u and v; we drew conclusions regarding continuity of u and v. Since the coefficients A and B depend upon u and v, we thus improved our information on the system (39.1), which led in turn to differentiability statements for u and v. We might feed back this new conclusion into Eq. (39.1) and thus improve continuously the statements regarding the regularity of u and v. A complete series development at infinity might thus be derived.

We shall follow a different procedure, based on the theory of linear differential equations with analytic coefficients, in order to obtain an infinite series development. Let us assume, for the sake of simplicity, that the constants $\varkappa$ and μ in Eq. (39.10) are not both zero, in other words, that $\Sigma^2 + \Gamma^2 \neq 0$. By the mapping

$$\zeta_1 = (x + i\beta y)^{-1}, \qquad \zeta_1 = \xi_1 + i\eta_1, \tag{41.1}$$

which was already used in Eq. (39.15), we transform a neighborhood of infinity into a neighborhood of the origin in the ζ_1-plane. Here u and v are continuously differentiable functions of ξ_1 and η_1, and as we showed in Eqs. (39.17), (39.18′) the Jacobian at the origin has the value:

$$\frac{\partial(u, v)}{\partial(\xi_1, \eta_1)} = \mu\lambda - \nu\varkappa = -\left(\sqrt{1 - M_0^2}\,\mu^2 + \frac{\varkappa^2}{\sqrt{1 - M_0^2}}\right) < 0. \tag{41.2}$$

Therefore we may assert that the exterior of a large circle $x^2 + y^2 = r_0^2$ in the (x, y)-plane is mapped bicontinuously upon a domain in the (u, v)-plane containing the point $(u_0, 0)$ in its interior.

The velocity potential $\varphi(x, y)$ and its conjugate stream function $\psi(x, y)$ are continuously differentiable functions of the new variables u, v, except at the point of singularity $(u_0, 0)$. As was shown in Chap. III, the functions φ and ψ satisfy in the hodograph plane a linear elliptic system of differential equations with coefficients which are well defined analytic functions of u and v. The particular solutions φ, ψ which lead to the asymptotic behavior (39.19) at the point of infinity in the (x, y)-plane must, of course, be singular at $(u_0, 0)$. We can remove these singularities, however, by the following artifice:

We consider the harmonic function

$$\varphi_0(x, y) = u_0 x + \frac{\Sigma_0}{4\pi\varrho_0} \log(x^2 + y^2) + \frac{\Gamma_0}{2\pi} \arctan\frac{y}{x} \tag{41.3}$$

which can be interpreted as the velocity potential of an incompressible fluid flow with density ϱ_0, free stream velocity u_0 in the x-direction, source strength Σ_0 and circulation Γ_0. If we put

$$\zeta = \log(u - iv)\,, \qquad u = \frac{\partial\varphi_0}{\partial x}\,, \qquad v = \frac{\partial\varphi_0}{\partial y}\,, \tag{41.4}$$

we can write $\varphi_0(x, y)$ in the logarithmic hodograph plane as the real part of an analytic function of ζ:

$$f(\zeta) = \frac{\frac{1}{\varrho_0}\Sigma_0 - i\Gamma_0}{2\pi}\left[\frac{1}{\zeta - \zeta_0} - \log(\zeta - \zeta_0) + r(\zeta)\right], \tag{41.5}$$

where $\zeta_0 = \log u_0$ and $r(\zeta)$ is regular analytic near $\zeta = \zeta_0$. We may use $f(\zeta)$ as a generating function in the Bergman or the Lighthill method in order to obtain a solution for the stream function equation in the pseudo-logarithmic λ, ϑ-plane.

On the other hand, the above solution ψ will satisfy also the stream function equation in the same λ, ϑ-plane. We can subtract from it a solution ψ_0 obtained from Eq. (41.3) by the above procedure and can choose Σ_0, ϱ_0 and Γ_0 in such a way as to make the difference function bounded even at the singular point. This difference function is a solution of a second order partial differential equation with analytic coefficients and must, therefore, be analytic even at the exceptional point; that is, it can be developed into a power series in λ and ϑ. Similarly, the difference between the original velocity potential φ and the auxiliary $\varphi_0(\lambda, \vartheta)$ must be analytic at the critical point.

Since the singularity functions $\varphi_0(\lambda, \vartheta)$ and $\psi_0(\lambda, \vartheta)$ can be constructed explicitly and can be developed into series around the singular point $(\lambda_0, 0)$ which corresponds to $(u_0, 0)$ we have established a series development for the functions $\varphi(\lambda, \vartheta)$ and $\psi(\lambda, \vartheta)$. By the transition formulas (23.6) we can finally return to the physical plane and obtain series developments for φ and ψ in terms of x and y. The rather elaborate calculations were carried out by LUDFORD [*12*], who obtained the following series development for $\varphi(x, y)$:

$$\left.\begin{aligned}\varphi(x, y) = u_0 r\cos\vartheta + \frac{\Sigma}{2\pi\varrho_0}\log r + \frac{\Gamma}{2\pi}\arctan(\beta\tan\vartheta) + \\ + \sum_{k,l=0}^{\infty} A_{kl}(\vartheta)(\log r/r)^k r^{-l}\end{aligned}\right\} \tag{41.6}$$

with $x = r\cos\vartheta$, $y = r\sin\vartheta$ and coefficients $A_{kl}(\vartheta)$ which have the period 2π in ϑ.

In the case that $\Sigma = \Gamma = 0$ a somewhat more involved analysis leads to the series development in polar coordinates

$$\varphi(x, y) = u_0 r\cos\vartheta + \sum_{k=0}^{\infty} A_k(\vartheta)\, r^{-k} \tag{41.7}$$

where the $A_k(\vartheta)$ are again periodic functions of the angle.

The series development for the velocity potential of a compressible fluid flow at infinity was first considered by LAMB [*14*] and BATEMAN [*15*]. See also [*13*]. It is important for the calculation of the lift and moment upon the obstacle. See [*36*] and Sect. 52.

42. Kutta-Joukowski condition and circulation. Let us suppose that the profile P considered in Sect. 40 is a smooth curve whose tangent varies Hölder-continuously with the arc length except at one point T. At this point we assume P to have a corner or cusp characterized by the external angle $\varkappa\pi$, $1\leqq\varkappa\leqq 2$. $\varkappa=1$ corresponds to a smooth point T, $\varkappa=2$ to a cusp point. T is called the trailing edge of the profile (or airfoil) P. If $\varkappa>1$, it is not possible to prescribe the free stream velocity u_0 and circulation Γ arbitrarily and to find a subsonic flow around the profile. The requirement of subsonic flow past P will, indeed, determine the circulation constant Γ in a unique way.

We shall say that a flow past the profile P satisfies the Kutta-Joukowsky condition at the trailing edge T if the flow is uniformly subsonic near the point T and if the speed at T is zero in the case of a proper corner ($\varkappa<2$).

We can then formulate the uniqueness theorem: *A subsonic flow past a profile P is uniquely determined by its free stream velocity and the Kutta-Joukowsky condition at the trailing edge T.*

In order to prove this result let us assume that satisfying the conditions stated there are two flows, described by the velocity potentials φ and φ_1 and by stream functions ψ and ψ_1. The circulations Γ and Γ_1 must be different, since otherwise the uniqueness theorem of Sect. 40 would suffice to ensure the identity of the two flows. We may suppose, therefore, that

$$\Gamma_1<\Gamma. \tag{42.1}$$

From the asymptotic formula of the type (39.24) for the velocity potential, we conclude easily

$$\tilde{\psi}(x,y)=\psi_1(x,y)-\psi(x,y)=-\frac{\Gamma_1-\Gamma}{2\pi}\varrho_0\log r+O(1)\quad\text{as}\quad r\to\infty \tag{42.2}$$

while on the boundary P of the flow region D obviously $\tilde{\psi}(x,y)=0$.

We introduce again the difference function

$$\tilde{\varphi}(x,y)=\varphi_1(x,y)-\varphi(x,y) \tag{42.3}$$

and obtain the differential equations (40.5) connecting $\tilde{\varphi}$ and $\tilde{\psi}$. If we eliminate $\tilde{\varphi}$ from the equations we obtain for $\tilde{\psi}$ a second order partial differential equation, and by the methods of Sect. 36 it is easy to show that $\tilde{\psi}$ cannot take a relative minimum at a finite interior point of D. Since $\tilde{\psi}\to+\infty$ if $r\to\infty$ and $\tilde{\psi}\to 0$ as one approaches P, it is clear that, therefore,

$$\tilde{\psi}(x,y)>0 \text{ in } D. \tag{42.4}$$

On the other hand, $\tilde{\psi}\equiv 0$ on P; the derivative of $\tilde{\psi}$ in the direction of the interior normal must, therefore, be non-negative and a more refined argument shows that at every point of P (even at T) we have [*4*]

$$\frac{\partial}{\partial n}\tilde{\psi}(x,y)>0. \tag{42.5}$$

In the case $1<\varkappa\leqq 2$, we can even assert that

$$\lim|\nabla\tilde{\psi}|=\infty \tag{42.5'}$$

if we approach the trailing edge. But the Kutta-Joukowsky condition demands

$$\nabla\tilde{\psi} = \nabla\psi_1 - \nabla\psi = 0 \tag{42.6}$$

at a corner point T (i.e. $1 \leqq \varkappa < 2$) while in the case of a cusp ($\varkappa = 2$) we have the requirement

$$|\nabla\tilde{\psi}| \leqq |\nabla\psi_1| + |\nabla\psi| = \text{finite}. \tag{42.6'}$$

In every case we obtain thus a contradiction between the formulas (42.5) and (42.6), and the assumption $\Gamma_1 < \Gamma$ is shown to be impossible. This proves the uniqueness theorem.

It is again possible to improve this result and to compare a given subsonic flow with possibly mixed flows. This is formulated in the theorem [*4*]:

A subsonic flow past a profile P is uniquely determined by its free stream velocity and the Kutta-Joukowsky condition at a trailing edge T, even among all possibly mixed flows, provided that the velocity components have bounded derivatives near the profile P.

The Kutta-Joukowsky condition is natural in the case of subsonic flows, for it seems physically evident that a corner or a cusp will generate supersonic speeds if this condition is not fulfilled. Finn and Gilbarg [*4*] proved indeed: If a subsonic flow is defined in the neighborhood of a corner T with $1 < \varkappa < 2$ and if its speed is continuous up to T, then T is a stagnation point of the flow.

We may apply the preceding reasoning to obtain a comparison theorem for subsonic flows. Let ψ and ψ_1 be the stream functions of two subsonic flows past a given smooth profile P; we suppose that both flows have the free stream velocity u_0 in the positive x-direction but that their circulations satisfy $\Gamma_1 < \Gamma$. We can then assert again that their difference function $\tilde{\psi} = \psi_1 - \psi$ satisfies the inequality (42.5); we have, therefore, at every point of P

$$\frac{\partial\psi_1}{\partial n} > \frac{\partial\psi}{\partial n}. \tag{42.7}$$

Since $\psi = \psi_1 = 0$ on P, the inequality (42.7) is equivalent to

$$|\nabla\psi_1| < |\nabla\psi| \quad \text{or} \quad |\nabla\psi_1| > |\nabla\psi|, \tag{42.8}$$

depending on the sense of the flow with respect to the profile P. The first inequality holds if the flow sense is positive (counterclockwise), the second in the opposite case.

By Eq. (35.1) we conclude

$$\varrho_1 q_1 < \varrho q \quad \text{or} \quad \varrho_1 q_1 > \varrho q \tag{42.9}$$

respectively. Finally, we recognize from Eqs. (35.8) and (35.9) that ϱq increases with q and therefore

$$q_1 < q \quad \text{or} \quad q_1 > q, \tag{42.10}$$

depending on the sense of motion with respect to P.

Thus we have proved: If the circulation Γ is increased, the speed at every point of the profile will increase wherever the motion is in the positive sense and will decrease at all points of negative flow sense. This result is in agreement with the intuitive definition of circulation; however, it is, not obvious a priori that the effect has the expected character at every point of the profile, as has been established by our argument.

43. Variational problem for subsonic flow without circulation. Let $x(s)$, $y(s)$ be a parametric representation of a closed curve P in the (x, y)-plane in terms of the arc length s; we assume $x(s)$ and $y(s)$ to have at least three continuous derivatives with respect to s. We conceive P to be the boundary of an object submerged in a two-dimensional flow filling the entire (x, y)-plane and satisfying the adiabatic equation (35.11) as pressure-density relation. We impress on the fluid a uniform velocity u_0 in the direction of the positive x-axis; this will create in the (x, y)-plane a flow characterized by a velocity potential $\varphi(x, y)$ satisfying the quasi-linear differential equation (35.6) and the boundary condition

$$\frac{\partial \varphi}{\partial n} = 0 \text{ on } P. \tag{43.1}$$

In the case of the adiabatic pressure-density relation (35.11), we have the speed-density relation (35.12′), which in turn determines the basic function $F(\varphi_x^2 + \varphi_y^2)$ occurring in the differential equation (35.6). It is immediate that Eq. (35.6) can now be written in form

$$(c^2 - \varphi_x^2)\,\varphi_{xx} - 2\,\varphi_x \varphi_y \varphi_{xy} + (c^2 - \varphi_y^2)\,\varphi_{yy} = 0, \tag{43.2}$$

where c is the local velocity of sound; hence $c^2 = p'(\varrho) = a\,\gamma \varrho^{\gamma-1}$, and by the Bernoulli equation (35.12):

$$\frac{1}{2}(\varphi_x^2 + \varphi_y^2) + \frac{c^2}{\gamma - 1} = \frac{1}{2}\,q_{\max}^2 = \frac{1}{2}\,u_0^2 + \frac{c_0^2}{\gamma - 1}. \tag{43.2′}$$

The first Eq. (43.2′) determines c as a function of the local speed; the second equation expresses $q_{\max}$ in terms of the free stream velocity u_0 and the velocity of sound c_0 at infinity. Since we assume there is no circulation Γ at infinity, we have for $\varphi(x, y)$ the asymptotic formula (39.24) near infinity:

$$\varphi(x, y) = u_0 x + O(r^{-\alpha}), \quad r^2 = x^2 + y^2. \tag{43.3}$$

We have already shown that conditions (43.1) to (43.3) determine the function $\varphi(x, y)$ uniquely provided that the flow is subsonic in the exterior of P. Now arises the fundamental question whether such a flow does exist. It was first shown by FRANKL and KELDYSH (1934) that for sufficiently small Mach number at infinity $M_0 = u_0/c_0$ such a flow does indeed exist [*16*]. In 1952, Shiffman proved the following general theorem [*17*], [*18*]:

Each given profile P has a critical number $C(P) < 1$ such that for every Mach number $M_0 = u_0/c_0$ at infinity with $M_0 < C(P)$ there exists a completely subsonic flow past P with this Mach number at infinity and with no circulation. Let $M_{\max}(M_0)$ denote the maximum of the local Mach number q/c for the flow characterized by M_0; then $M_{\max}$ will range over all values $0 \leqq M_{\max} < 1$ if M_0 varies over $0 < M_0 < C(P)$.

In other words: There is a critical constant $C(P)$ such that for $M_0 < C(P)$ the local Mach numbers stay below 1, while somewhere one comes arbitrarily near to sonic speed if M_0 approaches the critical constant. If therefore, one insures, by choosing $M_0 < C(P)$ that the flow will stay subsonic outside of P, one is certain of the existence of a flow as characterized above. The uniqueness of the flow is, of course, guaranteed by the results of Sect. 40.

The existence proof of Shiffman is based on the direct methods of the calculus of variations. It was already pointed out in Sect. 35 that the partial differential equation (35.6) for the velocity potential may be considered as the Euler-Lagrange equation for the variational problem (35.5); it is, therefore, suggestive to attack the existence problem of the velocity potential with those tools which have been prepared in connection with the Dirichlet problems of the theory of partial differential equations. There are, however, various additional difficulties inherent in our specific problem:

a) We have to consider an infinite flow domain, and the integral (35.5) does not converge.

b) The variational problem is a regular variational problem only as long as the flow remains subsonic; we do not know a priori when the flow considered will remain subsonic, and it is difficult to speak about the subsonic character of a flow whose existence has first to be established.

It is advisable to consider the stream function $\psi(x, y)$ rather than the velocity potential $\varphi(x, y)$, since the boundary condition $\psi = 0$ on P is much easier to handle than the requirement (43.1) for the normal derivative of φ. Therefore we recast the variational problem into a form involving ψ.

We start from the fundamental differential relations (35.1) between φ and ψ and find

$$\psi_x^2 + \psi_y^2 = \varrho^2 \cdot (\varphi_x^2 + \varphi_y^2) = \varrho^2 q^2. \tag{43.4}$$

Because of the known speed density relation the right term is a well defined function of ϱ; we can solve with respect to ϱ and obtain

$$\frac{1}{\varrho} = f'(\psi_x^2 + \psi_y^2). \tag{43.5}$$

With this notation the partial differential equation (35.2′) for the stream function becomes

$$\frac{\partial}{\partial x}\left(\frac{\partial f}{\partial \psi_x}\right) + \frac{\partial}{\partial y}\left(\frac{\partial f}{\partial \psi_y}\right) = 0, \tag{43.6}$$

which is obviously the Euler-Lagrange equation for the variational problem

$$\delta \iint f(\psi_x^2 + \psi_y^2)\, dx\, dy = 0. \tag{43.7}$$

Let us study the function $f(z)$ a little more in detail. We have in view of Eqs. (43.4) and (43.5)

$$\frac{d}{dq^2} f(\varrho^2 q^2) = \frac{1}{\varrho} \frac{d}{dq^2} (\varrho^2 q^2) = \varrho + 2q^2 \frac{d\varrho}{dq^2}. \tag{43.8}$$

We recognize, therefore, from Eqs. (35.8) and (35.10) that

$$\frac{d}{dq^2} f(\varrho^2 q^2) = \varrho\left(1 - \frac{q^2}{c^2}\right) \tag{43.9}$$

is positive for subsonic flow. Consider next the combination

$$\alpha(q^2) = 2\left[\varrho q^2 - F(q^2)\right] \tag{43.10}$$

where F is the function defined by Eq. (35.3). Clearly

$$\frac{d\alpha}{dq^2} = \varrho + 2q^2 \frac{d\varrho}{dq^2}. \tag{43.10′}$$

Hence, comparing Eqs. (43.8) and (43.10′), we see that we can put

$$f(\psi_x^2 + \psi_y^2) = 2\left[\varrho q^2 - F(q^2)\right]. \tag{43.11}$$

We come now to the decisive question of the regularity of the variational problem (43.7); a variational problem is called regular if its corresponding Euler-Lagrange equation is of elliptic type. Most methods of the calculus of variations apply only in the case of a regular variational problem. The ellipticity of the differential equation (43.6) is characterized by the inequality

$$\frac{\partial^2 f}{\partial \psi_x^2}\,\frac{\partial^2 f}{\partial \psi_y^2} - \left(\frac{\partial^2 f}{\partial \psi_x \partial \psi_y}\right)^2 > 0 \tag{43.12}$$

which in view of the special dependence of f on $z = \psi_x^2 + \psi_y^2$ becomes:

$$2z f'(z)\, f''(z) + f'(z)^2 = \frac{d}{dz}\left(z f'(z)^2\right) > 0. \tag{43.12'}$$

By the definitions (43.5) and because of Eq. (43.4) we have

$$z f'(z)^2 = q^2, \qquad z = \varrho^2 q^2, \tag{43.13}$$

and Eq. (43.12') may be expressed in the form

$$\frac{d}{dz} q^2 = \left[\frac{d}{dq^2}(\varrho^2 q^2)\right]^{-1} = \varrho^{-2}\left(1 - \frac{q^2}{c^2}\right)^{-1} > 0 \tag{43.14}$$

where we have used Eqs. (35.8) and (35.10). We find, therefore, as expected, that Eq. (43.6) is elliptic and the variational problem (43.7) is regular as long as the flow is subsonic.

Given a free stream velocity u_0 we can always choose our units in such a way that $q_{\max} = 1$. In this case, it follows from Eq. (35.13) that the speed belonging to the Mach number 1 is given by

$$q_s = \sqrt{\frac{\gamma - 1}{\gamma + 1}}. \tag{43.15}$$

Since $\varrho^2 q^2 = z$ is a monotonic function of q as long as $q < c$, we recognize that the flow will be subsonic as long as

$$z = \psi_x^2 + \psi_y^2 < q_s^2 \cdot \varrho(q_s^2) = \frac{\gamma - 1}{\gamma + 1}\, 2F'\left(\frac{\gamma - 1}{\gamma + 1}\right) = b_0. \tag{43.16}$$

The knowledge of the function $f(z)$ in the interval $0 \leqq z < b_0$ is sufficient to establish the mathematical theory of the subsonic flow of a gas with adiabatic pressure-density relation. It is well determined beyond this interval, but it will then describe supersonic motion and give rise to a non-regular variational problem (43.7). The basic idea in SHIFFMAN'S existence proof is now to replace the adiabatic equation by another one which gives rise to a function $h(z)$ which coincides with the original function $f(z)$ in an interval $0 \leqq z \leqq b_1 < b_0$ but is defined for all values $z > b_1$ in such a way that it continues to lead to a regular variational problem (43.7), that is, to an elliptic partial differential equation (43.6). This change in the equation of state will not be felt so long as the fluid motion stays under the Mach number M_1 corresponding to $z = b_1$; on the other hand, we can bring b_1 as near as we wish to b_0 and hence choose M_1 arbitrarily near to one. In other words, our change of $f(z)$ into $h(z)$ is without any consequence as far as subsonic motion is concerned. But we can now develop an elegant theory of the variational problem (43.7) based on the fact that it will be regular throughout. In this way the particular difficulty (b) mentioned above is overcome. The idea of changing the equation of state of the gas considered in a convenient way was first used by VASZONYI [*19*] and utilized also in BERS' existence proof [*7*]. See Sects. 48, 49.

The actual construction of the function $h(z)$ is as follows: We choose $zh'(z)^2 = zf'(z)^2$ in the interval $0 \leq z \leq b_1$; then we continue $zh'(z)^2 = H(z)$ as a monotonically increasing function of z and let it become equal to $a^2 z$ for $z > b_2$. Clearly

$$h(z) = \int \sqrt{\frac{H(z)}{z}}\, dz \tag{43.17}$$

will be well defined and for $z > b_2$ will coincide with the linear function $az + b$. Since

$$\frac{d}{dz}\left[z\, h'(z)^2\right] = H'(z) > 0, \tag{43.17'}$$

we may characterize $h(z)$ by the fact that $\sqrt{z}\, h'(z)$ increases with z and that $h(z)$ is, therefore, an increasing convex function of $\sqrt{z}$. We can obviously choose $h(z)$ in such a way that it possesses everywhere at least three continuous derivatives.

Finally, we have to deal with the difficulty (a) which is due to the infinite flow domain D. For this purpose, we have to investigate the character of ψ near the point at infinity in the (x, y)-plane. Since at infinity $\varphi_x = u_0$, $\varphi_y = 0$ we have $q_0 = u_0$, $\varrho_0 = \varrho(q_0^2)$ and by Eq. (35.1)

$$\psi_y = \varrho_0 u_0 = t_0, \quad \psi_x = 0 \quad \text{at infinity}. \tag{43.18}$$

The argument z of $h(z)$ near infinity will tend to

$$z_0 = \varrho_0^2 q_0^2 = t_0^2. \tag{43.18'}$$

We consider then the following variational problem:

$$\delta \iint_D \left[h(\psi_x^2 + \psi_y^2) - h(t_0^2) - 2t_0 h'(t_0^2)(\psi_y - t_0)\right] dx\, dy = 0. \tag{43.19}$$

The new problem leads to the same Euler-Lagrange equation as the original problem (43.7). The additional terms, however, ensure the convergence of the integral if a suitably restricted class of competing functions for the variational problem (43.19) is selected.

44. SHIFFMAN's existence proof for subsonic flows past a profile. Let $\mathfrak{C}$ be the class of all functions $\psi(x, y)$ which are continuously differentiable in the infinite domain D, which vanish on the boundary P of D and for which $\psi_x \to 0$, $\psi_y \to t_0$ as $x^2 + y^2 \to \infty$; the convergence shall be fast enough such that

$$\iint_D \left[\psi_x^2 + (\psi_y - t_0)^2\right] dx\, dy < \infty. \tag{44.1}$$

The class $\mathfrak{C}$ is obviously not empty.

We want to show that the integral (43.19) converges for all functions $\psi(x, y)$ of the class $\mathfrak{C}$. We introduce the function

$$\tilde{\psi}(x, y) = t_0 y + [\psi(x, y) - t_0 y]\, \tau \tag{44.2}$$

which depends on the parameter τ, $0 \leq \tau \leq 1$. The function

$$J_\tau(x, y) = h(\tilde{\psi}_x^2 + \tilde{\psi}_y^2) - h(t_0^2) - 2t_0 h'(t_0^2)(\tilde{\psi}_y - t_0) \tag{44.3}$$

coincides for $\tau = 1$ with the integrand of Eq. (43.19). Since we have by definition (44.2)

$$\tilde{\psi}_x^2 + \tilde{\psi}_y^2 = t_0^2 + 2t_0(\psi_y - t_0)\,\tau + \left[\psi_x^2 + (\psi_y - t_0)^2\right]\tau^2 \tag{44.4}$$

it is easily seen that $J_0(x, y) = 0$ and $\frac{d}{d\tau} J_\tau(x, y)|_{\tau=0} = 0$. Hence, we can put the integrand of Eq. (43.19) into the form:

$$J_1(x, y) = \int_0^1 (1-\tau) \frac{d^2 J_\tau}{d\tau^2} d\tau = \int_0^1 (1-\tau) \frac{d^2}{d\tau^2} h(\tilde{\psi}_x^2 + \tilde{\psi}_y^2) \, d\tau. \tag{44.5}$$

Finally, we have with $\tilde{h} = h(\tilde{\psi}_x^2 + \tilde{\psi}_y^2]$

$$\left.\begin{aligned} \frac{d^2}{d\tau^2} \tilde{h} = [\tilde{h}'' 4\tilde{\psi}_x^2 + 2\tilde{h}'] \psi_x^2 + 8\tilde{h}'' \cdot \tilde{\psi}_x \tilde{\psi}_y \psi_x (\psi_y - t_0) + \\ + [\tilde{h}'' 4\tilde{\psi}_y^2 + 2\tilde{h}'] (\psi_y - t_0)^2. \end{aligned}\right\} \tag{44.6}$$

If we insert Eq. (44.6) into Eq. (44.5) and use the properties of $h(z)$ as described in its construction in Sect. 43, we find that there exist positive constants k and K such that the integrand $J_1(x, y)$ of Eq. (43.19) satisfies inequality

$$k[\psi_x^2 + (\psi_y - t_0)^2] \leq J_1(x, y) \leq K[\psi_x^2 + (\psi_y - t_0)^2]. \tag{44.7}$$

We have now verified that the integral

$$D[\psi] = \iint_D [h(\psi_x^2 + \psi_y^2) - h(t_0^2) - 2t_0 h'(t_0^2)(\psi_y - t_0)] \, dx \, dy \tag{44.8}$$

converges for all functions of the class $\mathfrak{C}$ and has a non-negative lower bound. Using the classical results of the calculus of variations for double integrals which were obtained in connection with the theory of minimal surfaces (HAAR [*20*], TONELLI [*21*], HOPF [*22*], MORREY [*5*]), one can show that there exists a function $\psi(x, y)$ in the class $\mathfrak{C}$ which minimizes the integral (44.8). The minimum function has continuous first and second derivatives in D and satisfies, therefore, the Euler-Lagrange equation (43.6). It can even be shown that the minimum function has continuous first derivatives in the closed flow region $D+P$. We shall show below that the minimum function is uniquely determined by the variational problem.

The minimum function depends still on the two parameters t_0 and b_1; the latter number denotes the lowest value of z for which the function $f(z)$ is modified into $h(z)$. By Eq. (43.18), the parameter t_0 is an elementary function of the Mach number at infinity M_0. It can be shown that ψ and its first derivatives depend continuously on t_0 (and therefore on the Mach number M_0) in the closed region $D+P$.

Let us select now a fixed value b_1; we write

$$z(t_0, b_1) = \max_D [\psi_x^2 + \psi_y^2], \tag{44.9}$$

where ψ is the minimum function belonging to t_0 and b_1. In view of the preceding we know that $z(t_0, b_1)$ depends continuously on t_0. Since $t_0 = 0$ obviously admits the minimum function $\psi \equiv 0$, we know in view of the uniqueness of the minimum function that $z(0, b_1) = 0$. On the other hand, we can choose so high a velocity at infinity that M_0 is larger than the Mach number M_1 which corresponds to $z = b_1$ and for the value t_0 of this flow clearly $z(t_0, b_1) > b_1$. Hence, in view of the continuity of $z(t_0, b_1)$, we can assert that there exists an interval $0 \leq t_0 < t_1(b_1)$ such that

$$z(t_0, b_1) < b_1 \quad \text{for} \quad 0 \leq t_0 < b_1, \quad \text{while} \quad z(t_1, b_1) = b_1. \tag{44.10}$$

As long as $t_0 < t_1(b_1)$ the modification of the function $f(z)$ does not affect the flow obtained and the corresponding minimum function ψ of the integral (44.8) is the correct stream function of the hydrodynamical problem considered. ψ does not depend on the modification $h(z)$.

If $h^*(z)$ is a modification of $f(z)$ beginning at the point b_1^*, it may also be considered as a modification $h(z)$ from $b_1 < b_1^*$ on. The interval of admissible t_0 belonging to b_1^*, therefore, cannot be smaller than that belonging to b_1. Hence $t_1(b_1^*) \geq t_1(b_1)$; the function $t_1(b)$ increases monotonically. There exists, therefore, the limit

$$\lim_{b_1 \to b_0} t_1(b_1) = t_c, \tag{44.11}$$

which corresponds to the Mach number C at infinity. Since

$$\lim_{b_1 \to b_0} z(t_1, b_1) = b_0, \tag{44.11'}$$

we see that the maximum Mach numbers in the flows must converge to the Mach number 1 which corresponds to b_0 [see Eqs. (43.15) and (43.16)]. This proves all the statements made in the fundamental existence theorem in Sect. 43.

45. Uniqueness of the minimum function. One fundamental link in our above existence proof is the uniqueness of the minimum function for given parameters t_0 and b_1. This uniqueness proof provides also a uniqueness theorem for a subsonic flow with prescribed free stream velocity u_0 and without circulation around a given profile P, provided that the additional condition (44.1) is fulfilled. It is somewhat weaker than the result of Sect. 40; but since it illustrates the role of the integral (44.8) in the theory we shall sketch the reasoning of this variational uniqueness proof.

We suppose that $\psi(x, y)$ is one stationary function for the variational problem (43.19). Let $\zeta(x, y)$ be continuously differentiable in $D+P$, vanish on P and have a finite Dirichlet integral

$$\iint_D |\nabla\zeta|^2\, dx\, dy < \infty\,. \tag{45.1}$$

We introduce now the function $\psi(x, y) + \zeta(x, y)$ as argument function into the integral (44.8). In order to estimate the change of $D[\psi]$, we define in analogy to our procedure in Sect. 44 the linear interpolation function

$$\psi^*(x, y) = \psi(x, y) + \tau\zeta(x, y) \tag{45.2}$$

and can then write

$$\left.\begin{aligned} h\big((\psi_x+\zeta_x)^2+(\psi_y+\zeta_y)^2\big) &= h(\psi_x^2+\psi_y^2) + 2h'(\psi_x^2+\psi_y^2)(\psi_x\zeta_x+\psi_y\zeta_y) + \\ &\quad + \int_0^1 (1-\tau)\, \frac{d^2 h(\psi_x^{*2}+\psi_y^{*2})}{d\tau^2}\, d\tau. \end{aligned}\right\} \tag{45.3}$$

In view of Eq. (45.3) we can now calculate

$$\left.\begin{aligned} D[\psi+\zeta] &= D[\psi] + 2\iint_D [h'(\psi_x^2+\psi_y^2)(\psi_x\zeta_x+\psi_y\zeta_y) - t_0 h'(t_0^2)\zeta_y]\, dx\, dy + \\ &\quad + \iint_D \left\{\int_0^1 (1-\tau)\, \frac{d^2 h^*}{d\tau^2}\, d\tau\right\} dx\, dy = D[\psi] + 2\,\delta D + Q. \end{aligned}\right\} \tag{45.4}$$

The last integral in Eq. (45.4) can be estimated easily by means of a formula analogous to Eq. (44.6); we have only to replace ψ_x and $\psi_y - t_0$ there by ζ_x and ζ_y. It is then seen that with the same constants k and K as in Eq. (44.7)

$$k\iint_D [\zeta_x^2+\zeta_y^2]\, dx\, dy \leq Q \leq K\iint_D [\zeta_x^2+\zeta_y^2]\, dx\, dy. \tag{45.5}$$

The integral δD vanishes obviously because of the stationariness of $\psi(x, y)$. Thus we have proved:

$$D[\psi+\zeta] \geq D[\psi] + k \iint_D [\zeta_x^2 + \zeta_y^2]\, dx\, dy. \tag{45.6}$$

From this inequality follows immediately that ψ yields an absolute minimum for the integral $D[\psi]$ within the class $\mathfrak{C}$ and, in particular, that $\psi(x, y)$ is uniquely determined by its minimum property.

46. Variational problem for subsonic flows with circulation. In the case of circulation, it is easily seen from Eqs. (35.1) and (39.24) that a stream function $\psi(x, y)$ must have near infinity the same asymptotic behavior as

$$S_0(x, y) = \varrho_0 u_0 y - \frac{\Gamma \varrho_0}{2\pi} \log r. \tag{46.1}$$

In order to obtain a convergent integral in the variational problem, we choose an arbitrary function $S(x, y)$ with the asymptotic form (46.1) and consider the difference expression

$$h(\psi_x^2 + \psi_y^2) - h(S_x^2 + S_y^2) - A(x, y)(\psi_x - S_x) - B(x, y)(\psi_y - S_y), \tag{46.2}$$

which leads to the Euler-Lagrange equation

$$\frac{\partial}{\partial x}\left(\frac{\partial h}{\partial \psi_x}\right) + \frac{\partial}{\partial y}\left(\frac{\partial h}{\partial \psi_y}\right) - \left(\frac{\partial A}{\partial x} + \frac{\partial B}{\partial y}\right) = 0. \tag{46.3}$$

If we can choose A and B in such a way that

$$\frac{\partial A}{\partial x} + \frac{\partial B}{\partial y} = 0 \tag{46.4}$$

and that the integral of Eq. (46.2) extended over the infinite region D converges, we have a suitable integrand for the variational problem.

The most natural choice for $A(x, y)$ and $B(x, y)$ would be

$$A_0(x, y) = \frac{\partial}{\partial S_x} h(S_x^2 + S_y^2), \qquad B_0(x, y) = \frac{\partial}{\partial S_y} h(S_x^2 + S_y^2) \tag{46.5}$$

since in this case the difference (46.2) would be at least quadratic in the difference terms $\psi_x - S_x$, $\psi_y - S_y$. However, these terms will not satisfy Eq. (46.4). We observe that $S_x \to 0$, $S_y \to \varrho_0 u_0 = t_0$ as we approach infinity. It is, therefore, advisable to replace Eq. (46.5) by the linear approximation near infinity

$$A(x, y) = 2h'(t_0^2)\, S_x, \quad B(x, y) = 2t_0 h'(t_0^2) + [4h''(t_0^2)\, t_0^2 + 2h'(t_0^2)] \cdot (S_y - t_0). \tag{46.6}$$

If we wish that Eq. (46.4) be satisfied, we have to require

$$h'(t_0^2)\, S_{xx} + [2h''(t_0^2)\, t_0^2 + h'(t_0^2)]\, S_{yy} = 0. \tag{46.7}$$

In order to simplify Eq. (46.7) we replace $h(z)$ by $f(z)$ which is admissible in the only case of interest, that when $t_0 < b_1$. From Eqs. (43.5), (43.13) and (43.14) we obtain easily

$$1 + 2z \frac{f''(z)}{f'(z)} = \left(1 - \frac{q^2}{c^2}\right)^{-1}. \tag{46.8}$$

Thus the differential equation (46.7) for the singularity function $S(x, y)$ becomes

$$\beta^2 S_{xx} + S_{yy} = 0, \quad \beta = \sqrt{1 - M_0^2}. \tag{46.9}$$

We can choose the following integral of Eq. (46.9), which has the same asymptotic behavior as S_0:

$$S(x, y) = \varrho_0 u_0 y - \frac{\Gamma \varrho_0}{2\pi} \log \sqrt{x^2 + \beta^2 y^2}. \tag{46.10}$$

From Eq. (46.6) we then obtain expressions $A(x, y)$ and $B(x, y)$ which satisfy Eq. (46.4) and lead to a convergent integral for Eq. (46.2) if we restrict ourselves to a suitable class of functions ψ. This class is defined as follows:

$\mathfrak{C}_\Gamma$ is the class of all functions $\psi(x, y)$ which are continuously differentiable in D, vanish on the boundary P of D and at infinity behave like $S(x, y)$ in the following sense:

$$\iint_D [(\psi_x - S_x)^2 + (\psi_y - S_y)^2]\, dx\, dy < \infty. \tag{46.11}$$

We can now formulate the following minimum problem: To determine the functions $\psi(x, y)$ of class $\mathfrak{C}_\Gamma$ that minimize the integral

$$D_\Gamma[\psi] = \iint_D [h(\psi_x^2 + \psi_y^2) - h(S_x^2 + S_y^2) - A(\psi_x - S_x) - B(\psi_y - S_y)]\, dx\, dy. \tag{46.12}$$

It can be shown again that there exists a unique solution $\psi(x, y)$ which is continuously differentiable in $D + P$ and satisfies the Euler-Lagrange equation

$$\frac{\partial}{\partial x}\left(\frac{\partial h}{\partial \psi_x}\right) + \frac{\partial}{\partial y}\left(\frac{\partial h}{\partial \psi_y}\right) = 0. \tag{46.13}$$

Using the same arguments as in Sect. 44 we can then prove the following existence theorem:

Let c_0 be the local velocity of sound at infinity, u_0 the speed in the x-direction, $v_0 = 0$; consider all points $\xi = M_0 = \frac{u_0}{c_0}$, $\eta = \frac{\Gamma}{c_0}$. There exists a region in the (ξ, η)-plane including the origin such that for each point in this region there exists a unique subsonic flow around P with these values of u_0, Γ. If M_1 is the maximum local Mach number throughout the flow, $M_1(\xi, \eta)$ will approach the value 1 as the point ξ, η tends to the boundary of the region.

Finally, it should be remarked that we make little use of the adiabatic pressure-density relation and that an analogous theory could be established for other functional forms $p = p(\varrho)$, provided only that $p'(\varrho) = c^2 > 0$.

47. The Janzen-Rayleigh iteration procedure. The most natural approach to the boundary value problem for the non-linear differential equation (43.2) of the velocity potential is the method of successive approximations. For this purpose we put the Eq. (43.2) into the form

$$\Delta\varphi = \varphi_{xx} + \varphi_{yy} = \frac{1}{c^2}[\varphi_x^2 \varphi_{xx} + 2\varphi_x \varphi_y \varphi_{xy} + \varphi_y^2 \varphi_{yy}], \quad \Delta = \frac{\partial^2}{\partial x^2} + \frac{\partial^2}{\partial y^2}, \tag{47.1}$$

where c^2 is a simple function of $q^2 = \varphi_x^2 + \varphi_y^2$. Indeed, in view of Eq. (43.2′)

$$c^2 = c_0^2 + \frac{\gamma - 1}{2}(u_0^2 - q^2). \tag{47.2}$$

We are to find a solution $\varphi(x, y)$ of (47.1) behaving like $u_0 x$ near infinity and having a vanishing normal derivative on the boundary P of the given profile.

In the case of an incompressible fluid, where $c = \infty$, the right hand side of Eq. (47.1) vanishes, and we face a boundary value problem of the Neumann type for harmonic functions. This is by no means an elementary problem for an arbitrarily given boundary P; but since it is a linear problem, we can consider it as solvable. It is our aim to reduce the solution of (47.1) to a succession of such linear boundary value problems.

If the maximum Mach number q/c in the flow is small, the coefficients of the second derivatives of φ will be much smaller on the right hand side of (47.1) than on the left. Therefore we can set up the following recursive procedure:

(a) We determine the function $\varphi_0(x, y)$ which satisfies $\Delta\varphi=0$, has the required behavior at infinity and vanishing normal derivative on P.

(b) We denote the right hand side of Eq. (47.1) by $R[\varphi]$ and consider the inhomogeneous differential equations

$$\Delta\varphi_n(x, y) = R[\varphi_{n-1}], \qquad n = 1, 2, \ldots, \tag{47.3}$$

with the side conditions that each $\varphi_n(x, y)$ behaves like $u_0 x$ at infinity and has a vanishing normal derivative at P. The Eq. (47.3) with these side conditions determines $\varphi_n(x, y)$ in a unique way if $\varphi_{n-1}(x, y)$ is given. We obtain thus an iterative procedure which converges in many cases to the actual solution of the desired boundary value problem. This procedure is due to JANZEN [*23*] and Lord RAYLEIGH [*24*]; it has been applied by many investigators in order to improve a known solution for an incompressible fluid flow around a given profile into a solution for a compressible fluid flow with a small maximum Mach number [*25*] to [*29*]. Most authors were satisfied with a single step in the iteration procedure, that is, they stopped with $\varphi_1(x, y)$. G. I. TAYLOR [*30*], [*31*] devised an analogue computor for carrying out the iteration step (47.3) by means of electric currents in a conducting layer of variable resistance. The investigations of FRANKL and KELDYSH [*16*] show that to a given smooth profile P there exists a constant $c(P)$ such that the Janzen-Rayleigh iterative procedure does indeed converge for $u_0 < c(P)$. In this sense, the method of successive approximations yields a constructive existence proof for a compressible fluid flow without circulation around a given profile P, provided only that the free stream velocity u_0 be small enough.

The recursive procedure (47.3) can be reduced to integrations if one knows the harmonic NEUMANN's function $N(z, \zeta)$ of the profile P. It is defined as the velocity potential at the point $z = x + iy$ of an incompressible fluid flow around P, due to a source of strength 1 at the point $\zeta = \xi + i\eta$ and a sink of strength 1 at infinity. Thus, $N(z, \zeta)$ is characterized by the three requirements:

1. $N(z, \zeta)$ is harmonic in $z = x + iy$, except at $z = \zeta$ and $z = \infty$.
2. $N(z, \zeta) + \frac{1}{2\pi}\log|z-\zeta|$ is harmonic at ζ and at infinity.
3. $N(z, \zeta)$ has a vanishing normal derivative on P.

Since these conditions determine $N(z, \zeta)$ only up to an additive constant we may also require that $N(z, \zeta) + \frac{1}{2\pi}\log|z-\zeta|$ vanishes for $z = \infty$. This gives a unique NEUMANN's function.

Now let $u(z)$ be twice continuously differentiable in the closed flow region $D + P$, vanish at infinity and have derivatives which vanish to the second order at infinity. Then GREEN's formula applied to $u(z)$ and $N(z, \zeta)$ yields immediately

$$u(\zeta) = -\int_P \frac{\partial u(z)}{\partial \nu} N(z, \zeta)\, ds - \iint_D \Delta u(z)\, N(z, \zeta)\, dx\, dy \tag{47.4}$$

where ν denotes the normal at z on P showing into the flow region D. This general formula may now be applied to

$$u(z) = \varphi_n(x, y) - u_0 x. \tag{47.5}$$

Since

$$\Delta u = R[\varphi_{n-1}], \qquad \frac{\partial u}{\partial \nu} = -u_0 \cos(\nu, x) \tag{47.5'}$$

are known, Eq. (47.4) leads to the determination of $\varphi_n(x, y)$ in terms of $\varphi_{n-1}(x, y)$. This formalism shows clearly that the Janzen-Rayleigh method reduces the solution of the non-linear boundary value problem to the determination of the harmonic NEUMANN's function of P and to an iterated integration procedure.

The Janzen-Rayleigh procedure becomes relatively simple in the case of a circular boundary P. Using polar coordinates, we can carry out the successive iterations by elementary processes and to every degree desired. The method was indeed first applied to this particular case and also extended to problems of flow around a sphere in space. In this case, too, the use of polar coordinates and the radial symmetry simplify matters considerably.

The Janzen-Rayleigh method is a particular case of a series development of the solution in terms of a smallness parameter (in this case the Mach number M) discussed generally in Sect. 10. Because of its particularly simple structure it can be made rigorous for small values of M and leads, in principle, to existence proofs for flows around profiles as will be seen in the next sections.

An analogous iteration procedure can be applied to the stream function instead of the velocity potential [*40*]. In this case, the recursion terms are more complicated but the boundary conditions on P become much easier to handle.

48. The boundary value problem for the velocity potential and a fixed-point problem in function space. The preceding method of successive approximation, though rather involved, can in principle be carried out to any desired degree of precision, provided only that the procedure does converge. The numerical evaluation is possible since the non-linear equation (47.1) has been reduced to the classical Laplace equation which is, by far, the most studied and best known partial differential equation of elliptic type. We may, however, connect Eq. (47.1) with other linear elliptic differential equations; we shall then lose much in constructiveness of the procedure but obtain a better theoretical insight into the existence problem for the solution.

Let $\mathfrak{C}$ be the class of all functions $\chi(x, y)$ which are continuously differentiable in the closed flow region $D+P$, which behave at infinity asymptotically like $u_0 x$ (that is, satisfy $\lim(\chi - u_0 x) = 0$ at infinity), and which have a vanishing normal derivative on P. With each function χ in $\mathfrak{C}$ we connect the coefficient functions

$$A[\chi] = 1 - \frac{1}{c^2}\chi_x^2, \qquad B[\chi] = -\frac{1}{c^2}\chi_x\chi_y, \qquad C[\chi] = 1 - \frac{1}{c^2}\chi_y^2 \tag{48.1}$$

where

$$c(\chi)^2 = c_0^2 + \frac{\gamma-1}{2}(u_0^2 - \chi_x^2 - \chi_y^2). \tag{48.1'}$$

There is exactly one harmonic function in $\mathfrak{C}$; we denote it by $\Phi_0(x, y)$. We solve then the sequence of linear partial differential equations

$$A[\Phi_{n-1}]\,\Phi_{xx} + 2B[\Phi_{n-1}]\,\Phi_{xy} + C[\Phi_{n-1}]\,\Phi_{yy} = 0 \tag{48.2}$$

and denote the solution of Eq. (48.2) in $\mathfrak{C}$ by $\Phi_n(x, y)$. This leads to a sequence of approximation functions $\Phi_n(x, y)$ which is analogous to that in Sect. 47 given by the Janzen-Rayleigh method.

Consider the partial differential equation

$$A[\chi]\,\Phi_{xx} + 2B[\chi]\,\Phi_{xy} + C[\chi]\,\Phi_{yy} = 0. \tag{48.3}$$

In order to find a solution of Eq. (48.3) in $\mathfrak{C}$ we have to solve a boundary value problem of the Neumann type for the linear equation (48.3) and we are sure of the existence of such a solution if Eq. (48.3) is elliptic in the entire domain D. This means that the inequality

$$A\,C - B^2 = 1 - \frac{q^2}{c^2} > 0, \qquad q^2 = \chi_x^2 + \chi_y^2 \tag{48.4}$$

is fulfilled in D, which leads by Eqs. (43.2′) and (48.1′) to the inequality

$$q^2 < \frac{\gamma - 1}{\gamma + 1}\, q_{\max}^2 = q_s^2. \tag{48.5}$$

The iteration (48.2) might come to a stop when a function $\Phi_n(x, y)$ is reached for which $(\nabla\Phi_n)^2 > q_s^2$. This possibility would decrease considerably the chances of an existence proof. We can sidestep this difficulty by the same artifice which was used in Sect. 43 in connection with the Shiffman existence proof. We replace the speed-density relation $\varrho(q^2)$ for a real gas by a new relation $\varrho = \varrho^*(q^2)$ which coincides with $\varrho(q^2)$ for $q \le q_1 < q_s$ where q_1 can be as near to q_s as we wish. Beyond q_1 we modify the function $\varrho(q)$ in such a way that

$$1 - \frac{q^2}{c^2} \ge \alpha(q_1) > 0. \tag{48.6}$$

The differential equation (35.6) for the velocity potential always has the form (43.2) independently of the particular equation of state. In fact, in view of Eqs. (35.3) and (35.9), we have generally

$$F''(q^2) = -\frac{\varrho}{4c^2}, \tag{48.7}$$

and from this it is easy to compute

$$F_{uu} = \varrho\left(1 - \frac{u^2}{c^2}\right), \qquad F_{uv} = -\varrho\,\frac{uv}{c^2}, \qquad F_{vv} = \varrho\left(1 - \frac{v^2}{c^2}\right). \tag{48.8}$$

We now connect every function $\chi(x, y)$ of the class $\mathfrak{C}$ with the coefficient functions (48.1) but now define $c(\chi)$ by means of the new speed-density relation and in view of Eq. (35.9′) by

$$c(\chi)^2 = -\frac{\varrho(q^2)}{2\varrho'(q^2)}, \qquad q^2 = (\nabla\chi)^2. \tag{48.9}$$

We are now sure that the differential equation (48.3) will be elliptic in the entire exterior D of the profile P. In view of Eq. (48.6), we can even assert the inequality

$$0 < \frac{(A + C)^2}{4(A\,C - B^2)} = \frac{\left(2 - \frac{q^2}{c^2}\right)^2}{4\left(1 - \frac{q^2}{c^2}\right)} \le \alpha(q_1)^{-1}. \tag{48.10}$$

We may express the inequality (48.10) in the language of the general theory of elliptic differential equations by stating that Eq. (48.3) is a uniformly elliptic equation in the domain D.

If we choose an arbitrary function χ in $\mathfrak{C}$, the differential equation (48.3) will determine a unique function in $\mathfrak{C}$ which satisfies it. The existence and uniqueness of such a function $\Phi(x, y)$ is a consequence of the classical boundary value theorems for linear partial differential equations of elliptic type. We may consider this relation between $\chi(x, y)$ and $\Phi(x, y)$ as a transformation of the class $\mathfrak{C}$ into itself. The existence problem for a compressible fluid flow past the given profile P is equivalent to the question of whether the transformation of $\mathfrak{C}$ into itself $\chi \rightarrow \Phi$ has elements which remain fixed; in fact, $\Phi \rightarrow \Phi$ simply means that Φ has the right asymptotic behavior at infinity, has vanishing normal derivatives on P and, finally, satisfies the equation

$$A[\Phi]\,\Phi_{xx} + 2B[\Phi]\,\Phi_{xy} + C[\Phi]\,\Phi_{yy} = 0 \tag{48.3'}$$

which is precisely the partial differential equation (35.6), (43.2) for the velocity potential.

We have thus formulated the problem of a compressible fluid flow without circulation around a given profile P as the problem of a fixed point in the mapping of a certain function space into itself. The theory of such fixed points is an important branch of modern topology [*32*] to [*34*]. Bers applied this theory [*7*], [*8*] in the above context in order to give an alternative proof for SHIFFMAN's existence theorem which was formulated in Sect. 43. We shall sketch his method in the next section.

49. BERS' existence proof for subsonic flow past a profile. Since BERS' method of reasoning is typical for a large class of modern existence proofs, we will present here the chain of arguments without giving proofs for the constituent statements. The ennumeration of the intermediate steps will bring out clearly the leading ideas and analytical tools necessary in the topological approach [*7*].

a) We were dealing with functions $\chi(x, y)$ of the class $\mathfrak{C}$ and their gradient fields which can be expressed in the complex form

$$w(z) = \chi_x - i\chi_y, \qquad z = x + iy. \tag{49.1}$$

These complex gradients are particular cases of general complex-valued bounded and continuous functions $W(z)$ in $D+P$. Let $\mathfrak{B}$ be the class of all such functions; clearly, $\mathfrak{B}$ is a linear space. We introduce into $\mathfrak{B}$ a metric by associating with every $W(z)$ in $\mathfrak{B}$ the norm

$$\|W\| = \text{least upper bound of } |W(z)| \text{ in } D+P. \tag{49.2}$$

If $W_1(z)$ and $W_2(z)$ are two elements of $\mathfrak{B}$, we may call $\|W_1 - W_2\|$ the distance between these two elements. If we have a sequence of elements W_n in $\mathfrak{B}$, we may now say that W_n converges to W in $\mathfrak{B}$ if $\|W_n - W\|$ converges to zero as $n \rightarrow \infty$. It is easily seen that the closure of $\mathfrak{B}$ under this topology satisfies all axioms of a Banach space.

b) For each $W(z)$ in $\mathfrak{B}$, we can define a partial differential equation

$$\tilde{A}[W]\,\chi_{xx} + 2\tilde{B}[W]\,\chi_{xy} + \tilde{C}[W]\,\chi_{yy} = 0 \tag{49.3}$$

with

$$\tilde{A}[W] = 1 - \frac{U^2}{c^2}, \qquad \tilde{B}[W] = -\frac{UV}{c^2}, \qquad \tilde{C}(W) = 1 - \frac{V^2}{c^2} \tag{49.3'}$$

where $W = U - iV$ and c^2 is given by (48.9) with $q^2 = |W|^2$. Obviously, this equation is a generalization of Eq. (48.3). Indeed, comparing Eq. (48.1) with Eq. (49.3') we have

$$A[\chi] = \tilde{A}[w], \qquad B[\chi] = \tilde{B}[w], \qquad C[\chi] = \tilde{C}[w]. \tag{49.3''}$$

By our choice of the equation of state, we can assert that

$$0 < \frac{(\tilde{A}+\tilde{C})^2}{4(\tilde{A}\tilde{C}-\tilde{B}^2)} \leqq \alpha(q_1)^{-1}; \tag{49.4}$$

that is, all partial differential equations (49.3) are uniformly elliptic in $D+P$.

c) We have assumed that P is a closed smooth curve in the $z=x+iy$ plane. Let $\zeta=\zeta(z)$ be that univalent function in the exterior D of P which maps D conformally onto the exterior $|\zeta|>1$ of the unit circle in the ζ-plane in such a way that the points at infinity correspond and that $\zeta'(\infty)>0$. This mapping function is continuous in the closed region $D+P$. It plays a central role in the theory of an incompressible fluid flow around P and will even be of importance in the present problem of a compressible fluid. This is clearly shown in the following theorem:

If $W(z)$ lies in the Banach space $\mathfrak{B}$ and if $\chi(x, y)$ is a solution of the partial differential equation (49.3) *of the class $\mathfrak{C}$, defined in Sect. 48, then its complex gradient. $w(x)$ will satisfy the inequality*

$$|w(z_1)-w(z_2)| \leqq K u_0 |\zeta(z_1)^{-1}-\zeta(z_2)^{-1}|^{\lambda}, \quad 0<\lambda<1 \tag{49.5}$$

for any two points z_1 and z_2 in $D+P$. The constants K and λ depend only on the profile P and the constant $\alpha(q_1)$ defined in Eq. (48.6).

In other words: If a function of the class $\mathfrak{C}$ satisfies a partial differential equation (49.3), its complex gradient is Hölder-continuous in the closed region $|\zeta| \geqq 1$ and the Hölder constants are independent of the coefficient function $W(z)$.

d) To each element W of the Banach space $\mathfrak{B}$ belongs a solution $\chi(x, y)$ of Eq. (49.3) belonging to the class $\mathfrak{C}$. It is uniquely determined, and so is its complex gradient $w(z)$. Thus, through the differential equation (49.3) with the side conditions characteristic for the class $\mathfrak{C}$, we can establish a well-defined transformation

$$W \to w = TW \tag{49.6}$$

of all elements of $\mathfrak{B}$ into a distinguished subset of the same space.

We can prove the following theorem:

If a sequence of elements W_n of $\mathfrak{B}$ converges to a limit W, then the corresponding complex gradients w_n will converge likewise, and we have

$$\lim_{n\to\infty} w_n = \lim_{n\to\infty} TW_n = w = TW = T \lim_{n\to\infty} W_n. \tag{49.7}$$

In other words: The transformation (49.6) is continuous in $\mathfrak{B}$.

Now let $\mathfrak{B}_0$ be the subset of $\mathfrak{B}$ consisting of all elements W for which

$$\lim_{z\to\infty} W(z) = u_0 \tag{49.8}$$

holds and

$$|W(z_1)-W(z_2)| \leqq K u_0 |\zeta(z_1)^{-1}-\zeta(z_2)^{-1}|^{\lambda}. \tag{49.9}$$

The set $\mathfrak{B}_0$ is a convex subset of $\mathfrak{B}$; in fact, if $W_0(z)$ and $W_1(z)$ belong to $\mathfrak{B}_0$, then all linear combinations

$$W_t(z) = t W_1(z) + (1-t) W_0(z), \quad 0 \leqq t \leqq 1 \tag{49.10}$$

will likewise lie in $\mathfrak{B}_0$. It is easily seen that $\mathfrak{B}_0$ is bounded, closed and compact.

The complex-valued functions $W(z)$ of the subset $\mathfrak{B}_0$ of the Banach space have also image elements TW, and they lie likewise in $\mathfrak{B}_0$. Thus, the mapping $w=TW$ of the entire Banach space carries the subset $\mathfrak{B}_0$ into itself.

e) There exists a classical theorem in the topology of finite dimensional spaces: If a bounded and closed convex set is mapped continuously into itself, there exists at least one fixed point, that is, a point which goes into itself under the mapping.

This is the celebrated Brouwer fixed point theorem for which numerous and simple proofs are available. This theorem was extended to more general spaces and in particular to function spaces by BIRKHOFF and KELLOG [34] and, in greatest generality, by SCHAUDER [32], [33] who utilized it systematically in existence problems of analysis. The fixed point theorem needed for our purposes states:

If a closed and bounded subset of a Banach space is mapped continuously into a compact and convex subset of itself, there exists at least one fixed element of the mapping.

In view of our preceding considerations this theorem can be applied immediately to the set $\mathfrak{B}_0$, which may also be conceived as a compact subset of itself. Hence there must exist an element $W(z)$ such that it corresponds to itself in the mapping $w = TW$. Hence $W(z)$ is the complex gradient of a function $\Phi(x, y)$ lying in the class $\mathfrak{C}$ and satisfying the partial differential equation

$$\left(1 - \frac{\Phi_x^2}{c^2}\right)\Phi_{xx} - 2\frac{\Phi_x \Phi_y}{c^2}\Phi_{xy} + \left(1 - \frac{\Phi_y^2}{c^2}\right)\Phi_{yy} = 0. \tag{49.11}$$

This proves the existence of a solution of the boundary value problem for the non-linear differential equation (49.11) with respect to the given profile P.

In the sketch of the above existence proof we have restricted ourselves for sake of simplicity to the problem of the flow without circulation around a given smooth profile P. It is possible to extend this proof without much modification to the case that P is smooth and the value Γ of the circulation is prescribed or that P has a trailing edge T and the Kutta-Joukowski condition with respect to T is required.

The existence proofs described are given at first under the assumption of the modified equation of state. However, if the maximum speed of the flow obtained remains below that speed for which the modification begins, one obtains an existence statement for the real fluid. All theorems obtained contain, therefore, a maximum admissible free stream velocity, in the same manner as SHIFFMAN'S theorem of Sect. 43.

The topological methods in the existence theory of compressible fluid flows are flexible enough to lead to existence proofs for free boundaries and wakes behind obstacles [35].

50. Conservation laws and estimates for the velocity field. If we consider a flow without circulation around a profile P which has at infinity the velocity u_0 in direction of the positive x-axis, there exists a maximum speed $u_{\max}$ in the flow which depends on u_0. We call the largest free stream Mach number at infinity which leads to a purely subsonic flow around P the critical free stream Mach number of the profile. SHIFFMAN'S existence theorem shows clearly the significance of this quantity, and it is desirable to give estimates for it in terms of simple geometric measures for the profile P.

In the case of incompressible fluid flow the numerous distortion theorems of conformal mapping can be applied in order to obtain various inequalities for the velocity potential and the stream function. LOEWNER [37], [38] has developed a useful formalism for deriving analogous estimates in the theory of a compressible fluid flow. His procedure exhibits, moreover, the importance of general conservation laws in the treatment of the partial differential equations of the flow and is, therefore, of interest in itself.

The differential equations of a steady irrotational flow may be written in the form of a system

$$u_y - v_x = 0, \qquad (\varrho u)_x + (\varrho v)_y = 0. \tag{50.1}$$

Both equations have the following typical form

$$\frac{\partial}{\partial x}\xi(u, v) + \frac{\partial}{\partial y}\eta(u, v) = 0 \tag{50.1'}$$

where ξ and η are functions of the velocity components. Various important conservation laws of fluid dynamics appear in this general form; the second equation (50.1) expresses the conservation of matter while the law of conservation of momentum leads to the equations

$$\frac{\partial}{\partial x}(\varrho u^2 + p) + \frac{\partial}{\partial y}(\varrho u v) = 0, \qquad \frac{\partial}{\partial x}(\varrho u v) + \frac{\partial}{\partial y}(\varrho v^2 + p) = 0 \tag{50.2}$$

which are again of the general type (50.1'). Mathematically, the formula (50.1') is a divergence law. It can, therefore, be integrated over an arbitrary region and can be reduced by integration by parts to an identity involving boundary terms only. It leads to laws of motion in integral instead of differential form.

It is of interest to determine the most general form for ξ and η which satisfy the conservation law (50.1') as a consequence of the differential system (50.1). Carrying out explicitly the differentiations in (50.1) and (50.1'), we find

$$\left.\begin{aligned} (\varrho + 2u^2\varrho')\,u_x + 2\varrho' u v\, v_x + 2\varrho' u v\, u_y + (\varrho + 2v^2\varrho')\,v_y &= 0\\ - v_x + \qquad u_y \qquad &= 0\\ \xi_u u_x \qquad + \xi_v v_x + \eta_u u_y \quad + \eta_v v_y &= 0 \end{aligned}\right\} \tag{50.3}$$

with $\varrho = \varrho(q^2)$ and $\varrho' = d\varrho/dq^2$. We require that the first two equations imply the third independently of the particular solution u, v. The coefficient matrix of this system must have the rank 2 at most. It is easily seen that the corresponding determinant relations reduce to the two differential equations

$$(\varrho + 2v^2\varrho')\,\xi_u = (\varrho + 2u^2\varrho')\,\eta_v, \quad (\varrho + 2v^2\varrho')(\xi_v + \eta_u) = 4\varrho' u v \eta_v. \tag{50.4}$$

We know already two particular solutions of Eqs. (50.4), namely

$$\xi = -v, \quad \eta = u \quad \text{and} \quad \xi = \varrho u, \quad \eta = \varrho v. \tag{50.4'}$$

We set up the general solution of Eqs. (50.4) in the form

$$\xi = \lambda(-v) + \mu(\varrho u), \qquad \eta = \lambda u + \mu(\varrho v). \tag{50.5}$$

If we introduce polar coordinates q, ϑ in the hodograph plane, we find for $\lambda(q, \vartheta)$ and $\mu(q, \vartheta)$ the system of differential equations

$$\mu_q = \frac{\varrho + 2\varrho' q^2}{\varrho^2 q}\lambda_\vartheta, \qquad \mu_\vartheta = -\frac{q}{\varrho}\lambda_q. \tag{50.6}$$

We observe that Eqs. (35.8) and (35.10) imply the identity

$$\varrho + 2q^2\varrho' = \frac{\partial}{\partial q}(\varrho q) = \varrho\left(1 - \frac{q^2}{c^2}\right) \tag{50.7}$$

The differential system (50.6) therefore, becomes identical with the system of differential equations (23.8) connecting the velocity potential and the stream function in the hodograph plane. We have to put

$$\lambda(q,\vartheta) = -\psi_0(q,\vartheta), \qquad \mu(q,\vartheta) = \varphi_0(q,\vartheta). \tag{50.5'}$$

Every particular solution φ_0, ψ_0 of the hodograph differential system leads thus not only to a special flow pattern in the physical plane, but yields also a universal conservation law (50.1′) with

$$\xi = \psi_0 v + \varphi_0 \varrho u, \qquad \eta = -\psi_0 u + \varphi_0 \varrho v \tag{50.5''}$$

for all possible flow patterns. The freedom obtained in expressing the basic differential equations of fluid motion in the plane through various conservation laws will now be utilized in order to obtain useful inequalities for all possible flow patterns.

By a somewhat lengthy but straightforward calculation one can derive from Eqs. (50.5) and (50.6) the identity

$$\begin{vmatrix} d(\varrho u) & d(\varrho v) \\ d\xi & d\eta \end{vmatrix} = \frac{\partial}{\partial q}(\lambda q)\left[(\varrho + 2q^2\varrho')\,dq^2 + \varrho q^2\,d\vartheta^2\right] \tag{50.8}$$

valid for every vector ξ, η of the form (50.5) and satisfying a conservation law (50.1′) and for every displacement $dq, d\vartheta$ in the hodograph plane. Observe that the second factor on the right hand side of Eq. (50.8) is positive in the entire subsonic region and that the sign of the determinant depends, therefore, upon $(\lambda q)_q$.

The importance of the identity (50.8) is that it provides a differential inequality for velocity fields (u, v) in a region where there is known a particular $\lambda(q, \vartheta)$ such that $(\lambda q)_q$ is positive. We shall soon construct such test functions; but we wish to show first how in many cases the differential inequality (50.8) can be transformed into a functional inequality.

Suppose, for example, that there exists a function $\lambda(q, \vartheta)$ which is a solution of the system (50.6), defined in an angular sector of the subsonic circle $|q| < q_s$ and which satisfies there the inequality $(\lambda q)_q \geqq 0$. We introduce the variables $t_1 = \varrho u$, $t_2 = \varrho v$, which form a coordinate system in the sector. If two arbitrary points t_1, t_2 and t_1^*, t_2^* are given in the sector the entire line

$$\tau_1 = t_1 + l(t_1^* - t_1), \qquad \tau_2 = t_2 + l(t_2^* - t_2), \qquad 0 \leqq l \leqq 1 \tag{50.9}$$

will lie in this convex region. Consider the vector (50.5) which belongs to λ and the determinant

$$D(l) = \begin{vmatrix} \tau_1 - t_1, & \tau_2 - t_2 \\ \xi(\tau_1,\tau_2) - \xi(t_1,t_2) & \eta(\tau_1,\tau_2) - \eta(t_1,t_2) \end{vmatrix} \tag{50.10}$$

which satisfies the identity

$$\frac{d}{dl} D(l) = \frac{1}{l} D(l) + l \begin{vmatrix} t_1^* - t_1 & t_2^* - t_2 \\ \dfrac{d}{dl}\xi & \dfrac{d}{dl}\eta \end{vmatrix}. \tag{50.11}$$

The last determinant is non-negative for all values of the parameter l since it is proportional to the determinant (50.8) when we choose in the latter the displacement $(dq, d\vartheta)$ which corresponds to the direction from (t_1, t_2) to (t_1^*, t_2^*). For $l = 0$ we have obviously $D(0) = 0$ and for small values of l we shall have $D(l) \geqq 0$ since the differential inequality (50.8) will decide the sign. Hence, we have everywhere

$$\frac{d}{dl} D(l) \geqq 0 \quad \text{for} \quad 0 \leqq l \leqq 1. \tag{50.12}$$

Integrating from 0 to 1, we arrive at the final inequality

$$\begin{vmatrix} (\varrho u)^* - \varrho u & (\varrho v)^* - \varrho v \\ \xi^* - \xi & \eta^* - \eta \end{vmatrix} \geqq 0 \tag{50.13}$$

where $\xi, \eta, \varrho u$ and ϱv are taken at the two points u, v and u^*, v^* of the sector, respectively. Thus, every $\lambda(q, \vartheta)$ of the type described gives rise to inequalities for the solutions u, v of the basic differential equations (50.1).

We introduce now the functions

$$\psi(x, y) = \int [-\varrho v\, dx + \varrho u\, dy], \qquad \chi(x, y) = \int [-\eta\, dx + \xi\, dy] \tag{50.14}$$

and observe that the conservation laws represent precisely the integrability conditions for the integrands; ψ is the stream function of the flow. We can then write Ineq. (50.13) in the form

$$\begin{vmatrix} \psi_x^* - \psi_x & \psi_y^* - \psi_y \\ \chi_x^* - \chi_x & \chi_y^* - \chi_y \end{vmatrix} \geqq 0. \tag{50.13'}$$

A mapping $f = f(x, y)$, $g = g(x, y)$ is called strongly elliptic if for every pair of points x, y and x^*, y^* the inequality

$$\begin{vmatrix} f_x^* - f_x & f_y^* - f_y \\ g_x^* - g_x & g_y^* - g_y \end{vmatrix} \geqq 0 \tag{50.13''}$$

is fulfilled. Strongly elliptic mappings have numerous significant properties and were extensively studied by LAVRENTIEFF [*39*]. The following applications to fluid dynamics are due to the fact that ψ, χ provide a strongly elliptic mapping.

Let us suppose now that we are dealing with a flow which satisfies the following conditions:

(a) It is uniform at infinity in the direction of the positive x-axis and free of circulation.

(b) Its image domain in the hodograph plane lies in the sector where the inequality (50.13) is valid.

(c) The velocity components have continuous boundary values along the profile P.

We let the point x^*, y^* move to infinity and integrate the inequality (50.13') with respect to x, y over the exterior D of the profile P. We introduce the auxiliary functions

$$\left.\begin{aligned} \tilde\psi &= \psi - \varrho_0 u_0 y, \\ \tilde\chi &= \chi + \eta_0 x - \xi_0 y, \quad \xi_0 = \xi(u_0, 0), \quad \eta_0 = \eta(u_0, 0), \end{aligned}\right\} \tag{50.15}$$

where u_0 is the free stream velocity and ϱ_0 the corresponding density. We can then write

$$\iint_D \begin{vmatrix} \psi_x & \psi_y - \varrho_0 u_0 \\ \chi_x - \chi_x^\infty & \chi_y - \chi_y^\infty \end{vmatrix} dx\, dy = \iint_D d\tilde\psi\, d\tilde\chi \geqq 0. \tag{50.15'}$$

It is easy to see that the integral converges in spite of the infinite domain of integration. We can integrate the second integral in Eq. (50.15') by parts and obtain an integral over the boundary P only:

$$-\int_P \tilde{\psi}\, d\tilde{\chi} \geqq 0. \tag{50.16}$$

We return finally to the original functions ψ and χ; we observe that the stream function ψ vanishes on P and that $\varrho_0 u_0$ is positive. Hence we obtain

$$\int_P y\, d(\chi + \eta_0 x - \xi_0 y) \geqq 0. \tag{50.17}$$

Since we have obviously

$$\int_P y\, dy = 0, \qquad -\int_P y\, dx = A_P = \text{area of profile } P, \tag{50.18}$$

Ineq. (50.17) leads to the final inequality

$$\int_P y\, d\chi + \chi_x^\infty A_P \geqq 0. \tag{50.19}$$

Integrating by parts over P, we can bring inequality (50.19) into the equivalent but more convenient form

$$\int_P \chi\, dy \leqq \chi_x^\infty A_P. \tag{50.19'}$$

Every function $\chi(x, y)$ derived from a vector ξ, η with conservation law and belonging to a $\lambda(q\,\vartheta)$ with non-negative $(\lambda q)_q$ leads thus to an inequality connecting values on the boundary P with values at infinity.

As a first and simplest application we consider $\xi = -v$, $\eta = u$, which corresponds in view of Eq. (50.5) to $\lambda = 1$. Hence $(\lambda q)_q = 1$ is positive in the entire subsonic circle and the inequality (50.19') can be asserted for every subsonic flow without circulation around a smooth profile. In this particular case, χ coincides with $-\varphi(x, y)$ and hence Ineq. (50.19') becomes:

$$\int_P \varphi\, dy \geqq u_0 A_P, \tag{50.20}$$

an interesting inequality for the free stream velocity in terms of the values of the velocity potential on the profile.

51. Estimates for fluid pressure and free stream Mach number. In order to apply the inequality (50.19') systematically, we have to obtain test functions $\chi(x, y)$ whose corresponding λ-factor satisfies the inequality

$$\frac{d}{dq}(\lambda q) \geqq 0 \tag{51.1}$$

in a sector of the subsonic circle, say for $|\vartheta| \leqq \frac{\pi}{2\beta}$, $\beta > \frac{1}{2}$. For this purpose, we solve the system (50.6) by separation of variables:

$$\lambda = A(q)\cos\beta\vartheta, \qquad \mu = B(q)\sin\beta\vartheta. \tag{51.2}$$

We obtain the system of equations between $A(q)$ and $B(q)$:

$$B'(q) = -\frac{(\varrho q)_q}{\varrho^2 q}\beta A(q), \qquad A'(q) = -\frac{\varrho}{q}\beta B(q), \qquad \varrho = \varrho(q^2). \tag{51.3}$$

The point $q = 0$ is a singular point for this system of ordinary differential equations. If we assume the normalization $\varrho(0) = 1$, it is easily seen that there exists a solution of Eq. (51.3) of the form:

$$A(q) = q^{\beta} A_1(q), \qquad B(q) = q^{\beta} B_1(q), \tag{51.4}$$

where $A_1(q)$ and $B_1(q)$ are regular power series in q. They do not vanish at $q=0$ and we may assume in view of the Eqs. (51.3) that

$$A_1(0) = -B_1(0) = 1. \tag{51.4'}$$

Since the factor $\cos\beta\vartheta$ is non-negative in the sector considered, we have only to show that

$$\frac{d}{dq}(q A(q)) = A(q) + q A'(q) = A(q) - \beta \varrho B(q) \geqq 0 \tag{51.5}$$

for subsonic values of q. We prove this statement by considering the functions

$$U(q) = A(q) - \beta \varrho B(q), \qquad V(q) = \beta A(q) - \varrho B(q) \tag{51.6}$$

which by virtue of Eqs. (51.3) and (50.7) satisfy the differential system

$$U'(q) = \frac{\beta}{\varrho q} (\varrho q)_q V(q), \qquad V'(q) = \frac{\beta}{q} U(q) + \frac{2 q \varrho'}{\varrho} V(q). \tag{51.7}$$

In view of Eq. (51.4) and our assumption (51.4') we can assert that U and V are zero for $q=0$ and become positive for small positive values of q. As long as $V(q)>0$, $U(q)$ must increase for $q<c$. Let q_1 be the least positive value for which $V(q)$ changes its sign; since $V(q_1)=0$, $U(q_1)>0$, we deduce from the second equation (51.7) that $V'(q_1)>0$. But $V(q)$ was supposed to change at q_1 from positive to negative values, that is, to decrease. We have obtained thus a contradiction: $V(q)$ is positive for all subsonic values of q and, consequently, $U(q)>0$ for $0\leqq q\leqq q_s$. This proves the inequality (51.5) and the set of functions (51.2) provides us with an admissible test function χ for the sector $q\leqq q_s$, $|\vartheta|\leqq\frac{\pi}{2\beta}$ in the inequality (50.19').

We observe that in the particular case $\beta=1$ we have $U\equiv V$, and we derive from the first equation (51.7)

$$U(q) = \varrho q. \tag{51.8}$$

The inequality (51.5) expresses in this case the known fact that ϱq increases with subsonic q. We can compute $A(q)$ from Eqs. (51.5), (51.6) and (51.8) by integration. Using Eq. (35.9'), we find

$$q A(q) = \int_0^q \varrho q\, dq = -\int_0^q c^2\, d\varrho = p(0) - p(q). \tag{51.9}$$

Thus, in the particular case $\beta=1$, $A(q)$ is closely related to the fluid pressure.

Since $A(q)$ occurs in various inequalities it is useful to state the following monotony properties:

(a) $A(q)$ is positive and increases monotonically with subsonic q.

(b) $A_1(q) = q^{-\beta} A(q)$ decreases monotonically with subsonic q.

In order to prove (a) we have only to consider the differential system (51.3) and to observe that by our assumption (51.4') $A(q)$ and $-B(q)$ are positive for small positive values of q. From Eq. (51.3) follows indeed that both functions must increase as long as $(\varrho q)_q>0$, that is, for all subsonic q-values.

To prove (b), we simply compute for Eq. (51.3)

$$A_1'(q) = -\beta q^{-\beta-1} C(q), \qquad C(q) = A(q) + \varrho B(q), \tag{51.10}$$

and we have to prove the inequality $C(q)>0$ for subsonic q. But from Eq. (51.3) again, it follows at once that

$$C'(q) = -\frac{\beta}{q}C(q) + \left(\frac{-\varrho'}{\varrho}\right)V(q) = -\frac{\beta}{q}C(q) + P(q). \tag{51.11}$$

It was already shown that $V(q)$ is positive for all values $q \leqq q_s$ and since $\varrho' < 0$ the term $P(q)$ in Eq. (51.11) is positive. We can write

$$C(q) = q^{-\beta}\int_0^q r^\beta P(r)\,dr \tag{51.12}$$

which shows the positive character of $C(q)$ and prove, therefore, the statement (b).

We consider now the flow around a profile P which is symmetric with respect to the x-axis. We suppose that the part of P above the x-axis is described by a single-valued non-negative function $y=f(x)$ over an interval $a \leqq x \leqq b$. We require $f(a)=f(b)=0$ and that the tangent angle against the x-axis be less or equal to $\frac{\pi}{2\beta}$, that is

$$|f'(x)| \leq \tan\frac{\pi}{2\beta}. \tag{51.13}$$

These assumptions guarantee that the flow pattern has an image in the hodograph plane which lies entirely in the angular space $|\vartheta| \leqq \frac{\pi}{2\beta}$. This is an easy consequence of the maximum principle for the stream function $\psi(x, y)$. In fact, let H be the upper half of the flow region whose boundary consists in part of the x-axis and in part of the upper half of P. Since $\psi(x, y)$ vanishes obviously on this boundary and is positive near infinity, we have $\psi > 0$ inside of H. Now let (a, b) with $b > 0$ be an arbitrary fixed vector forming with the x-axis an angle larger than $\frac{\pi}{2\beta}$. If we attach this vector to any boundary point of H, it will point into the region H and, hence, we have in view of Eq. (35.1)

$$\frac{\partial\psi}{\partial x}a + \frac{\partial\psi}{\partial y}b = \varrho(u\,b - v\,a) > 0. \tag{51.14}$$

Since the left hand side of Eq. (51.14) is defined in H and satisfies the minimum principle, we can extend the inequality from the boundary into the domain. Since (a, b) can be any vector showing into the sector $\pi\left(1-\frac{1}{2\beta}\right) > |\vartheta| > \frac{\pi}{2\beta}$, it follows that the velocity vector must lie in the angular space $|\vartheta| \leqq \frac{\pi}{2\beta}$ which proves the assertion.

We can now apply the inequalities (50.19) and (50.19′) to flows around profiles P with the preceding geometric specifications and with the particular test functions $A(q)$ discussed above. If we use the definition (50.5) of ξ, η, the expression (35.1) for u, v and the fact that $\psi=0$ on P, we can reduce (50.19) to the simple form

$$\int_P y(-\eta\,dx + \xi\,dy) = -\int_P \lambda\,y\,d\varphi \geqq -\chi_x^\infty A_P \tag{51.15}$$

where the integration is to be extended around P in the positive (counter-clockwise) sense. Because of the particular choice (51.2) of λ, we can put the inequality (51.15) into the form

$$\int_P y\,A(q)\cos\beta\vartheta\,d\varphi \leqq -A(u_0)\,u_0\cdot A_P, \tag{51.15'}$$

where u_0 is the free stream velocity.

If we make the special choice $\beta=1$ and use the identity (51.9) which is valid in this case, we obtain after an easy calculation

$$\frac{1}{A_P}\int\limits_P p(-y\,dx) \leqq p(u_0), \tag{51.16}$$

which leads to an estimate for the fluid pressure at infinity by an average over the pressures on the profile P. Observe that $(-y\,dx)\geqq 0$ in our sense of integration. The estimate (51.16) is valid for every profile P symmetric with respect to the x-axis which is cut by each line parallel to the y-axis in one straight line segment, if it is cut at all.

Let us consider now a profile for which $\beta>1$. We transform inequality (51.15′) by introducing x as the variable of integration and obtain easily (since $dx/ds=\cos\vartheta$)

$$2\int\limits_a^b f(x)\,A(q)\,q\,\frac{\cos\beta\vartheta}{\cos\vartheta}\,dx \geqq A(u_0)\,u_0\cdot A_P. \tag{51.17}$$

This gives an estimate for the free stream velocity u_0 in terms of the speed on the profile. If q_m denotes the maximum speed of the flow, we can use the monotonic character of $A(q)$ and derive

$$A(q_m)\,q_m\cdot\gamma \geqq A(u_0)\,u_0 \tag{51.18}$$

where

$$\gamma=\left(\int\limits_a^b f(x)\,\frac{\cos\beta\vartheta}{\cos\vartheta}\,dx\right)\left(\int\limits_a^b f(x)\,dx\right)^{-1} \tag{51.19}$$

is a purely geometric constant. The estimate (51.18) holds for all values $q_m<q_s$, and hence we have in

$$A(q_s)\,q_s\cdot\gamma \geqq A(u_{cr})\,u_{cr} \tag{51.18′}$$

an estimate for the critical free stream Mach number M_{cr} of the profile P.

Observe now that because of the monotony property (b) of the test function $A(q)$, we have

$$\frac{A(q_s)}{q_s^\beta} \leqq \frac{A(u_{cr})}{u_{cr}^\beta}. \tag{51.20}$$

From the inequalities (51.18′) and (51.20) we obtain finally

$$\frac{u_{cr}}{q_s} \leqq \gamma^{\frac{1}{\beta+1}}. \tag{51.21}$$

This estimate is particularly remarkable since it does not depend on the particular form of the equation of state [represented by the function $A(q)$] but only on the geometry of the profile P. Its importance in connection with the preceding existence theorems is obvious.

52. Aerodynamical forces and moments. We can apply the conservation laws of Sect. 50 in order to determine the forces and moments exerted by the flow of a compressible fluid upon a given profile P. Let C be a closed curve surrounding P and denote by D the ring domain bounded by the curves C and P. We integrate the differential equations (50.2) over the domain D and obtain by means of GREEN'S identity:

$$\int\limits_{C+P}[p\cos(\nu,x)+\varrho(\boldsymbol{q}\cdot\boldsymbol{\nu})\,u]\,dx=\int\limits_{C+P}[p\cos(\nu,y)+\varrho(\boldsymbol{q}\cdot\boldsymbol{\nu})\,v]\,dx=0. \tag{52.1}$$

Here $\boldsymbol{\nu}$ denotes the inner normal on the boundary $C+P$ with respect to the domain D. We use now the fact that the flow direction is tangential along the profile P and hence

$$\boldsymbol{q}\cdot\boldsymbol{\nu}=0 \qquad \text{on } P. \tag{52.2}$$

On the other hand, the vector

$$\boldsymbol{F}=-\oint_P p\,\boldsymbol{\nu}\,ds \tag{52.3}$$

represents obviously the total force due to the fluid pressure upon the obstacle. Hence, we obtain from Eq. (35.1) the fundamental equation

$$\boldsymbol{F}=\oint_C [p\,\boldsymbol{\nu}+\varrho(\boldsymbol{q}\cdot\boldsymbol{\nu})\,\boldsymbol{q}]\,ds. \tag{52.4}$$

We may choose any closed curve C surrounding the given profile P for the calculation of the force vector $\boldsymbol{F}$. C is called a control contour for the calculation; it may be chosen conveniently in order to simplify the computation.

In a similar fashion, we may express the moment of the pressure forces with respect to a given point, say the origin. We start with the conservation law

$$\frac{\partial}{\partial x}[(p+\varrho u^2)\,y-\varrho\,u\,v\,x]+\frac{\partial}{\partial y}[-(p+\varrho v^2)\,x+\varrho\,u\,v\,y]=0 \tag{52.5}$$

which is an immediate consequence of the two differential equations (50.2). We integrate Eq. (52.5) over the ring domain D and again apply GREEN's identity. We obtain

$$\int_{C+P}[p(\boldsymbol{\nu}\times\boldsymbol{r})+\varrho(\boldsymbol{q}\cdot\boldsymbol{\nu})\,(\boldsymbol{q}\times\boldsymbol{r})]\,ds=0 \tag{52.6}$$

where $\boldsymbol{a}\times\boldsymbol{b}$ denotes the vector product of $\boldsymbol{a}$ and $\boldsymbol{b}$ and $\boldsymbol{r}=(x,y)$ is the radius vector from the origin.

We again make use of Eq. (52.2) and observe that

$$\boldsymbol{M}=-\oint_P p(\boldsymbol{\nu}\times\boldsymbol{r})\,ds \tag{52.7}$$

represents the moment of the pressure forces with respect to the origin. Hence, we obtain finally:

$$\boldsymbol{M}=\oint_C[p(\boldsymbol{\nu}\times\boldsymbol{r})+\varrho(\boldsymbol{q}\cdot\boldsymbol{\nu})\,(\boldsymbol{q}\times\boldsymbol{r})]\,ds. \tag{52.8}$$

We have thus expressed the aerodynamical force and moment of the flow upon the profils by means of integrals over an arbitrary control contour.

We shall apply the formulas (52.4) and (52.8) to the particular case that the flow past P is due to a subsonic doublet at infinity. That is, we suppose that the flow considered is uniform and subsonic at infinity. Let its uniform stream velocity be directed in the positive x-direction and have the magnitude u_0. Applying the theorem at the end of Sect. 39 and the asymptotic formula (39.24), we find that the velocity field near infinity has the form

$$u=u_0-\frac{\beta\Gamma}{2\pi}\frac{y}{x^2+\beta^2y^2}+o(r^{-1}),\qquad v=\frac{\beta\Gamma}{2\pi}\frac{x}{x^2+\beta^2y^2}+o(r^{-1}). \tag{52.9}$$

We choose a circle C_R with sufficiently large radius R around the origin as the control surface. We calculate from Eqs. (52.4) and (52.9) the force vector $\boldsymbol{F}$; sending R to infinity, we obtain after an obvious but somewhat lengthy calculation:

$$\boldsymbol{F}=(0,-\varrho_0u_0\Gamma). \tag{52.10}$$

Here, ϱ_0 denotes the density of the fluid at infinity, corresponding to the speed u_0 of the uniform flow.

We have thus extended the classical Kutta-Joukowski formula from the theory of an incompressible fluid to that of a compressible one. There exists a lifting force normal to the incident uniform flow; no drag in the flow direction can arise even for a compressible fluid as long as it has no viscosity or vorticity.

For the moment M, we can simplify Eq. (52.8) to the limit formula

$$M = \lim_{R\to\infty} \Big[R \int_{C_R} \varrho\, u\, v\, ds\Big]. \tag{52.11}$$

It appears that more terms are needed in the asymptotic development of the velocity potential near infinity than are provided by Eq. (39.24). One has to use the series development (41.6), and it can be shown [4] that the decisive term is the coefficient $A_{01}(\vartheta)$ in this series.

The preceding considerations show again the importance of the study of the velocity potential $\Phi(x, y)$ at infinity and of the asymptotic formulas (39.24) and (41.6).

V. The analytical theory of two-dimensional transonic flows.

53. Transonic flows and differential equations of mixed type. Let us consider the steady motion of a two-dimensional compressible fluid around a smooth profile P. Suppose that the flow is uniform near infinity and has there the velocity u_0 in the direction of the positive x-axis. Since there are necessarily stagnation points on P, it is clear that the flow must be either subsonic or of mixed character, that is, partially subsonic and partially supersonic. Flows which contain subsonic and supersonic regions are called transonic flows. This chapter will deal with questions of existence and uniqueness of transonic flows around given profiles.

For the sake of definiteness and simplicity we consider the particular case that the free stream velocity is subsonic. In this case, the supersonic regions will consist of finite pockets in the flow each of which is bounded by sonic line arcs or parts of the boundary of P. We may exclude the possibility of supersonic enclosures which are entirely bounded by a sonic line. In fact, we can start to follow a Mach line from some point of the sonic boundary into the supersonic region. As was stated in Sect. 34 the speed of the flow is a monotonic function along a Mach line and therefore we cannot return to the sonic boundary again. If we exclude the possibility of points with maximum (escape) velocity, the Mach curves have to end, therefore, on the profile P. Thus, each supersonic pocket is attached to the boundary P of the flow region.

It is well-known that in transonic flows shock waves will tend to appear, that is, lines across which the entropy changes suddenly and across which the pressure-density relation jumps discontinuously. We wish to investigate here, however, isentropic flows with a fixed pressure-density relation. Such transonic flows have been constructed explicitly for various special profiles. However, very serious arguments have been brought forth which throw doubt on the existence of regular transonic solutions for arbitrary profiles P and make it even less probable that these solutions depend continuously on the physical data. Therefore we shall discuss transonic flow problems not so much for their physical significance but rather in order to understand the high probability of shock phenomena in transonic flows. The mathematical tools developed may be of considerable value also in a detailed study of the various shock possibilities.

The question as to the existence of a continuous potential flow around a profile which is uniform at infinity and contains finite supersonic enclosures is known as the *Taylor problem* [1] to [3].

Since we assumed the flow to be uniform at infinity, the Helmholtz-Kelvin circulation theorem (1.16) guarantees that the flow is irrotational throughout and that we can introduce a velocity potential $\varphi(x, y)$ satisfying the differential equation

$$(c^2 - \varphi_x^2)\,\varphi_{xx} - 2\varphi_x\varphi_y\varphi_{xy} + (c^2 - \varphi_y^2)\,\varphi_{yy} = 0. \tag{53.1}$$

On the boundary P we require

$$\frac{\partial\varphi}{\partial n} = 0 \tag{53.2}$$

and the asymptotic behavior of φ at infinity is determined by our assumption:

$$\varphi(x, y) = u_0\, x + O\left(\frac{1}{r}\right), \qquad r^2 = x^2 + y^2. \tag{53.3}$$

The analytical theory of the flow has now to deal with the existence and uniqueness of a solution for the quasi-linear differential equation (53.1) with the side conditions (53.2) and (53.3). The difficulty of the problem becomes evident when we consider the discriminant of the equation

$$(c^2 - \varphi_x^2)\,(c^2 - \varphi_y^2) - \varphi_x^2\,\varphi_y^2 = c^2\,(c^2 - q^2). \tag{53.4}$$

We see that the equation is elliptic for subsonic flow and hyperbolic for supersonic flow with an abrupt change of type across the sonic boundary line of the supersonic enclosure.

Partial differential equations of elliptic-hyperbolic type, also called equations of mixed type, present considerable analytical difficulties even if they are linear [4], [5]. The best known linear equation of mixed type

$$y\,\varphi_{xx} + \varphi_{yy} = 0 \tag{53.5}$$

is called the Tricomi equation and was first carefully investigated in 1923 [6] to [13]. As was shown in (31.31′) this equation occurs, indeed, in the theory of the stream function in the hodograph plane if a gas with a proper equation of state is considered.

Very little is known in the theory of quasi-linear differential equations of mixed type. It is easy to see, however, that at least the uniqueness theory for such equations can be reduced to the theory of linear mixed equations. Let

$$L[\varphi] = a\,\varphi_{xx} + 2b\,\varphi_{xy} + c\,\varphi_{yy} = 0 \tag{53.1′}$$

be a quasi-linear partial differential equation whose coefficients are continuously differentiable functions of x, y and of φ_x, φ_y. If we wish to stress the fact that the coefficient a, for example, has been computed for the argument function φ, we shall denote it by $a[\varphi]$. Now let $\varphi(x, y)$ and $\tilde{\varphi}(x, y)$ be two solutions of (53.1); we wish to show that

$$\omega(x, y) = \varphi(x, y) - \tilde{\varphi}(x, y) \tag{53.6}$$

satisfies a simple linear differential equation. For this purpose, we write

$$\left.\begin{aligned} L[\varphi] - L[\tilde{\varphi}] = a[\varphi]\,\omega_{xx} + 2b[\varphi]\,\omega_{xy} + c[\varphi]\,\omega_{yy} + (a[\varphi] - a[\tilde{\varphi}])\,\tilde{\varphi}_{xx} + {}\\ + 2(b[\varphi] - b[\tilde{\varphi}])\,\tilde{\varphi}_{xy} + (c[\varphi] - c[\tilde{\varphi}])\,\tilde{\varphi}_{yy} \end{aligned}\right\} \tag{53.7}$$

and by means of the mean value theorem obtain

$$a[\varphi] - a[\tilde{\varphi}] = \alpha_0(x, y)\,\omega_x + \beta_0(x, y)\,\omega_y. \tag{53.8}$$

Combining Eqs. (53.7), (53.8) and the fact that φ and $\tilde{\varphi}$ are solutions of Eq. (53.1') we obtain

$$a[\varphi]\,\omega_{xx} + 2b[\varphi]\,\omega_{xy} + c[\varphi]\,\omega_{yy} + \alpha\,\omega_x + \beta\,\omega_y = 0. \tag{53.9}$$

This is a linear differential equation for the difference function $\omega(x, y)$ which has the same discriminant as the original equation (53.1') and is, therefore, of the same type. For such boundary conditions on $\varphi(x, y)$ as would lead to boundary conditions for $\omega(x, y)$ implying $\omega \equiv 0$ by means of the theory for *linear* mixed equations, we can then assert the uniqueness of the solution.

The preceeding artifice explains the important role of the theory of linear equations of mixed type in the transonic flow problems which are of non-linear character.

54. A maximum principle and a uniqueness theorem for equations of mixed type. In order to clarify the determination of a solution by boundary conditions in the case of a linear differential equation of mixed type, we shall study the particular mixed equation

$$L[\varphi] = K(y)\,\varphi_{xx} + \varphi_{yy} = 0 \tag{54.1}$$

where $K(y)$ is a monotonically increasing function of y; we assume $K(0) = 0$ and, hence $K(y) > 0$ for $y > 0$ and $K(y) < 0$ for $y < 0$. The differential equation (54.1) is a generalization of TRICOMI's equation (53.5) but still simple enough to permit an easy treatment [*14*] to [*16*]. It is of immediate physical importance since the differential equation system (23.8) in the hodograph plane leads to a differential equation of the type (54.1) for the stream function. Obviously $L[\varphi]$ is elliptic in the upper half-plane $y > 0$ and hyperbolic in the lower.

The differential equation (54.1) has for $y < 0$ real characteristics satisfying the ordinary differential equations

$$\frac{dy}{dx} = l(y)^{-1}, \tag{54.2}$$

$$\frac{dy}{dx} = -\,l(y)^{-1}, \tag{54.2'}$$

respectively, with

$$l(y) = \sqrt{-K(y)}. \tag{54.2''}$$

Through each point of the lower half-plane passes a curve of each family of characteristics. Let $\xi(x, y) = \text{const}$ and $\eta(x, y) = \text{const}$ be the equations of the integrals of Eqs. (54.2) and (54.2'), respectively. We may then introduce ξ and η as new coordinates in the half-plane $y < 0$.

We have the differentiation rules

$$\frac{\partial}{\partial\xi} = \frac{\partial}{\partial x} + l(y)^{-1}\frac{\partial}{\partial y}, \qquad \frac{\partial}{\partial\eta} = \frac{\partial}{\partial x} - l(y)^{-1}\frac{\partial}{\partial y}, \tag{54.3}$$

$$\frac{\partial}{\partial x} = \frac{1}{2}\left(\frac{\partial}{\partial\xi} + \frac{\partial}{\partial\eta}\right), \qquad \frac{\partial}{\partial y} = \frac{1}{2}\,l(y)\left(\frac{\partial}{\partial\xi} - \frac{\partial}{\partial\eta}\right). \tag{54.3'}$$

On the characteristic $\eta = \text{const}$ by Eqs. (54.3') we have $\frac{\partial\xi}{\partial y} > 0$, which shows that ξ increases along the characteristic if we ascend towards $y = 0$. Similarly, we have on $\xi = \text{const}$ $\frac{\partial\eta}{\partial y} < 0$; that is, η increases if we descend.

We calculate from Eqs. (54.3′) the new form of the differential operator L:

$$L[f] = K f_{\xi\eta} - \tfrac{1}{8}(K_\xi - K_\eta)(f_\xi - f_\eta). \tag{54.4}$$

Since K depends only on y, we have $K_x = \frac{1}{2}(K_\xi + K_\eta) = 0$ and we can simplify Eq. (54.4) to

$$L[f] = K f_{\xi\eta} - \frac{1}{4} K_\xi (f_\xi - f_\eta) = -l^{\frac{3}{2}} \left\{ \frac{\partial}{\partial\xi}(l^{\frac{1}{2}} f_\eta) - \frac{\partial}{\partial\xi}(l^{\frac{1}{2}}) f_\xi \right\}. \tag{54.4′}$$

A solution $\varphi(x, y)$ of Eq. (54.1), therefore, satisfies the equation in characteristic coordinates

$$\frac{\partial}{\partial\xi}(l^{\frac{1}{2}} \varphi_\eta) = \frac{\partial}{\partial\xi}(l^{\frac{1}{2}}) \cdot \varphi_\xi. \tag{54.4″}$$

Let $\eta = \text{const}$ be a characteristic of the equation and let $A \equiv (\xi_1, \eta)$, $B \equiv (\xi_2, \eta)$ be two points upon it. We integrate the identity (54.4″) between A and B with respect to ξ and, using one integration by parts, we obtain

$$[l^{\frac{1}{2}} \varphi_\eta]_A^B = (\varphi(B) - \varphi(A)) \left(\frac{\partial}{\partial\xi} l^{\frac{1}{2}} \right)_A + \int_A^B [\varphi(B) - \varphi(R)] \frac{\partial^2}{\partial\xi^2}(l^{\frac{1}{2}})\, d\xi; \tag{54.5}$$

here R denotes the variable point of integration (ξ, η). The identity (54.5) is valid for all solutions of the differential equation (54.1) and plays the role of the Green's identity in the case of elliptic equations.

In order to utilize the identity (54.5) we have to express $(l^{\frac{1}{2}})_{\xi\xi}$ in simpler form. We have (with $l' = dl/dy$)

$$\frac{\partial^2}{\partial\xi^2}(l^{\frac{1}{2}}) = l^{-\frac{7}{2}}\left(l'' l - \frac{3}{2} l'^2\right) = -l^{-\frac{7}{2}} l'(y)^2 \left\{ \left(\frac{l}{l'}\right)' + \frac{1}{2} \right\}. \tag{54.6}$$

Since $l(y)$ is positive for $y < 0$ we can assert that the expression (54.6) is negative if

$$\left(\frac{l}{l'}\right)' + \frac{1}{2} = 2\left(\frac{K}{K'}\right)' + \frac{1}{2} > 0, \quad \text{i.e. } 5K'^2 \geqq 4KK''. \tag{54.7}$$

The inequality (54.7) holds for the Tricomi equation where $K = y$ and for the case studied by Gellerstedt [33] with $K = y^n$. We shall make from now the assumption that Ineq. (54.7) is fulfilled.

We remark finally that in view of the monotonicity of $K(y)$

$$\frac{\partial}{\partial\xi} l^{\frac{1}{2}} < 0. \tag{54.8}$$

We are now able to apply the fundamental identity (54.5). We start with a characteristic triangle D which lies in the lower half-plane and is bounded by the segment $x_\alpha \leqq x \leqq x_\beta$ of the x-axis, by the characteristic $\eta = \alpha$ through x_α and by the characteristic $\xi = \beta$ through x_β. Let $\varphi(x, y)$ be a solution of Eq. (54.1) in the domain D. We state the maximum principle:

If the solution $\varphi(x, y)$ of Eq. (54.1) is non-decreasing with y on the characteristic boundary $\xi = \beta$, it must take its maximum on the horizontal boundary $y = 0$.

This is a typical maximum result for a hyperbolic differential equation. We observe the great difference in the role played by the various boundary components and the monotonicity assumption on the solution, necessary for the possibility of a maximum statement at all.

In order to prove the maximum principle, let us suppose that $\varphi(x, y)$ takes its maximum at a point B inside of the domain D or on the characteristic $\eta = \alpha$

(Fig. 13). For otherwise it would have to be located on the characteristic $\xi = \beta$ and because of the monotonicity of φ along this curve on the horizontal boundary; in this case the theorem would be true. We follow the characteristic $\eta = \gamma$ through B in the direction of decreasing values of y until we meet the boundary arc $\xi = \beta$ at the point A. We apply the identity (54.5) to these two points. Since $\varphi(B)$ is the maximum value of φ in D and since we have the inequalities

$$\frac{\partial}{\partial \xi}(l^{\frac{1}{2}}) < 0, \qquad \frac{\partial^2}{\partial \xi^2}(l^{\frac{1}{2}}) < 0 \tag{54.6'}$$

we conclude that the right side in Eq. (54.5) is negative. If B lies inside of D the maximum assumption ensures $\varphi_\eta(B) = 0$; if B lies on the characteristic arc $\eta = \alpha$, we can assert because of the maximum property $\varphi_\eta(B) > 0$. Thus, in any case, Eq. (54.5) reduces to

$$\varphi_\eta(A) \geqq 0. \tag{54.9}$$

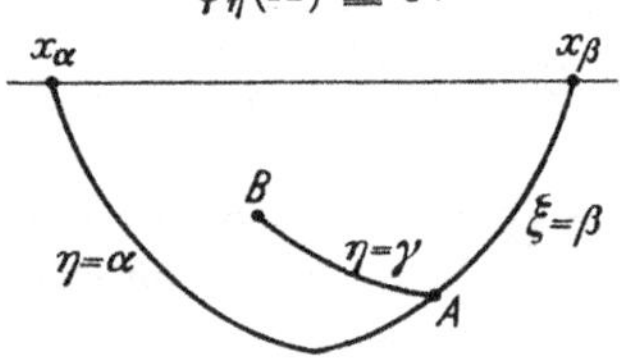

Fig. 13.

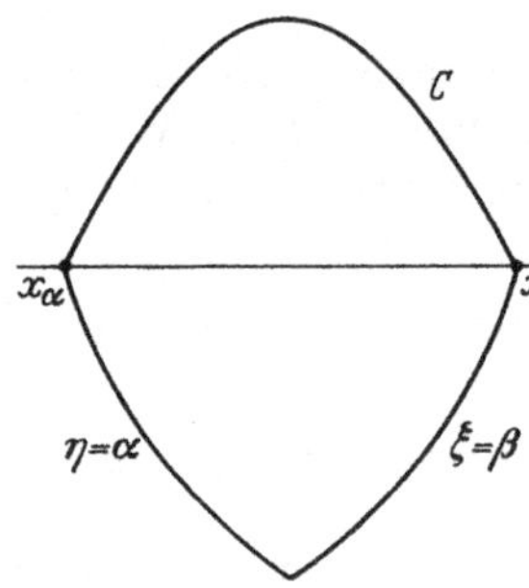

Fig. 14.

On the other hand, η decreases with increasing y and the monotony of φ on the characteristic $\xi = \beta$ implies there

$$\varphi_\eta \leqq 0. \tag{54.10}$$

From inequalities (54.9) and (54.10) we can conclude $\varphi_\eta(A) = 0$; but the equality can hold only if $\varphi(B) = \varphi(A)$, as follows from our estimation of the identity (54.5). Hence, $\varphi(A)$ is also the maximum value and because of the monotony of φ on the characteristic arc the maximum must be taken also on the horizontal boundary. Thus, the above maximum principle is completely proved.

By a finer argument it can be shown that if $\varphi(x, y)$ takes its maximum at an interior point of the horizontal segment $x_\alpha \leqq x \leqq x_\beta$ then at this maximum point must hold [*18*]

$$\varphi_y > 0. \tag{54.11}$$

So far, we are dealing with a hyperbolic differential equation since we restricted ourselves to the region $y < 0$. We consider now a domain D (Fig. 14) which is bounded by an arc C in the upper half plane which connects the points x_α and x_β, $x_\alpha < x_\beta$, on the x-axis and by the two characteristics $\eta = \alpha$ and $\xi = \beta$ which start at the points $(x_\alpha, 0)$ and $(x_\beta, 0)$, respectively. Let $\varphi(x, y)$ be a solution of (54.1) in D and let us make the further assumption that $\varphi(x, y)$ is non-decreasing with y on the characteristic $\xi = \beta$. We have then the theorem:

The solution φ takes its maximum on the arc C.

Indeed, the preceding maximum principle guarantees that the maximum of φ in the lower half of D is attained somewhere on the segment $x_\alpha \leqq x \leqq x_\beta$ of the x-axis. Since in the upper half of D φ satisfies an elliptic differential equation, it must take its maximum there either on C or on the segment $\langle x_\alpha, x_\beta \rangle$. In the first case our statement is proved; in the second case there would be a point inside the segment where φ would attain its absolute maximum in D. This would, of course, imply $\varphi_y = 0$ while in view of Ineq. (54.11) this is seen to be impossible. Hence the second possibility is ruled out, and the maximum principle for the mixed equation (54.1) is proved.

An immediate and important consequence of the preceding maximum principle is the following uniqueness theorem for the mixed differential equation (54.1):

A solution $\varphi(x, y)$ of the mixed equation (54.1) in the domain D is uniquely determined by its values on the arc C and on the characteristic $\xi = \beta$.

We have here a fundamental difference between the possible boundary value problems for elliptic and for mixed differential equations. In the case of a mixed differential equation it is not possible to prescribe arbitrarily the boundary values of a solution along the entire boundary. TRICOMI was the first to point out this fact and to discuss the particular boundary value problem for which the preceding theorem guarantees the uniqueness of the solution. He gave also an existence proof for such a solution in the particular case (53.5).

The preceding uniqueness proof for a class of mixed equations (54.1) is due to GERMAIN and BADER [*17*]. Their result has been extended to large classes of hyperbolic and mixed equations by AGMON, NIRENBERG and PROTTER [*18*]. The main restriction of the above method is the inequality (54.7) assumed for the coefficient K. This inequality is not fulfilled for the differential equation of the stream function of a real gas in arbitrary domains of the hodograph plane.

55. Uniqueness theory for the Tricomi and Frankl boundary value problems. In view of the somewhat unsatisfactory uniqueness theorem of the preceding section it seems desirable to consider alternative and more flexible methods for establishing uniqueness theorems for mixed equations. We shall sketch such a procedure which goes back to FRANKL [*19*] to [*21*] and has been clarified by FRIEDRICHS and applied systematically by PROTTER [*22*] and MORAWETZ [*25*] to numerous problems. It has a considerable analogy to corresponding uniqueness proofs for elliptic and for hyperbolic equations since it is likewise based on a positive-definite integral corresponding to the classical energy integrals.

The basic idea is to start with the identity

$$\iint_D (a\,\varphi + b\,\varphi_x + c\,\varphi_y)\,(K\,\varphi_{xx} + \varphi_{yy})\,dx\,dy = 0 \tag{55.1}$$

which holds for every solution $\varphi(x, y)$ of Eq. (54.1) independently of the choice of the coefficient functions a, b, c. Integrating by parts, one can transform Eq. (55.1) into the sum of a surface integral over D and a line integral over the boundary C of D. Both integrands are quadratic and homogeneous in φ, φ_x and φ_y. By a proper choice of a, b, c one tries to make the integrand in the surface integral positive-definite, so that it can vanish only if $\varphi \equiv 0$. On the other hand, one looks for a linear boundary condition on φ such as to ensure that the line integral be positive. This condition will depend, in general, also on the geometry of the boundary curve C, and geometric restrictions on the domains D considered may become necessary. If φ is a solution in D and satisfies such a boundary condition, it must be identically zero in that domain. It is easy to recast such results in the form of uniqueness theorems for certain boundary value problems.

We illustrate this general programm in the particular case $a = 0$. It is easy to verify the identities

$$\left.\begin{aligned}(x\,\varphi_x + y\,\varphi_y)\,(K\,\varphi_{xx} + \varphi_{yy}) &= (\tfrac{1}{2} K\,x\,\varphi_x^2 + K\,y\,\varphi_x\,\varphi_y - \tfrac{1}{2}\,x\,\varphi_y^2)_x - \\ &\quad - (\tfrac{1}{2} K\,y\,\varphi_x^2 - x\,\varphi_x\,\varphi_y - \tfrac{1}{2}\,y\,\varphi_y^2)_y + \tfrac{1}{2}\,y\,K'(y)\,\varphi_x^2\end{aligned}\right\} \tag{55.2}$$

and

$$x\varphi_x(K\varphi_{xx}+\varphi_{yy}) = (\tfrac{1}{2}Kx\varphi_x^2 - \tfrac{1}{2}x\varphi_y^2)_x + (x\varphi_x\varphi_y)_y + (-\tfrac{1}{2}K\varphi_x^2 + \tfrac{1}{2}\varphi_y^2). \quad (55.3)$$

Observe that when we integrate the right hand sides in Eqs. (55.2) and (55.3) the first two terms will give rise to line integrals over the boundary by virtue of GREEN'S formulas. The third term in each expression gives a surface integral; since $\frac{1}{2}yK'\varphi_x^2$ is positive for $y>0$ and $-\frac{1}{2}K\varphi_x^2+\frac{1}{2}\varphi_y^2$ is positive for $y<0$, it is natural to consider the integral (55.1) with the choice

$$a=0, \qquad b=x, \qquad c=\begin{cases} y & \text{for } y\geqq 0 \\ 0 & \text{for } y\leqq 0. \end{cases} \quad (55.4)$$

We are sure that the surface integrals are positive, and we have now to concentrate on the boundary conditions and the geometry of the boundary curves.

Following MORAWETZ we consider the following domain D (Fig. 15). In the upper half plane D is bounded by a curve C_0 which intersects the x-axis at the points $x_\alpha<0$ and $x_\beta>0$. In the lower half-plane, D is bounded by two characteristic arcs, say $\xi=0$ and $\eta=0$ which issue from the point $x=y=0$ and by two arcs C_1 and C_2 which run from x_α to the ξ-characteristic and from x_β to the η-characteristic, respectively. We suppose that C_0 is star-shaped with respect to the origin 0; that means that every point on C_0 can be seen from 0 and leads to the analytic condition

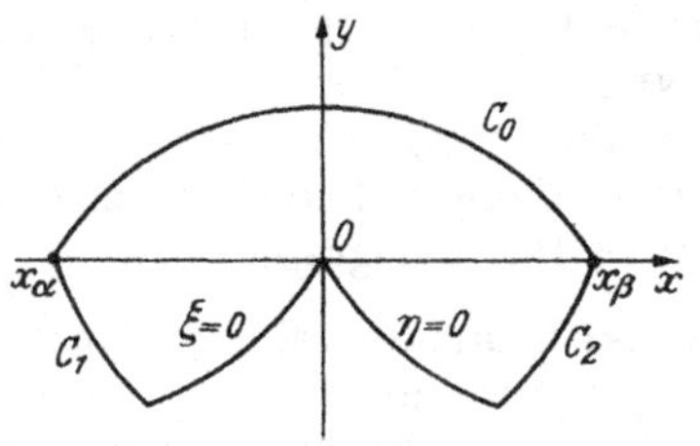

Fig. 15.

$$x\,dy - y\,dx \geqq 0 \quad (55.5)$$

if C_0 is followed in the positive sense. The curves C_1 and C_2 are monotonic curves, and we assume that C_1 is intersected at most once by every curve $\xi=$ const, and the same is assumed for C_2 with respect to curves $\eta=$ const. This is ensured by assuming

$$0\geqq y'\geqq -l^{-1} \text{ on } C_1, \qquad 0\leqq y'\leqq l^{-1} \text{ on } C_2, \qquad l=\sqrt{-K}. \quad (55.5')$$

Having thus characterized the domain D, we can prove the following uniqueness theorem:

A solution $\varphi(x, y)$ of Eq. (54.1) is uniquely determined in D by its boundary values on the curves C_0, C_1 and C_2.

The significance of this result is two-fold. We consider in the hyperbolic half-plane arcs of the boundary which are not necessarily characteristics. We do not need any inequalities on the coefficient $K(y)$ except the obvious monotonicity requirement. On the other hand, there appears the rather restrictive condition (55.5) on the elliptic part C_0 of the boundary, unnecessary in our preceding approach.

In order to prove the theorem, it is sufficient to show that if a solution $\varphi(x, y)$ in D has vanishing boundary values on the arcs C_0, C_1 and C_2 it must be identically zero in D. For this purpose, we compute the identity (55.1) for the domain D considered, the solution φ in question and the coefficients (55.4). Let D_U and D_L denote the parts of D in the upper and lower half-plane; we have then

$$\left.\begin{aligned} 0 &= \iint_{D_U} \tfrac{1}{2}yK'(y)\varphi_x^2\,dx\,dy + \iint_{D_L}[-\tfrac{1}{2}K\varphi_x^2+\tfrac{1}{2}\varphi_y^2]\,dx\,dy + \\ &\quad + \int_C\{-[b\varphi_x\varphi_y - \tfrac{1}{2}c(K\varphi_x^2-\varphi_y^2)]\,dx + [\tfrac{1}{2}b(K\varphi_x^2-\varphi_y^2)+Kc\varphi_x\varphi_y]\,dy\} \end{aligned}\right\} \quad (55.6)$$

where the boundary integral over C is extended in the positive sense.

We consider first the part of the boundary integral which is extended over the arcs C_0, C_1, C_2. Here we have by assumption $\varphi \equiv 0$ and hence

$$\varphi_x\,dx + \varphi_y\,dy = 0. \tag{55.7}$$

By Eq. (55.7) we can eliminate φ_y from the integrand in Eq. (55.6) and find easily for the contribution of these arcs to the line integral:

$$J_1 = \int\limits_{C_0+C_1+C_2} \frac{1}{2}\,\varphi_x^2 \left[K + \left(\frac{dx}{dy}\right)^2\right] (b\,dy - c\,dx). \tag{55.8}$$

It follows at once from our assumptions (55.5) and (55.5′) on the geometry of the C-curves that the integrand is non-negative and hence $J_1 \geq 0$.

We come finally to the characteristic part of the boundary $\xi = 0$ and $\eta = 0$. We have by Eq. (54.2)

$$\frac{dy}{dx} = l^{-1} \quad \text{on} \quad \xi = 0, \qquad \frac{dy}{dx} = -\,l^{-1} \quad \text{on} \quad \eta = 0. \tag{55.9}$$

Therefore it is easy to see that the contribution of the integrals along the characteristics is

$$J_2 = \int\limits_{\xi=0} \left(-\tfrac{1}{2}\,x\right) \left(l^{\frac{1}{2}}\,\varphi_x + l^{-\frac{1}{2}}\,\varphi_y\right)^2 dx + \int\limits_{\eta=0} \tfrac{1}{2}\,x \left(l^{\frac{1}{2}}\,\varphi_x - l^{-\frac{1}{2}}\,\varphi_y\right)^2 dx, \tag{55.10}$$

and it is obvious that this expression is non-negative.

We have shown that Eq. (55.6) is the sum of four non-negative terms; since the sum vanishes, each of these terms must vanish separately. From the surface integrals we can conclude that $\varphi_x \equiv 0$ in D; since φ vanishes on the arcs C_0, C_1 and C_2 this implies $\varphi \equiv 0$ in D. The asserted uniqueness theorem is therefore proved.

The following terminology has become customary: The boundary value problem for a mixed partial differential equation in a domain whose hyperbolic boundary consists of characteristic arcs only is called a *Tricomi* boundary value problem. If the hyperbolic boundary consist in part or entirely of non-characteristic arcs, we call the boundary value problem a *Frankl* boundary value problem. In the latter case, generally, one will have to restrict the class of admissible hyperbolic arcs; for example, one may require that such arc cuts every curve of one family of characteristics at most once. The Tricomi problems appear as particular or limit cases of Frankl problems. The significance of the more general Frankl problem in fluid dynamics will appear in the sequel. In Sect. 54 we solved the uniqueness question for an important class of Tricomi problems, while in the present section we gave a uniqueness theorem for a Frankl problem.

Frankl [*20*] considered the Tricomi problem of a domain D bounded by an elliptic arc C and two characteristics; a solution of Eq. (54.1) in D is to be found with prescribed boundary values on the arc C and on one characteristic. He showed that if the coefficient function $K(y)$ satisfies for $y<0$ the differential inequality

$$F(y) = 2\left(\frac{K}{K'}\right)' + 1 > 0, \quad \text{i.e. } 3K'^2 > 2K\,K'', \tag{55.11}$$

then the above Tricomi problem can have at most one solution. Since the inequality (55.11) is less restrictive than inequality (54.7), Frankl's result is stronger than the theorem given in Sect. 54 for the same type of domain D. But it is still not strong enough to cover the case of a real gas in the entire hodograph plane.

PROTTER [22] used the method described in this section to prove the stronger result: *There exists a constant $d_0 < 0$ such that if $F(y) \geq d_0$ in D the above Tricomi problem has a unique solution.*

The main significance of this result is to show that the inequality (55.11) in FRANKL'S result is a consequence of his method of proof and not inherent in the nature of the problem. The condition $F(y) \geq d_0$ still means a restriction on the hyperbolic part of the domain considered. It is, therefore, interesting that by the same method the following result can be proved:

There exists a constant $d_1 > 0$ such that if the ordinate y on the elliptic boundary curve C always satisfies the inequality $y < d_1$, one can assert the uniqueness of the Tricomi problem for every monotonic coefficient function $K(y)$ and every admissible domain D.

Thus, we have removed the restriction on $F(y)$ on the hyperbolic part of D but have introduced instead a geometric condition for the elliptic part. PROTTER showed also that his two uniqueness theorems hold in the case that the hyperbolic boundary of D consists of four characteristic arcs and when the boundary values for the solution are prescribed on the elliptic boundary and the two outer characteristic arcs.

The great variety of overlapping results in the uniqueness theory of the mixed equation (54.1) suggests the possibility of a unifying and very general uniqueness theorem. At the present stage the theory even of the particular linear equation (54.1) is not too satisfactory.

It should further be remarked that while Eq. (54.1) is the general type of the stream function equation in the hodograph plane, the uniqueness theory for the non-linear differential equation (53.1) in the physical plane leads to the more general linear mixed equation (53.9). TRICOMI has shown that the most general mixed equation

$$a\,\omega_{xx} + 2b\,\omega_{xy} + c\,\omega_{yy} + d\omega_x + e\,\omega_y + f\,\omega = 0 \tag{55.12}$$

whose coefficients depend on x and y can be transformed by a change of variables into the form

$$y\,\omega_{xx} + \omega_{yy} + \alpha\,\omega_x + \beta\,\omega_y + \gamma\,\omega = 0. \tag{55.13}$$

Thus, the second derivatives in the normal form coincide with the type of Eq. (54.1). However, the presence of the lower derivatives unfortunately invalidates the conclusions drawn above. It is conjectured, but not yet proved, that for this more general case uniqueness theorems similar to those for in the case of Eq. (54.1) will hold.

56. The existence problem for transonic flows. In this section we wish to show the close connection between the uniqueness theory for linear partial differential equations of elliptic type and the existence problem for transonic flows around arbitrary profiles P. We return to the situation discussed in Sect. 53 of a flow around P which is subsonic and uniform at infinity and has supersonic enclosures attached to the profile P (Fig. 16). We make the assumption that P is a smooth convex curve which is symmetric with respect to the x-axis; we can therefore restrict our attention to the half-plane $y \geq 0$. We suppose further that there exists only one supersonic pocket in this half-plane; it will be bounded by an arc CD of the profile P and by the curve Γ in the flow along which $q = q_s$ and which connects C with D. We choose now a point E on Γ and draw the two Mach lines starting from E, which will cut out from $\widehat{CD}$ a subarc $\Lambda = \widehat{FG}$ on P.

Let us consider the velocity potential $\varphi(x, y)$ in the domain D_Λ obtained from the upper half-plane $y > 0$ by removal of the profile P and the curvilinear triangle $\Delta(E, F, G)$. $\varphi(x, y)$ will be a solution of the partial differential equation (53.1) in D_Δ; it will have at infinity the asymptotic form (53.3) and it will have a vanishing normal derivative on the boundary of D_Δ except for the Mach arcs $\widehat{EF}$ and $\widehat{EG}$. We observe that Eq. (53.1) is elliptic in D_Δ outside of the arc Γ and hyperbolic inside of Γ; the Mach lines are obviously the characteristics of the equation there; see Sect. 5.

We wish to show now that the considerations of Sects. 54 and 55 make it highly probable that $\varphi(x, y)$ is uniquely determined in D_Δ by the data described. In other words, though we do not make any assumption on φ on the characteristic part of the boundary, the conditions on the remainder of the boundary suffice to determine φ in a unique way. Indeed, suppose there were a second function $\tilde{\varphi}(x, y)$ with the same properties. The difference function $\omega = \varphi - \tilde{\varphi}$

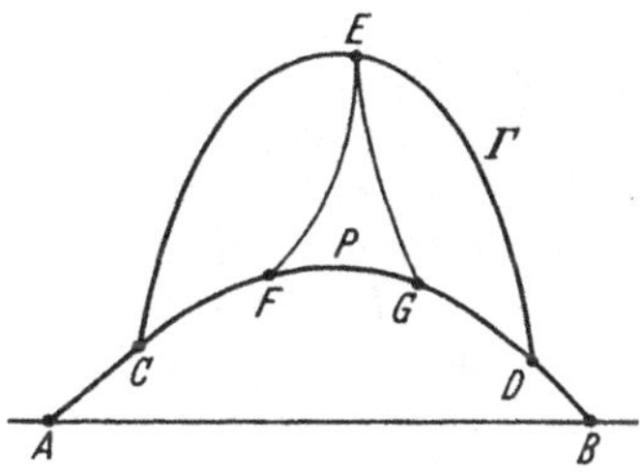

Fig. 16.

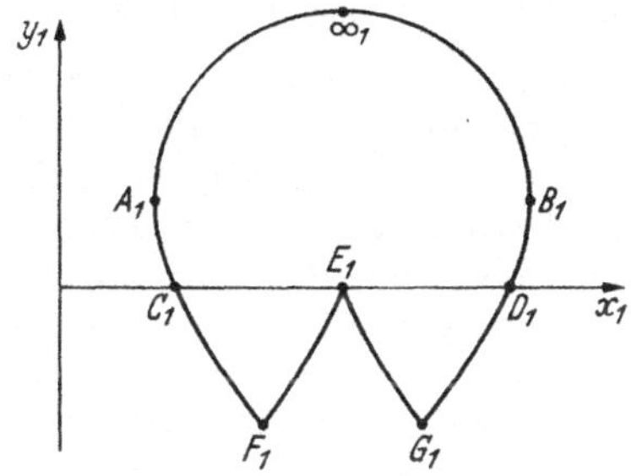

Fig. 17.

would be regular at infinity and would satisfy the linear partial differential equation (53.9) and the linear boundary condition $\partial\omega/\partial n = 0$ on the non-characteristic part of the boundary.

As stated in Sect. 55, we can introduce new variables x_1, y_1 such that Eq. (53.9) appears in the normal form

$$y_1 \frac{\partial^2 \omega}{\partial x_1^2} + \frac{\partial^2 \omega}{\partial y_1^2} + \alpha \frac{\partial \omega}{\partial x_1} + \beta \frac{\partial \omega}{\partial y_1} + \gamma \omega = 0. \tag{56.1}$$

In the new (x_1, y_1)-plane the image of the domain D (Fig. 17) appears bounded by an elliptic arc which corresponds to the line $\widehat{DB\infty AC}$ on the boundary of D_Δ; by two non-characteristic arcs $\widehat{C_1F_1}$ and $\widehat{D_1G_1}$ which correspond to the arcs $\widehat{CF}$ and $\widehat{DG}$ of P in the supersonic region, and by the two characteristics $\widehat{F_1E_1}$ and $\widehat{G_1E_1}$ which correspond to the analogous Mach lines in the (x, y)-plane. The boundary condition for ω has the form

$$a_1 \frac{\partial \omega}{\partial x_1} + a_2 \frac{\partial \omega}{\partial y_1} = 0 \tag{56.2}$$

and is prescribed along the non-characteristic boundary.

The difference function $\omega(x_1, y_1)$, therefore, satisfies in the (x_1, y_1)-plane a Frankl problem very similar to that discussed by MORAWETZ [*24*] and sketched in Sect. 55. The Eq. (56.1), however, is more general than Eq. (54.1), for which our preceding treatment holds; the boundary condition $\varphi = 0$ on the non-characteristic part of the boundary has been replaced by the linear homogeneous boundary condition (56.2). The great analogy between the problems makes it very probable that $\omega \equiv$ const under our assumptions and that $\varphi(x, y)$ is determined up to an additive constant by the data on the boundary, even without the cut-out segment $\widehat{FG}$ of P. Once it is shown that $\varphi(x, y)$ is determined in D_A, it is easily seen that it is determined in the same way inside of the curvilinear triangle $\Delta(E, F, G)$ and, therefore, in the entire flow region.

Suppose now that we know the actual flow around a given profile P of the described kind. If we should deform the profile P in an arbitrary way along the supersonic arc $\widehat{FG}$ we should not change the determining data with respect to the velocity potential $\varphi(x, y)$; hence the same function φ would be the velocity potential for all profiles P^* obtained from P by arbitrary deformation along the arc $\widehat{FG}$. But by the very definition of $\varphi(x, y)$ we must have

$$\frac{\partial \varphi}{\partial n} = 0 \qquad \text{on } P^*. \tag{56.3}$$

This condition, in general, will not be fulfilled for the fixed function φ along the arbitrary arc $\widehat{TG}$. The only way out of this dilemma is the following:

Either the conjectured uniqueness theorem for linear mixed equations is not true *or* there does not exist a solution of the fluid dynamical flow problem past profiles P^* which are arbitrarily near to a profile P with a solution $\varphi(x, y)$.

This highly suggestive argument against the existence of transonic flows around arbitrary profiles P is due to C. G. GARDNER [*14*]. Its basic idea is to replace the non-linear mixed equation (56.1) for the velocity potential by the linear equation (56.9) for the difference function $\omega(x, y)$. However, at the present stage of our knowledge on linear mixed equations even the linear equation is too general to allow us to make the argument conclusive. It is natural to try to obtain simpler linear mixed equations by comparing infinitesimally near profiles only. This has been done successfully by MORAWETZ [*25*] and we proceed now to sketch her argument.

Since we are going to deal with perturbation arguments we wish first to study the following question. Suppose that we have a pair of functions $\varphi(x, y; \tau)$ and $\psi(x, y; \tau)$ which depend on a real parameter τ and satisfy for fixed value of τ the differential system

$$u = \frac{\partial \varphi}{\partial x} = \frac{1}{\varrho}\frac{\partial \psi}{\partial y}, \qquad v = \frac{\partial \varphi}{\partial y} = -\frac{1}{\varrho}\frac{\partial \psi}{\partial x} \tag{56.4}$$

which describes the relation between velocity potential and stream function. We denote by

$$\omega(x, y) = \frac{\partial \varphi(x, y; \tau)}{\partial \tau}\bigg|_{\tau=0}, \qquad \omega^* = \frac{\partial \psi(x, y; \tau)}{\partial \tau}\bigg|_{\tau=0} \tag{56.5}$$

the first partial derivatives at $\tau = 0$, i.e. the first variations of φ and ψ. It is easy to obtain new partial differential equations for ω and ω^* by differentiating Eqs. (56.4). MANWELL [*26*] pointed out, however, that these functions satisfy particularly simple differential equations in the hodograph plane; we shall now derive these equations.

We consider the functions $\varphi(x, y; \tau)$ and $\psi(x, y; \tau)$ in dependence on the variables u, v in the hodograph plane. We have then

$$\varphi(x, y; \tau) = \tilde{\varphi}(u, v; \tau), \qquad \psi(x, y; \tau) = \tilde{\psi}(u, v; \tau). \tag{56.6}$$

Observe that x and y are functions of u and v which in view of Eq. (23.5) depend in turn on τ. Let

$$\frac{\partial x(u, v; \tau)}{\partial \tau}\bigg|_{\tau=0} = \xi(u, v), \qquad \frac{\partial y(u, v; \tau)}{\partial \tau}\bigg|_{\tau=0} = \eta(u, v), \tag{56.7}$$

$$\tilde{\varphi}_\tau(u, v; 0) = \Omega(u, v), \qquad \tilde{\psi}_\tau(u, v; 0) = \Omega^*(u, v). \tag{56.7'}$$

If q and ϑ are the polar coordinates in the hodograph plane, by differentiating Eq. (23.6) with respect to τ and putting $\tau=0$ we derive the relations

$$\left.\begin{aligned} d\xi &= \left(\frac{\cos\vartheta}{q}\Omega_q - \frac{\sin\vartheta}{\varrho q}\Omega_q^*\right)dq + \left(\frac{\cos\vartheta}{q}\Omega_\vartheta - \frac{\sin\vartheta}{\varrho q}\Omega_\vartheta^*\right)d\vartheta\,,\\ d\eta &= \left(\frac{\sin\vartheta}{q}\Omega_q + \frac{\cos\vartheta}{\varrho q}\Omega_q^*\right)dq + \left(\frac{\sin\vartheta}{q}\Omega_\vartheta + \frac{\cos\vartheta}{\varrho q}\Omega_\vartheta^*\right)d\vartheta\,. \end{aligned}\right\} \tag{56.8}$$

On the other hand, differentiating Eq. (56.6) with respect to τ and putting $\tau=0$, we find by use of Eq. (56.4) for $\tilde\omega(u,v)=\omega(x,y)$ and $\tilde\omega^*(u,v)=\tilde\omega^*(x,y)$ the relations

$$\tilde\omega(u,v) = \Omega(u,v) - u\,\xi - v\,\eta; \qquad \tilde\omega^* = \Omega^* + \varrho\,v\,\xi - \varrho\,u\,\eta. \tag{56.9}$$

Differentiating Eq. (56.9) with respect to q and ϑ and using Eq. (56.8), we show easily that

$$\left.\begin{aligned} \tilde\omega_q &= -\frac{1}{q}(u\,\xi + v\,\eta), & \tilde\omega_\vartheta &= v\,\xi - u\,\eta,\\ \tilde\omega_q^* &= \frac{1}{q}\frac{d}{dq}(\varrho\,q)\,(v\,\xi - u\,\eta), & \tilde\omega_\vartheta^* &= \varrho\,(u\,\xi + v\,\eta)\,. \end{aligned}\right\} \tag{56.10}$$

Eliminating the first variations ξ, η we obtain the simple system

$$\tilde\omega_q^* = \frac{\varrho}{q}\left(1 - \frac{q^2}{c^2}\right)\tilde\omega_\vartheta, \qquad \tilde\omega_\vartheta^* = -\,\varrho\,q\,\tilde\omega_q. \tag{56.11}$$

We have used here the identity (23.9).

It is quite remarkable that the first variations $\omega(x,y)$ and $\omega^*(x,y)$ of the velocity potential and the stream function satisfy in the hodograph variables such a simple differential system very similar, moreover, to the system (23.8) for the functions themselves.

We introduce next the new variable

$$\sigma = -\int\limits_{q_s}^{q} \frac{dq}{\varrho q}\,, \tag{56.12}$$

which represents just a change of scale for the velocity. For small values of q we have $\sigma \sim -\log q$, showing the analogy between σ and the variable $-\log q$ which is convenient in the incompressible case. The sonic velocity q_s corresponds to $\sigma=0$. We have now the system

$$\tilde\omega_\sigma^* = -\,K(\sigma)\tilde\omega_\vartheta, \qquad \tilde\omega_\vartheta^* = \tilde\omega_\sigma \tag{56.11'}$$

with

$$K(\sigma) = \varrho^2\left(1 - \frac{q^2}{c^2}\right). \tag{56.13}$$

It is easily seen that

$$\sigma\,K(\sigma) \geq 0, \qquad \frac{dK}{d\sigma} > 0 \quad \text{for} \quad \sigma \geq 0; \tag{56.14}$$

that is, K is monotonic when $\sigma>0$, and yet in a little region $\sigma \leq 0$.

In the new variables σ, ϑ of the hodograph plane we have the following partial differential equation for the first variation $\tilde\omega$ of the velocity potential φ:

$$K(\sigma)\,\tilde\omega_{\vartheta\vartheta} + \tilde\omega_{\sigma\sigma} = 0. \tag{56.15}$$

It is elliptic in the subsonic part of the hodograph plane and hyperbolic in the supersonic half.

Eq. (56.15) holds for every type of variation of the velocity potential. We describe now a particular variation which is connected with the existence problem for transonic flows. We use again the symmetric profile P described at the beginning of the section and can restrict ourselves, therefore, to the upper half-plane $y>0$. We assume that the flow region D_0 in this half-plane has a one-to-one image Δ_0 in the hodograph plane; there are many explicitly known flow patterns of this kind. Next, we cut from the boundary arc $\widehat{AB}$ of P the subarc $\widehat{FG}$ by two Mach lines from a point E on the sonic line Γ. We replace the arc $\widehat{FG}$ by a sequence of continuously varying arcs which depend on a parameter τ. Let D_τ be the flow region corresponding to the parameter value τ and $\varphi(x, y; \tau)$ the corresponding velocity potential.

The first variation $\omega(x, y)$ of $\varphi(x, y; \tau)$ will be defined in the hodograph image Δ_0; it will satisfy the partial differential equation (56.15), and on the boundary of D_0 it will satisfy the condition

$$\frac{\partial \omega}{\partial n} = 0, \tag{56.16}$$

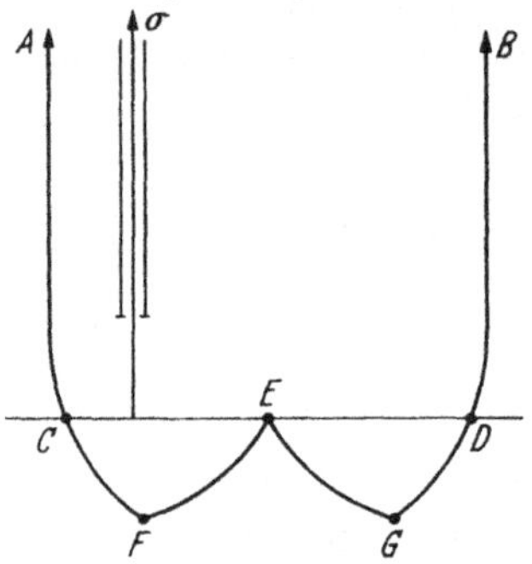

Fig. 18.

except for the critical arc $\widehat{FG}$ on P. Since we wish to deal with ω in the hodograph plane, we have to translate Eq. (56.16) in terms of the hodograph variables. Obviously, we can characterize Eq. (56.16) also by $\omega^* = 0$ on the boundary part considered. That is, by virtue of Eq. (56.11'),

$$\left.\begin{aligned} d\tilde{\omega}^* &= \tilde{\omega}^*_\sigma d\sigma + \tilde{\omega}^*_\vartheta d\vartheta \\ &= -K\tilde{\omega}_\vartheta d\sigma + \tilde{\omega}_\sigma d\vartheta = 0. \end{aligned}\right\} \tag{56.17}$$

Finally let D_Λ be the domain obtained from D_0 by removal of the curvilinear triangle $\Delta(E, F, G)$. The image Δ_Λ of D_Λ in the hodograph plane (Fig. 18) is bounded in the elliptic half-plane $\sigma>0$ by a piece of the σ-axis whose two edges correspond to the two pieces of the x-axis in the boundary of D_Λ and by two arcs $\widehat{AC}$ and $\widehat{BD}$ corresponding to the subsonic arcs on P denoted by the same letters. In the hyperbolic half- plane $\sigma<0$, Δ_Λ is bounded by the two non-characteristic arcs $\widehat{CF}$ and $\widehat{DG}$ and the two characteristics $\widehat{EF}$ and $\widehat{EG}$ of the differential equation (56.15) which correspond to the Mach lines in the physical plane. The stagnation points A, B of the flow correspond, of course, to the value $\sigma = +\infty$.

In the domain Δ_Λ the function $\tilde{\omega}(\sigma, \vartheta)$ satisfies the differential equation (56.15) with the homogeneous boundary conditions (56.17), except on the characteristic arcs $\widehat{EF}$ and $\widehat{EG}$. In order to make the boundary value problem for the mixed equation (56.15) a definite one, one must also prescribe the asymptotic behavior of $\tilde{\omega}$ for $\sigma = +\infty$. This behavior is entirely determined by the properties of the velocity potential at the stagnation points A and B.

Using the energy-integral method described in Sect. 55, MORAWETZ showed that the preceding boundary value problem admits only the solution $\tilde{\omega} = \text{const}$. Since $\varphi(x, y; \tau)$ is only determined up to an additive constant which may depend on τ, we may even assume that $\tilde{\omega} = 0$. Thus, the velocity potential $\varphi(x, y)$ of the transonic flow around P has the variation zero if the arc $\widehat{FG}$ of its supersonic boundary is arbitrarily deformed. This is impossible since on this variable arc the velocity potential must satisfy $\partial\varphi/\partial n = 0$ and must, therefore, adjust itself to the directional changes on this part of the boundary.

We have thus arrived at a contradictory result; it is based on the implicit assumption that the velocity potential $\varphi(x, y; \tau)$ depends differentiably on the parameter τ of the boundary deformation. We have proved in all rigor that the velocity potential does not depend differentiably on the supersonic part of the profile P. This result makes it highly probable that in the transonic case the entire boundary value problem for the velocity potential is incorrectly posed and has, in general, no solution.

MORAWETZ [*25*] has also shown that if the profile P has a continuous transonic solution φ for the parameter value $\tau=0$ and if we even admit weak shock discontinuities for the solutions with $\tau \neq 0$ we cannot have a first variation ω in the family of solutions. Thus even a slight change of the profile seems to engender strong shock discontinuities for the flow pattern.

It should be remarked that the preceding considerations make strong use of the fact that the supersonic enclosure is bounded by sonic arcs and boundary arcs only. The situation may be different for transonic flows whose supersonic regions are infinite or bounded in part by shock lines (see, for example, [*27*], [*28*]).

57. Arguments against the existence of transonic flows for arbitrary profiles. Various intuitive and qualitative arguments have been given for the nonexistence of continuous transonic flow around arbitrary profiles. One argument refers to the different dependence on boundary data of solutions of elliptic and hyperbolic equations. Consider, for example, the equation

$$\varepsilon\, \varphi_{yy} + \varphi_{xx} = 0, \qquad \varepsilon = \begin{Bmatrix} +1 & \text{for} & y > 0 \\ -1 & \text{for} & y < 0 \end{Bmatrix} \tag{57.1}$$

which is of the general type (54.1) but has a discontinuous coefficient $K(y)$. We can write down a great number of explicit solutions of Eq. (57.1); for example

$$\varphi_n(x, y) = \frac{1}{n} \sin n x \left(e^{\sqrt{\varepsilon}\, n y} + e^{-\sqrt{\varepsilon}\, n y}\right). \tag{57.2}$$

These solutions are continuous across the x-axis and obviously are bounded for $y<0$. If we let n tend to infinity, φ_n will converge to zero uniformly in the hyperbolic half-plane $y<0$ while it will converge to infinity almost everywhere in the upper half-plane. If Eq. (57.1) were the differential equation of some physical situation we could describe this result as follows: Any small perturbation $\varphi_n(x, y)$ coming from the hyperbolic region would be indefinitely magnified in the elliptic half-plane. The smaller the perturbation, the stronger will be its influence for $y>0$. Clearly, such a physical theory would be incorrectly formulated and the differential equation (57.1) would be useless for a description of the phenomenon.

The actual differential equation of the stream function in the hodograph plane has a continuous coefficient $K(y)$ but in common with (57.1) it has the fact that the sign of $K(y)$ changes across the x-axis. BUSEMANN [*29*] has discussed the asymptotic character of solutions of this equation analogous to Eq. (57.2) and has shown for them precisely the same behavior. He has pointed out that mathematically the transonic flow problem represents the fusion of an elliptic and a hyperbolic solution along the sonic line; the continuity requirement on this solution leads to an intial value problem for the elliptic solution which is incorrectly posed to the extent that the solution does not depend continuously on the data. This is indeed a strong argument against the possibility of arbitrary transonic flows [*30*].

In order to describe a more geometric consideration we express the equations of the steady irrotational and two-dimensional flow in a particular set of variables.

Let q and ϑ denote the speed and directional angle of the flow and let $\partial/\partial s$ and $\partial/\partial n$ denote differentiations along the arc length s and the normal of the stream line through the point considered. Clearly we have

$$\frac{\partial}{\partial s} = \cos\vartheta \frac{\partial}{\partial x} + \sin\vartheta \frac{\partial}{\partial y}, \qquad \frac{\partial}{\partial n} = -\sin\vartheta \frac{\partial}{\partial x} + \cos\vartheta \frac{\partial}{\partial y}. \tag{57.3}$$

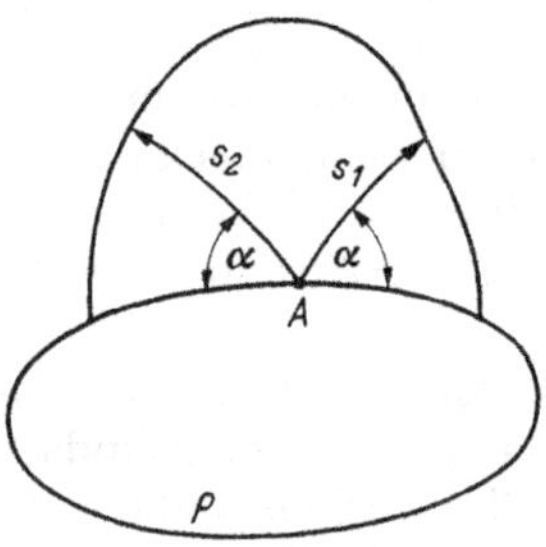

Fig. 19.

If $\varphi(x, y)$ is the velocity potential of the flow, we have

$$q^2 = \varphi_x^2 + \varphi_y^2, \qquad \vartheta = \operatorname{arc\,tan} \frac{\varphi_y}{\varphi_x} \tag{57.4}$$

and hence we can compute easily

$$\left.\begin{aligned} q^2 \frac{\partial q}{\partial s} &= \varphi_{xx}\varphi_x^2 + 2\varphi_{xy}\varphi_x\varphi_y + \varphi_{yy}\varphi_y^2, \\ q^2 \frac{\partial q}{\partial n} &= (\varphi_x^2 - \varphi_y^2)\,\varphi_{xy} + \varphi_x\varphi_y(\varphi_{yy} - \varphi_{xx}) = q^3 \frac{\partial\vartheta}{\partial s}, \\ q^3 \frac{\partial\vartheta}{\partial n} &= \varphi_{xx}\varphi_y^2 - 2\varphi_{xy}\varphi_x\varphi_y + \varphi_{yy}\varphi_x^2. \end{aligned}\right\} \tag{57.5}$$

We can now use the differential equation (53.1) satisfied by the velocity potential in order to simplify the first and last equation. We find

$$q^2 \frac{\partial q}{\partial s} = c^2(\varphi_{xx} + \varphi_{yy}), \qquad q^3 \frac{\partial\vartheta}{\partial n} = (q^2 - c^2)(\varphi_{xx} + \varphi_{yy}). \tag{57.5'}$$

We arrive thus finally at the system of differential equations between q and ϑ

$$\frac{\partial q}{\partial n} = q \frac{\partial\vartheta}{\partial s}, \qquad (M^2 - 1)\frac{\partial q}{\partial s} = q \frac{\partial\vartheta}{\partial n}; \qquad M = \frac{q}{c}. \tag{57.6}$$

M is a known function of q; for supersonic motion we have $M > 1$ and putting

$$\sin\alpha = \frac{1}{M}, \qquad 0 \leqq \alpha \leqq \frac{\pi}{2}, \tag{57.7}$$

we know that $\vartheta + \alpha$ and $\vartheta - \alpha$ are the directions of the two Mach lines through the point considered.

Now let us consider again a flow around a profile P with a supersonic enclosure (Fig. 19). Let A be a point on a supersonic arc of P; there are two Mach lines starting from A and running to the sonic boundary of the enclosure. Let s_1 and s_2 be the arc length along the Mach lines counted from A; by the rules of directional differentiation we can then compute

$$\frac{\partial q}{\partial s_1} = \frac{\partial q}{\partial s}\cos\alpha + \frac{\partial q}{\partial n}\sin\alpha, \qquad \frac{\partial q}{\partial s_2} = -\frac{\partial q}{\partial s}\cos\alpha + \frac{\partial q}{\partial n}\sin\alpha. \tag{57.8}$$

Since the speed is monotonic along a Mach line and since we are running on each from supersonic to sonic values we have clearly $\frac{\partial q}{\partial s_i} < 0$ $(i = 1, 2)$. Since the first terms on the right hand sides of the above two equations change sign, the second terms must be the decisive ones and, we have the inequality:

$$\left|\frac{\partial q}{\partial n}\right| > \cot\alpha \left|\frac{\partial q}{\partial s}\right|. \tag{57.9}$$

We use now the first equation (57.6); since the stream line follows the profile P at A, we have there

$$\frac{\partial\vartheta}{\partial s} = \varkappa, \qquad \varkappa = \text{curvature of } P \text{ at } A, \tag{57.10}$$

and Ineq. (57.9) becomes therefore

$$|\varkappa| > \cot\alpha \left|\frac{\partial q}{\partial s}\right| q^{-1}. \tag{57.11}$$

Inequality (57.11) is a geometric restriction for arcs on the profile P which can possibly bound a supersonic enclosure. If we were to replace a very short arc of P by a straight segment along such a boundary arc, no corresponding transonic flow could exist in vicw of the inequality (57.11). This simple argument shows again the improbability of transonic flow patterns past arbitrary profiles [*31*], [*32*].

It is hard to evaluate the actual physical significance of the preceding mathematical arguments. Clearly the fiction of an ideal non-viscous fluid flow must break down near the boundary of the profile and a thin viscous layer is built up along P. Its presence makes the concepts of non-analytic curves or infinitesimal variations rather vague. However, it seems probable that rapid changes of flow pattern with the data and instability phenomena are to be expected in actual transonic flow. The entire Taylor problem is still a fruitful subject for further mathematical and experimental research.

58. The idealized fluid of Tomotika-Tamada. Nozzle flow. The question of explicit solutions for transonic flows is naturally an important one. While the Taylor problem discussed so far seems to yield a negative answer in the general case, special solutions are important as starting solutions which may be adjusted to more general situations by the addition of shock lines. Transonic solutions in nozzle flows are not excluded by our general mathematical considerations and will in fact be developed in this section. A successful approach to transonic flows which avoids excessive analytical complications rests on the replacement of a real fluid by an idealized fluid whose behavior is similar enough to be physically significant and whose equation of state leads to simple differential equations [*33*], [*37*].

We consider the hodograph variables q, ϑ as functions of the velocity potential φ and the stream function ψ and transform the basic equations in the hodograph plane (23.8) into the form

$$\vartheta_\varphi = \frac{\varrho}{q}\cdot q_\psi\,, \qquad \vartheta_\psi = \frac{1}{q\,\varrho}\,(M^2-1)\,q_\varphi\,. \tag{58.1}$$

We used here only the law of inverse differentiation and the identity (23.9). The Eqs. (58.1) can be simplified by introducing the new speed variable

$$w = \int\limits_{q_s}^{q} \frac{\varrho}{q}\,dq\,, \qquad q_s = \text{sonic speed}. \tag{58.2}$$

Then Eqs. (58.1) become

$$\vartheta_\varphi = w_\psi\,, \qquad \vartheta_\psi = \frac{1}{\varrho^2}\,(M^2-1)\,w_\varphi\,. \tag{58.1'}$$

The term $\varrho^{-2}(M^2-1)$ is obviously a function of w. From Eq. (24.2) we conclude

$$M^2 - 1 = \frac{\frac{\gamma+1}{\gamma-1}\,q^2 - q_1^2}{q_1^2 - q^2}\,, \qquad q_s = \sqrt{\frac{\gamma-1}{\gamma+1}}\,q_1\,, \tag{58.3}$$

and it is easy to see that near $w=0$, i.e. near the sonic speed q_s,

$$\frac{1}{\varrho^2}\,(M^2-1) = \frac{1}{\varrho_s^3}\,(\gamma+1)\,w + 0(w^2)\,, \tag{58.3'}$$

where ϱ_s is the density of the fluid at the sonic speed q_s.

If we are interested only in nearly uniform transonic flow where the speed never deviates much from q_s and where w stays small, we may neglect the higher order terms of w in Eq. (58.3′) and obtain the differential system,

$$\vartheta_\varphi = w_\psi, \quad \vartheta_\psi = 2k w w_\varphi, \quad k = \frac{\gamma+1}{2\varrho_s^3}. \tag{58.4}$$

This system has been discussed extensively by TOMOTIKA and TAMADA [33]. It may be interpreted also as the correct differential system for a gas with a changed equation of state. Near the sonic speed the real gas and the hypothetical one have a very similar behavior.

If we eliminate ϑ from the system (58.4), we obtain the non-linear partial differential equation

$$w_{\psi\psi} = k(w^2)_{\varphi\varphi}. \tag{58.5}$$

This equation is simple enough to lead to various interesting particular solutions. TOMOTIKA and TAMADA set up for example:

(a) $w = f(\varphi + \lambda\psi)$; (b) $w = F(\varphi) + G(\psi)$; (c) $w = F(\varphi)\cdot G(\psi)$;

(d) $w = F(\psi) + G(\psi)\,\varphi^2$; (e) $w = \left(Z(\varphi+\psi^2) + 2\psi^2\right)k^{-1}$.

In each of these cases an ordinary differential equation for the functions is obtained and the ensuing flow pattern is described. We discuss the interesting pattern arising from the set-up (e). Inserting this expression for w into Eq. (58.5) we obtain for $Z(u)$ the differential equation

$$\frac{d}{du}\left(Z\frac{dZ}{du}\right) - \frac{dZ}{du} - 2 = 0 \tag{58.6}$$

and since φ is determined only up to an additive constant we may consider the integral of Eq. (58.6),

$$Z\frac{dZ}{du} - Z - 2u = 0, \tag{58.7}$$

without loss of generality. The differential equation (58.7) has the integrating factor

$$m(Z,u) = (Z-2u)^{-1}(Z+u)^{-1} \tag{58.7'}$$

and hence, the integral

$$(Z-2u)^2(Z+u) = 2a^3, \tag{58.8}$$

where a is a constant of integration. By definition, we have

$$w = [Z(u) + 2\psi^2]\,k^{-1}, \quad u = \varphi + \psi^2, \tag{58.9}$$

and the Eqs. (58.4) can be satisfied by choosing

$$\vartheta = \frac{2}{k}\psi\left[Z(u) + 2\varphi + \frac{2}{3}\psi^2\right]. \tag{58.10}$$

The Eqs. (58.9) and (58.10) describe the flow completely; we see that everything is symmetric with respect to the axis $\psi=0$. If we choose the region

$$|\psi| \leqq \psi_0 \tag{58.11}$$

we obtain a flow bounded by two rigid walls corresponding to $\psi = \pm\psi_0$ with the straight stream line $\psi=0$. This may be interpreted as a flow through a Laval nozzle with the axis corresponding to $\psi=0$.

In order to investigate the possible flows of the idealized fluid through the nozzles obtained it is important to understand the (Z, u)-relation. Observe that for $a=0$ Eq. (58.8) leads to the two straight lines $Z=2u$ and $Z=-u$ (Fig. 20). For $a<0$ the curve $Z(u)$ consists of two branches each of which has these lines as asymptotes. One branch lies entirely in the half-plane $Z<0$ while the other lies in the half-plane $u<0$.

We observe that by Eq. (58.9) $w=Z(\varphi)\,k^{-1}$ determines the velocity distribution along the nozzle axis $\psi=0$. Therefore, if we choose, the first branch of the $Z(u)$ curve where Z is negative throughout, we see that w is negative on the nozzle axis. This means by our definition (58.2) of the speed variable that the velocity is subsonic all along the axis. If the nozzle is

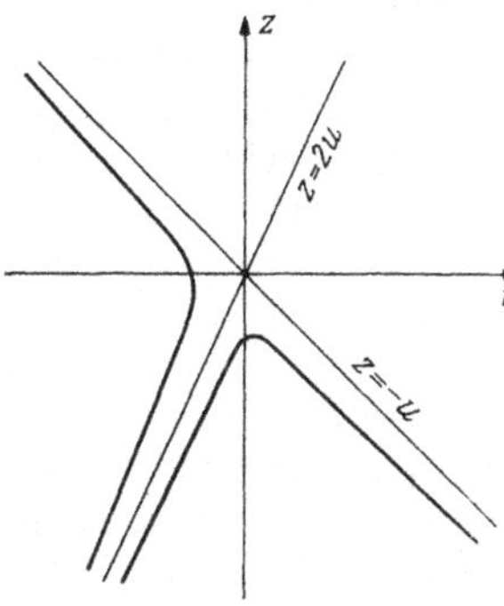

Fig. 20.

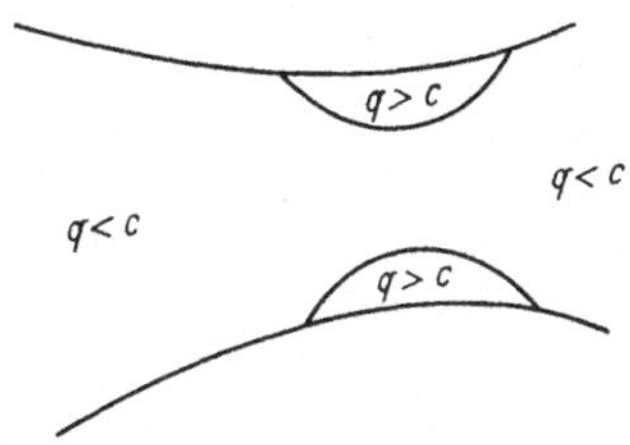

Fig. 21.

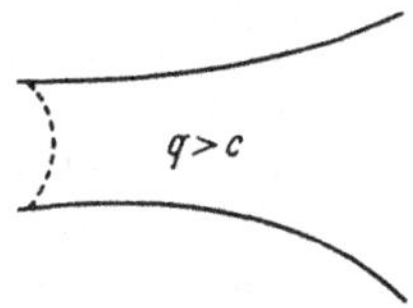

Fig. 22. $Z>0$.

wide enough, ψ may become so large that finite pockets with positive w-values, that is, supersonic speed, may arise. Tomotika and Tamada call this type of flow the "Taylor type" of transonic flow through a nozzle (Fig. 21).

If we wish to use the branch of the curve $Z(u)$ which lies in the half-plane $u<0$, we have to consider its two halves with $Z>0$ and $Z<0$ separately. Indeed, according to Eq. (58.7) we have $Z'(u)=\infty$ for $Z=0$, and we can pass from the upper half to the lower only by crossing the u-axis. Observe that by Eqs. (58.1) and (58.2)

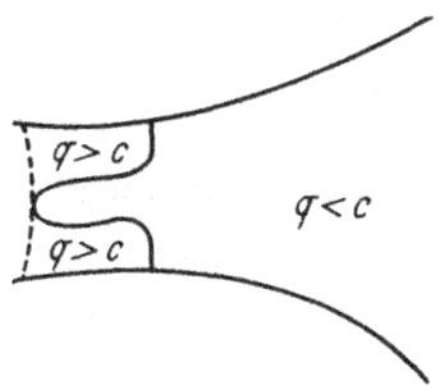

Fig. 23. $Z<0$.

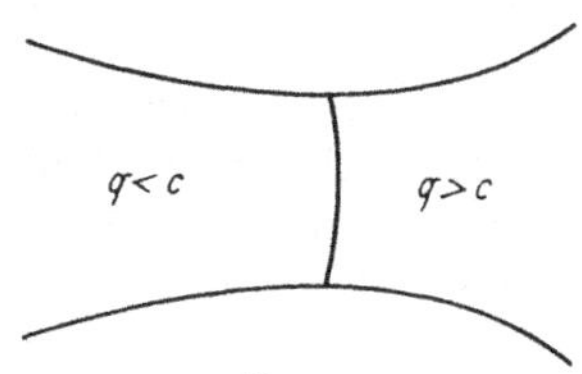

Fig. 24.

$$\left.\begin{aligned}\frac{\partial(q,\vartheta)}{\partial(x,y)} &= \frac{\partial(q,\vartheta)}{\partial(\varphi,\psi)}\cdot\frac{\partial(\varphi,\psi)}{\partial(x,y)} = \frac{q}{\varrho}\left[(M^2-1)\,q_\varphi^2 - q_\varrho^2 q_\psi^2\right]\\ &= \frac{q^3}{\varrho^3}\left[(M^2-1)\,w_\varphi^2 - q_\varrho^2 w_\psi^2\right].\end{aligned}\right\} \tag{58.12}$$

For $Z=0$ we have obviously infinite values for w_φ and w_ψ, and the transition from the (x, y)-plane to the hodograph plane breacks down. At the point $Z=0$ the flow reaches a limit line as defined in Sect. 33.

If we use the arc $Z>0$, $u<0$, in view of the velocity formulas (58.2) and (58.9) we shall obtain an entirely supersonic motion bounded on one side by a limit line and extending on the other to infinity (Fig. 22). If we use the lower half $Z<0$ of this branch, however, we will find (Fig. 23) a flow which is in its larger part subsonic but which becomes supersonic in a region adjacent to the limit line which bounds it to one side. The limit line, which is essentially a mathematical singularity due to the Jacobian of the mapping onto the hodograph plane, may be interpreted as the location of a shock discontinuity in the flow.

There is a simple flow pattern in a Laval nozzle which was studied by Meyer [34] and which consists of an infinite supersonic half of the nozzle and an infinite subsonic part; it may be called "Meyer's type" of transonic nozzle flow (Fig. 24).

It may be composed from the elements corresponding to the parts $Z<0$ and $Z>0$ in the above solution based on the second branch of the $Z(u)$-curve.

A particularly interesting problem in the flow through a Laval nozzle is the transition from the Taylor type to the Meyer type flow. If we consider various solutions belonging to different choices of the constant of integration a in Eq. (58.8) we may investigate this problem as follows. We start with $a<0$ and let a converge to zero. The two branches of the curve $Z(u)$ approach each other indefinitely and fuse into the asymptotes for $a=0$. This permits us to study the transition of type in detail. It is true that the form of the nozzle itself will also change with a since the stream lines $\psi=\pm\psi_0$ have an image in the physical plane which depends on the function $Z(u)$. Fortunately, it can be shown that the nozzle form is practically unchanged under the a-variation, so that the transition may be considered as taking place in a fixed nozzle, being due only to a change of the rate of flow in the nozzle.

The transition may be described as follows: We start with a Taylor type nozzle flow. If we increase the rate of flow the supersonic pockets will increase and come nearer to each other. They will finally meet on the axis; this is the moment when for the first time sonic speed occurs on the nozzle axis. If the rate of flow continues to grow the pattern changes to MEYER'S type and shock waves will appear. This change goes on continuously in the upstream half of the nozzle and is naturally abrupt in the downstream region (see also [1], [35], [36]).

The preceding theory holds precisely only for the idealized fluid, but it is surely very indicative for the behavior of the flow of a real fluid through a Laval nozzle.

The elementary theory of transonic flow of a real gas through a nozzle is based on the method of series developments [34] to [36]. We suppose the nozzle to be symmetric with respect to the x-axis and assume for the x-component of the velocity on the nozzle axis the power series development

$$u(x,0)=u_0(x)=q_s(1+ax+bx^2+\cdots). \tag{58.13}$$

We have obviously chosen the origin $x=y=0$ as that point on the axis where the subsonic motion changes into supersonic motion. For reason of symmetry, the y-component of the velocity must vanish on the nozzle axis:

$$v(x,0)\equiv 0. \tag{58.13'}$$

The velocity potential $\varphi(x,y)$ of the flow satisfies the second order partial differential equation (53.1) with the initial conditions on the x-axis

$$\varphi_x(x,0)=u_0(x), \qquad \varphi_y(x,0)=0. \tag{58.14}$$

The coefficient of φ_{yy} in the Eq. (53.1) is c^2-v^2 and in view of Eq. (58.13') is non-zero on the x-axis. Hence the Cauchy-Kovalevski theory of partial differential equations guarantees the possibility of developing φ into a power series of y

$$\varphi(x,y)=\sum_{\nu=0}^{\infty}A_\nu(x)\,y^\nu, \tag{58.15}$$

where the $A_\nu(x)$ are analytic functions of x. φ is determined except for an additive constant and the $A_\nu(x)$ may be computed successively by means of the initial data and the differential equation. In our particular problem, the symmetry of the data leads to the identical vanishing of all A_ν with odd subscript such that only even powers of y will occur in Eq. (58.15).

The usual way of applying the series method to the nozzle flow is the indirect approach of starting with an arbitrary but fixed series (58.13) and determining the nozzle walls as a pair of symmetric stream lines of the ensuing flow. The limitation of the method lies in the lack of knowledge about the range of convergence of the series (58.15).

An easy and straightforward calculation leads to the following series for the velocity potential:

$$\left.\begin{aligned}\varphi(x, y) = q_s\Big[\Big(x + \frac{a}{2}\, x^2 + \cdots\Big) + \Big(\frac{\gamma+1}{2}\, a^2 x + \cdots\Big) y^2 + \\ + \Big(\frac{(\gamma+1)^2}{24}\, a^3 + \cdots\Big) y^4 + \cdots\Big].\end{aligned}\right\} \tag{58.16}$$

The velocity field (58.13), (58.13') on the nozzle axis leads thus to the velocity field in the large

$$\left.\begin{aligned}u &= q_s\Big[1 + a x + \frac{\gamma+1}{2}\, a^2 y^2 + \cdots\Big],\\ v &= q_s\Big[(\gamma+1)\, a^2 x y + \frac{(\gamma+1)^2}{6}\, a^3 y^3 + \cdots\Big].\end{aligned}\right\} \tag{58.17}$$

In order to discuss some qualitative aspects of nozzle flow, in Eq. (58.17) we shall neglect all terms beyond those explicitly shown. We suppose that we have selected a stream line sufficiently near to the axis as nozzle wall. Let x_0 be the abscissa of its throat and let $2d$ be the narrowest nozzle diameter, attained at that point. Clearly, the velocity of the flow will be horizontal at the throat point (x_0, d), and we find thus from Eq. (58.17) the relation

$$x_0 = -\frac{\gamma+1}{6}\, a\, d^2. \tag{58.18}$$

This result shows that the sonic point on the nozzle axis (i.e. $x=0$) lies downstream from the throat.

We can also calculate with the same degree of approximation the equation of the sonic line, along which $u^2+v^2=q_s^2$. We find easily

$$x + \frac{\gamma+1}{2}\, a y^2 = 0, \tag{58.19}$$

i.e. the line of transition from subsonic to supersonic flow is a parabola lying toward the subsonic side of its vertex.

The series method allows numerous applications and has been used extensively. It can be extended immediately to three-dimensional flows through nozzles with axial symmetry. While it is most useful for qualitative insight, it must be applied with caution because of the difficult convergence problems involved. For finer investigations the theory of an idealized transonic gas flow is mathematically preferable and it is in good agreement with the physical evidence as long as the deviation from sonic speed is not too large.

59. Solutions for TRICOMI's equation. Particular solutions for TAYLOR's problem. We return to the mathematical theory of the idealized gas of TOMOTIKA-TAMADA. While the non-linear system (58.4) provides us with a number of interesting particular solutions, the natural approach to a general solution theory is through the linear system of equations which we obtain when we consider φ and ψ as functions of w and ϑ. Indeed, Eqs. (58.4) transform into

$$\varphi_w = 2k\, w\, \psi_\vartheta, \qquad \varphi_\vartheta = \psi_w. \tag{59.1}$$

Eliminating φ we obtain the simple partial differential equation for ψ:

$$\psi_{ww} - 2k\, w\, \psi_{\vartheta\vartheta} = 0, \tag{59.2}$$

which reduces by the substitution

$$z = (2k)^{\frac{1}{3}} w \tag{59.3}$$

exactly to the Tricomi equation in the hodograph variables z, ϑ:

$$\psi_{zz} - z\psi_{\vartheta\vartheta} = 0. \tag{59.2'}$$

Particular solutions of Eq. (59.2') can be obtained by separation of variables. In fact,

$$\psi = Z_\nu(z)\, e^{i\nu\vartheta} \tag{59.4}$$

leads to the ordinary differential equation for $Z_\nu(z)$

$$Z_\nu'' + \nu^2 z Z_\nu = 0 \tag{59.5}$$

with the solution

$$Z_\nu(z) = z^{\frac{1}{2}} J_{\pm\frac{1}{3}}\left(\tfrac{2}{3}\nu z^{\frac{3}{2}}\right) \tag{59.6}$$

where $J_\alpha(t)$ is the Bessel function of order α.

A more ingenious assumption for $\psi(z, \vartheta)$ is

$$\psi(z, \vartheta) = z^\alpha\, Y_\alpha\left(\tfrac{9}{4}\vartheta^2 z^{-3}\right), \tag{59.7}$$

which transforms Eq. (59.2') into the ordinary differential equation for $Y_\alpha(t)$:

$$t(1-t)\, Y_\alpha''(t) + \left[\tfrac{1}{2} - \left(\tfrac{4}{3} - \tfrac{2}{3}\alpha\right)t\right] Y_\alpha'(t) - \tfrac{1}{9}\alpha(\alpha-1)\, Y_\alpha(t) = 0. \tag{59.8}$$

This equation is a special case of the hypergeometric differential equation, and since we can choose α arbitrarily we obtain a great number of particular solutions with various kinds of multivaluedness and singularities. For a discussion of the set of solutions (59.7) see [*42*]. We content ourselves with mentioning the case $\alpha = \frac{1}{2}$, where the solutions reduce to simple algebraic expressions. It is easily seen that

$$\psi(z, \vartheta) = \left[\sqrt{\vartheta^2 - \tfrac{4}{9} z^3} \pm \vartheta\right]^{\frac{1}{3}} \tag{59.9}$$

are solutions of Eq. (59.2') and of the form (59.7). Since the coefficients of Eq. (59.2') are independent of ϑ, we can obtain a more general solution by putting $\vartheta - i\lambda$ instead of ϑ. ψ will now become complex-valued; but since Eq. (59.2') is linear, the real and imaginary parts of ψ will represent real solutions of Eq. (59.2').

TOMOTIKA and TAMADA discussed in detail the solution

$$\left.\begin{aligned} \psi(w, \vartheta) = &\left[\sqrt{(\vartheta - i\lambda)^2 - \frac{8k}{9} w^3} - (\vartheta - i\lambda)\right]^{\frac{1}{3}} - \\ &- \left[\sqrt{(\vartheta - i\lambda)^2 - \frac{8k}{9} w^3} + (\vartheta - i\lambda)\right]^{\frac{1}{3}}. \end{aligned}\right\} \tag{59.10}$$

The point $\vartheta = 0$, $w = -\left[(9/8k)\,\lambda^2\right]^{\frac{1}{3}}$ is a branch of order $\frac{1}{2}$. The solutions Re $\{\psi\}$ and Im $\{\partial\psi/\partial\vartheta\}$ will have this point as branch point of order $\frac{1}{2}$ and $-\frac{1}{2}$, respectively. The second solution has this critical point as a source singularity and the authors show that linear combinations of these two particular solutions lead to flow patterns of the following kind: The flow is subsonic and uniform at infinity; it encloses a profile P and has supersonic regions adjacent to P. In order to obtain a result of physical interest one has to restrict the hodograph image of the flow to a neighborhood of the sonic line $w = 0$, for only there do the idealized and the real fluid behave sufficiently alike. The profiles obtained, therefore, are very slender and with sharp leading and trailing edges so that no stagnation points will occur. The flow patterns obtained are of interest in connection with the Taylor problem discussed above. We see that there exists a great variety of continuous transonic flows, uniform and subsonic at infinity, around profiles of various shapes.

From the fact that Eq. (59.10) is a solution of Eq. (59.2) we can easily draw the conclusion that the Tricomi equation

$$\psi_{zz} + z\,\psi_{\vartheta\vartheta} = 0 \tag{59.2''}$$

has a particular solution of the implicit form

$$\tfrac{1}{3}\psi^3 + z\,\psi = \vartheta. \tag{59.11}$$

MARTIN and THICKSTUN [38] obtained this result by considering in Eq. (59.2″) ψ and z as the independent variables and by looking for a solution of the form $\vartheta = A(\psi) + z\,B(\psi)$. Their method leads to various other particular solutions for the Tricomi equation.

Further interesting solutions of Eq. (59.2′) can be obtained from Eq. (59.9) by differentiating these solutions with respect to ϑ. If we use the notation

$$r = \sqrt{\vartheta^2 - \tfrac{4}{9} z^3} \tag{59.12}$$

we obtain from Eq. (59.9) by differentiating twice with respect to ϑ the new solutions

$$\psi_1(z,\vartheta) = r^{-\frac{5}{3}}\left(1 + \frac{\vartheta}{r}\right)^{\frac{1}{3}}\left(1 - \frac{3\vartheta}{r}\right), \qquad \psi_2(z,\vartheta) = r^{-\frac{5}{3}}\left(1 - \frac{\vartheta}{r}\right)^{\frac{1}{3}}\left(1 + \frac{3\vartheta}{r}\right). \tag{59.13}$$

These solutions have singularities along the lines $r = 0$ and in particular at the point $\vartheta = 0$, $z = 0$. They were used by GUDERLEY and YOSHIHARA [39], [40] to construct a flow coming from infinity with sonic speed and meeting the nose of a wedge-shaped obstacle.

A transonic flow past an obstacle with the free stream Mach number 1 is particularly puzzling from the theoretical point of view. By Eq. (22.15′) we have

$$\frac{d}{dq}(\varrho\,q) = \varrho\,(1 - M^2), \tag{59.14}$$

which shows that the flow density $\varrho\,q$ has a maximum for $M = 1$. Thus the fluid comes from infinity with the highest flow density possible and this quantity is reduced by the presence of the obstacle. On the other hand, the flow has to bypass the obstacle and carry away all the fluid streaming in from infinity. This seems impossible, and the existence of a steady flow of this kind is improbable. The result of GUDERLEY and YOSHIHARA will describe only a part of the flow, while the downstream part is separated from this solution by a line of discontinuity.

60. Explicit solutions for another idealized fluid. The system of differential equations (59.1) holds for the velocity potential and the stream function of a fluid with a very special equation of state which was introduced for the sake of mathematical convenience. We always have, however, the system

$$\varphi_w = \frac{1}{\varrho^2}(M^2 - 1)\,\psi_\vartheta, \qquad \varphi_\vartheta = \psi_w, \tag{60.1}$$

which is obtained from the general system (58.1′) by a simple interchange of dependent and independent variables. The function $\frac{1}{\varrho^2}(M^2 - 1)$ depends on w in a way determined by the equation of state of the fluid considered. In the preceding idealization, we took its value as $k\,w$ since this is the best linear approximation to the actual function at the point $w = 0$.

We shall now consider a new hypothetical fluid whose behavior approximates that of an adiabatic gas over a much wider speed range and allows even the treatment of a flow with stagnation points. Following TOMOTIKA and TAMADA, we assume

$$\frac{1}{\varrho^2}(M^2-1) = a\,(e^{2\alpha w}-1) \tag{60.2}$$

and try to adjust the constants a and α as adequately as possible. This assumption leads to an equation of state which is more realistic from a physical point of view and whose mathematical theory is still simple enough, as will be shown below.

In order to match the behavior of an adiabatic gas at the stagnation points $w=-\infty$ and at the sonic points $w=0$, we require

$$a = \varrho_0^{-2} \qquad 2a\,\alpha = (\gamma+1)\,\varrho_s^{-3} \tag{60.3}$$

where ϱ_0 and ϱ_s denote the density of the fluid at rest and for sonic speed, respectively. For $p=\varrho^\gamma$, that is, for the equation of state of the adiabatic gas, we compute from Eqs. (24.2) and (24.3)

$$\frac{\varrho_s}{\varrho_0} = \left(\frac{2}{\gamma+1}\right)^{\frac{1}{\gamma-1}}. \tag{60.4}$$

If we choose our units in such a way that $\varrho_s=1$, we find therefore

$$a = \left(\frac{\gamma+1}{2}\right)^{-\frac{2}{\gamma-1}}, \qquad \alpha = \left(\frac{\gamma+1}{2}\right)^{\frac{\gamma+1}{\gamma-1}}. \tag{60.3'}$$

It can be shown that the function (60.2) coincides very satisfactorily with the correct function $\frac{1}{\varrho^2}(M^2-1)$ over the entire subsonic region and even in a small supersonic interval.

Before discussing the mathematical theory of the new fluid we mention briefly its speed-density relation, from which the pressure-density relation can be found by the Bernoulli formula. We have by Eq. (23.9)

$$\frac{1}{\varrho^2}(M^2-1) = \frac{q^2}{\varrho}\left(\frac{1}{\varrho q}\right)' \tag{60.5}$$

where the dash denotes differentiation with respect to q. Hence, Eqs. (58.2) and (60.2) lead to the second order differential equation

$$2\alpha\frac{\varrho}{q}\left\{a + \frac{q^2}{\varrho}\left(\frac{1}{\varrho q}\right)'\right\} = \left\{\frac{q^2}{\varrho}\left(\frac{1}{\varrho q}\right)'\right\}' \tag{60.6}$$

for the unknown function $\varrho(q)$, from which it can be easily calculated.

Using Eq. (60.2) and eliminating φ from the system (60.1), we obtain the following differential equation for the stream function in the hodograph variables

$$\psi_{ww} + a(1-e^{2\alpha w})\,\psi_{\vartheta\vartheta} = 0. \tag{60.7}$$

Introducing the new speed variable

$$t = e^{\alpha w} = \exp\left\{\alpha\int\limits_{q_s}^{q}\frac{\varrho}{q}\,dq\right\} \tag{60.8}$$

and the new angular variable

$$\beta = \frac{\alpha}{\sqrt{a}}\,\vartheta, \tag{60.9}$$

we can express Eq. (60.7) in the standard form

$$L[\psi] = t^2 \psi_{tt} + t \psi_t + (1 - t^2) \psi_{\beta\beta} = 0. \tag{60.10}$$

Observe that the sonic speed corresponds now to $t=1$ and that Eq. (60.10) is still of the same mixed type as Eq. (60.7).

In order to obtain a large number of solutions for Eq. (60.10), we remark that the functions of the form

$$G(\tau \sin t - t - \beta) = F(\tau, \beta; t) \tag{60.11}$$

with arbitrary twice continuously differentiable G satisfy the partial differential equation

$$\tau^2 F_{\tau\tau} + \tau F_\tau + (1 - \tau^2) F_{\beta\beta} = 2 F_{t\beta} - F_{tt} = \frac{\partial}{\partial t} [2 F_\beta - F_t]. \tag{60.12}$$

Now let C be a closed curve in the complex t-plane and consider the integral

$$\psi(\tau, \beta) = \oint_C G[\tau \sin t - t - \beta]\, dt \tag{60.13}$$

depending on the arbitrary analytic function G. Clearly, in view of Eq. (60.12)

$$L[\psi] = \oint_C d[2 F_\beta - F_t] = 0, \tag{60.13'}$$

provided that G' is a single-valued function of its argument. Thus, every function G which leads to a non-vanishing contour integral (60.13) yields a solution for the partial differential equation (60.10).

As an illustrative example which will also be applied soon we consider the particular choice

$$G(z) = \frac{\lambda}{2\pi i} (\lambda - e^z)^{-1} \tag{60.14}$$

and take C as a small circle in the t-plane around a pole of $G(\tau \sin t - t - \beta)$. Such a pole will occur for fixed real values of τ, β and λ if $t = t_0$ with

$$\tau \sin t_0 - t_0 - \beta = \log \lambda \tag{60.14'}$$

and the residue of the integral (60.13) at that point will be

$$\psi(\tau, \beta) = \frac{1}{1 - \tau \cos t_0}. \tag{60.15}$$

By our general result this function of τ and β must be a solution of the differential equation (60.10). t_0 is, of course, to be understood as a function of τ and β by means of the implicit equation (60.14').

The particular solution (60.15) has a singularity for the point $\tau_\infty, \beta_\infty$ for which the two equations

$$1 - \tau_\infty \cos t_0 = 0, \qquad \tau_\infty \sin t_0 - t_0 - \beta_\infty - \log \lambda = 0 \tag{60.16}$$

have a common root; since this means that $\tau_\infty \sin t - t - \beta_\infty - \log \lambda = 0$ has a double root in this case, we see easily that the singular point of ψ corresponds to a branch point in the (τ, β)-plane with the order $-\frac{1}{2}$. We have thus found a solution on a two-sheeted Riemann surface over the hodograph plane with a source singularity at the point $\tau_\infty, \beta_\infty$. We know already from the theory of an incompressible fluid flow around a profile that such singularities are needed to describe flows which are uniform at infinity and pass along a contour P.

Since the coefficients of the Eq. (60.10) do not contain the variable β we obtain a new solution in the form

$$\psi^*(\tau,\beta) = -\int^{\beta}\psi(\tau,\gamma)\,d\gamma = -\int^{t_0}\frac{1}{1-\tau\cos t}\cdot\frac{d\gamma}{dt}\,dt = \int^{t_0}dt = t_0(\tau,\beta). \qquad (60.17)$$

Here we made use of the defining relation between t and γ:

$$\tau\sin t - t - \gamma = \log\lambda, \qquad \frac{d\gamma}{dt} = \tau\cos t - 1. \qquad (60.17')$$

Clearly, $\psi^*(\tau,\beta)$ is a solution of Eq. (60.10) which has at $\tau_\infty, \beta_\infty$ a branch point of order $\frac{1}{2}$. We can express this solution also in a different form by using the identity (60.14'). Since $t_0(\tau,\beta)$ is a solution and since also β itself is a trivial solution of Eq. (60.10), clearly

$$\psi_1(\tau,\beta) = \tau\sin t_0 \qquad (60.17'')$$

is also a solution of this equation. It is, of course, easy to verify directly in these particular cases that the functions (60.15) and (60.17) are solutions of Eq. (60.10).

Just as was done in the preceding section for another hypothetical fluid, one can form a linear combination of the two solutions $\psi(\tau,\beta)$ and $\psi_1(\tau,\beta)$ which represents the stream function of a flow around an obstacle which is uniform at infinity and has supersonic regions adjacent to the obstacle. TOMOTIKA and TAMADA made detailed calculations of the flow around various profiles. They can obtain increasing free stream Mach numbers until it happens for the first time that a limit line touches the profile after which the hodograph method breaks down. The authors compare their results with wind-tunnel measurements of LIEPMANN [*41*]; the profiles obtained in their theory coincide well with those considered by LIEPMANN (biconvex circular arc profiles) and the velocity distributions along the profile measured and calculated appear in satisfactory agreement. The shock wave observed is located at the profile point where the limit line touches first.

There is no doubt that a certain discrepancy exists between the special theory and experimental evidence, on the one hand, and between the general mathematical non-existence results discussed in Sects. 56 and 57. The particular solutions obtained by GOLDSTEIN, LIGHTHILL, CHERRY, TOMOTIKA and TAMADA seem to be stable and experimentally reproducible. Why this is so remains a challenging problem for the mathematical investigator.

General references and textbooks.

1. ACKERET, J.: Gasdynamik. In Handbuch der Physik, Bd. 7. Berlin 1927.
2. BUSEMAN, A.: Gasdynamik. In Handbuch der Experimentalphysik, Bd. 4. Leipzig 1931.
3. COURANT, R., and K.O. FRIEDRICHS: Supersonic flow and shock waves. New York 1948.
4. DURAND, W. J., ed.: Aerodynamic Theory, Vol. 3. Berlin 1935.
5. GOLDSTEIN, S., ed.: Modern developments in fluid dynamics, Vol. 1/2. Oxford 1938.
6. HOWARTH, L., ed.: Modern developments in fluid dynamics, Vol. 1/2. Oxford 1953.
7. KIBEL, I. A., N. B. KOCHIN and N. V. ROSE: Theoretical Hydrodynamics, 3rd ed. Moscow-Leningrad 1948 [Russian]. German translation: Berlin 1955.
8. LAMB, H.: Hydrodynamics, 6th ed. Cambridge 1932.
9. LIEPMAN, H. W., and A. E. PUCKETT: Introduction to aerodynamics of a compressible fluid. New York 1947.
10. MILNE-THOMPSON, L. M.: Theoretical Hydrodynamics, 2nd ed. New York 1950.
11. MILNE-THOMPSON, L. M.: Theoretical Aerodynamics. London 1938.
12. OSWATITSCH, K.: Gasdynamik. Wien 1952.
13. PRANDTL, L., u. O. TIETJENS: Hydro- und Aeromechanik. Berlin 1929.

14. ROBINSON, A., and J. A. LAURMANN: Wing Theory. Cambridge 1956.
15. SAUER, R.: Einführung in die theoretische Gasdynamik, 2. Aufl. Berlin-Göttingen-Heidelberg 1951.
6. SAUER, R.: Méthodes mathématiques de la théorie des ecóulements gaseux. Paris 1951.
17. SOMMERFELD, A.: Mechanik der deformierbaren Medien. Wiesbaden 1947.
18. WARD, G. N.: Linearized theory of steady high-speed flow. Cambridge 1955.

Bibliography: Chapter I (pp. 2—26).

[1] HOWARTH, L.: The equations of flow in gases. Mod. developments in fluid dynamics, Vol. 1, pp. 34—70. Oxford 1953.
[2] CROCCO, L.: Eine neue Stromfunktion für die Erforschung der Bewegung der Gase mit Rotation. Z. angew. Math. Mech. **17**, 1—7 (1937).
[3] ST. VENANT, B. DE, et L. WANTZEL: Mémoire et expériences sur l'écoulement déterminé par des différences de pression considérables. J. École polytech. **27**, 85—122 (1839).
[4] TAYLOR, G. I.: Recent work on the flow of compressible fluids. J. Lond. Math. Soc. **5**, 224—240 (1930).
[5] TAYLOR, G. I.: Strömung um einen Körper in einer kompressiblen Flüssigkeit. Z. angew. Math. Mech. **10**, 334—345 (1930).
[6] BATEMAN, H.: Irrotational motion of a compressible inviscid fluid. Proc. Nat. Acad. Sci. U.S.A. **16**, 816—825 (1930).
[7] RINGLEB, F.: Exakte Lösungen der Differentialgleichung einer adiabatischen Strömung. Z. angew. Math. Mech. **20**, 185—198 (1940).
[8] OPATOWSKI, I.: Two-dimensional compressible flows. Proc. Symp. Appl. Math. **1**, 87—93 (1949).
[9] BICKLEY, W. G.: Some exact solutions of the equations of steady homentropic flow of an inviscid gas. Mod. developments in fluid dynamics, Vol. I, pp. 158—189. Oxford 1953.
[10] COURANT, R., u. D. HILBERT: Methoden der mathematischen Physik, Vol. II. Berlin 1937.
[11] FRANK, P., u. R. v. MISES: Die Differential- und Integralgleichungen der Mechanik und Physik. New York 1943.
[12] WEBSTER, A., u. G. SZEGÖ: Partielle Differentialgleichungen der mathematischen Physik. Leipzig u. Berlin 1930.
[13] LEVI-CIVITA, T.: Caratteristiche e propagazione ondosa. Bologna 1931.
[14] LEWY, H.: Über das Anfangswertproblem einer hyperbolischen nichtlinearen partiellen Differentialgleichung. Math. Ann. **98**, 179—191 (1928).
[15] MASSAU, M.: Mémoire sur l'intégration graphique des équations aux dérivées partielles. Gand 1900—1903.
[16] PRANDTL, L., u. A. BUSEMANN: Näherungsverfahren zur zeichnerischen Ermittlung von ebenen Strömungen mit Überschallgeschwindigkeit. Stodola-Festschrift, pp. 499—509. Zürich 1929.
[17] GUDERLEY, G.: Erweiterung der Charakteristiken-Methode. Ber. Lilienthalges **139** (II), 15—23 (1941).
[18] OSWATITSCH, K.: Über die Charakteristikenverfahren der Hydrodynamik. Z. angew. Math. Mech. **25**, 196—208 (1947); **27**, 264—270 (1947).
[19] MEYER, R. E.: The method of characteristics. Modern developments in fluid dynamics, Vol. I, pp. 71—104. Oxford 1953.
[20] MEYER, R. E.: The method of characteristics for problems of compressible flow involving two independent variables. Quart. J. Mech. Appl. Math. **1**, 196—219, 451—469 (1948).
[21] MEYER, TH.: Über zweidimensionale Bewegungsvorgänge in einem Gas, das mit Überschallgeschwindigkeit strömt. Thesis, Göttingen 1908. Forschungsh. VDI **62** (1908).
[22] TAYLOR, G. I., and J. C. MACCOLL: Air pressure on a cone moving at high speeds. Proc. Roy. Soc. Lond., Ser. A **139**, 278—297, 298—311 (1933).
[23] BUSEMANN, A.: Die achsensymmetrische kegelige Überschallströmung. Luft.-Forsch. **19**, 137—144 (1942).
[24] BUSEMANN, A.: Drücke auf kegelförmige Spitzen bei Bewegung mit Überschallgeschwindigkeit. Z. angew. Math. Mech. **9**, 496—498 (1929).
[25] Tables of supersonic flow around cones. Cambridge 1947.

Bibliography: Chapter II (pp. 26—56).

[1] PRANDTL, L.: Über Strömungen, deren Geschwindigkeiten mit der Schallgeschwindigkeit vergleichbar sind. J. Aeronaut. Res. Inst. Tokyo **65**, 14 (1930).

[2] PRANDTL, L.: Allgemeine Überlegungen über die Strömung zusammendrückbarer Flüssigkeiten. Atti Accad. d'Italia (5th Volta Congress), Rome 1935, pp. 169—197.
[3] PRANDTL, L.: Theorie des Flugzeugtragflügels im zusammendrückbaren Medium. Luftf.-Forsch. **13**, 313—319 (1936).
[4] GLAUERT, H.: The effect of compressibility on the lift of an aerofoil. Proc. Roy. Soc. Lond., Ser. A **118**, 113—119 (1928).
[5] PISTOLESI, E.: La portanza alle alte velocità inferiori quella del suono. Atti. Accad. d'Italia (5th Volta Congress), Rome 1935, pp. 283—326.
[6] GOLDSTEIN, S., and A.D. YOUNG: The linear perturbation theory of compressible flow with application to wind tunnel interference. Rep. Memor. Aero. Res. Coun., London, No. 1909 (1943).
[7] TSIEN, H.S., and L. LEES: The Prandtl-Glauert approximation for subsonic flows of a compressible fluid. J. Aeronaut. Sci. **12**, 173—187 (1945).
[8] LEES, L.: A discussion of the application of the Prandtl-Glauert method to subsonic compressible flow over a slender body of revolution. NACA Tech. Notes, No. 1127 (1946).
[9] WARD, G.N.: Approximate Methods. Modern Developments in Fluid Dynamics, Vol. I, pp. 267—324. Oxford 1953.
[10] WARD, G.N.: Linearized theory of steady high speed flow. Cambridge 1955.
[11] LAITONE, E.V.: Subsonic flow about a body of revolution. Quart. Appl. Math. **5**, 227—231 (1947).
[12] LAITONE, E.V.: The linearized subsonic and supersonic flow about inclined slender bodies of revolution. J. Aeronaut. Sci. **14**, 631—642 (1947).
[13] SEARS, W.R.: On compressible flow about bodies of revolution. Quart. Appl. Math. **4**, 191—193 (1946); **5**, 89—91 (1947).
[14] YOUNG, A.D., and S. KIRKBY: Application of the linear perturbation theory to compressible flow about bodies of revolution. Rep. Memor. Aero. Res. Coun., London, No. 2624 (1947).
[15] HANTZSCHE, W., u. H. WENDT: Der Kompressibilitätseinfluß für dünne wenig gekrümmte Profile bei Unterschallgeschwindigkeit. Z. angew. Math. Mech. **22**, 72—86 (1942).
[16] HANTZSCHE, W.: Die Prandtl-Glauertsche Näherung als Grundlage für ein Iterationsverfahren zur Berechnung kompressibler Unterschallströmungen. Z. angew. Math. Mech. **23**, 185—199 (1943).
[17] SCHMEIDLER, W.: Integralgleichungen mit Anwendungen in Physik und Technik. Leipzig 1950.
[18] SCHRÖDER, K.: Über eine Integralgleichung erster Art der Tragflügeltheorie. Sitzgsber. Preuß. Akad. Wiss., Phys.-math. Kl. **1938**, 345—362.
[19] SÖHNGEN, H.: Die Lösung der Integralgleichung ... und deren Anwendung in der Tragflügeltheorie. Math. Z. **45**, 245—264 (1939).
[20] MACH, E., u. P. SALCHER: Photographische Fixierung der durch Projektile in der Luft eingeleiteten Vorgänge. Sitzgsber. Akad. Wiss. Wien IIa **95**, 764—780 (1887).
[21] MACH, E., u. L. MACH: Weitere ballistisch-photographische Versuche. Sitzgsber. Akad. Wiss. Wien IIa **98**, 1310—1326 (1889).
[22] MACH, L.: Weitere Versuche über Projektile. Sitzgsber. Akad. Wiss. Wien IIa **105**, 605—633 (1896).
[23] COURANT, R., u. D. HILBERT: Methoden der mathematischen Physik, Vol. II. Berlin 1937.
[24] SAUER, R.: Anfangswertprobleme bei partiellen Differentialgleichungen. Berlin 1954.
[25] HADAMARD, J.: Le problème de Cauchy. Paris 1932.
[26] HEASLET, M.A., and H. LOMAX: The use of source-sink and doublet distributions to the solution of arbitrary boundary value problems in supersonic flow. NACA Techn. Note, No. 1515 (1948).
[27] LOMAX, H., M.A. HEASLET and F.B. FULLER: Integrals and integral equations in linearized wing theory. NACA Rep. No. 1054 (1951).
[28] ROBINSON, A.: On source and vortex distributions in the linearized theory of steady supersonic flow. Quart. J. Mech. Appl. Math. **1**, 408—432 (1948).
[29] PUCKETT, A.E.: Supersonic wave drag of thin airfoils. J. Aeronaut. Sci. **13**, 475—484 (1946).
[30] PUCKETT, A.E., and H.L. STEWART: Aerodynamic performance of delta wings at supersonic speeds. J. Aeronaut. Sci. **14**, 567—568 (1947).
[31] EVVARD, J.C.: Use of source distributions for evaluating theoretical aero-dynamics of thin finite wings at supersonic speeds. NACA Rep. No. 951 (1950).
[32] HAYES, W.D., and H.A. LINSTONE: A development of EVVARD's supersonic wing theory. Rep. North Amer. Aviation Inc., No. AL-746.
[33] WARD, G.N.: Supersonic flow past thin wings. I. General theory. Quart. J. Mech. Appl. Math. **2**, 136—152 (1949).

[34] ACKERET, J.: Über Luftkräfte auf Flügel, die mit größerer als Schallgeschwindigkeit bewegt werden. Z. Flugtechn. **16**, 72—74 (1925).
[35] ACKERET, J.: Über Luftkräfte bei sehr großen Geschwindigkeiten, insbesondere bei ebenen Strömungen. Helv. phys. Acta **1**, 301—322 (1928).
[36] KÁRMÁN, T. VON: The problem of resistance in compressible fluids. Atti Accad. d'Italia (5th Volta Congress), Rome 1935, pp. 222—277.
[37] SCHLICHTING, F.: Tragflügeltheorie bei Überschallgeschwindigkeit. Luftf.-Forsch. **13**, 320—335 (1936).
[38] TAYLOR, G.I.: Applications to aeronautics of ACKERET's theory of aerofoils moving at speeds greater than that of sound. Rep. Memor. Aero. Res. Coun., London, No. 1467 (1932).
[39] BUSEMANN, A.: Aerodynamischer Auftrieb bei Überschallgeschwindigkeit. Atti Accad. d'Italia (5th Volta Congress), Rome 1935, pp. 328—360.
[40] LIGHTHILL, M. J.: Two-dimensional supersonic aerofoil theory. Rep. Memor. Aero. Res. Coun., London, No. 1929 (1944).
[41] LIGHTHILL, M. J.: The condition behind the trailing edge of the supersonic aerofoil. Rep. Memor. Aero. Res. Coun., London, No. 1930 (1944).
[42] KÁRMÁN, T. VON: Supersonic aerodynamics—principles and applications. J. Aeronaut. Sci. **14**, 373—409 (1947).
[43] BUSEMANN, A.: Infinitesimale kegelige Überschallströmung. Luftf.-Forsch. **7**B, 105—120 (1943).
[44] HAYES, W.D.: Linearized conical supersonic flow. Rep. North Amer. Aviation Inc., No. NA-46-818 (1946).
[45] STEWART, H.J.: The lift of a delta-wing at supersonic speeds. Quart. Appl. Math. **4**, 246—254 (1946).
[46] GOLDSTEIN, S., and G.N. WARD: The linearized theory of conical fields in supersonic flows, with applications to plane aerofoils. Aeronaut. Quart. **2**, 39—84 (1950).
[47] LAGERSTROM, P.A.: Linearized supersonic theory of conical wings. NACA Techn. Note No. 1685 (1948).
[48] BEHRBOHM, H., et K. OSWATITSCH: Corps coniques plats dans un écoulement supersonique. Centre d'Études supérieures de Mécanique, Bull. 10—12 (1950/51).
[49] GERMAIN, P.: La théorie des mouvements homogènes et son application au calcul de certaines ailes delta en régime supersonique. Rech. aéronaut. **7**, 3—16 (1949).
[50] LOMAX, H., and M.A. HEASLET: Generalized conical-flow fields in supersonic theory. NACA Techn. Note No. 2497 (1951).
[51] HAYES, W.D., R.C. ROBERTS and N. HAASER: Generalized linearized conical flow. NACA Techn. Note No. 2667 (1952).
[52] ROBINSON, A.: Aerofoil theory of a flat delta wing at supersonic speeds. Rep. Memor. Aero. Res. Coun., London, No. 2548 (1946).
[53] GÖRTLER, H.: Gasströmungen mit Übergang von Unterschall- zu Überschallgeschwindigkeiten. Z. angew. Math. Mech. **20**, 254—262 (1940).
[54] FERRARI, C.: Campi di correnti ipersonora attorno a solidi di rivoluzioni. Aerotecnica **17**, 507—518 (1937).
[55] KÁRMÁN, TH. V.: Druckverteilung an Luftschiffkörpern. Abh. Aerodyn. Inst. Aachen 1927, H. 6, 1—17.
[56] KÁRMÁN, TH. V., and N.B. MOORE: Resistance of slender bodies moving at supersonic velocities. Trans. Amer. Soc. Engrs. **54**, 303—310 (1932).
[57] TSIEN, H.S.: Supersonic flow over an inclined body of revolution. J. Aeronaut. Sci. **5**, 480—483 (1938).
[58] LIGHTHILL, M. J.: Supersonic flow past bodies of revolution. Rep. Memor. Aero. Res. Coun., London, No. 2003 (1945).
[59] DYKE, M.D. VAN: First and second order theory of supersonic flow past bodies of revolution. J. Aeronaut. Sci. **18**, 161—178 (1951).
[60] LIGHTHILL, M. J.: A new approach to thin aerofoil theory. Aeronaut. Quart. **3**, 193—210 (1951).
[61] FRANKL, F.I., u. M. KELDYSH: Die äußere Neumannsche Aufgabe für nicht-lineare elliptische Differentialgleichungen mit Anwendung auf die Theorie der Flügel im kompressiblen Gas. Bull. Akad. Sci. USSR. **12**, 561—601 (1934).

Bibliography: Chapter III (pp. 56 — 92).

[1] SAUER, R.: Linearverbindung kompressibler ebener Strömungsfelder. Z. angew. Math. Mech. **21**, 313—315 (1941).
[2] MOLENBROEK, P.: Über einige Bewegungen eines Gases mit Annahme eines Geschwindigkeitspotentials. Arch. Math. Phys. II **9**, 157—195 (1890).

[3] CHAPLYGIN, C.T.: On gaseous jets. Ann. Math. Phys. Sect. Sci., Imperial Univ. Moscow **21**, 1—121 (1904). English translation: NACA Techn. Note No. 1063 (1944).
[4] RIABOUCHINSKY, D.: Mouvement d'un fluide incompressible autour d'un obstacle. C. R. Acad. Sci., Paris **194**, 1215—1217 (1932).
[5] DEMTSCHENKO, B.: Sur les mouvements lents des fluides compressibles. C. R. Acad. Sci., Paris **194**, 1218—1220 (1932).
[6] DEMTSCHENKO, B.: Variation de la résistance aux faibles vitesses sous l'influence de la compressibilité. C. R. Acad. Sci., Paris **194**, 1720—1723 (1932).
[7] DEMTSCHENKO, B.: Sur la relation entre la dynamique des fluides compressibles et celle des fluides incompressibles. Publ. Math. Univ. Belgrade **2**, 85—105 (1933).
[8] MISES, R. v., and M. SCHIFFER: On BERGMAN'S integration method in two dimensional compressible flow. Adv. Appl. Mech. **1**, 249—285 (1948).
[9] LIGHTHILL, M. J.: The hodograph transformation in transonic flow. I. Symmetric channels. Proc. Roy. Soc. Lond., Ser. A **191**, 323—341 (1947).
[10] LIGHTHILL, M. J.: The hodograph transformation in transonic flow. II. Auxiliary theorems on the hypergeometric functions $\psi_n(\tau)$. Proc. Roy. Soc. Lond., Ser. A **191**, 341—351 (1947).
[11] LIGHTHILL, M. J.: The hodograph transformation in transonic flow. III. Flow round a body. Proc. Roy. Soc. Lond. A **191**, 352—369 (1947).
[12] FERGUSON, D. F., and M. J. LIGHTHILL: The hodograph transformation in transonic flow. IV. Tables. Proc. Roy. Soc. Lond., Ser. A **192**, 135—142 (1948).
[13] LIGHTHILL, M. J.: The hodograph transformation. Modern developments in fluid dynamics, Vol. I. Oxford 1953.
[14] BERGMAN, S.: On two-dimensional flows of compressible fluids. NACA Techn. Note No. 972 (1945).
[15] BERGMAN, S.: Operator methods in the theory of compressible fluids. Proc. Symp. Appl. Math. **1**, 19—40 (1949).
[16] BERGMAN, S.: Operatorenmethode der Gasdynamik. Z. angew. Math. Mech. **32**, 33—45 (1952).
[17] CHERRY, T. M.: Relation between BERGMAN'S and CHAPLYGIN'S methods of solving the hodograph equation. Quart. Appl. Math. **9**, 92—94 (1951).
[18] NEMÉNYI, P. F.: Recent developments in inverse and semi-inverse methods in the mechanics of continua. Adv. Appl. Mech. **2**, 123—151 (1951).
[19] GOLDSTEIN, S., M. J. LIGHTHILL and J. W. CRAGGS: On the hodograph transformation for high speed flow. Quart. J. Mech. Appl. Math. **1**, 344—357 (1948).
[20] LIGHTHILL, M. J.: On the hodograph transformation for high speed flow. II. A flow with circulation. Quart. J. Mech. Appl. Math. **1**, 442—450 (1948).
[21] TRICOMI, F.: Sulle equazione lineari alle derivati partiali di 2 ordine di tipo misto. Atti Accad. Lincei, Ser. V, **14**, 133—247 (1923).
[22] TSIEN, H. S.: Two-dimensional subsonic flow of compressible fluids. J. Aeronaut. Sci. **6**, 399—407 (1940).
[23] KÁRMÁN, T. v.: Compressibility effects in aerodynamics. J. Aeronaut. Sci. **8**, 337—356 (1941).
[24] GERMAIN, P.: Fluides compressibles. Étude directe du cas simplifié de CHAPLYGIN. C. R. Acad. Sci., Paris **223**, 532—534 (1946).
[25] TOLLMIEN, W.: Zum Übergang von Unterschall- in Überschallströmungen. Z. angew. Math. Mech. **17**, 117—136 (1937).
[26] TOLLMIEN, W.: Grenzlinien adiabatischer Potentialströmung. Z. angew. Math. Mech. **21**, 140—151, 308 (1941).
[27] RINGLEB, F.: Exakte Lösungen der Differentialgleichungen einer adiabatischen Gasströmung. Z. angew. Math. Mech. **20**, 185—198 (1940).
[28] GUDERLEY, G.: Rückkehrkanten in ebener kompressibler Potentialströmung. Z. angew. Math. Mech. **22**, 121—126 (1942).
[29] CRAGGS, J. W.: The breakdown of the hodograph transformation for irrotational compressible fluid flow in two dimensions. Proc. Cambridge Phil. Soc. **44**, 360—379 (1948).
[30] BUSEMANN, A.: Hodographenmethode der Gasdynamik. Z. angew. Math. Mech. **17**, 73—79 (1937).
[31] JACOB, C.: Étude d'un jet gaseux. Bull. Sci. Ecole Polyt. Timisoara **7**, 47—59, 224—244 (1937).
[32] CHERRY, T. M.: Flow of a compressible fluid about a cylinder. Proc. Roy. Soc. Lond., Ser. A **192**, 45—79 (1947).
[33] CHERRY, T. M.: Flow of a compressible fluid about a cylinder. II. Flow with circulation. Proc. Roy. Soc. Lond., Ser. **196**, 1—32 (1949).
[34] CHERRY, T. M.: Numerical solutions for transonic flow. Proc. Roy. Soc. Lond., Ser. A **196**, 32—36 (1949).

[35] Garrick, J. E., and C. Kaplan: On the flow of a compressible fluid by the hodograph method. NACA Rep. No. 789/790 (1944).

[36] Christianovich, S. A.: Gas flow over bodies at high subsonic velocity. Trudy Aero. Inst., No. 481 (1940).

[37] Bers, L.: On a method of constructing two-dimensional subsonic compressible flows around closed profiles. NACA Techn. Note, No. 969 (1945).

[38] Bergman, S.: Two-dimensional transonic flow patterns. Amer. J. Math. **70**, 856–891 (1948).

[39] Bers, L., and A. Gelbart: On a class of differential equations in mechanics of continua. Quart. Appl. Math. **1**, 168–188 (1943).

[40] Bers, L., and A. Gelbart: On a class of functions defined by partial differential equations. Trans. Amer. Math. Soc. **56**, 67–93 (1944).

[41] Preiswerk, E.: Anwendung gasdynamischer Methoden auf Wasserströmungen mit freier Oberfläche. Mitt. Inst. Aerodyn., E.T.H. Zürich, Nr. 7 (1938).

[42] Prandtl, L., u. A. Busemann: Näherungsverfahren zur zeichnerischen Ermittlung von ebenen Strömungen mit Überschallgeschwindigkeit. Stodola-Festschrift, pp. 499–509. Zürich 1929.

[43] Meyer, R. E.: The method of characteristics. Modern developments in fluid dynamics, Vol. I. Oxford 1953.

[44] Holt, M.: The numerical method of characteristics for supersonic flows with axial symmetry. Quart. J. Mech. Appl. Math. **2**, 473–478 (1949).

[45] Guderley, G.: Erweiterung der Charakteristiken-Methode. Ber. Lilienthalges **139**, 15–23 (1941).

Bibliography: Chapter IV (pp. 92–130).

[1] Lichtenstein, L.: Neuere Entwicklung der Theorie partieller Differentialgleichungen zweiter Ordnung vom elliptischen Typus. Enz. Math. Wissensch., Vol. II, Part 3.2, No. 12. 1924.

[2] Miranda, C.: Equazioni alle derivata parziali di tipo ellittico. Berlin-Göttingen-Heidelberg 1954.

[3] Hopf, E.: Elementare Betrachtungen über die Lösungen partieller Differentialgleichungen zweiter Ordnung vom elliptischen Typus. Sitzgsber. preuß. Akad. Wiss. **19**, 147–152 (1927).

[4] Finn, R., and D. Gilbarg: Asymptotic behavior and uniqueness of plane subsonic flows. Stanford University, Technical Report No. 225(11), No. 50.

[5] Morrey, C. B.: Multiple integral problems in the calculus of variations and related topics. Univ. California Publ. Math. **1**, 1–130 (1943).

[6] Morrey, C. B.: On the solutions of quasi-linear elliptic partial differential equations. Trans. Amer. Math. Soc. **43**, 126–166 (1938).

[7] Bers, L.: Existence and uniqueness of a subsonic flow past a given profile. Comm. Pure Appl. Math. **7**, 441–504 (1954).

[8] Bers, L.: Results and conjectures in the mathematical theory of subsonic and transonic gas flows. Comm. Pure Appl. Math. **7**, 79–104 (1954).

[9] Korn, A.: Zwei Anwendungen der Methode der sukzessiven Annäherungen. Schwarz-Festschrift, pp. 215–229. Berlin 1914.

[10] Korn, A.: Über Minimalflächen, deren Randkurven wenig von ebenen Kurven abweichen. Abh. preuß. Akad. Wiss. Berlin **1909**, Anhang II.

[11] Lichtenstein, L.: Zur Theorie der Konformen Abbildung. Konforme Abbildung nichtanalytischer singularitäten-freier Flächenstücke auf ebene Gebiete. Bull. Acad. Sci. Cracovie A 1916, 192–217.

[12] Ludford, G.: The behavior at infinity of the potential function of a two-dimensional subsonic compressible flow. J. Math. Phys. **30**, 117–130 (1951/52).

[13] Imai, I.: On the asymptotic behavior of compressible fluid flow at a great distance from a cylinder in the absence of circulation. J. Phys. Soc. Japan **8**, 537–544 (1953).

[14] Lamb, H.: On the flow of a compressible fluid past an obstacle. Rep. Memor. Aero. Res. Coun., London, No. 1156 (1928).

[15] Bateman, H.: The lift and drag functions for an elastic fluid in two-dimensional irrotational flow. Proc. Nat. Acad. Sci. **24**, 246–251 (1938).

[16] Frankl, F. I., u. M. Keldysh: Die äußere Neumannsche Aufgabe für nichtlineare elliptische Differentialgleichungen mit Anwendung auf die Theorie der Flügel im kompressiblen Gas. Bull. Acad. Sci. USSR. **12**, 561–601 (1934).

[17] Shiffman, M.: On the existence of subsonic flows of a compressible fluid. Proc. Nat. Acad. Sci. **38**, 434–438 (1952).

[18] Shiffman, M.: On the existence of subsonic flows of a compressible fluid. J. Rat. Mech. a. Analysis **1**, 605–652 (1952).

[19] VASZONYI, A.: An existence theorem in the theory of compressible fluids. Abstract. Bull. Amer. Math. Soc. **50**, 673 (1944).
[20] HAAR, A.: Über das Plateausche Problem. Math. Ann. **97**, 124—158 (1927).
[21] TONNELLI, L.: L'estremo assoluto degli integrali doppi. Ann. Scu. Norm. Sup. Pisa (2) **2**, 89—130 (1933).
[22] HOPF, E.: Zum analytischen Charakter der Lösungen regulärer zweidimensionaler Variationsprobleme. Math. Z. **30**, 404—413 (1929).
[23] JANZEN, O.: Beitrag zu einer Theorie der stationären Strömung kompressibler Flüssigkeiten. Phys. Z. **14**, 639—643 (1913).
[24] Lord RAYLEIGH: On the flow of compressible fluid past an obstacle. Phil. Mag. (6) **32**, 1—6 (1916).
[25] ESER, F.: Zur Strömung kompressibler Flüssigkeiten um feste Körper mit Unterschallgeschwindigkeit. Luftf.-Forsch. **207**, 220—230 (1943).
[26] KRAHN, E.: Die Janzen-Rayleigh zweite Näherung der kompressiblen Strömung um ein beliebiges Profil. Z. angew. Math. Mech. **23**, 33—35 (1943).
[27] KRAHN, E.: Berechnung der zweiten Näherung der kompressiblen Strömung um ein Profil nach JANZEN-RAYLEIGH. Luftf.-Forsch. **20**, 147—151 (1943).
[28] EHLERS, F.E.: Methods of linearization in compressible flow. I. Janzen-Rayleigh method. Wright Field Rep. No. F-Tr-1180A-ND.
[29] WENDT, H.: Die Janzen-Rayleighsche Näherung zur Berechnung von Unterschallströmungen. Sitzgsber. Heidelberg. Akad. Wiss., Math.-naturwiss. Kl. 1948.
[30] TAYLOR, G.I., and C.F. SHARMAN: A mechanical method for solving problems of flow in compressible fluids. Proc. Roy. Soc. Lond., Ser. A **121**, 194—217 (1928).
[31] TAYLOR, G.I.: Strömung um einen Körper in einer kompressiblen Flüssigkeit. Z. angew. Math. Mech. **10**, 334—345 (1930).
[32] LERAY, J., et J. SCHAUDER: Topologie et équations fonctionelles. Ann. Sci. Ecole Norm. Sup. Paris **51**, 45—78 (1934).
[33] SCHAUDER, J.: Der Fixpunktsatz in Funktionalräumen. Studia Math. **2**, 171—180 (1930).
[34] BIRKHOFF, G.D., and O.D. KELLOGG: Invariant points in function space. Trans. Amer. Math. Soc. **23**, 96—115 (1922).
[35] BERG, P.W.: On the existence of Helmholtz flows of a compressible fluid. Thesis, New York University 1953.
[36] GLAUERT, H.: The effect of compressibility on the lift of an aerofoil. Aero. Res. Coun., London, Techn. Rep. No. 1135 (1927/28).
[37] LOEWNER, C.: Conservation laws in compressible fluid flow and associated mappings. J. Rat. Mech. a. Analysis **2**, 537—561 (1953).
[38] LOEWNER, C.: Some bounds for the critical free stream Mach number of a compressible flow around an obstacle. Studies in Math. and Mech. presented to R.v. MISES, pp. 177—183. New York 1954.
[39] LAVRENTIEFF, M.: The general problem of quasi-conformal mappings of plane regions. Math. Sbornik, N.S. **21** (63), 285—320 (1947). Engl. Translation: Amer. Math. Soc. Translation No. 46.
[40] GOLDSTEIN, S., and M.J. LIGHTHILL: Two-dimensional compressible flow past a solid body in unlimited fluid or symmetrically placed in a channel. Phil. Mag. (7) **35**, 549—568 (1944).

Bibliography: Chapter V (pp. 130 — 154).

[1] TAYLOR, G.I.: The flow of air at high speeds past curved surfaces. Rep. Memor. Aero. Res. Coun., London, No. 1381 (1930).
[2] TAYLOR, G.I.: Strömung um einen Körper in kompressibler Flüssigkeit. Z. angew. Math. Mech. **10**, 334—345 (1930).
[3] TAYLOR, G.I.: Recent work on the flow of compressible fluids. J. Lond. Math. Soc. **5**, 224—240 (1930).
[4] GERMAIN, P.: Remarks on the theory of partial differential equations of mixed type and application to the study of transonic flow. Comm. Pure Appl. Math. **7**, 117—143 (1954).
[5] GERMAIN, P.: Introduction à l'étude mathématique des écoulements transoniques. Rech. aéronaut. 1951, No. 22, 7—20.
[6] TRICOMI, F.: Sulle equazioni lineari alle derivati parziali de 2° ordine di tipo misto. Atti Accad. naz. Lincei, Ser. V, **14**, 134—247 (1923).
[7] TRICOMI, F.: Ancora sull'equazione $y z_{xx} + z_{yy} = 0$. Atti Accad. naz. Lincei, Ser. VI, **6**, 567—571 (1927).
[8] TRICOMI, F.: Ulteriori ricerche sull'equazione $y z_{xx} + z_{yy} = 0$. Rend. Circ. Mat. Palermo **52**, 63—90 (1928).
[9] GERMAIN, P.: Nouvelles solutions de l'équation de TRICOMI. C. R. Acad. Sci., Paris **231**, 1116—1118 (1950).

[10] GERMAIN, P., et R. BADER: Sur le problème de TRICOMI: C. R. Acad. Sci., Paris **232**, 463–465 (1951).
[11] GERMAIN, P., et R. BADER: Application de la solution fondamentale à certains problèmes relatifs à l'équation de TRICOMI. C. R. Acad. Sci., Paris **231**, 1203–1205 (1950).
[12] WEINSTEIN, A.: On TRICOMI's equation and generalized axially symmetric potential theory. Bull. Acad. Roy. Belg., Cl. Sci. (5) **37**, 348–358 (1951).
[13] WEINSTEIN, A.: The singular solutions and the Cauchy problem for generalized Tricomi equations. Comm. Pure Appl. Math. **7**, 105–116 (1954).
[14] BERS, L.: Results and conjectures in the mathematical theory of subsonic and transonic gas flows. Comm. Pure Appl. Math. **7**, 79–104 (1954).
[15] GERMAIN, P., et R. BADER: Problème de DIRICHLET pour une équation du type mixte. C. R. Acad. Sci., Paris **230**, 1824–1826 (1950).
[16] GERMAIN, P., et R. BADER: Sur quelques problèmes aux limites, singuliers, pour une équation hyperbolique. C. R. Acad. Sci., Paris **231**, 268–270 (1950).
[17] GERMAIN, P., et R. BADER: Sur quelques problèmes relatifs à l'équation de type mixte de TRICOMI. Office nationale d'étude et de recherche aéronautique, Publ. No. 54 (1952).
[18] AGMON, S., L. NIRENBERG and M.H. PROTTER: A maximum principle for a class of equations of hyperbolic and mixed type. Comm. Pure Appl. Math. **6**, 455–470 (1953).
[19] FRANKL, F.I.: On CAUCHY's problem for equations of mixed elliptic-hyperbolic type with initial data on the transition line. Izvest. Akad. Nauk USSR. **8**, 195–224 (1944).
[20] FRANKL, F.I.: On the problems of CHAPLYGIN for mixed sub- and supersonic flows. Izvest. Akad. Nauk USSR. **9**, 121–143 (1945).
[21] FRANKL, F.I.: On a new boundary value problem for the equation $y z_{xx} + z_{yy} = 0$. Učenye Zap. Mosk. Gos. Univ. **152**, 99–116 (1951).
[22] PROTTER, M.H.: Uniqueness theorems for the Tricomi equation. J. Rat. Mech. a. Analysis **2**, 107–114 (1953).
[23] GELLERSTEDT, S.: Sur un problème aux limites pour une équation linéaire aux dérivéés partielles du second ordre de type mixte. Thesis, Uppsala 1935.
[24] MORAWETZ, C.S.: A uniqueness theorem for the Frankl problem. Comm. Pure Appl. Math. **7**, 697–703 (1954).
[25] MORAWETZ, C.S.: On the non-existence of continuous transonic flows past profiles. Comm. Pure Appl. Math. **9**, 45–68 (1956).
[26] MANWELL, A.R.: The variation of compressible flows. Quart. J. Appl. Math. **7**, 40–50 (1954).
[27] FRANKL, F.I.: On the theory of the Laval nozzle. Izvest. Akad. Nauk USSR. **9**, 387–422 (1945).
[28] FALKOVICH, S.V.: On the theory of the Laval nozzle. Prikl. Mat. Mekh. **10**, 503–512 (1946).
[29] BUSEMANN, A.: The non-existence of transonic flows. Proc. Symp. Appl. Math. **4**, 29–39 (1953).
[30] GUDERLEY, G.: On the presence of shocks in mixed subsonic-supersonic flow patterns. Adv. Appl. Mech. **3**, 145–184 (1953).
[31] NIKOLSKY, A.A., and G.I. TAGANOV: Gas motion in a locally supersonic region and conditions of breakdown of potential flow. Prikl. Mat. Mekh. **10**, 481–502 (1946).
[32] MISES, R. VON: Discussion on transonic flow. Comm. Pure Appl. Math. **7**, 145–148 (1954).
[33] TOMOTIKA, S., and K. TAMADA: Studies on two-dimensional transonic flows of a compressible fluid. I–III. Quart. Appl. Math. **7**, 381–397 (1949); **8**, 127–136 (1950); **9**, 129–147 (1951).
[34] MEYER, TH.: Über zweidimensionale Bewegungsvorgänge in einem Gas, das mit Schallgeschwindigkeit strömt. Thesis, Göttingen 1908. Forschungsh. VDI **62** (1908).
[35] GÖRTLER, H.: Zum Übergang von Unterschall- zu Überschallgeschwindigkeiten in Düsen. Z. angew. Math. Mech. **19**, 325–337 (1939).
[36] OSWATITSCH, K., u. W. ROTHSTEIN: Das Strömungsfeld in einer Laval-Düse. Jb. dtsch. Luftf. 1942, 91–102.
[37] GERMAIN, P.: Application de l'approximation homographique à l'étude des écoulements transoniques. C. R. Acad. Sci., Paris **232**, 1811–1813 (1951).
[38] MARTIN, M.H., and W.R. THICKSTUN: An example of transonic flow for the Tricomi gas. Proc. Symp. Appl. Math. **4**, 61–73 (1953).
[39] GUDERLEY, G., and H. YOSHIHARA: An axial-symmetric transonic flow pattern. Quart. Appl. Math. **8**, 333–339 (1951).
[40] GUDERLEY, G., and H. YOSHIHARA: The flow over a wedge profile at Mach number one. J. Aeronaut. Sci. **17**, 723–735 (1950).
[41] LIEPMANN, H.W.: The interaction between boundary layer and shock waves in transonic flow. J. Aeronaut. Sci. **13**, 623–637 (1946).
[42] GUDERLEY, G.: On the development of solutions of TRICOMI's differential equation in the vicinity of the origin. J. Rat. Mech. a. Analysis **5**, 747–790 (1956)

Théorie des ondes de choc.

Par

HENRI CABANNES.

Avec 59 Figures.

A. Les équations des phénomènes de choc.

a) Introduction.

1. Définition des ondes de choc. Le mouvement d'un solide dans un fluide compressible initialement au repos donne naissance à un mouvement du fluide. En chaque point, à chaque instant, on peut définir un vecteur vitesse. Dans une première étude des phénomènes aérodynamiques, on suppose que le champ des vitesses ainsi défini est un champ continu par rapport aux variables d'espace et par rapport au temps. Cependant, pour une étude plus précise, il est nécessaire de supposer que la vitesse du fluide et par suite la masse spécifique et la pression peuvent subir des discontinuités. Deux cas sont possibles; dans le premier, les molécules fluides affectées par la discontinuité sont toujours les mêmes au cours du temps; on dit que la discontinuité est stationnaire. Dans un second cas, la discontinuité affecte à chaque instant des molécules différentes, elle se propage dans le fluide; le lieu des points en lesquels la vitesse, la masse spécifique et la pression sont discontinues est appelé onde de choc. Chaque fois qu'un solide se déplace à une vitesse supérieure à la célérité du son, il donne naissance à plusieurs ondes de choc; ces ondes produisent les détonations que l'on entend au passage d'un avion supersonique. L'étude des ondes de choc à laquelle est consacré ce chapitre est l'un des problèmes posés par le vol supersonique. La conquête des vitesses supersoniques constitue la prochaine étape dans le développement de l'industrie aéronautique.

Les premiers travaux relatifs à la propagation des discontinuités dans les fluides datent du siècle dernier; ils ont pour auteurs principaux RIEMANN [*1*], RANKINE [*2*], HUGONIOT [*3*] et HADAMARD [*4*]. L'étude des ondes de choc fut reprise vers 1930. Différentes questions furent résolues de façon systématique et la solution fut poussée jusqu'à la construction de tables détaillées et précises. COURANT et FRIEDRICHS rédigèrent en 1948 un ouvrage de synthèse remarquable [*5*]. Le présent chapitre se propose d'indiquer, dans l'état actuel de nos connaissances, comment on peut déterminer les ondes de choc engendrées par un obstacle donné animé d'un mouvement connu. Dans un souci d'unité, nous avons renoncé à faire un exposé encyclopédique; certaines questions ne sont pas traitées. Nous avons voulu faire ressortir les méthodes et les résultats; pour cela, nous avons négligé le détail des calculs compliqués. Nous avons également fait un choix dans la bibliographie.

Dans les années qui viennent, l'étude théorique des ondes de choc semble devoir subit un certain ralentissement. La plupart des problèmes de caractère local ont été résolus; on pourra perfectionner les démonstrations et compléter les résultats, mais cela ne constituera pas un progrès fondamental. Par contre, lorsque la théorie des équations aux dérivées partielles, sous l'influence de la topologie, aura fait des progrès suffisants, on pourra s'attaquer aux problèmes de caractère global avec quelque espoir de succès. On pourra alors reprendre la plupart des problèmes étudiés et aller beaucoup plus loin dans leur solution.

Nous avons divisé ce chapitre en trois parties. Dans la première, nous établissons les équations qui sont à la base de la théorie. Dans la partie B, nous étudions les ondes de choc dans les mouvements stationnaires; l'obstacle est placé dans un fluide animé à l'infini d'une translation uniforme et nous supposons qu'un écoulement permanent a pu s'établir; en chaque point, la vitesse est indépendante du temps. Dans la partie C, nous étudions la propagation des ondes de choc.

2. Rappel des équations des écoulements continus. L'espace est rapporté à trois axes de coordonnées rectangulaires fixes $Oxyz$; le temps est désigné par t. Nous désignons par u, v, w les composantes cartésiennes de la vitesse du fluide sur les trois axes précédents, par ϱ et p la masse spécifique et la pression. Ces cinq fonctions dépendent de quatre variables.

Nous considérons une portion du fluide constamment formée des mêmes élements de matière, à l'instant t, elle occupe un volume V, limité par une surface fermée S. Nous désignons par α, β, γ les cosinus directeurs en un point de la surface S de la normale orientée vers l'extérieur du volume V. Nous négligerons la viscosité du fluide, c'est-à-dire que nous supposerons les pressions normales aux surfaces sur lesquelles elles s'exercent; nous supposons également que les forces extérieures sont négligeables. En désignant par $d\tau$ et ds les éléments de volume et de surface, le théorème des quantités de mouvement se traduit par les trois équations suivantes:

$$\left.\begin{aligned} \frac{d}{dt}\iiint\limits_V \varrho u\, d\tau + \iint\limits_S \alpha p\, ds = 0, \\ \frac{d}{dt}\iiint\limits_V \varrho v\, d\tau + \iint\limits_S \beta p\, ds = 0, \\ \frac{d}{dt}\iiint\limits_V \varrho w\, d\tau + \iint\limits_S \gamma p\, ds = 0. \end{aligned}\right\} \tag{2.1}$$

On admet que la masse de fluide contenue dans le volume V conserve, au cours du temps, une valeur constante; cette hypothèse se traduit par l'équation de la conservation de la masse

$$\frac{d}{dt}\iiint\limits_V \varrho\, d\tau = 0. \tag{2.2}$$

Nous écrivons pour terminer l'équation de la conservation de l'énergie; e désignant l'énergie interne par unité de masse, cette équation est la suivante:

$$\frac{d}{dt}\iiint\limits_V \left(\varrho\frac{u^2+v^2+w^2}{2} + e\right) d\tau = -\iint\limits_S p\, q_n\, ds. \tag{2.3}$$

Nous supposons que l'écoulement est adiabatique; c'est-à-dire que nous négligeons les échanges de chaleur entre les divers éléments fluides. On établit en thermodynamique la relation suivante: $p = (\gamma - 1)\, e$ où γ représente la rapport des chaleurs spécifiques à pression constante et à volume constant.

Pour evaluer les premiers membres des équations (2.1) à (2.3) nous utilisons la formule (2.4) que l'on établit en mathématiques et qui exprime la dérivée par rapport au temps d'une intégrale multiple étendue à un volume variable:

$$\frac{d}{dt}\iiint\limits_{(D)} \psi\, d\tau = \iiint\limits_{(D)} \frac{d\psi}{\partial t}\, d\tau + \iint\limits_{(F)} \psi\, q_n\, ds. \tag{2.4}$$

Dans cette formule ψ désigne une fonction continue, et dérivable par rapport au temps; la surface fermée (F) est la frontière du volume (D) et q_n désigne la vitesse de déplacement de cette frontière; le segment de normale déterminé par les deux positions infiniment voisines qu'occupe la surface (F) aux instants t et $t+dt$ est en chaque point $q_n\, dt$.

Nous appliquerons la formule (2.4) en prenant pour domaine d'intégration le volume V précédemment défini et nous supposerons que les molécules gazeuses ne se mélangent pas, si bien que ce sont toujours les mêmes molécules qui constituent la surface S; dans ces conditions on a $q_n = \alpha u + \beta v + \gamma w$. Si nous nous plaçons dans le cas où la fonction ψ possède des dérivées partielles premières continues par rapport aux variables d'espace et si nous supposons que les fonctions u, v et w possèdent également des dérivées partielles premières continues par rapport à ces mêmes variables, on peut utiliser la formule d'Ostrogradsky pour évaluer le dernier terme du second membre de l'équation (2.4); on obtient ainsi la formule suivante:

$$\frac{d}{dt}\iiint\limits_{V} \psi\, d\tau = \iiint\limits_{V} \left\{\frac{\partial\psi}{\partial t} + \frac{\partial(\psi u)}{\partial x} + \frac{\partial(\psi v)}{\partial y} + \frac{\partial(\psi w)}{\partial z}\right\} d\tau. \tag{2.5}$$

En utilisant cette formule pour transformer les équations (2.1) à (2.3) et en faisant tendre ensuite le volume V vers zéro, on obtient les équations du mouvement qui peuvent être écrites sous la forme suivante:

$$\left.\begin{aligned}
\frac{\partial u}{\partial t} + u\frac{\partial u}{\partial x} + v\frac{\partial u}{\partial y} + w\frac{\partial u}{\partial z} + \frac{1}{\varrho}\frac{\partial p}{\partial x} &= 0,\\
\frac{\partial v}{\partial t} + u\frac{\partial v}{\partial x} + v\frac{\partial v}{\partial y} + w\frac{\partial v}{\partial z} + \frac{1}{\varrho}\frac{\partial p}{\partial y} &= 0,\\
\frac{\partial w}{\partial t} + u\frac{\partial w}{\partial x} + v\frac{\partial w}{\partial y} + w\frac{\partial w}{\partial z} + \frac{1}{\varrho}\frac{\partial p}{\partial z} &= 0,
\end{aligned}\right\} \tag{2.6}$$

$$\frac{\partial\varrho}{\partial t} + \frac{\partial(\varrho u)}{\partial x} + \frac{\partial(\varrho v)}{\partial y} + \frac{\partial(\varrho w)}{\partial z} = 0, \tag{2.7}$$

$$\frac{\partial}{\partial t}\left(\frac{p}{\varrho^\gamma}\right) + u\frac{\partial}{\partial x}\left(\frac{p}{\varrho^\gamma}\right) + v\frac{\partial}{\partial y}\left(\frac{p}{\varrho^\gamma}\right) + w\frac{\partial}{\partial z}\left(\frac{p}{\varrho^\gamma}\right) = 0. \tag{2.8}$$

La dernière équation indique que, pour un élément fluide, le produit $p\varrho^{-\gamma}$ possède une valeur constante au cours du temps. Lorsque, en outre, cette valeur est la même pour tous les éléments fluides, l'écoulement est dit isentropique. Nous désignons par $\boldsymbol{V}$ le vecteur de composantes u, v, w et par q le module de ce vecteur. Les équations (2.6) peuvent s'écrire sous la forme vectorielle suivante:

$$\frac{\partial \boldsymbol{V}}{\partial t} + 2\operatorname{rot}\boldsymbol{V}\times\boldsymbol{V} + \operatorname{grad}\frac{V^2}{2} + \frac{1}{\varrho}\operatorname{grad} p = 0. \tag{2.9}$$

Lorsque le vecteur $(\partial\boldsymbol{V}/\partial t)$ est nul, l'écoulement est dit stationnaire ou permanent. Supposant qu'il en soit ainsi, nous projetons l'égalité (2.9) sur la vitesse. On obtient la relation suivante, appelée relation de BERNOULLI:

$$dp + \varrho\, q\, dq = 0. \tag{2.10}$$

La relation de BERNOULLI est valable pour tout déplacement le long d'une ligne de courant. Dans un tel déplacement, le produit $p\varrho^{-\gamma}$ demeure constant et la relation (2.10) peut être intégrée sous la forme suivante:

$$\frac{2\gamma}{\gamma-1}\,\frac{p}{\varrho}+q^2=\text{constante}. \tag{2.11}$$

La constante qui figure au second membre et dont la valeur dépend de la ligne de courant envisagée est désignée par q_m. Nous désignons par ϱ_0 et p_0 les valeurs de la masse spécifique et de la pression lorsque la vitesse est nulle; on déduit de l'équation (2.1) les relations suivantes, valables en tous les points d'une même ligne de courant:

$$\frac{\varrho}{\varrho_0}=\left(1-\frac{q^2}{q_m^2}\right)^{\frac{1}{\gamma-1}},\qquad \frac{p}{p_0}=\left(1-\frac{q^2}{q_m^2}\right)^{\frac{\gamma}{\gamma-1}}. \tag{2.12}$$

q_m représente la vitesse limite que peut atteindre le fluide sur une ligne de courant déterminée.

Dans l'étude des écoulements stationnaires, nous utiliserons également un système de coordonnées sphériques. Nous posons $x=r\cos\vartheta$, $y=r\sin\vartheta\cos\varphi$, $z=r\sin\vartheta\sin\varphi$ et nous désignons par u, v, w les composantes du vecteur vitesse sur les directions correspondant aux variations de r, ϑ et φ respectivement. Les équations (2.6) à (2.8) s'écrivent de la façon suivante:

$$\left.\begin{aligned}
&u\frac{\partial u}{\partial r}+\frac{v}{r}\frac{\partial u}{\partial\vartheta}+\frac{w}{r\sin\vartheta}\frac{\partial u}{\partial\varphi}+\frac{1}{\varrho}\frac{\partial p}{\partial r}-\frac{v^2+w^2}{r}=0,\\
&u\frac{\partial v}{\partial r}+\frac{v}{r}\frac{\partial v}{\partial\vartheta}+\frac{w}{r\sin\vartheta}\frac{\partial v}{\partial\varphi}+\frac{1}{\varrho r}\frac{\partial p}{\partial\vartheta}+\frac{u\,v-w^2\cot\vartheta}{r}=0,\\
&u\frac{\partial w}{\partial r}+\frac{v}{r}\frac{\partial w}{\partial\vartheta}+\frac{w}{r\sin\vartheta}\frac{\partial w}{\partial\varphi}+\frac{1}{\varrho r\sin\vartheta}\frac{\partial p}{\partial\varphi}+\frac{u\,w+v\,w\cot\vartheta}{r}=0,
\end{aligned}\right\} \tag{2.13}$$

$$\frac{\partial}{\partial r}(r^2\varrho\,u\sin\vartheta)+\frac{\partial}{\partial\vartheta}(r\,\varrho\,v\sin\vartheta)+\frac{\partial}{\partial\varphi}(r\,\varrho\,w)=0, \tag{2.14}$$

$$u\frac{\partial}{\partial r}\left(\frac{p}{\varrho^\gamma}\right)+\frac{v}{r}\frac{\partial}{\partial\vartheta}\left(\frac{p}{\varrho^\gamma}\right)+\frac{w}{r\sin\vartheta}\frac{\partial}{\partial\varphi}\left(\frac{p}{\varrho^\gamma}\right)=0. \tag{2.15}$$

b) Démonstration des équations des phénomènes de choc.

3. Formule préliminaire. Pour établir les équations du mouvement, nous avons utilisé la formule (2.4) qui exprime la dérivée par rapport au temps d'une intégrale de volume. Nous allons établir une formule analogue, valable dans le cas où la fonction ψ subit des discontinuités à l'intérieur du volume V. Nous désignons par Σ une certaine surface, appelée onde de choc, sur laquelle la vitesse du fluide, la masse spécifique et la pression sont discontinues. La surface Σ varie au cours du temps; son équation cartésienne est désignée par $f(x, y, z, t)=0$. La distance dn, normale à Σ, qui sépare cette surface de la position Σ' qu'elle occupe à l'instant $t+dt$ possède la valeur suivante:

$$dn=\frac{f_x\,dx+f_y\,dy+f_z\,dz}{\sqrt{f_x^2+f_y^2+f_z^2}}=\frac{-f_t}{\sqrt{f_x^2+f_y^2+f_z^2}}\,dt. \tag{3.1}$$

Le quotient $U=(dn/dt)$ est appelé vitesse de déplacement de l'onde de choc. Nous supposons qu'une partie de la surface Σ est intérieure au volume V; ce volume est alors partagé en deux volumes V_1 et V_2 limités par les portions S_1 et S_2 de la surface S: Fig. 1a. Considérons le premier membre de l'équation (2.2); nous désignons par ϱ_1 la valeur de la masse spécifique dans le volume V_1 et par ϱ_2

sa valeur dans le volume V_2. Le volume balayé par l'aire $d\sigma$ de la surface Σ pendant le temps dt est $U\,dt\,d\sigma$; le signe de la fonction $f(x, y, z, t)$ est choisi de façon que la vitesse U soit positive lorsque la surface Σ se déplace de la région 2 vers la région 1. L'augmentation de masse due au déplacement de l'aire $d\sigma$ pendant le temps dt est $\varrho_2 U\,dt\,d\sigma - \varrho_1 U\,dt\,d\sigma$. On peut donc écrire la formule suivante:

$$\frac{d}{dt}\iiint\limits_V \varrho\,d\tau = \iiint\limits_V \frac{\partial\varrho}{\partial t}\,d\tau + \iint\limits_S \varrho\,q_n\,ds + \iint\limits_\Sigma (\varrho_2 - \varrho_1)\,U\,d\sigma. \tag{3.2}$$

La formule (3.2) demeure valable lorsqu'on remplace la masse spécifique une fonction ψ arbitraire.

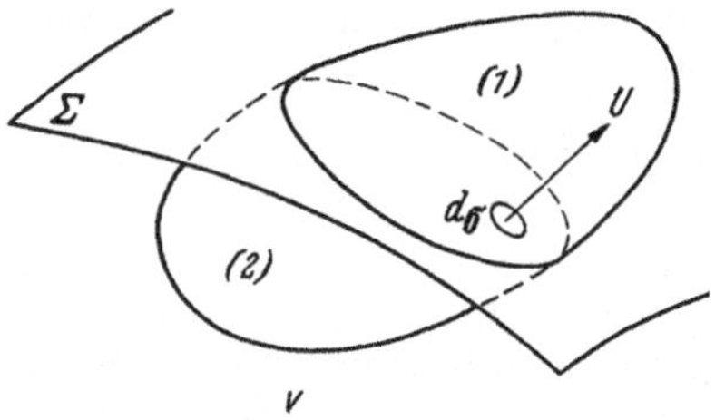

Fig. 1 a.

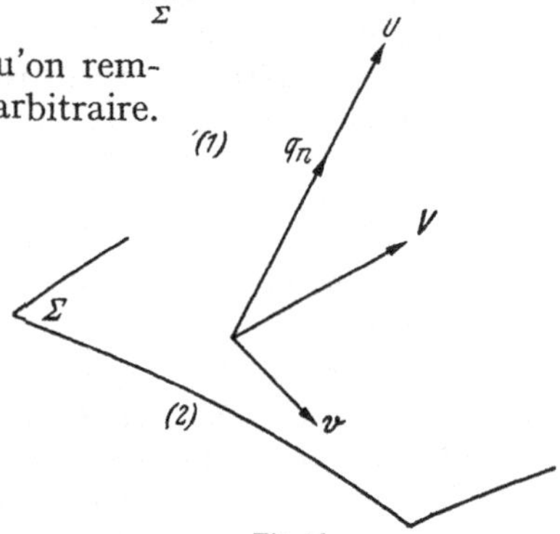

Fig. 1 b.

Nous supposons que la dérivée $(\partial\psi/\partial t)$ est bornée; la surface Σ étant fixe $(dt=0)$, nous faisons tendre les deux portions S_1 et S_2 de la surface S vers la surface Σ. On obtient la formule suivante:

$$\operatorname*{Limite}_{S\to\Sigma} \frac{d}{dt}\iiint\limits_V \psi\,d\tau = \iint\limits_\Sigma \psi_2\{(U - q_{n_2}) - \psi_1(U - q_{n_1})\}\,d\sigma. \tag{3.3}$$

Les indices 1 et 2 repèrent chacune des deux faces de la surface Σ; la vitesse U et les vitesses normales q_{n_1} et q_{n_2} sont comptées positivement de la région (2) vers la région (1). De même que nous avons utilisé la formule (2.4) pour établir les équations du mouvement, nous allons utiliser la formule (3.3) pour établir les équations des chocs. Dans les deux cas, nous partons des équations (2.1), (2.2) et (2.5) qui traduisent le théorème des quantités de mouvement, la conservation de la masse et la conservation de l'énergie.

4. Equations des phénomènes de choc. Nous supposons que les fonctions u, v, w, ϱ et p sont discontinues sur la surface Σ; nous désignons par $\boldsymbol{n}$ le vecteur normal unitaire de cette surface orienté de la région (2) vers la région (1). Nous faisons tendre le volume V vers zéro de façon qu'il soit toujours traversé par la surface Σ et nous appliquons la formule (3.3) aux équations (2.1), (2.2) et (2.5); nous obtenons les relations suivantes:

$$\left.\begin{aligned} \varrho_2 \boldsymbol{V}_2(q_{n_2} - U) + \boldsymbol{n}\,p_2 &= \varrho_1 \boldsymbol{V}_1(q_{n_1} - U) + \boldsymbol{n}\,p_1, \\ \varrho_2(q_{n_2} - U) &= \varrho_1(q_{n_1} - U), \\ \left(\frac{\varrho_2 q_2^2}{2} + \frac{p_2}{\gamma-1}\right)(q_{n_2} - U) + p_2 q_{n_2} &= \left(\frac{\varrho_1 q_1^2}{2} + \frac{p_1}{\gamma-1}\right)(q_{n_1} - U) + p_1 q_{n_1}. \end{aligned}\right\} \tag{4.1}$$

Nous désignons par $\boldsymbol{v}$ la projection du vecteur $\boldsymbol{V}$ sur le plan tangent à Σ et nous projetons la première des équations (4.1) sur la normale et sur le plan tangent à Σ; nous obtenons les équations suivantes:

$$\left.\begin{aligned} \varrho_2 q_{n_2}(q_{n_2} - U) + p_2 &= \varrho_1 q_{n_1}(q_{n_1} - U) + p_1 \\ \boldsymbol{v}_2 &= \boldsymbol{v}_1. \end{aligned}\right\} \tag{4.2}$$

Lorsque l'état du fluide avant le choc et la vitesse de déplacement U sont connus, les discontinuités subies par les cinq fonctions u, v, w, ϱ et p sont déterminées par les équations (4.1). En particulier la discontinuité du vecteur vitesse est normale à l'onde de choc. Les équations (4.1) se simplifient si on les résout par rapport aux inconnues q_{n_2}, ϱ_2 et p_2. Dans ce but, nous introduisons la célérité du son avant le choc c_1; elle peut être définie pour les écoulements adiabatiques par la relation $\gamma p_1 = \varrho_1 c_1^2$. Nous poserons $\mu^2 = (\gamma - 1)/(\gamma + 1)$. Nous appellerons

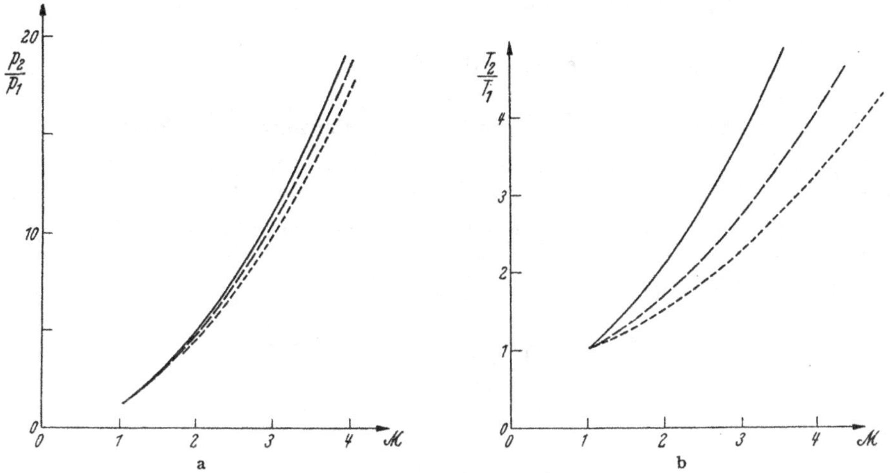

Fig. 2a et b. (a) Rapport des pressions. (b) Rapport des températures. ——— gaz monoatomique, — — — gaz diatomique, - - - - gaz triatomique.

nombre de Mach normal $\mathcal{M}$ le quotient de la vitesse de déplacement de l'onde par rapport au fluide par la célérité du son avant le choc. Les équations (4.1) admettent la solution suivante:

$$\frac{q_{n_2} - q_{n_1}}{c_1} = (1 - \mu^2)\left(\mathcal{M} - \frac{1}{\mathcal{M}}\right), \tag{4.3}$$

$$\frac{\varrho_1}{\varrho_2} = \frac{1 - \mu^2}{\mathcal{M}^2} + \mu^2, \tag{4.4}$$

$$\frac{p_2}{p_1} = (1 + \mu^2)\mathcal{M}^2 - \mu^2 \tag{4.5}$$

avec

$$\mathcal{M} = \frac{U - q_{n_1}}{c_1}. \tag{4.6}$$

Le second principe de la thermodynamique exige que la variation de l'entropie S d'un élement fluide soit positive ou nulle. Désignant par c_v la chaleur spécifique à volume constant, on peut écrire

$$S = c_v \log (p\, \varrho^{-\gamma}) + \text{constante}, \tag{4.7}$$

$$\left\{(1 + \mu^2)\mathcal{M}^2 - \mu^2\right\}\left\{\frac{1 - \mu^2}{\mathcal{M}^2} + \mu^2\right\}^{\gamma} \geqq 1. \tag{4.8}$$

Le rapport γ des chaleurs spécifiques étant supérieur à l'unité, on déduit de l'inégalité (4.8) que le nombre de Mach normal $\mathcal{M}$ est en valeur absolue supérieur ou égal à l'unité. Il en résulte qu'un choc est toujours accompagné d'une compression, d'une élévation de température et d'une augmentation de mass spécifique. Les variations du rapport des pressions et du rapport des températures en fonction du nombre de Mach $\mathcal{M}$ sont représentées sur les Figs. 2;

dans le cas des gaz monoatomiques, $\gamma=\frac{5}{3}$; diatomiques, $\gamma=\frac{7}{5}$ et triatomiques, $\gamma=\frac{4}{3}$. Les équations du choc permettent de calculer la discontinuité subie par les deux membres de l'équation (2.11):

$$\left(\frac{2\gamma}{\gamma-1}\frac{p_2}{\varrho_2}+q_2^2\right)-\left(\frac{2\gamma}{\gamma-1}\frac{p_1}{\varrho_1}+q_1^2\right)=2U(q_{n_2}-q_{n_1}). \tag{4.9}$$

Dans un écoulement stationnaire, l'onde de choc est immobile et la vitesse U est égale à zéro; la vitesse limite q_m définie par l'équation (2.11) ne subit pas de discontinuité à la traversée de l'onde de choc.

c) Chocs stationnaires dans les écoulements uniformes.

5. Equations en coordonnées cartésiennes. L'écoulement avant le choc est stationnaire, uniforme et homogène; la vitesse q_1 est parallèle à l'axe des abscisses. Nous désignons par $x=f(y,z)$ l'équation de l'onde de choc. Les composantes cartésiennes de la vitesse du fluide immédiatement après le choc s'expriment de la façon suivante: (où $f_y=\partial f/\partial y$ et $f_z=\partial f/\partial z$)

$$\left.\begin{aligned} u&=q_1-\frac{1-\mu^2}{\sqrt{1+f_y^2+f_y^2}}\,c_1\left(\mathcal{M}-\frac{1}{\mathcal{M}}\right),\\ v&=(q_1-u)\frac{\partial f}{\partial y},\qquad w=(q_1-u)\frac{\partial f}{\partial z},\end{aligned}\right\} \tag{5.1}$$

avec

$$\mathcal{M}=\frac{q_1}{c_1\sqrt{1+f_y^2+f_y^2}}. \tag{5.2}$$

Le nombre de Mach normal $\mathcal{M}$ est supérieur ou égal à l'unité en valeur absolue; on en déduit l'inégalité $q_1\geqq c_1$. Il ne peut donc pas y avoir d'onde de choc dans un écoulement stationnaire subsonique. Le cas limite dans lequel la vitesse du fluide est égale à la célérité du son est appelé écoulement critique. Désignant par une astérisque les valeurs correspondantes des diverses grandeurs, nous déduisons des formules (2.11) et (2.2) les expressions suivantes:

$$\left.\begin{aligned} q^{*2}&=c_0^2\frac{2}{\gamma-1}=q_m^2\frac{\gamma-1}{\gamma+1},\\ \varrho^*&=\varrho_0\left(\frac{2}{\gamma+1}\right)^{\frac{1}{\gamma-1}},\qquad p^*=p_0\left(\frac{2}{\gamma+1}\right)^{\frac{\gamma}{\gamma-1}}.\end{aligned}\right\} \tag{5.3}$$

6. Equations intrinsèques. Nous envisageons un écoulement stationnaire de révolution autour de Ox; nous supposons en outre que la vitesse du fluide en un point M est située dans le plan MOx. Nous désignons par Or la direction perpendiculaire à Ox dans un demi-plan méridien, par u et v les composantes de la vitesse sur les directions Ox et Or.

α) Equations du mouvement. Les équations cartésiennes du mouvement s'écrivent de la façon suivante:

$$\left.\begin{aligned} u\frac{\partial u}{\partial x}+v\frac{\partial u}{\partial r}+\frac{1}{\varrho}\frac{\partial p}{\partial x}&=0,\\ u\frac{\partial v}{\partial x}+v\frac{\partial v}{\partial r}+\frac{1}{\varrho}\frac{\partial p}{\partial r}&=0,\\ \frac{\partial}{\partial x}(r\varrho u)+\frac{\partial}{\partial r}(r\varrho v)&=0,\\ u\frac{\partial}{\partial x}(p\varrho^{-\gamma})+v\frac{\partial}{\partial r}(p\varrho^{-\gamma})&=0.\end{aligned}\right\} \tag{6.1}$$

En un point A, nous désignons par $\boldsymbol{t}$ le vecteur unitaire colinéaire à la vitesse du fluide et de même sens, par $\boldsymbol{n}$ le vecteur unitaire directement perpendiculaire. Nous considérons dans le plan méridien de A un point B voisin; le vecteur $\boldsymbol{AB}$ a pour composantes ds et dn sur les directions des vecteurs $\boldsymbol{t}$ et $\boldsymbol{n}$. Nous introduisons les coordonnées polaires q et ϑ du vecteur vitesse: $u = q\cos\vartheta$, $v = q\sin\vartheta$. Les variations subies par l'angle ϑ et la pression p dans le déplacement élémentaire $\boldsymbol{AB}$ sont des fonctions linéaires et homogènes des différentielles ds et dn:

$$\left.\begin{aligned} d\vartheta &= \frac{\partial\vartheta}{\partial s}\,ds + \frac{\partial\vartheta}{\partial n}\,dn, \\ dp &= \frac{\partial p}{\partial s}\,ds + \frac{\partial p}{\partial n}\,dn \end{aligned}\right\} \tag{6.2}$$

avec

$$\boldsymbol{AB} = \boldsymbol{t}\,ds + \boldsymbol{n}\,dn. \tag{6.3}$$

Les quatre coefficients différentiels ainsi définis seront notés comme des dérivées partielles bien que les coordonnées intrinsèques ne constituent pas de vraies variables. Les coordonnées intrinsèques et les coordonnées cartésiennes sont liées par les relations (6.4); en effectuant le changement de variables dans les équations (6.1), on obtient les équations (6.5) et (6.6):

$$\left.\begin{aligned} ds &= dx\cos\vartheta + dr\sin\vartheta, \\ dn &= -\,dx\sin\vartheta + dr\cos\vartheta, \end{aligned}\right\} \tag{6.4}$$

$$\left.\begin{aligned} \varrho q^2 \frac{\partial\vartheta}{\partial s} &= -\frac{\partial p}{\partial n}, \\ \varrho q^2 \left(\frac{\partial\vartheta}{\partial n} + \frac{\sin\vartheta}{r}\right) &= -\frac{\partial(\varrho q)}{\partial s}, \end{aligned}\right\} \tag{6.5}$$

$$\left.\begin{aligned} \frac{\partial p}{\partial s} + \varrho q \frac{\partial q}{\partial s} &= 0, \\ \frac{\partial}{\partial s}(p\varrho^{-\gamma}) &= 0. \end{aligned}\right\} \tag{6.6}$$

Les équations (6.6) traduisent respectivement la relation de Bernoulli et le caractère adiabatique de l'écoulement. En introduisant la célérité du son c, on peut évaluer le second membre de la seconde des équations (6.5) et écrire les équations des écoulements de révolution sous la forme définite suivante:

$$\left.\begin{aligned} \varrho q^2 \frac{\partial\vartheta}{\partial s} &= -\frac{\partial p}{\partial n}, \\ \varrho q^2 \left(\frac{\partial\vartheta}{\partial n} + \frac{\sin\vartheta}{r}\right) &= \left(1 - \frac{q^2}{c^2}\right)\frac{\partial p}{\partial s}. \end{aligned}\right\} \tag{6.7}$$

En supprimant le terme $(\sin\vartheta/r)$, on obtient les équations relatives au cas des écoulements plans.

β) Equations du choc. Les équations du choc sont les mêmes pour les écoulements plans ou les écoulements de révolution. Nous supposons l'écoulement avant le choc uniforme, de vitesse q_1 parallèle à Ox. Après le choc, nous désignons toujours par q et ϑ les coordonnées polaires de la vitesse, par β l'angle que fait avec Ox la tangente à la méridienne de l'onde de choc. Nous pouvons écrire

$$\left.\begin{aligned} q_{n_1} &= -\,q_1\sin\beta, & q_n &= -\,q\sin(\beta-\vartheta), \\ q_{t_1} &= q_1\cos\beta, & q_t &= q\cos(\beta-\vartheta), \end{aligned}\right\} \tag{6.8}$$

$$\mathcal{M} = \frac{U - q_{n_1}}{c_1} = \frac{q_1}{c_1}\sin\beta. \tag{6.9}$$

Le quotient $M=(q_1/c_1)$ est appelé nombre de Mach de l'écoulement amont; il est lié au nombre de Mach normal par la relation $M\sin\beta=\mathcal{M}$. En écrivant l'équation (4.3) et la continuité de la vitesse tangentielle sur l'onde de choc,

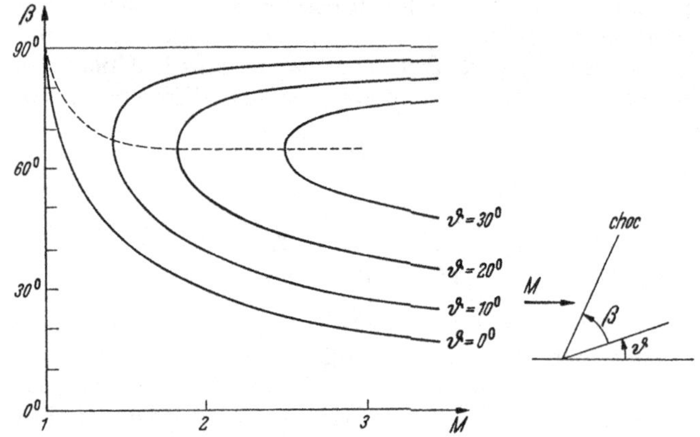

Fig. 3. Direction de la vitesse après le choc.

nous obtenons les équations (6.10); ensuite, par division membre à membre, les équations (6.11) ou (6.12):

$$\left.\begin{aligned} q\sin(\beta-\vartheta) &= q_1\sin\beta-(1-\mu^2)\,c_1\left(\mathcal{M}-\frac{1}{\mathcal{M}}\right),\\ q\cos(\beta-\vartheta) &= q_1\cos\beta, \end{aligned}\right\} \tag{6.10}$$

$$\tan(\beta-\vartheta)=\frac{\mu^2 M^2\sin^2\beta+1-\mu^2}{M^2\sin\beta\cos\beta}, \tag{6.11}$$

$$\tan(\beta-\vartheta)=\mu^2\,\frac{q_m^2-q_1^2\cos\beta}{q_1^2\sin\beta\cos\beta}. \tag{6.12}$$

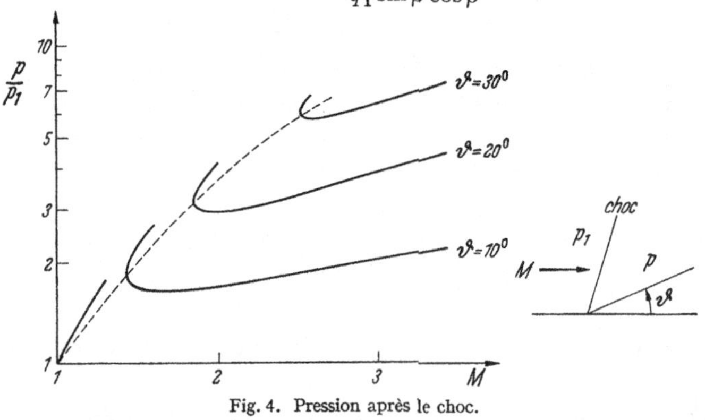

Fig. 4. Pression après le choc.

La masse spécifique ϱ, la pression p et l'angle de Mach α défini par la relation $c=q\sin\alpha$ s'expriment ensuite en fonction des angles β et ϑ par les formules suivantes:

$$\varrho=\varrho_1\,\frac{\tan\beta}{\tan(\beta-\vartheta)}, \tag{6.13}$$

$$p=p_1\,\frac{\tan\beta-\mu^2\tan(\beta-\vartheta)}{\tan(\beta-\vartheta)-\mu^2\tan\beta}, \tag{6.14}$$

$$\frac{\sin^2\alpha}{\sin^2(\beta-\vartheta)}=\frac{1}{1-\mu^2}\left\{\frac{\tan\beta}{\tan(\beta-\vartheta)}-\mu^2\right\}. \tag{6.15}$$

Les équations (6.11) et (6.14) déterminent pour chaque valeur de ϑ les variations de l'angle β et de la pression p en fonction du nombre de Mach M; ces variations sont représentées respectivement sur les Figs. 3 et 4, pour laquelle on a adopté la valeur numérique $\gamma = 1{,}4$. Il résulte enfin de l'équation (6.11) et de l'inégalité $\mathcal{M}^2 \geqq 1$ que les deux membres de l'équation (6.15) sont supérieurs ou égaux à l'unité.

Il sera commode pour l'étude des écoulements par la méthode des caractéristiques d'introduire l'angle auxiliaire ω défini par la formule (6.16) et de définir ensuite par les formules (6.17) les variables α at β appelées variables caractéristiques:

$$\omega = \int\limits_{q_1}^{q} \left(\frac{q^2}{c^2} - 1\right)^{\frac{1}{2}} \frac{dq}{q}, \tag{6.16}$$

$$\alpha = \omega + \vartheta, \qquad \beta = \omega - \vartheta. \tag{6.17}$$

Lorsque le choc est faible, l'angle ϑ est un infiniment petit; si q_1 désigne la grandeur de la vitesse avant le choc, l'angle ω est un infiniment petit équivalent soit à $+\vartheta$, soit à $-\vartheta$, et l'une des deux variables caractéristiques est un infiniment petit du troisiéme ordre par rapport à ϑ.

B. Les ondes de choc dans les écoulements stationnaires.

Dans l'étude d'un phénomène aérodynamique, la logique voudrait que l'on considère pour commencer le départ d'un obstacle dans un fluide initialement au repos. Le mouvement de l'obstacle donne naissance à un mouvement du fluide qu'il conviendrait d'étudier; lorsque le mouvement de l'obstacle tend vers une translation uniforme, on peut admettre que le mouvement du fluide tend vers un état stationnaire. L'étude des mouvements stationnaires devrait donc se faire en dernier; cependant, cette étude est la plus simple par suite de la diminution du nombre des variables et les résultats obtenus sont plus nombreux et peuvent être présentés pour cette raison en un ensemble cohérent. Nous consacrons donc la partie B à une étude systématique des chocs stationnaires, tandis que, dans la partie C, nous nous contenterons d'étudier quelques problèmes de chocs non stationnaires.

Le fluide est animé à l'infini d'un mouvement rectiligne uniforme. Les diverses sections de cette partie B correspondent à des valeurs de plus en plus grandes de la vitesse à l'infini. Lorsque cette vitesse à l'infini est faible, l'écoulement est continu; pour une certaine valeur (subsonique), des ondes de choc apparaissent au voisinage de l'obstacle; ce phénomène est étudié dans la partie I. Lorsque la vitesse à l'infini est faiblement supersonique, la première des ondes de choc ou onde de tête est détachée de l'obstacle; cette étude fait l'objet de la partie II. Lorsque la vitesse à l'infini est encore plus grande, l'onde de tête peut être attachée si l'obstacle présente une pointe; les ondes attachées sont étudiées dans la partie III pour les écoulements plans et dans la partie IV pour les écoulements à trois dimensions.

I. La formation des ondes de choc.

7. Détermination d'une loi de compressibilité approchée. Nous envisageons un écoulement plan uniforme à l'infini. Nous supposons que le fluide est homogène à l'infini, si bien que les constantes ϱ_0, p_0 et q_m figurant dans les formules (2.12) possèdent la même valeur sur toutes les lignes de courant. Les forces extérieures étant négligées, tous les mouvements du fluide ont lieu avec potentiel des vitesses[1]. Désignant par φ ce potentiel nous avons $d\varphi = u\,dx + v\,dy$; d'après l'équation (2.7), il existe une fonction ψ appelée fonction de courant telle que $\varrho^* d\psi = -\varrho v\,dx + \varrho u\,dy$. En introduisant les coordonnées polaires de la vitesse, on peut écrire

$$\left.\begin{aligned} q\,dx &= \cos\vartheta\,d\varphi - \sin\vartheta\,\frac{\varrho^*}{\varrho}\,d\psi,\\ q\,dy &= \sin\vartheta\,d\varphi + \cos\vartheta\,\frac{\varrho^*}{\varrho}\,d\psi.\end{aligned}\right\} \tag{7.1}$$

On a également les égalités $d\varphi = q\,ds$, $\varrho^* d\psi = \varrho\,q\,dn$, si bien que nous pouvons introduire les variables φ et ψ dans les équations intrinsèques des écoulements plans. En permutant le rôle des fonctions et des variables, nous obtenons le système suivant:

$$\left.\begin{aligned} \frac{\partial\varphi}{\partial\vartheta} &= q\,\frac{\varrho^*}{\varrho}\,\frac{\partial\psi}{\partial q},\\ \frac{\partial\varphi}{\partial q} &= -\frac{\varrho^*}{\varrho q}\left(1-\frac{q^2}{c^2}\right)\frac{\partial\psi}{\partial\vartheta}.\end{aligned}\right\} \tag{7.2}$$

En introduisant la variable w et la fonction $X(w)$ définies par le système (7.3), on peut écrire le système (7.2) sous la forme simplifiée (7.4).

$$\left.\begin{aligned} w &= \int\limits_{q^*}^{q}\frac{\varrho}{\varrho^*}\,\frac{dq}{q},\\ X(w) &= \frac{\varrho^{*2}}{\varrho^2}\left(1-\frac{q^2}{c^2}\right),\end{aligned}\right\} \tag{7.3}$$

$$\frac{\partial\varphi}{\partial\vartheta} = \frac{\partial\psi}{\partial w}, \qquad \frac{\partial\varphi}{\partial w} = -X(w)\,\frac{\partial\psi}{\partial\vartheta}. \tag{7.4}$$

La fonction $X(w)$ est déterminée par la loi de compressibilité. Réciproquement, le choix de la function $X(w)$ détermine une loi de compressibilité; pour expliciter cette dernière, il suffit d'intégrer le système (7.5) équivalent au système (7.3) et d'y ajouter l'équation (7.6), conséquence de la relation de BERNOULLI, valable dans tout le fluide puisque les quantités ϱ_0, p_0 et q_m ont la même valeur sur toutes les lignes de courant.

$$\left.\begin{aligned} \frac{d}{dw}\left(\frac{\varrho^* q^*}{\varrho q}\right) &= -X(w)\,\frac{q^*}{q},\\ \frac{d}{dw}\left(\frac{q^*}{q}\right) &= -\frac{\varrho^* q^*}{\varrho q},\end{aligned}\right\} \tag{7.5}$$

$$p = p^* - q^*\int\limits_0^w q^2\,dw. \tag{7.6}$$

Avec TOMOTIKA et TAMADA [6], nous adopterons la fonction $X(w)$ définie par l'équation suivante:

$$\left.\begin{aligned} &X(w) = a\,(1 - e^{2\alpha w}),\\ &a = \left(\frac{2}{\gamma+1}\right)^{\frac{2}{\gamma-1}}, \qquad \alpha = \left(\frac{\gamma+1}{2}\right)^{\frac{\gamma+1}{\gamma-1}}.\end{aligned}\right\} \tag{7.7}$$

[1] Ce résultat est la conséquence d'un théorème dû à LAGRANGE dont la démonstration déborde le cadre de ce chapitre.

Avec le choix précédent et les conditions initiales $q=q^*$, $\varrho=\varrho^*$ pour $w=0$, les équations (7.5) définissent une loi de compressibilité $\varrho(q)$ représentée sur la Fig. 5. Cette loi de compressibilité et la loi de compressibilité adiabatique sont tangentes lorsque la vitesse du fluide est nulle; elles sont osculatrices lorsque la vitesse du fluide est sonique. Nous considérons désormais un fluide fictif obéissant à la loi de compressibilité définie par l'équation (7.7); ce fluide se comportera comme un fluide réel pour les vitesses subsoniques, ainsi que pour les vitesses faiblement supersoniques.

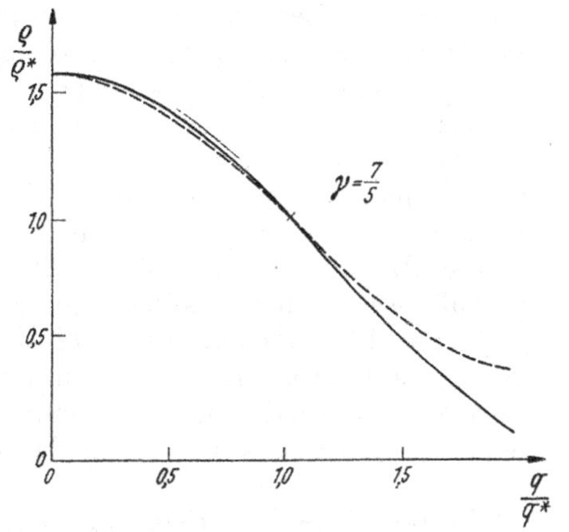

Fig. 5. Lois de compressibilité.

8. Ecoulement uniforme autour d'un obstacle. Le changement de variables $\alpha w=\log\tau$ et $\beta\sqrt{a}=\alpha\vartheta$ transforme le système (7.4) en le système suivant [6]

$$\left.\begin{aligned}\frac{\partial\varphi}{\partial\beta}&=\sqrt{a}\,\tau\,\frac{\partial\psi}{\partial\tau},\\ \frac{\partial\varphi}{\partial\tau}&=\sqrt{a}\left(\tau-\frac{1}{\tau}\right)\frac{\partial\psi}{\partial\beta}.\end{aligned}\right\}\tag{8.1}$$

λ désignant un paramètre réel, nous introduisons la famille de fonctions $\omega=r+is$ définies par l'équation suivante:

$$\beta-i\log\lambda+\tau\sin\omega-\omega=0,\tag{8.2}$$

$$\left.\begin{aligned}\tau&=\frac{s+\log\lambda}{\cos r\,\mathrm{Sin}\,s},\\ \beta&=r-(s+\log\lambda)\tan r\,\mathrm{Cot}\,s.\end{aligned}\right\}\tag{8.3}$$

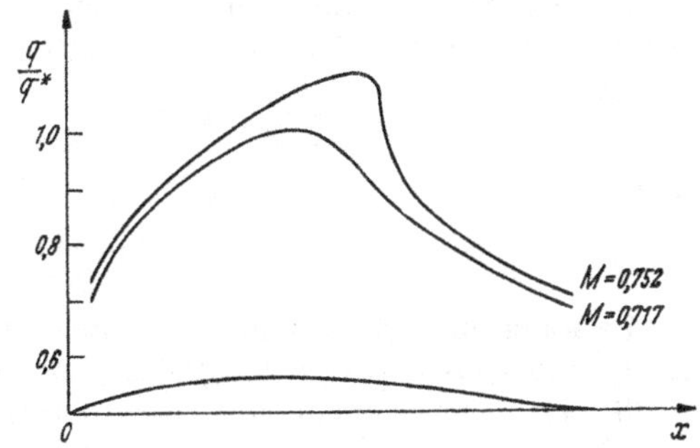

Fig. 6. Répartition des vitesses sur un profil.

Le système (8.1) admet la solution particulière $\psi=\omega$, $\varphi=-\sqrt{a}\,(\tau\cos\omega+\log\tau)$. D'une façon plus générale, nous considérons la famille de solutions constitutées par les parties réelles des fonctions suivantes:

$$\left.\begin{aligned}\psi&=\frac{1-\lambda^2}{1+\lambda^2}\,\tau\sin\omega-\frac{i}{1-\tau\cos\omega},\\ \varphi&=\sqrt{a}\,\tau\left\{\frac{\lambda^2-1}{\lambda^2+1}\left(\frac{\tau}{2}+\cos\omega\right)+\frac{i\sin\omega}{1-\tau\cos\omega}\right\}.\end{aligned}\right\}\tag{8.4}$$

Les formules (8.4) déterminent les valeurs des fonctions ψ et φ dans le plan de l'hodographe; on passe au plan physique par intégration du système (7.1). Lorsqu'on a $r=0$, $\mathrm{Tan}\,s=s+\log\lambda$, la fonction ψ est indéterminée tandis que la fonction φ est infinie. Les vitesses à l'infini sur toutes les lignes de courant sont donc équipollentes; ce qui signifie que l'écoulement est uniforme à l'infini:

$$\beta_\infty=0,\qquad \tau_\infty=\frac{1}{\mathrm{Cos}\,s_\infty}.\tag{8.5}$$

La vitesse à l'infini est subsonique si λ est plus petit que l'unité, ce que nous supposerons. Les quantités β et ψ sont nulles quand on a $r=0$; la ligne de courant $\psi=0$ est donc décomposée; elle comprend l'axe des abscisses et une courbe qui constitue le profil autour duquel s'écoule le fluide. On déduit des équations (7.1) les équations paramétriques du profil ainsi que la répartition des vitesses sur celui-ci. A chaque valeur de λ, correspond un profil différent et un nombre de Mach à l'infini M différent. La forme du profil varie peu avec λ, si bien que nous pouvons supposer confondus les profils correspondant à des valeurs voisines de λ; par contre, les variations de la répartition des vitesses ne sont pas négligeables. Deux cas sont représentés sur la Fig. 6.

Dans le premier cas $M = M_1 = 0{,}717$, la vitesse maximum sur le profil est exactement sonique; lorsque M est inférieur à M_1, l'écoulement est partout subsonique. Dans le second cas $M = M_2 = 0{,}752$, la courbe de répartition des vitesses sur le profil possède une tangente perpendiculaire à l'axe des abscisses; lorsque M est supérieur à M_2, la vitesse sur le profil n'est plus fonction univalente de l'absisse curviligne, si bien que la solution continue ne possède pas de réalité physique; l'écoulement comporte donc des ondes de choc. Celles-ci apparaîtront dès que le nombre de Mach à l'infini depassera une certaine valeur critique. La présence d'ondes de choc étant impossible dans une région subsonique, le nombre de Mach critique admet pour bornes inférieure et supérieure les valeurs M_1 et M_2.

Le problème de la détermination de l'ecoulement avec ondes de choc autour d'un obstacle n'est pas résolu. On a pu traiter quelques cas simples et démontrer de nombreuses propriétés relatives aux ondes de choc. L'objet de ce chapitre est d'indiquer brièvement ces divers résultats.

9. Exemple d'écoulement avec choc. Les équations des écoulements plans permanents isentropiques (7.2) admettent la solution suivante:

$$\psi = e^{n\vartheta} A(q), \qquad \varphi = \frac{e^{n\vartheta}}{n} \frac{\varrho^*}{\varrho} q A'(q) \tag{9.1}$$

avec

$$\tan\alpha(q) = \frac{n}{q} \frac{A(q)}{A'(q)}. \tag{9.2}$$

La fonction $\alpha(q)$ est solution d'une certaine équation différentielle de RICCATI, dont toutes les courbes intégrales concourent au point $q = q_m$, $n \tan\alpha = 1$ ([7], p. 28—31). Les écoulements correspondants ont été étudiés par TOLLMIEN [8]. Les lignes d'égale vitesse sont des spirales logarithmiques qui coupent les lignes de courant sous l'angle α. Nous plaçons une onde de choc sur l'une de ces spirales; le nombre de Mach normal possède la même valeur en tous les points de l'onde, si bien que l'écoulement après le choc est encore isentropique. Les conditions du choc prouvent que cet écoulement est encore défini par les formules (9.1) et (9.2) dans lesquelles les fonctions $A(q)$ et $\alpha(q)$ doivent être remplacées par deux autres fonctions $B(q)$ et $\beta(q)$; la fonction $\beta(q)$ est une seconde solution de l'équation différentielle de RICCATI vérifiée par la fonction $\alpha(q)$.

II. Les ondes de choc détachées.

10. Détermination de l'écoulement derrière une onde donnée. Un solide mobile dans un fluide compressible est animé d'une translation uniforme. Lorsque la vitesse du solide est faible, le mouvement du fluide est continu; lorsque la vitesse du solide dépasse la valeur critique, l'écoulement comporte des ondes de choc. Lorsque la vitesse du solide est supersonique, les perturbations engendrées ne peuvent se propager jusqu'à l'infini amont et il existe une région de l'espace dans laquelle le fluide est au repos; cette région est limitée par une onde de choc qui, pour les vitesses faiblement supersoniques, est détachée de l'obstacle.

Nous nous plaçons dans le cas des écoulements de révolution et nous raisonnons dans des axes liés à l'obstacle: Ox axe de révolution, Or axe perpendiculaire. ϱ_1, p_1 et M désignent la masse spécifique, la pression et le nombre de Mach dans l'écoulement amont; u et v désignent les composantes de la vitesse du fluide sur Ox et Or; les équations du mouvement s'écrivent sous la forme (6.1). Nous supposons que l'onde de choc est une courbe analytique, convexe vers l'amont et dont nous écrivons l'équation sous la forme suivante:

$$x = x(r) = \frac{r^2}{2R} + \sum_{j=2}^{\infty} \lambda_j \frac{r^{2j}}{2j\,R^{2j-1}}. \tag{10.1}$$

Sur la courbe $x = x(r)$, les fonctions u, v, ϱ et p vérifient les conditions du choc (4.4), (4.5) et (5.1); au-delà, elles vérifient les équations de l'écoulement. La détermination de l'écoulement est donc un problème de CAUCHY. En vertu de la troisième des équations (6.1), il existe une fonction de courant $\psi(x, r)$ vérifiant les relations (10.3):

$$\psi(x, r) = \sum_{j=1}^{\infty} r^{2j}\, \psi_{2j}(x), \tag{10.2}$$

$$\frac{\partial \psi}{\partial r} = r\varrho u, \qquad \frac{\partial \psi}{\partial x} = -r\varrho v. \tag{10.3}$$

Toute fonction proportionnelle à $\psi(x, r)$ joue le rôle de fonction de courant; nous pouvons donc choisir $R^2\psi_2(0) = 1$. Les conditions du choc et les équations du mouvement permettent de calculer les valeurs pour $x = 0$ de toutes les dérivées $\psi_{2j}^{(n)}(x)$ [*9*]. Les résultats se présentent sous la forme suivante:

$$\begin{aligned}
R^{2j+n}\psi_{2j}^{(n)}(0) &= f(M) && 2 < 2j + n \leqq 4\\
&= f(M) + \lambda_2 g(M) && 4 < 2j + n \leqq 6\\
&= f(M) + \lambda_2 g(M) + \lambda_3 h(M) && 2j + n = 7\\
&= f(M) + \lambda_2 g(M) + \lambda_3 h(M) + \lambda_2^2 k(M) && 2j + n = 8\\
&\ \vdots
\end{aligned}$$

Les fonctions $f(M)$, $g(M)$ ... dépendent des indices j et n et du nombre de Mach M. Si on considère seulement les valeurs des dérivées $\psi_{2j}^{(n)}(0)$ pour lesquelles on a $2j + n \leqq 6$, ces fonctions sont au nombre de 13; pour $2j + n \leqq 8$, elles sont au nombre de 38.

La fonction $\psi(x, r)$ demeure constante sur les surfaces de courant; lorsque la valeur constante est nulle, la surface de courant comprend l'axe de révolution et une surface (Σ). Supposant que cette surface rencontre l'axe de révolution en un point S, son équation peut être écrite sous la forme (10.4). Les quantités h, $\mathscr{R}$ et les paramètres σ_j se déterminent en écrivant que la fonction de courant est identiquement nulle sur la surface (Σ).

$$x = h + \frac{r^2}{2\mathscr{R}} + \sum_{j=2}^{\infty} \sigma_j \frac{r^{2j}}{2j\,\mathscr{R}^{2j-1}}, \tag{10.4}$$

$$\left.\begin{aligned}
\psi_2(h) &= 0\\
\psi_2'(h) + 2\mathscr{R}\psi_4(h) &= 0\\
&\ \vdots
\end{aligned}\right\} \tag{10.5}$$

Pour expliciter les résultats, nous faisons l'approximation qui consiste à supposer nulles les quantités $\psi_{2j}^{(n)}(0)$ pour lesquelles la somme $2j + n$ dépasse une valeur donnée $N + 2$. Dans l'approximation d'ordre quatre ($N = 4$), la première équation (10.5) ne possède pas de racine réelle lorsque le paramètre λ_2 est supérieur à une certaine valeur maximum fonction de M, soit $\lambda^*(M)$. Lorsque λ_2 est supérieur à ce maximum, la surface (Σ) ne rencontre pas l'axe de révolution; il existe alors une surface de courant qui possède un cercle double sur lequel le fluide est au repos. L'une des nappes de cette surface constitue l'obstacle, l'autre nappe sépare le fluide en deux régions; les molécules fluides situées dans l'une des régions contournent l'obstacle, les molécules fluides situées dans l'autre région pénètrent à l'intérieur. Deux exemples sont représentés sur les Figs. 7 et 8.

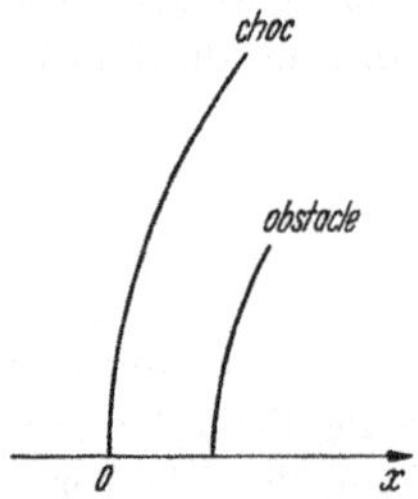

Fig. 7. $\lambda_2 < \lambda^*(M)$.

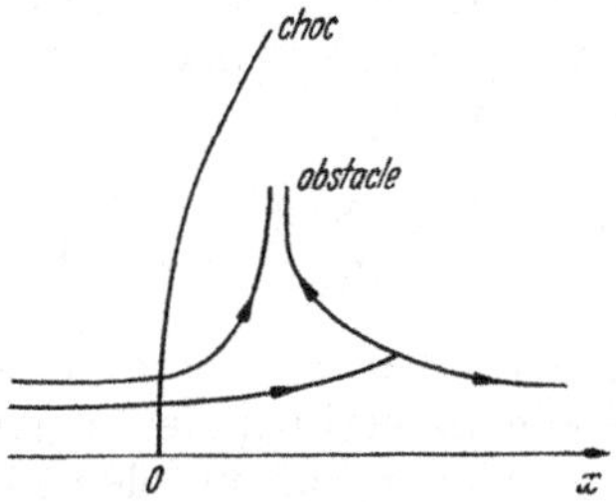

Fig. 8. $\lambda_2 > \lambda^*(M)$.

Ecoulement derrière une onde de choc détachée.

On peut traiter par la même méthode le cas des mouvements à trois dimensions. La vitesse q_1 avant le choc étant parallèle à l'axe des abscisses, on peut mettre

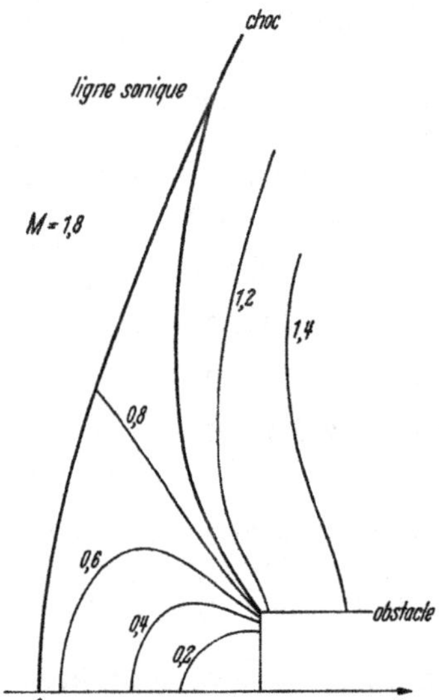

Fig. 9. Ecoulement plan.

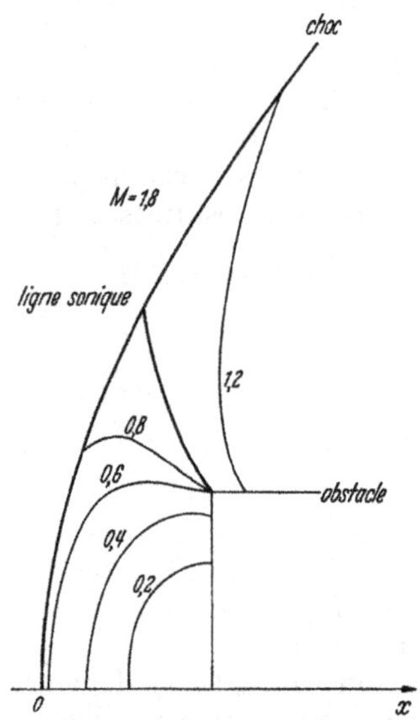

Fig. 10. Ecoulement de révolution.

l'équation de l'onde de choc sous la forme (10.6); la valeur du produit ϱq sur l'axe est donnée par la relation (10.7).

$$x = \frac{y^2}{2R_1} + \frac{z^2}{2R_2} + \cdots, \tag{10.6}$$

$$\varrho q = \varrho_1 q_1 (1 + a\,x + b\,x^2 + \cdots), \tag{10.7}$$

$$\left.\begin{aligned} a &= -\left(\frac{q_1^2}{q^{*2}} - 1\right)\left(\frac{1}{R_1} + \frac{1}{R_2}\right), \\ 2b &= \left\{\frac{q_1^4}{q^{*4}} - 3\,(2-\mu^2)\,\frac{q_1^2}{q^{*2}} + 2\right\}\left(\frac{1}{R_1} + \frac{1}{R_2}\right)^2, \\ &\quad - 2\left\{\frac{q_1^4}{q^{*4}} - 2\,(2-\mu^2)\,\frac{q_1^2}{q^{*2}} + 1\right\}\frac{1}{R_1 R_2}. \end{aligned}\right\} \tag{10.8}$$

Des calculs plus précis ont été effectués par MITCHELL en utilisant les méthodes de relaxation à partir d'une onde de choc donnée par photographie [*10*], [*11*]. La précision dans les résultats est plus grande, mais l'étude systématique de la correspondance entre l'onde de choc et l'obstacle n'est plus possible. Deux exemples sont représentés sur les Figs. 9 et 10.

11. Détermination de l'onde devant un obstacle donné. Pour déterminer l'onde de choc détachée devant un obstacle donné, il suffit théoriquement d'adopter un ordre d'approximation donné et de résoudre les équations (10.5) qui sont alors algébriques en y considérant comme inconnus les quantités h, R et les paramètres λ_j. Malheureusement, l'approximation d'ordre quatre semble insuffisante et les calculs correspondant à l'approximation d'ordre six sont excessivement longs. De nombreuses méthodes approchées ont été proposées, notamment par DUNGUNDJI [*12*], MOECKEL [*13*], SHU [*14*] et HIDA [*15*].

Dans toutes ces méthodes, on étudie l'onde de choc seulement au voisinage de son sommet; on a eu le droit d'agir ainsi pour les raisons suivantes. Le choc étant normal sur l'axe Ox, il existe, derrière l'onde, une région subsonique limitée par l'arc SI de l'obstacle, l'arc OA de l'onde et la ligne sonique IA: région (1) sur la Fig. 11. Les caractéristiques rencontrant la ligne sonique couvrent une région (2) limitée par la caractéristique IB. Une perturbation issue d'un point de l'arc SI se transmet dans les régions (1) et (2); une perturbation issue d'un autre point de l'obstacle n'atteint aucune de ces deux régions. La région (2), région supersonique influencée par la région subsonique, est appelée région transsonique; la caractéristique IB est appelée frontière transsonique. L'arc OB de l'onde de choc ne dépend que de l'arc SI de l'obstacle.

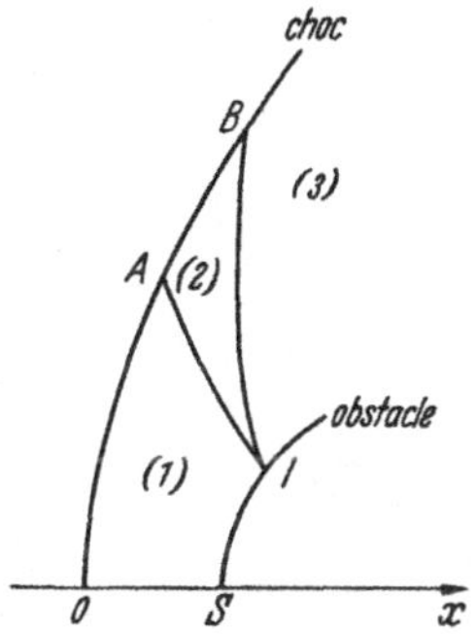

Fig. 11.

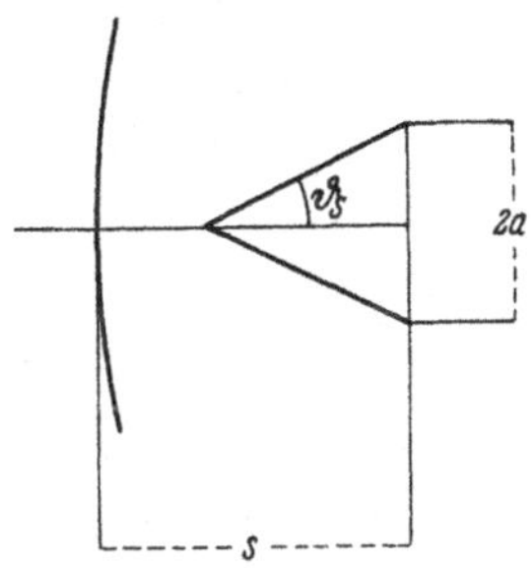

Fig. 12.

Lorsque l'obstacle est terminé par un cône de révolution (ou par un dièdre) de demi-angle au sommet ϑ_s, nous désignons par s la distance du sommet de l'onde au plan de la base et par a le rayon de la base: Fig. 12. MOECKEL admet que le rapport s/a est indépendant de l'angle ϑ_s et que la méridienne de l'onde de choc est une hyperbole. Les asymptotes devant être parallèles aux directions de Mach de l'écoulement amont, l'équation de l'onde est la suivante:

$$x^2 - (M^2 - 1)\,r^2 + 2(M^2 - 1)\,R\,x = 0. \tag{11.1}$$

SHU explicite la valeur de la fonction de courant dans la région subsonique; les équations du mouvement sont vérifiées de façon approchée et les conditions du choc de façon exacte. Cependant aucune des méthodes proposées jusqu'à l'heure actuelle ne peut être considérée comme satisfaisante; le problème de l'onde de choc détachée demeure posé.

12. Résultats expérimentaux. Des expériences ont été effectuées par HEBERLE, WOOD et GOODERUM [*16*]. Dans le cas de la sphère, les résultats sont les suivants:

M	$h/\mathscr{R}$	$\mathscr{R}/R$	λ_2	M	$h/\mathscr{R}$	$\mathscr{R}/R$	λ_2
1,17	1,477	0,109	−1,37	1,62	0,541	0.348	−0,31
1,30	0,961	0,197	−0,88	1,81	0,438	0,424	−0,19
1,37	0,817	0.224	−0,67				

La valeur du paramètre λ_2 diffère peu de celle que l'on déduit de l'équation (11.1); on peut donc admettre la relation $1+2(M^2-1)\,\lambda_2=0$.

Dans le cas d'un obstacle terminé par un cône, le rapport s/a varie peu avec l'angle ϑ_s; ses variations en fonction du nombre de Mach M sont représentées sur la Fig. 13. Dans le cas d'un obstacle terminé par un dièdre (écoulement plan), la variation du rapport s/a en fonction de l'angle ϑ_s a été étudié par JOHNSTON [*17*]. Les résultats sont représentés sur la Fig. 14; pour $\vartheta_s \geqq \vartheta_d = 29°30'$, l'onde de choc cesse d'être détachée. La variation du rapport s/a est importante surtout au voisinage de la valeur limite ϑ_d. Il est probable que le même phénomène se retrouve pour les écoulements de révolution, si bien que la Fig. 13 et l'hypothèse de MOECKEL sont valables seulement pour les valeurs de ϑ_s qui sont nettement supérieures à ϑ_d.

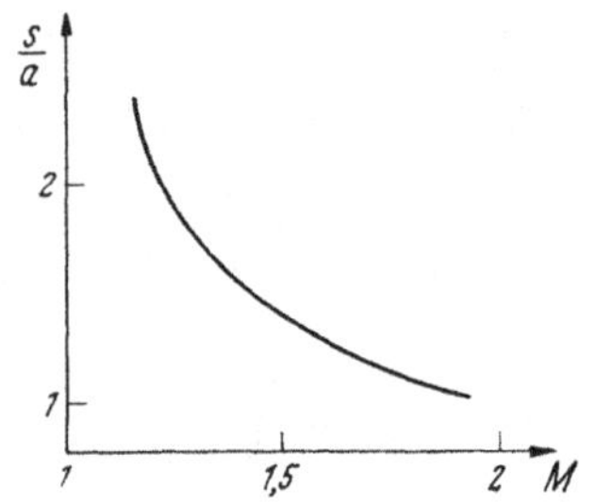

Fig. 13. Onde détachée devant un cône.

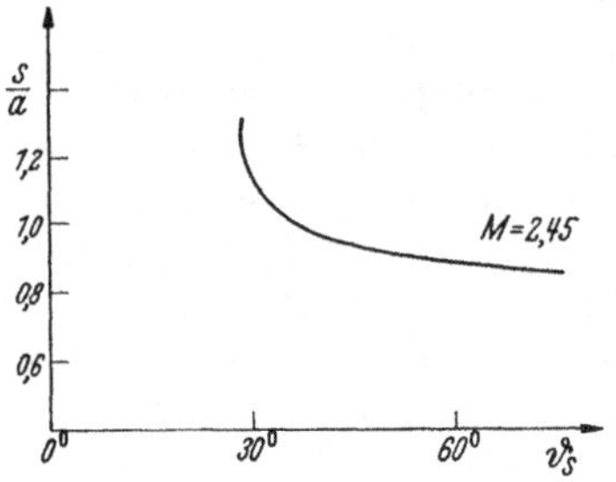

Fig. 14. Onde détachée devant un dièdre.

13. Branches infinies de l'onde de choc. Les directions asymptotiques de l'onde de choc sont les directions de Mach de l'écoulement amont, sinon la variation d'entropie sur l'onde de choc serait infinie, ainsi que l'énergie nécessaire pour engendrer le mouvement. Nous désignons par β_1 l'angle de Mach amont avant le choc (nous posons $\alpha=\cot\beta_1$). Nous nous plaçons dans le cas des écoulements de révolution et nous écrivons l'équation de l'onde de choc sous la forme (13.1). A l'infini, le quotient $G(r)/r$ est nul et le nombre de Mach normal $\mathscr{M}$ possède la valeur approchée $1+\sin\beta_1\cos\beta_1 G'(r)$; les équations du choc s'écrivent sous la forme (13.2).

$$x=\alpha r-G(r), \tag{13.1}$$

$$\left.\begin{aligned} u &= q_1-2(1-\mu^2)\,q_1\sin^3\beta_1\cos\,\beta_1 G'(r)+\cdots,\\ v &= \phantom{q_1-{}}2(1-\mu^2)\,q_1\sin^2\beta_1\cos^2\beta_1\,G'(r)+\cdots. \end{aligned}\right\} \tag{13.2}$$

Derrière l'onde, les fonctions u, v vérifient les équations de l'écoulement (6.1). Ces équations, linéarisées au voisinage de la solution représentée par l'écoulement amont, admettent la solution suivante:

$$\left.\begin{aligned} u &= q_1(1+\delta)=q_1-q_1\int\limits_0^{x-\alpha r}\frac{f(t)\,dt}{\sqrt{(x-t)^2-\alpha^2 r^2}},\\ v &= \frac{q_1}{r}\int\limits_0^{x-\alpha r}\frac{(x-t)\,f(t)\,dt}{\sqrt{(x-t)^2-\alpha^2 r^2}}. \end{aligned}\right\} \tag{13.3}$$

La fonction $f(t)$ est arbitraire. Les caractéristiques de la solution linéarisée ont pour équation: $x-\alpha r=\text{const}$. Nous désignons par $y(x,r)=\text{const}$ les caractéristiques de la solution réelle et nous admettons avec WITHAM [*18*] que les formules (13.3) représentent l'écoulement avec une meilleure approximation lorsque on remplace $x-\alpha r$ par $y(x,r)$. La caractéristique qui rencontre l'onde de choc à l'infini correspond à une certaine valeur y que nous supposons bornée. Nous remplaçons dans les relations (13.3) x par $y-\alpha r$ et nous négligeons y devant αr; on obtient ainsi les formules plus simples suivantes:

$$\delta=-\frac{F(y)}{\sqrt{2\alpha}}\,r^{-\frac{1}{2}},\qquad v=-\alpha q_1\,\delta. \tag{13.4}$$

La fonction $F(y)$ ne dépend que de la fonction $f(t)$; elle est donc également arbitraire. Pour déterminer la fonction $y(x,r)$, nous écrivons l'équation différentielle des caractéristiques $dr=\tan(\beta-\vartheta)\,dx$ où β désigne l'angle de Mach et ϑ l'angle de la vitesse avec Ox. On calcule l'angle de Mach en écrivant la relation de BERNOULLI, ensuite on intègre l'équation des caractéristiques le long des courbes $y=\text{const}$; on écrit ainsi les relations successives suivantes:

$$c^2=c_1^2-(\gamma-1)\,q_1^2\,\delta, \tag{13.5}$$

$$\beta=\beta_1-\left(1+\frac{\gamma-1}{2\sin^2\beta_1}\right)\tan\beta_1\,\delta, \tag{13.6}$$

$$\frac{dx}{dr}=\alpha-\frac{1}{1-\mu^2}\,\frac{F(y)}{\sqrt{2\alpha}}\,\frac{r^{-\frac{1}{2}}}{\sin^3\beta_1\cos\beta_1}, \tag{13.7}$$

$$x=\alpha r-\frac{2}{1-\mu^2}\,\frac{F(y)}{\sqrt{2\alpha}}\,\frac{r^{\frac{1}{2}}}{\sin^3\beta_1\cos\beta_1}+\text{const}. \tag{13.8}$$

Les caractéristiques représentées par l'équation (13.8) diffèrent de celles que l'on obtient dans la théorie linéarisée par le terme en $r^{\frac{1}{2}}$, qui est négligeable devant αr, mais infiniment grand avec r. Sur l'onde de choc, r et y sont liés par la relation (13.9) obtenue en éliminant x entre l'équation de l'onde et l'équation des caractéristiques; mais, sur l'onde de choc, r et y sont également liés par la relation (13.10) qui exprime que les fonctions u et v déduites des formules (13.4) vérifient les conditions du choc (13.2).

$$G(r)=\frac{2}{1-\mu^2}\,\frac{F(y)}{\sqrt{2\alpha}}\,\frac{r^{\frac{1}{2}}}{\sin^3\beta_1\cos\beta_1}+\text{const}, \tag{13.9}$$

$$G'(r)=\frac{1}{2(1-\mu^2)}\,\frac{F(y)}{\sqrt{2\alpha}}\,\frac{r^{-\frac{1}{2}}}{\sin^3\beta_1\cos\beta_1}. \tag{13.10}$$

Les relations (13.9) et (13.10) devant être compatibles, la fonction $G(r)$ vérifie l'équation différentielle $4rG'(r)-G(r)=\text{const}$. L'équation de l'onde de choc au voisinage de l'infini peut donc être écrite sous la forme suivante, dans laquelle A et B désignent deux constantes:

$$x=\alpha r-A\,r^n+B+\cdots,\qquad n=\tfrac{1}{4}. \tag{13.11}$$

Le problème analogue relatif au cas des écoulements plans a été résolu par OSWATITSCH [*19*]. L'écoulement après le choc est assimilé à un écoulement par ondes simples; la formule (13.11) dans laquelle on remplace r par y demeure valable, mais avec la valeur $n=\frac{1}{2}$.

III. Les ondes de choc attachées dans les écoulements plans.

a) Etude des chocs uniformes.

14. Ecoulement derrière une onde rectiligne. Nous envisageons un obstacle plan admettant un axe de symétrie et animé d'une translation uniforme supersonique parallèle à cet axe. La région amont dans laquelle le fluide est au repos est limitée par une onde de choc; celle-ci est détachée lorsque l'obstacle n'admet pas de pointe. Nous supposons que l'obstacle admet sur l'axe de symétrie une pointe de demi-angle ϑ_s; lorsque ϑ_s est supérieur à la valeur limite $\vartheta_m = \arcsin(1/\gamma)$, l'onde est encore détachée. Lorsque ϑ_s est inférieur à ϑ_m, l'onde est détachée pour les vitesses faiblement supersoniques; pour les vitesses suffisamment grandes, elle est attachée à la pointe de l'obstacle. A l'infini, les ondes attachées et détachées se comportent de la même façon; nous allons donc étudier les ondes de choc attachées au voisinage de la pointe.

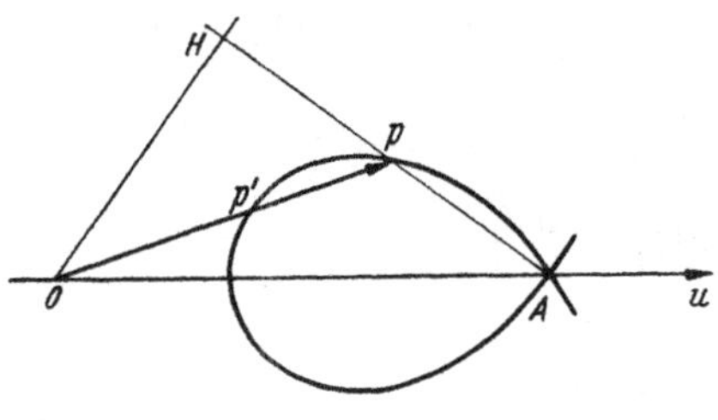

Fig. 15. Polaire de choc.

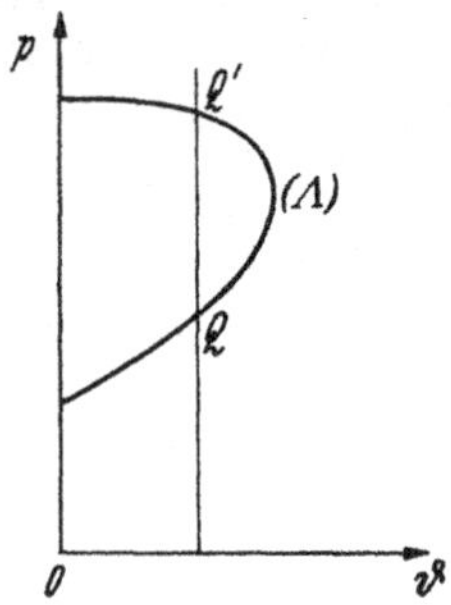

Fig. 16. Polaire de choc.

Nous supposons pour commencer une onde de choc rectiligne, faisant un angle β avec la direction Ox de l'écoulement amont. L'écoulement après le choc est uniforme; les valeurs des grandeurs q, ϑ, ϱ et p relatives à cet écoulement sont données par les équations du choc (6.10), (6.13) et (6.14); ces valeurs sont des fonctions de l'angle β et du nombre de Mach amont M. Il est commode de traduire les équations du choc par des graphiques. Dans le plan de l'hodographe, on trace, à partir d'un point fixe O, deux vecteurs $\boldsymbol{OA}$ et $\boldsymbol{OP}$ équipollents à la vitesse avant et après le choc; lorsque β varie, M étant fixe, le point P décrit une courbe algébrique (P) appelée polaire de choc: Fig. 15. Lorsque M varie, la courbe (P) se déforme et on obtient la famille des polaires de choc; l'onde est perpendiculaire à la discontinuité de vitesse, c'est-à-dire au vecteur $\boldsymbol{AP}$. On peut également tracer le point Q dont les coordonnées sont la déviation ϑ du fluide à travers le choc et la pression finale p: Fig. 16. Lorsque β varie, M étant fixe, le point Q décrit une courbe (Λ).

La déviation ϑ est déterminée par l'équation (6.11), qui est de la forme $F(\beta, \vartheta, M) = 0$. Le nombre de Mach M étant donné, nous faisons décroître β à partir de la valeur $(\pi/2)$; l'étude de l'équation (6.11) ou l'examen de la Fig. 15 montre que ϑ commence par croître à partir de zéro. Pour une certaine valeur β_0, la déviation ϑ est maximum; à ce moment, la dérivée $(\partial F/\partial \beta)$ est nulle; la fonction $\beta_0(M)$ est déterminée par l'équation suivante:

$$(1+\mu^2)\, M^4 \sin^4\beta + (2 - 2\mu^2 - M^2)\, M^2 \sin^2\beta - (1-\mu^2) - M^2 = 0. \quad (14.1)$$

Lorsque β continue à décroître, ϑ diminue jusqu'à la valeur zéro; à ce moment, β est égal à l'angle de Mach $\arcsin(1/M)$ et le nombre de Mach normal $\mathcal{M}$ est égal à l'unité; β ne peut être inférieur à l'angle de Mach, car $\mathcal{M}$ serait alors inférieur à 1, ce qui est en contradiction avec le second principe de la thermodynamique. La fonction $\beta_0(M)$ est représentée par la courbe en pointillé sur la Fig. 3; lorsque M est infini, on a $\cot\beta_0=\mu$.

Le nombre de Mach M_2 après le choc est déterminé par l'équation (6.15) dans laquelle on doit remplacer $\sin\alpha$ par $1/M_2$. Dans les conditions précédentes, M_2 croît de la valeur subsonique qui correspond au choc normal jusqu'à la valeur M. M_2 est égal à l'unité pour une certaine valeur β^*, déterminée par l'équation suivante:

$$(1+\mu^2)\,M^4\sin^4\beta+(1-2\mu^2-M^2)\,M^2\sin^2\beta-(1-\mu^2)=0. \tag{14.2}$$

Dans l'intervalle $\arcsin(1/M)\leq\beta\leq(\pi/2)$, le premier membre de l'équation (14.2) est une fonction croissante de β, négative pour $\beta=\arcsin(1/M)$, positive

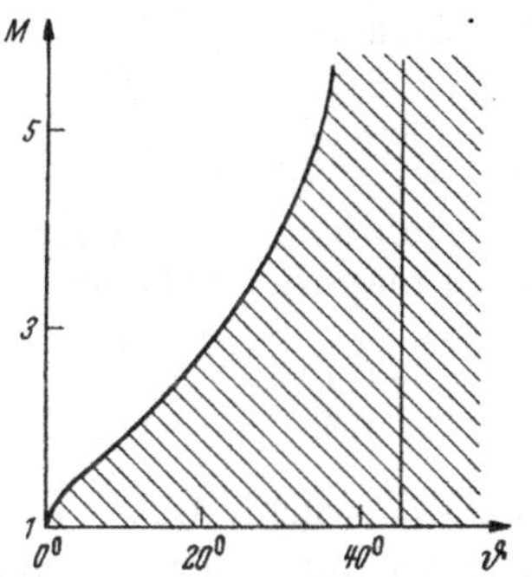

Fig. 17.

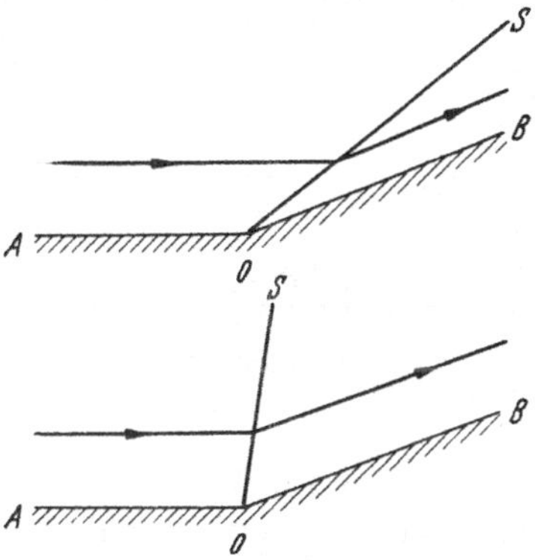

Fig. 18.

pour $\beta=\beta_0(M)$. L'angle $\beta^*(M)$ est donc inférieur à l'angle $\beta_0(M)$, ce qui signifie encore que M_2 est inférieur à 1 pour $\beta=\beta_0(M)$.

La valeur maximum de la déviation s'obtient en éliminant β entre les équations (6.11) et (14.1). On définit ainsi une fonction croissante $\vartheta_m(M)$ représentée sur la Fig. 17; lorsque M est infini, on a $\gamma\sin\vartheta_m=1$, soit $2\mu\tan\vartheta_m=1-\mu^2$. Pour les gaz diatomiques ($\gamma=1{,}4$), la déviation ne peut dépasser la valeur 45°35′05″.

15. Ecoulement supersonique dans un angle. Nous considérons un angle AOB, $(\boldsymbol{AO},\boldsymbol{OB})=\vartheta$; un écoulement supersonique uniforme parallèle au côté AO vient heurter le côté OB; nous cherchons à déterminer une onde de choc rectiligne OS telle que l'écoulement aval soit parallèle à OB. Nous supposons $2\mu\tan\vartheta$ inférieur à $1-\mu^2$.

La direction β de l'onde de choc est déterminée par l'équation (6.11), à laquelle nous joignons la condition que le produit $M\sin\beta$ soit supérieur à l'unité. L'angle ϑ étant donné, nous faisons croître le nombre de Mach M à partir de l'unité; l'étude de l'équation (6.11) ou l'éxamen de la Fig. 3 montre qu'il n'existe aucune valeur réelle de β lorsque M est un peu supérieur à 1. Pour une certaine valeur M_0 l'équation (6.11) admet une racine double en β; à ce moment, la dérivée $(\partial F/\partial\beta)$ est nulle. La racine double est donc égale à β_0 et la fonction $M_0(\vartheta)$ s'obtient en éliminant β entre les équations (6.11) et (14.1). Il en résulte que les fonctions $M_0(\vartheta)$ et $\vartheta_m(M)$ sont inverses l'une de l'autre, si bien que la Fig. 17 représente également la fonction $M_0(\vartheta)$; pour les couples de valeurs de ϑ et M représentées par un point situé dans la région hachurée de la Fig. 17, l'équation (6.11) ne possède pas de solution en β et le problème posé est impossible. Lorsque M est supérieur à $M_0(\vartheta)$, il existe deux solutions β_1 et β_2 (nous désignerons par β_1 la plus petite des deux valeurs de β). Le problème posé possède deux solutions représentées sur les Fig. 18; l'augmentation d'entropie est plus petite dans la première solution; c'est cette solution, correspondant à β_1, que l'on observe dans les expériences. L'angle β est une fonction décroissante du nombre de Mach M.

Le nombre de Mach M est déterminée par l'équation (6.15); par dérivation, on obtient l'équation suivante:

$$2\frac{\partial \log M_2}{\partial \beta} + M_2^2 \sin(\beta - \vartheta)\cos(\beta - \vartheta) + 2\cot 2(\beta - \vartheta) + 1 = 0. \qquad (15.1)$$

La valeur qui correspond à la première solution est donc une fonction croissante de M. Lorsqu'on a $M = M_0(\vartheta)$, l'angle β est égal à β_0 et M_2 est inférieur à 1; lorsque M est infini, les équations (6.11) et (6.15) s'écrivent de la façon suivante:

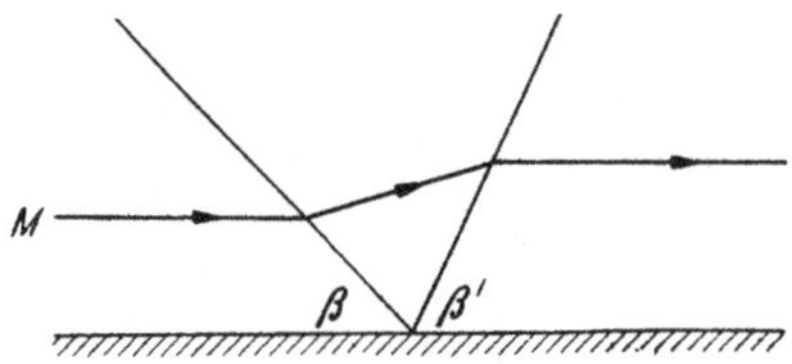

Fig. 19. Réflexion des ondes de choc.

$$\tan(\beta - \vartheta) = \mu^2 \tan\beta, \qquad (15.2)$$

$$M_2^2 = \frac{\mu^2}{(1+\mu^2)\sin^2(\beta - \vartheta)}. \qquad (15.3)$$

$\tan\beta_1$, déterminée par une équation du second degré, est majorée par la demi-somme des racines

$$\tan\beta_1 \leqq \frac{1-\mu^2}{2\mu^2}\cot\vartheta \leqq \frac{1}{\mu}. \qquad (15.4)$$

Il en résulte que M_2 est supérieur à 1. Le nombre de Mach M_2 est donc égal à l'unité pour une certaine valeur M^*, obtenue en éliminant β entre les équations (6.11) et (14.2). On a l'inégalité suivante:

$$M_0(\vartheta) < M^*(\vartheta). \qquad (15.5)$$

Par exemple, pour $\vartheta = 10°$, $M_0 = 1{,}4210$, $M^* = 1{,}4363$.

On étudie de la même façon le problème de la réflexion des ondes de choc: Fig. 19. Une onde de choc rectiligne vient rencontrer sous l'angle β une paroi rectiligne; on cherche s'il est possible de déterminer une onde de choc réfléchie faisant un angle β' avec la paroi, de telle façon que les équations du choc soient vérifiées sur chacune des deux ondes et que la direction de l'écoulement après le second choc coïncide avec la direction de l'écoulement avant le premier choc. Ce problème coïncide avec celui de la rencontre d'une onde plane et d'un dièdre que nous traiterons dans les Sects. 41 à 45. L'angle de réflexion β' est fonction de l'angle d'incidence β et du nombre de Mach incident M.

b) Ecoulement autour d'un obstacle terminé par un dièdre.

16. Cas où la vitesse aval est subsonique. Dans la pratique, on ne rencontre pas d'angle dont les côtés soient infinis, mais des obstacles terminés par des segments rectilignes. Nous envisageons un tel obstacle, symétrique par rapport à Ox et terminé par un segment rectiligne OI qui fait l'angle ϑ_s avec Ox: Fig. 20. Lorsque M est inférieur à $M_0(\vartheta_s)$, l'onde de choc est détachée; nous choisissons donc M supérieur à $M_0(\vartheta_s)$. Nous supposons que l'onde de choc est rectiligne; elle fait avec Ox l'angle β_1 calculé précédemment et son équation est $x = y\cot\beta_1$. En I, l'écoulement est perturbé; la perturbation atteint l'onde de choc en un point au delà duquel l'onde cesse d'être rectiligne. Lorsque M est compris entre $M_0(\vartheta_s)$ et $M^*(\vartheta_s)$, l'écoulement aval est subsonique, la perturbation se transmet dans tout le fluide et en particulier atteint le point O. L'onde ne possède alors aucun segment rectiligne; nous écrirons son équation sous la forme suivante:

$$x = y \cot \beta_1 - G(y). \tag{16.1}$$

Le terme $G(y)$ représente l'influence de la perturbation issue de I. Nous linéarisons les équations du mouvement au voisinage de la solution qui correspondrait au choc rectiligne de direction β_1. Les coordonnées intrinsèques peuvent alors être considérées comme des vraies variables et, en introduisant de nouvelles coordonnées ξ et η, les équations (6.7) s'écrivent de la façon suivante:

$$\left.\begin{aligned} \varrho q^2 \frac{\partial \vartheta}{\partial \xi} &= -\sqrt{1-\frac{q^2}{c^2}}\,\frac{\partial p}{\partial \eta}, \\ \varrho q^2 \frac{\partial \vartheta}{\partial \eta} &= \sqrt{1-\frac{q^2}{c^2}}\,\frac{\partial p}{\partial \xi}, \end{aligned}\right\} \tag{16.2}$$

$$d\xi = \frac{ds}{\sqrt{1-\frac{q^2}{c^2}}}, \qquad d\eta = dn. \tag{16.3}$$

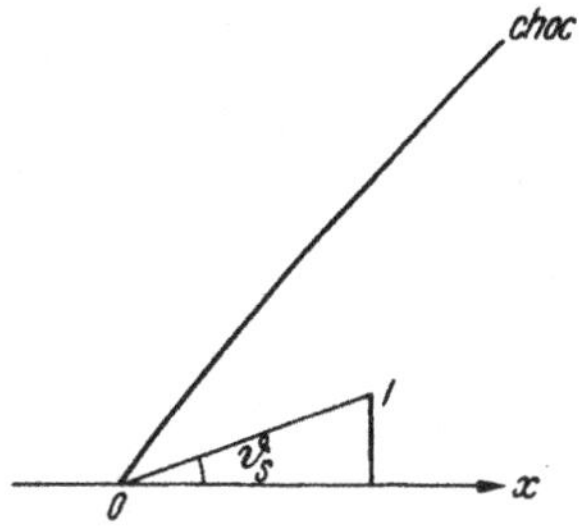

Fig. 20.

Dans le système (16.2), les quantités q, ϱ et c possèdent leur valeur au point O après le choc. Au point O, on a $\vartheta = \vartheta_s$ et $p = \bar{p}$; au voisinage de ce point, les expressions de ϑ et p sont les suivantes:

$$\left.\begin{aligned} \vartheta &= \vartheta_s + \sqrt{1-\frac{q^2}{c^2}}\, Q(\xi, \eta), \\ p &= \bar{p} + \varrho q^2\, P(\xi, \eta) \end{aligned}\right\} \tag{16.4}$$

où P et Q désignent la partie réelle et la partie imaginaire d'une fonction arbitraire $F(\zeta)$ de la variable complexe ζ ($\zeta = \xi + i\eta$); les formules (16.4) expriment en effet la solution la plus générale du système (16.2). Il faut choisir la fonction $F(\zeta)$ de façon que la condition sur l'obstacle et les équations du choc soient vérifiées. Sur l'obstacle, dn est nul; sur l'onde de choc, on a $ds = dn \tan(\beta - \vartheta)$; ainsi les équations de ces deux courbes dans le plan ξ, η sont les suivantes:

$$\text{pour l'obstacle} \qquad \eta = 0, \tag{16.5}$$

$$\text{pour le choc} \qquad \frac{\eta}{\xi} = \sqrt{1-\frac{q^2}{c^2}}\, \tan(\beta - \vartheta) = \tan \varphi. \tag{16.6}$$

Nous avons désigné par $\tan \varphi$ la valeur commune des deux membres de l'équation (16.6). Nous écrivons ensuite les équations du choc au voisinage du point O; pour cela, M étant fixe, nous différentions les équations (6.11) et (6.14); on obtient

$$d\vartheta = f_0(M, \vartheta)\, d\beta, \qquad dp = \frac{\varrho q^2}{1-\frac{q^2}{c^2}}\, g_0(M, \vartheta)\, d\vartheta, \tag{16.7}$$

$$\left.\begin{aligned} f_0 &= 1 - 2\mu^2 \cos^2(\beta - \vartheta) - \sin(\beta - \vartheta)\cos(\beta - \vartheta)\,(\tan\beta - \cot\beta), \\ g_0 &= 2(1-\mu^2)\left(1 - \frac{q^2}{c^2}\right) \sin(\beta - \vartheta)\cos(\beta - \vartheta), \end{aligned}\right\} \tag{16.8}$$

$$d\beta = \sin^2 \beta_1\, G'(y). \tag{16.9}$$

Pour exprimer simplement les conditions sur l'obstacle et sur l'onde, nous posons $\zeta = r\, e^{i\omega}$. L'obstacle correspond à $\omega = 0$, l'onde à $\omega = \varphi$; les conditions s'écrivent

$$\text{pour } \omega = 0, \qquad Q = 0 \tag{16.10}$$

$$\text{pour } \omega = \varphi \qquad \left.\begin{aligned} \sqrt{1-\frac{q^2}{c^2}}\, Q &= f_0\, d\beta, \\ \left(1-\frac{q^2}{c^2}\right) P &= g_0\, d\beta. \end{aligned}\right\} \tag{16.11}$$

Il est possible de satisfaire à toutes ces conditions en prenant $F(\zeta) = B_1 \zeta^m$, où B_1 et m sont deux constantes réelles [20]. La condition (16.10) est vérifiée. Sur l'onde de choc, les fonctions P et Q sont proportionnelles à r^m, donc à y^m, si bien que les deux équations (16.11) peuvent être écrites sous la forme suivante:

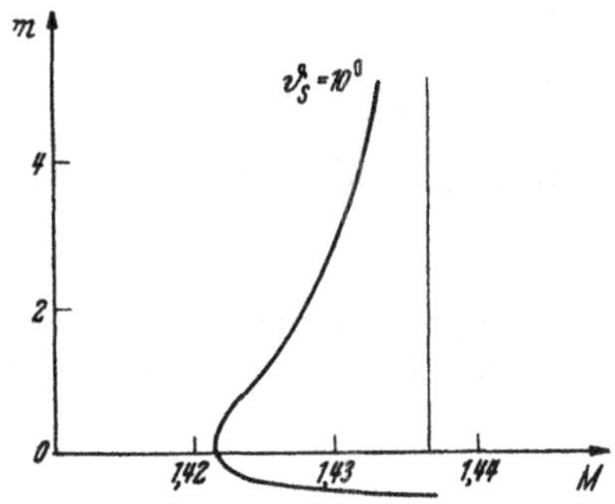

Fig. 21.

$$\tan m\varphi = \frac{f_0}{g_0}\sqrt{1-\frac{q^2}{c^2}}, \tag{16.12}$$

$$B_2\, y^m = G'(y). \tag{16.13}$$

L'équation (16.12) détermine l'exposant m et l'équation (16.13) la fonction $G(y)$. On obtient $G(y) = A\, y^{m+1}$ où A est une constante. L'exposant m qui dépend des quantités M et ϑ est une fonction croissante de M. Pour $\vartheta_s = 10°$ et $\gamma = 1{,}4$, les résultats numériques (Fig. 21) sont les suivants [21]:

β	M	m
67°	1,4211	0,135
66°	1,4226	0,554
65°	1,4255	1,270
64°	1,4297	2,857
63°	1,4356	14,704

L'onde de choc est représentée par l'équation (16.14); lorsque l'écoulement aval devient sonique, le nombre de Mach atteint la valeur M^* et la singularité disparait

$$x = y \cot\beta_1 - A\, y^{m+1} + \cdots. \tag{16.14}$$

17. Cas où la vitesse aval est supersonique. Lorsque le nombre de Mach M est supérieur à $M^*(\vartheta_s)$, l'écoulement après le choc est supersonique. La perturbation issue du point I se propage le long d'une caractéristique rectiligne et rencontre l'onde de choc en un point A: Fig. 22. En ce point, la courbure du choc est discontinue; la valeur R_A du rayon de courbure de l'onde immédiatement après le point A est proportionnelle à OI.

Nous supposons que l'obstacle possède en I une tangente discontinue, si bien que le fluide subit une détente par ondes simples concourantes limitée à la seconde caractéristique issue de A, soit (γ). D'après les formules relatives aux détentes, le rayon de courbure $\mathcal{R}_A$ de (γ) au point A est donné par la relation suivante:

$$\frac{IA}{\mathcal{R}_A} = \sin 2\alpha \{(1-\mu^2)\cos 2\alpha - \mu^2\}. \tag{17.1}$$

Nous désignons par σ l'abscisse curviligne sur (γ); la tangente fait avec Ox l'angle $\vartheta - \alpha$; $d\sigma = \mathcal{R}_A(d\vartheta - d\alpha)$. Pour tout déplacement sur une caractéristique, on a

$$(1-\mu^2)\cos^2\alpha\, d\alpha = (\sin^2\alpha + \mu^2\cos^2\alpha)\, d\vartheta, \tag{17.2}$$

$$\frac{d\vartheta}{d\sigma} = \frac{\partial\vartheta}{\partial s}\cos\alpha - \frac{\partial\vartheta}{\partial n}\sin\alpha = (1-\mu^2)\frac{\cos^2\alpha\sin 2\alpha}{IA}. \tag{17.3}$$

Nous calculerons dans la section suivante les valeurs des deux coefficients différentiels $(\partial\vartheta/\partial s)$ et $(\partial\vartheta/\partial n)$. La formule (17.3) détermine ensuite la courbure

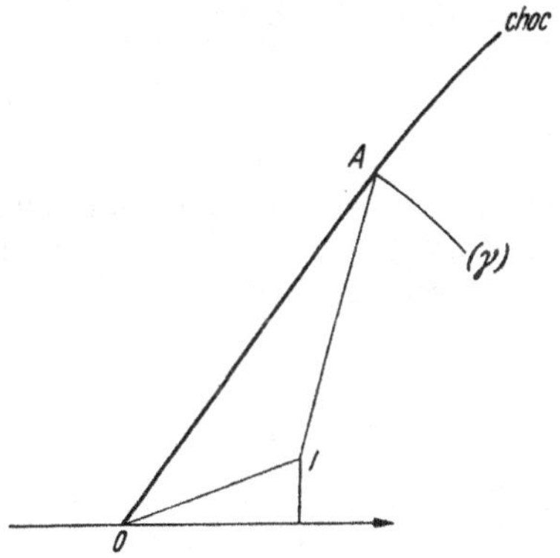

Fig. 22.

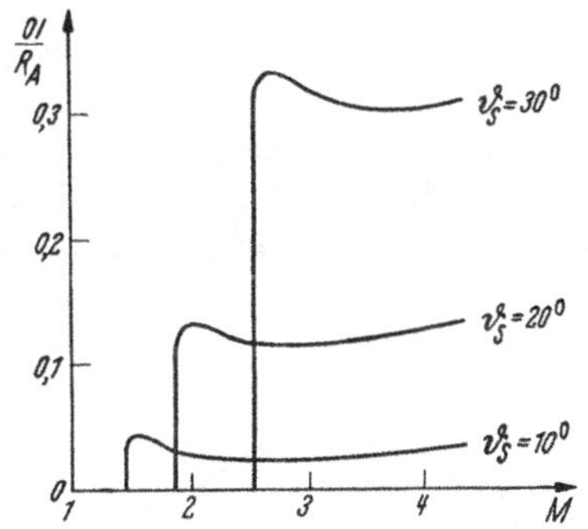

Fig. 23.

du choc au point A; les résultats pour $\gamma = 1{,}4$ sont indiqués sur la Fig. 23 ([7], p. 55); ils s'expriment par l'équation suivante:

$$\frac{OI}{R_A} = 2(1-\mu^2)\frac{\cos^2\alpha\sin^2(\beta-\vartheta-\alpha)}{(f_0 - g_0\tan\alpha)\sin(\beta-\vartheta)}. \tag{17.4}$$

c) Ecoulement autour d'une ogive.

18. Courbure du choc à la pointe. Nous désignons par ϑ_s l'angle que fait avec l'écoulement amont la tangente à l'obstacle au point O et par $\mathcal{R}$ le rayon de courbure en ce point; $\mathcal{R}$ est supposé fini et non nul. Le nombre de Mach amont M est donné. Puisque l'obstacle est une ligne de courant particulière, on doit avoir en O $\vartheta = \vartheta_s$, et l'équation (6.11) détermine la valeur en ce point de l'angle β que fait avec l'écoulement amont la tangente à l'onde de choc. L'abscisse curviligne sur l'onde de choc est désignée par σ. Les variations de la déviation ϑ et de la pression p pour tout déplacement sur l'onde de choc peuvent s'exprimer à l'aide des coefficients différentiels $(\partial\vartheta/\vartheta s)$, $(\partial\vartheta/\partial n)$, $(\partial p/\partial s)$ et $(\partial p/\partial n)$; les relations (16.7) peuvent s'écrire de la façon suivante:

$$\left.\begin{aligned} \frac{\partial\vartheta}{\partial s}\cos(\beta-\vartheta) + \frac{\partial\vartheta}{\partial n}\sin(\beta-\vartheta) &= f_0\frac{d\beta}{d\sigma},\\ \frac{\partial p}{\partial s}\cos(\beta-\vartheta) + \frac{\partial p}{\partial n}\sin(\beta-\vartheta) &= \frac{\varrho q^2}{1-\frac{q^2}{c^2}} g_0\frac{d\beta}{d\sigma}. \end{aligned}\right\} \tag{18.1}$$

Les équations (18.1) et les équations intrinsèques du mouvement, équations (6.7), permettent de calculer la valeur de chacun des quatre coefficients différentiels. On obtient en particulier

$$\left.\begin{aligned}\left\{1-\frac{q^2}{c^2}\sin^2(\beta-\vartheta)\right\}\frac{\partial\vartheta}{\partial s}&=f_1\frac{d\beta}{d\sigma},\\ \left\{1-\frac{q^2}{c^2}\sin^2(\beta-\vartheta)\right\}\frac{\partial\vartheta}{\partial n}&=g_1\frac{d\beta}{d\sigma},\end{aligned}\right\}\tag{18.2}$$

$$\left.\begin{aligned}f_1&=\cos(\beta-\vartheta)\,f_0-\sin(\beta-\vartheta)\,g_0,\\ g_1&=\left(1-\frac{q^2}{c^2}\right)\sin(\beta-\vartheta)\,f_0+\cos(\beta-\vartheta)\,g_0.\end{aligned}\right\}\tag{18.3}$$

Au point O, on a $\vartheta=\vartheta_s$ et $(\partial\vartheta/\partial s)=\mathcal{R}^{-1}$; la première des équations (18.2) détermine la valeur R du rayon de courbure de l'onde de choc:

$$\frac{R}{\mathcal{R}}=f_1\,\frac{\sin^2\alpha}{\sin^2\alpha-\sin^2(\beta-\vartheta)}\,.\tag{18.4}$$

Le rapport des courbures est représenté sur la Fig. 24; son signe est celui de f_1, d'après la relation (6.15). Lorsqu'on a $M=M_0(\vartheta)$, la fonction f_0 est nulle et la fonction g_0 est négative; la fonction f_1 est alors négative [22], [23].

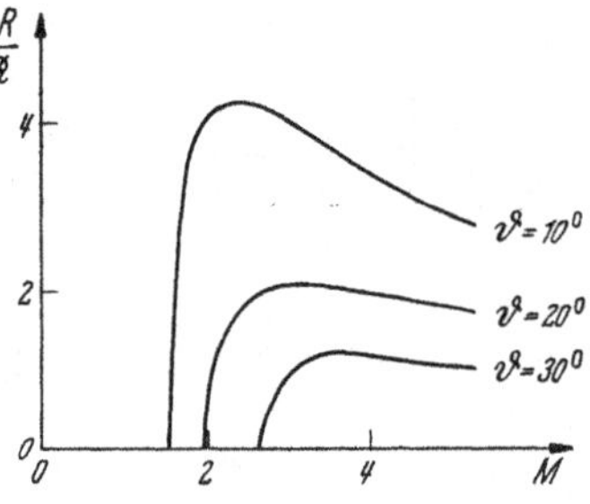

Fig. 24. Courbure des chocs.

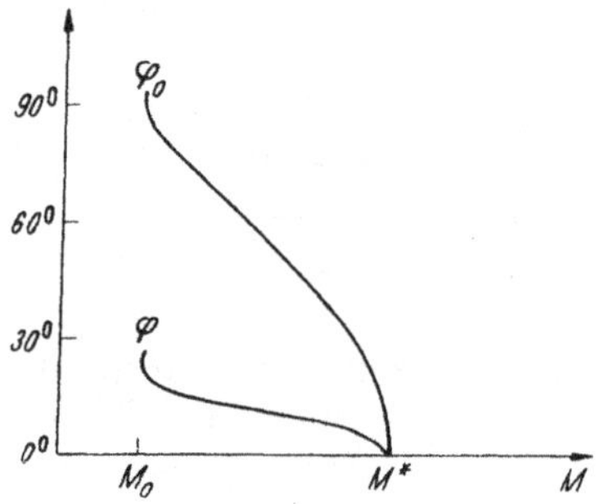

Fig. 25.

19. Cas d'un obstacle terminé par un point d'inflexion. Lorsque l'obstacle est terminé par un point d'inflexion, la courbure $(\partial\vartheta/\partial s)$ est nulle au point O; il en résulte que la courbure $(d\beta/d\sigma)$ et tous les coefficients différentiels du premier ordre sont nuls. D'une façon générale, nous supposons que l'on a au point O les relations suivantes:

$$\frac{\partial^i\vartheta}{\partial s^i}=0\quad\text{pour } i<r,\qquad \frac{\partial^r\vartheta}{\partial s^r}\neq 0.\tag{19.1}$$

En différentiant les équations intrinsèques du mouvement, puis les équations (18.2), on obtient six relations entre les huit coefficients différentiels du second ordre des quantités ϑ et p. On a en outre la relation suivante:

$$\frac{\partial}{\partial n}\left(\frac{\partial}{\partial s}\right)-\frac{\partial}{\partial s}\left(\frac{\partial}{\partial n}\right)=\frac{\partial\vartheta}{\partial s}\cdot\frac{\partial}{\partial s}-\frac{\partial\vartheta}{\partial n}\cdot\frac{\partial}{\partial n}\,.\tag{19.2}$$

On peut donc calculer tous les coefficients différentiels du second ordre. Comme l'un d'eux est nul, ils sont tous nuls. Il en est de même pour tous les coefficients d'ordre i et toutes les dérivées $(d^i\beta/d\sigma^i)$ pour lesquels on a $i<r$. Après $r-1$

différentiations des équations intrinsèques et des équations (18.2), on obtient les résultats suivants:

$$\left.\begin{aligned}\left\{1-\frac{q^2}{c^2}\sin^2(\beta-\vartheta)\right\}^r\frac{\partial^r\vartheta}{\partial s^r}&=f_r\frac{d^r\beta}{d\sigma^r},\\ \left\{1-\frac{q^2}{c^2}\sin^2(\beta-\vartheta)\right\}^r\frac{\partial}{\partial n}\left(\frac{\partial^{r-1}\vartheta}{\partial s^{r-1}}\right)&=g_r\frac{d^r\beta}{d\sigma^r},\end{aligned}\right\}\tag{19.3}$$

$$\left.\begin{aligned}f_r&=\cos(\beta-\vartheta)f_{r-1}-\sin(\beta-\vartheta)g_{r-1},\\ g_r&=\left(1-\frac{q^2}{c^2}\right)\sin(\beta-\vartheta)f_{r-1}+\cos(\beta-\vartheta)g_{r-1}.\end{aligned}\right\}\tag{19.4}$$

Au point O, le contact de l'onde de choc avec sa tangente est du même ordre que le contact de l'obstacle avec sa tangente. Le sens de la concavité du choc dépend du signe de la fonction f_r. Nous introduisons le nombre complexe

$$z_r=f_r+i\frac{g_r}{\sqrt{1-\frac{q^2}{c^2}}}.\tag{19.5}$$

L'argument du nombre complexe z_0 est l'angle $\varphi_0=(\pi/2)-m\varphi$ défini par la formule (16.12). Le nombre complexe z_r a pour argument $\varphi_0+r\varphi$ où φ est l'angle défini par la formule (16.6). Lorsque le nombre de Mach M croît de $M_0(\vartheta_s)$ jusqu'à $M^*(\vartheta_s)$, φ_0 décroît de $(\pi/2)$ à zéro, φ décroît également jusqu'à zéro: Fig. 25. L'argument de z_r est donc une fonction décroissante de M; elle est égale à $(\pi/2)$ lorsque $m=r$, elle est nulle pour M^*. Nous désignons par $M_r(\vartheta_s)$ la valeur de M pour laquelle on a $m=r$. La fonction f_r est sûrement positive pour $M_r(\vartheta_s)<M\leqq M^*(\vartheta_s)$. Les nombres de Mach M_r vérifient les relations suivantes:

$$M_0(\vartheta_s)<M_1(\vartheta_s)<\cdots,\qquad <M_r(\vartheta_s)<\cdots,\tag{19.6}$$

$$\operatorname*{Limite}_{r\to\infty}M_r(\vartheta_s)=M^*(\vartheta_s).\tag{19.7}$$

Lorsque le nombre de Mach M est égal à $M^*(\vartheta_s)$, toutes les fonctions g_r sont nulles, tandis que les fonctions f_r vérifient la relation suivante:

$$f_r=\cos^r(\beta-\vartheta)f_0.\tag{19.8}$$

Lorsque le nombre de Mach M est supérieur à $M^*(\vartheta_s)$, la fonction f_0 est positive et la fonction g_0 négative; il en résulte, d'après les formules (19.4), que toutes les fonctions f_r sont positives et toutes les fonctions g_r négatives. L'angle de Mach α étant réel, les formules (19.4) s'explicitent de la façon suivante:

$$\left.\begin{aligned}f_r\cos\alpha-g_r\sin\alpha&=\left\{\frac{\sin(\alpha+\beta-\vartheta)}{\sin\alpha}\right\}^r(f_0\cos\alpha-g_0\sin\alpha),\\ f_r\cos\alpha+g_r\sin\alpha&=\left\{\frac{\sin(\alpha-\beta+\vartheta)}{\sin\alpha}\right\}^r(f_0\cos\alpha+g_0\sin\alpha).\end{aligned}\right\}\tag{19.9}$$

Pour les valeurs infiniment petites de l'angle ϑ_s on obtient la formule (19.10) sur laquelle on retrouve les résultats contenus dans les formules (19.6) et (19.7).

$$M_r(\vartheta_s)=1+\frac{2+t}{2}\left(\frac{\gamma+1}{2}\,\frac{\vartheta_s}{1+t}\right)^{\frac{2}{3}}+\cdots,\tag{19.10}$$

avec

$$t=\tan^2\frac{\pi}{2r+4}.\tag{19.11}$$

IV. Les ondes de choc attachées dans les écoulements à trois dimensions.

a) Ecoulements coniques de révolution.

20. Etude des écoulements coniques de révolution. Nous envisageons un écoulement supersonique uniforme, parallèle à Ox; le nombre de Mach est M. Nous considérons une onde de choc constituée par un cône de révolution d'axe Ox,

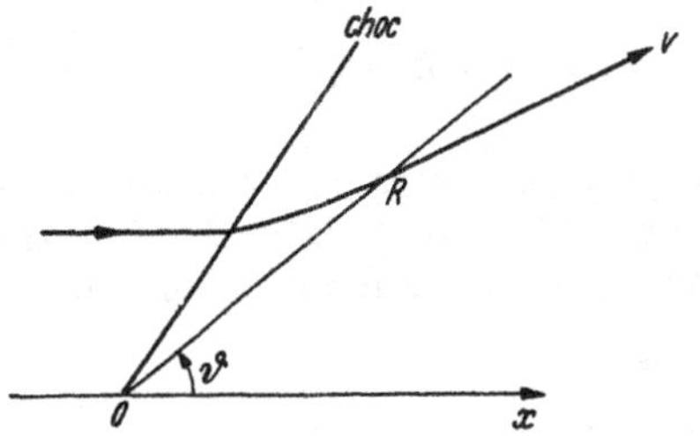

Fig. 26. Ecoulement conique de révolution.

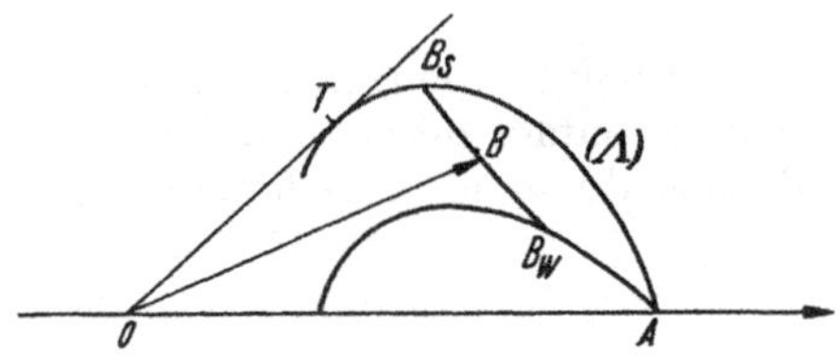

Fig. 27.

de demi-angle au sommet ϑ_w. Pour étudier le mouvement du fluide après le chcc, nous utilisons les équations (2.13) à (2.15); u, v, w sont les composantes de la vitesse sur les directions associées aux coordonnées sphériques r, ϑ, φ.

Sur l'onde de choc sont vérifiées les équations du choc, qui déterminent pour $\vartheta = \vartheta_w$ les valeurs des fonctions u, v, w, ϱ et p. Ces valeurs sont indépendantes

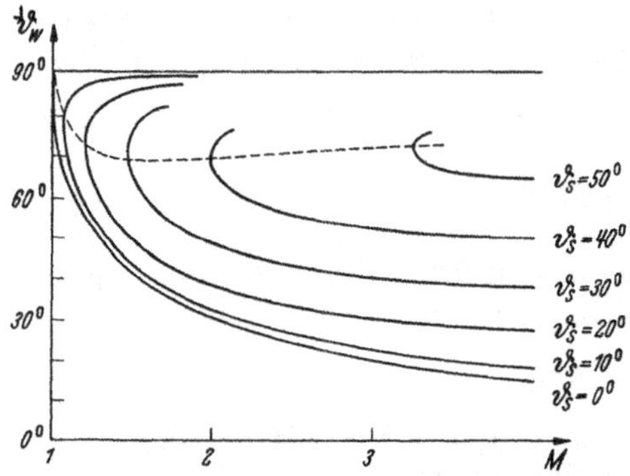

Fig. 28. Direction de l'onde de choc devant un cône

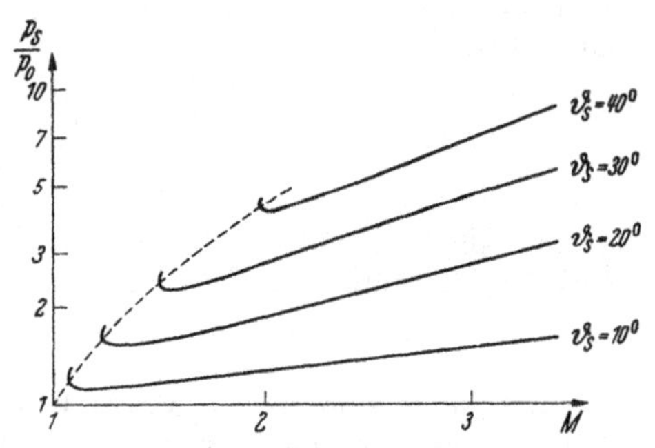

Fig. 29. Pression sur le cône.

des variables r et φ; en outre, w est nul. Comme les équations (2.13) à (2.15) possèdent des solutions présentant les mêmes caractères, on peut affirmer que, dans l'écoulement après le choc, w demeurera nul, tandis que u, v, ϱ et p dépendront de la seule variable ϑ. Un tel écoulement est appelé écoulement conique de révolution; en vertu de l'équation (2.15), il est isentropique. Les équations (2.13) et (2.14) s'écrivent de la façon suivante:

$$\left.\begin{aligned} \frac{du}{d\vartheta} - v &= 0\,, \\ \frac{dv}{d\vartheta} + u + \frac{1}{\varrho v}\,\frac{dp}{d\vartheta} &= 0\,, \\ \frac{d}{d\vartheta}(\varrho v \sin\vartheta) + 2\varrho u \sin\vartheta &= 0\,. \end{aligned}\right\} \tag{20.1}$$

Ces équations ont été intégrées numériquement [*24*] à [*26*]. En un point P, la vitesse du fluide est un vecteur $\boldsymbol{V}(\vartheta)$: Fig. 26. Nous menons par O un vecteur $\boldsymbol{OA}$ équipollent à la vitesse avant le choc et un vecteur $\boldsymbol{OB}$ équipollent à $\boldsymbol{V}(\vartheta)$: Fig. 27. Lorsque l'angle ϑ est égal à β, le point B est en B_w; lorsque ϑ varie, le point B décrit l'hodographe du mouvement. Pour une certaine valeur $\vartheta = \vartheta_s$, la composante $v(\vartheta)$ est nulle, si bien que le vecteur $\boldsymbol{V}$ est tangent au cône de révolution d'axe Ox, de demi-angle ϑ_s; ce cône constitue donc l'obstacle contourné par le fluide. Nous désignons par B_s la position correspondante du point B; on a $(OA, OB_s) = \vartheta_s$.

Les quantités ϑ_w, ϑ_s et M sont liées par une relation qui ne peut pas être exprimée en termes finis; cette relation est représentée sur la Fig. 28, homologue de la Fig. 3. Les valeurs du rapport des pressions p_s sur le cône et p_0 avant le choc sont représentées sur la Fig. 29, homologue de la Fig. 4.

Le nombre de Mach amont étant donné, nous faisons décroître l'angle ϑ_w de $(\pi/2)$ jusqu'à $\arcsin(1/M)$. Le point B_w décrit la polaire de choc, tandis que le point B_s décrit une courbe (Λ). La valeur maximum de la déviation finale ϑ_s s'obtient en menant de O la tangente OT à la courbe (Λ); on définit ainsi une fonction croissante $\overline{\vartheta}_s(M)$ représentée sur la Fig. 30, homologue de la Fig. 17. Lorsque M est infini, on obtient, pour les gaz diatomiques $(\gamma = 1{,}4)$, $\overline{\vartheta}_s = 57°32'$; a déviation finale ne peut dépasser cette valeur.

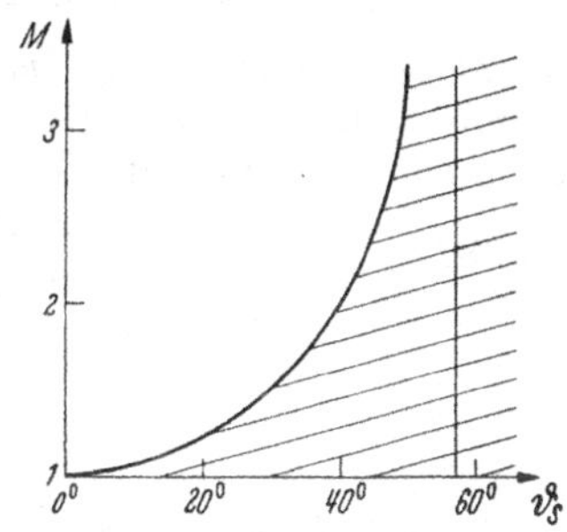

Fig. 30.

b) Ecoulements de révolution.

21. Onde attachée devant un cône de révolution d'axe parallèle à la vitesse. Nous considérons un obstacle constitué par un cône de révolution d'axe Ox, de demi-angle au sommet ϑ_s. Un écoulement supersonique uniforme parallèle à Ox, de nombre de Mach M vient heurter la surface du cône. Si ϑ_s est inférieur à l'angle limite ($57°32'$ pour $\gamma = 1{,}4$), on peut trouver une onde attachée constituée par un deuxième cône de révolution d'axe Ox et tel que l'écoulement après le choc contourne l'obstacle donné, chaque fois que le nombre de Mach M suffisamment grand. Le demi-angle au sommet ϑ_w de l'onde de choc est déterminé par les courbes tracées sur la Fig. 28.

L'angle ϑ_s étant donné, nous faisons croître M à partir de l'unité. Lorsque M est peu supérieur à 1, il n'existe aucune valeur de ϑ_w; pour une certaine valeur $M_0(\vartheta_s)$, on obtient deux valeurs confondues; on obtient ensuite deux valeurs distinctes dont la plus petite possède seule une réalité physique. La fonction $M_0(\vartheta_s)$, inverse de la fonction $\overline{\vartheta}_s(M)$, est représentée sur la Fig. 30. Pour un couple de valeurs ϑ_s et M dont le point figuratif est situé dans la région hachurée, le problème posé est impossible. Le nombre de Mach M_2 après le choc est une fonction croissante de ϑ: $M_2(\vartheta_s) < M_2(\vartheta_w)$. Les deux valeurs $M_2(\vartheta_s)$ et $M_2(\vartheta_w)$ qui correspondent au choc réel augmentent avec M et sont inférieures à 1 pour $M = M_0(\vartheta_s)$. Il existe donc une première valeur particulière $M^*(\vartheta_s)$ pour laquelle la vitesse aval sur l'onde est sonique et une deuxième valeur particulière $M^{**}(\vartheta_s)$ pour laquelle la vitesse aval sur l'obstacle est sonique. L'écoulement après le choc est subsonique pour $M_0 \leqq M < M^*$, il est supersonique pour $M^{**} < M$. Dans le cas intermédiaire, la vitesse est supersonique sur l'onde de choc et subsonique sur l'obstacle. Les trois nombres de Mach M_0, M^*, M^{**} et la différence $M^{**} - M_0$

augmentent indéfiniment avec ϑ_s. Pour $\gamma = 1,4$, M^{**} est infini pour $\vartheta_s = 55°41'$ et M_0 pour $\vartheta_s = 57°32'$; les valeurs numériques sont les suivantes:

ϑ_s	M_0	M^{**}
15°	1,1191	1,2183
30°	1,4805	1,6711
45°	2,3721	2,7786
55°	5,4820	11,7040

Dans la pratique, on ne rencontre pas de cône infini, mais des obstacles terminés par un cône. Nous envisageons un tel obstacle admettant Ox comme axe de révolution; la méridienne est terminée par un segment rectiligne OI qui fait l'angle ϑ_s avec Ox. Lorsque M est inférieur à $M_0(\vartheta_s)$, l'onde de choc est détachée. Nous supposerons qu'il n'en est pas ainsi.

α) La vitesse sur le cône est subsonique. Lorsque le nombre de Mach M est compris entre $M_0(\vartheta_s)$ et $M^{**}(\vartheta_s)$, la perturbation issue du point I atteint tous les points de l'onde de choc. La méridienne de l'onde de choc ne peut posséder aucun segment rectiligne; nous écrirons son équation sous la forme suivante:

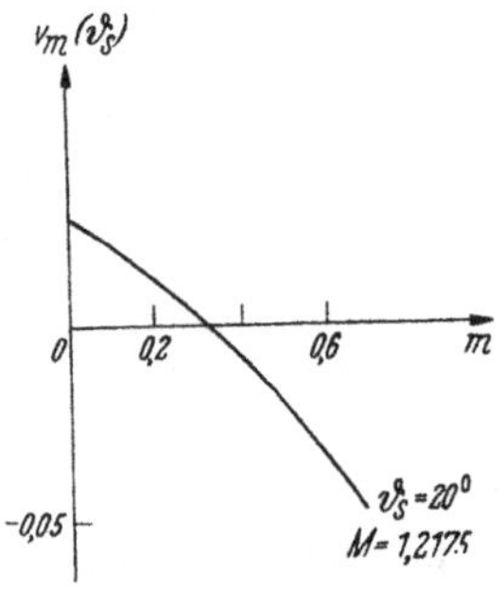

Fig. 31.

$$\vartheta = \vartheta_w - G(r) \tag{21.1}$$

où l'angle ϑ_w est celui qui correspond à l'écoulement conique. Nous cherchons à déterminer l'écoulement après le choc sous la forme suivante:

$$\left.\begin{aligned} u(r,\vartheta) &= u_0(\vartheta) + 2A\,r^m u_m(\vartheta) + \cdots, \\ v(r,\vartheta) &= v_0(\vartheta) + 2A\,r^m v_m(\vartheta) + \cdots, \\ \varrho(r,\vartheta) &= \varrho_0(\vartheta) + 2A\,r^m \varrho_m(\vartheta) + \cdots, \\ p(r,\vartheta) &= p_0(\vartheta) + 2A\,r^m p_m(\vartheta) + \cdots. \end{aligned}\right\} \tag{21.2}$$

Nous portons les fonctions ainsi écrites dans les équations du mouvement (2.13) a (2.15), w étant nul; nous identifions suivant les puissances croissantes de la variable r. Les fonctions d'indice zéro vérifient le système (20.1); les fonctions d'indice m vérifient le système suivant:

$$\left.\begin{aligned} u'_m v_0 + m u_m v_0 + v_m(u'_0 - 2v_0) + m\frac{p_m}{\varrho_0} &= 0, \\ v'_m v_0 + u_m v_0 + v_m\{v'_0 + (m+1)u_0\} + \frac{p'_0}{\varrho_0}\left(\frac{p'_m}{p'_0} - \frac{\varrho_m}{\varrho_0}\right) &= 0, \\ \{\varrho_m v_0 + \varrho_0 v_m)\sin\vartheta\}' + (m+2)(\varrho_m v_0 + \varrho_0 u_m)\sin\vartheta &= 0, \\ m u_0\left(\frac{p_m}{p_0} - \gamma\frac{\varrho_m}{\varrho_0}\right) + v_0\left(\frac{p_m}{p_0} - \gamma\frac{\varrho_m}{\varrho_0}\right)' &= 0. \end{aligned}\right\} \tag{21.3}$$

Les équations du choc déterminent les valeurs sur l'onde de choc des fonctions d'indice zéro et d'indice m; il est nécessaire que la fonction $G(r)$ soit proportionnelle à l'origine à r^m; nous poserons $G(r) = -A\,r^m + \cdots$. Les fonctions d'indice zéro représentent l'écoulement conique de révolution, qui correspond aux valeurs ϑ_s et M. Pour chaque valeur de m, les équations (21.3) peuvent être intégrées numériquement. La condition sur l'obstacle exige que la fonction $v(r,\vartheta_s)$ soit identiquement nulle. L'angle ϑ_w ayant été choisi de façon à annuler la quantité $v_0(\vartheta_s)$, il reste à satisfaire la condition $v_m(\vartheta_s) = 0$. Lorsque ϑ_s et M sont donnés, la quantité $v_m(\vartheta_s)$ est une fonction de l'exposant m représentée sur la Fig. 31; la valeur de cet exposant est donc déterminée. Pour $\vartheta_s = 20°$ et $\gamma = 1{,}4$, les valeurs numériques sont les suivantes [27]:

M	1,212	1,218	1,225	1,343
m	0,00	0,33	1,00	infini.

β) La vitesse sur la cône est supersonique. Lorsque le nombre de Mach M est supérieur à $M^{**}(\vartheta_s)$, la perturbation issue du point I se propage le long d'une caractéristique (γ) de l'écoulement conique: Fig. 32; elle atteint l'onde de choc en un point A. La méridienne de l'onde de choc possède un segment rectiligne OA; on peut calculer le rayon de courbure R_A de cette méridienne immédiatement

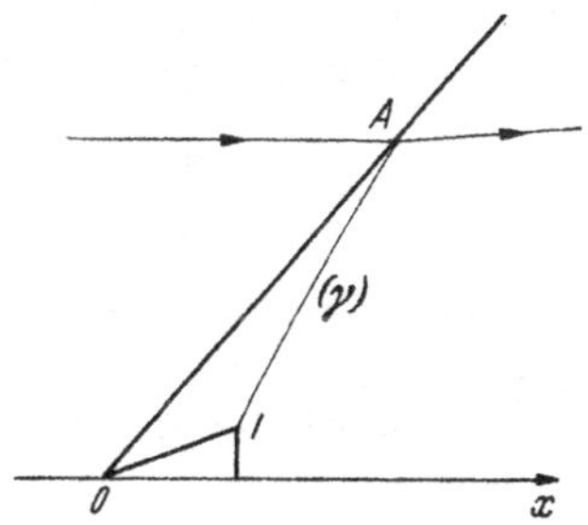

Fig. 32.

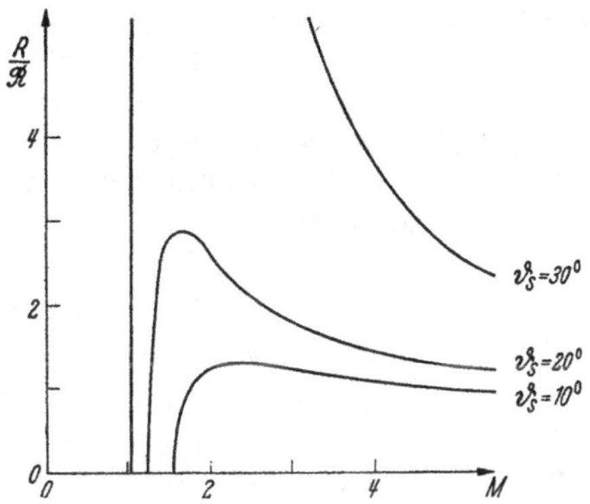

Fig. 33.

après le point A [27]. Dans ce but, on exprime la courbure des lignes de courant en chaque point de (γ) en fonction de l'abscisse curviligne sur cette courbe; cette fonction vérifie une équation différentielle de RICCATI. L'obstacle est une ligne de courant particulière; on en déduit la courbure de la ligne de courant issue du point A; ensuite, par des calculs analogues à ceux qui ont été faits dans le cas des écoulements plans, on détermine la valeur du quotient OI/R_A. L'écoulement conique de révolution est réalisé dans la région située en avant de la caractéristique (γ).

22. Onde de choc attachée à la pointe d'une ogive. Nous nous proposons d'étudier l'écoulement derrière une onde de choc de révolution représentée au voisinage du point O par une équation de la forme

$$\vartheta = \vartheta_w + \frac{r}{2R} + \cdots. \tag{22.1}$$

Nous cherchons à vérifier les équations du mouvement à l'aide de fonctions représentées par les formules (21.2); nous prenons $2A = R^{-1}$. Les équations du choc exigent que l'on ait $m = 1$, et déterminent les valeurs sur l'onde des fonctions d'indice zéro et des fonctions d'indice 1. Les fonctions d'indice zéro représentent toujours les écoulements coniques de révolution, tandis que les fonctions d'indice 1 vérifient le système (21.3), dans lequel m est égal à l'unité; on peut donc déterminer ces fonctions par intégration numérique.

Nous déterminons l'obstacle par son équation au voisinage du point O

$$\vartheta = \vartheta_s + \frac{r}{2\mathcal{R}} + \cdots. \tag{22.2}$$

En écrivant que la fonction $\vartheta(r)$ ainsi définie vérifie l'équation différentielle des lignes de courant $v\,dr - r\,u\,d\vartheta = 0$ et en tenant compte des équations (20.1), on obtient les relations suivantes:

$$v_0(\vartheta_s) = 0, \tag{22.3}$$

$$2\frac{v_1(\vartheta_s)}{R} - 3\frac{u_0(\vartheta_s)}{\mathcal{R}} = 0. \tag{22.4}$$

L'angle ϑ_s déterminé par la relation (22.3) est celui que l'on obtient dans la théorie des écoulements coniques de révolution. L'équation (22.4) détermine le rapport des courbures à la pointe de l'onde de choc et de l'obstacle. Les variations de ce rapport en fonction de M pour diverses valeurs de ϑ_s sont indiquées sur la Fig. 33, homologue de la Fig. 24 [*28*] et [*29*]. Lorsqu'on a $M = M_0(\vartheta_s)$, le quotient $R/\mathcal{R}$ est négatif; il est nul pour une certaine valeur $M_1(\vartheta_s)$, inférieure à $M^{**}(\vartheta_s)$.

Les divers résultats qui avaient été obtenus pour les écoulements plans se retrouvent pour les écoulements de révolution. Dans ce dernier cas, il n'est pas possible d'obtenir des formules explicites; des intégrations numériques doivent chaque fois être effectuées.

c) Ecoulements coniques.

23. Onde attachée devant un cône de révolution. Nous reprenons le problème traité dans la Sect. 21, en supposant maintenant que l'axe Δ du cône soit situé dans le plan xOy et fasse l'angle ε avec Ox. En utilisant les coordonnées sphériques r, ϑ, φ, l'équation du cône s'écrit

$$\cos\vartheta\cos\varepsilon + \sin\vartheta\sin\varepsilon\cos\varphi = \cos\vartheta_s, \tag{23.1}$$

$$\vartheta = \vartheta_s + \varepsilon\cos\varphi - \frac{\varepsilon^2}{2}\cot\vartheta_s\sin^2\varphi + \cdots. \tag{23.2}$$

Nous supposons l'angle ε petit et nous utilisons la formule (23.2) en négligeant les termes d'ordre supérieur à deux. Les données sont indépendantes de la variable r; la solution possède la même propriété. Les fonctions inconnues u, v, w, ϱ et p dépendent des variables ϑ et φ et du paramètre ε. Nous les développons suivant les puissances de ε; les coefficients qui admettent la période $\varphi = 2\pi$ peuvent ensuite être développées en série de FOURIER. La composante tangentielle de la vitesse sur l'obstacle doit être nulle; cela exige, compte tenu des équations du mouvement (2.13) à (2.15), que la solution possède la forme suivante [*30*]:

$$\left.\begin{aligned}
u &= \bar{u} + \varepsilon\,x\cos\varphi + \varepsilon^2(u_0 + u_2\cos 2\varphi) + \cdots,\\
v &= \bar{v} + \varepsilon\,y\cos\varphi + \varepsilon^2(v_0 + v_2\cos 2\varphi) + \cdots,\\
w &= \phantom{\bar{v} +{}} \varepsilon\,z\cos\varphi + \varepsilon^2\phantom{(v_0 +{}} w_2\cos 2\varphi + \cdots,\\
\varrho &= \bar{\varrho} + \varepsilon\,\xi\cos\varphi + \varepsilon^2(\varrho_0 + \varrho_2\cos 2\varphi) + \cdots,\\
p &= \bar{p} + \varepsilon\,\eta\cos\varphi + \varepsilon^2(p_0 + p_2\cos 2\varphi) + \cdots.
\end{aligned}\right\} \tag{23.3}$$

Sur l'onde de choc doivent être vérifiées les équations du choc; compte tenu des équations (23.3), l'équation de l'onde de choc possède la forme suivante:

$$\vartheta = \vartheta_w + \varepsilon\,\alpha\cos\varphi + \varepsilon^2(\beta_0 + \beta_2\cos 2\varphi) + \cdots. \tag{23.4}$$

Les équations du mouvement et les conditions du choc permettent, au moyen d'intégrations numériques, de calculer toutes les fonctions inconnues $\bar{u}, x, u_0 \ldots p_0, p_2$ qui dépendent de la seule variable ϑ et d'en déduire les valeurs des quantités

ϑ_w, α, β_0 et β_2. Les calculs ont été effectués sous la direction de Z. KOPAL [*31*]; ROBERTS et RILEY ont signalé que les valeurs de z sont inexactes; les valeurs correctes s'obtiennent en changeant le signe de la quantité $x - z \sin\vartheta$, laquelle est nulle sur l'onde de choc [*34*]. Pour un cône de demi-angle au sommet $\vartheta_s = 20°$, on obtient les résultats suivants (en adoptant la valeur $\gamma = 1{,}405$):

M	ϑ_w	α	β_0	β_2
1,2175	68°,507	0,4748	2,66	0,04
1,3144	58°,043	0,4251	0,72	0,11
1,6531	44°,424	0,4880	0,50	0,22
2,1297	36°,124	0,6017	0,43	0,31
2,8387	30°,435	0,7288	0,29	0,40

Les valeurs de la pression sur le cône sont données par la dernière des equations (23.3) dans laquelle on fait $\vartheta = \vartheta_s$; la concordance avec les résultats expérimentaux [*32*] est très satisfaisante. Si on néglige les termes en ε^2, l'onde de choc est un cône de révolution admettant le même angle au sommet que dans l'écoulement conique de révolution; son axe fait avec Ox l'angle $\alpha\varepsilon$. Dans l'approximation suivante, l'onde de choc est un cône admettant pour section droite une courbe de degré six, voisine d'une ellipse de faible excentricité.

HOLT a montré que l'entropie demeure constante sur l'obstacle et que les cônes sur lesquels l'entropie est constante possèdent une génératrice commune située dans le plan de symétrie de l'écoulement [*35*]. Cela sugère de choisir l'entropie comme variable pour étudier le cas où l'angle ε est petit.

24. Ecoulement derrière une onde de choc conique. Nous envisageons, dans un écoulement supersonique uniforme parallèle à Ox, une onde de choc constituée par un cône de sommet O; nous supposerons que le cône possède les plans xOy et xOz comme plane de symétrie, mais la théorie est générale. Utilisant les coordonnées sphériques, nous écrirons l'équation de l'onde de choc sous la forme suivante:

$$\vartheta = \vartheta_w + a_2 \cos 2\varphi + a_4 \cos 4\varphi + \cdots. \tag{24.1}$$

D'après les équations du choc, les valeurs sur l'onde des fonctions inconnues sont indépendantes de r; comme les équations du mouvement possèdent des solutions indépendantes de r, l'écoulement après le choc dépend uniquement des variables ϑ et φ. Un tel écoulement est appelé écoulement conique. L'écoulement étant symétrique par rapport aux plans xOy et xOz, nous posons

$$\left.\begin{aligned} u(\vartheta,\varphi) &= u_0(\vartheta) + u_2(\vartheta)\cos 2\varphi + u_4(\vartheta)\cos 4\varphi + \cdots,\\ v(\vartheta,\varphi) &= v_0(\vartheta) + v_2(\vartheta)\cos 2\varphi + v_4(\vartheta)\cos 4\varphi + \cdots,\\ w(\vartheta,\varphi) &= \qquad\quad w_2(\vartheta)\sin 2\varphi + w_4(\vartheta)\sin 4\varphi + \cdots,\\ \varrho(\vartheta,\varphi) &= \varrho_0(\vartheta) + \varrho_2(\vartheta)\cos 2\varphi + \varrho_4(\vartheta)\cos 4\varphi + \cdots,\\ p(\vartheta,\varphi) &= p_0(\vartheta) + p_2(\vartheta)\cos 2\varphi + p_4(\vartheta)\cos 4\varphi + \cdots. \end{aligned}\right\} \tag{24.2}$$

Afin de pouvoir se ramener à un problème portant uniquement sur des équations différentielles, il est nécessaire de faire des approximations. Nous supposons que l'onde de choc est voisine d'un cône de révolution: les coefficients $a_2, a_4, \ldots$ sont petits. Nous admettons que l'écoulement est voisin d'un écoulement conique de révolution; les fonctions d'indice supérieur ou égal à 2 sont petites. Les fonctions d'indice zéro vérifient le système (20.1), tandis que les fonctions d'indice 2 vérifient le système suivant:

$$\left.\begin{aligned} u_2' - v_2 &= 0, \\ v_0 v_2' + u_2 v_0 + v_2(u_0 + v_0') - \frac{\varrho_2 p_0'}{\varrho_0^2} + \frac{p_2'}{\varrho_0} &= 0, \\ v_0 w_2' + w_2(u_0 + v_0 \cot\vartheta) - \frac{2p_2}{\varrho_0 \sin\vartheta} &= 0, \\ \{(\varrho_0 v_2 + \varrho_2 v_0)\sin\vartheta\}' - 2(\varrho_0 u_2 + \varrho_2 u_0)\sin\vartheta - 2\varrho_0 w_2 &= 0, \\ \left(\frac{p_2}{p_0} - \gamma\frac{\varrho_2}{\varrho_0}\right)' &= 0. \end{aligned}\right\} \tag{24.3}$$

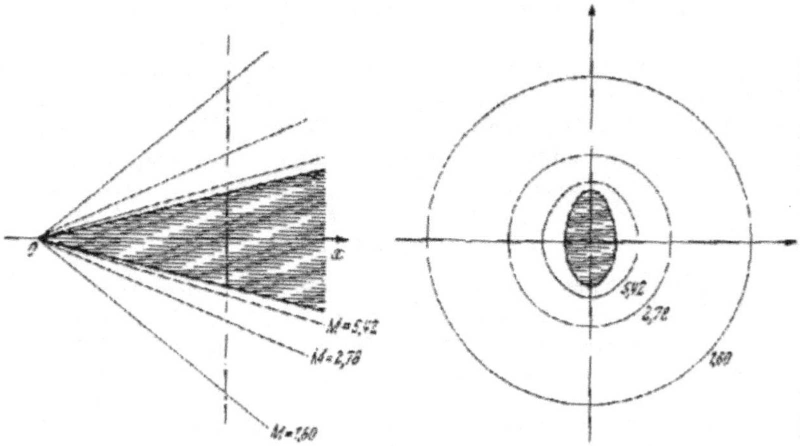

Fig. 34. L'onde de choc devant un cône donné, pour trois valeurs du nombre de Mach, M.

Les équations du choc déterminent les valeurs des fonctions inconnues pour $\vartheta = \vartheta_w$; on peut calculer ces fonctions par intégration numérique. L'obstable est le cône déterminé par les conditions suivantes:

$$\left.\begin{aligned} \vartheta = \vartheta_s + b_2 \cos 2\varphi &+ \\ + b_4 \cos 4\varphi &+ \cdots, \end{aligned}\right\} \tag{24.4}$$

$$\frac{dr}{u} = \frac{r\,d\vartheta}{v} = \frac{r\sin\vartheta\,d\varphi}{w}. \tag{24.5}$$

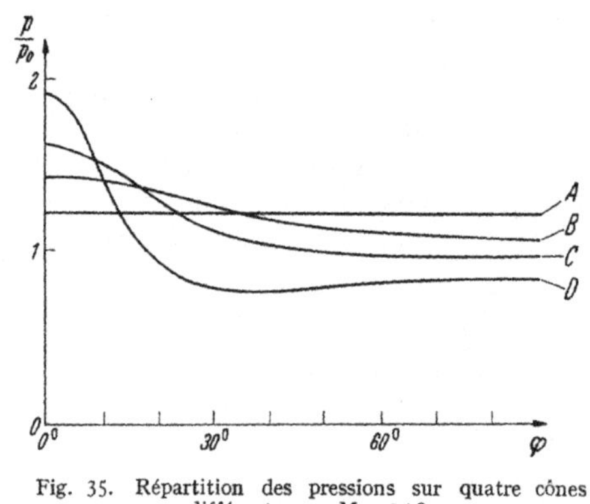

Fig. 35. Répartition des pressions sur quatre cônes différents pour $M = 2.78$.

Sur l'obstacle, r est arbitraire, tandis que ϑ est une fonction de φ; en écrivant que le système (24.5) (équations différentielles des lignes de courant) est satisfait pour tout déplacement tangent à la surface (24.4), on calcule les quantités ϑ_s, b_2, b_4 .... De nombreux exemples ont été traités: FERRI [36]. La Fig. 34 représente l'onde de choc devant un cône donné, pour trois valeurs différentes du nombre de Mach. La Fig. 35 représente, pour une même valeur du nombre de Mach, la répartition des pressions sur quatre cônes différents; la pression est maximum sur la génératrice la plus éloignée de l'axe de symétrie de cône; les cônes sont définis par l'équation suivante:

$$\frac{x^2}{h^2} = \frac{y^2}{a^2} + \frac{z^2}{b^2}. \tag{24.6}$$

Courbe	A	B	C	D
a/b	1,00	0,66	0,50	0,33
$h/\sqrt{\pi\, a\, b}$	3,21	3,15	3,10	2,96

C. Les ondes de choc dans les écoulements non stationnaires.

25. Préliminaires. Mouvement d'un point dans un fluide compressible. Comme nous l'avons indiqué, la détermination des chocs non stationnaires est beaucoup plus compliquée que celle des chocs stationnaires. Les résultats les plus nombreux sont relatifs aux ondes de choc planes; la partie I est consacrée à la détermination de ces ondes et la partie II à l'étude du problème de la rencontre d'une onde plane et d'un dièdre. Le cas des ondes de choc sphériques est également assez simple, puisque le nombre des variables se reduit à deux; les ondes de choc sphériques apparaissent dans les explosions; nous les étudions dans la partie III. Dans le cas général, les résultats sont assez fragmentaires; dans la partie IV, nous avons signalé quelques problèmes réunis sous le titre général: étude des translations rectilignes de vitesse variable.

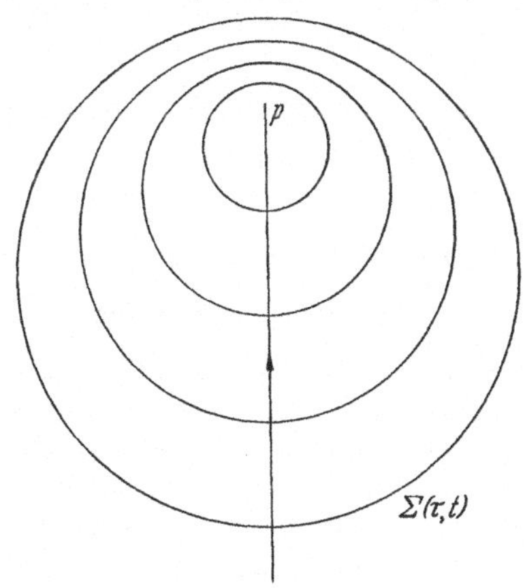

Fig. 36.

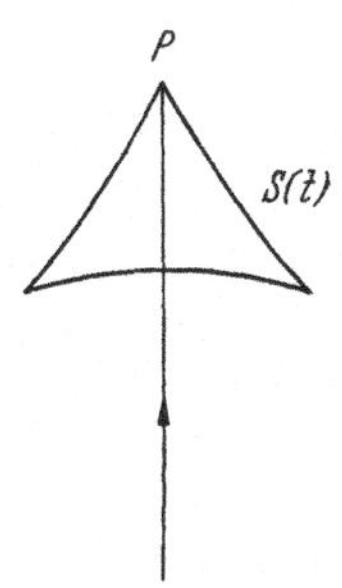

Fig. 37.

En dehors de ces divers cas, il est encore possible de simplifier l'étude des ondes de choc non stationnaires; il suffit de négliger les dimensions de l'obstacle et de l'assimiler à un point matériel P; il est naturel d'admettre qu'à une grande distance les résultats obtenus traduiront les phénomènes réels avec une bonne approximation. La théorie est élémentaire et nous l'exposerons rapidement à titre d'introduction à l'étude des ondes de choc dans les écoulements non stationnaires. Nous supposons que le point P se déplace dans un fluide homogène au repos; il occupe à chaque instant une position connue $P(t)$ et les perturbations émises se propagent à la vitesse constante c_0. La perturbation émise à l'instant τ se trouve à l'instant t sur la sphère $\Sigma(\tau, t)$ qui a pour centre le point $P(\tau)$ et pour rayon la longueur $c_0(t-\tau)$. L'enveloppe $S(t)$ de la famille des sphères Σ obtenue en faisant varier τ constitue un lieu d'accumulation des perturbations; c'est l'onde de choc à l'instant t. Au cours du temps cette onde de choc se déforme; chaque fois qu'une nappe de la surface $S(t)$ passera en un point fixe O, un observateur placé en O percevra une détonation provoquée par la discontinuité de pression qui accompagnera l'onde de choc.

Pour que les sphères $\Sigma(\tau, t)$ possèdent une enveloppe, il est nécessaire et suffisant que deux sphères voisines correspondant aux instants τ et $\tau + d\tau$ se coupent en des points réels; cela exige que la vitesse du point P à l'instant τ soit supersonique. Un avion volant à une vitesse subsonique ne provoquera jamais de détonation: Fig. 36. Au contraire un avion volant à une vitesse supersonique provoquera en général deux détonations: Fig. 37; il se peut également que l'on perçoive un nombre de détonations supérieur à deux pour certains mouvements du point P. Les détonations pourront être particulièrement violentes si l'onde de choc se réduit à un point; il est simple de déterminer les mouvements du point P pour lesquels cela se produit à un certain moment.

Supposons pour fixer les idées que le point P décrive l'axe des abscisses suivant la loi $x = \xi(t)$; nous désignons par r la distance à l'axe Ox. L'onde de choc $S(t)$ est définie en fonction du paramètre τ par les équations suivantes:

$$\left.\begin{aligned} x &= \xi(\tau) - \xi(t) + c_0(t-\tau)\sin\alpha(\tau), \\ r &= c_0(t-\tau)\cos\alpha(\tau) \end{aligned}\right\} \tag{25.1}$$

où on a posé $c_0 = \xi' \sin\alpha$. Le rayon de courbure R de l'onde de choc en son sommet est déterminé par la relation suivante:

$$\frac{R(t)\,\xi''(t)}{c_0^2} = -M(M^2-1), \tag{25.2}$$

où M désigne le nombre de Mach ($\xi' = c_0/M$). Ces phénomènes ont été étudiés par Esclangon [*37*] et Lilley [*38*].

I. Formation et propagation des ondes de choc planes.

a) Etude des chocs uniformes.

26. Choc engendré par un piston animé d'une vitesse constante. Les ondes de choc planes se rencontrent essentiellement dans les mouvements rectilignes; les fonctions inconnues dépendent de deux variables x et t. Le premier problème qui se pose est le suivant: un fluide homogène est au repos à la pression p_0, la célérité du son a pour valeur c_0; un piston, initialement immobile, se déplace à partir de l'instant $t=0$ vers le fluide avec une vitesse constante; on demande de trouver le mouvement du fluide. Dès l'instant initial, un choc se produit sur le piston; il convient d'étudier la propagation de ce choc.

La solution de ce problème est immédiate. Si un choc se propage avec la vitesse U dans le fluide au repos, l'état du fluide après le choc est déterminé par les équations (4.3) et (4.5)

$$\frac{u}{c_0} = (1-\mu^2)\left(\mathcal{M} - \frac{1}{\mathcal{M}}\right), \qquad \frac{p}{p_0} = (1+\mu^2)\,\mathcal{M}^2 - \mu^2, \tag{26.1}$$

avec $c_0\mathcal{M} = U$. Il suffit de choisir $\mathcal{M}$ de façon que la vitesse u soit égale à la vitesse du piston. Les équations (26.1) déterminent une relation entre la pression et la vitesse du fluide; dans les compressions par ondes simples ces deux grandeurs sont liées par la relation suivante ([*5*], p. 95):

$$\frac{p}{p_0} = \left\{1 + \frac{\gamma-1}{2}\,\frac{u}{c_0}\right\}^{\frac{2\gamma}{\gamma-1}}. \tag{26.2}$$

Les Figs. 38 représentent les courbes qui traduisent les deux relations dans le plan u, p. Ces courbes sont osculatrices sur l'axe des ordonnées. La différence Δp entre la pression après le choc et la pression correspondant aux ondes simples vérifie la relation (26.3). En vertu des équations (4.4), (4.5) et (4.7), la variation d'entropie sur l'onde de choc vérifie la relation (26.4)

$$\frac{\Delta p}{p_0} \sim \left(\gamma - \frac{5}{3}\right)\gamma(\gamma+1)\left(\frac{u}{c_0}\right)^3 + \cdots, \tag{26.3}$$

$$\frac{S_2 - S_1}{c_v} \sim \frac{16}{3}\,\frac{\gamma(\gamma-1)}{(\gamma+1)^3}\,(\mathcal{M}-1)^3 + \cdots. \tag{26.4}$$

On en déduit que l'écoulement derrière une onde de choc de faible intensité ($\mathcal{M}$ voisin de l'unité) peut être assimilé à une compression isentropique par ondes simples concourantes.

Lorsque le piston n'est plus une surface plane, l'onde de choc est déformée; le cas d'un piston constitué par un cylindre dont la section droite est une sinusoide

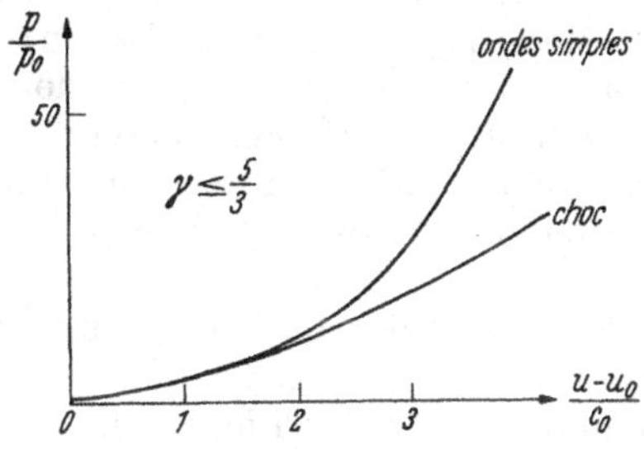

Fig. 38a.

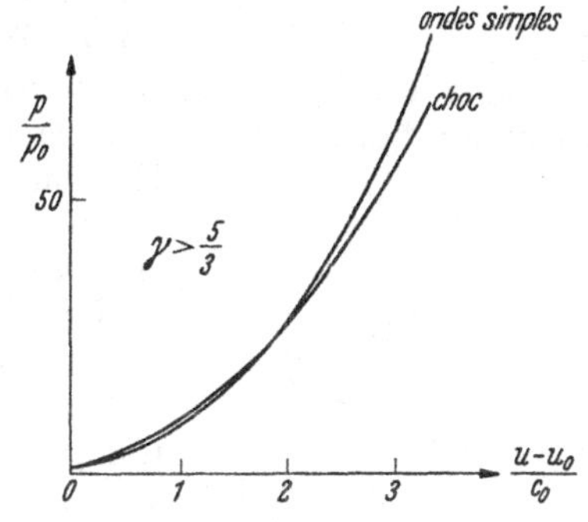

Fig. 38b.

a été étudié par FREEMAN [39] à l'aide des équations linéarisées. L'influence de la déformation du piston sur la déformation de l'onde de choc dépend du nombre de Mach; dans le cas d'un gaz diatomique ($\gamma = 1{,}4$), cette influence est minimum pour $\mathcal{M} = 1{,}14$; on peut dire que la stabilité du choc est alors maximum.

27. Problème du tube de choc. Un second procédé permettant d'obtenir des chocs uniformes est utilisé dans le tube de choc. Un tube illimité dans les deux sens est séparé en deux parties par une membrane. Des deux côtés de la membrane, le fluide est homogène et au repos; à droite (Fig. 39), région (1): la pression est p_1 et la célérité du son c_1; à gauche, région (2): la pression est p_2 et la célérité du son c_2. La pression p_2 est supposée supérieure à p_1; la témperature a la même valeur des deux côtés. A l'instant $t = 0$, on supprime la membrane et on demande de déterminer le mouvement du fluide.

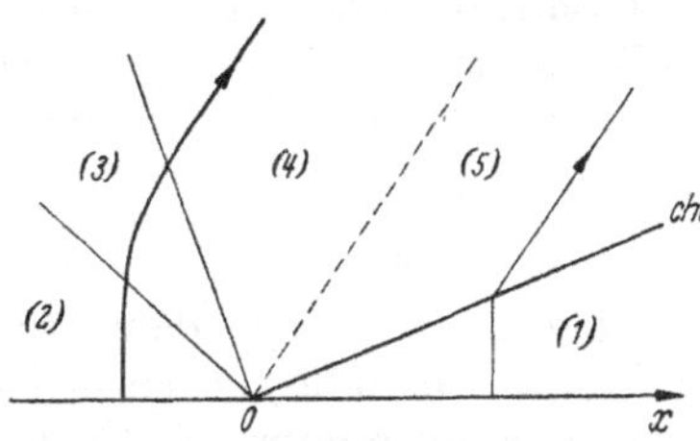

Fig. 39. Tube de choc.

On peut traiter ce problème en imaginant un piston qui coïncide initialement avec la membrane et qui se déplace ensuite vers la droite avec la vitesse constante u_3. La pression à droite du piston est donnée par les formules (26.1), la pression à gauche par la formule (26.2) dans laquelle on change u en $-u$. Pour une certaine valeur de la vitesse u_3, ces deux pressions sont égales et on peut supprimer le piston sans modifier l'écoulement; une discontinuité de masse spécifique et de température se propage à la place du piston. L'écoulement dans le plan x, t est représenté sur la Fig. 39.

La relation entre le nombre de Mach $\mathcal{M}$ de l'onde de choc et le rapport des pressions est donnée par la formule (27.1); les valeurs M_5 et M_4 des nombres de Mach des écoulements uniformes dans les régions (5) et (4) sont données par les formules (27.2) et (27.3):

$$\frac{p_2}{p_1} = \frac{2\gamma_1 \mathcal{M}^2 - \gamma_1 + 1}{\gamma_1 + 1} \left\{1 - \sqrt{\frac{\gamma_1}{\gamma_2} \frac{\gamma_2 - 1}{\gamma_1 + 1}} \left(\mathcal{M} - \frac{1}{\mathcal{M}}\right)\right\}^{-\frac{2\gamma_2}{\gamma_2 - 1}}, \tag{27.1}$$

$$M_5 = \frac{\frac{2}{\gamma_1 - 1}(\mathcal{M}^2 - 1)}{\sqrt{\frac{2\gamma_1}{\gamma_1 - 1}\mathcal{M}^2 - 1}\sqrt{\mathcal{M}^2 + \frac{2}{\gamma_1 - 1}}}, \tag{27.2}$$

$$M_4 = \frac{\frac{2}{\gamma_1 - 1}\left(\mathcal{M} - \frac{1}{\mathcal{M}}\right)}{\sqrt{\frac{\gamma_2}{\gamma_1}\frac{\gamma_1 + 1}{\gamma_2 - 1}} - \mathcal{M} + \frac{1}{\mathcal{M}}}. \tag{27.3}$$

Lorsque le rapport des pressions augmente indéfiniment, le nombre de Mach M_5 demeure borné, tandis que le nombre de Mach M_4 augmente indéfiniment. En outre le rapport des pressions étant donné, les nombres de Mach M_4 et M_5 augmentent avec le rapport γ_2/γ_1; pour cette raison on utilise parfois des gaz différents de part et d'autre de la membrane.

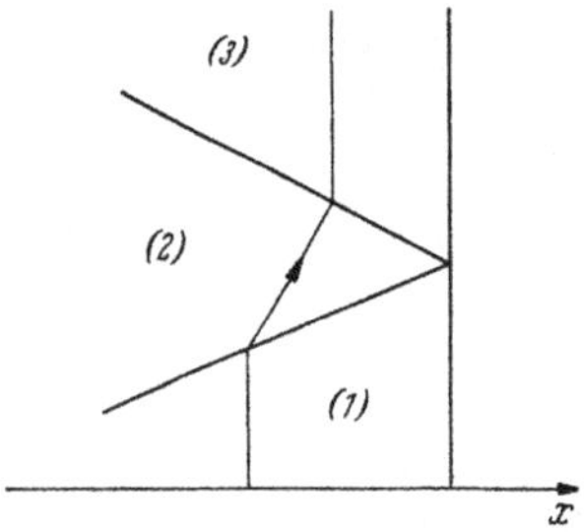

Fig. 40. Réflexion du choc.

Dans la pratique, le tube de choc n'est pas illimité mais possède des parois; l'onde de choc se réfléchit sur la paroi de droite: Fig. 40. L'écoulement après le choc incident est uniforme; la vitesse, la pression et la célérité du son ont pour valeurs u_2, p_2 et c_2:

$$\left.\begin{aligned} \frac{u_2}{c_1} &= (1 - \mu^2)\left(\mathcal{M} - \frac{1}{\mathcal{M}}\right), \qquad \frac{p_2}{p_1} = (1 + \mu^2)\mathcal{M}^2 - \mu^2, \\ \frac{c_2}{c_1} &= \sqrt{\{(1 + \mu^2)\mathcal{M}^2 - \mu^2\}\left\{\frac{1 - \mu^2}{\mathcal{M}^2} + \mu^2\right\}}. \end{aligned}\right\} \tag{27.4}$$

L'onde de choc se réfléchit et repart avec la vitesse constante U' et le nombre de Mach $\mathcal{M}' = (U' - u_2)/c_2$. L'écoulement après le choc réfléchi est encore uniforme, la vitesse et la pression ont pour valeur u_3 et p_3

$$\frac{u_3 - u_2}{c_2} = (1 - \mu^2)\left(\mathcal{M}' - \frac{1}{\mathcal{M}'}\right), \qquad \frac{p_3}{p_2} = (1 + \mu^2)\mathcal{M}'^2 - \mu^2. \tag{27.5}$$

Pour que le fluide demeure en contact avec la paroi, il est nécessaire que la vitesse u_3 soit nulle; on en déduit, en fonction du nombre de Mach incident $\mathcal{M}$, les valeurs du nombre de Mach réfléchi et du rapport des pressions

$$\mathcal{M}' = -\sqrt{\frac{2\gamma_1 \mathcal{M}^2 - \gamma_1 + 1}{(\gamma_1 - 1)\mathcal{M}^2 + 2}}, \tag{27.6}$$

$$\frac{p_3}{p_1} = \mathcal{M}'^2 \left\{\frac{3\gamma_1 - 1}{\gamma_1 + 1}\mathcal{M}^2 - \frac{\gamma_1 - 1}{\gamma_1 + 1}\right\}. \tag{27.7}$$

28. Choc engendré par des ondes simples concourantes. Nous signalons un dernier exemple dans lequel le mouvement du piston donne naissance à un choc uniforme. Nous supposons que le piston soit animé du mouvement défini par l'équation suivante:

$$1 - \frac{x}{x_1} = -\frac{2}{\gamma - 1}\left(1 - \frac{t}{t_1}\right) + \frac{\gamma + 1}{\gamma - 1}\left(1 - \frac{t}{t_1}\right)^{\frac{2}{\gamma + 1}} \qquad t \geqq 0, \tag{28.1}$$

dans laquelle x_1 et t_1 désignent deux constantes. Si le piston se déplace dans un fluide au repos, région (1), son mouvement donne naissance à une compression par ondes simples concourantes au point I de coordonnées x_1 et t_1 ([5], p. 105) Lorsque la vitesse du piston atteint une certaine valeur u_3, nous supposons que son mouvement se poursuit à la vitesse constante u_3. La trajectoire du piston comprend un arc de courbe OA et une demi-droite. La compression par ondes simples a lieu dans la région (2), comprise entre l'arc OA et les caractéristiques rectilignes OI et AI: Fig. 42. Au delà de la caractéristique AI, l'écoulement est uniforme: région (3); le problème consiste à raccorder les régions (1) et (3) en satisfaisant partout les équations du mouvement ou les conditions du choc. Dans ce but, nous considérons un choc uniforme issu du point I; derrière ce choc, se trouve une région d'écoulement uniforme: région (6). Dans le plan u, p, Fig. 41, la région (1) est représentée par un point sur l'axe des pressions; la région

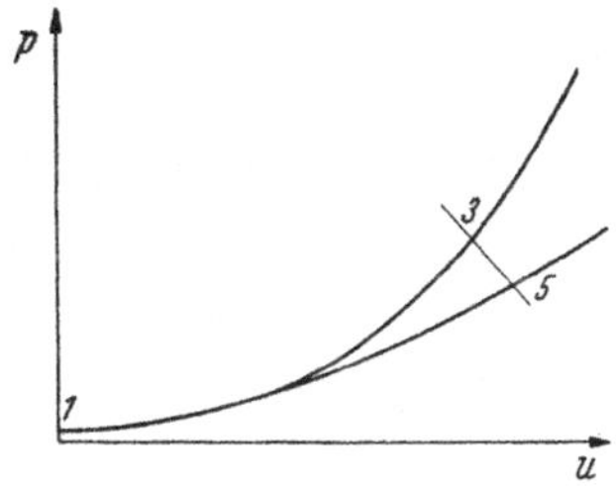

Fig. 41.

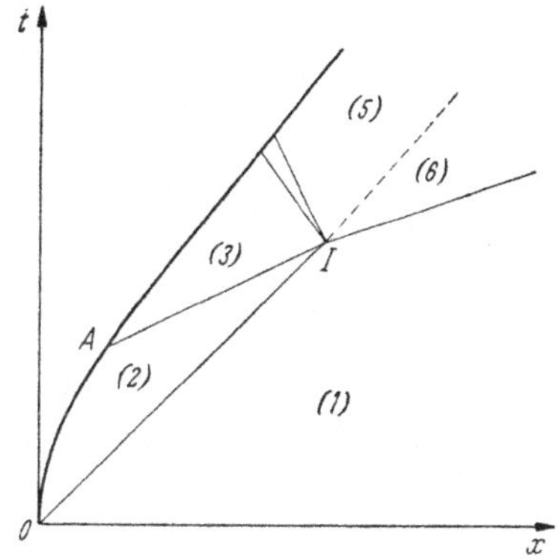

Fig. 42.

(2), par un arc de courbe; la région (3), par un point. La région (6) est également représentée par un point situé sur la courbe qui traduit les équations du choc. Lorsque le point qui représente la région (3) est au-dessus de cette courbe, on peut imaginer, dans une région (4), une détente par ondes simples concourantes en I, se terminant lorsque la courbe de détente issue du point (3) rencontre la courbe de choc issue du point I; au delà de la détente, se trouve une région d'écoulement uniforme: région (5). Les régions (5) et (6) sont représentées par le même point dans le plan u, p, si bien que toutes les conditions sont satisfaites.

L'examen des Fig. 38 nous permet d'énoncer les conclusions suivantes:

Lorsqu'on a $\gamma \leqq \frac{5}{3}$, une compression par ondes simples concourantes donne naissance à un choc et à une détente. Lorsqu'on a $\gamma > \frac{5}{3}$, la conclusion est la même si la vitesse du piston est suffisamment grande; si la vitesse du piston est faible, la compression par ondes simples concourantes donne naissance à deux chocs.

Il est possible de déterminer la valeur du nombre de Mach du premier choc en fonction de la vitesse finale du piston. Par exemple, pour $\gamma = 3$, on obtient

$$\frac{u_3}{c_1} = \frac{\mathcal{M}^2 - 1}{2\mathcal{M}} + \eta\left(\frac{\mathcal{M}^2 - 1}{2\mathcal{M}} + 1\right), \tag{28.2}$$

$$\eta = \frac{\left(\frac{3\mathcal{M}^2 - 1}{2}\right)^{\frac{1}{3}} - \left(\frac{\mathcal{M}^2 - 1}{2\mathcal{M}} + 1\right)^{\frac{1}{3}}}{\left(\frac{3\mathcal{M}^2 - 1}{2}\right)^{\frac{1}{3}} + \left(\frac{\mathcal{M}^2 - 1}{2\mathcal{M}} + 1\right)^{\frac{1}{3}}}. \tag{28.3}$$

b) Exemples de chocs non uniformes.

29. Cas d'intégrabilité des équations du problème. L'écoulement derrière une onde de choc non uniforme se propageant dans un fluide au repos n'est plus isentropique; c'est une des raisons pour lesquelles l'étude des écoulements comportant une onde de choc est compliquée. Il existe cependant un cas remarquable, signalé par Stanjukovich [*40*], dans lequel on sait intégrer les équations des mouvements rectilignes non isentropiques d'un fluide compressible. Ces équations sont les suivantes:

$$\left.\begin{aligned} \varrho\left(\frac{\partial u}{\partial t}+u\frac{\partial u}{\partial x}\right)+\frac{\partial p}{\partial x}&=0,\\ \frac{\partial\varrho}{\partial t}+\frac{\partial(\varrho u)}{\partial x}&=0,\\ \frac{\partial}{\partial t}\left(p\varrho^{-\frac{1}{\gamma}}\right)+u\frac{\partial}{\partial x}\left(p\varrho^{-\frac{1}{\gamma}}\right)&=0.\end{aligned}\right\}\tag{29.1}$$

La seconde équation équivaut à l'existence d'une fonction ψ (fonction trajectoire) telle que $d\psi=\varrho\,dx-\varrho\,u\,dt$. Si on prend ψ et t comme variables indépendantes, les équations (29.1) s'écrivent sous la forme (29.2) où la fonction $\delta(\psi)$ est arbitraire positive. On en déduit que la fonction $p(\psi,t)$ vérifie l'équation (29.3)

$$\frac{\partial u}{\partial t}+\frac{\partial p}{\partial\psi}=0,\qquad \frac{\partial\varrho}{\partial t}+\varrho^2\frac{\partial u}{\partial\psi}=0,\qquad \frac{1}{\varrho}=\delta(\psi)\,p^{-\frac{1}{\gamma}},\tag{29.2}$$

$$E\equiv\frac{\partial^2 p}{\partial\psi^2}-\delta(\psi)\frac{\partial^2}{\partial t^2}\left(p^{-\frac{1}{\gamma}}\right)=0.\tag{29.3}$$

Lorsque la fonction $\delta(\psi)$ est constante, l'écoulement est isentropique et on sait former l'intégrale générale de l'équation précédente pour toutes les valeurs entières du rapport $(\gamma-3)/(2\gamma-2)$. Si nous posons $p_1=(p/\psi)$ et $\psi_1=(1/\psi)$, l'équation $E=0$ prend la forme (29.4) qui est du même type:

$$E_1\equiv\frac{\partial^2 p_1}{\partial\psi_1^2}-\delta_1(\psi_1)\frac{\partial^2}{\partial t^2}\left(p^{-\frac{1}{\gamma}}\right)=0,\qquad \delta_1=\delta(\psi)\cdot\psi^{3-\frac{1}{\gamma}}.\tag{29.4}$$

Lorsque la fonction $\delta_1(\psi_1)$ est constante, on sait intégrer l'équation $E_1=0$ donc l'équation $E=0$. Lorsque une onde de choc d'équation $c_0 t=F(x)$ se propage dans un fluide au repos dans lequel la pression, la masse spécifique et la célérité du son ont pour valeurs p_0, ϱ_0 et c_0, on a, en vertu des équations du choc,

$$\delta(\psi)=\frac{p_0^{1/\gamma}}{\varrho_0}\left\{\frac{2}{\gamma+1}F'^2\left(\frac{\psi}{\varrho_0}\right)+\frac{\gamma-1}{\gamma+1}\right\}\left\{\frac{2\gamma}{\gamma+1}F'^{-2}\left(\frac{\psi}{\varrho_0}\right)-\frac{\gamma-1}{\gamma+1}\right\}^{1/\gamma}.\tag{29.5}$$

Il est toujours possible de choisir la trajectoire de choc de façon que l'on ait $\delta_1=\text{const}$; x et t s'expriment en fonction du paramètre $\mathcal{M}=F'^{-1}(x)$ par les équations

$$\left.\begin{aligned} x^{1-3\gamma}&=\left\{\frac{2}{\gamma+1}\mathcal{M}^{-2}+\frac{\gamma-1}{\gamma+1}\right\}^{\gamma}\left\{\frac{2\gamma}{\gamma+1}\mathcal{M}^2-\frac{\gamma-1}{\gamma+1}\right\},\\ c_0 t&=\int^{\mathcal{M}}\frac{dx}{\mathcal{M}}.\end{aligned}\right\}\tag{29.6}$$

L'intensité du choc diminue avec le temps et s'annule au bout d'un temps fini. Dans ce cas particulier, on peut déterminer explicitement l'écoulement après le choc, si bien que l'on obtient une solution exacte comportant une onde de choc.

30. Interaction d'une compression et d'une détente. La compression d'un fluide donne naissance en général à une onde de choc; nous supposons une compression réalisée par l'intermédiaire d'un piston animé d'un mouvement connu $x = X(t)$; à l'instant initial, le fluide est au repos et la vitesse du piston est nulle. La fonction $X(t)$ est supposée croissante et trois fois dérivable. Le mouvement engendré par le piston est une compression par ondes simples définie par les formules suivantes ([5], p. 92).

$$x - X(\vartheta) = \left\{c_0 + \frac{\gamma+1}{2} X'(\vartheta)\right\}(t - \vartheta), \tag{30.1}$$

$$u = X'(\vartheta), \qquad c = c_0 + \frac{\gamma-1}{2} u. \tag{30.2}$$

Les caractéristiques rectilignes définies par l'équation (30.1) possèdent une enveloppe au voisinage de laquelle les fonctions qui définissent l'écoulement ne sont plus uniformes, si bien que le principe d'impénétrabilité de la matière n'est pas respecté. Un choc se produit au plus tard à l'instant t_c égal à la valeur minimum du temps sur l'enveloppe. La valeur de t_c est donnée par la formule (30.3) dans laquelle ϑ est la plus petite racine positive de l'équation (30.4) ou zéro si cette équation ne possède pas de racine positive.

$$t_c = \vartheta + \frac{c_0 + \frac{\gamma-1}{2} X'(\vartheta)}{\frac{\gamma+1}{2} X''(\vartheta)}, \tag{30.3}$$

$$c_0 + \frac{\gamma-1}{2} X'(\vartheta) = \gamma \frac{X''^2(\vartheta)}{X'''(\vartheta)}. \tag{30.4}$$

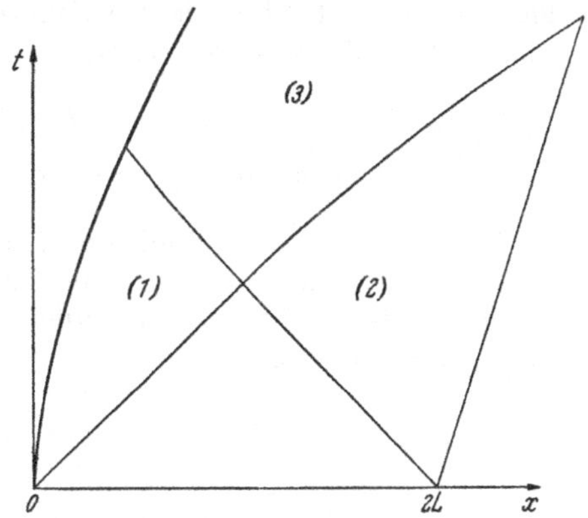

Fig. 43.

Un procédé qui permet de retarder l'apparition du choc consiste à détendre le fluide. Nous allons examiner ce problème dans le cas particulier suivant étudié par P. M. Stocker [41]. Un fluide homogène est au repos entre deux pistons P_1 et P_2, placés aux points d'abscisses 0 et $2L$; à partir de l'instant $t = 0$, on communique aux pistons les mouvements suivants: (P_1), soit $x = c_0 \eta t^2$, et (P_2), soit $x = 2L + c_0 t$, où η désigne une constante positive et c_0 la célérité du son dans le fluide au repos. Le premier piston engendre une compression par ondes simples: région (1); le second, une détende par ondes simples concourantes: région (2). Ces deux écoulements sont définis par les relations suivantes:

$$\left.\begin{aligned} (1)\quad & x - c_0 \eta \vartheta^2 = c_0 \{1 + (\gamma+1)\eta\}(t - \vartheta), \\ & u = 2 c_0 \eta \vartheta \qquad c = c_0 + \frac{\gamma-1}{2} u, \end{aligned}\right\} \tag{30.5}$$

$$(2)\quad x - 2L = (u - c)\, t, \qquad c = c_0 - \frac{\gamma-1}{2} u. \tag{30.6}$$

Nous supposons que l'ordonnée du point A, point de concours des caractéristiques rectilignes issues des positions initiales des pistons, est inférieure à l'instant tc qui correspond à la région (1), c'est-à-dire $\lambda = \{1/(\gamma+1)\}\{c_0/(\eta L)\} > 1$. Dans la région (3), adjacente aux régions (1) et (2) le long des caractéristiques curvilignes issues du point A, l'écoulement est isentropique. La solution générale des équations des mouvements isentropiques s'exprime à l'aide d'une seule fonction $w(r, s)$ par les formules (30.7); dans le cas d'un gaz monoatomique ($\gamma = \frac{5}{3}$), la fonction $w(r, s)$ s'exprime par la relation (30.8), dans laquelle $X(r)$ est une fonction arbitraire de r et $Y(s)$ une fonction arbitraire de s[1].

$$\left.\begin{aligned} u &= \frac{r+s}{2}, \qquad c = -\frac{\gamma-1}{4}(r-s), \\ 2cx &= (u-c)\frac{\partial w}{\partial r} - (u-c)\frac{\partial w}{\partial s}, \\ 2ct &= \frac{\partial w}{\partial r} - \frac{\partial w}{\partial s}, \end{aligned}\right\} \tag{30.7}$$

$$w(r, s) = \frac{X(r) - Y(s)}{r-s}. \tag{30.8}$$

En écrivant que les fonctions u et c sont continues sur les caractéristiques curvilignes issues de A, on obtient les expressions des fonctions $X(r)$ et $Y(s)$ qui déterminent l'écoulement dans la région (3).

$$\left.\begin{aligned} X(r) &= 2L(r^2 - 9c_0^2), \\ 144\eta c_0 Y(s) &= -(s-3c_0)^2(s+3c_0)(5s+3c_0). \end{aligned}\right\} \tag{30.9}$$

Les fonctions qui définissent l'écoulement dans la région (3) cessent d'être uniformes lorsque le déterminant fonctionnel des fonctions u, c des variables x, t et nul. Cette condition définit une ligne (ligne limite) sur laquelle la valeur minimum du temps est

$$t_s = \frac{3}{8}\frac{1}{\eta}\frac{(1+\lambda)^2}{4\lambda} \geqq \frac{3}{8}\frac{1}{\eta} = t_c. \tag{30.10}$$

Cette dernière inégalité prouve que la détente retarde l'apparition du choc. Stocker a traité également le cas où le piston P_1 possède une accélération proportionnelle au temps.

31. Choc se propageant dans une région de densité variable. Certains auteurs ont envisagé des ondes de choc planes qui se propageraient dans un fluide au repos non homogène. C. W. Jones a montré que l'on peut satisfaire les équations (29.1) à l'aide des fonctions u, p, ϱ qui possèdent la forme particulière indiquée dans les formules suivantes:

$$\left.\begin{aligned} &\tau = \alpha t, \qquad \xi = \beta x e^{-\alpha t}, \\ &\beta u = \alpha e^{\tau} U(\xi), \qquad p = p_0 P(\xi), \qquad \alpha^2 \varrho = p_0 \beta^2 e^{-2\tau} \Omega(\xi), \end{aligned}\right\} \tag{31.1}$$

dans lesquelles α, β et p_0 désignent trois constantes [42]. Les trois fonctions $U(\xi)$, $P(\xi)$ et $\Omega(\xi)$ vérifient un système de trois équations différentielles ordinaires. L'introduction des fonctions Y et Z définies par les formules (31.2) permet de se ramener à l'équation différentielle unique (31.3) et de ramener à des quadratures la détermination de toutes les fonctions inconnues.

$$Y = -\frac{U}{U-\xi}, \qquad Z = \gamma\frac{PY^2}{\Omega U^2}, \tag{31.2}$$

$$(\gamma+1)\frac{dY}{dZ} = \frac{Y+Z}{Y}. \tag{31.3}$$

[1] Dans le cas général, γ étant arbitraire, la fonction $w(r, s)$ peut être exprimée au moyen des fonctions hypergéometriques: cf. Chap. V.

Nous considérons un fluide au repos, à pression constante p_0, dont la masse spécifique est de la forme $\varrho = \sigma_0 x^{-2}$, où σ_0 désigne une constante. Lorsqu'une onde de choc se déplace dans ce fluide avec un nombre de Mach constant, l'écoulement après le choc est du type précédent. On obtient ainsi en solution exacte un écoulement comportant une onde de choc.

c) La formation des chocs dans les tuyères.

32. Etude qualitative. L'étude de la formation des ondes de choc est étroitement liée à l'étude de la stabilité des écoulements. Un écoulement est stable lorsqu'une modification infiniment petite des conditions aux limites et des conditions initiales entraîne une modification infiniment petite pour chacune des fonctions qui caractérisent l'écoulement. R. E. MEYER a étudie un cas simple dans lequel une telle modification donne naissance à un choc au bout d'un temps fini [*43*]; l'écoulement est instable et par suite ne possède pas de réalité physique.

Nous considérons une tuyère plane dont la section droite $Q(x)$ varie de façon suffisamment lente pour que l'écoulement puisse être considéré comme rectiligne. Nous supposons qu'un régime permanent sonique dans le col de la tuyère soit établi. Il existe des solutions dans lesquelles la vitesse est subsonique avant le col et supersonique après; l'accélération étant positive, ces écoulements sont dits accélérés. A toute solution des équations du mouvement, correspond une nouvelle solution dans laquelle la vitesse en chaque point possède la valeur opposée, tandis que la pression et la masse spécifique demeurent inchangées. Il existe donc des écoulements continus dans lesquels la vitesse est supersonique avant le col et subsonique après; ces écoulements sont dits retardés.

Une petite perturbation provoquée en un point du fluide se propage suivant deux ondes; pour un observateur lié au fluide, celles-ci se propagent avec la célérité du son: l'une vers l'aval, appelée onde progressive, l'autre vers l'amont, appellée onde régressive. Les ondes progressives créées à la sortie ne pénètrent pas dans la tuyère; celles qui sont créées à l'entrée traversent la tuyère et la quittent au bout d'un temps fini. Les ondes régressives penètrent dans la tuyère seulement dans le cas des écoulements retardés; elle ne peuvent pas dépasser la ligne sonique et vont s'accumuler au col de la tuyère. Ces considérations qualitatives laissent prévoir que les écoulements retardés sont instables, tandis que les écoulements accélérés sont stables. Nous allons démontrer ces résultats en étudiant de façon quantitative la propagation des ondes.

33. Etude quantitative. Nous supposons les mouvements isentropiques; en introduisant les opérateurs qui sont définis par les formules (33.1) et qui représentent les différentiations suivant chacune des caractéristiques et en utilisant les fonctions r et s définies par les formules (30.7), les équations du mouvement s'écrivent sous la forme (33.2).

$$\frac{D_+}{Dt} = \frac{\partial}{\partial t} + (u+c)\frac{\partial}{\partial x}, \qquad \frac{D_-}{Dt} = \frac{\partial}{\partial t} + (u-c)\frac{\partial}{\partial x}, \tag{33.1}$$

$$\frac{D_- r}{Dt} = c\,u\,H, \qquad \frac{D_- s}{Dt} = -\,c\,u\,H, \tag{33.2}$$

$$H = \frac{d\log Q}{dx}. \tag{33.3}$$

Nous supposons qu'un écoulement permanent retardé est établi dans la tuyère. La vitesse U, la célérité du son C et le nombre de Mach M sont des fonctions de l'abscisse x. Une perturbation créée à la sortie se propage vers l'amont suivant une onde régressive qui sépare le fluide en deux régions; à droite de l'onde, l'écoulement est perturbé; à gauche, l'écoulement est permanent. Nous posons $\sigma = D_+ r/Dt$ et nous désignons par Σ la valeur de σ dans l'écoulement permanent; on déduit des équations (33.2) que la fonction σ vérifie le long d'une onde régressive l'équation différentielle (33.4).

$$\frac{D_- \sigma}{Dt} = -\frac{\gamma+1}{8C}\sigma\{\sigma + F(x)\} + G(x), \tag{33.4}$$

$$2\Sigma + F = \frac{4CU'}{\gamma+1}\left(3 + \frac{1}{M}\right)\left(1 + \frac{\gamma-1}{2}M^2\right). \tag{33.5}$$

Nous renvoyons le lecteur au mémoire de MEYER pour le détail des calculs. L'équation (33.4) est du type de RICCATI et possède la solution particulière Σ. En posant $\sigma = \Sigma + z^{-1}$, on en déduit l'expression de la fonction z.

$$\left.\begin{aligned} \frac{z}{K} &= \frac{z_0}{K_0} + \frac{\gamma+1}{8}\int_0^t \frac{dt}{KC}, \\ \frac{K}{K_0} &= \exp\left\{\frac{\gamma+1}{8}\int_0^t \frac{2\Sigma+F}{C}\,dt\right\}. \end{aligned}\right\} \tag{33.6}$$

Les intégrales sont calculées le long de l'onde régressive. Si la valeur initiale z_0 est positive, la fonction z ne s'annule jamais, tandis qu'elle s'annule au bout d'un temps fini si z_0 est négatif. On déduit des équations du mouvement les relations (33.7) qui indiquent qu'une valeur négative de z_0 correspond à une augmentation de la pression à la sortie et qu'une racine de la fonction z correspond à une accélération infinie, c'est-à-dire à la formation d'une onde de choc. Nous venons donc de prouver qu'une augmentation de pression à la sortie engendre une onde de choc au bout d'un temps fini, d'où il résulte que les écoulements retardés sont instables.

$$\frac{\partial p}{\partial t} = \varrho\,\frac{U-c}{4z}, \qquad \left(\frac{\partial}{\partial t} + u\frac{\partial}{\partial x}\right)(u - U) = \frac{1}{4z}. \tag{33.7}$$

Les ondes progressives s'étudient de manière analogue. Seules les ondes progressives qui se forment à l'entrée pénètrent dans la tuyère et il est nécessaire de supposer celle-ci infinie vers l'aval si l'on veut préciser les résultats. On forme une équation différentielle de RICCATI pour déterminer les variations de la quantité $\tau = D_- s/Dt$ le long d'une onde progressive. Les résultats sont moins simples; seules les augmentations de pression à l'entrée de la tuyère peuvent éventuellement donner naissance à un choc; cependant, contrairement au cas précédent, ce choc ne se formera pas toujours. Pour qu'il puisse se former, il est nécessaire que la tuyère à l'infini diverge moins rapidement qu'un cône, c'est-à-dire que le produit $x^2\,Q(x)$ demeure borné; il est en outre nécessaire que l'augmentation de pression ne soit pas trop petite.

Dans le cas des écoulements accélérés, la stabilité s'établit par la même méthode.

II. Rencontre d'une onde de choc plane et d'un dièdre.

a) Cas de la réflexion régulière.

34. Lois de la réflexion régulière. Un problème important relatif aux ondes de choc planes est celui de la réflexion sur un obstacle. Ce problème se rencontre dans le vol supersonique; en effet, lorsqu'un avion se déplace à une vitesse supersonique, il est accompagné d'une onde de choc qui, dans une petite portion d'espace, éloignée de l'avion, peut toujours être assimilée à un plan. Au cours du mouvement, cette onde se réfléchit sur les obstacles qu'elle rencontre; nous allons étudier cette réflexion dans le cas d'un obstacle constitué par un dièdre.

Nous désignons par A et A' les points de rencontre de l'onde incidente avec les côtés du dièdre; Fig. 44. Nous supposons la bissectrice Ox du dièdre perpendiculaire à l'onde incidente; $(Ox, OA) = \psi$. Lorsque le nombre de Mach $\mathcal{M}$ de l'onde incidente est égal à l'unité, l'onde réfléchie est une enveloppe de cercles; elle comprend l'arc $R'R$ d'un cercle centré au sommet O du dièdre et les segments de tangentes AR et $A'R'$ au cercle précédent. Les ondes incidente et réfléchie partagent le fluide en trois régions (0), (1) et (2). Lorsque le choc incident n'est

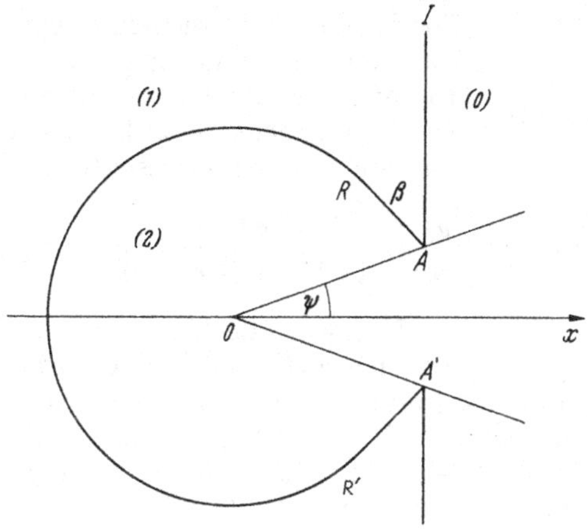

Fig. 44.

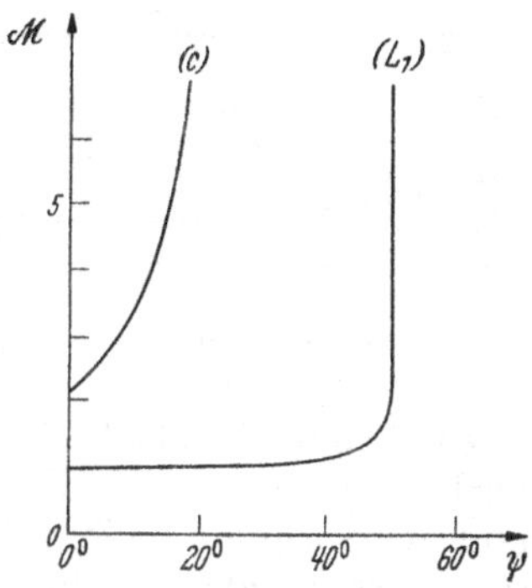

Fig. 45.

pas nul, il est encore possible de satisfaire à toutes les conditions du problème avec une onde réfléchie qui possède encore un segment rectiligne AR. Nous désignons par β l'angle de réflexion; $\beta = (AI, AR)$; nous posons $\omega = (AR, AO)$.

Les équations de choc (4.3) à (4.5) déterminent l'état du fluide dans la région (1), c'est-à-dire u_1, ϱ_1, p_1 et c_1 en fonction de ϱ_0, p_0, c_0 et $\mathcal{M}$. L'onde incidente se déplace avec la vitesse $U = \mathcal{M} c_0$; le point A décrit le côté du dièdre avec la vitesse $U/\cos\psi$ et le nombre de Mach $\mathcal{M}_1$ de la partie rectiligne de l'onde réfléchie possède la valeur suivante:

$$\mathcal{M}_1 = \frac{U(\cos\beta + \tan\psi \sin\beta) - u_1 \cos\beta}{c_1}. \tag{34.1}$$

Les équations du choc déterminent ensuite l'état du fluide dans la région (2); la vitesse doit être parallèle au côté du dièdre. On en déduit (pour $\gamma = 1{,}4$) la relation suivante qui détermine l'angle β en fonction des données $\mathcal{M}$ et ψ.

$$A\mathcal{M}^4 + 2B\mathcal{M}^2 = 5\sin^2\beta\{1 + 6\cot\beta\cot(\beta - \psi)\}, \tag{34.2}$$

$$\left.\begin{aligned} A &= 7 + \sin\beta\cos\beta\{6\cot(\beta-\psi)\tan\psi - 1\}(\cot\beta + 6\tan\psi), \\ B &= 12\{1 + \cot(\beta-\psi)\tan\psi\} - 5\sin^2\beta\{6\cot(\beta-\psi)\tan\psi - 1\}. \end{aligned}\right\} \tag{34.3}$$

Les lois de la réflexion régulière sont contenues dans l'équation (34.2) qui fut discutée par POLACHEK et SEEGER [44]. Les résultats sont indiqués sur la Fig. 46. Les asymptotes correspondent aux racines de la fonction A; ces racines sont réelles pour ψ supérieur à ψ_1 ($\psi_1 = 50{,}^\circ 030$). Une des courbes possède un point double; elle correspond à $\psi = \psi_2$ ($\psi_2 = 50{,}^\circ 769$). Suivant les valeurs de $\mathcal{M}$ et ψ, il existe zéro, une ou deux valeurs de β (ou ω). Dans le cas limite, l'équation (34.2) possède une racine double en β; $\mathcal{M}$ est alors une certaine fonction $\overline{\mathcal{M}}(\psi)$, représentée par la courbe (L_1) sur la Fig. 45; à gauche de la courbe (L_1), la réflexion régulière est impossible. Dans le cas où le problème possède deux solutions, seule la plus grande valeur de β (ou la plus petite valeur de ω) possède une réalité physique; ainsi, dans le cas particulier $\psi=\psi_2$, l'angle de réflexion β est indépendant du nombre de Mach $\mathcal{M}$ aussi longtemps que celui-ci demeure inférieur à la valeur 2,48.

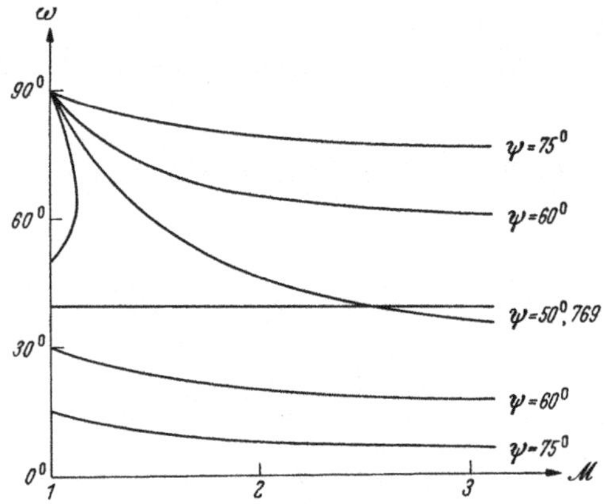

Fig. 46. Lois de la réflexion régulière.

Nous terminons l'étude des lois de la réflexion régulière en démontrant ave Borg [45] que l'onde de choc réfléchie régulière est toujours une onde détachée. En effet, le nombre de Mach u_1/c_1 de l'écoulement dans la region (1) est une fonction croissante de

$$\frac{u_1}{c_1}=\frac{5(\mathcal{M}^2-1)}{\sqrt{(\mathcal{M}^2+5)(7\mathcal{M}^2-1)}}. \tag{34.4}$$

L'onde réfléchie ne peut être attachée en O que si le rapport u_1/c_1, donc le nombre de Mach $\mathcal{M}$, est suffisamment grand. Le cas limite correspond à la courbe (C) tracée sur la Fig. 45; la comparaison des courbes (C) et (L_1) établit le résultat puisque l'onde réfléchie est détachée à droite de la courbe (C). En particulier, l'onde réfléchie est toujours détachée pour $\psi > 20°44'$.

b) Cas de la réflexion de Mach.

35. Equations de la réflexion de Mach. Lorsque la réflexion régulière ne se produit pas, l'expérience fait apparaître une réflexion analogue à celle qui est représentée sur la Fig. 47 et appelée réflexion de Mach. En un point A viennent concourir une ligne de discontinuité stationnaire (L) et trois ondes de choc: l'onde incidente AI, l'onde réfléchie AR, l'onde de Mach AM perpendiculaire au côté du dièdre; le fluide est divisé en quatre régions (0), (1), (2) et (3). Nous posons $\alpha=(Ox, OA)$, nous désignons encore par β l'angle de réflexion: $\beta=(AI, AR)$; la ligne de discontinuité (L), supposée rectiligne, fait avec Ox l'angle $\psi+(\pi/2)-\gamma$. Nous supposons l'indice adiabatique égal à 1,4.

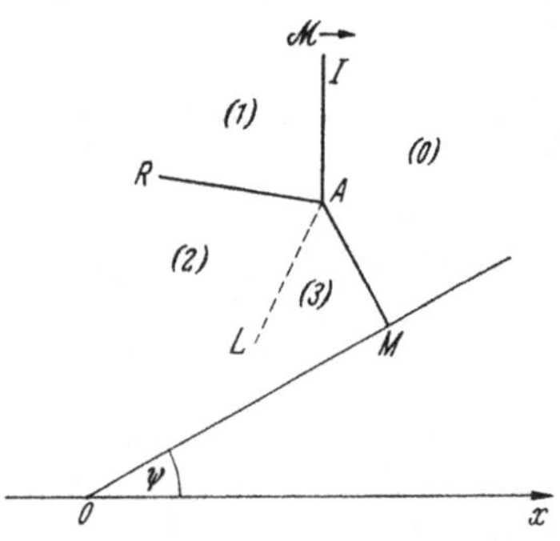

Fig. 47. Réflexion de Mach.

Les conditions sur l'onde de Mach sont les suivantes:

$$q_{n_3}=\frac{5}{6}c_0\left(\mathcal{M}_3-\frac{1}{\mathcal{M}_3}\right),\qquad q_{t_3}=0, \tag{35.1}$$

$$p_3=p_0\frac{7\mathcal{M}_3^2-1}{6}, \tag{35.2}$$

avec

$$\mathcal{M}_3=\mathcal{M}(\cos\psi+\sin\psi\tan\alpha). \tag{35.3}$$

On en déduit les composantes u et v de la vitesse dans la région (3): $u=q_{n_3}\cos\psi$, $v=q_{n_3}\sin\psi$.

Les conditions sur l'onde incidente (4.3) à (4.5) déterminent l'état du fluide dans la région (1). Les conditions sur l'onde réfléchie s'obtiennent en raisonnant comme dans le cas de la réflexion régulière, mais le point A décrit maintenant la demi-droite qui fait l'angle α avec Ox.

$$\mathcal{M}_1 = \frac{U(\cos\beta + \tan\alpha\sin\beta) - u_1\cos\beta}{c_1}, \tag{35.4}$$

$$\left.\begin{aligned} q_{n_2} &= u_1\cos\beta + \frac{5}{6}c_1\left(\mathcal{M}_1 - \frac{1}{\mathcal{M}_1}\right), \\ q_{t_2} &= u_1\sin\beta, \\ p_2 &= p_1\frac{7\mathcal{M}_1^2 - 1}{6}. \end{aligned}\right\} \tag{35.5}$$

On en déduit les valeurs des composantes u_2 et v_2 de la vitesse du fluide dans la région (2).

L'équation de la ligne de discontinuité (L) est la suivante:

$$(x - Ut) = \tan(\gamma - \psi)(y - Ut\tan\alpha). \tag{35.6}$$

Les composantes de la vitesse d'un point qui décrit cette ligne (L) verifient l'équation obtenue en dérivant par rapport au temps; mais les molécules fluides décrivant la ligne (L) appartiennent les unes à la région (2), les autres à la région (3); on en déduit la relation (35.7) et l'équation (35.8) qui traduit l'égalité des pressions de part et d'autre de (L).

$$\frac{u_2 - U}{v_2 - U\tan\alpha} = \frac{u_3 - U}{v_3 - U\tan\alpha}, \tag{35.7}$$

$$p_2 = p_3. \tag{35.8}$$

Nous avons ainsi obtenu toutes les équations de la réflexion de Mach; elles peuvent être résumées par les huit équations suivantes:

$$\left.\begin{aligned} \mathcal{M}_1 &= \frac{(\mathcal{M}^2 + 5)\cos\beta + 6\mathcal{M}^2\sin\beta\tan\alpha}{\sqrt{(7\mathcal{M}^2 - 1)(\mathcal{M}^2 + 5)}}, \\ \mathcal{M}_3 &= \mathcal{M}(\cos\psi + \sin\psi\tan\alpha), \\ 6(7\mathcal{M}_3^2 - 1) &= (7\mathcal{M}^2 - 1)(7\mathcal{M}_1^2 - 1), \\ v_2 &= \frac{5}{6}c_0\frac{\sqrt{(7\mathcal{M}^2 - 1)(\mathcal{M}^2 + 5)}}{6\mathcal{M}}\left(\mathcal{M}_1 - \frac{1}{\mathcal{M}_1}\right)\sin\beta, \\ u_2 &= \frac{5}{6}c_0\left(\mathcal{M} - \frac{1}{\mathcal{M}}\right) + v_2\cot\beta, \\ v_3 &= \frac{5}{6}c_0\left(\mathcal{M}_3 - \frac{1}{\mathcal{M}_3}\right)\sin\psi, \\ u_3 &= v_3\cot\psi, \\ u_2v_3 - u_3v_2 &= c_0\mathcal{M}\{\tan\alpha(u_2 - u_3) - (v_2 - v_3)\}. \end{aligned}\right\} \tag{35.9}$$

La célérité du son c_0, l'angle ψ et le nombre de Mach $\mathcal{M}$ sont donnés; les huit équations (35.9) contiennent huit inconnues. Nous allons étudier les valeurs des angles α et β.

36. Etude des cas particuliers. Les équations (35.9) se simplifient dans les cas particuliers; nous supposons d'abord que l'angle ψ est infiniment petit. Les angles

α et β s'expriment alors en fonction de $\mathcal{M}$ par les équations suivantes:

$$6\mathcal{M}^2 \frac{3\mathcal{M}^2-1}{(\mathcal{M}^2+5)^2} \tan^4\alpha - 2\frac{5\mathcal{M}^2+1}{\mathcal{M}^2+5}\tan^2\alpha + \frac{7\mathcal{M}^2-1}{6\mathcal{M}^2} = 0, \tag{36.1}$$

$$\frac{\mathcal{M}^2}{\mathcal{M}^2-1}\tan\alpha\tan\beta = \frac{\tan^2\alpha + \frac{\mathcal{M}^2+5}{6\mathcal{M}^2}}{6\frac{\mathcal{M}^2+1}{\mathcal{M}^2+5}\tan^2\alpha + 2\frac{4-\mathcal{M}^2}{3\mathcal{M}^2}}. \tag{36.2}$$

On étudie de façon analogue le système (35.9) dans le cas des chocs infiniment faibles ($\mathcal{M}=1$) ou infiniment forts ($\mathcal{M}$ infini), ainsi que dans les cas de la réflexion orthogonale ($\beta=\pi/2$) ou de la réflexion stationnaire ($\alpha=\psi$). La réflexion stationnaire constitue un cas limite puisque, pour α inférieur à ψ, la réflexion de Mach est impossible. Si on ajoute la condition $\alpha=\psi$ au système (35.9), on obtient la relation suivante entre les données $\mathcal{M}$ et ψ [44]:

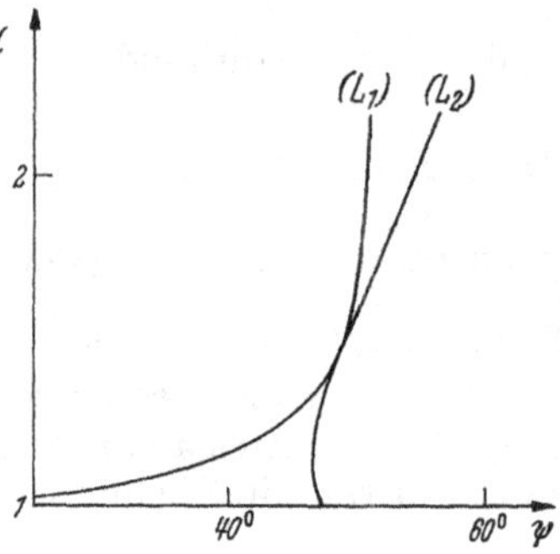

Fig. 48.

$$\left.\begin{aligned}&\frac{5(7\mathcal{M}^2-1)}{6(\mathcal{M}^2+5)}x^2 - \\ &\quad - \frac{187\mathcal{M}^4-326\mathcal{M}^2+175}{6(\mathcal{M}^2+5)^2}x - 6 = 0,\end{aligned}\right\} \tag{36.3}$$

avec

$$x = \frac{6\mathcal{M}^2}{7\mathcal{M}^2-1}\tan^2\psi. \tag{36.4}$$

Les équations précédentes définissent une courbe (L_2), représentée sur la Fig. 48. La réflexion stationnaire est un cas particulier de la réflexion régulière; la courbe (L_2) est donc située à droite de la courbe (L_1). Polachek a démontré que les deux courbes sont tangentes en un point d'ordonnée $\mathcal{M}=1{,}54$.

37. Lois de la réflexion de Mach. Les lois de la réflexion de Mach sont contenues dans le système (35.9); on le résout numériquement par interpolation des résultats

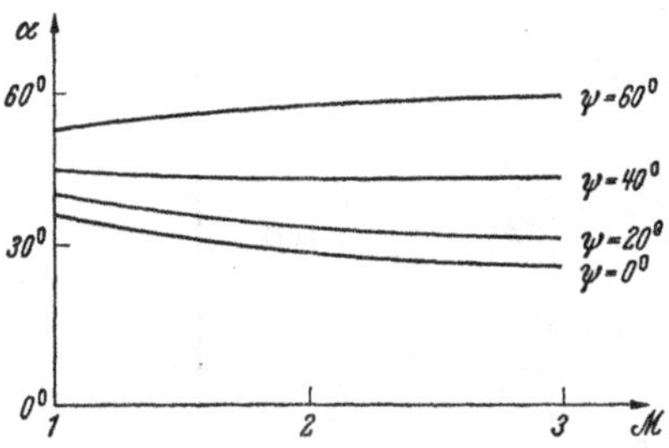

Fig. 49. Lois de la réflexion de Mach.

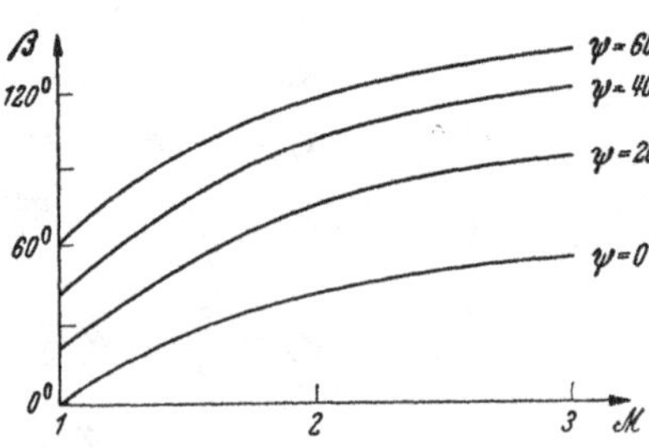

Fig. 50. Lois de la réflexion de Mach.

obtenus dans les divers cas particuliers précédents. Les valeurs des fonctions $\alpha(\mathcal{M},\psi)$ et $\beta(\mathcal{M},\psi)$ sont représentées sur les Figs. 49 et 50. A chaque couple de valeurs $\mathcal{M}$ et ψ correspondent deux solutions; seule la solution qui correspond à la plus grande valeur de β est observée; c'est cette solution qui est représentée sur les Figs. 49 et 50. Cette solution ne possède de réalité physique que si la valeur obtenue pour α est supérieure à ψ; il faut pour cela que le point de coordonnées $\mathcal{M}$, ψ sur la Fig. 48 soit situé à gauche de la courbe (L_2). Cette courbe est asymptote à la droite $\psi=68{,}^\circ23$; donc, pour les valeurs supérieures de ψ, la réflexion de Mach n'est jamais possible. On peut énoncer le résultat suivant:

à gauche de (L_2), la réflexion est du type de Mach,

à droite de (L_2), la réflexion est régulière.

Entre les deux courbes, les deux types de réflexion sont possibles. Le point de contact sépare la région comprise entre (L_1) et (L_2) en deux sous-régions. Pour des raisons de continuité que nous allons préciser, la réflexion est regulière dans la sous-région inférieure ($\mathcal{M}<1{,}54$); elle est du type de Mach dans la sous-région supérieure.

Lorsque le nombre de Mach $\mathcal{M}$ est égal à l'unité, l'augmentation d'entropie est nulle dans la réflexion régulière, positive dans la réflexion de Mach; on observe donc la réflexion régulière. Nous avons tracé sur la Fig. 51 les variations de l'angle β en fonction de $\mathcal{M}$ pour la valeur $\psi=40°$. Le point de coordonnées $\mathcal{M}$ et β se trouve en A pour $\mathcal{M}=1$ et décrit l'arc AB lorsque $\mathcal{M}$ augmente; il atteint le point B pour la valeur $\mathcal{M}=\bar{\mathcal{M}}(\psi)$ et passe ensuite à la position C, situé sur la courbe relative à la solution réelle de la réflexion de Mach; il décrit ensuite la branche CD. Pour la valeur $\bar{\mathcal{M}}(\psi)$, l'écoulement est instable. L'instabilité disparaît lorsque la branche $A'D$ coupe l'arc AB, c'est-à-dire pour $\mathcal{M}$ supérieur à 1,54.

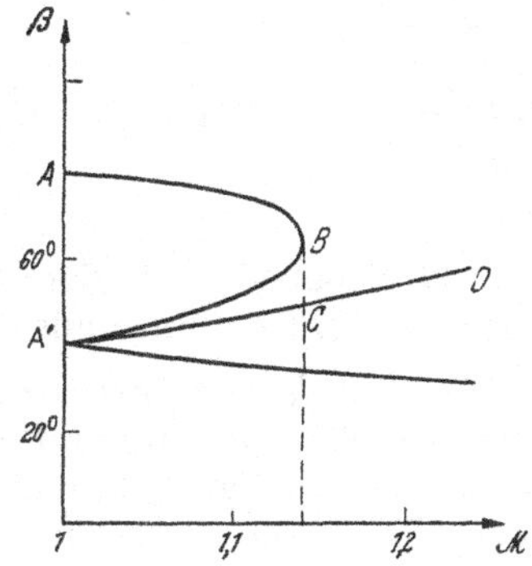

Fig. 51.

c) Etude de l'écoulement après le choc.

38. Ecoulements plans pseudo-stationnaires. Dans les équations des écoulements plans, nous effectuons le changement de variables $x=\xi t$, $y=\eta t$, $t=\zeta$; nous exprimons les inconnues en fonction des variables ξ, η, ζ et nous cherchons s'il existe des solutions indépendantes de ζ. La réponse est affirmative; les mouvements correspondants sont appelés écoulements plans pseudo-stationnaires. Ces écoulements sont régis par les équations suivantes:

$$\left.\begin{aligned}
&(u-\xi)\frac{\partial u}{\partial \xi}+(v-\eta)\frac{\partial u}{\partial \eta}+\frac{1}{\varrho}\frac{\partial p}{\partial \xi}=0,\\
&(u-\xi)\frac{\partial v}{\partial \xi}+(v-\eta)\frac{\partial v}{\partial \eta}+\frac{1}{\varrho}\frac{\partial p}{\partial \eta}=0,\\
&(u-\xi)\frac{\partial \varrho}{\partial \xi}+(v-\eta)\frac{\partial \varrho}{\partial \eta}+\varrho\left(\frac{\partial u}{\partial \xi}+\frac{\partial v}{\partial \eta}\right)=0,\\
&(u-\xi)\frac{\partial}{\partial \xi}\left(\frac{p}{\varrho^{\gamma}}\right)+(v-\eta)\frac{\partial}{\partial \eta}\left(\frac{p}{\varrho^{\gamma}}\right)=0.
\end{aligned}\right\}\tag{38.1}$$

La célérité du son étant désignée par c ($\gamma p=\varrho c^2$), le système (38.1) est elliptique pour $(u-\xi)^2+(v-\eta)^2<c^2$; il est hyperbolique dans le cas contraire. On appelle pseudo-lignes de courant les lignes du plan ξ, η qui sont en chaque point tangentes au vecteur dont les composantes sont $u-\xi$ et $v-\eta$. Nous posons $u-\xi=Q\cos\Theta$ et $v-\eta=Q\sin\Theta$. Lorsque le système est hyperbolique, les caractéristiques sont définies par les équations suivantes:

$$\left.\begin{aligned}
d\eta&=\tan(\Theta\pm M)\,d\xi,\\
c&=Q\sin M.
\end{aligned}\right\}\tag{38.2}$$

Un écoulement uniforme défini par les valeurs constantes u_1, v_1, ϱ_1 et p_1 est un écoulement plan pseudo-stationnaire. La ligne parabolique est le cercle centré au point P_1 de coordonnées u_1, v_1 et dont le rayon est égal à la célérité du son c_1. Les pseudo-lignes de courant sont les droites issues du point P_1 et les caractéristiques les tangentes au cercle parabolique [*46*]; l'équation différentielle des caractéristiques s'écrit en effet sous la forme (38.3), tandis que les tangentes au cercle parabolique peuvent être représentées en fonction d'un paramètre α par l'équation (38.4).

$$d\xi^2\{(\eta-v_1)^2-c_1^2\}-2d\xi\,d\eta\,(\xi-u_1)(\eta-v_1)+d\eta^2\{(\xi-u_1)^2-c_1^2\}=0, \tag{38.3}$$

$$(\xi-u_1)\cos\alpha+(\eta-v_1)\sin\alpha-c_1=0. \tag{38.4}$$

39. Ecoulement derrière l'onde de choc réfléchie. Nous revenons au problème de la rencontre d'une onde de choc plane et d'un dièdre.

α) Cas de la réflexion régulière. Nous supposons que l'onde réfléchie est une demi-droite issue du point A: Fig. 52. Dans la région (0), le fluide est au repos et les pseudo-lignes de courant sont les droites passant par

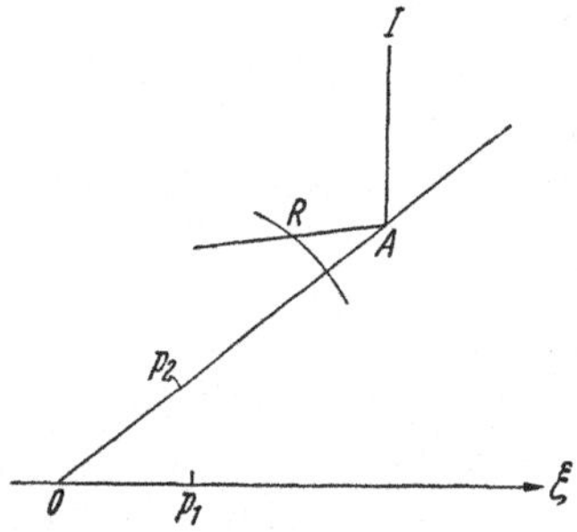

Fig. 52.

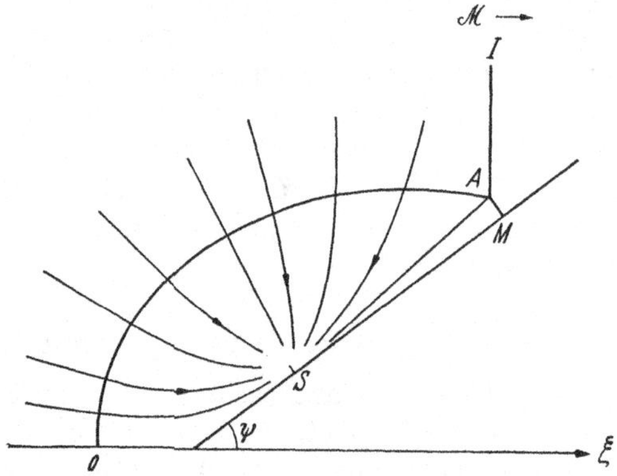

Fig. 53.

l'origine; l'écoulement est uniforme dans la région (1) et les pseudo-lignes de courant sont les droites qui passent par le point P_1 d'abscisse u_1 sur $O\xi$; dans la région (2), les pseudolignes de courant sont les droites qui passent par le point P_2 sur OA. Le vecteur $P_1\,P_2$ qui représente la discontinuité de vitesse est perpendiculaire à l'onde réfléchie AR. La ligne parabolique dans la région (2) est le cercle de centre P_2 de rayon c_2; en vertu des diverses équations des chocs, le point O est toujours intérieur à ce cercle; il appartient donc à la région elliptique. La présence de la pointe du dièdre ne permet pas que l'écoulement soit uniforme dans toute la région (2). Les perturbations issues de O se transmettent dans toute la région elliptique, atteignent le cercle parabolique et se transmettent ensuite dans une partie de la région hyperbolique le long des caractéristiques. Le cercle parabolique coupe toujours l'onde réfléchie; nous désignons par R le point de rencontre le plus proche de A. Le point A est intérieur ou extérieur au cercle parabolique suivant la valeur de $\mathcal{M}$. Pour $\psi=40°$, on a $\overline{\mathcal{M}}(\psi)=1{,}135$; le point A est extérieur au cercle parabolique pour $\mathcal{M}<1{,}128$; il est intérieur au cercle parabolique dans le cas contraire. Dans le cas $\mathcal{M}<1{,}128$, il existe dans le région (2) des points inaccessibles aux perturbations issues de O; en ces points, l'écoulement est uniforme. On admet que la région d'écoulement uniforme est la plus grande possible; elle est alors limitée par le cercle parabolique qui constitue une caractéristique particulière. L'onde réfléchie possède une portion rectiligne.

β) Cas de la réflexion de Mach. Les considérations précédentes demeurent valables. Si l'onde réfléchie est rectiligne, les écoulements dans les régions (1), (2) et (3) sont uniformes et les pseudo-lignes de courant concourent en des points P_1, P_2, P_3. P_1 est sur $O\xi$, P_3 sur le côté du dièdre. P_2, P_3 et A sont alignés sur la ligne de discontinuité (L). L'onde réfléchie peut encore, suivant la valeur de $\mathcal{M}$, posséder ou ne pas posséder de segment rectiligne.

Ludloff et Friedmann ont calculé l'écoulement après le choc réfléchie en supposant connue (d'après une photographie) l'onde réfléchie. Ils ont résolu numériquement un problème de Cauchy [*47*]; les pseudo-lignes de courant sont représentées sur la Fig. 53; on a $\psi = 38°$, $\mathcal{M} = 1{,}59$. Dans un ordre d'idées différentes, le problème de la rencontre d'une onde de choc plane et d'un dièdre a été résolu par Lighthill, dans le cas des dièdres dont le demi-angle ψ est voisin de zéro ou voisin de $(\pi/2)$ [*48*]. Des calculs ont également été effectués par Kofink [*49*].

III. Etude des ondes de choc sphériques.

40. Ondes se propageant avec une vitesse constante. A côté des mouvements rectilignes, on rencontre d'autres exemples de mouvements non stationnaires dans lesquels le nombre des variables se réduit à deux. C'est le cas en particulier des phénomènes d'explosion. Lorsqu'une explosion se produit dans un fluide homogène au repos, on peut considérer que les phénomènes présentent la symétrie sphérique. O étant le centre de l'explosion, la vitesse du fluide, la pression et la masse spécifique dépendent seulement du temps t et de la distance $OM = r$. La vitesse en un point M est portée par la droite OM; nous désignons par u sa valeur. Les équations du mouvement sont les suivantes:

$$\left.\begin{aligned} \varrho(u_t + u\,u_r) + p_r &= 0,\\ r^2\varrho_t + (r^2\varrho\,u)_r &= 0,\\ (p\,\varrho^{-\gamma})_t + u(p\,\varrho^{-\gamma})_r &= 0. \end{aligned}\right\} \tag{40.1}$$

Nous supposons qu'une onde de choc se propage dans un fluide au repos et nous écrivons son équation sous la forme $r = R(t)$. Les équations du choc s'écrivent de la façon suivante:

$$\left.\begin{aligned} u\{R(t), t\} &= (1-\mu^2)\left\{R'(t) - \frac{c_0}{R'(t)}\right\},\\ \frac{\varrho_0}{\varrho\{R(t), t\}} &= (1-\mu^2)\left\{\frac{c_0}{R'(t)}\right\}^2 + \mu^2,\\ \frac{p\{R(t), t\}}{p_0} &= (1+\mu^2)\left\{\frac{R'(t)}{c_0}\right\}^2 - \mu^2. \end{aligned}\right\} \tag{40.2}$$

Nous supposons que le mouvement du fluide est provoqué par une sphère en expansion, d'équation $r = \alpha(t)$. Comme les molécules fluides coincident avec la surface de la sphère, on a sur celle-ci $u\{\alpha(t), t\} \equiv \alpha'(t)$.

Nous cherchons s'il est possible que les trois fonctions inconnues dépendent de la seule variable $\xi\,(\xi = r/t)$. Les équations (40.1) prennent la forme suivante:

$$\left.\begin{aligned} (u-\xi)\,u' + p'\,\varrho^{-1} &= 0,\\ (u-\xi)\,\varrho' + \varrho\,u' + 2\varrho\,u\,\xi^{-1} &= 0,\\ (u-\xi)\,(p\,\varrho^{-\gamma})' &= 0. \end{aligned}\right\} \tag{40.3}$$

Les équations différentielles (40.3) définissent des écoulements isentropiques. Les ondes de choc susceptibles d'engendrer un écoulement du type précédent se propagent nécessairement avec une vitesse constante U. Pour chaque valeur du nombre de Mach $\mathcal{M}$ ($\mathcal{M}=U/c_0$), on obtient une solution. Les différentes solutions sont représentées sur la Fig. 54 ($\gamma=1{,}4$); u est une fonction de ξ. La fonction $\alpha(t)$ est de la forme $\bar{u}t$; la condition sur la sphère exige que la vitesse de propagation u soit solution de l'équation $u(\bar{u})=\bar{u}$. Résoudre cette équation revient à couper les courbes qui représentent les variations des fonctions $u(\xi)$ par la droite $u=\xi$. Ainsi une sphère en expansion uniforme dans un gaz au repos engendre une onde de choc sphérique qui se propage avec une vitesse constante. La correspondance entre le nombre de Mach $\mathcal{M}$ et la vitesse $\bar{u}$ de la sphère est indiquée sur la Fig. 55, sur laquelle est représentée également la courbe analogue relative aux ondes de choc planes et définie par la première des équations (25.1). Ces résultats, ainsi que ceux établis dans la section suivante, sont dûs à G. I. TAYLOR [50].

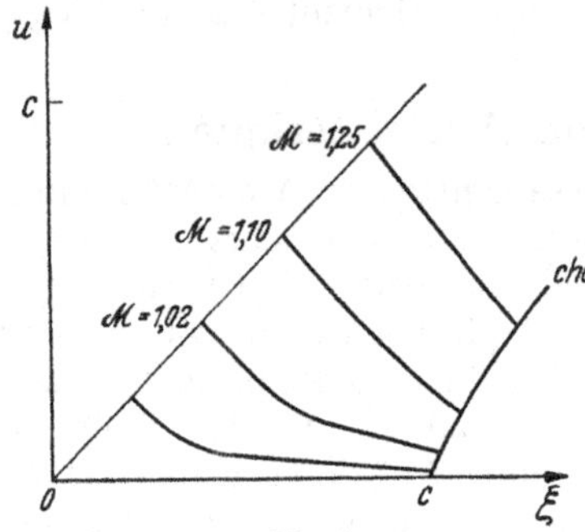

Fig. 54.

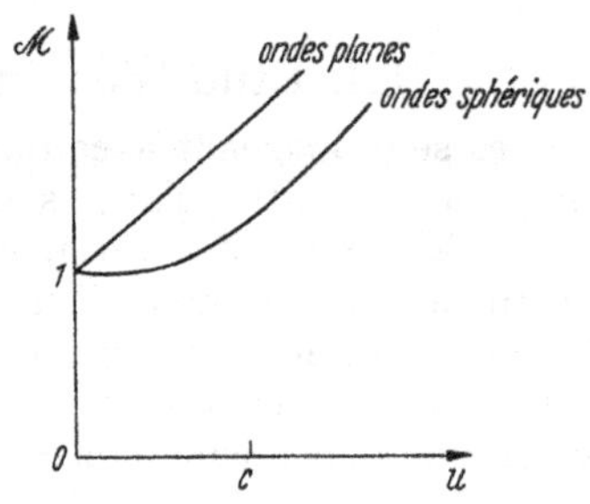

Fig. 55.

41. Ondes transportant une énergie constante. La solution que nous venons de construire ne peut correspondre à un cas se présentant dans la pratique; il est nécessaire en effet que la célérité du choc tende vers la célérité du son lorsque le temps augmente indéfiniment, sinon l'énergie mise en jeu pour engendrer le mouvement ne serait pas bornée. On obtient une meilleure approximation des phénomènes réels en cherchant une solution dans laquelle la somme de l'énergie cinétique et de l'énergie interne du gaz demeure constante au cours du temps. Cette somme est égale au premier membre de la formule (2.3), soit

$$E=\int_0^{R(t)}\left(\varrho\frac{u^2}{2}+\frac{p}{\gamma-1}\right)4\pi r^2\,dr. \tag{41.1}$$

Il est possible de satisfaire de façon approchée cette condition de la façon suivante. Nous imposons à la fonction $R(t)$ de satisfaire la condition $R^\alpha(dR/dT)=A$, où α et A désignent des constantes. Nous cherchons ensuite à determiner les fonctions inconnues sous la forme suivante:

$$\left.\begin{aligned} u(r,t)&=A\,\frac{\Phi(\eta)}{R^\alpha},\\ p(r,t)&=A^2\,\frac{\varrho_0 f(\eta)}{R^{2\alpha}},\\ \varrho(r,t)&=\varrho_0\psi(\eta).\end{aligned}\right\} \tag{41.2}$$

La variable η désigne le quotient $r/R(t)$. L'intégrale (41.1) est proportionnelle à $R^{2\alpha-3}$; elle est indépendante du temps si nous avons soin de choisir $\alpha=\frac{3}{2}$, ce que nous ferons. Lorsqu'on porte les expressions (41.2) dans le système (40.1), on obtient, pour les fonctions $\Phi(\eta)$, $f(\eta)$ et $\psi(\eta)$, les trois équations différentielles suivantes:

$$\left.\begin{aligned} -\left(\frac{3}{2}\Phi+\eta\Phi'\right)+\Phi\Phi'+\frac{f'}{\psi}&=0,\\ (\Phi-\eta)\psi'+\psi\Phi'+2\psi\frac{\Phi}{\eta}&=0,\\ -3f+(\Phi-\eta)\left(f'-\gamma f\frac{\psi'}{\psi}\right)&=0.\end{aligned}\right\}\tag{41.3}$$

En combinant la troisième équation avec la seconde, puis avec la première, on obtient les deux intégrales premières (41.4). La constante b doit être nulle, si on impose $\Phi(0)=0$. La fonction $\Phi(\eta)$ est alors solution d'une équation différentielle homogène dont l'intégrale générale s'écrit sous la forme (41.5).

$$\left.\begin{aligned} \eta(\eta-\Phi)f&=a\psi^{\gamma-1},\\ f\Phi-(\eta-\Phi)\left(\frac{f}{\gamma-1}+\frac{\psi\Phi^2}{2}\right)&=b\eta^{-2},\end{aligned}\right\}\tag{41.4}$$

$$\frac{\eta}{\eta_1}=\frac{\left|\frac{\Phi}{\eta}-\frac{1}{\gamma}\right|^{\frac{\gamma-1}{2\gamma+1}}}{\left|\frac{\Phi}{\eta}\right|^{\frac{2}{5}}\left|\frac{\Phi}{\eta}-\frac{5}{3\gamma-1}\right|^{\frac{13\gamma^2-7\gamma+12}{5(2\gamma+1)(3\gamma-1)}}}.\tag{41.5}$$

Ces trois équations obtenues par R. LATTER [51] définissent l'intégrale générale du système (41.3). Il n'est malheureusement pas possible, à l'aide des fonctions de la forme (41.2), de satisfaire les équations de choc; cependant ces équations peuvent être satisfaites si on suppose le choc violent ($\mathcal{M}$ infini); elles s'écrivent alors sous la forme suivante:

$$\left.\begin{aligned} \Phi(1)=\frac{2}{\gamma+1},\qquad f(1)&=\frac{2}{\gamma+1},\\ \psi(1)&=\frac{\gamma+1}{\gamma-1}.\end{aligned}\right\}\tag{41.6}$$

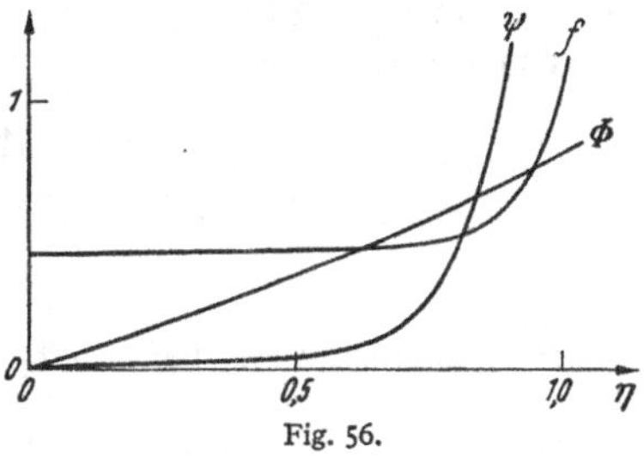

Fig. 56.

Les valeurs des trois fonctions $\Phi(\eta)$, $f(\eta)$ et $\psi(\eta)$ sont représentées sur la Fig. 56 pour un gaz diatomique ($\gamma=\frac{7}{5}$). L'intégrale E possède la valeur constante $33{,}5\,\varrho_0 A$. Cette théorie concorde très bien avec l'expérience, car l'examen des photographies des explosions atomiques prouve que le produit R^5t^{-2} demeure constant au début de l'explosion, c'est-à-dire tant que le choc demeure violent; de la valeur de la constante, on peut déduire l'énergie E mise en jeu.

42. Dégénérescence de l'onde de choc. L'énergie mise en jeu pour provoquer le choc étant bornée, le nombre de Mach $\mathcal{M}$ de l'onde de choc a pour limite l'unité lorsque le temps augmente indéfiniment. La variation d'entropie ΔS sur l'onde de choc est un infiniment petit d'ordre 3 par rapport à $\mathcal{M}-1$:

$$\Delta S\sim\frac{16}{3}c_v\frac{\gamma(\gamma-1)}{(\gamma+1)^2}(\mathcal{M}-1)^3.\tag{42.1}$$

On peut donc supposer que l'écoulement après le choc est isentropique; en outre il existe une fonction potentielle $\Phi(r,t)$ telle que $u=\partial\Phi/\partial r$. D'après l'équation (2.9), dans laquelle rot $\boldsymbol{V}=0$, l'expression (42.2) est une fonction du temps que l'on peut toujours supposer nulle par modification de la fonction Φ. En introduisant la célérité du son, on en déduit l'équation (42.3).

$$\frac{\partial \Phi}{\partial t} + \frac{V^2}{2} + \int \frac{dp}{\varrho} = 0, \tag{42.2}$$

$$c^2 = c_0^2 - (\gamma - 1)\left\{\frac{\partial \Phi}{\partial t} + \frac{1}{2}\left(\frac{\partial \Phi}{\partial r}\right)^2\right\}. \tag{42.3}$$

En écrivant l'équation de la conservation de la masse, on obtient, pour la fonction $\Phi(r, t)$, l'équation (42.4). Dans la théorie linéarisée, cette équation prend la forme (42.5) et la fonction Φ la forme (42.6).

$$c^2\left(\frac{\partial^2 \Phi}{\partial r^2} + \frac{2}{r}\frac{\partial \Phi}{\partial r}\right) = \frac{\partial^2 \Phi}{\partial t^2} + 2\frac{\partial \Phi}{\partial r}\frac{\partial^2 \Phi}{\partial r\,\partial t} + \left(\frac{\partial \Phi}{\partial r}\right)^2 \frac{\partial^2 \Phi}{\partial r^2}, \tag{42.4}$$

$$\frac{\partial^2 \Phi}{\partial r^2} + \frac{2}{r}\frac{\partial \Phi}{\partial r} = \frac{1}{c_0^2}\frac{\partial^2 \Phi}{\partial t^2}, \tag{42.5}$$

$$\Phi(r, t) = \frac{f(c_0 t - r)}{r}. \tag{42.6}$$

Le produit $r\Phi(r, t)$ demeure constant sur chaque caractéristique de la solution linéarisée. En raisonnant comme dans la Sect. 13, on désigne par $z(r, t) = \text{const}$ l'équation des caractéristiques de la solution réelle, et on cherche à déterminer la fonction $\Phi(r, t)$ en écrivant que le produit $r\Phi$ demeure constant sur chaque caractéristique de la solution réelle. Les calculs ont été effectués par WHITHAM [*52*]; on obtient, pour les grandes valeurs du temps, l'équation de l'onde de choc sous la forme (42.7) et la valeur en un point donné de la pression après le choc sous la forme (42.8):

$$c_0 t = r - A(\log r)^{\frac{1}{2}} - B - C(\log r)^{-\frac{1}{2}} + \cdots, \tag{42.7}$$

$$\frac{p}{p_0} = \frac{2\gamma}{\gamma + 1}\,\frac{r - c_0 t - B}{r \log r} + O\left(r^{-1}\log^{-\frac{3}{2}}\right). \tag{42.8}$$

A, B et C désignent des constantes.

IV. Etude des translations rectilignes de vitesse variable.

43. Départ d'un obstacle dans un fluide au repos. A côté des problèmes que nous venons d'indiquer, un certain nombre de questions ont été résolues; à défaut d'une étude générale qui n'est pas encore au point, nous nous proposons de terminer ce chapitre relatif aux ondes de choc par l'étude des translations rectilignes de vitesse variable. Nous envisageons, pour commencer, le cas d'un obstacle initialement immobile dans un fluide homogène et au repos. Nous supposons que l'obstacle admet un axe de révolution Ox et qu'il est animé, à partir de l'instant $t=0$, d'une translation rectiligne parallèle à Ox. La vitesse initiale de l'obstacle n'étant pas nulle, une onde de choc prend naissance instantanément; l'équation de l'obstacle se présente sous la forme $x = f(r) + \xi(t)$, et celle de l'onde de choc, sous la forme $x = F(r, t)$; la variable r représente la distance à l'axe de révolution.

On ne sait pas déterminer la fonction $F(r, t)$, mais on peut calculer la valeur de toutes les dérivées $(\partial^n F/\partial t^n)(r, 0)$. La position initiale de l'onde coïncide avec celle de l'obstacle; on a $F(r, 0) = f(r) + \xi(0)$; nous écrivons que les équations du choc sont vérifiées sur la surface $x = F(r, t)$, que les équations du mouvement sont vérifiées dans la région qui sépare l'onde et l'obstacle et que, sur l'obstacle, la vitesse relative du fluide par rapport à l'obstacle ne possède pas de composante normale. Les quantités $\bar{c}$, $\bar{\varrho}$ et $\bar{p}$ désignent la célérité du son, la masse spécifique et la pression dans le fluide au repos; nous définissons les angles φ et β par les formules (43.1). Le nombre de Mach normal $\mathcal{M}$ est une fonction des variables r et t, définie par la formule (43.2); nous désignons par $\mathcal{M}_0$ sa valeur pour $t=0$.

$$\cot\varphi = f'(r), \quad \cot\beta = \frac{\partial F}{\partial r}, \tag{43.1}$$

$$\mathcal{M}\bar{c} = \sin\beta\,\frac{\partial F}{\partial t}. \tag{43.2}$$

La condition sur l'obstacle s'écrit sous la forme (43.3); en comparant avec les équations du choc (Sect. 4) à l'instant initial, on en déduit la valeur de la vitesse initiale de l'onde de choc $(\partial F/\partial t)(r, 0)$:

$$u\sin\varphi - v\cos\varphi = \xi'(t)\sin\varphi, \tag{43.3}$$

$$\frac{\partial F}{\partial t}(r,0) = \frac{\xi'(0)}{2(1-\mu^2)} + \sqrt{\left\{\frac{\xi'(0)}{2(1-\mu^2)}\right\}^2 + \bar{c}^2\{1+f'^2(r)\}}. \tag{43.4}$$

Pour calculer la valeur de l'accélération initiale de l'onde de choc, nous dérivons par rapport à r et t la condition sur l'obstacle et les quatre équations du mouvement. En nous plaçant à l'instant initial, nous obtenons quatorze équations linéaires qui nous permettent de calculer les douze dérivées partielles des quatre fonctions u, v, p et ϱ par rapport aux trois variables x, y, t ainsi que les deux derivées secondes $(\partial^2 F/\partial r\,\partial t)$ et $(\partial^2 F/\partial t^2)$ [*48*]. On obtient les résultats suivants:

$$A\frac{\partial^2 F}{\partial t^2}(r,0) = \frac{\gamma+1}{2}\xi''(0) + (B\sin^2\varphi + C\cos^2\varphi)\bar{c}^2 f''(r) + B\bar{c}^2\frac{\cot\varphi}{r}, \tag{43.5}$$

$$\left.\begin{aligned} A &= \frac{2(2\gamma-1)\mathcal{M}_0^4 + (\gamma+5)\mathcal{M}_0^2 - \gamma + 1}{(\gamma+1)\mathcal{M}_0^2(\mathcal{M}_0^2-1)}, \\ B &= \frac{2\gamma\mathcal{M}_0^2 - \gamma + 1}{\gamma+1}\,\frac{(\gamma-1)\mathcal{M}_0^2 + 2}{(\gamma+1)\mathcal{M}_0^2}, \\ C &= \frac{2\gamma - (\gamma-1)\mathcal{M}_0^2}{(\gamma+1)\mathcal{M}_0^2}. \end{aligned}\right\} \tag{43.6}$$

La formule (43.5) détermine la valeur de l'accélération initiale de l'onde de choc en chacun de ses points. Cette accélération est infinie sur l'axe de révolution chaque fois que l'obstacle possède une pointe; dans le cas contraire, le second membre de l'équation (39.7) se réduit à $\frac{\gamma+1}{2}\xi''(0) + 2B\bar{c}^2 f''(0)$ pour $r=0$. En supprimant le dernier terme du second membre de l'équation (43.5), on obtient la valeur de l'accélération initiale du choc relative au cas des écoulements plans; en supposant en outre que la fonction $f(r)$ est constante, on obtient la valeur relative au cas des écoulements rectilignes.

Au voisinage de l'instant initial, l'onde de choc peut être représentée par l'équation suivante:

$$F(r,t) = f(r) + t\frac{\partial F}{\partial t}(r,0) + \frac{t^2}{2}\frac{\partial^2 F}{\partial t^2}(r,0). \tag{43.7}$$

Lorsque l'obstacle est terminé par une pointe, l'étude de l'onde de choc présente plus de difficultés. Le cas d'un dièdre a été traité par SAKURAI [*54*] à l'aide de méthodes approchées; la théorie présente de nombreuses analogies avec le problème de la rencontre d'une onde de choc plane et d'un dièdre que nous avons étudié précédemment.

44. Courbure des chocs attachés: écoulements plans. La solution du problème que nous venons de résoudre présente un caractère local; la fonction $x=F(r,t)$

qui définit l'onde de choc a été déterminée au voisinage de la valeur $t=0$, la variable r étant arbitraire. En adoptant toujours ce point de vue local, il est naturel de chercher des cas dans lesquels cette même fonction puisse être déterminée au voisinage de la valeur $r=0$, tandis que la variable t est arbitraire. Cela est possible lorsque l'onde de choc est attachée à la pointe de l'obstacle.

Nous envisageons d'abord le cas des écoulements plans. Un obstacle présentant une pointe sur l'axe Ox est défini par l'équation $x=f(y)+\xi(t)$; nous supposons qu'une onde de choc soit attachée à la pointe de l'obstacle et nous écrivons son èquation sous la forme $x=F(y,t)$. On a $F(0,t)=f(0)+\xi(t)$. Nous allons calculer ensuite les valeurs des deux fonctions suivantes:

$$\frac{\partial F}{\partial y}(0,t)=F_1(t), \qquad \frac{\partial^2 F}{\partial y^2}(0,t)=F_2(t). \tag{44.1}$$

Nous définissons les angles ω et β par les formules (44.2); la condition sur l'obstacle s'écrit sous la forme (44.3).

$$f'(0)=\cot\omega, \qquad \frac{\partial F}{\partial y}(0,t)=\cot\beta, \tag{44.2}$$

$$u(f+\xi,y,t)-f'(y)\,v(f+\xi,y,t)=\xi'(t). \tag{44.3}$$

Les valeurs des fonctions F_1 et F_2 s'obtiennent ensuite en raisonnant comme dans la section précédente [55]. Ces valeurs font intervenir le nombre de Mach normal $\mathcal{M}(y,t)$, dont nous désignons par $\mathcal{M}_0$ la valeur à la pointe. Dans les formules (44.5) et (44.6), on a supposé $\gamma=1,4$.

$$\left\{\frac{\bar c}{\xi'(t)}\right\}^2=\frac{1}{1+F_1^2}-\frac{1}{(1-\mu^2)\{1+f'(0)F_1\}}, \tag{44.4}$$

$$f''(0)-\left(A-\frac{B}{F_1^2}\right)F_2+\frac{3}{5}\,\frac{\mathcal{M}_0^4}{\mathcal{M}_0^2-1}\left(C-\frac{D}{F_1^2}\right)F_1'\,\frac{1+F_1^2}{\xi' F_1}=0, \tag{44.5}$$

$$\left.\begin{aligned}
A&=\frac{6\mathcal{M}_0^4}{25}\,\frac{9\mathcal{M}_0^4+16\mathcal{M}_0^2-1}{(\mathcal{M}_0^2-1)^4},\\
B&=\frac{9\mathcal{M}_0^2}{25}\,\frac{(\mathcal{M}_0^2+1)(\mathcal{M}_0^2+5)}{(\mathcal{M}_0^2-1)^3},\\
C&=\mathcal{M}_0^2\,\frac{3\mathcal{M}_0^6-5\mathcal{M}_0^4-43\mathcal{M}_0^2-3}{5(\mathcal{M}_0^2-1)^3},\\
D&=\frac{(\mathcal{M}_0^2+5)(\mathcal{M}_0^2+9)}{10\mathcal{M}_0^2(\mathcal{M}_0^2-1)}.
\end{aligned}\right\} \tag{44.6}$$

La formule (44.4) est une relation entre les angles β et ω; elle a déjà été résolue: Fig. 3. La formule (44.5) détermine la courbure du choc à la pointe. Nous désignons par R et $\mathcal{R}$ les rayons de courbure à la pointe de l'onde et de l'obstacle; dans le cas stationnaire ($F_1'=0$), la formule (44.5) s'écrit sous la forme suivante qui constitue une nouvelle façon d'écrire la formule (18.4):

$$\frac{R}{\mathcal{R}}=\frac{\sin^3\omega}{\sin^3\beta}\{A(\mathcal{M}_0^2)-B(\mathcal{M}_0^2)\tan^2\beta\}. \tag{44.7}$$

Nous désignons par $\lambda(\omega,M)$ la valeur commune des deux membres; dans le cas général, l'équation (44.5) s'écrit sous la forme suivante:

$$\frac{R(t)}{\mathcal{R}}=\frac{\lambda(\omega,M)}{1+\dfrac{\mathcal{R}\,\xi''(t)}{\bar c^2}\,\mu(\omega,M)}, \tag{44.8}$$

$$\mu=\frac{6}{5}\,\frac{\sin^3\omega}{\cos^2\beta}\,\frac{C-D\tan^2\beta}{\dfrac{\mathcal{M}_0^2+1}{\mathcal{M}_0^2-1}-\dfrac{\mathcal{M}_0^2+5}{6\mathcal{M}_0^2}\tan^2\beta}\cdot\frac{1}{(\mathcal{M}_0^2-1)^2}. \tag{44.9}$$

Les valeurs de la fonction λ sont représentées sur la Fig. 24 (avec $\vartheta_s = \pi - \omega$); les valeurs de la fonction μ sont représentées sur la Fig. 57. Comme la fonction λ, la fonction μ s'annule pour une certaine valeur $\overline{M}(\omega)$ du nombre de Mach $(M = \xi'(t)/\bar{c})$. La fonction $\overline{M}$ est représentée sur la Fig. 58; la valeur $\overline{M}$ est toujours supérieure à la valeur M_1 pour laquelle la fonction λ est nulle.

La formule (44.8) prouve que la courbure du choc est une fonction linéaire de la courbure et de l'accélération de l'obstacle. Dans le cas d'un obstacle terminé par un dièdre, le rapport $R(t)\,\xi''(t)/c^2$ est égal au quotient des fonctions λ et μ; lorsque l'angle ϑ_s $(\vartheta_s = \pi - \omega)$ est nul, on retrouve la formule (25.2).

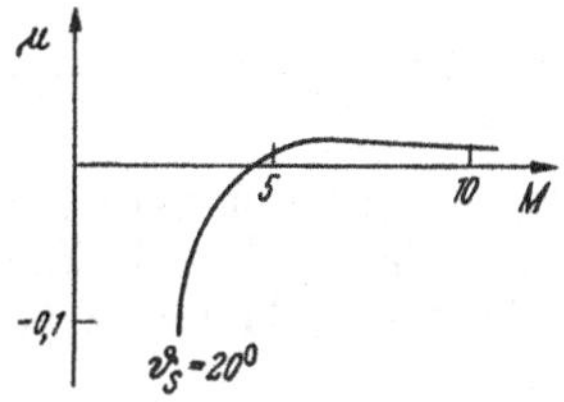

Fig. 57.

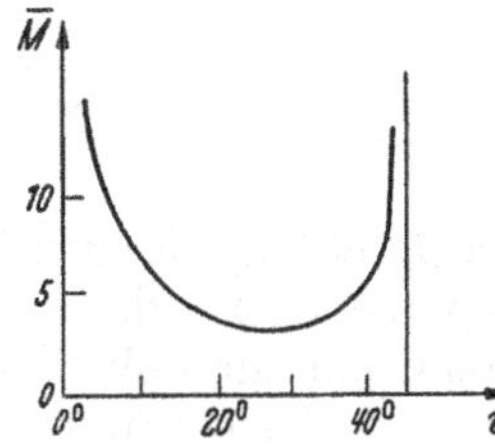

Fig. 58.

45. Courbure des ondes de choc attachées: écoulements de révolution. Nous reprenons le problème précédent en nous plaçant dans le cas des écoulements de révolution. Nous raisonnons dans un système d'axes liés à l'obstacle ce système est animé d'une translation rectiligne, parallèle à l'axe de révolution. En chaque point, la force d'inertie d'entraînement s'exerçant sur la masse unité est $-\xi''(t)$. Nous utiliserons les coordonnées sphériques r et ϑ, l'origine étant le sommet commun S de l'onde et de l'obstacle; u et v étant les composantes de la vitesse sur les directions associées à ces coordonnées, nous posons

$$\left.\begin{aligned} u(r,\vartheta,t) &= \bar{c}\,\{u_0(\vartheta,t) + r\,u_1(\vartheta,t) + \cdots\},\\ v(r,\vartheta,t) &= \bar{c}\,\{v_0(\vartheta,t) + r\,v_1(\vartheta,t) + \cdots\},\\ \varrho(r,\vartheta,t) &= \bar{\varrho}\,\{\varrho_0(\vartheta,t) + r\,\varrho_1(\vartheta,t) + \cdots\},\\ p(r,\vartheta,t) &= \bar{p}\,\{p_0(\vartheta,t) + r\,p_1(\vartheta,t) + \cdots\}. \end{aligned}\right\} \tag{45.1}$$

Les fonctions d'indice zéro vérifient toujours le système (20.1), le temps jouant le rôle de paramètre; les fonctions d'indice 1 vérifient le système suivant:

$$\left.\begin{aligned} &\frac{\partial u_1}{\partial t} v_0 + u_1 u_0 - v_1 v_0 + \frac{1}{\gamma}\frac{p_1}{p_0} = -\frac{1}{\bar{c}}\frac{\partial u_0}{\partial t} - \cos\vartheta\,\frac{\xi''(t)}{\bar{c}^2},\\ &\frac{\partial v_1}{\partial t} v_0 + u_1 v_0 + v_1\left(\frac{\partial v_0}{\partial\vartheta} + 2u_0\right) + \frac{1}{\gamma}\left(\frac{1}{\varrho_0}\frac{\partial p_1}{\partial\vartheta} - \frac{\varrho_1}{\varrho_0^2}\frac{\partial p_1}{\partial\vartheta}\right)\\ &\qquad\qquad = -\frac{1}{\bar{c}}\frac{\partial v_0}{\partial t} + \sin\vartheta\,\frac{\xi''(t)}{\bar{c}^2},\\ &\frac{\partial}{\partial\vartheta}\{(\varrho_1 v_0 + \varrho_0 v_1)\sin\vartheta\} + 3(\varrho_1 u_0 + \varrho_0 u_1)\sin\vartheta = -\frac{\sin\vartheta}{\bar{c}}\frac{\partial\varrho_0}{\partial t},\\ &u_0\left(\frac{p_1}{p_0} - \gamma\frac{\varrho_1}{\varrho_0}\right) + v_0\frac{\partial}{\partial\vartheta}\left(\frac{p_1}{p_0} - \gamma\frac{\varrho_1}{\varrho_0}\right) = -\frac{1}{\bar{c}}\frac{\partial\log p_0}{\partial t} + \frac{\gamma}{\bar{c}}\frac{\partial\log\varrho_0}{\partial t}. \end{aligned}\right\} \tag{45.2}$$

Les fonctions d'indice zéro ne dépendent du temps que par l'intermédiaire du nombre de Mach $M(M = -\xi'(t)/\bar{c})$. Les seconds membres des équations (45.2) sont donc proportionnels à l'accélération $\xi''(t)$. D'après les équations du choc, les fonctions d'indice 1 peuvent alors être écrites sous la forme suivante:

$$u_1(\vartheta,t) = \frac{u_1^*(\vartheta,t)}{R(t)} + \xi''(t)\,\frac{u_1^{**}(\vartheta,t)}{\bar{c}^2}. \tag{45.3}$$

Une représentation analogue est valable pour les trois autres fonctions d'indice 1. Les fonctions u_1^*, v_1^*, ϱ_1^*, p_1^*, appelées fonctions stationnaires d'indice 1, vérifient le système (45.2), dans lequel les seconds membres sont nuls; ce sont les fonctions définies dans la Sect. 22. Les fonctions u_1^{**}, v_1^{**}, ϱ_1^{**} et p_1^{**} se calculent ensuite par intégration numérique à partir des conditions initiales déterminées par les équations du choc [*55*]. Dans un calcul approché, on peut se contenter de calculer la valeur sur l'onde des dérivées par rapport à ϑ des fonctions u_1^{**}, v_1^{**}, ϱ_1^{**} et p_1^{**}; on suppose ensuite ces fonctions linéaires.

$$\frac{\partial v_1^{**}}{\partial \vartheta} = \frac{5}{6} \frac{(3\mathcal{M}_0^2+1)}{\mathcal{M}(\mathcal{M}_0^2-1)} \sin\vartheta_w + \frac{5(21\mathcal{M}_0^4+2\mathcal{M}_0^2+1)}{36\mathcal{M}_0^3(\mathcal{M}_0^2-1)} \cos\vartheta_w\, \xi' \frac{\partial \vartheta_w}{\partial \xi'}, \tag{45.4}$$

$$v_1^{**}(\vartheta, t) = v_1^{**}(\vartheta_w, t) + (\vartheta - \vartheta_w) \frac{\partial v_1^{**}}{\partial \vartheta}(\vartheta_w, t) \tag{45.5}$$

où ϑ_w et $\mathcal{M}_0$ désignent l'angle que fait avec l'axe de révolution la tangente au sommet de l'onde de choc et le nombre de Mach normal ($\mathcal{M}_0 = M \sin\vartheta_w$). La condition sur l'obstacle s'écrit sous la forme (45.6); on en déduit la valeur du rapport des courbures $R/\mathcal{R}$.

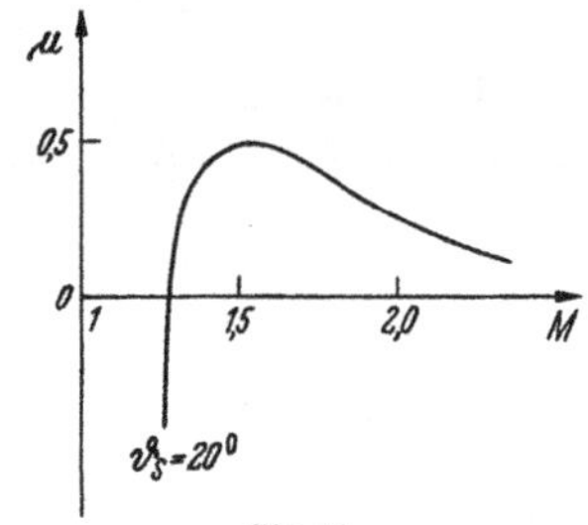

Fig. 59.

$$\frac{1}{2\mathcal{R}} \frac{\partial}{\partial \vartheta} v_0(\vartheta_s, t) + v_1(\vartheta_s, t) = \frac{u_0(\vartheta_s, t)}{2\mathcal{R}}, \tag{45.6}$$

$$\frac{R(t)}{\mathcal{R}} = \frac{\lambda(\vartheta_s, M)}{1 + \frac{\mathcal{R}\xi''}{\bar{c}^2}\mu(\vartheta_s, M)}, \tag{45.7}$$

$$\left.\begin{aligned} \lambda(\vartheta_s, M) &= \frac{2}{3} \frac{v_1^*(\vartheta_s, t)}{u_0(\vartheta_s, t)}, \\ \mu(\vartheta_s, M) &= -\frac{2}{3} \frac{v_1^{**}(\vartheta_s, t)}{u_0(\vartheta_s, t)}. \end{aligned}\right\} \tag{45.8}$$

Les fonctions d'indice zéro ont été calculées [*26*], ainsi que les fonctions v_1^* [*28*], [*29*]. En utilisant l'expression approchée (45.5), on obtient pour la fonction $\mu(\vartheta_s, M)$ des valeurs représentées sur la Fig. 59. Les valeurs de la fonction $\lambda(\vartheta_s, M)$ sont représentées sur la Fig. 33.

Le même problème a été étudié expérimentalement par MAIR [*56*]; mais les mesures ne sont pas assez précises pour permettre une vérification quantitative des résultats que nous venons d'indiquer.

46. Conclusion. Nous venons d'exposer les principaux résultats relatifs aux ondes de choc en aérodynamique. Toutes les méthodes utilisées consistent à chercher les obstacles qui correspondent à certaines ondes de choc, et à choisir une onde pour laquelle l'obstacle obtenu se rapproche le plus possible d'un obstacle déterminé. Comme nous l'avons dit, les progrès que l'on pourra obtenir en travaillant dans cette direction porteront surtout sur des questions de détail. Il serait souhaitable d'orienter les recherches relatives aux ondes de choc dans une direction entièrement différente. Dans cet ordre d'idées, je terminerai en montrant comment, dans le cas d'une équation très simple, on peut déterminer directement les ondes de choc éventuelles; on pourra se reporter également à la référence [*57*].

Les problèmes posés par l'aérodynamique sont des problèmes aux limites pour un système d'équations aux dérivées partielles. Nous considérons le cas d'une seule équation; une certaine fonction inconnue $u(t, x)$ dépendant de deux variables doit satisfaire l'équation (46.1) et la condition aux limites (46.2) dans laquelle la fonction $g(x)$ est supposée uniforme et douée d'une derivée seconde continue.

$$\frac{\partial u}{\partial t} + 2u \frac{\partial u}{\partial x} = 0, \tag{46.1}$$

$$u(0, x) = g(x). \tag{46.2}$$

La recherche de la fonction $u(t, x)$ est un problème de CAUCHY, dont la solution est donnée par la formule suivante:

$$u = g(x - 2ut). \tag{46.3}$$

Cette solution peut être représentée dans le plan x, t par la famille de droites sur lesquelles u demeure constant; cette famille admet une enveloppe (E). Dans les problèmes de mécanique, on s'intéresse seulement aux valeurs positives de la variable t. Lorsque la fonction $g(x)$ est croissante, l'enveloppe (E) ne possède aucun point dans le demi-plan $t > 0$ et la fonction $u(t, x)$, définie par la formule (46.3), est uniforme dans ce demi-plan.

Nous supposons que la fonction $g(x)$ est décroissante; la fonction $u(t, x)$ n'est plus uniforme au voisinage de (E). Un choix doit être fait entre les déterminations, si bien que la solution $u(t, x)$ est discontinue sur une certaine ligne (S) d'équation $x = \xi(t)$, jouant le rôle d'une onde de choc. D'après la formule (2.4) dans laquelle nous remplaçons ψ par u, l'équation (46.1) exprime qu'une certaine intégrale $\int u(t, x)\, dx$ demeure constante au cours du temps. Donc, si la fonction $u(t, x)$ est discontinue sur la ligne (S), les valeurs u_1 et u_2 immédiatement avant et immédiatement après (S) doivent vérifier, en vertu de la relation (3.3), l'équation suivante:

$$u_1 + u_2 = \xi'(t). \tag{46.4}$$

Le problème consiste donc à associer à la solution (46.3) une ligne (S) le long de laquelle la condition (46.4) soit vérifiée. Dans ce but, nous associons à l'équation (46.1) la fonction $F(t, x, u)$, définie de la façon suivante:

$$F(t, x, u) = -ux + u^2 t + f(u) \tag{46.5}$$

dans laquelle la fonction $f(u)$ est arbitraire mais possède un nombre suffisant de dérivées.

En changeant, si cela est nécessaire, F en $-F$, on peut interpréter les solutions de l'équation (46.1) comme les valeurs de u qui rendent la quantité F minimum. Lorsque F possède plusieurs minimum, la fonction $u(t, x)$ ainsi définie est multiforme; si nous convenons de choisir la valeur de u qui rend F minimum strict, on obtient le résultat remarquable suivant: la fonction $u(t, x)$ vérifie non seulement l'équation (46.1) quand elle est dérivable, mais encore la condition (46.4) quand elle est discontinue. Les valeurs t et x pour lesquelles F possède deux minimum égaux vérifient en effet les conditions suivantes:

$$\left.\begin{aligned} x - 2u_1 t = f'(u_1), \qquad x - 2u_2 t = f'(u_2), \\ -u_1 x + u_1^2 t + f(u_1) = -u_2 x + u_2^2 t + f(u_2). \end{aligned}\right\} \tag{46.6}$$

Ces équations déterminent dans le plan x, t une courbe le long de laquelle la condition (46.4) est vérifiée; cette courbe est donc l'onde de choc (S). Elle coïncide avec le lieu des points doubles des intégrales de l'équation différentielle $dx = u(t, x)\, dt$.

Ces considérations sont élémentaires; il semble cependant qu'une fois généralisées elles puissent servir de point de départ à de nouvelles recherches relatives à la détermination des ondes de choc en aérodynamique.

Annexe. Propagation des chocs dans les gaz ionisés.

Des travaux récents ont été consacrés à la dynamique des fluides compressibles doués de conductivité électrique; c'est un cas qui se présente dans les phénomènes d'ionisation. Dans les lignes qui suivent, les équations générales du mouvement sont établies, et nous en déduisons les résultats relatifs à la propagation des ondes sonores. Dans les cas extrêmes d'une conductivité électrique nulle (aérodynamique classique) ou d'une conductivité électrique infinie (conducteurs parfaits), des ondes de choc peuvent se propager au sein du fluide.

La viscosité et la conductivité thermique sont négligées tandis que la conductivité électrique est prise en considération. Le fluide (compressible) est plongé dans un champ électrique $\boldsymbol{E}$ et un champ magnétique $\boldsymbol{H}$. La perméabilité magnétique μ, la conductivité électrique σ et les chaleurs spécifiques sont supposées constantes; le gaz obéit à la loi des gaz parfaits $p = R\varrho T$. Toutes les molécules gazeuses étant supposées chargées électriquement, le fluide est parcouru par un courant électrique de densité $\boldsymbol{J} = \sigma(\boldsymbol{E} + \boldsymbol{V}\times\mu\boldsymbol{H})$. Les formules (1) et (2) traduisent respectivement les lois de FARADAY et D'AMPÈRE; les intégrales simples sont étendues à une courbe fermée d'élément linéaire $\boldsymbol{dl}$ et les intégrales doubles à une portion de surface de normale unitaire $\boldsymbol{n}$ et s'appuyant sur la courbe précédente.

$$\int \boldsymbol{E}\cdot \boldsymbol{dl} = -\iint \boldsymbol{n}\cdot\frac{\partial\mu\boldsymbol{H}}{\partial t}\,ds, \tag{1}$$

$$\int \boldsymbol{H}\cdot \boldsymbol{dl} = 4\pi\iint \boldsymbol{n}\cdot\boldsymbol{J}\,ds. \tag{2}$$

Les formules (3) à (5) traduisent respectivement l'équation de la dynamique, les principes de conservation de la masse et de l'énergie. Les intégrales triples sont étendues à un volume variable constamment formé des mêmes molécules, les intégrales doubles sont étendues à la surface de ce volume; cette surface, de normale extérieure unitaire $\boldsymbol{n}$, est supposée également toujours formée des mêmes molécules.

$$\frac{d}{dt}\iiint \varrho\boldsymbol{V}\,d\tau + \iint \boldsymbol{n}p\,ds = \iiint (\boldsymbol{J}\times\mu\boldsymbol{H})\,d\tau, \tag{3}$$

$$\frac{d}{dt}\iiint \varrho\,d\tau = 0, \tag{4}$$

$$\frac{d}{dt}\iiint \Lambda\,d\tau + \iint \boldsymbol{n}p\cdot\boldsymbol{V}\,ds = \iiint (\boldsymbol{J}\times\mu\boldsymbol{H})\cdot\boldsymbol{V}\,d\tau + \iiint \sigma^{-1}\boldsymbol{J}^2\,d\tau \tag{5}$$

avec

$$\Lambda = \varrho\left(c_v T + \frac{V^2}{2}\right).$$

Lorsque toutes les fonctions inconnues admettent des dérivées partielles premières continues, on déduit des relations précédentes les équations aux dérivées partielles du mouvement, qui, lorsque σ est infini, peuvent être écrites sous la forme suivante:

$$\left.\begin{aligned}
&\frac{\partial\boldsymbol{H}}{\partial t} = \operatorname{rot}(\boldsymbol{V}\times\boldsymbol{H}),\\
&\varrho\left(\frac{\partial\boldsymbol{V}}{\partial t} + \operatorname{rot}\boldsymbol{V}\times\boldsymbol{V} + \operatorname{grad}\frac{V^2}{2}\right) + \operatorname{grad}p = \frac{\operatorname{rot}\boldsymbol{H}\times\mu\boldsymbol{H}}{4\pi},\\
&\frac{\partial\varrho}{\partial t} + \operatorname{div}\varrho\boldsymbol{V} = 0,\\
&\frac{\partial}{\partial t}(p\varrho^{-\gamma}) + \boldsymbol{V}\cdot\operatorname{grad}(p\varrho^{-\gamma}) = 0.
\end{aligned}\right\} \tag{6}$$

Le système (6) est un système de huit équations du premier ordre de type hyperbolique normal. De la détermination des variétés caractéristiques on déduit les vitesses de déplacement U des ondes sonores [63]. A chaque instant, en chaque point, à chaque direction (de vecteur unitaire $\boldsymbol{\delta}$) correspondent huit vitesses de propagation $U-V_n (V_n=\boldsymbol{\delta}\cdot\boldsymbol{V})$; ces vitesses sont deux à deux opposées et deux d'entre elles sont nulles. On obtient donc pour $(U-V_n)^2$ les trois valeurs données par les formules (7) dans lesquelles ϑ désigne l'angle des directions $\boldsymbol{H}$ et $\boldsymbol{\delta}$.

$$\left.\begin{aligned}(U_1-V_n)^2&=\frac{\gamma p}{2\varrho}+\frac{\mu H^2}{8\pi\varrho}-\left\{\left(\frac{\gamma p}{2\varrho}+\frac{\mu H^2}{8\pi\varrho}\right)^2-\frac{\gamma p}{\varrho}\frac{\mu H^2}{4\pi\varrho}\cos^2\vartheta\right\}^{\frac{1}{2}},\\(U_2-V_n)^2&=\frac{\mu H^2}{4\pi\varrho}\cos^2\vartheta,\\(U_3-V_n)^2&=\frac{\gamma p}{2\varrho}+\frac{\mu H^2}{8\pi\varrho}+\left\{\left(\frac{\gamma p}{2\varrho}+\frac{\mu H^2}{8\pi\varrho}\right)^2-\frac{\gamma p}{\varrho}\frac{\mu H^2}{4\pi\varrho}\cos^2\vartheta\right\}^{\frac{1}{2}}.\end{aligned}\right\}\quad(7)$$

Lorsque les fonctions inconnues sont discontinues sur une surface (onde de choc), les discontinuitées doivent vérifier un certain nombre de conditions. Nous désignons par $\boldsymbol{n}$ le vecteur unitaire normal à l'onde de choc, par U la vitesse de déplacement du choc, et nous posons $H_n=\boldsymbol{n}\cdot\boldsymbol{H}$ et $V_n=\boldsymbol{n}\cdot\boldsymbol{V}$. On déduit des équations (1), (3) et (4) que les quantités suivantes sont continues à la traversée de l'onde de choc:

$$\left.\begin{aligned}&\boldsymbol{n}\times\boldsymbol{E}-U\mu\boldsymbol{H},\\&(V_n-U)\,\varrho\boldsymbol{V}+\boldsymbol{\pi},\\&(V_n-U)\,\varrho\end{aligned}\right\}\quad(8)$$

avec

$$\boldsymbol{\pi}=\boldsymbol{n}p+\boldsymbol{n}\frac{\mu H^2}{8\pi}-\frac{\mu H_n\boldsymbol{H}}{4\pi}.$$

Lorsque σ est infini, on peut éliminer le champ électrique à l'aide de l'équation (2), qui s'écrit alors $\boldsymbol{E}=-\boldsymbol{V}\times\mu\boldsymbol{H}$; on déduit ensuite de l'équation (5) que la quantité suivante est également continue à la traversée de l'onde de choc:

$$(V_n-U)\left(\varrho\frac{V^2}{2}+\frac{p}{\gamma-1}+\frac{\mu H^2}{8\pi}\right)+\boldsymbol{\pi}\cdot\boldsymbol{V}.\quad(9)$$

Les résultats précédents ont été appliqués au problème de l'onde de choc stationnaire attachée à la pointe d'un dièdre [64].

Bibliographie.

Articles de caractère général.

[1] RIEMANN, B.: Über die Fortpflanzung ebener Luftwellen von endlicher Schwingungsweite. Abh. Ges. Wiss. Göttingen, Math.-phys. Kl. **8**, 43 (1860).

[2] RANKINE, W. J. M.: On the thermodynamic theory of waves of finite longitudinal disturbance. Trans. Roy. Soc. Lond. **160**, 277—288 (1870).

[3] HUGONIOT, H.: Sur la propagation du mouvement dans les corps et spécialement dans les gaz parfaits. J. Ecole polytech. **58**, 1—125 (1889).

[4] HADAMARD, J.: Leçons sur la propagation des ondes et les équations de l'hydrodynamique. Paris: Hermann 1903; réimprimé par Chelsea Publishing Company, New-York 1949.

[5] COURANT, R., and K. O. FRIEDRICHS: Supersonic Flow and Shock Waves. New-York: Interscience Publ. 1948. — Handbook of Supersonic Aerodynamics. Navord Report 1488. Bureau of Ordnance, 5 volumes, 1950—1955. — Graphs for use in calculations of compressible airflow. Oxford: Clarendon Press 1954.

Ecoulements stationnaires.

[6] TOMOTIKA, S., and K. TAMADA: Studies on two-dimensional transonic flows of compressible fluide. Part III. Quart. Appl. Math. **9**, No. 2, 129—147 (1951).
[7] CABANNES, H.: Contribution à l'étude théorique des fluides compressibles. Ann. Ecole Normale Supérieure **63**, fasc. 1 (1952).
[8] TOLLMIEN, W.: Zum Übergang von Unterschall- in Überschallströmungen. Z. angew. Math. Mech. **17**, 117—136 (1937).

Ondes de choc détachées.

[9] CABANNES, H.: Détermination théorique de l'écoulement d'un fluide derrière une onde de choc détachée O.N.E.R.A. Note technique No. 5, 1951. — Rech. Aéronaut. **43**, 1—5 (1956).
[10] MITCHELL, A. R.: Application of relaxation to the rotational field of flow behind a bow shock wave. Quart. J. Mech. Appl. Math. **4**, 371—383 (1951).
[11] MITCHELL, A. R., and F. McCALL: The rotational field behind a bow shock wave in axially symmetric flow using relaxation methods. Proc. Roy. Soc. Edinburgh **53**, 371—381 (1952).
[12] DUGUNDJI: An investigation of detached shock in front of a body of revolution. J. Aeronaut. Sci. **15**, 699—705 (1948).
[13] MOECKEL, W. E.: Approximate method for predicting form and location of detached shock waves ahead of plane or axially symmetric bodies. N.A.C.A. Technical Note No. 1921, 1949.
[14] SHU, S. S.: On two-dimensional flow after stationary shock (with special reference to the problem of detached shock waves) N.A.C.A. Technical Note No. 2364, 1951.
[15] HIDA, K.: An approximate study on the detached shock wave in front of a circular cylinder and a sphere. J. Phys. Soc. Japan **8**, 740—745 (1953).
[16] HEBERLE, J., G. WOOD and P. GOODERUM: Data on shape and location of detached shock waves on cones and spheres. N.A.C.A. Technical Note No. 2000, 1950.
[17] JOHNSTON, G. W.: An investigation of the flow about cones and wedges at and beyond the critical angle. J. Aeronaut. Sci. **20**, 378—382 (1953).
[18] WHITHAM, G. B.: The flow pattern of a supersonic projectile. Comm. Pure Appl. Math. **5**, 301—348 (1952).
[19] OSWATITSCH, K.: Der Verdichtungsstoß bei der stationären Umströmung flacher Profile. Z. angew. Math. Mech. **29**, 129—141 (1949).

Ondes de choc attachées: écoulements plans.

[20] GUDERLEY, K.: Considerations of the structure of mixed subsonic-supersonic patterns. Technical report Project H.A. 21, 9, 1947.
[21] CABANNES, H.: Sur l'onde de choc attachée lorsque la vitesse aval à la pointe de l'obstacle est subsonique. C. R. Acad. Sci., Paris **239**, 1830—1832 (1950) et Publications Scientifiques et Techniques du Ministère de l'Air **250**, 181—195 (1952).
[22] CROCCO, L.: Singolarità della corrente gassosa iperacustica nell'interno di una prora a diedro. Atti del I. Congresso dell'Unione Matematica Italiana, pp. 597—615, 1937.
[23] THOMAS, T. Y.: Calculation of the curvature of attached shock waves. J. Math. Phys. **27**, 279—297 (1949).

Ondes de choc attachées: écoulement à trois dimensions.

[24] BOURQUARD, F.: Mém. Artillerie Française **2**, 135 (1932).
[25] TAYLOR, G. I., and J. W. MACCOLL: The air pressure on a cone moving at high speeds. Proc. Roy. Soc. Lond., Ser. A **139**, 278—311 (1933).
[26] ZDENEK, KOPAL: Tables of supersonic flow around cones. Massachusetts Institute of Technology. Technical Report No. 1, 1947.
[27] CABANNES, H.: Etude de l'onde de choc attachée dans les écoulements de révolution. Rech. Aéronaut. **27**, 7—16 (1952).
[28] SHEN, S. F., and C. C. LIN: On the attached curved shock in front of a sharpnosed axially symmetrical body placed in a uniform stream. N.A.C.A. Technical Note No.2505, 1951.
[29] CABANNES, H.: Etude de l'onde de choc attachée dans les écoulements de révolution. Rech. Aéronaut. **24**, 17—23 (1951).
[30] STONE, A. H.: On supersonic flow past a slightly yaving cone. J. Math. Phys. **27**, 64—81 (1948); **30**, 200—213 (1952); **31**, 300 (1953).

[31] KOPAL, ZDENEK: Tables of supersonic flow around yaving cones. Massachusetts Institute of Technology. Technical Report No. 3, 1947. — Tables of supersonic flow around cones of large yaw. Massachusetts Institute of Technology. Technical Report No. 5, 1949.

[32] HOLT, M., and J. BLACKIE: Experiments on circular cones at yaw in supersonic flow. J. Aeronaut. Sci. **23**, 931—936 (1956).

[33] FERRI, A.: Supersonic flow around circular cones at angles of attack. N.A.C.A. Technical Note No. 2236, 1950.

[34] ROBERTS, R., and J. RILEY: A guide to the use of the M.I.T. Cone Tables. J. Aeronaut. Sci. **21**, 336—342 (1954).

[35] HOLT, M.: A vortical singularity in conical flow. Quart. J. Mech. Appl. Math. **7**, 438—445 (1954).

[36] FERRI, A., N. NESS and T. KAPLITA: Supersonic flow over conical bodies without axial symmetry. J. Aeronaut. Sci. **20**, 563—571 (1953).

Ecoulement non stationnaires.

[37] ESCLANGON, E.: Ondes et détonations balistiques engendrées par les avions et les projectiles. Mém. Artillerie Française **27**, 4ème fascicule (1953).

[38] LILLEY, G. M., R. WESTLEY, A. H. YOTES and J. R. BUSING: The supersonic bang. J. Aeronaut. Soc. 1953.

Ondes de choc planes.

[39] FREEMAN, N. C.: A theory of the stability of plane shock waves. Proc. Roy. Soc. Lond., Ser. A **228**, 341—362 (1955).

[40] STANJUKOVICH: Dokl. Akad. Nauk. SSSR. **96**, 441—444 (1954).

[41] STOCKER, P. M.: On the problem of interaction of plane waves of finite amplitude involving retardation of shock-formation by an expansion wave. Quart. J. Mech. Appl. Math. **4**, 170—181 (1951).

[42] JONES, C. W.: On the propagation of shock waves in regions of non-uniform density. Proc. Roy. Soc. Lond., Ser. A **226**, 82—99 (1955).

[43] MEYER, R. E.: On waves of finite amplitude in ducts. Quart. Mech. Appl. Math. **5**, 257—299 (1952).

Rencontre d'une onde plane et d'un dièdre.

[44] POLACHEK, H., and R. J. SEEGER: On shock wave phenomena. Proc. Symp. Appl. Math. **1**, 119—144 (1948).

[45] BORG, S. F.: On unsteady nonlinearized conical flow. J. Aeronaut. Sci. **19**, 85—92 (1950).

[46] JONES, MARTIN and THORNHILL: A note on the pseudo-stationary flow behind a strong shock diffracted or reflected at a corner. Proc. Roy. Soc. Lond., Ser. A **209**, 238—248 (1951).

[47] LUDLOFF, H. E., and M. B. FRIEDMANN: Mach reflection of shocks at arbitrary incidence. J. Appl. Phys. **14**, 1247—1248 (1950).

[48] LIGHTHILL, M. J.: The diffraction of blast. Proc. Roy. Soc. Lond., Ser. A **198**, 454—470 (1949); **200**, 554—565 (1950).

[49] KOFINK, W., u. VOLLMER: Der gegabelte Verdichtungsstoß in Luft. Z. angew. Math. Mech. **33**, 73—88 (1953).

Ondes de choc sphériques.

[50] TAYLOR, G. I.: The air wave surrounding an expanding sphere. Proc. Roy. Soc. Lond., Ser. A **186**, 273—293 (1946). — The formation of a blast by a very intense explosion. Proc. Roy. Soc. Lond., Ser. A **201**, 159—189 (1950).

[51] LATTER, R.: Similarity solution for a spherical shock wave. J. Appl. Phys. **26**, 954—960 (1955).

[52] WHITHAM, G. B.: The propagation of spherical blast. Proc. Roy. Soc. Lond., Ser. A **203**, 571—581 (1950).

Translations rectilignes de vitesse variable.

[53] CABANNES, H.: Etude du départ d'un obstacle dans un fluide au repos. Rech. Aéronaut. **36**, 7—12 (1953).

[54] SAKURAI, A.: The flow due to impulsive motion of a wedge and its similarity to the diffraction of shock waves. J. Phys. Soc. Japan **10**, 221—228 (1955).

[55] CABANNES, H.: Influende des accélérations sur la courbure des chocs. Rech. Aéronaut. **39**, 3—13 (1954).

[56] MAIR, A. W.: Experiments on separation of boundary layers on probes in front of blunt-nosed bodies in a supersonic air stream. Phil. Mag., Sér. VII **43**, 695—716 (1952).

[57] GERMAIN, P., et R. BADER: Unicité des écoulements avec choc dans la mécanique de Burgers. Jb. wiss. Ges. Luftfahrt 144—148 (1953).

Propagation des chocs dans les gaz ionisés.

[58] ALFVÉN, H.: Cosmological Electrodynamics. Oxford: Clarendon Press 1950.

[59] COWLING, T. G.: Magnetohydrodynamics. New York: Interscience Publ. 1957.

[60] HOFFMANN, F., and E. TELLER: Magnetohydrodynamic shocks. Phys. Rev. **80**, 692—703 (1950).

[61] LUNDQUIST, S.: Studies in magnetohydrodynamics. Ark. Fysik **5**, 297 et suivantes (1952).

[62] LUST, R.: Magnetohydrodynamische Stoßwellen in einem Plasma unendlicher Leitfähigkeit. Z. Naturforsch. **8**, 277—284 (1953).

[63] FRIEDRICHS, K. O.: Nonlinear wave motion in magneto-hydrodynamics. Los Alamos Scientific Laboratory report (1954).

[64] CABANNES, H.: Sur les mouvements d'un fluide compressible doué de conductivité électrique. C. R. Acad. Sci., Paris **245**, 1379—1382 (1957).

Theory of Characteristics of Inviscid Gas Dynamics.

By

R. E. MEYER.

With 44 Figures.

A. Introduction.

1. Scope of the theory. In the largest and most important class of fluid motions, viscosity plays only a restricted role, and a very useful approximation to the real behavior of the fluids is obtained from a study of their motion in the limit of zero viscosity. Such motions are governed by the "inviscid equations", that is, the Navier-Stokes equations[1] truncated by omission of the terms proportional to the coefficients of viscosity and heat conductivity and their derivatives.

It is well known, however, that by no means all solutions of these truncated equations represent motions which are the limits of real fluid motions. To obtain predictions of any relevance from the inviscid equations, it is necessary to supplement them by rules distinguishing the solutions which do represent the limit of real fluid motions. Such rules are well known for some classes of motions, but their discussion is outside the scope of the theory of characteristics.

In some cases, the limiting motions are represented by solutions of the inviscid equations for boundary conditions related in a straightforward manner to the geometry of the immersed bodies. In other cases, however, the relation may not be so very direct, nor even properly understood yet, as for motions in which boundary layer separation occurs at points not clearly distinguished in a geometrical sense. It is convenient, nevertheless, to discuss the boundary conditions in terms of body geometry or of other concepts naturally associated with the inviscid equations, even though these be fictitious in the context of real fluid behaviour, and it is in this sense that the present article is to be interpreted. This manner of discussion remains useful also when fluid motions with large but finite Reynolds number are studied; the inviscid equations then approximate the fluid mechanism closely in the bulk of the motion, and the analysis of that part only is considered here.

In the motion of compressible gases, viscosity plays a crucial role not only in the boundary layers and their wakes, but also in narrow layers in the gas which are called shock waves, and a particular concern of characteristics theory is to elucidate the non-linear inviscid mechanism which generates the shocks.

The inviscid gas motions may be divided into three classes, according to whether the fluid mechanism is of the equilibrium or radiation type, or mixed (in the sense that each type occurs in part of the field). The theory of characteristics is essentially concerned with the radiation type, i.e. that represented by hyperbolic differential equations, and thus with the unsteady motions and the steady supersonic flows of compressible fluid. From the discussion of the steady flows are

[1] Cf. SERRIN's article in Vol. VIII, Part 1 of this Encyclopedia, Sects. 58 to 64.

excluded those in which the velocity, pressure etc. depend only on one space coordinate and which thus satisfy ordinary differential equations.

To the flows of mixed type, i.e. the steady flows containing both subsonic and supersonic regions, the theory applies only in a restricted sense. The local mechanism in the supersonic regions is of course the same as that in entirely supersonic flows, but when such regions are embedded in subsonic regions, the structure of the surrounding subsonic flow so dominates that of the supersonic flow that only pieces of information of restricted usefulness can be obtained from the supersonic equations by themselves. On the other hand, when the supersonic regions are not altogether embedded, for instance, in nozzles or many flows with supersonic regions of infinite extent, then the major part of the supersonic flow can be shown to be identical with part of a flow that is entirely supersonic (even if not physically realizable in its entirety). Such regions can be discussed usefully in terms of boundary conditions natural for the hyperbolic equations, even though the real boundary conditions may be prescribed in a subsonic region, and it is in this sense that characteristics theory contributes to the understanding of the flow.

The discussion of mixed flows in the first article of this volume is based on the Hodograph Method, which provides an approach to the solution of the exact inviscid equations applicable to both subsonic and supersonic flows and thus seems indispensable for the study of mixed flows. On the other hand, it is at present restricted to two-dimensional homentropic flow, and great difficulties have been encountered in its adaptation to physical boundary conditions, especially those encountered in flows with large supersonic regions[1]. The usefulness of the theory of characteristics as a complementary tool, restricted as it is to purely supersonic flow regions, arises from its success in providing methods which are well suited to the physical boundary conditions and may be generalized to flows that are not two-dimensional and homentropic.

2. Scope of the article. Within the theory of inviscid gas motion with a mechanism of the radiation type, a strong distinction has grown up between the approach aiming at the solution of the exact equations and that approximating them by the classical wave equation, with a similar approximation for the boundary conditions. The concept of characteristics plays a role in both approaches, but in the approximate linearized theory the characteristics are simple and the stress has been on the adaptation of the classical representations of the solution to the boundary conditions arising from aeronautical problems, with the primary aim of obtaining expressions for the aerodynamic forces. This theory is the subject of Division II of SCHIFFER'S article, and of TIMMAN'S article, in this volume. The theory of the exact equations, on the other hand, is dominated entirely by the concept of characteristics, and the application of this concept to the study of the motion is the subject of the present article.

The body of theory developed to-date to fill the place just sketched is still too large for a single article. An important part of the theory is concerned with the construction, by direct numerical integration, of particular solutions for numerically defined physical boundary conditions. This Method of Characteristics has been developed to a high degree, but it has already been described in considerable detail in a number of books[2] and is therefore omitted here, apart from a brief indication of the principle of the approach (Sect. 17).

---

[1] A notable exception is the solution of the nozzle problem given by T.M. CHERRY, Phil. Trans. Roy. Soc. Lond. A **245**, 583 (1953).

[2] [*3*], [*4*], [*7*], [*8*], R. SAUER: Einführung in die theoretische Gasdynamik. Berlin 1951.

The present account, then, is concerned with the analytical theory of characteristics, and in particular, with the approach aiming at the construction of solutions for physical boundary conditions. For the sake of homogeneity, no attempt is made to survey the fruits of the indirect approach of exploring particular, exact solutions of the equations of motion without regard to boundary conditions.

Since a direct numerical method and an approximate linearized theory exist, a particular aim of the analytical theory is to provide a general understanding of the "non-linear" features of the exact solutions. In fact, some understanding of the structure of the equations is necessary before useful explicit representations of the solutions can be found, and the present article is concerned as much with the qualitative results of the theory as the quantitative ones. A similar aim has been pursued by COURANT and FRIEDRICHS [2], and since the lucidity of their account can handly be surpassed, no attempt is made here to duplicate it. Only a brief discussion of the uniqueness theorems and the simple wave solutions is therefore given below[1], and the reader is referred to [2] for full detail. The classical treatment of uniqueness concentrates on the Cauchy problem, but the physicist is more often led to different formulations, and the sketch (Sect. 5) of the classical picture is therefore complemented by a further discussion (Sect. 18). The primary subject of the present article, however, is the theory of structure and representation of the solutions, which has advanced notably since 1948.

The unsteady one-dimensional homentropic motion and the steady two-dimensional supersonic irrotational homentropic flow provide the best examples for the presentation of the theory, since it has been developed most completely for these two classes of motion. Their structure is almost identical, so that the discussion of only one class would be mathematically sufficient. On the other hand, both have independent physical importance, and the plan adopted here is to give an introduction to the theory in Division BI, couched in physical language as far as possible without too much detriment to precision, and to complete the account of the more abstract parts of the theory in Division CI. The concepts and results[2] described in either Division apply to both classes of motion, but are at most alluded to briefly in the other Division.

The general treatment of shock waves and their interaction with the rest of the gas motion is the subject of the preceding article, but the explanation of shock formation by the "non-linear" focusing mechanism of the gas motion is discussed here. The general theory of this effect (Sect. 23) is based on the modern theory of flow structure. An introductory example, however, is provided by the "piston problem" (Sects. 4, 7, 11), and for this problem the theory has advanced to the stage where entropy variation can be accounted for to a fair degree of approximation (Sects. 11, 12). The analogous problem of the interaction between a near-simple wave and a weak shock in steady supersonic flow is taken up more briefly in Sect. 27.

The uniform first-order approximations for essentially non-linear stability problems (cf. CABANNES's article in this volume, Sects. 32, 33) and for the analysis of flow fields inadequately described even to this order by linearized theory, are an important recent chapter of characteristics theory. A brief explanation of the reason for such short-comings of linearized theory is given in Sect. 15; methods for overcoming them have been developed in particular for axially symmetrical supersonic flow, but lack of time has prevented the inclusion of more than the roughest outline (Sect. 32) of a unified theory.

[1] Complementing the brief mathematical discussion in SCHIFFER's article, Sects. 5 to 7.

[2] With the obvious exception of the details of Sects. 3, 8, 9, 14, 24, 25.

The theory of characteristics for flow problems with more than two independent variables has not advanced sufficiently beyond the state reported in [9] to justify a detailed account here, but a few references are collected in the appendix.

B. One-dimensional unsteady motion.

I. Homentropic motion.

a) Characteristic equations and simple waves.

3. Characteristic equations. The one-dimensional motion of a compressible inviscid gas is governed[1] by the equation of continuity,

$$\frac{\partial \varrho}{\partial t} + \frac{\partial (\varrho u)}{\partial x} = 0, \tag{3.1}$$

and the equation of motion,

$$\frac{\partial u}{\partial t} + u \frac{\partial u}{\partial x} = -\frac{1}{\varrho} \frac{\partial p}{\partial x}, \tag{3.2}$$

where t denotes the time, x the position, ϱ the density, u the velocity and p the pressure. A homentropic motion is defined as one in which the entropy is constant and uniform, so that the pressure is a function of the density, which is known from the thermodynamical equation of state. Any discontinuities of u, ϱ and p as functions of x and t can be shown[2] from the conservation principles to be either contact surfaces, which are convected with the motion of the gas, or shock waves. It will be assumed in the following that the velocity, pressure, etc., are continuous, except where shock waves are explicitly referred to.

The system (3.1), (3.2) is hyperbolic and possesses two families of characteristics[3]. This is most easily seen as follows. The quantity

$$a = \sqrt{\frac{dp}{d\varrho}} \tag{3.3}$$

is called the (local) speed of sound[4], and if (3.1) be multiplied by (a/ϱ), (3.1) and (3.2) become

$$\frac{\partial \omega'}{\partial t} + u \frac{\partial \omega'}{\partial x} + a \frac{\partial u}{\partial x} = 0, \tag{3.4}$$

$$\frac{\partial u}{\partial t} + u \frac{\partial u}{\partial x} + a \frac{\partial \omega'}{\partial x} = 0, \tag{3.5}$$

respectively, where

$$\omega' = \int_0^{\varrho} \frac{a}{\varrho} \, d\varrho. \tag{3.6}$$

Addition and subtraction of (3.4) and (3.5) yields

$$\frac{\partial r}{\partial t} + (u + a) \frac{\partial r}{\partial x} = 0, \tag{3.7}$$

$$\frac{\partial s}{\partial t} + (u - a) \frac{\partial s}{\partial x} = 0, \tag{3.8}$$

---

[1] Cf. Sects. 9 and 10 of OSWATITSCH's article in Vol. VIII, Part 1 of this Encyclopedia.
[2] Cf. Sects. 3 and 4 of CABANNES' article in this volume.
[3] Cf. Sects. 5 and 6 of SCHIFFER's article in this volume.
[4] The positive root is always understood.

where

$$r = \tfrac{1}{2}(\omega' + u), \qquad s = \tfrac{1}{2}(\omega' - u) \tag{3.9}$$

are variables first introduced by RIEMANN [1]. Since each of Eqs. (3.9), (3.10) involves a partial derivative in only one direction in the x, t-plane, these equations are characteristic. They show, in fact, that (3.1) and (3.2), for homentropic motion, are equivalent to the statements:

$$\left.\begin{array}{r} r = \text{const on the lines of local slope } \dfrac{dx}{dt} = u + a \\ (\text{"advancing Mach}^1 \text{ lines"}), \end{array}\right\} \tag{3.10}$$

$$\left.\begin{array}{r} s = \text{const on the lines of local slope } \dfrac{dx}{dt} = u - a \\ (\text{"receding Mach lines"}). \end{array}\right\} \tag{3.11}$$

The Mach lines are therefore characteristic lines in the sense of the mathematical theory. They are the paths traced out in the x, t-plane by observers traveling, with respect to the moving fluid, with the local speed of sound in the directions of x increasing and decreasing, respectively.

For simplicity, the following discussion is restricted to the case of a perfect gas with constant specific heat, for which the equation of state is

$$p\,\varrho^{-\gamma} \exp(-S/c_v) = \text{const}, \tag{3.12}$$

where $\gamma = c_p/c_v$. Then from (3.3) and (3.6),

$$a^2 = \gamma \frac{p}{\varrho}, \tag{3.13}$$

$$\omega' = \frac{2a}{\gamma - 1}, \tag{3.14}$$

and by (3.9),

$$r = \frac{a}{\gamma - 1} + \frac{u}{2}, \qquad s = \frac{a}{\gamma - 1} - \frac{u}{2}, \tag{3.15}$$

or

$$u = r - s, \qquad 2a = (\gamma - 1)(r + s). \tag{3.16}$$

The qualitative results are valid, and the quantitative results can be modified on the lines of Sect. 24, for any gas for which $(\partial p/\partial \varrho)_S$ and $(\partial^2 p/\partial (1/\varrho)^2)_S$ are positive.

The equation of continuity for a two-dimensional gas motion with radial symmetry is

$$x \frac{\partial \varrho}{\partial t} + \frac{\partial}{\partial x}(\varrho\, u\, x) = 0,$$

and that for a spherically symmetrical motion is

$$x^2 \frac{\partial \varrho}{\partial t} + \frac{\partial}{\partial x}(\varrho\, u\, x^2) = 0,$$

if x denotes the distance from the centre of symmetry and u, the velocity, which is assumed to be purely radial. For unsteady flow of gas through ducts of slightly varying cross-section, a useful physical approximation is often obtained by assuming that the velocity, density and pressure are constant over any cross-section normal to the axis of the duct, and the equation of continuity is then

$$Q \frac{\partial \varrho}{\partial t} + \frac{\partial}{\partial x}(\varrho\, u\, Q) = 0,$$

[1] After ERNST MACH, 1838–1916.

where Q denotes the area of cross-section, x the distance along the axis and u the axial velocity. In all three cases, the dynamical equation of motion is (3.2), the equation of continuity may be written in the form

$$\frac{\partial \varrho}{\partial t} + \frac{\partial}{\partial x}(\varrho u) = -\varrho u h(x)$$

where $h = x^{-1}$ for the plane radial motion, $h = 2x^{-1}$ for the spherical motion and $h = d(\log Q)/dx$ for the flow in ducts, and the introduction of the Riemann variables (3.9) leads to

$$\left.\begin{aligned} \frac{\partial r}{\partial t} + (u+a)\frac{\partial r}{\partial x} &= -\frac{1}{2} a u h, \\ \frac{\partial s}{\partial t} + (u-a)\frac{\partial s}{\partial x} &= -\frac{1}{2} a u h, \end{aligned}\right\} \tag{3.17}$$

instead of (3.7), (3.8). Since these equations are inhomogeneous, they do not possess particular solutions of the simple wave type (see below).

The equations of the plane radial motion are seen to stand in the same relation to those of the one-dimensional motion as the equations of steady, axially symmetrical, supersonic flow (Sect. 28) stand to those of steady, two-dimensional supersonic flow (Sect. 14), and the properties of the radial motion can thus be inferred directly from the properties of the steady axially symmetrical flow (Sects. 28 to 32). The equations of the spherically symmetrical motion are similar to those of plane radial motion, but the structure of the solutions shows some significant differences[1]. In both the cases of cylindrically and spherically symmetrical motion, the physical interest centres on the propagation of shock waves, and the case of spherical waves is treated in Sects. 40 to 42 of CABANNES' article in this volume. The application of the wave front concept (Sect. 6) to the study of the stability of steady shock-free flow in ducts of slightly varying cross-section is discussed in Sects. 32 to 33 of the article by CABANNES.

4. Simple waves. A region in the x, t-plane in which either r or s is constant is called a simple wave, and any region adjacent to a region of uniform motion must be of that type[2]. For this reason, the simple wave solutions are of particular importance for the description of the initial phases of the motion in shock tubes (cf. Sect. 27 of CABANNES's articles in this volume).

By (3.10), (3.11) both r and s are constant along each of the Mach lines of *one* family in a simple wave, and it follows from (3.16), (3.13) and (3.12) that the velocity and all the thermodynamical variables are constant along each Mach line of that family, and by (3.10) or (3.11), these Mach lines are straight. The detailed properties of the simple wave solutions are therefore easily deduced from the initial and boundary conditions.

As an example, consider the gas motion due to a piston advancing with finite acceleration into a gas initially at rest and of uniform density. The discussion is made easier by reference to the x, t-plane (Fig. 1), in which the origin may be chosen at the initial position of the piston and the instant when it begins to advance. The successive positions of the piston are represented by a curve $x = X_P(t)$.

Since the gas is initially at rest, $r = s = \text{const} = a_0/(\gamma - 1)$, by (3.15), for $t = 0$ and all $x \geqq 0$, and by (3.10), (3.11), r and s must retain this value—and the gas must therefore remain at rest—at all points (x, t) below, and on, the advancing Mach line OA through the origin. The same does not hold for points above this Mach line, which therefore marks the successive positions of the front of the pressure wave generated by the motion of the piston. By (3.10) the front travels with the velocity $dx/dt = a_0$ into the gas at rest, and the equation of the Mach line OA is

$$x = X_f(t) = a_0 t. \tag{4.1}$$

[1] Y.W. CHEN: Comm. Pure Appl. Math. **6**, 179 (1953) and G.B. WHITHAM: Proc. Roy. Soc. Lond., Ser. A **203**, 571 (1950). Of course, some such differences of structure are evident already in the solutions of the linear wave equation; cf. H. LAMB: Hydrodynamics, 6th ed., § 302. Cambridge 1932.

[2] A proof is given in Sect. 16.

The receding Mach lines through all points between the Mach line OA and the piston curve (Fig. 1) still meet the axis $t=0$ at $x>0$, so by (3.11),

$$s = \frac{a_0}{(\gamma - 1)} = \text{const} = s_0 \tag{4.2}$$

in this region, and it is an "advancing" simple wave. It will be noted that (4.2) and (3.15) imply a direct relationship,

$$a = a_0 + \tfrac{1}{2}(\gamma - 1)\, u \tag{4.3}$$

between velocity and speed of sound, and hence by (3.13) and (3.12) also a direct relationship between velocity and pressure. The advancing Mach lines in this simple wave carry constant values of r, u, a etc., by (3.10), (3.16), and thus form a one-parameter family of straight lines. The most convenient parameter is usually r, in terms of which the velocity and speed of sound are given by

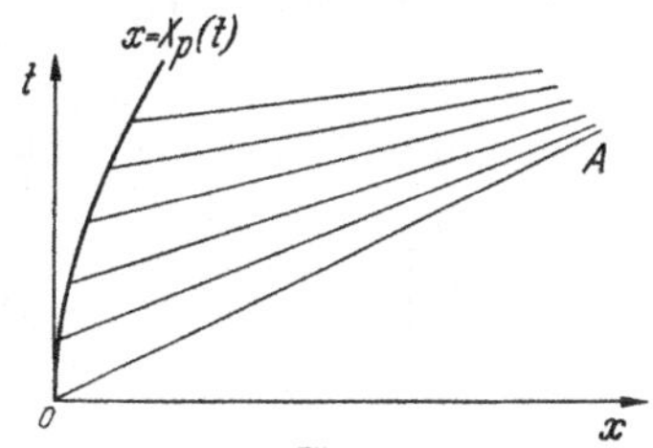

Fig. 1.

$$u = r - s_0, \qquad a = \tfrac{1}{2}(\gamma - 1)(r + s_0), \tag{4.4}$$

by (3.16) and (4.2). On the piston path,

$$u = \frac{dX_P}{dt}, \tag{4.5}$$

so $x = x_P(r)$ and $t = t_P(r)$ are known there when $X_P(t)$ is given, and by (3.10) and (4.4), the equation of the advancing Mach lines is

$$x(r) - x_P(r) = (u + a)\,[t(r) - t_P(r)]. \tag{4.6}$$

Clearly, this exact solution of the equations of motion in the simple wave region is not the analytic continuation of the solution in the region of rest, for all the partial derivatives of velocity and pressure vanish identically in the region of rest, but not all vanish in the simple wave. The physical significance of the wave front OA is thus reflected in the mathematical property that it carries a break in the analytic homogeneity of the solution—for instance, a discontinuity in the fluid acceleration or in a higher derivative of the velocity. This situation is typical of physical mechanisms of the radiation type, in contrast to those of the equilibrium or elliptic type.

b) Structure of the motion.

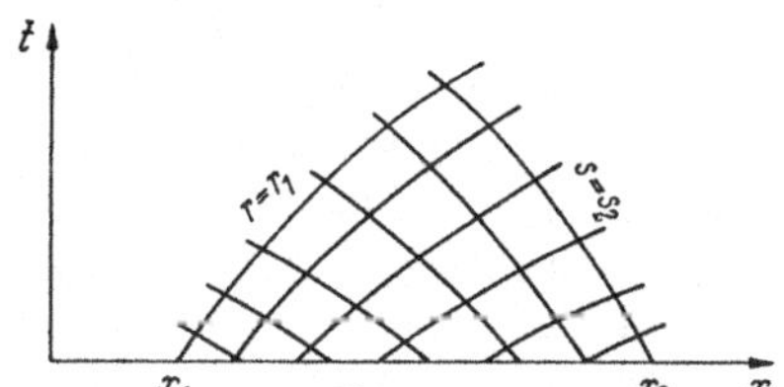

Fig. 2.

5. Uniqueness. The characteristic equations (3.10), (3.11) state that any value of r is propagated with velocity $(u+a)$, and any value of s, with velocity $(u-a)$. These velocities, in turn, depend linearly on r and s, by (3.16), and $a>0$. Therefore, if r and s are given initially as strictly monotonic functions of x, then the value r_1 of r which is taken initially at any position $x=x_1$ (Fig. 2) will in due course meet with all the values of s taken initially at positions $x>x_1$, but never with values of s taken initially at $x<x_1$. Similarly, the value s_2 of s taken initially at any position $x=x_2$ will meet all the values of r taken initially at $x<x_2$, but never those taken initially at $x>x_2$. Hence, if $x_2>x_1$, the time and position at which r_1 meets with s_2 can depend only on the initial values of r and s for $x_1 \leqq x \leqq x_2$, but *not* on the initial values for $x<x_1$ and $x>x_2$.

This argument, due to RIEMANN [1], shows the physical reason underlying the uniqueness theorem. Its conclusion can be sharpened, by the help of RIEMANN's explicit representation of the solution of the equations of motion (Sect. 9), to the theorem that the time and position at which r_1 and s_2 meet depends on *all* the initial values of r and s for $x_1 \leqq x \leqq x_2$ and *only* on these[1].

The prescription of initial conditions is not the only method, however, by which the solution of the equations of motion can be determined. RIEMANN's argument can be applied similarly to the case where r and s are prescribed arbitrarily along any line segment[2] C in the x, t-plane. This more general formulation is called Cauchy initial value problem. RIEMANN's argument can, moreover, be applied equally well to both sides of C in the x, t-plane (Fig. 3). It indicates, and the explicit representation of the solution (Sect. 9) confirms, that the distribution of r and s on the line segment C determines the solution uniquely in the curvilateral

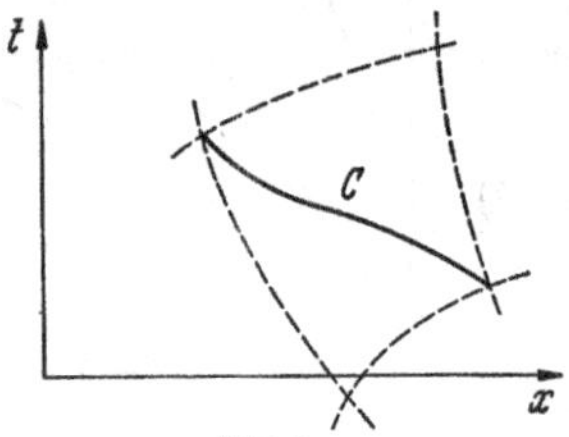

Fig. 3.

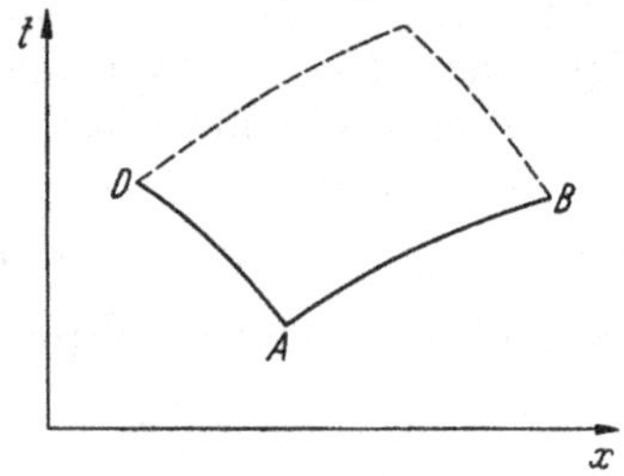

Fig. 4.

quadrangle formed by the Mach lines through the end points of C (Fig. 3), but not outside this Mach quadrangle.

An important special case arises when boundary conditions are given on two Mach line segments AB and AD (Fig. 4), i.e. when x and t are given as functions of s on the advancing Mach line segment AB, subject to (3.10), and as functions of r on the receding Mach line segment AD, subject to (3.11). The solution is then uniquely determined in, and only in, the Mach quadrangle of which AB and AD form two sides (and RIEMANN's explicit representation is also given in Sect. 9 below). But note that data on a single Mach line do not determine a solution, nor data on two Mach line segments which do not have a point in common.

The Cauchy and characteristic boundary value problems are not, however, the most frequent formulations arising from physical problems. For instance, in the piston problem of the preceding section, the piston curve (Fig. 1) is non-characteristic and the velocity is prescribed on it, but the pressure is not known on it to begin with, so that r and s are not both prescribed and it is not a Cauchy line. Still other formulations are possible, and many different cases arise, the discussion of which is deferred to Sect. 18 below. But in any case, if the initial or boundary conditions determine the solution in any finite region R of the x, t-plane (Fig. 5), then the boundaries of R are composed of Cauchy or characteristic

[1] The theorem holds for initial data of a generality sufficient to cover any physical situation, but it should be understood that it relates to the mathematical solution of the equations of motion and does not ensure that the solution it predicts must remain physically realisable (Sect. 7).

[2] Such a line segment is usually called "non-characteristic", in order to distinguish explicitly the special case of a line segment on which both of (3.10), or both of (3.11), are satisfied. The classical restrictions [2] on non-characteristic line segments are often violated in physical problems, but the conditions that the line segment be not closed and that $dr/dt = 0$ on it where $dx/dt = u + a$, and $ds/dt = 0$ where $dx/dt = u - a$, are necessary for a solution that is physically realisable in at least a small neighbourhood on both sides of C (Sects. 7, 9, 22).

line segments, or both, with proper data, so that the solution is, in fact, determined throughout the (smallest) characteristic quadrangle enclosing R (Fig. 5).

Outside this quadrangle Q, however, the solution may be continued in an infinite number of ways. One particular continuation is such that s is constant on the advancing Mach lines, and r on the receding Mach lines, until they enter Q (Fig. 6). The motion in any finite region of the x, t-plane, if it is neither uniform

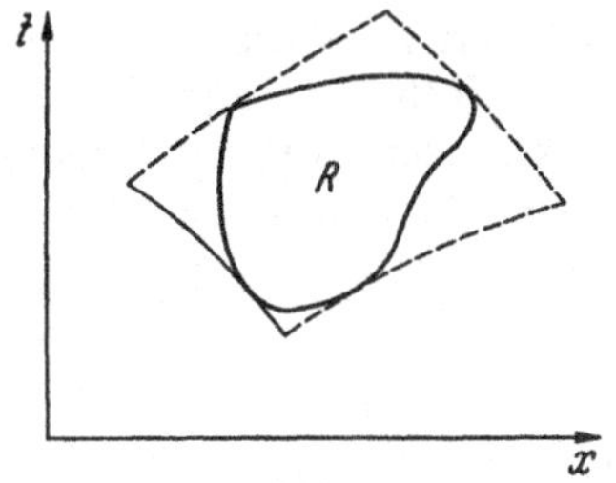

Fig. 5.

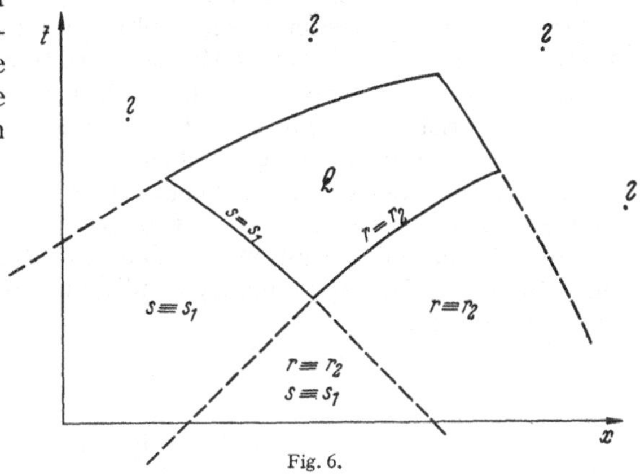

Fig. 6.

nor of the simple wave type, may therefore always be regarded as the interaction of two simple waves. This particular continuation implies, by (3.10), (3.11) that the region between the two simple waves (Fig. 6) is one of uniform motion, but in the rest of the x, t-plane the solution can still be continued in an infinite number of ways.

For the unsteady motion of compressible gas, much in contrast to the motion of incompressible fluid, it is thus necessary to consider, in the first place, only the finite regions in the x, t-plane in which the solution is determined by any particular initial or boundary conditions on finite line segments.

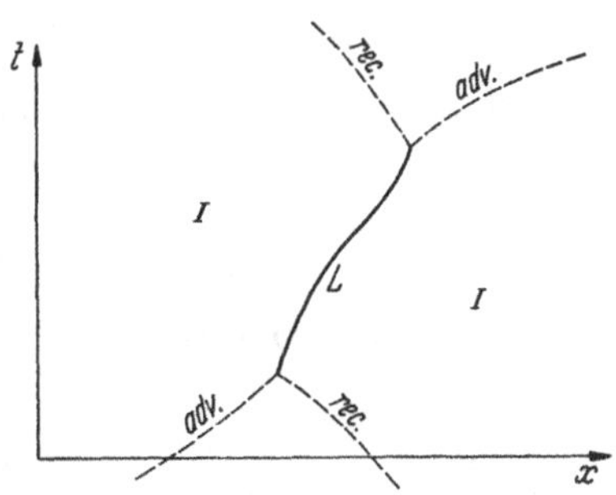

Fig. 7.

The fundamental mosaic character of the motion can be illustrated further by the concept of the "region of influence"[1] of a line segment, defined as that part of the x, t-plane in which any initial or boundary conditions prescribed on the line segment can contribute towards the determination of the solution. The great usefulness of the concept lies in the corollary that those initial or boundary conditions can have no influence whatsoever on the solution in the rest of the plane. It can be shown[2] that, if the solution is determined at a point in the x, t-plane, then both the Mach lines through that point must meet a line on which initial or boundary conditions are prescribed. It follows that the region of influence I of a non-characteristic line segment L (Fig. 7) consists of all the points connected to L by Mach lines. Similarly, that of a Mach line segment consists of the strip bounded by the Mach lines of the other family through the end points of the segment.

The line segment is seen to divide its region of influence in two parts, which are distinguished by the sense of time for Mach line segments and also for many non-characteristic segments[3], and it may be natural to reserve the name region of influence for the part representing the future. The other part is then called the "domain of dependence" of the line segment because the motion on the line segment depends on the earlier motion in that domain, but not

[1] The concept was pioneered by Courant, but the definition chosen here, on account of its broader usefulness in the discussion of physical problems, differs from the original one [2].

[2] R.E. Meyer: Z. angew. Math. Phys. 9b, 454 (1958).

[3] Alternatively, the enquiry into the nature of the motion may proceed in some definite sense distinguishing the two parts.

on the earlier motion outside it. When this distinction can be made, the concepts remain definite even if the line segment is shrunk to a point—its region of influence and domain of dependence are the Mach wedges of the future and past, respectively.

6. Wave fronts. The uniqueness theorems show that a complete discussion of the motion is possible only in the finite regions of the x, t-plane where it is determined by any particular set of physical initial or boundary conditions. None the less, certain properties of the motion remain invariant, however it be continued beyond the finite region in which it is first defined. For instance, r and s are characteristic invariants, and some other such properties will be discussed in the present and following sections.

The uniqueness theorems suggest that the characteristic variables r and s be taken as independent variables, since the regions of influence, and the Mach quadrangles in which solutions are determined, are fixed regions in the r, s-plane directly defined by the boundary conditions. Moreover, the non-linear equations (3.7), (3.8) become

$$\frac{\partial x}{\partial s} - (u+a)\frac{\partial t}{\partial s} = 0, \qquad (6.1)$$

$$\frac{\partial x}{\partial r} - (u-a)\frac{\partial t}{\partial r} = 0, \qquad (6.2)$$

which are linear, by (3.16). Further differentiation gives

$$\frac{\partial^2 t}{\partial r\,\partial s} + \frac{n}{r+s}\left(\frac{\partial t}{\partial r} + \frac{\partial t}{\partial s}\right) = 0, \qquad n = \frac{\gamma+1}{2(\gamma-1)}. \qquad (6.3)$$

Thus

$$U = (r+s)^n \frac{\partial t}{\partial r} \qquad (6.4)$$

and

$$V = (r+s)^n \frac{\partial t}{\partial s} \qquad (6.5)$$

satisfy

$$\frac{\partial U}{\partial s} = -n(r+s)^{-1} V \qquad (6.6)$$

and

$$\frac{\partial V}{\partial r} = -n(r+s)^{-1} U. \qquad (6.7)$$

Division of (6.6) by (6.5) gives the rate of change of U along advancing Mach lines,

$$\left(\frac{dU}{dt}\right)_{\text{adv}} = -n(r+s)^{n-1} \qquad (6.8)$$

and since $(r+s)$ is a continuous function of x and t, U is a continuously differentiable function of time along advancing Mach lines, and similarly for V along receding Mach lines. Discontinuities of U and V may, however, occur *across* the respective Mach lines, and if the jump of a quantity be denoted by a square bracket, any discontinuity of U satisfies

$$[U] = \text{const on advancing Mach lines}, \qquad (6.9)$$

by (6.8)[1]. Thus $[U]$ is also invariant.

The wave front concept is not restricted to breaks in the analytic homogeneity of the motion characterised by discontinuities of the fluid acceleration, and the

[1] To any statement concerning U there corresponds a dual statement concerning V, with the roles of the characteristics families interchanged, and vice versa. For brevity, the dual statements will now be omitted.

definition of the square bracket symbol can be extended accordingly and given more physical meaning. If the solution of the equations of motion is known in some region Q, it may be possible to continue it "analytically" into an adjacent region S (Fig. 8). More precisely, Q may be imagined restricted in extent so that the derivatives of all orders of velocity and pressure with respect to x and t are continuous in Q[1]. Then r and s can be represented near any point in Q by a pair of double power series in x and t, and if a pair of such series converges also in S, functions $r(x, t)$ and $s(x, t)$ satisfying the equations of motion in S are thereby defined. This solution in S may be regarded as representing a natural continuation of the pressure wave (or interaction of pressure waves) in Q. If boundary conditions are prescribed so as to determine the solution in S, and if the analytically continued solution does not satisfy these boundary conditions, then the proper solution in S may be regarded as representing the interaction of the naturally continued wave with a new wave introduced by the new analytic element in the boundary condition for the region S. The difference between the proper solution and the analytically continued solution at any point in S may be distinguished by a square bracket and regarded as the contribution of the new wave to the solution in S.

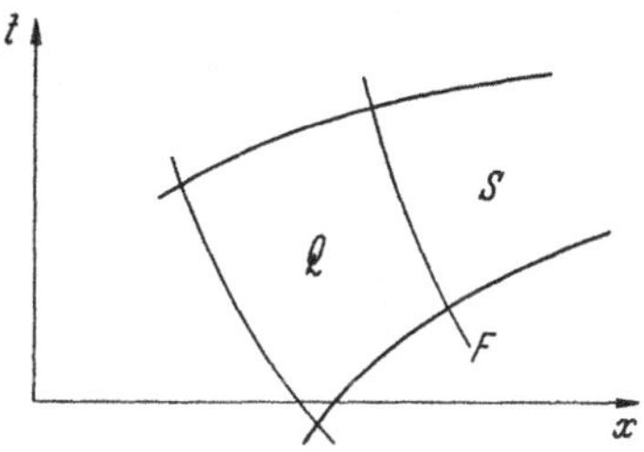

Fig. 8.

The border between Q and S may be called a wave front, if it carries such a break in the analytic homogeneity of the motion (cf. also Sect. 4). It must be a Mach line, since the solution is uniquely determined throughout a Mach quadrangle (Sect. 5), if anywhere. Assume it to be an advancing Mach line[2], $r = r_0$. With r and s adopted as independent variables, it is convenient to interpret the differences to be distinguished by the square bracket as those at a point in the characteristic r, s-plane, rather than the x, t-plane of Fig. 8. Then from (6.7), for sufficiently small $|r - r_0|$,

$$[V] = O\big([U]\,(r - r_0)/(r_0 + s)\big)$$

only, and to the first approximation, $[U]$ is a function of r only. Thus (6.9) gives the contribution of the new wave in a strip of Mach lines sufficiently close to the wave front F. Observe that this strip need not be narrow in the x, t-plane; indeed, this type of wave front approximation provides a powerful method of obtaining information on problems for which a complete, exact solution is difficult to obtain (cf. Sects. 32 to 33 of CABANNES' article in this volume and Sect. 32 below).

The wave front concept also explains the use of the name "speed of sound", in the present context, for the quantity a defined by (3.3): it is the speed of propagation of a wave front with respect to the moving gas, by the uniqueness theorem and (3.10), (3.11). Conversely, the wave front concept affords a precise physical definition, which also lends itself to direct substantiation by measurement[3], of the speeds of sound in more general gas motions.

7. Branch and limit lines. The change from x and t to r and s as independent variables raises the mathematical question whether it is always permissible, that is, whether the transformation from the "physical" x, t-plane to the characteristic r, s-plane represented by a solution of the equations of motion is always 1-1 and regular. The answer is, no. The general study of this question is deferred to

[1] The assumption of piecewise continuity of all such derivatives is implicit in the point of view taken in this article. It appears sufficient to cover all physical situations.

[2] Cf. footnote 1, p. 234.

[3] I. I. GLASS: Inst. Aerophys. Univ. Toronto, UTIA Rep. 9 (1952).

Sects. 21, 22. For the present, it may suffice to note that the singularities of the transformation from the physical plane to the characteristic plane, which are called branch-type singularities, affect only the use of the mathematical device of employing r and s as independent variables. The results given in Sect. 9 show that the device need not be abandoned, except in simple waves, where $r \equiv$ const or $s \equiv$ const so that the transformation is altogether degenerate and unnecessary. Even then, it can be shown (Sect. 19) that—though some of the Eqs. (6.1) to (6.7) may be quite meaningless as they stand—any meaningful relation obtained from them by algebraic manipulation is valid. For instance, in the simple wave $s \equiv s_0$ of the piston problem (Sect. 4) Eqs. (6.5) and (6.6) are meaningless, but (6.8) is meaningful and correct.

The singularities of the inverse transformation are called limit-type singularities and pose a more serious problem. They indicate a situation where the solution of the equations of motion for well-posed boundary conditions is degenerate in the physical x, t-plane and hence, not entirely admissible as representing a gas

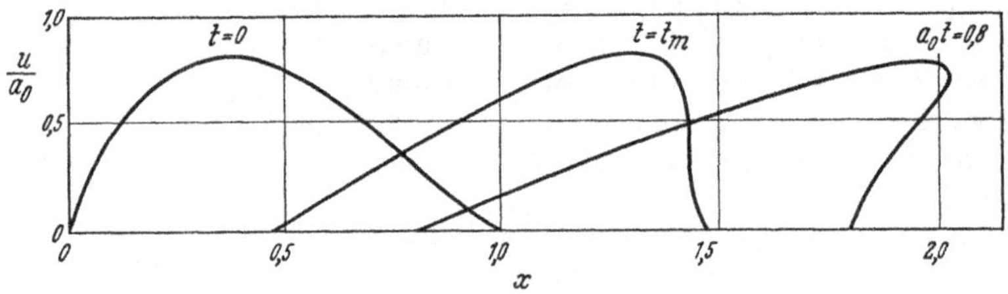

Fig. 9. Distribution of velocity in an advancing simple wave with initial velocity $u = a_0(5x - 9x^2 + 4x^3)$ for $0 \leqq x \leqq 1$, for $\gamma = 1.4$; $a_0 t_m = \frac{10}{21}$, $x_m = 1.44$ (from [3]).

motion. It can be shown[1], moreover, that no solution is free of such singularities at all times and positions, unless it represents a purely uniform motion. The general discussion is deferred to Sect. 22, but it is important here to illustrate the physical situation at the instance of the piston problem of Sect. 4.

In the simple wave region $s \equiv s_0 = a_0/(\gamma - 1)$ between the wave front OA and the piston curve $x = X_P(t)$ (Fig. 1), the velocity enforced by the piston at any time upon the gas in immediate contact with it remains constant along the advancing Mach lines, that is, the velocity u is propagated into the gas with velocity

$$u + a = a_0 + \tfrac{1}{2}(\gamma + 1)\, u, \tag{7.1}$$

by (3.16), which increases as u increases. The "wave profile", i.e. the curve of u vs. x for fixed t, therefore becomes progressively steeper and ultimately "breaks" (Fig. 9). In geometrical terms, the inverse slope (7.1) of the straight advancing Mach lines in the x, t-plane increases with increasing u, so the Mach lines converge (Fig. 1) and ultimately intersect. A little geometrical imagination indicates that the pattern of Fig. 1 thus develops into one where, at any point x, t in a certain region towards the upper right, three[2] distinct advancing Mach lines intersect. The exact solution of the piston problem obtained in Sect. 4 thus predicts three distinct velocities, and similarly three distinct pressures etc., at the same position and time. This must be preceded, however, by the intersection of neighbouring straight Mach lines, to form an envelope ("limit line"). Since u is constant on the straight Mach lines, the normal distance between neighbouring ones is inversely proportional to the fluid acceleration and to the pressure gradient. Both are

[1] G. S. S. LUDFORD: Proc. Cambridge Phil. Soc. **48**, 499 (1952).
[2] One of which may belong to the region of uniform flow.

therefore singular on the envelope, so that neglect of viscosity and heat condictivity ceases to be justifiable and the equations of motion (3.1,) (3.2) fail to apply to any real gas.

To determine in more detail the extent to which the solution of Sect. 4 can represent physical reality, note that the fluid acceleration,

$$b = \frac{Du}{Dt} = \frac{\partial u}{\partial t} + u\frac{\partial u}{\partial x} = \frac{1}{2}\left(\frac{\partial u}{\partial t} + (u-a)\frac{\partial u}{\partial x}\right),$$

since $u = \text{const}$ on advancing Mach lines, and thus by (3.16),

$$2b = \frac{\partial r}{\partial t} + (u-a)\frac{\partial r}{\partial x},$$

since $s \equiv \text{const}$. This expression, on the other hand, is the reciprocal of the derivative of t with respect to r in the receding Mach direction, $\partial t/\partial r$, as it occurs in (6.2), and hence from (6.4),

$$U = \tfrac{1}{2}(r+s)^n b^{-1}. \tag{7.2}$$

But as noted before, (6.8) is valid, whence $(dU/dt)_{\text{adv}}$ is constant on advancing Mach lines and U vanishes, and b becomes singular, at the time

$$\left.\begin{aligned} t_L(r) &= t_P(r) + n^{-1}(r+s_0)^{1-n}\, U_P(r), \\ &= t_P(r) + \frac{2}{\gamma+1}\,\frac{a(r)}{b_P(r)}, \end{aligned}\right\} \tag{7.3}$$

by (3.16) and (7.2), where b_P and U_P denote respectively the fluid acceleration and U on the piston curve (Fig. 1). The corresponding position is

$$x_L(r) = x_P(r) + (u+a)\,(t_L(r) - t_P(r)). \tag{7.4}$$

Since the piston is assumed to advance into the gas, $u > 0$ and thus $(u+a) > 0$, and the line of singularity of the acceleration occurs in the region of the x, t-plane representing gas motion if $t_L(r) > t_P(r)$, i.e. if the piston acceleration $b_P(r) > 0$. From (7.3) and (4.3), the lowest value, t_m, which the expression

$$\chi \equiv t + \frac{2a}{(\gamma+1)\,b} = t + \frac{2a_0 + (\gamma-1)\,u}{(\gamma+1)\,b}$$

takes on the piston curve when $b > 0$ therefore represents an upper bound for the time during which the motion can remain homentropic. If the acceleration of the piston never exceeds its initial value, b_0, the lowest value of χ occurs for $t = 0$, so that the acceleration first becomes infinite at the front of the pressure wave generated by the piston, at the time

$$t_m = \frac{2a_0}{(\gamma+1)\,b_0}. \tag{7.5}$$

The further development of the motion is discussed in Sects. 11, 12.

c) Solution of the general wave-interaction problem.

8. Solution for certain values of γ[1]. It was noted by RIEMANN [*1*] that the function $w(r, s)$ which may, by (6.1), (6.2) and (3.16), be defined by

$$\frac{\partial w}{\partial r} = x - (u+a)\,t, \qquad \frac{\partial w}{\partial s} = -x + (u-a)\,t, \tag{8.1}$$

satisfies the equation

$$\frac{\partial^2 w}{\partial r\,\partial s} + \frac{k}{r+s}\left(\frac{\partial w}{\partial r} + \frac{\partial w}{\partial s}\right) = 0, \qquad k = \frac{3-\gamma}{2(\gamma-1)}, \tag{8.2}$$

[1] A.E.H. LOVE and F.B. PIDDUCK: Phil. Trans. Roy. Soc. Lond. A **222**, 167 (1922).

which is slightly simpler than the equation (6.3) for $t(r, s)$, since $k=n-1$. Now,

$$t=-(\gamma-1)^{-1}(r+s)^{-1}\left(\frac{\partial w}{\partial r}+\frac{\partial w}{\partial s}\right), \tag{8.3}$$

by (8.1) and (3.16), and comparison of (6.3) and (8.2) shows that $w_{m+1}=(r+s)^{-1}(\partial w_m/\partial r+\partial w_m/\partial s)$ satisfies (8.2) with $k=m+1$, if w_m satisfies it with $k=m$. But for $k=0$ it is the wave equation, the general solution of which has the representation, $w_0=F(r)+G(s)$, where F and G are arbitrary functions, so for $k=1$ a representation is $w_1=(r+s)^{-1}(F'(r)+G'(s))$, and for k equal to any positive integer m it is

$$w=\left\{\frac{1}{r+s}\left(\frac{\partial}{\partial r}+\frac{\partial}{\partial s}\right)\right\}^{m-1}\left(\frac{f(r)+g(s)}{r+s}\right), \tag{8.4}$$

where f and g are arbitrary functions depending on the boundary conditions. The values $m=1$ and 2 correspond to $\gamma=\frac{5}{3}$ and $\frac{7}{5}$, respectively. A convenient method of determining f and g from characteristic boundary conditions for larger values of m has been given by TAUB[1]. For an application of (8.4), cf. Sect. 30 of CABANNES' article in this volume.

Of course, the solution (8.4) is, in the first place, a mathematical solution and it remains physically acceptable without modification only as long as it predicts a finite acceleration. Even at later times, however, it remains partly acceptable and its complete knowledge is useful for the determination of the necessary modification (Sect. 23). The same holds for the solution presented in the next section.

9. RIEMANN's solution [*1*]. The function $w(r', s')$ satisfies (8.2),

$$L(w)\equiv\frac{\partial^2 w}{\partial r'\,\partial s'}+\lambda\left(\frac{\partial w}{\partial r'}+\frac{\partial w}{\partial s'}\right)=0, \qquad (r'+s')\,\lambda=k=\text{const},$$

and if $R(r', s'; r, s)$ is the function satisfying

$$M(R)\equiv\frac{\partial^2 R}{\partial r'\,\partial s'}-\frac{\partial(\lambda R)}{\partial r'}-\frac{\partial(\lambda R)}{\partial s'}=0 \tag{9.1}$$

with boundary conditions

$$\frac{\partial R}{\partial s'}=\lambda R \quad \text{when } r'=r, \tag{9.2}$$

$$\frac{\partial R}{\partial r'}=\lambda R \quad \text{when } s'=s \tag{9.3}$$

and

$$R(r, s; r, s)=1, \tag{9.4}$$

then

$$0=RL(w)-wM(R)=\frac{\partial}{\partial r'}\left\{R\left(\frac{\partial w}{\partial s'}+\lambda w\right)\right\}-\frac{\partial}{\partial s'}\left\{w\left(\frac{\partial R}{\partial r'}-\lambda R\right)\right\}. \tag{9.5}$$

When this identity is integrated over a rectangle $APBD$ in the r', s'-plane (Fig. 10), P being taken as the point with coordinates $r'=r$, $s'=s$, it yields, by (9.3),

$$0=\int_D^B w\left(\frac{\partial R}{\partial r'}-\lambda R\right)dr'+\int_B^P-\int_D^A R\left(\frac{\partial w}{\partial s'}+\lambda w\right)ds',$$

and the first two integrals may be integrated by parts so that the equation becomes

$$0=(Rw)\Big|_D^P-\int_D^B R\left(\frac{\partial w}{\partial r'}+\lambda w\right)dr'-\int_B^P w\left(\frac{\partial R}{\partial s'}-\lambda R\right)ds'-\int_D^A R\left(\frac{\partial w}{\partial s'}+\lambda w\right)ds',$$

[1] H. TAUB: Ann. Math. (2) **47**, 811 (1946).

whence by (9.4) and (9.2),

$$w(P) = R(D)\,w(D) + \int_D^B R\left(\frac{\partial w}{\partial r'} + \lambda w\right) dr' + \int_D^A R\left(\frac{\partial w}{\partial s'} + \lambda w\right) ds', \qquad (9.6)$$

where $\lambda = k(r'+s')^{-1}$. The solution w at a general point P with characteristic coordinates r, s is thus obtained by quadrature from the "Riemann function" $R(r', s'; r, s)$ and the distribution of w on any two characteristic segments DA and DB forming a rectangle with the characteristics through P (Fig. 10). That distribution of w, in turn, is obtained by quadrature from those of x and t, by (8.1).

Similarly, when the identity (9.5) is integrated over the triangle ABP (Fig. 10) the base of which is a non-characteristic[1] curve C, it yields, by (9.3),

$$0 = \int_A^B w\left(\frac{\partial R}{\partial r'} - \lambda R\right) dr' + \int_B^P - \int_B^A R\left(\frac{\partial w}{\partial s'} + \lambda w\right) ds',$$

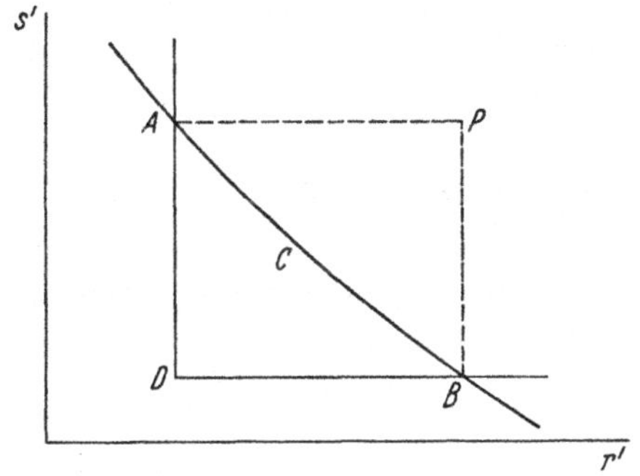

Fig. 10.

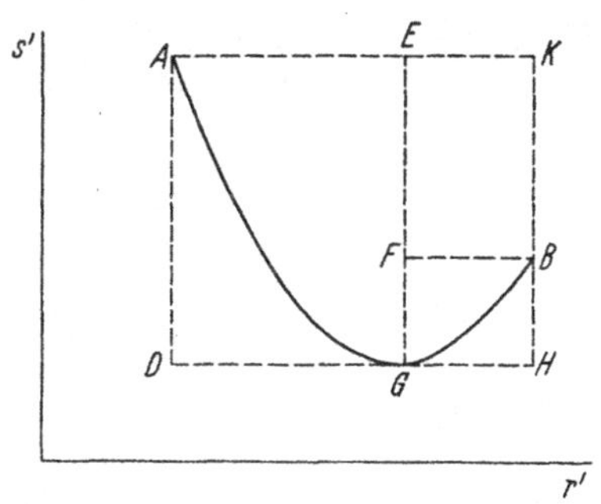

Fig. 11.

where the first and last integrals are taken along C. The first two integrals may again be integrated by parts, and the arbitrary constant $w(A)$ chosen equal to zero, so that the equation becomes

$$w(P) = \int_A^B \left\{\left(\frac{\partial w}{\partial r'} + \lambda w\right) R\,dr' + \left(\frac{\partial R}{\partial s'} - \lambda R\right) w\,ds'\right\}, \qquad (9.7)$$

by (9.4) and (9.2), where again $\lambda = k(r'+s')^{-1}$ and the integral is to be taken along C (Fig. 10). This representation furnishes the solution at the general point P by quadrature from the Riemann function and the distributions of w and its first derivatives on any non-characteristic curve segment C spanning a sector between the characteristics through P. The strict initial value problem is the case where $t=0$ on C so that, by (8.1) and (3.16), $\partial w/\partial r' = x$ and the value of w at any point Q on C is

$$w = \int_A^Q x\,(dr' - ds') = \int_A^Q x\,du. \qquad (9.8)$$

The formulae (9.6), (9.7) are independent of the orientation of A, B, and P in the r', s'-plane, provided A is taken as the point where $s'=s$, and B as that where $r'=r$.

In the evaluation of the integrals, some attention must be given to branch lines (Sect. 21). For instance, the distribution of w and its first derivatives on the non-characteristic segment C as shown in Fig. 11 determine w at all points in the rectangles $ADGE$ and $BFGH$ by (9.7),

[1] Cf. footnote 2, p. 232.

and then by (9.6) at all points in the rectangle $EGHK$. Of course, two different values of w are thereby obtained at each point in $BFGH$, but as long as it predicts a finite acceleration, such a solution is entirely acceptable and indeed, it arises in many physical problems. The solution may be regarded as having two sheets in $BFGH$, connected along the fold GH, and if (9.7) is interpreted accordingly, it represents the solution on both sheets, no recourse to (9.6) being necessary. Alternatively, the characteristic plane may be unfolded by a transformation to different characteristic variables[1].

If r and s are differentiable with respect to the arc length on the boundary line segments in the x, t-plane, the conditions of footnote 2, p. 232, suffice to exclude limit points on the boundary; in some contexts, it is desirable to admit them (Sect. 27 of CABANNES' article in this volume, and Sect. 22 of this paper).

The Riemann function satisfying (9.1) to (9.4) is [*1*]

$$R(r', s'; r, s) = \left(\frac{r'+s'}{r+s}\right)^k F\left(1-k, k; 1; \frac{-(r-r')(s-s')}{(r+s)(r'+s')}\right), \tag{9.9}$$

where F denotes the hypergeometric function and, from (8.2), $k=(3-\gamma)/(2\gamma-2)$. When k is any positive integer, the hypergeometric series terminates and the representation (8.4) is recovered.

The representations (9.6) to (9.8) exhibit explicitly many of the general properties of the solutions discussed in Sects. 5 to 7. They also show that the solution depends continuously on the initial and boundary conditions.

II. Motion with entropy variation.

10. Characteristic equations. When viscous dissipation and heat conduction are negligible, the specific entropy S of a fluid element can change during its motion only due to the addition of heat, and by the first and second laws of thermodynamics [*3*],

$$T\left(\frac{\partial S}{\partial t} + u\frac{\partial S}{\partial x}\right) = H, \tag{10.1}$$

the local rate of heat addition per unit mass of fluid. The equations of continuity and momentum for one-dimensional gas motion are still (3.1), (3.2),

$$\frac{\partial \varrho}{\partial t} + \frac{\partial(\varrho u)}{\partial x} = 0, \tag{10.2}$$

$$\frac{\partial u}{\partial t} + u\frac{\partial u}{\partial x} = -\frac{1}{\varrho}\frac{\partial p}{\partial x}, \tag{10.3}$$

and it will be assumed here that a single equation of state,

$$p = p(\varrho, S), \tag{10.4}$$

is adequate to describe the gas throughout the process.

From (10.1), (10.2) and (10.4),

$$\frac{\partial p}{\partial t} + u\frac{\partial p}{\partial x} + \varrho a^2\frac{\partial u}{\partial x} = \left(\frac{H}{T}\right)\frac{\partial p}{\partial S},$$

where $a^2 \equiv \partial p/\partial\varrho$, and this may be combined with (10.3) to give

$$\frac{\partial u}{\partial t} + (u+a)\frac{\partial u}{\partial x} + \frac{1}{a\varrho}\left(\frac{\partial p}{\partial t} + (u+a)\frac{\partial p}{\partial x}\right) = \frac{H}{a\varrho T}\frac{\partial p}{\partial S}, \tag{10.5}$$

$$\frac{\partial u}{\partial t} + (u-a)\frac{\partial u}{\partial x} - \frac{1}{a\varrho}\left(\frac{\partial p}{\partial t} + (u-a)\frac{\partial p}{\partial x}\right) = -\frac{H}{a\varrho T}\frac{\partial p}{\partial S}. \tag{10.6}$$

[1] G. S. S. LUDFORD: Proc. Cambridge Phil. Soc. **48**, 499 (1952) and J. Rational Mech. Anal. **3**, 77 (1954).

These equations are characteristic, since they involve only derivatives along advancing and receding Mach lines [cf. Eqs. (3.10), (3.11)], respectively. Their relation to the equations of homentropic motion, (3.7), (3.8), is made clearer by introducing again

$$\omega'(\varrho, S) = \int_0^{\varrho} a(\varrho', S)\frac{d\varrho'}{\varrho'} \tag{10.7}$$

and

$$r = \tfrac{1}{2}(\omega' + u), \qquad s = \tfrac{1}{2}(\omega' - u) \tag{10.8}$$

so that (10.5), (10.6) become

$$\frac{\partial r}{\partial t} + (u+a)\frac{\partial r}{\partial x} + \frac{1}{2}\left(\frac{1}{a\varrho}\frac{\partial p}{\partial S} - \frac{\partial \omega'}{\partial S}\right)\left(\frac{\partial S}{\partial t} + (u+a)\frac{\partial S}{\partial x}\right) = \frac{H}{2a\varrho T}\frac{\partial p}{\partial S}, \tag{10.9}$$

$$\frac{\partial s}{\partial t} + (u-a)\frac{\partial s}{\partial x} + \frac{1}{2}\left(\frac{1}{a\varrho}\frac{\partial p}{\partial S} - \frac{\partial \omega'}{\partial S}\right)\left(\frac{\partial S}{\partial t} + (u-a)\frac{\partial S}{\partial x}\right) = \frac{H}{2a\varrho T}\frac{\partial p}{\partial S}, \tag{10.10}$$

or by (10.1),

$$\frac{\partial r}{\partial t} + (u+a)\frac{\partial r}{\partial x} = \frac{1}{2}\left(a\frac{\partial \omega'}{\partial S} - \frac{1}{\varrho}\frac{\partial p}{\partial S}\right)\frac{\partial S}{\partial x} + \frac{1}{2}\frac{H}{T}\frac{\partial \omega'}{\partial S},$$

$$\frac{\partial s}{\partial t} + (u-a)\frac{\partial s}{\partial x} = \frac{1}{2}\left(\frac{1}{\varrho}\frac{\partial p}{\partial S} - a\frac{\partial \omega'}{\partial S}\right)\frac{\partial S}{\partial x} + \frac{1}{2}\frac{H}{T}\frac{\partial \omega'}{\partial S}.$$

The equations are thus formally similar to the characteristic Eqs. (3.17) of spherical waves etc., and being inhomogeneous, possess no particular solutions of the simple wave type (Sect. 4). But they differ fundamentally from (3.17) in not being the only characteristic equations of the motion. Eq. (10.1) is also characteristic, since it involves only the derivative along a particle path. The motion is thus seen to be generated not only by the propagation of pressure waves, but also by the convection of entropy.

For simplicity, the following discussion is again restricted to the case of a perfect gas with constant specific heats, for which the equation of state is (3.12). It follows that

$$\frac{\partial p}{\partial S} = \frac{p}{c_v} \tag{10.11}$$

and

$$\varrho T = \frac{p}{c_p - c_v}, \tag{10.12}$$

and by (3.13), the Eqs. (10.5), (10.6) may be written

$$\frac{\partial u}{\partial t} + (u+a)\frac{\partial u}{\partial x} + \frac{a}{\gamma p}\left(\frac{\partial p}{\partial t} + (u+a)\frac{\partial p}{\partial x}\right) = (\gamma - 1)\frac{H}{a}, \tag{10.13}$$

$$\frac{\partial u}{\partial t} + (u-a)\frac{\partial u}{\partial x} - \frac{a}{\gamma p}\left(\frac{\partial p}{\partial t} + (u-a)\frac{\partial p}{\partial x}\right) = -(\gamma - 1)\frac{H}{a}, \tag{10.14}$$

while the Eqs. (10.9), (10.10) become, by (3.12), (3.13),

$$\frac{\partial r}{\partial t} + (u+a)\frac{\partial r}{\partial x} - \frac{a}{2(\gamma-1)c_p}\left(\frac{\partial S}{\partial t} + (u+a)\frac{\partial S}{\partial x}\right) = \frac{(\gamma-1)H}{2a}, \tag{10.15}$$

$$\frac{\partial s}{\partial t} + (u-a)\frac{\partial s}{\partial x} - \frac{a}{2(\gamma-1)c_p}\left(\frac{\partial S}{\partial t} + (u-a)\frac{\partial S}{\partial x}\right) = \frac{(\gamma-1)H}{2a}, \tag{10.16}$$

with r and s related to u and a by (3.15), (3.16). Eqs. (10.15), (10.16), or (10.13), (10.14), together with (10.1), the relations (3.12), (3.13), (3.16), and appropriate boundary conditions are a complete set of equations determining the motion.

11. Shock formation in a simple wave. One of the most important physical effects of which the explanation has been sought by the theory of characteristics is the formation of shock waves in an initially continuous gas motion. Since the thickness of a shock front is in general only of the order of the molecular mean free path, the explanation has been sought by representing the shock as a discontinuity of pressure, velocity etc., satisfying the shock equations[1] which express the conservation principles of mass, momentum and energy.

It should be noted that a theory based on the equations of inviscid gas motion and the shock equations cannot give a completely satisfactory explanation of shock formation, since it predicts—as will be seen here and in Sect. 23—a singularity of the pressure gradient at the point and time of shock birth. Moreover, the Navier-Stokes equations show [3] that the thickness of a shock front becomes large compared with the mean free path as the shock strength tends to zero. The very birth of the shock thus represents a singular perturbation problem, and this has not yet been solved. On the other hand, the growth and propagation of the shock predicted by the theory available at present agree with experimental observation[2], for weak shocks, and the finer details of the actual birth of the shock seem of less practical importance.

The problem which has been found susceptible to quantitative analysis in considerable detail is that of shock formation in a simple wave. It is of particular importance since the results apply more generally to the propagation of weak shocks, and to their interaction with the gas motion, in many practical problems. Moreover, the theory described here may be transcribed directly into the notation of steady, two-dimensional, hypersonic flow[3]. The corresponding theory for steady two-dimensional supersonic flow is described more briefly in Sect. 27, and the qualitative theory of shock formation in a general (in contrast to simple wave) steady supersonic flow is presented in Sect. 23.

The theory is founded on the observation that the increase of entropy due to a weak shock is much smaller than the increase of pressure etc. If a shock advances into gas with relative velocity $a_0 M$, where a_0 denotes the speed of sound in the gas just before the shock passes, and if

$$\varepsilon = M^2 - 1 \ll 1,$$

the Riemann variable $r = a/(\gamma-1) + u/2$ just after the passage of the shock is given by[4]

$$\delta \equiv \frac{r - r_0}{a_0} = (\gamma+1)^{-1}\,\varepsilon\left(2 - \varepsilon + \frac{7}{8}\,\varepsilon^2\right) + O(\varepsilon^4). \tag{11.1}$$

if r_0 is its value just before the shock passes. But the other Riemann variable, $s = a/(\gamma-1) - u/2$, and the entropy are similarly given by

$$\frac{s - s_0}{a_0} = \frac{\varepsilon^3}{8(\gamma+1)} + O(\varepsilon^4) = \left(\frac{\gamma+1}{8}\right)^2 \delta^3 + O(\delta^4), \tag{11.2}$$

and

$$\frac{S - S_0}{c_v} = \frac{2\gamma(\gamma-1)}{3(\gamma+1)^2}\,\varepsilon^3 + O(\varepsilon^4) = \frac{\gamma(\gamma^2-1)}{12}\,\delta^3 + O(\delta^4). \tag{11.3}$$ [4]

---

[1] Cf. Sect. 4 of CABANNES' article in this volume.

[2] R.A. SHUNK: Inst. Res. Lehigh Univ., Tech. Rep. 11 (1958).

[3] W.D. HAYES: Quart. Appl. Math. **5**, 105 (1947).

[4] If the shock recedes into gas, $M < 0$, $u < 0$ and it is found that $(s - s_0) = O(\varepsilon)$, $(r - r_0)/a_0 = O(\varepsilon^3)$, $(S - S_0)/c_v = O(\varepsilon^3)$. It may be worth noting also that the coefficient of ε^3 in (11.3) never exceeds the value 5/48, which it takes for $\gamma = \frac{5}{3}$, and tends to zero with $(\gamma - 1)$.

It is thus plausible that a useful approximation can be obtained by neglecting the changes of both S and s across a weak advancing shock. To this approximation[1], an advancing shock formed in an advancing simple wave $s \equiv s_0$ does not change the simple wave character of the motion. Moreover, the approximation can be refined so that the effect of entropy changes is accounted for to a first, and even to a second, approximation.

Without loss of generality[2], the shock may be assumed to be formed in an advancing simple wave which, in turn, may be regarded as generated by the motion of a piston, as in Sect. 4. Without serious loss of physical generality, moreover, the piston acceleration may be assumed never to exceed its initial value, so that the first limit point occurs on the very front of the wave (Sect. 7). That the shock must start at this first limit point, on the basis of the inviscid-discontinuous model, can be deduced from the homentropic approximation as follows.

The solution of the equations of continuous motion in the simple wave $s \equiv s_0$ was seen in Sect. 4 to be completely determined when the piston curve, $x = x_P(r)$ and $t = t_P(r)$, is prescribed. If r and t are taken as (non-characteristic) independent variables in this simple wave, (4.6) gives the position as

$$x(t, r) = x_P(r) + (u + a)(t - t_p(r)), \tag{11.4}$$

whence

$$\frac{\partial x}{\partial t} = u + a, \qquad \frac{\partial x}{\partial r} = \frac{\gamma+1}{2}(t - t_P(r)) - \frac{a}{b_P(r)}, \tag{11.5}$$

by (3.16) and since $dx_P/dt_P = u$ and $dt_P/dr = dt_P/du = b_P^{-1}$, for $s \equiv \text{const}$, where $b_P(r)$ is the piston acceleration. The values of x and t on the shock, $x_s(r)$ and $t_s(r)$, are related by

$$\frac{dx_s}{dr} = \frac{\partial x}{\partial t}\frac{dt_s}{dr} + \frac{\partial x}{\partial r} = a_0 M \frac{dt_s}{dr} \tag{11.6}$$

and by (11.5), (3.16), (11.1),

$$\frac{\partial x}{\partial t} - a_0 M = u + a - a_0 M = \frac{\gamma+1}{4} a_0 \delta + O(\delta^2), \tag{11.7}$$

since $r_0 = s_0$ in the gas at rest, by (3.16). From (11.5) to (11.7), if $b_0 = b_P(r_0)$,

$$\frac{\delta}{2}\frac{dt_s}{d\delta} + t_s = t_P + \frac{2a}{(\gamma+1)\,b_P} = \frac{2a_0}{(\gamma+1)\,b_0} + o(1),$$

since $t_p(r_0) = 0$, $dt_p/d\delta = a_0/b_P$, and so

$$t_s = \frac{2a_0}{(\gamma+1)\,b_0} + o(1) = t_m + o(1), \tag{11.8}$$

by (7.5). It follows from (11.4) that also $x_s = a_0 t_m + o(1)$. (For a more general proof, cf. Sect. 23.)

Since the shock travels into gas with a relative velocity at least equal to the local speed of sound[3], it cannot be left behind by the wave front and must advance into the gas at rest. The relative shock velocity, $a_0 M$, therefore coincides with the absolute shock velocity, and the local values distinguished by a suffix 0 in (11.1) to (11.3) are identical with those taken in the gas at rest, as in Sects. 4, 7. In the x, t-diagram (Fig. 12) the shock thus appears as the continuation AB, curving slightly downward, of the straight wave front OA of Fig. 1. Between OAB and the piston curve OC, three regions may be distinguished. Region I is that covered by receding Mach lines not crossing the shock, and since particle paths traversing this region do not cross the shock either, both s and S retain their undisturbed values and the motion in I is represented exactly by the simple

[1] Due to K.O. Friedrichs: Comm. Pure Appl. Math. **1**, 211 (1948).
[2] Cf. footnote 1, p. 234.
[3] Cf. Sect. 4 of Cabannes' article in this volume.

wave solution of Sect. 4. Region II ($CDAE$) is that covered by particle paths that have not, and receding Mach lines that have, crossed the shock. The entropy therefore retains its undisturbed value exactly in II, but this region is not exactly a simple wave. Region III, finally, is that in which both s and S differ from their undisturbed values.

It may be noted that any particle path in Fig. 12[1] can be interpreted as a possible piston curve, the particle paths in region III corresponding to piston motions starting from rest with non-zero velocity. The solution for one piston motion may thus be transformed by a simple shift of the origin into the solution for any one of a family of piston motions. If the velocity is continuous on the particle path identified as piston curve (as in Fig. 12), the initial acceleration must be positive, by (11.8), for a shock to be formed at a time $t_s > 0$. It is admissible, however, that the acceleration becomes negative later, and solutions are

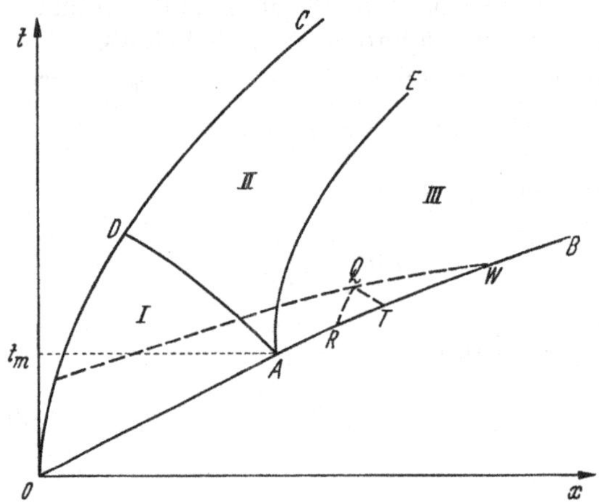

Fig. 12.

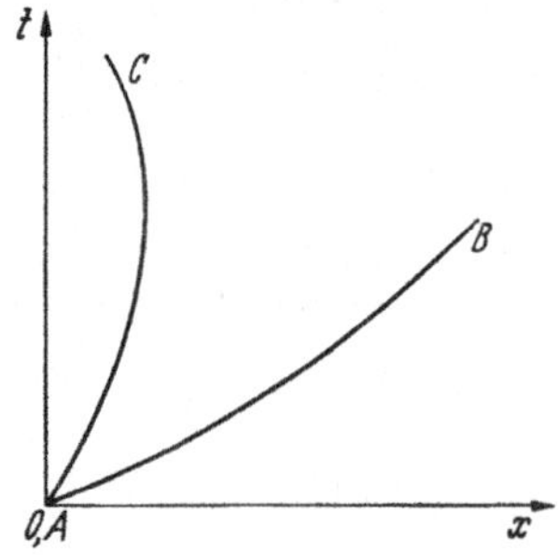

Fig. 13.

then obtained also for piston motions with impulsive start and subsequent deceleration (Fig. 13) so that the shock, formed with non-zero initial strength, decays in time. This case is of particular interest because of its close analogy to the decaying nose-shock of an aerofoil in steady supersonic flight (Sect. 27).

The basic features of the method by which the rise of entropy across the shock can be accounted for approximately, and some of the important results, were first discovered by PILLOW[2]. The crucial fact is that the variation of s *along particle paths*, even in region III, is so small compared with the variation of r that the error incurred by neglecting it is smaller than that inherent in FRIEDRICHS' homentropic approximation.

More precisely, if Q is any point in the x, t-plane (Fig. 12) between the piston curve and shock curve, and if R and W are the respective points where the particle path and advancing Mach line through Q meet the shock curve, then the difference between the values of s at Q and R is of the order of the difference between the corresponding values of r, multiplied by the cube of the highest shock strength occurring between R and W. A complete proof of this result has been published only for the case of a uniformly accelerated piston, but it can be extended to the case of non-uniform acceleration. It may be mentioned that the proof rests on two main facts. First, the difference between the shock velocity and the velocity of propagation $(u+a)$ of advancing signals in the gas is small of the first order in the shock strength δ, by (11.7). Secondly,

$$\frac{a}{\gamma p} = \frac{d\overline{\omega}}{dp} \exp\left\{\frac{S-S_0}{2c_p}\right\}, \tag{11.9}$$

[1] Including those obtained when the simple wave is continued analytically beyond the piston curve of Fig. 12 to smaller values of x and t.

[2] A.F. PILLOW: Proc. Cambridge Phil. Soc. **45**, 558 (1949).

if

$$\overline{\omega} = \frac{2}{\gamma - 1} a_0 \left(\frac{p}{p_0}\right)^{\frac{\gamma-1}{2\gamma}}, \tag{11.10}$$

so that the characteristic Eqs. (10.13), (10.14) may be written

$$\frac{\partial \overline{r}}{\partial t} + (u+a)\frac{\partial \overline{r}}{\partial x} = \left[1 - \exp\frac{S - S_0}{2c_p}\right]\left(\frac{\partial \overline{\omega}}{\partial t} + (u+a)\frac{\partial \overline{\omega}}{\partial x}\right),$$

$$\frac{\partial \overline{s}}{\partial t} + (u-a)\frac{\partial \overline{s}}{\partial x} = \left[\exp\frac{S - S_0}{2c_p} - 1\right]\left(\frac{\partial \overline{\omega}}{\partial t} + (u-a)\frac{\partial \overline{\omega}}{\partial x}\right),$$

where $\overline{r} = \frac{1}{2}(\overline{\omega} + u)$, $\overline{s} = \frac{1}{2}(\overline{\omega} - u)$, and therefore $\overline{r}$ and $\overline{s}$ are constant along the respective Mach lines to a higher approximation than r and s, when $(S - S_0)/c_v \ll 1$.[1]

Another way of looking at it[2] is to note that the change in s, from Q to R, is the difference of the changes from R to T (Fig. 12) and from T to Q, T being the point where the receding Mach line through Q meets the shock. On the shock, ds/dS is given by the shock equations and from (11.2), (11.3), $ds/dS > 0$. On the receding Mach line, $ds/dS > 0$ again, by (10.16), and since S takes the same value at Q and R, by (10.1), the change in s, from Q to R, must be markedly smaller than either of the changes in s from R to T and from T to Q, unless γ is close to unity. To formulate this in more physical terms, note that the shock advances into gas at rest, so that any change in S along it is due to the interaction of the shock with the advancing pressure wave generated by the piston and $(c_v/s)\, ds/dS$ on the shock can be interpreted as a "reflection coefficient" giving the strength of the receding pressure wave generated by this interaction. Of course, an entropy wave is also generated by this interaction, and as the receding pressure wave is propagated into the gas behind the shock, it interacts both with that entropy wave and with the advancing pressure wave. The interaction with the advancing pressure wave, in the absence of the entropy wave, would be characterised by the constancy of s on receding Mach lines, by (3.11), and $(c_v/s)\, ds/dS$ on receding Mach lines can therefore be interpreted as a "reflection coefficient" characterising the interaction of the receding pressure wave with the entropy wave. The signs of the two reflection coefficients show that the interaction processes counteract each other as regards the variation of s on particle paths produced by them.

A more refined approximation than the homentropic one can therefore be obtained by taking s to be a function only of the entropy,

$$s = s(S), \tag{11.11}$$

which is given by the shock equations.

12. Shock-expansion theory. Eq. (11.11) opens the way for a simple approximate theory[3] of the motion due to an advancing piston. By (3.16) and (10.15), (10.16), with $H = 0$,

$$\frac{\partial u}{\partial t} + (u+a)\frac{\partial u}{\partial x} = \frac{2a}{(\gamma-1)\,c_p}\left(\frac{\partial S}{\partial t} + u\frac{\partial S}{\partial x}\right) - 2\left(\frac{\partial s}{\partial t} + u\frac{\partial s}{\partial x}\right),$$

and by (11.11) and (10.1), u is constant along advancing Mach lines! The same follows for p, by (10.13), so that the advancing Mach lines remain isovels and isobars, on PILLOW's approximation, even if they are not lines of constant density, nor straight, as on FRIEDRICHS' simple wave approximation.

The mathematical simplification resulting from (11.11) is that it includes (10.16) (with $H = 0$) as one of its consequences, to PILLOW's approximation, so that only two characteristic equations, (10.1) and (10.15), remain to be satisfied. A convenient approach is to choose as independent variables the velocity u

[1] M. J. LIGHTHILL: Phil. Mag. (7) **41**, 1101 (1950), where this argument is developed for a gas with general equation of state.

[2] M. J. LIGHTHILL: Phil. Mag. 7, **40**, 214 (1949) and [4]; J. J. MAHONY: J. Aeronaut. Sci. **22**, 673 (1955); W.D. HAYES and R.F. PROBSTEIN: Hypersonic Theory. New York 1959.

[3] R.E. MEYER: Quart. Appl. Math. **14**, 433 (1957).

and the mass ψ defined by

$$\frac{\partial \psi}{\partial x} = \varrho, \qquad \frac{\partial \psi}{\partial t} = -\varrho u. \tag{12.1}$$

They are characteristic variables, since $u = \text{const}$ on advancing Mach lines and $\psi = \text{const}$ on particle paths. Then

$$p = p(u), \quad S = S(\psi), \quad s = s(\psi) \tag{12.2}$$

and the characteristic Eqs. (10.15) and (10.1) become

$$\frac{\partial x}{\partial \psi} = (u + a)\frac{\partial t}{\partial \psi}, \tag{12.3}$$

$$\frac{\partial x}{\partial u} = u\frac{\partial t}{\partial u}. \tag{12.4}$$

On advancing Mach lines, $d\psi = (\partial\psi/\partial t + (u+a)\ \partial\psi/\partial x)\ dt = a\varrho\ dt$, by (12.1), so

$$\frac{\partial t}{\partial \psi} = \frac{1}{a\varrho} = \frac{d\bar{\omega}}{dp}\exp\left(\frac{S - S_0}{2c_p}\right), \tag{12.5}$$

by (3.13) and (11.9). On the piston, $s = \text{const} = s_P$, by (11.11) and (10.1), so that the relation (12.2) between pressure and velocity can be written more explicitly in the form

$$u = \frac{2a}{\gamma - 1} - 2s_P = \bar{\omega}(p)\exp\left(\frac{S_P - S_0}{2c_p}\right) - 2s_P \tag{12.6}$$

by (3.16), (3.13), (3.12) and (11.10). On the piston, moreover, $(\partial t/\partial u)^{-1} = b_p(u)$, the piston acceleration. Thus if the second derivative, $\partial^2 t/\partial\psi\,\partial u$ is computed from (12.5), by the help of (12.2) and (12.6), and then integrated along an advancing Mach line from the piston to an arbitrary particle path, the expression

$$\frac{\partial t}{\partial u} \equiv t_u(\psi, u) = \frac{1}{b_P} + \left(\frac{d}{d\bar{\omega}}\left(\frac{d\bar{\omega}}{dp}\right)\right)\int_0^{\psi}\exp\left(\frac{S(\psi') - S_P}{2c_p}\right)d\psi' \tag{12.7}$$

for $\partial t/\partial u$ as a function of ψ and u is obtained. On the shock, $dx/dt = a_0 M$, so the slope of the shock curve, $\psi = \psi_s(u)$, in the ψ, u-plane is

$$\frac{d\psi_s}{du} = \frac{a_0 M\dfrac{\partial t}{\partial u} - \dfrac{\partial x}{\partial u}}{\dfrac{\partial x}{\partial \psi} - a_0 M\dfrac{\partial t}{\partial \psi}} = \frac{a_0 M - u}{u + a - a_0 M}\,a\varrho\, t_u(\psi_s, u), \tag{12.8}$$

by (12.3)to (12.5). Since S is known as a function $S(u)$ on the shock and a, ϱ, $\bar{\omega}$ and M are similarly known, from the shock equations, (12.6) to (12.8) together represent an ordinary differential equation for

$$f(u) = \int_{\alpha}^{u}\exp\left(\frac{S(u) - S_P}{2c_p}\right)\frac{d\psi_s}{du}\,du \tag{12.9}$$

the solution of which is

$$\left.\begin{aligned} f(u) &= \frac{1}{g}\left(f_0 + \int_{\alpha}^{u} g\,h\,du,\right), \quad h(u) = \frac{(a_0 M - u)\,a\varrho}{(u + a - a_0 M)\,b_P}\exp\left(\frac{S(u) - S_P}{2c_p}\right), \\ \frac{1}{g}\frac{dg}{du} &= -b_P h\frac{d}{d\bar{\omega}}\left(\frac{d\bar{\omega}}{dp}\right) = \frac{\gamma+1}{2\gamma}\frac{b_P h}{p}, \quad g(\alpha) = 1 \end{aligned}\right\} \tag{12.10}$$

by (11.10). The position x_s and time t_s on the shock are given by

$$\varrho_0 x_s = \psi_s = \varrho_0 x_m + \int_{f_0}^{f} \exp\left(\frac{S_P - S(u)}{2c_p}\right) df,$$

$$t_s = t_m + \frac{1}{a_0 \varrho_0} \int_{f_0}^{f} \frac{1}{M} \exp\left(\frac{S_P - S(u)}{2c_p}\right) df,$$

by (12.1) and (12.9). The constants are $\alpha = 0$, $S_P = S_0$ and $t_m = x_m/a_0 = f_0/(a_0 \varrho_0) = 2(\gamma+1)^{-1} a_0/b_0$, when the initial velocity of the piston is zero and its initial acceleration, $b_0 > 0$; and $S_P = S(\alpha)$, $x_m = t_m = f_0 = 0$, when the initial velocity of the piston is $\alpha > 0$. The solution must be modified if $\alpha = 0$ and the piston acceleration is initially an increasing function of time so that the shock does not start at the head of the pressure wave. A second approximation extending the theory beyond the validity of (11.11) has been found by STOCKER[1].

A simple, exact solution—on the basis of the inviscid-discontinuous model—can be given for the special case of shock formation by a centred simple wave of compression (Sect. 28 of CABANNES' article in this volume). The analogous steady flow solution is of interest in connection with the design of supersonic engine intakes (Fig. 30).

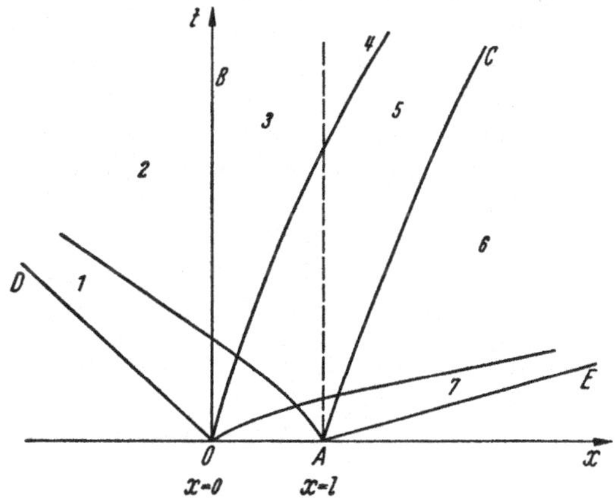

Fig. 14.

13. Heat addition. The Eqs. (10.1), (10.13) and (10.14) may be employed to study the waves generated in a perfect gas by the unsteady addition of heat, provided the motion remains substantially[2] one-dimensional. An instructive example is that of heat addition to an initially uniform subsonic flow in a straight duct, with the heat added only from $t=0$ onward at a constant and uniform rate in a finite segment $0 \leqq x \leqq l$ of the duct. The x, t-diagram of Fig. 14 shows the Mach lines and particle paths through the points (0, 0) and $(l, 0)$, which are the wave fronts during the initial phase of the motion. The closed regions, formed by them and the lines $x=0$ and $x=l$, represent wave interactions and the open, numbered regions represent the waves which are formed by these interactions. Since the entropy is convected with the gas motion, by (10.1), it differs from its initial value S_0 only in $BOAC$. The uniqueness theorem for the initial value problem of homentropic motion (Sect. 5) shows that the gas retains its initial, uniform motion to the left of OD and to the right of AE, and (3.10), (3.11) then show that the regions 1, 2 and 6, 7 are receding and advancing simple waves, respectively (Sect. 4).

Some useful information can be obtained simply by tracing the sense of variation of u, p and S along Mach lines and particle paths, by the help of (10.1) and (10.13), (10.14). It is particularly noteworthy that the gas approaching the heating region is pre-compressed and preheated by the receding simple wave.

[1] P.M. STOCKER: unpublished.

[2] The wide scope of one-dimensional models for the treatment of flow processes in ducted machinery is discussed by CROCCO in [7].

To see this, integrate (10.13) and (10.14) formally along the respective Mach lines from $t=0$ to the same point on the t-axis (Fig. 14), and subtract to eliminate u; the result is

$$\int_{p_0}^{p(0,t)} \left[\left(\frac{a}{\gamma p}\right)_{\rm rec} + \left(\frac{a}{\gamma p}\right)_{\rm adv}\right] dp = (\gamma - 1) H \int_{t_1}^{t} \left(\frac{1}{a}\right)_{\rm rec} dt$$

where the suffix indicates the Mach line on which the integrand is to be evaluated, p_0 denotes the initial pressure and t_1, the time at which the receding Mach line enters the heating region, $0 \leqq x \leqq l$. Since the integrands are positive and $t > t_1$, it follows that $p(0,t) > p_0$ for $t > 0$, if heat is added; and from (3.12) and (10.12), $T(0,t) > T_0$, since $S(0,t) = S_0$. It is seen similarly that the advancing simple wave is a wave of compression, and that both exhibit a tendency to form shocks. To study the motion beyond the first few milliseconds, however, it is necessary to take account of the conditions at the inlet and outlet of the duct and of their interaction with the waves.

A detailed prediction of the motion for any particular numerical case can be obtained—if the rate H of heat addition is a given function of x, t, ϱ and S—by the numerical step-by-step integration [7] of equations (10.1), (10.13) and (10.14) It is difficult, however, to secure in this way any reliable information regarding the ultimate nature of the motion. An approximate, analytical solution which illustrates the most important physical effects and, in particular, the asymptotic behaviour of the motion has been worked out by STOCKER[1].

He assumed the rate of heat addition to be so small that a useful approximation is obtained when Eqs. (10.1), (10.13) and (10.14) are linearized by replacing the variable coefficients by their values in the initial uniform flow. They then integrate to

$$\frac{S}{c_p} - \frac{\lambda t}{a_0} = \text{const on particle paths, } x - u_0 t = \text{const,} \tag{13.1}$$

$$2r' - \lambda t = \text{const on advancing Mach lines, } x - (u_0 + a_0)\, t = \text{const} \tag{13.2}$$

$$2s' - \lambda t = \text{const on receding Mach lines, } x - (u_0 - a_0)\, t = \text{const,} \tag{13.3}$$

respectively, where

$$\left.\begin{aligned} \lambda &= \frac{(\gamma-1)\,H}{a_0} = \text{const} \quad \text{for} \quad 0 \leqq x \leqq l, \quad t \geqq 0, \\ \lambda &= 0 \qquad\qquad\qquad\qquad \text{otherwise}, \end{aligned}\right\} \tag{13.4}$$

and

$$r' = \frac{a_0 p}{2\gamma p_0} + \frac{u}{2}, \qquad s' = \frac{a_0 p}{2\gamma p_0} - \frac{u}{2}, \tag{13.5}$$

and a suffix 0 denotes values taken in the initial uniform flow. The boundary conditions assumed by STOCKER were that the pressure remains constant at the outlet of the duct and the stagnation temperature, $T_H = T + u^2/(2c_p)$, and stagnation pressure, $p\,(T_H/T)^{\gamma/(\gamma-1)}$, remain constant at the inlet, corresponding approximately to the conditions for a duct connecting two large reservoirs.

From (13.5),

$$r' + s' = r'_0 + s'_0 \quad \text{at the outlet,} \tag{13.6}$$

while at the inlet,

$$(\gamma-1)\,(p-p_0)/(\gamma p_0) = (T - T_0)/T_0 = -u_0\,(u-u_0)/(c_p T_0) = -(\gamma-1)\,u_0\,(u-u_0)/a_0^2$$

to the first order, so that

$$r' - r'_0 = \frac{M_0 - 1}{M_0 + 1}\,(s' - s'_0) \quad \text{at the inlet,} \tag{13.7}$$

---

[1] P.M. STOCKER: Proc. Cambridge Phil. Soc. **48**, 482 (1952).

by (13.5), where $M_0 = u_0/a_0 < 1$ is the Mach number of the initial flow. By (13.2), the outlet value of r' exceeds the inlet value on the same advancing Mach line by $\frac{1}{2}\lambda l/(a_0 + u_0)$, and by (13.3), the inlet value of s' exceeds the outlet value on the same receding Mach line by $\frac{1}{2}\lambda l/(a_0 - u_0)$. Thus, if $s' = s'_n$ at the inlet at time t, then by (13.6), (13.7) and (13.2), (13.3), s' equals

$$s'_{n+1} = s'_0 + \frac{1 - M_0}{1 + M_0}(s'_n - s'_0) - \frac{1}{2}\lambda l\left(\frac{1}{a_0 + u_0} - \frac{1}{a_0 - u_0}\right)$$

at the inlet at time

$$t + \frac{L}{a_0 + u_0} + \frac{L}{a_0 - u_0}$$

(Fig. 15), where L is the length of the duct. If

$$s'_* = s'_0 + \tfrac{1}{2}\lambda l/(a_0 - u_0),$$

it follows that

$$\nu = (s'_{n+1} - s'_*)/(s'_n - s'_*) = (1 - M_0)/(1 + M_0) < 1,$$

and s'_* is therefore the asymptotic value of s' at the inlet. Moreover, the number of reflections at inlet and outlet (Fig. 15) required to reduce $s' - s'_*$ by a factor ε is $\log \varepsilon/\log \nu$ (or the first integer greater than this number) and the time required is therefore

$$2\,(L/a_0)\,(1 - M_0^2)^{-1}\log \varepsilon/\log \nu,$$

approximately. The asymptotic values of r' and s' at other stations in the duct are found similarly[1], and the approach is the same. The final steady motion is, of course, that obtained from the equations of steady flow with heat addition when the same approximations are made.

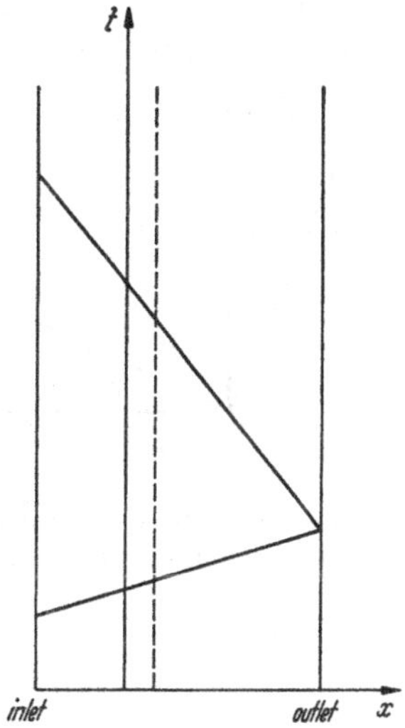

Fig. 15.

When the initial Mach number is close to unity, even small changes in u and a may lead to marked fractional changes in $(a - u)$ and hence also in the increase of s' along receding Mach line segments in the heating region. The accuracy of the approximation may then be improved greatly by accounting for terms $O(Hla_0^{-3}(1 - M_0)^{-1})$ in the coefficients of (10.1) and (10.13), (10.14), while still neglecting terms $O(Hla_0^{-3})$ in comparison with unity. The Eqs. (13.1), (13.2) remain unaltered, but (13.3) becomes

$$2s' - \lambda t = \text{const} \quad \text{on receding Mach lines,} \quad dx/dt = u - a.$$

The solution[1] provides a useful approximation as long as Hla_0^{-3} is fairly small, even if $Hl > (1 - M_0)\,a_0^3$.

C. Steady two-dimensional supersonic flow.

I. Homentropic irrotational flow.

a) Characteristic equations and simple waves.

14. Characteristic equations. The equations governing the steady, two-dimensional irrotational homentropic flow of a gas[2] are the equation of continuity,

$$\frac{\partial(\varrho u)}{\partial x} + \frac{\partial(\varrho v)}{\partial y} = 0, \tag{14.1}$$

and that for the absence of vorticity,

$$\frac{\partial u}{\partial y} - \frac{\partial v}{\partial x} = 0, \tag{14.2}$$

together with the equation of state of the gas, $p = p(\varrho, S)$, and the statement that the specific entropy S is uniform; x and y denote Cartesian coordinates in

[1] Cf. footnote 1, p. 248.

[2] Cf. Sects. 9 to 11 of OSWATITSCH's article in Vol. VIII, Part 1, of this Encyclopedia.

the flow plane, u and v the corresponding velocity components, p and ϱ the pressure and density. By CROCCO's theorem[1], the stagnation enthalpy

$$h + \tfrac{1}{2} q^2 = \text{const}, \tag{14.3}$$

where h denotes the specific enthalpy and $q = \sqrt{u^2 + v^2}$, the velocity magnitude. It will be assumed that $(\partial p/\partial \varrho)_S$ and $\partial^2 p/\partial (1/\varrho)^2)_S$ are both positive and bounded, as they are for most gases, and also that u, v, ϱ, etc., are continuous functions of x and y, except where shock waves are referred to explicitly.

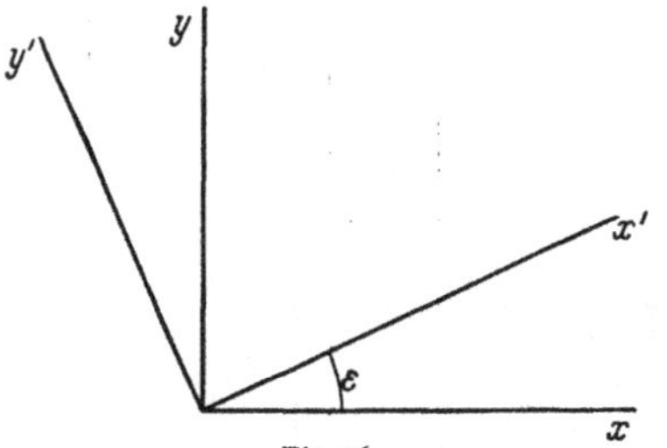

Fig. 16.

When the velocity magnitude q exceeds the speed of sound

$$a \equiv \sqrt{(\partial p/\partial \varrho)_S}, \tag{14.4}$$

the governing equations are a hyperbolic system and posses characteristics (Sect. 5 of SCHIFFER's article in this volume). That is, for any point in the flow plane where $q > a$, a coordinate system x', y' (Fig. 16) can be found such that the governing equations imply a partial differential equation containing derivatives with respect to only one of the independent variables. In fact, it is easy to verify that they imply

$$\frac{\partial \theta}{\partial x'} + \frac{1}{q} \cot \mu \frac{\partial q}{\partial x'} = 0 \quad \text{when} \quad \varepsilon = \theta - \mu, \tag{14.5}$$

where

$$\theta = \arc \tan \frac{v}{u} \tag{14.6}$$

is the stream direction and[2]

$$\mu = \arc \sin \frac{a}{q} \tag{14.7}$$

is the "Mach angle". Similarly,

$$\frac{\partial \theta}{\partial x'} - \frac{1}{q} \cot \mu \frac{\partial q}{\partial x'} = 0 \quad \text{when} \quad \varepsilon = \theta + \mu, \tag{14.8}$$

and the verification also shows that a system of governing equations equivalent to the original one is obtained when (14.1), (14.2) are replaced by (14.5), (14.8).

There are thus two characteristic directions at any point, and these define two families of characteristic curves, again called Mach lines, which cover the supersonic region of flow. On any one Mach line the relation between θ and q cannot be arbitrary, but must satisfy (14.5) or (14.8) at every point, and these equations therefore represent ordinary differential equations on the individual Mach lines. On the other hand, (14.5) specifies only one relation between two derivatives and thus permits the occurrence of discontinuities of the derivatives (Sect. 20); and similarly for (14.8). Altogether, the equations show a very close resemblance to those of Division BI.

The form of (14.5) and (14.8) suggests the introduction of new dependent variables. Let the Prandtl angle[3] ω be defined by

$$q \frac{\partial \omega}{\partial q} = \cot \mu \quad \text{and} \quad \omega = 0 \quad \text{when} \quad q = a, \tag{14.9}$$

[1] Cf. Sect. 25 ibidem.

[2] Use is made of $\varrho (\partial h/\partial \varrho)_S = (\partial p/\partial \varrho)_S$, which can be deduced from the second law of thermodynamics.

[3] After LUDWIG PRANDTL, 1875–1953.

and let

$$\alpha = \theta + \omega, \quad \beta = \theta - \omega. \tag{14.10}$$

Then (14.5) and (14.8) are equivalent to

$$\frac{dy}{dx} = \tan(\theta - \mu) \text{ on the lines } \alpha = \text{const} \quad (\text{"plus" Mach lines}) \tag{14.11}$$

and

$$\frac{dy}{dx} = \tan(\theta + \mu) \text{ on the lines } \beta = \text{const} \quad (\text{"minus" Mach lines}), \tag{14.12}$$

respectively, which are called the *characteristic equations* of the flow.

The relations between ω, q and the thermodynamic variables depend on the equation of state. For a perfect gas with constant specific heat, it is (3.12)

$$p\,\varrho^{-\gamma} \exp\left(\frac{-S}{c_v}\right) = \text{const}, \quad \gamma = \frac{c_p}{c_v}, \tag{14.13}$$

so that

$$a^2 = \frac{\gamma p}{\varrho} = \gamma(c_p - c_v)\,T = (\gamma - 1)\,h. \tag{14.15}$$

The constancy of stagnation enthalpy may therefore be expressed by

$$\frac{q^2}{2} + \frac{a^2}{\gamma - 1} = \text{const}, \tag{14.16}$$

the constancy of entropy by

$$a^2 \propto \varrho^{\gamma - 1}, \tag{14.17}$$

and the Prandtl angle may be expressed in terms of the Mach angle by [3]

$$\omega = \mu + 2n \operatorname{arc\,cot}(2n \tan\mu) - \frac{\pi}{2}, \quad n = \frac{\gamma + 1}{2(\gamma - 1)}, \tag{14.18}$$

whence it is seen to be related directly to the (local) Mach number

$$M = \frac{q}{a} = \frac{1}{\sin\mu} \tag{14.19}$$

(of which it is a monotonically increasing function), and also to q, a, p, ϱ etc. by (14.19) and (14.15) to (14.17); tables of these homentropic relations are found e.g. in [8][1].

15. Uniform first-order approximation. In a region of uniform flow, α and β are both constant, by (14.10), (14.9) and (14.7). Flows which are near-uniform in the sense that the velocity components, and hence also α and β, differ little from constants are of particular importance in engineering applications, since they are associated with small drag[2].

If

$$|\alpha - \alpha_1| < \tau \quad \text{and} \quad |\beta - \beta_1| < \tau, \quad \text{with} \quad \tau \ll 1, \tag{15.1}$$

where α_1, β_1 and τ are constants[3], and if the reference direction from which the stream direction θ is measured is chosen so that $\alpha_1 + \beta_1 = 0$, then a first approxi-

[1] Where the Prandtl angle is denoted by ν.

[2] Cf. Sect. 40 of OSWATITSCH's article in Volume VIII, Part 2, of this Encyclopedia.

[3] The uniqueness theorems (Sect. 18) show, moreover, that this condition is satified throughout any region if, and only if, it is satisfied by the boundary conditions determining the flow in that region.

mation for the Mach lines is, by (14.11), (14.12),

$$y + x \tan \mu_1 = \text{const} \quad \text{and} \quad y - x \tan \mu_1 = \text{const}, \tag{15.2}$$

and an approximate general solution of (14.11), (14.12) is therefore

$$\alpha = F(x + b y), \qquad \beta = G(x - b y), \tag{15.3}$$

where $b = \cot \mu_1 = \sqrt{M_1^2 - 1}$, by (14.19), and F and G are arbitrary functions. This is the solution of linearized theory (Sect. 18 of SCHIFFER'S article in this volume); indeed, any function $W(\alpha, \beta)$ is linear in $(\alpha - \alpha_1)$ and $(\beta - \beta_1)$, if terms $O(\tau^2)$ are neglected, and satisfies

$$b^2 \frac{\partial^2 W}{\partial x^2} - \frac{\partial^2 W}{\partial y^2} = 0$$

by (15.3).

While this approximation has proved extraordinarily fruitful, it is seen not to be entirely consistent—terms $O(\tau)$ being neglected against unity in (15.2), but not in (15.3)—and it may fail to be of *uniform* quality, even if the "thin-body" assumptions (Sect. 17 of SCHIFFER'S article in this volume) are satisfied. If terms $O(\tau^2)$ only are neglected in a consistent manner, the Mach line slopes become

$$\frac{dy}{dx} = -b^{-1} + d(\alpha - \alpha_1) + (c - d)(\beta - \beta_1) \tag{15.4}$$

and

$$\frac{dy}{dx} = b^{-1} + (c - d)(\alpha - \alpha_1) + d(\beta - \beta_1), \tag{15.5}$$

respectively, with $c = (b^2 + 1)/b^2 = M_1^2/(M_1^2 - 1)$ and $d = (\gamma + 1)\, c^2/4$, by (14.9) to (14.12), (14.16) and (14.19). On the plus Mach lines, e.g., $\alpha = \text{const}$ and (15.4) shows that such a Mach line may diverge by an arbitrarily large distance from the straight line given by (15.2), however small τ, provided the Mach line is followed over a sufficient distance. Moreover, while the partial derivatives of α and β with respect to x and y must be $O(\tau)$ on the average, by (15.1), if a typical boundary dimension is chosen as unit of length, it does not follow that these derivatives are uniformly $O(\tau)$. If $\partial\alpha/\partial x$ and $\partial\alpha/\partial y$ are sufficiently large near some point, then the slope (15.4) changes very rapidly with distance along any line crossing the plus Mach lines and a strong *local* distortion of the pattern (15.2) results. That such effects are neglected by the linearized theory accounts for its failure to predict, for instance, shock formation.

It is plausible that a more uniform approximation is obtained by approximating the Mach line slopes in (14.11), (14.12) by (15.4), (15.5). This hypothesis, due to WHITHAM[1], is confirmed by the exact solution (Sect. 24).

16. Simple waves. The characteristic Eqs. (14.11), (14.12) are almost identical with those of the one-dimensional unsteady homentropic motion, (3.10), (3.11), and much of the discussion of Division BI could be repeated literally. For instance, any region in the flow plane throughout which either α or β is constant is again called a simple wave (cf. Sect. 4), and again, in the absence of shocks any region adjacent to a region of supersonic flow must be a simple wave.

The proof [2] is instructive. Let $\alpha \equiv \alpha_1$, $\beta \equiv \beta_1$ in the region of uniform flow. If the border of this region was not a Mach line, then Mach lines of both families would cross it, by (14.11), (14.12), and $\alpha \equiv \alpha_1$, $\beta \equiv \beta_1$ also beyond it, so that it could not be the border. On the other hand, if the border is a plus Mach line, then minus Mach lines cross it, and $\beta \equiv \beta_1$ in the adjacent region, and if the border is a minus Mach line, $\alpha \equiv \alpha_1$ in the adjacent region.

[1] G.B. WHITHAM: Comm. Pure Appl. Math. 5, 338 (1952).

A steady-flow example of simple waves is found in supersonic wind-tunnel nozzles, which are designed to produce a uniform flow in the "test-section" (Fig. 17). It follows that the region just upstream of the test-section is a simple wave.

Since one of α, β is identically constant in a simple wave, both are constant on any individual Mach line of one of the families, by (14.11), (14.12). It follows, as in Sect. 4, that all the physical variables are constant on these Mach lines, and that they are straight. For instance, in a region where $\alpha \equiv$ const ("plus"

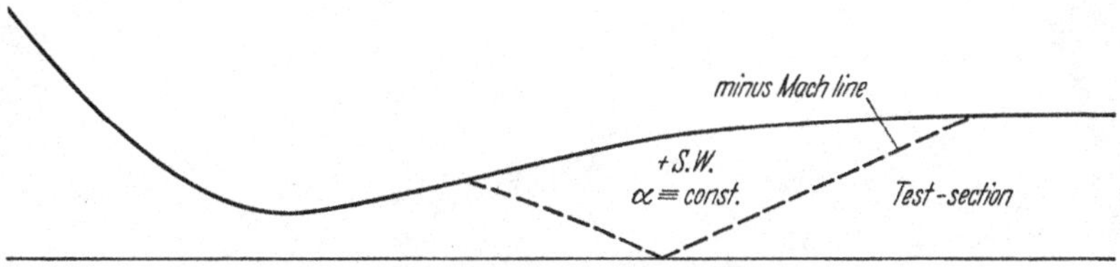

Fig. 17. Simple wave region in a nozzle (sense of flow from left to right).

simple wave) the minus Mach lines are isoclines, isobars, etc., and straight. Note also the direct relationship between stream direction and pressure in a simple wave implied by (14.10) and

$$d p = -\varrho q\, d q = -\varrho q^2 \tan \mu\, d\omega, \tag{16.1}$$

which follows directly from (14.9), the constancy of entropy and stagnation pressure and the second law[1].

The most important engineering application of simple waves is to the supersonic flow past a sharp-nosed aerofoil (Fig. 18) with attached shocks (Sects. 14, 15, 17 and 18 of Cabannes' article in this volume) and uniform flow upstream. If the aerofoil is "thin", the shocks are weak, and an approximate description of the flow may be obtained by neglecting the entropy rise across them, which is only of the third order in the shock strength (Sect. 6 of Cabannes' article in this volume), and treating the flow as homentropic throughout[2]. The jump of one of the characteristic variables, α, β, across the shock is negligible, to the same approximation (Sect. 6 of Cabannes' article in this volume). For the shocks on the upper side of the aerofoil (Fig. 18), it is the jump in α that may be neglected, so that $\alpha \equiv \alpha_1$ above the aerofoil, if the suffix 1 is used to denote values in the undisturbed stream ahead of the nose-shocks. If θ_1 is chosen zero, $\alpha_1 = \omega_1$, by (14.10), and may be found from M_1 by the homentropic tables [8]. From the known distribution of the inclination θ of the aerofoil surface, the pressure distribution on the upper side may then be read off the tables, since $\omega = \theta - \alpha_1$, by (14.10). If μ is similarly read off, the distribution of minus Mach line slopes follows from (14.12), and since these Mach lines are isobars etc., and straight, the solution is also found away from the aerofoil surface; for the shape of the shocks cf. [2]. Below the aerofoil, $\beta \equiv \beta_1$ and the solution may be computed similarly.

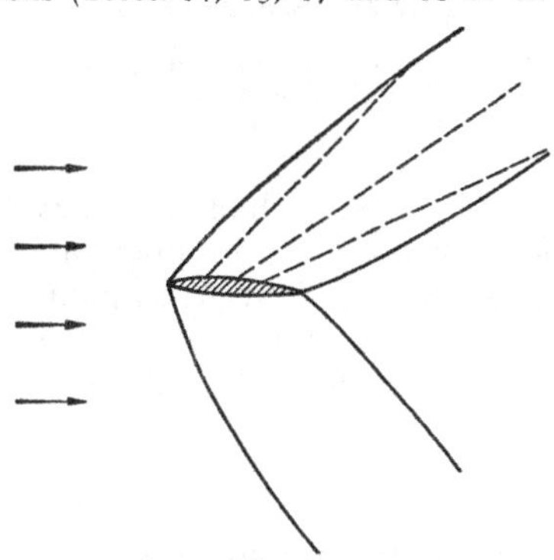

Fig. 18.

As noted in Sect. 7, the straight Mach lines in a simple wave diverge or converge, and shock formation may occur (Sect. 11). An important special case arises when all the straight Mach lines pass through a single point, the "centre" of the wave. At such a point α (or β) takes a whole range of values, and so do

[1] Cf. footnote 2, p. 250.

[2] This approximation, due to Ackeret and Busemann, ranks between the cruder one of linearized theory (Sect. 18 of Cabannes' article) and the more refined one of shock-expansion theory (Sect. 27).

therefore pressure and stream direction, and the stream lines has a corner. When the pressure rises along the stream lines in the sense of flow, a shock is formed (Sect. 12). When it falls (Fig. 19), the centred simple wave is called a Prandtl-Meyer expansion (Sect. 7 of SCHIFFER's article in this volume)[1].

Fig. 19. Schlieren photograph of supersonic flow past a double-wedge aerofoil at $M = 2.4$. Sense of flow from right to left. The light wedge emanating from the "roof-top" is the image of a simple wave centred there. (Courtesy T. A. D'EWS THOMSON, W. and A. BENNETT Supersonic Laboratory, University of Sydney.)

17. The method of characteristics. The characteristic Eqs. (14.11), (14.12) lend themselves to numerical step-by-step integration. If α, β take known values α_1, β_1 and α_2, β_2 at given points x_1, y_1 and x_2, y_2, respectively, they take the values α_1, β_2 at the point of intersection, x_3, y_3, of the plus Mach line through x_1, y_1 and minus Mach line through x_2, y_2 (Fig. 20), and

$$\frac{y_3 - y_1}{x_3 - x_1} = t_+, \qquad \frac{y_3 - y_2}{x_3 - x_2} = t_- . \tag{17.1}$$

by (14.11), (14.12), where t_+, t_- denote suitable mean values of $\tan(\theta - \mu)$ and $\tan(\theta + \mu)$ on the respective Mach line segments.

Assume now that α and β are prescribed as function of x, y along a non-characteristic[2] line C (Cauchy problem, cf. Sect. 5). Then a set of points A, B, D, E, etc. can be chosen close together on C (Fig. 21), and any pair of adjacent points of the set can be employed to compute α, β, x and y at a third point, by (17.1). With α, β, x, y thus determined at a new set of points F, G, H, etc. (Fig. 21), the procedure can be repeated to compute them at K, L, etc. In this way, a network of points can be built up, step by step, at each of which α, β, x

[1] Even though a whole range of values of α be taken at the wave centre, each corresponds to a definite plus Mach direction, and thus a continuous correspondence is defined between the directions of approach to the centre and the limiting values of α. It is in this sense that the assumed continuity of velocity, pressure etc. in terms of x and y has to be interpreted at a wave centre connected with fall of pressure along stream lines.

[2] Cf. footnote 2, p. 232.

and y are computed, and at intermediate points they can be found by interpolation. The values of all the other variables can then be found from (14.10) and (14.13) to (14.19) or the homentropic tables.

The method of computation is sufficiently flexible to cope with all initial and boundary value problems and can, moreover, be extended to more general types of flow, like those of Divisions C II, D below, for which no alternate analytical methods are available yet. Detailed and extensive descriptions will be found in [3], [4], [8].

Attention may also be drawn to the similarity and contrast with the relaxation method for elliptic differential equations. Both methods employ difference equations and a network, but whereas the relaxation procedure is repetitive, the characteristics procedure is progressive. From the way in which the network is built up (Fig. 21), it can be seen that the effect of a computing error is felt

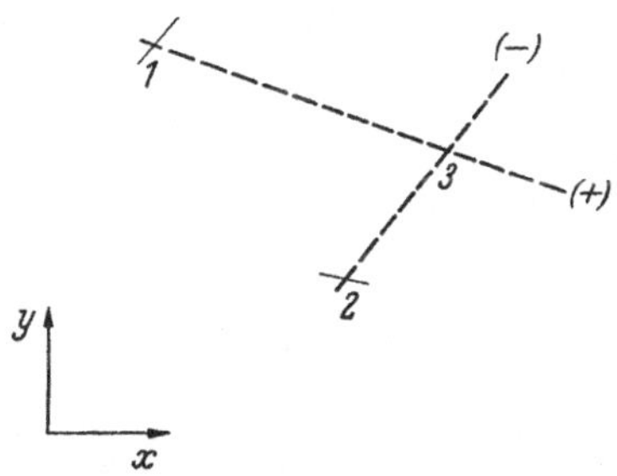

Fig. 20.

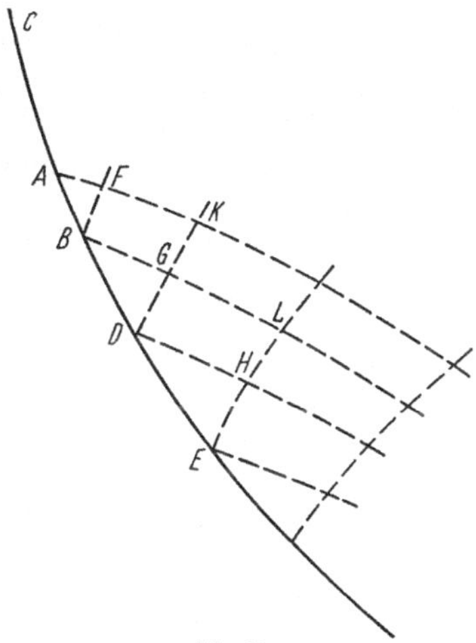

Fig. 21.

only in the region of influence (Sect. 5) of the point where the error is introduced. On the other hand, errors can be greatly inflated during their "propagation" along Mach lines, and the problem of accuracy is therefore of great importance in characteristics computations. A comprehensive analysis for the characteristics method applied to (14.11), (14.12) was made by HALL[1], so as to determine the best compromise between the demands of accuracy and economy of labour.

b) Structure of the flow.

18. Uniqueness. The method of characteristics can also be extended to provide strict existence and uniqueness proofs, and those for the CAUCHY and characteristic boundary value problems were first given in this manner[2]. More convenient proofs are now available by the method of Sect. 24. The results are strictly analogous to those of Sect. 5. If α and β are prescribed along a non-characteristic[3] line segment C in the flow plane (Cauchy problem), they are uniquely determined by (14.11), (14.12) at all points in the curvilateral quadrangle formed by the Mach lines through the end points of C (Fig. 3). If they are prescribed, subject to (14.11), (14.12), on two Mach line segments in the flow plane which have a point in common but do not belong to the same family (characteristic boundary value problem), then they are uniquely determined in the Mach quadrangle of which the Mach line segments form two sides (Fig. 4).

[1] M. G. HALL: Quart. J. Mech. Appl. Math. **9**, 320 (1956).
[2] H. LEWY: Math. Ann. **97**, 179 (1928).
[3] Cf. footnote 2, p. 232.

The concepts of region of influence and domain of dependence also apply in the same way as in Sect. 5, and so does the whole discussion expounding the fundamental mosaic character of the motion. Much in contrast to steady subsonic flow, a complete discussion of steady supersonic flow is possible only in the finite regions of the flow plane where the flow is determined by any particular set of physical boundary conditions. Its continuation beyond such regions is arbitrary.

As noted in Sect. 5, however, the CAUCHY and characteristic boundary value problems are not the only formulations arising from physical problems. A third basic problem is the "mixed" one where x, y, α and β [subject to (14.11) or (14.12)] are given on a Mach line segment AB, and x, y and one relation between α and β are given on a non-characteristic line segment AC. A typical example is that of the piston problem of Sect. 4, or the analoguous steady-flow problem where the shape of a stremline segment, but not the pressure distribution on it, and x, y, α and $\beta(=\text{const})$ on a minus Mach line are given (Fig. 22). The knowledge of the

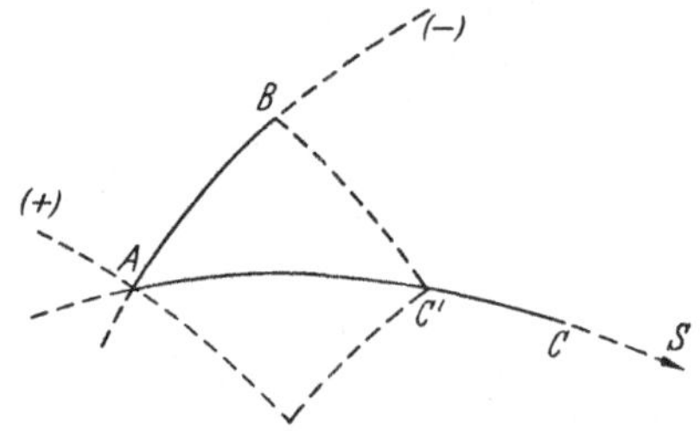

Fig. 22.

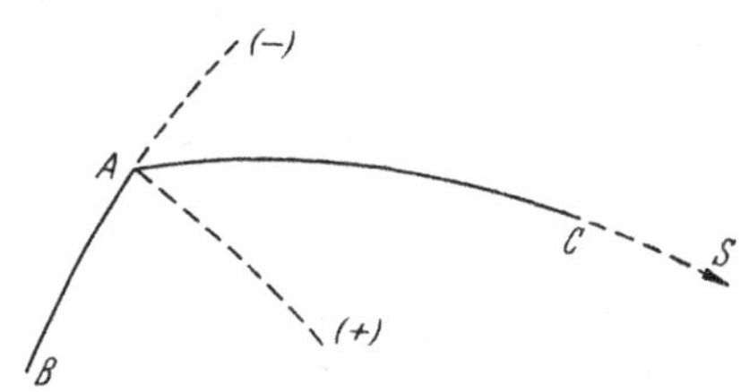

Fig. 23.

shape of the streamline segment implies knowledge of the stream direction, $\theta=\frac{1}{2}(\alpha+\beta)$, at every point, but since the pressure, and hence $\omega=\frac{1}{2}(\alpha-\beta)$, is unknown on the stremline, it is not a Cauchy line.

Another typical example is that of a jet boundary AC, in the place of the streamline of given shape. On purely inviscid theory, the pressure on a jet boundary is constant, and if it is given, $\omega=\frac{1}{2}(\alpha-\beta)$ is known, but $\theta=\frac{1}{2}(\alpha+\beta)$ is not known a priori.

The proof of existence and uniqueness of the solution of this mixed problem has been given by the integral-equation method (Sect. 25). It establishes that the solution is uniquely determined in the Mach quadrangle of which the Mach line segment forms the whole of one side and the non-characteristic segment forms the whole of one "diagonal" (Fig. 22). Note that this implies that the two segments have one end-point in common and, moreover, that the other Mach direction at that point must not lie inside the angle which the two segments form there (as in Fig. 23, where AB and AC cannot form side and diagonal, respectively, of the same Mach quadrangle).

It would appear that most physical problems are reducible to a set of problems of the three basic types. This is best illustrated by a few examples. If data are given on a streamline segment AC and an isobar segment AD (Fig. 24) including between them an angle greater than the local Mach angle, such that the data on one of the two segments formulate a Cauchy problem, and the data on the other provide only one relation between α and β, then the solution of the Cauchy problem furnishes data on the Mach line between the two segments, which reduce the rest of the problem to one of the mixed type. On the other hand, if the data on neither the isobar nor the streamline segment are sufficient to formulate a Cauchy problem, the solution is undetermined; and if the data on both line segments formulate Cauchy problems, the complete problem is over-determined, since each of the two Cauchy problems determines x, y, α and β on the Mach line between the non-characteristic line segments. Conditions are similar when boundary values are given on two other non-characteristic line segments that have (just) one end-point in common and form an angle there which includes (just) one of the local Mach directions.

When the mixed problem arises as sketched in Fig. 25, the basic theorem establishing uniqueness in the Mach quadrangle of which AB' forms one side

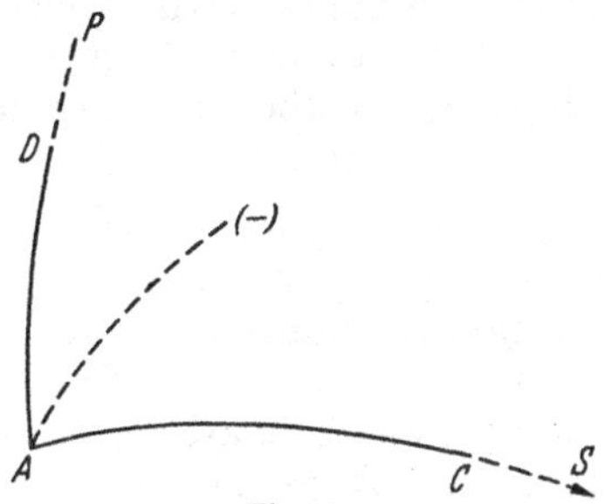

Fig. 24.

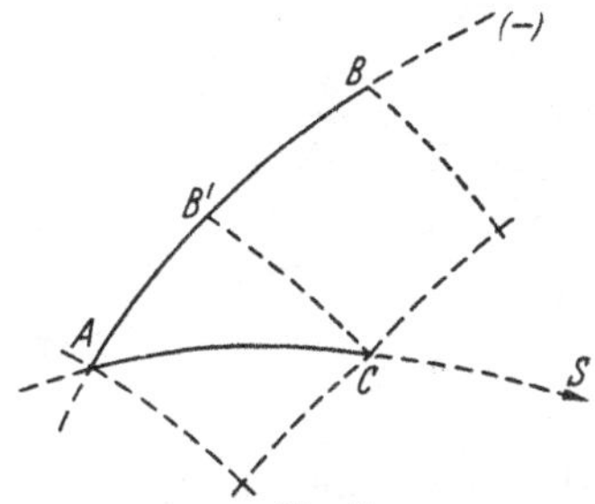

Fig. 25.

implies the unique determination of the solution on the Mach line $B'C$. The solution there, together with the data originally prescribed on $B'B$, then formulates a characteristic boundary value problem for the Mach quadrangle of which $B'B$ forms one side. In the case of Fig. 22, however, the solution is not determined in any region beyond the Mach quadrangle indicated, since the solution determined on the Mach line BC', together with the prescribed data on $C'C$, only formulates a problem of the type of Fig. 23. That such a problem is indeterminate when only one relation between α and β on the ordinary line AC (Fig. 23) is prescribed, follows from the fact that a determinate problem results for a region including both AB and AC when a second such relation is added, so that the data on AC present a Cauchy problem; its uniqueness implies the formulation of a characteristic boundary value problem for an adjacent Mach quadrangle of which AB is one side.

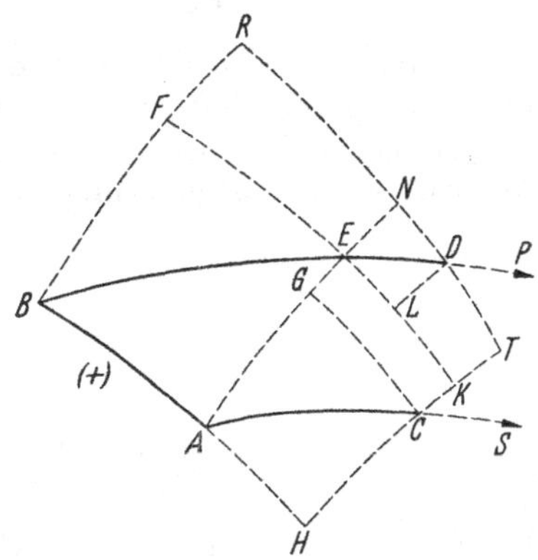

Fig. 26.

An example of boundary conditions on a line consisting of more than two distinct segments is that where the pressure on a segment BD of a jet boundary and the shape of (but not pressure on) another streamline segment AC are given (Fig. 26), together with x, y, α and β [subject to (14.11) or (14.12)] on the Mach line segment AB connecting them. Here a succession of mixed and characteristic boundary value problems arises for the regions indicated, starting with $ABFE$, and ending with $EFRN$ and $DLKT$, so that the solution is finally determined in the whole of the Mach quadrangle $BHTR$, but not beyond it.

Not all problems, however, can be analysed in terms of only the basic three. They do not help to decide, for instance, whether a unique solution exists when data are prescribed on two non-characteristic line segments meeting at an angle not including a local Mach direction; nor is the numerical method of characteristics (Sect. 17) applicable, in view of the special Mach line pattern involved (Fig. 27). This problem has received scant attention, since it does not appear to arise often in physics[1].

It may be noted also that the discussion of the present section has been based on the flow plane because it is not necessary that its mapping into the characteristic plane be regular. On the other hand, the questions discussed here refer only to the mathematical solutions of the equations of continuous motion, which may be physically acceptable only in part. In fact, it will be seen in Sect. 22 that the uniform flow is the only supersonic solution which can be so acceptable throughout the whole of the flow plane. This result, in turn, has led to a generalisation[2] of the existence and uniqueness theory sketched here.

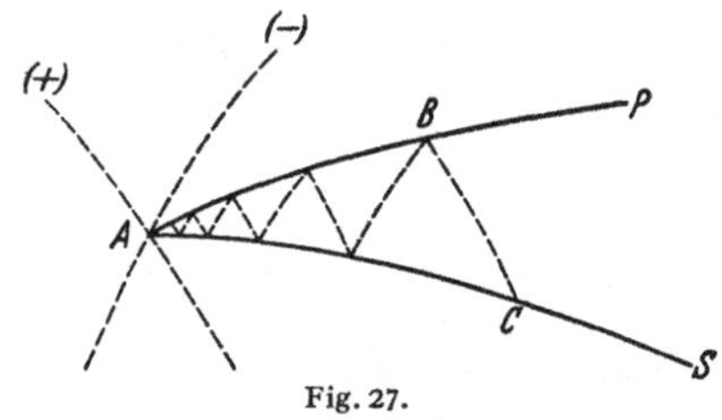

Fig. 27.

19. Focusing equations. For the further study of steady supersonic flow it is convenient to take as independent variables the characteristic parameters α, β of (14.11), (14.12). The new coordinate system will then be based, not on the shapes of boundaries in the flow plane, but on the Mach lines[3] distinguished by the differential equations representing the mechanism of the flow. The characteristic Eqs. (14.11), (14.12) become

$$\frac{\partial x}{\partial \beta} = h_\beta \cos(\theta - \mu), \qquad \frac{\partial y}{\partial \beta} = h_\beta \sin(\theta - \mu), \tag{19.1}$$

$$\frac{\partial x}{\partial \alpha} = h_\alpha \cos(\theta + \mu), \qquad \frac{\partial y}{\partial \alpha} = h_\alpha \sin(\theta + \mu), \tag{19.2}$$

respectively. In this form, the equations are linear, but contain a new pair of unknown parameters, $h_\alpha(\alpha, \beta)$ and $h_\beta(\alpha, \beta)$, which need be introduced, if only for dimensional reasons. Cross-differentiation of (19.1), (19.2) yields the differential equations governing the variation of h_α and h_β,

$$\frac{\partial h_\beta}{\partial \alpha} = m(\omega)\,(h_\beta \cos 2\mu - h_\alpha), \tag{19.3}$$

$$\frac{\partial h_\alpha}{\partial \beta} = m(\omega)\,(h_\beta - h_\alpha \cos 2\mu), \tag{19.4}$$

where

$$m(\omega) = \frac{1 - d\mu/d\omega}{2\sin 2\mu} = \frac{\sec^2\mu}{2\sin 2\mu}\left(1 + a\frac{da}{dh}\right), \tag{19.5}$$

by (14.7), (14.9) and (14.3); by the help of the second law (cf. footnote 2, p. 250), this may also be written

$$m = \frac{1}{8\varrho^3 a^2 \sin\mu \cos^3\mu} \, \frac{d^2 p}{d(1/\varrho)^2},$$

which is positive and bounded, for $0 < \mu < \pi/2$, for the gases considered in this article. For the perfect gas with constant specific heats,

$$m = \frac{\gamma + 1}{8 \sin\mu \cos^3\mu}, \tag{19.5a}$$

by (19.5) and (14.18).

[1] But cf. M.H. PROTTER: Pacific J. Math. **4**, 99 (1954) for the solution of a related differential equation with such boundary conditions.

[2] R.E. MEYER: to be published.

[3] Note that the Mach directions (14.11) and (14.12) are distinct at any point where $\pi/2 > \mu > 0$, i.e. $1 < M < \infty$, by (14.19). The usefulness of the transformation for transonic and hypersonic flows is therefore limited, and it will be assumed in the following that $\pi/2 > \mu > 0$.

It will emerge that there are great advantages in regarding the "length parameters" h_α and h_β as the fundamental dependent variables, rather than x and y (which are obtained from h_α and h_β by quadrature). From the physical point of view, this approach means that the primary aim will be to determine the velocity gradient as a function of the velocity. Indeed, the components of the pressure gradient in the Mach directions [in the sense making an acute angle with the stream direction (Fig. 28)] are

$$\frac{\partial p}{h_\beta \partial \beta} = \frac{\varrho q^2 \tan\mu}{2 h_\beta}, \qquad \frac{\partial p}{h_\alpha \partial \alpha} = \frac{-\varrho q^2 \tan\mu}{2 h_\alpha}, \tag{19.6}$$

by (16.1), (14.10), and (19.1), (19.2), the streamline curvature is

$$\varkappa_s = \frac{1}{4\cos\mu}\left(\frac{1}{h_\alpha} + \frac{1}{h_\beta}\right), \tag{19.7}$$

and similar expressions are obtained for other derivatives. From a mathematical point of view, the transformation casts the flow problem into that of a Riemannian geometry with unknown metric tensor, to be determined from (19.3), (19.4) and the boundary conditions. The components of the metric tensor are $g_{11} = h_\alpha^2$ etc., but direct use of h_α and h_β is greatly preferable, since the Riemannian geometry is singular. It is a fundamental feature of the problem that (19.1), (19.2) define not only the magnitudes but also the *signs* of h_α and h_β, since an orientation in the flow plane is distinguished by the stream direction (Fig. 28).

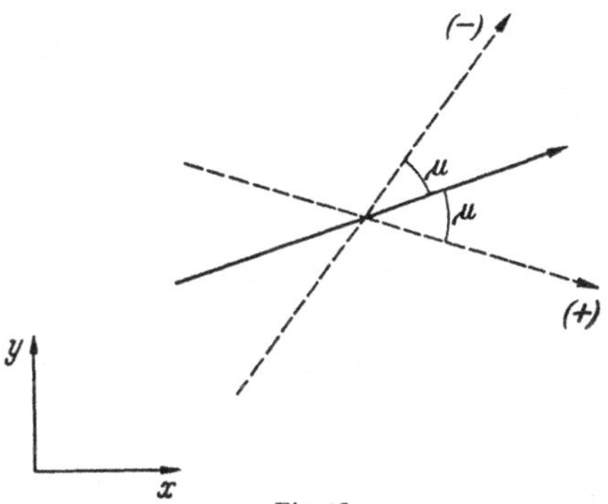

Fig. 28.

The variation of h_α and h_β describes the convergence or divergence of the Mach lines in the flow plane, and because of the analogy with optics in a medium of variable refractive index, (19.3), (19.4) are called focusing equations. They can be simplified further by the introduction of

$$U = h_\alpha / f(\mu), \qquad V = h_\beta / f(\mu), \tag{19.8}$$

where

$$f(\mu) = \exp\left(2 \int^{\omega} m \cos 2\mu \, d\omega\right). \tag{19.9}$$

[For a perfect gas with constant specific heat, by (19.5), (19.5a),

$$f = \left[\left\{\frac{(\gamma - \cos 2\mu)^\gamma}{(\sin\mu)^{\gamma+1}}\right\}^{\frac{1}{\gamma-1}} \frac{1}{\cos\mu}\right]^{\frac{1}{2}},$$

which is plotted in Fig. 29.] Then

$$\frac{\partial U}{\partial \beta} = m V, \tag{19.10}$$

$$\frac{\partial V}{\partial \alpha} = -m U. \tag{19.11}$$

As they stand, Eqs. (19.1) to (19.11) are valid only when the transformation from the physical x, y-plane to the characteristic α, β-plane is regular. This is not usually so, and the singularities of the transformation and its inverse will be discussed in Sects. 21, 22. Moreover, it follows from the definition of characteristics [and more specifically from (14.11), (14.12) in our case] that, if β is a characteristic variable, then so is any function $\beta'(\beta)$, and the Eqs. (19.10), (19.11) have the further defect of not being invariant under a transformation from one such variable to another. The two defects are related, for a change from β to $\beta'(\beta) \neq \beta$ changes h_β in (19.1), and thus also V, by (19.8); the regularity of the transformation depends on the existence of h_α and h_β and therefore, on the choice of characteristic variables. A partially invariant form of (19.10) is obtained, however, by dividing it by V so that it reads

$$\partial U / h_\beta \, \partial \beta = m/f, \tag{19.12}$$

by (19.8). Since $\partial / h_\beta \, \partial\beta$ has the invariant meaning of differentiation with respect to length, in the flow plane, in the plus Mach direction taken in the sense making an acute angle with the local stream direction (Fig. 28), (19.12) is valid [5] provided only U exists, even if U^{-1} and V or V^{-1} do not. Similarly,

$$\partial V / h_\alpha \, \partial \alpha = - m/f, \tag{19.13}$$

provided only that V exists, and four more equations can be obtained [5] to form a set complete in the sense that, whichever pair out of U, V, U^{-1} and V^{-1} exist, the set furnishes a pair of valid equations for them. It thus provides a basis for the discussion of the singularities of the transformation from the flow plane to the characteristic plane and of the inverse transformation even in those physical problems where the singularities defeat more conventional methods.

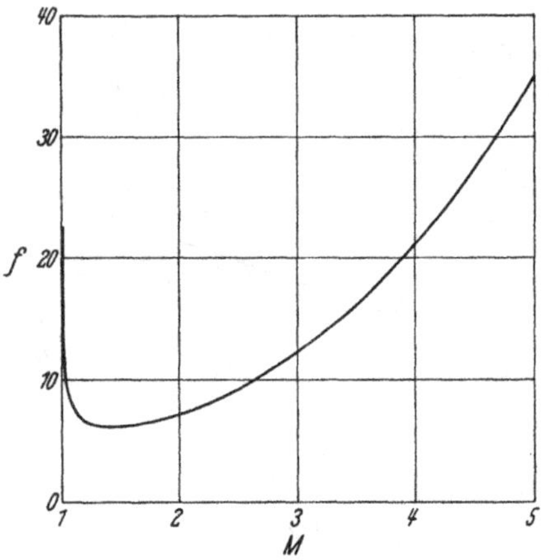

Fig. 29. The propagation function f vs. Mach number for the perfect gas with $\gamma = 1.4$.

20. Wave fronts. Eqs. (19.10) to (19.12) are analogous to (6.6) to (6.8), and the results of Sect. 6 apply therefore also to steady, supersonic flow. Thus, since $(f/m) > 0$ for $0 < \mu < \pi/2$, U is a continuously differentiable function of length along plus Mach lines, and similarly for V along minus Mach lines. On the other hand, discontinuities of U or V may occur *across* the respective Mach lines, and if the jump of a quantity is again denoted by a square bracket,

$$[U] = \text{const on a plus Mach line,} \tag{20.1}$$

by (19.12)[1]. This result is based only on the differential equations representing the mechanism of the flow, independently of any boundary conditions. $[U]$ is therefore again an *invariant under continuation*, in the sense that, if a discontinuity of U exists in a region where the solution is determined by any particular set of boundary conditions, then (20.1) remains valid also beyond that region, however the solution be continued there. In conjunction with the uniqueness theorems (Sect. 18), the invariance of $[U]$ also implies that a discontinuity of U can occur only if it is introduced either directly by the boundary conditions, or indirectly by reflection of a discontinuity of V from a boundary on which only a relation between U and V is prescribed.

[1] Cf. footnote 1, p. 234.

The same result holds for $[\partial U/\partial\alpha]$, when U itself is continuous, and similarly for higher derivatives of all orders [5]. A more significant generalisation, however, is that explained in Sect. 6, where $[U]$ is generalized to represent the contribution of the new wave in regions where such a wave is introduced by a break in the analytic homogeneity of boundary conditions. Then (20.1) remains valid to a first approximation, for $|\alpha-\alpha_f|\ll 1$, near a wave front $\alpha=\alpha_f$ on which the analytic homogeneity of the solution suffers a break.

21. Branch lines. A line in the flow plane at which the mapping of this plane into the characteristic α, β-plane has a fold[1] is called branch line. It must[2] be a plus (minus) Mach line which is either a wave front across which $U(V)$ changes sign or a line on which $U^{-1}\equiv 0$ $(V^{-1}\equiv 0)$, and in any case, it cannot end in a supersonic region.

The branch property is therefore another property of characteristics invariant under continuation. On the other hand, it is not invariant under change of characteristic variables, and this implies that, if $U(\alpha,\beta)$ is singular, then its singular part U^* must be a function of α only[3]. The invariance under continuation and the uniqueness theorems again imply that a branch line can occur only in either of two ways. It may be introduced directly by the boundary conditions, in which case $U^*(\alpha)$ is simply *equal to the singular part of the boundary conditions*. Or it may be introduced indirectly by reflection of a branch line from a boundary on which only a relation between U and V is prescribed, in which case $U^*(\alpha-\alpha_h) = k V^*(\beta-\beta_h)$, where k is a constant depending on that relation and α_h, β_h are the respective branch values of α, β (and when one of these is known, the other can be deduced directly from the boundary conditions, cf. Sect. 18). If these facts are appreciated, branch type singularities are easily dealt with[4], and this is the main reason for the usefulness of α, β as independent variables despite the very frequent occurrence of branch lines in physical problems.

Some of the local properties of branch lines are discussed in Sect. 33 of SCHIFFER's article in this volume, and references will be found there. Other properties are easily deduced from (19.1) to (19.9). For instance, across a branch line, the derivatives of pressure and stream direction in the *other* Mach direction change sign (or they vanish at it), by (19.6) and (14.10). In the characteristic plane, the streamlines are tangent to branch lines $U^{-1}=0$ or $V^{-1}=0$, since the differential equation of the streamlines is

$$U\,d\alpha = V\,d\beta, \tag{21.1}$$

by (19.1), (19.2) and (19.8). The same is true in the hodograph plane, since the transformation between these two planes is $1-1$, for $0<\mu<\pi/2$, by (14.9), (14.10).

A plus simple wave ($\alpha\equiv$const) is covered by branch lines $U^{-1}=0$, by (19.2) and (19.8)[3], and a region of uniform flow is covered by branch lines of both families. Except in such regions, x and y are continuous functions of α, and by (19.2), any singularity of U must be *integrable*, as a function of α.

22. Limit lines. A line in a characteristic plane is called *singular* if the mapping of that plane into the flow plane has a fold at the line such that a point (x, y) exists at which the solution predicts two different velocities. Since the velocity components are continuous functions of both α, β and x, y, it follows from (19.1), (19.2) and (19.8), (19.9) that U or V must change sign at a singular line that is

[1] This word is meant to include limiting cases which are local degenerations of the mapping, rather than distinct folds. The definition thus differs from that of Sect. 33 in SCHIFFER's article by including folds due to discontinuities.

[2] R.E. MEYER: Z. angew. Math. Phys. 9b, 454 (1958).

[3] Cf. footnote 1, p. 234. For more detail on singular parts, cf. [6].

[4] Also in Hodograph Theory, Sect. 33 of SCHIFFER's article in this volume.

not the image of a branch line. And conversely, a line in the α, β-plane *across* which U or V changes sign must be singular[1].

The image of a singular line in the flow plane is called *limit line*, since it represents a bound for the extent of the region in which the solution can correspond to physical reality. A limit line is, by its definition in the flow plane, invariant under change of characteristic variables; in contrast to a wave front or branch line, it is therefore an essential singularity of the characteristic mapping. It can be shown[2], moreover, that *any Mach line which is neither a branch line, nor meets a limit line, must be isolated, if the supersonic region is extended far enough in the flow plane.* This does not imply that a limit line must occur in every real flow, but rather, the view familiar from classical hydrodynamics is taken that it is profitable not to restrict attention to the part of the solution which is intended to be realised, if an understanding of its structure is desired. The theorem may, in fact, be called a supersonic Liouville theorem, for the uniform flow is the only solution such that all the Mach lines are branch lines, and the theorem therefore has the corollary that *the uniform flow is the only solution which is supersonic and free of limit lines throughout the whole of the flow plane.* In this sense, the limit type singularities of supersonic flow are seen to be an analogue of the more familiar singularities which define the structure of potential flow of incompressible fluid. The existence of a supersonic analogue of the Liouville theorem is all the more remarkable since any solution is, in the first place, determined only in a finite region, and its continuation beyond that region is arbitrary. The theorem holds, however the continuation be chosen, as long as it leaves the flow supersonic.

The basic idea of the proof is to exploit the monotoneity property expressed by (19.12), (19.13), the righthand sides of which are of definite sign, by (19.9) and (19.5). In the absence of limit lines, α and β are single-valued functions of x, y and the Mach direction (14.11) is uniquely defined at every point. Thus, if it is impossible to follow a plus Mach line M_+ over an indefinite distance in the flow plane in both the sense making an acute, and in that making an obtuse, angle with the local stream direction, the Mach line must turn back on itself; but this implies a change of sign of V, by (19.1) and (19.8), (19.9). On the other hand, if it is possible to follow M_+ indefinitely in both senses, without approaching sonic or vacuum conditions, and if M_+ is not a branch line (so that U exists), then (19.12) implies that U must change sign on M_+. To tighten this argument to a proof[2] requires much discussion, especially of the vacuum singularity of (14.11), (14.12). Moreover, it follows from the theorem itself that multi-valued solutions must be considered from the start, in contrast to the simpler point of view taken so far in this article.

Any one singular line may be composed of segments on which $U=0$ (or $V=0$) and of wave front segments across which U (or V) changes sign discontinuously. Indeed, if a singular line $U=0$ in the α, β-plane meets a wave front across which U jumps by a finite amount, that wave front must possess a singular segment, by (20.1). It is a little confusing that a discontinuous change of sign of U, say[3], can be associated with either a limit type or a branch type fold, but the distinction between the two types is not difficult, since all branch lines can be recognised by the help of the boundary conditions (Sect. 21). Of course, a wave front segment may be both a branch and limit segment[4]. Moreover, a

[1] The Jacobian $\partial(x, y)/\partial(q, \theta) = 4\,U\,V q^{-1} f^2 \cos^2\mu$, and since $f\cos\mu > 0$ and bounded, for $0 < \mu < \pi/2$, the limit lines and branch lines discussed here include all those of SCHIFFER's Sect. 33.

[2] R.E. MEYER: to be published.

[3] Cf. footnote 1, p. 234.

[4] As an example, consider a simple wave $\beta \equiv$ const ($V^{-1} \equiv 0$) adjacent to a region of uniform flow. The Mach line M_+ forming the border is a branch line, since it belongs to the uniform region. At the same time, it belongs to the family of straight Mach lines in the simple wave. These form an envelope, at any rate if extended far enough in the flow plane, which (19.2) shows to be a line $U=0$. If the streamline curvature is discontinuous at M_+, (19.7) to (19.9) and (19.12) show that a point $U=0$ occurs also on M_+, and part of M_+ must therefore be a limit line as well. The pattern is similar to that of Fig. 31.

singular segment $U=0$ may coincide with a coordinate line segment $\beta=$const in the characteristic plane, so that the limit segment degenerates to a point, the "centre" of a wave, by (19.2), as e.g. in a Prandtl-Meyer expansion (cf. Sect. 16, Fig. 19 and Sect. 7 of SCHIFFER's article in this volume) or in Fig. 30. Clearly, quite complicated patterns may occur, especially in the flow plane, when limit lines, wave fronts and branch lines, of both families, occur together in a solution[1]. In any case, however, the properties of the mapping are described by the signs of U and V.

The local properties of a limit line at which U changes sign are that the streamlines and minus Mach lines return on themselves, by (19.1), (19.2), since θ and μ are continuous in the flow plane; note that the continuity of θ implies that the sense of the stream is not reversed at such a limit point. If $U=0$ on the limit line, the streamlines and minus Mach lines are cusped in the conventional sense [cf. e.g. (19.7)] and the limit line is an envelope of the plus Mach lines, since it has their slope at every point, by (19.2), while the singular line touches coordinate lines $\alpha=$const only at exceptional points. Conversely, an envelope of plus Mach lines can be shown to be a limit line $U=0$.

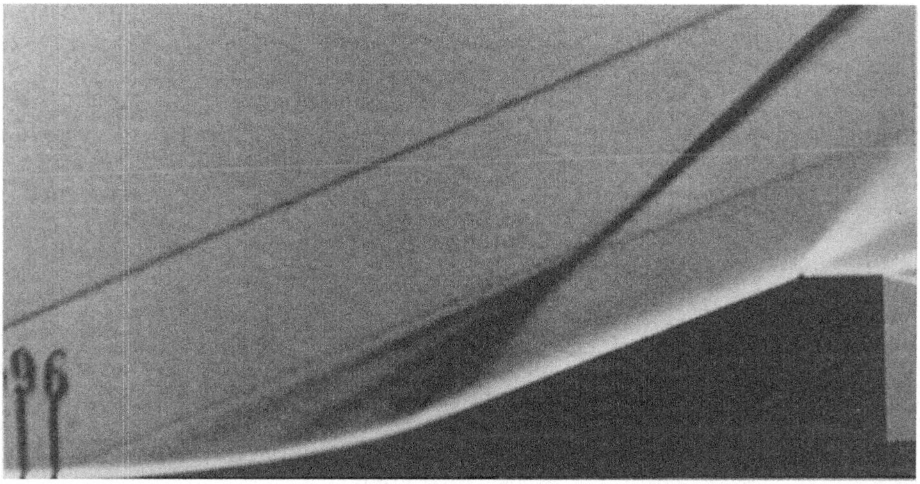

Fig. 30. Schlieren photograph of a centred simple wave of compression at $M=2.75$. Sense of flow from left to right. The dark wedge-shaped region is the image of the compression, continued upwards as a shock; the fainter dark line sloping up to the right from the centre of compression is the image of a vortex sheet. The light wedges are images of centred simple waves of expansion [Courtesy Dr. G. DROUGGE, Aeron. Res. Inst., Sweden; cf. A.R.I. Sweden Rep. 46 (1953)].

As noted in Sect. 33 of SCHIFFER's article and Sect. 7 of this paper, the pressure gradient is infinite at a point where $U=0$ or $V=0$, by (19.6), and hence, the neglect of viscosity and heat conduction in the equations of motion (14.1), (14.2) cannot be strictly justified. The influence of viscosity is relatively slight, however, when the pressure falls in the stream direction, e.g. near the centre of a Prandtl-Meyer expansion (Fig. 19). The occurrence of a limit line with the pressure rising in the stream direction indicates the presence of a shock wave in the real flow, but the influence of viscosity still remains relatively slight, if the range of variation of α and β is small (Sect. 23).

To bring some order into the picture, from a practical point of view, it is useful to note that there are, just as in classical hydrodynamics, two main types of boundary conditions by which a solution can be determined. First, a solution

[1] For some examples found in physical problems, cf. [5], P.M. STOCKER and R.E. MEYER: Proc. Cambridge Phil. Soc. **47**, 518 (1951); J.J. MAHONY: Phil. Trans. Roy. Soc. Lond. A **248**, 499 (1956).

can be defined by its very singularities. For instance, a characteristic segment of singular line corresponding to a wave centre may be prescribed, as is effectively done in the flow around a corner (Sect. 7 of SCHIFFER's article in this volume and Sect. 16, Fig. 19 of this paper). Such a singularity may also be prescribed in the middle of the flow (Fig. 30), as in some cases of supersonic intake design.

Secondly, a solution may be defined by non-singular boundary conditions that can be physically realized (at least, approximately), and no limit line can then meet the boundary. It is not difficult to see that a limit line starting or ending inside a supersonic region must do so at a highly singular point of the mapping, if at all, and no example appears to have been found to-date. If it be assumed that this does not occur, it follows as in Sect. 33 of SCHIFFER's article in this volume that the limit lines occurring in solutions for non-singular boundary conditions must be *cusped*. A limit line at which U changes sign has a cusp where the singular line touches, but does not cross, a coordinate line $\beta = \text{const}$, by (19.1), (19.2). If U is continuous, the cusp is a conventional one and the pattern in the flow plane is as shown in Fig. 31, with the image of the characteristic plane covering a sector of the flow plane *three-fold*. Observe that the branch of the limit line first encountered when the streamlines are followed from upstream is OR, not OQ. If U is discontinuous at the cusp-point O, one of the branches OR, OQ is a wave front (as in the piston problem of Sect. 7), but the pattern is similar [5], and this three-fold pattern is indeed typical for regions in the flow plane where a limit line approaches closest to a non-singular boundary.

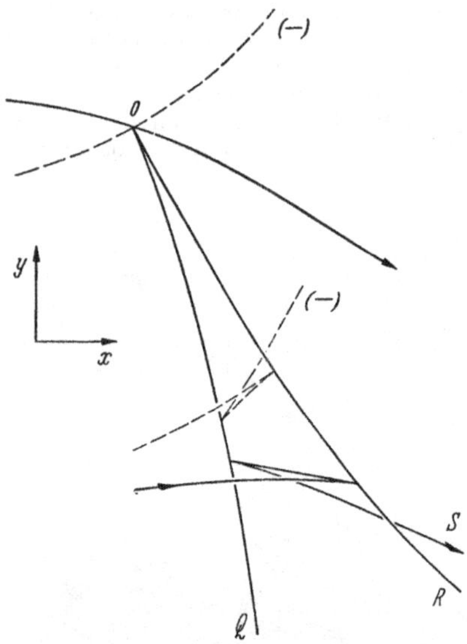

Fig. 31. Pattern with cusped limit line ROQ in the flow plane. The streamlines are denoted by S; the sense of flow is towards the right on all their segments. Minus Mach lines are shown dashed, the plus Mach lines enveloped by the limit line are not shown.

The pattern is encountered most commonly when boundary conditions are prescribed so as to define a flow with pressure rising in the stream direction, and then a shock wave starts at the cusp point of the limit line (Sect. 23), at any rate on the basis of the inviscid-discontinuous model of the flow (Sect. 11). To determine the location of the cusp point, it is in general necessary to compute the solution by numerical or analytical means (Sects. 17, 24 and 25). An indication can, however, be obtained in advance, if the signs of U and V on the boundary are known, e.g. from (19.6) to (19.8). Then (19.12), (19.13) show which sense along the Mach lines leads toward limit lines.

The three-fold pattern can also occur, however, when the boundary conditions are intended to define a monotonically accelerating flow, and a physical solution is then impossible, even if account is taken of viscosity and heat conductivity. For instance, in the design of supersonic nozzles (Fig. 17) it is common to define the flow by prescribing a monotonic distribution of pressure along the straight streamline. By (19.6), (19.7), $U = -V > 0$ on that stream line and (19.13) shows that a limit line is approached in the part of the nozzle shown in Fig. 17 as the plus Mach lines are followed away from the axis. If the prescribed rate of expansion on the straight streamline is too strong, the intended boundary streamline (Fig. 17) meets the limit line and is not realizable. This example illustrates a third type of boundary conditions by which a solution may be determined, viz. boundary conditions which are proper from the point of view of the mathematical existence and uniqueness theorems, and sufficiently regular to appear physically "reasonable", but which cannot themselves be realized *directly* by physical means. They ensure the existence of a physical solution only in a small neighborhood of the boundary.

23. Shock formation. On the basis of the results of the last three sections, it is possible to take up the problem of shock formation in a more general manner than in Sect. 11, where the discussion was restricted to simple waves. As in that section, the shock will be regarded as a line of discontinuity of velocity, pressure etc. governed by the shock equations (Sects. 4 to 6 of CABANNES' article in this

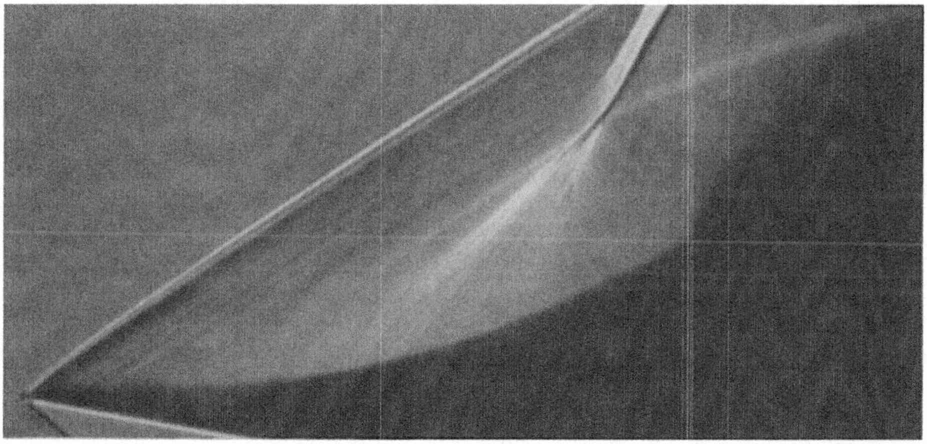

Fig. 32. Schlieren photograph of shock formation in a simple wave of compression. Images of the straight minus Mach lines can be discerned. The wave is near-centred, like that of Fig. 30, and a definite vortex sheet is seen to trail from the centre, but it can also be seen that part of the shock formation takes place gradually, at some distance from the centre. Sense of flow from left to right. [Courtesy Dr. N. H. JOHANNESEN, University of Manchester: cf. Phil. Mag. (7) **43**, 567 (1952).]

volume), embedded in a continuous flow of inviscid gas. It is true also for steady flow that the entropy rise across a weak shock is of the third order in the shock strength (Sect. 4 of CABANNES), and attention will be restricted here to the *homentropic approximation* neglecting those entropy differences altogether.

Two further comments are necessary for steady flow. Since the theory of characteristics applies only to strictly supersonic flow, it will be assumed that the shock is oblique and weak enough for the flow on both sides to be supersonic. This is justified for an initial segment, at least, of any shock starting in the middle of the flow with zero strength (Fig. 32). Secondly, important viscous effects must be anticipated when strong adverse pressure gradients occur on a solid boundary. As explained in Sect. 1, the inviscid solution approximating the flow outside the boundary layer may not correspond to the purely geometric boundary, but whatever its real boundary conditions, the solution is best discussed in terms of the concepts of Sect. 5 and 18. Moreover, even when shocks start in the immediate vicinity of a solid wall, they do so in the outer part of the boundary layer and the effect of vorticity and viscosity on the local process of shock formation is a relatively minor one.

Consider then a region of supersonic flow, away from solid surfaces, in which a shock starts with zero strength at a point O (Fig. 33). If the streamlines are deflected clockwise at the shock, as indicated by the arrow, the Mach line pattern must be as indicated by the dashed lines, since on the upstream side of a shock the local Mach angle is less than the angle which the shock makes with the stream direction, while on the downstream side it is greater [3].

On the homentropic approximation, the α, β-plane (Fig. 34) is a characteristic plane both for the region I ($AOBCD$) upstream of the shock (Fig. 33) and the region II ($AOBEF$) downstream. Since pressure and stream direction are discontinuous across the shock, each point P of the shock (Fig. 33) has two images P_1, P_2 (Fig. 34) (the point O where the shock strength vanishes, is an exception). Let suffices 1, 2 denote respectively the upstream and downstream sides of the shock. Then it can be shown (Sect. 6 of CABANNES' article in this volume) from the shock equations that the jump of one of the characteristic

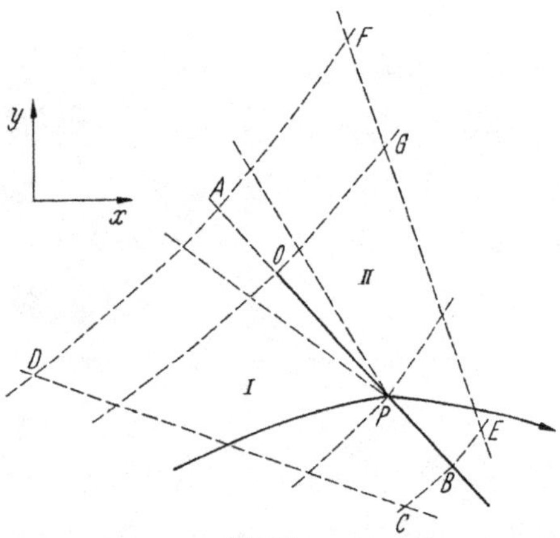

Fig. 33.

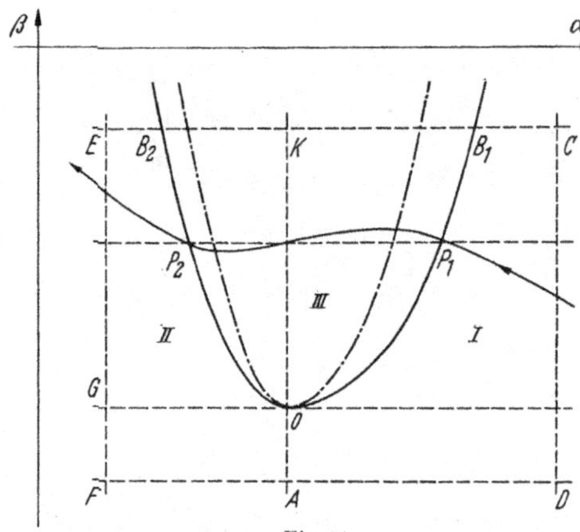

Fig. 34.

variables is of the order of the entropy rise; with clockwise stream deflection, i.e. $\delta \equiv \theta_2 - \theta_1 < 0$, it is found that

$$\beta_2 - \beta_1 = O(\delta^3), \quad \alpha_2 - \alpha_1 = 2\delta + O(\delta^3), \tag{23.1}$$

at any point P of the shock. Observe that a region III appears in the characteristic plane (Fig. 34) which is not the image of any region of the flow plane (Fig. 33)!

Fig. 34 is drawn with the shock curve approximating a parabola, near O; it will be seen below that this is a possible case, even if not the only one. Fig. 33 is sketched with the local Mach angle increasing along the plus Mach line DC, from D to C, so β must also increase (Fig. 34), by (19.5) and (14.10). Fig. 34 is also drawn on the assumption that no branch line occurs in the region under consideration, since that would complicate the figure, without changing the results.

Consider now the question of uniqueness. Assume that proper characteristic boundary conditions are prescribed on the Mach lines DC and DF. Then (Sect. 18) a unique solution Σ of the equations of continuous homentropic flow is determined in the characteristic rectangle $CDFE$ of the α, β-plane (Fig. 34). In the flow plane (Fig. 33) the same boundary conditions determine a unique solution in the regions $CDAOB$ (in view of the inequality between Mach angle and shock angle mentioned above) and $OAFG$, and this must be part of the same solution Σ. If—as is generally accepted because it is true in the simple-wave case (Sects. 11, 12) and agrees with the experience of the numerical method of characteristics [3]—the boundary conditions on DC and DF also determine uniquely the position of the shock segment OB in the flow plane, then it can be shown[1] that the solution in region II of the flow plane is identical with another part of the *same* solution Σ,

[1] J. J. MAHONY: Phil. Trans. Roy. Soc. Lond. A **248**, 499 (1956).

on the homentropic approximation. The solution downstream of the shock is therefore also determined by the equations of continuous motion and the boundary conditions, independently of the shock equations. What depends on the shock equations is the extent of the region III (Fig. 34) which is missing in the flow plane, and in particular, the position of the lines OPB and $B_1P_1OP_2B_2$ representing the shock in the two planes.

Since the points P_1, P_2 (Fig. 34) are images of the same point P (Fig. 33),

$$\left(\int_{\alpha_1}^{\alpha_2} h_\alpha \, d\alpha\right)_{\beta=\beta_1} = 0, \tag{23.2}$$

by (19.2), and h_α must change sign on P_1P_2, whence U must also change sign, by (19.8), (19.9). Assume first that AOK is not a branch line; then P can be chosen so that P_1 and P_2 are not separated by a branch line, and U must change sign at a singular point. Moreover, since the flow with shock is to be one that can be realized physically, there can be no limit lines in regions I and II, and U must therefore have the same sign at P_1 and P_2. Since P is an arbitrary point of the shock, it follows that there must be an even number of singular lines in III across which U changes sign[1] and two[2] of which join at O.

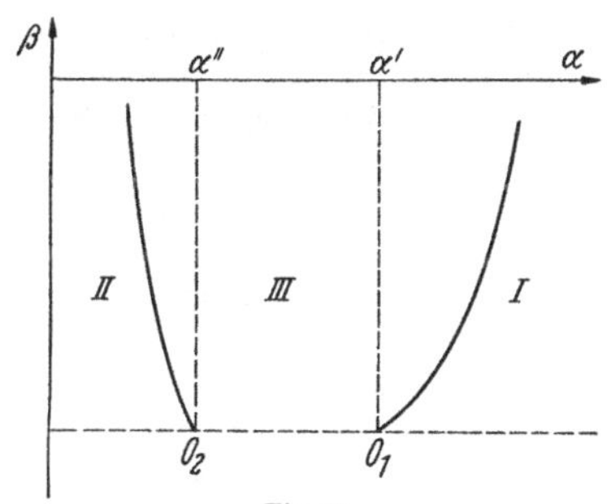

Fig. 35.

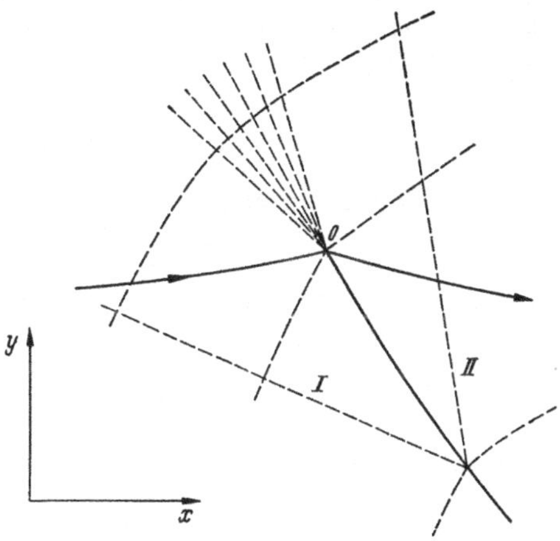

Fig. 36.

Hence, if the whole of the characteristic plane (Fig. 34) is mapped back onto the flow plane, the three-fold pattern of Fig. 31 must result! The shock (Fig. 33) represents a cut from the top sheet of this pattern to the bottom sheet.

Similarly, a shock with anti-clockwise stream deflection is associated with a cusped limit line at which V changes sign.

MAHONY[3] succeeded also in showing how a shock can be fitted as a cut into any solution of the equations of continuous motion with a cusped limit line (of only one of the characteristics families) associated with rise of pressure in the stream direction. Three main cases arise in which the local properties of the mapping differ. If the limit line has an ordinary cusp at O, the singular line is of parabolic shape there, touching a line $\beta = \text{const}$, and so is the shock (Fig. 34). If one branch of the limit line at O is a wave front, the singular line has a discontinuity of slope there, and so has the shock. Finally, the limit point O may be a wave centre the image of which is a segment of singular line coinciding with a characteristic segment $\beta = \text{const}$ (Fig. 35); the shock then starts at O with non-zero strength $\delta = (\alpha'' - \alpha')/2$ and the stream line through O has a corner (Fig. 36).

[1] If AOK is a branch line, $\alpha = \alpha_b$, U must have different sign at P_1 and P_2 and the same result follows, since the integral in (23.2) must be taken from α_1 to α_b and back to α_2.

[2] It can be shown that more than two singular line segments cannot join at O.

[3] Cf. footnote 1, p. 266.

In all three cases, the part of the solution containing the point O need not be realised, some streamline crossing the shock (such as the streamline shown in Figs. 33, 34) being identified with the boundary of the flow. The more complex problem of the reflection of a weak shock from an (ideal) jet boundary was also solved by MAHONY[1].

MAHONY's solution is derived from the homentropic approximation to the equations of motion and shock equations. Much thought has been given to the question to what degree this solution approximates the solution of the full inviscid equations of motion and shock equations. PILLOW's proof and shock-expansion theory (Sects. 11, 12 and 27) do not cover the general case of shock formation in a region of wave interaction, and the most advanced result for it is perhaps MAHONY's[1] that, if the homentropic solution with shock provides any approximation at all, then the error in the velocity field it predicts is indeed only of the order of the entropy rise across the shock; the error in the shock shape predicted is of the same order, provided the analysis is suitably arranged; but the error in the pressure gradients predicted down-stream of the shock is of the order of the shock strength δ itself, and the shape of succeeding shocks in the flow would not be predicted with the same accuracy as that of the first.

c) Analytical solution methods.

24. Riemann representation. Since (19.10), (19.11) are linear, several solution methods are available, and the choice between them must be made according to the boundary conditions and the information sought. The most important method is RIEMANN's. With

$$U+V=Z, \quad U-V=Y, \tag{24.1}$$

(19.10), (19.11) become

$$\frac{\partial^2 Z}{\partial\alpha\,\partial\beta}+\varphi(\alpha-\beta)\,Z=0, \tag{24.2}$$

$$\frac{\partial^2 Y}{\partial\alpha\,\partial\beta}+\psi(\alpha-\beta)\,Y=0, \tag{24.3}$$

where

$$\varphi(\alpha-\beta)=m^2-\tfrac{1}{2}\,dm/d\omega, \tag{24.4}$$

$$\psi(\alpha-\beta)=m^2+\tfrac{1}{2}\,dm/d\omega, \tag{24.5}$$

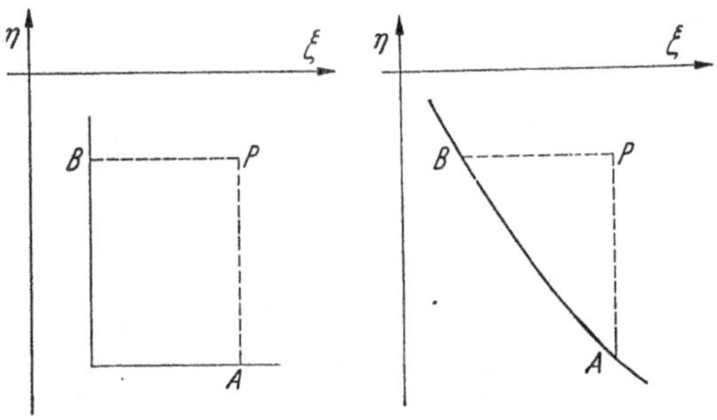

Fig. 37, 38. Characteristic plane.

and the method of Sect. 9 can be applied to express the solution of the CAUCHY and characteristic boundary value problems for either equation in terms of its Riemann function. That of (24.2), $R(\xi,\eta;\alpha,\beta)$, satisfies

$$\frac{\partial^2 R}{\partial\xi\,\partial\eta}+\varphi(\xi-\eta)\,R=0, \tag{24.6}$$

$$R(\alpha,\eta;\alpha,\beta)=R(\xi,\beta;\alpha,\beta)=1, \tag{24.7}$$

and yields the representation

$$Z(P)=Z(A)+\int_A^B\left(Z\,\frac{\partial R}{\partial\xi}\,d\xi+R\,\frac{\partial Z}{\partial\eta}\,d\eta\right), \tag{24.8}$$

where P is the point with characteristic coordinates α,β which appear as parameters in the Riemann function and the integral is taken along the boundary, from the point A where $\xi=\alpha$ to the point B where $\eta=\beta$ (Figs. 37, 38). A representation of U and V is obtained by a further quadrature, by (19.10), (19.11) and (24.1). Alternatively, (24.8) and the analogous representation of Y in terms

[1] Cf. footnote 1, p. 266.

of its Riemann function $S(\xi, \eta; \alpha, \beta)$ may be used to obtain the more symmetrical representation

$$\left.\begin{aligned} 2U(P) - 2U(A) = \int_A^B \Big\{ \Big[m(R-S) + \frac{\partial(R+S)}{\partial\xi} \Big] U\, d\xi + \\ + \Big[m(R+S) - \frac{\partial(R-S)}{\partial\eta} \Big] V\, d\eta \Big\}, \end{aligned}\right\} \tag{24.9}$$

$$\left.\begin{aligned} 2V(P) - 2V(B) = \int_A^B \Big\{ \Big[m(R+S) + \frac{\partial(R-S)}{\partial\xi} \Big] U\, d\xi + \\ + \Big[m(R-S) - \frac{\partial(R+S)}{\partial\eta} \Big] V\, d\eta \Big\}, \end{aligned}\right\} \tag{24.10}$$

which exhibits explicitly many of the general properties of the solutions discussed in Sects. 5, 18, 20, 21. The representations (24.8) and (24.9), (24.10) also show that the solution depends continuously on the boundary conditions. The solution of the third basic boundary value problem (Sects. 18, 25) cannot be represented similarly explicitly in terms of the boundary conditions and Riemann functions.

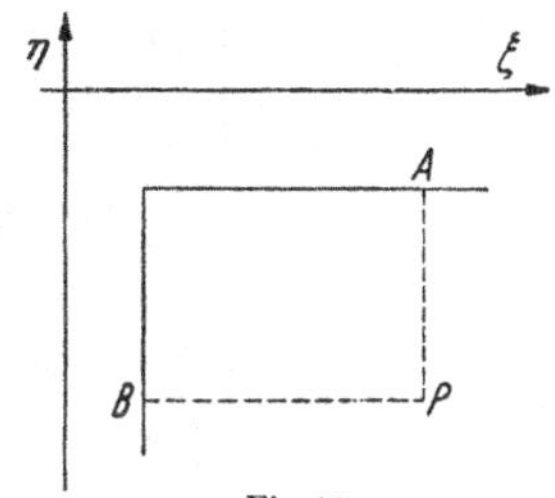

Fig. 39.

The formulae (24.8) to (24.10) are valid also when the orientation in the characteristic plane is different, e.g. as in Fig. 39, provided A is still taken to be the point on the boundary where $\xi = \alpha$, and B that where $\eta = \beta$. When branch lines are present, the path of integration has to be suitably interpreted, so that no relevant part of the boundary is omitted; alternatively, the integrals may be taken in the flow plane, or the formulae may be applied separately to each sheet of the α, β-plane.

It remains to consider the construction of the Riemann functions. Observe, first of all, that a trivial approximation is obtained by putting

$$R \equiv S \equiv 1 \tag{24.11}$$

to satisfy (24.7), without any attempt at satisfying (24.6). The more accurate computation of R and S is complicated somewhat by the awkward nature of the dependence of φ and ψ on $\xi - \eta$, given by (24.4), (24.5), (19.5), (14.18) and (14.10), and illustrated in Fig. 40 (for tables for $\gamma = 1.3$ and 1.4 cf. [6]). Both φ and ψ are singular at sonic speed and at vacuum speed, and it is assumed in the following that both $(M-1)$ and $1/M$ are bounded away from zero[1]. To construct the Riemann functions it is then convenient to choose a base point (α_0, β_0), preferably within the region of the characteristic plane to be covered by the solution, to consider the Riemann functions as functions of the variables

$$\alpha' = \alpha - \alpha_0, \quad \beta' = \beta - \beta_0,$$
$$\xi' = \xi - \alpha_0, \quad \eta' = \eta - \beta_0,$$

and to write (24.6) in the form

$$\frac{\partial^2 R}{\partial\xi'\,\partial\eta'} + \varphi_0 R = \varphi_1(\xi' - \eta')\, R, \tag{24.12}$$

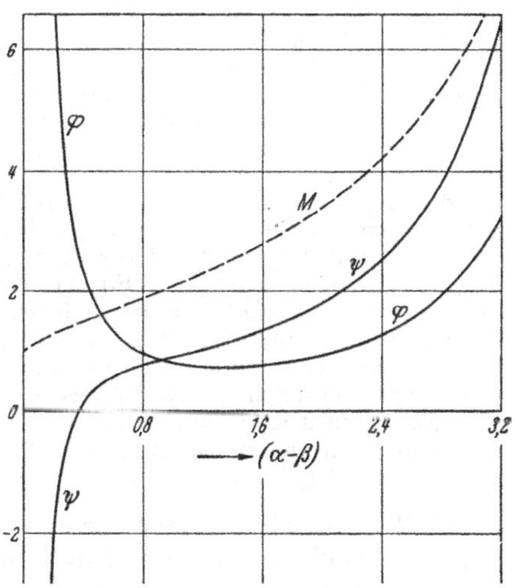

Fig. 40. Functions $\varphi(\alpha-\beta)$ and $\psi(\alpha-\beta)$ for the perfect gas with $\gamma = 1.4$.

where

$$\varphi_0 = \varphi(\alpha_0 - \beta_0), \quad \varphi_1(\xi' - \eta') = \varphi_0 - \varphi(\xi' - \eta' + \alpha_0 - \beta_0).$$

[1] For first approximations for the excluded cases cf. [6].

If the right-hand side of (24.12) were known, it would be the telegraph equation with Riemann function [9]

$$R_0(w, z; \xi', \eta') = J_0 [2 \sqrt{\varphi_0 (w - \xi')(z - \eta')}], \tag{24.13}$$

where J_0 denotes the Bessel function of order zero, and the Riemann representation of its solution would be

$$R(\xi', \eta'; \alpha', \beta') = R_0(\xi', \eta'; \alpha', \beta') + \\ + \int_{\alpha'}^{\xi'} \int_{\beta'}^{\eta'} \varphi_1 (w - z) R_0 (w, z; \xi', \eta') R(w, z; \alpha', \beta') \, dw \, dz.$$

This integral equation for R can be solved by iteration, starting from $R = R_0$, the convergence of which can be proved by PICARD's method [9], provided the dashed variables are not too large in absolute magnitude. The same method yields S, if φ is replaced by ψ. For most

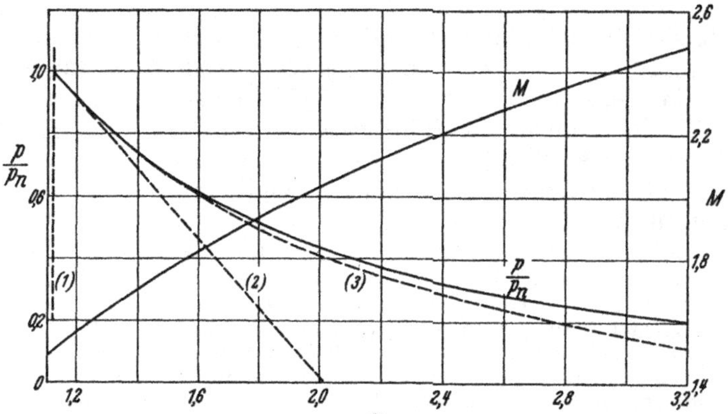

Fig. 41. Pressure and Mach number distributions on the axis of symmetry of the first wave-interaction region in an ideal jet of perfect gas, with $\gamma = 1.3$, issuing from a perfect nozzle for $M = 1.5$. Distance x measured from nozzle exit in multiples of half the distance between the lips. (From [6]).

practical problems, however, the first approximation to R represented by (24.13) and the analogous approximation to S obtained by replacing φ_0 by $\psi(\alpha_0 - \beta_0)$, will be found sufficiently accurate. For higher approximations cf. [6].

The application of RIEMANN's method is illustrated by Fig. 41, which shows the initial fall of pressure along the axis of an ideal jet issuing from a perfect nozzle (for details cf. [6]). On the approximation of linearized theory[1], the pressure falls discontinuously as shown by the dashed curve (1). The trivial approximation of characteristics theory can be interpreted in two ways. If the solution is expanded systematically in powers of $p/p_n - 1$, where p_n is the pressure in the mouth of the nozzle, then the approximation corresponding to (24.11) is given by the tangent (2) in Fig. 41. On the other hand, if (24.11) is regarded as the first of a sequence of successive approximations to the Riemann functions and is accordingly employed in conjunction with the *exact* representation of the solution given by (24.9), (24.10) together with the formulae of Sects. 14 and 19, then the curve (3) is obtained! The approximation to the solution similarly obtained from the first approximation, (24.13), to the Riemann function is indistinguishable from the exact solution in Fig. 41. The example suggests the conjecture that the transformation leading from the equations of motion to (24.9), (24.10) accounts exactly for some of the important non-linear elements common to all solutions.

25. The mixed boundary value problem. The Cauchy and characteristic boundary value problems are associated entirely with "firm" boundaries, in the sense that the whole boundary is a known curve in both the flow plane and the characteristic plane, since the data prescribe all of x, y, α and β at every point of the boundary[2]. For the mixed problem (Sect. 18) part of the boundary is

[1] L. PRANDTL: Phys. Z. **5**, 599 (1904).

[2] On a characteristic boundary segment only one of x and y, or a single relation between them will in general be prescribed explicitly. Since x and y are also related by (14.11) or (14.12), however, a quadrature suffices to determine both.

characteristic and similarly firm, but the other part is noncharacteristic with data less complete than for a Cauchy boundary and hence, is "floating" in the flow plane or the characteristic plane[1]. For instance, if the non-characteristic boundary segment of a mixed problem is a boundary segment of an ideal jet, the pressure is a given constant on it, and so is $(\alpha-\beta)$, but it is a floating boundary in the flow plane. A case of particular importance is that where the non-characteristic boundary segment of a mixed problem is a streamline of given shape on which the pressure distribution is not given to begin with, since this formulation corresponds to the "direct" problem of aerofoil theory.

RIEMANN'S formulae (24.8) or (24.9), (24.10) can be applied also to the mixed problems, by supposing the non-characteristic boundary segment was firm and then using the formulae to "predict" the given data on the characteristic segment. What is obtained in this way is not a representation of the solution, however, but an *integral equation* (in the case of a boundary segment floating in the flow plane) or *integro-differential equation* (in the case of a boundary segment floating in the characteristic plane) the solution of which furnishes just the information required to make the floating boundary segment firm. In many cases, this is precisely the information sought, and in any case, a renewed application of the Riemann representation furnishes the complete solution.

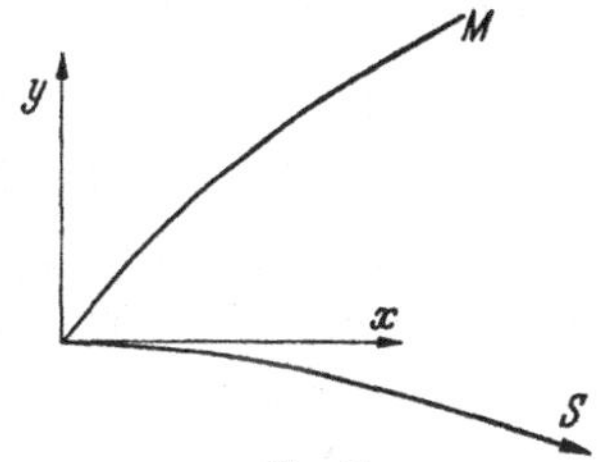

Fig. 42.

As an example, consider the reflection of a given simple wave $\beta\equiv\beta_0$ from a curved streamline S of given shape. This problem arises, for instance, in the calculation of the flow on the "upper" side of an aerofoil immersed in a non-uniform, homentropic stream (Fig. 42). Its solution predicts the exact flow field when the incidence is such that a nose shock is not formed on the upper side. When a nose shock is present, but the incident stream is only slightly non-uniform, the solution predicts the flow to the second order[2]. The boundary conditions are that $\beta=\beta_0$ and x and y are known as functions of α on the minus Mach line M (Fig. 42) which terminates the simple wave; it follows from (19.2) and (19.8) that $U(\alpha,\beta_0)=U_M(\alpha)$ is also known there. On the streamline, x, y and $\theta=(\alpha+\beta)/2$ are given as functions of the arc length s, measured from the common point of the lines M and S (Fig. 42), and U, V, α and β are related by (21.1). The immediate problem is to determine α as a function of s, $\alpha(s)$, on the streamline, whence $\omega(s)=\alpha(s)-\theta(s)$ and the pressure distribution on the streamline follow by (14.10) and the homentropic relations. Since $\beta(s)=2\theta(s)-\alpha(s)$, the streamline then becomes a Cauchy boundary, and the complete solution can be found by RIEMANN'S method (Sect. 24), if required.

Now apply (24.9), with P chosen on the Mach line M, and the streamline S chosen as boundary (Fig. 42). V may be eliminated by (21.1), and (19.1), (19.2) and (21.1) may be employed to express U on the streamline in terms of $\alpha(s)$ by

$$U(\alpha(s),\, 2\theta(s)-\alpha(s))=g(\alpha(s)-\theta(s))\,\frac{ds}{d\alpha}, \tag{25.1}$$

where

$$\frac{1}{g(\omega)}=2f\cos\mu, \tag{25.2}$$

μ being understood as the function of ω given by (14.18). Then (24.9) reads

$$\left.\begin{aligned} U_M(\alpha(s)) &= g(\alpha(s)-\theta(s))\,\frac{ds}{d\alpha}-\int\limits_0^s\Big\{m(\alpha(s')-\theta(s'))\,R+\\ &\quad+\frac{1}{2}\left(\frac{\partial R}{\partial\xi}+\frac{\partial S}{\partial\xi}-\frac{\partial R}{\partial\eta}+\frac{\partial S}{\partial\eta}\right)\Big\}\,g(\alpha(s')-\theta(s'))\,ds',\end{aligned}\right\} \tag{25.3}$$

[1] The hodograph plane is equivalent to the characteristic plane, for the purpose of the present discussion, since the transformation between the two planes is 1 — 1, for $1<M<\infty$, and given explicitly by (14.9), (14.10) and (14.18).

[2] In the thickness ratio and incidence.

where the argument and parameter points of the Riemann functions are $(\xi, \eta) = (\alpha(s'), 2\theta(s') - \alpha((s')))$ and $(\alpha, \beta) = (\alpha(s), \beta_0)$, respectively, and this is the integro-differential equation for $\alpha(s)$. For its solution by expansion in powers of a slenderness parameter cf. [6].

The integral equation, and the Riemann functions altogether, can be avoided, for instance, by expanding U and V in double power series with respect to α, and β to be determined directly from (19.10), (19.11) and the boundary conditions. Of course, many different series may be required when a number of wave fronts are present, but on the other hand, very few terms of each series are sometimes needed [6]. The double power series approach may also be used [6] to reduce the mixed boundary value problem to one of inverting a sequence of matrices, and this formulation may be practical when a solution of very high accuracy is sought and machines are available. Convergence of such series can be proved by deriving them from an iteration, so that some form of PICARD's method [9] can be applied, provided the range of variation of α and β is suitably restricted.

II. Flow with entropy variation.

26. Characteristic equations. As noted in Sect. 10, the specific entropy S of an inviscid gas can change during its motion only due to the addition of heat, which will be excluded in the following. In steady flow, the entropy then remains constant along any one streamline,

$$S = \text{const on the lines of local slope } dy/dx = \tan\theta. \tag{26.1}$$

Similarly, the stagnation enthalpy remains constant on streamlines[1]. The limit of zero viscosity and heat conductivity, however, can rarely be approached uniformly throughout a flow field. In particular, the effect of dissipation and heat conduction remains in general appreciable where the gas passes through a shock. If the shock thickness may be neglected, the stagnation enthalpy remains constant on streamlines[2], and it will be assumed in the following that it is uniform throughout the flow. The entropy, on the other hand, suffers a discontinuous rise across the shock, the magnitude of which is given by the shock equations (Sect. 4 of CABANNES' article in this volume), and if the shock is not plane, the entropy cannot be uniform on both sides of the shock. CROCCO's Theorem[1] shows that a variation of S from streamline to streamline implies the presence of vorticity, and thus (14.2) fails. The equations governing the flow of a perfect gas with constant specific heats are therefore the equation of continuity (14.1),

$$\frac{\partial(\varrho u)}{\partial x} + \frac{\partial(\varrho v)}{\partial y} = 0, \tag{26.2}$$

the Eulerian equations of motion,

$$\varrho u \frac{\partial u}{\partial x} + \varrho v \frac{\partial u}{\partial y} = -\frac{\partial p}{\partial x}, \tag{26.3}$$

$$\varrho u \frac{\partial v}{\partial x} + \varrho v \frac{\partial v}{\partial y} = -\frac{\partial p}{\partial y}, \tag{26.4}$$

the adiabatic relation (14.16),

$$\frac{1}{2}(u^2 + v^2) + \frac{a^2}{\gamma - 1} = \text{const}, \tag{26.5}$$

[1] Cf. Sect. 25 of OSWATITSCH's article in Vol. VIII, Part 1, of this Encyclopedia.
[2] Sect. 4 of CABANNES' article in this volume.

and the equation of state (3.12),

$$p\varrho^{-\gamma}\exp(-S/c_v) = \text{const}. \tag{26.6}$$

These equations are seen to imply (26.1), which has the form of a characteristic equation. A systematic search for further characteristics (Sect. 5 of SCHIFFER'S article in this volume) shows that there are just two more families, when the velocity magnitude q exceeds the speed of sound a, and leads to the additional characteristic equations

$$\left.\begin{array}{l} d\theta = \dfrac{dp}{\varrho q^2 \tan\mu} \quad \text{on the lines of local slope } \dfrac{dy}{dx} = \tan(\theta-\mu) \\ \qquad\qquad (\text{"plus" Mach lines}) \end{array}\right\} \tag{26.7}$$

and

$$\left.\begin{array}{l} d\theta = \dfrac{-dp}{\varrho q^2 \tan\mu} \quad \text{on the lines of local slope } \dfrac{dy}{dx} = \tan(\theta+\mu) \\ \qquad\qquad (\text{"minus" Mach lines}), \end{array}\right\} \tag{26.8}$$

where μ is again the Mach angle defined by (14.7). It is easy to verify that (26.1) and (26.7), (26.8), together with (26.5), (26.6) are equivalent to the system (26.2) to (26.6).

The relation of (26.7), (26.8) to the characteristic Eqs. (14.11), (14.12) of homentropic flow becomes clearer when it is noted that, by (14.15) and (14.7), differentiation of (26.5), (26.6) leads to

$$\gamma(\gamma-1)\,a^{-2}q\,dq = \gamma\varrho^{-1}d\varrho - \gamma p^{-1}dp = -(\gamma-1)\,p^{-1}dp - c_v^{-1}dS,$$

i.e.

$$\lambda^{-1}d\omega + (\gamma-1)\,p^{-1}dp + c_v^{-1}dS = 0, \tag{26.9}$$

where

$$\lambda = \sin\mu\cos\mu/[\gamma(\gamma-1)] \tag{26.10}$$

and ω is again the Prandtl angle defined by (14.9). By (14.15) and (14.7), moreover,

$$\varrho q^2\tan\mu = \gamma p/(\sin\mu\cos\mu),$$

so that (26.7), (26.8) can, by (26.9) and (14.10), be written

$$d\alpha = -\lambda c_v^{-1}dS \quad \text{when} \quad \frac{dy}{dx} = \tan(\theta-\mu), \tag{26.11}$$

$$d\beta = \lambda c_v^{-1}dS \quad \text{when} \quad \frac{dy}{dx} = \tan(\theta+\mu). \tag{26.12}$$

It is worth noting that these equations admit a simple particular solution such that θ and p, but not S, are uniform in a region. Such a region may be called one of *pure shear flow*, since the variation of S implies also a variation of q from streamline to streamline, by (26.5), (26.6) and (14.15).

A step-by-step computation of the solution at a network of points can be carried out in a manner similar to that described in Sect. 17, except that use has to be made also of (26.1) at every step, since (26.5), (26.6) and (26.11), (26.12) are not now a complete set of equations. A procedure is described e.g. in [*3*]; the problem of accuracy and economy has received little attention yet.

In most physical problems, the variation of entropy is due to the presence of shocks the positions and shapes of which are not known a priori in the flow plane. The shock equations then play the role of floating boundary conditions in the computation, and it is on account of this complication that the question of existence and uniqueness has not yet been discussed to the same degree as for homentropic flow. It is easy to see from the shock equations, however, that they do not determine the shock, if only the flow field upstream of the shock is known [4], and the calculation of the shock is therefore linked intimately with that of the rotational flow field downstream.

27. Shock-expansion theory. The problem which has received most attention is that of the *weak shear flow* encountered when a uniform stream has passed through a curved shock producing entropy variations which are appreciable, but not sufficient to demand a treatment to more than an approximation. This situation occurs, in particular, in the supersonic flow past an aerofoil of moderately small thickness and incidence (Fig. 18) with a shock attached to the sharp nose. The common approach is based, as in Sect. 11, on the observation that the rise of entropy across the shock is much smaller than the rise of pressure and the stream deflection. Indeed, for sufficiently weak shocks a homentropic approximation suffices (Sect. 16). Moreover, it would appear that PILLOW'S proof (Sect. 11) can be extended to the case of steady two-dimensional flow[1], and thus the shock-expansion theory of Sect. 12 can also be so extended. On the upper side of the aerofoil (Fig. 18), the stream deflection at the shock is anti-clockwise so that the change of α across the shock is of the order of the entropy rise [cf. (23.1)] and the relation corresponding to (11.11) is

$$\alpha = \alpha(S). \tag{27.1}$$

This implies, first, that α is constant on streamlines, and in particular, on the upper surface of the aerofoil. This explains, at least in part, the long-standing success of the engineering conjecture[2] that the surface pressure and surface slope are related by (16.1) and (14.10), with $\alpha = \text{const}$. Secondly, it follows from the characteristic equations, in a manner similar to that described in Sect. 12, that the minus Mach lines are isoclines and isobars of the flow above the aerofoil (Fig. 18), even if not isovels nor straight (as on the simple-wave approximation of Sect. 16). Finally, (27.1) encompasses (26.11), to the shock-expansion approximation, and the remaining two characteristic Eqs. (26.1), (26.12), together with the auxiliary Eqs. (26.5), (26.6) and (27.1), can be solved in a manner similar to that described in Sect. 12[3].

D. Steady axially symmetrical supersonic flow.

28. Characteristic equations. The equations governing steady, homentropic and irrotational flow (to which the discussion will be restricted) with axial symmetry and with velocity everywhere parallel to the Meridian planes differ from those of Sect. 14 only in the equation of continuity, which is now[4]

$$r\frac{\partial(\varrho u)}{\partial x} + \frac{\partial(r\varrho v)}{\partial r} = 0, \tag{28.1}$$

if r, u, v denote respectively the distance from the axis of symmetry and the axial and radial velocity components. Since (28.1) differs from (14.1) only by a term not involving derivatives, the characteristic ("Mach") directions in the

[1] A sketch of such an extension has been given by J. P. GUIRAUD: C. R. Acad. Sci., Paris **245**, 1778 (1957).

[2] Due to P. S. EPSTEIN: Proc. Nat. Acad. Sci. Wash. **17**, 532 (1931).

[3] R. E. MEYER: Quart. Appl. Math. **14**, 433 (1957). A thorough discussion from an engineering point of view will be found in W. D. HAYES and R. P. PROBSTEIN: Hypersonic Theory. New York 1959.

[4] Cf. Sect. 12 of SERRIN's article in Vol. VIII, Part 1 of this Encyclopedia.

Meridian plane are the same as in two-dimensional flow, but when the equations are referred to local (Cartesian) coordinate systems based on those directions, they are found to imply[1]

$$\frac{d\alpha}{dr} = \frac{\sin\theta \sin\mu}{r\sin(\theta-\mu)} \quad \text{when} \quad \frac{dr}{dx} = \tan(\theta-\mu) \quad \text{(plus Mach lines)}, \tag{28.2}$$

$$\frac{d\beta}{dr} = \frac{-\sin\theta \sin\mu}{r\sin(\theta+\mu)} \quad \text{when} \quad \frac{dr}{dx} = \tan(\theta+\mu) \quad \text{(minus Mach lines)}, \tag{28.3}$$

where the symbols have the same meaning as in Sect. 14. The inhomogeneous form of the left-hand pair of Eqs. (28.2), (28.3) shows that simple wave regions (throughout which $\alpha \equiv \text{const}$ or $\beta \equiv \text{const}$) do not exist.

The step-by-step computation of a solution at a network of points, for numerically prescribed boundary conditions, can be carried out by an extension [*3*] of the process sketched in Sect. 17. For the computation of the flow outside a body of revolution it is advantageous, however, to replace (28.2), (28.3) by OSWATITSCH's equations[2]. The question of economy of labor has been studied by W. J. TURNER who found that any mean values occurring in the difference equations [such as t_+, t_- in (17.1)] are best approximated by rough extrapolation from those computed in earlier steps; where this is impossible near the boundaries of the network, the method of HOLT[3] may be used. The limiting form [*3*] of (28.2), (28.3) near the axis of symmetry was derived by GOLDSTEIN on the assumption that u is a differentiable function of x; when it is not, a modification of the equations to take full account of CHEN's results (Sect. 31) is advantageous.

When $r \neq 0$, (28.2), (28.3) have a similar structure as (14.11), (14.12), and the results regarding existence and uniqueness described in Sects. 5 and 18 apply also to axially symmetrical flow. In particular, as in two-dimensional homentropic flow, if the solution is determined in a region, then the extent of this region is such that every Mach line in it meets the line on which boundary conditions are prescribed. On the axis, $\theta \equiv 0$; the singularity of (28.2), (28.3) which occurs there requires special investigation (Sect. 31).

29. Focusing equations. As in two dimensions, the study of the flow is greatly assisted by the use of characteristic independent variables. If ξ and η are constant on plus and minus Mach lines, respectively, the characteristic Eqs. (28.2), (28.3) may be written

$$\frac{\partial x}{\partial \eta} = h_\eta \cos(\theta-\mu), \quad \frac{\partial r}{\partial \eta} = h_\eta \sin(\theta-\mu), \quad \frac{\partial \alpha}{\partial \eta} = h_\eta r^{-1} \sin\theta \sin\mu, \tag{29.1}$$

$$\frac{\partial x}{\partial \xi} = h_\xi \cos(\theta+\mu), \quad \frac{\partial r}{\partial \xi} = h_\xi \sin(\theta+\mu), \quad \frac{\partial \beta}{\partial \xi} = - h_\xi r^{-1} \sin\theta \sin\mu, \tag{29.2}$$

where h_ξ, h_η are parameters analogous to h_α, h_β of Sect. 19. In contrast to the two-dimensional flow, the characteristics are not fixed lines in the hodograph plane, and it is now necessary to distinguish not only the flow plane (i.e. the x, r-plane) from the characteristic ξ, η-plane, but also the hodograph plane, which is here represented by the α, β-plane[4].

---

[1] Sect. 49, *ibid.*

[2] K. OSWATITSCH: Öst. Ing.-Arch. **10**, 359 (1956).

[3] M. HOLT: Quart. J. Mech. Appl. Math. **2**, 473 (1949).

[4] The relation between q, θ and α, β is still given by (14.9), (14.10), (14.7) and (19.5), and more explicitly by (14.10), (14.16) and (14.18), (14.19) for a perfect gas.

If f denotes again the propagation function (19.9) and

$$h_\xi = f\,A(\xi,\eta)\,U(\xi,\eta), \qquad h_\eta = f\,B(\xi,\eta)\,V(\xi,\eta), \tag{29.3}$$

$$\sigma = r^{\frac{1}{2}}\frac{\partial\alpha}{\partial\xi}, \qquad \tau = r^{\frac{1}{2}}\frac{\partial\beta}{\partial\eta}, \tag{29.4}$$

with

$$A^{-1}\frac{\partial A}{\partial\eta} = (m - \cot\mu)\frac{\partial\alpha}{\partial\eta}, \qquad B^{-1}\frac{\partial B}{\partial\xi} = (\cot\mu - m)\frac{\partial\beta}{\partial\xi}, \tag{29.5}$$

then by equating the respective mixed second derivatives of r and x with respect to ξ and η and making use of (29.1) to (29.4), (19.9), (19.5) and (14.10),

$$\frac{\partial U}{h_\eta\,\partial\eta} = \frac{m\,\sigma}{f\,A}r^{-\frac{1}{2}}, \qquad \frac{\partial(A\,\sigma)}{h_\eta\,\partial\eta} = A\left(\frac{3}{2}r^{-\frac{1}{2}}\sin(\theta+\mu) + \frac{m\,\tau}{h_\eta}\right)\frac{\partial\beta}{\partial\xi}, \tag{29.6}$$

$$\frac{\partial V}{h_\xi\,\partial\xi} = -\frac{m\,\tau}{f\,B}r^{-\frac{1}{2}}, \qquad \frac{\partial(B\,\tau)}{h_\xi\,\partial\xi} = B\left(\frac{3}{2}r^{-\frac{1}{2}}\sin(\theta-\mu) - \frac{m\,\sigma}{h_\xi}\right)\frac{\partial\alpha}{\partial\eta}, \tag{29.7}$$

are obtained, which correspond to (19.12), (19.13).

Since $\partial\beta/\partial\xi$ may be expressed in terms of U by (29.2), (29.3), $A\sigma$ and U are seen to satisfy a homogeneous system of first-order differential equations on any one plus Mach line. Thus, if a finite discontinuity of σ or U occurs at any point P where $r \neq 0$, then there is no point on the plus Mach line through P at which neither σ nor U is discontinuous, and similarly for τ and V on minus Mach lines. This result can be generalized, on the lines of Sect. 6, to show again that the wave front property is invariant under continuation. The reflection of wave fronts at the axis $r=0$, however, is of a curious nature (Sect. 31).

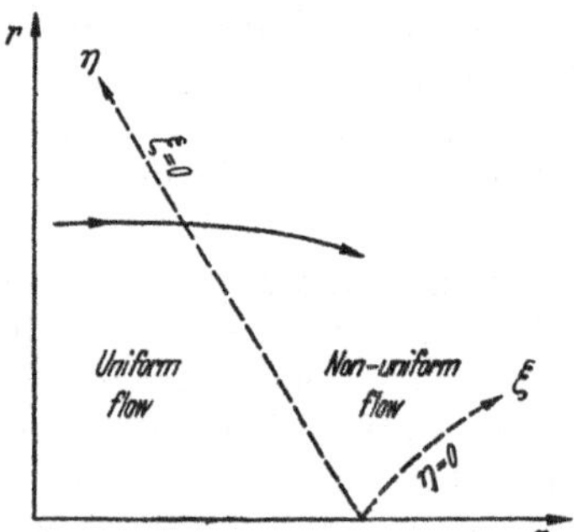

Fig. 43.

30. A wave front. The understanding of the features distinguishing the axiallysymmetrical solutions of the equations of motion from the two-dimensional ones has been accelerated by a study[1] of the wave fronts forming the borders of regions of uniform flow. In such regions α and β are constant, and it follows from (28.2), (28.3) that $\theta \equiv 0$, i.e. the flow is a purely axial one. With velocity and pressure assumed continuous functions of x and r, α and β retain their constant values also on the wave front, which must be a Mach line, since only the Mach directions are characteristic. For definiteness, assume it to be a plus Mach line (Fig. 43), say $\xi=0$. From (29.6) and (29.2), since $\theta=0$,

$$\sigma = \text{const on } \xi = 0, \tag{30.1}$$

and since m, f and A are all constant on the wave front, by (19.5), (19.9), (29.5) and (14.10), Eqs. (29.6) and (29.1) yield

$$U = -\frac{2m\,\sigma}{f A\sin\mu}r^{\frac{1}{2}} + c \qquad \text{on } \xi = 0. \tag{30.2}$$

Since even $\partial\beta/h_\xi\,\partial\xi = 0$ on $\xi=0$, by (29.2) the pressure gradient on the wave front is proportional to $\Pi = (\partial\alpha/\partial\xi)/(\partial r/\partial\xi)$, and by (29.4), (29.2), (29.3) and (30.2),

$$\Pi^{-1} = r^{\frac{1}{2}}(c_1 - 2m\,r^{\frac{1}{2}}), \tag{30.3}$$

[1] R.E. MEYER: Quart. J. Mech. Appl. Math. **1**, 451 (1948).

where c_1 is a constant of integration and m is given by (19.5). In particular,

$$\text{as } r \to 0, \quad \Pi = O(r^{-\frac{1}{2}}), \tag{30.4}$$

but

$$\text{as } r \to \infty, \quad \Pi \sim -2 m r^{-1}, \tag{30.5}$$

if $\Pi \neq 0$ on the wave front. The first of these two results demonstrates the existence of a marked *radial focusing effect*. [The second is analogous to the result obtained in two-dimensional flow from (19.6), (19.8) and (19.12). Moreover, if $c_1 \geqq 0$, a limit point with singularity of the pressure gradient occurs at $r^{\frac{1}{2}} = c_1/(2m)$, again in analogy with conditions in two-dimensional flow.]

31. The singularity at the axis. The singularity indicated by (30.4) was first studied by the help of Linearized Theory, which was expected to apply because the radial velocity component vanishes at the axis. The results, however, were patently unsatisfactory, and the nature of the singularity was elucidated only by CHEN's[1] searching investigation of the existence of continuous solutions of the non-linear equations near the axis.

For definiteness, let ξ and η be chosen so that $\xi + \eta = 0$ is the image of the axis $r = 0$ and

$$\frac{\partial x}{\partial \xi} = 2 \cot \mu \quad \text{on} \quad \xi + \eta = 0. \tag{31.1}$$

Then from (29.1) to (29.3), since also $\theta = 0$ on the axis,

$$A U = -B V = (f \sin \mu)^{-1} \quad \text{on} \quad \xi + \eta = 0, \tag{31.2}$$

and if

$$r(\xi, \eta) = (\xi + \eta)\big(1 + w(\xi, \eta)\big), \tag{31.3}$$

(29.1), (29.2) and (31.2) show that

$$w(\xi, -\xi) = 0, \tag{31.4}$$

and thus the last of Eqs. (29.1) may be written

$$\frac{\partial \alpha}{\partial \eta} = -\theta/(\xi + \eta) + a_1(\xi, \eta), \tag{31.5}$$

where $a_1 = (\xi + \eta)^{-1}\theta[1 + \theta^{-1} \sin\theta (1 + w)^{-1} f B V \sin \mu]$ becomes negligible, as $(\xi + \eta) \to 0$, compared with $\theta/(\xi + \eta)$. Similarly,

$$\frac{\partial \beta}{\partial \xi} = -\theta/(\xi + \eta) + a_2(\xi, \eta), \tag{31.6}$$

and a further differentiation leads, by (14.10), to

$$\frac{\partial^2 \theta}{\partial \xi \, \partial \eta} + \frac{1}{2}(\xi + \eta)^{-1}\left(\frac{\partial \theta}{\partial \xi} + \frac{\partial \theta}{\partial \eta}\right) - (\xi + \eta)^{-2}\theta = \frac{1}{2}\left(\frac{\partial a_1}{\partial \xi} + \frac{\partial a_2}{\partial \eta}\right), \tag{31.7}$$

$$\frac{\partial^2 \omega}{\partial \xi \, \partial \eta} + \frac{1}{2}(\xi + \eta)^{-1}\left(\frac{\partial \omega}{\partial \xi} + \frac{\partial \omega}{\partial \eta}\right) = (\xi + \eta)^{-1}(a_1 - a_2) + \frac{1}{2}\left(\frac{\partial a_1}{\partial \xi} - \frac{\partial a_2}{\partial \eta}\right), \tag{31.8}$$

and closer inspection confirms that every term on the right-hand sides of these two equations becomes negligible, as $(\xi + \eta) \to 0$, compared with at least one of the terms occurring on the respective left-hand sides. The differential operators on the left-hand sides, moreover, are linear and closely related to that of (8.2), so that they possess Riemann functions similar to (9.9). Eqs. (31.7), (31.8) may

[1] Y.W. CHEN: Comm. Pure Appl. Math. **5**, 57 (1952); **6**, 179 (1953).

therefore be transformed into integral equations, for any given characteristic or Cauchy boundary conditions, by supposing the right-hand sides to be known functions of ξ and η and formulating the Riemann representation (Sect. 9 or [9]) of the solution of these "inhomogeneous" equations. The integral equations, in turn, can be solved by iteration[1], starting from the first approximation furnished by the solution of (31.7), (31.8) with $a_1 \equiv a_2 \equiv 0$. It is significant that this approach, in addition to splitting each of the focusing Eqs. (29.6), (29.7) into a major, linear part and a minor, non-linear part, also leads to the conclusion that there should be a definite order of attack, the right-hand pair of (29.6), (29.7) [which is equivalent to the pair (31.7), (31.8)] being treated first, at each stage of iteration, and the left-hand pair thereafter. This means that, at each stage, the characteristic pattern in the α, β-plane (or hodograph plane) is worked out first, and then used to work out the Mach line pattern in the flow plane (i.e. meridian plane).

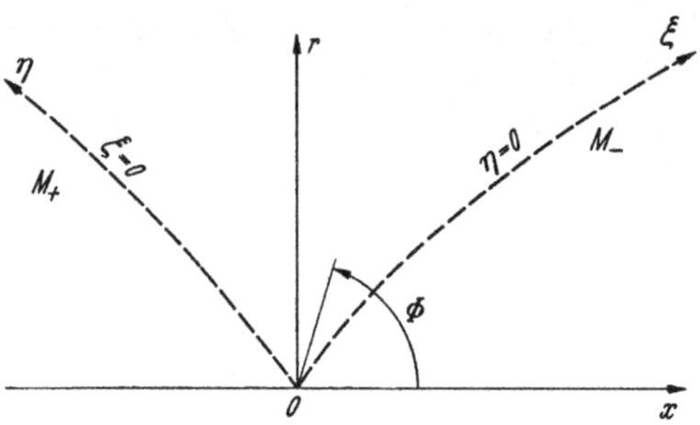

Fig. 44.

By estimating the Riemann functions sufficiently closely, CHEN deduced the following results regarding the existence and uniqueness of solutions in a sufficiently restricted neighbourhood of the axis.

Regular solutions. If the radial velocity component v is prescribed as a function $v(r)$ on a Mach line M_-, (Fig. 44) such that

$$v(r) = O(r) \quad \text{on } M_-, \tag{31.9}$$

then the pressure and velocity are continuously differentiable functions of position between M_- and the half-axis $r=0$, $x>0$, on which the Mach number

$$M(x) = M(0) + O(x), \quad x > 0. \tag{31.10}$$

Conversely, the boundary condition (31.10) on the half axis $r=0$, $x>0$, implies (31.9). Moreover, if v is also prescribed on the plus Mach line M_+ meeting M_- at the axis (Fig. 44) such that

$$v(r) = O(r) \quad \text{on } M_+, \tag{31.11}$$

then the solution is not only similarly determined between M_+ and the half-axis $r=0$, $x<0$, but also in the sector between M_+ and M_-, and pressure and velocity are uniformly differentiable functions of position near O (Fig. 44). Indeed, not only are U, V, σ, τ (Sect. 29) uniformly bounded[2], but also $\sigma \to 0$, $\tau \to 0$ as $r \to 0$.

The same results hold if the order terms in (31.9) to (31.11) are r^ε, x^ε and r^δ, respectively, with $\varepsilon > \frac{1}{2}$, $\delta > \frac{1}{2}$, except that pressure and velocity are not uniformly differentiable, if ε, δ are not both ≥ 1. A local singularity of derivatives then occurs at O, being bridged in the flow so that, if ζ denotes the lesser of ε, δ, the quantities $r^{-\zeta}\theta$, $r^{1-\zeta}\,\partial\theta/\partial r$ and $r^{1-\zeta}\,\partial\theta/\partial x$ tend to continuous functions of the angle φ (Fig. 44) as $(x^2+r^2) \to 0$, and the Mach number distribution and its derivatives behave similarly.

Singular solutions. If the boundary condition

$$v = v(r) = k\,r^{\frac{1}{2}} + o(r^{\frac{1}{2}}), \quad k = \text{const}, \quad \text{on } M_- \tag{31.12}$$

[1] Cf. footnote 1, p. 277. To complement (31.7), (31.8), CHEN used, rather than the first pair of (29.6), (29.7), the equivalent second-order equation for $(1+w)$, also split into its major linear and minor non-linear part.

[2] That A, B must be bounded even at the axis $r=0$, $\theta=0$ for all solutions of the equations of motion follows from (29.5) and (31.5), (31.6).

(Fig. 44) is prescribed, then the pressure and velocity are continuous between M_- and the half-axis $r=0$, $x>0$, on which the Mach number

$$M(x) = M(0) + k_1 x^{\frac{1}{2}} + o(x^{\frac{1}{2}}), \qquad \text{for } x>0, \tag{31.13}$$

with $k_1/k = \text{const} > 0$. Conversely, the boundary condition (31.13) implies (31.12). Moreover, if the additional boundary condition (31.11) is prescribed, the solution is determined also in the sector between M_+ and M_- (Fig. 44) and pressure and velocity are uniformly continuous, but their gradients become $O(r^{-\frac{1}{2}})$ as M_+ is approached. As noted before, (31.11) implies a regular solution between M_+ and the half axis $r=0$, $x<0$ (Fig. 44). The Mach line M_+ is therefore a *wave front* with a finite discontinuity of σ, as in the example of Sect. 30. In the sector between M_+ and the right-hand half-axis ($r=0$, $x>0$) U, V, σ and τ tend to non-zero, finite values depending continuously on the angle φ (Fig. 44), except on M_-, where τ and V are singular!

To understand the nature of this singularity on the "reflected" Mach line M_- it is necessary to consider the characteristic plane. To form a more definite picture of its relation to the flow plane, we may complete the choice of ξ and η by putting $\xi=0$ on M_+, $\eta=0$ on M_-. [Since the distribution of Mach number is continuous near O (Fig. 44), it follows from (31.1) and (14.7) that

$$\partial r/\partial\eta \to 1 \quad \text{on } M_+, \qquad \partial r/\partial\xi \to 1 \quad \text{on } M_-, \qquad \text{as } r\to 0, \tag{31.14}$$

so that we may think of ξ and η as being equal to r, to a first approximation, on M_- and M_+, respectively.] Then CHEN's result is

$$\tau = O\left(\log\frac{\eta}{\xi}\right), \qquad V = O(\xi^{\frac{1}{2}}\tau) \tag{31.15}$$

as $\eta/\xi\to 0$ and $\xi\to 0$.

These properties are already exhibited by the first approximation to the solution, i.e. they follow from (31.7), (31.8) with $a_1\equiv a_2\equiv 0$. It is worth noting that these truncated equations become identical with the equations of Linearized Theory (Sects. 15, 21 of SCHIFFER's article in this volume) for v or θ and u or ω, if ξ is replaced by $(br+x)$, and η by $(br-x)$, with $b^2=M_0^2-1$, and it is with this mistaken interpretation that (31.15) was first discovered. Indeed, the singularity of V shows that the Mach line pattern in the meridian plane of the flow must show a strong local distortion near M_- (Fig. 44), and hence it is a serious error to confuse (ξ,η) with $(br+x, br-x)$. Observe, in particular, that the distortion so matches the singularity of τ that

$$\frac{\partial\beta/\partial\eta}{\partial r/\partial\eta} = O(r^{-1}), \qquad \text{on } M_-$$

by (31.15), (29.4), (29.1), (29.3), (31.14) and since B is bounded[1], and hence the physical velocity and pressure gradients are *continuous* on M_- for $r>0$, becoming singular (like r^{-1}) only as $r\to 0$.

None the less, it turns out that the solution is only a formal, mathematical one, when $k<0$ in (31.12), i.e. when the pressure rises in the stream direction. An inspection[2] of the signs of the leading terms shows that V then changes sign near M_-, i.e. a limit line[3] occurs at which the mapping of the α,β- and ξ,η-planes into the x,r-plane is folded and the pressure gradient is singular, so that shock formation must be anticipated. When the pressure falls in the

[1] Cf. footnote 2, p. 278.

[2] Y.W. CHEN: Comm. Pure Appl. Math. **6**, 179 (1953).

[3] With two branches, one on each side of M_-, merging at O (Fig. 44).

stream direction ($k>0$), on the other hand, the mappings are regular and the singular solution is physically acceptable. It predicts that the discontinuity of the pressure gradient across M_+ is reflected neither as a discontinuity, nor as a singularity, but as a remarkably rapid variation of the pressure near M_-, with a strong but quite local singularity of the pressure and velocity gradients at O (Fig. 44)[1].

32. Uniform first-order theory. It has been noted in Sect. 15, and again in the preceding section, that the Linearized Theory (Sects. 15 to 21 of SCHIFFER'S article in this volume), as conventionally formulated, does not provide an entirely consistent approach and cannot be expected in general to provide an approximation of uniform quality. An approximation free of these defects is obviously desirable for a proper understanding of the flow pattern, and in particular, for the prediction of shock formation and boundary layer separation[2].

Two approaches to a uniform first-order theory have been developed simultaneously. The first, formulated on a heuristic basis by WHITHAM[3], starts from the idea (Sect. 15) that the approximation of Linearised Theory may be inadequate, even though θ and $\omega-\omega_1$ be uniformly small[4], where ω_1 is a constant, because the Mach line pattern is inadequately approximated. Another approximation can be obtained by quadrature from the second pair of Eqs. (28.2), (28.3), taking θ and μ to be the functions of x and r which Linearized Theory predicts them to be. This approximation formulates a mapping of the characteristic plane into the flow plane, represented by two functions $x=X(\xi,\eta)$, $r=R(\xi,\eta)$. WHITHAM'S hypothesis is essentially that the perturbation potential φ predicted by Linearized Theory as a function $\varphi=\varphi(br+x, br-x)$, with $b^2 = M_1^2-1$, is such that the three functions

$$\varphi=\varphi(\xi,\eta), \qquad x=X(\xi,\eta), \qquad r=R(\xi,\eta) \tag{32.1}$$

eepresent a uniform first order approximation to the solution of the non-linear rquations of motion when $|\theta|$ and $|\omega-\omega_1|$ are uniformly small.

This turns out to be correct to an extent covering a large class of the most important physical problems. For a justification of the hypothesis on the basis of the equations of motion, however, it is necessary to confirm that the assumption of small θ and $\omega-\omega_1$ can serve to split the characteristic Eqs. (29.1), (29.2) into a major and minor part—in the manner pioneered by CHEN (Sect. 31)—such that the major part of the second-order equations for θ and ω in terms of ξ and η is identical with the equations of Linearized Theory for θ and ω in terms of $br+x$ and $br-x$. It is also necessary to re-examine the boundary conditions in the light of the characteristic equations, to check whether their conventional linearization is uniformly admissible.

The second approach, first developed for an unsteady flow problem[5], is based on the wave front principle of Sect. 6. The variation along a wave front of discontinuities of the velocity derivatives may be obtained from the focusing equa-

---

[1] Of course, an analoguous solution is obtained when the roles of M_+ and M_-, and of the half-axes $x>0$ and $x<0$, are interchanged.

[2] It may be premature to pin down the term "uniform first-order approximation". What seems primarily understood, at the present stage of gas dynamics, in view of the results noted in Sects. 7, 22, 23, is an approximation predicting not only the velocity and pressure, but also their first derivatives and the position of weak shocks, everywhere with a relative error only of the order of the slenderness parameter characterising the boundary conditions.

[3] G.B. WHITHAM: Comm. Pure Appl. Math. 5, 301 (1952).

[4] That this holds throughout the flow, if on the boundary, does not follow directly from the uniqueness theorems, as for the two-dimensional flow (footnote 3, p. 251).

[5] R.E. MEYER: Quart. J. Mech. Appl. Math. 5, 257 (1952).

tions, (29.6), (29.7). It may be interpreted, by appeal to continuity as in Sect. 6, as a general first approximation valid in a sufficiently narrow strip of Mach lines adjacent to the wave front, even if no discontinuity of velocity derivatives occurs. When θ and $\omega-\omega_1$ are small on the boundary, and only wave fronts of one characteristics family are present, "sufficiently narrow" can be shown to imply no restriction on the extent of the strip in the flow plane, and the uniformity of the approximation is guaranteed automatically by the fact that it is based directly on the non-linear equations of motion and boundary conditions.

The exclusion of wave fronts of the second characteristics family restricts the approximation, to begin with, to problems involving single progressive waves[1], rather than wave-interactions. The restriction was removed, however, by MAHONY'S[2] extension of the approximation to the flow in ducts, which led him back to precisely the system of approximate characteristic equations shown by CHEN to provide the uniform first-order approximation near the axis (Sect. 31). The two approaches are therefore seen to be closely related, and a definitive uniform first-order theory seems within sight.

The contributions of the uniform first-order theory to the understanding of the physical features of the flow lie in three main areas. Whitham[3] studied the asymptotic pattern far from a body of revolution with pointed nose in a uniform incident stream and predicted the number of shocks formed and their asymptotic shape and decay, relating these features directly to the cross-sectional area distribution of the body.

The second area is that of the flow pattern near bodies of revolution which are slender in the sense that $\theta \ll 1$, but such that the curvature of their meridian contour is not uniformly small. The assumption of conventional Linearized Theory (Sect. 10 of SCHIFFER'S article in this volume) that the velocity derivatives satisfy a condition similar to (15.1) cannot then be satisfied even near the body. The difficulty was first overcome by LIGHTHILL'S[4] elegant direct extension of Linearized Theory. A point of discontinuity of slope of the body contour is a wave centre, as in two-dimensional flow (Sects. 16, 22, 23). Much in contrast to two-dimensional flow, however, such a wave centre is followed immediately by a striking reversal of the pressure gradient. In the case of a cylindrical body with a cone as nose, for instance, the expansion centred at the "shoulder" where cone and cylinder join, is followed immediately by a rapid recompression[5], which decays only relatively slowly with distance along the cylinder.

The third area is that of the flow inside ducts, where the uniform first-order theory serves to bridge the gap between the boundary conditions (which must be given on lines at some distance from the axis, if they are to be realized directly by physical means) and CHEN'S results (Sect. 31) on the nature of the flow in the immediate neighbourhood of the axis. The illuminating example studied by MAHONY[2] is that of an expansion, with $\theta \ll 1$, centred at the wall of the duct.

---

[1] I.e. waves of the general type of which the simple waves of unsteady one-dimensional and steady two-dimensional flow and the waves studied in shock-expansion theory are examples. WHITHAM'S original study, and original hypothesis, were similarly restricted.

[2] J. J. MAHONY: Phil. Trans. Roy. Soc. Lond. A **251**, 1 (1958).

[3] Cf. footnote 3, p. 280.

[4] M. J. LIGHTHILL: Quart. J. Mech. Appl. Math. **1**, 90 (1948).

[5] For the exact limiting value of the pressure gradient as a centre of expansion is approached, cf. N.H. JOHANNESEN and R.E. MEYER: Aeronaut. Quart. **2**, 127 (1950). The recompression also leads, in general, to the formation of a "shoulder shock" at some distance from the wave centre; cf. ibid. and footnote 3, p. 280.

Such an expansion must be bounded by two wave fronts, the "head" and "tail" fronts, separating it respectively from the flow upstream and downstream. The head front is very similar to the wave front of Sect. 30, with $c_1 < 0$, and causes precisely the pattern near the axis described by CHEN'S singular solution, with $k > 0$ (Sect. 31). The tail front is also similar to the wave front of Sect. 30, but the discontinuity of the pressure gradient is an adverse one, because of the marked recompression which must follow a centred expansion in axially symmetrical flow. No limit point occurs on the tail front[1], but the pattern it causes near the axis is that of CHEN'S singular solution with $k < 0$, so that its reflection from the axis involves a limit line. The real reflection of the tail front from the axis must therefore be expected to be a shock.

Appendix.

Some references for problems with 3 or 4 independent variables.

1. Characteristic equations.

DOLPH, C.L., and N. COBURN: Proc. First Sympos. Appl. Math. 1947, Amer. Math. Soc., 55 (1948), for steady homentropic irrotational supersonic flow.

HOLT, M.: J. Fluid Mech. **1**, 409 (1956), for steady rotational supersonic flow.

THORNHILL, C.K.: Aeron. Res. Coun. Rep. & Mem. 2615 (1948), for unsteady motion in two dimensions.

BRUHN, G., and W. HAACK: Z. angew. Math. Phys. **9b**, 173 (1958), for unsteady motion in three dimensions.

2. Perturbation methods.

FERRI, A. [*4*].

HOLT, M.: J. Aero/Space Sci. **26**, 787 (1959).

3. Wave front theory.

SCHIFFER, M.M., Sect. 9 of his article in this volume.

THOMAS, T.Y.: J. Math. Mech. **6**, 455 (1957).

4. Generalised simple wave theory.

GIESE, J.: Quart. Appl. Math. **9**, 237 (1951), and Aberdeen Prov. Ground Ball. Res. Lab. Rep. 894 (1954).

NAYLOR, D.: J. Rational. Mech. Anal. **5**, 687 (1956) and J. Math. Mech. **7**, 705 (1958).

References.

[*1*] RIEMANN, B.: Ges. Werke, p. 144—167. Leipzig 1876.

[*2*] COURANT, R., and K.O. FRIEDRICHS: Supersonic Flow and Shock Waves. New York 1948.

[*3*] HOWARTH, L. (Ed.): Modern Developments in Fluid Dynamics, High-Speed Flow, Vol. I. Oxford 1953.

[*4*] SEARS, W.R. (Ed.): General Theory of High Speed Aerodynamics. Princeton 1956.

[*5*] MEYER, R.E.: Phil. Trans. Roy. Soc. Lond. A **242**, 153 (1949).

[*6*] MAHONY, J.J., and R.E. MEYER: Phil. Trans. Roy. Soc. Lond. A **248**, 467 (1956).

[*7*] EMMONS, H.W. (Ed.): Fundamentals of Gas Dynamics. Princeton 1958.

[*8*] FERRI, A.: Elements of Aerodynamics of Supersonic Flow. New York 1949.

[*9*] COURANT, R., and D. HILBERT: Methoden der Mathematischen Physik. Berlin 1937.

---

[1] Because $\theta \ll 1$. For the case of a stronger centred expansion, cf. footnote 5, p. 281.

Linearized Theory of Unsteady Flow of a Compressible Fluid.

By

R. TIMMAN.

With 5 Figures.

I. Formulation of the problem.

1. Introduction. The theory of the unsteady motion of a body in a compressible fluid has been the subject of numerous researches in the past twenty years. This is caused by its technical importance for the explanation and prevention of flutter in aeroplanes.

Since the theory, at least in the linearized form, has now reached a stage where it can be considered complete, a unified account of this linearized theory will be given. After the fundamental equations have been derived, it is shown that exact analytical solutions can be given in all cases where the acceleration potential and a GREEN'S function of the second kind for the surface of the aerofoil can be expressed analytically.

In all other cases an integral equation can be derived from KIRCHHOFF'S formula, a consequence of the wave equations. For practical purposes, this integral equation can be solved by numerical methods, not described here.

This formulation of the problem has made it inevitable that many original papers which approached it differently could not be represented in this article. A more complete review of the literature on the subject is given by VAN DE VOOREN in Vol. V of the "Advances in applied Mechanics".

2. Basic equations. The exact mathematical theory of unsteady gas flow is extremely complicated, in particular if viscosity effects are included. This introductory section derives the equations for unsteady flow of a non-viscous gas which does not conduct heat. We employ rectangular coordinates x, y and z; the unit vectors are $\boldsymbol{i}$, $\boldsymbol{j}$ and $\boldsymbol{k}$. The velocity vector is denoted by $\boldsymbol{v}$; pressure, temperature and density by p, T and ϱ; entropy by S. The basic equations are

$$\frac{d\boldsymbol{v}}{dt} = \boldsymbol{v}_t + (\boldsymbol{v}\cdot\nabla)\,\boldsymbol{v} = -\frac{1}{\varrho}\operatorname{grad} p \qquad \text{(momentum equation)}, \qquad (2.1)$$

$$\varrho_t + \operatorname{div}(\varrho\boldsymbol{v}) = 0 \qquad \text{(continuity equation)}, \qquad (2.2)$$

$$p = RT\varrho \qquad \text{(equation of state)}, \qquad (2.3)$$

$$S_t + (\boldsymbol{v}\cdot\nabla)\,S = 0 \qquad \text{(energy equation)}. \qquad (2.4)$$

The entropy S of a fluid particle is defined by the differential relation

$$T\,ds = de + p\,d\frac{1}{\varrho} \qquad (2.5)$$

where

$$e = c_v\,T \qquad (2.6)$$

is the specific internal energy of the ideal gas.

We consider the case where the flow at infinity is uniform with velocity $U\boldsymbol{i}$ in the direction of the positive x axis. At infinity then, all particles start with the same entropy. Since Eq. (2.4) states that (in the absence of viscosity and heat conduction) the entropy of a particle is preserved during its motion, it will be constant over all space, at least for subsonic flow. Thus the flow is *homentropic*. In supersonic flow discontinuities may appear in the form of shock waves. It is well known that the change of entropy across a shock is of the third order in the shock-strength; hence, in the linearized approximation, to be introduced in Sect. 3, any resulting entropy gradient can be neglected even in supersonic flow. Thus throughout the flow the pressure p is a function of ϱ only. For an ideal gas this function takes the form of POISSON's law

$$p = C \cdot \varrho^{\gamma} \tag{2.7}$$

where $\gamma = C_p/C_v$ is POISSON's constant.

Then there is an acceleration potential φ satisfying

$$\frac{d\boldsymbol{v}}{dt} = -\frac{1}{\varrho}\operatorname{grad} p = \operatorname{grad}\varphi; \tag{2.8}$$

in fact

$$\varphi = -\int \frac{dp}{\varrho} = -\frac{\gamma}{\gamma-1}\cdot\frac{p}{\varrho} + f(t), \tag{2.9}$$

where $f(t)$ is an arbitrary function of the time which has no effect on $\boldsymbol{v}$. Since p and ϱ are supposed uniform at infinity, by putting $f(t) = 0$ we have $\varphi = \text{const}$ at infinity.

The basic equations then can be written in the form:

$$\boldsymbol{v}_t + \tfrac{1}{2}\operatorname{grad}(\boldsymbol{v}\cdot\boldsymbol{v}) - \boldsymbol{v}\times\operatorname{rot}\boldsymbol{v} = -\operatorname{grad}\varphi. \tag{2.10}$$

Defining the vorticity γ by

$$\gamma = \operatorname{rot}\boldsymbol{v}, \tag{2.11}$$

by taking the curl of both sides of (2.8) we obtain

$$\gamma_t = \operatorname{rot}(\boldsymbol{v}\times\gamma). \tag{2.12}$$

Now

$$\operatorname{rot}(\boldsymbol{v}\times\gamma) = \gamma\cdot\nabla\boldsymbol{v} - \boldsymbol{v}\cdot\nabla\gamma + \boldsymbol{v}\cdot\operatorname{div}\gamma - \gamma\cdot\operatorname{div}\boldsymbol{v};$$

by use of the continuity equation (2.2) it follows that

$$\frac{d\frac{\gamma}{\varrho}}{dt} = \left(\frac{1}{\varrho}\gamma\cdot\nabla\right)\boldsymbol{v}. \tag{2.13}$$

This equation is known as HELMHOLTZ' equation.

Its physical interpretation is that the vortices are carried along with the flow. In fact, consider an element of a vortex-line (which has the vorticity as a tangent)

$$\xi = \frac{\varepsilon}{\varrho}\gamma. \tag{2.14}$$

Then the rate of increase of vector ξ follows from the difference in velocity at both ends of the vector:

$$\frac{d\xi}{dt} = (\xi\cdot\nabla)\,\boldsymbol{v} = \varepsilon\left(\frac{\gamma}{\varrho}\cdot\nabla\right)\boldsymbol{v},$$

or, according to (2.13)

$$\frac{d}{dt}\left\{\xi - \varepsilon\frac{\gamma}{\varrho}\right\} = 0.$$

If for a fluid particle $\gamma = 0$ at infinity, γ remains zero for that particle: a flow which starts as irrotational remains irrotational.

For such a flow $\operatorname{rot} \boldsymbol{v} = 0$, and there exists a velocity potential Φ, so that

$$\boldsymbol{v} = \operatorname{grad} \Phi. \tag{2.15}$$

In this case (2.10) may be integrated to yield

$$\Phi_t + \tfrac{1}{2}(\boldsymbol{v}\cdot\boldsymbol{v}) + \varphi = g(t). \tag{2.16}$$

For flow uniform at infinity, $g(t) = 0$.

The velocity of sound is introduced by

$$c^2 = \frac{dp}{d\varrho} = \gamma \frac{p}{\varrho}. \tag{2.17}$$

Obviously

$$c^2 = -(\gamma - 1)\,\varphi, \tag{2.18}$$

and (2.16) takes the form

$$\Phi_t + \frac{1}{2}(\boldsymbol{v}\cdot\boldsymbol{v}) - \frac{c^2}{\gamma - 1} = 0, \tag{2.19}$$

known as BERNOULLI's law for non-steady flow. The partial differential equation satisfied by Φ is found by substituting the equations of motion in the continuity equation:

$$\frac{p_t}{\varrho} + c^2 \operatorname{div} \boldsymbol{v} + \frac{1}{\varrho}\boldsymbol{v}\cdot\operatorname{grad} p = 0$$

or

$$-\varphi_t + c^2 \operatorname{div}\boldsymbol{v} + \boldsymbol{v}\cdot\operatorname{grad} p = 0.$$

The result is

$$-\Phi_{tt} + c^2 \operatorname{div}\boldsymbol{v} - \boldsymbol{v}\cdot\operatorname{grad}\Phi_t - \boldsymbol{v}\cdot\tfrac{1}{2}\operatorname{grad}(\boldsymbol{v}\cdot\boldsymbol{v}) - \tfrac{1}{2}(\boldsymbol{v}\cdot\boldsymbol{v})_t = 0, \tag{2.20}$$

where $\boldsymbol{v}$ and c^2 are given by (2.15), (2.18) and (2.16).

3. The linearized equations. We consider a thin wing, subject to a uniform translational motion with velocity U. On this main motion is superposed a time-dependent motion with a velocity which can be assumed small with respect U, so that at any moment the tangent plane of the wing (except near the leading edge) makes a small angle with the main velocity U. This additional motion induces an additional velocity field $\boldsymbol{v}$, which also we consider as small with respect to U.

We first derive the equation in a coordinate system x, y, z fixed in space for a small disturbance with velocity potential Φ. Substituting in (2.20) and neglecting second order terms then yields the common wave equation.

$$\Phi_{xx} + \Phi_{yy} + \Phi_{zz} - \frac{1}{c^2}\Phi_{tt} = 0, \tag{3.1}$$

where c can be considered as a constant velocity.

The linearized form of BERNOULLI's equation is

$$\varphi = -\Phi_t. \tag{3.2}$$

Obviously, the acceleration potential φ also satisfies the wave equation

$$\varphi_{xx} + \varphi_{yy} + \varphi_{zz} - \frac{1}{c^2}\varphi_{tt} = 0. \tag{3.3}$$

If the disturbance field caused by the aerofoil moves with a velocity U in the direction of the negative x axis, a coordinate system can be introduced which is fixed to this field.

A simple translation, however, changes the form of Eq. (3.1), but, a Lorentz transformation leaves the wave equation invariant [3]:

$$\left.\begin{aligned} \xi &= \frac{1}{\beta}(x+Ut), & x &= \frac{1}{\beta}(\xi - U\tau),\\ \eta &= y, & y &= \eta,\\ \zeta &= z, & z &= \zeta,\\ \tau &= \frac{1}{\beta}\left(t + M\frac{x}{c}\right), & t &= \frac{1}{\beta}\left(\tau - \frac{M}{c}\xi\right), \end{aligned}\right\} \tag{3.4}$$

where

$$M = \frac{U}{c}, \qquad \beta = \sqrt{1-M^2}.$$

Then φ and Φ still satisfy the wave equation

$$\varphi_{\xi\xi} + \varphi_{\eta\eta} + \varphi_{\zeta\zeta} - \frac{1}{c^2}\varphi_{\tau\tau} = 0, \tag{3.5}$$

and the relation between the two potentials is

$$\varphi = -\frac{1}{\beta}(\Phi_\tau + U\Phi_\xi). \tag{3.6}$$

Contrary to the behaviour of Lorentz transformations in special relativity, τ and ξ are not the time and length variables observed in a system fixed to the aerofoil.

This is obtained by a simple Galilean transformation

$$\left.\begin{aligned} x' &= x + Ut = \beta\xi, & x &= x' - Ut',\\ y' &= y = \eta, & y &= y',\\ z' &= z = \zeta, & z &= z',\\ t' &= t = \frac{1}{\beta}\left(\tau - \frac{M}{c}\xi\right). & t &= t'. \end{aligned}\right\} \tag{3.7}$$

Substitution gives the equation

$$\beta^2\varphi_{x'x'} + \varphi_{y'y'} + \varphi_{z'z'} - 2\frac{M}{c}\varphi_{x't'} - \frac{1}{c^2}\varphi_{t't'} = 0. \tag{3.8}$$

In most cases, however, Eq. (3.5) is preferable.

For a supersonic translation velocity $U>c$ the transformation formulas become complex.

In this case, we write $\beta = \sqrt{M^2-1}$ and introduce

$$\left.\begin{aligned} \xi &= \frac{1}{\beta}(x+Ut), & x &= \frac{1}{\beta}(\xi - M\tau),\\ \eta &= y, & y &= \eta,\\ \zeta &= z, & z &= \zeta,\\ \tau &= \frac{1}{\beta}(ct - Mx), & t &= \frac{1}{\beta c}(\tau - M\xi). \end{aligned}\right\} \tag{3.9}$$

Then the equation again transforms into the wave equation

$$-\varphi_{\xi\xi} + \varphi_{\eta\eta} + \varphi_{\zeta\zeta} + \varphi_{\tau\tau} = 0 \tag{3.10}$$

where ξ is the time-like variable, while τ is a space-like variable.

In linearized approximation φ gives the pressure disturbance

$$p = -\varrho_0\varphi, \tag{3.11}$$

where ϱ_0 is the undisturbed value of the density.

4. The influence of viscosity, Kutta-Joukowski condition. In a theory taking no account at all of effects of viscosity, the boundary condition for the flow about a body is simply that the normal velocity component of the fluid relative to the body at the surface vanishes. Viscosity, however, changes this picture greatly, and a non-viscous theory, which does not take account of these changes, is completely unrealistic.

The proper boundary condition in a viscous fluid is that the fluid adheres to the bounding surface. Both the normal and the tangential velocity relative to the body must vanish. Since, at a small distance from the surface, the velocity has already reached a value of the order of the free-stream value, the influence of viscosity is restricted to a small *boundary layer* with strong vorticity near the surface. For a body of large thickness this boundary layer will separate; a large region of strong vorticity extends behind the body, forming a flow pattern completely different from the corresponding potential flow. For thin wings, however, the vortex layer along the surface is also thin. The vortices are carried along with the flow and form a thin vortex wake behind the wing. The strength of this wake can be determined, approximately, by the condition of KUTTA-JOUKOWSKI.

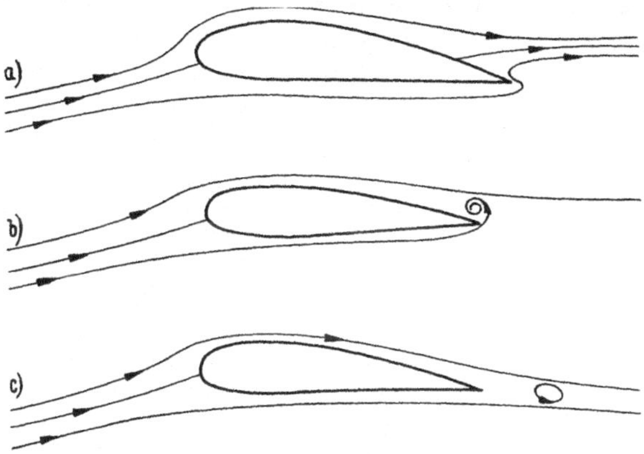

Fig. 1a—c. a Initial flow pattern. b Formation of a vortex. c Carrying away of vortex.

In order to understand this condition, we consider the case of a two-dimensional flow about a wing, starting from rest into a translational motion at a small angle of incidence.

In the first stage of the motion a boundary layer extends along the contour, but outside this boundary layer a quasi-potential flow is formed, which has a stagnation point near the leading edge at the lower side of the wing and near the (sharp) trailing edge at the upper side.

Since the trailing edge is sharp, the velocity is high and the pressure is low, but at the stagnation point the pressure is high. Therefore, between the trailing edge and the stagnation point there is a large adverse pressure gradient. This causes a separation of the boundary layer, and a concentrated vortex is formed. The induced velocity of this vortex causes a circulatory flow in the opposite direction along the aerofoil, and the rear stagnation point is shifted towards the trailing edge. At first all vorticity is in the boundary layer and the total circulation is zero. But the trailing edge vortex separates from the aerofoil, and a resultant circulation results with strength equal and opposite to the strength of the trailing edge vortex. This process continues till the pressure difference at the trailing edge is extenuated, or, what amounts to the same, till the rear stagnation point is at the trailing edge. This condition is known as the condition of KUTTA-JOUKOWSKI.

In a general unsteady motion, at each change of the motion of the wing vortices are shed off at the trailing edge and are carried away with the fluid motion relative to the aerofoil. Hence there extends behind the aerofoil a vortex wake, the strength of which is determined by the Kutta condition of zero pressure difference at the trailing edge.

5. Boundary conditions. As was mentioned above, the normal component of the velocity of the fluid at the surface of the body must be equal to the velocity component of the surface itself in the normal direction.

We formulate this condition in a coordinate system x, y, z which partakes in the translational velocity of the wing.

The equation of the wing is expressed in the form

$$\{z - \varepsilon f(x, y, t)\}^2 = \varepsilon^2 g(x, y, t), \tag{5.1}$$

where $g(x, y, t) > 0$ in the projection of the wing on the x, y plane. $g = 0$ corresponds to the edge of the wing.

$z = \varepsilon f(x, y, t)$ denotes a "mean surface" and $\varepsilon \sqrt{g}$ a thickness distribution.

For a thin wing ε is a small parameter. The direction of the normal is given by the vector

$$\left\{\varepsilon\left(f_x \pm \frac{g_x}{2\sqrt{g}}\right),\ \varepsilon\left(f_y \pm \frac{g_y}{2\sqrt{g}}\right),\ -1\right\} = (\varepsilon f_x, \varepsilon f_y, -1) \pm \frac{\varepsilon}{2\sqrt{g}}(g_x, g_y, 0),$$

where the upper sign refers to the upper side of the wing, the lower sign to the lower side.

On the surface the dominant vector is the vector $(0, 0, -1)$, except near the edge, where $\sqrt{g}$ is very small and the vector has the direction of $(g_x, g_y, 0)$.

We now derive an expression for the normal vector in terms of the δ-function (distribution) which gives this behaviour in the limit as $\varepsilon \to 0$. In order to achieve this, we consider the flux through the surface of an arbitrary "test vector" $\boldsymbol{v} = (u, v, w)$

$$\tilde{\Phi} = \iint_S (\boldsymbol{v} \cdot \boldsymbol{n})\, dS$$

where $\boldsymbol{n}$ is the unit outward normal.

We consider three regions: the inner part of the upper surface S^+, the inner part of the lower surface S^- and the edge region E, defined by $0 \leqq g \leqq \varepsilon^2$.

The contribution of the surface regions to the integral is

$$\iint_{S^+}\left[-\varepsilon\left(f_x + \frac{g_x}{2\sqrt{g}}\right)u - \varepsilon\left(f_y + \frac{g_y}{2\sqrt{g}}\right)v + w\right]dx\,dy -$$

$$-\iint_{S^-}\left[-\varepsilon\left(f_x - \frac{g_x}{2\sqrt{g}}\right)u - \varepsilon\left(f_y - \frac{g_y}{2\sqrt{g}}\right)v + w\right]dx\,dy.$$

Along the edge $g = 0$ in the x, y plane we introduce a local orthogonal coordinate system S along the contour and ν normal to the contour. Consider now a vertical section through a point S of the edge and the normal ν (Fig. 2).

In this plane the flux through the section is

$$\tilde{\Phi} = \int (\boldsymbol{v} \cdot \boldsymbol{n})\, d\sigma,$$

where σ is the arc length of the section. Along the part where $\sqrt{g} < \varepsilon$, the normal vector $\boldsymbol{n}$ is approximated by the vector

$$\boldsymbol{\nu} = (g_x, g_y, 0)\,(g_x^2 + g_y^2)^{-\frac{1}{2}}. \tag{5.2}$$

Hence, approximately, the total flux through the edge, divided by ε, is

$$\left.\begin{aligned}\frac{\Phi E}{\varepsilon} &= \oint [(\boldsymbol{v}\cdot\boldsymbol{\nu})\,(2\sqrt{g})]_{\nu=0}\,ds = \oint [(\boldsymbol{v}\cdot\boldsymbol{\nu})\,2\sqrt{g(\nu)}\,]_{\nu=0}\,ds\\ &= \oint \int_{\nu} (\boldsymbol{v}\cdot\boldsymbol{\nu})\,\delta(\nu)\,2\sqrt{g(\nu)}\,d\nu\cdot ds.\end{aligned}\right\}\tag{5.3}$$

The δ function is defined by the functional relation

$$\int_0^\infty \delta(\nu)\,h(\nu)\,d\nu = h(0).$$

If $\boldsymbol{v}$ is regular for $\nu=0$ the contribution will vanish, but if $\boldsymbol{v}$ is singular, so that as $\varepsilon\to 0$ the limit of $(\boldsymbol{v}\cdot\boldsymbol{\nu})\sqrt{g(\nu)}$ as $\nu\to 0$ is finite, a contribution remains.

Hence we write for the direction vector of the normal

$$\left\{\varepsilon\left(f_x\pm\frac{g_x}{2\sqrt{g}}\right),\ \varepsilon\left(f_y\pm\frac{g_y}{2\sqrt{g}}\right),\ -1\right\}_{S\pm}+2\varepsilon\,\boldsymbol{\nu}\,2\sqrt{g(\nu)}\,\delta(\nu).$$

We now can write down the boundary conditions.

In general, we do not consider pulsating motions. This means that the thickness distribution g is independent of the time. The only case we consider is that of a local rotation about a horizontal axis and a vertical translation.

In this case the displacement of the projection of a point of the surface on the x, y plane only is of second order in the amplitudes. Then each point is characterized by its projection (x, y) on the horizontal plane, and the velocity of a point of the aerofoil is $(0, 0, \dot{z}) = (0, 0, \varepsilon f_t)$.

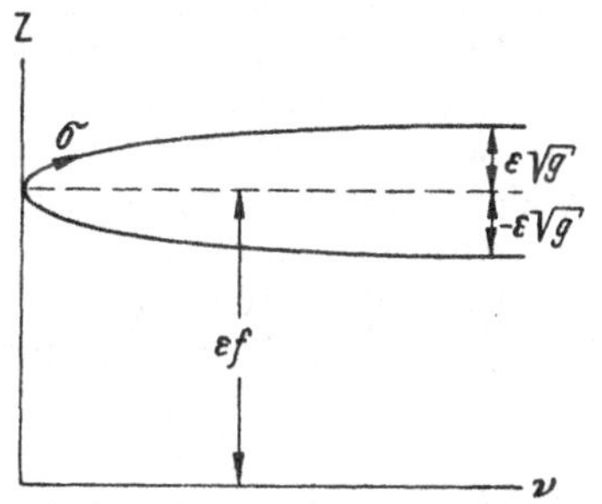

Fig. 2. Cross section normal to the edge.

The total velocity potential Φ_{tot}, in our reference system x, y, z, t [which in Sect. 3 was denoted by x', y', z', t', Eq. (3.7)], is composed of a main part and a disturbance potential Φ,

$$\Phi_{\text{tot}} = U x + \varepsilon\Phi.\tag{5.4}$$

The boundary condition is now

$$\left.\begin{aligned}&\left\{\varepsilon(U+\varepsilon\Phi_x)\left(f_x\pm\frac{g_x}{2\sqrt{g}}\right)+\varepsilon^2\Phi_y\left(f_y\pm\frac{g_y}{2\sqrt{g}}\right)-\varepsilon\Phi_z\right\}_{S\pm}+\\ &\quad+2\varepsilon\{U g_x+\varepsilon(\Phi_x g_x+\Phi_y g_y)\}\,(g_x^2+g_y^2)^{-\frac12}\,\delta(\nu)\sqrt{g(\nu)}=-\varepsilon f_t,\end{aligned}\right\}\tag{5.5}$$

from which we find that to the first order

$$\begin{aligned}\Phi_z &= f_t+U f_x\pm\left[\frac{1}{2\sqrt{g}}\{U f_x\mp\varepsilon(\Phi_x g_x+\Phi_y g_y)\}\right]_{S\pm}-\\ &\quad-2\,[U g_x(g_x^2+g_y^2)^{-\frac12}+\varepsilon\,\text{grad}\,\Phi\cdot\boldsymbol{\nu}]\,\delta(\nu)\sqrt{g(\nu)}\,.\end{aligned}$$

We separate the potential into a steady and an unsteady part and remark that g does not depend on the time.

As we are only interested in the non-steady part, which from now on we denote by Φ, we obtain

$$\Phi_z = w(x, y, t)_{S\pm} - 2\varepsilon\,\delta(\nu)\,\sqrt{g(\nu)}\,(\operatorname{grad}\Phi\cdot\boldsymbol{\nu}), \tag{5.6}$$

where

$$w(x, y, t) = f_t + U f_x$$

is a function determined by the motion of the aerofoil. In the linearized approximation, we apply this boundary condition to the area $g>0$ of the x, y plane instead of the surface.

Obviously Φ_z has the same form for the upper and lower sides of the wing, and Φ_z must be an even function of z.

Then Φ must be an odd function of z.

$$\Phi(x, y, +z) = -\Phi(x, y, -z). \tag{5.7}$$

The wake behind the aerofoil is a region of vorticity of small thickness. This means that at the upper and lower sides the tangential velocities are different. In the linearized approximation we let the thickness go to zero, and the projection of the vortex region on the x, y plane becomes a surface of discontinuity for the tangential velocities and, consequently, for Φ.

We have seen that Φ is an odd function of z.

This means that in the x, y plane $\Phi=0$ except on the aerofoil ($g>0$) and in the wake, where Φ is discontinuous, but at opposite points Φ assumes values which are equal and opposite, $\Phi^+ = -\Phi^-$.

However, the pressure is continuous in the wake.

Instead of the pressure, we consider the acceleration potential φ, whence the pressure difference from the undisturbed flow is easily calculated [Eq. (3.11)]:

$$p = -\varrho_0\,\varphi,$$

where the suffix 0 denotes undisturbed values.

In the linearized approximation the acceleration potential φ is related to the velocity potential Φ,

$$\varphi = \Phi_t + U\Phi_x,$$

and we see that φ also is an odd function of z.

This means that $\varphi=0$ everywhere in the x, y plane except on the aerofoil. The boundary conditions for φ are simpler than those for Φ. Moreover, φ is related to the pressure, which is the function most important for the applications, since it determines the aerodynamical force. The normal derivative of φ is given by

$$\varphi_{\boldsymbol{n}} = \varphi_z + 2\varepsilon\,\delta(\nu)\left\{\sqrt{g}\,\frac{\partial\varphi}{\partial\boldsymbol{\nu}}\right\} = \Phi_{tz} + U\Phi_{zz} + 2\varepsilon\,\delta(\nu)\left\{\sqrt{g}\,\frac{\partial\varphi}{\partial\boldsymbol{\nu}}\right\}$$

or

$$\varphi_{\boldsymbol{n}} = w_t + U w_x + 2\varepsilon\,\delta(\nu)\left\{\sqrt{g}\,\frac{\partial\varphi}{\partial\boldsymbol{\nu}}\right\}.$$

Hence, on the aerofoil the normal derivative of the acceleration potential is given, but along the edge a distribution of singularities of an unknown intensity must be added. Here the condition of KUTTA-JOUKOWSKI becomes of primary importance. It requires that the pressure remains finite at the trailing edge, and consequently $\sqrt{g}\,\boldsymbol{\nu}\cdot\operatorname{grad}\varphi$ also remains finite. The contour (Fig. 3) is divided into a leading edge L and a trailing edge T, the boundary points being given by the two tangents parallel to the x axis. (It is assumed for simplicity that only two such tangents exist.) Only at the leading edge is a singularity required. This singularity can be explained by the following physical model. At the lower side of an aerofoil at incidence a stagnation point is formed near the leading edge; at the upper side is a region of high velocity.

This means that at the lower side a high pressure appears, but at the upper side, a low pressure.

Consequently, a pressure dipole is formed along the contour. For a rounded leading edge where this occurs, we may put $g=\alpha\nu$, and the arc length σ at the leading edge is approximately

$$\sigma \approx \varepsilon \sqrt{g(\nu)} \approx \varepsilon \sqrt{\alpha \nu}.$$

The dipole moment is obtained by differentiating the pressure φ with respect to σ, which is the coordinate along the dipole axis:

$$\frac{\partial \varphi}{\partial \sigma} = \frac{\partial \varphi}{\partial \nu} \frac{2\sqrt{\nu}}{\varepsilon \sqrt{\alpha}} = \frac{\partial \varphi}{\partial \nu} \frac{2\sqrt{g(\nu)}}{\varepsilon \alpha},$$

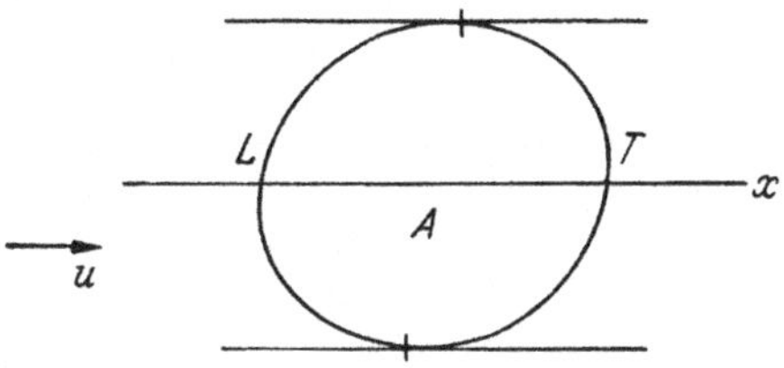

Fig. 3. Plan form of aerofoil.

which explains the term.

The basic problem to be solved is now put in terms of the acceleration potential φ.

φ must satisfy Eq. (3.8),

$$\beta^2 \varphi_{xx} + \varphi_{yy} + \varphi_{zz} - 2\frac{M}{c}\varphi_{xt} - \frac{1}{c^2}\varphi_{tt} = 0.$$

In the plane $z=0$ for a certain region A the condition

$$\frac{\partial \varphi}{\partial z} = w_t + U w_x + \varepsilon\,\delta(\nu)\left\{\sqrt{g}\,\frac{\partial \varphi}{\partial \nu}\right\}$$

is satisfied; outside A, $\varphi=0$.

The additional term is added only for a part L of the contour. The unknown value of $\partial\varphi/\partial\nu$ must be determined afterwards by a separate consideration.

II. Explicit solutions.

a) Subsonic case.

6. If a Green's function of the second kind exists for the equation satisfied by φ and for the region A in the x, y plane, the solution can be expressed explicitly in terms of this function.

This is the case if the differential equation involves only differentiations with respect to the space variables, as happens both for flow of an incompressible fluid and for harmonic oscillations in a compressible fluid. In the latter case the time factor $e^{i\nu t}$ can be cancelled and the equation is essentially the Helmholtz equation.

If the linear equation for the potential is

$$L(\varphi) = 0, \tag{6.1}$$

Green's function of the second kind for a closed surface S is defined as the solution $G(\boldsymbol{r}, \boldsymbol{r}_p)$ of the adjoint non-homogeneous equation

$$M\big(G(\boldsymbol{r}, \boldsymbol{r}_p)\big) = \delta(\boldsymbol{r} - \boldsymbol{r}_p) \tag{6.2}$$

satisfying on S the boundary condition

$$G_n = 0. \tag{6.3}$$

Application of GREEN's identity then gives for a point p inside S

$$\varphi_p = \iint \varphi_n G_p \, dS. \tag{6.4}$$

Here the region is bounded by the aerofoil and a large sphere, on which the potential vanishes. On the surface (the two sides of the aerofoil and the edge) we have for the acceleration potential the boundary condition

$$\varphi_n = w_t + U w_x + 2\varepsilon\, \delta(\nu) \left\{ \sqrt{g(\nu)}\, \frac{\partial \varphi}{\partial \nu} \right\}, \tag{6.5}$$

where the singular term must be added along the leading edge L. Substitution in (6.4) gives

$$\varphi_p = \iint_A (w_t + U w_x)\, G_p \, dx\, dy + \int_L \frac{\partial \varphi}{\partial \nu} \left\{ \sqrt{\nu}\, \frac{\partial G_p}{\partial \nu} \right\}_{\nu=0} ds, \tag{6.6}$$

where the function $\varepsilon \dfrac{\partial \varphi}{\partial \nu} = \lambda(s, t)$ along the leading edge is unknown. We can derive an integral equation for this unknown function from the normal velocity on the aerofoil surface, which is our primary datum. The velocity potential follows from an integration of the first order partial differential equation

$$\varphi = \Phi_t + U\Phi_x,$$

viz.

$$\left.\begin{aligned}
\Phi(x_p, y_p, z_p, t) &= \frac{1}{U} \int_{-\infty}^{x_p} \varphi\left(x', y_p, z_p, t - \frac{x_p - x'}{U}\right) dx' \\
&= \frac{1}{U} \int_{-\infty}^{x_p} dx' \iint_A \left\{ w_t\left(x, y, t - \frac{x_p - x'}{U}\right) + U w_x\left(x, y, t - \frac{x_p - x'}{U}\right) \right\} \times \\
&\quad \times G(x, y, 0; x'_p, y_p, z_p)\, dx\, dy + \frac{1}{U} \int_{-\infty}^{x_p} dx' \int_L \lambda\left(s, t - \frac{x_p - x'}{U}\right) \times \\
&\quad \times \left\{ \sqrt{\nu}\, \frac{\partial G_p}{\partial \nu} (x(s), y(s), x', y, z_p) \right\} ds.
\end{aligned}\right\} \tag{6.7}$$

The first integral can be reduced by partial integration, since

$$w_t\left(x, y, t - \frac{x_p - x'}{U}\right) = -U w_{x'}\left(x, y, t - \frac{x_p - x'}{U}\right). \tag{6.8}$$

This yields

$$\left.\begin{aligned}
\Phi(x_p, y_p, z_p, t) = \iint_A & w(x, y, t)\, G(x, y, 0; x_p, y_p, z_p)\, dx\, dy + \\
&+ \int_{-\infty}^{x_p} dx' \iint_A \{ -w\, G_{x'} + w_x G \}\, dx\, dy + \\
&+ \frac{1}{U} \int_{-\infty}^{x_p} \int_L \lambda\left(s, t - \frac{x_p - x'}{U}\right) \left\{ \sqrt{\nu}\, \frac{\partial G}{\partial \nu} \right\}_{\nu=0} ds\, dx'.
\end{aligned}\right\} \tag{6.9}$$

The second integral, which is a surface integral, can be reduced to an integral along the complete edge.

In order to achieve this reduction, we observe that if

$$\frac{\partial \varphi}{\partial n} = w = \frac{\partial \varphi}{\partial z} + \varepsilon\, \delta(\nu) \left\{ \sqrt{g}\, \frac{\partial \varphi}{\partial \nu} \right\}, \tag{6.10}$$

then

$$\frac{\partial}{\partial n}\frac{\partial \varphi}{\partial x} = \frac{\partial^2 \varphi}{\partial x\, \partial z} + \varepsilon\, \delta(\nu) \left\{\sqrt{g}\, \frac{\partial^2 \varphi}{\partial x\, \partial \nu}\right\}, \tag{6.11}$$

but

$$\frac{\partial}{\partial x}\frac{\partial \varphi}{\partial n} = \frac{\partial^2 \varphi}{\partial x\, \partial z} + \varepsilon\, \delta(\nu) \sqrt{g}\, \frac{\partial^2 \varphi}{\partial x\, \partial \nu} + \varepsilon \frac{\partial \varphi}{\partial \nu}\frac{\partial}{\partial x}\{\delta(\nu)\sqrt{g}\} = \frac{\partial w}{\partial x}. \tag{6.12}$$

The potential having the normal derivative given by (6.10) is

$$\varphi_p' = \iint_A w\, G_p\, dx\, dy \tag{6.13}$$

and

$$\frac{\partial \varphi}{\partial x_p} = \iint_A w \frac{\partial G}{\partial x_p}\, dx\, dy. \tag{6.14}$$

But on the other hand

$$\left.\begin{aligned} \frac{\partial \varphi}{\partial x_p} &= \iint_A \frac{\partial}{\partial n}\left(\frac{\partial \varphi}{\partial x}\right) G\, dx\, dy \\ &= \iint_A \frac{\partial w}{\partial x} G\, dx\, dy - \varepsilon \iint_A \frac{\partial \varphi}{\partial \nu}\frac{\partial}{\partial x}\{\delta(\nu)\sqrt{g}\}\, G\, dx\, dy \\ &= \iint_A \frac{\partial w}{\partial x} G\, dx\, dy + \varepsilon \iint \delta(\nu)\sqrt{g}\, \frac{\partial}{\partial x}\left(G\, \frac{\partial \varphi}{\partial \nu}\right) dx\, dy \\ &= \iint_A \frac{\partial w}{\partial x} G\, dx\, dy + \varepsilon \int_{L+T} \left\{\sqrt{g(\nu)}\, \frac{\partial G}{\partial x}\right\}_{\nu=0} \frac{\partial \varphi}{\partial \nu}\, ds + \\ &\quad + \varepsilon \int_{L+T} \{\sqrt{g(\nu)}\, G_{\nu\pm 0}\} \frac{\partial^2 \varphi}{\partial x\, \partial \nu}\, ds. \end{aligned}\right\} \tag{6.15}$$

If G is such that $\sqrt{g(\nu)}\, G$ vanishes at the edge but $\sqrt{g(\nu)}\, \frac{\partial G}{\partial x}$ remains, we obtain the identity:

$$\iint_A \{w\, G_{x_p} - w_x G\}\, dx\, dy = \varepsilon \int_{L+T} \left(\sqrt{g(\nu)}\, \frac{\partial G}{\partial x}\, \frac{\partial \varphi}{\partial \nu}\right)_{\nu=0} ds. \tag{6.16}$$

With this result the formula for the velocity potential takes the form:

$$\left.\begin{aligned} \Phi_p = \tilde{\Phi}(x_p, y_p, z_p) &- \varepsilon \int_{-\infty}^{x_p}\int_{L+T} \left[\sqrt{\nu}\, \frac{\partial G}{\partial x}\right]_{\nu=0} \left(\frac{\partial \varphi}{\partial \nu}\right) dx'\, ds + \\ &+ \frac{1}{U}\int_{-\infty}^{x_p}\int_{L} \lambda\left(s, t - \frac{x_p - x'}{U}\right)\left\{\sqrt{\nu}\, \frac{\partial G}{\partial \nu}\right\}_{\nu=0} ds\, dx', \end{aligned}\right\} \tag{6.17}$$

where $\tilde{\Phi}_p$ is given by

$$\tilde{\Phi}_p = \iint_A w\, G\, dx\, dy. \tag{6.18}$$

Differentiation with respect to z_p and putting $z_p = 0$ gives for a point on the aerofoil

$$\left.\begin{aligned} \left(\frac{\partial \Phi_p}{\partial z_p}\right)_0 = w = \left(\frac{\partial \tilde{\Phi}_p}{\partial z_p}\right)_0 &- \varepsilon \frac{\partial}{\partial z_p}\int_{-\infty}^{x_p}\int_{L+T} \left(\frac{\partial \Phi}{\partial \nu}\right)_{\nu=0} ds \left[\sqrt{\nu}\, \frac{\partial G}{\partial x}\right]_{\nu=0} dx' + \\ &+ \frac{1}{U}\frac{\partial}{\partial z_p}\int_{-\infty}^{x_p}\int_{L} \lambda\left(s, t + \frac{x' - x_p}{U}\right)\left\{\sqrt{\nu}\, \frac{\partial G}{\partial \nu}\right\}_{\nu=0} ds\, dx'. \end{aligned}\right\} \tag{6.19}$$

But $\left(\frac{\partial \tilde{\Phi}_p}{\partial z_p}\right)_0 = w$ from (6.18), and we obtain the following integral equation for $\lambda(s,t)$; where (6.13) is replaced by (6.18).

$$\left.\begin{aligned} 0 = & \frac{\partial}{\partial z_p} \int_{-\infty}^{x_p} \int_L \left\{ \sqrt{\nu}\, \frac{\partial G}{\partial \nu}\, \frac{\partial \nu}{\partial x} \left(\varepsilon \frac{\partial \Phi}{\partial \nu} \right) + \frac{\lambda}{U} \right\} ds\, dx' + \\ & + \frac{\partial}{\partial z_p} \int_{-\infty}^{x_p} \int_L \left(\sqrt{\nu}\, \frac{\partial G}{\partial \nu}\, \frac{\partial \nu}{\partial x}\, \varepsilon \frac{\partial \Phi}{\partial \nu} \right)_{\nu=0} ds\, dx'. \end{aligned}\right\} \quad (6.20)$$

This equation can be solved by numerical methods.

b) Applications.

7. Flow of an incompressible fluid. The preceding equations can be applied in all cases where GREEN's function of the second kind exists. The first example is the case of an incompressible fluid, where φ satisfies LAPLACE's equation,

$$\varphi_{xx} + \varphi_{yy} + \varphi_{zz} = 0. \quad (7.1)$$

α) Two-dimensional flow. The first and now classical papers by KÜSSNER [2], CIGALA [5] and THEODORSEN [1] are on the two-dimensional problem. Because of its relative simplicity, we shall now treat this case fully by the method developed above.

The flow is independent of y; the aerofoil extends along the segment $-1 \leqq x \leqq +1$ of the x-axis. We introduce elliptic coordinates

$$x = \cosh\mu \cos\vartheta, \qquad 0 \leqq \vartheta < 2\pi,$$
$$z = \sinh\mu \sin\vartheta, \qquad \mu \geqq 0.$$

The segment is given by $\mu = 0$.

GREEN's function for the equation

$$\varphi_{\mu\mu} + \varphi_{\vartheta\vartheta} = 0$$

is easily found to be

$$\left.\begin{aligned} & G(\mu, \vartheta; \mu_p, \vartheta_p) \\ & = \frac{1}{2\pi} \ln \{\cosh(\mu - \mu_p) - \cos(\vartheta - \vartheta_p)\}\{\cosh(\mu + \mu_p) - \cos(\vartheta - \vartheta_p)\}. \end{aligned}\right\} \quad (7.2)$$

Near the leading edge $\vartheta = \pi$ the coordinate ν is

$$\nu = 1 + x = 1 + \cos\vartheta, \quad (7.3)$$

and the singular potential is

$$\left(\sqrt{\nu}\, \frac{\partial G}{\partial \nu}\right)_{\nu=0} = + \frac{1}{2\pi\sqrt{2}}\, \frac{\sin\vartheta_p}{\cosh\mu_p + \cos\vartheta_p}. \quad (7.4)$$

With this result we substitute for the acceleration potential

$$\varphi_p = \frac{1}{\pi} \int_0^{2\pi} (w_t + U w_x) \ln\{\cosh\mu_p - \cos(\vartheta - \vartheta_p)\} \sin\vartheta\, d\vartheta + \lambda(t)\, \frac{\sin\vartheta_p}{\cosh\mu_p + \cos\vartheta_p}, \quad (7.5)$$

where the function $\lambda(t)$ must be determined from the integral equation.

$$\left.\begin{aligned} 0 = \frac{\partial}{\partial z_p}\int\limits_{-\infty}^{x_p} dx' \left\{\lambda\left(t + \frac{x' - x_p}{U}\right) + \varepsilon\left(\frac{\partial \Phi}{\partial x}\right)_{-1}\right\}\left(\sqrt{v}\,\frac{\partial G}{\partial v}\right)_{x=-1} + \\ + \frac{\partial}{\partial z_p}\int\limits_{-\infty}^{x_p} dx'\,\varepsilon\left(\frac{\partial \Phi}{\partial x}\right)_{-1}\left(\sqrt{v}\,\frac{\partial G}{\partial v}\right)_{x=+1}. \end{aligned}\right\} \tag{7.6}$$

Here Φ is the regular velocity potential, which is given by

$$\Phi_p = \int\limits_0^{2\pi} w(\vartheta, t)\, G_p \sin\vartheta\, d\vartheta. \tag{7.7}$$

After some calculations we find that

$$\left.\begin{aligned} \lim_{\varepsilon\to 0}\left\{\varepsilon\left(\frac{\partial \Phi}{\partial x}\right)_{-1}\right\} = \frac{1}{\pi}\int\limits_0^{2\pi} w(\vartheta, t)\,\frac{\sin^2\vartheta\, d\vartheta}{1+\cos\vartheta} = \frac{1}{\pi}\int\limits_0^{2\pi} w(\vartheta, t)\,(1-\cos\vartheta)\, d\vartheta, \\ \lim_{\varepsilon\to 0}\left\{\varepsilon\left(\frac{\partial \Phi}{\partial x}\right)_{+1}\right\} = \frac{1}{\pi}\int\limits_0^{2\pi} w(\vartheta, t)\,\frac{\sin^2\vartheta\, d\vartheta}{1-\cos\vartheta} = \frac{1}{\pi}\int\limits_0^{2\pi} w(\vartheta, t)\,(1+\cos\vartheta)\, d\vartheta. \end{aligned}\right\} \tag{7.8}$$

If we assume that the given velocity $w(\vartheta, t)$ is expanded in a Fourier cosine series,

$$w(\vartheta, t) = a_n(t) + 2\sum_{n=1}^{\infty} a_n(t)\cos n\vartheta, \tag{7.9}$$

where the coefficients $a_n(t)$ are given functions of t, we obtain the following equation for the determination of $\lambda(t)$

$$\left.\begin{aligned} &\int\limits_{-\infty}^{-1} \lambda\left(t + \frac{1+x'}{U}\right) d\sqrt{\frac{x'-1}{x'+1}} \\ &\quad = \int\limits_{-\infty}^{-1}\left\{a_0\left(t + \frac{1+x'}{U}\right) - a_1\left(t + \frac{1+x'}{U}\right)\right\} d\sqrt{\frac{x'-1}{x'+1}} + \\ &\qquad + \int\limits_{-\infty}^{-1} (a_0 + a_1)\, d\sqrt{\frac{x'+1}{x'-1}}\,. \end{aligned}\right\} \tag{7.10}$$ [1]

We solve this equation for the case of a non-steady motion which starts from rest at $t = 0$. Then, for $t < 0$ a_0, a_1 and λ are all zero. Putting

$$1 + x' = -\sigma, \qquad \tau = U t,$$

we transform the equation into

$$\left.\begin{aligned} \int\limits_0^{\tau} \lambda(\tau-\sigma)\, d\sqrt{\frac{2+\sigma}{\sigma}} = \int\limits_0^{\tau}\{a_0(\tau-\sigma) - a_1(\tau-\sigma)\}\, d\sqrt{\frac{2+\sigma}{\sigma}} + \\ + \int\limits_0^{\tau}\{a_0(\tau-\sigma) - a_1(\tau-\sigma)\}\, d\sqrt{\frac{\sigma}{2+\sigma}}\,. \end{aligned}\right\} \tag{7.11}$$

[1] This equation holds in the sense that for the divergent integrals the finite values in the sense of HADAMARD [7], [8] must be taken.

We introduce the Laplace transforms,

$$\Lambda(p) = \int_0^\infty \lambda(\tau)\, e^{-p\tau}\, d\tau, \qquad A_i(p) = \int_0^\infty a_i(\tau)\, e^{-p\tau}\, d\tau \qquad (i = 0, 1), \tag{7.12}$$

and remark that

$$\left.\begin{aligned} \int_0^\infty e^{-p\sigma}\sqrt{\frac{2+\sigma}{\sigma}}\, d\sigma &= e^p\{K_1(p) + K_0(p)\},\\ \int_0^\infty e^{-p\sigma}\sqrt{\frac{\sigma}{2+\sigma}}\, d\sigma &= e^p\{K_1(p) - K_0(p)\}, \end{aligned}\right\} \tag{7.13}$$

where $K_0(p)$ and $K_1(p)$ are the Hankel functions of imaginary argument. From the convolution theorem the transformed function $\Lambda(p)$ is seen to be equal to

$$\Lambda(p) = 2\{A_0 \cdot C(p) + A_1(1 - C(p)\} \tag{7.14}$$

where

$$C(p) = \frac{K_1(p)}{K_0(p) + K_1(p)} = \frac{1}{1 + K_1(p)/K_0(p)}. \tag{7.15}$$

The significance of the function $C(p)$ is obvious if we assume that $w(x, t)$ has the form

$$w(x, t) = 1 \cdot \delta(t). \tag{7.16}$$

i.e. a unit impulse of a constant velocity at $t=0$. Then

$$A_0(p) = 1, \qquad A_i(p) = 0, \qquad i > 1, \tag{7.17}$$

and

$$\Lambda(p) = 2C(p). \tag{7.18}$$

The acceleration potential on the surface of the wing is

$$\varphi_p = \frac{1}{\pi}\delta'(t) + \lambda(t) \cdot \tan\frac{1}{2}\vartheta, \tag{7.19}$$

where, by the inversion formula for Laplace transforms,

$$\lambda(t) = \frac{1}{\pi i}\int_{\gamma-i\infty}^{\gamma+i\infty} \frac{e^{p\tau} K_1(p)}{K_0(p) + K_1(p)}\, dp, \tag{7.20}$$

γ being chosen so that the contour lies to the right of all poles of the integrand. Since the pressure on the aerofoil is proportional to φ, we find that the total force becomes

$$K = \varrho U \int_0^{2\pi} \varphi(\vartheta) \sin\vartheta\, d\vartheta = 2\varrho U [\delta'(t) + \lambda(t)]. \tag{7.21}$$

Hence, apart from the $\delta'(t)$ function, $\lambda(t)$ is the response to the input (7.16) (KÜSSNER [9], SEARS [10]).

For steady harmonic oscillations

$$w(\vartheta, t) = w(\vartheta) \cdot e^{i\nu\tau}.$$

We easily find that

$$\lambda(t) = \frac{H_1^{(2)}(\omega)}{H_1^{(2)}(\omega) + i H_0^{(2)}(\omega)} \cdot e^{i\nu\tau} = C(i\omega)\, e^{i\nu\tau}$$

where $\omega = \dfrac{\nu l}{U}$.

Other cases which have been investigated are the asymptotic behaviour of harmonic oscillation starting at $t=0$. VAN DE VOOREN [11], TIMMAN [6] or the classical case of a sudden start of a uniform motion at $t=0$ (WAGNER [12]). In the first case it is shown that the harmonic case results if ν is real.

The same method applies also to two-dimensional flow of an incompressible fluid in more complicated configurations (e.g. wind-tunnel wall corrections [13]) where GREEN'S function is given by analytical expressions.

β) Circular wings. Three-dimensional flow can be attacked by the analytical method only for configurations in which the potential equation can be separated. It is well known [14] that this is the case only for coordinate systems consisting of confocal quadratic surfaces. The only sensible planforms of this kind are the ellipse or the circle. In the first case, Lamé functions are involved. As a comprehensive account of the properties of these functions is not available, only the case of the circular planform has been considered in the literature. Here the case of harmonic oscillations is treated by SCHADE [15] and VAN SPIEGEL [16], [16a].

SCHADE solves the boundary value problem for the velocity potential. In this case the pressure becomes singular at the leading edge as well as at the trailing edge. In order to suppress these singularities a compensation calculus has to be performed. The approach by VAN SPIEGEL is about the same as that developed here. In both methods, GREEN'S function is found by the introduction of confocal coordinates

$$\left.\begin{aligned} x &= \sqrt{1+\eta^2}\sqrt{1-\mu^2}\cos\vartheta, && \eta \geqq 0, \\ y &= \sqrt{1+\eta^2}\sqrt{1-\mu^2}\sin\vartheta, && -1 \leqq \mu \leqq +1, \\ z &= \mu\eta, && 0 \leqq \vartheta < 2\pi. \end{aligned}\right\} \tag{7.22}$$

The wing is represented by $\eta=0$, the outer part of the x-y plane by $\mu=0$. GREEN'S function of the second kind has the form:

$$\left.\begin{aligned} &G(\eta,\mu,\vartheta,0,\mu_1,\vartheta_1) \\ &= \sum_{n=1}^{\infty}\sum_{m=0}^{n-1}(2n+1)\frac{\varepsilon_m}{\pi i}\frac{(n-m)!}{(n+m)!}P_n^m(\mu)P_n^m(\mu_1)\frac{Q_n^m(i\eta)}{Q_n^{m\prime}(i0)}\cos m(\vartheta-\vartheta_1) \end{aligned}\right\} \tag{7.23}$$

with $n+m$ even, $\varepsilon_m=\frac{1}{2}$ for $m=0$, $\varepsilon_m=1$ for $m\geqq 1$, P_μ^m and Q_n^m the Legendre and associated Legendre functions. The singular term, which arises from the singularity at the leading edge, is

$$\int_{\pi/r}^{3\pi/r}\lambda(\vartheta_1)\,G_{\mu_1}(\eta,\mu,\vartheta;0,0_1\vartheta_1)\,d\vartheta_1,$$

where G_{μ_1} can be expressed in finite form:

$$G_{\mu_1}(\eta,\mu,\vartheta;0,0_1\vartheta_1) = \frac{\mu}{\pi^2}\cdot\frac{1}{1+\eta^2-2\sqrt{1+\eta^2}\sqrt{1-\mu^2}\cos(\vartheta-\vartheta_1)+1-\mu^2}. \tag{7.24}$$

The integral equation for $\lambda(\vartheta_1)$ is an equation of the first kind; it may be solved by a numerical method [expansion in a series of terms in $\cos(2\eta+1)\,\vartheta_1$].

KÜSSNER [17] uses a different method. The numerical results of SCHADE and VAN SPIEGEL for the steady case are in good agreement but differ from KÜSSNER'S.

8. Compressible subsonic flow. In subsonic flow of a compressible fluid we consider only the problem of harmonic vibrations. The velocity and acceleration potential then contain a time factor $e^{i\nu t}$. In the transformed coordinates we can represent the acceleration potential as

$$\varphi(\xi, \eta, \zeta)\, e^{ik\tau}$$

where

$$k = \frac{\nu}{\beta}. \tag{8.1}$$

Plainly $\varphi\, e^{ik\tau}$ satisfies HELMHOLTZ's equation:

$$\varphi_{\xi\xi} + \varphi_{\eta\eta} + \varphi_{\zeta\zeta} + k^2 \varphi = 0. \tag{8.2}$$

The acceleration potential in the original coordinates (x, y, z, t) moving with the aerofoil is related to this function through

$$\varphi(x, y, z, t) = \tilde{\varphi}(x, y, z)\, e^{i\nu t} = \varphi(\xi, \eta, \zeta) \cdot e^{i\frac{\nu}{\beta}\tau - i\frac{\nu M}{\beta}\xi}. \tag{8.3}$$

The only case which has been solved up to now is that of two-dimensional flow (F. HASKIND [*10*], REISSNER [*19*], BILLINGTON [*20*], TIMMAN [*21*], TIMMAN and v. D. VOOREN [*22*]). The aerofoil extends over the segment $-1 < x < +1$, which corresponds to $-\frac{1}{\beta} < \xi < \frac{1}{\beta}$.

The boundary value problem for φ is the outer Neumann problem. On the segment $-\frac{1}{\beta} < \xi < +\frac{1}{\beta}$ the normal derivative is given. Here, however, an additional condition at infinity must be added. The waves produced by the aerofoil are outgoing waves, and this leads to SOMMERFELD's condition that at large distances they must behave as $r^{-\frac{1}{2}}\, e^{-ik(r-c\tau)}$.

For the solution of the problem we introduce elliptic coordinates

$$\left.\begin{aligned} \xi &= \frac{1}{\beta} \cosh\mu \cos\vartheta, \\ \eta &= \frac{1}{\beta} \sinh\mu \sin\vartheta. \end{aligned}\right\} \tag{8.4}$$

Then HELMHOLTZ's equation takes the form

$$\varphi_{\mu\mu} + \varphi_{\eta\eta} + 2k^2 (\cosh 2\mu - \cos 2\vartheta)\, \varphi = 0, \tag{8.5}$$

which is satisfied by the products

$$\mathrm{Ne}_n^{(2)}(\mu)\, \mathrm{se}_n(\vartheta).$$

Where $\mathrm{se}_n(\vartheta)$ and $\mathrm{Ne}_n^{(2)}(\mu)$ are Mathieu functions[1] and associated Mathieu functions. These satisfy the equations

$$\frac{d^2\, \mathrm{se}_n(\vartheta)}{d\vartheta^2} + (\lambda_n - 2k^2 \cos 2\vartheta)\, \mathrm{se}_n(\vartheta) = 0, \tag{8.6}$$

$$\frac{d^2\, \mathrm{Ne}_n(\mu)}{d\mu^2} + (\lambda_n + 2k^2 \cosh 2\mu)\, \mathrm{Ne}_n(\mu) = 0, \tag{8.7}$$

where λ_n is the proper value of order μ of the Mathieu equation $(\mu = 1, 2, \ldots)$. The functions $\mathrm{Ne}_n^{(2)}(\mu)$ are the solutions of (8.7) which have the asymptotic behaviour required by the Sommerfeld condition (MACLACHLAN [*23*], MEIXNER and SCHÄFKE [*24*]). The functions $\mathrm{se}_n(\vartheta)$ form an orthogonal set.

For $\xi = 0$ the normal derivative of our function φ is given, i.e.

$$\left(\frac{\partial\varphi}{\partial\mu}\right)_{\mu=0} = \sin\tilde{w}(\vartheta) = \sum_{n=1}^{\infty} a_n\, \mathrm{se}_n\, \vartheta, \tag{8.8}$$

[1] For details on Mathieu functions see J. MEIXNER, this Encyclopedia, Vol. I, pp. 208—216.

where the a_n are the coefficients in the expansion of $\widetilde{w}(\vartheta)\sin\vartheta$.

$$a_n = \frac{2}{\pi}\int_0^{\pi} \widetilde{w}(\vartheta)\sin\vartheta\,\mathrm{se}_n(\vartheta)\,d\vartheta. \tag{8.9}$$

Assuming for φ the expansion

$$\varphi = \sum_{n=1}^{\infty} \alpha_n\,\mathrm{Ne}_n(\mu)\,\mathrm{se}_n(\vartheta), \tag{8.10}$$

we see that

$$\left(\frac{\partial\varphi}{\partial\mu}\right)_{\mu=0} = \sum_{n=1}^{\infty} \alpha_n\,\mathrm{Ne}_n'(0)\,\mathrm{se}_n(\vartheta), \tag{8.11}$$

and hence by equating of corresponding coefficients we obtain

$$\left.\begin{aligned}\varphi &= \sum_{n=1}^{\infty} a_n \frac{\mathrm{Ne}_n(\mu)}{\mathrm{Ne}_n'(0)}\,\mathrm{se}_n(\vartheta)\\ &= \frac{2}{\pi}\sum_{n=1}^{\infty}\int_0^{2\pi} \widetilde{w}(\vartheta_1)\cdot\frac{\mathrm{Ne}_n(\mu)}{\mathrm{Ne}_n'(0)}\,\mathrm{se}_n(\vartheta)\,\mathrm{se}_n(\vartheta_1)\sin\vartheta_1\,\mathrm{d}\vartheta_1.\end{aligned}\right\} \tag{8.12}$$

Hence we see that GREEN's function is given by the series expansion

$$G(\mu,\vartheta;0,\vartheta_1) = \frac{2}{\pi}\sum_{n=1}^{\infty}\frac{\mathrm{Ne}_n^{(2)}(\mu)}{\mathrm{Ne}_n'(0)}\,\mathrm{se}_n(\vartheta)\,\mathrm{se}_n(\vartheta_1). \tag{8.13}$$

For the calculations of the integrals needed in the determination of the coefficient λ of the singular potential, some analytical difficulties occasioned by the singularities of the integrals must be surmounted. This can be achieved by splitting off the singular expressions for the incompressible case, which have the same analytical behaviour. For large values of $k = \nu/\beta$, i.e. either high frequencies or transonic flow, the method is very cumbersome. Here asymptotic methods can be used [25]. In the three-dimensional case the problem for the elliptic wing can be solved in terms of Lamé wave functions; for an axis ratio of β, only spheroidal wave functions are involved. No explicit analytic solution has been obtained.

III. The method of integral equations.

A more general method in subsonic and supersonic flow is the method of integral equations for the unknown pressure distribution (acceleration potential).

This method is applicable also in cases where GREEN's function does not exist, or if it exists, cannot be expressed by analytic functions. Such a case is furnished by the problem of unsteady motion of wings of arbitrary planform.

If the integral equation is derived, it must be solved by numerical methods. The derivation of the integral equation is somewhat different in the subsonic and supersonic cases. In both cases the transformed coordinates ξ, η, ζ, τ are used.

9. Subsonic flow. If the main flow is subsonic, the equation satisfied by φ is

$$\varphi_{\xi\xi} + \varphi_{\eta\eta} + \varphi_{\zeta\zeta} - \frac{1}{c^2}\varphi_{\tau\tau} = 0. \tag{9.1}$$

The aerofoil is represented by a region A in the ξ, η plane, on which the normal derivative of the velocity potential $\Phi \left(\frac{\partial \Phi}{\partial \zeta}\right)_{\zeta=0} = w$ is given. The derivation of the integral equation is based on KIRCHHOFF's theorem (e.g. [26], p. 37).

Let $\varphi(\xi, \eta, \zeta, \tau)$ be a solution of the partial differential equation (9.1) having partial derivatives of the first and second orders that are continuous within and on a closed surface S, and let $P(\xi_p, \eta_p, \zeta_p)$ be a point within S. Then

$$\varphi(\xi_p, \eta_p, \zeta_p) = \frac{1}{4\pi} \iint_S \left\{[\varphi] \frac{\partial}{\partial n}\left(\frac{1}{\varrho}\right) - \frac{1}{c\varrho} \frac{\partial \varrho}{\partial n}\left[\frac{\partial \varphi}{\partial t}\right] - \frac{1}{\varrho}\left[\frac{\partial \varphi}{\partial n}\right]\right\} dS, \tag{9.2}$$

where φ is the distance from P to a point (ξ, η, ζ) of S.

$$\varrho = \sqrt{(\xi - \xi_p)^2 + (\eta - \eta_p)^2 + (\zeta - \zeta_p)^2}, \tag{9.3}$$

$\partial/\partial n$ denotes differentiation along the inward normal to S, and square brackets denoted "retarded" values, i.e., for any function $U(\xi, \eta, \zeta, \tau)$

$$[U] = U\left(\xi, \eta, \zeta, \tau - \frac{\varrho}{c}\right). \tag{9.4}$$

N.B. In the third term of the integrand the normal differentiation is to be performed before the retarded value is taken.

If P lies outside the surface S, the value of the integral is zero. The aerofoil occupies a part A of the ξ, η plane. We now take for the surface S the closed surface formed by a hemisphere with large radius R and centre at the origin in the upper half space $\zeta > 0$ and by its circle of intersection with the ξ, η plane. If φ goes to zero sufficiently rapidly with R, the integral along the spherical surface vanishes, and we have for a point P with $\xi_p > 0$:

$$\varphi_p = \frac{1}{4\pi} \int_{-\infty}^{+\infty}\int_{-\infty}^{+\infty} \left\{[\varphi]^+ \frac{\partial}{\partial \zeta}\left(\frac{1}{\varrho}\right) - \frac{1}{c\varrho} \frac{\partial \varrho}{\partial \zeta}\left[\frac{\partial \varphi}{\partial \tau}\right]^+ - \frac{1}{\varrho}\left[\frac{\partial \varphi}{\partial \zeta}\right]^+\right\} d\xi\, d\eta \tag{9.5}$$

since $\frac{\partial}{\partial n} = +\frac{\partial}{\partial \zeta}$. The $+$-sign means that the limit of the quantities at the upper side of the ξ, η plane must be taken. Since P lies in the upper half space, the result of the same reasoning when applied to a hemisphere in this half space is

$$0 = \frac{1}{4\pi} \int_{-\infty}^{+\infty}\int_{-\infty}^{+\infty} \left\{[\varphi]^- \frac{\partial}{\partial \zeta}\left(\frac{1}{\varrho}\right) - \frac{1}{c\varrho} \frac{\partial \varrho}{\partial \zeta}\left[\frac{\partial \varphi}{\partial \tau}\right]^- - \frac{1}{\varrho}\left[\frac{\partial \varphi}{\partial \zeta}\right]^-\right\} d\xi\, d\eta \tag{9.6}$$

where the $-$-sign refers to limit values from below. We now remark that φ is an odd function of ζ, hence

$$\left[\frac{\partial \varphi}{\partial \zeta}\right]^+ = \left[\frac{\partial \varphi}{\partial \zeta}\right]^- \text{ and } [\varphi]^+ = -[\varphi]^-, \quad \left[\frac{\partial \varphi}{\partial \tau}\right]^+ = -\left[\frac{\partial \varphi}{\partial \tau}\right]^-.$$

Obviously $[\varphi]^+$ and $\left[\frac{\partial \varphi}{\partial \tau}\right]^-$ vanish outside the aerofoil A. Addition of (9.5) and (9.6) eliminates $[\partial\varphi/\partial\zeta]$ and gives

$$\varphi_p = \frac{1}{2\pi} \iint_A \left\{[\varphi] \frac{\partial}{\partial \zeta}\left(\frac{1}{\varrho}\right) - \frac{1}{c\varrho} \frac{\partial \varrho}{\partial \zeta}\left[\frac{\partial \varphi}{\partial \tau}\right]\right\} d\xi\, d\eta. \tag{9.7}$$

For the evaluation of the unknown functions $[\varphi]$ and $[\partial\varphi/\partial\tau]$ the velocity potential $\Phi(\xi_p, \eta_p, \zeta_p, \tau_p)$ must be calculated.

From the differential relation

$$\Phi(\xi_p, \eta_p, \zeta_p, \tau_p) = \beta \int_{-\infty}^{\xi_p} \varphi\left(\xi', \eta_p, \zeta_p, \tau - \frac{\xi_p - \xi'}{U}\right) d\xi', \tag{9.8}$$

for a point on the aerofoil, where w is given, differentiation with respect to ζ_p yields

$$\left.\begin{aligned} w(\xi_p, \eta_p, 0, \tau_p) = \beta \lim_{\zeta_p \to 0} \frac{\partial}{\partial \zeta_p} \int_{-\infty}^{\xi_\varrho} \iint_A \left\{ \varphi\left(\xi, \eta, \tau - \frac{\varrho'}{c} - \frac{\xi_p - \xi'}{U}\right) \frac{\partial}{\partial \xi}\left(\frac{1}{\varrho}\right) - \right. \\ \left. - \frac{1}{c\varrho'} \frac{\partial \varrho'}{\partial \zeta} \frac{\partial}{\partial \tau} \varphi\left(\xi, \eta, \tau - \frac{\varrho'}{c} - \frac{\xi_p - \xi'}{U}\right)\right\} d\xi\, d\eta\, d\xi' \end{aligned}\right\} \tag{9.9}$$

where

$$\varrho' = \sqrt{(\xi - \xi')^2 + (\eta - \eta_p)^2 + \zeta_p^2}. \tag{9.10}$$

In this general form the equation has not been solved; only the case of harmonic vibrations has been considered. We put

$$\varphi(\xi, \eta, \tau) = \tilde{\varphi}(\xi, \eta)\, e^{i\nu\beta\tau} \tag{9.11}$$

and, for the given function,

$$w(\xi, \eta, \tau) = \tilde{w}(\xi, \eta)\, e^{i\nu\beta\tau} \tag{9.12}$$

The integral equation takes the form

$$\left.\begin{aligned} \tilde{w}(\xi_p, \eta_p) = \beta \lim_{\zeta_p \to 0} \frac{\partial}{\partial \zeta_p} \int_{-\infty}^{\xi_p} \iint_A \tilde{\varphi}(\xi\, \eta), \left[\frac{\partial}{\partial \zeta} \frac{1}{\varrho'} \exp\left(-\frac{i\nu\beta\varrho'}{c}\right)\right] \times \\ \times \exp\left(-\frac{i\nu\beta(\xi' - \xi)}{U}\right) d\xi\, d\eta\, d\xi'. \end{aligned}\right\} \tag{9.13}$$

The kernel

$$\left.\begin{aligned} K(\xi, \eta; \xi_p, \eta_p) = \lim_{\zeta_p \to 0} \frac{\delta}{\delta \zeta_p} \int_{-\infty}^{\xi_p} \exp\left\{-\frac{i\nu\beta(\xi' - \xi)}{U}\right\} \times \\ \times \frac{\partial}{\partial \zeta}\left\{\frac{1}{\varrho'} \exp\left(-\frac{i\nu\beta\varrho'}{c}\right)\right\} d\xi' \end{aligned}\right\} \tag{9.14}$$

is singular with a singularity of the dipole type. Tables of this kernel have been computed [27]. The two-dimensional integral equation has been solved completely by numerical methods. In this case $\tilde{w}$ and $\tilde{\varphi}$ are independent on η. Then the kernel is obtained by integrating with respect to η':

$$K(\xi, \xi_p) = \lim_{\zeta_p \to 0} \frac{\partial}{\partial \zeta_p} \int_{-\infty}^{\xi_p} \exp \frac{ik}{M}(\xi - \xi') \left\{\frac{\partial}{\partial \zeta} \int_{-\infty}^{+\infty} \frac{e^{-ik\varrho'}}{\varrho'} d\eta\right\} d\xi'. \tag{9.15}$$

The inner integral can be expressed into Hankel functions

$$\int_{-\infty}^{+\infty} \frac{e^{-ik\varrho'}}{\varrho'} d\eta = -\pi i H_0^{(2)}\left(k\sqrt{(\xi - \xi')^2 + (\zeta - \zeta_p)^2}\right), \tag{9.16}$$

which gives

$$K(\xi, \xi_p) = \pi i k \int_{-\infty}^{\xi_p} e^{\frac{ik}{M}(\xi' - \xi)} \frac{H_1^{(2)}(k|\xi - \xi'|)}{|\xi - \xi'|} d\xi'. \tag{9.17}$$

The corresponding singular integral equation

$$w(\xi_p) = \int_{-1}^{+1} \varphi(\xi) K(\xi, \xi_p) \, d\xi \tag{9.18}$$

is known as POSSIO's equation [28]. Numerical solutions have been given by SCHADE [29], DIETZE [30] and FETTIS [31].

10. Supersonic flow. *α) The integral equation.* In supersonic flow the potentials Φ and φ are solutions of

$$-\Phi_{\xi\xi} + \Phi_{\eta\eta} + \Phi_{\zeta\zeta} + \Phi_{\tau\tau} = 0. \tag{10.1}$$

The derivative $\Phi_\zeta = w(\xi, \eta, \tau)$ is given on the fixed region A in the plane $\zeta = 0$ in the ξ, η, ζ space, or, in the four-dimensional ξ, η, ζ, τ space, on a cylinder with A as base and the τ axis as axis. Contrary to the behaviour in subsonic flow, for most planforms in supersonic flow the vortex wake behind the aerofoil does not influence the pressure distribution on the aerofoil. (This is the case only for what are called subsonic trailing edges.) For this reason in this section we consider only the velocity potential Φ.

If this potential is found, the pressure distribution follows in linearized approximation from

$$\varphi = \frac{1}{\beta} \{c\,\Phi_\tau + U\,\Phi_\xi\}. \tag{10.2}$$

As in subsonic flow, the integral equation derives from KIRCHHOFF'S equation. Here, however, ξ is a time-like variable, while τ is a space-like variable.

The boundary values of Φ_ζ, for each value of ξ, are given on the intersection of the plane $\zeta = 0$ in the subspace η, ζ, τ with the cylinder. Hence, we apply KIRCHHOFF'S formula to the upper half space $\zeta > 0$ and a point $P(\xi_p; \eta_p, \zeta_p, \tau_p)$:

$$\Phi(\xi_p; \eta_p, \zeta_p, \tau_p) = \frac{1}{4\pi} \iint \left\{ [\Phi]^+ \frac{\partial}{\partial \zeta} \frac{1}{\varrho} - \frac{1}{\varrho} \frac{\partial \varrho}{\partial \zeta} \left[\frac{\partial \Phi}{\partial \xi_p}\right]^+ - \frac{1}{\varrho} \left[\frac{\partial \Phi}{\partial \xi}\right]^+ \right\} d\tau \, d\eta, \tag{10.3}$$

where brackets here denote "retarded" values:

$$[\Phi] = \Phi(\xi_p - \varrho, \eta, \zeta, \tau), \tag{10.4}$$

and

$$\varrho = \sqrt{(\eta_p - \eta)^2 + (\zeta_p - \zeta)^2 + (\tau_p - \tau)^2}.$$

(Again the retarded value in the last term must be inserted after the differentiation with respect to ζ.)

Performing the differentiations gives

$$\Phi_p = \frac{1}{4\pi} \iint \left\{ [\Phi]^+ \frac{\zeta_p - \zeta}{\varrho^3} + \frac{\zeta_p - \zeta}{\varrho^2} \left[\frac{\partial \Phi}{\partial \xi_p}\right]^+ - \frac{1}{\varrho} \left[\frac{\partial \Phi}{\partial \zeta}\right]^+ \right\} d\tau \, d\eta. \tag{10.5}$$

We transform this formula into a formula in the ξ, η, ζ space, which is more accessible to physical interpretation. Introducing a new variable of integration

$$\xi' = \xi_p - \varrho = \xi_p - \sqrt{(\eta_p - \eta)^2 + (\zeta_p - \zeta)^2 + (\tau_p - \tau)^2} \tag{10.6}$$

instead of τ, we have

$$\tau = \tau_\varrho \pm \varrho' \tag{10.7}$$

where

$$\varrho' = \sqrt{(\xi_p - \xi')^2 - (\eta_p - \eta)^2 - (\zeta_p - \zeta)^2}. \tag{10.8}$$

Since the cylinder in the η, ζ, τ space extends over all values of τ, both signs enter in the equation, and the points where $\varrho' > 0$ are mapped on two points on the η, τ plane. The contributions of both points must be taken into account.

The transformed values are

$$\left[\frac{\partial \Phi}{\partial \zeta}\right] = \frac{\partial^*}{\partial \zeta} \Phi(\xi_p - \varrho, \eta, \zeta, \tau) = \frac{\partial^*}{\partial \zeta} \Phi(\xi', \eta, \zeta, \tau_p \pm \varrho'), \tag{10.9}$$

where the asterisk indicates that ϱ and ϱ' must be differentiated with respect to ζ.

Further

$$\left.\begin{aligned} \left[\frac{\partial \Phi}{\partial \xi_p}\right] &= \frac{\partial}{\partial \xi_p} \Phi(\xi_p - \varrho, \eta, \zeta, \tau) = \frac{\partial^*}{\partial \xi'} \Phi(\xi', \eta, \zeta, \tau_p \pm \varrho') \\ &= \frac{\partial}{\partial \xi'} \Phi(\xi', \eta, \zeta, \tau_p \pm \varrho') \pm \frac{\partial \Phi}{\partial \tau_p} (\xi', \eta, \zeta, \tau_p \pm \varrho') \frac{\partial \varrho'}{\partial \xi'}. \end{aligned}\right\} \tag{10.10}$$

As

$$\frac{d\tau}{d\xi'} = \pm \frac{\xi' - \xi_p}{\varrho'} = \mp \frac{\varrho}{\varrho'} \tag{10.11}$$

we have

$$\left.\begin{aligned} \Phi(\xi_p, \eta_p, \zeta_p, \tau_p) = &\frac{-1}{4\pi} \iint^* \Big\{ [\Phi]_1^+ \frac{\zeta_p - \zeta}{\varrho' \varrho^2} + \frac{\zeta_p - \zeta}{\varrho \varrho'} \frac{\partial}{\partial \xi'} [\Phi]_1^+ + \\ &+ \frac{(\zeta_p - \zeta)}{\varrho \varrho'} \frac{(\xi' - \xi_p)}{\varrho'} \left[\frac{\partial \Phi}{\partial \tau}\right]_1^+ - \frac{1}{\varrho'} \left[\frac{\partial \Phi}{\partial \zeta}\right]_1^+ \Big\} d\xi' \, d\eta + \\ &+ \frac{1}{4\pi} \iint^* \Big\{ [\Phi]_2^+ \frac{\zeta_p - \zeta}{\varrho' \varrho^2} + \frac{\zeta_p - \zeta}{\varrho \varrho'} \frac{\partial}{\partial \xi'} [\Phi]_2^+ + \\ &+ \frac{(\zeta_p - \zeta)}{\varrho \varrho'} \frac{\xi' - \xi_p}{\varrho'} \left[\frac{\partial \Phi}{\tau \partial}\right]_2^+ - \frac{1}{\varrho'} \left[\frac{\partial \Phi}{\partial \zeta}\right]_2^+ \Big\} d\xi' \, d\eta, \end{aligned}\right\} \tag{10.12}$$

where the index 1, 2 means that the last argument is $\tau_p \pm \varrho'$. The region of integration is the forward part of the intersection of the cone

$$(\xi_p - \xi')^2 - (\eta_p - \eta)^2 - \zeta_p^2 \geqq 0,$$

with the ξ', η plane, since $\xi_p > \xi'$.

At the boundary the integrand becomes infinite, and the integral has to be considered as a finite part in the sense of HADAMARD [7] and RIESZ [8].

We transform the second term in the integrals by partial integration with respect to ξ' and remark that the integrated terms do not contribute anything, the potential vanishes for $\xi' = -\infty$ and the boundary does not contribute to the finite part. This gives

$$\left.\begin{aligned} \Phi(\xi_p, \eta_p, \zeta_p, \tau_p) = &\frac{-1}{4\pi} \iint^* \Big\{ [\Phi]_2 - [\Phi]_1 \Big\} \frac{\zeta_p - \zeta}{\varrho'^3} - \frac{\zeta_p - \zeta}{\varrho'^2} \left\{ \left[\frac{\partial \Phi}{\partial \tau}\right]_2 - \left[\frac{\partial \Phi}{\partial \tau}\right]_1 \right\} - \\ &- \frac{1}{\varrho'} \left\{ \left[\frac{\partial \Phi}{\partial \zeta}\right]_2 - \left[\frac{\partial \Phi}{\partial \zeta}\right]_1 \right\} \Big\} d\xi' \, d\eta \\ = &\frac{1}{4\pi} \iint^* \Big\{ \{[\Phi]_2 - [\Phi]_1\} \left(\frac{\partial}{\partial \zeta} \frac{1}{\varrho'}\right) + \left\{ \left[\frac{\partial \Phi}{\partial \tau}\right]_2 - \left[\frac{\partial \Phi}{\partial \tau}\right]_1 \right\} \frac{1}{\varrho'} \frac{\partial \varrho'}{\partial \zeta} + \\ &+ \frac{1}{\varrho'} \left\{ \left[\frac{\partial \Phi}{\partial \zeta}\right]_2 - \left[\frac{\partial \Phi}{\partial \zeta}\right]_1 \right\} \Big\} d\xi' \, d\eta. \end{aligned}\right\} \tag{10.13}$$

Application of the theorem to the lower half space, since $\zeta_p > 0$, gives the same equation with left hand side zero, but here for Φ and its derivatives the lower limiting values must be taken.

As Φ is an odd function of ζ, $\Phi=0$ everywhere outside the aerofoil, and $\Phi^+ = -\Phi^-$, while $\partial\Phi^+/\partial\zeta = \partial\Phi^-/\partial\zeta$.

Addition of the two equations then gives:

$$\Phi(\xi_p, \eta_p, \zeta_p, \tau_p) = \frac{1}{2\pi} \iint \frac{1}{\varrho'} \left\{ \left[\frac{\partial\Phi}{\partial\zeta}\right]_2 - \left[\frac{\partial\Phi}{\partial\zeta}\right]_1 \right\} d\xi' \, d\eta . \tag{10.14}$$

This formula has been found by HEASLET and LOMAX [32] by the very elegant reasoning of acoustical planforms, considering at the point P the origin of those disturbances in the plane which can reach P.

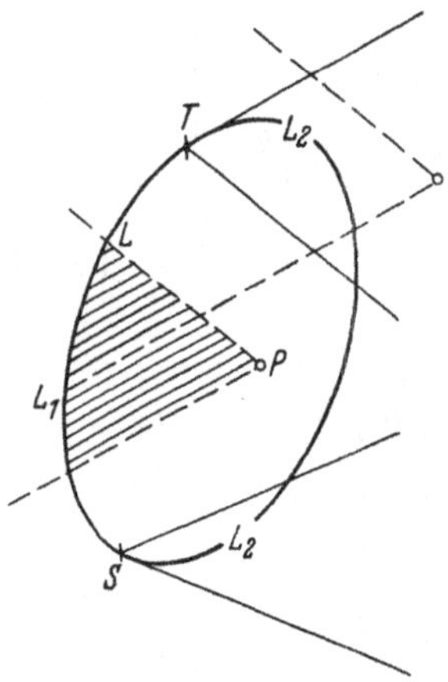

Fig. 4. Integration regions for supersonic flow.

On the aerofoil $\partial\Phi/\partial\zeta$ is given. Hence we see that if the region of integration, i.e. the forward Mach cone, contains only a part of the aerofoil, (10.14) immediately gives the value of Φ_p. In order to find the region for which this is true, we have to consider the backward Mach cones, starting from points at the leading edge of the aerofoil. The leading edge consists of two parts, the supersonic part which lies ahead of the Mach cones of its points, and the subsonic part L_2, which lies inside its Mach cones. The Mach lines starting from the separation point T and S mark the part of the aerofoil where Φ is immediately given by (10.14); for the points in the side regions the integrand is partly unknown, and an integral equation must be solved first in order to obtain these unknown values. This integral equation is obtained by requiring that in these regions Φ must be zero.

Another integral equation can be obtained if we remark that Φ and Φ_τ are different from zero only on the aerofoil. Addition of the formula (10.13) and its counterpart now gives

$$\Phi_p = \frac{1}{2\pi} \iint^{+} \left\{ \{[\Phi]_2 - [\Phi]_1\} \frac{\partial}{\partial\zeta} \frac{1}{\varrho'} + \left\{ \left[\frac{\partial\Phi}{\partial\tau}\right]_2 - \left[\frac{\partial\Phi}{\partial\tau}\right]_1 \right\} \frac{1}{\varrho'} \frac{\partial\varrho'}{\partial\zeta} \right\} d\xi' \, d\varrho . \tag{10.15}$$

The region of integration here is the part of the aerofoil in the forward Mach cone of the point P.

By differentiation with respect to ζ_p for $\zeta_p=0$ an integral equation can be obtained. For arbitrary non-steady motion this equation has not been solved.

β) A solution method for the integral equation. As has been remarked the integral equation in $2+1$ variables corresponding to (10.14) can be solved, but this method fails in the case of $3+1$ variables. Here GARDNER [33] has given a method for the reduction of the problem to two successive integral equations of the Evvard type. We give here a slightly different description of the method, starting from KIRCHHOFF's formula

$$\Phi(\xi_p, \eta_p, \zeta_p, \tau_p) = \frac{1}{2\pi} \iint \frac{1}{\varrho} \frac{\partial\Phi(\xi_p - \varrho, \eta, 0, \tau)}{\partial\zeta} \, d\tau \, d\eta , \tag{10.16}$$

which follows from (10.5) if the function Φ is an odd function of ζ.

We first derive an integral representation for the function $\psi(\xi_p - |\varrho|)$, considered as a function of two variables ξ_p and ϱ, which is seen to satisfy the equation

$$\psi_{\xi\xi} - \psi_{\varrho\varrho} = 0 .$$

In order to derive this integral representation, we use HADAMARD's method of descent and consider instead $\psi(\xi_p - |\varrho|)$ as a solution of the equation

$$\psi_{\xi\xi} - \psi_{\varrho\varrho} - \psi_{\sigma\sigma} \mp 0 \tag{10.17}$$

which is independent of the additional variable σ.

For $\varrho = 0$ we have the boundary value $\psi(\xi)$, and, further, $\psi(\xi - |\varrho|)$ is an even function of ϱ.

Then we have the integral representation

$$f(\xi_p - |\varrho|) = \iint^{*} f(\xi)\, d\xi\, d\sigma \left\{\frac{\partial}{\partial r} \frac{1}{\sqrt{(\xi_p - \xi)^2 - \sigma^2 - (\varrho - r)^2}}\right\}_{r=0}, \tag{10.18}$$

which follows from (10.15) for functions which are independent of the time.

Substitution gives the integral equation

$$\left.\begin{aligned} \Phi(\xi_p, \eta_p, \zeta_p, \tau_p) &= \frac{1}{4\pi^2} \iiint^{*} \frac{\partial \Phi}{\partial \zeta}(\xi, \eta, 0, \tau) \times \\ &\times \frac{1}{\varrho}\left[\frac{\partial}{\partial r} \frac{1}{\sqrt{(\xi_p - \xi)^2 - \sigma^2 - (\varrho - r)^2}}\right]_{r=0} d\xi\, d\eta\, d\tau\, d\sigma \end{aligned}\right\} \tag{10.19}$$

where

$$\varrho^2 = (\eta_p - \eta)^2 + (\zeta_p - 0)^2 + (\tau_p - \tau)^2, \tag{10.20}$$

or, after performing the differentiation:

$$\left.\begin{aligned} \Phi(\xi_p, \eta_p, \zeta_p, \tau_p) &= \frac{1}{4\pi^2} \iiint^{*} \frac{\partial \Phi}{\partial \zeta}(\xi, \eta, 0, \tau) \times \\ &\times \{(\xi_p - \xi)^2 - (\eta_p - \eta)^2 - \zeta_p^2 - (\tau_p - \tau)^2 - \sigma^2\}^{-\frac{3}{2}}\, d\xi\, d\eta\, d\tau\, d\sigma. \end{aligned}\right\} \tag{10.21}$$

The region of integration is the part of the space $\zeta = 0$ inside the characteristic cone.

We obtain GARDNER's set of equations by substitution of a new variable of integration λ instead of σ, defined by

$$\lambda^2 = -\sigma^2 + (\tau_p - \tau)^2 + (\xi_p - \xi)^2. \tag{10.22}$$

The equation transforms into

$$\left.\begin{aligned} \Phi(\xi_p, \eta_p, \zeta_p, \tau_p) &= \frac{1}{4\pi^2} \iiint^{*} \frac{\partial \Phi}{\partial \zeta}(\xi, \eta, 0, \tau) \times \\ &\times \frac{d\xi\, d\tau}{\sqrt{(\xi_p - \xi)^2 - (\tau_p - \tau)^2 - \lambda^2}} \frac{\lambda\, d\lambda\, d\eta}{\sqrt{\lambda^2 - (\eta_p - \eta)^2 - \zeta_p^2}}. \end{aligned}\right\} \tag{10.23}$$

We introduce a function

$$\psi(\xi_p, \tau_p, \eta, \lambda) = \frac{1}{2\pi} \iint \frac{\partial \Phi}{\partial \zeta}(\xi, \eta, 0, \tau) \frac{d\xi\, d\tau}{\sqrt{(\xi_p - \xi)^2 - (\tau_p - \tau)^2 - \lambda^2}}. \tag{10.24}$$

Then

$$\Phi(\xi_p, \eta_p, 0_p, \tau_p) = \frac{1}{2\pi} \left\{\frac{\partial}{\partial \lambda_1} \iint \frac{\psi(\xi_p, \tau_p, \eta, \lambda)}{\sqrt{(\lambda - \lambda_1)^2 - (\eta_p - \eta)^2}}\, d\eta\, d\lambda\right\}_{\lambda_1 = 0}, \tag{10.25}$$

and we have the two integral equations.

The method has been applied to the case of a rectangular wing tip travelling at supersonic speed [*34*].

The tip extends along the quadrant $\xi > 0$, $\eta > 0$ of the ξ, η plane. This means that $\Phi = 0$ for $\eta < 0$, $\partial\Phi/\partial\zeta$ is given on the quadrant and $\partial\Phi/\partial r$ is unknown.

In Eq. (10.24) $\partial\Phi/\partial\zeta$ is known for all values of ξ and τ and $\eta > 0$ in the region of integration, and we see that ψ is determined for $\eta > 0$.

For Eq. (10.25) we introduce the function

$$\chi(\xi_p, \tau_p, \eta_p, \lambda_1) = \frac{1}{2\pi} \iint \frac{\psi(\xi_p, \tau_p, \eta, \lambda)\, d\eta\, d\lambda}{\sqrt{(\lambda - \lambda_1)^2 - (\eta_p - \eta)^2}}$$

and remark that $\Phi = \left(\frac{\partial \chi}{\partial \lambda_1}\right)_{\lambda_1 = 0}$. For $\eta < 0$ we put the additional condition on χ: $\chi = 0$.

For the determination of ψ this gives again an equation of the Evvard type which can be solved. Then χ and Φ can easily be calculated.

γ) Harmonic oscillations. For harmonic oscillations Φ contains the time factor $e^{i\nu\tau}$. We can write

$$\Phi(x, y, z, t) = \Psi(\xi, \eta, \zeta)\, e^{\frac{i\nu\tau}{\beta}}. \tag{10.26}$$

The integral equation (10.14) becomes

$$\Psi(\xi_p, \eta_p, \zeta_p) = \frac{-1}{\pi} \iint \frac{\partial\psi}{\partial\zeta} \frac{\cos\frac{\nu\varrho'}{\beta}}{\varrho'}\, d\xi'\, d\eta. \tag{10.27}$$

For two-dimensional flow ζ is independent of η, and the integration with respect to η can be performed.

The leading edge is now supersonic, and we have

$$\Psi(\xi_p, \eta_p, \zeta_p) = -\frac{1}{\pi} \int_{\xi'=0}^{\xi'=\xi_p-\zeta_p} \psi_\zeta\, J_0\left\{\frac{\nu}{\beta}\sqrt{(\xi_p - \xi')^2 - \zeta_p^2}\right\} d\xi'. \tag{10.28}$$

On the aerofoil ψ_ζ is given, and we see that the two-dimensional case is completely solved by (10.28).

The difficulties encountered in the case of subsonic leading edges with the integral equation (10.27) have already been described. A numerical method has been presented in [*35*]. For a rectangular wing tip the solution has been given by GOODMAN [*34*].

The second integral equation, which involves only integrations over parts of the wing, follows from (10.15):

$$\psi(\xi_p, \eta_p, \zeta_p) = \frac{-1}{\pi} \iint \psi(\xi', \eta, 0) \left\{\frac{\partial}{\partial\zeta} \frac{\cos\frac{\nu\varrho'}{\beta}}{\varrho'}\right\} d\xi'\, d\eta. \tag{10.29}$$

The same relation holds for the acceleration potential, which immediately yields the pressure distribution.

For this reason WATKINS and BERMAN [36] derive for this potential an integral equation which is completely analogous to the corresponding equation in the subsonic case. If the kernel is tabulated, a numerical solution method can be given.

IV. Reciprocity relations.

11. For several purposes the "reciprocity relations" are important. They are extensions of the well known Helmholtz reciprocity relations in acoustics [37].

They follow mathematically from GREEN'S identity. We base our consideration on the wave equation in the transformed coordinates ξ, η, ζ, τ.

For a subsonic flow this equation is

$$\Phi_{\xi\xi} + \Phi_{\eta\eta} + \Phi_{\zeta\zeta} - \frac{1}{c^2}\Phi_{\tau\tau} = 0. \tag{11.1}$$

For two solutions Φ and $\bar{\Phi}$ of this equation, substitution into GREEN'S identity gives

$$\iiint \frac{1}{c^2}\{\Phi_{\tau\tau}\bar{\Phi} - \Phi\bar{\Phi}_{\tau\tau}\}\,d\xi\,d\eta\,d\zeta = \int\limits_{-\infty}^{+\infty}\!\!\int \left(\Phi\frac{\partial\bar{\Phi}}{\partial\zeta} - \bar{\Phi}\frac{\partial\Phi}{\partial\zeta}\right)d\xi\,d\eta, \tag{11.2}$$

if, as in previous sections, the theorem is applied at first to a hemisphere in the half space $\zeta > 0$, the radius of which is extended to infinity.

We seek to eliminate the volume integral in this identity and take for the first solution Φ the velocity potential $\Phi(\xi, \eta, \zeta, \tau)$ corresponding to a flow starting from rest at $\tau = 0$ and having a main flow with velocity U in the direction of the positive x axis.

For the second solution $\bar{\varphi}$ we take the acceleration potential $\bar{\varphi}$ corresponding to a "reverse flow" with main velocity U in the direction of the negative x axes. The time variable is here $\tau' = T - \tau$, and the flow is supposed to start from rest at the time $\tau' = 0$. The relation between acceleration potential φ and velocity potential Φ for the first flow is given by Eq. (10.2),

$$\varphi = \frac{1}{\beta}\{c\,\Phi_\tau + U\,\Phi_\xi\}, \tag{11.3}$$

while for the second flow this relation is

$$\bar{\varphi} = \frac{1}{\beta}\{c\,\bar{\Phi}_{\tau'} - U\,\bar{\Phi}_\xi\} = -\frac{1}{\beta}\{c\,\bar{\Phi}_\tau + U\,\bar{\Phi}_\xi\}. \tag{11.4}$$

By partial integration of the volume integral in (11.2) with respect to τ we see that

$$\int\limits_0^T \{\Phi_{\tau\tau}\bar{\varphi} - \Phi\bar{\varphi}_{\tau\tau}\}\,d\tau = [\Phi_\tau\bar{\varphi} - \Phi\bar{\varphi}_\tau]_0^T - \int\limits_0^T \{\Phi_\tau\bar{\varphi}_\tau - \Phi_\tau\bar{\varphi}_\tau\}\,d\tau = 0, \tag{11.5}$$

and the relation becomes

$$\int\limits_0^T \int\limits_{-\infty}^{+\infty} \int\limits_{-\infty}^{+\infty} \left\{\Phi\frac{\partial\bar{\varphi}}{\partial\zeta} - \bar{\varphi}\frac{\partial\Phi}{\partial\zeta}\right\}d\xi\,d\eta\,d\tau = 0, \tag{11.6}$$

where the integration is extended over the whole ξ, η plane. We remark that for the first flow the wake W extends behind the aerofoil A, starting at the trailing edge T, and is bounded by two lines parallel to the ξ axis in the direction of the positive ξ axis; for the second flow the wake extends along the negative ξ axis and starts at the leading edge L, which here takes the role of the trailing edge.

We remark that φ and $\bar{\varphi}$ are different from zero only at the aerofoil A, Φ is different from zero at $A+W$, $\bar{\Phi}$ is different from zero at $A+\bar{W}$.

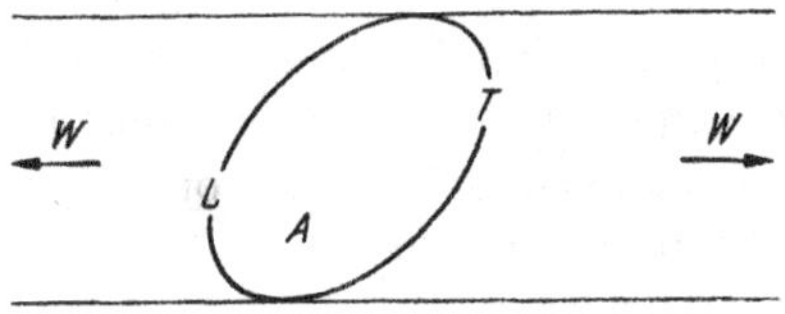

Fig. 5. Reciprocity relations.

Hence the second integral is

$$\int_0^T \iint_A \bar{\varphi}\,\frac{\partial \Phi}{\partial \zeta}\,d\xi\,d\eta\,d\tau$$

and the first integral is

$$\int_0^T \iint_{A+W} \Phi\,\frac{\partial \bar{\varphi}}{\partial \zeta}\,d\xi\,d\eta\,d\tau .$$

We now transform this integral. Since

$$\frac{\partial \bar{\varphi}}{\partial \zeta} = -\frac{1}{\beta}\{c\,\bar{\Phi}_{\tau\zeta} + U\,\bar{\Phi}_{\xi\xi}\},$$

the result is

$$+\int_0^T \iint_{A+W} \Phi\,\frac{\partial \bar{\varphi}}{\partial \zeta}\,d\xi\,d\eta\,d\tau = -\frac{c}{\beta}\int_0^T \iint_{A+W} \Phi\,\bar{\Phi}_{\tau\zeta}\,d\xi\,d\eta\,d\tau - \frac{U}{\beta}\int_0^T \iint_{A+W} \Phi\,\bar{\Phi}_{\xi\xi}\,d\xi\,d\eta\,d\tau .$$

The first integral is transformed by partial integration with respect to τ,

$$\int_0^T \iint_{A+W} \Phi\,\bar{\Phi}_{\tau\zeta}\,d\xi\,d\eta\,d\tau = -\int_0^T \iint_{A+W} \Phi_\tau\,\bar{\Phi}_{\tau\zeta}\,d\xi\,d\eta\,d\tau ,$$

the second by partial integration with respect to ξ,

$$\int_0^T \iint_{A+W} \Phi\,\bar{\Phi}_{\xi\xi}\,d\xi\,d\eta\,d\tau = \int_0^T d\tau \int d\eta\,[\Phi\,\bar{\Phi}_\zeta]_L^\infty - \int_0^T \iint_{A+W} \Phi_\xi\,\bar{\Phi}_\zeta\,d\xi\,d\eta\,d\tau .$$

Remark that at the trailing edge the singularity is eliminated by the Kutta condition, so that the partial integration does not leave terms arising from this trailing edge. The integrated terms vanish since Φ is zero at the leading edge. In this way we have reduced the first integral to

$$\int_0^T \iint_{A+W} \left\{\frac{c}{\beta}\,\Phi_\tau + \frac{U}{\beta}\,\Phi_\xi\right\} \bar{\Phi}_\zeta\,d\xi\,d\eta\,d\tau$$

or, taking account of (11.3) and of the fact that $\varphi = 0$ on W,

$$\int_0^T \iint_A \varphi\,\bar{\Phi}_\zeta\,d\xi\,d\eta\,d\tau .$$

In this way we have obtained the relation

$$\int_0^T \iint_A (\Pi\,\bar{W} - \bar{\Pi}\,W)\,d\xi\,d\eta\,d\tau = 0, \tag{11.7}$$

where $W = \partial\Phi/\partial\zeta$, $\bar{W} = \partial\bar{\Phi}/\partial\zeta$, and the pressure distributions $\Pi(\xi, \eta, \tau)$ and $\bar{\Pi}(\xi, \eta, \tau')$ are proportional to the acceleration potentials $\Pi = \varrho\,\varphi$, $\Pi = \varrho\,\bar{\Pi}$. This is the general form of the reciprocity relation for non-steady flows.

For oscillatory motion the direct flow has the time factor $e^{i\nu\tau}$; the reverse flow, $e^{-i\nu\tau}$. For this case the theorem has been given by FLAX *[38]*, the general case for non-steady flow by HEASLET and SPREITER *[39]*. The same relations hold in supersonic main flow.

References.

[1] THEODORSEN, TH.: General theory of aerodynamic instability and the mechanism of flutter. NACA Techn. Rep. No. 496, 1935.

[2] KÜSSNER, H. G.: Zusammenfassender Bericht über den instationären Auftrieb von Flügeln. Luftf.-Forsch. **13**, 410—424 (1936).

[3] KÜSSNER, H. G.: Allgemeine Tragflächentheorie. Luftf.-Forschg. **17**, 370—378 (1940).

[4] KÜSSNER, H. G.: A review of the two-dimensional problem of unsteady lifting surface theory during the last thirty years. University of Maryland Lecture Series, No. 23, 1953.

[5] CIGALA, P.: Lo stato attuale delle ricerche sul moto instazionario di una superficie portante. L'Aerotecnica **21**, 671—685, 759—777 (1941).

[6] TIMMAN, R.: La théorie des profils minces en écoulement non stationnaire en fluide incompressible ou compressible. N.L.L. (Nat. Luchtv. Lab.) Publ. No. MP 69.

[7] HADAMARD, J.: Lectures on CAUCHY's problem. Yale 1923.

[8] RIESZ, M.: Acta math. **81**, 1—223 (1949).

[9] KÜSSNER, H. G.: Das zweidimensionale Problem der beliebig bewegten Tragfläche unter Berücksichtigung von Partialbewegungen der Flüssigkeit. Luftf.-Forschg. **17**, 355—363 (1940).

[10] SEARS, W. R.: Operational methods in the theory of airfoils in non-uniform motion. J. Franklin Inst. **230**, 95—111 (1940).

[11] v. D. VOOREN, A. I.: Generalization of the Theodorsen function to stable oscillations. J. Aeronaut. Sci. **19**, 209—211 (1952).

[12] WAGNER, H.: Über die Entstehung des dynamischen Auftriebs von Tragflügeln. Z. angew. Math. Mech. **5**, 17 (1925).

[13] TIMMAN, R.: The oscillating airfoil between two parallel walls. Appl. Sci. Res. A **3**, 31—57 (1951).

[14] MORSE and FESHBACH: Methods of theoretical physics. New York, 1953.

[15] SCHADE, TH.: Theorie der schwingenden kreisförmigen Tragfläche auf potentialtheoretischer Grundlage. Luftf.-Forschg. **17**, 387—400 (1940).

[16] VAN SPIEGEL, E.: Theory of the circular wing in steady incompressible flow. NLL Report F. 189, 1956.

[16a] VAN SPIEGEL: Boundary value problems in lifting surface theory. Thesis: Delft, 1959.

[17] KÜSSNER, H. G.: A general method for solving problems of the unsteady lifting surface theory in the subsonic range. J. Aeronaut. Sci. **21**, 17—27 (1954).

[18] HASKIND, M. D.: Oscillations of a wing in a subsonic gas flow. Prikl. Mat. i Mehk. (Moscow) **11**, 129—146 (1947). Also available as Translation No. A 9-T-22. Air Mat. Command and Brown Univ.

[19] REISSNER, E.: On the application of Mathieu functions in the theory of subsonic compressible flow past oscillating airfoils. NACA Techn. Note 2363, 1951.

[20] BILLINGTON, A. E.: Harmonic oscillations of an aerofoil in subsonic flow. ARL Report A 65, 1949.

[21] TIMMAN, R.: Beschouwingen over de luchtkrachten op trillende vliegtuigvleugels. Thesis, Technological University Delft 1946.

[22] TIMMAN, R., and A. I. v. D. VOOREN: Theory of the oscillating wing with aerodynamically balanced control surface in a two-dimensional subsonic compressible flow. NLL Report F. 54, 1949.

[23] McLACHLAN, N. W.: Theory and application of Mathieu functions. Oxford: Clarendon Press 1947.

[24] MEIXNER, J., u. F. W. SCHÄFKE: Mathieusche Funktionen und Sphäroidfunktionen. Springer, 1953.

[25] BURGER, A. P.: On the asymptotic solution of wave propagation and oscillation problems. NLL Report F. 157, 1955.

[26] BAKER, B. B., and E. T. COPSON: The Mathematical theory of HUYGENS' principle. 2nd edition, Oxford: Clarendon Press 1939.

[27] WATKINS, CH. E., H. L. RUNYAN and D. S. WOOLSON: On the kernel function of the integral equation relating the lift and downwash distributions of oscillating finite wings in subsonic flow. NACA Techn. Note 3131, 1954.

[28] POSSIO, C.: L'azione aerodinamica sul profilo oscillante in un fluido compressibile a velocità iposonora. L'Aerotecnica **18**, 441—458 (1938).

[29] SCHADE, W.: The numerical solution of POSSIO's integral equation for an oscillating aerofoil in a two-dimensional subsonic stream. A.R.C. Report 9506, 1946.

[30] DIETZE, F.: Die Luftkräfte des harmonisch schwingenden Flügels im kompressibelen Medium bei Unterschallgeschwindigkeit. D.V.L. Forschungsber. 1733, 1943.

[31] FETTIS, H.E.: An approximate method for the calculation of non-stationary air forces at subsonic speeds. WADC Techn. Rep. 52—56, 1952.

[32] HEASLET, M., and H. LOMAX: Supersonic and transonic small perturbation theory in High Speed Aerodynamics and Jet Propulsion, Vol. VI: General theory of high speed aerodynamics (W.R. SEARS, ed.). Princeton: University Press 1954.

[33] GARDNER, C.: Time dependent linearized supersonic flow past planar wings. Comm. Pure Appl. Math. **3**, 33—38 (1950).

[34] GOODMAN, TH.R.: The quarter infinite wing oscillating at supersonic speeds. Quart. Appl. Math. **10**, 189—192 (1952).

[35] PINES, S., J. DUGUNDJI and J. NEURINGER: Aerodynamic flutter derivatives for a flexible wing with supersonic and subsonic edges. J. Aeronaut. Sci. **22**, 693—700 (1955).

[36] WATKINS, CH.E., and J.H. HERMAN: Velocity potential and air forces associated with a triangular wing in supersonic flow with subsonic leading edges, and deforming harmonically according to a general quadratic equation. NACA Techn. Note No. 3009, 1953.

[37] RAYLEIGH, Lord.: The theory of sound. Dover Publications 2nd ed. 1945.

[38] FLAX, A.H.: Reverse flow and variational theorems for lifting surfaces in non-stationary compressible flow. Cornell Aero. Lab. Cal-42, 1952.

[39] HEASLET, M.A., and J.R. SPREITER: Reciprocity relations in aerodynamics. NACA Techn. Note No. 2700, 1952.

Jets and Cavities.

By

David Gilbarg.

With 60 Figures.

1. Introduction. The problems of free streamline theory have long attracted both hydrodynamicists and mathematicians, the former because of the many applications to jet, cavity, and wake phenomena, and the latter because of the unusual and exciting mathematical features of this branch of potential theory. The different interests of these two groups have resulted in a parallel development of the field, emphasizing on the one hand particular solutions and technical applications, and on the other hand general theory and results of a more qualitative nature. These two trends are reflected in the following exposition in that Chap. II and V are concerned with special flows and numerical methods, and Chap. III und IV with the general theory. While this is not an altogether natural division from the physical point of view, we have adopted it as a convenient means of presenting the various methods and of tracing the development of the ideas. In devoting a separate chapter to qualitative features (Chap. III), we hope to direct attention to an elegant part of the general theory which should be useful to those with theoretical and applied interests alike.

The emphasis in this survey is on the basic models of cavity and jet flow and the mathematical methods that have been devised for their study. The discussion centers almost entirely on the exact theory, which is now in a relatively mature (but by no means final) state. From limitation of space and the essentially different nature of the material, a number of allied topics which might reasonably be included here have been touched on only perfunctorily or not at all. Among the most important are compressible fluid jets, and wakes in real fluids. For these and other subjects related to this work we refer to the literature. In particular, the reader is directed to Birkhoff and Zarantonello [*1*][1] for topics not covered here, for references to numerous applications, and also for a more detailed account of many parts of this article.

In recent years a large portion of the literature on free surface phenomena has appeared in the form of laboratory reports with only limited circulation. We have attempted where possible to present these results, although a complete exposition is of course impossible.

I. Physical and mathematical foundations.

2. Experimental background. We review here some of the observations and experimental data concerning water entry, cavitation, wakes, and jet flow, that seem most pertinent in motivating and appraising the theory. For discussion of equipment, accuracy of measurement, etc., we refer the reader to the original sources.

[1] I am grateful to these authors for having provided me with a pre-publication copy of their valuable work.

α) *Cavity flows*[1]. Although one of the first applications of the concept of streamline of discontinuity was to the theory of wakes, the assumptions of the theory are better realized in the cavity phenomena associated with high speed underwater motion. These occur, for example, when a missile enters water from the atmosphere at moderate or high speed, its inertia creating behind it an air filled cavity; or, if the body is completely immersed and moves with such high speed that the local pressures descend to the vapor pressure of the liquid, the fluid "boils" and forms a vapor cavity behind the body. The practical importance of underwater motion at high speeds has led in recent years to extensive research on the properties of these cavity flows.

The phenomena accompanying water entry[2] include a surface impact phase, in which the fluid has not yet detached from the body; the splash; cavity formation and growth; subsurface and surface closure of the cavity; jet formation at each closure; and decay of the cavity. Of particular interest for the later discussion are the jets which form at the rear of the closed cavity and which can be observed (in high speed photographs) traveling down its length, often overtaking the missile and sometimes affecting its motion. The scaling of water entry phenomena is a problem complicated by several factors, including the influence of atmospheric density and, sometimes, surface tension and viscosity[3].

Water entry cavities, because of their non-stationary gravity-dependent character, are yet too complex for quantitative theoretical treatment. However, water tunnel experiments make it possible to observe steady-state cavities behind obstacles under conditions very close to those assumed in the theory.

Letting p_∞ denote the pressure in the undisturbed stream, and p_c the cavity pressure, we define the *cavitation parameter* σ by

$$\sigma = \frac{p_\infty - p_c}{\frac{1}{2}\varrho U^2}, \tag{2.1}$$

where U is the velocity of the stream and ϱ the fluid density. Assuming ideal fluid theory in the absence of gravity, one sees that σ has the significance of being the only non-dimensional similarity parameter associated with the flow. That is, two cavity flows under geometrically similar conditions will be geometrically and dynamically similar if and only if they have equal values of σ, and any non-dimensional quantity depending only on p_∞, p_c, ϱ and U (such as the drag coefficient of a fixed obstacle), is a function of σ alone. The free stream values p_∞ and U are experimentally assigned, while p_c is the (known) vapor pressure of the liquid[4]. The cavitation number is therefore an experimentally controlled variable and can be considered a datum in any flow problem.

A number of investigators[5] have carried out water tunnel studies of forces and cavity dimensions as functions of σ. The results of drag measurements on cones and discs in three different tunnels are shown in Fig. 1. At least for small

[1] For a survey of the experimental work on cavitation, see P. EISENBERG: David Taylor Model Basin Reports 712 (1950) and 842 (1953); also Forschungsh. Schiffstechn. **3**, 111–124 (1953); **4**, 155–168 (1953); **5**, 201–212 (1954). An early report on cavitation phenomena is contained in J. ACKERET, Techn. Mech. Thermodyn. **1** (1930).

[2] D. GILBARG and R. ANDERSON: J. Appl. Phys. **19**, 127–139 (1948). — A. WORTHINGTON: A study of splashes. New York: Longmans Green 1908. These contain numerous photographs of the phenomena involved.

[3] Cf. GILBARG and ANDERSON: Preceding footnote; also H. WAYLAND: Naval Ord. Test Station, NAVORD Rep. 978, 1947.

[4] In water tunnel experiments with air-filled cavities the cavity pressure must be determined by direct measurement.

[5] H. REICHARDT: Min. Aircraft Prod. Rep. and Transl. 766 (1946). — P. EISENBERG: Footnote 1. — M. PLESSET and B. PERRY: Mémoire sur la mécanique des fluides offerts à

σ, the observations show that the drag coefficient is given in good approximation by the formula

$$C_D(\sigma) = C_D(0)(1+\sigma). \tag{2.2}$$

This formula, first observed by REICHARDT, is asymptotically valid according to both plane and axially symmetric theory (cf. Sects. 7, 8). The experimental

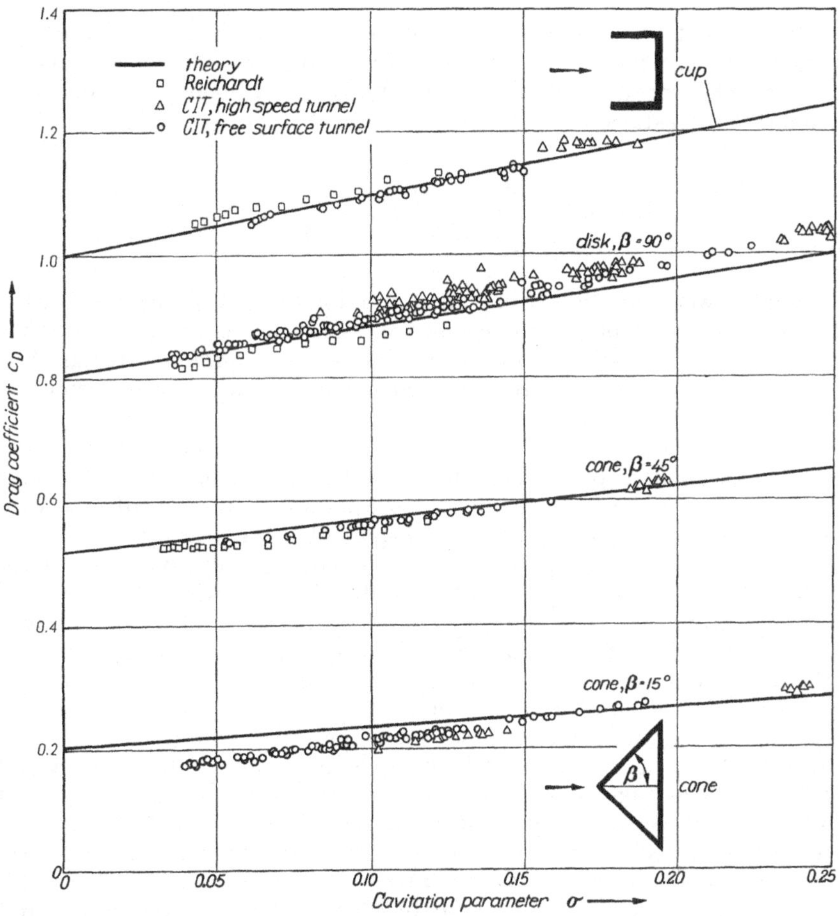

Fig. 1. After M. PLESSET and B. PERRY: Mémoires sur la mécanique des fluides offerts à M. DIMITRI RIABOUCHINSKY, pp. 251—261. The theoretical curve is duscussed in Sect. 22.

data gives some indication that the slopes of the C_D curves increase in the σ range shown in Fig. 1. This is in contrast with the two-dimensional theory, which predicts negligible deviation from (2.2) up to $\sigma=1$ in symmetric flow[1],

M. Dimitri Riabouchinsky, pp. 251—261. Publ. Sci. Tech. Min. de l'Air, Paris 1954. — *Added in proof:* Experiments on cavitating flat plate and circular-arc hydrofoils have been carried out by B. PARKIN [J. Ship. Res. **1**, 35—46 (1958)]. His results have been compared with two-dimensional theory by T. Y. WU [J. Math.-Phys. **35**, 236—265 (1956)], who finds excellent agreement, over all angles of incidence, between the experimental lift and drag values and those obtained by calculation (cf. Sect. 11).

[1] Formula (2.2) ceases to be accurate at oblique incidence and for very fine wedges; cf. WU, preceding footnote.

but is in agreement with GARABEDIAN'S calculations on axially symmetric cavities[1] (cf. Sect. 45).

While (2.2) seems to be accurate for sharp-edged bodies such as cones, measurements indicate that the relation $C_D(\sigma) = C_D(0)(1 + \alpha\sigma)$ is more appropriate for smooth obstacles. EISENBERG and POND[2] find that this formula provides a good fit to their data with $\alpha = 2.02$ for the sphere and $\alpha = 3.65$ for the 2-caliber ogive.

Measurements of cavity dimensions as functions of σ have also been reported[3]. Observations by REICHARDT on cavities with $\sigma < 0.12$ fit closely the following formulas for cavity width d and length h, in units of body diameters, for arbitrary axially symmetric obstacles,

$$\left.\begin{aligned} d &= \left(\frac{C_D}{\sigma - 0.132\sigma^{\frac{8}{7}}}\right)^{\frac{1}{2}}, \\ \frac{h}{d} &= \frac{\sigma + 0.008}{\sigma(0.066 + 1.70\sigma)}. \end{aligned}\right\} \tag{2.3}$$

These formulas were obtained from an approximate theory due to MÜNZNER and REICHARDT[4] based on a source-sink representation of cavity flow. They may be compared with the theoretically derived asymptotic formulas (8.14), (8.15). REICHARDT'S formulas are in good agreement with measurements on spheres and 45° cones[5] up to $\sigma = 0.3$.

A vital feature of cavity mechanics is the re-entrant jet, which can be observed visually in larger cavities, projecting inward from the rear and striking the cavity wall. Fig. 2 shows such a jet in the cavity behind a disc at two different values of σ. These jets are the steady state analogues of those already mentioned in connection with water entry cavities. In large cavities (small cavitation number) the jet is relatively weak and is swept away by the flow without affecting the overall cavity structure. The cavity in these flows is clear and transparent. For larger values of the cavitation number[6], the disturbance created by the jet is filling the cavity and deforming the wall may be sufficient to destroy the predominantly steady state character of the flow, giving rise to a cavity which, when viewed by high speed photography, is seen to be pulsating and irregularly shaped, but which takes on a smooth steady opaque appearance when observed by eye[7]. High speed motion pictures of small cavities formed on an obstacle consisting of a hemisphere with a cylindrical afterbody show a periodic phenomenon characterized by three stages[8]: (1) formation of a cavity which grows until it achieves a steady condition; (2) the re-entrant jet fluid, moving upstream from the rear stagnation region, enters and partly fills the cavity; (3) the cavity breaks away from the body and is transported downstream while a new cavity begins to form immediately. The connection between the jet and breakoff of the cavity

---

[1] P. GARABEDIAN: Bull. Amer. Math. Soc. **62**, 219–235 (1956).

[2] P. EISENBERG and H. POND: David Taylor Model Basin Rep. No. 668 (1948); see also EISENBERG, footnote 1, p. 312.

[3] REICHARDT: Footnote 5, p. 312. — EISENBERG and POND: Preceding footnote. — M. SELF and J. RIPKEN: St. Anthony Falls Hydraulic Lab., Univ. of Minn., Rep. No. 47, 1955.

[4] H. MÜNZNER and H. REICHARDT: Armament Res. Est. Transl. No. 1/50 (1950). — German original UM 6616 (1944).

[5] Cf. SELF and RIPKEN: footnote 3.

[6] These values lie approximately in the range $0.1 \leqq \sigma \leqq 0.25$ depending on the shape of the obstacle and probably on the experimental arrangement.

[7] See, for example, the photographs in EISENBERG, footnote 1, p. 312.

[8] R. KNAPP: Trans. Amer. Soc. Mech. Eng. **77**, 1045–1054 (1955); similar observations are reported by SELF and RIPKEN, footnote 3. KNAPP discusses the connection with cavitation damage.

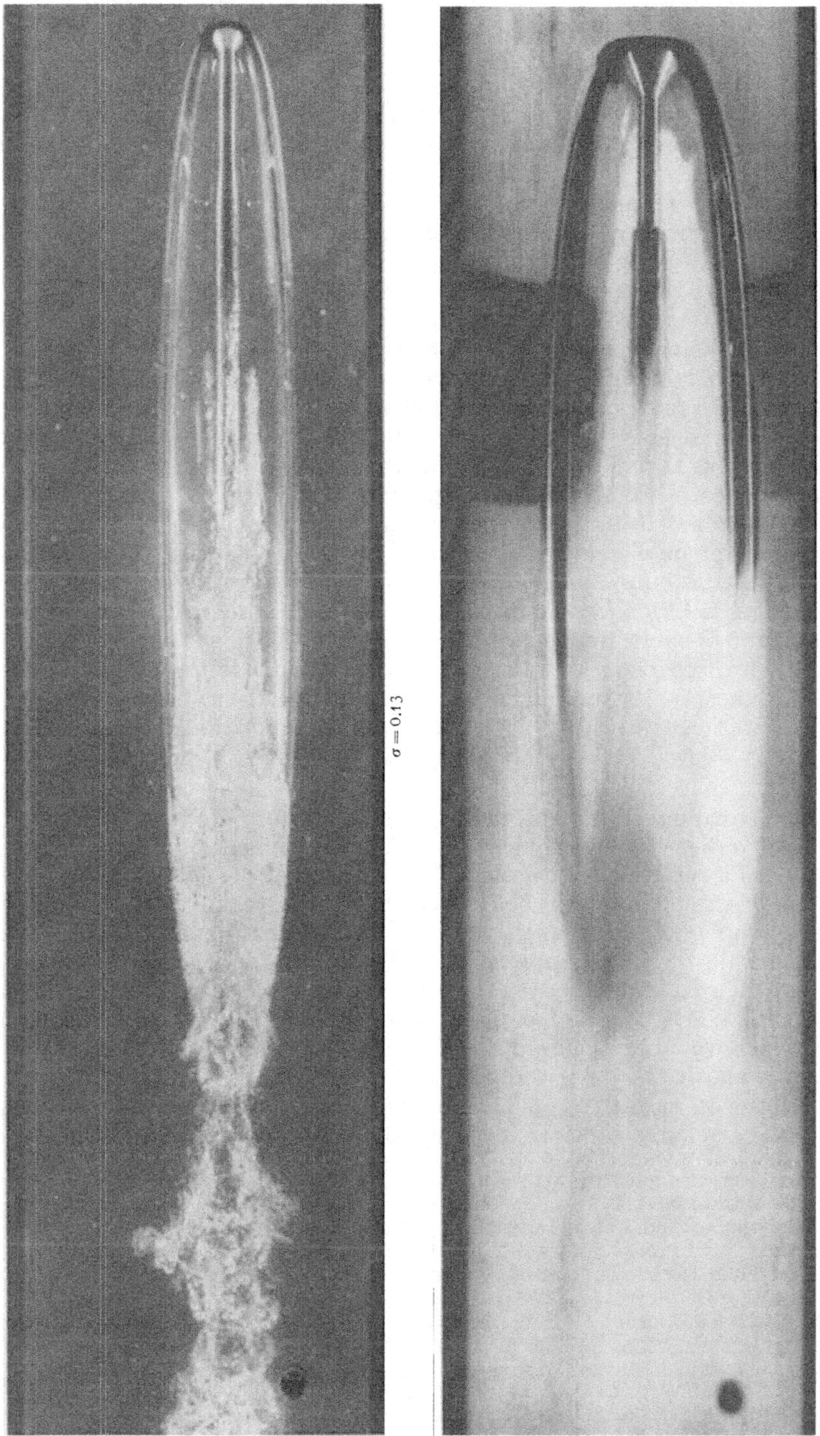

$\sigma = 0.13$ $\sigma = 0.23$

Fig. 2a and b. Water tunnel cavities behind a circular disc. Courtesy of the Hydrodynamics Laboratory, California Institute of Technology.

seems conclusive, although the periodicity is possibly a feature of the experimental setup.

It is of interest to know the extent to which factors disregarded in ideal fluid theory affect the results of water tunnel measurements. Calculations on widely contrasting cavity models (cf. Sects. 7, 8, 11) indicate that the turbulent mixing at the rear of the cavity and its attendant wake have little if any effect on drag and cavity dimensions. The measured cavity pressure is slightly (at most a few percent) in excess of vapor pressure when the cavity is wholly gaseous[1], but may appear to differ from it appreciably for larger cavitation numbers when the cavity is a liquid-vapor mixture[2] (presumably because of the re-entrant jet). In the former case the cavity pressure is sensibly constant over the entire length. Under the conditions of most water tunnel testing (i.e., small scale models and $\sigma<1$), viscosity is on the whole a minor factor. However, the cavity inflection and finite angle of separation often observed[3] on smooth bodies may be a boundary layer effect, and skin friction probably influences drag measurements on slender obstacles. Experiments on cylinders[4] indicate the existence of a "critical" cavitation number, equal to about 1.5 for the cylinder, at which value the drag coefficient varies markedly with changes in the REYNOLDS' number. Similar REYNOLDS' number effects are visible in water entry phenomena[5]. The influence of gravity on drag measurements is not clear, but the asymmetry of many cavities due to buoyancy is quite apparent. There is strong evidence also, in larger cavities, of trailing hollow core vortex pairs resulting from the circulation set up by the hydrostatic pressure gradients[6]. The influence of all these factors on measurements and on comparison with theory has yet to be made precise, but it would appear that the usual assumptions of free streamline theory are fulfilled on the whole in cavitation experiments and that theory and experiment should not differ by more than a few percent where gross quantities (such as drag and cavity dimensions) are concerned.

The free streamline theory provides a less satisfactory description of the wake behind bluff bodies. Photographic studies of wakes behind discs and other blunt bodies[7] reveal a surface of discontinuity similar to that postulated by the HELMHOLTZ theory extending a few body diameters downstream, but thereafter the surface rolls up into eddies and loses its identity in a more or less turbulent wake. The wake itself is composed of eddies whose motion creates an underpressure at the rear of the body. This explains the fact that the observed drag coefficient, $C_D=2.13$ for the flat plate[8], exceeds even the theoretical maximum of unity which would be achieved if the stagnation pressure acted over the front of the plate, and if the wake were at static pressure as required by the classical theory. However, measurements on the rear of bluff bodies reveal that the base underpressure is fairly constant between the separation points of the discon-

[1] In the tests conducted by SELF and RIPKEN, the cavity and vapor pressures were essentially equal up to $\sigma=0.15$.

[2] Cf. EISENBERG and POND: Footnote 2, p. 314; also EISENBERG: Footnote 1, p. 312.

[3] EISENBERG and POND: Footnote 2, p. 314; see also H. WAYLAND and J. WHITE: Heat Transfer and Fluid Mechanics Institute. Amer. Soc. Mech. Eng. **1949**, 51–64.

[4] V. A. KONSTANTINOV: Izv. Akad. Nauk. SSSR., OTN **1946**, 1355–1373. — David Taylor Model Basin Transl. 233, 1950. — E. MARTYRER: Hydromechanische Probleme des Schiffsantriebs, p. 268–286, Edit. by G. KEMPF and E. FOERSTER. Hamburg 1932.

[5] Cf. WAYLAND and WHITE: Footnote 3.

[6] R. N. COX and W. A. CLAYDEN: Proc. Symp. on Cavitation in Hydrodynamics, Nat. Phys. Lab., 1956. This phenomenon is visible in motion pictures and often by eye; indications of it are present in Fig. 2a.

[7] Cf. S. GOLDSTEIN (ed.): Modern developments in fluid dynamics, Vol. 2. Oxford 1938.

[8] A. FAGE and F. JOHANSEN: Proc. Roy. Soc. Lond., Ser. A **116**, 170–197 (1927).

tinuity surfaces[1]. This suggests that a wake pressure p_w be associated with each flow past a bluff body, and that a *wake parameter*

$$\sigma_w = \frac{p_\infty - p_w}{\frac{1}{2}\varrho U^2} \tag{2.4}$$

which plays the same role as the cavitation parameter (2.1), be defined for the flow. This idea has been made the basis of a modified wake theory (cf. Sect. 11) which proves to be in good quantitative agreement with pressure and drag measurements. It should be emphasized, however, that unlike the cavitation number, the wake parameter is a quantity which is not known *a priori*, and must be empirically determined in each case.

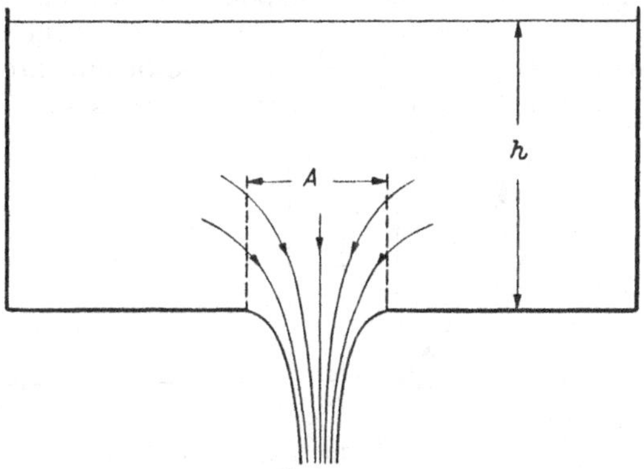

Fig. 3a.

β) Jet flows. The problem of jet efflux from an orifice is one of the oldest in hydrodynamics and the first to be treated by the HELMHOLTZ free streamline theory. Of particular importance for engineering applications is the *discharge coefficient* C_d, which is defined in terms of the discharge Q per unit time, the pressure P, and the cross-sectional area A of the orifice, by the formula,

$$Q = C_d A \sqrt{2P/\varrho} \tag{2.5}$$

where ϱ is the fluid density. Two methods of measuring C_d have been most frequently adopted. In the first the liquid issues from an orifice in a large vessel under the influence of gravity (Fig. 3a), while in the second it is forced out of a nozzle or pipe under high pressure (Fig. 3b). In the former case, $P = \varrho g h$, where h is the height of the fluid surface above the orifice, and in the latter P is the nozzle pressure.

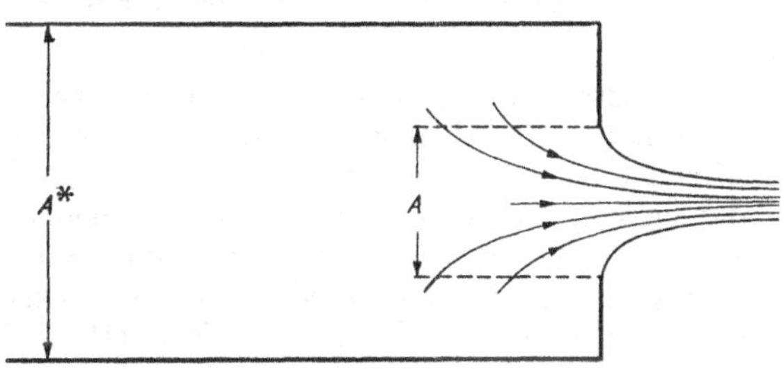

Fig. 3b.

The *contraction coefficient* C_c of the flow is defined as the ratio of the limiting cross-sectional area of the jet to that ofthe orifice. In the theory which neglects gravity, the limit shape of the jet is achieved at infinity, but in real flows the area is essentially constant several orifice diameters downstream, after which the effects of gravity and surface tension become significant[2]. The contraction coefficient may be measured in terms of the discharge coefficient, in flow out of a pipe of cross-sectional area A^* (Fig. 3b), by use of the formula

$$C_d = \frac{C_c}{\sqrt{1 - (C_c A/A^*)^2}}\,. \tag{2.6}$$

[1] Cf. GOLDSTEIN (footnote 7, p. 316), p. 421.

[2] If the orifice is non-circular the cross-section of the jet varies in shape periodically along its length. For a description of the motion and discussion of stability, see RAYLEIGH: Theory of Sound, Vol. 2, 356ff. New York: Dover Publications 1945.

In case of flow out of a large vessel, when $A/A^* \approx 0$, C_c and C_d become identical.

The discharge coefficient for jet flow out of a circular orifice at the end of a pipe or in a large plane wall has been measured many times[1]. These experiments yield values in the neighborhood of $C_c = 0.61$ for efflux from an infinite plane wall. This figure is of particular interest because it agrees almost exactly with that given by the corresponding two-dimensional theory, namely, $\pi/(\pi+2) = 0.611$ (cf. Sect. 10), and suggests that the theoretical values of the plane and axially symmetric contraction coefficients are perhaps identical. The calculational evidence for and against this view is discussed at length in Sects. 43, 46, 47.

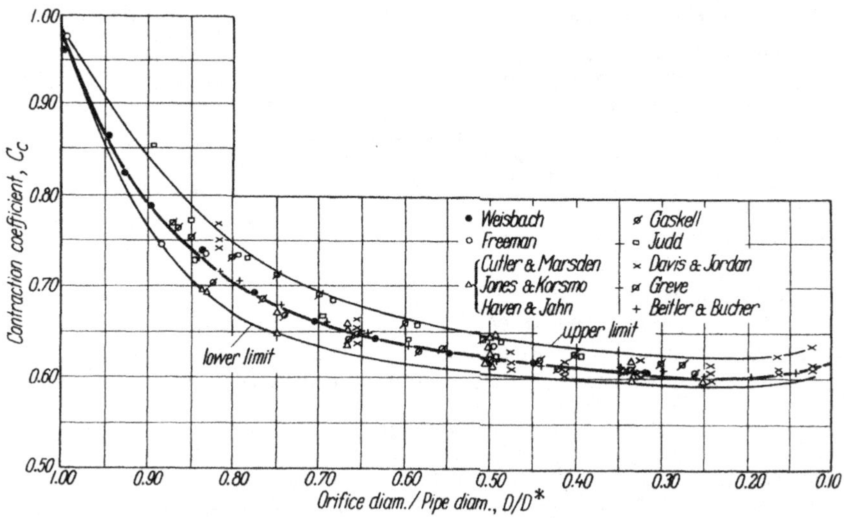

Fig. 4. Contraction coefficients of pipe orifices. After W. LANSFORD: Civ. Engng., Lond. 4, 245—247 (1934).

Because of the apparent agreement between the observed value 0.61 and the two-dimensional theory, it is all the more important to examine the possible influence on the measurement of those factors, such as gravity, surface tension, and viscosity, that are ignored in the theory. Without attempting to evaluate the various effects, we call attention to the following experimental results. (1) LANSFORD[2] has averaged the data of eight different investigators to obtain a curve of C_c vs. the diameter ratio D/D^* for flow out of pipe orifices. As shown in Fig. 4, this curve achieves a minimum of 0.60 at $D/D^* = 0.23$ and rises to 0.61 at $D/D^* = 0.12$, which is the last recorded value. However, the theory (ignoring gravity, etc.) shows, entirely rigorously, that C_c must be a monotone increasing function of D/D^*, and hence cannot have a minimum (cf. Sect. 28). (2) In measurements of flow out of a large vessel under gravity several investigators[2] have found discharge coefficients for flow through a long rectangular orifice a few percent in excess of the theoretical value 0.61. If this data for plane flow is indicative of a similar excess in the axially symmetric case, it would

[1] For example, see the data compiled by W. LANSFORD: Civ. Engng., Lond. **4**, 245—247 (1934). — R. BAZIN: Mém. prés. par div. savants Acad. Sci., Paris (2) **32**, 1—63 (1902). — H. SMITH: Hydraulics. New York 1886. — For this data see also H. W. KING: Handbook of Hydraulics, 3rd ed. New York-Toronto-London: McGraw-Hill 1939. — P. FORCHHEIMER: Hydraulik, 3rd ed. Leipzig: Teubner 1930. — F. MEDAUGH and G. JOHNSON: Civ. Engng., Lond. **10**, 422—424 (1940). — For additional references, see [*1*], Chap. 10.

[2] See preceding footnote.

appear that the correct theoretical value of C_c for the circular orifice is somewhat smaller than the one given by experiment. (3) The experiments and compilations of H. SMITH, STRICKLAND, and MEDAUGH and JOHNSON[1] reveal a noticeable effect of (presumably) both viscosity and gravity on the measurement of C_c. Their observed values decrease monotonically with an increase of the orifice diameter and the pressure head. Typically, for an orifice of 1 in. diameter, C_c decreases from 0.603 when $h=2$ ft to 0.595 when $h=35$ ft; and for fixed $h=35$ ft, C_c has the values 0.607, 0.60, and 0.595, corresponding to $D=\frac{1}{4}$, $\frac{1}{2}$, and 1 in. MEDAUGH and JOHNSON conclude (by extrapolation) that $C_c=0.588$ for perfect contraction under very high head. SMITH's measurements on mercury show a similar monotonic dependence of C_c on h, with his extrapolated limit value of C_c lying between 0.58 and 0.59.

The preceding remarks have dealt only with liquid jets. Gas jets exhibit qualitatively different effects of considerable interest, but are unfortunately beyond the scope of this article. For their discussion we refer to the literature[2].

3. Mathematical foundations. For flows in which the free boundary is an interface between liquid and gas, as in cavitation, it is reasonable to assume viscosity a minor factor and therefore the motion irrotational. This hypothesis is retained also when the free surface is not a liquid-gas interface, as in the theory of wakes, but in situations which lack a well-defined surface of discontinuity the description provided by ideal fluid theory often proves unrealistic (cf. Sect. 2).

Under the assumption of irrotationality, the flow of incompressible fluids is governed by LAPLACE's equation for the velocity potential φ,

$$\Delta\varphi = \varphi_{xx} + \varphi_{yy} + \varphi_{zz} = 0. \tag{3.1}$$

Our considerations here will be confined almost entirely to plane and axially symmetric flows. These can be also described in terms of a stream function ψ which is conjugate harmonic to φ in the case of plane flows, but for axially symmetric flows satisfies the equation

$$\psi_{xx} + \psi_{yy} - \frac{1}{y}\psi_y = 0, \tag{3.2}$$

where x is the axial and y the radial coordinate. The potential and stream function of a plane flow together determine the *complex potential* $f(z)=\varphi+i\psi$, which is an analytic function of $z=x+iy$ and has the important porperty that its derivative,

$$w = \frac{df}{dz} = u - iv = q\,e^{-i\vartheta}, \tag{3.3}$$

is the conjugate of the velocity, where q is the flow speed, ϑ is the inclination of the velocity vector, and u, v are the x, y velocity components. The velocity components of an axially symmetric flow are given by

$$u = \varphi_x = \frac{1}{y}\psi_y, \qquad v = \varphi_y = -\frac{1}{y}\psi_x. \tag{3.4}$$

In both plane and axially symmetric flows the condition of irrotationality takes the frequently useful intrinsic form

$$\frac{\partial q}{\partial n} + \varkappa q = 0. \tag{3.5}$$

[1] See footnote 1, p. 318.

[2] E.g., S. I. PAI: Fluid dynamics of jets. New York: Van Nostrand 1954. — A. OUDART: L'étude des jets et la mécanique théorique des fluides. Publ. Sci. Tech. Min. de l'Air 234, Paris 1949; and accompanying references.

Here n denotes the normal to a streamline and $\varkappa$ its curvature; $\varkappa$ is taken to be positive when the streamline is convex to the region on the side of the positive normal. From Eq. (3.5) one concludes that the normal acceleration on a streamline is in the same direction as the curvature. A similar result holds on stream surfaces in three-space.

Boundary conditions. In the boundary value problems under consideration the flow boundaries will be either rigid, in which case they are known data of the problem, or they will be free boundaries (free surfaces, free streamlines), in which case their shape is unknown beforehand. On rigid boundaries it is assumed, as is customary in ideal fluid theory, that the motion is tangential to the surface. For steady motions this implies that the normal component of the fluid velocity is zero, or equivalently, that

$$\psi = \text{const}, \quad \frac{\partial \varphi}{\partial n} = 0 \tag{3.6}$$

on the boundary.

On the free boundaries two conditions must be satisfied which are presumably sufficient, in conjunction with the other data, to balance the incomplete knowledge of the boundary and to render the flow problem determinate. The first condition, kinematic in nature, states that the free boundary is a material surface: particles initially on the surface remaining thereon. If the surface is described by the equation $F(x, y, z, t) = 0$ (where F might be the pressure, for example), then the velocity vector $\boldsymbol{q}$ and the shape are connected by the relation

$$F_t + \boldsymbol{q} \cdot \operatorname{grad} F = 0. \tag{3.7}$$

For steady flows ($F_t = 0$) this implies simply that the free boundary is a stream surface, so that this boundary condition reduces again to (3.6).

The second boundary condition, which really characterizes the free boundary problem, states that pressure is constant on the free surface. BERNOULLI's equation allows this condition to be converted into one containing only the kinematic quantities. Thus, in an unsteady motion under the influence of gravity BERNOULLI's equation states that

$$p + \frac{1}{2} q^2 + g y + \frac{\partial \varphi}{\partial t} = c(t) \tag{3.7a}$$

throughout the fluid (assumed to be of unit density), $c(t)$ being independent of the space variables. If, in addition, flow conditions are constant at some fixed point in the fluid, usually at infinity, then $c(t)$ is constant also with respect to time, and on a free boundary we have

$$\frac{1}{2} q^2 + g y + \frac{\partial \varphi}{\partial t} = \text{const}. \tag{3.8}$$

This would be the free surface condition, for example, on the cavity walls in the water entry of a missile. Further specializations yield the following conditions on the free boundary:

$$\tfrac{1}{2} q^2 + g y = \text{const} \quad \text{—steady flow with gravity,} \tag{3.9}$$

$$\frac{1}{2} q^2 + \frac{\partial \varphi}{\partial t} = \text{const} \quad \text{—unsteady flow without gravity,} \tag{3.10}$$

$$q = \text{const} \quad \text{—steady flow without gravity.} \tag{3.11}$$

The condition (3.9) is appropriate, for example, in the problem of steady flow over a weir, or discharge of a heavy jet from an orifice; Eq. (3.10) is the free surface condition in the problems of surface impact and of unsteady cavity flows without gravity. However, by all means the greatest body of theory has been developed for free surface problems in which the constant speed condition (3.11) is imposed on the free boundary, and most of the discussion here will be devoted to this class of problems.

From the mathematical point of view, the fact that the free boundary problem is properly set can be inferred from general heuristic considerations. The two boundary conditions satisfied on the free surface can be looked upon as CAUCHY data for LAPLACE'S equation [or Eq. (3.2)] on the unknown free boundary. As is well known, these conditions cannot be assigned on arbitrary curves in the case of elliptic equations, and their fulfillment here is expressed in the analyticity of the free surface. Since the flow must also satisfy a condition on the fixed boundary, the possible forms that may be assumed by the free surface are further limited. The fact that a unique solution is indeed determined in general is the problem of the existence and uniqueness theory of Chap. IV.

Notation and conventions. The notation of this section will be followed throughout the sequel. In symmetric flows the axis of symmetry, unless otherwise noted, will be assumed to coincide with the x-axis. The fluid density and the constant speed on free streamlines are taken to be unity, except when specifically given other values. In the figures that accompany the text the same letters will be used to identify corresponding points in different planes, and the notation z_A, ζ_A, etc. will be used for the z and ζ coordinates of A, etc.

4. Physical models[1]. We turn now to a preliminary survey and formulation of the principal boundary value problems of this article. We shall be concerned here partly with the mathematical statement of the various problems, but more with their physical content and motivation. The details of solution and general theory are presented in later sections.

α) Cavities and wakes. Among the first applications of the Helmholtz concept of streamline of discontinuity[2] was to the model of the infinite wake[3]. Proposed as a means of circumventing the D'ALEMBERT paradox within the realm of classical hydrodynamics, this model postulates that in the motion of a body through an ideal fluid, a streamsurface of discontinuity forms behind the body, extending to infinity and enclosing within it a wake of constant pressure. Although it was recognized early[4] that this picture corresponds only roughly to observed fact, the infinite wake and its complete mathematical treatment by the hodograph method proved a great stimulus to the further elaboration of the theory of fluid resistance. Only recently has it become apparent that the Helmholtz concept, while yielding a poor account of the wake, is quite successful as a description of cavity phenomena[5]. The stability and permanence of the liquid-gas interface[6], which are lacking in the wake, are major factors contributing to the success of

[1] Cf. D. GILBARG: Proc. Symposium on Naval Hydrodynamics, 281–295, Nat. Res. Counc. Publ. 515, 1957. — G. BIRKHOFF: Hydrodynamics. Princeton, N. J.: Princeton University Press 1950.

[2] H. HELMHOLTZ: Mber. Akad. Wiss. Berlin **1868**, 215–228. — Wiss. Abh. **1**, 154.

[3] G. KIRCHHOFF: J. reine angew. Math. **70**, 289–298 (1869). — Ges. Abh. 416.

[4] Lord RAYLEIGH: Phil. Mag., Dec. **1876**. — Papers **1**, 287.

[5] Among the earliest work was that of A. BETZ and E. PETERSOHN: Ing.-Arch. **2**, 190–211 (1931). See also the articles on cavitation in: Hydromechanische Probleme des Schiffsantriebs. Edit. by G. KEMPF and E. FOERSTER, Hamburg 1932.

[6] According to theory (see [*1*], Chap. 11) the interface is actually unstable, but very weakly so because of the large relative density of liquid and gas. This is in marked contrast with the wake, where the surface of discontinuity separates two fluids of equal density and hence is strongly unstable.

free streamline theory both in cavitation and in the theory of jets. For sake of a fixed terminology which is at the same time more in keeping with the observed facts, we shall henceforth refer to "cavities" rather than to "wakes" in describing free surface flows.

β) Infinite cavity. To study this model in more detail let us consider a steady plane infinite stream of unit velocity in the x-direction impinging upon a fixed obstacle C. It is supposed that free streamlines γ_1, γ_2 detach from points A and B on the body and extend to infinity downstream in the free stream direction (Fig. 5), so that the flow region is bounded by $AB+\gamma_1+\gamma_2$. The mathematical description of the flow is given by a complex potential $f(z)=\varphi+i\psi$ with the property that $f(z)$ maps the flow region into the full f plane slit along the positive real axis, and such that $|f'(z)|=1$ on the free streamlines.

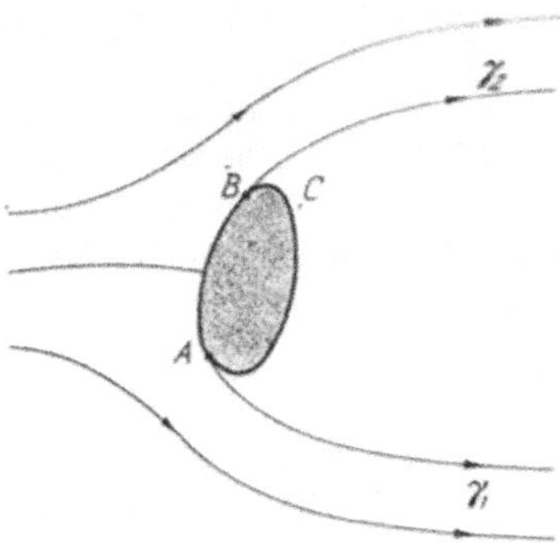

Fig. 5. Infinite cavity.

To this characterization of the cavity flow are sometimes added that following physical conditions emphasized by BRILLOUIN[1]:

(i) *the free streamlines are simple curves which do not intersect the obstacle or each other;*

(ii) *the maximum flow speed is achieved on the free boundary.*

The first condition is imposed for evident reasons of continuity, while the second expresses the basic physical assumption that the fluid cannot withstand negative pressures, or, stated in terms of cavitation, that the fluid cavitates when the pressure descends below vapor pressure (which is equal to static pressure in the infinite cavity model).

Incorporation of the Brillouin conditions into the mathematical theory involves serious difficulties. At the point of detachment the curvature of the free streamline is in general infinite; however, it may also be finite, in which case it has the same value as the curvature of the fixed boundary at detachment (cf. Sect. 24). Consider then the situation at the point of detachment B, which we assume is an interior point of C. If the free streamline curvature at B is infinite and convex to the flow, then clearly condition (i) is violated. On the other hand, if the curvature is infinite and concave to the flow, then the flow speed increases from B into the fluid, and hence condition (ii) is violated. Thus, either the first or second Brillouin condition will be contradicted at an interior separation point if the free streamline curvature is infinite there. From this it would appear that the natural requirement at detachment is finiteness of the free streamline curvature. However, it is simple to exhibit bodies for which the Brillouin conditions cannot both be satisfied in infinite cavity flow (cf. Sect. 6), and LERAY[2] has even constructed examples of flows past convex bodies in which the curvature requirement is met, but *neither* Brillouin condition is satisfied near detachment. Thus far it has been possible to satisfy both Brillouin conditions simultaneously only by restricting greatly the class of obstacles under consideration (cf. Sect. 24). It is likely that they can be satisfied in general only by allowing multiple free streamlines and re-entrant flows[3]. A theory of this type has yet to be developed.

[1] M. BRILLOUIN: Ann. chim. phys. **23**, 145–230 (1911).

[2] J. LERAY: Volume du Jubilé de M. BRILLOUIN, pp. 246–257. Paris: Gauthier-Villars 1935.

[3] Cf. Sect. 6, and GILBARG, footnote 1, p. 321.

For purposes of mathematical convenience it has been the practice to formulate the infinite cavity problem without regard to the Brillouin conditions, and then to discuss the latter separately. In the *problem of fixed detachment*[1] the obstacle and the points of detachment are assumed known. This form of the cavity problem is appropriate for bodies with sharp edges. However, when the separation points are not prescribed *a priori* by the physical situation, for example when the body is smooth, another condition must be imposed to fix their position. In the *problem of smooth detachment*[2] the curvature of the free streamlines is required to be finite at separation. One must then determine the flow, including the points of detachment and the free streamlines, so that the smooth separation condition is satisfied as well as the usual boundary conditions. The precise mathematical formulation of these problems appear in the later sections. In these forms the cavity problem enjoys a relatively complete mathematical theory, but from the physical standpoint the correctly set problem for general obstacles, inclusive of the Brillouin conditions, has still to be formulated.

γ) Cavities with positive cavitation number. From the point of view of cavitation, the infinite cavity model and a zero cavitation number are unrealistic. In real fluids the cavity pressure is less than the static pressure and the cavity dimensions are finite; it is the description of these flows that is the central problem of the theory of cavitation. It is of practical importance to know the various physical quantities (drag, cavity dimensions, etc) in their dependence on the cavitation parameter.

Assuming steady plane flow and the Brillouin conditions, one may investigate the influence of the positive cavitation number on the cavity shape. From the convexity of the free streamlines, which is a consequence of the second Brillouin condition and of Eq. (3.5), it follows that these curves must either intersect one another, extend to infinity downstream without intersection, or turn back to form a re-entrant jet. The first possibility is excluded for reasons of continuity, while the second can be proved self-contradictory[3] (cf. Sect. 25). The remaining alternative is the existence of a re-entrant jet which eventually strikes the body or the cavity wall. It follows that *within the framework of the theory of irrotational non self-intersecting flows, a steady cavitating flow with positive cavitation number cannot exist.*

To allow further development of the theory of steady state cavities one may proceed in either of two directions: by permitting suitable doubly covered flow regions (self-intersecting flows), or by introducing an artificial model of the cavity. In the former case we are led to the re-entrant jet model and in the latter to the Riabouchinsky cavity and similar constructs. It is one of the remarkable features of the theory that whichever viewpoint is taken, the quantitative results are essentially the same over a large and physically significant range of cavitation numbers, and are in substantial agreement with observation (cf., Sects. 8, 45).

δ) Re-entrant jet model[4]. We have seen that steady flow with positive cavitation number can exist in theory only when the free streamlines form a re-entrant jet. We consider now the mathematical idealization obtained by allowing the jet to "pass through" the obstacle, forming a second sheet of the flow plane

[1] French: Problème du sillage.

[2] French: Détachement en proue.

[3] If doubly covered regions are admitted into consideration, the first alternative can be proved mathematically untenable for the same reasons as the second.

[4] See references Sect. 7.

(Fig. 6). The validity of this model from the experimental side has already been discussed in Sect. 2, and there seems little doubt, despite the non-physical feature of the second flow plane, that it provides the "correct" description of cavitation phenomena.

To state the flow problem in mathematical terms, we consider the re-entrant jet cavity illustrated in Fig. 6, in which there is a single jet extending in a fixed direction to infinity, C, and a single interior stagnation point. B, at which the incident flow separates into two parts, consisting of the streamlines that contribute to the jet and of those that go downstream with the main flow. The complex potential $f(z)$ is many-valued and its expansion at infinity I is of the form[1]

$$f(z) = \varkappa \log z + U z + P(1/z), \quad \varkappa \neq 0, \tag{4.1}$$

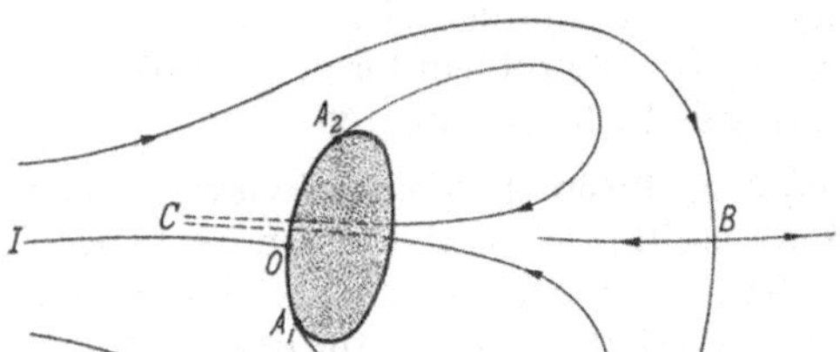

Fig. 6. Re-entrant jet cavity.

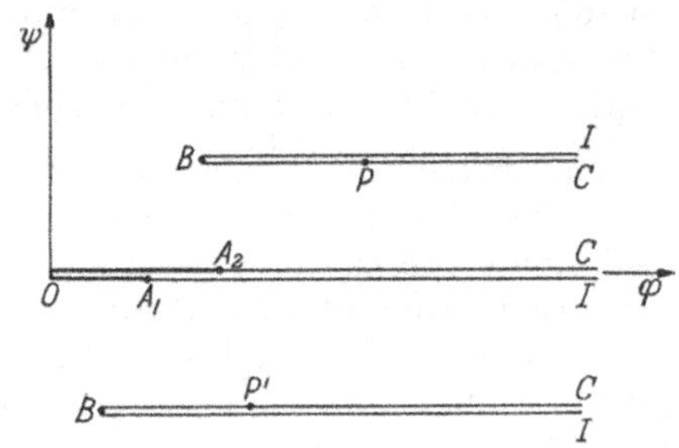

Fig. 7.

where U is the free stream velocity, and $\varkappa$ is a complex constant which describes the circulation about the cavity and the flux through the jet. At the stagnation point B, $f(z)$ is of the form[2]

$$f = f_B + (z - z_B)^2 P(z - z_B), \quad P(0) \neq 0, \tag{4.2}$$

and we may assume that elsewhere in the flow $f(z)$ is regular with non-vanishing derivative. A single branch of the Riemann surface determined by $f(z)$ is obtained by cutting the z-plane along the streamline IBC which joins the branch point I to be boundary (Fig. 7). The horizontal slits IBC in the upper and lower parts of the f-plane are images of the upper and lower sides of the cut IBC, andhence points on corresponding edges of each slit at equal distances from f_B (e.g., P, P' in Fig. 7) can be identified, since they are images of the same z value. The complete Riemann surface over the f plane consists of infinitely many such congruent sheets joined pairwise along BI and BC. The flow past the obstacle $A_1 0 A_2$ is described by a potential $f(z)$ such that (i) $z(f)$ takes the portion $A_1 0 A_2$ of the real, f-axis into the fixed boundary; (ii) $|dz/df| = 1$ on $A_1 C$ and $A_2 C$; (iii) $z(f_1) = z(f_2)$ if f_1 and f_2 are identified points on the slits IBC in the upper and lower half planes, (iv) $dz/df \to 1/U > 1$ as $f \to f_I$. The explicit solution of the re-entrant jet problem is discussed in Sect. 7 and the general theory in Chap. III and IV.

ε) *Artificial cavity models.* Whereas the re-entrant jet model lightens the physical demands on the flow by allowing self-intersections, a second approach to the finite cavity problem is to construct simple cavity models incorporating reasonable but artificial restrictions on the flow. The best known example of this type is the *Riabouchinsky cavity*[2]. In this model let AB be the given fixed

[1] The notation $P(z - z_0)$ and $P(1/z)$ will be used to represent functions regular at $z = z_0$ and $z = \infty$.

[2] See references Sect. 8.

boundary and $A'B'$ its mirror image with respect to a vertical line. A uniform flow impinging on AB describes a Riabouchinsky cavity if AB and $A'B'$ are both streamlines and if the end points A, A' and B, B' are joined by free streamlines of the flow (Fig. 8). The region enclosed by the fixed and free boundaries is the cavity. The cavitation number, or equivalently, the speed on the free streamlines, will of course depend on the distance between AB and its image $A'B'$; and conversely, if the cavitation number is prescribed, the separation of AB and $A'B'$ is an unknown of the problem, as is the length of the re-entrant jet cavity.

Clearly the Riabouchinsky cavity is only one of an infinite family of possible artificial models in each of which the cavity is closed by some fixed curve. At the extreme opposite end of the family is the model introduced by JOUKOWSKY, ROSAKO and EPPLER[1] in which the rear of the cavity is specified by two horizontal hHlf-lines $A'I$, $B'I$, the free streamlines extending from A to A' and B

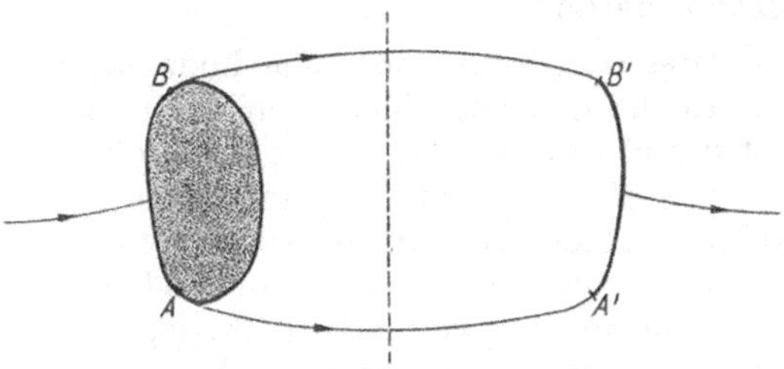

Fig. 8. Riabouchinsky cavity.

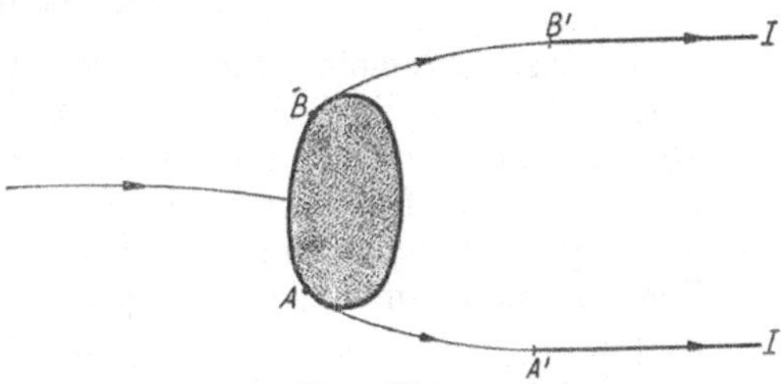

Fig. 9. Wake model.

to B' (Fig. 9). The position of A' and B' of course depends on the cavitation number. Since this cavity is infinite in extent it is more appropriate as a description of a wake than of a cavity, and has been discussed by ROSHKO and EPPLER from this point of view.

The cavity problems formulated above in terms of plane flows can be stated, after simple modifications, for axially symmetric and spatial flows, which may also be unsteady and include gravity. However, aside from a few special results a substantial theory has been developed only for steady plane and axially symmetric flows satisfying the constant speed condition on the free boundary.

ζ) Jet flows. The jet problem is concerned with the discharge of a fluid from an orifice (in a fixed vessel or container) into an atmosphere at constant pressure. The quantities of principal interest in application are the discharge and contraction coefficients (Sect. 2). For purposes of theory the convenient idealization assumes that the container is infinite in extent and completely filled with fluid, and that the jet emerges under such high pressure that the effect of gravity can be ignored.

In steady two-dimensional flow from a vessel let free streamlines γ_1, γ_2 detach from the edges of the orifice (slot) and extend to infinity with the same limiting direction and unit velocity. The boundary value problem seeks the curves γ_1 and γ_2 and a complex potential $f(z) = \varphi + i\psi$ which maps the flow plane onto an infinite horizontal strip of width h in the f-plane, and such that $|f'(z)| = 1$ on γ_1, γ_2. The quantity h is the mass flow per unit time, and is also the jet width (since the density of the fluid is unity). In general it is an unknown of the problem, although, as in the flow from the Borda mouthpiece (Sect. 10), it may

[1] A. ROSHKO: NACA TN 3168 (1954). — R. EPPLER: J. Rat. Mech. a. Analysis **3**, 591 to 644 (1954). — N. JOUKOWSKY: Coll. Works, Vol. 2, No. 3; Rec. Math. **25** (1890). — Cf. Sect. 11 for further discussion of this model

sometimes be determined *a priori*. In this form the theory of plane jets has been well explored, although important problems remain unsolved, particularly as concern asymmetric flows.

The modifications required for axially symmetric flows are evident: the governing equation in the meridian plane is (3.2), the free surface condition is $y^{-1}\,\partial\psi/\partial n = 1$, while $\psi = 0$ on the axis of symmetry and $\psi = \frac{1}{2}h^2$ on the bounding surface (where h is now the half-width of the jet).

If gravity is taken into account and the jet is allowed to fall freely, it is a consequence of the free surface condition (3.9) that the jet speed increases without limit, and conservation of mass requires that the width becomes arbitrarily fine. Observation, however, shows that the jet eventually breaks up into droplets[1].

II. Particular flows.

a) The hodograph method.

5. Basic principles; the Schwarz-Christoffel transformation. The hodograph method[2], which first brought prominence to the free boundary problem in the 19th century, remains even now the most useful technique available for explicit construction of free surface flows. The essence of the method is to introduce the velocity as a new variable and then to exploit the fact that constant speed streamlines, unknown in shape in the physical plane, become known curves in the hodograph, or velocity, plane. For plane flows of incompressible fluids the hodograph variable is $w = f'(z) = q\,\mathrm{e}^{-i\vartheta}$, although it is often more convenient to work with the logarithmic hodograph variables

$$\omega = i \log \frac{df}{dz} = \vartheta + i \log q \tag{5.1}$$

or

$$\zeta = \log \frac{dz}{df} = \log \frac{1}{q} + i\,\vartheta. \tag{5.2}$$

It is clear that any flow region bounded entirely by polygonal streamlines and constant speed free streamlines is mapped into a region of the w plane bounded by radial segments ($\vartheta = \text{const}$) and circular arcs ($q = \text{const}$), and is mapped into a polygonal domain in the ω and ζ planes. Hence the hodograph image of the flow is a known region.

Since the plane of the complex potential $f = \varphi + i\,\psi$ is already a known polygonal region except for the specific location of certain vertices the flow problem is essentially solved if the mapping between the f plane and the hodograph plane is known, with appropriate boundary correspondence, for then $f(z)$ appears as the solution of a differential equation

$$\frac{df}{dz} = w(f), \tag{5.3}$$

and the flow region is determined by quadrature, $z = \int df/w(f)$. Thus the flow problem is reduced to that of conformal mapping between *known* regions, and the difficulty raised by the unknown free boundary has been transformed away. This is the basic achievement of the hodograph method.

---

[1] Lord RAYLEIGH: Theory of Sound, Vol. 2, pp. 351–365. New York: Dover Publications 1945. Cf. also J. KELLER and I. KOLODNER: J. Appl. Phys. **25**, 918–921 (1954).

[2] The hodograph method began with the basic work of G. KIRCHHOFF, J. reine angew. Math. **70**, 289–298 (1869), and was systematized into essentially its present form by M. PLANCK, Wied. Ann. **21** (1884), and J. MICHELL, Phil. Trans. Roy. Soc. Lond., Ser. A **181**, 389–431 (1890).

In the solution of problems involving polygonal boundaries the required mappings can always be exhibited in closed form, except perhaps for the determination of certain parameters. The principal tool is the Schwarz-Christoffel formula which, through explicit mappings on a half plane, expresses Eq. (5.3) in parametric form, $\omega = \omega(t)$, $f = f(t)$.

The Schwarz-Christoffel transformation[1] in its usual form asserts that a simple plane polygon P in the z plane, with vertices A_k having interior angles $\alpha_k \pi$, $k = 1, \ldots, n$, is mapped conformally from a half-plane by the formula,

$$z(t) = A \int^{t} \prod_{k=1}^{n} (t - a_k)^{\alpha_k - 1} dt + B, \qquad \operatorname{Im} t \geqq 0 \tag{5.4}$$

ehere A, B are complex constants, and the a_k are points on the real t-axis whose images are the respective vertices A_k. In this formula three of the constants a_k can be chosen arbitrarily, and if one is placed at infinity the corresponding factor does not appear in the integrand.

A more general form of the Schwarz-Christoffel transformation is sometimes useful in the free boundary problem, and can be stated as follows[2]: Let $z(t)$ map the upper half t-plane on a Riemann surface with closed polygonal boundary[3]. In the neighborhood of interior points $t = b_j$ $(j = 1, \ldots, m)$, let

$$z'(t) = (t - b_j)^{\beta_j} P(t - b_j), \qquad P(0) \neq 0, \tag{5.5}$$

where $\beta_j \neq 0$, and at all other interior points let $z(t)$ be regular and $z'(t) \neq 0$. If the boundary has interior angles $\alpha_k \pi$ $(k = 1, \ldots, n)$ at vertices corresponding to $t = a_k$ (a_k real), then $z(t)$ is of the form

$$z(t) = A \int^{t} \left\{ \prod_{j=1}^{m} \left[(t - b_j)^{\beta_j} (t - \bar{b}_j)^{\bar{\beta}_j}\right] \prod_{k=1}^{n} (t - a_k)^{\alpha_k - 1} \right\} dt + B \tag{5.6}$$

where A and B are complex constants. If the Riemann surface is given, then the precise values of A, B, a_k, b_j are to be determined from the known positions of the vertices and the known values of α_k, β_j. As in the simpler formula (5.4), three of the constants a_k can be chosen arbitrarily, and if one is chosen at infinity the corresponding factor does not appear in the integrand.

In the usual case of simple polygons without winding points and not containing the point at infinity in the interior, the exponents β_j are all zero, and (5.6) reduces to the classical formula (5.4). In general the polygon may contain winding points ($\beta_j =$ integer $\neq -1, -2$) and linear boundary identifications if the mapping (5.6) is many valued[4] (β_j non-integral).

It must be emphasized that the Schwarz-Christoffel formula, while exceedingly elegant, nevertheless contains the unknown parameters A, B, a_k, b_j, whose effective determination is a very laborious computational problem in general[5]. Thus,

[1] See, e.g., Z. NEHARI: Conformal Mapping: New York-Toronto-London: McGraw-Hill 1952.

[2] D. GILBARG: Proc. Nat. Acad. Sci. U.S.A. **35**, 609–612 (1949).

[3] The Riemann surface is here considered over the z-plane. If $z(t)$ is many-valued under analytic continuation, then we define a single-valued branch by introducing suitable cuts in the z-plane. The images of opposite edges of the cut are identified on the image Riemann surface. The surface is considered to be defined when the boundary, the location and type of interior winding points, identifications, and connections between the sheets, are known. The general Riemann mapping theorem asserts the existence of a mapping from the half-plane onto an arbitrary given Riemann surface.

[4] This occurs, for example, in the mapping on the f-plane of the re-entrant jet flow (Fig. 7); see Sect. 7 for the mapping formula.

[5] Cf. BIRKHOFF and ZARANTONELLO [*1*], Chap. 9.

it is only for the simplest problems, usually involving no more than three vertices, that the mapping (5.4) or (5.6) is really "explicit".

The usual procedure in solving free boundary problems by the hodograph method is to start with a form of the hodograph plane suggested by the particular flow problem and then to check that the analytic solution thus obtained has the properties demanded of the flow. This is the procedure we shall follow here. However, it is also possible, as BIRKHOFF and ZARANTONELLO [1] have done, to follow a systematic course whereby the form of the hodograph plane is rigorously deduced from the given data and the flow then determined.

The literature of hydrodynamics abounds with a rich assortment of flows calculated by the hodograph method[1]. However, in the following sections we shall limit ourselves to just a few cases of interest, laying particular stress on the cavity and jet flows with which the larger part of this article is concerned.

6. The infinite cavity. We consider first the classical KIRCHHOFF[2] example of a symmetric flow of unit velocity past a vertical flat plate, with free stream-

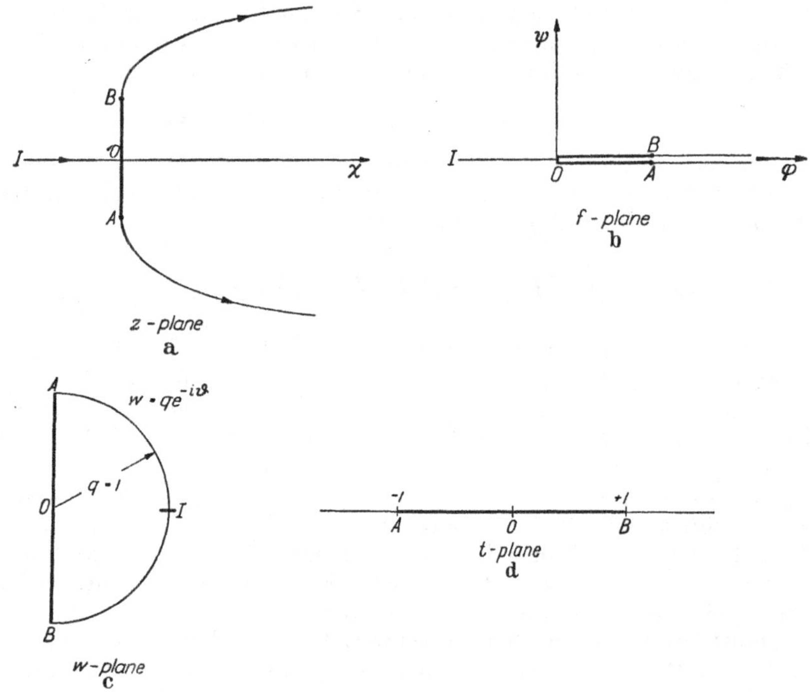

Fig. 10a—d.

line detachment at the ends. The physical plane and its images in the f and w planes are illustrated in Fig. 10. The mappings on the half-plane $\operatorname{Im} t > 0$ are given by

$$f = k t^2, \qquad \frac{i}{2}\left(w - \frac{1}{w}\right) = \frac{1}{t}, \tag{6.1}$$

[1] See, for example, BIRKHOFF and ZARANTONELLO [1]; CISOTTI [2]; B. DEMTCHENKO: Problèmes mixtes harmoniques en hydrodynamique des fluides parfaits. Paris: Gauthier-Villars 1933; G. GREENHILL: British Aero. Res. Counc. Reps. and Memos. 19, 1910; H. VILLAT [4].

[2] G. KIRCHHOFF: J. reine angew. Math. **70**, 289—298 (1869). — Ges. Abh. 416.

where $t = +1, -1$ correspond to the end points of the plate. Hence

$$w = \frac{df}{dz} = i\left(-\frac{1}{t} + \sqrt{\frac{1}{t^2} - 1}\right), \tag{6.2}$$

the branch of the square root being chosen to make $w = 0$ at $t = 0$. From

$$z = \int \frac{dz}{df}\, df = 2ik \int_0^t \left(1 + \sqrt{1 - t^2}\right) dt, \tag{6.3}$$

we have $k = l/(4+\pi)$, where l is the length of the plate. The upper free streamline is therefore given in parametric form by

$$\left.\begin{aligned} x &= \frac{l}{4+\pi}\left[t\sqrt{t^2-1} - \log\left(t + \sqrt{t^2-1}\right)\right], \\ y &= \frac{l}{2} + \frac{2l}{4+\pi}(t-1), \quad t \geq 1. \end{aligned}\right\} \tag{6.4}$$

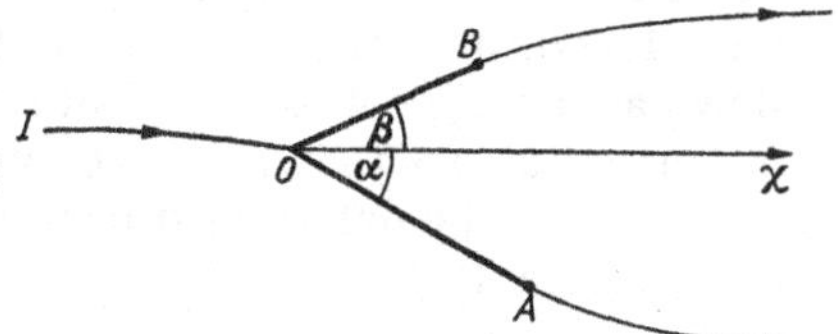

a

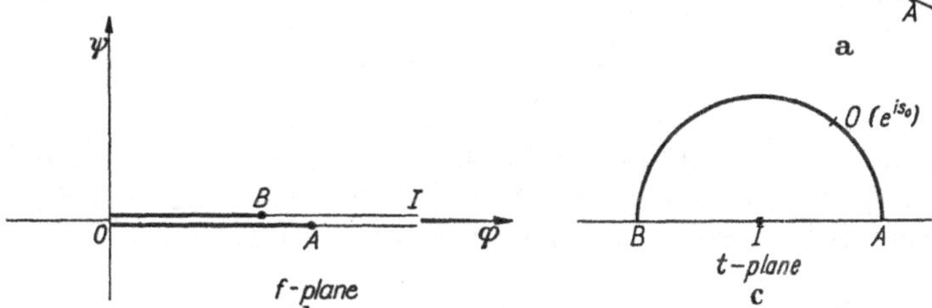

b c

Fig. 11 a—c.

The force on the plate is calculated by integrating the pressure difference on the two sides. If the cavity pressure is p_c, we have for the drag

$$\left.\begin{aligned} D &= \int_{-l/2}^{+l/2} (p - p_c)\, dy = -\frac{i}{2} \int_{-il/2}^{+il/2} (1 + w^2)\, dz \\ &= \frac{l}{2} + \frac{l}{4+\pi} \int_{-1}^{+1} \left(-1 + \sqrt{1-t^2}\right) dt = \frac{\pi l}{4+\pi}. \end{aligned}\right\} \tag{6.5}$$

When the free stream speed is equal to U and the density equal to ϱ, the expression for the drag becomes

$$D = \frac{2\pi}{4+\pi} \cdot \frac{1}{2} \varrho U^2 l \tag{6.6}$$

and the value of the drag coefficient is

$$C_{\mathrm{D}} = \frac{2\pi}{4+\pi} \approx 0.88. \tag{6.7}$$

The large difference between this result and the observed value $C_{\mathrm{D}} \approx 2$ for a flat plate in non-cavitating flow (cf. Sect. 2) demonstrates clearly the inadequacy of the Kirchhoff model as a theory of the wake.

To describe the flow past a wedge having a stagnation point at the vertex (Fig. 11) it is convenient to introduce as parameter plane the semi-circle $|t| < 1$, $\operatorname{Im} t > 0$. The mapping between the f and t planes which takes the points of detachment A, B, and the point at infinity respectively into $t = +1, -1$, and 0, is given by

$$f = M\left[\cos s_0 - \tfrac{1}{2}(t + 1/t)\right]^2, \tag{6.8}$$

where

$$M = \frac{1}{4} (\sqrt{f_A} + \sqrt{f_B})^2, \qquad \cos s_0 = \frac{\sqrt{f_B} - \sqrt{f_A}}{\sqrt{f_B} + \sqrt{f_A}}. \tag{6.9}$$

Defining the function $\omega = \vartheta + i \log q$ as in Eq. (5.1), one sees by inspection that

$$\omega(t) = \beta + i \frac{(\alpha+\beta)}{\pi} \log \frac{t - e^{i s_0}}{t\, e^{i s_0} - 1} \tag{6.10}$$

has the desired behavior on the boundary. For the logarithm term is real on the real axis and (if the branch is chosen to be equal to 0 at $t = -1$) has its real part equal to $-\pi$ on AO and to O on OB. Thus, $q = 1$ on the real axis and $\vartheta = -\alpha, \beta$ on OA, OB, from which it follows that the formulas (6.8), (6.10) describe a cavity flow past a wedge whose sides are inclined at angles $-\alpha$ and β. If the free stream is horizontal, so that $\omega(0) = 0$, we have $s_0 = \pi\beta/(\alpha+\beta)$. The remaining parameter M can be determined from the given dimensions of the obstacle by means of the formula

$$z(t) = \int_{e^{i s_0}}^{t} e^{i\omega(t)} \frac{df}{dt} dt. \tag{6.11}$$

Since there are two conditions for the determination of this one quantity, it is evident that the length of only one of the faces can be assigned arbitrarily. This loss of freedom is the result of having previously fixed the position of the stagnation point.

In the special case that $\alpha + \beta = \pi$ the wedge degenerates into a flat plate inclined at an angle β. For a plate of length l in a stream of velocity U and density ϱ one finds the expression

$$F = \frac{\pi \sin\beta}{4 + \pi \sin\beta} \varrho U^2 l \tag{6.12}$$

for the total force, which acts in a direction perpendicular to the plate[1].

The example of the wedge serves to illustrate the variety of mathematical possibilities occurring in the cavity problem, and exhibits some of the difficulties in arriving at a formulation whose solutions are physically acceptable. If the free streamlines are required to detach from the ends, then there are two possibilities: (i) the dividing streamline meets the body at the vertex, in which case, as we have seen, the ratio of the lengths cannot be arbitrary; (ii) the dividing streamline meets the body at a point other than the vertex, in which case the velocity is infinite at the corner, and the flow is physically unacceptable. If we widen the possibilities by allowing the free streamline to detach from the vertex, then several additional solutions present themselves; (iii) the solution is simply the flow past an inclined plate, so that the free streamline detaching from the vertex may intersect the second side of the wedge if it is sufficiently long; (iv) the free streamline detaching from the vertex turns back, forming a re-entrant jet which passes through the cavity wall (Fig. 12), while the part of the flow not included in the jet forms an infinite cavity with detachment from the ends[2]. If the possibilities are broadened still further to allow the speed on the free streamline to differ from the value at infinity, then additional *families* of solutions enter consideration[3]. One of these is illustrated in Fig. 13, which

[1] Cf. MILNE-THOMSON [3], Chap. 12.

[2] Calculations based on this model have been carried out by A. COX and W. CLAYDEN, J. Fluid Mech. **3**, 615–637 (1958); see also R. COX and J. MACCOLL: Proc. Symposium on Naval Hydrodynamics, 215-232, Nat. Res. Counc. Publ. 515, 1957.

[3] Cf. VILLAT [4]. — G. JAFFÉ: Phys. Z. **6**, 129–133 (1922).

shows a free streamline detaching from the vertex and meeting the second side tangentially, while an infinite cavity is formed by the free streamlines detaching from the end points. In this flow the speed on the finite free stramline must be less than at infinity.

In case (iv) the flow speed on all the free streamlines is the same as at infinity and is the maximum speed of the flow. Except for the feature of the two-sheeted flow plane, this model satisfies the Brillouin conditions and, in light of the experimental evidence[1] (cf. Sect. 2), would appear to be the "proper" physical solution whenever flows of type (i) and (iii) exist only with self-intersections. The possibilities discussed above remain essentially unchanged if the vertex is slightly rounded, thereby showing the existence of smooth convex bodies without physically acceptable solutions of the Kirchhoff type.

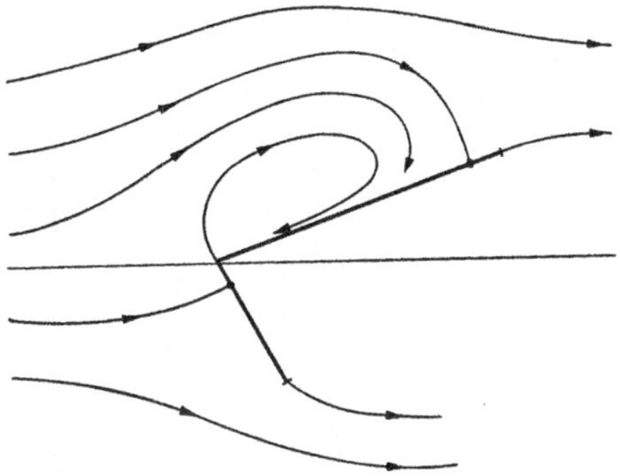

Fig. 12.

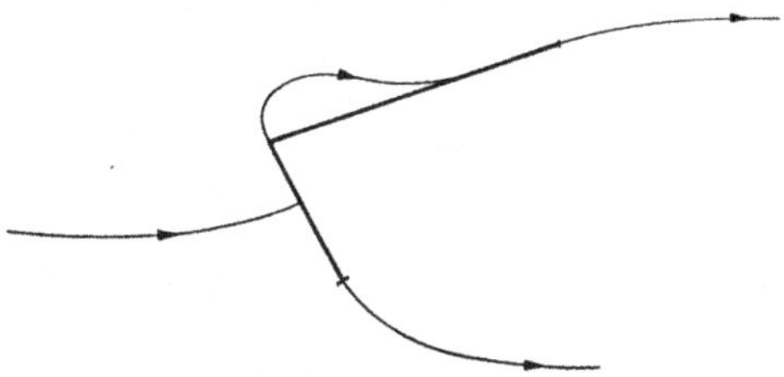

Fig. 13.

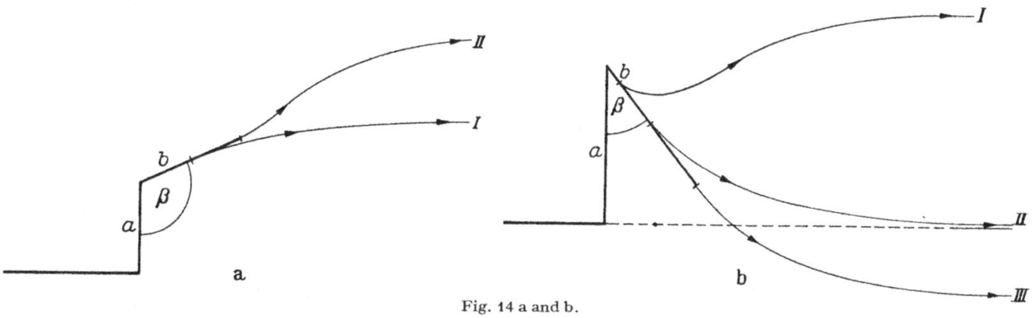

Fig. 14 a and b.

Additional qualitative features of interest are present in the symmetric infinite cavity flow past a polygonal obstacle of three faces, detachment occurring at the end points[2]. The possibilities are illustrated in Fig. 14 (which shows only the upper half of the flow plane). When the angle β between the faces exceeds $\pi/2$ (Fig. 14a), the cavity may be convex (I) or the free streamline may have an inflection point (II), according to the size of the ratio b/a of the lengths of the faces. At $\beta=\pi/2$ only the inflection point case can occur. If $\beta<\pi/2$ (Fig. 14b), there are three possibilities, depending on the ratio b/a. For smaller b/a the free streamline has an inflection point (I), and for larger b/a the free streamline is entirely concave, extending below the axis of symmetry (III). The entire flow is of course doubly sheeted in the latter case. At an intermediate value of b/a, separating the two cases I and III, the free streamline is concave and is asymptotic to the axis of symmetry. In particular, if $\beta=0$, the

[1] Wedge flows of this type have been observed experimentally; cf. Cox and Clayden, footnote 2, p. 330.

[2] C. Schmieden: Ing.-Arch. **3**, 356—370 (1932). In this flow the velocity is of course infinite at the vertices.

obstacle degenerates into a single flat plate and there are the above three indicated possibilities of detachment from the back of the plate. The critical ratio $(a-b)/a$ corresponding to the cavity of zero asymptotic width is $4\log(2+\sqrt{3})/(3+4\log 2) \approx 0.913$. We shall see (Sect. 32) that the preceding behavior is typical for curved obstacles as well.

7. The re-entrant jet cavity[1]. The general features of the re-entrant jet cavity have been discussed in Sect. 3. The simplest example, which we now calculate explicitly, is that produced in symmetric flow against a vertical flat

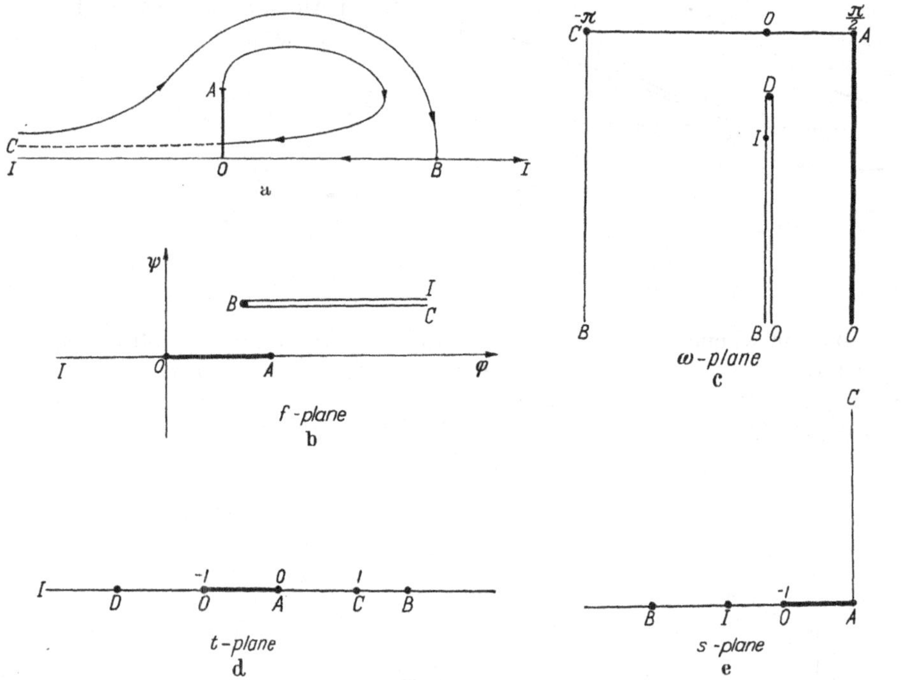

Fig. 15a—e.

plate[2]. It suffices to consider the flow in the upper half plane, which is shown with its various image planes in Fig. 15. The Schwarz-Christoffel formula (5.4) gives at once,

$$f'(t) = K_1 \frac{(t-t_B)}{(t-t_I)^2(t-1)}, \tag{7.1}$$

$$\omega'(t) = K_2 \frac{t-t_D}{\sqrt{t(t-1)(t-t_B)(t+1)}}. \tag{7.2}$$

The substitution $t = s^2/(s^2-2)$ (see Fig. 15e) leads to certain simplifications. Eqs. (7.1), (7.2) become

$$f'(s) = L_1 \frac{s(s^2-s_B^2)}{(s^2-s_I^2)^2}, \tag{7.3}$$

$$\omega'(s) = L_2 \frac{s^2-s_D^2}{(s^2-1)(s^2-s_B^2)^2}. \tag{7.4}$$

---

[1] This model seems to have been first proposed by H. WAGNER (unpublished) and later, independently, by G. KREISEL, Admiralty Res. Lab. Rep. No. R1/H/36 (1946), and D. EFROS, Dokl. Akad. Nauk. SSSR. **51**, 267—270 (1946).

[2] Cf. D. GILBARG and D. ROCK: Naval Ord. Lab. Memo. 8718 (1946); M. GUREVICH: Izv. Akad. Nauk. SSSR, OTN **1947**, 143—150 (David Taylor Model Basin Transl. 224); D. GILBARG and J. SERRIN: J. Math. Phys. **29**, 1—12 (1950); E. ARNOFF: Naval Ordnance Test Station NAVORD Report 1298 (1951).

These can both be integrated in closed form, but only the expression for $\omega(s)$ is ever needed. Observing that the residues of $\omega'(s)$ at $s=s_B, -s_B, -1, +1$ are respectively $i, -i, i/2, -i/2$, we obtain from (7.4),

$$\omega(s) = i \log L_3 \left(\frac{s+1}{s-1}\right)^{\frac{1}{2}} \left(\frac{s-s_B}{s+s_B}\right); \tag{7.5}$$

and since $\omega(s) \to -\pi$ as $s \to i\infty$, it follows that $L_3 = -1$, so that

$$\omega(s) = i \log \left(\frac{s+1}{s-1}\right)^{\frac{1}{2}} \left(\frac{s_B - s}{s_B + s}\right). \tag{7.6}$$

Setting $b = s_B$ and $h = s_I$, we infer from the fact that $\omega(h) = i \log U$, where U is the free stream velocity, the relation

$$U = \left(\frac{h+1}{h-1}\right)^{\frac{1}{2}} \left(\frac{b-h}{b+h}\right). \tag{7.7}$$

Also, since the residue of dz/ds is zero at $s=h$ (in order that the flow plane be simply covered at infinity), it follows that

$$b = -2h^2 - h + 2, \tag{7.8}$$

and thus the cavitation number is expressed in terms of h by the formula

$$\sigma = \frac{1}{U^2} - 1 = \frac{-4h^3 + h^2 + 4h - 2}{(h^2 + h - 1)^2}. \tag{7.9}$$

The value $\sigma = 0$ corresponds to $h = -\infty$.

The force on the plate is calculated by integrating $p - p_c$ over the front surface, disregarding the impact of the jet against the back of the plate. This gives for the value of the drag coefficient,

$$C_D(\sigma) = (1+\sigma) \left[1 - \frac{\int\limits_{-1}^{0} \left(\frac{1+s}{1-s}\right)^{\frac{1}{2}} \left(\frac{s-b}{s^2-h^2}\right)^2 s\, ds}{\int\limits_{-1}^{0} \left(\frac{1-s}{1+s}\right)^{\frac{1}{2}} \left(\frac{s+b}{s^2-h^2}\right)^2 s\, ds} \right]. \tag{7.10}$$

It is apparent from (7.8) and (7.9) that $[(s \pm b)/(s^2 - h^2)]^2 = 4 - \sigma + O(\sigma^2)$ as $\sigma \to 0$, so that the expression in brackets in (7.10) approaches $2\pi/(4+\pi)$. Thus we arrive at the asymptotic formula, valid for small σ[1],

$$C_D(\sigma) \sim \frac{2\pi}{4+\pi} (1+\sigma). \tag{7.11}$$

A more precise estimate is[2]

$$C_D(\sigma) \sim \frac{2\pi}{4+\pi} \left[1 + \sigma + \frac{\sigma^2}{8(\pi+4)}\right]. \tag{7.12}$$

Calculations show that the formula (7.11) underestimates within 0.8% at $\sigma = 1$ and improves in accuracy as $\sigma \to 0$[3]. (7.11) is a special case of the approximate formula

$$C_D(\sigma) \sim C_D(0)\,(1+\sigma), \tag{7.13}$$

[1] The notation $\sim$ is taken to mean that the error term is of higher order in σ.

[2] Cf. T. Wu: J. Math. Phys. **35**, 236–265 (1956).

[3] Cf. references to Gilbarg-Rock, Gurevich and Arnoff in footnote 2, p. 332. The formula (7.11) was first stated by K. Zoller, Deutsche Luftfahrtforschung, UM 4518, 1943, for the Riabouchinsky model, and later, independently, by Gilbarg and Rock for both the re-entrant jet and Riabouchinsky models. The cavitation model of A. Betz, Proc. Third. Intern. Congr. Appl. Mech. Stockholm 1930, yields a similar result. H. Reichardt [Min. Aircraft Prod. Rep. and Transl. 766 (1946)] inferred formula (7.11) from a similarity hypothesis on the pressure distribution (cf. [*1*], p. 18).

which seems to be valid for arbitrary bodies with fixed points of detachment, and for other cavity models as well (cf. Sects. 8, 11).

If the flat plate is replaced by a symmetric wedge of semiangle $\alpha\pi$, then formulas (7.7), (7.10) remain unchanged except that the exponent $\frac{1}{2}$ is replaced by α, and Eq. (7.8) becomes $\alpha b = -h^2 - \alpha h + 1$ [1]. It follows from momentum considerations[2], and also analytically, that the drag on an arbitrary obstacle of diameter l in symmetric flow and the jet width J are related by the formula [cf. Eq. (7.16)]

$$C_D(\sigma) = 2\frac{J}{l}\left(1 + \sigma + \sqrt{1+\sigma}\right). \tag{7.14}$$

With Eq. (7.13) this gives the asymptotic relation

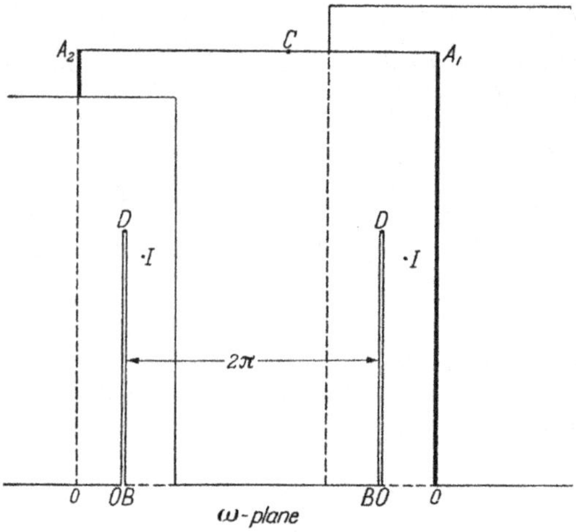

Fig. 16.

$$\frac{J}{l} \sim \frac{C_D(0)}{2} \frac{\sqrt{1+\sigma}}{1+\sqrt{1+\sigma}}. \tag{7.15}$$

For small σ, $J/l \approx C_D(0)/4$.

The formula (7.14) is a special case of a more general expression for the force on an obstacle in re-entrant jet flow. If V is the cavity speed, γ the direction of the jet, and $\varkappa$ the circulation about the cavity, then the formulas for the drag and lift in a fluid of density ϱ become[3] (cf. Sect. 23)

$$\left.\begin{aligned} D &= \varrho J V(U - V\cos\gamma), \\ L &= -\varrho U(U J \sin\gamma + \varkappa). \end{aligned}\right\} \tag{7.16}$$

For asymmetric bodies[4] the mathematical details of the re-entrant jet cavity differ from the preceding in that both the f and ω images of the flow plane ecome Riemann surfaces. The mapping of the f-plane (Fig. 7) on a half-plane, $\operatorname{Im} t > 0$, is easily derived by means of the generalized Schwarz-Christoffel formula (5.6). Since $f'(t)$ has a simple zero at $t = t_B$ and a pole of second order at $t = t_I$, but is otherwise regular with non-vanishing derivative, it follows that

$$f'(t) = M\frac{(t-t_B)(t-\bar{t}_B)(t-t_0)}{(t-t_I)^2(t-\bar{t}_I)^2}, \qquad M \text{ real}. \tag{7.17}$$

In case the obstacle is a wedge of angle $\alpha\pi$ and the vertex O is a stagnation point, the Riemann surface in the ω plane is polygonal (Fig. 16), and by a similar application of Eq. (5.6) one finds

$$\omega'(t) = K\frac{(t-t_D)(t-\bar{t}_D)}{(t-t_B)(t-\bar{t}_B)(t-t_0)(t^2-1)^{\frac{1}{2}}}, \qquad K \text{ real}. \tag{7.18}$$

[1] Calculations on flows past symmetric wedges are presented by ARNOFF, footnote 2, p. 332.

[2] G. BIRKHOFF: Revista de Ciencias, Lima **50**, 105–116 (1948); this treats the axially symmetric case.

[3] GILBARG and SERRIN: Footnote 2, p. 332.

[4] Cf. D. EFROS: Dokl. Akad. Nauk SSSR. **51**, 267–270 (1946); **60**, 29–31 (1948). — M. SHIFFMAN: Comm. Pure Appl. Math. **2**, 1–11 (1949). — GILBARG and SERRIN: Footnote 2, p. 332.

In the ζ semi-circle defined by the mapping $t = -\frac{1}{2}(\zeta + 1/\zeta)$ the expression for ω takes the simple form,

$$\omega(\zeta) = i \log\left(\frac{\zeta - e^{i s_0}}{\zeta - e^{-i s_0}}\right)^{\alpha} \left(\frac{\zeta - \zeta_B}{\zeta - \bar{\zeta}_B}\right)\left(\frac{\zeta - 1/\bar{\zeta}_B}{\zeta - 1/\zeta_B}\right) + \text{const}, \tag{7.19}$$

where $\zeta = e^{i s_0}$ is the image of the vertex O, and the constant is real. The formula (7.19) can be verified by inspection; it can also be derived from Eq. (7.18) by integration, or by direct application of the Schwarz principle of reflection.

8. The Riabouchinsky cavity[1]. In this cavity model the physically objectionable double-sheeted flow plane of the re-entrant jet cavity is avoided by artifically closing the cavity at its rear. This is accomplished by introducing into the flow as a fixed boundary the mirror image of the obstacle in a plane perpendicular to the flow direction. The principal advantage of this model is its mathematical simplicity. At the same time it yields values for the important physical flow quantities that are almost indistinguishable from those provided by the re-entrant jet cavity.

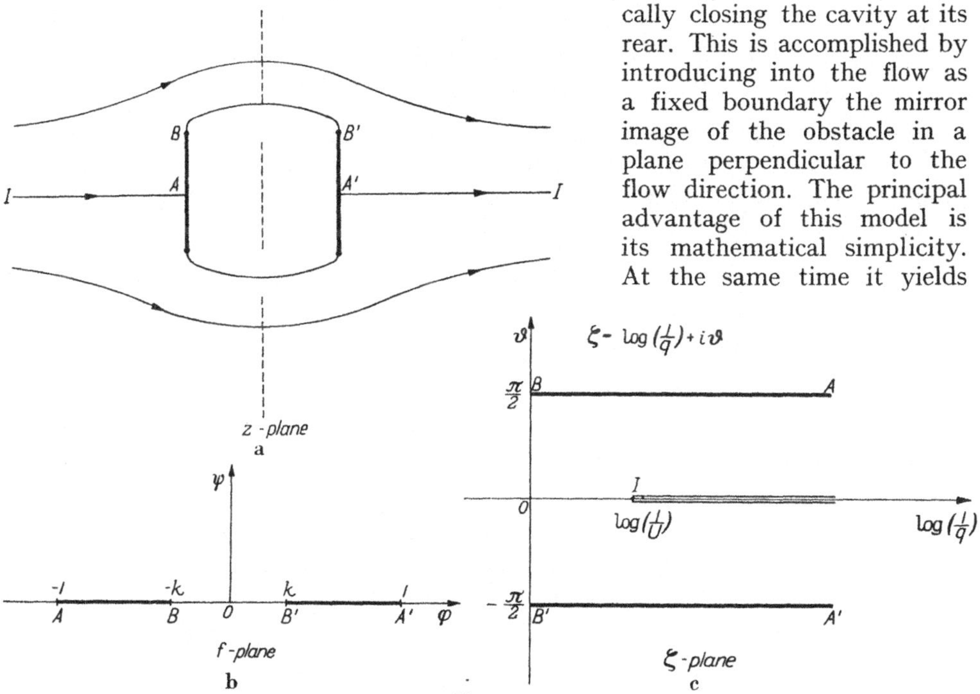

Fig. 17a—c.

Consider the symmetric Riabouchinsky cavity behind a flat plate, shown in Fig. 17 with the various image planes corresponding to the upper half of the flow. We assume for convenience that the flow is so normalized that $f = -1, +1$ at the front and rear stagnation points A, A'. We obtain from the Schwarz-Christoffel transformation

$$\frac{d\zeta}{df} = \frac{M}{(f^2 - 1)\sqrt{f^2 - k^2}}, \qquad (M = \text{const}), \tag{8.1}$$

which yields, after integration and making the appropriate boundary correspondence,

$$\frac{dz}{df} = \frac{\sqrt{f^2 - k^2} + f\sqrt{1 - k^2}}{k\sqrt{f^2 - 1}}. \tag{8.2}$$

[1] D. Riabouchinsky: Proc. London Math. Soc. (2) **19**, 206—215 (1920); **25**, 185—194 (19.26). Apparently the first application of this model to cavities rather than wakes was made by F. Weinig: Hydromechanische Probleme des Schiffsantriebs, ed. by G. Kempf and E. Foerster. Hamburg 1932.

This formula determines the flow completely in terms of the parameter k. Since the left member becomes $1/U = \sqrt{1+\sigma}$ as $f \to \infty$, the cavitation number and k are related by the formula,

$$\sigma = 2\,\frac{\sqrt{1-k^2}}{k^2}\left(1+\sqrt{1-k^2}\right) = \frac{2k'}{1-k'}, \tag{8.3}$$

where

$$k' = \sqrt{1-k^2}. \tag{8.4}$$

Introducing the complete elliptic integrals of the first and second kind,

$$E(k) = \int_0^1 \sqrt{\frac{1-k^2 t^2}{1-t^2}}\,dt, \qquad K(k) = \int_0^1 \frac{dt}{\sqrt{(1-t^2)(1-k^2 t^2)}}, \tag{8.5}$$

and designating by E' and K' the same functions with argument k', we are able by means of Eq. (8.2) to represent the various flow quantities of interest by the following explicit formulas:

the ratio of cavity width d to plate length l,

$$\frac{d}{l} = \frac{k' + E' - k^2 K'}{k'^2 + E' - k^2 K'}; \tag{8.6}$$

the ratio of cavity length h to plate length,

$$\frac{h}{l} = \frac{E - k'^2 K}{k'^2 + E' - k^2 K'}; \tag{8.7}$$

the drag coefficient[1],

$$C_{\mathrm{D}} = 2(1+\sigma)\left(1 - \frac{k'^2}{k'^2 + E' - k^2 K'}\right). \tag{8.8}$$

The dependence of these quantities on the cavitation number in the physically important range of small σ is easily determined from the properties of elliptic integrals. From Eqs. (8.6) to (8.8) and the asymptotic relations for $k' \to 0$[2],

$$K' \sim \frac{\pi}{2}\left(1 + \frac{k'^2}{4}\right), \qquad E' \sim \frac{\pi}{2}\left(1 - \frac{k'^2}{4}\right)$$

$$E \sim 1 + \frac{1}{2}\left(\log\frac{4}{k'} - \frac{1}{2}\right)k'^2, \qquad K \sim \log\frac{4}{k'} + \left(\log\frac{4}{k'} - 1\right)\frac{k'^2}{4},$$

we obtain the following asymptotic formulas for small σ[3]:

$$\frac{d}{l} \sim \frac{4}{4+\pi}\left(\frac{2+\sigma}{\sigma} + \frac{\pi}{4}\right), \tag{8.9}$$

$$\frac{h}{l} \sim \frac{4}{4+\pi}\left[\left(\frac{2+\sigma}{\sigma}\right)^2 - \frac{1}{2}\log 4\left(\frac{2+\sigma}{\sigma}\right) - \frac{1}{4}\right] \tag{8.10}$$

$$C_{\mathrm{D}}(\sigma) \sim \frac{2\pi}{4+\pi}(1+\sigma). \tag{8.11}$$

---

[1] The integration of $p - p_c$ is of course taken over the forward plate alone.

[2] E. JAHNKE and F. EMDE: Tables of Functions, 3rd ed. Leipzig u. Berlin: Teubner 1938; reprinted New York: Dover Publ. 1943.

[3] DUE to K. ZOLLER: Deutsche Luftfahrtforschung, UM 4518, 1943; this reference has (incorrectly) $+\frac{3}{4}$ rather than $-\frac{1}{4}$ in Eq. (8.10).

Formula (8.9) is accurate within 0.6% for $\sigma<3$, that for h/l is too high by less than 0.7% at $\sigma=1$, and the expression (8.11) for C_D underestimates by less than 0.8% at $\sigma=1$ [1]. All three formulas increase in accuracy for smaller σ. The more precise asymptotic relation [2]

$$C_D(\sigma) \sim \frac{2\pi}{4+\pi}\left[1+\sigma+\frac{\sigma^2}{8(\pi+4)}\right], \tag{8.12}$$

[which is the same as Eq. (7.12) for the re-entrant jet] shows that $C_D/(1+\sigma)$ is constant within a positive term of the second order in σ [3].

It is of interest to compare the asymptotic formulas (8.9) to (8.11) with those derived for the general axially symmetric Riabouchinsky cavity behind a fixed arc of diameter l. Namely [4],

$$C_D(\sigma) \sim C_D(0)\,(1+\sigma) \tag{8.13}$$

and

$$\frac{d^2}{l^2} \sim \frac{C_D(\sigma)}{\sigma}, \qquad \frac{h^2}{l^2} \sim \frac{C_D(\sigma)}{\sigma^2}\log\frac{1}{\sigma}. \tag{8.14}$$

The error term in Eq. (8.13) is positive. We note that the ratio of dimensions is asymptotcally given by

$$\frac{h^2}{d^2} \sim \frac{1}{\sigma}\log\frac{1}{\sigma} \tag{8.15}$$

as compared with the formula

$$\frac{h}{d} \sim \frac{2}{\sigma} \tag{8.16}$$

in the plane case. From this it is apparent that for a given cavitation number the plane cavity has a much larger length-width ratio than the three dimensional one. On the other hand, the three dimensional cavity is the flatter of the two (as measured by d/l).

It is a significant fact that despite the great conceptual difference between the re-entrant jet and Riabouchinsky cavities, the two models give almost identical results as regards gross behavior [5]. Thus the two models differ in their values of the drag coefficient for the flat plate by about 0.1% at $\sigma=2$, and show almost as close agreement in the drag coefficient of wedges. With respect to cavity dimensions, the agreement between the two models is still quite good, although not so spectacularly as for the drag coefficient. Thus it would seem that the gross physical quantities associated with the cavity, such as dimensions and drag, are not sensitive to the choice of cavity model, or, in other words, to the conditions at the rear of the cavity. This view is given additional support by calculations on other cavity models which yield essentially the same results as obtained from the RIABOUCHINSKY and re-entrant jet cavities (cf. Sect. 11).

---

[1] Cf. remarks appended to M. GUREVICH: David Taylor Model Basin Transl. 224 (1948).

[2] Cf. T. WU: J. Math. Phys. **35**, 236–265 (1956).

[3] For calculations on wedges see M. PLESSET and P. SHAFFER: Rev. Mod. Phys. **20**, 228–231 (1948); J. Appl. Phys. **19**, 934–939 (1948); flows in channels are considered by G. BIRKHOFF, M. PLESSET and N. SIMMONS: Quart. Appl. Math. **9**, 413–421 (1952).

[4] Cf. P. GARABEDIAN: Pac. J. Math. **6**, 611–684 (1956); the derivations are convincing but have still to be made rigorous. See also H. REICHARDT: Ministry Aircraft Prod. Rep. and Transl. 766 (1946), and [*1*], Chap. 10.

[5] D. GILBARG and D. ROCK: Naval Ordnance Lab. Memo 8718, 1946. — M. GUREVICH: Footnote 1. — E. ARNOFF: Naval Ordnance Test Station, NAVORD Report 1298, 1951.

9. The cusped cavity. We consider now a model of the finite cavity characterized by a cusp at its trailing edge[1] (Fig. 18). It is clear from Eq. (3.5) that the maximum speed of the flow cannot be achieved on the free streamlines of a cusped cavity, and one can show for symmetric flows that the cavity speed is less than the free stream value (cf. Sect. 27). This model is therefore unsuited to the theory of cavitation, and is of questionable physical significance altogether. It is evident from the D'ALEMBERT paradox that the drag on an obstacle with an attached cusped cavity must be zero.

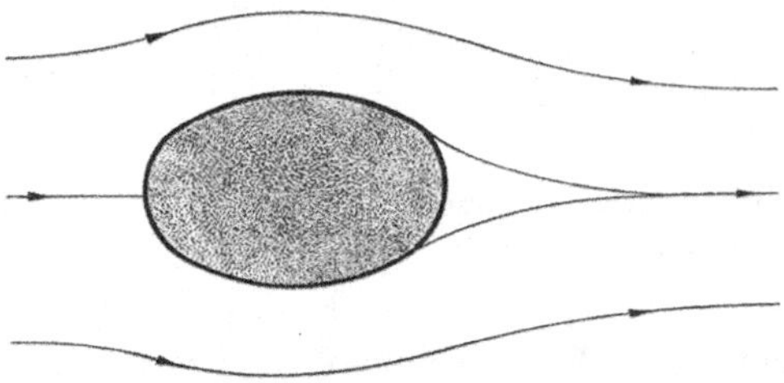

Fig. 18.

We illustrate the cusped cavity in the symmetric flow past a flat plate[2] (Fig. 19) with unit free stream velocity and cavity speed V. The connection

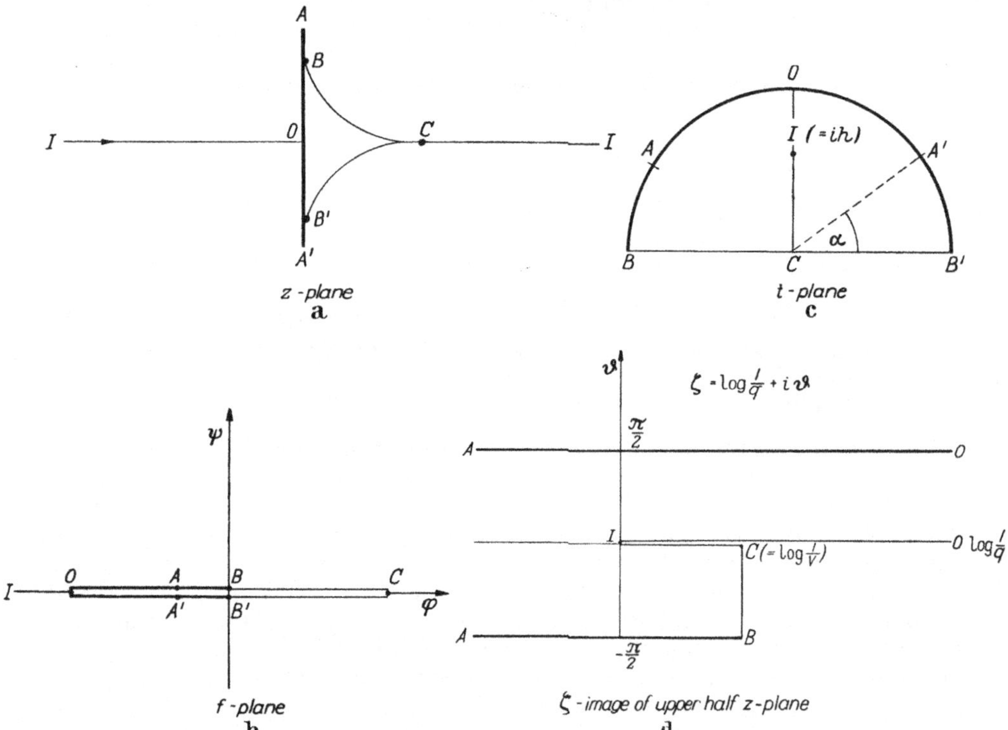

Fig. 19a–d.

[1] The cusped cavity was proposed by M. BRILLOUIN and rejected by him as a finite wake model [Ann. chim. phys. **23**, 145–230 (1911)] because it violates the maximum speed condition on the free boundary. H. VILLAT: Ann. Fac. Sci. Toulouse (3) **5**, 375–404 (1913), conjectured the non-existence of these flows in general after proving it for flow past a wedge (with detachment from the endpoints). Calculations of cusped cavity flows past polygonal obstacles were carried out by U. CISOTTI: Ann. Scuola Norm. Sup. Pisa (2) **1**, 101–112 (1932), and M. KOLSCHER: Luftf.-Forschg. **17**, 154–160 (1940). An explicit example of such a flow past a curved obstacle was constructed by M. LIGHTHILL: Aeronaut. Res. Council Rep. and Memo. 2328 (1945). General theory and numerical calculations will be discussed in later sections.

[2] Cf. KOLSCHER (preceding footnote).

between the image planes is given by the formulas,

$$f = K \frac{(1+t^2)^2}{(t^2+h^2)(h^2t^2+1)}, \qquad K>0,$$

$$\frac{dz}{df} = e^{\zeta} = \frac{1}{V} \frac{(i+t)(t-e^{i\alpha})(t+e^{-i\alpha})}{(i-t)(t-e^{-i\alpha})(t+e^{i\alpha})}.$$

Since $(dz/df)_{t=ih}=1$, we have

$$V = \frac{(1+h)(1-2h\sin\alpha+h^2)}{(1-h)(1+2h\sin\alpha+h^2)}$$

and from the fact that the residue of dz/dt is zero at $t=ih$ (where z has a simple pole) it follows

$$h^2 - 2h^2 \frac{1+2\sin\alpha-2\sin^2\alpha}{2\sin\alpha-1} + 1 = 0.$$

The latter equation has exactly one real root h, $0<h<1$, for $\pi/6<\alpha<\pi/2$. The cavity speed V decreases from 1 to 0 as α varies between $\pi/6$ and $\pi/2$. The position of the cusp x_C and of the point of detachment of the free streamline y_B are derived from

$$z(t) = \int_i^t \frac{dz}{df}\frac{df}{dt}\,dt$$

which gives

$$x_C = x(0) = c(\alpha)\left[\frac{2-\sin\alpha}{2\sin\alpha-1} - \frac{\pi-2\alpha}{2\cos\alpha}\right],$$

$$y_B = y(-1) = c(\alpha)\left[\frac{1-2\sin\alpha}{\sin\alpha(2-\sin\alpha)} + \frac{1}{\cos\alpha}\log\frac{\sin\alpha}{1-\cos\alpha}\right],$$

where

$$c(\alpha) = \frac{\sin 2\alpha}{\cos^2\alpha - 2\sin\alpha\log\sin\alpha}.$$

As $\alpha\to\pi/2$ the cavity shrinks to zero and the flow becomes the classical flow around the flat plate. As $\alpha\to\pi/6$ the cusp moves down stream towards infinity, the point of detachment B approaches

$$y_B = \frac{4\log(2+\sqrt{3})}{3+4\log 2} \approx 0.913,$$

and the flow approaches the infinite cavity of zero asymptotic width described in Sect. 6. The points on the plate above this value are those from which detach infinite cavities with inflected free streamlines (as in Fig. 14b), while ordinates below this value determine double sheeted flows. Hence the separation points of the cusped cavities coincide with those of the two-sheeted infinite cavity flows. This proves to be typical of the general situation (cf. Sect. 32).

The several cavity models considered in this and the preceding sections exhibit some of the possibilities of cavitational flow past a single flat plate. These are only a few of the totality of possible configurations. If we restrict considerations to single sheeted uniform flows with bounded velocity past a single flat plate, with finitely many free streamlines, the entire set of possibilities (of which there are nine) has been described by ZARANTONELLO[1].

10. Jet flows. Among the first achievements of the hodograph method was the solution, for plane flows, of the classical problem of jet efflux from an orifice. However, fully one hundred years before the development of the hodograph

[1] E. ZARANTONELLO: J. Math. pures appl. (9) **33**, 29–80 (1954).

method BORDA[1] succeeded in determining exactly the contraction coefficient of the jet issuing from what is now known as the Borda mouthpiece. This problem concerns the motion of a jet of fluid escaping from a large vessel through a long inwardly projecting channel. The idealization of the problem considers an infinite straight channel in an unbounded fluid, disregarding gravity, and leads to a flow illustrated in Fig. 20, where BIB' is the channel, BB' the orifice, and J the jet. It is noteworthy that BORDA's method applies equally well to two and three dimensional flows, and also to compressible fluids. His argument was based on momentum considerations[2], and is essentially the following (in its modern form).

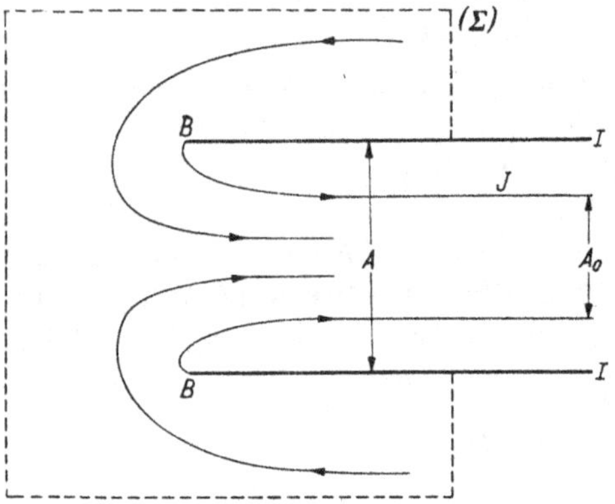

Fig. 20.

If Σ denotes a large control surface surrounding the channel, and A, A_0 are the respective cross-sectional areas of the channel and jet, the momentum principle gives the equality

$$\int\limits_{\Sigma+A_0} \varrho u V_n dS = - \int\limits_{\Sigma+A_0+J} p n_x dS \quad (10.1)$$

where the left member is the x component of momentum flux through the surface $\Sigma + A_0$, and the right member is the x component of force on the fluid surface $\Sigma + A_0 + J$. The pressure integral over the horizontal channel vanishes and hence does not appear in the right member. (The argument works for precisely this reason.) Let the stagnation pressure be p_∞, that in the channel p_0 and let u_0 be the jet speed. Assuming that the velocity attenuates at infinity so rapidly that the momentum integrals over Σ vanish as Σ becomes infinitely large[3], we obtain from Eq. (10.1), for incompressible fluids,

$$\varrho u_0^2 A_0 = p_\infty A - p_0 A_0 - p_0 (A - A_0) = (p_\infty - p_0) A,$$

whence, using BERNOULLI's equation, $p_\infty = p_0 + \frac{1}{2} \varrho u_0^2$, we have

$$C_c = A_0/A = \tfrac{1}{2}. \quad (10.2)$$

The same argument can be used to show that $C_c \geq \frac{1}{2}$ in general[4]. If the fluid is contained in a vessel which is parallel to the channel and has a cross-sectional area A^*, the relation $A^*(A - 2A_0) + A_0^2 = 0$ holds, and hence

$$C_c = \frac{1}{(1 + \sqrt{1 - A/A^*})}.$$

---

[1] J. BORDA: Mém. Acad. Sci. Paris **1766**, 519–607.

[2] A number of interesting and useful results, some approximate and some exact, are derivable entirely from conservation principles. See, for example, BIRKHOFF and ZARANTONELLO [*1*], Chap. 1, in particular, the applications to lined hollow charges and wall effects.

[3] To make the argument rigorous the assumed asymptotic behavior of the velocity must be proved a priori. This assumption is implicit in the usual demonstration, in which the surface Σ is the containing vessel and the fluid velocity is presumed to vanish on it.

[4] Cf. H. LAMB: Hydrodynamics, 6th ed., pp. 24–25. C_c can be less than $\frac{1}{2}$ for an inwardly projecting diverging mouthpiece. — F. KÖTTER: Arch. Math. Phys. (2) **5**, 392–417 (1887), has shown by a momentum argument that $0.54 < C_c < 0.71$ for jet efflux from a circular orifice in a plane wall, on the assumption that the flow at the orifice section is directed entirely into the jet.

For a compressible fluid obeying the pressure-density relation $p \propto \varrho^{\gamma}$, one derives in the same manner

$$C_c = \frac{\gamma - 1}{2\gamma} \frac{p_{\infty}/p_0 - 1}{(p_{\infty}/p_0)^{\gamma/(\gamma-1)} - 1}.$$

These values for the contraction coefficient are among the few exact results known concerning spatial free surface flows of either compressible or incompressible fluids. For a complete description of the flows in closed form one must again specialize to the plane incompressible case.

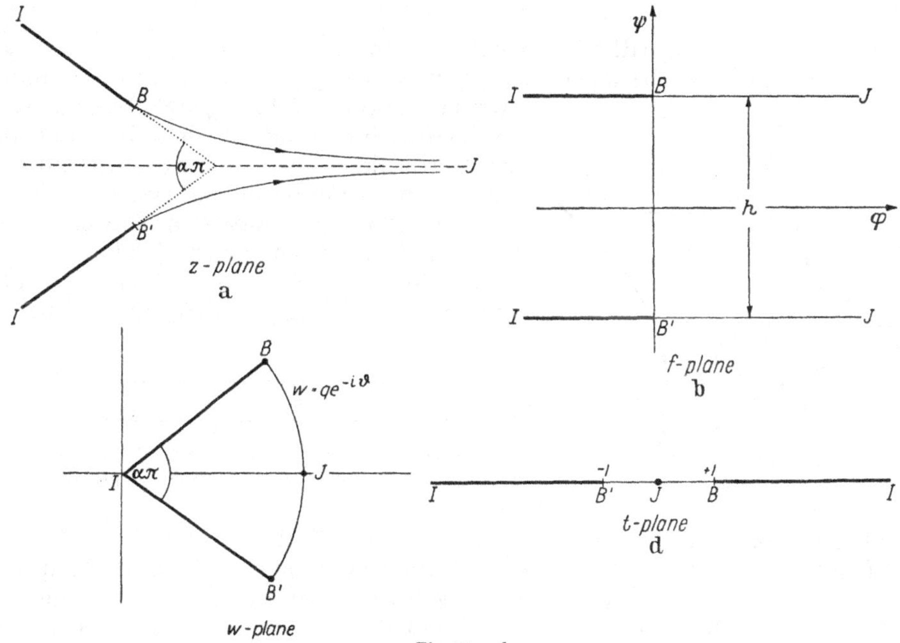

Fig. 21a–d.

Consider now the efflux of a plane jet from an aperture in a vessel bounded by semi-infinite straight walls, as in Fig. 21, assuming the flow to be symmetric with respect to the x-axis. If the angle between the walls is $\alpha\pi$, then for $\alpha = 2$ we have once more the flow from the Borda mouth-piece, while for $\alpha = 1$ the flow describes a jet issuing from slot in a plane wall. If the speed on the free streamlines is taken equal to unity, the difference h between the values of the stream function on the upper and lower free streamlines is equal to the jet width at infinity. The flow plane and its images are shown in Fig. 21. The mappings between the planes are given by the formulas,

$$\left.\begin{aligned} f &= \frac{h}{\pi}\left(i\frac{\pi}{2} - \log t\right), \\ w &= f'(z) = \left(\sqrt{1-t^2} + i t\right)^{\alpha}. \end{aligned}\right\} \tag{10.3}$$

The shape of the free streamlines is given by

$$z(t) = -\frac{h}{\pi} \int\limits_1^t \frac{\left(\sqrt{1-\tau^2} - i\tau\right)^{\alpha}}{\tau}\, d\tau, \quad 0 < t < 1. \tag{10.4}$$

In the special case $\alpha = 1$, this yields the value h/π for the change in ordinate along BJ, and hence the value

$$C_c = \frac{\pi}{\pi + 2} \approx 0.611 \tag{10.5}$$

for the contraction coefficient of a jet issuing from a slot in a plane wall[1,2]. In the general case one obtains

$$\frac{1}{C_c} = 2 - \frac{1}{\pi} \sin \frac{\alpha \pi}{2} \left[\Psi\left(\frac{1}{2} + \frac{\alpha}{4}\right) - \Psi\left(\frac{\alpha}{4}\right) - \frac{2}{\alpha} \right], \tag{10.6}$$

where $\Psi(x)$ is the logarithmic derivative of the gamma function. The contraction coefficient varies between the extreme values $\frac{1}{2}$ and 1 as α decreases from 2 (Borda mouthpiece) to 0 (parallel flow). That the increase in C_c is monotonic—also for axially symmetric flows—can be proved without calculation from the results of Sect. 28.

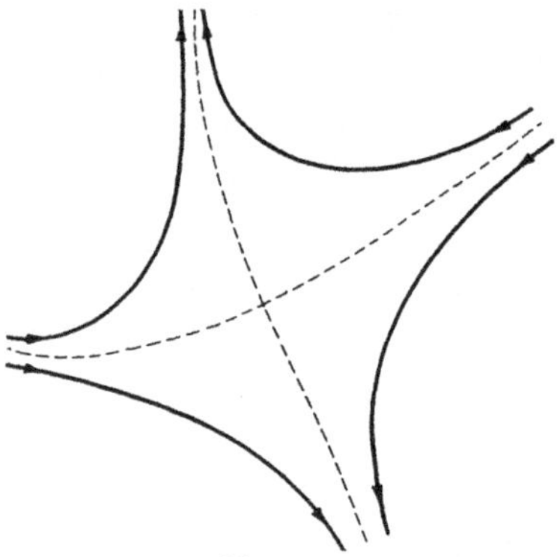

Fig. 22.

The following comparison between theoretical and experimental values[3] indicates fairly good agreement, with the observed values running slightly higher than the calculated ones (cf. Sect. 2).

α	0.5	1	1.5	2
C_c exp.	0.753	0.632	0.557	0.541
C_c theor.	0.746	0.611	0.537	0.500

The motion produced by the confluence and branching of two jets[4] (Fig. 22) is of special interest because of the indeterminacy it exhibits. The resultant flow and, in particular, the asymptotic jet widths and directions, are not uniquely determined by the asymptotic properties of the incident jets. This would seem to imply that the history of the motion, before it assumes its steady character, is essential to the final configuration.

In principle, the hodograph method allows us to describe by explicit formulas jet flows from arbitrary polygonal vessels[5]. However, these formulas generally contain undetermined parameters which are connected with the shape of the fixed boundary through complicated transcendental equations. Whether a flow is actually realized by suitable values of the free parameters is a problem of great complexity belonging properly to the existence theory considered in Sect. 38.

11. Wake; planing; wall effects. This section presents briefly several additional flows of physical interest that have been treated by the hodograph method.

[1] The solution for $\alpha = 1$ was first obtained by G. KIRCHHOFF: Ges. Abh. 416; that for the Borda mouthpiece ($\alpha = 2$) is due to H. HELMHOLTZ: Wiss. Abh. **1**, 154.

[2] The contraction coefficients for jet flow from various tunnels and channels of finite width have been calculated by R. VON MISES: Z.VDI **61**, 447—452, 469—473, 493—498 (1917).

[3] The experimental data is due to WEISBACH, for which see v. MISES, preceding footnote. For other relevant data see H.W. KING: Handbook of Hydraulics, 3rd ed. New York-Toronto-London: McGraw-Hill 1939.

[4] See, for example, BIRKHOFF and ZARANTONELLO [*1*] and MILNE-THOMSON [*3*].

[5] Extensive references and a wide assortment of jet flows can be found in BIRKHOFF and ZARANTONELLO [*1*] and CISOTTI [*2*].

α) The underpressure wake. The wake model of ROSHKO and EPPLER[1] improves on the classical Kirchhoff model by postulating a fixed base pressure which is chosen to agree with observed values (cf. Sect. 2). In this model it is supposed that the wake parameter σ_w (2.4) is given, and that free streamlines, with flow speed $V = U\sqrt{1+\sigma_w}$ (U = incident velocity), detach from the obstacle and extend down stream until they join smoothly to horizontal half-lines (Fig. 9). The symmetric flow past a flat plate and the corresponding image planes are shown in Fig. 23. The mapping between the z and f planes is given by

$$z = i\,\frac{k^2+1}{2k}\left\{\sqrt{f(1-f)} + \arctan\sqrt{\frac{f}{1-f}} + \frac{1}{\alpha}\sqrt{f(\alpha^2-f)} + \alpha\arctan\sqrt{\frac{f}{\alpha^2-f}}\right\},$$

where

$$k^2 = 1+\sigma_w \quad \text{and} \quad \alpha = \frac{k^2+1}{k^2-1}.$$

The value of the wake parameter required to yield the observed drag coefficient $C_D = 2.13$ is $\sigma_w = 1.37$, and for this value the calculated pressure distribution

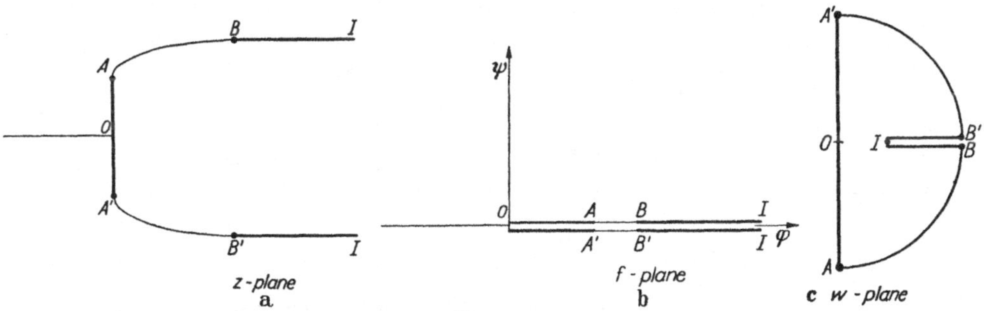

Fig. 23a—c.

over the plate is in excellent agreement with the measured ones[2]. Calculations on a circular cylinder when *smooth separation* is assumed show much less satisfactory agreement with observed pressure distributions (in fully wetted flow).

This model gives values for $C_D(\sigma)$ close to those calculated for the re-entrant jet and Riabouchinsky flows. The asymptotic dependence of C_D on σ for the flat plate is expressed in the formula,

$$C_D \sim \frac{2\pi}{4+\pi}\left[1+\sigma+\frac{\sigma^2}{6(4+\pi)}\right],$$

which may be compared with Eqs. (7.12) and (8.12) for the other models[3].

β) Planing on a water surface[4]. The simplest two-dimensional planing motion in the absence of gravity is that produced by a semi-infinite flat plate on the

[1] A. ROSHKO: NACA Tech. Note 3168 (1954). — E. EPPLER: J. Rat. Mech. a. Analysis **3**, 591—644 (1954). The latter has a full discussion of the general theory.

[2] See references in the preceding footnote for comparison with experiment.

[3] Cf. T. WU: J. Math. Phys. **35**, 236—265 (1956); this reference contains general theory and calculations on oblique incidence based on this model; the results show excellent agreement with experiments on cavitating hydrofoils. For other calculations see the references in footnote 1.

[4] Cf. H. WAGNER: Z. angew. math. Mech. **12**, 193—215 (1932). — A. E. GREEN: Proc. Cambridge Phil. Soc. **31**, 589 (1935); **32**, 67—85, 248—252 (1936). — E. COOPER: Naval Ordnance Test Station, NAVORD Report No. 1154 (1949).

surface of an infinitely deep sea (Fig. 24). This model has the fluid dividing at a stagnation point on the plate, the upper portion forming a jet or splash, while the lower part is bounded by a free streamline detaching from the trailing edge of the plate and extending to infinity downstream where, as the theory shows, it lies infinitely below the undisturbed surface. The depth of immersion of the plate is therefore undefined. This divergence remains in the solution even when the plate is finite (Fig. 25). When the depth of the fluid is finite a solution is obtained for all depths of immersion of the trailing edge, and also for a range of positions of the trailing edge above the undisturbed stream surface. However, in the limiting case as the depth approaches infinity the downstream free surface is again infinitely below the undisturbed stream. With the introduction of gravity[1] one easily infers that both this divergence and the infinite splash

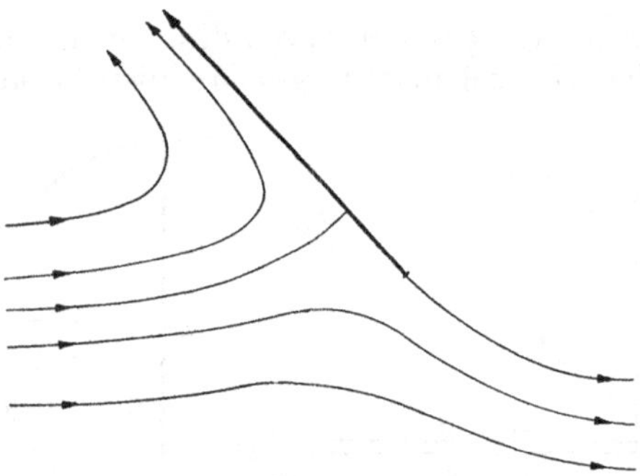

Fig. 24.

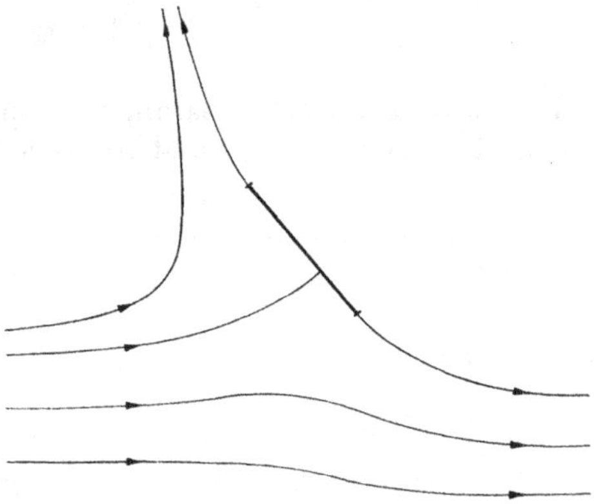

Fig. 25.

disappear. Calculations of the forces in the gravity-free case have been mady by GREEN[2] and several other gliding configurations have been considered be COOPER[2].

γ) Wall effects[3]. The estimation of wall interference is of evident importance in cavitation water tunnel testing. In the two-dimensional case a conservative analysis can be achieved by the hodograph method using flat plate obstacles. Comparison of closed and free jet tunnels shows the former to be somewhat less suitable at low cavitation numbers because of a blockage effect. Since the infinite cavity in a closed tunnel must have a cavity speed in excess of the upstream value it is clear that the minimum attainable cavitation number (for a given obstacle) exceeds zero. For example, in the flow past a plate of width $^1/_{20}$ that of the channel the cavitation number cannot descend below 0.6, and, to achieve a value of σ as low as 0.05 for a body having a drag coefficient less than unity, the ratio of channel to body width must exceed 400[4]. BIRKHOFF, PLESSET, and SIMMONS are thus led to the concept of *blockage ratio* (in analogy with wind tunnel phenomena near sonic speed) for the minimum channel-body ratio admitting a given cavitation number.

[1] Relaxation calculations of planing in a gravity field have been carried out by R. SOUTHWELL and G. VAISEY: Phil. Trans. Roy. Soc. Lond., Ser. A **240**, 117–161 (1948).

[2] See footnote 4, p. 393.

[3] G. BIRKHOFF, M. PLESSET and N. SIMMONS: Quart. Appl. Math. **8**, 161–168 (1950); **9**, 413–421 (1952). Cf. also A. ARMSTRONG: Arm. Res. Est. Rep. No. 12/51 (1951). — A. ARMSTRONG and K. TADMAN: Arm. Res. Est. Rep. Nos. 3/52 and 7/52 (these two papers deal with axially symmetric tunnels). — H. COHEN and Y. TU: Proc. 9th Intern. Cong. Appl. Mech. Brussels 1956. — H. COHEN and R. GILBERT: J. Appl. Mech. **24**, 170–176 (1957).

[4] Numerically, the limitation is of course much less severe in the three-dimensional case.

Calculations show that the drag coefficient in a closed tunnel is numerically very close to that in the infinite stream in the attainable range of cavitation numbers, and hence the wall effect on measurements can be expected to be small.

12. Compressible fluids. In the theory of compressible as well as of incompressible fluids the hodograph transformation is the principal device for the construction of exact solutions. Although the ramifications of the hodograph method are extensive and occupy a large portion of the literature on compressible fluids, in this section we can only summarize briefly the basic principles as they relate to the free boundary problem[1].

The usefulness of the hodograph mapping for compressible fluids stems from the fact that the nonlinear equations describing a steady, irrotational, and isentropic motion in the physical plane become linear in the hodograph variables. In the case of free surface flows past polygonal boundaries the hodograph image is again a region bounded by radial segments and circular arcs, so that the flow problem becomes one of solving linear partial differential equations in (essentially) known regions[2]. However, in contrast to the incompressible case, potential theory and conformal mapping are no longer the appropriate mathematical tools.

The equations connecting the potential and stream functions in the physical plane are

$$\varrho\varphi_x = \psi_y, \quad \varrho\varphi_y = -\psi_x, \quad \varrho = \varrho(q^2) = \varrho(\varphi_x^2 + \varphi_y^2), \tag{12.1}$$

where $\varrho(q^2)$ is a known function determined from BERNOULLI's equation, $\int \frac{dp}{\varrho} + \frac{1}{2}q^2 = \text{const}$, when the pressure-density relation $p = p(\varrho)$ is given. For polytropic flow, $p \propto \varrho^\gamma$, we have, if ϱ is normalized to equal unity when $q = 0$,

$$\varrho = \left(1 - \frac{\gamma - 1}{2} q^2\right)^{\frac{1}{\gamma - 1}}. \tag{12.2}$$

In the hodograph variables (q, ϑ) Eqs. (12.1) become

$$\varphi_q = -\frac{(1 - M^2)}{\varrho q}\psi_\vartheta, \quad \varphi_\vartheta = \frac{q}{\varrho}\psi_q, \tag{12.3}$$

where M is the Mach number q/c, c being the speed of sound, $c = (dp/d\varrho)^{\frac{1}{2}} = (-\varrho/2\varrho')^{\frac{1}{2}}$. The flow is subsonic, sonic, or supersonic, at a point according as $M < 1$, $= 1$, or > 1. The equations satisfied by φ and ψ are obtained from (12.3) by cross differentiation. Thus,

$$q^2\psi_{qq} + q\left(1 + \frac{q^2}{c^2}\right)\psi_q + \left(1 - \frac{q^2}{c^2}\right)\psi_{\vartheta\vartheta} = 0, \tag{12.4}$$

and similarly for φ. The connection between the physical and hodograph variables is given by

$$\left.\begin{aligned} dz &= \frac{e^{i\vartheta}}{q}\left(d\varphi + \frac{i}{\varrho}d\psi\right) \\ &= \frac{e^{i\vartheta}}{\varrho q}\left[(q\psi_q + i\psi_\vartheta)\,d\vartheta + \left(-\frac{1 - M^2}{q}\psi_\vartheta + i\psi_q\right)dq\right], \end{aligned}\right\} \tag{12.5}$$

and a similar relation holds in terms of φ.

[1] For a more complete discussion and extensive references see A. OUDART: L'étude des jets et la mécanique theorique des fluides. Publ. Sci. Tech. du Min. de l'Air 234, Paris 1949. — S. PAI: Fluid dynamics of jets. New York: Van Nostrand 1954. — BIRKHOFF and ZARANTONELLO [*1*], Chap. 8. — R. HOWARTH (ed.): Modern developments in fluid dynamics. High speed flow. Oxford 1953. — M. SCHIFFER: Encyclopedia of Physics, this volume.

[2] A treatment analogous to that based on the Schwarz-Christoffel transformation is discussed by S. BERGMAN: J. Rat. Mech. a. Analysis **4**, 883—905 (1955).

Various procedures for solving Eq. (12.4) have been developed[1], starting with the method of separation of variables used by CHAPLYGIN[2] in his basic memoir on gas jets. Briefly, this method proceeds as follows. Assuming polytropic flow, we introduce the notation

$$\tau = \frac{\gamma - 1}{2} q^2, \qquad \beta = \frac{1}{\gamma - 1},$$

in terms of which Eq. (12.4) becomes

$$\frac{\partial}{\partial\tau}\left[2\tau(1-\tau)^{-\beta}\frac{\partial\psi}{\partial\tau}\right] + \frac{1-(2\beta+1)\tau}{2\tau(1-\tau)^{\beta+1}}\frac{\partial^2\psi}{\partial\vartheta^2} = 0. \tag{12.6}$$

Separable solutions of this equation are

$$\psi = \tau^{\lambda_n} y_n(\tau)\sin(2\lambda_n\vartheta + \mu_n),$$

where λ_n, μ_n are constants, and $y_n(\tau)$ is a hypergeometric function

$$y_n(\tau) = F(a_n, b_n, c_n; \tau),$$

in which

$$c_n = 2\lambda_n + 1, \qquad a_n + b_n = 2\lambda_n - \beta, \qquad a_n b_n = -\beta\lambda_n(2\lambda_n + 1).$$

Thus, any expansion

$$A + B\vartheta + \sum_{n=1}^{\infty} B_n \tau^{\lambda_n} y_n(\tau)\sin(2\lambda_n\vartheta + \mu_n) \tag{12.7}$$

represents a formal solution of Eq. (12.6).

These expansions become useful by virtue of a correspondence principle, established by CHAPLYGIN, which relates solutions of analogous problems for potential flows of compressible and incompressible fluids. Namely, suppose the stream function ψ_i for an incompressible fluid has a convergent expansion in the hodograph plane,

$$\psi_i = A + B\vartheta + \sum_{n=1}^{\infty} B_n\left(\frac{q^2}{q_0^2}\right)^{\lambda_n}\sin(2\lambda_n\vartheta + \mu_n), \tag{12.8}$$

where $q<q_0$ throughout the flow, and q_0 is subsonic. Then the correspondence principle asserts that the expansion

$$\psi = A + B\vartheta + \sum_{n=1}^{\infty} B_n\left(\frac{\tau}{\tau_0}\right)^{\lambda_n}\frac{y_n(\tau)}{y_n(\tau_0)}\sin(2\lambda_n\vartheta + \mu_n) \tag{12.9}$$

describes a subsonic solution of Eq. (12.6). In particular, if Eq. (12.8) is the stream function of a free surface flow having $q=q_0$ on the free boundary and $\vartheta=\vartheta_0$ on a fixed boundary, then Eq. (12.9) describes a subsonic free surface flow having $q=q_0$ as free streamline and $\vartheta=\vartheta_0$ as fixed boundary. This follows readily from the fact that (1), ψ and ψ_i reduce to the same (and therefore constant) value for $q=q_0$ in Eq. (12.8) and for $\tau=\tau_0$ in Eq. (12.9); and (2), since ψ_i is constant on $\vartheta=\vartheta_0$ the coefficients B_n must all be zero in Eq. (12.8), and hence $\psi=A+B\vartheta_0=$ const on $\vartheta=\vartheta_0$. The physical plane of the subsonic flow is given by the mapping (12.5).

Applications of this principle to various types of jet flows appear in the literature. In general such applications involve considerable effort, both because the expansions (12.8) are not easy to obtain and because long calculations are

[1] See references, footnote 1, p. 345.

[2] S. CHAPLYGIN: Ann. Sci. Imp. Univ., Math.-Phys. Cl. **1904**, 1–121; translated in NACA Tech. Note 1063 (1944).

required in applying Eq. (12.9). This method has been made more practical in recent contributions[1].

A reduction of the flow problem to conformal mapping is made possible by an approximation initiated by CHAPLYGIN and later extended by others[2]. We observe that if the pressure-density relation satisfied by the gas is

$$p = a - \frac{b}{\varrho}, \tag{12.10}$$

then we have, after suitable choice of units, $q^2 = 1/\varrho^2 - 1$ and $\varrho^2 = 1 - M^2$. It follows that the substitution, $\lambda = \int (\sqrt{1-M^2}/q)\, dq$, or equivalently, $q = -$ cosech λ, takes the system (12.3) into the Cauchy-Riemann equations,

$$\varphi_\lambda = -\psi_\vartheta, \qquad \varphi_\vartheta = \psi_\lambda.$$

Thus, $f = \varphi + i\psi$ is a complex analytic function of the hodograph variable $\omega = \vartheta + i\lambda$. Hence, for a gas satisfying Eq. (12.10) a free surface flow with polygonal fixed boundaries is described in the ω plane by a polygon which is the conformal image of the plane of the complex potential f, just as in the case of incompressible fluids. However, the connection with the physical plane is no longer conformal, but is given by Eq. (12.5); specifically,

$$dx = -\operatorname{Im}(\sin\omega\, df), \qquad dy = \operatorname{Re}(\cos\omega\, df).$$

The equation for the potential in the physical plane becomes the equation for minimal surfaces:

$$(1+\varphi_y^2)\,\varphi_{xx} - 2\varphi_x\varphi_y\varphi_{xy} + (1+\varphi_x^2)\,\varphi_{yy} = 0.$$

The Chaplygin approximation consists in replacing the adiabatic of the actual gas by a straight line (12.10) tangent to it at some convenient point, corresponding for example to the state of the fluid at infinity or at stagnation. It is easy to see that much of the theory of the free boundary problem for incompressible fluids can be carried over to subsonic flows under this approximation[3].

For further discussion of free surface phenomena in compressible fluids we refer to the literature[4].

b) The method of reflection

13. Basic principles. Another procedure for explicit construction of free surface flows past polygonal boundaries, without use of the hodograph mapping, has been devised by SHIFFMAN[5]. The basic idea of this method is to continue the flow analytically across the free streamline in the physical plane, thereby obtaining a new flow region having essentially known polygonal boundaries. The free streamline itself is contained in the interior of the new enlarged flow and is therefore eliminated from direct consideration. This approach results in explicit formulas $z = z(t)$, $f = f(t)$, describing the flow, and is therefore formally related

[1] See references, footnote 1, p. 345.

[2] CHAPLYGIN: Footnote 2, p. 346. — S. TSIEN: J. Aeron. Sci. **6**, 399—407 (1939).

[3] Cf. BIRKHOFF and ZARANTONELLO [*1*], Chap. 8. — OUDART: Footnote 1, p. 345, Chap. 5. — L. C. WOODS: Quart. J. Math. Mech. **7**, 263—282 (1954). — Proc. Roy. Soc. Lond., Ser. A **227**, 367—386 (1955).

[4] C. JACOB [Acad. Roum. Bull. Sect. Sci. **28**, 637—641 (1946)], extends the Jantzen-Rayleigh method to subsonic free surface flows. Cavity flows at Mach number 1 have been considered by T. SAKURAI [J. Phys. Soc. Japan **11**, 710—715 (1956)]; J. HELLIWELL and A. MACKIE [J. Fluid Mech. **3**, 93—109 (1957)] and by others—for references see the preceding two papers. Supersonic jets are discussed in the references under footnote 1 on p. 345.

[5] M. SHIFFMAN: Comm. Pure Appl. Math. **1**, 89—99 (1948); **2**, 1—11 (1949).

to the hodograph method by a simple transformation. The point of view, however, is conceptually different.

Consider a flow region R adjacent to a free streamline Γ on which $|w|=1$, and let R_f and R_w be images of R in the f and w planes. The free streamline images Γ_f and Γ_w are arcs of $\psi=\text{const}=0$ and of $|w|=1$ respectively, and consequently the function $w(f)$ can be continued analytically by Schwarzian reflection across Γ_f. The relation $w=df/dz$ induces an analytic continuation of $z=z(f)$, which incidentally exhibits the analytic character of the free streamline in the physical plane. The explicit determination of the reflected domain defined by $z(f)$ is the essence of the present method.

Let us denote by z^*, f^*, and w^* the points corresponding to z, f, and w under continuation. We have from the Schwarz principle of reflection,

$$f^*=\bar{f}, \qquad w^*=1/\bar{w}, \tag{13.1}$$

where bars denote complex conjugates. The connection between z^* and z is exhibited in the fundamental differential relation,

$$dz^*=df^*/w^*=\bar{w}\,d\bar{f}=\bar{w}^2\,d\bar{z}=q^2\,dz, \tag{13.2}$$

in which the last equality follows from $w=q\,e^{-i\vartheta}$ and $dz=ds\,e^{i\vartheta}$. If the speed on the free streamline is V, then Eq. (13.2) is replaced by

$$dz^*=\frac{q^2}{V^2}\,dz. \tag{13.3}$$

Since a line $\psi=\text{const}$ maps into another such line under reflection in $\psi=0$, it is plain that continuation across Γ takes streamlines into streamlines, but in inverted order. Furthermore, Eq. (13.2) shows that the direction of the flow at corresponding points remains unchanged. This is summarized in the statement that *under analytic continuation across a free streamline the image of a streamline element is another streamline element in the same direction, the ordering of streamlines being inverted in the process.* In particular, straight streamlines map into parallel straight streamlines.

Further insight into the nature of the continued flow is derived by considering the images of distinguished points of hydrodynamical interest, such as sources, stagnation points, etc. If z_0 is any point (possibly infinity) at which the velocity has the expansion[1]

$$w=A(z-z_0)^m+\cdots, \qquad A\neq 0$$

then at the image points z_0^*, we infer from (13.1) and (13.2) the expansions (if $m\neq-\frac{1}{2}$)

$$z^*-z_0^*=\frac{\bar{A}^2}{2m+1}(\bar{z}-\bar{z}_0)^{2m+1}+\cdots, \tag{13.4}$$

$$w^*=B(z^*-z_0^*)^{-m/(2m+1)}+\cdots, \tag{13.5}$$

where B is a constant depending on $\bar{A}$ and m. The special case $m=-\frac{1}{2}$ gives rise to logarithmic terms.

From Eq. (13.5) we deduce the behavior of the continued flow in the following cases of interest. (i) If z_0 is a stagnation point ($m=1$), then its image is a three-sheeted branch point in the finite part of the plane. (ii) If z_0 is a finite simple source (vortex), corresponding to $m=-1$ and real (pure imaginary) A, the image is a simple source (vortex) of equal strength located at infinity, and conversely.

[1] The dots signify ascending powers of $z-z_0$ if z_0 is finite, descending if z_0 is the point at infinity. In the latter case $z-z_0$ is of course replaced by z in the expansions.

(iii) If the flow is parallel at infinity ($m=0$), the image is again a uniform flow in the same direction with speed magnified in the ratio $1/|A|^2$; the neighborhood of infinity is magnified in the ratio $|A|^2$; in particular, if $|A|=1$, i.e., if the limit speed is unity, then both magnifications also have this value. (iv) If z_0 is a corner of angle $\alpha\pi$ ($m=1/\alpha-1$), then the image is a finite point if $\alpha<2$ and the image flow is one about a corner of angle $(2-\alpha)\pi$; when the corner has angle 2π ($m=-\frac{1}{2}$) the image is a strip at infinity of width $2\pi|A|^2$ with zero velocity asymptotically. (v) The image of a free streamline γ on which $|w|=c$ is another free streamline, similar in shape to γ for all values of c, and a translation of it in the special case $c=1$.

14. Applications. As an application of these ideas we treat the problem of a jet issuing from a slot in a plane wall[1] (cf. Sect. 10). The flow in the upper half-plane before reflection is bounded by the axis of symmetry BC, the semiinfinite wall AB, and the unknown free streamline AC detaching from A and extending to the right to infinity (Fig. 26). According to the above theorem on the reflection of streamlines, BC maps into a parallel

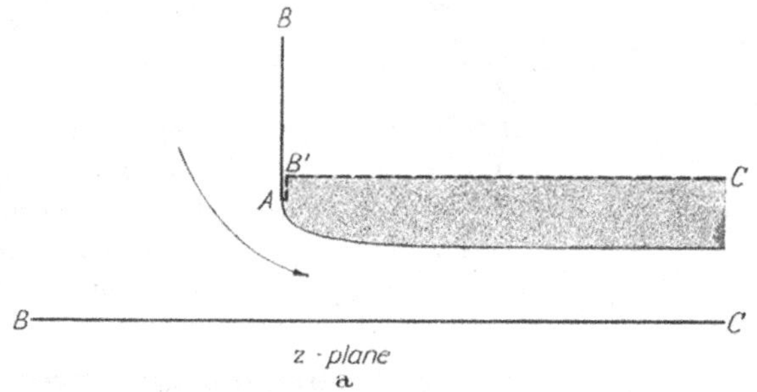

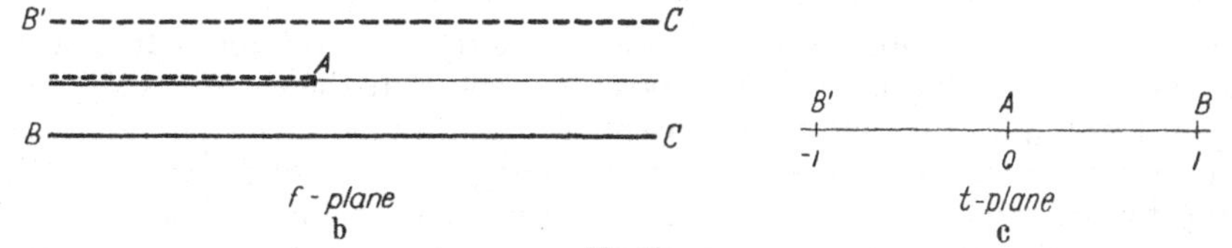

Fig. 26a–c.

line $B'C$ under reflection across AC, and AB into a segment AB' on AB; (it is clear that B' cannot lie below A on the extension of AB). The reflected flow region is the shaded portion of Fig. 26. By virtue of (iii) the vertical distance between BC and $B'C$ is twice the asymptotic width of the (half) jet (showing thereby that the contraction coefficient exceeds $\frac{1}{2}$).

The complete flow bounded by the polygon $BCB'AB$ has a source at B and an image source of equal strength at B' [by property (ii)]. The two impinging flows thus produced are separated by the free streamline AC. The mathematical description of the flow is obtained easily from the Schwarz-Christoffel transformation (5.4). Letting B', A, B, C correspond respectively to $-1, 0, 1, \infty$ in the t half-plane, we derive the mapping formulas

$$\frac{df}{dt}=\frac{Kt}{(t-1)(t+1)}, \tag{14.1}$$

$$\frac{dz}{dt}=\frac{Lt}{(t-1)^{\frac{3}{2}}(t+1)^{\frac{1}{2}}}. \tag{14.2}$$

The ratio of these expressions gives the same formula for $w=df/dz$ as obtained by the hodograph method in Sect. 10 after the connection between the two parameter planes is taken into account. Eqs. (14.1) and (14.2) furnish a complete description of the flow.

[1] Additional examples are treated by SHIFFMAN. See also MILNE-THOMSON [3].

The reflection formula (13.2) provides an elegant geometric interpretation of the force on an obstacle in cavity flow. The following remarks apply to all cavity models. Let the arc AOB be an obstacle in a flow with cavity speed V, free stream velocity U, and density ϱ. The resultant force, in complex form, is given by

$$F = -\frac{i}{2}\int_A^B (p - p_c)\,dz = -\frac{i}{2}\,\varrho V^2 \int_A^B \left(1 - \frac{q^2}{V^2}\right) dz.$$

Denote by A^* the image of A in the flow reflected across the free streamline BC (Fig. 27). Then by Eq. (13.3),

$$\left.\begin{aligned} F &= -\frac{i}{2}\,\varrho V^2 \int_A^B (dz - dz^*) \\ &= -\frac{i}{2}\,\varrho U^2 (1+\sigma)\,(z_{A*} - z_A). \end{aligned}\right\} \qquad (14.3)$$

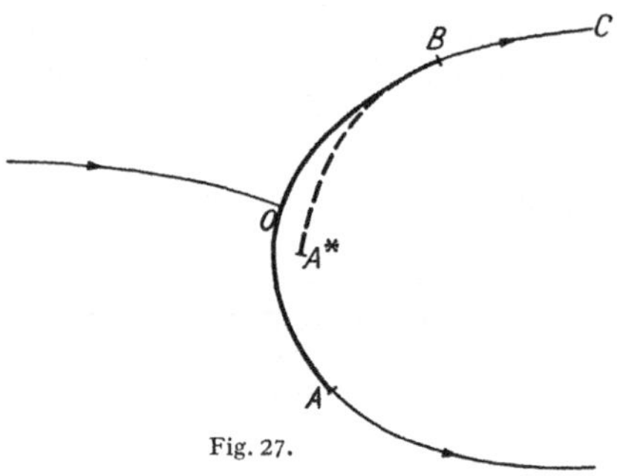

Fig. 27.

Thus the hydrodynamic force on AOB is equal to $1+\sigma$ times the force produced by the dynamic pressure $\frac{1}{2}\varrho U^2$ acting over the front face of the straight segment that joins A and its reflected image A^*.

c) Inverse and semi-inverse solutions.

When we turn to problems with axial symmetry or involving a free surface condition other than that of constant speed, we find that the theory is altogether lacking in techniques for determining explicitly flows past given boundaries. As in many other cases in classical physics, this suggests the use of *inverse methods* as a step toward understanding these more difficult problems. That is, we seek solutions of the flow equations satisfying the exact free surface condition without imposing other boundary conditions; or, we may satisfy certain fixed boundary conditions as well, but restrict the flow to have convenient mathematical properties (semi-inverse solutions). In this section we consider several techniques for the determination of inverse and semi-inverse solutions of free boundary problems[1]. While the physical significance of some of the flows obtained is slight, their importance resides in their being the first exact solutions in a difficult and unexplored area.

15. Free surface flows in a gravity field[2]. The inverse solutions of this and the following three sections are based on a technique whereby the real equations expressing the free surface conditions are continued analytically into the complex domain.

Let us consider steady plane free surface flows in the presence of gravity and choose a suitable origin of coordinates for which condition (3.9) on the free

[1] LEVI-CIVITA's well known inverse method for constructing plane jet and cavity flows satisfying the constant speed free surface condition is discussed in Sects. 23 and 33.

[2] The method presented here is based on the ideas of H. LEWY: Comm. Pure Appl. Math. **5**, 413–414 (1952); the treatment is that of M. VITOUSEK: Tech. Report No. 25 (1954), Appl. Math. Stat. Lab., Stanford Univ., and P. GARABEDIAN, E. MCLEOD, M. VITOUSEK: Amer. Math. Monthly **61**, 870–873 (1954). Apparently, inverse gravity-dependent solutions were first constructed, by C. SAUTREAUX, Ann. Sci. Ec. Norm. Sup. (3) **10** Suppl., 95–182 (1893) and J. de Math. (5) **7**, 125–159 (1901), for which see BIRKHOFF and ZARANTONELLO [*1*], Chap. 8. See also H. BLASIUS: Z. Math. Phys. **58**, 90–110 (1909); H. VILLAT: Ann. Sci. Ec. Norm. Sup. (3) **32**, 177–214 (1915); A. RICHARDSON: Phil. Mag. (6) **40**, 97–110 (1920); F. JOHN: Comm. Pure Appl. Math. **6**, 497–503 (1953) (cf. Sect. 18).

boundary becomes

$$\tfrac{1}{2}|f'(z)|^2 + g\,y = 0.$$

Introducing units in which $g = 1$ and recalling that f is real on the free surface S, we may write this equation for $y \neq 0$ in the form,

$$\frac{dz}{df}\cdot\frac{d\bar{z}}{df} + \frac{1}{2y} = 0 \quad \text{on} \quad S \tag{15.1}$$

or equivalently,

$$z'(z' - 2i\,y') + \frac{1}{2y} = 0 \quad \text{on} \quad S, \tag{15.2}$$

where the primes denote differentiation with respect to $f\,(y' = \partial y/\partial\varphi$ on $S)$. Suppose now that the free streamline $S = (x(\varphi), y(\varphi))$ is an analytic curve[1]. Then there is an analytic function $\lambda(f)$ coinciding with $y(\varphi)$ on S, and hence Eq. (15.2) can be continued as an analytic relation between $z(f)$ and $\lambda(f)$,

$$z'(z' - 2i\,\lambda') + \frac{1}{2\lambda} = 0. \tag{15.3}$$

This has the solution $z(f)$ given by

$$z(f) = i\,\lambda + i\int\sqrt{\lambda'^2 + \frac{1}{2\lambda}}\,df. \tag{15.4}$$

Since $\lambda(f) = y(f)$ on S, so that $z' - i\lambda'$ is real, it follows that

$$\lambda'^2 + \frac{1}{2\lambda} \leqq 0 \tag{15.5}$$

on the portion of the real f axis corresponding to S.

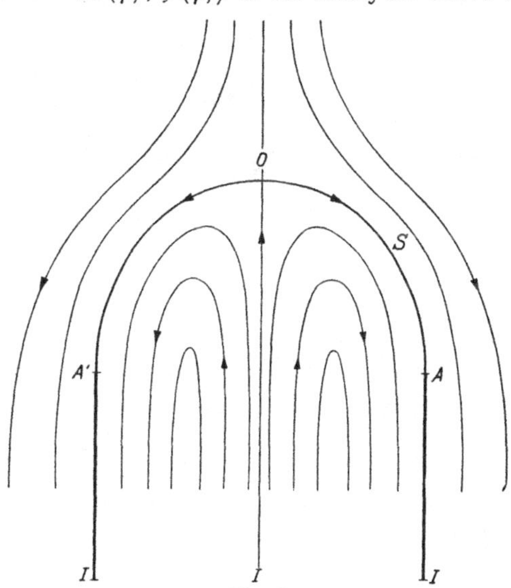

Fig. 28.

Conversely, *given any analytic function $\lambda(f)$ real on the real f axis and satisfying* Eq. (15.5), *then the flow $z(f)$ determined by* Eq. (15.4) *satisfies the free surface condition* (15.1) *and thus describes a free surface flow in the presence of gravity.* The free surface S corresponds only to that portion of the real f axis on which Eq. (15.5) is satisfied, while the image of the remainder is a streamline which can be considered the fixed boundary from which S detaches.

Several authors[2] have presented examples of these flows and have studied the properties of $\lambda(f)$ required to produce solutions of a given character; for instance, flows describing surface waves. As an illustration, let us set $\lambda = -f$, then Eq. (15.4) takes the form,

$$z = i\,f + i\int\sqrt{1 - \frac{1}{2f}}\,df.$$

As f varies from ∞ to $\frac{1}{2}$ the image in the flow plane (Fig. 28) is a vertical fixed boundary IA, at the upper end of which a cycloidal free boundary S detaches as f decreases from $\frac{1}{2}$ to 0; the free streamline meets another vertical streamline OI

[1] The analyticity of the free streamline has been proved independently by R. GERBER: C. R. Acad. Sci., Paris **233**, 1560—1562 (1951), and H. LEWY: Proc. Amer. Math. Soc. **3**, 11—113 (1952).

[2] Cf. references, footnote 2, p. 350, esp. VITOUSEK.

which is the image of the negative real axis. The resulting flow can be reflected in OI to give a single fountain-like flow bounded by the free surface AOA' and the fixed boundaries $A'I$, AI. By taking another branch of the square root we are able to get a flow which is uniform at infinity and bounded by the streamline $IAOA'I$. This can be considered the flow due to a semi-infinite bubble trapped within fixed plane walls rising at constant speed under gravity[1].

16. An example involving surface tension[2]. We consider now a symmetric plane flow past a constant pressure cavity or bubble, taking account of surface tension, but neglecting gravity. The novelty in this problem is that the free surface C of the fluid is no longer at constant pressure. Instead, the fluid pressure p at any point on C is connected to the bubble pressure p_0 through the relation

$$p - p_0 = T\varkappa \tag{16.1}$$

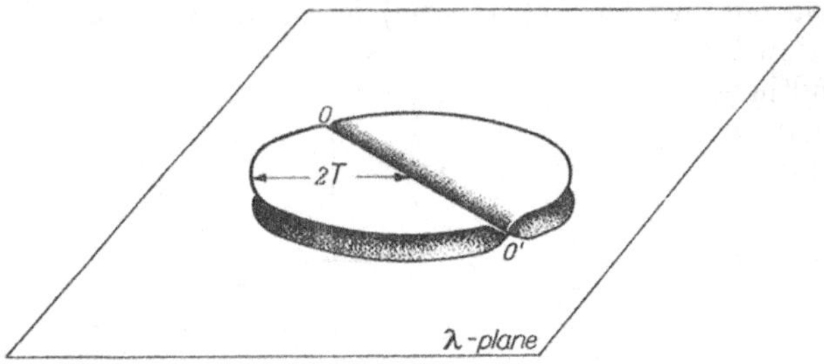

Fig. 29.

where T is the surface tension (assumed constant), and $\varkappa$ is the curvature of C at the point in question. From BERNOULLI's equation, assuming unit velocity at infinity, one obtains

$$\left.\begin{aligned} p_\infty - p_0 + \tfrac{1}{2}(1 - |f'(z)|^2) \\ = T\varkappa \quad \text{on } C. \end{aligned}\right\} \tag{16.2}$$

In the special case that $p_\infty - p_0 + \frac{1}{2} = 0$ the flow problem admits a complete solution, which we now determine.

From Eq. (16.2), which becomes

$$|f'(z)|^2 = -2T\varkappa, \tag{16.3}$$

we see that the bubble is convex. Letting s designate arc length along C, we have (on C)

$$f'^2 \mathrm{e}^{i\vartheta} = -2T\,\mathrm{e}^{-i\vartheta}\frac{d\vartheta}{ds},$$

so that, after integration with respect to s, we can write

$$\lambda(z) = \int^z f'(z)^2\,dz = -2iT\,\mathrm{e}^{-i\vartheta(z)}, \quad z \in C. \tag{16.4}$$

Hence the function $\lambda(z)$, defined and analytic in the flow region D, maps C on a circle of radius $2T$ which is described once in a counter-clockwise sense as C in turn is traversed clockwise. It follows by a standard counting argument that $\lambda(z)$ maps D on a double-sheeted Rieman surface covering the disc $|\lambda| < 2T$ exactly twice and all other points of the λ-plane once (Fig. 29). Let $t(z)$ map D conformally on the exterior of the unit circle so that infinity maps into infinity and the stagnation points 0, 0′ on C correspond to $t = \pm 1$. One checks that

$$\frac{\lambda}{2T} = t\,\frac{t^2 + A^2}{1 + A^2 t^2} \tag{16.5}$$

[1] The steady-state problem of semi-infinite bubbles rising under gravity has been studied by P. GARABEDIAN [Proc. Roy. Soc. Lond., Ser. A **241**, 423–431 (1957)], and by G. BIRKHOFF and D. CARTER [J. Math. Mech. **6**, 769–779 (1957)]. For other calculations of free surface flows in a gravity field, see [*1*], Chap. 8; R. SOUTHWELL and G. VAISEY: Phil. Trans. Roy. Soc. Lond., Ser. A **240**, 117–161 (1948); B. PERRY: Diss. Stanford Univ. 1957.

[2] E. MCLEOD: J. Rat. Mech. a. Analysis **4**, 557–567 (1955); see also P. GARABEDIAN, E. MCLEOD and M. VITOUSEK: Amer. Math. Monthly **61**, 8–10 (1954).

provides a mapping between the λ and t planes, and since f and λ have the same leading term in their expansion at infinity, we have

$$\frac{f}{2T} = t + \frac{1}{t}. \tag{16.6}$$

These formulas give

$$z(t) = \int \frac{f'(t)^2}{\lambda'(t)}\, dt = 2T \int \frac{(1 - 1/t^2)^2\, dt}{\frac{d}{dt}\left[t\left(\frac{t^2 + A^2}{1 + t^2 A^2}\right)\right]}.$$

Since $dz/dt \neq 0$, A is determined by the condition that the denominator vanishes at $t = \pm 1$. Thus $A^2 = -3$, and we get the solution,

$$z = 2T\left(t - \frac{2}{3t} - \frac{1}{27t^3}\right). \tag{16.7}$$

The bubble takes on an elongated shape, as illustrated in Fig. 30.

The same method can be used to treat the case of a bubble bounded in part by a fixed straight segment.

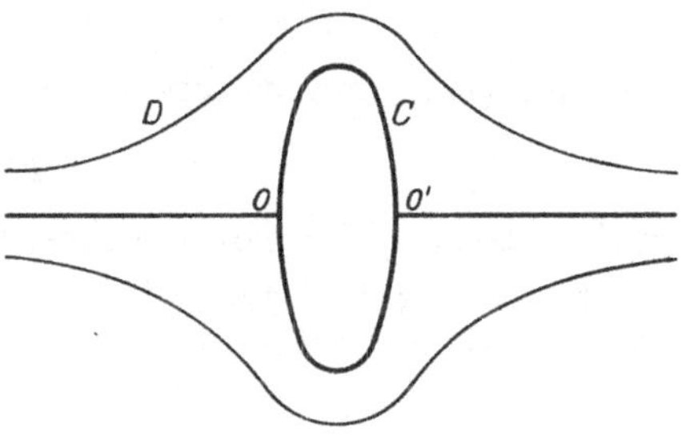

Fig. 30.

17. Axially symmetric flows. Despite the development of an extensive general theory of axially symmetric free surface flows, the only known explicit solutions are those provided by inverse methods. In this section we outline a procedure devised by GARABEDIAN[1] for constructing in closed form axially symmetric free surface flows with *prescribed free boundary*.

In the meridian z plane let

$$\Gamma\colon\ \bar{z} = g(z) \tag{17.1}$$

be the equation of a given analytic curve. This form of the equation can be derived from the usual representation in terms of x and y by the substitutions $x = (z + \bar{z})/2$, $y = (z - \bar{z})/2i$. The Cauchy-Kowalewski theorem insures the existence of a unique axially symmetric stream function ψ satisfying Eq. (3.2) and the boundary conditions

$$\psi = 0, \quad \frac{1}{y}\frac{\partial \psi}{\partial n} = 1 \quad \text{on } \Gamma, \tag{17.2}$$

but provides little specific information concerning the flow. This solution can be expressed in closed form by GARABEDIAN's formula[2],

$$\psi(z, \bar{z}) = \mathrm{Re}\left\{\frac{1}{2i}\int_{z_0}^{z} (z - g(t))^{\frac{1}{2}} (\bar{z} - t)^{\frac{1}{2}} F\left[\frac{(z - t)\,(\bar{z} - g(t))}{(\bar{z} - t)\,(z - g(t))}\right] g'(t)^{\frac{1}{2}}\, dt\right\}, \tag{17.3}$$

where z_0 is any point on Γ, and

$$F[\zeta] = F\left(-\frac{1}{2}, -\frac{1}{2}, 1; \zeta\right) = \sum_{m=0}^{\infty} \frac{[1 \cdot 3 \cdot 5 \ldots (2k - 3)]^2}{2^{2m}(m!)^2}\, \zeta^m$$

---

[1] P. GARABEDIAN: Studies in Mathematics and Mechanics presented to RICHARD VON MISES, pp. 149—159. New York: Academic Press 1954.

[2] Essentially this formula is contained in P. GARABEDIAN, H. LEWY and M. SCHIFFER: Ann. of Math. **56**, 560—602 (1952).

is the hypergeometric function (with parameters $-\frac{1}{2}$, $-\frac{1}{2}$, 1) satisfying the ordinary differential equation

$$\zeta(1-\zeta)\,F''(\zeta)+F'(\zeta)-F(\zeta)/4=0.$$

Before discussing the formula (17.3) in more detail, we remark that it is the analogue of the expression

$$f(z)=\int^{z} g'(t)^{\frac{1}{2}}\,dt \tag{17.4}$$

for the complex potential of the plane flow having Γ as its free boundary[1]. For, writing the free surface condition of the plane flow in the form

$$\frac{df}{dz}\cdot\frac{df}{d\bar z}=1 \quad \text{on} \quad \Gamma,$$

from Eq. (17.1) and analytic continuation we conclude that

$$f'(z)^2=g'(z)$$

in the interior of the flow, whence, Eq. (17.4) follows.

In deriving Eq. (17.3) it is convenient to introduce complex notation by means of the substitutions,

$$\frac{\partial}{\partial z}=\frac{1}{2}\left(\frac{\partial}{\partial x}-i\frac{\partial}{\partial y}\right),\qquad \frac{\partial}{\partial \bar z}=\frac{1}{2}\left(\frac{\partial}{\partial x}+i\frac{\partial}{\partial y}\right). \tag{17.5}$$

In this notation the Eq. (3.2) for the stream function takes the canonical hyperbolic form

$$L(\psi)=\psi_{z\bar z}+\frac{1}{2(z-\bar z)}\,\psi_z-\frac{1}{2(z-\bar z)}\,\psi_{\bar z}=0 \tag{17.6}$$

and suggests the adoption of hyperbolic methods in the complex domain. From this point of view we look on the boundary conditions (17.2) as Cauchy initial values on Γ for the solution of Eq. (17.6), and seek to determine the latter by RIEMANN's method of integration[2]. The Riemann function for Eq. (17.6) is given explicitly by

$$R(z,\bar z;t,\bar t)=\frac{(z-\bar t)^{\frac{1}{2}}(t-\bar z)^{\frac{1}{2}}}{t-\bar t}\,F\left[\frac{(z-t)\,(\bar z-\bar t)}{(z-\bar t)\,(\bar z-t)}\right],$$

where z and $\bar z$ are to be considered independent complex variables, as are t and $\bar t$. As is customary in the Riemann method, we may represent the solution ψ explicitly in terms of R and the boundary values (17.2), but now the arguments are to be points in the four dimensional space of the complex variables z and $\bar z$. The final result, after some manipulation, is the formula (17.3).

The solution (17.3) can of course be checked by direct computation. The equality $L(R)=0$ implies

$$L\left\{(\bar z-t)^{\frac{1}{2}}\,[z-g(t)]^{\frac{1}{2}}\,F\left[\frac{(z-t)\,((\bar z-g(t))}{(\bar z-t)\,(z-g(t))}\right]\right\}=0,$$

so that it suffices in evaluating $L(\psi)$ to consider only the contribution of the variable upper limit of integration in Eq. (17.3). These terms are

$$\operatorname{Im}\left\{\frac{\partial}{\partial z}\left[(\bar z-z)^{\frac{1}{2}}\,(z-g(z))^{\frac{1}{2}}\,g'(z)^{\frac{1}{2}}\right]+\frac{1}{2(z-\bar z)}\left[(\bar z-z)^{\frac{1}{2}}\,(z-g(z))^{\frac{1}{2}}\,g'(z)^{\frac{1}{2}}\right]\right\}=0,$$

[1] If the equation of the free streamline is given in intrinsic form, $s=f(\vartheta)$, then $f(\vartheta+i\log q)$ is the complex potential of the flow, expressed in terms of the hodograph variables.

[2] See, for example, R. COURANT and D. HILBERT: Methoden der Mathematischen Physik, Vol. 2, Chap. 5. Berlin: Springer 1938.

and hence it follows that Eq. (17.3) is a solution of $L(\psi)=0$. Since the integrand in Eq. (17.3) is real when the path of integration coincides with Γ, the boundary condition $\psi=0$ is seen to be satisfied. To check the constant speed condition, we use the relations

$$\frac{\partial}{\partial n}=i\left(\dot z\frac{\partial}{\partial z}-\dot{\bar z}\frac{\partial}{\partial \bar z}\right),\qquad g'(z)=\dot{\bar z}^2,$$

where the dot represents differentiation with respect to arc length along Γ. We obtain

$$\frac{\partial\psi}{\partial n}=\operatorname{Im}\left\{i\,\dot z\,(\bar z-z)^{\frac12}(z-\bar z)^{\frac12}g'(z)^{\frac12}+\frac12\int_{z_0}^{z}\frac{\partial}{\partial n}\left\{|\bar z-t|\,F\left[\frac{|z-t|^2}{|\bar z-t|^2}\right]\right\}|dt|\right\}$$
$$=\frac{|\bar z-z|}{2}=y.$$

This completes the direct verification that Eq. (17.3) describes an axially symmetric free surface flow.

Of course, the formula (17.3) provides no way of determining flows past prescribed barriers. Nevertheless, one can find conditions on $g(z)$ to insure

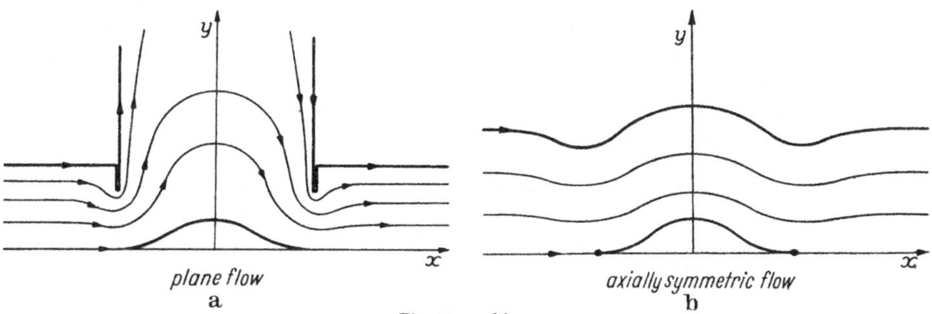

Fig. 31 a and b.

that the flow generated by Eq. (17.3) has certain desired properties. Thus, the condition that the axis of symmetry be a streamline ($\psi=0$) is

$$\operatorname{Im}\left[\int_{z_0}^{z}[x-g(t)]^{\frac12}(x-t)^{\frac12}g'(t)^{\frac12}\,dt\right]=0;$$

and the condition that a vertical line L which is tangent to the free boundary at z_0 be a streamline is that $\operatorname{Re}(g(z)-\bar z_0)=0$ on L.

The usefulness of Eq. (17.3) in generating flows in the large is limited primarily by the intrusion of singularities. These may occur at the singularities of $g(z)$, or on the axis of symmetry (when the factor $\bar z-t$ vanishes at the upper limit of integration), or at the zeros of the analytic function $z-g(z)$. It is natural to attempt to construct axially symmetric flows by means of Eq. (17.3), using the same free boundary as for a known plane flow. Thus, the axially symmetric flow obtained from the free streamline of the plane Riabouchinsky cavity between vertical plates (Sect. 8) proves to have the same fixed boundary and to be defined without singularities everywhere outside the cavity except on the axis of symmetry, where there is a distribution of sources and sinks. GARABEDIAN constructs an explicit example of a flow through a tube of varying cross-section past an isolated cavity having a cusp at both ends (on the axis of symmetry). This axially symmetric flow and the plane flow from which it is derived by the procedure of taking over the same free streamline are both illustrated in Fig. 31.

The analogue of Eq. (17.3) for generalized axially symmetric flows is

$$\psi(z, \bar{z}) = \operatorname{Re}\left\{\frac{1}{2i}\int_{z_0}^{z} [z - g(t)]^{\frac{\varepsilon}{2}} (\bar{z} - t)^{\frac{\varepsilon}{2}} F\left[\frac{(z-t)(\bar{z} - g(t))}{(\bar{z} - t)(z - g(t))}\right] g'(t)^{\frac{1}{2}} dt\right\}, \tag{17.7}$$

where $F[\zeta]$ is now the hypergeometric function $F\left(-\frac{\varepsilon}{2}, -\frac{\varepsilon}{2}; 1; \zeta\right)$. ψ is a solution of the equation

$$\psi_{xx} + \psi_{yy} - \frac{\varepsilon}{y}\psi_y = 0$$

for the axially symmetric stream function in a space of $\varepsilon + 2$ dimensions, and satisfies the free surface conditions $\psi = 0$, $\frac{1}{y^\varepsilon}\frac{\partial \psi}{\partial n} = 1$ on the curve decribed by Eq. (17.1).

18. Unsteady flows. Probably the most difficult of the free boundary problems to treat exactly are those involving unsteady flows[1]. It is therefore of interest that a general inverse method has been devised by F. JOHN[2] for constructing non-stationary plane free surface flows in a gravity field. Let $f(z, t)$ represent the time-dependent complex potential of a flow for which the curve $C_t: z = F(s, t)$ is a free boundary, s being a real Lagrangian coordinate of the surface particles. Thus the particle velocity and acceleration are given respectively by $F_t\,(=\bar{f}_z)$ and F_{tt}. The constant pressure condition on C_t states that the pressure gradient $F_{tt} + ig$, where g is the gravitational acceleration, must be normal to the surface. Since F_s is a vector tangential to C_t it follows that

$$F_{tt} + i g = i\, r(s, t)\, F_s \tag{18.1}$$

where $r(s, t)$ is a real function. The solutions of this parabolic equation, then, define the possible free surfaces. We observe further that on C_t,

$$f_s(s, t) = f_z(z, t)\, z_s(s, t) = \bar{F}_t(s, t)\, F_s(s, t), \tag{18.2}$$

$f(s, t)$ having been identified with $f[z(s, t), t]$. If now F and F_t are analytic in s, this relation can be extended into the complex s plane and written in the form

$$f_s(s, t) = \overline{F_t(\bar{s}, t)}\, F_s(s, t), \tag{18.3}$$

where both sides are analytic functions of s. Hence, if F and F_t are analytic in s, Eq. (18.3) determines f after quadrature as an analytic function of s and thus also of z. By reversing the process, we conclude that *any solution* $F(s, t)$ of Eq. (18.1), *analytic* (*with* F_t) *in* s, *determines by means of* Eq. (18.3) *a flow* $f(z, t)$ *having the curve* $z = F(s, t)$ *as free boundary.*

This provides a means of constructing free surface flows corresponding to convenient choices of $r(s, t)$. For example, if $r = R(t)$, where R is real, we obtain

---

[1] A perturbation procedure, starting with the exact boundary conditions, is given by N. CURLE: Proc. Roy. Soc. Lond., Ser. A **235**, 375–381, 382–395 (1956). Approximations based on linearization are due to L. C. WOODS: Proc. Roy. Soc. Lond., Ser. A **229**, 152–180 (1955); T. Y. WU: Cal. Inst. Tech., Hydrodynamics Lab. Rep. 85–6 (1957); B. PARKIN: same Lab., Rep. 85–2 (1957). Another type of approximation, in which the unsteady free boundary is assumed to be a slowly varying streamline, is found in D. GILBARG: Z. angew. Math. Phys. **3**, 34–42 (1952), and L. C. WOODS: Aeron. Res. Coun. Current Paper No. 149 (1954). Stability under small disturbances has been treated by C. ABLOW and W. HAYES: Tech. Rep. No. 1, 1951, Grad. Div. Appl. Math., Brown Univ.; J. L. FOX and G. W. MORGAN: Quart. Appl. Math. **11**, 439–456 (1954); N. CURLE: Proc. Roy. Soc. Lond., Ser. A **238**, 489–501 (1957). See also [*1*], Chap. 11.

[2] F. JOHN: Comm. Pure Appl. Math. **6**, 497–503 (1953).

by inspection solutions of the form,

$$F = -\frac{i}{2} g t^2 + \sum_\lambda h_\lambda(t) e^{i\lambda s}, \tag{18.4}$$

where h_λ is a solution of the ordinary differential equation $h_\lambda'' + \lambda R(t) h_\lambda = 0$.

When restricted to steady flows the preceding method leads to the same results as described in Sect. 15.

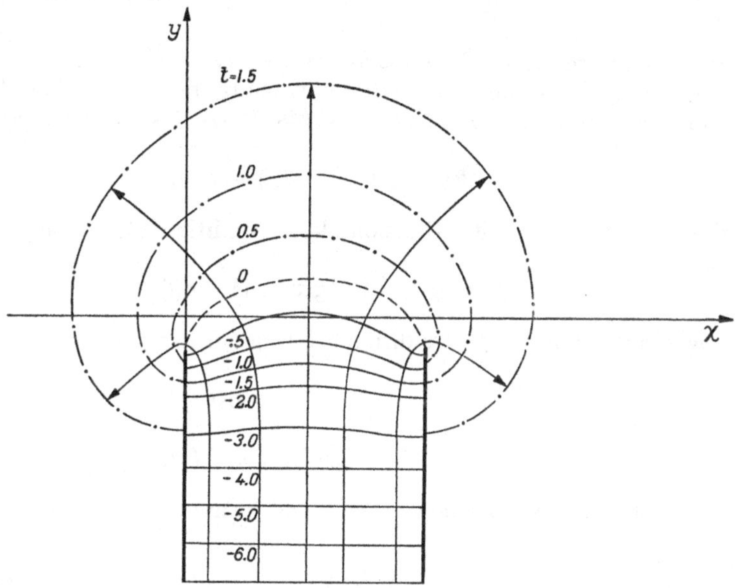

Fig. 32.

As an example, if we take $r \equiv 1$ in (18.1), we find the solution

$$z = F(s, t) = g \cdot (s + i t - i e^{t - is}),$$

which defines the complex potential,

$$f(s, t) = -i g^2 [(1 + e^{2t}) s - 2 e^t \sin s].$$

The free surface is a trochoid for negative t,

$$x = g(s - e^t \sin s), \qquad y = g(t - e^t \cos s),$$

but for $t > 0$ is self intersecting with branch points at $z = g(2n\pi - i)$, n = integer.

This flow is illustrated in Fig. 32. It shows a fluid rising between two vertical straight walls with uniform velocity, surmounted by a free surface which is essentially flat for large negative t. If the walls end exactly at two branch points $z = g(2n\pi - i)$ we find that the flow can be continued without self intersection for $t \geq 0$, describing an overflow of fluid into the region exterior to the two walls.

John has also constructed an example of an axially symmetric unsteady flow with free boundary[1]. Assuming a potential of the form

$$\varphi(x, y, z, t) = \frac{1}{A(t)} (r^2 - 2z^2), \qquad r^2 = x^2 + y^2,$$

[1] F. John: Rev. gén. Hydraul. 18, No. 71, 230–232 (1952).

he finds that the boundary conditions (3.7), (3.8) imply the relation,

$$A^6(2-A')(8+2A')^2=\text{const}$$

and that the free surfaces must be quadric. In case $-4<A'<2$, the free surface is an ellipsoid of revolution of constant volume which flattens out over the x, y plane as $t\to-\infty$ and becomes elongated about the z axis as $t\to+\infty$. In case $A'>2$ or $A'<-4$, the free surfaces may be hyperboloids of one or two sheets[1].

19. Accelerated flow of a constant shape cavity. In keeping with the inverse approach we seek to characterize those unsteady free surface flows having streamlines of constant shape, or in other words, flows described by potentials of the form

$$\Phi(x, y, z, t) = U(t)\,\varphi(x, y, z). \tag{19.1}$$

In the absence of gravity the boundary condition (3.7a) on the free surface becomes

$$U'(t)\,\varphi+\tfrac{1}{2}U^2\,|\text{grad}\,\varphi|^2=c(t), \tag{19.2}$$

from which it follows (e.g., by differentiating with respect to the space variables) that

$$U'/U^2=\text{const} \tag{19.3}$$

and

$$c(t)=\text{const}\;U^2. \tag{19.4}$$

Hence all such flows must be of the special form[2],

$$U(t)=\frac{U_0}{1-a_0U_0t} \tag{19.5}$$

where $U_0=U(0)$ and a_0 is the constant in (19.3). When the potential (19.1) describes a uniform flow at infinity, with values $U(t)$ and p_∞ for the velocity and pressure at infinity, and pressure p_0 on the free boundary, BERNOULLI's law gives

$$c(t)=\tfrac{1}{2}U^2+p_\infty-p_0=\tfrac{1}{2}U^2(1+\sigma), \tag{19.6}$$

σ being the cavitation number. Comparing this with Eq. (19.4), one sees that these flows have constant cavitation number, although in general $p_\infty-p_0$ will (unrealistically) be a function of time.

We now determine the class of symmetric plane flows past a flat plate trailing a finite cavity of constant shape[3] (Fig. 33). These flows have a complex potential $F(z, t)$ of the form

$$F(z,t)=U(t)\,f(z), \qquad f=\varphi+i\psi, \tag{19.7}$$

where $U(t)$ is the velocity at infinity. On the free streamlines we have from Eqs. (19.2) to (19.4)

$$a\varphi+|f'|^2=b, \qquad (a, b=\text{const}, \qquad f'=df/dz). \tag{19.8}$$

---

[1] Additional details and some generalizations are contained in the preceding reference.

[2] This result, for $c(t)=0$, has been derived by H. S. TAN: Quart. Appl. Math. **12**, 78–80 (1954).

[3] The first example of such a flow was exhibited by T. VON KÁRMÁN: Ann. Mat. pura appl. (4) **29**, 247–249 (1949). The entire set of flows was derived by D. GILBARG: Z. angew. Math. Phys. **3**, 34–42 (1952).

Limiting ourselves to the upper half flow plane, let us introduce the auxiliary variable ζ which, under the mapping

$$f = \alpha\left[1 - \frac{1}{2}\left(\zeta + \frac{1}{\zeta}\right)\right], \qquad \alpha = \text{const} > 0, \tag{19.9}$$

takes the half-plane $\psi \geqq 0$ into the semi-circle $|\zeta| \leqq 1$, $\operatorname{Im}\zeta \geqq 0$ with the correspondence indicated in Fig. 33. The image $\zeta = k$ of the stagnation point D will prove to be variable, the choice of k specifying a particular flow. On the real ζ axis the inclination ϑ of the velocity is zero for $-1 \leqq \zeta < k$, and equals $\pi/2$ for $k < \zeta \leqq 1$. Hence the function

$$g(\zeta) = \log\left[f'\left(\frac{1 - k\zeta}{k - \zeta}\right)^{\frac{1}{2}}\right] \tag{19.10}$$

is analytic in the ζ plane and real and continuous on the real axis. Therefore by reflection $g(\zeta)$ can be defined in the full circle $|\zeta| \leqq 1$. On $|\zeta| = 1$ we have from Eqs. (19.8) and (19.9)

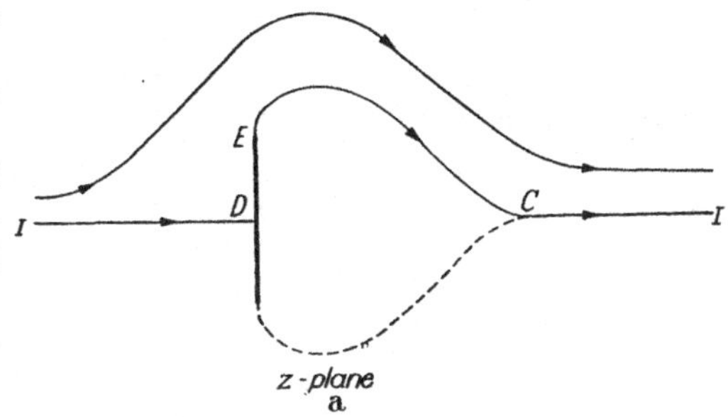

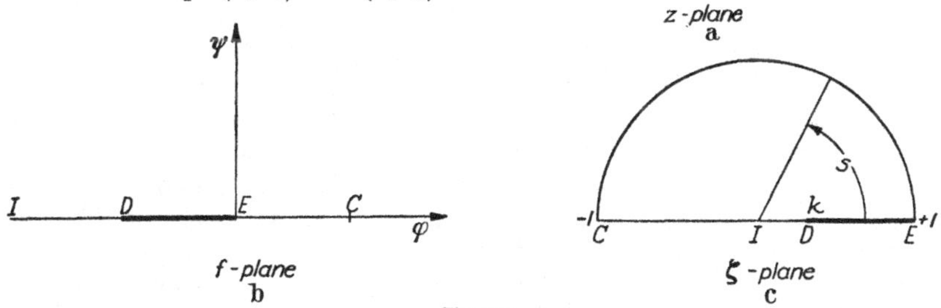

Fig. 33a–c.

$$\operatorname{Re} g(\zeta) = \log|f'| = \tfrac{1}{2}\log\left[A + B(\zeta + 1/\zeta)\right], \qquad (|\zeta| = 1), \tag{19.11}$$

where $A = b - a\alpha$ and $B = \frac{1}{2}\,a\alpha$ are constants. This determines $g(\zeta)$ in $|\zeta| \leqq 1$, and one finds either

$$\text{(i)} \qquad g(\zeta) = \frac{1}{2}\log\left\{\left|\frac{B}{\varrho}\right|(1 - \varrho\zeta)^2\right\} \tag{19.12}$$

or

$$\text{(ii)} \qquad g(\zeta) = \tfrac{1}{2}\log\{|B|\,(\zeta^2 - 2\zeta\cos\beta + 1)\}, \tag{19.13}$$

according as the quadratic equation $B(\zeta^2 + 1) + A\zeta = 0$ has a real root ϱ, $|\varrho| < 1$, or a root $e^{i\beta}$.

Considering the first case, we obtain from Eq. (19.10) and from the fact that $f' = 1$ at $\zeta = 0$, the equalities

$$f' = (1 - \varrho\zeta)\left(\frac{1 - \zeta/k}{1 - k\zeta}\right)^{\frac{1}{2}} \tag{19.14}$$

and

$$|B| = |\varrho|/k. \tag{19.15}$$

The condition that the flow plane be simply covered at infinity implies that the residue of $dz/d\zeta = 0$ at $\zeta = 0$; this yields the relation

$$2\varrho = k - \frac{1}{k}, \tag{19.16}$$

where $-1 \leqq \varrho < 0$, and $\sqrt{2}-1 \leqq k < 1$. The semi-width h of the plate fixes α in terms of k by means of the relation

$$\left.\begin{aligned} h i &= \int_k^1 \frac{dz}{df} \frac{df}{d\zeta} d\zeta = i\alpha \int_k^1 \frac{1-\zeta^2}{\zeta^2(1-\varrho\zeta)} \left(\frac{1-k\zeta}{\zeta/k-1}\right)^{\frac{1}{2}} d\zeta \\ &= i\alpha G(k), \quad \left(\varrho = \frac{1}{2}(k-1/k)\right) \end{aligned}\right\} \tag{19.17}$$

while by Eqs. (19.15), (19.16),

$$a = \frac{G(k)}{h}\left(\frac{1}{k^2}-1\right). \tag{19.18}$$

Hence in case (i) the constant shape cavities behind a flat plate of width $2h$ are described by a complex potential $F(z,t) = U(t) f(z)$, where

$$f = \frac{h}{G(k)}\left[1 - \frac{1}{2}(\zeta + 1/\zeta)\right],$$

$$\frac{df}{dz} = (1-\varrho\zeta)\left(\frac{1-\zeta/k}{1-k\zeta}\right)^{\frac{1}{2}},$$

and

$$U(t) = \frac{U_0}{1-\frac{1}{2} a U_0 t};$$

in these formulas, $\sqrt{2}-1 \leqq k < 1$, and ϱ, a are given by Eqs. (19.16), (19.18). The flow reduces to KARMAN's example[1] when $k = \sqrt{2}-1$. In this case the trailing edge of the cavity is blunt and is a stagnation point, but for all other values of k it is cusped and the velocity is non-zero. We remark that the case $\sigma = 0$ is not included among the flows, so that $p_\infty - p_0$ must be time-dependent.

The upper and lower streamlines of the cavities described by case (ii) intersect one another, the flows thus being doubly covered. Generalization to the case of polygonal obstacles, symmetric or asymmetric, involves no basic difficulty.

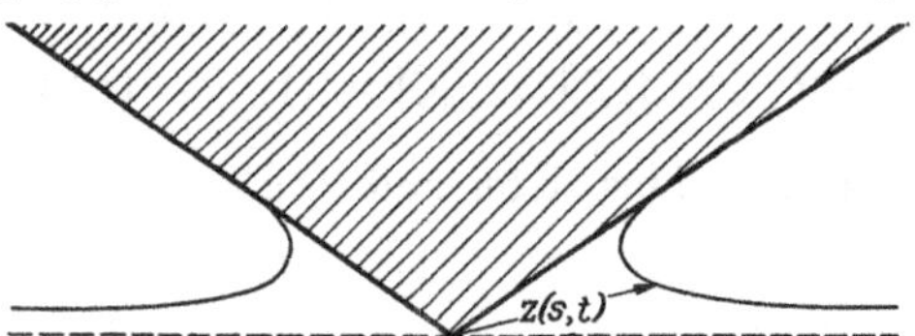

Fig. 34.

20. Surface impact of wedges and cones. The self-similar unsteady flows described by potentials of the form

$$\Phi(z,t) = t\,\varphi\left(\frac{z}{t}\right), \tag{20.1}$$

where z is a vector from the origin, enjoy the property that under similarity transformation ($z' = \lambda z$, $t' = \lambda t$) the velocity field is invariant and a free surface is taken into a free surface. These flows are therefore useful in describing the surface impact of wedges and cones[2] (Fig. 34).

Let s denote the Lagrangian coordinate of a free surface particle s units from the origin at $t=0$, and let $z(s,t)$ be the vector describing its position at time $t > 0$ (Fig. 34). The similarity assumption on the surface may be written in the form

$$\frac{z}{t} = g\left(\frac{s}{t}\right)$$

or equivalently,

$$\lambda z(s,t) = z(\lambda s, \lambda t),$$

[1] Cf. preceding footnote.

[2] The similarity hypothesis was introduced by H. WAGNER: Z. angew. Math. Mech. **12**, 193—215 (1932), to describe the vertical entry of wedges. It was later applied by M. SHIFFMAN and D. SPENCER: Comm. Pure Appl. Math. **4**, 379—417 (1951), to the vertical impact of cones, and by P. GARABEDIAN:: Comm. Pure Appl. Math. **6**, 157—167 (1953), to a type of oblique water entry of a wedge.

for any λ. From this follows

$$z = s z_s + t z_t. \tag{20.2}$$

At the same time, the condition

$$z_{tt} \cdot z_s = 0 \tag{20.3}$$

must be satisfied, this expressing the fact that the acceleration of a particle on a constant pressure surface is normal to the surface. Differentiating Eq. (20.2) with respect to s and t, and then applying Eq. (20.3), one obtains

$$\frac{\partial}{\partial s} |z_s|^2 = 0.$$

Hence,

$$|z_s| = \text{const} = 1, \tag{20.4}$$

the constant being equal to unity since the free surface is horizontal at infinity. Thus, the arc length from any free surface particle to the fixed boundary remains constant and is equal to the coordinate s.

Consider now the vertical entry of a cone into water[1], supposing that its velocity is unity. For definiteness we study the flow at $t=1$, the motion at all other times being given by the similarity law (20.1). One finds that the boundary conditions satisfied by the potential φ and stream function ψ in the meridian plane are

$$\varphi = \tfrac{1}{2}(s^2 - x^2 - y^2), \qquad d\psi = x(y\,dx - x\,dy) \tag{20.5}$$

on the free surface Γ, and

$$\psi = -\tfrac{1}{2} x^2 \tag{20.6}$$

on the wetted portion of the cone C; $y=0$ is the undisturbed surface. Introducing the ring source $V(x, y)$ and its stream function $S(x, y)$, and applying GREEN's theorem, one obtains in the manner of TREFFTZ (cf. Sect. 46) the following integral equation for φ on the boundary of the flow,

$$\varphi(p) = \frac{1}{\pi} \int_{C+\Gamma} [\varphi(q)\, dS(p, q) - V(p, q)\, d\psi(q)], \tag{20.7}$$

where φ, ψ satisfy the boundary condition (20.5), (20.6).

The problem set by Eqs. (20.5) to (20.7) must be handled numerically. The method adopted by SHIFFMAN and SPENCER[1] for the determination of Γ and of φ is based on an approximation process similar to that first employed by TREFFTZ in his calculation of the vena contracta (cf. Sect. 46). This consists in choosing a trial solution for the form of Γ, thereby determining φ on Γ and ψ on $C+\Gamma$ [by Eqs. (20.5), (20.6)]. Thus, Eq. (20.7) becomes a linear integral equation for φ on C, which can be solved by standard methods. Inserting these values into the right member of Eq. (20.7) and taking p on Γ, one obtains the error in satisfying the boundary condition (20.5) for φ. The successive approximations for Γ are chosen to reduce this error. A similar method is suggested by WAGNER[2] for the entry of wedges. SHIFFMAN and SPENCER present calculations, carried out by HILLMAN, on the entry of a 120° cone, and WAGNER reports results on a wedge of angle 144°. Further calculations on wedges have been carried out by PIERSON[3]. The details are evidently quite laborious.

[1] SHIFFMAN and SPENCER: Footnote 2, p. 360.
[2] H. WAGNER: Footnote 2, p. 360.
[3] J. PIERSON: Stevens Inst. Tech. Rep. No. 381 (1950).

A comparison of the calculated flows past the 120° wedge and cone appears in Fig. 35, which plots the dimensionless pressure and velocity as functions of position along the two bodies. It is especially interesting that the pressure minimum occurs at the vertex and that the maximum is situated almost at the extreme edge of the wetted surface. These facts had been observed by WAGNER.

To discuss impact forces it is useful to introduce the concept of *induced* or *virtual mass*[1] of the fluid. On vertical entry of a body with instantaneous velocity U let the vertical momentum possessed by the fluid be P; then the *momentum* virtual mass M is defined by the relation $P = MU$. Similarly, an *energy* virtual mass $\overline{M}$ can be defined by $E = \frac{1}{2}\overline{M}\,U^2$, where E is the kinetic energy of the fluid. In the case of constant speed entry one sees readily that $\overline{M} = 2M$. These quantities are expressed in terms of the boundary values of φ and ψ by the formulas (in axially symmetric flow)

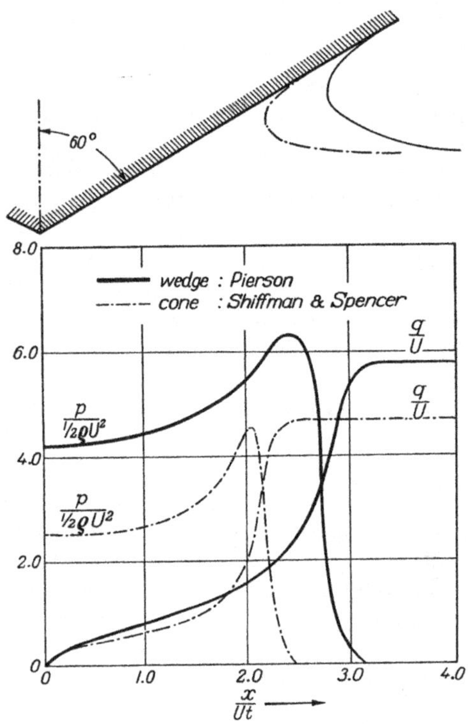

Fig. 35. Surface impact of wedge and cone. Adapted from R. N. Cox and J.W. MACCOLL: Proc. Symposium on Naval Hydrodynamics, 215-232, Nat. Res. Counc. Publ. 515, 1957.

$$\left.\begin{aligned} M &= 2\pi\,\varrho \int \varphi\, x\, dx, \\ \overline{M} &= 2\pi\,\varrho \int \varphi\, d\psi \end{aligned}\right\} \qquad (20.8)$$

where both integrals are taken over the flow boundary in the meridian plane.

Assuming constant velocity during entry, we have the expression

$$F = \frac{dM}{dY}\,U^2 = \frac{1}{2}\,\frac{d\overline{M}}{dY}\,U^2 \qquad (20.9)$$

for the force on the body, where Y is the depth of penetration below the undisturbed surface. Let a dimensionless coefficient k be defined in three-dimensional flow by the formula

$$M = k\,\varrho\, Y^3, \qquad (20.10)$$

and by $M = k\varrho Y^2$ in the plane case. In general k will depend only on the shape of the body and on Y. For a cone or wedge it is evident from similitude that k is a function only of the vertex angle. Inserting Eq. (20.10) into Eq. (20.9), one obtains $F = 3\,k\,\varrho\, U^4 t^2$ for the force on a cone at time t after the vertex first pierces the surface. An analogous formula, $F \propto U^3 t$, holds for wedges.

The coefficient k can be evaluated by means of Eqs. (20.8) and (20.10) when the shape of the free surface and the value of φ on the cone have been determined. For a cone of angle 120° SHIFFMAN and SPENCER calculate $k = 8.31$[2], which corresponds to a drag coefficient of 5.28 based on the cross-section at the undisturbed surface. They compare their results with the experiments of WATANABE[3],

---

[1] Cf. BIRKHOFF and ZARANTONELLO [1], Chap. 11, and SHIFFMAN and SPENCER, footnote 2, p. 360.

[2] WAGNER finds $F = 49.8\,\varrho\, U^3 t$ for a wedge of 144°.

[3] S. WATANABE: Inst. Phys. Chem. Res. Tokyo (Sci. Papers) **12**, 251–267 (1930); **14**, 153–168 (1930).

but since these were performed on cones of larger angle and show considerable variation, the comparison is inconclusive[1].

An exact solution describing oblique water entry of a wedge has been derived by GARABEDIAN[2] using a method of reflection similar to that in Sects. 15 to 17. However, this solution has the property that the constant surface pressures on the two sides of the wedge are unequal, thereby excluding the physically most important case[3].

All existing treatments of surface impact of curved bodies are based on approximate methods. One such method disregards the rise in the water surface and calculates the classical potential flow past the submerged portion of the body assuming the free surface is a plane equipotential[4] (Fig. 36). The impact force can be calculated from the virtual mass by Eq. (20.9). A refinement of the procedure to include a surface rise correction has been given by WAGNER and by SHIFFMAN and SPENCER[5]. A frequently used lower order of approximation replaces the immersed body by a simpler shape for which the classical potential flow is known, and bases the force calculations on the latter (still assuming the free surface is a plane equipotential). Thus, KÁRMÁN[6] takes the varying virtual mass of the flow to be that of the plane section of the body at the undisturbed surface[7], while others[8] replace the immersed body by an ellipsoid. These ideas have been extended to include oblique entry and other situations as well[9]. BIRKHOFF and ZARANTONELLO ([1], Chap. 11) assert that the experimental data are within $\pm 20\%$ of the theoretical values for the first quarter diameter of entry.

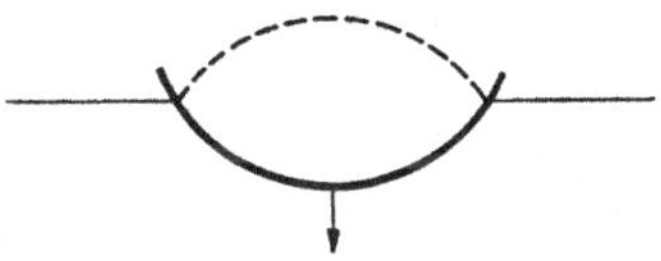

Fig. 36.

d) Approximate theories.

It is all too often the case that free surface problems arising in practice are beyond the scope of the exact theory. It is therefore natural, and sometimes a

[1] Futher measurements of surface impact are discussed in papers by S. WATANABE: Inst. Phys. Chem. Res. Tokyo (Sci. Papers) **21**, (1931); **23**, 118—135 (1933); **23**, 202—208 (1934); **23**, 249—255 (1934); R. KREPS: CAHI Rep. No. 438 (1939), translated in NACA Tech. Memo. 1046 (1943); A. WEIBLE: German Aviation Res. Rep. No. 4541, Naval Res. Lab. Transl. No. 286 (1952)—this paper reviews the experimental work of WATANABE and others, and compares their results with approximate theories of WAGNER (footnote 2, p. 360) and C. SCHMIEDEN: Ing.-Arch. **10**, 1—13 (1939). Additional references to experimental work are in BIRKHOFF and ZARANTONELLO [1], Chap. 11.

[2] P. GARABEDIAN: Footnote 2, p. 360.

[3] For other exact solutions related to water entry of wedges, see E. COOPER: Naval Ordnance Test Station, NAVORD Rep. No. 1154 (1949).

[4] See, for example, WAGNER: Footnote 2, p. 360, and M. SHIFFMAN and D. SPENCER: Quart. Appl. Math. **5**, 270—288 (1947).

[5] See also C. SCHMIEDEN: Footnote 1 and Z. angew. Math. Mech. **33**, 147—151 (1953).

[6] T. VON KARMAN: NACA Tech. Note 321 (1929).

[7] Accordingly, in the two dimensional case $\overline{M} = \frac{1}{2}\pi\varrho a^2$, and in axial symmetry $\overline{M} = \frac{4}{3}\varrho a^3$, where a is the radius of the surface section—cf. H. LAMB: Hydrodynamics, 6th ed., pp. 85, 139. New York: Macmillan 1932.

[8] E.g., SHIFFMAN and SPENCER: Footnote 2, p. 360 and L. TRILLING: J. Appl. Phys. **21**, 161—170 (1950); the latter treats oblique entry.

[9] L. SEDOV: CAHI Rep. No. 187 (1934). — M. LAVRENTIEFF, M. KELDYSH, A. MARKUSHEVITCH, L. SEDOFF and A. LOTOFF: Collected reports on the problem of impact against a water surface. CAHI Rep. No. 152 (1935). — L. TRILLING: preceding footnote. Several approximation schemes are reviewed by R. N. COX and J. W. MACCOLL: Proc. Symposium on Naval Hydrodynamics, 215—232, Nat. Res. Counc. Publ. 515, 1957; see also the references in the discussion of this article.

matter of practical necessity, to seek approximate solutions, giving up in the process some of the conditions of the correctly posed problem in exchange for ease of calculation and quick answers. The engineering literature contains a variety of such approximations[1], frequently not well founded in theory, but often useful because of good agreement with exact results and experimental observation. The methods discussed in the following sections are applicable in two important situations where solutions according to the exact theory are either unavailable or involve laborious calculations.

21. Linearized theory of plane cavity flow. It is reasonable to expect that a linearized theory will give accurate results for flows past slender bodies at small cavitation numbers. TULIN[2] has developed such a theory possessing the important practical feature that the flows under consideration can be completely described by quadratures.

Let a slender symmetric obstacle[3] be described in the upper halfplane by a function $y = Y(x)$ defined in $-a \leqq x \leqq 0$. It is assumed that a finite cavity with cavitation number $\sigma > 0$ and cavity speed unity forms behind the body (Fig. 37), and that its length measured from $x = 0$ is equal to b. (The fact that the cavity closes at the rear, as in Fig. 37, is consistent with the linearized theory, but in the exact theory such a model clearly cannot exist.)

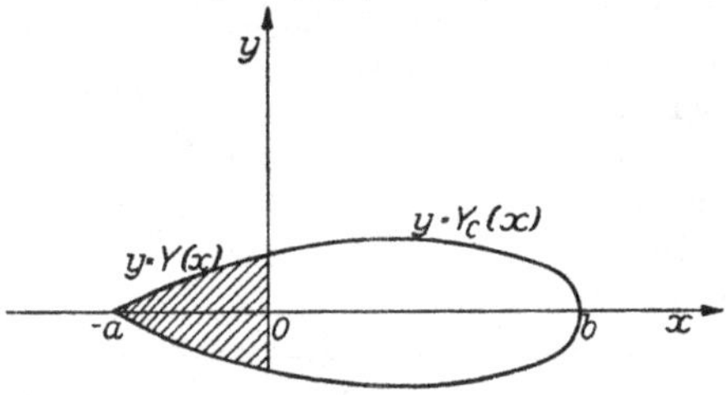

Fig. 37.

Let U denote the free stream velocity, (u, v) the perturbation velocity with respect to U, and φ the perturbation potential, so that grad $\varphi = (u, v)$. The basic assumptions of the linearized theory is that all quantities quadratic or of higher degree in u, v, etc., may be neglected in comparison with first degree terms. To this order of approximation, if the boundary conditions are carried over to $y = 0$, one obtains

$$\left.\begin{aligned} v(x, 0) &= Y'(x), & -a < x < 0, \\ u(x, 0) &= \tfrac{1}{2}\sigma U, & 0 < x < b. \end{aligned}\right\} \tag{21.1}$$

Thus, the potential φ is a solution of the boundary value problem:

$$\Delta\varphi = 0,$$

$$\varphi_y(x, 0) = \left\{\begin{matrix} 0, & -\infty < x < -a, & b < x < \infty \\ Y'(x), & -a < x < 0, & \end{matrix}\right\} \tag{21.2}$$

$$\left.\begin{aligned} &\varphi_x(x, 0) = \tfrac{1}{2}\sigma U, \quad 0 < x < b \\ &\varphi_x,\ \varphi_y \to 0 \quad \text{as} \quad (x, y) \to \infty. \end{aligned}\right\} \tag{21.3}$$

---

[1] For extensive references and discussion of various approximation methods, see BIRKHOFF and ZARANTONELLO [*1*].

[2] M. TULIN, (1), David Taylor Model Basin Rep. 834 (1953). Applications and extensions of this theory are contained in: M. TULIN (2), Proc. Symposium on Cavitation, National Phys. Lab., 1956. — A. ACOSTA: Hydrodynamics Lab., Cal. Inst. Tech., Rep. No. E-19 (1955). — H. COHFN and R. GILBERT: J. Appl. Mech. **24**, 170—176 (1957). — T.Y. WU: Hydrodynamics Lab., Cal. Inst. Tech. Rep. No. 21—22 (1956) and 85—5 (1957). — J. GEURST and R. TIMMAN: Proc. 9th Intern. Cong. Appl. Mech., Brussels 1956. — H. COHEN and Y. TU: Same Proceedings.

[3] The theory can be easily extended to asymmetric bodies; cf. TULIN (2) and WU, preceding footnote.

To these conditions must be added the requirement that the cavity be closed at $x = b$. This can be expressed in the form

$$\int_{-a}^{b} \varphi_y(x, 0)\, dx = 0, \tag{21.4}$$

since on the boundary $dy/dx = v(x, 0) = \varphi_y(x, 0)$.

We seek a solution of the form

$$\varphi(x, y) = \frac{1}{2\pi} \int_{-a}^{b} m(\xi) \log r \, d\xi, \qquad r = \sqrt{(x-\xi)^2 + y^2} \tag{21.5}$$

where $m(x)$ represents a suitable source distribution on the x-axis. It follows that

$$v(x, 0) = \varphi_y(x, 0) = m(x)/2, \qquad -a < x < b. \tag{21.6}$$

Use of this fact and Eqs. (21.2), (21.3) leads to the integral equation

$$\int_{0}^{b} \frac{m(\xi)}{x-\xi}\, d\xi = \pi \sigma U - 2 \int_{-a}^{0} \frac{Y'(\xi)}{x-\xi}\, d\xi, \tag{21.7}$$

which is well-known from thin airfoil theory. The required solution must be such that $v(x, 0)$, and therefore $m(x)$ itself, is continuous at the separation point $x = 0$. This solution is

$$m(x) = \left(\frac{x}{b-x}\right)^{\frac{1}{2}} \left[-\sigma U + \frac{2}{\pi} \int_{-a}^{0} Y'(\xi) \frac{\sqrt{b-\xi}}{(x-\xi)\sqrt{-\xi}}\, d\xi\right]. \tag{21.8}$$

From this formula all quantities of interest can be easily derived in closed form.

The cavity shape $Y_c(x)$ is expressed in terms of $m(x)$ by

$$Y_c(x) = Y(0) + \int_0^x Y_c'(\xi)\, d\xi = Y(0) + \tfrac{1}{2} \int_0^x m(\xi)\, d\xi. \tag{21.9}$$

The requirement (21.4) that the cavity be closed at $x = b$ together with the approximation $1/U = 1 + \frac{1}{2}\sigma$ yields the following relation between cavitation number and cavity length,

$$\frac{\sigma}{1 + \frac{1}{2}\sigma} = \frac{4}{\pi b} \int_{-a}^{0} Y'(\xi)\, d\xi. \tag{21.10}$$

Using this formula and (21.8), (21.9), we obtain for the shape of the cavity (in $0 < x < b$),

$$Y_c(x) = \frac{\sigma b}{2+\sigma} \sqrt{\frac{x}{b}\left(1 - \frac{x}{b}\right)} + \frac{2}{\pi} \int_{-a}^{0} Y'(\xi) \arctan \sqrt{\frac{\xi(b-x)}{x(\xi-b)}}\, d\xi. \tag{21.11}$$

For small σ this represents a body approximately elliptical in shape, of length b and length-width ratio $(2+\sigma)/\sigma$. The drag is expressed by the formula

$$\frac{D}{\frac{1}{2}\varrho U^2} = 2\sigma \sqrt{1+\sigma} \int_{-a}^{0} Y'(\xi) \sqrt{\frac{\xi}{\xi - b}}\, d\xi + \frac{2b}{\pi}(1+\sigma) \left[\int_{-a}^{0} \frac{Y'(\xi)\, d\xi}{\sqrt{\xi(\xi-b)}}\right]^2. \tag{21.12}$$

In the case of infinite cavity flow ($\sigma=0$, $b=\infty$) the asymptotic shape is given by

$$\lim_{x\to\infty}\frac{Y_c(x)}{\sqrt{x}}=\frac{2}{\pi}\int_{-a}^{0}Y'(\xi)\frac{d\xi}{\sqrt{-\xi}}. \tag{21.13}$$

This is connected with the drag of the obstacle by the relation

$$\frac{D}{\frac{1}{2}\varrho U^2}=\frac{\pi}{2}\left[\lim_{x\to\infty}\frac{Y_c(x)}{\sqrt{x}}\right]^2=\frac{2}{\pi}\left[\int_{-a}^{0}Y'(\xi)\frac{d\xi}{\sqrt{-\xi}}\right]^2. \tag{21.14}$$

The fact that the first equality holds also in the exact theory (cf. Sect. 25) is evidence in support of the linearized approximation.

Comparison of the drag coefficient computed in the linearized and exact theories for a 30° wedge in infinite cavity flow shows agreement within 15%, with rapid improvement for smaller angles. A comparison with results for the RIABOUCHINSKY cavity behind the same wedge shows agreement within 7% and 4% for the cavity length and cavity diameter when $\sigma=0.2$.

The linearized theory has played an essential part in the design of supercavitating propellers of high efficiency. This and other important applications of the theory are discussed in the Proceedings of the Second Symposium on Naval Hydrodynamics, Superintendent of Documents, Washington, D. C.

22. Approximate axially symmetric cavity flow[1]. The problems of axially symmetric cavitational flow are of great practical importance, and therefore, in the absence of exact solutions, considerable effort has been expended in seeking useful approximations.

Several attempts[2] have been made to describe such flows in terms of axial source distributions, but a common failing of these solutions, since they are analytic off the axis, is that they cannot provide an accurate description near detachment where the flow behavior may be critical. However, other types of distributed singularities show more promise of success. Thus, the use of axially distributed sources in conjunction with solutions that reproduce the singularity of the flow at detachment is part of an effective numerical procedure discussed in Sect. 45. Another approach is to consider the flow as generated by a distribution of ring vortices (vortex sheet) situated on the flow boundary, of strength equal to the flow speed[3]. By imposing the boundary conditions, one can reduce the flow problem directly to an integral equation involving both the free surface and the unknown velocity distribution on the given body, which must then be solved numerically[4].

The quite natural idea of applying exact results from the two dimensional theory to the problems of axially symmetric flow has motivated several related

[1] A survey of the material in this section can be found in R.N. COX and J.W. MACCOLL: Proc. Symposium on Naval Hydrodynamics, 215—232, Nat. Res. Counc. Publ. 515, 1957.

[2] E.g., W. BAUER: Ann. d. Phys. **82**, 1014—1016 (1927). — H. MÜNZNER and H. REICHARDT: Armament Res. Est. Transl. No. 1/50 (1950)—German original UM 6616 (1944). — A. ARMSTRONG and K. TADMAN: Armament Res. Est. Rep. No. 7/52 (1953).

[3] A. ARMSTRONG and J. DUNHAM: Armament Res. Est. Rep. No. 12/53 (1953); this is an adaptation of a method of calculating flows past bodies of revolution developed by F. VANDREY: Aeronaut. Res. Counc. FM 1665 (1951), and L. LANDWEBER: David Taylor Model Basin Rep. No. 761 (1951).

[4] This approach is quite similar to the one used by TREFFTZ for the calculation of the vena contracta (cf. Sect. 46).

approximation methods, which turn out to provide good agreement with observation. These methods are based on a correspondence principle which assigns the same velocity distribution to an obstacle in axially symmetric flow as in the corresponding plane flow. A first crude justification for this procedure is that the velocity distributions in the two cases agree at the center and at the point of detachment, and presumably do not differ too much in between.

Using this device, PLESSET and SHAFFER[1] have calculated the drag coefficient of cones in the Riabouchinsky model on the assumption that *for a given cavitation number* the flow speeds on the cone and wedge of the same angle are equal at equal distances from the vertex. In paricular, the drag coefficient of a circular disk in a flow of cavitation number σ is given by the formula

$$\frac{C_D}{1+\sigma} = -2\int (1+w^2)\, z\, dz = \frac{2}{(1+B)^2}\left(B^2 + \frac{1}{k^2} - \frac{1}{2}\frac{k'^2}{k^3}\log\frac{1+k}{1-k}\right), \tag{22.1}$$

where the integral is taken from the center to the edge of the disk, w is the function of z describing the plane flow in Sect. 8, k and k' are the parameters defined by Eqs. (8.3), (8.4), and $k^2 B = E - k'^2 K$. This approximation yields the value 0.805 for the drag coefficient of the disk in infinite cavity flow ($\sigma = 0$). Comparison with the observations recorded in Fig. 1 shows that the computed curve of $C_D(\sigma)$, which is essentially linear for $0 \leqq \sigma \leqq 0.5$, passes through the experimental points, but has a lower slope than the indicated experimental curve. The agreement with observation is best for the disk and stagnation cup, and poorest for cones of small angle at the lower values of σ.

In attempting to account for the latter discrepancy, ARMSTRONG[2] points out that the velocity gradients at the vertex are unequal for flows past wedges and cones of the same angle, the difference being greatest for the smaller angles. By expanding the flows in the neighborhood of the vertex, he finds that the leading terms can be made identical if a certain explicit relation holds between the cavitation numbers and the vertex angles in the two cases. He suggests therefore that the correspondence associating equal velocity distributions in plane and axially symmetric flows be set up between the appropriate cones and wedges at cavitation numbers satisfying this relation. The results of drag calculations on this basis show greatly improved agreement with observation at smaller angles for $\sigma \leqq 0.1$, which is the limit of the reported calculations.

FISHER[3] has suggested a different refinement of the method of correspondence. He observes that in the limit case of fully wetted non-cavitating flow ($\sigma \to \infty$), the velocity distributions on the plate and circular disk are identical when the incident velocity of the axially symmetric flow is $\pi/2$ times that of the plane flow[4]. He therefore suggests as an approximation that the velocity distribution of the plane flow at cavitation number σ_2 be transposed to the axially symmetric flow at cavitation number σ_3 when the incident velocity of the latter is $\pi/2$ times that of the plane flow. In other words, the velocity distributions in the two cases

[1] M. PLESSET and P. SHAFFER: Rev. Mod. Phys. **20**, 228—231 (1948), and J. Appl. Phys. **19**, 934—939 (1948). This correspondence was previously suggested by H. REICHARDT: Min. Aircraft Prod. Rep. and Transl. No. 766 (1946), and J. FISHER: Underwater Ballistics Res. Comm. Rep. No. 34 (1945).

[2] A. ARMSTRONG: Armament Res. Est. Rep. No. 21/54 (1954).

[3] FISHER: Footnote 1, see also ARMSTRONG and DUNHAM: Footnote 3 on p. 366.

[4] Such an equality holds also between the velocity distributions on ellipses and their ellipsoids of revolution in fully wetted flow when the ratio of the respective free stream velocities has a certain value depending on the eccentricity. This observation has been applied by A. ARMSTRONG and K. TADMAN, Armament Res. Est. Memo. 5/54 (1954), to cavity flows past ellipsoidal head forms.

are to be identified when the cavitation numbers are related by

$$1+\sigma_2=\frac{\pi^2}{4}(1+\sigma_3). \tag{22.2}$$

Hence, in this approximation scheme the values of C_D computed by PLESSET and SHAFFER at $\sigma=\sigma_2$ are to be assigned to $\sigma=\sigma_3$ determined by Eq. (22.2). This yields the figure $C_D(0)=0.825$, which proves to be in very good agreement with GARABEDIAN'S more systematic calculation of this quantity (cf. Sect. 44). However, the agreement is poorer for $\sigma>0$.

ARMSTRONG and DUNHAM[1] use still another, somewhat more involved, procedure in determining the connection between σ_2 and σ_3, and arrive at results comparable with those of FISHER. They propose not only that the velocity distribution over the obstacle be the same in both plane and axially symmetric flows, but also that the free streamline in the latter case be an appropriate affine transformation of the plane curve. This is in keeping with the fact that the cavity dimensions differ markedly between plane and axially symmetric flows for the same cavitation number. The known connection between the shape of the free streamline at detachment and the velocity distribution at the edge of the obstacle is used as a consistency condition in determining the choice of the affine transformation and of the cavitation number to appear in the transposed plane velocity distribution. Comparison with measured cavity dimensions shows good agreement up to $\sigma=0.4$.

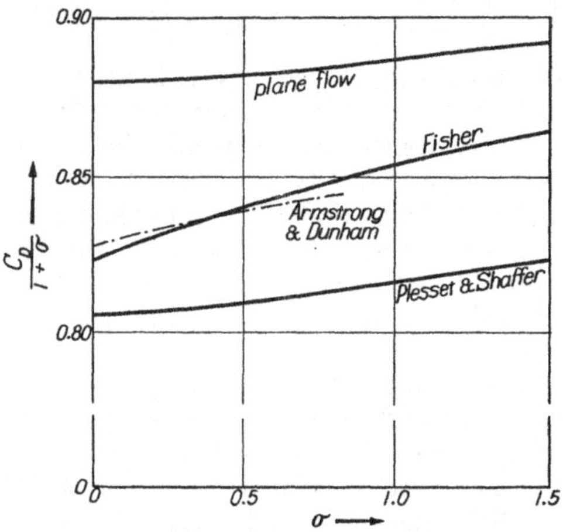

Fig. 38. Theoretical determinations of C_D for the disk. Adapted from A. ARMSTRONG and J. DUNHAM: Armament Res. Est. Rep. 12/53 (1953).

The comparative results of PLESSET and SHAFFER, FISHER, and ARMSTRONG and DUNHAM in estimating $C_D(\sigma)$ for the circular disk in the Riabouchinsky model are shown in Fig. 38. We emphasize that the method of computation of C_D is exactly the same for all three, being based on Eq. (22.1); the difference is only in the value of σ that is to correspond to k in Eq. (22.1)

Finally, we call attention once more to the important asymptotic formulas (8.14), (8.15) for the dimensions of the axially symmetric Riabouchinsky cavity. Their derivation, which is partly heuristic and not yet fully general, is based on comparison of the actual cavity flows with those past ellipsoids of revolution.

III. Qualitative theory.

This chapter is concerned with the general properties of plane and axially symmetric free surface flows that obey the condition of constant speed on the free boundary. For this class of flows the theory boasts a variety of interesting qualitative results which add greatly to understanding of the phenomena and are often useful in application. A similar body of results has not yet been developed for other classes of free boundary problems.

[1] ARMSTRONG and DUNHAM: Footnote 3 on p. 366.

23. LEVI-CIVITA's representation[1]. LEVI-CIVITA's treatment of the free boundary problem may be considered the beginning of the general theory for curved obstacles. This contribution is based on a convenient parametrization of the flow which removes the unknown free boundary from discussion and allows the general solution to be described in terms of an arbitrary analytic function in the unit circle.

We illustrate the Levi-Civita representation in the infinite cavity flow past a fixed curve AOB. Let the origin be located at the bifurcation point O on the dividing streamline and the separation points of the free streamlines at A and B

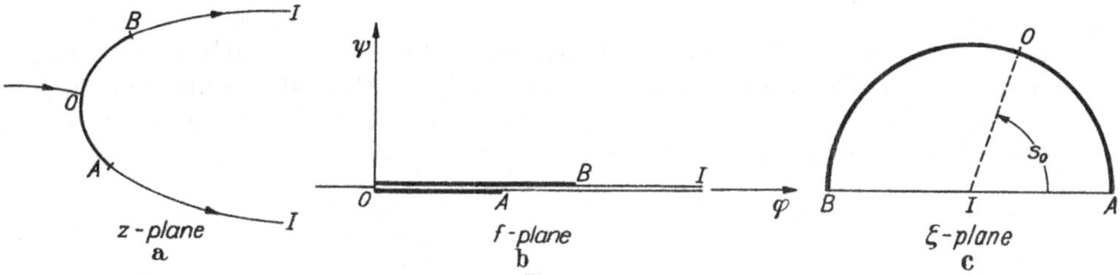

Fig. 39 a—c.

(Fig. 39). The complex potential $f(z)$ takes the flow region into the full f plane slit along the positive real axis. Introducing the variable $\zeta = \xi + i\eta$ and the semicircle

$$\Gamma: |\zeta| < 1, \quad \eta > 0, \tag{23.1}$$

one sees that the f plane is mapped conformally onto Γ by

$$f = M\left[\cos s_0 - \frac{1}{2}\left(\zeta + \frac{1}{\zeta}\right)\right]^2 \tag{23.2}$$

where

$$\cos s_0 = \frac{\sqrt{f_B} - \sqrt{f_A}}{\sqrt{f_B} + \sqrt{f_A}}, \qquad M = \tfrac{1}{4}\left(\sqrt{f_A} + \sqrt{f_B}\right)^2$$

and where $\zeta_A = 1$, $\zeta_B = -1$, $\zeta_I = 0$, and $\zeta_0 = e^{i s_0}$, under the boundary correspondence. The function $\omega(\zeta)$,

$$\omega = i \log f'(z) = \vartheta + i \log q = \vartheta + i\tau, \tag{23.3}$$

plays an important part in the sequel. It has the property of being real on the real axis since $q = 1$ on the free boundary[2], and vanishes at the origin because of the uniformity of the flow at infinity[3].

The various features of the flow are completely determined by the functions $f(\zeta)$ and $\omega(\zeta)$. In particular, the fixed boundary AOB is described by the curve

$$z(e^{is}) = \int_{e^{is_0}}^{e^{is}} e^{i\omega(\zeta)} f'(\zeta)\, d\zeta, \qquad 0 \leqq s \leqq \pi \tag{23.4}$$

and the free streamlines by

$$z(\xi) = \int_{e^{is_0}}^{\xi} e^{i\omega(\zeta)} f'(\zeta)\, d\zeta, \qquad -1 \leqq \xi \leqq +1. \tag{23.5}$$

[1] T. LEVI-CIVITA: Rend. Circ. Mat. Palermo **18**, 1—37 (1907).

[2] In flows having $q = V$ on the free boundary we define $\omega = \vartheta + i \log(q/V)$.

[3] It is assumed here that the velocity approaches a limit at infinity; this will be proved in Sect. 25.

This representation of the free boundary proves useful in deriving the asymptotic behavior of the free streamlines (Sect. 25) and in discussing the nature of the separation from the fixed boundary (Sect. 24).

We note that, conversely, any function $\omega(\zeta)$ analytic in Γ, real on the real axis, and satisfying $\omega(0)=0$, defines with $f(\zeta)$ (in which $0<s_0<\pi$) an infinite cavity past an arc described by Eq. (23.4). This is a means therefore of constructing inverse solutions of the flow problem[1] (see also Sect. 33). The solution for a prescribed obstacle requires the determination of $\omega(\zeta)$ and the parameters σ_0, M in Eq. (23.2) such that Eq. (23.4) is a parametric representation of the given curve.

The expression for the force on the obstacle takes a particularly elegant form in terms of $\omega(\zeta)$. We observe first that ω can be analytically continued by reflection into the full circle $|\zeta|<1$. Designating by X and Y the drag and lift respectively on the obstacle AOB, we find that

$$X+iY=-i\int_{AOB}(p-p_c)\,dz=-\frac{i}{2}\int_{AOB}(1-|f'(z)|^2)\,dz$$

$$=-\frac{i}{2}\left[\int_{AOB}e^{i\omega(\zeta)}f'(\zeta)\,d\zeta-\int_{AOB}\overline{f'(z)}\,f'(\zeta)\,d\zeta\right].$$

Since $\overline{f'(z)}=e^{i\omega(\bar\zeta)}$ and f is real on AOB, we may write

$$\int_{AOB}\overline{f'(z)}\,f'(\zeta)\,d\zeta=\int_{AO'B}e^{i\omega(\zeta)}f'(\zeta)\,d\zeta,$$

O' being the reflection of O in the lower half plane. Hence

$$X+iY=-\frac{i}{2}\oint e^{i\omega(\zeta)}f'(\zeta)\,d\zeta, \tag{23.6}$$

where $f'(\zeta)$ is given by Eq. (23.2) and the integral is taken counterclockwise about any simple circuit containing $\zeta=0$. Using the expansion

$$e^{i\omega(\zeta)}=1+i\zeta\,\omega'(0)+\frac{\zeta^2}{2}\left[i\,\omega''(0)-\omega'(0)^2\right]+\cdots,$$

we calculate the integral (23.6) from the residue of the integrand to obtain

$$X+iY=\frac{\pi}{4}M\left\{\omega'(0)^2+i\left[4\omega'(0)\cos s_0-\omega''(0)\right]\right\}. \tag{23.7}$$

Thus the force on the obstacle depends only on the first two coefficients in the expansion of $\omega(\zeta)$ and on the parameters s_0 and M[2]. If the fluid has density ϱ and free stream velocity U, then $\omega=i\log(f'(z)/U)$ and Eq. (23.7) is multiplied by ϱU.

The Levi-Civita parametrization is applicable to other cavity and jet models as well, the details differing of course with the various expressions for $f(\zeta)$ and sometimes because of the use of different image domains in the ζ-plane. The common feature in all the problems is the use of the flow function $\omega(\zeta)$ which can be continued by reflection across the free boundary.

In the re-entrant jet problem[3], as we have seen in Sect. 7, the f plane is a Riemann surface (Fig. 7) whose mapping into the semi-circle Γ is given by the

[1] A number of inverse solutions constructed by this method are contained in G. GREENHILL: Aeron. Res. Counc. Rep. and Memo. 19 (1910), Appendix (1916).

[2] An analogous formula can be derived for the moment acting on the obstacle; see S. BRODETSKY: Proc. Roy. Soc. Lond., Ser. A **102**, 361–372 (1922).

[3] For the following cf. D. GILBARG and J. SERRIN: J. Math. Phys. **29**, 1–12 (1950).

formula[1],

$$f'(\zeta) = M \frac{(1-\zeta^2)(\zeta - e^{is_0})(\zeta - e^{-is_0})(\zeta - \zeta_B)(\zeta - \bar{\zeta}_B)(\zeta - 1/\zeta_B)(\zeta - 1/\bar{\zeta}_B)}{\zeta(\zeta-\zeta_I)^2(\zeta-\bar{\zeta}_I)^2(\zeta - 1/\zeta_I)^2(\zeta - 1/\bar{\zeta}_I)^2}. \tag{23.8}$$

Here ζ_B, ζ_I in Γ are the images of the interior stagnation point B of the flow and of the point at infinity I, while $\zeta=0$ is the image of the jet at infinity (Fig. 40). From the discussion in Sects 4 and 7 one sees that $\omega(\zeta)$ has the following properties in Γ:

(i) $\tau = 0$ for real ζ;

(ii) at ζ_B $\omega(\zeta)$ has a singularity of the form,

$$\omega(\zeta) = i\log(\zeta - \zeta_B) + P(\zeta - \zeta_B),$$

and is regular elsewhere in Γ;

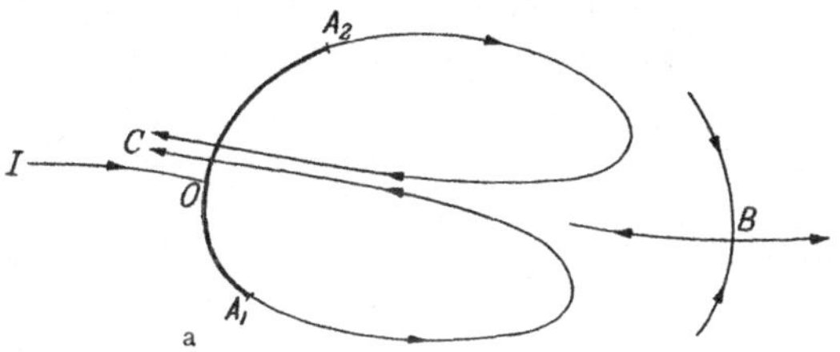

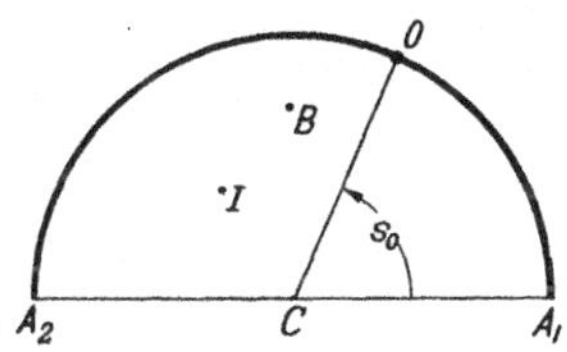

Fig. 40 a und b.

(iii) the neighborhood of the point at infinity z_I must be simply covered in the mapping $z(\zeta)$, that is,

$$\operatorname*{Res}_{\zeta=\zeta_I} \frac{dz}{d\zeta} = \frac{d}{d\zeta}\left[(\zeta-\zeta_I)^2 e^{i\omega(t)} \frac{df}{d\zeta}\right]_{\zeta=\zeta_I} = 0, \tag{23.9}$$

where $df/d\zeta$ is given by Eq. (23.8);

(iv) $\omega(\zeta_I) = i\log U$, U = free stream speed.

Conversely, any function $\omega(\zeta)$ having these properties describes with $f(\zeta)$ a re-entrant jet flow (with unit cavity speed) past an obstacle given by Eq. (23.4) and having the free streamlines (23.5)[2].

The jet width J and its limiting direction γ are given by

$$J = -\pi M, \quad \gamma = \omega(0). \tag{23.10}$$

The first of these formulas follows from the equality $J = -\frac{1}{2i}\oint f'(\zeta)\,d\zeta$, the integral being taken about a contour containing the origin and excluding other singular points. The force on the obstacle can be computed from the formula (23.6), where the integral is now taken a contour containing both $\zeta=0$ and $\zeta=\bar{\zeta}_I$, these being the only singularities at which the integrand has residues. Calculation of the residues gives the result, for circulation $\varkappa$ and cavity speed V,

$$X = \varrho V J(U - V\cos\gamma), \quad Y = -\varrho U(UJ\sin\gamma + \varkappa). \tag{23.11}$$

a) Geometric properties of free streamlines.

24. Free streamline behavior at detachment. Flows past curved boundaries present a new problem in that the points of detachment are not known *a priori* as they are for sharp edged bodies. We note that examples of the latter variety

[1] This follows from (7.17) after the substitution $t = -\frac{1}{2}(\zeta + 1/\zeta)$.

[2] The inverse problem for re-entrant jet flow has been treated by G. Pykhteev: Prikl. Mat. Mekh. **20**, 378—381 (1956).

considered in Sects. 6 to 10 have the property that the free streamline curvatures are infinite at separation. While this behavior is unobjectionable in flows with detachment from fixed endpoints, it violates the physical basis of the theory in flows past general curved obstacles (cf. Sect. 4). A natural criterion for selecting the separation point is provided by the basic result of VILLAT[1] which states that *the curvature of a free streamline at detachment is either infinite or equal to that of the fixed boundary at the point.*

To prove VILLAT's alternative, let a flow $f(z)=\varphi+i\psi$ be defined on one side of a streamline arc C, with $\psi=0$ on C and $\psi>0$ elsewhere. Let $A\,(f_A=0)$ divide C into two arcs γ_1 and γ_2 such that γ_1 is a free streamline (on which the flow speed equals unity) and γ_2 is some fixed curve. Consider now the function $\omega=\vartheta+i\tau$ in the auxiliary plane $t=\sqrt{f}$. In this plane the flow region is mapped into the first quadrant, with γ_1 and γ_2 corresponding to the real and imaginary axes, so that $\omega(t)$ can be defined by reflection in a neighborhood abutting on the imaginary axis and containing a segment of the positive real axis. On γ_1 the curvature is given by

$$\varkappa_1=\frac{d\omega}{df}=\frac{1}{2t}\frac{d\omega}{dt}; \tag{24.1}$$

and on γ_2,

$$\varkappa_2=\mathrm{Re}\left(e^{\tau}\frac{d\omega}{df}\right)=\mathrm{Re}\left(\frac{e^{\tau}}{2t}\frac{d\omega}{dt}\right). \tag{24.2}$$

Consider now the behavior of $\omega'(t)$ as $t\to 0$. Under suitable regularity assumptions on the fixed boundary, in particular if the curvature satisfies a Hölder condition[2] with respect to arc length, an expansion

$$\omega'(t)=\omega'(0)+2\varkappa t+o(t),\qquad \varkappa \text{ real}, \tag{24.3}$$

holds in a neighborhood $|t|\leqq t_0$, $\mathrm{Re}\,t\geqq 0$[3]. Then (24.1) and (24.2) become

$$\varkappa_1=\frac{\omega'(0)}{2t}+\varkappa+o(1), \tag{24.4}$$

$$\varkappa_2=\mathrm{Re}\left[e^{\tau}\left(\frac{\omega'(0)}{2t}+\varkappa+o(1)\right)\right]. \tag{24.5}$$

It is now evident that the curvature of the free streamline has an infinite limit at $t=0$ if $\omega'(0)\neq 0$ and a finite limit if $\omega'(0)=0$. In the latter case the two curvatures (24.4) and (24.5) are identical at A, equal to $\varkappa$, and the alternative is proved. In the former case the free streamline is curved towards the flow or away from it according as $\omega'(0)>0$ or $\omega'(0)<0$[4]. In terms of the Levi-Civita variable ζ the three types of detachment are similarly characterized; namely the free streamline has finite curvature equal to that of the body, infinite curvature concave to the flow, and infinite curvature convex to the flow according as $\omega'(\zeta)_{\zeta=\pm 1}=0,\ >0,\ <0$. Henceforth we adopt the terminology *smooth* detachment[5] (or separation) to describe the first of these cases.

[1] H. VILLAT: J. de Math. (6) **10**, 231–290 (1914).

[2] A function $f(x)$ is Hölder continuous in a set S if $|f(x_1)-f(x_2)|\leqq K|x_1-x_2|^{\alpha}$ for constants K, α, $0<\alpha\leqq 1$, and all $x_1, x_2\in S$.

[3] For a proof see J. LERAY: Comm. Math. Helv. **8**, 149–180 (1935); also BIRKHOFF and ZARANTONELLO, Chap. 4. The further behavior of ω at detachment has been discussed by LERAY. In the typical case, when neither the curvature of the body nor $\omega'(\zeta)$ vanishes at detachment, $d^3\omega/d\zeta^3$ has a logarithmic infinity. If either of these exceptions occurs, then the singularity appears (in general) in $d^5\omega/d\zeta^5$, etc.

[4] A similar analysis of detachment in the axially symmetric case has been given by A. ARMSTRONG: Armament Res. Est. Memo. 22/53 (1953).

[5] French: detachement en proue.

VILLAT's alternative leads to a natural and physically significant classification of free boundary problems according to the type of detachment: *The problem of fixed detachment*[1] seeks to determine a free surface flow past a given boundary in which the points of detachment are prescribed. The free streamline solutions of this problem will in general have infinite curvature at separation, as Eq. (24.4) shows. *The problem of smooth detachment* seeks a free surface flow past a given boundary with the property that the free streamline curvature at separation is equal to that of the fixed body. Since a given arc does not always admit a flow with smooth detachment, it is often preferable to treat the more general *prow problem*[2]. This seeks a flow either with smooth separation (of one or both free streamlines) or with detachment from the given endpoints; in the latter case the free streamline is to be convex to the flow at detachment. (The convexity is to insure that the physical requirements are satisfied at least locally; see below.)

Although it is possible that neglected physical effects, such as the boundary layer, play a part in separation[3], it is still of interest to explore the connection between the type of detachment and fulfillment of the Brillouin conditions (Sect. 4), where we regard the latter as the criterion of physical acceptability of a solution within the classical theory.

In certain cases it is possible to infer *a priori* that the Brillouin conditions are satisfied. For example, in any flow where the fixed boundary is concave to the fluid, the inward directed normal derivative of the speed at smooth points on the body must be positive [because of Eq. (3.5)], and hence the maximum speed occurs on the free surface. (This remains true for spatial as well as for plane flows.) Thus, any flow past a concave barrier automatically satisfies the Brillouin conditions.

The above assertion that the maximum speed occurs on the free surface is still correct when the barrier is not strictly concave, that is, when zero curvature is allowed. For, if to the contrary the maximum speed were achieved at a point P on the fixed boundary, let this point be chosen as origin of the coordinates, with the x-axis in the direction of the velocity and the y-axis normal to the surface at P directed into the fluid. Since the velocity component φ_x is a maximum at P and φ_x is harmonic, it follows that $\varphi_{xy}(P) < 0$. This contradicts the fact that the surface has non-negative curvature at P (with respect to the chosen coordinate system).

Evidently a *necessary* condition that a solution be physically acceptable in the sense of BRILLOUIN is that the detachment be smooth if it occurs at an interior point of the barrier, and that the free streamline be convex to the flow at separation if it occurs at an end point. The validity of the solutions of the problem of smooth separation in this extended sense (the prow problem) has been studied by LERAY[4] in connection with the infinite cavity. Plainly the Brillouin conditions can be satisfied only if $d\omega/d\zeta \leqq 0$ at detachment. Pursuing a step further the argument which leads to this inequality, LERAY establishes that when *either* Brillouin condition holds in a neighbohrood of smooth detachment, it implies the inequality at this point,

$$\frac{d}{d\zeta}\left(\frac{d\omega}{df}\right) \geqq 0. \tag{24.6}$$

Conversely, both Brillouin conditions are satisfied in the neighborhood of a point where the obstacle is convex and the detachment is smooth, provided the strict inequality holds in (24.6). LERAY offers a simple and illuminating example of

[1] French: problème du sillage.

[2] French: problème de la proue.

[3] The available experimental data on cavitation is in substantial agreement with the theory; cf. footnote 2, p. 420.

[4] J. LERAY: Volume du Jubile de M. BRILLOUIN, pp. 246—257. Paris: Gauthier-Villars 1935; see also VILLAT, footnote 1, p. 372.

a flow past a symmetric convex obstacle enjoying smooth detachment of the free streamlines without satisfying (24.6). Consequently, this solution fulfills neither of the Brillouin conditions even locally. It follows that smooth detachment is not sufficient in itself for physical acceptability of a solution, even for convex obstacles.

LERAY establishes that a solution of the problem of smooth detachment for the infinite cavity satisfies the Brillouin conditions in the following limited but important circumstances: the obstacle B_0C_0 consists of a convex arc B_0B_1, a concave arc B_1A, another concave arc AC_1, and another convex arc C_1C_0, it being assumed that the flow divides at A, and that the absolute curvature of the arcs B_0B_1 and C_0C_1 is non-decreasing when they are traversed from C_1 to C_0, and from B_1 to B_0 (Fig. 41). The arcs B_1A and AC_1 may contain straight segments, and each of the arcs B_0B_1, B_1C_1, and C_1C_0 may reduce to a point. These obstacles are called *accolades* by LERAY.

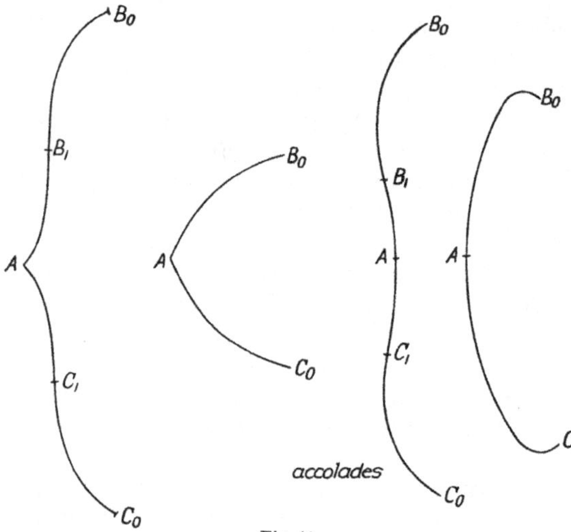

Fig. 41.

The proof of the above result can be seen in its essentials in the case of a symmetric obstacle whose concave portion B_1C_1 shrinks to the point A. We shall establish that the flow speed is non-decreasing on the fixed boundary and that the curvature of the free boundary decreases monotonically after detachment[1]. Consider the function

$$\frac{d\omega}{df} = \alpha + i\beta$$

in the quarter-circle, $|\zeta| \leqq 1$, $0 \leqq s = \arg\zeta \leqq \pi/2$, corresponding to the lower half of the flow. Measuring arc length l of a streamline in the direction of motion and denoting streamline curvature $d\vartheta/dl$ by $\varkappa$, we have on the segment I: $0 < \zeta \leqq 1$, corresponding to the free streamline,

$$\alpha = \varkappa, \quad \beta = 0,$$

and on the arc C: $\zeta = e^{is}$, $0 \leqq s < \pi/2$, corresponding to the wetted (half) obstacle,

$$\alpha = \frac{\varkappa}{q}, \quad \beta = -\frac{d}{dl}\left(\frac{1}{q}\right). \tag{24.7}$$

Let G be a component of the set on which $\beta < 0$, and let $R \subset G$ be a component of the subset on which $\alpha > 0$. Denote the boundary of R by R'; this set consists of points on $|\zeta| = 1$ or at which $\beta = 0$ or $\alpha = 0$. If R' is non-empty it must also include $\zeta = i$. For α is decreasing on the arcs where $\beta = 0$ as R' is traversed in the positive sense, while on C,

$$\frac{d\alpha}{ds} = \frac{1}{q}\frac{d\varkappa}{ds} - \beta\frac{dl}{ds} \leqq 0.$$

[1] This property is true also for the general *accolade*. It is easy to see in the latter case that smooth detachment can occur only on the convex portion of the obstacle.

Hence α is non-increasing along the entire length of R'; this can occur only if α has a discontinuity, which would have to be at $\zeta = i$.

The behavior of

$$\frac{d\omega}{df} = \frac{2\zeta^3 \frac{d\omega}{d\zeta}}{M(\zeta^2+1)(\zeta^2-1)}$$

at $\zeta = 0$ shows there must be a non-null set D on which $\alpha < 0$ and $\beta < 0$ having $\zeta = 0$ on its boundary. The components of D, for the same reason as the preceding, must have $\zeta = i$ on their boundaries. Hence a level line λ, on which $\alpha = \text{const} < 0$, extends from $\zeta = 0$ to $\zeta = i$, so that if R has points of I or C on its boundary, it is separated by λ from the vertical segment $(0, i)$. This would imply the existence of two level lines $\beta = 0$ issuing from $\zeta = i$, which is inconsistent with the expansion at $\zeta = i$,

$$\frac{d\omega}{df} = \frac{c}{(\zeta - i)^2} + \cdots .$$

It follows that any component R meets neither I nor C.

The desired results now follow easily. On C we have $\alpha > 0$ and hence, by the above, $\beta \geqq 0$. This establishes the monotonicity of the speed, and, as an immediate consequence, the convexity of the free streamline. From the latter we infer $\alpha > 0$ on I and, again by the preceding, $\beta > 0$ in the neighborhood of I. We conclude $d\alpha/ds > 0$ on I, or, in other words, the curvature increases as the free streamline is traversed toward the separation point.

25. Asymptotic behavior. We study next the behavior of the free streamlines at large distances from the fixed boundary. It suffices to consider only a local portion of the flow plane, which we shall suppose is a simply connected region R_z bounded by two free streamlines extending to infinity and by a simple arc joining the free streamlines. It is assumed further that the inclination of the velocity vector is bounded in R_z.

We remark first that the assumption of uniform flow at infinity made in the previous sections is a consequence of the above general hypotheses. For let $t(z)$ map R_z onto a neighborhood R_t on the upper side of the real t-axis, the point at infinity mapping into $t = 0$ and the two free streamlines into segments of the real axis. The function $\omega = \vartheta + i \log q$ is real on the real axis, and can therefore be continued analytically into the lower t-plane. It follows from the theorem on removable singularities for harmonic functions that ϑ is regular at $t = 0$, and hence so is ω. Thus, as asserted, the velocity vector has a limit at infinity[1].

Let us now specialize the flow in R_z to be of the type either of the infinite cavity or of the jet[2], and parallel to the x-axis at infinity. In the former case the image R_f in the f-plane is (by definition) a neighborhood of infinity slit along the positive real f-axis. The mapping $f = 1/t^2$ takes R_f into a region R_t in the upper z-plane such that the edges of the slit go into segments of the positive and negative real axis. The correspondence between the t and z-planes is given by

$$\left.\begin{aligned} z(t) &= \int e^{i\omega(t)} f'(t)\, dt \\ &= \frac{1}{t^2} + \frac{2i\omega'(0)}{t} + [\omega'(0)^2 - i\omega''(0)] \log t + P(t) \end{aligned}\right\} \qquad (25.1)$$

[1] Unless a bound is placed on the inclination, pathological flows are possible; for an example, see R. Finn: J. d'Analyse Math. **4**, 246—291 (1956).

[2] A more detailed classification of flows is treated in [*1*], Chap. 4.

where $P(t)$ represents a regular power series in t. Recalling that the derivatives $\omega'(0)$, $\omega''(0)$, ..., are real, and that $\omega(0)=0$, we have the following parametric representation of the free streamlines near $t=0$,

$$\left.\begin{aligned} x(t) &= \frac{1}{t^2} + \omega'(0)^2 \log|t| + P(t), \\ y(t) &= \frac{2\omega'(0)}{t} - \omega''(0)\log|t| + P(t), \end{aligned}\right\} \quad t \text{ real}. \tag{25.2}$$

From Eq. (25.2) it is apparent that if $\omega'(0) \neq 0$ the free streamlines are asymptotic to the parabolas

$$y = \pm 2\,\omega'(0)\sqrt{x}\,. \tag{25.3}$$

The higher order terms in the asymptotic development can be inferred from Eq. (25.2). In the exceptional case $\omega'(0)=0$, $\omega''(0)\neq 0$, the two free streamlines have the same logarithmic asymptotic behavior, $y \sim \omega''(0)\log x$. If, further, $\omega'(0)=0$ and the flow is symmetric with respect to the x-axis, then $\omega''(0)=0$ and the free streamlines have the asymptotic shape,

$$y^2 = \text{const}\; x^{-(2n+1)},$$

where n is a non-negative integer. Infinite cavities of this type with zero asymptotic width have already been met in Sect. 6.

In the Levi-Civita representation of the infinite cavity the expression (23.2) replaces $f(t)$ in Eq. (25.1), and the preceding results take essentially the same form. In place of Eq. (25.2) we have, for real ζ,

$$\left.\begin{aligned} x &= \frac{M}{4}\left[\frac{1}{\zeta^2} - \frac{4\cos s_0}{\xi} + \omega'(0)^2\log|\zeta| + \cdots\right] \\ y &= \frac{M}{4}\left[\frac{2\omega'(0)}{\zeta} + \big(4\omega'(0)\cos s_0 - \omega''(0)\big)\log|\zeta| + \cdots\right], \end{aligned}\right\} \tag{25.4}$$

and

$$y = \pm\omega'(0)\sqrt{M\,x} \tag{25.5}$$

is the asymptotic shape of the free streamlines, where $-$ and $+$ correspond to the upper and lower curves respectively. If $\omega'(0) > 0$ the upper and lower streamlines overlap. Let C denote the *shape factor* $|\omega'(0)|\sqrt{M}$ of the free streamlines. For symmetric flow one obtains from Eq. (25.4) [in which $s_0=\pi/2$, $\omega''(0)=0$] the more precise asymptotic formula,

$$y - y_0 = \pm C\,x^{\frac{1}{2}}\left(1 + \frac{C^2}{16}\,\frac{\log x}{x} + O\left(\frac{1}{x}\right)\right); \tag{25.5a}$$

here the sign is positive or negative according as the streamline lies eventually above or below the axis of symmetry.

The force formula (23.7) shows that the drag on an obstacle in infinite cavity flow is proportional to C^2,

$$D = \frac{\pi}{4}\,\varrho\, C^2\, U^2. \tag{25.6}$$

The degenerate cavity corresponding to the case $\omega'(0)=0$ has the property that its associated drag is zero. Hence the name "cavity of zero drag" has been applied to these flows[1].

[1] Cf. BIRKHOFF and ZARANTONELLO [1].

In the case of a jet flow the image R_f of the flow plane is assumed to cover a semi-infinite horizontal strip of width h in the f-plane extending to $f=+\infty$. The mapping $f-f_0=-\frac{h}{\pi}\log t$, where f_0 is a suitable constant, takes R_f into a neighborhood R_f in the upper t-plane such that the free streamlines correspond to segments of the real axis on opposite sides of the origin. The mapping between the t and z planes is given by

$$z(t)=\int e^{i\omega(t)}f'(t)\,dt=-\frac{h}{\pi}\left(\log t+P(t)\right), \tag{25.7}$$

from which it is apparent that the free streamlines approach their asymptotes logarithmically.

Determination of the asymptotic shape of axially symmetric cavities is considerably more difficult than in the plane case. LEVINSON[1] has shown under suitable regularity assumptions on the free boundary that the asymptotic shape in the meridian plane is given by the remarkable formula,

$$y=\frac{C\,x^{\frac{1}{2}}}{(\log x)^{\frac{1}{4}}}\left[1-\frac{1}{8}\,\frac{\log\log x}{\log x}+O\left(\frac{1}{\log x}\right)\right], \tag{25.8}$$

and from this has derived the following expression for cavity drag in a free stream of velocity U and density ϱ,

$$D=\frac{\pi}{8}\,\varrho\,C^4\,U^2. \tag{25.9}$$

Thus, according to Eq. (25.8) the shape of the axially symmetric cavity is flatter than that of the plane case by the factor $1/(\log x)^{\frac{1}{4}}$, and the drag is proportional to the fourth power of the shape factor C, in contrast with Eq. (25.6) for plane flows.

Since the proofs of Eqs. (25.8) and (25.9) are too lengthy for inclusion here, we limit ourselves to the following brief remarks concerning the details. The analysis centers around a singular non-linear integro-differential equation containing the free boundary and the potential on the fixed boundary as unknowns. This equation is similar to that appearing in the problem of water entry of a cone (Sect. 20) and to that considered by TREFFTZ (Sect. 46) in the axially symmetric jet problem. LEVINSON assumes first that the cavity is convex at infinity and is described for large x by a function $y=x^k g(x)$ where $0<k<1$ and

$$\lim_{x\to\infty} x\,\frac{g'(x)}{g(x)}=0 \tag{25.10}$$

[so that $g(x)$ has less than algebraic growth]. Estimates of the terms in the integral equation show that the possibilities $0<k<\frac{1}{2}$ and $\frac{1}{2}<k<1$ separately lead to contradiction. A contradiction follows also from the assumptions $k=\frac{1}{2}$, Eq. (25.10), and $\int^{\infty}(g^2(x)/x)\,dx<\infty$. In the remaining case, for which $k=\frac{1}{2}$ and $\int^{\infty}(g^2(x)/x)\,dx=\infty$, it follows that

$$\frac{1}{(\log x)^{\frac{1}{4}+\varepsilon}}<g(x)<\frac{1}{(\log x)^{\frac{1}{4}-\varepsilon}}$$

for any $\varepsilon>0$ and for large x. This suggests taking $g(x)$ in the form $g(x)=h(x)/(\log x)^{\frac{1}{4}}$, where $(\log x)^{-\varepsilon}<h(x)<(\log x)^{\varepsilon}$ for any $\varepsilon>0$. It is now assumed that $x\log x\,h'(x)/h(x)=O(1)$ for large x. This leads finally to the asymptotic formula (25.8). LEVINSON's argument has been pursued still further by SCHEID[2] who carried the asymptotic expansion to several additional terms.

The derivation of Eq. (25.9) is based on a standard momentum flux argument in which the following growth estimates for the velocity components play the essential part. For large y

[1] N. LEVINSON: Ann. of Math. **47**, 704–730 (1946).
[2] F. SCHEID: Amer. J. Math. **72**, 485–501 (1950).

and $x > y$

$$v = \frac{C^2}{2y(\log x)^{\frac{1}{2}}}(1 + o(1)),$$

$$u = 1 - \frac{C^2 \log y}{4x(\log x)^{\frac{3}{2}}} + O\left(\frac{y^2}{x^3(\log x)^{\frac{1}{2}}}\right) + o\left(\frac{1}{x(\log x)^{\frac{1}{2}}}\right).$$

For large y and $-y < x < y$,

$$v = O\left(\frac{1}{y(\log y)^{\frac{1}{2}}}\right), \qquad u = 1 + O\left(\frac{1}{y(\log y)^{\frac{1}{2}}}\right).$$

For large $|x|$ and $x < -y$,

$$v = O\left(\frac{y}{x^2(\log|x|)^{\frac{1}{2}}}\right), \qquad u = 1 + O\left(\frac{1}{|x|(\log|x|)^{\frac{1}{2}}}\right).$$

These growth estimates are derived by LEVINSON using Eq. (25.8). The formula (25.9) then follows by direct application of the momentum theorem to a region (in the meridian plane) bounded by a vertical and horizontal segment, the axis of symmetry, the obstacle, and the free streamline. It turns out that the entire value of the drag is contributed by the momentum flow through the region $x > y$ as $y \to \infty$.

The formula (25.8) was derived independently by GUREVICH[1], who showed that this is the asymptotic shape required of an axially symmetric half-body (not necessarily a cavity) in order that its drag be finite and non-zero. Specifically, GUREVICH considers axially symmetric flows whose potentials are represented in the form

$$\varphi(r, \vartheta) = \int_{-N}^{0} a(n)\,[r^n P_n(\cos\vartheta) - 1]\,dn + r\cos\vartheta, \qquad N > 0, \tag{25.11}$$

where r, ϑ are polar coordinates in the meridian plane, P_n is the Legendre function, and $a(n)$ is a distribution which is singular but of one sign at $n = 0$. The potential (25.11) describes a flow of unit speed past a half body having the asymptotic shape

$$r(1 - \cos\vartheta) \sim 2\int_{-\varepsilon}^{0} n\,a(n)\,r^n\,dn \tag{25.12}$$

if ε is sufficiently small. With Eqs. (25.11) and (25.12) the momentum theorem yields the result that the drag D is related to the shape by the equation

$$D = \lim_{\vartheta \to \infty} \frac{\pi}{2} r^2 (1 - \cos\vartheta)^2 \log\frac{2}{1 - \cos\vartheta},$$

whence

$$\frac{8D}{\pi} \sim \frac{y^4}{x^2}\log\frac{x^2}{y^2},$$

or

$$y \sim \frac{C x^{\frac{1}{2}}}{(\log x)^{\frac{1}{4}}}\left(1 - \frac{\log\log x}{8\log x}\right)$$

where $C = (8D/\pi)^{\frac{1}{4}}$.

26. Connection between the geometry of the fixed and free boundaries.

The behavior of the free streamlines in the large is of course dependent on the shape of the fixed boundary[2]. The results of this section show that this dependence can be described by the statement that *in a suitable sense the free streamlines reproduce the geometric properties of the fixed boundary.*

We consider first symmetric plane flows, including the infinite and cusped cavities, the Riabouchinsky cavity, and jet flows from an orifice. Suppose that

[1] M. GUREVICH: Prikl. Mat. Mekh. **11**, 97–104 (1947) [in Russian with English summary].

[2] The possibility of pathological behavior of the free streamlines (e.g. self-intersections) was observed by M. BRILLOUIN: Ann. Chem. Phys. **23**, 145–230 (1911).

on the fixed barrier Γ in the upper half-plane[1] the inclination ϑ satisfies the inequality

$$\alpha - \pi \leqq \vartheta \leqq \alpha, \quad 0 < \alpha \leqq \pi, \tag{26.1}$$

where α is some fixed angle. Such is the case, for example, when $\alpha = \pi/2$ and Γ can be described by a single-valued function $y = f(x)$. The condition (26.1) states that a line having inclination α can intersect Γ in at most a single point or segment. The harmonic function ϑ, which is defined by reflection in the semi-circle $|\xi| < 1$, $\operatorname{Re}\zeta < 0$, of the Levi-Civita representation, is equal to zero on the imaginary ζ-axis and satisfies (26.1) on the boundary. It follows by the maximum principle for harmonic functions that $\alpha - \pi < \vartheta < \alpha$ on the free streamline. In other words, any line inclined at angle α intersects both the fixed and free boundary at most once. In particular, a free streamline detaching from a monotonic curve must itself be strictly monotonic (in the same sense as the barrier).

The preceding result is also a consequence of the following general *single intersection theorem*, proved by SERRIN[2] for both axially symmetric and plane flows by comparison methods (cf. Sect. 27). In any infinite cavity, cusped cavity, or Riabouchinsky flow, let T denote the curve consisting of the obstacle and the upstream axis of symmetry; in the case of jet flow let T denote the wall of the vessel or channel[3]. *Then any straight line which does not cut T can intersect the free streamline detaching from T in at most one point.*

The effect of this theorem is to limit pathological behavior of the free streamlines to flows past barriers of more unusual shape[4]. A useful application is the following result. Let T either satisfy Ineq. (26.1) for some α or be starlike[5] with respect to a point exterior to the flow. Then the free streamline and the curve consisting of T plus the free streamline enjoy the same property.

Consider now the curvature properties of the free streamlines. We have already seen in Sect. 25 that whenever the fixed boundary is concave or nowhere convex to the flow the free surface is convex, and that for the *accolades* of LERAY the problem of smooth separation always has convex solutions with monotonically decreasing curvature. An interesting general result connecting the curvature properties of the fixed and free boundaries in (not necessarily symmetric) plane cavity and jet flow is contained in the following theorem: *The number of inflection points on the free boundary does not exceed the number on the fixed boundary plus the inflections at detachment*[6]. (An inflection is said to occur at detachment if the fixed and free boundaries have oppositely directed curvatures there.)

To carry through the proof in case of cavity flow, consider in the ζ-plane the set of points corresponding to inflection points of streamlines. Since $\varkappa/q = \operatorname{Re}(d\omega/df)$ is harmonic, where $\varkappa$ denotes streamline curvature, the level lines $\varkappa = 0$ must terminate on the boundary of the semi-circle C: $|\zeta| < 1$, $\operatorname{Im}\zeta > 0$.

---

[1] In the Riabouchinsky flow the barrier is understood to be the obstacle alone and does not include its reflected image. The following discussion of this flow concerns only the upper left half of the physical plane.

[2] J. SERRIN: J. Rat. Mech. a. Analysis 2, 563–575 (1953).

[3] We refer only to the upper half-plane in case of plane flow and to the meridian half-plane in case of axially symmetric flow; cf. footnote 1 with regard to Riabouchinsky flows.

[4] R. FINN: J. d'Analyse Math. 4, 246–291 (1956), discusses jet flows under various conditions of self-intersection of the boundary, and shows that the topology of the barrier is intimately connected with the possible existence or non-existence of the flows.

[5] A curve is *starlike* with respect to a point P if every ray from P intersects the curve in at most a single point or segment.

[6] J. SERRIN: Thesis, Indiana University 1951, Appendix A. — BIRKHOFF and ZARANTONELLO [*1*], Chap. 4; see also R. EPPLER: J. Rat. Mech. a. Analysis 3, 591–644 (1954).

Any such level line starting at an interior point of the free boundary must terminate either (i) on the free boundary; (ii) at $\zeta = 0$; (iii) on the fixed boundary; or (iv) at a detachment point ($\zeta = \pm 1$). The possibility (i) is excluded since otherwise the level line would bound with $\operatorname{Im}\zeta = 0$ a region on whose boundary either $\operatorname{Re}(d\omega/df) = 0$ or $\operatorname{Im}(d\omega/df) = 0$, from which would follow the absurdity $d\omega/df \equiv 0$. The same argument can be used to exclude (ii). This leaves only the possibilities (iii) and (iv). In the former case we observe that two level lines beginning on the free boundary cannot terminate at the same point on the fixed boundary, for the same reason that applied to (i)—unless the point in question is the bifurcation point. Since the expansion of $d\omega/df$ at the latter begins with a term $c(\zeta - e^{is_0})^{-2}$, there are at most three level lines $\varkappa = 0$ approaching $\zeta = e^{is_0}$ from C. At the same time, since the expansion of $d\omega/df$ at $\zeta = 0$ begins with a term in ζ^3, or of higher degree if $\omega'(0) = 0$, there are at least three level lines issuing from $\zeta = 0$ into C. These must terminate on the fixed boundary. Finally, the behavior of $d\omega/df$ at detachment [cf. Eq. (24.5)] shows that no more than one level line can issue from a separation point where the free streamline curvature is infinite. It is now apparent that for every inflection on the free boundary there is at least one on the fixed boundary provided the endpoints are included. This proves the theorem. Similar proofs apply to the other flow models.

We infer directly from the above result that a free streamline detaching from a convex barrier can have at most a single inflection point, and that one separating from a barrier with a finite number of inflections must have curvature of one sign sufficiently far downstream.

b) Comparison methods.

A fruitful source of qualitative theorems concerning free surface flows is the comparison method originated by LAVRENTIEFF[1] and later extended by GILBARG and SERRIN[1]. A distinctive feature of this method is its reliance on simple geometric considerations in the physical plane, and as a consequence the results obtained often generalize naturally to axially symmetric problems and to flows of compressible fluids.

27. Comparison theorems. In the following the flows under consideration will be described in terms of the stream function ψ. It will be convenient to treat plane and axially symmetric flows simultaneously, in which case the governing equation can be written as

$$\psi_{xx} + \psi_{yy} - \frac{\delta}{y}\psi_y = 0, \tag{27.1}$$

where $\delta = 0$ or 1 according as the flow is plane or axially symmetric[2]. Every solution of Eq. (27.1) satisfies the maximum principle; that is, it cannot take on an interior maximum or minimum unless it is constant. In the sequel a useful consequence of the maximum principle will be the following *boundary*

[1] The first applications of the comparison method—to plane free boundary problems—are in the basic memoir of M. LAVRENTIEFF: Mat. Sbornik **46**, 391—458 (1938) [Russian with French summary]. His work was simplified, further elaborated, and extended to axially symmetric flows and to flows of compressible fluids in the following series of papers: D. GILBARG: (1) J. Rat. Mech. a. Analysis **1**, 309—320 (1952). — (2) J. Rat. Mech. a. Analysis **2**, 233—251 (1953). — (3) D. GILBARG and M. SHIFFMAN: J. Rat. Mech. a. Analysis **3**, 209—230 (1954). — J. SERRIN: (1) Amer. J. Math. **74**, 492—506 (1952). — (2) J. Rat. Mech. a. Analysis **1**, 1—48 (1952). — (3) J. Rat. Mech. a. Analysis **1**, 563—572 (1952). — (4) J. Rat. Mech. a. Analysis **2**, 563—575 (1953). — (5) J. Math. Phys. **33**, 27—45 (1954).

[2] For extensions to compressible fluids see GILBARG (2) and SERRIN (5), preceding footnote.

point lemma[1], which is applicable to any elliptic equation of the form,

$$a u_{xx} + 2b u_{xy} + c u_{yy} + d u_x + e u_y = 0, \tag{27.2}$$

having bounded coefficients, and satisfying $ac - b^2 \geqq m > 0$ for some constant m. Let u be a solution of Eq. (27.2) in a neighborhood N having a regular boundary point[2] P and suppose that $u \geqq 0$ in N, $u(P) = 0$. Then the inward drawn normal derivative at P satisfies the inequality

$$\frac{\partial u}{\partial n}(P) \geqq 0, \tag{27.3}$$

where the equality holds if and only if $u \equiv 0$ in N. If the equation is Eq. (27.1) with $\delta = 1$ and P lies on the axis of symmetry ($y_P = 0$), the same result is true if (27.3) is replaced by $\frac{\partial^2 u}{\partial y^2}(P) \geqq 0$[3].

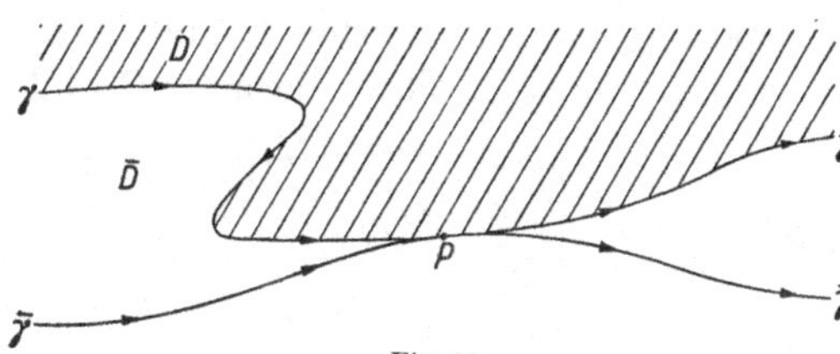

Fig. 42.

Comparison theorem 1[4]. Let D and $\bar{D}$ be flow regions for two (plane or axially symmetric) flows having uniform non-zero velocities q_0, $\bar{q}_0$ at infinity, where $\bar{q}_0 \geqq q_0$. Let D and $\bar{D}$ be bounded by the streamlines γ and $\bar{\gamma}$, extending to $x = \pm\infty$. If $D \subset \bar{D}$, and if γ and $\bar{\gamma}$ have a point P in common (Fig. 42), then *the respective flow speeds q and $\bar{q}$ satisfy at P the inequality*

$$\bar{q}(P) \geqq q(P), \tag{27.3}$$

and furthermore, when $\bar{q}(P) \neq 0$ the equality holds if and only if $D = \bar{D}$ and the two flows are identical.

Proof. To begin with let us suppose that $\bar{q}_0 > q_0$. It follows from the limit relation at infinity, $\frac{1}{y^\delta}\frac{\partial(\bar{\psi} - \psi)}{\partial y} \to \bar{q}_0 - q_0 > 0$, that the stream functions $\bar{\psi}, \psi$—which are assumed equal to zero on $\bar{\gamma}, \gamma$ and which are positive in $\bar{D}$, D—satisfy the inequality $\bar{\psi} - \psi > 0$ in D at points sufficiently far from the origin. At the same time, $\bar{\psi} - \psi \geqq 0$ on γ. We infer from the maximum principle that

$$\bar{\psi} - \psi > 0 \quad \text{in} \quad D. \tag{27.4}$$

Suppose now that $\bar{q}_0 = q_0$. Then the stream function $\alpha\bar{\psi}$, where $\alpha > 1$, describes a flow in $\bar{D}$ whose velocity vector is α times that of the flow determined by $\bar{\psi}$. Hence as in the preceding case, $\alpha\bar{\psi} - \psi > 0$ in D. Letting $\alpha \to 1$ we see that $\bar{\psi} - \psi \geqq 0$ in D, whence it follows that either $\bar{\psi} > \psi$ throughout D or $\bar{\psi} \equiv \psi$.

To complete the proof we may assume Ineq. (27.4) and as sume that $q(P) > 0$. If P is a regular boundary point of D, then the required result

$$\bar{q}(P) > q(P) \tag{27.5}$$

is an immediate consequence of Ineq. (27.4) and the boundary point lemma. If P is not a regular point, let C be an arc in D whose end points Q, Q' lie on γ on opposite sides of P and are regular boundary points of D; such an arc C

[1] E. Hopf: Proc. Amer. Math. Soc. **3**, 791—793 (1952).

[2] A boundary point P of a region R is said to be *regular* if there is a circle having P on its circumference and its interior entirely within R.

[3] This extension is due to Serrin (5), footnote 1, p. 380.

[4] Cf. Lavrentieff, Serrin (1), Gilbarg (1) to (3), footnote 1, p. 380.

can easily be found. For a suitable $\varepsilon > 0$ we must have from Ineq. (27.4), if neither Q nor $\bar{Q}$ lies on $\bar{\gamma}$, $\bar{\psi} - \psi - \varepsilon\psi \geqq 0$ on C; if either Q or Q' does lie on $\bar{\gamma}$, this result follows from an application of the boundary point lemma. The same inequality holds on γ and therefore also in the region bounded by C and γ. We conclude that

$$\bar{q}(P) \geqq (1+\varepsilon)\, q(P) > q(P),$$

and this proves the theorem.

The extension of the preceding result to flows bounded by two streamlines, as in a channel, is contained in

Comparison theorem 1'[1]. Let D and $\bar{D}$ be two (plane or axially symmetric) flow regions bounded by the pairs of streamlines γ_1, γ_2 and $\bar{\gamma}_1, \bar{\gamma}_2$ respectively. Suppose $\bar{\gamma}_1$ and γ_2 together bound a region containing γ_1 and $\bar{\gamma}_2$ in its closure (Fig. 43). Consider flows through D and $\bar{D}$ for which the total flux Q through

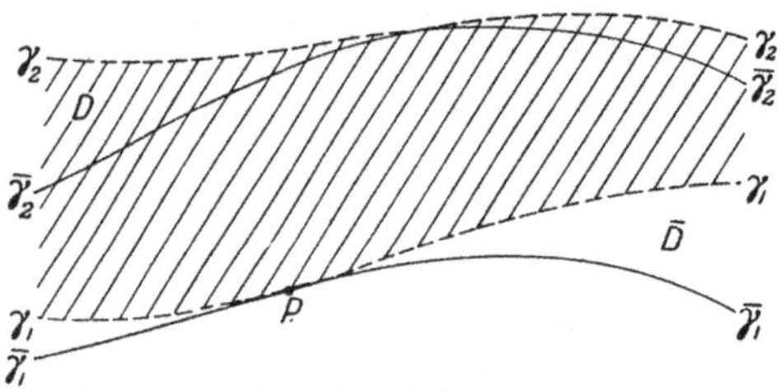

Fig. 43.

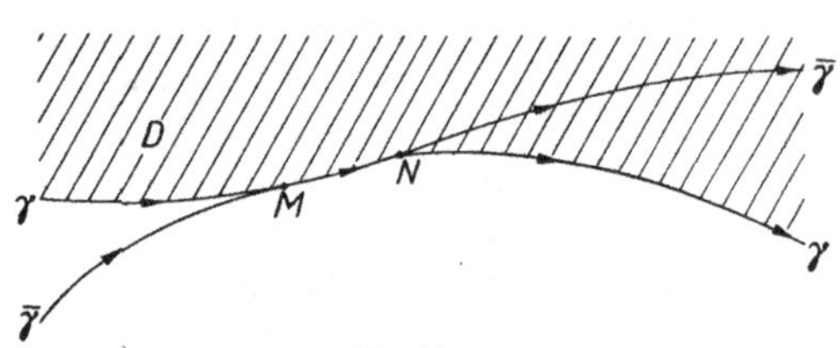

Fig. 44.

D and $\bar{Q}$ through $\bar{D}$ satisfy the inequality $\bar{Q} \geqq Q$, and for which the stream functions satisfy $0 \leqq \psi \leqq Q$ in D, $0 \leqq \bar{\psi} \leqq \bar{Q}$ in D. If P is a boundary point common to γ_1 and $\bar{\gamma}_1$ then the respective flow speeds obey at P the inequality $\bar{q}(P) \geqq q(P)$; and furthermore, when $\bar{q}(P) \neq 0$ the equality holds if and only if the two flows are identical and $D = \bar{D}$.

A deeper result is provided by the following

Comparison theorem 2[2]. Let two (plane or axially symmetric) flows be defined in regions D and $\bar{D}$ bounded by the streamlines γ and $\bar{\gamma}$ extending to infinity. Let γ and $\bar{\gamma}$ have an arc MN in common such that the direction of each flow on MN is from M to N. Furthermore, let the intersection of D and $\bar{D}$ be bounded by the single curve consisting of MN, the arc $(-\infty, M)$ of γ and $(N, +\infty)$ of $\bar{\gamma}$ (Fig. 44). Then *the flow speeds obey the inequality*

$$\frac{\bar{q}(M)}{\bar{q}(N)} \geqq \frac{q(M)}{q(N)}$$

provided $q(N)$, $\bar{q}(N) \neq 0$; *furthermore, when* $\bar{q}(M) \neq 0$ *the equality holds if and only if* $D = \bar{D}$ *and the two flows are similar.*

We outline the proof. Let us assume that $q(N)$, $\bar{q}(N) \neq 0$ and that $D \neq \bar{D}$. We may also assume without loss of generality that $q(N) = \bar{q}(N)$, for this can be achieved by multiplication of either flow by a suitable factor. It remains to show that $q(M) < \bar{q}(M)$, if $\bar{q}(M) \neq 0$. Consider the function $\Omega = \bar{\psi} - \psi$. Since $0 = \bar{q}(N) - q(N) = \frac{1}{y^\delta}\frac{\partial\Omega}{\partial n}(N)$, a level line $\Omega = 0$ issues from N into the intersection R of D and $\bar{D}$. The argument can be reduced to its essentials and a number of special

[1] Cf. GILBARG (2), footnote 1, p. 380.
[2] Cf. SERRIN (1), (3), (5), footnote 1, p. 380.

cases avoided if we disregard possible branchings of the level line C and if we suppose that γ and $\bar{\gamma}$ do not have a point in common except on MN. Under these assumptions C must extend to infinity, since $\Omega > 0$ on the arc $\gamma_M = (-\infty, M) \subset \gamma$, $\Omega < 0$ on the arc $\bar{\gamma}_N = (N, +\infty) \subset \bar{\gamma}$, and C obviously cannot return to MN. C devides the common region R into two parts: R_1, which is bounded by $\gamma_M + MN + C$, and R_2, bounded by $\bar{\gamma}_N + C$. It follows from the relation $\Omega \geqq 0$ on the boundary of R_1 and an application of the maximum principle[1] that $\Omega > 0$ in R_1. From this inequality and the boundary point lemma (as in the proof of the first comparison theorem) we conclude the desired result: $\bar{q}(M) - q(M) > 0$ provided $\bar{q}(M) \neq 0$.

A limiting form of the above result is the following interesting inflection point theorem[2]. Let R be a region bounded by a three times differentiable curve γ which extends from $-\infty$ to $+\infty$. Let P be an inflection point on γ, and suppose the arcs $(-\infty, P)$ and $(P, +\infty)$ of γ lie on opposite sides of the tangent line L at P (Fig. 45). Suppose there is a plane flow in R with γ as bounding streamline. If $q(s)$ denotes flow speed as function of arc lenth s, and if L is assigned the direction corresponding to increasing s at P, then

$$\frac{dq}{ds}(P) > 0 \quad \text{or} \quad \frac{dq}{ds}(P) < 0$$

according as L is directed into R or out of R at P.

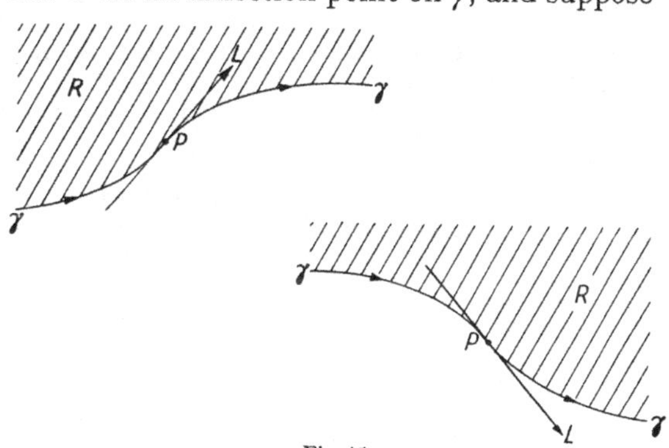

Fig. 45.

Several consequences of these theorems for plane symmetric and axially symmetric flows are more or less immediate: The upper free streamline in the solution of the infinite cavity problem either lies entirely above the axis of symmetry, or, if it once meets the axis of symmetry, remains below it for all larger values of the abscissa. The upper free streamlines of cusped cavity and re-entrant jet flows[3] are situated entirely above the axis of symmetry. A cusped cavity and non-overlapping infinite cavity cannot both detach from the same point on an obstacle[4]. If the free streamlines of a re-entrant jet and cusped cavity flow separate from the same point on a body, the former flow region contains the latter; this statement remains true with the (non-overlapping) infinite cavity in place of the cusped cavity. Another, more subtle, consequence of the comparison theorems is the *single intersection theorem* already mentioned in Sect. 26.

28. Monotonicity theorems. The preceding comparison theorems are often useful in deriving monotonicity properties of free surface flows. Consider, for example, the dependence of the contraction coefficient of a jet on the ratio of orifice to pipe diameter in flow through an orifice plate (cf. Sect. 2). It will

[1] This application involves several cases depending on the configuration of γ and $\bar{\gamma}$ at infinity; see, for example, SERRIN (5), footnote 1, p. 380.

[2] Cf. LAVRENTIEFF, SERRIN (3), footnote 1, p. 380.

[3] Although the hypotheses of the two comparison theorems do not, as stated, include the re-entrant jet flow, it is easy to see that the proofs are still applicable when $\bar{D}$ is the upper half of such a flow; cf. SERRIN (2), footnote 1, p. 380.

[4] The proofs of this and the following result presuppose that the obstacle is either starlike with respect to a point on the axis of symmetry or (in the case of plane flow) that (26.1) is satisfied.

be just as simple to treat the more general case in which the upper half[1] of the channel or vessel C is any curve that is concave or nowhere convex to the flow[2]. The convex free streamline Γ detaching from C at its end point $A=(0, a)$ extends downward to the right where it is asymptotic to $y=h$. Consider now the barrier

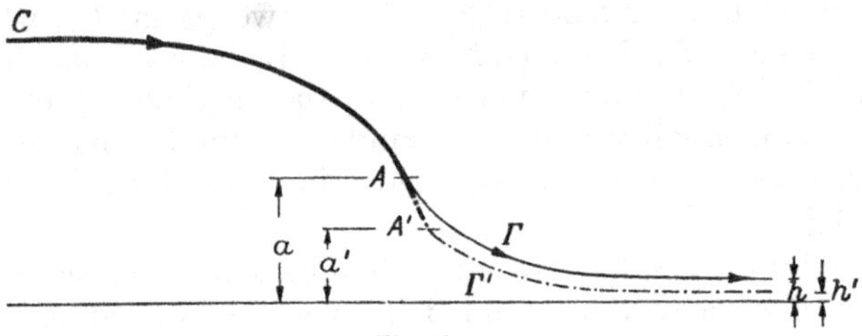

Fig. 46.

C' obtained by extending C in a straight segment from A to A', where the ordinate is a' (Fig. 46). Let the jet flow issuing from A' have asymptotic width h'. Then we assert $h/a > h'/a'$, or in other words, *the contraction coefficient decreases in passing from C to C'*.

To prove this result let 0 be the point where the extension of AA' meets the axis of symmetry. A suitable expansion with respect to 0 as center takes A' into A and the bounding streamline $C'+\Gamma'$ into a curve $C''+\Gamma''$ with asymptotic

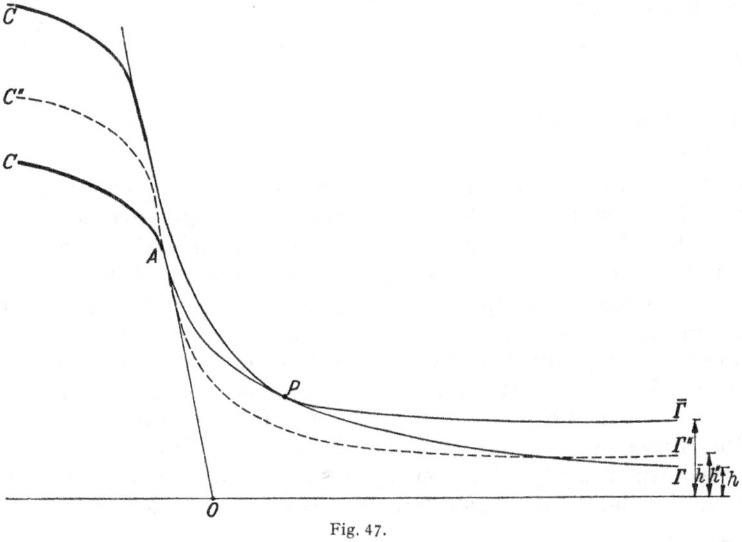

Fig. 47.

width equal to h'', while the flow bounded by $C'+\Gamma'$ is taken into a similar flow through $C''+\Gamma''$. C'' lies above C or may touch it. It suffices to prove that $h'<h$. Suppose to the contrary that $h''\geqq h$. Then another similarity expansion with respect to 0 as center takes $C''+\Gamma''$ into a curve $\overline{C}+\overline{\Gamma}$ which lies above $C+\Gamma$ but has at least one point of contact between Γ and $\overline{\Gamma}$, say at P (Fig. 47). The flow through $\overline{C}+\overline{\Gamma}$ may be assumed (after multiplication by a suitable factor) to have the same flux as that through $C+\Gamma$. The conditions of

[1] In this and the next section all statements apply equally to plane symmetric and axially symmetric flows.

[2] More general configurations can also be treated.

Comparison Theorem 1′ are now fulfilled, and it follows that $\bar{q}=\bar{q}(P)>q(P)=q$. However, this inequality is inconsistent with the assumption $\bar{h}\geqq h$ and the equality $\bar{h}\,\bar{q}=hq$ of the mass flows. This proves the monotonic character of the contraction coefficient.

This result is of particular interest because of its conflict with the experimental data compiled by LANSFORD (Fig. 4), which shows the contraction coefficient to be a decreasing function of the ratio of orifice to pipe diameter until this ratio is about 0.23, at which point the experimental values apparently begin to rise. According to the above monotonicity theorem the latter rise is inconsistent with the theory and is presumably due to effects other than those considered in the theory[1].

In a similar way one proves that the length and ratio of length to width of the Riabouchinsky cavity are monotonically decreasing functions of cavitation number. (The length is understood to be the distance between the front and rear stagnation points.)

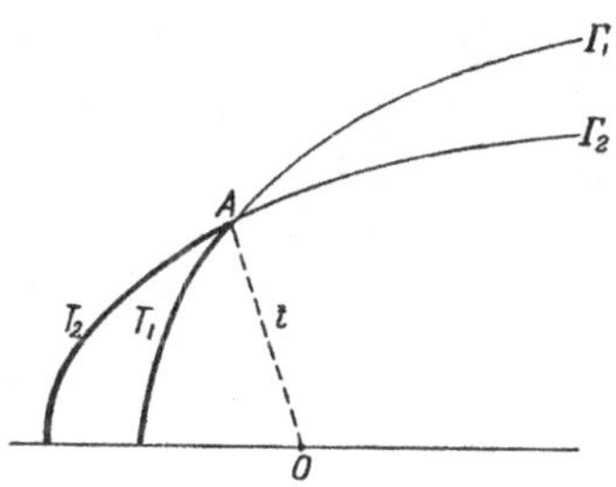

Fig. 48.

Another type of monotonicity theorem is the following, due to LAVRENTIEFF and SERRIN[2]. Let T_1 and T_2 be two obstacles in the cusped, infinite, or RIABOUCHINSKY, cavity problem, or two fixed walls in the jet problem. Suppose that T_1 and T_2 have the same end point A, and that T_1 (or T_2) is starlike with respect to a point O on the axis of symmetry. Let t denote the straight line determined by O and A, and suppose that T_2 and T_1 are both on the upstream side of t. Then *if T_2 lies above (or touches) T_1, the free streamline Γ_2 detaching from T_2 lies below the free streamline Γ_1 detaching from T_1* (Fig. 48).

We present the proof for the case that the two flows are of the infinite cavity type; the other cases can be treated in a similar way, although the Riabouchinsky cavity introduces a new feature. Consider first the relative positions of Γ_1 and Γ_2 at infinity. Letting $r=r_1(\vartheta)$ and $r=r_2(\vartheta)$ be the polar representations of Γ_1 and Γ_2 with respect to 0 as origin, we observe from the asymptotic formulas (25.5a), (25.8) that the ratio $r_1(\vartheta)/r_2(\vartheta)$ has the limit unity as $\vartheta\to 0$ if the two curves Γ_1, Γ_2 intersect infinitely often as they extend to infinity (i.e. as $\vartheta\to 0$). Suppose then, contrary to the theorem, that $\limsup\limits_{\vartheta\to 0} r_1(\vartheta)/r_2(\vartheta)\leqq 1$, as would be the case if Γ_2 did not lie below Γ_1 at infinity. It follows that a similarity contraction of the flow region R_1 (bounded by $T_1+\Gamma_1$) with respect to 0 as center takes the flow in R_1 into one in $\bar{R}_1$ bounded by $\bar{T}_1+\bar{\Gamma}_1$, such that velocities are unchanged at corresponding points, and furthermore, such that R_2 (bounded by $T_2+\Gamma_2$) $\subset \bar{R}_1$ (Fig. 49a). If the contraction ratio is taken equal to $1/\varrho$ where $\varrho=\max\limits_{\vartheta\leqq\vartheta_A} r_1(\vartheta)/r_2(\vartheta)\geqq 1$, then $\bar{\Gamma}_1$ has a point of contact P with Γ_2. Comparison Theorem 1, applied to the flows in R_2 and $\bar{R}_1$ (which are assumed to have the same free stream speed), asserts inequality of the two flow speeds at P and therefore on the free streamlines, which is a contradiction. We may assume, therefore, that

$$\liminf_{\vartheta\to 0} r_1(\vartheta)/r_2(\vartheta)>1\,. \tag{28.1}$$

[1] The monotonicity theorem and analysis of the experimental data are due to P. GARABEDIAN: Pac. J. Math. **6**, 611–684 (1956).

[2] M. LAVRENTIEFF: Mat. Sbornik **46**, 391–458 (1938). — J. SERRIN: J. Rat. Mech. a. Analysis **2**, 563–575 (1953).

A typical configuration, if the theorem were untrue, is shown in Fig. 49b and no essential loss of generality is involved in restricting discussion to this case. Let $\varrho' \leqq 1$ be the minimum value or $r_1(\vartheta)/r_2(\vartheta)$ for $\vartheta \leqq \vartheta_A$, which, by virtue of (28.1), is achieved for some value ϑ_N corresponding to the point N on Γ_2. Also, let $\varrho'' \geqq 1$ be the maximum of $r_1(\vartheta)/r_2(\vartheta)$ for $\vartheta_A \geqq \vartheta \geqq \vartheta_N$, achieved at some point M on Γ_2. Let R_1', R_1'' be obtained from R_1 under similarity transformations in the ratio $1/\varrho'$ and $1/\varrho''$ (Fig. 49b), and denote the images of Γ_1 by Γ_1', Γ_1''. Now construct the flow region R_3 bounded by the curves $(-\infty, M)$ of Γ_1'', MN of Γ_2 and $(N, +\infty)$ of Γ_1', and observe that $R_1'' \supset R_3 \supset R_1'$. From the first comparison theorem, using evident notation,

$$\left.\begin{aligned} q_3(M) &< q_1''(M) = U, \\ q_3(N) &> q_1'(N) = U, \end{aligned}\right\} \quad (28.2)$$

where U is the cavity speed of the flows in R_1, R_1', R_1''. Using Eqs. (28.2) and applying the second comparison theorem to the flow in R_2 and R_3, we obtain the contradiction $1 = q_2(M)/q_2(N) < q_3(M)/q_3(N) < 1$. *QED.*

An immediate consequence of the above theorem is the result that the contraction coefficient in flow out of a conical or wedge-shaped vessel in a monotonically decreasing function of the angular opening (cf. Sect. 10).

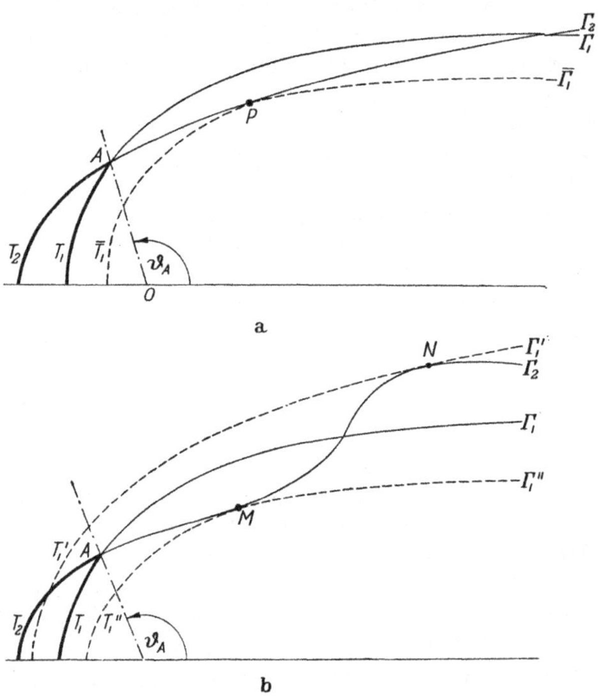

Fig. 49a and b.

Another type of monotonicity theorem has been obtained by GARABEDIAN[1] but does not seem to follow from comparison arguments alone. It states that in any free boundary problem formulated for the same fixed boundaries in both plane and axially symmetric flow, the region of flow expands with the dimension. We refer to the original work for the details.

29. Drag dependence on shape; body with minimum cavity drag. The preceding results are sometimes useful in comparing the cavity drags of different bodies. Since the approximate formula $C_D(\sigma) = C_D(0)(1+\sigma)$ is valid for small σ when the point of detachment is fixed, it sufficesi n this case to consider only the infinite cavity $(\sigma = 0)$.

Let T_1 and T_2 be given symmetric obstacles $(T_1 \neq T_2)$ having the same separation point, and let us suppose that T_2 lies above T_1 in the upper half plane. Then by the results of the preceding section, in particular by (28.1), it follows that the asymptotic width of the cavity—as measured by the shape factor C in (25.5a) or (25.8)—is greater for T_1 than for T_2. We conclude from (25.6), or from (25.9) in the case of axially symmetric flow, that *the cavity drag of T_1 exceeds that of T_2.*

[1] GARABEDIAN: Footnote 1, p. 385.

Consider now the problem of determining the symmetric obstacle of given dimensions whose infinite cavity drag is a minimum[1]. The obstacle is described by a simple curve defined for $0 \leq x \leq a$ with ordinates taking all values between 0 and b. The admissible class of obstacles will consist of curves of this type trailing an infinite cavity satsfying the Brillouin conditions, namely, (i) the free streamline may detach from C at any point provided it does not intersect C beyond the point of detachment; (ii) the maximum speed of the flow is achieved on the free streamline. A particular curve of this class is selected as follows. Let a Kirchhoff infinite cavity be determined by the vertical segment L: $x=0, 0 \leq y \leq h$, and let h be so chosen that the free streamline passes through the point (a, b). Denote by E the curve consisting of both the vertical segment and the arc of the free streamline joining it to (a, b). *The curve E achieves the minimum cavity drag among all admissible curves C*[2].

To establish the extremal property of E consider any other admissible curve C, and let its free streamline be Γ. If C lies below E, then the required inequality of the drag forces can be inferred as above. *If C lies above E somewhere in* $0 < x < a$, then there is a segment L': $x=0$, $0 \leq y \leq h'$, with $h' < h$, whose corresponding free streamline Γ' is tangent to C, say at P (Fig. 50). Plainly, Γ' must cross Γ, for otherwise Comparison Theorem 1 would provide a contradiction to (ii) at P. Hence the above result can again be applied, and we obtain the inequality, drag $C >$ drag $L' >$ drag E. *QED.*

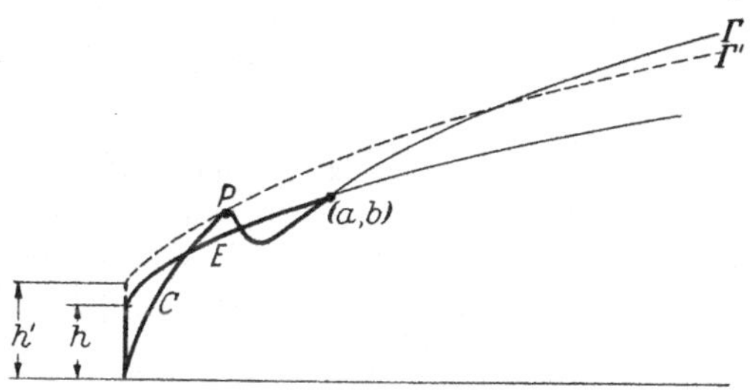

Fig. 50.

The same argument shows that the drag of the segment, $x=a$, $0 \leq y \leq b$, is the maximum for the curves under considerations.

c) Variational principles.

30. Principle of minimum virtual mass and kinetic energy. Most of the free surface flows discussed in this work can be characterized as solutions of variational problems. In this section we shall be concerned with one such extremal characterization bearing a close resemblance to HAMILTON's principle.

Consider a uniform flow in space past a finite body B, the free stream velocity being unity. The potential has an expansion at infinity of the form

$$\varphi = x + \frac{a\,x}{r^3} + \cdots. \tag{30.1}$$

A simple application of GREEN's theorem in the flow region R shows that the coefficient a is connected with the *virtual mass*

$$M = \iiint\limits_R \left[(\varphi_x - 1)^2 + \varphi_y^2 + \varphi_z^2\right] dx\,dy\,dz \tag{30.2}$$

[1] This problem was solved by M. LAVRENTIEFF: Mat. Sbornik **46**, 391—458 (1938) for plane flows and convex obstacles; the generalization discussed here, which holds also for axial symmetry and non-convex bodies, is due to J. SERRIN: J. Rat. Mech. a. Analysis **2**, 563—575 (1953).

[2] LAVRENTIEFF (preceding footnote) does not impose condition (ii) on his admissible class; on the other hand, he defines the drag of an obstacle to be the maximum drag for all possible points of detachment. The final result is the same.

and the volume V enclosed in B by the formula[1]

$$4\pi a = M + V. \tag{30.3}$$

(In the case of plane flow the factor 4 is replaced by 2.) Except for a factor of $\frac{1}{2}$ the virtual mass M is identical with the kinetic energy of the flow in a coordinate system moving with the free stream.

To derive the necessarity variational formulas let B be bounded by a fixed surface C and by a surface Γ which will be allowed to vary. The displacement of Γ into a neighboring surface $\overline{\Gamma}$ is assumed to be described by a shift $\delta n = \varepsilon\varrho\,(x, y, z)$ of each point on Γ in the direction of the outer normal. Let $\overline{\varphi}, \overline{a}$, etc., denote quantities associated with the flow bounded by $C + \overline{\Gamma}$. Our first objective is the basic formula for the variation in a,

$$\delta a = \frac{1}{4\pi}\iint\limits_{\Gamma} (\nabla\varphi)^2\,\delta n\, dS, \tag{30.4}$$

where dS is the surface element on Γ and $\delta a = \varepsilon a'(\varepsilon)_{\varepsilon=0}$. To obtain this formula we suppose for convenience that ϱ and Γ are analytic[2] so that φ and $\overline{\varphi}$ can be defined by analytic continuation in the exteriors of both B and $\overline{B}$. Let D and $\overline{D}$ be regions bounded by a sufficiently large sphere Σ and by B and $\overline{B}$ respectively. Then

$$\left.\begin{aligned}\iiint\limits_{\overline{D}} \nabla\varphi\cdot\nabla\overline{\varphi}\,dV - \iiint\limits_{D} \nabla\varphi\cdot\nabla\overline{\varphi}\,dV &= -\iint\limits_{\Gamma} \nabla\varphi\cdot\nabla\overline{\varphi}\,\delta n\,dS + o(\varepsilon)\\ &= -\iint\limits_{\Gamma} (\nabla\varphi)^2\,\delta n\,dS + o(\varepsilon);\end{aligned}\right\} \tag{30.5}$$

the latter equality follows from the fact that $\nabla\varphi$ and $\nabla\overline{\varphi}$ differ by $o(1)$ on Γ. Applying GREEN's identity and recalling that $\partial\varphi/\partial n = 0$ on $C + \Gamma$, $\partial\overline{\varphi}/\partial n = 0$ on $C + \overline{\Gamma}$, we obtain

$$\left.\begin{aligned}&\iiint\limits_{\overline{D}} \nabla\varphi\cdot\nabla\overline{\varphi}\,dx\,dy\,dz - \iiint\limits_{D} \nabla\varphi\cdot\nabla\overline{\varphi}\,dx\,dy\,dz\\ &\qquad = \iint\limits_{\Sigma}\left(\varphi\frac{\partial\overline{\varphi}}{\partial n} - \overline{\varphi}\frac{\partial\varphi}{\partial n}\right) dS = 4\pi(a - \overline{a}).\end{aligned}\right\} \tag{30.6}$$

Combining this result with Eq. (30.5), we arrive at Eq. (30.4).

We now pose the problem of minimizing the virtual mass (or kinetic energy) subject to the condition of fixed volume enclosed by B. Let us assume this problem has a solution. Since

$$\delta V = \iint\limits_{\Gamma} \delta n\,dS \tag{30.7}$$

it follows from Eqs. (30.3) and (30.4) that

$$\delta M = \iint\limits_{\Gamma} [(\nabla\varphi)^2 - 1]\,\delta n\,dS, \tag{30.8}$$

and therefore by the Lagrange multiplier rule there is a constant σ such that

$$0 = \delta M - \sigma\,\delta V = \iint\limits_{\Gamma} [(\nabla\varphi)^2 - 1 - \sigma]\,\delta n\,dS \tag{30.9}$$

---

[1] See, for example, P. GARABEDIAN and D.C. SPENCER: J. Rat. Mech. a. Analysis **1**, 359–409 (1952).

[2] The regularity properties required for the proof of Eq. (30.4) can be greatly lightened; cf. P. GARABEDIAN and M. SCHIFFER: J. d'Analyse Math. **2**, 281–368 (1953).

on the extremal surface Γ, whatever the variation δn. From the arbitrariness of δn one concludes in the usual way that

$$(\nabla\varphi)^2 = 1 + \sigma \quad \text{on} \quad \Gamma. \tag{30.10}$$

Γ is therefore a constant speed free surface in the flow past $C+\Gamma$. In other words, *the extremal body is a solution of the free boundary problem for the fixed boundary* C[1]. We note that the Lagrange multiplier is precisely the cavitation number of the flow.

A slightly different formulation seeks a solution of the problem

$$M - \sigma V = \text{minimum} \tag{30.11}$$

or

$$\delta(M - \sigma V) = 0 \tag{30.12}$$

for fixed $\sigma > 0$. It is clear from the preceding that on the extremal surface Γ the constant speed condition (30.10) is satisfied. The problem (30.12) can be viewed as a type of HAMILTON's principle in which $M/2$ is the kinetic energy of the flow, while the quantity $\sigma V/2$ is interpreted as the potential energy of the cavity, at pressure $\sigma/2$ and of volume V.

Similar variational problems can be formulated for other classes of free surface flows, including flows from a nozzle, and in a gravity field[2]. Of interest is the fact that the problem of the trailing vortex sheet, which is equivalent to the problem of three-dimensional lifting surfaces, can also be stated as a minimum problem for the virtual mass or kinetic energy[3]. The derivation of the infinite cavity problem from a minimum energy principle involves difficulties connected with the divergence of the integral (30.2) However, the variational problem can be made sensible by subtracting a suitable term describing the behavior of the flow at infinity and rendering the energy integral finite.

The variational procedure outlined above has an interesting application in the estimation of drag of the Riabouchinsky cavity[4]. Let C denote the fixed boundary consisting of the obstacle C_1 and its reflected image C_2, which we take to be cones whose axes coincide with the x-axis and whose vertices lie a distance h from the origin. The cones may degenerate into plane sections (which need not be circular). The free surface Γ spans the oppositely situated curves at the rim of the cones. Let us now expand the entire boundary $C+\Gamma$ by factor $1+\eta$, where η is small and positive. As in Eqs. (30.7) and (30.9), we obtain

$$\delta M - \sigma\,\delta V = \iint\limits_{C+\Gamma} \left[(\nabla\varphi)^2 - 1 - \sigma\right] \delta n \, dS. \tag{30.13}$$

[1] The present discussion is based on the work of GARABEDIAN and SPENCER, footnote 1, p. 388, who elaborated the ideas and founded an existence theory on this variational principle (cf. Sect. 35). The fact that stationary energy is equivalent to the free boundary condition was first pointed out by D. RIABOUCHINSKY: C. R. Acad. Sci., Paris **185**, 840—841 (1927). A formulation of the above minimum principle was achieved independently by M. SCHIFFER in 1947 (cf. M. BERGMAN and M. SCHIFFER: Kernel Functions and Elliptic Differential Equations in Mathematical Physics. New York: Academic Press 1953). A similar variational principle, stated for flows from a nozzle, was propounded by K. FRIEDRICHS: Math. Ann. **109**, 60—82 (1933—1934), who based a uniqueness theorem on the positive character of the second variation; (cf. Sect. 38). This formulation is essentially the same as in the classical Dirichlet principle, in which the competing functions are arbitrary and are not selected only from the more restricted class of harmonic functions, as above.

[2] Cf. GARABEDIAN and SPENCER, footnote 1, p. 388.

[3] Cf. GARABEDIAN and SCHIFFER, footnote 2, p. 388.

[4] For the following see P. GARABEDIAN: Pacif. J. Math. **6**, 611—684 (1956); another treatment, without use of variational methods, is presented by the same author in Bull. Amer. Math. Soc. **62**, 219—235 (1956).

Along Γ the integrand vanishes, while on C we have that $\delta n = \eta h \sin\vartheta$ where ϑ is the angle made by the elements of the cones with the x-axis. It follows that

$$\delta M - \sigma\,\delta V = \eta\, h \iint\limits_{C_1+C_2} \left[(\nabla\varphi)^2 - 1 - \sigma\right] dy\,dz. \tag{30.14}$$

At the same time we note that the drag D of the obstacle C_1 is given by the expression

$$D = \tfrac{1}{2} \iint\limits_{C_1} \left[1 + \sigma - (\nabla\varphi)^2\right] dy\,dz, \tag{30.15}$$

and therefore the two surface integrals in Eq. (30.14) must each have the value $-2D$. From Eq. (31.13) it follows that

$$\delta(M - \sigma V) = -\,4\eta\, h\, D. \tag{30.16}$$

But since V and M have the dimension of length cubed, and so increase under the magnification by the factor $(1+\eta)^3$, the left member of Eq. (30.16) is simply $3\eta(M - \sigma V)$. We conclude that

$$4h\,D = 3(\sigma V - M). \tag{30.17}$$

This formula can be used estimating the drag because of the minimum property (30.11) satisfied by $M - \sigma V$. For, if σ is known accurately, and if the cavity shape is known within an error δn in the displacement of the free surface, then by virtue of the minimum principle a lower bound for the drag D is given by Eq. (30.17) to within an error of higher order in δn. This fact is exploited in the calculations of axially symmetric cavities discussed in Sect. 45.

Another consequence of Eq. (30.17) is the somewhat paradoxical one that the Riabouchinsky cavity drag of an arbitrary flat plate depends only on the cavitation number σ and the cavity length $2h$, independently of the shape. For let C_1 and C_2 be plane sections which vary in shape while σ and h remain constant; then $\delta(M - \sigma V) = 0$, and the constancy of D follows from Eq. (30.17). The significance of this result is unclear, but it implies that the drag coefficient of a long, variable plate, whose dimensions are chosen so that σ and h remain constant, tends to zero as the length approaches infinity. This very unlikely proposition casts doubt on the existence of the three-dimensional Riabouchinsky cavity in general.

31. Minimax principle. A second type of extremal characterization of free surface flows is obtained from a minimax principle for the velocity[1]. Consider, for example, the plane or axially symmetric Riabouchinsky cavity generated by a curve C_1 in the upper half plane whose reflected image in the y-axis is C_2. Let Γ be the free streamline joining the endpoints of C_1 and C_2, and let the incident velocity be unity. Consider now, for any other are L joining the endpoints of C_1 and C_2, the flow of unit velocity past the curve $B = C_1 + L + C_2$, and denote by $v(L)$ the maximum speed achieved by this flow on L. The minimax principle asserts that *among all possible curves L the free streamline Γ is the one for which $v(L)$ is a minimum.* In other words, the speed v^* on the free streamline satisfies the condition,

$$\min_{L}\ \max_{P\in L} q(P) = v^* \tag{31.1}$$

where $q(P)$ is the speed at the point $P \in L$ in the flow past B; furthermore, Γ is the only curve having v^* as its maximum speed.

This extremal property of Γ is proved by a simple comparison argument. Under a suitable contraction (possibly the identity), B can be taken into a curve

[1] Cf. P. GARABEDIAN and D.C. SPENCER: J. Rat. Mech. a. Analysis 1, 359–409 (1952).

$B' = C_1' + L' + C_2'$ whose interior is contained within the cavity $C_1 + \Gamma + C_2$ such that L' has a point P of contact with Γ. The flow past B is taken into a similar flow past B' having the same velocity at corresponding points. From the first comparison theorem of Sect. 27, we conclude the inequality $q(P) > v^*$ unless $L' = \Gamma$, whence Eq. (31.1) follows immediately. It is clear from the argument that the cavity flow can be just as well characterized by a ***maxmin*** principle. This has been made the basis of an existence theory by GARABEDIAN and SPENCER. They show directly, without presupposing existence of the cavity flow, that the body having the maxmin property must solve the free boundary problem.

Other types of free surface flows, such as the cusped and infinite cavities, and jet flows from a nozzle, can also be described by suitable minimax principles.

A related extremal property is satisfied by the Riabouchinsky cavity C behind the flat plate or disc. Namely, among all bodies of the same length-width ratio as C in a flow with fixed incident velocity, C is the one having the smallest maximum speed on its boundary. In the case of compressible fluids this is equivalent to the characterization of the Riabouchinsky cavity behind the flat plate or disc as the body of given dimensions for which the critical Mach number is maximum[1]. These statements are proved easily by the comparison method.

IV. Existence and uniqueness theory.

The conditions that are proper to determine a unique and meaningful solution of the free boundary problem for curved barriers have been studied at length since BRILLOUIN and VILLAT emphasized the possibility of pathological behavior and indeterminacy of solutions. Probably because of the mathematical difficulties in taking account of the physical requirements (BRILLOUIN's conditions), the general existence and uniqueness theory has developed in such a way that most effort has been applied to the problem of fixed detachment and (somewhat less) to the problem of smooth detachment, with only incidental attention being given to the physical significance of the solutions. The theory of these problems is now well advanced, after having attracted some of the world's leading analysts and inspired many original and important ideas. In the following sections we trace the principal developments of the theory, but, because of the length and complexity of the proofs, we are limited in most cases to only a broad outline of the methods.

a) Existence theory.

32. Survey of the results. We first summarize the results achieved by several different approaches to the existence problem. Before entering into the details, we remark that while these results are of qualitative interest and often add to our insight concerning cavity and jet phenomena, all the general proofs are nonconstructive in character and have only limited usefulness in the calculation of flows.

Apparently the first existence theorem for curved boundaries was obtained by NEKRASOFF[2] for infinite cavity flow past circular arcs of small extent. He used the method of successive approximations, based directly on iteration of the functional equation of the flow (to be discussed in Sect. 33). This proof has been extended[3] to include other convex curved barriers and flow models, and is

[1] Cf. D. GILBARG and M. SHIFFMAN: J. Rat. Mech. a. Analysis **3**, 209–230 (1954).

[2] N. A. NEKRASOFF: Publ. Inst. Polytech. Ivanovo-Voszniesiensk, 1922; for related work see the references in BIRKHOFF and ZARANTONELLO [*1*], Chap. 7.

[3] Cf. BIRKHOFF and ZARANTONELLO [*1*], Chap. 7.

applicable as long as the parameter M in the functional equation is sufficiently small.

The first general attack on the existence problem was by the method of continuity, applied initially by WEINSTEIN, and elaborated by him and others[1]. In this method the existence of a jet flow from a given polygonal channel (or cavity flow past a polygonal obstacle) is inferred as the end result of a process of continuous deformation starting from a reference configuration for which a solution is known. The existence of the flow corresponding to a curved boundary is obtained by polygonal approximation. The earlier work using this method proved the existence of jet flows (with fixed detachment) from concave symmetric channels of finite width whose walls each turn through an angle less than π[2]. Later FINN extended this result to a wide class of concave symmetric channels without limitation, other than finiteness, on the total angular variation of the walls, and stated analogous results for cavity flows. He also proved the existence of flows through asymmetric concave channels having the following properties: the total angular change of the walls does not exceed π, and the separation points lie on opposite sides of some line parallel to the incident flow. The unknown direction of the jet seems to present a major difficulty in the solution of the asymmetric jet problem. The proofs of existence and uniqueness by the method of continuity are closely intertwined and are therefore discussed together in Sect. 38.

A fundamental advance, greatly broadening the scope of the existence theory, was provided by LERAY[3] with his application of the Leray-Schauder fixed point method. This approach, which is basically an extension of the method of continuity to function spaces, has contributed a variety of results to the free boundary problem. LERAY considered the infinite cavity problem for a class of obstacles (symmetric or asymmetric) intersected in at most a single point or segment by straight lines parallel to the incident flow. For these obstacles[4] he proved that the problem of fixed detachment and the prow problem (cf. Sect. 24) always have a solution.

Also using the fixed point method, SERRIN[5] extended the class of obstacles and the flow models covered by the existence theory. He treated symmetric flows past starlike obstacles[6] and later[7] showed that his results could be stated for any symmetric obstacle with the property that its upper half and the adjoining axis of symmetry lie on one side of a straight line through the endpoint.

To describe these results it is useful to formulate the ***mixed cavity problem***[8]. For a given point of detachment this problem seeks as solution either a cusped

[1] A. WEINSTEIN: Math. Z. **19**, 265–274 (1924); **31**, 424–433 (1929). — G. HAMEL: Proc. Second Int. Cong. Appl. Mech. Zurich 1926, p. 489. — H. WEYL: Nachr. Ges. Wiss. Göttingen **1927**, 227–237. — K. FRIEDRICHS: Math. Ann. **109**, 60–82 (1933/34). — J. LERAY and A. WEINSTEIN: C. R. Acad. Sci., Paris **198**, 430 (1934). — R. FINN: J. d'Analyse Math. **4**, 246–291 (1956). For other references and discussion see A. WEINSTEIN: Proc. Symposia Appl. Math., Vol. 1, Amer. Math. Soc., 1949.

[2] Cf. LERAY and WEINSTEIN, preceding footnote.

[3] J. LERAY: Comm. Math. Helv. **8**, 149–180 (1935).

[4] The following regularity properties are assumed: in the problem of fixed detachment the inclination $\Psi(l)$ ($l=$ arc length) obeys a Hölder condition, and in the problem of smooth detachment and the prow problem the curvature $\Psi'(l)$ obeys such a condition. The same regularity hypotheses are made in the later applications of the fixed point method.

[5] J. SERRIN: J. Rat. Mech. a. Analysis **1**, 1–48 (1952).

[6] In the following a plane obstacle C will be called *starlike* if it satisfies Eq. (26.1) or if the curve consisting of C plus the upstream axis of symmetry is starlike with respect to a point on or below the axis of symmetry (cf. Sect. 26, footnote 5, p. 379).

[7] J. SERRIN: J. Rat. Mech. a. Analysis **2**, 563–575 (1953).

[8] This is the *schlicht cavity problem* in SERRIN's terminology.

cavity or a non-overlapping infinite cavity. It is easily shown by the comparison method that both possibilities cannot occur simultaneously. SERRIN establishes that[1]: (1) the problem of fixed detachment for the infinite cavity always has a unique solution (which may, however, overlap the axis of symmetry); (2) the mixed cavity problem always has a unique solution, and this solution is a cusped cavity if and only if the corresponding infinite cavity for that point of detachment is overlapping; (3) on any closed starlike curve the solutions of the cusped cavity problem take on all cavity speeds between zero and the free stream velocity, and at least one of the solutions of the infinite cavity problem is a cavity of zero asymptotic width; (4) if the obstacle has a local maximum, there is at least one preceding point at which the corresponding infinite cavity detaches smoothly; (5) at a given point of detachment, if U is the cavity speed of the corresponding solution of the mixed cavity problem, then the re-entrant jet problem has a solution for all cavity speeds in excess of U and only for these values[2]; the re-entrant jet flow region contains in its interior the region belonging to the solution of the mixed cavity problem.

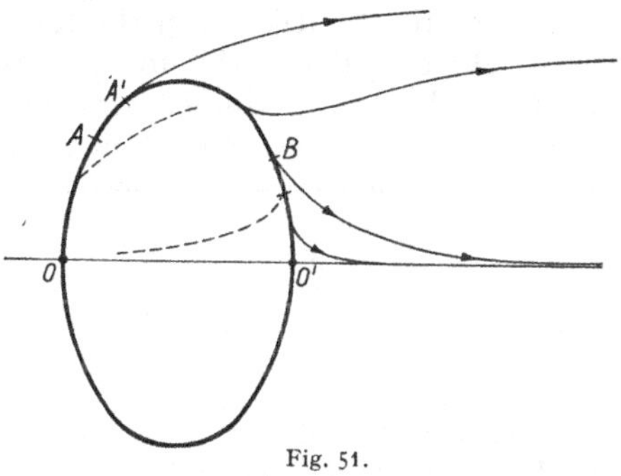

Fig. 51.

The various possibilities for flow past a convex closed obstacle can be summarized as follows[3] (see Fig. 51). There is an initial arc OA, starting at the stagnation point, such that the infinite cavity detaching from any point on OA is convex and has infinite curvature at separation; at A the detachment is smooth; there is an arc $A'B$ containing the maximum point of the obstacle such that the detaching free streamline from any interior point on $A'B$ has one inflection and stays above the axis of symmetry[4]; at B the free streamline is asymptotic to the axis; for each point on the arc BO' (where O' is the rear stagnation point) there is a unique solution of the cusped cavity problem, the value of the corresponding cavity speed decreasing monotonically to zero as the arc is traversed from B to O'; at each point on the obstacle the re-entrant jet problem has a solution related to the corresponding infinite or cusped cavity flow in accordance with (5) above.

BIRKHOFF and ZARANTONELLO [*1*] have achieved a relatively simple unified treatment of the existence problem for symmetric convex obstacles by the fixed point method. Their proof encompasses all the cavity models considered here and both fixed and smooth separation (cf. Sect. 34). The first existence proof for

[1] Of the following results those numbered (1) to (3) were first obtained by M. LAVRENTIEFF: Mat. Sbornik **46**, 391–458 (1938), but under more restrictive assumptions on the class of obstacles.

[2] In other words, the re-entrant jet flow may have a cavitation number *less than zero*.

[3] This description combines results due to LAVRENTIEFF, footnote 1, and SERRIN, footnote 5, p. 392.

[4] For the general convex obstacle there may be more than one point of smooth detachment, all of which lie on the arc AA', and furthermore, the corresponding free streamline may intersect the obstacle. However, if the curvature is non-decreasing on the face of the obstacle, there is exactly one point of smooth detachment (so that $A = A'$) and the corresponding flow satisfies the Brillouin conditions; (cf. Sect. 24 and J. LERAY: Volume du Jubilé de M. Brillouin, pp. 246–257. Paris: Gauthier-Villars 1935.) LAVRENTIEFF, by a simple comparison argument, has shown in general that there is a point on the face such that the corresponding free streamline does not intersect the body; the detachment will usually not be smooth, of course.

compressible fluids was given by BERG[1], who extended to subsonic infinite cavity flows essentially the same results as LERAY's (see above). This application of the fixed point method is based on function theoretic principles for elliptic partial differential equations and uses ideas developed by BERS[2] in his existence theory for subsonic flows past profiles. For other applications of the fixed point method we refer to the literature[3].

Variational methods have been the other major source of existence theorems for the free boundary problem. The basic idea of these methods is to start with an extremum characterization of the desired solution of a flow problem (cf. Sects. 30, 31), and then to establish the existence of a solution of the extremum problem. The first such contribution was made by LAVRENTIEFF[4], who characterized the free streamline as that curve (among a class of competing curves) on which the total variation in the flow speed is a minimum, which proves equal to zero. He considered the mixed cavity problem and the unrestricted infinite cavity problem for symmetric obstacles having (in the upper half plane) at most a single point or segment of intersection with any vertical line, and proved by arguments based on normal families and domain variation the existence of flows for which the zero minimum is actually achieved. A number of the preceding important results were established in this way for the first time.

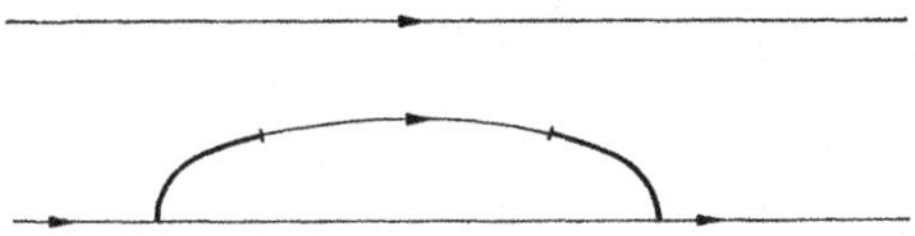

Fig. 52.

The variational principles of Sects. 30, 31 have been made the basis of an existence theory by GARABEDIAN and SPENCER[5]. Using the principle of minimum virtual mass, they prove the existence of the Riabouchinsky cavity behind a symmetric obstacle[6] at a fixed (but arbitrary) distance from its image[7], and a finite half-cavity in a straight channel in the presence of gravity (Fig. 52). The existence of other flows can be inferred as limiting cases when gravity is zero and when the image obstacle recedes to infinity. The same authors obtain the Riabouchinsky and cusped cavities behind convex bodies by the minimax principle of Sect. 31.

A basic contribution of the variational method was the first proof, by GARABEDIAN, LEWY, and SCHIFFER[8], of existence of axially symmetric cavity flows. Applying the principle of minimum virtual mass, they obtain first the Riabouchinsky cavity generated by a monotonic curve, and then the infinite cavity as a limiting case of this flow when the image obstacle moves off to infinity. Although the basic plan of the proof is the same for both plane and axially symmetric flows, fundamentally new ideas are required in the latter case to compensate for the loss of conventional complex variable methods.

[1] P. BERG: Diss. New York University 1953.

[2] L. BERS: Comm. Pure Appl. Math. **7**, 441–504 (1954).

[3] J. KRAVTCHENKO: J. Math. pures appl. **20**, 35–239 (1941) —cavity flows in a channel, extensive bibliography; Ann. Sci. Ec. Norm. Sup. **62**, 233–268 (1946); **63**, 161–184 (1947); A. OUDART: J. Math. pures appl. **22**, 245–320 (1943); **23**, 1–36 (1944); R. HURON: C. R. Acad. Sci., Paris **228**, 290, 357 (1949); P. THERON: C. R. Acad. Sci., Paris **228**, 1922 (1949); R. GERBER: J. Math. pures appl. (9) **34**, 185–299 (1955) — Flows in a gravity field.

[4] LAVRENTIEFF: Footnote 1, p. 393.

[5] P. GARABEDIAN and D. C. SPENCER: J. Rat. Mech. a. Analysis **1**, 359–409 (1952).

[6] It is assumed that the curve is analytic and starlike. GARABEDIAN and SPENCER assert that the former assumption can be removed by approximating smooth curves analytically (cf. preceding reference, p. 387).

[7] An alternative proof has been given by P. GARABEDIAN and H. ROYDEN: Proc. Nat. Acad. Sci. U.S.A. **38**, 57–61 (1952).

[8] P. GARABEDIAN, H. LEWY and M. SCHIFFER: Ann. of Math. **56**, 560–602 (1952).

33. The functional equations. In this section we formulate the flow problems in terms of functional equations that play a central role in the existence theory and in the calculation of flows past curved obstacles. To illustrate the principles we consider first the infinite cavity past a given smooth body with prescribed separation points (Fig. 39). Let the inclination Ψ of the tangent on the fixed boundary AOB be given as a function $\Psi(l)$, where l is arc length measured along AOB from A to B, and let $l=a$ at A, $=b$ at B. For the present it suffices to assume that $\Psi(l)$ is Hölder continuous in l for $a\leqq l\leqq b$. Assuming the flow to be given, let $l(s)$ provide the boundary correspondence between AOB and the semi-circle $\zeta=e^{is}$, $0\leqq s\leqq\pi$, in the Levi-Civita plane. The inclination ϑ of the velocity vector is related to Ψ on AOB by

$$\vartheta(e^{is})=\begin{cases}\Psi[l(s)]-\pi, & 0\leqq s<s_0\\ \Psi[l(s)], & s_0<s\leqq\pi.\end{cases} \tag{33.1}$$

Let us now define the function

$$\Omega(\zeta)=\omega(\zeta)-i\log\frac{\zeta-e^{is_0}}{\zeta\, e^{is_0}-1} \tag{33.2}$$

where as before $\omega=i\log df/dz=\vartheta+i\tau$. From the properties of the second term on the right (cf. Sect. 6) it follows that $\operatorname{Re}\Omega(e^{is})=\Psi[l(s)]$[1], and hence $\Omega(\zeta)$ is Hölder continuous in the closed disc $|\zeta|\leqq 1$[2]. Also, by the Schwarz-Poisson formula, and using the fact that $\Omega(\zeta)$ is real on the real axis, we have

$$\Omega(\zeta)=\Theta+iT=\frac{1}{\pi}\int\limits_0^\pi\Psi[l(s)]\frac{1-\zeta^2}{1-2\zeta\cos s+\zeta^2}\,ds, \tag{33.3}$$

from which $T(e^{is})$ can be expressed by the formula

$$T(e^{is})=\frac{\sin s}{\pi}\int\limits_0^\pi\frac{\Psi[l(s')]-\Psi[l(s)]}{\cos s'-\cos s}\,ds'. \tag{33.4}$$

$\Theta(e^{is})$ is the inclination of the obstacle, parametrized with respect to arc length on $|\zeta|=1$.

For a given obstacle [i.e., given $\Psi(l)$], the function Ω is completely determined if $l(s)$ is known. We now show that $\omega(\zeta)$ and $f(\zeta)$, and hence the entire flow, are also determined by $l(s)$. From $\omega(0)=0$ and Eq. (33.3) we infer that

$$s_0=\Theta(0)=\frac{1}{\pi}\int\limits_0^\pi\Psi[l(s)]\,ds \tag{33.5}$$

so it remains only to determine M in the expression (23.2) for $f(\zeta)$. Inserting

$$\frac{df}{d\zeta}=\frac{M}{2}\zeta^{-3}(\zeta^2-1)(\zeta-e^{is_0})(\zeta-e^{-is_0})$$

[1] If there is a corner of angle $\beta\pi$ at O, the logarithm term in Eq .(33.2) is multiplied by β; see [1] for the treatment of this more general case.

[2] The boundary mapping $z(e^{is})$, and hence $l(s)$, is differentiable since AOB has a Hölder continuously turning tangent (cf. G. Gategno and A. Ostrowski: Mém. Sci. Math. No.110). Thus $\Psi[l(s)]$ is Hölder continuous in s with the same exponent as $\Psi(l)$. It follows by a theorem of A. Korn, Abh. Kgl.-Preuss. Akad. Wiss., Berlin 1909, Anhang II [which is usually attributed to J. Privaloff, Bull. Soc. Math. France **44**, 100–103 (1916)], that $\Omega(\zeta)$ is also Hölder continuous in $|\zeta|\leqq 1$ with this exponent.

into the equality $dz/d\zeta = e^{i\omega(\zeta)}\, df/d\zeta$, we obtain[1]

$$\frac{dl}{ds} = 4M\, e^{-T(e^{is})} \sin^2 \frac{s+s_0}{2} \sin s, \tag{33.6}$$

or

$$l(s) = l(0) + 4M \int_0^s e^{-T(e^{is})} \sin^2 \frac{s+s_0}{2} \sin s\, ds. \tag{33.7}$$

The relation

$$l(\pi) - l(0) = b - a = 4M \int_0^\pi e^{-T(e^{is})} \sin^2 \frac{s+s_0}{2} \sin s\, ds \tag{33.8}$$

now determines M in terms of $l(s)$.

Conversely, if $T(e^{is})$, s_0, and M are defined in terms of $l(s)$ by Eqs. (33.4), (33.5), and (33.8), then the right member of Eq. (33.7) can be considered an operator $F[l, \Psi(l)]$ depending on $l(s)$ and $\Psi(l)$, and Eq. (33.7) is a functional equation

$$l = F[l, \Psi(l)] \tag{33.9}$$

for the determination of the boundary correspondence $l(s)$ in terms of the given function $\Psi(l)$. Suppose that $l(s)$ is any such solution for which $0 < s_0 < \pi$[2]. Then, as above, we may define an infinite cavity by the functions $f(\zeta)$ and $\omega(\zeta)$, and this flow has the given obstacle as a bounding streamline. Thus, *any solution* $l(s)$, $0 \leqq s \leqq \pi$, *of* Eq. (33.9) *for which* s_0 *satisfies the inequality* $0 < s_0 < \pi$, *determines an infinite cavity flow past the obstacle defined by* $\Psi(l)$.

The above considerations provide a means of constructing inverse and approximate solutions of the cavity problem. We notice first that any analytic function $\Omega(\zeta) = \Theta + iT$ defined in $|\zeta| < 1$, which is real on the real axis, and for which $0 < s_0 < \pi$ in Eq. (33.5), defines an infinite cavity flow past a body parametrized by its inclination $\Theta(e^{is})$ and arc length $l(s)$, the latter being given by Eq. (33.7). The parameter M can be chosen arbitrarily and essentially fixes the dimensions of the body. If the body is given, say by $\Psi = \Psi(l)$, then a flow past an approximating body can be found by making a reasonable choice of $l(s)$ to insert in Eq. (33.3). This defines $\Omega(\zeta)$ which, with Eqs. (33.5) and (33.8), determines the parameters s_0, M. The right member of Eq. (33.7) now defines a function $L(s)$, which in general is of course unequal to $l(s)$. The functions $\Theta(e^{is})$ and $L(s)$ describe a curve of the same length as the given one and varying through the same angles, but differing in actual shape. $\Omega(\zeta)$ determines an infinite cavity behind this approximating body.

In the *symmetric* cavity problem, $s_0 = \pi/2$, $l(\pi - s) = -l(s)$, and Eq. (33.7) can be written as

$$l(s) = \frac{b \int_{\pi/2}^{s} e^{-T(e^{is})} (1 + \sin s) \sin s\, ds}{\int_{\pi/2}^{\pi} e^{-T(e^{is})} (1 + \sin s) \sin s\, ds} = F[l, \Psi(l)]. \tag{33.10}$$

The functional equations associated with other free boundary problems can be derived in essentially the same way. We summarize the results for several of the flow models considered previously[3].

---

[1] This equation is essentially due to H. VILLAT: Ann. Sci. Ec. Norm. Sup. **28**, 203–240 (1911).

[2] This condition must be imposed if the flow is to contain a streamline that branches on the obstacle.

[3] Additional models are treated by BIRKHOFF and ZARANTONELLO [*1*], Chap. 6.

Re-entrant jet. Let $\zeta_B = b\,e^{i\beta}$, $\zeta_I = h\,e^{i\alpha}$ $(0 < h, b < 1)$ be the singularities of $\omega(\zeta)$ corresponding to the interior stagnation point B and to the point at infinity I (Fig. 40). Let the free stream velocity be U and the cavity speed unity. We now have

$$\omega(\zeta) = \Omega(\zeta) + i\log\frac{(\zeta-\zeta_B)(\zeta\bar{\zeta}_B-1)(\zeta-e^{is_0})}{(\zeta-\bar{\zeta}_B)(\zeta\zeta_B-1)(\zeta e^{is_0}-1)}, \tag{33.11}$$

where Ω is the analytic function in $|\zeta|<1$ given by Eq. (33.3); Ω is Hölder continuous in $|\zeta|\leqq 1$. Using the expression (23.8) for $df/d\zeta$ we get

$$\left.\begin{aligned} &\frac{dl}{ds} = 8M\,e^{-T(e^{is})}\,\nu(s;s_0,\zeta_I,\zeta_B),\\ &\text{where}\\ &\nu(s;s_0,\zeta_I,\zeta_B) = \frac{(1-2b\cos(s+\beta)+b^2)^2\sin^2\frac{s+s_0}{2}\sin s}{[(1-2h\cos(s-\alpha)+h^2)(1-2h\cos(s+\alpha)+h^2)]^2}.\end{aligned}\right\} \tag{33.12}$$

The determining equations for the parameters M, s_0, ζ_I, ζ_B in (33.12) are

$$\omega(\zeta_I) = i\log U, \tag{33.13}$$

$$\operatorname*{Res}_{\zeta=\zeta_I}\frac{dz}{d\zeta} = \frac{d}{d\zeta}\left[(\zeta-\zeta_I)^2 e^{i\omega(\zeta)}\frac{df}{d\zeta}\right]_{\zeta=\zeta_I} = 0, \tag{33.14}$$

$$a_2 - a_1 = 8M\int_0^\pi e^{-T(e^{is})}\,\nu(s;s_0,\zeta_I,\zeta_B)\,ds. \tag{33.15}$$

To these must be added one more condition expressing the freedom in choice of the circulation about the cavity. Denoting circulation by Γ, we may write this condition as

$$\operatorname{Im}\operatorname*{Res}_{\zeta=\zeta_I}\frac{df}{d\zeta} = \operatorname{Im}\frac{d}{d\zeta}\left[(\zeta-\zeta_I)^2\frac{df}{d\zeta}\right]_{\zeta=\zeta_I} = \Gamma \tag{33.16}$$

where $df/d\zeta$ is given by Eq. (23.8).

When $a_2 - a_1$, U, Γ, and $l(s)$ are known Eqs. (33.13) to (33.16) are two real and two complex equations for the same number of real and complex parameters s_0, M, ζ_B, ζ_I. We may consider Eqs. (33.12) to (33.16) as a system of functional equations for the determination of $l(s)$, ζ_I, ζ_B, s_0, and M when $\Psi(l)$ a_1, a_2, U, and Γ, are given. Any solution for which $0 < s_0 < \pi$, and $0 < h$, $b < 1$, $0 < \alpha$, $\beta < \pi$, provides a flow past the given obstacle with a re-entrant jet cavity having cavitation number $\sigma = 1/U^2 - 1$ and circulation Γ.

In the symmetric case we have $s_0 = \pi/2$, $\zeta_I = ih$, $\zeta_B = ib$, $\Gamma = 0$, $a_2 = -a_1 = a$, and the basic equations (33.12) to (33.16) reduce to

$$\left.\begin{aligned} &l(s) = \frac{a\int_{\pi/2}^{s} e^{-T(e^{is})}\,\nu(s;b,h)\,ds}{\int_{\pi/2}^{\pi} e^{-T(e^{is})}\,\nu(s;b,h)\,ds},\\ &\text{where}\\ &\nu(s;b,h) = \frac{(1+2b\sin s+b^2)^2(1+\sin s)\sin s}{(1+2h^2\cos 2s+h^4)^2};\end{aligned}\right\} \tag{33.17}$$

$$\frac{(h-b)(1-hb)(1-h)}{(h+b)(1+hb)(1+h)}\,e^{T(ih)} = U, \tag{33.18}$$

$$\Omega'(ih) + \frac{2}{h^2-1} + \frac{2b(1-h^2)}{h(h+b)(1+hb)} = 0. \tag{33.19}$$

Cusped cavity. In this model the ζ plane has the cusp C at the origin and the point at infinity as an interior point $\zeta_I = h\,e^{i\alpha}$ (Fig. 53). The mapping $f(\zeta)$ is given by[1]

$$\frac{df}{d\zeta} = M\frac{\zeta(\zeta^2-1)(\zeta-e^{is_0})(\zeta-e^{-is_0})}{(\zeta-\zeta_I)^2(\zeta-\bar{\zeta}_I)^2(\zeta\zeta_I-1)^2(\zeta\bar{\zeta}-1)^2}.$$

[1] This is easily derived from Eq. (5.6); see J. SERRIN: J. Rat. Mech. a. Analysis 1, 1–48 (1952) for details.

Since $\omega(\zeta)$ is regular in $|\zeta| < 1$, the relations (33.2) to (33.4) remain unchanged. The functional equation for $l(s)$ is now

$$\left.\begin{aligned} &\frac{dl}{ds} = 8M\,\mathrm{e}^{-T(\mathrm{e}^{is})}\,\nu(s;\,s_0,\,\zeta_I) \\ &\text{where} \\ &\nu(s;\,s_0,\,\zeta_I) = \frac{\sin^2\dfrac{s+s_0}{2}\,\sin s}{[(1-2h\cos(s-\alpha)+h^2)\,(1-2h\cos(s+\alpha)+h^2)]^2}\,. \end{aligned}\right\} \qquad (33.20)$$

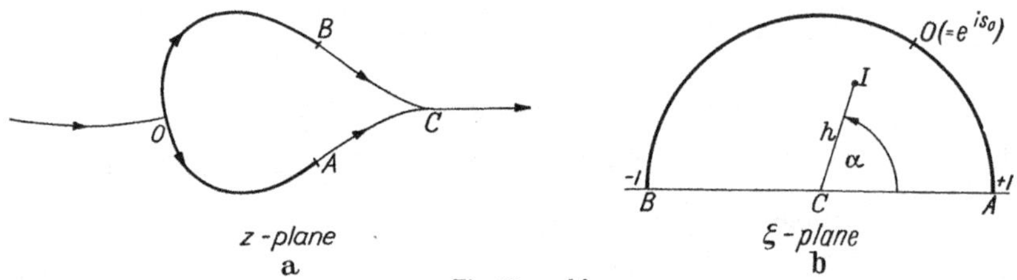

Fig. 53a and b.

The determining equations for the parameters s_0, M, ζ_I, and the speed U at infinity (which is an unknown if the cavity speed is fixed at unity), are again Eqs. (33.13) to (33.15), with obvious modifications for the present flow. When $l(s)$ and $b-a$ are given, these are five real equations for the five real constants s_0, M, h, α, and U. We observe that the circulation cannot be prescribed in this model, but rather is a derived quantity.

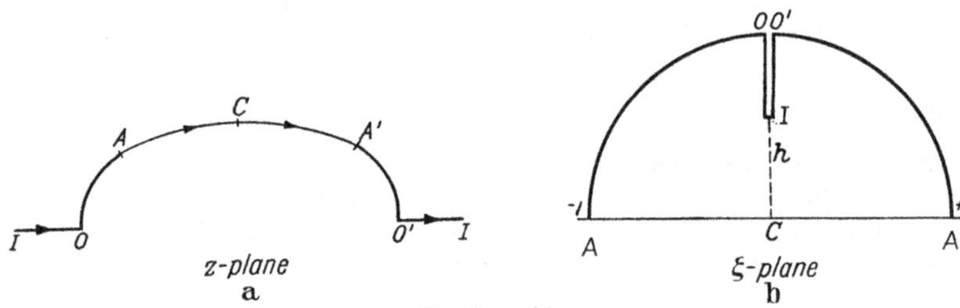

Fig. 54 a and b.

In the case of symmetric flow the basic equations (33.13) to (33.15), (33.20), become

$$\frac{1-h}{1+h}\,\mathrm{e}^{T(ih)} = U,$$

$$\Omega'(i\,h) + \frac{2}{h^2-1} = 0,$$

$$l(s) = \frac{b\int\limits_{\pi/2}^{s}\mathrm{e}^{-T(\mathrm{e}^{is})}\,\nu(s;\,h)\,ds}{\int\limits_{\pi/2}^{\pi}\mathrm{e}^{-T(\mathrm{e}^{is})}\,\nu(s;\,h)\,ds}, \qquad \text{where} \qquad \nu(s;\,h) = \frac{(1+\sin s)\sin s}{(1+2h^2\cos 2s+h^4)^2}\,.$$

Riabouchinsky cavity. The upper half of the flow plane and its image in the ζ plane are indicated in Fig. 54. Since ω is regular in $|\zeta| < 1$ (including the slit) we have Eqs. (33.2) to (33.4) as before, with $s_0 = \pi/2$. The functional equation for $l(s)$ takes the form

$$l(s) = \frac{b\int\limits_{\pi/2}^{s}\mathrm{e}^{-T(\mathrm{e}^{is})}\,\nu(s;\,h)\,ds}{\int\limits_{\pi/2}^{\pi}\mathrm{e}^{-T(\mathrm{e}^{is})}\,\nu(s;\,h)\,ds}, \qquad \text{where} \qquad \nu(s;\,h) = \frac{(1+\sin s)\sin s}{(1+2h^2\cos 2s+h^4)^{\frac{3}{2}}},$$

and

$$\omega(ih) = i\log U$$

is the determining equation for h when U is the free stream velocity.

In the preceding we have assumed as given the points of detachment and hence the corresponding arc lengths. In the problem of smooth detachment this knowledge is replaced by the conditions $\omega'(1)=0$ and $\omega'(-1)=0$ (cf. Sect. 24), and these in turn can be expressed in terms of the shape of the obstacle and $l(s)$ by the formulas

$$\left.\begin{aligned} A[l,\Psi(l)] &= \frac{1}{2\pi}\int_0^\pi \{\Psi[l(0)]-\Psi[l(s)]\}\frac{ds}{\sin^2\frac{s}{2}} - \cot\frac{s_0}{2} = \omega'(1) = 0,\\ B[l,\Psi(l)] &= \frac{1}{2\pi}\int_0^\pi \{\Psi[l(s)]-\Psi[l(\pi)]\}\frac{ds}{\cos^2\frac{s}{2}} - \tan\frac{s_0}{2} = \omega'(-1) = 0.\end{aligned}\right\} \qquad (33.21)$$

These conditions must be added to previous ones [Eqs. (33.5), (33.6), (33.8), in the infinite cavity problem] to complete the system of equations for the problem of smooth detachment. For the prow problem the equalities in the relations (33.21) are replaced by inequalities, $A[l,\Psi(l)]\leqq 0$, $B[l,\Psi(l)]\leqq 0$[1].

An alternative form of the functional equations, based on curvature rather than arc length, has proved useful in the treatment of convex bodies, both in the existence theory and in numerical calculations[2]. Let the inclination of the obstacle be described by the angle $\Phi=\Psi-a_0$, where $a_0=\Theta(0)$, and express the curvature in the form $K(\Phi)=-d\Phi/dl$ (whence $K>0$ if the body is convex to the flow). Then the Eqs. (33.6), (33.12), etc., for $l(s)$, which we write as

$$\frac{dl}{ds} = M\,e^{-T(e^{is})}\,\nu(s), \qquad (33.22)$$

can be transformed into

$$\lambda(s) = -\Phi'(s) = M\,e^{-T(e^{is})}\,K(\Phi(s))\,\nu(s) \qquad (33.23)$$

(with the obvious notation $\Phi(s)=\Psi[l(s)]-a_0=\Theta(e^{is})-\Theta(0)$).

This equation is convenient in that T and Φ can be expressed simply in terms of λ and can be formally eliminated. Indeed, one verifies that since $\Phi(s)$ has zero mean value in $(0,\pi)$,

$$\Phi(s) = \int_0^\pi J(s,\sigma)\,\lambda(\sigma)\,d\sigma \equiv I[\lambda] \qquad (33.24)$$

where

$$J(s,\sigma) = \frac{2}{\pi}\sum_{m=1}^{\infty}\frac{\cos m s\sin m\sigma}{m} = \begin{cases} -\sigma/\pi, & 0\leqq\sigma<s\\ \frac{1}{2}-\sigma/\pi, & s=\sigma\\ 1-\sigma/\pi, & s<\sigma\leqq\pi.\end{cases} \qquad (33.25)$$

Also, by a formula of Dini[3] which relates the boundary values of a harmonic function on a circle with its normal derivative ($\partial T/\partial n=-\lambda$),

$$T(e^{is}) = \int_0^\pi D(s,\sigma)\,\lambda(\sigma)\,d\sigma \equiv D[\lambda], \qquad (33.26)$$

where

$$D(s,\sigma) = \frac{2}{\pi}\sum_{m=1}^{\infty}\frac{\sin m s\sin m\sigma}{m} = \frac{1}{\pi}\log\left|\frac{\tan\frac{s}{2}+\tan\frac{\sigma}{2}}{\tan\frac{s}{2}-\tan\frac{\sigma}{2}}\right|. \qquad (33.27)$$

---

[1] Cf. J. Leray: Comm. Math. Helv. **8**, 149–180 (1935).

[2] For the following see [*1*], Chap. 6.

[3] M. Dini: Ann. di Mat. (2) **5**, 305–345 (1871).

We may therefore write the *curvature equation* (33.23) in the form

$$\lambda = M \nu \, \mathrm{e}^{-D[\lambda]} K(I[\lambda]). \tag{33.28}$$

By means of Eqs. (33.24) and (33.26) the various conditions at detachment can all be expressed directly in terms of λ by integral relations of the form

$$\int_0^\pi \lambda(s) f(s) \, ds = 1 \tag{33.29}$$

where f is a suitable function. Thus, in symmetric flow, $f = \text{const}$ corresponds to a fixed separation angle, and hence to fixed detachment, $f = \pi^{-1} \operatorname{cosec} s$ corresponds to smooth detachment, and $f = \pi^{-1} \sin s$ to a cavity of zero asymptotic width (cf. Sect. 40).

34. The fixed point method. The proofs of existence by the fixed point method have been based thus far on the Leray-Schauder[1] theory of functional equations the basic principles of which are summarized in the following statement due almost verbatim to LERAY[2]:

Consider an equation of the form

$$x = F(x) \tag{34.1}$$

where x is a point of normed, linear, and complete (Banach) space E, and F is a completely continuous[3] transformation defined on E. We associate with this equation the mapping

$$y = x - F(x). \tag{34.2}$$

Let D be a bounded domain in E whose boundary contains no solution of Eq. (34.1). The topological degree[4] at $y=0$ of the mapping (34.2) restricted to D is called the *total index* of the solutions of Eq. (34.1) in D. The total index remains constant when the mapping $F(x)$ is varied continously[5] without any solution of Eq. (34.1) reaching the boundary of D. Existence theorems can be inferred by the following procedure. The Eq. (34.1) is reduced continuously, without any of its solutions reaching the boundary of D, to a sufficiently simple equation $x = F_0(x)$ for which the total index of the solutions in D can be determined; if it differs from zero[6], the total index of Eq. (34.1) in D, being the same, must also be different from zero; hence D must contain at least one solution of Eq. (34.1).

We sketch the procedure in the specific case[7] of symmetric infinite cavity flow with fixed detachment from an arc OB (in the upper half plane). For definiteness it is assumed that OB is starlike (cf. footnote 6, p. 392) and that its inclination $\Psi(l)$, $0 \leqq l \leqq b$, is Hölder continuous in l. Let E denote the Banach space of continuously differentiable functions $l(s)$, $\pi/2 \leqq s \leqq \pi$, with the norm $\|l(s)\| = \max|l(s)| + \max|l'(s)|$. We have seen in the preceding section that the solution of the flow problem is equivalent to finding an element $l(s) \in E$

[1] J. LERAY and J. SCHAUDER: Ann. Sci. Ec. Norm. Sup. **51**, 45—78 (1934).

[2] J. LERAY: Comm. Math. Helv. **8**, 149—180 (1935).

[3] A mapping is completely continuous if it is continuous and takes every bounded set into a set with compact closure.

[4] The topological degree of a mapping at a point y is, roughly speaking, the signed multiplicity of the covering of a neighborhood of y under the mapping.

[5] More precisely, if F contains a parameter, $F \equiv F_\alpha(x)$, then F_α is continuous in α, uniformly with respect to any bounded set in E.

[6] This is the case, for example, if the mapping (34.2) is one-to-one in the neighborhood of a known unique solution, or, in particular, if $F_0(x)$ is the constant mapping.

[7] For complete details concerning this and other flow models treated by the fixed point method see the references in Sect. 32.

satisfying the equation

$$l(s) = F[l, \Psi(l)] = \frac{b \int\limits_{\pi/2}^{s} e^{-T(e^{is})} (1 + \sin s) \sin s \, ds}{\int\limits_{\pi/2}^{\pi} e^{-T(e^{is})} (1 + \sin s) \sin s \, ds}. \tag{34.3}$$

In order that F be defined for all $l(s) \in E$, as required by the Leray-Schauder theory, let $\Psi(l) = \Psi(b)$ for $l \geq b$, and we recall that $\Psi(-l) = -\Psi(l)$. Note that under this extended definition of F, any solution of Eq. (34.3) is still a solution of the flow problem. Let us introduce the family of obstacles

$$\Psi_k(l) = k\Psi(l) + (1-k)\frac{\pi}{2}, \quad 0 \leqq k \leqq 1, \tag{34.4}$$

and consider the corresponding one-parameter family of functional equations,

$$l = F[l, \Psi_k(l)], \quad 0 \leqq k \leqq 1. \tag{34.5}$$

This becomes Eq. (34.3) for the given obstacle when $k=1$, and reduces to the equation for the vertical flat plate when $k=0$.

One now verifies the hypotheses of the Leray-Schauder theorem: (i) $F[l, \Psi_k(l)]$ is completely continuous in E; (ii) F is continuous in k, uniformly with respect to any bounded set in E. These two properties of the transformation F are easily proved, the former by means of the Korn-Privaloff theorem[1], which insures that the functions $F[l(s), \Psi_k(l(s))]$ and their derivatives are equicontinuous. The principal effort here, as in most proofs by LERAY-SCHAUDER, is in establishing *a priori* bounds on the solutions of Eq. (34.5). This is expressed by the statement: (iii) all solutions of Eq. (34.5) are uniformly bounded in E; i.e., for some C all solutions $l(s)$ satisfy $\|l(s)\| \leqq C$. For the details of this vital step we refer to the literature[2]. At $k=0$ the operator F defines a fixed function; in other words, $F[l, \Psi_0(l)]$ is constant in E. Hence the total index of solutions of Eq. (34.5) at $k=0$ is unity. We conclude: the total index of solutions of Eq. (34.5) at $k=1$ is unity, and therefore the flow problem has at least one solution.

Various modifications are necessary to handle other flow problems or conditions of detachment. Thus, the solution of the symmetric re-entrant jet problem for a given cavitation number and prescribed point of detachment is described by a function $l(s)$ and two parameters h, b for which the relations (33.17) to (33.19) are satisfied. This system can be written as a functional equation $x = F(x, \Psi(l), U)$, where $x = (l(s), h, b)$ is a point of the product space $E \times I \times I$ ($I =$ unit interval). A similar consideration is involved in all problems containing parameters, including the problem of smooth detachment, which has the arc length at separation as an undetermined parameter. Establishment of the required *a priori* bounds on the parameters[3] seems to be the principal difficulty in several unsolved problems. Thus, in the problems of the asymmetric re-entrant jet and jet flow from a channel the parameters are connected with the unknown direction and width of the jets, and no general proofs have yet been obtained for the required bounds.

The unified treatment achieved by BIRKHOFF and ZARANTONELLO [*1*] is simpler in certain respects than the proofs based on the Villat equation (33.22). They consider the curvature equation (33.28) for symmetric convex obstacles,

[1] Cf. footnote 2, p. 395.

[2] Cf. LERAY: Footnote 2, p. 400. — J. SERRIN: J. Rat. Mech. a. Analysis **1**, 1—48 (1952).

[3] When the parameters represent points in the interior of the Levi-Civita plane, as in the examples of Sect. 33, these bounds assert that for all solutions the points stay at a fixed distance from the boundary.

subject to the side condition (33.29) which expresses the type of detachment. Using the latter, one can eliminate M from Eq. (33.28) to obtain

$$\lambda = \frac{\nu\, K(I\,[\lambda])\, \mathrm{e}^{-D[\lambda]}}{\int_0^\pi f\,\nu\, K(I\,[\lambda])\, \mathrm{e}^{-D[\lambda]}\, ds}. \tag{34.6}$$

If the flow problem is such that ν contains parameters, as in the re-entrant jet, cusped cavity, and Riabouchinsky cavity problems, these are to be fixed, so that the Eq. (34.6) is without free parameters.

In order to apply the Leray-Schauder theory to Eq. (34.6), let $K_\alpha(\Phi) = \alpha K(\Phi) + (1-\alpha)$ be the curvature function of a family of obstacles varying continuously between the given body at $\alpha = 1$ to the circular arc at $\alpha = 0$. The corresponding family of functional equations is

$$\lambda = F_\alpha(\lambda) \equiv \frac{\nu\, K_\alpha(I\,[\lambda])\, \mathrm{e}^{-D[\lambda]}}{\int_0^\pi f\,\nu\, K_\alpha(I\,[\lambda])\, \mathrm{e}^{-D[\lambda]}\, ds}. \tag{34.7}$$

The Banach space E on which F_α is defined is the set of Lebesgue square-integrable functions on the interval $(0, \pi)$, with the Hilbert norm $\|\lambda\| = \left(\int_0^\pi |\lambda|^2\, ds\right)^{\frac{1}{2}}$. It is assumed that (i) $\nu(s)$ is continuous on $0 \leqq s \leqq \pi$ and positive in the open interval $(0, \pi)$; (ii) $f(s)$ is positive and continuous in $(0, \pi)$; (iii) $\int_0^\pi f(s)\, \nu(s)\, ds = N < \infty$; (iv) $D\,[f\nu] \leqq C f$. These conditions are satisfied by the functions ν and f corresponding to the flow models and conditions of detachment discussed in Sect. 33 and throughout this work.

The proof that the solutions of Eq. (34.7) are unformly bounded in E is relatively simple and proceeds as follows. Let k be a positive constant such that $0 < k \leqq K_\alpha = k^{-1} < \infty$. We have

$$\int_0^\pi f\,\nu\, K_\alpha(I\,[\lambda])\, \mathrm{e}^{-D[\lambda]}\, ds \geqq k \int_0^\pi f\,\nu\, \mathrm{e}^{-D[\lambda]}\, ds. \tag{34.8}$$

By a special case of Jensen's inequality[1]

$$\int_0^\pi f\,\nu\, \mathrm{e}^{-D[\lambda]}\, ds \Big/ \int_0^\pi f\,\nu\, ds \geqq \exp\left[-\int_0^\pi f\, D\,\nu\,[\lambda]\, ds \Big/ \int_0^\pi f\,\nu\, ds\right].$$

Also,

$$\int_0^\pi f\,\nu\, D\,[\lambda]\, ds = \int_0^\pi \lambda\, D\,[f\,\nu]\, ds \leqq C \int_0^\pi \lambda\, f\, ds,$$

whence,

$$\int_0^\pi f\,\nu\, \mathrm{e}^{-D[\lambda]}\, ds \geqq N\, \mathrm{e}^{-(C/N)\int_0^\pi \lambda f\, ds}. \tag{34.9}$$

For a solution λ, $\int_0^\pi \lambda f\, ds = 1$, so from Eqs. (34.8) and (34.9) follows

$$\int_0^\pi f\,\nu\, K_\alpha(I\,[\lambda])\, \mathrm{e}^{-D[\lambda]}\, ds \geqq k\, N\, \mathrm{e}^{-C/N}.$$

Inserting this into Eq. (34.6), we obtain

$$\lambda \leqq \frac{\nu\, \mathrm{e}^{C/N}}{k^2\, N} \leqq A\,\nu,$$

[1] G. Hardy, J. Littlewood and G. Polya: Inequalities, p. 138. Cambridge: Cambridge University Press 1934.

and this gives the desired *a priori* bound on solutions of Eq. (34.7).

$$\|\lambda\| = A\ \|\nu\| = A'. \tag{34.10}$$

The conditions required for the Leray-Schauder theory, that F_α be completely continuous in E and continuous with respect to α, are easily verified. It remains only to show that the total index of solutions of $\lambda = F_0(\lambda)$ is unequal to zero. This is accomplished in two steps, the first of which is to prove existence of a unique solution of $\lambda = F_0(\lambda)$, or, in other words, the existence of a unique solution of the flow problem (34.7) for a circular arc obstacle. This, rather than the *a priori* bound (34.10), is the principal difficulty of the present method. Its solution is clearly provided by the following theorems: (1) For given $M \geqq 0$ and $\nu(s) > 0$ in $(0, \pi)$, the equation $\lambda = M\nu\, e^{-D[\lambda]}$ has one and only one solution $\lambda(s)$; (2) (JACOB's lemma[1]) if λ and $\lambda + \Delta\lambda$ are solutions corresponding to two different non-negative values M and $M + \Delta M$, then $0 < \frac{\Delta\lambda(s)}{\Delta M} < \nu(s)$ in $(0, \pi)$. The second step is to show the one-to-one character of the mapping $\eta = \lambda - F_0(\lambda)$ in the neighborhood of the solution of $\lambda = F_0(\lambda)$[2]. The proof of this result is elementary and similar to that of JACOB's lemma.

The following existence theorems are immediate consequences. For any symmetric closed convex obstacle there exist: (1) An infinite cavity for each point of detachment; (2) an infinite cavity with smooth separation; (3) a cavity of zero asymptotic width; (4) a one-parameter family of cusped cavities; (5) a one-parameter family of Riabouchinsky cavities for each point of detachment; (6) a one-parameter family of Riabouchinsky cavities with smooth separation; (7) a two-parameter family of re-entrant jets. The connection between the parameters in (4) to (7) and the natural physical data (e.g. cavitation number) appears in the side relations (33.18), (33.19), etc. The existence of a solution under prescribed conditions does not follow at once from the above results, and requires separate consideration. This has yet to be done in most cases.

35. The variational method. The variational principles of Sects. 30, 31 offer a conceptually appealing approach to the existence problem and have the additional merit of not being limited to plane flow. This section outlines the existence proofs for plane and axially symmetric Riabouchinsky cavities based on the principle of minimum virtual mass[3].

We consider first plane flows. Let the obstacle C_1 be an arc symmetric with respect to the x-axis and monotonically increasing in the second quadrant from the axis to its end point $(-k, h)$. Let C_2 be the image of C_1 in the y axis, and denote by C the closed curve obtained by joining the endpoints of C_1 and C_2 with horizontal straight segments (on $y = \pm h$). Now let Γ be any curve lying in the strip $|x| \leqq k$, and joining the upper endpoints of C_1 and C_2 without intersecting C. We designate by B the symmetric closed curve composed of C_1, C_2, Γ, and the image of the latter in the x-axis. Consider the symmetric flow with unit free stream velocity past B. The potential and stream functions of this flow have the expansions at infinity

$$\varphi = x + \frac{a\,x}{r^2} + \cdots, \qquad \psi = y - \frac{a\,y}{r^2} + \cdots, \tag{35.1}$$

[1] C. JACOB: Mathematica **11**, 149–175 (1935); this result implies that in the case of circular arc obstacles the separation angle increases monotonically and continuously with M.

[2] Cf. footnote 6, p. 400.

[3] For plane flow see P. GARABEDIAN and D. SPENCER: J. Rat. Mech. a. Analysis **1**, 359–409 (1952), and P. GARABEDIAN and H. ROYDEN: Proc. Nat. Acad. Sci. U.S.A. **38**, 57–61 (1952); for axially symmetric flow see P. GARABEDIAN, H. LEWY and M. SCHIFFER: Ann. of Math. **56**, 560–602 (1952).

where a is connected with the area A and virtual mass M of B by the formula (cf. Sect. 30),

$$a = \frac{1}{2\pi}(M + A). \tag{35.2}$$

For any fixed $\sigma > 0$ we now seek a curve Γ and body B for which

$$2\pi a - \sigma A = \text{minimum}. \tag{35.3}$$

We have seen a formal proof in Sect. 30 that the solution Γ must be a constant speed free streamline on those portions γ where it does not coincide with C or with $x = \pm k$[1]. This is the heuristic basis of the present method. The formal proof can be made rigorous provided γ is sufficiently smooth, in particular, if it is analytic. Our goal then is to establish the existence of an analytic solution of the extremum problem (35.3).

One observes first that a minimizing sequence of curves B_n, i.e., a sequence for which $2\pi a_n - \sigma A_n \to$ minimum, can be chosen to consist of doubly symmetric curves[2] which are monotonic in each quadrant. To see this fact, we introduce the notion of *symmetrization:* Let B be any simple closed curve. Then B' is said to be the (STEINER) symmetrization of B in the x-axis if it has the property that (1) any vertical line L intersects the interior of B' in a single segment bisected by the x-axis; and (2) the length of this segment is equal to the total length of segments in the intersection of L with the interior of B. Under symmetrization, area is unchanged and the virtual mass does not increase[3]. If then B_n is a minimizing sequence for the problem (35.3), the corresponding doubly symmetrized curves likewise form a minimizing sequence, and these curves are monotonic in each quadrant. We note that the fixed curves C_1, C_2 are unaffected by symmetrization in either axis.

Consider now any member B of a minimizing sequence of the above type, and let m be its maximum height above the x-axis. B contains the vertical segment of length $2m$ and must therefore [cf. (30.4)] have a larger co-efficient a; i.e., $a \geq m^2/2$, the latter being the value of a for the segment[4]. At the same time, the area A satisfies the inequality $A \leq 2ml$, where l is the fixed length of B. It follows that for the minimizing sequence the quantity $\pi m^2 - 2\sigma m l$ is bounded and hence the sequence of curves is bounded in height.

In a coordinate system rotated through 45° with respect to the original one the individual curves of the minimizing sequence all satisfy the same Hölder condition (with coefficient and exponent both unity), and therefore form a bounded equicontinuous family. Hence a minimizing subsequence can be chosen which converges to a symmetric rectifiable curve B. The corresponding stream functions ψ_n form a normal family since the sequence $\psi_n - y$ is uniformly bounded; in other words, the minimizing sequence can be selected so that $\psi_n \to \psi$, where ψ describes the flow past the limit curve B. One thus obtains $a_n \to a$, and since $A_n \to A$, it follows that B is a solution of the extremum problem (35.3).

If the extremal configuration B differs from C, it contains an arc γ in $|x| < k$ which is not coincident with $y = \pm h$. It remains to prove that γ is a free streamline on which $|\operatorname{grad} \varphi|^2 = 1 + \sigma$, and for this it suffices to know that γ is analytic,

[1] On C and on $x = \pm k$ the extremal curve cannot be varied freely and therefore the argument of Sect. 30 ceases to apply.

[2] I.e., symmetric in both axes.

[3] Cf. GARABEDIAN and SPENCER, footnote 3, p. 403. For additional properties of symmetrization see G. POLYA and G. SZEGÖ, Isoperimetric inequalities of mathematical physics. Princeton 1951.

[4] Cf. H. LAMB: Hydrodynamics, 6th ed., p. 85. New York: Macmillan 1932.

since then the formal analysis of Sect. 30 is applicable. The proof of this property of γ is the essential difficulty of the present argument, in both the plane and axially symmetric cases. In the plane problem, analyticity is proved by the method of *interior variations*, which has proved useful in other connections in conformal mapping[1]. This consists in applying a special type of analytic variation which takes the above extremal flow into a neighboring admissible flow of problem (35.3). The fact that B is extremal yields an analytic variational condition in the flow region which can be reflected across the rectifiable arc γ and as a consequence exhibits the analytic character of γ. The constant speed condition on γ can also be proved directly from this arguments without recourse to Sect. 30.

Another proof of analyticity of γ and of the constant speed condition is based on approximation of the extremal curve by extremal polygons[2]. This method avoids interior variations and seems both simpler and more elementary.

Assuming it has been proved that the extremal arc γ is a free streamline, we see that for any σ there is a solution Γ of Eqs. (35.3) consisting either of a monotonic free streamline γ and adjoining horizontal segments of $y=\pm h$, or of γ and connecting segments of $x=\pm k$ (Fig. 55).

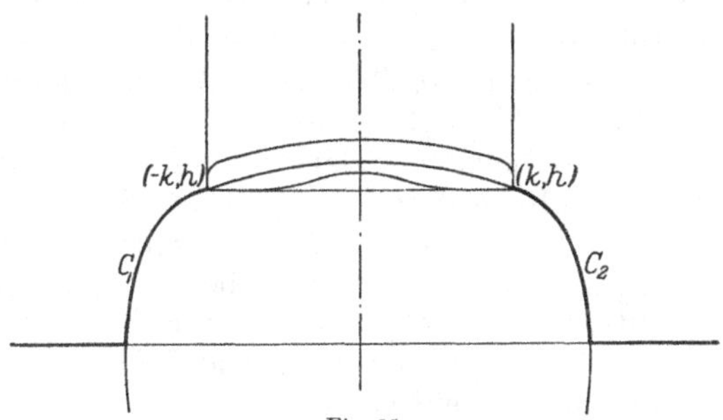

Fig. 55.

Consider now the dependence of the solutions on σ. Let D_1 and D_2 be extremal flow regions in the upper half plane corresponding to values σ_1 and σ_2 such that $\sigma_1<\sigma_2$. We assert that $D_1\supset D_2$. If this is not so, a downward translation of D_1 takes it into a position where $D_1\supset D_2$ and the free streamlines γ_1 and γ_2 have a common point of contact. However, this fact contradicts the first comparison theorem of Sect. 27, since the speeds on γ_1, γ_2 are unequal in the wrong order. If $\sigma_1=\sigma_2$, the same argument proves that the solution is unique. Thus the flow region D is monotonically decreasing with σ. The dependence on σ is furthermore continuous, since if $\sigma_n\to\sigma$, the extremal bodies B_n form a minimizing sequence for the problem (35.3)[3], and hence converge to the unique solution B. It is not difficult to see that for sufficiently large values of σ the extremal body contains segments of $x=\pm k$, and that for sufficiently small σ it contains segments of $y=\pm h$ (Fig. 55). Hence, by continuity, there is a value of σ such that the corresponding free streamline detaches from the endpoints of C_1 and C_2. The resulting flow describes the Riabouchinsky cavity behind the obstacle C_1 when its image is placed at C_2. From continuity considerations one establishes the existence of the cavity for prescribed cavitation number σ[4].

The guiding ideas of the above proof remain unchanged in the case of axially symmetric flows[5]. The potential and stream functions now have the expansions in the meridian plane,

$$\varphi=x+\frac{a\,x}{r^3}+\cdots,\qquad \psi=\frac{y^2}{2}-\frac{a\,y^2}{r^3}+\cdots$$

[1] Cf. GARABEDIAN and SPENCER, footnote 3, p. 403 and references therein.

[2] GARABEDIAN and ROYDEN: Footnote 3, p. 403.

[3] Namely, $2\pi a-\sigma_n A\geqq 2\pi a_n-\sigma_n A_n$, and letting $\sigma_n\to\sigma$ one obtains $\limsup_{n\to\infty}(2\pi a_n-\sigma A_n)\leqq 2\pi a-\sigma A$.

[4] Continuity of the velocity at detachment does not follow at once and must be established separately.

[5] For the following see GARABEDIAN, LEWY and SCHIFFER, footnote 3, p. 403.

and the variational problem characterizing the free surface flow is

$$4\pi a - \sigma V = \text{minimum}, \tag{35.4}$$

where V is the three dimensional volume enclosed by the body of revolution past which the flow is taken. Again a suitable symmetrization process allows an arbitrary minimizing sequence to be replaced by one consisting of monotonic curves; from this a subsequence can be chosen having as limit a body achieving the minimum (35.4). The extremal configuration again consists of a rectifiable monotonic curve γ and either horizontal or vertical segments (Fig. 55). These details do not differ greatly between the plane and axially symmetric cases, although the process of symmetrization is somewhat more complicated in the latter. The chief difference arises in proving that the curve is an analytic free streamline.

It is again possible to apply the method of interior variations, but in the present problem this leads to the not quite adequate result that the potential φ has continuous boundary values $\varphi = (1+\sigma)^{\frac{1}{2}} s$ on γ, where s is arc length. The needed differentiability, which implies the constant speed condition $|\operatorname{grad}\varphi|^2 = 1+\sigma$, is inferred from the analyticity of γ, which implies that of φ and ψ up to γ. The proof of analyticity is the essentially new feature in the treatment of the axially symmetric problem. It is based on the device, already encountered in Sect. 17, of analytic continuation of the stream function into the complex domain of the independent variables.

A few of the details can be presented as follows (cf. the paper of GARABEDIAN, LEWY, and SCHIFFER). We find for the equation $\bar{z} = f(z)$ of the free boundary the functional equations

$$f(t) = -\frac{F'(t)}{2\pi i\, R(t, g(t); t, \bar{z}_0)} - \int\limits_{z_0}^{t} \frac{R_t(z, g(z); t, \bar{z}_0)}{R(t, g(t); t, \bar{z}_0)} f(z)\, dz,$$

$$g(t) = \bar{z}_0 + \int\limits_{z_0}^{t} f(z)^2\, dz.$$

These are based essentially on Eq. (17.3), $R(z, \bar{z}; t, \bar{t})$ being the Riemann function defined in Sect. 17. It can be shown by the Picard method of successive approximations that there is a solution $g(z)$ of this equation which is analytic just inside the free boundary γ and reduces continuously to $\bar{z}$ on γ. Consider the mappings

$$\Phi = z + g(z), \qquad \Psi = z - g(z);$$

γ maps onto a segment of the real axis in the Φ-plane and onto a segment of the imaginary axis in the Ψ-plane. Hence by Schwarzian reflection Ψ is an analytic function of Φ on γ. Thus, $z = \frac{1}{2}(\Phi - \Psi)$ is also analytic as a function of the real argument Φ along the free boundary, and hence the free boundary is itself analytic.

The remainder of the argument leading to the Riabouchinsky cavity proceeds as in the plane case.

b) Uniqueness theory.

36. General remarks. That certain types of non-uniqueness are inherent in free surface phenomena is soon apparent. Thus, the steady flow resulting from the impact of two jets is not uniquely determined by the limiting velocity and direction of the impinging jets (cf. Sect. 10), and, as the examples of Sects. 6 to 9 show, a variety of different configurations with free boundaries are possible with the same barrier[1]. It is plain also that a continuum of flows of a given

[1] This type of multiplicity has been emphasized by VILLAT; see [4]; also G. JAFFÉ: Z. angew. Math. Mech. **1**, 398–410 (1921), and Phys. Z. **23**, 129–133 (1922).

model (e.g., infinite cavity) are possible if the points of detachment are unrestricted. Such a multiplicity can exist, further, even if the free streamlines are required not to intersect the body[1]. It is less clear that a multiplicity of solutions may exist for which the detachment is smooth, but examples of such non-uniqueness for cavity flows past even convex obstacles are known to exist[2].

The present status of the uniqueness theory suggests that the problem of fixed detachment, to which a major portion of the literature has been devoted, has only unique solutions. At least, uniqueness has been proved for this problem in the symmetric case under quite general hypotheses, for all flow models but the re-entrant jet cavity, and no counterexamples are yet known.

The solutions of the prow problem and the problem of smooth detachment have been proved unique for the symmetric *accolades* considered by LERAY (cf. Sect. 24).

The various proofs to date have been based on two conceptually different attacks—the method of continuity and the comparison method[3]. The early proofs of uniqueness were obtained by the method of continuity, first applied to the free boundary problem by WEINSTEIN, and extended by WEYL, HAMEL, FRIEDRICHS, LERAY, and FINN. The more recently developed comparison method, initiated by LAVRENTIEFF and elaborated by GILBARG and SERRIN, has the advantage of greater simplicity and fewer restrictions on the class of barriers, and furthermore applies as well to axially symmetric flows and to flows of compressible fluids. However, it has thus far been limited to symmetric flows, whereas the continuity method has had some success in treating problems with asymmetry, which seem to involve special difficulties. As far as symmetric flows past curved boundaries are concerned, the results obtained by the comparison method at present include those derived by the method of continuity. The contributions and principles of each approach are discussed in the following sections.

37. The comparison method. This method is illustrated here in the proof of uniqueness of the infinite cavity[4]. We consider first either plane symmetric or axially symmetric flows past an obstacle C (in the upper half or meridian plane) with free streamline detachment from a fixed endpoint A. The curve C is assumed to be starlike with respect to some point 0 on the downstream axis of symmetry[5]. Consider two such infinite cavity flows with free streamlines Γ and $\bar{\Gamma}$ detaching from A. The uniqueness theorem asserts that $\Gamma = \bar{\Gamma}$ and that the flows are identical.

Proof: The single intersection theorem (cf. Sect. 26) shows that the curves $C+\Gamma$ and $C+\bar{\Gamma}$ are also starlike with respect to 0. Let $r=r(\vartheta)$, $r=\bar{r}(\vartheta)$ be the polar representations of Γ and $\bar{\Gamma}$ with respect to 0 as origin. From the formulas (25.5a) and (25.8) for the asymptotic cavity shape it follows that $\lim\limits_{\vartheta\to 0} r(\vartheta)/\bar{r}(\vartheta)$ exists and is (say) greater than or equal to unity. As a consequence $0 < \min\limits_{0<\vartheta\leq\vartheta_A} r(\vartheta)/\bar{r}(\vartheta) = \varrho \leqq 1$, and the minimum is achieved for some $\vartheta > 0$. Thus a similarity transformation of the flow past $C+\Gamma$ in the ratio $1/\varrho$ takes $C+\Gamma$ into a curve $C'+\Gamma'$ and the flow region D into D' in such a way that Γ' and $\bar{\Gamma}$

[1] This follows immediately from the existence theory (Sect. 32). For explicit construction of such flows and additional theory, see S. BERGMAN: Z. angew. Math. Mech. **12**, 95—121 (1932).

[2] J. LERAY: Volume du Jubilé de M. Brillouin, pp. 246—257. Paris: Gauthier-Villars 1935.

[3] For references, see Sect. 32.

[4] Cf. M. LAVRENTIEFF: Mat. Sbornik **46**, 391—458 (1938). — J. SERRIN: Amer. J. Math. **74**, 492—506 (1952). — D. GILBARG: J. Rat. Mech. a. Analysis **1**, 309—320 (1952).

[5] In the case of plane flow C may be starlike in the more general sense of footnote 6, p. 392.

have a point P in common, and such that the flow region $\bar{D}$ (bounded by $C+\bar{\Gamma}$) contains the image region D' (Fig. 56). The flow can be defined in D' to have the same velocity at corresponding points in D. Comparison Theorem I of Sect. 27 now applies to the flows in D' and $\bar{D}$, and provides a contradiction at P unless $\Gamma=\Gamma'=\bar{\Gamma}$ and the flows are identical. This completes the proof.

Similar arguments prove uniqueness in the case of fixed detachment for plane and axially symmetric jet flows, cusped cavities, and Riabouchinsky cavities The proofs in these cases are fundamentally simpler than for the infinite cavity, since the asymptotic properties of the flow require no special consideration, as they do in the latter problem. The same method carries over to both plane and axially symmetric subsonic flows of compressible fluids[1]. Uniqueness has not yet been established for the reentrant jet cavity.

We turn now to the problem of smooth detachment for plane incompressible infinite cavity flows. Unless restrictions are placed on the class of obstacles

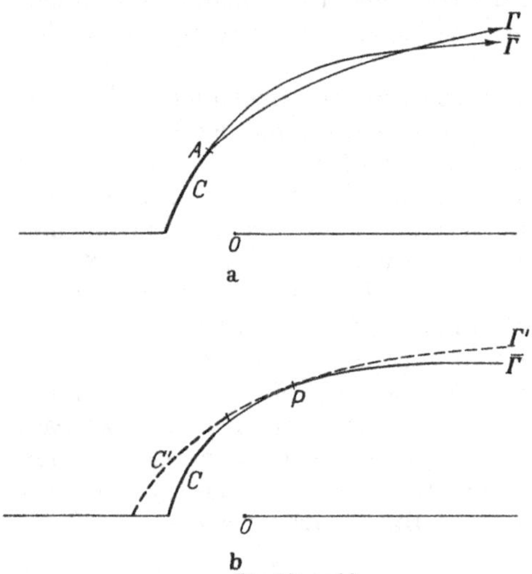

Fig. 56a and b.

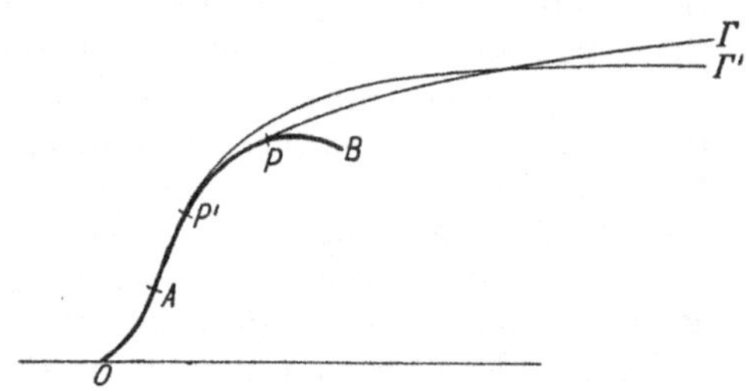

Fig. 57.

uniqueness does not necessarily hold for solutions of this problem and, as previously remarked, examples of non-uniqueness are known even for convex obstacles. However, for symmetric flows in which the obstacle is an *accolade*, consisting of a concave arc OA (which may degenerate into a single point) joined to a curve AB having non-decreasing curvature as a function of arc length (Fig. 41), the problem of smooth separation has only unique solutions. The first proof of this result was given by LERAY[2] using the method of continuity in function space (cf. Sect. 39). LERAY requires that any line parallel to the x-axis can intersect the obstacle in a single point or segment, and in the following proof by the comparison method we make the same assumption.

Proof: The results stated in Sect. 24 concerning the class of obstacles under consideration show that in smooth detachment from OAB the point of separation lies on the convex arc AB, that the curvature of the free streamline is a maximum at detachment (equal to that of the body) and decreases monotonically, and that the Brillouin conditions are satisfied by the flow. Suppose then that there are two distinct flows solving the problem of smooth detachment for the obstacle OAB, and let the assumed points of separation be P and P', with P' lying between A and P (Fig. 57). We denote the corresponding flow regions by

---

[1] D. GILBARG: J. Rat. Mech. a. Analysis **2**, 233—251 (1953). The results of this paper are strengthened by D. GILBARG and J. SERRIN: J. rational Mech. a. Analysis **4**, 169—175 (1955); also D. GILBARG: Math. Z. **72**, 165—174 (1959).

[2] J. LERAY: Comm. Math. Helv. **8**, 250—263 (1935).

D and D'. It may be assumed that the respective free streamline Γ and Γ' intersect, for otherwise $D' \subset D$ and Comparison Theorem I is contradicted at P'. From Sect. 28 we know that the curves $OAP+\Gamma$ and $OAP'+\Gamma'$ intersect precisely once, and that Γ lies above Γ' beyond this point (Fig. 57). It follows that a suitable horizontal displacement takes D into a region $\bar{D}$ contained in D' such that the streamline $OAP+\Gamma$ goes into another curve having contact with $OAP'+\Gamma'$ at a point C on Γ'. C cannot lie on the translated image of OAP since the curvature of the obstacle is nondecreasing from A to P, whereas the curvature of the free streamline *decreases* after separation at P'. Consequently C lies on both free streamlines Γ' and $\bar{\Gamma}$, and since $D' \supset \bar{D}$, we arrive again at a contradiction of the comparison theorem. A similar argument can be applied to other flow models as well.

38. The method of continuity-polygonal boundaries. Although the method of continuity, as an approach to uniqueness, has been largely superseded by the simpler comparison method, it is of continuing importance for its contributions to the existence theory and the uniqueness problem for asymmetric flows, and also because of the mathematical ideas originating from it.

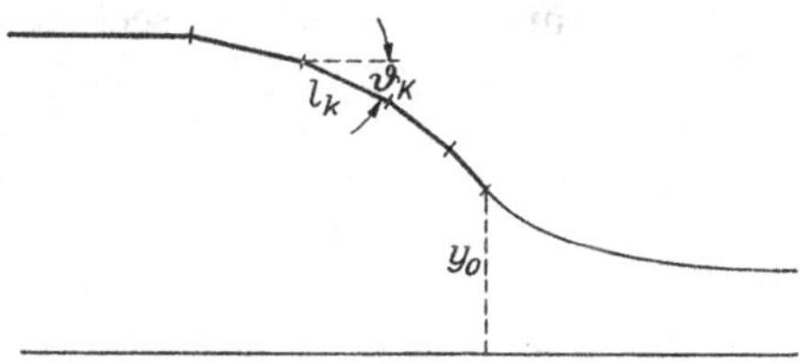

Fig. 58.

To illustrate the method we treat the problem of symmetric jet flow from a polygonal channel[1]. The same considerations, with minor modifications, are applicable to cavity flows past polygonal obstacles. Let the flux through the (upper half) channel be preassigned and equal to $\pi/2$. The speed of the jet, which we denote by $1/\mu$, is therefore an unknown of the problem. Let the channel wall start at $z = i y_0$ (the orifice), and consist of segments of length $l_1, l_2, \ldots, l_n, \infty$, having inclinations $\vartheta_1, \vartheta_2, \ldots, \vartheta_n, 0$. If the channel is concave to the flow, as we suppose in the following, then $\vartheta_k \leqq \vartheta_{k+1} \leqq 0$ (Fig. 58). The exterior angles are defined by $\beta_k \pi = \vartheta_{k+1} - \vartheta_k$ $(k = 1, \ldots, n)$. Thus a polygonal channel is represented by $2n+1$ *physical parameters*

$$y_0;\, l_1, \ldots, l_n;\ \ \beta_1, \ldots, \beta_n \tag{38.1}$$

such that

$$y_0, l_1, \ldots, l_n > 0, \quad \text{and} \quad 0 \leqq \beta_k < 1.$$

The jet flow through a channel is described in the Levi-Civita representation by the complex potential $f(\zeta)$ satisfying

$$\frac{df}{d\zeta} = \frac{1}{\zeta}\,\frac{\zeta^2-1}{\zeta^2+1} \tag{38.2}$$

and by the flow function

$$\omega(\zeta) = \vartheta + i \log \mu q = i \sum_{k=1}^{2n} \beta_k \log i \frac{\zeta - \zeta_k}{1 - \zeta \zeta_k}, \tag{38.3}$$

where the points $\zeta_k = e^{i s_k}$, $0 < s_1 < \cdots < s_n < \pi/2$, correspond to the vertices of the channel wall and $s_{2n-k} = \pi - s_k$. Conversely, to any $2n+1$ *flow parameters*

$$\mu;\, s_1, \ldots, s_n;\, \beta_1, \ldots, \beta_n \tag{38.4}$$

[1] The ensuing discussion essentially follows A. WEINSTEIN: Proc. Symp. Appl. Math. Vol. 1, Amer. Math. Soc. 1949, pp. 1–18.

for which $\mu>0$, $0<s<\cdots<s_n<\pi/2$, $0\leqq\beta_k<1$, there corresponds a jet flow through a channel in the physical plane, the transformation from the Levi-Civita variable ζ to the z plane being given by

$$z-i\,y_0=\int_{f_0}^{f}\frac{df}{w(f)}=\mu\int_{1}^{\zeta}e^{i\omega(\zeta)}\frac{\zeta^2-1}{\zeta^2+1}\,\frac{d\zeta}{\zeta}. \tag{38.5}$$

The existence of a flow through a given channel is therefore established by $2n+1$ flow parameters which, when inserted into Eqs. (38.3) and (38.5), yield the given boundary and with it the $2n+1$ physical parameters defining the channel. Uniqueness is proved when it is shown that only one set of flow parameters can correspond to the given physical parameters.

With these preliminaries behind us the basic idea of the method of continuity can be outlined simply. Since in this method the proof of uniqueness depends on an existence theory, we discuss both problems simultaneously. Let C be a given channel for which we seek to establish existence and uniqueness of a flow. We start from some initial channel C_0 for which a unique solution is known (C_0 might be the trivial straight channel for example), and deform it continuously along some path Γ joining C_0 to C in the $2n+1$ dimensional space E of physical parameters. Suppose it is known that (1) if a solution exists at some point P on Γ, then a neighborhood of P in E is mapped in a single-valued continuous manner onto a neighborhood of the image of P in the space E' of flow parameters; (2) throughout the deformation process along Γ the quantities s_k are bounded away from one another, from 0, and from $\pi/2$, and μ is bounded away from 0 and ∞. From (1) it follows that one can proceed along the path from C_0, describing a continuous family of solutions as the channel is deformed, while (2) insures that the deformation process does not lead to a degenerate solution and thus break down before reaching C. This establishes the existence of a flow through C. To infer uniqueness, we observe that if two distinct points C' and C'' in E' correspond to C, then for any path L joining C to C_0 in E, there correspond (by the existence theorem) two paths L', L'' in E' joining C' and C'' respectively to C_0', where C_0' describes the unique flow through C_0. As L' and L'' are traversed, starting from C' and C'', they must meet a first time (possibly at C_0'). However, this leads to an immediate contradiction of property (1) above.

Property (1) is established by showing that the Jacobian of the transformation from the flow parameters to the physical parameters is non-vanishing. This is a consequence of the theory of *infinitesimal uniqueness* to be discussed below. The second property is inferred from a theorem on *a priori* bounds on solutions which asserts that in any closed subregion of E the corresponding points in E' occupy a set in which μ, s_1, $s_{k+1}-s_k$ $(k=1,\ldots,n-1)$, and $\pi/2-s_n$ have positive upper and lower bounds. This type of uniform estimate is essential to all existence proofs in one form or other. It is easily proved for symmetric concave channels with prescribed bounds on the range of directions of the boundary segments[1], but it is a major source of difficulty in dealing with asymmetric channels[2].

To define the concepts used in the theory of infinitesimal uniqueness, let a given flow be imbedded in a continuously differentiable one-parameter family of flows described by the flow parameters $s_k(\varepsilon)$, $\beta_k(\varepsilon)$, $\mu(\varepsilon)$, and flow functions $f(z;\varepsilon)$, $\omega(\zeta;\varepsilon)$, the corresponding physical parameters being $y_0(\varepsilon)$, $l_k(\varepsilon)$, and $\beta_k(\varepsilon)$. The *variations* of these quantities with respect to the given flow (at $\varepsilon=0$)

[1] Cf. A. WEINSTEIN: Math. Z. **31**, 424–433 (1929).

[2] Cf. R. FINN: J. d'Analyse Math. **4**, 246–291 (1956).

are defined to be

$$\delta s_k = \varepsilon\, s_k'(\varepsilon)_{\varepsilon=0}, \qquad \text{etc.}, \qquad \delta f(z) = \varepsilon\, f_\varepsilon(z;\varepsilon)_{\varepsilon=0}, \qquad \delta\omega(\zeta) = \varepsilon\,\omega_\varepsilon(\zeta;\varepsilon)_{\varepsilon=0}. \tag{38.6}$$

The connection between the variations of the physical and flow parameters is given by the equations,

$$\left.\begin{aligned} \delta y_0 - \sum_{i=1}^{n} \frac{\partial y_0}{\partial \beta_i}\,\delta\beta_i &= \frac{\partial y_0}{\partial \mu}\,\delta\mu + \sum_{i=1}^{n}\frac{\partial y_0}{\partial s_i}\,\delta s_i \\ \delta l_k - \sum_{i=1}^{n} \frac{\partial l_k}{\partial \beta_i}\,\delta\beta_i &= \frac{\partial l_k}{\partial \mu}\,\delta\mu + \sum_{i=1}^{n}\frac{\partial l_k}{\partial s_i}\,\delta s_i, \qquad (k=1,\dots,n). \end{aligned}\right\} \tag{38.7}$$

From this system of linear equations for the variations $\delta\mu$, δs_k in terms of δy_0, δl_k it is apparent that the Jacobian

$$\partial(y_0; l_1, \dots, l_n; \beta_1 \dots \beta_n)/\partial(\mu; s_1, \dots, s_n; \beta_1 \dots \beta_n)$$

is non-vanishing—and hence property (1) holds—provided the relations

$$\delta y_0 = 0, \qquad \delta l_k = 0, \qquad \delta\beta_k = 0, \qquad k = 1, \dots, n \tag{38.8}$$

imply

$$\delta\mu = 0, \qquad \delta s_k = 0, \qquad k = 1, \dots, n. \tag{38.9}$$

The latter conclusion can be interpreted as stating there cannot be two distinct infinitesimally neighboring flows through the same channel. This is the property of *infinitesimal uniqueness*. It means that solutions cannot branch, and insures local existence and uniqueness.

Differentiating Eq. (38.5) with respect to ε, one obtains the general variational equation

$$\delta f = w \int_{f_c}^{f} \frac{\delta w}{w^2}\, df - i w\, \delta y_0. \tag{38.10}$$

In the present case this becomes

$$\delta f = \mathrm{e}^{-i\omega} \int_1^{\zeta} \mathrm{e}^{i\omega}\, \delta(-i\omega)\, \frac{\zeta^2-1}{\zeta^2+1}\, \frac{d\zeta}{\zeta} - \frac{\delta\mu}{\mu} \int_1^{\zeta} \mathrm{e}^{i\omega}\, \frac{\zeta^2-1}{\zeta^2+1}\, \frac{d\zeta}{\zeta} - \frac{i\,\delta y_0}{\mu},$$

where

$$\delta(-i\omega) = \sum_{k=1}^{2n} i\beta_k\, \frac{(\zeta^2-1)\,\zeta_k}{(\zeta-\zeta_k)\,(1-\zeta\zeta_k)}\, \delta s_k.$$

It is easily seen that $\delta f \equiv 0$ implies the relations (38.9). Hence the proof of infinitesimal uniqueness is equivalent to showing that $\delta f \equiv 0$ is a consequence of Eq. (38.8), and this is accomplished by considering the boundary value problem satisfied by the harmonic function $\delta\psi = \operatorname{Im} \delta f$.

We observe first that since the channel wall and flux are prescribed,

$$\delta\psi = 0 \quad \text{on the fixed boundary}. \tag{38.11}$$

The other boundary conditions are derived from Eq. (38.10), which can be written in the form,

$$\frac{d}{df}(\delta f) - \frac{d \log w}{df}\, \delta f = \delta \log w. \tag{38.12}$$

Taking the real part we obtain

$$\frac{\partial}{\partial n}(\delta\psi) - \varkappa\,\delta\psi = -\frac{\delta\mu}{\mu^2} \quad \text{on the free streamline}^1, \tag{38.13}$$

where $\varkappa = \mu^{-1}\,d\vartheta/d\varphi$ is the curvature of the free streamline, and n denotes the outward normal ($dn = \mu\,d\psi$). The condition of fixed detachment is expressed in terms of the imaginary part of Eq. (38.12) as

$$\frac{\partial}{\partial \varphi}(\delta\psi) = 0 \quad \text{at detachment}. \tag{38.14}$$

The theorem of infinitesimal uniqueness states that *a harmonic function* $\delta\psi$ *satisfying the boundary conditions* (38.11), (38.13), (38.14), *must be identically zero*. The proof of this result can be reduced to that for the special case $\delta\mu = 0$ (μ fixed), for which *it suffices to show that a harmonic function* $\delta\psi$ *satisfying* Eq. (38.11) *and* $\partial(\delta\psi)/\partial n - \varkappa\delta\psi = 0$ *on the free streamline vanishes identically*.

Infinitesimal uniqueness in this restricted sense has been the principal concern of most work based on the method of continuity. The main difficulty lies in showing that the quadratic integral,

$$D(u) = \iint_G (u_x^2 + u_y^2)\,dx\,dy - \int_\Sigma \varkappa u^2\,ds, \tag{38.15}$$

$$(G = \text{flow region}, \ \Sigma = \text{free streamline})$$

which vanishes for $u = \delta\psi$ (by GREEN'S theorem), is positive for all u that are not identically zero. For concave channels $\varkappa > 0$ on Σ, so that the positiveness of $D(u)$ is not at all evident.

The method of continuity as outlined above was first applied by WEINSTEIN[2], who proved infinitesimal uniqueness for concave channels under certain restrictions which were lightened successively by HAMEL, WEYL, FRIEDRICHS, and FINN[2]. These restrictions were stated in terms of the *total curvature*, defined as $\sum_k \beta_k \pi$ ($= -\vartheta_1$ for symmetric channels). FRIEDRICHS, whose work played an important part in later investigations, proved[3] infinitesimal uniqueness for concave channels of total curvature less than π by means of a simple lemma based on JACOBI'S transformation in the calculus of variations; this asserts, in the form needed here, that $D(u)$ is positive definite provided there is a harmonic function u unequal to zero in G and satisfying the natural boundary condition

$$\frac{\partial u}{\partial n} - \varkappa u = 0 \quad \text{on} \quad \Sigma.$$

Under the assumption of total curvature less than π there is a velocity component with the required properties and hence the definiteness of $D(u)$ follows.

We remark the FRIEDRICHS also succeeded in proving *absolute* uniqueness—without an existence theory, and for curved boundaries—in the case that μ is fixed and the point of detachment is allowed to vary. This he accomplished by showing that the actual flow minimizes an integral of the form $\iint (u_x^2 + u_y^2 + 1)\,dx\,dy$ whose second variation is essentially the positive definite expression (38.15). His argument carries over to axially symmetric flows as well.

---

[1] In cavity flow the right member of this equation is absent since μ is fixed. Variational formulas analogous to these will prove useful in Sect. 45 in the calculation of cavity flows.

[2] A. WEINSTEIN: Math. Z. **19**, 265–274 (1924). — G. HAMEL: Proc. Second Intern. Cong. Appl. Mech., Zürich 1926, p. 489. — H. WEYL: Nachr. Ges. Wiss. Göttingen **1927**, 227–237. — K. FRIEDRICHS: Math. Ann. **109**, 60–82 (1933/34). — R. FINN: Footnote 2, p. 410.

[3] This proof is valid for curved as well as for polygonal channels.

The most general results on polygonal boundaries were obtained about twenty years later by FINN[1], who proved infinitesimal uniqueness—among symmetric or *asymmetric* disturbances of the given symmetric flow—for concave channels, without any limitations on the total curvature, and proved the same result for convex obstacles in infinite cavity flow. FRIEDRICHS' result on infinitesimal uniqueness was used by LERAY and WEINSTEIN[2] to prove existence, and therefore uniqueness in the large, for concave polygonal channels of total curvature less than π. FINN extended this result to include a class of channels without restriction on the total curvature, and stated an analogous theorem for infinite cavity flow past convex obstacles. (The corresponding flows may be multiple-sheeted.) These results are of interest in showing that the magnitude of the angular variation of the fixed boundary has no importance as such in determining existence and uniqueness of free surface flows.

The preceding discussion has been concerned only with symmetric flows. The problem of asymmetric channel flows introduces a special difficulty in that asymptotic direction as well as the width of the jet is unknown beforehand. A first result on this problem was obtained by WEYL[3], who proved infinitesimal uniqueness for asymmetric channells having total curvature (of both walls) not in excess of π, under the hypothesis that the asymptotic width and direction of the jet were fixed. FINN removed the latter assumptions and proved infinitesimal uniqueness for the class of channels satisfying the following condition: all vectors from the origin, parallel to the boundary segments of the walls, and directed in the sense of motion along the segments toward the separation points, lie interior to a common half-plane. He proves the existence theorem, and hence uniqueness in the large, under the additional hypothesis that the channel is concave and that the separation points lie on opposite sides of some horizontal line; these conditions along with the one above imply that the total curvature is less than π.

39. The method of continuity-curved boundaries. The existence theorem for curved boundaries follows simply from the foregoing results by polygonal approximation[4]. However, uniqueness can obviously not be inferred from such a limiting process, but requires an appropriate generalization of the preceding finite parameter theory to one based on function space methods. Such an extension was provided by LERAY[5], who successfully applied the theory of the FRÉCHET differential and the methods developed by him jointly with SCHAUDER[5] to the problem of uniqueness of the infinite cavity.

We state briefly the basic principles as needed here. Let $F(x; k)$, $0 \leq k \leq 1$ be a one-parameter family of operators defined on a normed, linear, complete, (Banach) space E, with values in E. If at a fixed point $x \in E$ and for a fixed value k in the parameter interval I, there is a linear operator L such that

$$\|F(x+\delta x; k+\delta k) - F(x; k) - L(\delta x; \delta k)\| = o(\|\delta x\| + |\delta k|),$$

then L is said to be the (FRÉCHET) *differential of* F at the point $(x; k)$ of $E \times I$. Let x be a solution of the equation

$$x = F(x; k) \tag{39.1}$$

[1] FINN: Footnote 2, p. 410.

[2] J. LERAY and A. WEINSTEIN: C. R. Acad. Sci., Paris **198**, 430 (1934).

[3] WEYL: Footnote 2, p. 412.

[4] A. WEINSTEIN: Math. Z. **31**, 424—433 (1929). — J. LERAY and A. WEINSTEIN: C. R. Acad. Sci., Paris **198**, 430 (1934). — R. FINN: J. d'Analyse Math. **4**, 246—291 (1956).

[5] J. LERAY: Comm. Math. Helv. **8**, 250—263 (1935). — J. LERAY and J. SCHAUDER: Ann. Sci. Ec. Norm. Sup. (3) **51**, 45—78 (1934).

and suppose that the differential L is completely continuous at $(x; k)$. Under these assumptions, if the *equation of variations*

$$\delta x = L(\delta x; \delta k) \tag{39.2}$$

has a unique solution δx, then x is an isolated solution of Eq. (39.1) having the same topological index[1] as δx, and the Eq. (39.1) admits a solution x' for every k' sufficiently close to k, such that $x' \to x$ as $k' \to k$. This can be considered a theorem on *local existence and uniqueness* for Eq. (39.1).

Suppose now that the above hypotheses are fulfilled for all solutions $(x; k)$ $E \times I$, and that the solutions of Eq. (39.1) form a compact set in $E \times I$. This implies the existence of a one-parameter family of solutions $x(k)$, $0 \leqq k \leqq 1$, provided a single such solution exists. It is now possible to infer uniqueness at $k=1$ from the existence of a unique solution at $k=0$. Namely, any solution at $k=1$ can be joined to the given unique solution at $k=0$ by a family of solutions $x(k)$, all having the same (constant) index as the solutions of the corresponding equations of variation. Since it is a property of the total index [i.e., the sum of the indices of all solutions of Eq. (39.1) for a fixed k] to be constant with respect to k, the number of solutions of Eq. (39.1) at $k=1$ must be unity. This proves uniqueness.

In the application to the infinite cavity problem, the functional equation corresponding to Eq. (39.1) is the Villat equation (33.9)

$$l = F[l, \Psi_k(l), a_k, b_k] \tag{39.3}$$

for the boundary correspondence $l(s)$. Here $\Psi_k(l)$ describes a family of obstacles, and a_k, b_k the corresponding endpoints. The functions $l(s)$ are elements of the space E_α of Hölder continuous functions with Hölder exponent α in $0 \leqq s \leqq \pi$[2], and the curvatures $\Psi_k'(l)$ are assumed to satisfy a Hölder condition with exponent $>\frac{1}{2}$. Taking differentials and using Eqs. (33.4), (33.5), and (33.7), we obtain

$$\delta l(s) = \delta l(0) + 4M \int_0^s \mathrm{e}^{-T(\mathrm{e}^{is})} \sin^2 \frac{s+s_0}{2} \sin s \left(\frac{\delta M}{M} - \delta T(\mathrm{e}^{is}) + \cot \frac{s+s_0}{2}\, \delta s_0\right) ds \tag{39.4}$$

where

$$\delta T(\mathrm{e}^{is}) = \frac{\sin s}{\pi} \int_0^\pi \frac{\delta \Psi_k[l(s')] - \delta \Psi_k[l(s)]}{\cos s' - \cos s}\, ds',$$

$$\delta \Psi_k(l) = \Psi_k'(l)\, \delta l + \frac{\partial \Psi_k}{\partial k}\, \delta_k,$$

$$\delta s_0 = \frac{1}{\pi} \int_0^\pi \delta \Psi_k[l(s)]\, ds,$$

$$\delta b_k - \delta a_k = 4M \int_0^\pi \mathrm{e}^{-T(\mathrm{e}^{is})} \sin^2 \frac{s+s_0}{2} \sin s \left(\frac{\delta M}{M} - \delta T(\mathrm{e}^{is}) + \cot \frac{s+s_0}{2}\, \delta s_0\right) ds$$

and

$$\delta l(0) = \delta a_k = \frac{\partial a_k}{\partial k}\, \delta k, \qquad \delta l(\pi) = \delta b_k = \frac{\partial b_k}{\partial k}\, \delta k.$$

---

[1] The topological index of an isolated solution x of Eq. (39.1) is the degree of the mapping $y = x - F(x; k)$ at $y=0$, when the mapping is restricted to a neighborhood of x (cf. footnote 4, p. 400). A unique solution of Eq. (39.2) has index $+1$ or -1, and, assuming the same hypothesis are fulfilled for all k, the index remains constant as k varies.

[2] Let $H_\alpha = \text{l.u.b.}\ |l(s_2) - l(s_1)|/|s_2 - s_1|^\alpha$ for $s_1, s_2 \in [0, \pi]$; then $\|l\| = \max\limits_{0 \leqq s \leqq \pi} |l(s)| + H_\alpha$.

Inserting these relations into Eq. (39.4) we arrive at the equation of variations for Eq. (39.3). This can be expressed in the form

$$\delta l = dF[\delta l, \delta k; l, \Psi_k(l), a_k, b_k], \tag{39.5}$$

where the right member is linear and homogeneous in δl and δk.

One observes that the differential dF is completely continuous in the argument δl for every value of δk, and that Eq. (39.5) is an integral equation of Fredholm type for $\delta l(s)$. Hence Eq. (39.5) has a unique solution for each value of δk provided the equation obtained from Eq. (39.5) by setting $\delta k = 0$ has only the trivial solution $\delta l \equiv 0$. The proof that $\delta l \equiv 0$ (which is the heart of the uniqueness argument) is reduced, as in the theory of infinitesimal uniqueness for polygonal boundaries, to showing that a harmonic function solving a certain boundary value problem must vanish identically. This problem is basically the same as in the preceding section and is derived in the same manner by considering the variational equations satisfied by the flow functions. Using a modified form of the lemma of FRIEDRICHS (cf. Sect. 38, p. 412), LERAY supplies a proof for the class of obstacles described below. Uniqueness in the large follows by imbedding any given obstacle in the class in a family of curves $\Psi_k(l)$, every member of which is also in the class, and which reduces to a flat plate at $k = 0$. For the latter obstacle the right member of Eq. (39.5) is independent of δl, the topological index of the solution is (trivially) $+1$, and the solution of Eq. (39.5) is unique. The compactness of the solutions of Eq. (39.3) in $E_\alpha \times I$, which is all that remains to complete the proofs, is assured by the *a priori* bounds of the existence theory.

LERAY proves uniqueness of the infinite cavity for a class of obstacles having the property that a line parallel to the incident flow can intersect the barrier in at most a single point or a segment. The class of problems included in his proof are the following: the problem of fixed detachment for a convex obstacle (with or without symmetry); the symmetric problem of fixed detachment; the prow problem for a convex circular arc (with or without symmetry); the prow problem for symmetric *accolades* (cf. Sect. 24). The method is undoubtedly capable of wider application.

V. Numerical methods.

The hodograph method and the method of reflection are the only known systematic procedures for explicit construction of free surface flows past given barriers. Even in the case of polygonal boundaries the "explicit" solutions yielded by these methods contain undetermined parameters whose evaluation is generally a difficult computational problem[1]. For problems of plane flow past curved barriers, not to mention still more complex problems involving axial symmetry, gravity, and time-dependence, numerical methods seem to be the only means of deriving precise information based on the exact theory. The increasing emergence of the high speed computer as a working tool and the advanced state of the general theory make very probable the further development of numerical methods in the free boundary problem.

In the following sections we outline several methods of effective computation of free surface flows. It is impossible to discuss adequately here the practical aspects of the computations or the many details of numerical analysis bearing on the accuracy of the results. We therefore content ourselves with a description of the methods and results, and occasionally remark on the scope, limitations,

[1] See BIRKHOFF and ZARANTONELLO [*1*], Chap. 9. This reference contains a detailed discussion of the practical aspects of computation.

and accuracy of the various approaches. For more intimate details of the computations the reader is referred to the original sources.

At the present time none of the methods under discussion has been rigorously proved to converge[1], and theoretical estimates of accuracy are still lacking. The significance of the corresponding internal error estimates is therefore open to debate, especially in those cases where the procedure is not systematic. However, when the supporting evidence is convincing, these estimates are probably more realistic than the weak bounds that are usually provided by a general theory.

a) Plane flows past curved obstacles.

40. The discrete curvature equations; iteration. The first calculations of free surface flows past curved barriers were performed by Brodetsky[2], who successfully treated the plane infinite cavity flow past circular and elliptical arcs by a method of trigonometric interpolation that was later extended by Birkhoff, Zarantonello, *et al.*[3]. The work of the latter group is the basis of the following discussion.

As we have seen in Sect. 33, the solution of the free boundary problem for a curved barrier B can be described in terms of a regular function

$$\Theta + i\,T = \Omega(\zeta) = \sum_{m=0}^{\infty} a_m \zeta^m, \qquad a_m \text{ real} \tag{40.1}$$

defined in the unit circle $|\zeta| \leqq 1$, $\zeta = \varrho\, e^{is}$, where $\Theta(e^{is})$, $0 \leqq s \leqq \pi$, is the inclination of B with respect to the x-axis at the point corresponding to $\zeta = e^{is}$. The determination of $\Omega(\zeta)$ is reduced to the solution for $\lambda(s)$ of the functional equation (33.23)

$$\lambda(s) = M\,\nu(s)\,K(\Phi(s))\,e^{-T(e^{is})},$$

or Eq. (33.28),

$$\lambda(s) = M\,\nu(s)\,K(I[\lambda])\,e^{-D[\lambda]}, \tag{40.2}$$

where $K(\Phi)$ is the curvature of the given boundary, $\nu(s)$ is a known function — except possibly for parameters — depending on the type of flow under discussion, and where

$$\lambda(s) = \sum_{m=1}^{\infty} m\, a_m \sin m\, s = -\frac{d\Phi}{ds}, \tag{40.3}$$

$$\left.\begin{aligned} I[\lambda] &= \sum_{m=1}^{\infty} a_m \cos m\, s = \Phi(s), \\ &= \int_0^{\pi} J(s,\sigma)\,\lambda(\sigma)\,d\sigma, \qquad J(s,\sigma) = \begin{cases} -\sigma/\pi, & 0 \leqq \sigma < s \\ 1-\sigma/\pi, & s < \sigma \leqq \pi, \end{cases} \end{aligned}\right\} \tag{40.4}$$

$$\left.\begin{aligned} D[\lambda] &= \sum_{m=1}^{\infty} a_m \sin m\, s = T(e^{is}), \\ &= \int_0^{\pi} D(s,\sigma)\,\lambda(\sigma)\,d\sigma, \qquad D(s,\sigma) = \frac{1}{\pi}\log\left|\frac{\tan\frac{\sigma}{2} + \tan\frac{s}{2}}{\tan\frac{\sigma}{2} - \tan\frac{s}{2}}\right|. \end{aligned}\right\} \tag{40.5}$$

[1] Special instances of convergence have been proved, but not for the natural flow problems; these results are mentioned in the following section.

[2] S. Brodesky: Proc. Roy. Soc. Lond., Ser. A **102**, 542–553 (1923). — Scripta Univ. Bibl. Hieros. **1**, 1–14 (1923).

[3] G. Birkhoff, D. Young; E. Zarantonello: Proc. Symp. Appl. Math. **4**, 117–140 (1953). — G. Birkhoff, H. Goldstine and E. Zarantonello: Rend. Seminar. Mat. Univ. Politec. Torino **13**, 205–224 (1953–54); see also Birkhoff and Zarantonello [*1*], Chap. 9. Similar methods have been used by C. Schmieden: Ing.-Arch. **1**, 104–109 (1929); **3**, 356–370 (1932); **5**, 373–375 (1934); and R. Eppler: J. Rat. Mech. a. Analysis **3**, 591–644 (1954).

As in Sect. 33, Φ is the inclination of the obstacle, defined by $\Phi(s) = \Theta(e^{is}) - \Theta(0) = \Theta(e^{is}) - a_0$. In the symmetric case, when $a_0 = \pi/2$, Φ is simply the angle through which the upper arc has turned from the vertical in a clockwise direction. M and the parameters in $\nu(s)$ must ultimately be determined by side conditions. In certain problems such as that of the symmetric infinite cavity, M is the only unknown parameter, and this is determined by the separation condition. We shall discuss only such cases here.

The typical detachment conditions can also be expressed in terms of λ and the coefficients a_m. Thus, if the separation point, or equivalently, the angle at separation, is fixed—say $\Phi = \Phi_0$ at the point corresponding to $s = 0$—then by Eq. (40.4)

$$\Phi_0 = \int_0^{\pi} J(0, \sigma)\, \lambda(\sigma)\, d\sigma = \sum_{m=1}^{\infty} a_m. \tag{40.6}$$

Similarly, if the separation is smooth, then we must have $\Omega'(1) = \cot(s_0/2)$, whence from Eq. (40.5)

$$\cot\frac{s_0}{2} = \frac{1}{\pi}\int_0^{\pi} \lambda(s) \cot\frac{s}{2}\, ds = \sum_{m=1}^{\infty} m\, a_m; \tag{40.7}$$

in the symmetric case, for which $a_2 = a_4 = \cdots = 0$, this reduces to

$$1 = \frac{1}{\pi}\int_0^{\pi} \lambda(s) \operatorname{cosec} s\, ds = a_1 + 3a_3 + 5a_5 + \cdots. \tag{40.7a}$$

The discrete equations. The calculation, after these preliminaries, proceeds in two steps: first a reduction to discrete equations, and then a solution of the discrete system by iteration. Let us introduce the approximate solution,

$$\Omega(\zeta) = \sum_{m=0}^{N} a_m \zeta^m, \tag{40.8}$$

and define the corresponding trigonometric polynomials,

$$\lambda(s) = \sum_{1}^{N} m\, a_m \sin m s, \qquad I[\lambda] = \sum_{1}^{N} a_m \cos m s, \qquad D[\lambda] = \sum_{1}^{N} a_m \sin m s. \tag{40.9}$$

We now replace the Eq. (40.2) in the continuous variable s by the discrete system

$$\lambda_k = M\, \nu_k K(I[\lambda_k])\, e^{-D[\lambda_k]}, \qquad k = 1, \ldots, N, \tag{40.10}$$

where $\lambda_k = \lambda(s_k)$, $\nu_k = \nu(s_k)$. Let the N values of s_k be equidistant, with

$$s_k = \left(k - \frac{1}{2}\right)\frac{\pi}{N}, \qquad k = 1, \ldots, N. \tag{40.11}$$

The coefficients a_m can then be expressed in terms of the λ_k by the formulas

$$a_m = \begin{cases} \dfrac{2}{m N} \displaystyle\sum_{k=1}^{N} \lambda_k \sin m s_k, & m \neq N, \\[2ex] \dfrac{1}{m N} \displaystyle\sum_{k=1}^{N} \lambda_k \sin m s_k, & m = N. \end{cases} \tag{40.12}$$

When these are inserted into the polynomial approximations (40.9) for $I[\lambda]$ and $D[\lambda]$ one obtains

$$I[\lambda_k] = \sum_{j=1}^{N} J_{kj} \lambda_j, \qquad D[\lambda_k] = \sum_{j=1}^{N} D_{kj} \lambda_j, \tag{40.13}$$

Table 1. *Flows past*

	Φ_0	C_D	M	a_1	a_3	a_5
1. Fixed detachment . .	15°	0.81912	0.18196	0.26704	−0.00433	−0.00065
2. Fixed detachment . .	45°	0.67766	0.81727	0.77811	0.00746	0.00003
3. Smooth detachment .	55.04°	0.60838	1.13593	0.94277	0.01643	0.00123
4. Zero asymptotic width	124.21°	0	5.71464	2.00000	0.12518	0.02661
5. Cusped cavity	136.4°	0	14.04376	2.36906	−0.01893	0.02303

where (J_{kj}) and (D_{kj}) are matrices depending only on N and can therefore be tabulated. Thus the discrete problem corresponding to Eq. (40.2) reduces to solving the N Eqs. (40.10) for the N unknowns λ_k. To these equations must be added one side condition such as Eqs. (40.6) or (40.7) (cut off at a_N) for the determination of the separation parameter M. The solution thus obtained will describe a free surface flow past an obstacle having the same curvature as the given body at the N points $\Phi_k = I[\lambda_k]$, and satisfying the appropriate separation condition. ZARANTONELLO[1] has proved for convex obstacles ($K>0$) and for *fixed* M that the solutions of the discrete equations converge uniformly to the solution of Eq. (40.2) as $N\to\infty$ whenever that solution is unique. The computational scheme outlined above is essentially the one used by BRODETSKY.

Iteration. The discrete Eqs. (40.10) are now solved numerically by an iteration scheme. It turns out that a direct iteration is generally divergent unless M is sufficiently small, and hence a process of *averaged* iteration is suggested by BIRKHOFF *et al.* Let us define the transformation

$$S[\lambda] = \nu K(I[\lambda])\, e^{-D[\lambda]} \tag{40.14}$$

to be used in the iteration process. Consider first the case of prescribed M. In this case, for a suitable class of convex obstacles, iterations based on the transformation

$$M\, S_\varepsilon[\lambda] = \varepsilon M\, S[\lambda] + (1-\varepsilon)\,\lambda, \qquad 0<\varepsilon\leqq 1 \tag{40.15}$$

have been proved to converge uniformly to a solution of Eq. (40.10) if ε is sufficiently small[2]. The process based on Eq. (40.15) involves a weighted average between an iterate and its predecessor; its success is founded on the fact that the transformation (40.14) is order-inverting.

The case of greater interest is that in which M is unspecified and is to be determined from the detachment conditions. In the iteration process let

$$\lambda^{(r)} = \sum_{m=1}^{N} m\, a_m^{(r)} \sin m\, s \tag{40.16}$$

be the r-th iterate of λ and M_r the r-th iterate of M. If we set

$$g(\lambda) = \Phi_0 \Big/ \sum_1^N a_m \qquad \text{for fixed detachment,}$$

$$= 1 \Big/ \sum_1^N m\, a_m \qquad \text{for smooth detachment,}$$

[1] E. ZARANTONELLO: Collectanea Mathematica 5, 175—255 (1952). For sufficiently small fixed M, N. NEKRASOFF: Publ. Inst. Polytech. d'Ivanovo-Voszniesiensk 1922, and his followers have shown that direct iteration of the non-discretized equation (40.2) converges to a solution (cf. Sect. 32).

[2] E. ZARANTONELLO, preceding footnote; see also BIRKHOFF and ZARANTONELLO [*1*], Chap. 9.

circular arcs.

a_7	a_9	a_{11}	a_{13}	a_{15}	a_{17}	a_{19}	a_{21}	a_{23}
− 0.000 16	− 0.000 06							
− 0.000 10	− 0.000 05							
0.000 17	0.000 04							
0.008 58	0.003 49	0.001 67	0.000 89	0.000 53	0.000 35	0.000 24	0.000 18	0.000 16
0.003 93	0.002 12	0.000 94	0.000 52	0.000 31	0.000 20	0.000 14		

then the recursion formula for the averaged iteration scheme is defined by

$$\left.\begin{aligned}\bar{\lambda}^{(r)} &= M_r\, S[\lambda^{(r)}],\\ \lambda^{(r+1)} &= \varepsilon\,\bar{\lambda}^{(r)} + (1-\varepsilon)\,\lambda^{(r)}\end{aligned}\right\} \tag{40.17}$$

and

$$M_{r+1} = M_r + c\,\varepsilon\,(\overline{M}_r - M_r), \qquad 0 < \varepsilon \leqq 1, \tag{40.18}$$

where

$$\overline{M}_r = M_r\, g(\bar{\lambda}^{(r)})$$

and $0 \leqq c < 1$. This process has not been proved to converge, but in actual calculations by BIRKHOFF and his collaborators it has exhibited convergence in every case tried. This completes the outline of the computational scheme.

41. Results of calculations. Calculations of free surface flows based on the preceding scheme have been carried out only for symmetric obstacles ($a_{2\mu} = 0$, $\mu = 1, 2, \ldots$). The calculations by BRODETSKY of infinite cavity flows past circular and elliptical arcs used approximations (40.8) terminating at $N = 5$, or in other words, involving the only three coefficients a_1, a_3, a_5. Inasmuch as his flows were required to satisfy the smooth detachment condition (40.7a), he was left, effectively, with a two parameter family of curves to approximate the given obstacle. In the case of the circular arc he found the values

$$a_1 = 0.9415, \qquad a_3 = 0.0167, \qquad a_5 = 0.0017, \tag{41.1}$$

which describe a flow past an arc of half angle 55°. The drag coefficient based on the diameter was computed to be 0.50.

Systematic calculations of a variety of symmetric flows past convex and concave barriers have been carried out on a high speed computer by BIRKHOFF, GOLDSTINE, and ZARANTONELLO[1], using the averaged iteration process described above. The several different types of flows considered include the usual infinite cavity; the infinite cavity flow past an obstacle in a jet, in a channel, and in a jet from a nozzle; the cusped cavity; and the Riabouchinsky cavity. In addition to problems of fixed and smooth separation, they handle the side condition $a_1 = 2$ associated with the cavity of zero asymptotic width[2], and also that required for the cusped cavity. In all, about fifty different calculations were reported and tabulated, including a careful analysis of the practical aspects and of the accuracy of the results.

In these computations N was taken equal to 24, leaving ,after the detachment condition, an eleven parameter family of curves with which to approximate the given obstacle. We illustrate the results in Table 1 with the computed values for several flows past circular arcs of unit radius. (Here Φ_0 is equal to the central

[1] G. BIRKHOFF, H. GOLDSTINE and E. ZARANTONELLO: Rend. Seminar. Mat. Univ. Politec. Torino **13**, 205—224 (1953).

[2] The condition $\omega'(0) = 0$ characterizing this class of flows (cf. Sect. 25) is clearly equivalent to $a_1 = 2$.

angle subtended by the half arc.) The figures for the drag coefficient are based on the *wetted cross-section*, and are computed for the infinite cavities from the drag formula (cf. Sect. 23)[1]

$$D = \frac{\pi}{8} M (a_1 - 2)^2. \tag{41.2}$$

In particular, the value for C_D, when based on the diameter, is 0.50 in the flow with smooth detachment, which agrees with BRODETSKY's result. Comparison with Eq. (41.1) shows that the trinomial approximation is remarkably accurate. The available experimental data, although inconclusive, is in substantial agreement with the calculated value of C_D and of the angle of detachment[2].

The convergence of the coefficients indicated in Table 1 is more or less typical. In most of the calculations the ratio $r_m = |a_{m+1}/a_{m-1}|$ was less than 0.75 for all m, increasing with m in many cases (as expected from the theory), but behaving erratically in others. Typical values for c and ε in the iteration process (40.17), (40.18) were $c = 0.3$, $\varepsilon = 0.5$.

To estimate accuracy, the authors compute the relative error $|\tilde{K}(\Phi)/K(\Phi) - 1|$ between the prescribed curvature function $K(\Phi)$ and that given by the polynomial approximation $\tilde{K}(\Phi)$ at values of s intermediate between the s_k. By this measure the curvature values are usually accurate to better than 1%, and often to within 0.2%, except at the separation point $s = 0$. This reflects a corresponding accuracy in the shape of the fixed boundary, which is given in intrinsic form by $l(\Phi) = \int d\Phi/K(\Phi)$. At the point of detachment the errors in curvature are usually of the order 3 to 10% when the detachment is non-smooth, and less than 1% when smooth. What these estimates say with regard to the other physical quantities is not immediately clear. The drag, for example, which is given by the formula (41.2), is determined to within the accuracy of the parameters M and a_1, for which estimates are lacking. However, some indication is given by the close agreement between the values $a_1 = 0.9415$ and $a_1 = 0.9428$ obtained from the trinomial approximation and from the calculation for $N = 24$.

BIRKHOFF, GOLDSTINE, and ZARANTONELLO conclude from their analysis of the errors that the curvature Eq. (40.2) does not promote high accuracy when $K(\vartheta)$ varies by a large factor of say 10 or more; and also that the equidistant spacing (40.11), except in the case of smooth detachment, will yield inaccurate values of the curvature at separation unless $N \gg 24$.

[1] The present coefficient M is twice that in Eq. (23.7).

[2] Cavitation experiments on circular cylinders for $\sigma > 0.5$ have been performed by V. KONSTANTINOV: Izv. Akad. Nauk. SSSR., OTN **1946**, 1355–1373, David Taylor Model Basin Transl. 233 (1950), and by MARTYRER: Hydromechanische Probleme des Schiffsantriebs, Hamburg 1932, ed. by G. KEMPF and E. FOERSTER. Further experiments on circular cylinders, down to $\sigma < 0.3$, have been reported by R. WAID; Hydrodynamics Lab., Cal. Inst. Tech. Rep. E-73.6 (1957). The extrapolation of their drag data to $\sigma = 0$ gives a value for $C_D(0)$ in the neighborhood of 0.5. Konstantinov finds constant base pressure on the cylinder starting at about 60°, and sometimes forward of this point, for σ between 0.65 and 1. In the case of *fully wetted flow* the observed angle at which the base pressure becomes sensibly constant (boundary layer separation) is 20° or more farther downstream than the calculated 55°; cf. S. GOLDSTEIN (ed.), Modern developments in fluid dynamics, Vol. II, Oxford, p. 421. This suggests that smooth separation is not the appropriate condition to describe wakes behind bluff bodies; for other relevant calculations see A. ROSHKO: NACA TN 3168 (1954). The connection between the boundary layer and free streamline separation has been discussed theoretically by H. B. SQUIRE: Phil. Mag. **7** (17), 1150–1160 (1934); I. IMAI: J. Phys. Soc. Japan **8**, 399–402 (1953); and M. KAWAGUTI: J. Phys. Soc. Japan **8**, 403–406 (1953). C. SCHMIEDEN: Ann. d. Phys. (5) **2**, 350–356 (1929) finds that the experimental angle of detachment from a *sphere* in cavity flow is the same as the calculated value 55° for the cylinder, but this disagrees with the observations of E. HSU and B. PERRY: Hydrodynamics Lab., Cal. Inst. Tech., Rep. E-24.9 (1954).

b) Axially symmetric flows.

42. Dimensional perturbation. In this and the following sections we outline a systematic method of calculating axially symmetric flows which was developed by GARABEDIAN[1] and applied by him to several specific problems. His point of departure is the concept of axially symmetric free surface flow in a space of $\varepsilon+2$ dimensions. The stream function for such flows obeys the equation

$$\psi_{xx}+\psi_{yy}-\frac{\varepsilon}{y}\psi_y=0 \tag{42.1}$$

in the meridian plane, where x is the axial and y the radial variable. Let the meridian cross-section of the flow be denoted by $\Omega(\varepsilon)$ and its boundary by $\Gamma(\varepsilon)$. In the following $\Gamma(\varepsilon)$ will always be a streamline on which

$$\psi=0, \tag{42.1a}$$

and will consist of two parts, a fixed boundary Γ_1, and a free boundary $\Gamma_2(\varepsilon)$ on which the constant speed free surface condition

$$\frac{1}{y^\varepsilon}\frac{\partial\psi}{\partial n}=1 \tag{42.2}$$

is satisfied. The point of view of the method presented here is to investigate the dependence of ψ on ε as a means of acquiring information about the physically significant case $\varepsilon=1$ of three dimensional axially symmetric flow. The classical plane free boundary problem corresponds of course to $\varepsilon=0$.

We suppose that the solution is a regular function of ε in the half plane $\mathrm{Re}(\varepsilon)>-1$, as in other boundary value problems for Eq. (42.1)[2]. This defines the radius of convergence of the perturbation series

$$\psi(x,y;\varepsilon)=\psi_0(x,y)+\varepsilon\psi_1(x,y)+\varepsilon^2\psi_2(x,y)+\cdots. \tag{42.3}$$

We observe that an expansion in the variable

$$\delta=\frac{\varepsilon}{\varepsilon+2} \tag{42.4}$$

brings the entire regularity domain $\mathrm{Re}(\varepsilon)>-1$ into the circle of convergence $|\delta|<1$; and hence this is the appropriate variable in which to write the expansions that have to be calculated later at the pertinent value $\varepsilon=1$ or $\delta=\frac{1}{3}$.

One sees from Eq. (42.1) that the initial term ψ_0 is simply the stream function of the plane free boundary problem, and that the successive terms ψ_i satisfy the recursive system of Poisson equations,

$$\Delta\psi_i=\frac{1}{y}\frac{\partial\psi_{i-1}}{\partial y}. \tag{42.6}$$

To derive the boundary conditions for the ψ_i, we observe first that

$$\psi_i=0 \quad \text{on} \quad \Gamma_1, \quad i=0,1,\ldots. \tag{42.7}$$

The conditions on the free boundary are obtained by differentiating Eq. (42.2) with respect to ε. This yields, after some calculation, the boundary condition

$$\frac{\partial}{\partial n}\left(\frac{\partial\psi}{\partial\varepsilon}\right)+\varkappa\frac{\partial\psi}{\partial\varepsilon}=y^\varepsilon\log y \tag{42.8}$$

[1] P. GARABEDIAN: Pacif. J. Math. **6**, 611–684 (1956).

[2] Cf. A. WEINSTEIN: Bull. Amer. Math. Soc. **59**, 20–38 (1953), and see the discussion by GARABEDIAN, preceding footnote.

where $\varkappa = \varkappa(\varepsilon)$ is the curvature of $\Gamma_2(\varepsilon)$. Inserting $\varepsilon = 0$ into this relation, one obtains

$$\frac{\partial \psi_1}{\partial n} + \varkappa_0 \psi_1 = \log y \quad \text{on} \quad \Gamma_2(0), \tag{42.9}$$

$\varkappa_0$ being the curvature of the free streamline $\Gamma_2(0)$ for the plane problem. Further differentiations of Eq. (42.7) with respect to ε and specialization to $\varepsilon = 0$ yield a sequence of boundary conditions of the form

$$\frac{\partial \psi_i}{\partial n} + \varkappa_0 \psi_i = B_i, \tag{42.10}$$

where the B_i are known expressions involving only the earlier coefficients $\psi_0, \ldots, \psi_{i-1}$. Thus, all the solutions ψ_i are given recursively as solutions of the linear mixed boundary value problems (42.6), (42.7), and (42.10), defined in the (presumably) known region Ω_0 of the plane problem.

When Γ_1 is polygonal, this boundary value problem can be solved in closed form[1] by using the hodograph method to construct the Green's function of the homogeneous problem (42.6), (42.7), (42.10) in Ω_0, and then representing the desired solution in terms of the inhomogeneous data by means of Green's formula. Even in this case the labor involved in calculations beyond $i = 1$ proves to be forbidding, and hence Garabedian's results and the present discussion are limited to the first terms ψ_0 and ψ_1.

It should be remarked that in principle the foregoing perturbation scheme can be applied also to non-polygonal fixed boundaries, by starting with polygonal $\Gamma_1(0)$ and terminating at $\varepsilon = 1$ at the given curved boundary $\Gamma_1(1)$. This modification differs from the preceding only in making the boundary condition (42.7) along Γ_1 inhomogeneous.

To this outline of the general method we add a brief discussion of the following reduction to analytic functions which proves to be of great assistance in the calculations.

Let $\Gamma_2(\varepsilon)$ be represented in the form

$$\bar{z} = g(z; \varepsilon) \tag{42.11}$$

where g is an analytic function of its arguments. We define

$$W(z; \varepsilon) = \Phi + i\Psi = \int_{z_0}^{z} g'(t; \varepsilon)^{\frac{1}{2}}\, dt, \qquad z_0 \varepsilon \Gamma_2(\varepsilon), \tag{42.12}$$

which is simply the complex potential of a plane flow having $\Gamma_2(\varepsilon)$ as free boundary [cf. Eq. (17.4)]. Since the stream functions $\psi(x, y; \varepsilon)$ and $\Psi(x, y; \varepsilon)$ are formally connected by Eqs. (42.12) and (17.7), it is natural to consider the perturbation expansion

$$\Psi(x, y; \varepsilon) = \Psi_0(x, y) + \varepsilon \Psi_1(x, y) + \varepsilon^2 \Psi_2(x, y) + \cdots, \tag{42.13}$$

with the intention of relating the corresponding ψ_i and Ψ_i, especially since the latter are harmonic functions and might be simpler to handle than the ψ_i. If Γ_1 consists of vertical segments and z_0 is the point of detachment of $\Gamma_2(\varepsilon)$ from Γ_1, it follows from Eqs. (17.7) and (42.12) that the ψ_i and Ψ_i are connected by explicit formulas which reduce for $i = 0, 1$ to

$$\psi_0 = \Psi_0$$

---

[1] The condition $\frac{\partial G}{\partial n} + \varkappa_0 G = 0$ satisfied by the Green's function on $\Gamma_2(0)$ takes the simple form $\frac{\partial G}{\partial \tau} + G = 0$ in the logarithmic hodograph plane, $\omega = \vartheta + i\tau = i \log f_0'(z)$, where $f_0(z)$ is the complex potential of the flow determined by ψ_0. The function $G^* = \frac{\partial G}{\partial \tau} + G$ vanishes on the entire polygonal boundary in the ω plane, has a known singularity, and is therefore easily determined by the Schwarz-Christoffel mapping. G itself is then obtained by integration. For the details, see Garabedian, footnote 1, p. 421.

and

$$\psi_1 = \Psi_1 + \frac{1}{2}\,\mathrm{Im}\int\limits_{z_0}^{z}\left[\log\frac{(\bar{z}-t)\,(z-g(t;0))}{4}\right] W'(t;0)\,dt. \tag{42.14}$$

One can check that the equations satisfied by Ψ_1 in $\Omega(0)$ are therefore

$$\Psi_1 = 0 \quad \text{on} \quad \Gamma_1, \qquad \frac{\partial \Psi_1}{\partial n} + \varkappa_0 \Psi_1 = 0 \quad \text{on} \quad \Gamma_2(0), \tag{42.15}$$

$$\Psi_1 = -\tfrac{1}{2}\,\mathrm{Im}\int\limits_{z_0}^{x}\left[\log(x-t)\,(x-g(t;0))\right] W'(t;0)\,dt + \text{const} \tag{42.16}$$

on the x-axis, and

$$\Delta\Psi_1 = 0 \quad \text{in} \quad \Omega(0). \tag{42.17}$$

Disregarding its derivation, one can look upon Eq. (42.14) as just a substitution leading to these simpler equations for Ψ_1, which, if solved, yield ψ_1 in turn. In the problems at hand Ψ_1 proves to have an explicit representation in the hodograph plane that assists materially in the computations.

43. The vena contracta. We consider now the numerical determination of the contraction coefficient C_c of a jet issuing horizontally from a circular orifice in a plane vertical wall. In the meridian plane the flow region Ω is bounded by the x-axis, by the semi-infinite line Γ_1: $y \geq Y$, $x=0$, and by the free streamline Γ_2 which detaches from Γ_1 at $z_0 = iY$ and is asymptotic to the horizontal line $y=\bar{Y}$ at $x=+\infty$. The desired quantity is

$$C_c = \bar{Y}^2/Y^2. \tag{43.1}$$

This will be computed by an interpolation process involving the generalized contraction coefficient

$$C_c(\varepsilon) = (\bar{Y}(\varepsilon)/Y(\varepsilon))^{1+\varepsilon} \tag{43.2}$$

which reduces to Eq. (43.1) at $\varepsilon=1$ and to

$$C_c(0) = \frac{\pi}{\pi+2} \tag{43.3}$$

for plane flows (cf. Sect. 10).

The stream function $\psi(x, y; \varepsilon)$ of the generalized flow problem is a bounded solution of Eqs. (42.1), (42.1a), and (42.2) and satisfies the additional condition

$$\psi = \frac{\bar{Y}^{\varepsilon+1}}{1+\varepsilon} \quad \text{on} \quad y=0.$$

The interpolation scheme succeeds in this problem with relatively little calculation because of the fortunate circumstance that one can determine easily the ratio $\bar{Y}(\varepsilon)/Y(\varepsilon)$ in the limiting cases $\varepsilon=-1$ and $\varepsilon=\infty$.

Consider first the case $\varepsilon\to+\infty$. From the monotonicity properties of the free streamlines (cf. Sect. 28) it follows that the convex curve $\Gamma_2(\varepsilon)$ must rise as ε increases. At the same time the potential $\varphi(x, y; \varepsilon)$ satisfying

$$\varphi_{xx} + \varphi_{yy} + \frac{\varepsilon}{y}\,\varphi_y = 0 \tag{43.4}$$

can be expected, as in other singular perturbation problems of the same type to approach the solution of $\varphi_y=0$ in the interior of the jet as $\varepsilon\to\infty$, so that the limiting solution is a function of x alone, and hence defines a parallel flow. This gives the result

$$\bar{Y}(\infty)/Y(\infty) = 1. \tag{43.5}$$

The case $\varepsilon \to -1$ can be analyzed in terms of the source flow solution of Eq. (42.1),

$$\psi^*(\vartheta;\varepsilon) = \frac{\int\limits_{\pi/2}^{\vartheta} \sin^\varepsilon t\,dt}{(1+\varepsilon)\int\limits_{\pi/2}^{\pi} \sin^\varepsilon t\,dt},$$

where $\vartheta = \arctan(y/x)$. One observes that

$$\psi^*(x,y;-1) = \lim_{\varepsilon\to-1}\psi^* = \int\limits_{\pi/2}^{\vartheta}\frac{dt}{\sin t} = \log\frac{y}{x+\sqrt{x^2+y^2}},$$

so that the velocity on the y axis is

$$-y\frac{\partial\psi^*}{\partial x} = 1.$$

From this it follows that the entire y-axis is a free streamline, and, in particular, so is the portion $0 < y \leq Y$. It is resonable to infer that this degenerate jet flow is the limiting case of the *vena contracta* as $\varepsilon \to -1$, the jet width becoming vanishingly thin in the limit. Thus we have

$$\overline{Y}(-1)/Y(-1) = 0. \tag{43.6}$$

From the three data (43.3), (43.5), and (43.6) we now attempt a quadratic interpolation to determine $\overline{Y}(1)/Y(1)$, using the quantity $\delta = \varepsilon/(\varepsilon+2)$ as the appropriate interpolation variable. The quadratic expression

$$\frac{\overline{Y}}{Y} = \frac{\pi}{\pi+2}(1-\delta^2) + \frac{1}{2}(\delta+\delta^2) \tag{43.7}$$

fits all the data (43.3), (43.5), and (43.6), and gives at $\delta = \frac{1}{3}$ (which corresponds to $\varepsilon = 1$) the value 0.765 as the approximation for $\overline{Y}(1)/Y(1)$. This yields the preliminary estimate

$$C_c(1) = 0.586. \tag{43.8}$$

The next step is to refine this estimate by determining the derivative with respect to ε of $\overline{Y}/Y$ at $\varepsilon = 0$. To this end, we normalize the flows so that $\overline{Y} = 1$ for all ε. Hence it suffices to calculate $Y'(\varepsilon)$ at $\varepsilon = 0$. By analyzing the first order displacement of the free streamline with respect to ε, one finds that the required derivative can be represented in the logarithmic hodograph plane, $\omega = \vartheta + i\log q$, of the plane flow by the formula

$$\frac{dY}{d\varepsilon}(0) = \frac{\partial\psi_1}{\partial\vartheta} = \frac{\partial\Psi_1}{\partial\vartheta}, \tag{43.9}$$

where the right members are evaluated at $\omega = \pi/2$, corresponding to the point of detachment of the free streamline. The formula (43.9) is an expression of the first order tangential shift of the free boundary when the slope is vertical (and has the same significance when $\vartheta \neq \pi/2$ if Y denotes arc length). Since Ψ_1 can be found explicitly in the hodograph plane, there remains only the calculation of $\partial\Psi_1/\partial\vartheta$ in Eq. (43.9). This can be written in closed form as a complicated double integral which has to be evaluated numerically. The precise value obtained by GARABEDIAN is

$$\frac{dY}{d\varepsilon}(0) = -0.650544,$$

at least the first four figures of which he considers significant.

This gives for the slope of the $\overline{Y}/Y$ curve at $\varepsilon=0$ the value

$$\frac{d}{d\varepsilon}\left(\frac{\overline{Y}}{Y}\right)_{\varepsilon=0}=0.24287. \tag{43.10}$$

Using this additional information, we find a cubic polynomial in fitting the four data (43.3), (43.5), (43.6), and (43.10), namely,

$$\overline{Y}/Y=0.6110+0.4857\,\delta-0.1110\,\delta^2+0.0143\,\delta^3. \tag{43.11}$$

At $\delta=\frac{1}{3}$ this becomes

$$\overline{Y}(1)/Y(1)=0.7611, \tag{43.12}$$

whence our new approximation to the three dimensional contraction coefficient has the value

$$C_c(1)=0.5793\approx 0.58. \tag{43.13}$$

The earlier result (43.8), based on almost trivial calculations, is thus seen to be remarkably close to the final value 0.58, suggesting that the procedure is rapidly convergent. This is indicated also by the rapid decrease in the magnitude of the coefficients of Eq. (43.11).

The value 0.58 for the contraction coefficient differs somewhat from the commonly accepted figure of 0.61. The calculations by TREFFTZ and by SOUTHWELL and VAISEY leading to this value will be discussed in later sections. The pertinent experimental data was reviewed in Sect. 2, and it was noted there that several results displaying effects of gravity, viscosity, etc., point to a lower theoretical figure than 0.61.

44. The drag coefficient of the disk: Infinite cavity. We now apply the preceding method to the determination of the drag coefficient C_{D} of a circular disc with trailing infinity cavity. In the meridian plane the fixed boundary Γ_1 is now the segment $0\leqq y\leqq Y$ of the y-axis, and Γ_2 is the free streamline leaving the end point $z_0=iY$ of Γ_1 and proceeding to the right. As before we consider this flow in its dependence on the dimension parameter ε. The solution of the flow problem is a stream function $\psi(x, y;\varepsilon)$ which satisfies Eq. (42.1) and the usual boundary conditions (42.1a), (42.2), and also has the limit behavior

$$\psi_x/y^\varepsilon\to 0,\qquad \psi_y/y^\varepsilon\to 1 \tag{44.1}$$

at infinity.

The drag coefficient $C_{\mathrm{D}}[\varepsilon]$[1] is expressible in terms of $\psi(x, y;\varepsilon)$ by the formula

$$C_{\mathrm{D}}[\varepsilon]=\frac{1+\varepsilon}{Y^{1+\varepsilon}}\int\limits_0^Y(1-\psi_x^2/y^{2\varepsilon})\,y^\varepsilon\,dy. \tag{44.2}$$

The desired value $C_{\mathrm{D}}[1]$ will be determined by interpolation from calculated values of $C_{\mathrm{D}}[0]$, $C_{\mathrm{D}}[-1]$, and $C'_{\mathrm{D}}[0]$. We begin with the well known result (6.7)

$$C_{\mathrm{D}}[0]=\frac{2\pi}{4+\pi} \tag{44.3}$$

of the classical two-dimensional theory.

To evaluate $C_{\mathrm{D}}[-1]$ we observe that $-\psi_x$ is a solution of Eq. (42.1), and hence must have its maximum on the boundary of Ω or at infinity. The latter

[1] Bracket notation is used in this section to denote dependence of C_{D} on ε, while the notation $C_{\mathrm{D}}(\sigma)$ will be adopted in the sequel, when the cavitation number σ is the pertinent variable.

is excluded by the assumed asymptotic behavior of the flow (44.1). On the other hand, the maximum cannot occur at an interior point of Γ_1 since ψ_x can be continued by reflection across Γ_1 as an even function of x. It follows that $-\psi_x$ achieves its maximum on Γ_2; and furthermore, for $\varepsilon \leqq 0$ this must be at the point of detachment $z_0 = iY$, since Γ_2 is monotonic and $-\psi_x \leqq \partial\psi/\partial n = y^\varepsilon \leqq Y^\varepsilon$ along Γ_2. Hence we have the estimate $\psi_x^2 \leqq Y^{2\varepsilon}$ on Γ_1 for $\varepsilon \leqq 0$. When inserted into Eq. (44.2) this yields the inequality

$$C_{\mathrm{D}}[\varepsilon] \geqq 1 - \frac{1+\varepsilon}{1-\varepsilon} \quad \text{for} \quad \varepsilon \leqq 0. \tag{44.4}$$

At the same time, it is evident that $C_{\mathrm{D}}[\varepsilon] \leqq 1$ for all ε. Therefore, letting $\varepsilon \to -1$ in Eq. (44.4), we obtain in the limit,

$$C_{\mathrm{D}}[-1] = 1. \tag{44.5}$$

A similar *a priori* determination of $C_{\mathrm{D}}[\infty]$ is not known.

Finally, to calculate $C'_{\mathrm{D}}[0]$ we set $Y = 1$ in Eq. (44.2) and differentiate with respect to ε at $\varepsilon = 0$, getting

$$\left.\begin{aligned} C'_{\mathrm{D}}[0] &= \int_0^1 (\log y - 1)\left(\frac{\partial\psi_0}{\partial x}\right)^2 dy - 2\int_0^1 \frac{\partial\psi_0}{\partial x}\,\frac{\partial\psi_1}{\partial x}\,dy \\ &= \int_0^1 (\log 4y - 1)\left(\frac{\partial\psi_0}{\partial x}\right)^2 dy - 2\int_0^1 \frac{\partial\psi_0}{\partial x}\,\frac{\partial\Psi_1}{\partial x}\,dy - \\ &\quad - \operatorname{Im}\int_0^1 \frac{\partial\psi_0}{\partial x}\left\{\frac{\partial}{\partial x}\int_i^z [\log(\bar z - t)(z - g(t))]\,W'(t)\,dt\right\} dy \end{aligned}\right\} \tag{44.6}$$

where ψ_0 is the stream function of the plane flow and ψ_1, Ψ_1 are the first order perturbation terms in Eqs. (42.3), (42.13), related by (42.14). As stated before, Ψ_1 can be calculated explicitly in terms of ψ_0 by using an appropriate GREEN'S function constructed in the hodograph plane. The final lengthy integral formulas for $C'_{\mathrm{D}}[0]$ must be evaluated numerically, the result being

$$C'_{\mathrm{D}}[0] = -0.07419. \tag{44.7}$$

Fitting the three data (44.3), (44.5), and (44.7), to a quadratic interpolation formula in the variable δ, we obtain

$$C_{\mathrm{D}} = \frac{2\pi}{4+\pi} - 0.14838\,\delta - 0.02818\,\delta^2. \tag{44.8}$$

At $\delta = \frac{1}{3}$ this yields the estimate

$$C_{\mathrm{D}}[1] = 0.827 \tag{44.9}$$

for the drag coefficient of a circular disk in infinite cavity flow.

The accuracy of this value can perhaps be estimated by comparison with the errors in the calculation of the contraction coefficient. In the latter problem the value provided by the three data (43.3), (43.6), and (43.10) differs by about $\frac{1}{2}$% from the final result obtained by including Eq. (43.5). On this basis, we are led to predict that Eq. (44.9) is in error at most in the last figure.

The experimental values of $C_{\mathrm{D}}[1]$ are obtained by extrapolating the essentially linear curve of C_{D} versus cavitation number σ down to $\sigma = 0$ (Fig. 1). These values lie between the extremes of 0.79 (REICHARDT) and 0.83 (California Institute of Technology), so that the calculated figure 0.827 agrees with the largest observed values. However, the curve $C_D = 0.827\,(1+\sigma)$ passes through the center of the observed points at $\sigma = 0.12$ where the experimental data are most abundant.

A concluding remark about the perturbation method: It is plain that information about the degenerate cases $\varepsilon = -1$ and $\varepsilon = \infty$ contributed in an essential way to the final estimates of the contraction and drag coefficients. Without these data it would be necessary, for the same final accuracy, to compute additional derivatives at $\varepsilon = 0$. At present this seems to involve prohibitive labor, and even calculation of the first derivatives is a lengthy task. Thus the scope of the method may be limited by practical considerations to problems in which it is possible to obtain *a priori* information concerning either the degenerate cases or other values of ε. The following section describes a method that is not dependent on fortuitous circumstances, but it too involves considerable labor apparently inherent in these problems.

45. GARABEDIAN's iteration method: The finite cavity problem. In the problem of the finite cavity first order perturbation calculations yield results of relatively poor accuracy, at least compared with those obtained in the preceding two sections. This is a reflection of the fact that some flow quantities differ markedly in plane and axially symmetric cavities (cf. Sect. 8), whereas the quantities of interest in earlier calculations, the contraction and drag coefficients, change relatively little between the two and three dimensional cases. To cope with this situation GARABEDIAN[1] has devised an altogether different technique of successive approximation, which may be used either by itself or as an adjunct to the perturbation method. Here we outline his calculation of the Riabouchinsky flow past a circular disk.

We start with two vertical segments of length Y in the meridian plane, symmetrically located with respect to the y axis at a fixed distance $2X$ from one another. To obtain a first approximation, we study the dependence of the corresponding Riabouchinsky cavity on the dimension parameter ε. Let Γ_1, Γ_1' and $\Gamma_2(\varepsilon)$ designate respectively the nose, tail, and free boundary, of the flow, and let $d(\varepsilon)$ be the cavity width corresponding to the fixed obstacle of length Y and cavity length $2X$. The cavitation number $\sigma(\varepsilon)$ is also an unknown in this formulation of the problem.

The calculation is divided into two parts, the first of which determines $d(1)$ by interpolation from the computed values $d(0)$, $d'(0)$, and $d(\infty)$. We discuss this phase of the calculation only perfunctorily. The quantity $d(0)$ is given explicitly by the plane theory [cf. Eq. (8.6)], and $d'(0)$ can be derived by computations quite analogous to those undertaken for the derivatives $C_D'[0]$ and $Y'(0)$ in the two earlier problems. The plane problem is made specific by selection of the parameter k in Sect. 8. In the example treated by GARABEDIAN the value of k is taken to be 0.96, so that we have

$$X/Y = 6.2263\,, \quad d(0)/Y = 2.4339\,, \quad d'(0)/Y = -0.1747\,, \quad \sigma(0) = 0.778 \qquad (45.1)$$

The quantity $d(+\infty)$ can be estimated in much the same way as the ratio $\overline{Y}(\infty)/Y(\infty)$ in the problem of the contraction coefficient. Namely, we observe first that since the cavity flow speed is unity, the stream and potential functions have the following asymptotic bahavior at infinity

$$\psi \sim \frac{y^{1+\varepsilon}}{(1+\varepsilon)\,(1+\sigma)^{\frac{1}{2}}}\,, \qquad \varphi \sim \frac{x}{(1+\sigma)^{\frac{1}{2}}}\,. \qquad (45.2)$$

[1] P. GARABEDIAN: (1) Pacif. J. Math. **6**, 611–684 (1956). — (2) Bull. Amer. Math. Soc. **62**, 219–245 (1956).

From Eq. (43.4) we infer that on the limit of φ as $\varepsilon\to\infty$ is a function of x alone, whence

$$\lim_{\varepsilon\to\infty}\varphi(x, y; \varepsilon) = \frac{x}{(1+\sigma)^{\frac{1}{2}}}\,. \tag{45.3}$$

Since the convex curve Γ_2 descends monotonically as $\varepsilon\to\infty$, we conclude from Eq. (45.3) that the limiting position of Γ_2 is the horizontal line $y=Y$, so that

$$d(\infty)/Y = 1\,. \tag{45.4}$$

A quadratic interpolation formula fitting the data (45.1) and (45.4) is

$$d/Y = 2.4339 - 0.34094\,\delta - 1.0845\,\delta^2, \qquad \delta = \varepsilon/(\varepsilon+2)\,, \tag{45.5}$$

and this yields as a first approximation to the cavity width,

$$d(1)/Y = 2.197\,. \tag{45.6}$$

The magnitude of the δ^2 term in Eq. (45.5) is an indication that this estimate is not very accurate.

The next step in the calculation is a refinement of the preceding result by an iteration scheme. As we shall see, this procedure can be formulated as a general method applicable in principle to other free boundary problems, both plane and axially symmetric.

Suppose now that ψ is the exact solution of the cavity problem and that Γ_2^* is an approximation to the free stremline. We seek a boundary value problem satisfied by ψ in the approximating flow region Ω^* bounded by Γ_2^*, the fixed boundaries Γ_1, Γ_2', and the x-axis. The conditions fulfilled by ψ on all the boundary curves but Γ_2^* are the same as before. On Γ_2^* we can derive an approximate linear boundary condition as follows. From the irrotationality condition $\partial q/\partial n + \varkappa q = 0$ it follows that on any streamline $\psi=0$,

$$\frac{\partial}{\partial n}\left(q + \frac{\varkappa}{y}\psi\right) = 0\,. \tag{45.7}$$

In addition, on a free streamline, where both

$$q = \frac{1}{y}\frac{\partial\psi}{\partial n} = 1 \quad \text{and} \quad \psi = 0\,, \tag{45.8}$$

we have $q + \frac{\varkappa}{y}\psi = 1$, or

$$\frac{1}{y}\frac{\partial\psi}{\partial n} + \frac{\varkappa}{y}\psi = 1\,. \tag{45.9}$$

We conclude from Eq. (45.7) that the boundary condition (45.9) is fulfilled along the approximate free boundary Γ_2^* within an error of order $(\delta n)^2$ in the normal displacement δn between Γ_2 and Γ_2^*. The point of view of the iteration scheme is to attempt to solve the boundary value problem for a stream function ψ obeying the condition (45.9) on the known curve Γ_2^* of curvature $\varkappa$, and from this solution to determine as a new approximation to Γ_2 the curve Γ_2^{**} on which $\psi=0$. In actual computation the latter curve is obtained, within the same order of accuracy, by a normal displacement of Γ_2^* in the amount

$$\delta n = -\frac{\psi}{y}\,. \tag{45.9a}$$

We may then proceed to the next step in the iteration sequence, starting with Γ_2^{**} as the approximate free boundary. The above analysis shows that the solution ψ and its associated approximate free streamline $\psi = 0$ satisfy the correct boundary conditions (45.8) within an error of order $(\delta n)^2$ in the normal displacement δn between the successive approximate free streamlines. Thus, the error at each stage can be expected to shrink by a factor of order $(\delta n)^2$.

GARABEDIAN points out that this iteration scheme is formally analogous to NEWTON's method of determining roots of a polynomial and should therefore exhibit the same type of rapid geometric convergence to a solution.

The boundary conditions

$$\psi = 0 \quad \text{on } \Gamma_1, \Gamma_1', \qquad \psi(x, 0) = 0 \tag{45.10}$$

and

$$\frac{1}{y}\frac{\partial \psi}{\partial n} + \frac{\varkappa}{y}\psi = 1 \quad \text{on } \Gamma_2^* \tag{45.11}$$

are not sufficient to determine the stream function uniquely, a further condition being necessary to specify the value of σ appearing in the asymptotic relation

$$\psi \sim \frac{y^2}{2(1+\sigma)^{\frac{1}{2}}}. \tag{45.12}$$

The method of normalizing the problem is to take into account the nature of the free streamline detachment at the end points z_0, z_0' of Γ_1, Γ_1'. At these points the correct stream function has an expansion in positive powers of $(z-z_0)^{\frac{1}{2}}$, $(z-z_0')^{\frac{1}{2}}$, but because the velocity is bounded the expansion is lacking the terms $(z-z_0)^{\frac{1}{2}}$, $(z-z_0')^{\frac{1}{2}}$; hence

$$\psi = O(z-z_0) \text{ at } z_0, \qquad \psi = O(z-z_0') \text{ at } z_0'. \tag{45.13}$$

If these conditions are now imposed on the prospective solution ψ, they prove sufficient to determine it uniquely. The full set of conditions on the successive approximations ψ are therefore Eqs. (3.2), (45.10), (45.11), and (45.13).

We interpolate a remark here concerning application of this method to the problem of smooth separation from curved obstacles. In this case it is no longer possible to fix the point of detachment from one approximation to the next. However, one may compensate for this added freedom in the solution by imposing the condition of finite curvature at separation. This states formally that the term $(z-z_0)^{\frac{3}{2}}$, as well as $(z-z_0)^{\frac{1}{2}}$, is absent in the expansion of ψ at separation. Thus it is possible, in principle, to extend the present iteration scheme to the general free boundary problem for curved obstacles.

Returning to the finite cavity problem for the disk, let us take as the starting point of the iteration procedure the solution obtained previously from the perturbation method. More precisely, we adopt as the starting approximation Γ_2^* the curve into which the plane free boundary $\Gamma_2(0)$ is taken by an affine transformation that keeps the end points fixed and that moves the point $(0, d(0))$ into $(0, d(1))$.

The next and basic step in the procedure is the solution of the mixed linear boundary value problem (3.2), (45.10), (45.11), (45.13). This deceptively simple classical problem concentrates within itself the principal difficulties of the entire numerical calculation. GARABEDIAN's approach to this problem is as follows. He expands ψ as a linear combination of appropriate solutions (3.2) in which the coefficients are selected to provide a best least square fit of the boundary conditions at a certain number of points of Γ_1 and Γ_2^*. Specifically[1], in the problem at hand he selects twenty solutions of Eq. (3.2) for use in the expansion of

[1] The following calculations are described in reference (2), preceding footnote; earlier calculations by the same method are contained in reference (1), same footnote.

ψ and fits the boundary data (45.10), (45.11), in the least square sense at 62 boundary points of which 35 are situated on Γ_1, and 37 on half of Γ_2^*[1]. The twenty solutions consist of three source-sink flows distributed along the axis, the uniform flow solution y^2, eleven spherical harmonics, and five ellipsoidal harmonics with suitable square root singularities at the separation point z_0. The function of the latter solutions is to provide terms in the expansion of ψ that reproduce the required development in powers of $(z-z_0)^{\frac{1}{2}}$ at detachment. These explicit solutions, in other words, absorb the singular boundary behavior of the desired solution, and therefore avoid undue errors in calculation that might otherwise be promoted by the infinite curvature at detachment. As for the source-sink flows, these are well known means of generating flows about closed bodies, and are therefore naturally included in the basic set of solutions. The spherical harmonics serve to smooth out the solution between the interpolation points. This particular set of twenty functions was adopted after some experimentation, as was the location of the sources and the number and location of the interpolation points. The latter were naturally chosen to have greatest density near the point of separation.

The final results of the calculations are summarized in the values

$$\sigma = 0.2235, \qquad \frac{d}{Y} = 2.30, \qquad \frac{C_D}{1+\sigma} = 0.865. \tag{45.14}$$

The coefficient of the y^2 term in the expansion of ψ provides the value of σ [cf. (45.12)]. This figure is to be contrasted with $\sigma = 0.778$ in Eq. (45.1) for the plane flow having the same X/Y values. The drag coefficient was computed by numerical integration in the formula

$$\frac{C_D}{1+\sigma} = \frac{1}{Y^2}\int_0^Y (1-\psi_x^2/y^2)\, y\, dy. \tag{45.15}$$

The value of d/Y and the shape of the new approximate free streamline were obtained from Eq. (45.9a).

The figure $d/Y = 2.30$ at $\sigma = 0.2235$ is in close agreement with experiment[2], while 0.865 for the modified drag coefficient is about 2% above experimental values (cf. Fig. 1). GARABEDIAN attributes this discrepancy to a gravity effect whose relation to the pertinent mathematical quantities is similar here and in the problem of the vena contracta. However, the increase in $C_D/(1+\sigma)$ from 0.827 at $\sigma = 0$ to 0.865 at $\sigma = 0.2235$ agrees qualitatively with the observed behavior of the C_D curve, this increase apparently being much greater in the axially symmetric case than in plane flow.

The largest errors, as one might expect, appear in the velocity determinations, since these involve derivatives of the solution. In fact, the velocity data on the disk is even partly oscillatory, whereas it should be monotonic.

GARABEDIAN obtains an interesting independent check on the C_D value from calculations based on the variational formula (30.17), which we write here in the form

$$\tfrac{2}{3}\pi X Y^2 C_D = (1+\sigma)\, V - 4\pi a. \tag{45.16}$$

Because of the maximum property of the right member (cf. Sect. 30), this formula provides an accurate lower bound for C_D *when σ is given* and the cavity shape is

[1] It proves helpful to the accuracy to impose both of the conditions, $\psi = 0$ and $y^{-1}\,\partial\psi/\partial n = 1$ at detachment, this apparently being an effective way of formulating Eq. (45.13) numerically.

[2] Cf. P. EISENBERG and H. POND: David Taylor Model Basin Rep. 668 (1948).

approximately known. In the present problem σ is itself a computed quantity, so that Eq. (45.16) will not necessarily define a lower bound, although it is quite useful for comparison purposes. The magnitude V inserted in Eq. (45.16) for the computations was the volume of the cavity bounded by the two disks and the affinely transformed approximate free streamline Γ_2^*; the virtual mass a was derived from the classical potential flow past this same body. With these values of V and a, Eq. (45.16) provides the following relation between C_D and σ,

$$\frac{C_D}{1+\sigma} = 11.7737 - \frac{13.3469}{1+\sigma}. \tag{45.17}$$

This formula is plainly quite sensitive to small errors in σ and in the numerical coefficients. Fortunately, the latter quantities are determined very precisely (for the approximate cavity). An instructive comparison is obtained from an earlier, independent computation[1] which presumably gives a less accurate determination of V and a. This leads to the relation

$$\frac{C_D}{1+\sigma} = 11.1910 - \frac{12.6498}{1+\sigma}, \tag{45.18}$$

and, as one would expect from variational theory, the corresponding values of C_D are lower than those obtained from Eq. (45.17) (see below).

As a means of estimating accuracy, GARABEDIAN has carried out his calculations with 16, 18, and 19 solutions in the expansion of ψ. Listed below is a comparison in the three cases of the values of σ, of d/Y, of $C_D/(1+\sigma)$ computed by Eqs. (45.15), (45.17), and (45.18), and of the standard error η in the least square fit of the 62 boundary conditions.

	σ	d/Y	$C_D/(1+\sigma)$ by Eq. (45.15)	$C_D/(1+\sigma)$ by Eq. (45.17)	$C_D/(1+\sigma)$ by Eq. (45.18)	η
16	0.22882	2.354	0.8881	0.9121	0.8967	0.00832
18	0.22540	2.321	0.8697	0.8818	0.8680	0.00426
19	0.22454	2.312	0.8624	0.8742	0.8608	0.00302

The values appearing in Eq. (45.14) are GARABEDIAN'S best estimate of the actual solution, inferred largely by extrapolation of the above data. He considers a conservative estimate of the errors to be 1% in the values of σ and d/Y, and ± 0.015 in $C_D/(1+\sigma)$. He takes the view that the errors in solving the linear problem greatly overshadow those resulting from having stopped the iteration process at an early stage.

46. TREFFTZ' integral equation. The vena contracta. The calculation of axially symmetric free surface flows according to the exact theory was initiated by TREFFTZ[2] in a memoir which for many years was the principal contribution to the three-dimensional free boundary problem. This work had an added appeal in its agreement with observation and in its support of the hypothesis—suggested by the experiments and by the Borda mouthpiece—that the contraction coefficient is the same for corresponding plane and axially symmetric flows. However,

[1] Reference (1), footnote 1, p. 427.

[2] E. TREFFTZ: Z. Math. Phys. **64**, 34—61 (1916). The method of TREFFTZ has been applied by W. SCHACH: Ing.-Arch. **6**, 51—59 (1935), to the problem of a cylindrical jet impinging against a plane wall. Essentially the same method is used by M. SHIFFMAN and D. SPENCER (Sect. 20) in their calculation of the water entry of a cone. In the two-dimensional case the calculations of the water entry of a wedge by H. WAGNER (Sect. 20) and of the flow over a weir by A. LAUCK: Z. angew. Math. Mech. **5**, 1—16 (1925), are based on the same ideas.

this conjecture and TREFFTZ' generally accepted value $C_c = 0.61$ for flow out of a circular orifice in a plane wall is now cast into serious doubt by GARABEDIAN'S calculations and his analysis of the experimental data (Sect. 43).

To outline the method developed by TREFFTZ, let us designate by Γ_1 and Γ_2 the fixed and free boundaries in the meridian plane, the former consisting of the half-line $x=0$, $1 \leqq y < \infty$, and the latter extending downward and to the right to $x=+\infty$, where it is asymptotic to the line $y=b$. Let us also introduce an equipotential cross-section Γ_3 of the jet, which will taken so far down stream that the velocity is effectively constant $(=1)$ on it. The flow region bounded by $\Gamma_1, \Gamma_2, \Gamma_3$ will be denoted by Ω.

From GREEN'S representation for the potential φ in the three dimensional flow region we have for points s *on the boundary* of the flow,

$$2\pi\varphi(s) = \int \varphi(t) \frac{\partial}{\partial n_t}\left(\frac{1}{r}\right) dS - \int \frac{1}{r} \frac{\partial \varphi}{\partial n_t} dS \tag{46.1}$$

where $r = r_{st}$ is as usual the distance between s and the variable of integration t. Using the boundary conditions, and writing $dS = y(t)\, dt\, d\vartheta_t$ for the element of area, where dt is the element of arc length on the meridian curve, we can express Eq. (46.1) in terms of line integrals. Namely,

$$\pi\varphi(s) = \int\limits_{\Gamma_1+\Gamma_2} \varphi(t)\, y(t) \frac{\partial V}{\partial n_t}(s,t)\, dt - \int\limits_{\Gamma_3} \frac{\partial \varphi}{\partial n_t} y(t)\, V(s,t)\, dt \tag{46.2}$$

where

$$V(s,t) = \frac{1}{2}\int\limits_0^{2\pi} \frac{d\vartheta_t}{r_{st}}\,, \quad r_{st} = [(x_s - x_t)^2 + y_s^2 + y_t^2 - 2 y_s y_t \cos(\vartheta_s - \vartheta_t)]^{\frac{1}{2}}. \tag{46.3}$$

The ring source $V(s,t)$ and its normal derivative can be expressed in terms of elliptic integrals as follows. Letting

$$r_1^2 = (x_s - x_t)^2 + (y_s - y_t)^2, \quad r_2^2 = (x_s - x_t)^2 + (y_s + y_t)^2$$

and defining

$$\varrho(s,t) = \frac{1}{2}(r_1 + r_2), \quad \xi(s,t) = \frac{r_2 - r_1}{r_2 + r_1},$$

one obtains after appropriate transformation of Eq. (46.3),

$$V(s,t) = -\frac{2K(\xi)}{\varrho(s,t)}, \quad K(\xi) = \int\limits_0^{\pi/2} \frac{dt}{(1-\xi^2\sin^2 t)^{\frac{1}{2}}},$$

also,

$$y(t)\frac{\partial V}{\partial n_t} = \frac{K(\xi)}{\varrho} x'(t) + \frac{\varrho G(\xi)}{y(s)} \frac{\partial \alpha(s,t)}{\partial t}$$

where

$$G(\xi) = E(\xi) - \tfrac{1}{2}(1-\xi^2) K(\xi), \quad E(\xi) = \int\limits_0^{\pi/2} (1-\xi^2\sin^2 t)^{\frac{1}{2}} dt,$$

and where $\alpha(s,t)$ is the angle subtended at t by the segment between s and its mirror image in the x-axis.

Since $\partial\varphi/\partial n = 1$ on Γ_3, the second integral in Eq. (46.2) is a fixed function of s, which we designate by $f(s)$. If Γ_2 is assumed to be known $\varphi(s)$ is then a solution of the linear integral equation,

$$\pi\varphi(s) = \int\limits_{\Gamma_1+\Gamma_2} K(s,t)\,\varphi(t)\, dt - f(s), \quad K(s,t) = y(t)\frac{\partial V(s,t)}{\partial n_t}. \tag{46.4}$$

The shape of Γ_2 of course appears implicitly in the kernel $K(s,t)$. In the problem at hand we seek to determine Γ_2, and hence $K(s,t)$, so that the solution φ has the derivative $\partial\varphi/\partial l = 1$ with respect to arc length along Γ_2. It is simple to verify that this determines completely the solution of the flow problem.

On the exact free boundary Γ_2 the potential is given by

$$\varphi(t) = \varphi_0 + l(t) \tag{46.5}$$

where φ_0 is the value of φ at the separation point (0, 1); if t is a measure of arc length from this point, then $l(t) = t$. Inserting Eq. (46.5) into Eq. (46.4) and observing that $K(s, t) = 0$ when s and t are both on Γ_1, we find for points on the fixed boundary Γ_1 that $\varphi(s)$ is given by a quadrature along Γ_2:

$$\pi\varphi(s) = \int_{\Gamma_2} l(t)\, K(s, t)\, dt - f(s). \tag{46.6}$$

This forms the basis of TREFFTZ's scheme of successive approximations, which proceeds as follows. Let an approximate free boundary Γ_2^* be chosen. This defines the kernel $K(s, t)$ by Eq. (46.4) and then a first approximation $\varphi_1(s)$ to the potential by the formulas

$$\begin{aligned} \varphi_1(s) &= \varphi_0 + l(s), \qquad s \in \Gamma_2^* \\ \pi\varphi_1(s) &= \int_{\Gamma_2^*} l(t)\, K(s, t)\, dt - f(s), \qquad s \in \Gamma_1. \end{aligned}$$

The approximation φ_1 is now inserted into the integral in Eq. (46.4), yielding a new function $\varphi_2(s)$ for points s on Γ_2^*. The difference $\varphi_2 - \varphi_1$ along this curve is then used as a measure of the closeness of Γ_2^* to the correct free streamline, and as a basis for improvement of the approximation. For the exact solution, of course, we must have $\varphi_1 \equiv \varphi_2$. An interesting feature of this method is that the calculations are confined to the boundary and do not require construction of the flow in the large.

TREFFTZ suggests an heuristic procedure for improvement of the approximation Γ_2^*. Namely, at points where $\varphi_2 < \varphi_1$ (points of smaller velocity, essentially) the curve is to be pushed downward, presumably increasing the velocity, and at points where $\varphi_2 > \varphi_1$ it is to be displaced away the axis. The magnitude of the displacement and the relative importance to be assigned points along the curve are a matter of judgment. In this form the procedure is evidently not systematic and depends ultimately on the skill of the computer[1].

The calculations described by TREFFTZ were carried out on two initial curves that were affine transforms of the plane free streamline. The ordinates of the asymptotes in the two cases were $b = 0.75$ and $b = 0.80$, corresponding therefore to initial estimates of 0.56 and 0.64 of the contraction coefficient. The results of the two computations are shown in the following tables, the first for $b = 0.75$ and the second for $b = 0.80$.

$x(s)$	$y(s)$	$l(s)$	$\varphi_1(s)$	$\varphi_2(s)$	$\varphi_1 - \varphi_2$
0.0	1.0000	0.000	0.362		
0.1	0.9076	0.140	0.503	0.500	0.002
0.2	0.8662	0.248	0.610	0.602	0.008
0.5	0.8014	0.555	0.917	0.904	0.013
1.0	0.7634	1.057	1.419	1.407	0.012
2.5	0.7503	2.557	2.919	2.917	0.002

$x(s)$	$y(s)$	$l(s)$	$\varphi_1(s)$	$\varphi_2(s)$	$\varphi_1 - \varphi_2$
0.0	1.0000	0.000	0.408		
0.1	0.9261	0.127	0.535	0.553	−0.018
0.2	0.8930	0.233	0.641	0.655	−0.014
0.5	0.8411	0.538	0.946	0.955	−0.009
1.0	0.8107	1.039	1.447	1.454	−0.007
2.5	0.8002	2.539	2.947	2.055	−0.008

[1] A. ARMSTRONG and J. DUNHAM: Armament Res. Lab. Rep. 12/53 (1953), base a method of calculating the axially symmetric cavity on the same type of integral equation as used by TREFFTZ (cf. Sect. 22). These authors outline a systematic iteration scheme but do not perform an actual calculation.

The next approximation to the free streamline was obtained by modifying the initial curves more or less in accordance with the scheme outlined above. The final results are in the table below:

$x(s)$	$y(s)$	$l(s)$	$\varphi_1(s)$	$\varphi_2(s)$	$\varphi_1-\varphi_2$	$x(s)$	$y(s)$	$l(s)$	$\varphi_1(s)$	$\varphi_2(s)$	$\varphi_1-\varphi_2$
0.0	1.000	0.000	0.385			1.0	0.790	1.052	1.437	1.434	+0.003
0.1	0.913	0.138	0.523	0.525	−0.002	1.75	0.782	1.802			
0.2	0.875	0.245	0.630	0.630	0.000	2.5	0.780	2.552	2.937	2.934	+0.003
0.5	0.819	0.551	0.936	0.939	−0.003						

The value $b=0.78$ gives the well known result $C_c=0.61$. TREFFTZ estimates the error by observing that a change of 0.05 in the asymptote is accompanied by an average difference of 0.020 between the $\varphi_1-\varphi_2$ values of the starting approximations, and hence concludes that the extremes of 0.003 and -0.003 in the $\varphi_1-\varphi_2$ values of the final approximation correspond to errors of ± 0.0075 in the asymptote. Thus he estimates $0.7725 \leqq b \leqq 0.7875$, and accordingly, $0.60 \leqq C_c \leqq 0.62$. TREFFTZ completes his calculations by constructing the entire flow field.

Although the preceding method is of great interest and potentially quite useful, a critical comment is in order concerning the details. The suggested criterion for accuracy of approximation, the magnitude of $\varphi_1-\varphi_2$, is essentially a measure of how well the constant speed condition is fulfilled on the approximate free boundary. It is to be expected that when a small number of boundary points are used in the calculation, as in the present case, important local properties of the flow are not reflected in the computed solution; a particular instance is the infinite curvature at detachment. A small error in the shape here can be magnified into a relatively large error in the contraction coefficient without inducing significant velocity variations on the free boundary except in the immediate vicinity of separation. The small magnitude of $\varphi_1-\varphi_2$ in the final approximation above, based on only six boundary points, therefore may be quite deceptive. We shall see by example in the following section that extreme refinement is essential to accuracy.

47. The relaxation method. The general idea of the relaxation method is well known[1] and requires only brief mention here: The flow field is divided into a rectangular network of appropriate fineness, and the governing equation—in the present case the equation for the stream function—is written in finite difference form in terms of the function values at the mesh points. Starting from a reasonable trial solution, the values at the mesh points are successively corrected until the solution is stabilized and the errors in the boundary conditions are within the limits considered acceptable in the particular problem. Additional accuracy can generally be introduced in any part of the flow by refining the network there.

The free boundary problems are among the most difficult computationally to have been treated by relaxation methods. The procedure followed in this class of problems is illustrated in the calculation by SOUTHWELL and VAISEY[2] of the axially symmetric jet flow from an orifice plate at the end of a cylindrical

[1] See, e.g., R. SOUTHWELL: Relaxation Methods in Theoretical Physics. Oxford 1946.
[2] R. SOUTHWELL and G. VAISEY: Phil. Trans. Roy. Soc. Lond., Ser. A **240**, 117—161 (1948).

tube of diameter six times that of the orifice. The finite difference approximation to the Eq. (3.2) for the stream function is

$$\psi_1 + \psi_2 + \psi_3 + \psi_4 - 4\psi_0 - \frac{a}{2y_0}(\psi_2 - \psi_4) = 0 \qquad (47.1)$$

where the mesh pattern of dimension a is as shown in Fig. 59. The principal difficulty in the problem is that the shape and asymptotic width of the jet have to be adjusted, essentially by trial and error, in successive steps of the computation.

To begin with, values are permanently assigned the stream function on the axis and on the jet, and trial selections are made of the shape and asymptotic speed V of the jet. The computation process is aimed at reducing the error

$$\eta = \frac{1}{y}\frac{\partial \psi}{\partial n} - V \qquad (47.2)$$

in the constant speed condition along the free streamline, by successive modification of both V and the shape of the jet. The principle followed by SOUTHWELL and VAISEY in making corrections is to increase V to attain an overall decrease in η, and to enlarge the jet radius at any point where a decrease in velocity and hence in η, relative to neighboring points is desired. This heuristic device is similar to that employed by TREFFTZ in his calculation and is apparently quite effective in practice.

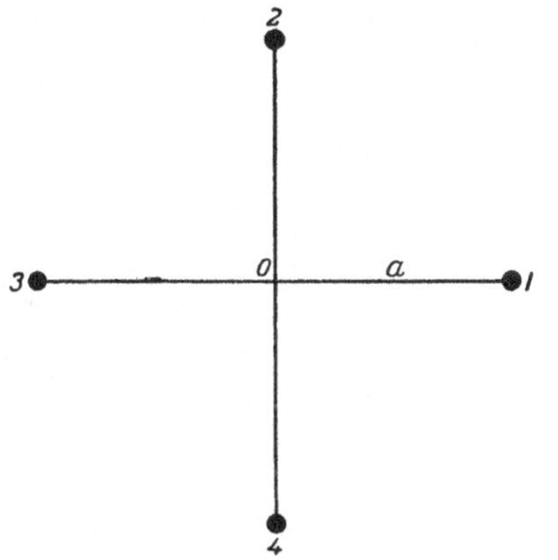

Fig. 59.

In order that η be kept uniformly small it is imperative that the mesh size be drastically refined in the neighborhood of the separation point, for here the curvature of the free streamline becomes infinite, and consequently a small error in shape can reflect itself in relatively large errors in the downstream position of the free boundary without incurring siginificant velocity fluctuations except near detachment. SOUTHWELL and VAISEY refine their network carefully, using mesh sides graded from $\frac{1}{2}$ at points remote from detachment to $\frac{1}{64}$ in a neighborhood of this point. Their grid consists of about 2000 points, which is substantially more than employed by others in similar computations. The neighborhood of detachment is shown in Fig. 60a, and a larger section in Fig. 60b. We observe that despite the refinement near separation, the free streamline makes an angle of about 30° with the plate, and that the largest velocity fluctuations occur near detachment.

The final solution yielded a contraction coefficient of 0.608, in agreement with TREFFTZ' result. The average error, as measured by $|\eta|/V$, was about 0.004, and the maximum variation in η/V over the entire length of the free streamline was about ± 0.012. However, it is not clear what accuracy this implies in the computed value of the contraction coefficient, since small velocity variations may be accompanied by relatively large displacements of the free streamline. The same question arises with regard to all calculations of free surface flows. The magnitude of truncation error in the finite difference approximation also remains to be clarified. It should be remarked that we are in the need here of an order of accuracy at least capable of choosing between GARABEDIAN's value 0.761 for the asymptotic width and the TREFFTZ-SOUTHWELL-VAISEY figure 0.78.

SOUTHWELL and VAISEY perform calculations on a number of other problems of interest, including the axially symmetric flow from the Borda mouthpiece and from an orifice under the influence of gravity; the plane flow of a free jet falling under gravity, as in a waterfall; the motion of a two-dimensional planing surface,

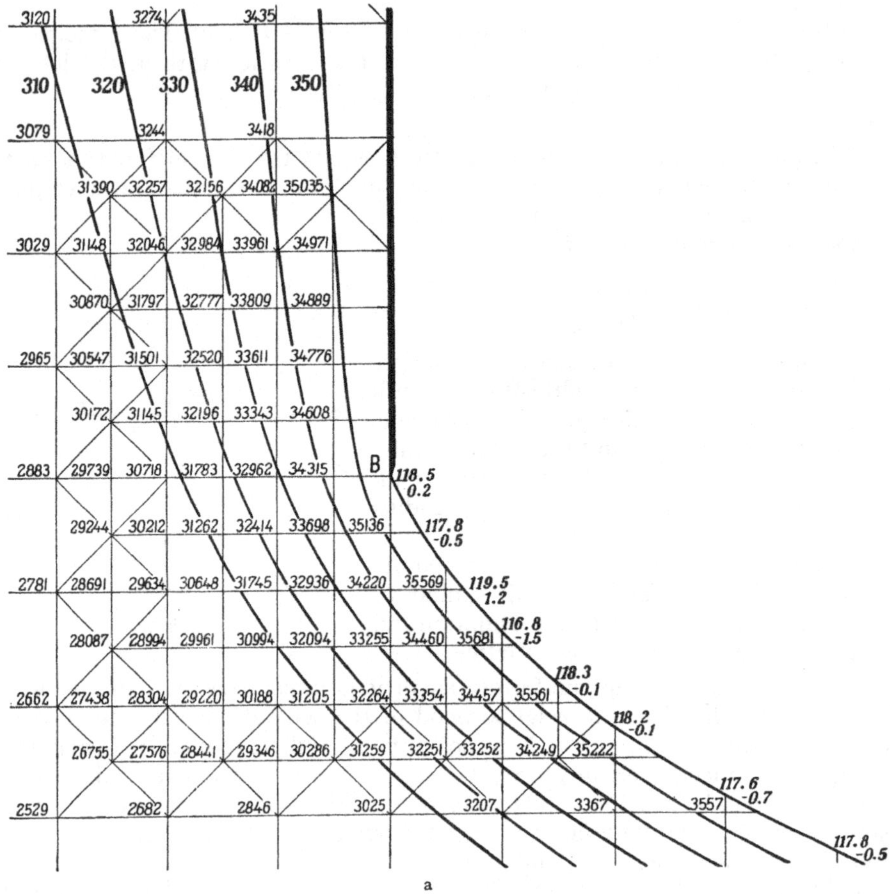

Fig. 60a and b. Calculation of flow through a circular orifice. After R. SOUTHWELL and G. VAISEY: Phil. Trans. Roy. Soc. Lond., Ser. A 240, 117–161 (1948).

also including gravity; and finally, the cusped cavity behind a cylinder and sphere. For a detailed discussion of these computations we refer to the original work.

Relaxation calculations on the efflux of axially symmetric jets from pipe orifices have also been carried out by ROUSE and ABUL-FETOUH[1]. They too obtain equality between the contraction coefficients of corresponding plane and axially symmetric flows, for all ratios of orifice to pipe cross-section.

In connection with our previous remarks concerning the accuracy of the relaxation method, it is instructive to consider a set of calculations by YOUNG

[1] H. ROUSE and A. ABUL-FETOUH: J. Appl. Mech. 17, 421–426 (1950); these authors also discuss the electrical analogy technique for calculating jet flows. This method has also been applied by P. MARCHET and M. MALAVARD: Direction Aerodynamique, Laboratoire d'Analogies Electriques, Centre Nat. Réch. Sci., Paris 1949. Cf. also M. MALAVARD: J. Aeronaut Sci. 24, 321–331 (1957).

et al.[1] on several plane and axially symmetric Riabouchinsky cavities, executed on a high speed electronic computer using a grid of 1145 points. Of particular interest in this regard is a comparison between a computed and exact solution

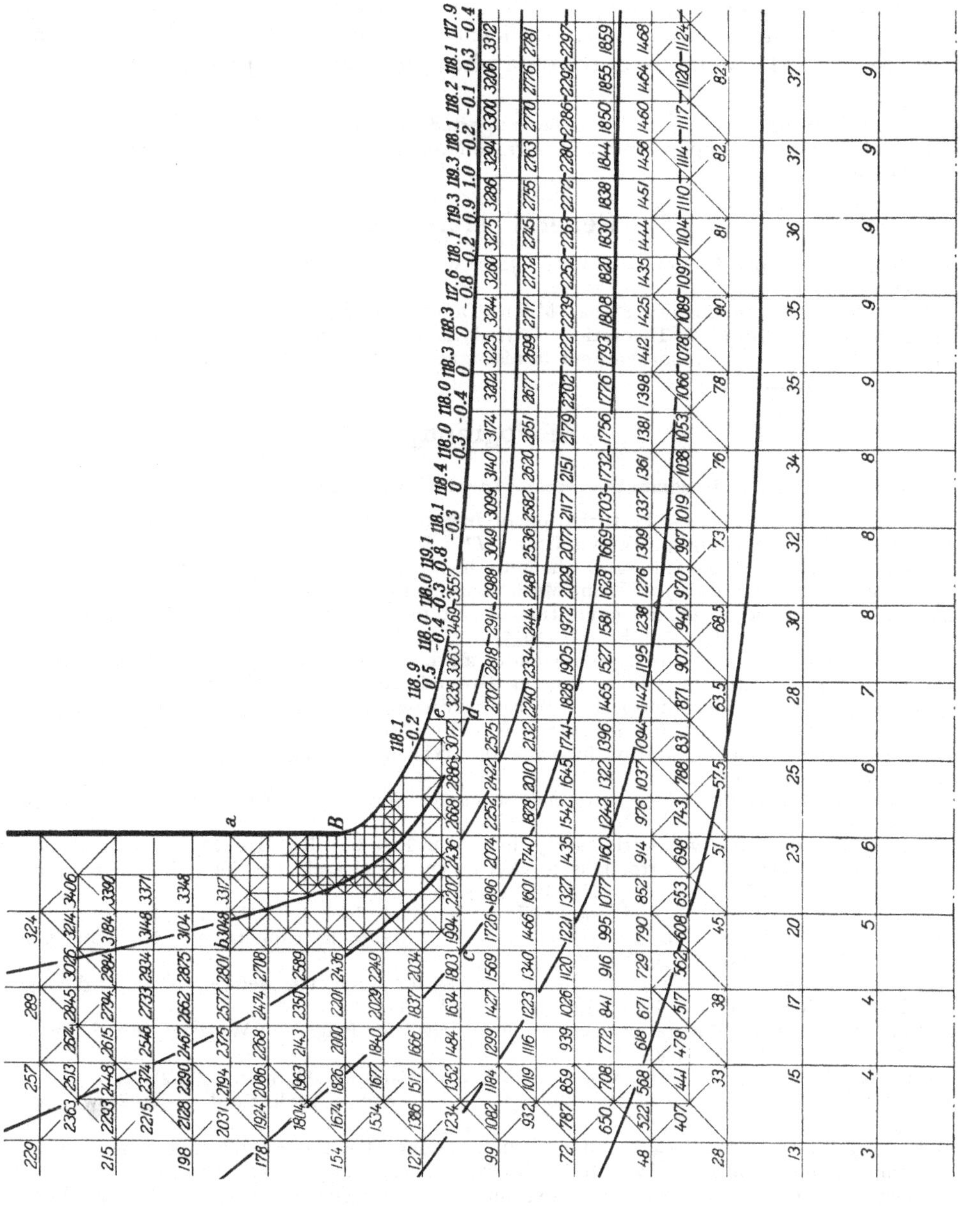

Fig. 60 b.

in the plane case, in which the velocity fluctuation of the former was less than $\pm 0.7\%$ of the average velocity, and yet the maximum ordinate of the free streamline above the separation point differed by about 10% in the two solutions. Further calculations by ARMS and GATES[1], based on the exact analytic solution

[1] D. YOUNG, L. GATES, R. ARMS and D. ELIEZER: U.S. Naval Proving Ground Rep. No. 1413 (1955). — R. ARMS and L. GATES: Rep. No. 1533 (1957).

as the trial boundary, yielded a 5% variation in the computed speed with an average value about 2% greater than the exact one. As they observe, this seems to imply relatively large truncation error, especially near the separation point where the fluctuations are largest. While the relaxation method is undoubtedly capable of great accuracy with sufficient refinement, these results show the need for critical evaluation of the data when the method is applied to free boundary problems.

Acknowledgment. This work has been supported in part by the Office of Naval Research. The author expresses appreciation to Mr. G. MARGULIES for his assistance in preparing the figures.

General references.

[*1*] G. BIRKHOFF and E. ZARANTONELLO: Jets, wakes and cavities. New York: Academic Press 1957.
[*2*] U. CISOTTI: Idromeccanica Piana. Milan: Tamburini 1921.
[*3*] L. M. MILNE-THOMSON: Theoretical Hydrodynamics, 3rd ed. New York: Macmillan 1956.
[*4*] H. VILLAT: Apercus théoriques sur la résistance des fluides. Collection Scientia. Paris: Gauthier-Villars 1920.

Bibliography.

Numbers in italics designate the sections in which the references appear.

ABLOW, C., and W. HAYES: Perturbation of free surface flows. Tech. Rep. No. 1 (1951), Grad. Div. Appl. Math., Brown Univ.; *11*.
ACKERET, J.: Experimentelle und theoretische Untersuchungen über Hohlraumbildung (Kavitation im Wasser). Techn. Mech. Thermodyn. **1** (1930); *2*.
ACOSTA, A.: A note on partial cavitation of flat plate hydrofoils. Hydrodynamics Lab., Cal. Inst. Tech., Rep. No. E-19.9 (1955); *21*.
ARMS, R., and L. GATES: The computation of an axially symmetric free boundary problem on NORC-Part II. U.S. Naval Proving Grounds Rep. No. 1533 (1957); *47*.
ARMSTRONG, A.: Boundary corrections to cavity dimensions and drag coefficients of a two-dimensional plate in a free-jet. Arm. Res. Est. Rep. No. 12/51 (1951); *11*.
— Abrupt and smooth separation in plane and axisymmetric cavity flow. Arm. Res. Est. Memo. 22/53 (1953); *24*.
— Drag coefficients of wedges and cones in cavity flow. Arm. Res. Est. Rep. No. 21/54 (1954); *22*.
—, and J. DUNHAM: Axisymmetric cavity flow. Arm. Res. Est. Rep. No. 12/53 (1953); *22*, *46*.
—, and K. TADMAN: Wall corrections to cavities in closed and open jet axially symmetric tunnels. Arm. Res. Est. Rep. No. 3/52 (1952); *11*.
— — Axisymmetric cavity flow about ellipsoids. Arm. Res. Est. Memo. 5/54 (1954); *22*.
— — Wall corrections to axially symmetric cavities in circular tunnels and jets. Arm. Res. Est. Rep. No. 7/52 (1952); *11*, *22*.
ARNOFF, E.: Re-entrant jet theory and cavity drag for symmetric wedges. Naval Ordnance Test Station NAVORD Rep. 1298 (1951); *7*, *8*.
BAUER, W.: Über das Widerstandsgesetz schnell bewegter Kugeln in Wasser. Ann. d. Phys. **82**, 1014—1016 (1927); *22*.
BAZIN, R.: Mem. pres. par div. savants Acad. Sci., Paris (2) **32**, 1—63 (1902); *2*.
BERG, P.: On the existence of Helmholtz flows in a compressible fluid. Diss. New York Univ. 1953; *32*.
BERGMAN, S.: On representation of stream functions of subsonic and supersonic flow of compressible fluids. J. Rat. Mech. Analysis **4**, 883—905 (1955); *12*.
— Mehrdeutige Lösungen bei Potentialströmungen mit freien Grenzen. Z. angew. Math. Mech. **12**, 95—121 (1932); *32*.
—, and M. SCHIFFER: Kernel functions and elliptic differential equations in mathematical physics. New York: Academic Press 1953; *30*.
BERS, L.: Existence and uniqueness of a subsonic flow past a given profile. Comm. Pure Appl. Math. **7**, 441—504 (1954); *32*.
BETZ, A.: Einfluß der Kavitation auf die Leistung von Schiffsschrauben. Proc. Third Intern. Congr. Appl. Mech. Stockholm 1930; *7*.
—, and E. PETERSOHN: Anwendung der Theorie der freien Strahlen. Ing.-Arch. **2**, 190—211 (1931); *4*.

BIRKHOFF, G.: Hydrodynamics. Princeton, N. J.: Princeton University Press 1950; *4*.
— Remarks on streamlines of discontinuity. Rev. Ciencias, Lima **50**, 105—116 (1948); *7*.
—, and D. CARTER: Rising plane bubbles. J. Math. Mech. **6**, 769—779 (1957); *15*.
— H. GOLDSTINE and E. ZARANTONELLO: Calculation of plane cavity flows past curved obstacles. Rend. Seminar. Mat. Univ. Politec, Torino **13**, 205—224 (1953/54); *40, 41*.
— M. PLESSET and N. SIMMONS: Wall effects in cavity flow I, II. Quart. Appl. Math. **8**, 161—168 (1950); **9**, 413—421 (1952); *8, 11*.
— D. YOUNG and E. ZARANTONELLO: Numerical methods in conformal mapping. Proc. Symp. Appl. Math. Vol. 4, pp. 117—140, Amer. Math. Soc. 1953; *40*.
BLASIUS, H.: Funktionentheoretische Methoden in der Hydrodynamik. Z. Math. Phys. **58**, 90—110 (1909); *15*.
BORDA, J.: Mémoire sur l'écoulement des fluides par les orifices des vases. Mem. Acad. Sci. Paris **1766**, 519—607; *10*.
BRILLOUIN, M.: Les surfaces de glissement de HELMHOLTZ et la résistance des fluides. Ann. chim. Phys, **23**, 145—230 (1911); *4, 9, 26*.
BRODETSKY, S.: The line of action of the resultant pressure in discontinuous fluid motion. Proc. Roy. Soc. Lond., Ser. A **102**, 361—372 (1922); *23*.
— Discontinuous fluid motion past circular and elliptic cylinders. Proc. Roy. Soc. Lond., Ser. A **102**, 542—553 (1923); *40*.
— Fluid motion past circular barriers. Bibl. Hieros. **1**, 1—14 (1923); *40*.
CHAPLYGIN, S.: On gas jets. Ann. Sci. Imp. Univ., math.-phys. Cl. **1904**, 1—121; translated in NACA Tech. Note 1063 (1944); *12*.
CISOTTI, U.: Scie limitate. Ann. Scuola Norm. Sup. Pisa (2) **1**, 101—112 (1932); *9*.
COHEN, H., and R. GILBERT: Two-dimensional, steady, cavity flow about slender bodies in channels of finite breadth. J. Appl. Mech. **24**, 170—176 (1957); *21*.
—, and Y. TU: A comparison of wall effects on supercavitating flows past symmetric bodies in solid wall channels and jets. Proc. 9th Intern. Cong. Appl. Mech., Brussels 1956; *11, 21*.
COOPER, E.: Theory of water entry of missiles with flat noses. Naval Ordnance Test Station NAVORD Rep. No. 1154 (1949); *11, 20*.
COURANT, R., and D. HILBERT: Methoden der Mathematischen Physik, Vol. 2. Berlin: Springer 1938; *17*.
COX, A., and W. CLAYDEN: Cavitating flow about a wedge at incidence. J. Fluid Mech. **3**, 615—637 (1958); *6*.
COX, R., and W. CLAYDEN: Air entrainment at the rear of a steady cavity. Symp. on Cavitation in Hydrodynamics. Nat. Phys. Lab. Her Majesty's Stationery Office, 1956; *2*.
—, and J. MACCOLL: Recent contributions to basic hydroballistics. Proc. Symp. on Naval Hydrodynamics, pp. 215—232, Nat. Res. Counc. Publ. 515 (1957); *6, 20, 22*.
CURLE, N.: On hydrodynamic stability in unlimited fields of viscous flow. Proc. Roy. Soc. Lond., Ser. A **238**, 489—501 (1957); *18*.
— Unsteady two-dimensional flows with free boundaries. Proc. Roy. Soc. Lond., Ser. A **235**, 275—381, 382—395 (1956); *18*.
DEMTCHENKO, B.: Problèmes mixtes harmoniques en hydrodynamique des fluides parfaits. Paris: Gauthier-Villars 1933; *5*.
DINI, M.: Sull'equazione $\Delta^2 u = 0$. Ann. di Mat. (2) **5**, 305—345 (1871); *33*.
EFROS, D.: Hydrodynamical theory of two-dimensional flow with cavitation. Dokl. Akad. Nauk SSSR. **51**, 267—270 (1946); *7*.
— Calculation of the hydrodynamic forces acting on a cavitation contour in plane flow. Dokl. Akad. Nauk SSSR. **60**, 29—31 (1948) [in Russian]; *7*.
EISENBERG, P.: On the mechanism and prevention of cavitation. David Taylor Model Basin Rep. 712 (1950) — Addendum, Rep. 842 (1953); same as: Kavitation. Forsch.-H. Schiffstechnik **3**, 111—124 (1953); **4**, 155—168 (1953); **5**, 201—212 (1954); *2*.
—, and H. POND: Water tunnel investigations of steady state cavities. David Taylor Model Basin Rep. 668 (1948); *2, 45*.
EPPLER, R.: Beiträge zu Theorie und Anwendung der unstetigen Strömungen. J. Rat. Mech. Analysis **3**, 591—644 (1954); *4, 11, 26, 40*.
FAGE, A., and F. JOHANSEN: On the flow of air behind an inclined flat plate of infinite space. Proc. Roy. Soc. Lond., Ser. A **116**, 170—197 (1920); *2*.
FINN, R.: Some theorems on discontinuous plane fluid motions. J. d'Analyse Math. **4**, 246 to 291 (1956); *25, 26, 32, 38, 39*.
FISHER, J.: Discontinuous fluid flow with cavity formation. Underwater Ballistics Res. Comm. Rep. No. 34 (1945); *22*.
FORCHHEIMER, P.: Hydraulik, 3rd ed. Leipzig: Teubner 1930; *2, 10*.
FOX, J., and G. MORGAN: On the stability of some flows of an ideal fluid with free surfaces. Quart. Appl. Math. **11**, 439—456 (1954); *18*.

Friedrichs, K.: Über ein Minimumproblem für Potentialströmungen mit freiem Rande. Math. Ann. **109**, 60—82 (1933/34); *30, 32, 38.*

Garabedian, P.: Oblique water entry of a wedge. Comm. Pure Appl. Math. **6**, 157—167 (1953); *20.*

— An example of axially symmetric flow with a free surface. Studies in Mathematics and Mechanics presented to Richard von Mises, pp. 149—159. New York: Academic Press 1954; *17.*

— The mathematical theory of three-dimensional cavities and jets. Bull. Amer. Math. Soc. **62**, 219—235 (1956); *2, 30, 45.*

— Calculation of axially symmetric cavities and jets. Pacific J. Math. **6**, 611—684 (1956); *8, 28, 30, 42, 45.*

— On steady-state bubbles generated by Taylor instability. Proc. Roy. Soc. Lond., Ser. A **241**, 423—431 (1957); *15.*

— H. Lewy and M. Schiffer: Axially symmetric cavitational flow. Ann. of Math. **56**, 560—602 (1952); *17, 32, 35.*

— E. McLeod and M. Vitousek: Recent advances at Stanford in the application of conformal mapping to hydrodynamics. Amer. Math. Monthly **61**, No. 7, Part II, 8—10 (1954); *15, 16.*

—, and H. Royden: A remark on cavitational flow. Proc. Nat. Acad. Sci. U.S.A. **38**, 57—61 (1952); *32, 35.*

—, and M. Schiffer: Convexity of domain functionals. J. d'Analyse Math. **2**, 281—368 (1953); *30.*

—, and D. C. Spencer: Extremal methods in cavitational flow. J. Rat. Mech. Analysis **1**, 359—409 (1952); *30, 31, 32, 35.*

Gategno, G., and A. Ostrowski: Représentation conforme à la frontière; domaines particulières. Mem. Sci. Math. No. 110; *33.*

Gerber, R.: Sur l'existence des écoulements irrotationels, plans, periodiques, d'un liquide pesant incompressible. C.R. Acad. Sci., Paris **233**, 1560—1562 (1951),; *15.*

— Sur les solutions exacts des équations du mouvement avec surface libre d'un liquide pesant. J. Math. pures appl. (9) **34**, 185—299 (1955); *32.*

Geurst, J., and R. Timman: Linearized theory of two-dimensional cavitational flow around a wing section. Proc. 9th Intern. Cong. Appl. Mech., Brussels 1956; *21.*

Gilbarg, D.: Unsteady flow with free boundaries. Z. angew. Math. Phys. **3**, 34—42 (1952); *18.*

— Uniqueness of axially symmetric flows with free boundaries. J. Rat. Mech. Analysis **1**, 309—320 (1952); *26, 37.*

— Comparison methods in the theory of subsonic flows. J. Rat. Mech. Analysis **2**, 233—251 (1953); *26, 27, 37.*

— Free streamline theory and steady state cavitation. Proc. Symp. on Naval Hydrodynamics, pp. 281—295, Nat. Res. Counc. Publ. 515 (1957); *4.*

—, and R. Anderson: Influence of atmospheric pressure on the phenomena accompanying the entry of spheres into water. J. Appl. Phys. **19**, 127—139 (1948); *2.*

—, and D. Rock: On two theories of plane potential flows with finite cavities. Naval Ord. Lab. Memo. 8718 (1946); *7, 8.*

—, and J. Serrin: Free boundaries and jets in the theory of cavitation. J. Math. Phys. **29**, 1—12 (1950); *7, 23.*

— — Uniqueness of axially symmetric subsonic flow past a finite body. J. Rat. Mech. Analysis **4**, 169—175 (1955); *37.*

—, and M. Shiffman: On bodies achieving extreme values of the critical Mach number, I. J. Rat. Mech. Analysis **3**, 209—230 (1954); *26, 31.*

Goldstein, S. (ed.): Modern developments in fluid dynamics, Vol. 2. Oxford 1938; *2, 41.*

Green, A. E.: The gliding of a plate on a stream of finite depth, I, II. Proc. Cambridge Phil. Soc. **31**, 589 (1935); **32**, 67—85 (1956); *11.*

— Note on the gliding of a plate on the surface of a stream. Proc. Cambridge Phil. Soc. **32**, 248—252 (1936); *11.*

Greenhill, G.: Theory of a streamline past a plane barrier. British Aero. Res. Counc. Reps, and Memos. **19** (1910) — Appendix (1916); *5, 23.*

Gurevich, M.: Some remarks on stationary schemes for cavitational flow about a flat plate. Izv. Akad. Nauk. SSSR. OTN **1947**, 143—150; David Taylor Model Basin Transl. 224 (1948); *7, 8.*

— Flow past an axisymmetric semi-body of finite drag. Prikl. Mat. Mekh. **11**, 97—104 (1947) [in Russian with English summary]; *25.*

Hamel, G.: Ein hydrodynamischer Unitätssatz. Proc. 2nd Int. Congr. Appl. Mech., Zurich 1926, p. 489; *32, 38.*

HARDY, G., J. LITTLEWOOD and G. POLYA: Inequalities. Cambridge; Cambridge University Press 1934; *34*.

HELLIWELL, J., and A. MACKIE: Two-dimensional subsonic and sonic flow past thin bodies. J. Fluid Mech. **3**, 93—109 (1957); *12*.

HELMHOLTZ, H.: Über diskontinuierliche Flüssigkeitsbewegungen. Mber. Akad. Wiss. Berlin **1868**, 215—228; Wiss. Abh. 1, 154; *4, 10*.

HOPF, E.: A remark on linear elliptic differential equations of the second order. Proc. Amer. Math. Soc. **3**, 791—793 (1952); *27*.

HOWARTH, L. (ed.): Modern developments in fluid dynamics. High-speed flow. Oxford 1953; *12*.

HSU, E., and B. PERRY: Water tunnel experiments on spheres in cavity flow. Hydrodynamics Lab., Cal. Inst. Tech. Rep. E-24.9 (1954); *41*.

HURON, R.: Sur l'unicité des solutions du problème de représentation conforme de HELMHOLTZ. C.R. Acad. Sci., Paris **228**, 290, 357 (1949); *32*.

IMAI, I.: Discontinuous potential flows as the limiting form of the viscous flow for vanishing viscosity. J. Phys. Soc. Japan **8**, 399—402 (1953); *41*.

JACOB, C.: Sur la détermination des fonctions harmoniques par certaines conditions aux limites. Mathematica **11**, 149—175 (1935); *34*.

— Sur la méthode approchée de M. LAMLA en dynamique des fluides compressibles. Acad. Roum. Bull. Sect. Sci. **28**, 637—641 (1946); *12*.

JAFFÉ, G.: Unstetige und mehrdeutige Lösungen der hydrodynamischen Gleichungen. Z. angew. Math. Mech. **1**, 398—410 (1920); *36*.

— Über mehrdeutige Lösungen der hydrodynamischen Gleichungen. Phys. Z. **23**, 129—133 (1922); *6, 36*.

JAHNKE, E., and F. EMDE: Tables of functions, 3rd ed. Leipzig: Teubner 1938; reprinted New York: Dover Publ. 1943; *8*.

JOHN, F.: An example of a transient three-dimensional flow with a free boundary. Rev. Gen. Hydraul. **18**, No. 71, 230—232 (1952); *18*.

— Two-dimensional potential flows with a free boundary. Comm. Pure Appl. Math. **6**, 497—503 (1953); *15, 18*.

JOUKOWSKY, N.: A modification of KIRCHHOFF's method of determining two-dimensional motions of a fluid given constant velocity along an unknown streamline. Coll. Works **2**, No. 3; also: Rec. Math. **25** (1890); *4*.

KÁRMÁN, T. VON: Accelerated flow of an incompressible fluid with wake formation. Ann. Mat. pura appl. (4) **29**, 247—249 (1949); *19*.

— The impact on seaplane floats during landing. NACA Tech. Note 321 (1929); *20*.

KAWAGUTI, M.: Discontinuous flow past a circular cylinder. J. Phys. Soc. Japan **8**, 403—406 (1953),; *41*.

KELLER, J., and I. KOLODNER: Instability of liquid surfaces and the formation of drops. J. Appl. Phys. **25**, 918—921 (1954); *4*.

KEMPF, G., u. E. FOERSTER (ed.): Hydromechanische Probleme des Schiffsantriebs. Hamburg 1932; *2, 4, 8, 41*.

KING, H. W.: Handbook of hydraulics, 3rd ed. New York-Toronto-London: McGraw-Hill 1939; *2, 10*.

KIRCHHOFF, G.: Zur Theorie freier Flüssigkeitsstrahlen. J. reine angew. Math. **70**, 289—298 (1869); Ges. Abh. 416; *4, 5, 6, 10*.

KNAPP, R.: Recent investigations of the mechanics of cavitation and cavitation damage. Trans. Amer. Soc. Mech. Engrs. **77**, 1045—1054 (1955); *2*.

KÖTTER, F.: Über die Contractio Venae bei spaltförmigen und kreisförmigen Öffnungen. Arch. für Math. Phys. (2) **5**, 392—417 (1887); *10*.

KOLSCHER, M.: Unstetige Strömungen mit endlichen Totwasser. Luftf.-Forschg. **17**, 154 to 160 (1940); *9*.

KONSTANTINOV, V. A.: Influence of Reynolds number on cavitational flow. Izv. Akad. Nauk. SSSR., OTN **1946**, 1355—1373; David Taylor Model Basin Transl. 233 (1950); *2, 41*.

KORN, A.: Über Minimalflächen, deren Randkurven wenig von ebenen Kurven abweichen. Abh. Kgl.-Preuss. Akad. Wiss., Berlin 1909, Anhang II; *33*.

KRAVTCHENKO, J.: Sur l'existence des solutions du problème de représentation conforme de HELMHOLTZ. Ann. Sci. Ec. Norm. Sup. **62**, 233—268 (1946); **63**, 161—184 (1947); *32*.

— Sur le problème de représentation conforme de HELMHOLTZ. J. Math. pures appl. **20**, 35—239 (1941); *32*.

KREISEL, G.: Cavitation with finite cavitation numbers. Admiralty Res. Lab. Rep. No. R1/H/36 (1946); *7*.

KREPS, R.: Experimental investigation of impact in landing on water. CAHI Rep. No. 438 (1939); translated in NACA Tech. Memo. 1046 (1943); *20*.

Lamb, H.: Hydrodynamics, 6th ed. Cambridge: Cambridge University Press 1932; *10, 20, 35.*
Landweber, L.: The axially symmetric potential flow about elongated bodies of revolution. David Taylor Model Basin Rep. No. 761 (1951); *22.*
Lansford, W.: Discharge coefficients for pipe orifices. Civ. Engng., Lond. **4**, 245—247 (1934); *2.*
Lauck, A.: Der Überfall über ein Wehr. Z. angew. Math. Mech. **5**, 1—16 (1925); *46.*
Lavrentieff, M.: Sur certaines propriétés des fonctions univalentes et leurs applications à la théorie des sillages. Mat. Sbornik **46**, 391—458 (1938) [Russian with French summary]; *26, 28, 29, 32, 37.*
— M. Keldysh, A. Markuschevitch, L. Sedoff and A. Lotoff: Collected reports on the problem of impact against a water surface. CAHI Rep. No. 152 (1935); *20.*
Leray, J.: Les problèmes de représentation conforme de Helmholtz; théorie des sillages et des proues, I, II. Comment Math. Helv. **8**, 149—180, 250—263 (1935); *24, 32, 33, 34, 37, 39.*
— Sur la validité des solutions du problème de la proue. Volume de jubilé de M. Brillouin, pp. 246—257. Paris: Gauthier-Villars 1935; *4, 24, 32, 36.*
—, et J. Schauder: Topologie et équations fonctionnelles. Ann. Sci. Ec. Norm. Sup. **51**, 45—78 (1934); *34, 39.*
—, et A. Weinstein: Sur un problème de représentation conforme pose par la théorie de Helmholtz. C. R. Acad. Sci., Paris **198**, 430 (1934); *32, 38, 39.*
Levi-Civita, T.: Scie e leggi de resistenza. Rend. Circ. Mat. Palermo **18**, 1—37 (1907); *23.*
Levinson, N.: On the asymptotic shape of the cavity behind an axially symmetric nose moving through an ideal fluid. Ann. of Math. **47**, 704—730 (1946); *25.*
Lewy, H.: On steady free surface flow in a gravity field. Comm. Pure Appl. Math. **5**, 413—414 (1952); *15.*
— A note on harmonic functions and a hydrodynamical application. Proc. Amer. Math. Soc. **3**, 11—113 (1952); *15.*
Lighthill, M.: A note on cusped cavities. Aeronaut. Res. Council Rep. and Memo. 2328 (1945); *9.*
Malavard, M.: Recent developments in the method of rheoelectric analogy applied to aerodynamics. J. Aeronaut. Sci. **24**, 321—331 (1957); *47.*
Marchet, P., and M. Malavard: Determination des lignes de jet dans les mouvements plans et de revolution. Direction Aerodynamique, Laboratoire d'Analogies Electriques, Centre Nat. Rech. Sci., Paris 1949; *47.*
Martyrer, E.: Kraftmessungen an Widerstandkörpern und Flügelprofilen im Wasserstrom bei Kavitation. Hydromechanische Probleme des Schiffsantriebs, Hamburg, 1932, ed. G. Kempf and E. Foerster; *2, 41.*
McLeod, E.: The explicit solution of a free boundary problem involving surface tension. J. Rat. Mech. Analysis **4**, 557—567 (1955); *16.*
Medaugh, F., and G. Johnson: Investigation of the discharge and coefficients of small circular orifices. Civ. Engng., Lond. **10**, 422—424 (1940).
Michell, J.: On the theory of free streamlines. Phil. Trans. Roy. Soc. Lond., Ser. A **181**, 389—431 (1890); *5.*
Mises, R. von: Berechnung von Ausfluß und Überfallzahlen. Z. VDI **61**, 447—452, 469—473, 493—498 (1917); *10.*
Münzner, H., and H. Reichardt: Rotationally symmetrical source-sink bodies with predominately constant pressure distribution. Arm. Res. Est. Transl. No. 1/50 (1950)—German original UM 6616 (1944); *2, 22.*
Nehari, Z.: Conformal mapping. New York-Toronto-London: McGraw-Hill 1952; *5.*
Nekrasoff, N. A.: Sur la mouvement discontinu à deux dimensions de fluide autour d'un obstacle en forme d'arc de cercle. Publ. Inst. Polytech. Ivanovo-Voszniesiensk **1922**; *32, 40.*
Oudart, A.: L'étude des jets et la mécanique théorique des fluides. Publ. Sci. Tech. Min. de l'air 234, Paris 1949; *3, 12.*
— Sur le schema de Helmholtz-Kirchhoff. J. Math. pures appl. **22**, 245—320 (1943); **23**, 1—36 (1944); *32.*
Pai, S.I.: Fluid dynamics of jets. New York: Van Nostrand 1954; *2, 12.*
Parkin, B.: Experiments on circular-arc and flat-plate hydrofoils. J. Ship Res. **1**, 34—56 (1958); *2.*
— Fully cavitating hydrofoils in nonsteady motion. Hydrodynamics Lab., Cal. Inst. Tech. Rep. 85-2 (1957); *18.*
Perry, B.: Methods for calculating the effect of gravity on two-dimensional free surface flows. Diss. Stanford U. 1957; *15.*
Pierson, J.: Stevens Inst. Tech. Rep. No. 381 (1950); *20.*
Planck, M.: Zur Theorie der Flüssigkeitsstrahlen. Wied. Ann. (2) **21**, 499—509 (1884); *5.*

PLESSET, M., et B. PERRY: On application of free streamline theory to cavity flows. Mémoires sur la mécanique des fluides offerts a M. DIMITRI RIABOUCHINSKY, pp. 251—261, Publ. Sci. Tech. Min. de l'Air, Paris 1954; *2.*

—, and P. SHAFFER: Cavity drag in two and three dimensions. J. Appl. Phys. **19**, 934—939 (1948); *8, 22.*

— — Drag in cavitating flow. Rev. Mod. Phys. **20**, 228—231 (1948); *8, 22.*

POLYA, G., and G. SZEGO: Isoperimetric inequalities of mathematical physics. Princeton 1951; *35.*

PRIVALOFF, J.: Sur les fonctions conjuguées. Bull. Soc. Math. France **44**, 100—103 (1916); *33.*

PYKHTEEV, G.: Solution of the inverse problem of a plane cavitational flow along an arc of a curve. Prikl. Mat. Mekh. **20**, 378—381 (1956); *23.*

Lord RAYLEIGH: On the resistance of fluids. Phil. Mag., Dec. **1876**; Papers 1, 287; *4.*

— Theory of sound. New York: Dover Publications 1945; *2, 4.*

REICHARDT, H.: The laws of cavitation bubbles at axially symmetric bodies in a flow. Min. Aircraft Prod. Rep. and Transl. 766 (1946)—German original Göttingen 1945; *2, 22.*

RIABOUCHINSKY, D.: On steady fluid motion with free surface. Proc. Lond. Math. Soc. (2) **19**, 206—215 (1920); *8.*

— On some cases of two-dimensional fluid motions. Proc. Lond. Math. Soc. (2) **25**, 185—194 (1926); *8.*

— Sur un problème de variation. C. R. Acad. Sci., Paris **185**, 840—841 (1927); *30.*

RICHARDSON, A.: Stationary waves in water. Phil. Mag. (6) **40**, 97—110 (1920); *15.*

ROSHKO, A.: A new hodograph for free streamline theory. NACA TN 3168 (1954); *4, 11, 41.*

ROUSE, H., and A. ABUL-FETOUH: Characteristics of irrotational flow through axially symmetric orifices. J. Appl. Mech. **17**, 421—426 (1950); *47.*

SAKURAI, T.: The flow past a flat plate accompanied with an unsymmetric dead air at Mach number 1. J. Phys. Soc. Japan **11**, 710—715 (1956); *12.*

SAUTREAUX, C.: Question d'hydrodynamique. Ann. Sci. Ec. Norm. Sup. (2) **10** Suppl., 95—182 (1893); *15.*

— Mouvement d'un liquide parfait soumis à la pesanteur. Determination des lignes de courant. J. de Math. (5) **7**, 125—159 (1901); *15.*

SCHACH, W.: Umlenkung eines kreisförmigen Flüssigkeitsstrahles an einer ebenen Platte senkrecht zur Strömungsrichtung. Ing.-Arch. **6**, 51—59 (1935); *46.*

SCHEID, F.: On the asymptotic shape of the cavity behind an axially symmetric nose moving through an ideal fluid. Amer. J. Math. **72**, 485—501 (1950); *25.*

SCHIFFER, M.: Analytical theory of subsonic and supersonic flows. Encyclopedia of Physics, Vol. IX; *12.*

SCHMIEDEN, C.: Die unstetige Strömung um einen Kreiszylinder. Ing.-Arch. **1**, 104—109 (1929); *40.*

— Über die Eindeutigkeit der Lösungen in der Theorie der unstetigen Strömungen. Ing.-Arch. **3**, 356—370 (1932); *6, 40.*

— Über die Eindeutigkeit der Lösungen der unstetigen Strömungen. Ing.-Arch. **5**, 373—375 (1934); *40.*

— Über die Hohlraumbildung in der Flüssigkeit. Ann. d. Phys. (5) **2**, 350—356 (1929); *41.*

— Der Aufschlag von Rotationskörpern auf eine Wasseroberfläche. Z. angew. Math. Mech. **33**, 147—151 (1953); *20.*

— Über den Landestoß von Flugzeugschwimmern. Ing.-Arch. **10**, 1—13 (1939); *20.*

SEDOV, L.: Der Stoß eines starren Körpers der auf Oberfläche einer inkompressiblen Flüssigkeit schwimmt. CAHI Rep. No. 187 (1934); *20.*

SELF, M., and J. RIPKEN: Steady-state cavity studies in a free-jet water tunnel. St. Anthony Falls Hydraulic Lab., Univ. of Minn. Rep. No. 47, 1955; *2.*

SERRIN, J.: Comparison theorems for subsonic flows. J. Math. Phys. **33**, 27—45 (1954); *26, 37.*

— Existence and uniqueness of the flows solving four free boundary problems. Thesis, Indiana University 1951, Appendix A.; *26.*

— Existence theorems for some hydrodynamical free boundary problems. J. Rat. Mech. Analysis **1**, 1—48 (1952); *26, 32, 33, 34.*

— Uniqueness theorems for two free boundary problems. Amer. J. Math. **74**, 492—506 (1952); *26, 37.*

— Two hydrodynamic comparison theorems. J. Rat. Mech. Analysis **1**, 563—572 (1952); *26.*

— On plane and axially symmetric free boundary problems. J. Rat. Mech. Analysis **2**, 563—575 (1953); *26, 28, 29, 32.*

SHIFFMAN, M.: On free boundaries of an ideal fluid. I, II. Comm. Pure Appl. Math. **1**, 89—99 (1948); 2, 1—11 (1949); *3, 7, 13.*

—, and D. SPENCER: The flow of an ideal incompressible fluid about a lens. Quart. Appl. Math. **5**, 270—288 (1947); *20.*

— — The force of impact on a cone striking a water surface. Comm. Pure Appl. Math. **4**, 379—417 (1951); *20.*

SMITH, H.: Hydraulics. London-New York: John Wiley 1886; *2.*
SOUTHWELL, R.: Relaxation methods in theoretical physics. Oxford 1946; *47.*
—, and G. VAISEY: Fluid motions characterized by "free" streamlines. Phil. Trans. Roy. Soc. Lond., Ser. A **240**, 117—161 (1948); *10, 11, 15, 47.*
SQUIRE, H. B.: On the laminar flow of a viscous fluid with vanishing viscosity. Phil. Mag. **7** (17), 1150—1160 (1934); *41.*
TAN, H. S.: A unique law for ideal incompressible flow with preserved pattern of finite separation. Quart. Appl. Math. **12**, 78—80 (1954); *19.*
THERON, P.: Sur un théorème d'existence des mouvements des fluides plans avec sillage. C. R. Acad. Sci., Paris **228**, 1922 (1949); *32.*
TREFFTZ, E.: Über die Kontraktion kreisförmiger Flüssigkeitsstrahlen. Z. Math. Phys. **64**, 34—61 (1916); *46.*
TRILLING, L.: The impact of a body on a water surface at an arbitrary angle. J. Appl. Phys. **21**, 161—170 (1950); *20.*
TSIEN, S.: Two-dimensional subsonic flow of compressible fluids. J. Aeronaut. Sci. **6**, 399 to 407 (1939); *12.*
TULIN, M.: Steady two-dimensional cavity flows about slender bodies. David Taylor Model Basin Rep. 834 (1953); *21.*
— Supercavitating flow past foils and struts. Symposium on Cavitation in Hydrodynamics, National Phys. Lab., Her Majesty's Stationery Office, 1956; *21.*
VANDREY, F.: A direct iteration method for the calculation of the velocity distribution of bodies of revolution and symmetrical profiles. Aeronaut. Res. Counc. FM 1665 (1951); *22.*
VILLAT, H.: Sur le changement d'orientation d'un obstacle dans un courant de fluide. Ann. Fac. Sci. Toulouse (3) **5**, 375—404 (1913); *9.*
— Sur la résistance des fluides. Ann. Sci. Ec. Norm. Sup. **28**, 203—240 (1911); *33.*
— Sur la validité des solutions de certains problèmes d'hydrodynamique. J. de Math. (6) **10**, 231—290 (1914); *24.*
— L'écoulement des fluides pesants. Ann. Sci. Ec. Norm. Sup. (3) **32**, 177—214 (1915); *15.*
VITOUSEK, M.: Some flows in a gravity field satisfying the exact free surface condition. Tech. Rep. No. 25 (1954), Appl. Math. Stat. Lab., Stanford Univ.; *15.*
WAGNER, H.: Über Stoß- und Gleitvorgänge an den Oberflächen von Flüssigkeiten. Z. angew. Math. Mech. **12**, 193—215 (1932); *11, 20.*
WAID, R.: Water tunnel investigations of two-dimensional cavities, Hydrodynamics Lab., Cal. Inst. Tech. Rep. E-73.6 (1957); *41.*
WATANABE, S.: Resistance of impact on water surface, Parts I, II, IV, V, VI. Inst. Phys. Chem. Res. Tokyo (Sci. Papers), I-cone: **12**, 251—267 (1930); II-cone: **14**, 153—168 (1930); IV-circular plane: **23**, 118—135 (1933); V-sphere: **23**, 202—208 (1934); VI-cylinder: **23**, 249—255 (1934); *20.*
WAYLAND, H.: Scale factors in water entry. Naval Ord. Test Station, NAVORD Rep. 978 (1947); *2.*
—, and J. WHITE: Boundary layer effects on spinning spheres. Heat Transfer and Fluid Mechanics Institute. Amer. Soc. Mech. Eng. **1949**, 51—64; *2.*
WEIBLE, A.: The penetration resistance of bodies with axial head forms at perpendicular impact on water. German Aviation Res. Rep. No. 4541; Naval Res. Lab. Transl. No. 286 (1952); *20.*
WEINIG, F.: Die Ausdehnung des Kavitationsgebietes. Hydromechanische Probleme des Schiffsantriebs, ed. by G. KEMPF and E. FOERSTER. Hamburg 1932; *8.*
WEINSTEIN, A.: Ein Hydrodynamischer Unitätssatz. Math. Z. **19**, 265—274 (1924); *32, 38.*
— Generalized axially symmetric potential theory. Bull. Amer. Math. Soc. **59**, 20—38 (1953); *42.*
— Non-linear problems in the theory of fluid motions with free boundaries. Proc. Symposia Appl. Math., Vol. 1, pp. 1—18, Amer. Math. Soc. 1949; *32, 38.*
— Zur Theorie der Flüssigkeitsstrahlen. Math. Z. **31**, 424—433 (1929); *32, 38, 39.*
WEYL, H.: Strahlbildung nach der Kontinuitätsmethode behandelt. Nachr. Ges. Wiss. Göttingen **1927**, 227—237; *32, 38.*
WOODS, L.: Compressible subsonic flow in two-dimensional channels with mixed boundary conditions. Quart. J. Math. Mech. **7**, 263—282 (1954); *13.*
— Unsteady cavitating flow past curved obstacles. Aeron. Res. Coun. Current Paper No. 149 (1954) *18.*
— Two-dimensional flow of a compressible fluid past given curved obstacles with infinite wakes. Proc. Roy. Soc. Lond., Ser. A **227**, 367—386 (1955); *13.*
— Unsteady plane flow past curved obstacles with infinite wakes. Proc. Roy. Soc. Lond., Ser. A **229**, 152—180 (1955); *18.*
WORTHINGTON, A.: A study of splashes. New York: Longmans Green 1908; *2.*

Wu, T.: A note on the linear and non-linear theories for fully cavitated hydrofoils. Hydrodynamics Lab., Cal. Inst. Tech. Rep. 21—22 (1956); *21.*

— A free streamline theory for two-dimensional fully cavitated hydrofoils. J. Math. Phys. **35**, 236—265 (1956); *2, 7, 8, 11.*

— A linearized theory for nonsteady cavity flows. Hydrodynamics Lab., Cal. Inst. Tech. Rep. 85-6 (1957); *18.*

— A simple method for calculating the drag in the linear theory of cavity flows. Hydrodynamic Lab., Cal. Inst. Tech. Rep. 85-5 (1957); *21.*

Young, D., L. Gates, R. Arms and D. Eliezer: The computation of an axially symmetric free boundary problem on NORC. U.S. Naval Proving Ground Rep. No. 1413 (1955); *47.*

Zarantonello, E.: A constructive theory for the equations of flows with free boundaries. Collectanea Mathematica **5**, 175—225 (1952); *40.*

— Parallel cavity flows past a plate. J. Math. pures appl. (9) **33**, 29—80 (1954); *9.*

Zoller, K.: Widerstand einer ebenen Platte mit Totwassergebiet. Dtsch. Luftf.-Forschg. UM 4518 (1943); *7, 8.*

Surface Waves.

By

JOHN V. WEHAUSEN and EDMUND V. LAITONE [1]

With 56 Figures.

A. Introduction.

The various problems of fluid motion treated in this article have in common the property that the fluid is subject to a gravitational force. In addition, in almost all cases they also have in common the presence of surfaces separating two fluids of different densities or, if only one fluid is present, of so-called free surfaces. However, not all fluid flows falling into this category are treated here: tidal motion is treated in Vol. XLVIII in the article by A. DEFANT. The observed properties of ocean waves and their generation by wind are treated in the article by H. U. ROLL, also in Vol. XLVIII. Closely related problems concerning flows with free surfaces are treated in the article by D. GILBARG in this volume.

The subject of water waves engaged many of the mathematicians and mathematical physicists of the last century. Moreover, the last several years have brought a renewed interest in the theory of water waves. In addition to this extensive literature on theoretical aspects of the subject, there have also been many experimental investigations, usually carried out by hydraulic engineers. Hydraulic engineers have also produced an extensive literature, both theoretical and experimental, on open channel flow, flow over weirs and through sluice-gates, etc.; included is a considerable literature on numerical and graphical methods of solving the equations involved. Oceanographers have produced their own literature, usually emphasizing different aspects of the subject. The theory of ship waves has produced its own literature.

All this material is pertinent to this article. Clearly some selection has to be made. We have followed roughly the following rules: Fundamental results are derived in full. The treatments of various special problems are selected so as to exemplify particular methods, other methods being mentioned only by literature citation. Experimental results are not usually reproduced, but references are given. Numerical methods of solving equations are not treated at all. The more special problems of hydraulic engineering are also not treated. Geophysical aspects which are omitted have already been mentioned.

Several excellent expositions of the theory of waves or of various parts of it already exist. We mention the following[2]: LAMB [1932, Chaps. VIII (pp. 250 to 362) and IX (pp. 363—475)]; BASSET [1888, Chap. XVII (pp. 144—187)]; WIEN [1900, Chap. V (pp. 166—224)]; KOCHIN, KIBEL', and ROZE [1948, Chap. 8 (pp. 394—526)]; MILNE-THOMSON [1956, Chap. XIV (pp. 374—431)]; AIRY (1845); BOUASSE (1924); AUERBACH (1931); THORADE (1931); SRETENSKII (1936); KHRISTIANOVICH (1938); KEULEGAN (1950); ECKART (1951); and STOKER (1957). The last cited book by STOKER gives an up-to-date account of much of the

[1] Chaps. A, B, C, D, F, G were prepared by J. V. WEHAUSEN, Chap. E by E. V. LAITONE. The former is much indebted to the Office of Naval Research, U.S. Navy, for support during the preparation of his chapters.

[2] References are collected at the end and identified in the text by author and date.

fundamental theory. For observation of waves of many kinds, CORNISH (1910, 1934) und MICHE (1954) should be consulted. SHULEIKIN [1953, part 3 (pp. 213 to 292)] contains a general discussion of topics of interest in oceanography. RUSSELL and MACMILLAN (1952) give a rather nontechnical discussion of ocean waves. A volume published by the Society of Naval Architects of Japan (Zôsen Kyôkai) contains expository papers on various aspects of water-wave theory related to ships [see MARUO (1957), JINNAKA (1957), NISHIYAMA (1957), BESSHO (1957), and INUI (1957)].

For extensive bibliographies one should consult THORADE (1931, pp. 195 to 211); SRETENSKII (1936, pp. 294—303); KAMPÉ DE FÉRIET (1932, pp. 225—229); and STOKER (1957, pp. 545—560). SRETENSKII (1950, 1951) in a survey of the accomplishments of the USSR during the years 1917—1947 has given a rather complete bibliography of Russian papers during those years. TAKAO INUI (1954) has included a valuable bibliography of Japanese papers in a survey of Japanese contributions to the theory of ship waves. An interesting early history of the subject may be found in a paper by ST. VENANT and FLAMANT (1887). The treatise by the WEBER brothers (1825) is still of interest for its content, and especially for its many references to and summaries of the early papers on water waves. The section on waves in the article on hydrodynamics by LOVE (1914), as modified by APPELL, BEGHIN and VILLAT, in the *Encyclopédie des sciences mathématiques* gives brief indications of the contents of many of the papers published up to about 1912.

B. Mathematical formulation.

1. Coordinate systems and conventions. In the mathematical description of waves one may, as in fluid mechanics in general, describe the motion by describing either the paths of individual fluid particles ("Lagrangian" description) or the velocity (and acceleration) field in the region occupied by fluid at a given moment ("Eulerian" description). Generally, but not always, the Eulerian description will be used.

Rectangular coordinates may be used conveniently for almost all problems. The y-axis will be taken directed oppositely to the force of gravity, the x-axis and z-axis so as to form a right-handed system (i.e., if the y-axis is toward the top of the page and the x-axis is toward the right, the z-axis will point toward the reader). This is a somewhat unconventional choice for the z-axis, but has the obvious advantage that in two-dimensional problems one can delete z-dependent terms from the equations, have conventional (x, y) coordinates, and set $z = x + iy$ without ambiguity when complex-variable methods are convenient.

It seems hardly worth while to try to formulate rules concerning when a moving coordinate system is preferable to a fixed one. However, use of a moving coordinate system is clearly convenient in those cases where it allows one to formulate a problem in a time-independent manner.

The following well-established convention with regard to use of certain letters will be adhered to. The components of the velocity vector $\boldsymbol{v}$ will be denoted by u, v, w the pressure by p and the density by ϱ. The coefficient of viscosity of the fluid will be denoted by μ, the coefficient of kinematic viscosity, μ/ϱ, by ν. The acceleration resulting from gravity is denoted by g.

In the Eulerian formulation one seeks $\boldsymbol{v}, p$ and ϱ as functions of x, y, z, t i.e., at any instant t one seeks a vector function and two scalar functions defined on the region occupied by fluid at that instant. In the Lagrangian system one focuses attention on the trajectories of individual particles in the fluid: if a, b, c

are the coordinates of a particle at time $t=0$, then one seeks the position $x(a, b, c, t)$, $y(a, b, c, t)$, $z(a, b, c, t)$ of this point at a later time t. One may pass from one system to the other by means of the equations

$$\frac{dx}{dt}=u(x, y, z, t), \qquad \frac{dy}{dt}=v(x, y, z, t), \qquad \frac{dz}{dt}=w(x, y, z, t) \tag{1.1}$$

with $x=a$, $y=b$, $z=c$ at $t=0$ as initial conditions.

2. Equations of motion. Derivations of the fundamental equations describing fluid motion are available in many places (e.g., Vol. VIII, Part 1 of this Encyclopedia). The equations are reproduced here for convenience of reference.

The equation of continuity in Eulerian coordinates is

$$\frac{\partial \varrho}{\partial t}+\frac{\partial(\varrho u)}{\partial x}+\frac{\partial(\varrho v)}{\partial y}+\frac{\partial(\varrho w)}{\partial z}=0. \tag{2.1}$$

If the fluid is incompressible, but not necessarily homogeneous, $d\varrho/dt=0$ (but not necessarily $\partial\varrho/\partial t=0$) and Eq. (2.1) becomes

$$\frac{\partial u}{\partial x}+\frac{\partial v}{\partial y}+\frac{\partial w}{\partial z}=0. \tag{2.2}$$

In Lagrangian coordinates this may be written

$$\varrho(x, y, z, t)\, D=\varrho(a, b, c, 0) \tag{2.3}$$

where

$$D=\begin{vmatrix} \frac{\partial x}{\partial a} & \frac{\partial x}{\partial b} & \frac{\partial x}{\partial c} \\ \frac{\partial y}{\partial a} & \frac{\partial y}{\partial b} & \frac{\partial y}{\partial c} \\ \frac{\partial z}{\partial a} & \frac{\partial z}{\partial b} & \frac{\partial z}{\partial z} \end{vmatrix}.$$

For an incompressible fluid $\varrho(x, y, z, t)=\varrho(a, b, c, 0)$ and (2.3) becomes

$$D=1. \tag{2.4}$$

The dynamical equations take different forms according as one does or does not try to take account of viscosity. The Navier-Stokes equations for the motion of an incompressible viscous fluid, when the only external force is that of gravity, are as follows in Eulerian coordinates:

$$\left.\begin{aligned} \frac{\partial u}{\partial t}+u\frac{\partial u}{\partial x}+v\frac{\partial u}{\partial y}+w\frac{\partial u}{\partial z} &= -\frac{1}{\varrho}\frac{\partial p}{\partial x}+\frac{\mu}{\varrho}\Delta u, \\ \frac{\partial v}{\partial t}+u\frac{\partial v}{\partial x}+v\frac{\partial v}{\partial y}+w\frac{\partial v}{\partial z} &= -g-\frac{1}{\varrho}\frac{\partial p}{\partial y}+\frac{\mu}{\varrho}\Delta v, \\ \frac{\partial w}{\partial t}+u\frac{\partial w}{\partial x}+v\frac{\partial w}{\partial y}+w\frac{\partial w}{\partial z} &= -\frac{1}{\varrho}\frac{\partial p}{\partial z}+\frac{\mu}{\varrho}\Delta w. \end{aligned}\right\} \tag{2.5}$$

If viscosity is neglected, the last two terms on the right side of the equations are to be deleted and one obtains the equations for an "ideal" fluid:

$$\left.\begin{aligned} \frac{\partial u}{\partial t}+u\frac{\partial u}{\partial x}+v\frac{\partial u}{\partial y}+w\frac{\partial u}{\partial z} &= -\frac{1}{\varrho}\frac{\partial p}{\partial x}, \\ \frac{\partial v}{\partial t}+u\frac{\partial v}{\partial x}+v\frac{\partial v}{\partial y}+w\frac{\partial v}{\partial z} &= -g-\frac{1}{\varrho}\frac{\partial p}{\partial y}, \\ \frac{\partial w}{\partial t}+u\frac{\partial w}{\partial x}+v\frac{\partial w}{\partial y}+w\frac{\partial w}{\partial z} &= -\frac{1}{\varrho}\frac{\partial p}{\partial z}. \end{aligned}\right\} \tag{2.6}$$

In Lagrangian coordinates the latter equations become:

$$\left.\begin{aligned}\frac{\partial^2 x}{\partial t^2}\frac{\partial x}{\partial a}+\left(g+\frac{\partial^2 y}{\partial t^2}\right)\frac{\partial y}{\partial a}+\frac{\partial^2 z}{\partial t^2}\frac{\partial z}{\partial a}&=-\frac{1}{\varrho}\frac{\partial p}{\partial a},\\ \frac{\partial^2 x}{\partial t^2}\frac{\partial x}{\partial b}+\left(g+\frac{\partial^2 y}{\partial t^2}\right)\frac{\partial y}{\partial b}+\frac{\partial^2 z}{\partial t^2}\frac{\partial z}{\partial b}&=-\frac{1}{\varrho}\frac{\partial p}{\partial b},\\ \frac{\partial^2 x}{\partial t^2}\frac{\partial x}{\partial c}+\left(g+\frac{\partial^2 y}{\partial t^2}\right)\frac{\partial y}{\partial c}+\frac{\partial^2 z}{\partial t^2}\frac{\partial z}{\partial c}&=-\frac{1}{\varrho}\frac{\partial p}{\partial c}.\end{aligned}\right\} \tag{2.7}$$

The equations of two-dimensional motion result if one deletes all terms containing z, w, and c.

The motion is called irrotational if it satisfies the additional equations

$$\frac{\partial w}{\partial y}-\frac{\partial v}{\partial z}=0,\qquad \frac{\partial u}{\partial z}-\frac{\partial w}{\partial x}=0,\qquad \frac{\partial v}{\partial x}-\frac{\partial u}{\partial y}=0, \tag{2.8}$$

or, in two-dimensional motion,

$$\frac{\partial v}{\partial x}-\frac{\partial u}{\partial y}=0. \tag{2.8'}$$

In the case of irrotational motion there exists a potential function $\Phi(x, y, z, t)$ such that

$$u=\frac{\partial\Phi}{\partial x},\qquad v=\frac{\partial\Phi}{\partial y},\qquad w=\frac{\partial\Phi}{\partial z}. \tag{2.9}$$

It is a classical theorem of hydrodynamics [cf. LAMB (1932, §§ 17, 33)] that, if the motion of an inviscid fluid with $\varrho=\varrho(p)$ is irrotational at any instant, it is so thereafter. In particular, a motion started from rest is irrotational.

If $\varrho=\varrho(p)$ is the equation of state, the following integral of the equations of motion exists for irrotational motion:

$$\frac{\partial\Phi}{\partial t}+\frac{1}{2}(u^2+v^2+w^2)+gy+P=A(t) \tag{2.10}$$

where

$$P=\int_{p_0}^{p}\varrho^{-1}\,dp$$

and $A(t)$ is an arbitrary function of t. If the fluid is incompressible, the usual case in this article, ϱ is independent of p and the integral becomes:

$$\frac{\partial\Phi}{\partial t}+\frac{1}{2}+(u^2+v^2+w^2)+gy+\frac{p-p_0}{\varrho}=A(t). \tag{2.10'}$$

In this case one obtains also from (2.2) and (2.9)

$$\Delta\Phi=\frac{\partial^2\Phi}{\partial x^2}+\frac{\partial^2\Phi}{\partial y^2}+\frac{\partial^2\Phi}{\partial z^2}=0. \tag{2.11}$$

Even if the motion is not irrotational, there still exists an integral like (2.10) if the motion is steady, the so-called Bernoulli integral:

$$\tfrac{1}{2}(u^2+v^2+w^2)+gy+P=C. \tag{2.10''}$$

Here C is constant along a single streamline:

$$\frac{dx}{dt}=u(x,y,z),\qquad \frac{dy}{dt}=v(x,y,z),\qquad \frac{dz}{dt}=w(x,y,z),$$

but may vary from one streamline to another.

There will be occasion in the following to treat problems in moving coordinate systems. Let $Oxyz$ be a fixed coordinate system and $\bar{O}\bar{x}\bar{y}\bar{z}$ be a system moving with respect to $Oxyz$ but without rotation. Let $\boldsymbol{v}_0$ be the vector $\frac{d}{dt}O\bar{O}$, the velocity of a particle referred to $Oxyz$ be $\boldsymbol{v}$ and to $\bar{O}\bar{x}\bar{y}\bar{z}$ be $\bar{\boldsymbol{v}}$. Then $\boldsymbol{v} = \bar{\boldsymbol{v}} + \boldsymbol{v}_0$. We shall generally want either to describe the absolute motion $\boldsymbol{v}$ with respect to the moving coordinate system $\bar{O}\bar{x}\bar{y}\bar{z}$ or the relative motion $\bar{\boldsymbol{v}}$ with respect to this coordinate system. In either case the continuity equation remains the same in form

$$\frac{\partial\varrho}{\partial t} + \frac{\partial(\varrho u)}{\partial\bar{x}} + \frac{\partial(\varrho v)}{\partial\bar{y}} + \frac{\partial(\varrho w)}{\partial\bar{z}} = 0, \qquad \varrho = \varrho(\bar{x},\bar{y},\bar{z},t) \tag{2.12}$$

or

$$\frac{\partial\varrho}{\partial t} + \frac{\partial(\varrho \bar{u})}{\partial\bar{x}} + \frac{\partial(\varrho \bar{v})}{\partial\bar{y}} + \frac{\partial(\varrho \bar{w})}{\partial\bar{z}} = 0, \qquad \varrho = \varrho(\bar{x},\bar{y},\bar{z},t). \tag{2.13}$$

The dynamical equations for an ideal fluid for the absolute motion described in the moving coordinate system are:

$$\left.\begin{aligned}
\frac{\partial u}{\partial t} + (u-u_0)\frac{\partial u}{\partial\bar{x}} + (v-v_0)\frac{\partial u}{\partial\bar{y}} + (w-w_0)\frac{\partial u}{\partial\bar{z}} &= - \frac{1}{\varrho}\frac{\partial p}{\partial\bar{x}},\\
\frac{\partial v}{\partial t} + (u-u_0)\frac{\partial v}{\partial\bar{x}} + (v-v_0)\frac{\partial v}{\partial\bar{y}} + (w-w_0)\frac{\partial v}{\partial\bar{z}} &= -g - \frac{1}{\varrho}\frac{\partial p}{\partial\bar{y}},\\
\frac{\partial w}{\partial t} + (u-u_0)\frac{\partial w}{\partial\bar{x}} + (v-v_0)\frac{\partial w}{\partial\bar{y}} + (w-w_0)\frac{\partial w}{\partial\bar{z}} &= - \frac{1}{\varrho}\frac{\partial p}{\partial\bar{z}}.
\end{aligned}\right\} \tag{2.14}$$

The dynamical equations for the relative motion are:

$$\left.\begin{aligned}
\frac{\partial\bar{u}}{\partial t} + \bar{u}\frac{\partial\bar{u}}{\partial\bar{x}} + \bar{v}\frac{\partial\bar{u}}{\partial\bar{y}} + \bar{w}\frac{\partial\bar{u}}{\partial\bar{z}} &= - \frac{1}{\varrho}\frac{\partial p}{\partial\bar{x}} - \dot{u}_0,\\
\frac{\partial\bar{v}}{\partial t} + \bar{u}\frac{\partial\bar{v}}{\partial\bar{x}} + \bar{v}\frac{\partial\bar{v}}{\partial\bar{y}} + \bar{w}\frac{\partial\bar{v}}{\partial\bar{z}} &= -g - \frac{1}{\varrho}\frac{\partial p}{\partial\bar{y}} - \dot{v}_0,\\
\frac{\partial\bar{w}}{\partial t} + \bar{u}\frac{\partial\bar{w}}{\partial\bar{x}} + \bar{v}\frac{\partial\bar{w}}{\partial\bar{y}} + \bar{w}\frac{\partial\bar{w}}{\partial\bar{z}} &= - \frac{1}{\varrho}\frac{\partial p}{\partial\bar{z}} - \dot{w}_0.
\end{aligned}\right\} \tag{2.15}$$

Let us suppose that the motion is irrotational and let $\Phi(x, y, z, t)$ be the velocity potential for the absolute motion in the fixed coordinate system. Let

$$\Phi(x,y,z,t) = \Phi\left(\bar{x} + \int^t u_0\,dt, \bar{y} + \int^t v_0\,dt, \bar{z} + \int^t w_0\,dt, t\right) = \bar{\Phi}(\bar{x},\bar{y},\bar{z},t).$$

Then $\bar{\Phi}$ is the velocity potential for the absolute motion in the moving coordinate system:

$$\frac{\partial\bar{\Phi}}{\partial\bar{x}} = u, \qquad \frac{\partial\bar{\Phi}}{\partial\bar{y}} = v, \qquad \frac{\partial\bar{\Phi}}{\partial\bar{z}} = w.$$

The integral (2.10) becomes:

$$\frac{\partial\bar{\Phi}}{\partial t} + \frac{1}{2}\left[(u-u_0)^2 + (v-v_0)^2 + (w-w_0)^2\right] + g\bar{y} + P = \bar{A}(t), \tag{2.16}$$

where $\bar{A}(t) = A(t) + \frac{1}{2}(u_0^2 + v_0^2 + w_0^2) - g\int^t v_0\,dt$. If one defines $\bar{\bar{\Phi}}$ by

$$\bar{\bar{\Phi}}(\bar{x},\bar{y},\bar{z},t) = \bar{\Phi}(\bar{x},\bar{y},\bar{z},t) - u_0\bar{x} - v_0\bar{y} - w_0\bar{z},$$

then $\bar{\bar{\Phi}}$ is the velocity potential for the relative motion:

$$\frac{\partial\bar{\bar{\Phi}}}{\partial\bar{x}} = \bar{u}, \qquad \frac{\partial\bar{\bar{\Phi}}}{\partial\bar{y}} = \bar{v}, \qquad \frac{\partial\bar{\bar{\Phi}}}{\partial\bar{z}} = \bar{w},$$

and the integral (2.10) may be written:

$$\frac{\partial\bar{\bar{\Phi}}}{\partial t}+\dot{u}_0\bar{x}+\dot{v}_0\bar{y}+\dot{w}_0\bar{z}+\frac{1}{2}(\bar{u}^2+\bar{v}^2+\bar{w}^2)+g\bar{y}+P=\bar{A}(t). \tag{2.17}$$

The more general equations when the system $\bar{O}\bar{x}\bar{y}\bar{z}$ is also rotating will not be necessary for this article.

3. Boundary conditions at an interface. Let us now suppose that we are given two immiscible fluids with a common boundary surface, $S(t)$. The one fluid, with density ϱ_1 and viscosity μ_1, will occupy region $R_1(t)$; the other, with density ϱ_2 and viscosity μ_2, the region $R_2(t)$. Let $F(x, y, z, t)=0$ describe the surface $S(t)$; we assume $F_x^2+F_y^2+F_z^2>0$ (where $F_x=\partial F/\partial x$, etc.).

The first condition which the surface $S(t)$ must satisfy is a kinematic one. As the surface moves, the velocity of a point (x, y, z) on the surface in the direction of the normal to the surface is given by $-F_t/\sqrt{F_x^2+F_y^2+F_z^2}$. Here one takes the normal in the direction (F_x, F_y, F_z). A particle of fluid at the same point of the surface at that instant will have a velocity component in the direction of the surface normal given by $\frac{uF_x+vF_y+wF_z}{\sqrt{F_x^2+F_y^2+F_z^2}}=v_n$. For $S(t)$ to be a bounding surface means, of course, that there can be no transfer of matter across the surface. Consequently the following equation must be satisfied:

$$uF_x+vF_y+wF_z=-F_t, \tag{3.1}$$

where we have used the assumption $F_x^2+F_y^2+F_z^2>0$ in dropping the denominators. If one defines the "material derivative" by the equation

$$\frac{DF}{Dt}=uF_x+vF_y+wF_z+F_t,$$

then (3.1) is the same as

$$\frac{DF}{Dt}=0. \tag{3.1'}$$

This condition must be satisfied by any bounding surface, whether an interface or a rigid boundary[1].

There are further dynamical conditions to be satisfied at an interface. Let us first consider the general case of viscous fluids with surface tension at the interface. The following assumptions are made:

1. The effect of surface tension as one passes through the interface is to produce a discontinuity in the normal stress proportional to the mean curvature of the boundary surface.

2. For viscous fluids the tangential stress must be continuous as one passes through the interface.

3. For viscous fluids the tangential component of the velocity must be continuous as one passes through the interface.

In order to formulate these statements in mathematical language, we introduce the following notation. Let $g(x, y, z)$ be some function defined in both R_1 and R_2 and let (x_0, y_0, z_0) be a point of the interface S. Assuming that the following limit exists, we shall write

$$g_i(x_0, y_0, z_0)=\lim g(x, y, z) \quad \text{as} \quad (x, y, z)\to(x_0, y_0, z_0), (x, y, z) \text{ in } R_i,$$

[1] For further discussion of this condition see C. Truesdell: Bull. Tech. Univ. Istanbul **3** (1950), No. 1, 71–78 (1951); L. Lichtenstein: Grundlagen der Hydromechanik, pp. 159 to 170, 234ff. Berlin: Springer 1929.

and

$$[g(x_0, y_0, z_0)] = g_2(x_0, y_0, z_0) - g_1(x_0, y_0, z_0).$$

Let the components of the stress tensor be denoted by

$$\begin{matrix} \sigma_{xx} & \sigma_{xy} & \sigma_{xz} \\ \sigma_{yx} & \sigma_{yy} & \sigma_{yz} \\ \sigma_{zx} & \sigma_{zy} & \sigma_{zz}. \end{matrix}$$

Consider an element of area of the surface S at a point (x, y, z) of S. Let the unit normal vector to S at (x, y, z) be (l, m, n). Then the surface element will have associated with it the stress vector with components:

$$\sigma_{xx} l + \sigma_{xy} m + \sigma_{xz} n, \quad \sigma_{yx} l + \sigma_{yy} m + \sigma_{yz} n, \quad \sigma_{zx} l + \sigma_{zy} m + \sigma_{zz} n.$$

Let R_1 and R_2 be the principal radii of curvature of S at (x, y, z). Then 1. and 2. are combined in the one equation

$$\left.\begin{aligned} [\sigma_{xx} l + \sigma_{xy} m + \sigma_{xz} n] &= T(R_1^{-1} + R_2^{-1})\, l, \\ [\sigma_{yx} l + \sigma_{yy} m + \sigma_{yz} n] &= T(R_1^{-1} + R_2^{-1})\, m, \\ [\sigma_{zx} l + \sigma_{zy} m + \sigma_{zz} n] &= T(R_1^{-1} + R_2^{-1})\, n, \end{aligned}\right\} \tag{3.2}$$

where T is a constant of proportionality depending upon the two fluids (and their temperatures, but this will not be considered here). T is called the coefficient of surface tension[1].

The kinematic condition imposed in (3.1) is clearly equivalent to continuity of the normal component of the velocity as one passes through S. Consequently, the condition 3. above may be combined with this to give

$$u_1 = u_2, \quad v_1 = v_2, \quad w_1 = w_2. \tag{3.3}$$

In the linearized theory of viscosity the stress tensor for an incompressible fluid is given by

$$\left.\begin{matrix} p - 2\mu u_x & -\mu(u_y + v_x) & -\mu(u_z + w_x) \\ -\mu(v_x + u_y) & p - 2\mu v_y & -\mu(v_z + w_y) \\ -\mu(w_x + u_z) & \mu(w_y + v_z) & p - 2\mu w_z. \end{matrix}\right\} \tag{3.4}$$

The geometric quantity $R_1^{-1} + R_2^{-1}$ is given by the formula[2]

$$\left.\begin{aligned} \frac{1}{R_1} + \frac{1}{R_2} &= -\frac{\partial}{\partial x} \frac{F_x}{\sqrt{F_x^2 + F_y^2 + F_z^2}} - \frac{\partial}{\partial y} \frac{F_y}{\sqrt{F_x^2 + F_y^2 + F_z^2}} - \frac{\partial}{\partial z} \frac{F_z}{\sqrt{F_x^2 + F_y^2 + F_z^2}} \\ &= -\frac{F_{xx}(F_y^2 + F_z^2) + F_{yy}(F_x^2 + F_z^2) + F_{zz}(F_x^2 + F_y^2) - 2(F_{xy} F_x F_y + F_{yz} F_y F_z + F_{zx} F_z F_x)}{[F_x^2 + F_y^2 + F_z^2]^{\frac{3}{2}}}. \end{aligned}\right\} \tag{3.5}$$

The sign is so selected that, if it is positive, the direction of increase of the normal component of the stress vector at the interface is in the direction

$$(l, m, n) = \left(\frac{F_x}{\sqrt{F_x^2 + F_y^2 + F_z^2}}, \frac{F_y}{\sqrt{F_x^2 + F_y^2 + F_z^2}}, \frac{F_z}{\sqrt{F_x^2 + F_y^2 + F_z^2}}\right). \tag{3.6}$$

---

[1] For an air-water interface $T = 72.8$ dynes/cm at 20° C, for mercury-air interface $T = 485$ dynes/cm at 20° C, for a mercury-water interface $T = 412$ dynes/cm, for benzene-air $T = 28.9$ dynes/cm at 20° C, for liquid helium-helium vapor $T = 0.24$ dynes/cm at $-270°$ C.

[2] See, e.g., A. Duschek and W. Mayer: Lehrbuch der Differentialgeometrie, Vol. I, pp. 150–152. Leipzig u. Berlin: Teubner 1930.

In the case of a surface given by $y=\eta(x, z)$ Eq. (3.5) becomes

$$\frac{1}{R_1}+\frac{1}{R_2}=\frac{\eta_{xx}(1+\eta_z^2)+\eta_{zz}(1+\eta_x^2)-2\eta_{xz}\eta_x\eta_z}{(1+\eta_x^2+\eta_z^2)^{\frac{3}{2}}}, \tag{3.5'}$$

where the direction of increase is upwards. In the case of two-dimensional motion this simplifies further to the well-known formula

$$\frac{\eta_{xx}}{(1+\eta_x^2)^{\frac{3}{2}}}. \tag{3.5''}$$

If one now substitutes (3.4) to (3.6) in (3.2), one obtains the general boundary condition at the interface. The result is unwieldy in its general form[1].

If the interface is given by $y=\eta(x, z)$, the boundary condition becomes

$$\left.\begin{aligned}
[p]\,\eta_x-\{2[\mu u_x]\,\eta_x-[\mu(u_y+v_x)]+[\mu(u_z+w_x)]\,\eta_z\} &= T(R_1^{-1}+R_2^{-1})\,\eta_x,\\
[p]+\{[\mu(v_x+u_y)]\,\eta_x-2[\mu v_y)+[\mu(v_z+w_y)]\,\eta_z\} &= T(R_1^{-1}+R_2^{-1}),\\
[p]\,\eta_z-\{[\mu(w_x+u_z)]\,\eta_x-[\mu(w_y+v_z)]+2[\mu w_z]\,\eta_z\} &= T(R_1^{-1}+R_2^{-1})\,\eta_z
\end{aligned}\right\} \tag{3.7}$$

with $R_1^{-1}+R_2^{-1}$ given by (3.5'). Here fluid$_1$ is the lower and fluid$_2$ the upper fluid. For two-dimensional motion the equations take the following form:

$$\left.\begin{aligned}
[p]\,\eta'(x)-\{2[\mu u_x]\,\eta'(x)-[\mu(u_y+v_x)]\} &= T\frac{\eta''(x)}{(1+\eta'(x)^2)^{\frac{3}{2}}}\,\eta'(x),\\
[p]+\{[\mu(u_y+v_x)]\,\eta'(x)-2[\mu v_y]\} &= T\frac{\eta''(x)}{(1+\eta'(x)^2)^{\frac{3}{2}}}.
\end{aligned}\right\} \tag{3.8}$$

One may also write this condition in terms of the components of the stress vector normal and tangential to the interface:

$$\left.\begin{aligned}
[p]-2\frac{[\mu u_x]\,\eta'^2-[\mu(u_y+v_x)]\,\eta'+[\mu v_y]}{(1+\eta'^2)^{\frac{1}{2}}} &= T\frac{\eta''(x)}{(1+\eta'^2)^{\frac{3}{2}}},\\
\frac{2[\mu(u_x-v_y)]\,\eta'+[\mu(u_y+v_x)]\,(\eta'^2-1)}{(1+\eta'^2)^{\frac{1}{2}}} &= 0.
\end{aligned}\right\} \tag{3.8'}$$

If surface tension is to be neglected, one obtains the resulting boundary condition by setting $T=0$ in the various equations above. In this case, Eq. (3.2) simply states the continuity of the stress vector as one passes through the interface.

If viscosity is neglected, but not necessarily surface tension, the condition on the stress vector becomes simply

$$[p]=T(R_1^{-1}+R_2^{-1}), \tag{3.9}$$

where, of course, the mean curvature is still given by (3.5). The other boundary condition (3.3) changes more drastically upon neglecting viscosity: Condition 3. stating the continuity of the tangential component of velocity is abandoned. The continuity of the normal component, i.e. (3.1), is still retained, of course.

---

[1] In tensor notation the condition is somewhat more perspicuous:

$$\{[p]\,\delta_{ij}-[\mu(u_{i,j}+u_{j,i})]\}\frac{F_{,j}}{(F_{,k}F_{,k})^{\frac{1}{2}}}=T\frac{F_{,r}F_{,s}F_{,rs}-F_{,r}F_{,r}F_{,ss}}{(F_{,k}F_{,k})^{\frac{3}{2}}}\cdot\frac{F_{,i}}{(F_{,k}F_{,k})^{\frac{1}{2}}},$$

where $(x_1, x_2, x_3)=(x, y, z)$, $(u_1, u_2, u_3)=(u, v, w)$ and $F_{,i}=\partial F/\partial x_i$. We have refrained from using tensor notation because its particular advantages cannot in general be exploited here.

Condition 2. concerning the tangential stress is satisfied vacuously for an inviscid fluid.

So far we have considered the boundary condition at an interface between two fluids. If the second fluid is absent, the boundary surface for the first fluid is called a "free surface". Usually the pressure above a free surface is assumed to be some given function, say $p_2(x, y, z, t)$, of position and time; in most cases it is taken to be a constant, either an assumed atmospheric pressure or zero. The boundary conditions concerning the stress vector at a free surface are slight modifications of those for an interface, and can be obtained by setting $\mu_2=0$, $\lambda_2=0$. The result is again somewhat unwieldy in its complete form[1]. For an incompressible fluid it is:

$$\left.\begin{aligned}(\bar{p}-p)\,F_x+\mu\,\{2u_x F_x+(u_y+v_x)\,F_y+(u_z+w_x)\,F_z\}&=T(R_1^{-1}+R_2^{-1})\,F_x,\\(\bar{p}-p)\,F_y+\mu\,\{(v_x+u_y)\,F_x+2v_y F_y+(v_z+w_y)\,F_z\}&=T(R_1^{-1}+R_2^{-1})\,F_y,\\(\bar{p}-p)\,F_z+\mu\,\{(w_x+u_z)\,F_x+(w_y+v_z)\,F_y+2v_z F_z\}&=T(R_1^{-1}+R_2^{-1})\,F_z.\end{aligned}\right\}\tag{3.10}$$

Here we have written $\bar{p}$ for p_2, and μ for μ_1; $\bar{p}, p, u_x, \ldots$ are to be evaluated at $F(x, y, z, t)=0$.

The case with which we shall be chiefly concerned is that of an inviscid fluid without surface tension and with $\bar{p}(x, y, z, t)=p_0$, a constant. In this case the boundary condition reduces to the single equation

$$p(x, y, z, t)=p_0 \tag{3.11}$$

on $F(x, y, z, t)=0$. If the motion is irrotational and incompressible, one may determine p explicitly from (2.10′) so that (3.10) becomes

$$\Phi_t+\tfrac{1}{2}(u^2+v^2+w^2)+g y=A(t) \tag{3.11′}$$

to be satisfied on $F(x, y, z, t)=0$.

In the case of steady motion of an incompressible fluid, the Bernoulli integral (2.10″) still exists even if the motion is rotational. Consequently, in certain two-dimensional problems of steady motion in which the free surface is a streamline one continues to have a boundary condition like (3.10″):

$$\frac{1}{2}(u^2+v^2)+g y+\frac{p_0}{\varrho}=C, \tag{3.11″}$$

to be satisfied on $F(x, y)=0$.

4. Boundary conditions on rigid surfaces. Let the equation of the rigid surface be given by the equation $G(x, y, z, t)=0$. Then in the case of an inviscid fluid the condition to be satisfied on $G=0$ is the same as the kinematic condition (3.1):

$$u\,G_x+v\,G_y+w\,G_z=-\,G_t, \tag{4.1}$$

i.e., the component of velocity of the fluid normal to the surface must equal the velocity of the rigid surface in the direction of its normal.

If the fluid is viscous, it must stick to a solid boundary and move with it without slippage. An equation of the form $G(x, y, z, t)=0$ is not suitable for

[1] In tensor notation it may be written:

$$\{(\bar{p}-p)\,\delta_{ij}+\mu\,(u_{i,j}+u_{j,i})\}\frac{F_{,j}}{(F_{,k}F_{,k})^{\frac{1}{2}}}=T\,\frac{F_{,r}F_{,s}F_{,rs}-F_{,r}F_{,r}F_{,ss}}{(F_{,k}F_{,k})^{\frac{3}{2}}}\,\frac{F_{,j}}{(F_{,k}F_{,k})^{\frac{1}{2}}}\,.$$

Here we have written $\bar{p}$ for p_2 and λ, μ for λ_1, μ_1. All variable quantities in the braces are, of course, to be evaluated at the free surface $F=0$.

formulating this statement in equations (e.g., $x^2+y^2+z^2=a^2$ does not distinguish between a rotating and a stationary sphere). Let the surface be given in parametric coordinates by: $x=X(r,s,t)$, $y=Y(r,s,t)$, $z=Z(r,s,t)$, where a given point on the surface corresponds to a given pair of values (r,s). Then the condition for viscous fluids may be written:

$$u=\frac{\partial X}{\partial t},\qquad v=\frac{\partial Y}{\partial t},\qquad w=\frac{\partial Z}{\partial t}. \tag{4.2}$$

If a solid boundary penetrates the free surface (or an interface) of a viscous fluid, there will be some difference in treatment of the boundary condition according as the fluid wets the surface or not. In the case of mercury sloshing in a clean glass basin, the fluid pulls free of the surface as it moves up and down, whereas water in the same basin will continue to adhere to any part of the walls already wetted. Furthermore, if surface tension is taken into account, the angle of contact of the free surface with the solid surface will enter into the boundary condition; in the first case mentioned above the angle may vary according as the liquid is rising or falling along the wall[1]. Although attempts to prove very general existence theorems for fluid motion would presumably take such complications into account, they are usually neglected in most solutions of special problems, there being indeed little choice in the matter.

5. Other types of boundary surfaces. Geophysical problems sometimes suggest situations in which there is an interface between a fluid and an elastic medium. This may occur, for example, in the study of the effect of ocean waves on the ocean floor, as in LONGUET-HIGGINS' (1950) theory of microseisms. Other possibilities are suggested by wave motion on a body of water covered with an ice sheet or at an interface between two fluids separated by an elastic membrane or plate. In one series of investigations the ice sheet has been assumed broken into pieces small with respect to the prevalent wave lengths. In this case only the density of the ice layer enters into the modified boundary condition [see PETERS (1950), KELLER and GOLDSTEIN (1953), KELLER and WEITZ (1953), SHAPIRO and SIMPSON (1953)]. Waves in a thin plate over an infinitely deep fluid have been considered briefly by LANDAU and LIFSHITS (1953, pp. 762—763), but with neglect of gravity. GREENHILL (1887, p. 68; 1916) included gravity.

The kinematic boundary condition (3.1) must always hold. The dynamical conditions will depend upon the nature of the assumptions. The matter will not be further considered here.

C. Preliminary remarks and developments.

6. Classification of problems. Most of the theory of water waves is concerned either with elucidating some general aspects of wave motion or with predicting the behavior of waves in the presence of some special configuration of interest to oceanographers, hydraulic engineers, or ship designers. Unfortunately, even some of the apparently simplest problems have proved too difficult to solve in their most complete formulation. Approximations have been necessary, and in many cases the problems which have been solved are those which could be solved by the approximate methods in use. An examination of the theory also shows that many of the concepts and definitions are almost inextricably bound up with these methods of approximation, following rather than preceding the making of the approximation.

[1] See, e.g., R. S. BURDON: Surface tension and the spreading of liquids, pp. 76—82. Cambridge 1949.

The nature of the approximations used in treating a particular problem provides a natural way of classifying it. First there are the assumptions concerning the properties of the fluid: viscous or inviscid, compressible or incompressible, surface tension or not. Although assuming the fluid to be inviscid, incompressible, and without surface tension simplifies the equations, they are still not easily manageable, even for the simplest kinds of problems. Other approximations of a different nature are required. These are in a sense mathematical approximations. Their physical significance is not in restricting the nature of the fluid but in restricting the character of the waves and the boundary configuration. The kind of mathematical approximation used provides another means of classifying problems, and is the principal one which will be used in this article. There are two principal methods of approximation, explained below in Sect. 10, the infinitesimal-wave approximation and the shallow-water approximation. Thus, the development of these two approximate theories and of the exact theory constitutes the bulk of this article.

7. Progressive waves and wave velocity. Standing waves. It will be convenient to call any motion of a fluid in a gravitational field with a free surface or an interface a *wave motion*.

If the velocity components, pressure, and free surface or interface may be expressed in the form

$$\boldsymbol{v}=\boldsymbol{v}(x-ct,y,z),\qquad p=p(x-ct,y,z),\qquad y=\eta(x-ct,z),$$

respectively, then the wave motion will be said to be a *progressive wave* travelling in the direction Ox. In this case a change to a moving coordinate system with $x'=x-ct$, $y'=y$, $z'=z$ reduces the motion to steady motion with respect to the moving coordinate system. With respect to the fixed coordinate system the profile of the free surface or interface is being transported without change of form in the direction Ox with velocity c. It might seem reasonable therefore to call c the velocity of propagation of the progressive wave.

However, STOKES (1849; or 1880, pp. 202ff.) has pointed out that the velocity of propagation of the profile of the free surface does not by itself give a useful definition of wave velocity. Let the fluid be inviscid, either infinitely deep or with a horizontal bottom, and unlimited otherwise. Now let the whole fluid in the progressive wave described above be transported with velocity C (positive or negative) in the direction Ox. Then the motion will still be consistent with the laws of fluid mechanics, the various parts of the fluid will move the same relatively to each other, but the velocity of propagation of the profile will be arbitrary, depending upon the choice of C. What is required for a useful definition of wave velocity is the velocity of propagation of the profile with respect to a coordinate system fixed in some sense in the fluid.

In the case of an infinitely deep fluid, if the axes may be chosen so that as $y\to-\infty$ the velocity relative to these axes vanishes, then one may reasonably measure the profile velocity with respect to these. If the motion far ahead or far behind the disturbance approaches a uniform velocity (possibly zero), then axes moving with the fluid with this velocity may be used. When the disturbance does not behave thus (as in the case of periodic waves) and when the depth is finite, there is no longer an obvious way to select a set of reference axes.

In order to put the problem somewhat differently, let us assume that the wave motion is given as a steady motion with velocity field $\boldsymbol{v}(x,y)$ and free surface $y=\eta(x)$. We wish to find a moving coordinate system $x'=x-u_0t$, $y'=y$, so that in some sense the relative motion vanishes on the average. We now have the free surface given by $y' = \eta(x' + u_0 t)$ and the relative velocity by $\boldsymbol{v}'(x' + u_0 t, y') = \boldsymbol{v}(x' + u_0 t, y') - u_0 \boldsymbol{i}$. How is u_0 to be chosen? STOKES made two suggestions. One is to define it by the equation

$$\left.\begin{aligned} &\lim_{a\to-\infty,\, b\to\infty} \frac{1}{b-a} \int_a^b dx' \int_{-h}^{\eta(x'+u_0 t)} u'(x' + u_0 t, y')\, dy' \\ &\quad = \lim_{a\to-\infty,\, b\to\infty} \frac{1}{b-a} \int_a^b dx \int_{-h}^{\eta(x)} [u(x, y) - u_0]\, dy = 0, \end{aligned}\right\} \tag{7.1}$$

where $y = -h$ is the equation for the bottom. In case the motion is periodic, with period λ, the defining equation may be written

$$\int_0^\lambda dx \int_{-h}^{\eta(x)} [u(x, y) - u_0)]\, dy = 0. \tag{7.2}$$

If one notes that the mean depth is given by

$$\lim \frac{1}{b-a} \int_a^b [\eta(x) + h]\, dx \quad \text{or} \quad \frac{1}{\lambda} \int_0^\lambda [\eta(x) + h]\, dx,$$

then one sees that, with h' as mean depth,

$$u_0 h' = Q \tag{7.3}$$

where Q is the average discharge rate per unit width. u_0 is thus defined so that the average discharge rate with respect to the (x', y') coordinate system is zero. u_0 is usually denoted by c'.

STOKES' other suggestion was to define u_0 by

$$\lim_{T\to\infty} \frac{1}{T} \int_0^T u'(x' + u_0 t, y')\, dt = 0 \tag{7.4}$$

or

$$\begin{aligned} u_0 &= \lim_{T\to\infty} \frac{1}{T} \int_0^T u(x' + u_0 t, y')\, dt \\ &= \lim_{T\to\infty} \frac{1}{u_0 T} \int_{x'}^{x'+u_0 T} u(x, y)\, dx = \lim_{a\to\infty} \frac{1}{a} \int_{x'}^{x'+a} u(x, y)\, dx. \end{aligned}$$

If u is periodic in x with period λ, one may write

$$u_0 = \frac{1}{\lambda} \int_{x'}^{x'+\lambda} u(x, y)\, dx. \tag{7.5}$$

In either case, for the definition to be useful u_0 must be independent of x' and y. If u is bounded, it follows easily that $\partial u_0/\partial x' = 0$ for both cases. If the motion is irrotational, $u_y = v_x$ and it follows again that $\partial u_0/\partial y = 0$ if v is bounded. Wave velocity defined in this manner is usually denoted by c. For the two special cases considered earlier, the two definitions coincide.

The definition of wave velocity in cases where the motion cannot be reduced to a steady motion is no longer straightforward. In many cases of interest, the asymptotic behavior of the motion for large positive or negative x allows one to define a wave velocity in a manner similar to that above. In more complicated wave motions one may simply follow the motion of some special phase of the profile, say a crest. This provides, for example, a definition of phase velocity for a cylindrical wave.

A general definition of *standing wave* is somewhat more awkward to formulate than that for a progressive wave. For the case of a plane wave, the free surface $y=\eta(x,t)$ must be periodic in each of x and t, with wave length λ and period τ, say. In addition, the curves in the (x,t)-plane represented by $\eta(x,t)=0$, where $y=0$ is the undisturbed surface, must consist of two sets of curves oscillating about the lines $x=\frac{1}{2}n\lambda$ and $t=\frac{1}{2}n\tau$, $n=0, \pm 1, \ldots$. For progressive waves the curves $\eta(x,t)=0$ consist of a single set of straight lines, all parallel to $x-ct=0$. The prototype for the standing wave is the surface defined by, say, $y=\sin 2\pi x/\lambda \times \cos 2\pi t/\tau$. However, as shown by both Penney and Price (1952b) and by Sekerzh-Zenkovich (1947), neither set of curves $\eta(x,t)=0$ consists of straight lines, or even fixed curves, for standing waves of finite amplitude.

There remains the problem of establishing that progressive and standing waves exist under suitable boundary conditions. For the exact boundary conditions for a perfect fluid, the existence of progressive waves was first established by Levi-Civita (1925) and Nekrasov (1921, 1922). The existence of standing waves satisfying the exact boundary conditions is apparently an open question.

8. Energy. Let $T(t)$ be a region occupied by a perfect fluid with a boundary $S(t)$ represented by

$$F(x,y,z,t)=0,$$

the representation being chosen so that (F_x, F_y, F_z) is in the direction of the exterior normal. The surface $S(t)$ moves independently of the motion of the fluid. It is assumed that $T(t)$ contains no singularities of $\boldsymbol{v}$ and that surface tension does not act upon the surface $S(t)$ at any time. The energy of the fluid contained in $T(t)$ is given by

$$E=\iiint\limits_{T(t)}\left[\tfrac{1}{2}\varrho(u^2+v^2+w^2)+\varrho g y\right]d\tau. \tag{8.1}$$

For irrotational motion of an inviscid incompressible fluid, one may use (2.10′) and express E by

$$E=\iiint\limits_{T(t)}\left[-p-\varrho\frac{\partial\Phi}{\partial t}\right]d\tau.$$

[Here Φ has been redefined so that $A(t)$ may be set equal to zero.] One may now compute dE/dt by using the general formula:

$$\frac{d}{dt}\iiint\limits_{T(t)} f(x,y,z,t)\,d\tau=\iiint\limits_{T(t)} f_t(x,y,z,t)\,d\tau+\iint\limits_{S(t)} f(x,y,z,t)\frac{-F_t}{\sqrt{F_x^2+F_y^2+F_z^2}}\,d\sigma.$$

One finds [cf. F. John (1949, p. 19ff.), which we follow closely here]:

$$\begin{aligned}\frac{dE}{dt}&=\iiint\limits_{T(t)}\varrho\,\operatorname{grad}\Phi\cdot\operatorname{grad}\Phi_t\,d\tau+\iint\limits_{S(t)}[\varrho\Phi_t+p]\frac{F_t}{\sqrt{F_x^2+F_y^2+F_z^2}}\,d\sigma\\&=\iint\limits_{S(t)}\varrho\Phi_t\frac{\partial\Phi}{\partial n}\,d\sigma+\iint\limits_{S(t)}[\varrho\Phi_t+p]\frac{F_t}{\sqrt{F_x^2+F_y^2+F_z^2}}\,d\sigma\end{aligned}$$

by Green's Theorem and the equation of continuity. Finally,

$$\frac{dE}{dt}=\iint\limits_{S(t)}\left\{\varrho\Phi_t\left[\frac{\partial\Phi}{\partial n}+\frac{F_t}{\sqrt{F_x^2+F_y^2+F_z^2}}\right]+p\frac{F_t}{\sqrt{F_x^2+F_y^2+F_z^2}}\right\}d\sigma. \tag{8.2}$$

We recall that $-F_t/\sqrt{F_x^2+F_y^2+F_z^2}$ is the velocity of $S(t)$ in the direction of the exterior normal. Two cases are of special interest. If $S(t)$ is a "physical" boundary, i.e., one moving with the fluid, then the first summand vanishes and one finds

$$\frac{dE}{dt} = -\iint\limits_{S(t)} p \frac{\partial \Phi}{\partial n} d\sigma \tag{8.3}$$

[cf. LAMB, Hydrodynamics, p. 9, Eq. (5)]. If $S(t)$ is a fixed "geometrical" boundary, then $F_t = 0$ and one gets

$$\frac{dE}{dt} = \iint\limits_{S(t)} \varrho\, \Phi_t \frac{\partial \Phi}{\partial n} d\sigma. \tag{8.4}$$

If one considers any portion of $S(t)$, then the integral of (8.2) taken over this portion and with a minus sign gives the rate of flow of energy through this portion of $S(t)$. In case a part of $S(t)$ is a physical boundary which is fixed, $\partial\Phi/\partial n = 0$ and the flow through this part is zero. The same conclusion holds for any portion of $S(t)$ that is a free surface, for then $p = 0$.

If one has a progressive wave moving to the right with $\Phi(x, y, z, t) = \varphi(x - ct, y, z)$ and takes S as a region in the fixed plane $x = x_0$, then the rate of flow of energy through S in the positive direction is given by

$$\iint\limits_S \varrho\, c\, \varphi_x^2(x_0 - c\,t, y, z)\, dy\, dz \geqq 0, \tag{8.5}$$

i.e., energy always flows in the direction of the wave.

In cases where one is dealing with waves generated by moving bodies, it is frequently possible to choose the region T so that no energy is lost from it, the latter being true only as an average if the motion is periodic in time. As an example, consider a body moving steadily with velocity c in the x-direction in an infinite ocean with horizontal bottom. In addition to the boundary conditions on the body, free surface, and bottom, we assume that the motion vanishes (in the limit) far ahead and to the sides of the body. The surface $S(t)$ may then be chosen as a plane $M: x - ct - a = 0$ far ahead, another plane $N: -(x - ct) + b = 0$ behind the body, planes R and L: $z = \pm a$ on either side, and the bottom H, the wetted surface of the body B, and the part of the free surface F included between the body and the planes. The energy within this region is clearly constant, and one easily obtains, with $\Phi(x, y, z, t) = \varphi(x - ct, y, z)$:

$$0 = -\iint\limits_B p \frac{\partial \varphi}{\partial n} d\sigma - \iint\limits_{M+N+R+L} \varrho\, c \frac{\partial \varphi}{\partial x} \frac{\partial \varphi}{\partial n} d\sigma + c \iint\limits_M \left[\frac{1}{2} \varrho (\varphi_x^2 + \varphi_y^2 + \varphi_z^2) + \varrho g y\right] d\sigma -$$
$$- c \iint\limits_N \left[\frac{1}{2} \varrho (\varphi_x^2 + \varphi_y^2 + \varphi_z^2) + \varrho g y\right] d\sigma.$$

Since on B one has $\partial\varphi/\partial n = c \cos(n, x)$, one finds for the first integral, remembering that $\boldsymbol{n}$ points into the body,

$$-\iint\limits_B p \frac{\partial \varphi}{\partial n} d\sigma = -c \iint\limits_B p \cos(n, x)\, d\sigma = R\, c,$$

where R is the force on the body. The parts of the second integral over M, R, L vanish as $a \to \infty$ and similarly for the first summand in the third integral. The terms in $\varrho g y$ give

$$\int_{-a}^{a} dz \int_{-h}^{\eta(a,z)} \varrho g y\, dy - \int_{-a}^{a} dz \int_{-h}^{\eta(b,z)} \varrho g y\, dy = \int_{-a}^{a} \tfrac{1}{2} \varrho g \left[\eta^2(a,z) - \eta^2(b,z)\right] dz$$

which, as $a \to \infty$, converges to

$$-\tfrac{1}{2} \varrho g \int_{-\infty}^{\infty} \eta^2(b,z)\, dz.$$

One obtains finally

$$\left.\begin{aligned} R = \tfrac{1}{2} \varrho \int_{-\infty}^{\infty} dz \int_{-h}^{\eta(b,z)} \left[-\varphi_x^2(b,y,z) + \varphi_y^2(b,y,z) + \varphi_z^2(b,y,z)\right] dy + \\ + \tfrac{1}{2} g \varrho \int_{-\infty}^{\infty} \eta^2(b,z)\, dz. \end{aligned}\right\} \tag{8.6}$$

This exact formula for resistance will be put into a different form later after linearization of the boundary conditions. Although the plane $x - ct = b$ may be taken at any distance behind the body without destroying the validity of (8.6), it is usually convenient to take it so far behind that asymptotic expressions for φ can be used.

If in (8.1) a part of the surface $S(t)$, say $S_1(t)$, is an interface with another fluid with surface tension acting, then the energy is given by

$$E = \iiint_{T(t)} \left[\tfrac{1}{2} \varrho (u^2 + v^2 + w^2) + \varrho g y\right] d\tau + T \iint_{S_1(t)} d\sigma. \tag{8.7}$$

Let $S_1(t)$ be bounded by the curve $C(t)$ given parametrically by $x(s,t)$, $y(s,t)$, $z(s,t)$ and let $S(t) = S_1(t) + S_2(t)$. Then the formula analogous to (8.2) is

$$\left.\begin{aligned} \frac{dE}{dt} = \iint_{S(t)} \varrho \Phi_t \left[\Phi_n + \frac{F_t}{\sqrt{F_x^2 + F_y^2 + F_z^2}}\right] d\sigma + \iint_{S_2(t)} p \frac{F_t}{\sqrt{F_x^2 + F_y^2 + F_z^2}} d\sigma + \\ + \iint_{S_1(t)} \left[p + T\left(\frac{1}{R_1} + \frac{1}{R_2}\right)\right] \frac{F_t}{\sqrt{F_x^2 + F_y^2 + F_z^2}} d\sigma + T \int_{C(t)} \begin{vmatrix} F_x & F_y & F_z \\ x_t & y_t & z_t \\ x_s & y_s & z_s \end{vmatrix} \frac{d_s}{\sqrt{F_x^2 + F_y^2 + F_z^2}}. \end{aligned}\right\} \tag{8.8}$$

If $S_1(t)$ is a free surface, then the boundary condition

$$p_0 - p = T\left(\frac{1}{R_1} + \frac{1}{R_2}\right)$$

where p_0 is an assumed constant pressure implies that there is no flux of energy through S_1.

If the motion is two-dimensional, with S_1 given by $y = \eta(x,t)$, $x_1(t) \leqq x \leqq x_2(t)$, then (8.7) becomes

$$E(t) = \iint_{T(t)} \left[\tfrac{1}{2} \varrho (u^2 + v^2) + \varrho g y\right] d\sigma + T \int_{X_1(t)}^{x_2(t)} ds \tag{8.9}$$

and (8.8) becomes

$$\left.\begin{aligned} \frac{dE}{dt} = \int_{S_2} \varrho \Phi_t \left[\Phi_n + \frac{F_t}{\sqrt{F_x^2 + F_y^2}}\right] ds + \int_{S_2} p \frac{F_t}{\sqrt{F_x^2 + F_y^2}} ds - \\ - \int_{x_1(t)}^{x_2(t)} \left[p + \frac{T \eta_{xx}}{[1 + \eta_x^2]^{\frac{3}{2}}}\right] \eta_t\, dx + T \frac{\eta_x \eta_t}{\sqrt{1 + \eta_x^2}} \Bigg|_{x_1}^{x_2} + T \sqrt{1 + \eta_x^2}\, x'(t) \Bigg|_{x_1}^{x_2}. \end{aligned}\right\} \tag{8.10}$$

If S_1 is a free surface, the integral over S_1 may be dropped by suitably redefining p.

9. Momentum. Expressions for rate of change of momentum may be derived which are analogous to those for rate of change of energy. With

$$\boldsymbol{M} = \iiint_{T(t)} \varrho\, \boldsymbol{v}\, d\tau, \tag{9.1}$$

and otherwise the same notation as in Sect. 8, one finds

$$\left.\begin{aligned}\frac{d\boldsymbol{M}}{dt} &= \iint_S \varrho \left\{\Phi_t \boldsymbol{n} + \operatorname{grad}\Phi \frac{-F_t}{\sqrt{F_x^2+F_y^2+F_z^2}}\right\} d\sigma,\\ &= -\iint_S \left\{(p+\varrho g y)\,\boldsymbol{n} + \varrho\left[\boldsymbol{v}\cdot\boldsymbol{n} + \frac{F_t}{\sqrt{F_x^2+F_y^2+F_z^2}}\right]\boldsymbol{v}\right\} d\sigma,\\ &= \iint_S \varrho \left\{\left(\Phi_t + \frac{1}{2} v^2\right)\boldsymbol{n} - \left[\boldsymbol{v}\cdot\boldsymbol{n} + \frac{F_t}{\sqrt{F_x^2+F_y^2+F_z^2}}\right]\boldsymbol{v}\right\} d\sigma.\end{aligned}\right\} \tag{9.2}$$

Here the first line of (9.2) is derived by a direct computation of $d\boldsymbol{M}/dt$ with $\boldsymbol{v} = \operatorname{grad}\Phi$; the second is derived analogously to (8.2); the third follows directly by use of (2.10′). Comparison of lines one and three gives the known relation (LEVI-CIVITA):

$$\iint_S \tfrac{1}{2} v^2 \boldsymbol{n}\, d\sigma = \iint_S (\boldsymbol{v}\cdot\boldsymbol{n})\, \boldsymbol{v}\, d\sigma. \tag{9.3}$$

Note that in (9.2) and (9.3) $S(t)$ may move in an arbitrary manner as long as the region $T(t)$ contains no singularities and only fluid. If the boundary is physical, the terms in square brackets vanish in (9.2); if the boundary is fixed, then $F_t = 0$.

Let $S_0(t)$ be a physical boundary, possibly the surface of a solid body, and $S(t)$ a closed surface containing it. Applying (9.2) to the region of fluid bounded jointly by S_0 and S, one finds

$$\left.\begin{aligned}F_0 &= \iint_{S_0} (p + \varrho g y)\, \boldsymbol{n}\, d\sigma\\ &= -\iint_{S_0} \varrho (\Phi_t \boldsymbol{n} + \boldsymbol{v}\cdot\boldsymbol{n}\,\boldsymbol{v})\, d\sigma + \iint_S \varrho \left(\tfrac{1}{2} v^2 \boldsymbol{n} - \boldsymbol{v}\cdot\boldsymbol{n}\,\boldsymbol{v}\right) d\sigma.\end{aligned}\right\} \tag{9.4}$$

Here F_0 is the hydrodynamic orce fon S_0 and does not include the hydrostatic force.

If singularities are allowed in the region occupied by fluid, they may be enclosed in spheres of small radius and the formula (9.4) applied to the remaining fluid, with S modified to include the spherical surfaces. If the singularities are isolated sources of strengths m_i at the points $\boldsymbol{a}_i$, then by shrinking the spheres about the singularities in a customary fashion [cf. MILNE-THOMSON (1956, pp. 448 to 450)], one obtains the following modification of (9.4):

$$\boldsymbol{F}_0 = -\iint_{S_0} \varrho (\Phi_t \boldsymbol{n} + \boldsymbol{v}\cdot\boldsymbol{n}\,\boldsymbol{v})\, d\sigma + \sum 4\pi \varrho\, m_i \boldsymbol{v}_i + \iint_S \varrho \left(\tfrac{1}{2} v^2 \boldsymbol{n} - \boldsymbol{v}\cdot\boldsymbol{n}\,\boldsymbol{v}\right) d\sigma, \tag{9.5}$$

where $\boldsymbol{v}_i$ is the velocity at the point $\boldsymbol{a}_i$ when the source at that point is removed. Other modifications may be derived for other types of singularities.

If the velocity field is such that $r^{1+\varepsilon} v \to 0$ as $r = \sqrt{x^2+y^2+z^2} \to \infty$ for some $\varepsilon > 0$, then the last integral in (9.4) or (9.5) will vanish as S is expanded to infinity, provided the latter can be done without destroying the validity of the formula. In the case of a body moving in a fluid with a free surface, one cannot expand in all directions and must include the contribution of the last integral over the free surface. However, the formulas are still useful in computing the force on an obstacle resulting from waves.

10. Expansion of solutions in powers of a parameter. In their exact form even the simplest problems with surface waves are difficult to solve. If one neglects viscosity and assumes irrotational motion, the problem is reduced to finding solutions of LAPLACE'S equation, which is at least linear in the unknown. However, the problem is still difficult because of the nonlinear boundary condition at the free surface or interface. This lack of linearity deprives one, for example, of the mathematical tool of superposition of solutions; expansion in eigenfunctions or use of GREEN'S functions is not possible.

In order to be able to treat special problems, the equations are approximated by ones which are more tractable. The two principal methods of approximation may each be treated as a perturbation procedure. As was mentioned in Sect. 7, this procedure is not concerned with the assumptions about the nature of the fluid, for example, whether or not viscosity is neglected, but rather with the nature of the motion and its generation. An advantage in using the perturbation procedure is that the assumptions about the motion are displayed in such a way that it is clear how to obtain approximations of higher order. The method has been applied to water-wave problems by SEKERZH-ZENKOVICH (1947, 1951, 1952), K. FRIEDRICHS (1948), KELLER (1948), F. JOHN (1949), LONGUET-HIGGINS (1953b), PETERS and STOKER (1957), and others. As used here the method is purely formal, the nature of the convergence of the perturbation series, whether it be uniform, pointwise, asymptotic or what not, being left open. However, for each method of approximation it is possible to point to several cases in which convergence has been proved: for the infinitesimal-wave approximation, LEVI-CIVITA'S (1925), STRUIK'S (1926) and NEKRASOV'S (1921, 1928) proofs of the existence of a periodic wave of permanent type; and for the shallow-water approximation, FRIEDRICHS and HYERS' (1954) proof of the existence of a solitary wave and LITTMAN'S (1957) proof of the existence of cnoidal waves.

To a certain extent the two methods of approximation have different aims. The infinitesimal-wave approximation fits into a general scheme for approximating nonlinear equations and boundary conditions by linear ones [see SOURIAU (1952) for a discussion]. To apply it, one must know a particular exact solution to start with. In addition, one must be able to select a dimensionless parameter (or parameters), say ε, which helps to determine the exact physical problem and is such that the solutions to the exact problems associated with each value of ε approach (in some sense) the known exact solution when $\varepsilon \to 0$. It is then assumed that the various functions entering into the problem may be expanded into power series in ε. The series are substituted into the equations and boundary conditions and grouped according to powers of ε. The coefficients of each power then yield a sequence of equations and boundary conditions, the coefficients of ε giving the first-order theory, those of ε^2 the second-order theory, etc. As an exact initial solution it is usually most convenient to take either a state of rest or of uniform motion. Various choices of ε will be made in the applications later.

The shallow-water approximation differs in that a change of variable involving the expansion parameter is made initially. This introduces ε into the exact equations. When the power series expansions are introduced into the equations, the resulting equations of the sequence are linear in quantities of the same order, but the equations are too degenerate to determine all these quantities without recourse to the equations of next higher order. This leads to nonlinear equations for the desired functions, but ones of a type which have been intensively investigated. In this case the procedure is perhaps artificial in that the perturbation scheme is devised to lead to a special set of equations for a first-order theory, derived originally by quite different reasoning. However, in doing this it makes clear the nature of the approximation and gives a systematic procedure for finding higher-order approximations. It is instructive, in this connection, to read the usual derivation as given, for example, in LAMB (1932, pp. 254—256) or STOKER (1957, pp. 22—25) (who also gives the one given here).

α) The infinitesimal-wave approximation. We shall derive the equations of motion and the free-surface or interface boundary conditions for this linearized theory without identifying explicitly the parameter ε used in the expansions. Later on, when specific choices are made, the boundary conditions on certain geometric boundaries associated with the choice of ε will be modified to conform with the linearization.

Consider two incompressible viscous fluids in contact along an interface represented by $y=\eta(x,z,t)$. Quantities referring to the upper fluid have subscript 2, those to the lower fluid subscript 1; the coefficient of surface tension is T. Assume the following expansions in the parameter ε:

$$\left.\begin{aligned} \boldsymbol{v}_i(x,y,z,t,\varepsilon) &= \varepsilon\,\boldsymbol{v}_i^{(1)} + \varepsilon^2\,\boldsymbol{v}_i^{(2)} + \cdots, \\ p_i(x,y,\zeta,t,\varepsilon) &= p_i^{(0)} + \varepsilon\,p_i^{(1)} + \varepsilon^2\,p_i^{(2)} + \cdots, \\ \eta(x,z,t,\varepsilon) &= \varepsilon\,\eta^{(1)} + \varepsilon^2\,\eta^{(2)} + \cdots. \end{aligned}\right\} \tag{10.1}$$

Substitute these expansions in Eqs. (2.2), (2.5), (3.1), (3.3), and (3.7), remembering in addition that formal expansions of the following sort, for example, hold:

$$\begin{aligned} u_1(x,\eta(x,z,t),z,t) &= u_1(x,0,z,t) + \eta\,u_{1y}(x,0,z,t) + \cdots \\ &= \varepsilon\,u_1^{(1)}(x,0,z,t) + \varepsilon^2\left[u_1^{(2)}(x,0,z,t) + \eta^{(1)}u_{1y}^{(1)}(x,0,z,t)\right] + \cdots. \end{aligned}$$

Collecting first the terms independent of ε, one finds from (2.5) and (3.3)

$$\operatorname{grad}\,(p_i^{(0)} + \varrho_i\,g\,y) = 0, \qquad p_2^{(0)}(x,0,z,t) = p_1^{(0)}(x,0,z,t). \tag{10.2}$$

Collecting the coefficients of the first power of ε, one finds

$$\left.\begin{aligned} &\frac{\partial u_i^{(1)}}{\partial x} + \frac{\partial v_i^{(1)}}{\partial y} + \frac{\partial w_i^{(1)}}{\partial z} = 0, \qquad i = 1, 2, \\ &\frac{\partial \boldsymbol{v}_i^{(1)}}{\partial t} = -\frac{1}{\varrho_i}\operatorname{grad} p_i^{(1)} + \nu_i\,\Delta\,\boldsymbol{v}_i^{(1)}, \qquad i = 1, 2, \\ &u_1^{(1)}(x,0,z,t) = u_2^{(1)}(x,0,z,t), \\ &v_1^{(1)}(x,0,z,t) = v_2^{(1)}(x,0,z,t) = \eta_t^{(1)}(x,z,t), \\ &w_1^{(1)}(x,0,z,t) = w_2^{(1)}(x,0,z,t), \\ &\mu_1(u_{1y}^{(1)}(x,0,z,t) + v_{1x}^{(1)}) = \mu_2(u_{2y}^{(1)} + v_{2x}^{(1)}), \\ &p_2^{(1)}(x,0,z,t) - p_1^{(1)} - (\varrho_2 - \varrho_1)\,g\,\eta^{(1)} - 2(\mu_2\,v_{2y}^{(1)} - \mu_1\,v_{1y}^{(1)}) = T(\eta_{xx}^{(1)} + \eta_{zz}^{(1)}), \\ &\mu_1(w_{1y}^{(1)}(x,0,z,t) + v_{1z}^{(1)}) = \mu_2(w_{2y}^{(1)} + v_{2z}^{(1)}). \end{aligned}\right\} \tag{10.3}$$

If the upper fluid is replaced by a given atmospheric pressure distribution $\bar{p}(x,z,t)$, then the equations for the lower fluid become (after dropping the subscripts)

$$\left.\begin{aligned} &\operatorname{grad}\,(p^{(0)} + \varrho\,g\,y) = 0, \qquad p^{(0)}(x,0,z,t) = \bar{p}^{(0)}(x,z,t), \\ &\frac{\partial u^{(1)}}{\partial x} + \frac{\partial v^{(1)}}{\partial y} + \frac{\partial w^{(1)}}{\partial z} = 0, \\ &\frac{\partial \boldsymbol{v}^{(1)}}{\partial t} = -\frac{1}{\varrho}\operatorname{grad} p^{(1)} + \nu\,\Delta\,\boldsymbol{v}^{(1)}, \\ &\eta_t^{(1)}(x,z,t) = v^{(1)}(x,0,z,t), \\ &u_y^{(1)}(x,0,z,t) + v_x^{(1)} = w_y^{(1)} + v_z^{(1)} = 0, \\ &p^{(1)}(x,0,z,t) - \varrho\,g\,\eta^{(1)} - 2\mu\,v_y^{(1)} = -T(\eta_{xx}^{(1)} + \eta_{zz}^{(1)}) + \bar{p}^{(1)}(x,z,t). \end{aligned}\right\} \tag{10.4}$$

For convenience we have assumed above that the expansion for η starts with $\varepsilon\eta^{(1)}$. If we had assumed instead $\eta=\eta^{(0)}+\varepsilon\eta^{(1)}+\cdots$, we would have found from (3.1) and (3.7) the equations

$$\eta_t^{(0)}=\eta_x^{(0)}=\eta_z^{(0)}=0$$

and, hence, $\eta^{(0)}=$const. The zero values of y in the boundary conditions would then be replaced by this constant. Taking the constant equal to zero means that we have taken the undisturbed interface as (x, z)-plane.

The equations above give the linearized equations of motion and boundary conditions at the interface or free surface. If one now proceeds, as we shall not do for this case, to collect coefficients of ε^2, one may obtain the differential equations and boundary conditions for the second-order corrections to be added to the solutions of the linearized equations, and so forth for higher-order corrections. In general the resulting equations are too unwieldy to be useful.

A special case of the linearized equations which is of particular interest is irrotational flow of a perfect fluid. There is then a velocity potential Φ which we assume has the following expansion:

$$\Phi(x, y, z, t, \varepsilon) = \varepsilon\,\Phi^{(1)} + \varepsilon^2\,\Phi^{(2)} + \cdots. \tag{10.5}$$

Condition (2.11) becomes

$$\Delta\Phi^{(i)} = 0, \qquad i = 1, 2, \ldots. \tag{10.6}$$

Let there be two superposed fluids with velocity potentials Φ_1 and Φ_2 describing the motion in each; otherwise the same notation as above. Then condition (3.1) at the interface gives the linearized condition

$$\eta_t^{(1)}(x, z, t) = \Phi_{1y}^{(1)}(x, 0, z, t) = \Phi_{2y}^{(1)}(x, 0, z, t) \tag{10.7}$$

and condition (3.9), together with (2.10′), gives

$$-\varrho_2\,\Phi_{2t}^{(1)}(x, 0, z, t) + \varrho_1\,\Phi_{1t}^{(1)}(x, 0, z, t) + (\varrho_1-\varrho_2)\,g\,\eta^{(1)}(x, z, t) = T(\eta_{xx}^{(1)}+\eta_{zz}^{(1)}). \tag{10.8}$$

The further special case when both the upper fluid and surface tension are missing will be dealt with so often later on that we repeat the boundary conditions for it. We allow, however, a pressure distribution on the free surface, $\bar{p}(x, z, t) = \varepsilon\bar{p}^{(1)}+\varepsilon^2\bar{p}^{(2)}+\cdots$. The first-order boundary conditions are

$$\left.\begin{aligned}&\eta_t^{(1)}(x, z, t) - \Phi_y^{(1)}(x, 0, z, t) = 0,\\ &g\,\eta^{(1)}(x, z, t) + \Phi_t(x, 0, z, t) + \varrho^{-1}\,\bar{p}^{(1)}(x, z, t) = 0.\end{aligned}\right\} \tag{10.9}$$

Eliminating $\eta^{(1)}$ between the last two equations, one gets

$$g\,\Phi_y^{(1)}(x, 0, z, t) + \Phi_{tt}^{(1)}(x, 0, z, t) + \varrho^{-1}\,\bar{p}_t^{(1)}(x, z, t) = 0. \tag{10.10}$$

The boundary conditions for the second-order corrections are not too long to write down:

$$\left.\begin{aligned}&\eta_t^{(2)}(x, z, t) - \Phi_y^{(2)}(x, 0, z, t) = \eta^{(1)}\,\Phi_{yy}^{(1)} - \eta_x^{(1)}\,\Phi_x^{(1)} - \eta_z^{(1)}\,\Phi_z^{(1)},\\ &g\,\eta^{(2)}(x, z, t) + \Phi_t^{(2)}(x, 0, z, t) + \varrho^{-1}\,\bar{p}^{(2)}(x, z, t) = -\eta^{(1)}\,\Phi_{ty}^{(1)} - \tfrac{1}{2}(\operatorname{grad}\Phi^{(1)})^2.\end{aligned}\right\} \tag{10.11}$$

Eliminating $\eta^{(1)}$ and $\eta^{(2)}$ from (10.11), one finds a counterpart to (10.10):

$$\left.\begin{aligned}g\,\Phi_y^{(2)}(x, 0, z, t) + \Phi_{tt}^{(2)} + \varrho^{-1}\,\bar{p}_t^{(2)} = -\frac{\partial}{\partial t}(\operatorname{grad}\Phi^{(1)})^2 +\\ + (\Phi_t^{(1)} + \varrho^{-1}\,\bar{p}^{(1)})\left(\Phi_{yy}^{(1)} + \frac{1}{g}\,\Phi_{tty}^{(1)}\right) - \varrho^{-1}\,(\Phi_x^{(1)}\,\bar{p}_x^{(1)} + \Phi_z^{(1)}\,\bar{p}_z^{(1)}).\end{aligned}\right\} \tag{10.12}$$

Under certain circumstances the next-to-last term will vanish. The boundary conditions for higher-order corrections will not be worked out in detail. However, they are of the form

$$\left.\begin{aligned} g\,\Phi_y^{(i)}(x,0,z,t) + \Phi_{tt}^{(i)} + \varrho^{-1} p_t^{(i)} &= A_i\{\Phi^{(1)},\ldots,\Phi^{(i-1)},\bar{p}^1,\ldots,\bar{p}^{(i-1)}\},\\ g\,\eta^{(i)}(x,z,t) + \Phi_t^{(i)}(x,0,z,t) + \varrho^{-1}\bar{p}^{(i)} &= B_i\{\Phi^{(1)},\ldots,\Phi^{(i-1)},\bar{p}^{(1)},\ldots,\bar{p}^{(i-1)}\},\end{aligned}\right\}\qquad(10.13)$$

where A_i and B_i are functionals of the functions in brackets, in this case complicated polynomials of the functions and their derivatives evaluated at $y=0$.

It is useful to have the form of the linearized boundary conditions when certain additional assumptions are made.

First, let us suppose that the $(\bar{x},\bar{y},\bar{z})$-coordinate system is moving with velocity $c(t)$ in the x-direction with respect to the fixed (x,y,z)-coordinate system. Then, from the equation following (2.15) with $\bar{y}=y$, $\bar{z}=z$

$$\Phi_t(x,y,z,t) = \bar{\Phi}_t - c\,\bar{\Phi}_{\bar{x}},\qquad \Phi_{tt} = \bar{\Phi}_{tt} - 2c\,\bar{\Phi}_{t\bar{x}} + c^2\,\bar{\Phi}_{\bar{x}\bar{x}} - \dot{c}\,\bar{\Phi}_{\bar{x}},$$

and the boundary conditions become

$$\left.\begin{aligned} g\,\bar{\eta}^{(1)}(\bar{x},\bar{z},t) + \bar{\Phi}_t^{(1)}(\bar{x},0,\bar{z},t) - c\,\bar{\Phi}_{\bar{x}}^{(1)}(\bar{x},0,\bar{z},t) + \varrho^{-1}\bar{p}^{(1)}(\bar{x},\bar{z},t) &= 0,\\ \bar{\Phi}_{tt}^{(1)}(\bar{x},0,\bar{z},t) - 2c\,\bar{\Phi}_{t\bar{x}}^{(1)} + c^2\,\bar{\Phi}_{\bar{x}\bar{x}}^{(1)} - \dot{c}\,\bar{\Phi}_{\bar{x}}^{(1)} + g\,\bar{\Phi}_{\bar{y}}^{(1)} + \varrho^{-1}\bar{p}_t^{(1)} - c\,\varrho^{-1}\bar{p}_{\bar{x}}^{(1)} &= 0.\end{aligned}\right\}\qquad(10.14)$$

If c is constant and the motion is steady in the moving coordinate system,

$$\Phi(x,y,z,t) = \varphi(x-ct,y,z) = \varphi(\bar{x},\bar{y},\bar{z})$$

and the linearized boundary conditions are

$$\left.\begin{aligned} g\,\bar{\eta}^1(\bar{x},\bar{z}) - c\,\varphi_{\bar{x}}^{(1)}(\bar{x},0,\bar{z}) + \varrho^{-1}\bar{p}^1(\bar{x},\bar{z}) &= 0,\\ g\,\varphi_{\bar{y}}^{(1)}(\bar{x},0,\bar{z}) + c^2\,\varphi_{\bar{x}\bar{x}}^{(1)}(\bar{x},0,\bar{z}) - c\,\varrho^{-1}\bar{p}_{\bar{x}}^{(1)}(\bar{x},\bar{z}) &= 0.\end{aligned}\right\}\qquad(10.15)$$

If the motion is steady with respect to a moving coordinate system, one may impose a uniform flow in the opposite direction and then treat the problem as a steady one in an absolute coordinate system, but carrying out the perturbation about the uniform flow. We illustrate this for the case of two-dimensional irrotational flow. Let $\varphi(x,y)$ and $\psi(x,y)$ be the velocity potential and stream function, respectively, and assume expansions of the form

$$\left.\begin{aligned} \varphi(x,y) &= -cx + \varepsilon\,\varphi^{(1)}(x,y) + \varepsilon^2\,\varphi^{(2)} + \cdots,\\ \psi(x,y) &= -cy + \varepsilon\,\psi^{(1)}(x,y) + \varepsilon^2\,\psi^{(2)} + \cdots,\\ \eta(x) &= \qquad\ \ \varepsilon\,\eta^{(1)}(x) \quad\ + \varepsilon^2\,\eta^{(2)} + \cdots.\end{aligned}\right\}\qquad(10.16)$$

The differential equations $\Delta\varphi=0$, $\Delta\psi=0$, $\varphi_x=\psi_y$, $\varphi_y=-\psi_x$ become

$$\Delta\varphi^{(i)}=0,\qquad \Delta\psi^{(i)}=0,\qquad \varphi_x^{(i)}=\psi_y^{(i)},\qquad \varphi_y^{(i)}=-\psi_x^{(i)}.\qquad(10.17)$$

The kinematic condition (3.1) is replaced by

$$\psi\big(x,\eta(x)\big)=0.$$

Substituting the expansions (10.16) in this equation and in (3.11′), one finds from the coefficients of ε

$$\left.\begin{aligned} -c\,\eta^{(1)}(x) + \psi^{(1)}(x,0) &= 0,\\ g\,\eta^{(1)}(x) - c\,\varphi_x^{(1)}(x,0) + \varrho^{-1}\bar{p}^{(1)}(x) &= 0.\end{aligned}\right\}\qquad(10.18)$$

Eliminating $\eta^{(1)}$ and using the third of Eqs. (10.17), one gets

$$g\,\psi^{(1)}(x,0) - c^2\psi_y^{(1)}(x,0) + c\,\varrho^{-1}\bar{p}'^{(1)}(x) = 0. \tag{10.19}$$

Collecting the coefficients of ε^2, one obtains after some manipulation

$$\left.\begin{aligned} g\,\psi^{(2)}(x,0) - c^2\psi_y^{(2)} &= \frac{1}{c}\,\psi^{(1)}\left[c^2\psi_{yy}^{(1)} - g\,\psi_y^{(1)}\right] - \frac{1}{2}\,c\left[\psi_x^{(1)\,2} + \psi_y^{(1)\,2}\right],\\ c\,\eta^{(2)}(x) &= \psi^{(2)}(x,0) + \frac{1}{c}\,\psi^{(1)}\psi_y^{(1)}; \end{aligned}\right\} \tag{10.20}$$

here we have assumed for simplicity that $\bar{p}=0$.

β) The shallow-water approximation. This approximation has been widely used by hydraulic engineers in the study of open-channel flow and, in a further simplification, is used for the theory of tides. In deriving the equations from the exact ones we shall follow the method of Friedrichs (1948) and Keller (1948). However, a somewhat different approach to this approximation due to Ursell (1953) is also instructive. Although it is possible to carry through the derivation while taking account of surface tension, this will not be done here. It will be assumed to start with that there are two perfect, incompressible fluids with an interface $y=\eta(x,z,t)$; the bottom fluid is bounded below by a rigid surface $y=-h(x,z)$. Variables pertaining to the lower fluid have subscript 1, those pertaining to the upper fluid subscript 2. The motion will be assumed irrotational.

Before making an expansion in powers of a parameter, it is essential to make a change of variable in which vertical and horizontal distances are stretched by different amounts. Let m be a scale for horizontal measurement and n one for vertical measurement. Define $\varepsilon = n^2/m^2$. Introduce new variables, $\bar{x}, \bar{y}, \bar{z}, \bar{t}$, by the equations

$$\bar{x} = x\sqrt{\varepsilon},\quad \bar{y} = y,\quad \bar{z} = z\sqrt{\varepsilon},\quad \bar{t} = t\sqrt{\varepsilon},\quad \bar{u} = u,\quad \bar{v} = v\sqrt{\varepsilon},\quad \bar{w} = w,\quad \bar{p} = p. \tag{10.21}$$

Eqs. (2.2), (2.6), (2.8), (3.1), (3.9), and (4.1) (with $T=0$) become:

$$\left.\begin{aligned} &\varepsilon\left(\frac{\partial\bar{u}}{\partial\bar{x}} + \frac{\partial\bar{w}}{\partial\bar{z}}\right) + \frac{\partial\bar{v}}{\partial\bar{y}} = 0,\\ &\varepsilon\left(\frac{\partial\bar{u}}{\partial t} + \bar{u}\frac{\partial\bar{u}}{\partial\bar{x}} + \bar{w}\frac{\partial\bar{u}}{\partial\bar{z}} + \frac{1}{\varrho}\frac{\partial\bar{p}}{\partial\bar{x}}\right) + \bar{v}\frac{\partial\bar{u}}{\partial\bar{y}} = 0,\\ &\varepsilon\left(\frac{\partial\bar{v}}{\partial t} + \bar{u}\frac{\partial\bar{v}}{\partial\bar{x}} + \bar{w}\frac{\partial\bar{v}}{\partial\bar{z}} + g + \frac{1}{\varrho}\frac{\partial\bar{p}}{\partial y}\right) + \bar{v}\frac{\partial\bar{v}}{\partial\bar{y}} = 0,\\ &\varepsilon\left(\frac{\partial\bar{w}}{\partial t} + \bar{u}\frac{\partial\bar{w}}{\partial\bar{x}} + \bar{w}\frac{\partial\bar{w}}{\partial\bar{z}} + \frac{1}{\varrho}\frac{\partial\bar{p}}{\partial\bar{z}}\right) + \bar{v}\frac{\partial\bar{w}}{\partial\bar{y}} = 0,\\ &\frac{\partial\bar{w}}{\partial\bar{y}} = \frac{\partial\bar{v}}{\partial\bar{z}},\qquad \frac{\partial\bar{u}}{\partial\bar{z}} = \frac{\partial\bar{w}}{\partial\bar{x}},\qquad \frac{\partial\bar{v}}{\partial\bar{x}} = \frac{\partial\bar{u}}{\partial\bar{y}},\\ &\varepsilon\left(\bar{u}\frac{\partial\bar{\eta}}{\partial\bar{x}} + \bar{w}\frac{\partial\bar{\eta}}{\partial\bar{z}} + \frac{\partial\bar{\eta}}{\partial\bar{t}}\right) - \bar{v} = 0 \text{ for } \bar{y} = \bar{\eta}(\bar{x},\bar{z},\bar{t}),\\ &\varepsilon\left(\bar{u}\frac{\partial\bar{h}}{\partial\bar{x}} + \bar{w}\frac{\partial\bar{h}}{\partial\bar{z}}\right) + \bar{v} = 0 \qquad \text{for } \bar{y} = -\bar{h}(\bar{x},\bar{z},\bar{t}),\\ &\bar{p}_2(\bar{x},\bar{\eta},\bar{z},\bar{t}) = \bar{p}_1(\bar{x},\bar{\eta},\bar{z},\bar{t}), \end{aligned}\right\} \tag{10.22}$$

where $\bar{u}, \bar{v}, \bar{w}, \bar{p}, \varrho$ possess suppressed subscripts 1 and 2 for the lower and upper fluids respectively, except in the last equation.

Now assume expansions of the form

$$\left.\begin{aligned} \bar{\boldsymbol{v}}_i &= \boldsymbol{v}_i^{(0)} + \varepsilon\, \boldsymbol{v}_i^{(1)} + \varepsilon^2 \boldsymbol{v}_i^{(2)} + \cdots, \qquad i = 1, 2,\\ \bar{p}_i &= p_i^{(0)} + \varepsilon\, p_i^{(1)} + \varepsilon^2 p_i^{(2)} + \cdots, \qquad i = 1, 2,\\ \bar{\eta} &= \eta^{(0)} + \varepsilon\, \eta^{(1)} + \varepsilon^2 \eta^{(2)} + \cdots, \end{aligned}\right\} \tag{10.23}$$

substitute in the Eqs. (10.22) and collect according to powers of ε. (We shall henceforth suppress the bars on $\bar{x}, \bar{y}, \bar{z}, \bar{t}, \bar{\eta}$). The terms independent of ε give the equations

$$\left.\begin{aligned} &v_y^{(0)} = 0,\\ &v^{(0)} u_y^{(0)} = 0, \quad v^{(0)} v_y^{(0)} = 0, \quad v^{(0)} w_y^{(0)} = 0,\\ &w_y^{(0)} = v_z^{(0)}, \quad u_z^{(0)} = w_x^{(0)}, \quad v_x^{(0)} = u_y^{(0)},\\ &v^{(0)}(x, \eta^{(0)}, z, t) = 0, \quad v_1^{(0)}(x, -h, z, t) = 0,\\ &p_2^{(0)}(x, \eta^{(0)}, z, t) = p_1^{(0)}(x, \eta^{(0)}, z, t). \end{aligned}\right\} \tag{10.24}$$

The first and fourth equations give

$$v^{(0)}(x, y, z, t) \equiv 0. \tag{10.25}$$

The third then states that

$$u_y^{(0)} = w_y^{(0)} = 0 \quad \text{or} \quad u^{(0)} = u^{(0)}(x, z, t), \quad w^{(0)} = w^{(0)}(x, z, t). \tag{10.26}$$

The terms which are coefficients of ε give, after making use of (10.25) and (10.26),

$$\left.\begin{aligned} &u_x^{(0)} + w_z^{(0)} + v_y^{(1)} = 0,\\ &u_t^{(0)} + u^{(0)} u_x^{(0)} + w^{(0)} u_z^{(0)} + p_x^{(0)}/\varrho = 0,\\ &g + p_y^{(0)}/\varrho = 0,\\ &w_t^{(0)} + u^{(0)} w_x^{(0)} + w^{(0)} w_z^{(0)} + p_z^{(0)}/\varrho = 0,\\ &u^{(0)} \eta_x^{(0)} + w^{(0)} \eta_z^{(0)} + \eta_t^{(0)} - v^{(1)} = 0 \quad \text{for } y = \eta^{(0)}(x, z, t),\\ &u_1^{(0)} h_x + w_1^{(0)} h_y + v_1^{(1)} = 0 \quad \text{for } y = -h(x, z). \end{aligned}\right\} \tag{10.27}$$

(The equations deriving from irrotationality and the continuity of pressure will be brought in later.) The first and last two equations of (10.27) together with (10.26) give

$$\left.\begin{aligned} v_1^{(1)} &= -y\,(u_{1x}^{(0)} + w_{1z}^{(0)}) - (u_1^{(0)} h)_x - (w_1^{(0)} h)_z,\\ v_2^{(1)} &= -y\,(u_{2x}^{(0)} + w_{2z}^{(0)}) + (u_2^{(0)} \eta^{(0)})_x + (w_2^{(0)} \eta^{(0)})_z - (u_1^{(0)} \eta^{(0)})_x - (w_1^{(0)} \eta^{(0)})_z +\\ &\qquad - (u_1^{(0)} h)_x - (w_1^{(0)} h)_z. \end{aligned}\right\} \tag{10.28}$$

The third equation of (10.27) gives

$$p^{(0)} = -\varrho g y + f(x, z, t).$$

In order to evaluate f, further information is necessary. Here are two cases of interest. 1. If the upper fluid is absent, the condition $p^{(0)}(x, \eta^{(0)}, z, t) = 0$ gives

$$p^{(0)} = -\varrho g y + \varrho g \eta^{(0)}(x, z, t). \tag{10.29}$$

2. If the upper fluid is unbounded above, then, up to an additive constant,

$$\left.\begin{aligned} p_1^{(0)} &= -\varrho_1 g y + (\varrho_1 - \varrho_2) g \eta^{(0)} + k,\\ p_2^{(0)} &= -\varrho_2 g y + k. \end{aligned}\right\} \tag{10.30}$$

If the upper fluid is bounded above by a free surface $y = d(x, z, t) = d^{(0)} + \varepsilon d^{(1)} + \cdots$, then one may satisfy the boundary conditions $p_2^{(0)}(x, d^{(0)}, z, t) = 0$, $p_2^{(0)}(x, \eta^{(0)}, z, t) = p_1^{(0)}(x, \eta^{(0)}, z, t)$ with

$$\left.\begin{aligned} p_1^{(0)} &= -\varrho_1 g(y - \eta^{(0)}) + \varrho_2 g(d^{(0)} - \eta^{(0)}), \\ p_2^{(0)} &= -\varrho_2 g(y - d^{(0)}). \end{aligned}\right\} \tag{10.31}$$

It is clear from the form of $p^{(0)}$ why the shallow-water approximation is sometimes called the hydrostatic approximation.

The usual equations for the first approximation to the shallow-water theory are those in which only the lower fluid is present. They may now be obtained by substituting (10.29) in the second and fourth equations in (10.27) and (10.28) in the fifth equation. They are (10.25), (10.29), and

$$\left.\begin{aligned} u_t^{(0)} + u^{(0)} u_x^{(0)} + w^{(0)} u_z^{(0)} + g\eta_x^{(0)} &= 0, \\ w_t^{(0)} + u^{(0)} w_x^{(0)} + w^{(0)} w_z^{(0)} + g\eta_z^{(0)} &= 0, \\ \eta_t^{(0)} + [u^{(0)}(\eta^{(0)} + h)]_x + [w^{(0)}(\eta^{(0)} + h)]_z &= 0. \end{aligned}\right\} \tag{10.32}$$

If one now collects the coefficients of ε^2 and the remaining coefficients of ε, one finds after some reduction

$$\left.\begin{aligned} & u_x^{(1)} + w_z^{(1)} + v_y^{(2)} = 0, \\ & u_t^{(1)} + u^{(1)} u_x^{(0)} + u^{(0)} u_x^{(1)} + w^{(1)} u_z^{(0)} + w^{(0)} u_z^{(1)} + v^{(1)} u_y^{(1)} + p_x^{(1)}/\varrho = 0, \\ & v_t^{(1)} + u^{(0)} v_x^{(1)} + w^{(0)} v_z^{(1)} + v^{(1)} v_y^{(1)} + p_y^{(1)}/\varrho = 0, \\ & w_t^{(1)} + u^{(1)} w_x^{(0)} + u^{(0)} w_x^{(1)} + w^{(1)} w_z^{(0)} + w^{(0)} w_z^{(1)} + v^{(1)} w_y^{(1)} + p_z^{(1)}/\varrho = 0, \\ & w_y^{(1)} = v_z^{(1)}, \quad u_z^{(1)} = w_x^{(1)}, \quad v_x^{(1)} = u_y^{(1)}, \\ & u^{(0)} \eta_x^{(1)} + u^{(1)} \eta_x^{(0)} + w^{(0)} \eta_z^{(1)} + w^{(1)} \eta_z^{(0)} + \eta_t^{(1)} - \eta^{(1)} v_y^{(1)} - v^{(2)} = 0 \\ & \qquad\qquad \text{for} \quad y = \eta^{(0)}(x, z, t), \\ & u_1^{(1)} h_x + w_1^{(1)} h_z + v_1^{(2)} = 0 \quad \text{for} \quad y = -h(x, z), \\ & p_2^{(1)} - p_1^{(1)} + \eta^{(1)}(p_{2y}^{(0)} - p_{1y}^{(0)}) = 0 \quad \text{for} \quad y = \eta^{(0)}(x, z, t). \end{aligned}\right\} \tag{10.33}$$

Some relations can be derived immediately from these equations. For the sake of brevity we introduce the following functions:

$$\begin{aligned} & A_i(x, z, t) = u_{ix}^{(0)} + w_{iz}^{(0)}, \qquad C_i(x, z, t) = (u_i^{(0)} \eta^{(0)})_x + (w_i^{(0)} \eta^{(0)})_z, \quad i = 1, 2, \\ & B_1(x, z, t) = -(u_1^{(0)} h)_x - (w_1^{(0)} h)_z, \qquad B_2 = C_2 - C_1 + B_1. \end{aligned}$$

Eqs. (10.28) may then be written

$$v_i^{(1)} = -y A_i + B_i, \qquad i = 1, 2. \tag{10.28'}$$

Then the fifth, first, and third equations of (10.33) give

$$\left.\begin{aligned} u^{(1)} &= -\tfrac{1}{2} y^2 A_x + y B_x + r(x, z, t), \\ w^{(1)} &= -\tfrac{1}{2} y^2 A_z + y B_z + s(x, z, t), \\ r_z &= s_x, \\ v^{(2)} &= \tfrac{1}{6} y^3 (A_{xx} + A_{zz}) - \tfrac{1}{2} y^2 (B_{xx} + B_{zz}) - y(r_x + s_z) + l(x, z, t), \\ p^{(1)}/\varrho &= \tfrac{1}{2} y^2 [A^2 + u^{(0)} A_x + w^{(0)} A_z + A_t] + \\ &\quad + y[A B + u^{(0)} B_x + w^{(0)} B_z + B_t] + q(x, z, t), \end{aligned}\right\} \tag{10.34}$$

where we have suppressed the subscripts indicating the fluid. The rest of the equations and the boundary conditions are still available to determine the unknown functions. We carry this out only for the case the upper fluid is missing. Then the last condition in (10.32) becomes $p^{(1)}(x, \eta^{(0)}, z, t) = \varrho g \eta^{(1)}$, which allows one to determine $q(x, z, t)$ after $\eta^{(1)}$ is found. The next-to-the-last condition in (10.32) determines $l(x, z, t)$. The equations for r, s and $\eta^{(1)}$ are

$$\left.\begin{aligned}
&u^{(0)} r_x + w^{(0)} r_z + r_t + u^{(0)}_x r + u^{(0)}_z s = - q_x - B\,B_x,\\
&u^{(0)} s_x + w^{(0)} s_z + s_t + w^{(0)}_x r + w^{(0)}_z s = - q_z - B\,B_z,\\
&u^{(0)} \eta^{(1)}_x + w^{(0)} \eta^{(1)}_z + \eta^{(1)}_t + A\,\eta^{(1)} = \left[v^{(2)} - u^{(1)} \eta^{(0)}_x - w^{(1)} \eta^{(0)}_z\right]_{y=\eta^{(0)}},\\
&r_z = s_x,
\end{aligned}\right\}\qquad(10.35)$$

where

$$\begin{aligned}
q(x, z, t) &= g\,\eta^{(1)} - \tfrac{1}{2} \eta^{(0)2} \left[A^2 + u^{(0)} A_x + w^{(0)} A_z + A_t\right] - \\
&\quad - \eta^{(0)} \left[A\,B + u^{(0)} B_x + w^{(0)} B_z + B_t\right],\\
l(x, z, t) &= - \left[u^{(1)} h_x + w^{(1)} h_z\right]_{y=-h} - \tfrac{1}{6} \eta^{(0)3} \left[A_{xx} + A_{zz}\right] + \\
&\quad + \tfrac{1}{2} \eta^{(0)2} \left[B_{xx} + B_{zz}\right] - \eta^{(0)} (r_x + s_z).
\end{aligned}$$

The solutions to these equations give the second-order corrections to the first-order shallow-water theory.

The equations resulting from the coefficients of ε^3 have been given by KELLER (1948) for two dimensions, but will not be reproduced here.

The Eqs. (10.32) for the first-order theory are nonlinear. In the theory of tides and seiches it is customary to simplify further by linearizing them in a manner similar to that used in deriving the equations for the infinitesimal-wave theory. Let $y = 0$ be the surface of the undisturbed water and assume that one may make further expansions in a small parameter α: $u^{(0)} = \alpha u^{(01)} + \cdots$, $w^{(0)} = \alpha w^{(01)} + \cdots$, $\eta^{(0)} = \alpha \eta^{(01)}, \ldots$. After some easy manipulations one finds for the linearized approximation to (10.32) the equations

$$\left.\begin{aligned}
u^{(01)}_t + g\,\eta^{(01)}_x = 0, \qquad w^{(01)}_t + g\,\eta^{(01)}_z = 0,\\
\eta^{(01)}_{tt} - g\left[\eta^{(01)}_x h\right]_x - g\left[\eta^{(01)}_z h\right]_z = 0.
\end{aligned}\right\}\qquad(10.36)$$

If the bottom is flat, the equation for $\eta^{(01)}$ becomes the simple wave equation.

D. Theory of infinitesimal waves.

This chapter will deal with special solutions of the linearized equations derived in Sect. 10α. This approximate theory has been very fruitful in its application to problems with various boundary configurations; the linear character of both the equations and boundary conditions allows one to use easily found simple solutions to construct other solutions satisfying special boundary conditions. The derivation of the equations in Sect. 10α suggests the limitations of their use in physical problems: If L and V are a typical length and velocity associated with the physical problem, then, when the perturbation parameter ε is small, the surface elevation and velocities (or their deviation from a uniform flow) should be small with respect to L and V respectively. The smallness may not be uniform, but the quantities in question should approach zero point-wise with ε except at singular points.

11. The fundamental equations. With few exceptions, this chapter will be concerned with the solution of a problem in potential theory. Let the (x, z)-plane be at the undisturbed free surface. We shall be seeking a function $\Phi(x, y, z, t)$, the velocity potential of the motion, satisfying the conditions [cf. Eq. (10.10)]

$$\left.\begin{aligned}&\Delta\Phi = \Phi_{xx} + \Phi_{yy} + \Phi_{zz} = 0\\&\Phi_{tt}(x,0,z,t) + g\,\Phi_y(x,0,z,t) = -\varrho^{-1}\bar{p}_t(x,z,t),\\&\Phi_n = V_n \quad \text{on solid boundaries},\end{aligned}\right\} \tag{11.1}$$

where $\Delta\Phi = 0$ is to be satisfied at all nonsingular points of the fluid in the region $y<0$ and V_n is the normal velocity of the solid boundary at a given point. $\bar{p}(x,z,t)$ is a given pressure distribution on the free surface; in many problems it will be 0. The form of the free surface is given by:

$$\eta(x,z,t) = -\frac{1}{g}\,\Phi_t(x,0,z,t) - \frac{1}{\varrho g}\,\bar{p}(x,z,t). \tag{11.2}$$

Two special cases occur frequently. If the motion is steady in a coordinate system moving with constant velocity c in the x-direction, then with x, y, z as moving coordinates, the free-surface boundary condition and equation of the surface are given by [cf. Eq. (10.15)]

$$\left.\begin{aligned}&\varphi_y(x,0,z) + \frac{c^2}{g}\,\varphi_{xx}(x,0,z) = \frac{c}{\varrho g}\,\bar{p}_x(x,z),\\&\eta(x,z) = \frac{c}{g}\,\varphi_x(x,0,z) - \frac{1}{\varrho g}\,\bar{p}(x,z).\end{aligned}\right\} \tag{11.3}$$

If Φ and $\bar{p}$ are harmonic functions of the time, i.e.

$$\Phi(x,y,z,t) = \varphi_1(x,y,z)\cos\sigma t + \varphi_2(x,y,z)\sin\sigma t = \operatorname{Re}\varphi(x,y,z)\,e^{-i\sigma t},$$

where

$$\varphi(x,y,z) = \varphi_1(x,y,z) + i\,\varphi_2(x,y,z)$$

and similarly for $\bar{p}$, then the free-surface condition and equation of the surface become

$$\left.\begin{aligned}&\varphi_{1y}(x,0,z) - \frac{\sigma^2}{g}\,\varphi_1(x,0,z) = -\frac{\sigma}{\varrho g}\,\bar{p}_2(x,z),\\&\varphi_{2y}(x,0,z) - \frac{\sigma^2}{g}\,\varphi_2(x,0,z) = \frac{\sigma}{\varrho g}\,\bar{p}_1(x,z),\\&\eta(x,z,t) = \frac{\sigma}{g}\,[\varphi_1(x,0,z)\sin\sigma t - \varphi_2(x,0,z)\cos\sigma t] -\\&\qquad - \frac{1}{\varrho g}\,[\bar{p}_1(x,z)\cos\sigma t + \bar{p}_2(x,z)\sin\sigma t].\end{aligned}\right\} \tag{11.4}$$

In the few cases where we consider superposed fluids, viscous fluids or surface tension, we shall refer back to Sect. 10 for the equations.

Use of complex variables. For two-dimensional irrotational motion, it is frequently advantageous to use complex variables. Let

$$z = x + i\,y, \qquad f(z,t) = \Phi(x,y,t) + i\,\Psi(x,y,t),$$

where Φ and Ψ are velocity potential and stream function, respectively. (It should be clear from context whether z is being used for $x+iy$ or one of the horizontal coordinates.) Since the equations relating Φ and Ψ,

$$\Phi_x = \Psi_y, \qquad \Phi_y = -\Psi_x,$$

are just the Cauchy-Riemann equations, the function $f(z,t)$ is an analytic function of z for all points z for which Φ_x and Φ_y exist. $f(z,t)$ will be called the "complex potential". The "complex velocity" is given by

$$w(z, t) = f'(z, t) = u - i v.$$

The boundary condition at the free surface in (11.1) can be expressed in the following equation in $f(z, t)$:

$$\mathrm{Re}\left\{i g f'(z, t) + \frac{d^2}{dt^2} f(z, t)\right\} = -\frac{1}{\varrho} p_t(x, t) \quad \text{for } y = 0. \tag{11.5}$$

The first equation of (11.3) becomes

$$\mathrm{Re}\{i g f'(z) + c^2 f''(z)\} = \frac{c}{\varrho} p'(x) \quad \text{for } y = 0. \tag{11.6}$$

However, Eq. (10.19) shows that this may also be taken in the form

$$\mathrm{Re}\{i g f(z) + c^2 f'(z)\} = \frac{c}{\varrho} p(x) \quad \text{for } y = 0.$$

If one may express $f(z, t) = f_1(z) \cos \sigma t + f_2(z) \sin \sigma t$, then the first of Eqs. (11.4) becomes

$$\mathrm{Re}\{i g f_k'(z) - \sigma^2 f_k(z)\} = (-1)^k \frac{\sigma}{\varrho} p_{k-(-1)^k}(x) \quad \text{for } y = 0,\ k = 1, 2. \tag{11.7}$$

We note that in order to express $f(z, t)$ in a manner analogous to that used for Φ immediately preceding (11.4) one must introduce a second complex unit j which does not "interact" with i. Thus let $f(z) = f_1(z) + j f_2(z)$. Then $f(z, t) = \mathrm{Re}_j f(z)\, e^{-j\sigma t}$.

If $f(z)$ is an analytic function satisfying any one of the conditions (11.5) to (11.7) with $p \equiv 0$, then $f^{(n)}(z)$ will also satisfy it.

12. Other boundary conditions. The boundary conditions given in Sect. 11 will not ordinarily be sufficient to ensure a unique solution to the problems in which the fluid occupies an unbounded region. An additional condition at infinity must be imposed upon the potential function. In certain cases the proper additional condition is fairly clear from the physical problem. For example, for a body moving steadily in an infinite ocean undisturbed except for the body, it seems reasonable to impose the condition that the fluid motion vanish far ahead of and far below the body. For the fluid motion produced by a stationary but steadily oscillating body, it seems reasonable to impose vanishing of the motion far below the body, but outgoing waves at infinity on all sides, if the body does not extend to infinity in some horizontal direction, the so-called "radiation condition".

If the body is not bounded in a horizontal direction, one may easily see that the radiation condition stated above cannot be expected to be satisfied. For example, suppose that waves are being generated by some type of oscillation of a vertical half-plane, say $z = 0$, $x > 0$, in which the oscillatory motion of the half-plane is independent of x. Then one will expect the generated waves to behave like outgoing plane waves from the two sides of the plane as $x \to \infty$; these will not satisfy the radiation condition in the direction Ox. On the other hand, one might expect that the influence of the edge at $x = 0$ would show up as waves satisfying the radiation condition. The formulation of proper boundary conditions in situations of this sort has been discussed by Peters and Stoker (1954); see also Stoker (1956, 1957, p. 109ff).

In diffraction problems one customarily prescribes the form of an incoming wave and then seeks the scattered wave. The preceding remarks concerning the boundary conditions for waves generated by an oscillating body apply also to the scattered wave.

In more complicated physical situations it is not always clear what boundary conditions should be imposed at infinity, and errors have been made. For example, for a body which is both oscillating with a fixed frequency σ and moving with a steady average velocity c, one might reasonably expect no motion far ahead if c is large, but a radiation condition if c is small. However, the formulation of the boundary condition cannot be completed until the problem is partly solved.

The proper formulation of the boundary conditions at infinity can frequently be obtained by a method recommended by HAVELOCK (1917, 1949a) and used also by BRARD (1948a, b), STOKER (1953, 1954), STOKER and PETERS (1957), DE PRIMA and WU (1957), WU (1957) and others. It consists in formulating an initial-value problem for which the desired steady-state problem is the limit as $t \to \infty$. For the initial-value problem the boundary condition at infinity is that the fluid motion vanishes everywhere. However, even though this procedure may produce the desired solution, it is not always obvious what boundary conditions at infinity in the steady-state problem would have produced it.

13. Some mathematical solutions. Some of the mathematical solutions to be derived in this section will provide solutions, without further modification, to certain physical problems; others, although apparently not acceptable physically, will provide fundamental solutions which can be used in constructing solutions to other more complicated physical problems. In all cases the fluid is assumed unbounded in a horizontal direction and either infinitely deep or with a horizontal bottom $y = -h$; the pressure on the free surface is taken to be zero everywhere. The solutions without singularities are obtained by the method of separation of variables, and are all harmonic in t. It will not be necessary to carry along the subscripts of (11.4).

α) Separation of the y-variable. Assume that one may express φ by

$$\varphi(x, y, z) = Y(y)\, \varphi(x, z).$$

Then $\Delta_3\varphi = \varphi_{xx} + \varphi_{yy} + \varphi_{zz} = 0$ becomes, after separation,

$$\Delta_2\varphi + A\,\varphi = 0, \qquad Y'' - A\,Y = 0.$$

The two cases $A = m^2 > 0$ and $A = -m^2 < 0$ lead to different solutions.

$A > 0$. In this case $\varphi(x, z)$ satisfies the wave equation

$$\Delta_2\varphi + m^2\,\varphi = 0$$

and Y is given by

$$Y = a\,e^{my} + b\,e^{-my}.$$

If the fluid is infinitely deep and $\varphi_y(x, y, z)$ is to remain bounded as $y \to -\infty$, one must have $b = 0$. Then condition (11.4) requires

$$m = \frac{\sigma^2}{g}$$

and $\varphi(x, y, z)$ is of the form

$$\varphi(x, y, z) = e^{\sigma^2 y/g}\, \varphi(x, z). \tag{13.1}$$

If the fluid is of finite depth h, the boundary condition $\varphi_y(x, -h, z) = 0$ requires Y to take the form

$$Y = a \cosh m\,(y + h)$$

and condition (11.4) becomes

$$m \tanh m h = \frac{\sigma^2}{g},$$

an equation with two real solutions, say $\pm m_0$. In this case, one has

$$\varphi(x, y, z) = \cosh m_0 (y + h)\, \varphi(x, z). \tag{13.2}$$

We note that, if $h_1 < h_2$ then $\sigma^2/g < m_0^{(2)} < m_0^{(1)}$. Also $m_0/h^{\frac{1}{2}} \to \sigma/g^{\frac{1}{2}}$ as $h \to 0$ and $m_0 \to \sigma^2/g$ as $h \to \infty$.

$A < 0$. In this case $\varphi(x, z)$ satisfies

$$\Delta_2 \varphi - m^2 \varphi = 0$$

and Y is given by

$$Y = a \cos m y + b \sin m y.$$

Condition (11.4) restricts Y further to

$$Y = C\left(m \cos m y + \frac{\sigma^2}{g} \sin m y\right).$$

If the fluid is infinitely deep, requiring φ_y to remain bounded imposes no further restriction. If the fluid is of depth h, then $\varphi_y(x, -h, z) = 0$ requires m to satisfy the equation

$$m \tan m h = -\frac{\sigma^2}{g},$$

an equation with an infinite number of real solutions, $\pm m_1, \pm m_2, \ldots$. In this latter case one may conveniently take Y in the form

$$Y = C \cos m (y + h).$$

The roots m_k satisfy $\frac{1}{2}(2k - 1)\, \pi/h < m_k < k\pi/h$. For fixed h, $m_k h \to k\pi$ as $k \to \infty$; for fixed k, $m_k h \to k\pi$ as $h \to 0$, and $m_k h \to \frac{1}{2}(2k - 1)\, \pi$ as $h \to \infty$.

For these two cases one finds then for $\varphi(x, y, z)$ the forms:

infinite depth:

$$\varphi(x, y, z) = C\left(m \cos m y + \frac{\sigma^2}{g} \sin m y\right) \varphi(x, z); \tag{13.3}$$

finite depth:

$$\varphi(x, y, z) = C \cos m_i (y + h)\, \varphi(x, z). \tag{13.4}$$

β) Further separation of variables. We now assume $\varphi(x, z) = X(x)\, Z(z)$ and substitute in each of the two equations for φ given above.

$A > 0$. In this case substitution in $\Delta \varphi + m^2 \varphi = 0$ gives

$$X'' + (m^2 - k^2)\, X = 0, \qquad Z'' + k^2 Z = 0.$$

(The equations obtained by replacing k^2 by $-k^2$ will give the solution obtained below for $A < 0$, with x and z interchanged.) The solution for Z is

$$Z = f \cos k z + g \sin k z = B \cos(k z + \gamma).$$

The solution for X depends upon the sign of $m^2 - k^2$:

$$k^2 < m^2: \quad X = c \cos x \sqrt{m^2 - k^2} + d \sin x \sqrt{m^2 - k^2};$$

$$k^2 = m^2: \quad X = c\, x + d;$$

$$k^2 > m^2: \quad X = c\, e^{x\sqrt{k^2 - m^2}} + d\, e^{-x\sqrt{k^2 - m^2}}.$$

$A<0$. Substitution in $\Delta_2\varphi - m^2\varphi = 0$ gives

$$X'' - (k^2+m^2)\,X = 0, \qquad Z'' + k^2 Z = 0,$$

which gives Z as above and

$$X = c\,\mathrm{e}^{x\sqrt{k^2+m^2}} + d\,\mathrm{e}^{-x\sqrt{k^2+m^2}}.$$

(Substituting $-k^2$ for k^2 would give the solutions corresponding to $A>0$ with x and z interchanged.) We may accumulate the preceding results to obtain the following fundamental solutions:

for infinite depth:

$$\left.\begin{aligned}
&\mathrm{e}^{\nu y}\left(a\cos x\sqrt{\nu^2-k^2}+b\sin x\sqrt{\nu^2-k^2}\right)\cos(kz+\gamma)\cos(\sigma t+\tau), \quad k^2<\nu^2,\\
&\mathrm{e}^{\nu y}(a x+b)\cos(kz+\gamma)\cos(\sigma t+\tau), \quad k^2=\nu^2,\\
&\mathrm{e}^{\nu y}\left(a\,\mathrm{e}^{x\sqrt{k^2-\nu^2}}+b\,\mathrm{e}^{-x\sqrt{k^2-\nu^2}}\right)\cos(kz+\gamma)\cos(\sigma t+\tau), \quad k^2>\nu^2,\\
&(m\cos m y+\nu\sin m y)\left(a\,\mathrm{e}^{x\sqrt{k^2+m^2}}+b\,\mathrm{e}^{-x\sqrt{k^2+m^2}}\right)\cos(kz+\gamma)\cos(\sigma t+\tau),
\end{aligned}\right\} \quad (13.5)$$

where $\nu=\sigma^2/g$;

for finite depth:

$$\left.\begin{aligned}
&\cosh m_0(y+h)\left(a\cos x\sqrt{m_0^2-k^2}+b\sin x\sqrt{m_0^2-k^2}\right)\times\\
&\qquad\qquad\times\cos(kz+\gamma)\cos(\sigma t+\tau), \quad k^2<m_0^2,\\
&\cosh m_0(y+h)(a x+b)\cos(kz+\gamma)\cos(\sigma t+\tau), \quad k^2=m_0^2,\\
&\cosh m_0(y+h)\left(a\,\mathrm{e}^{x\sqrt{k^2-m_0^2}}+b\,\mathrm{e}^{-x\sqrt{k^2-m_0^2}}\right)\cos(kz+\gamma)\cos(\sigma t+\tau), \quad k^2>m_0^2,\\
&\cos m_i(y+h)\left(a\,\mathrm{e}^{x\sqrt{k^2+m_i^2}}+b\,\mathrm{e}^{-x\sqrt{k^2+m_i^2}}\right)\cos(kz+\gamma)\cos(\sigma t+\tau),
\end{aligned}\right\} \quad (13.6)$$

where

$$m_0\tanh m_0 h - \frac{\sigma^2}{g} = 0 \quad\text{and}\quad m_i\tan m_i h + \frac{\sigma^2}{g} = 0.$$

The corresponding solutions for two dimensions may be obtained by setting $k=0$ and deleting the second and third equations in each group.

For either set of solutions only the first in each is bounded for all values of the variables for which $y\leqq 0$ or $-h\leqq y\leqq 0$. For two-dimensional motion it has been shown by WEINSTEIN (1927, 1949) that the only function harmonic in $-h<y<0$ and satisfying (11.4) and $\varphi_y(x,-h)=0$ for which both φ and φ_y are bounded in $-h\leqq y\leqq 0$ is $\varphi = A\cosh m(y+h)\sin(mx+\alpha)$. KELDYSH (1935) and STOKER (1947, pp. 7—9) have proved a similar theorem for the lower half-plane: If φ and $\varphi_x^2+\varphi_y^2$ are bounded for $y\leqq 0$ as $x^2+y^2\to\infty$, the only φ satisfying (11.4) and harmonic everywhere in the half-plane $y\leqq 0$ is $A\,\mathrm{e}^{ky}\sin(kx+\alpha)$. WEINSTEIN's theorem has been generalized by JOHN (1950, p. 59) to three dimensions: If $\varphi(x,y,z)$ satisfies (11.4), $\varphi_y(x,-h,z)=0$,

$$\lim_{R\to\infty}\varphi(R\cos\alpha, y, R\sin\alpha)\,R^{-\frac{1}{2}}\,\mathrm{e}^{-m_1 R} = 0$$

and is harmonic everywhere in $-h\leqq y\leqq 0$, then $\varphi(x,y,z)$ is of the form (13.2) with $\varphi(x,z)$ an everywhere regular solution of

$$\Delta_2\varphi + m_0^2\varphi = 0.$$

The condition at infinity is necessary, as the solution derived below in (13.8),

$$\varphi = I_0(m_1 R)\cos m_1(y+h),$$

shows. The corresponding theorem for infinite depth was proved by KOCHIN (1940).

The equations for $\varphi(x, y)$ may also be separated in polar coordinates (R, α), $x = R\cos\alpha$, $z = R\sin\alpha$. We give only the solutions:

infinite depth:

$$\left.\begin{array}{l} e^{\nu y}\,[A\,J_n(\nu R) + B\,Y_n(\nu R)]\cos(n\alpha + d)\cos(\sigma t + \tau), \\ (m\cos m y + \nu\sin m y)\,[A\,I_n(m R) + B\,K_n(m R)]\cos(n\alpha + d)\cos(\sigma t + \tau), \end{array}\right\} \quad (13.7)$$

where $\nu = \sigma^2/g$ and n is an integer;

finite depth:

$$\left.\begin{array}{l} \cosh m_0(y + h)\,[A\,J_n(m_0 R) + B\,Y_n(m_0 R)]\cos(n\alpha + \delta)\cos(\sigma t + \tau), \\ \cos m_i(y + h)\,[A\,I_n(m_i R) + B\,K_n(m_i R)]\cos(n\alpha + \delta)\cos(\sigma t + \tau), \quad i \geqq 1, \end{array}\right\} \quad (13.8)$$

where $m_0\tanh m_0 h - \sigma^2/g = 0$, $m_i\tan m_i h + \sigma^2/g = 0$ and n is an integer. Here J_n, Y_n, I_n, K_n are Bessel functions (we use WATSON'S notation). Y_n and K_n are both singular at $R = 0$ but approach zero as $R \to \infty$; J_n and I_n are both finite at $R = 0$; J_n approaches zero as $R \to \infty$, I_n increases exponentially.

γ) Singular solutions. In this section we shall find solutions of the problems set in Sect. 11 which have singularities of simple type at a single point. We shall indicate proofs only for the case of simple sources, i.e. singularities of the for $[(x-\alpha)^2 + (y-b)^2 + (z-c)^2]^{-\frac{1}{2}}$ or $\log[(x-a)^2 + (z-b)^2]^{\frac{1}{2}}$. We shall consider first the case of a stationary source of pulsating strength, then the case of a moving source. Three-dimensional problems are treated first.

Source of pulsating strength in three dimensions. Let (a, b, c) be in the lower half-space. We wish to find a function

$$\Phi(x, y, z, t) = \varphi_1(x, y, z)\cos\sigma t + \varphi_2(x, y, z)\sin\sigma t$$

defined for $y \leqq 0$ except at (a, b, c) and satisfying

$$\left.\begin{array}{l} 1.\ \Delta\varphi_i = 0 \text{ except at } (a, b, c),\ i = 1, 2, \\ 2.\ \varphi_{iy}(x, 0, z) - \nu\varphi_i(x, 0, z) = 0,\ i = 1, 2,\ \nu = \dfrac{\sigma^2}{g}, \\ 3.\ \Phi(x, y, z, t) = r^{-1}\cos\sigma t + \Phi_0(x, y, z, t), \\ \quad \text{where } \Phi_0 \text{ is harmonic in the whole region } y < 0, \\ 4.\ \lim\limits_{y \to -\infty} \operatorname{grad}\varphi_i = 0,\ i = 1, 2, \\ 5.\ \lim\limits_{R \to \infty} \sqrt{R}\left(\dfrac{\partial\varphi_1}{\partial R} + \nu\varphi_2\right) = 0,\ \lim\limits_{R \to \infty} \sqrt{R}\left(\dfrac{\partial\varphi_2}{\partial R} - \nu\varphi_1\right) = 0. \end{array}\right\} \quad (13.9)$$

Here $r^2 = (x-a)^2 + (y-b)^2 + (z-c)^2$ and $R^2 = (x-a)^2 + (z-c)^2$. Condition 5, usually called the "radiation condition", requires the waves at infinity to ben progressing outwards and imposes a uniqueness which would not otherwise be present. However, other such conditions could be imposed.

We assume that a solution Φ can be found in the form

$$\Phi(x, y, z, t) = [r^{-1} + \varphi_0(x, y, z)]\cos\sigma t + \varphi_2(x, y, z)\sin\sigma t. \quad (13.10)$$

φ_2 will be determined at the end so as to satisfy 5. Denote the double Fourier transform in x and z of φ by $\tilde{\varphi}$:

$$\varphi(x, y, z) = \frac{1}{2\pi}\int\limits_0^\infty\int\limits_{-\pi}^{\pi} \tilde{\varphi}(k, \vartheta, y)\, e^{ik(x\cos\vartheta + z\sin\vartheta)}\, d\vartheta\, dk.$$

Then condition 1 applied to φ_0 becomes after transforming

$$\tilde{\varphi}_{0yy} - k^2 \tilde{\varphi}_0 = 0$$

or

$$\tilde{\varphi}_0 = A_0(k, \vartheta)\, e^{yk} \tag{13.11}$$

where we have used 4. to discard the other solution. From the known integral

$$(x^2 + y^2 + z^2)^{-\frac{1}{2}} = \frac{1}{2\pi} \int_0^\infty \int_{-\pi}^{\pi} e^{-k|y|}\, e^{ik(x\cos\vartheta + z\sin\vartheta)}\, d\vartheta\, dk \tag{13.12}$$

one may compute

$$\widetilde{r^{-1}} = e^{-k|y-b|}\, e^{-ik(a\cos\vartheta + c\sin\vartheta)}. \tag{13.13}$$

Substituting $\tilde{\varphi}_0 + \widetilde{r^{-1}}$ in the transform of condition 2 gives

$$A_0(k, \vartheta) = \frac{k+\nu}{k-\nu}\, e^{kb}\, e^{-ik(a\cos\vartheta + c\sin\vartheta)}. \tag{13.14}$$

We now have, formally,

$$\varphi_0(x, y, z) = \frac{1}{2\pi} \int_0^\infty \int_{-\pi}^{\pi} \frac{k+\nu}{k-\nu}\, e^{k(y+b)}\, e^{ik[(x-a)\cos\vartheta + (z-c)\sin\vartheta]}\, d\vartheta\, dk.$$

Since the integrand has a singularity at $k = \nu$, the integral is not meaningful without further definition. We shall interpret the integral as a Cauchy principal value. Then

$$\left.\begin{aligned} \varphi_1(x, y, z) &= \frac{1}{r} + \frac{1}{2\pi}\, \mathrm{PV} \int_0^\infty \int_{-\pi}^{\pi} \frac{k+\nu}{k-\nu}\, e^{k(y+b)}\, e^{ik[(x-a)\cos\vartheta + (z-c)\sin\vartheta]}\, d\vartheta\, dk, \\ &= \frac{1}{r} + \frac{1}{r_1} + \frac{\nu}{\pi}\, \mathrm{PV} \int_0^\infty \int_{-\pi}^{\pi} \frac{1}{k-\nu}\, e^{k(y+b)}\, e^{ik[(x-a)\cos\vartheta + (z-c)\sin\vartheta]}\, d\vartheta\, dk, \end{aligned}\right\} \tag{13.15}$$

where $r_1^2 = (x-a)^2 + (y+b)^2 + (z-c)^2$. The second equation may be derived easily from the first one by use of (13.12) suitably modified. φ_1 satisfies 1., 2. and 4.; φ_0 is harmonic in the whole region.

In order to satisfy 5. we shall first find the asymptotic form of φ_1 for large R. With polar coordinates

$$x - a = R\cos\alpha, \qquad z - c = R\sin\alpha,$$

one may write (13.15) as

$$\begin{aligned} \varphi_1(x, y, z) = \varphi_1(R, \alpha, y) &= \frac{1}{r} + \frac{1}{r_1} + \frac{\nu}{\pi}\, \mathrm{PV} \int_0^\infty \int_{-\pi}^{\pi} \frac{1}{k-\nu}\, e^{k(y+b)}\, e^{ikR\cos(\vartheta-\alpha)}\, d\vartheta\, dk, \\ &= \frac{1}{r} + \frac{1}{r_1} + \frac{4\nu}{\pi}\, \mathrm{PV} \int_0^\infty \int_0^{\frac{1}{2}\pi} \frac{1}{k-\nu}\, e^{k(y+b)} \cos(kR\cos\vartheta)\, d\vartheta\, dk, \\ &= \frac{1}{r} + \frac{1}{r_1} + \frac{4\nu}{\pi}\, \mathrm{PV} \int_0^\infty \int_0^1 \frac{1}{\sqrt{1-\lambda^2}}\, \frac{1}{k-\nu}\, e^{k(y+b)} \cos Rk\lambda\, d\lambda\, dk, \\ &= \frac{1}{r} + \frac{1}{r_1} + 2\nu\, \mathrm{PV} \int_0^\infty \frac{1}{k-\nu}\, e^{k(y+b)}\, J_0(kR)\, dk. \end{aligned}$$

In the next to the last equation we shall change the order of integration, write

$$\cos R\lambda k = \cos R\lambda\nu\cos R\lambda(k-\nu) - \sin R\lambda\nu\sin R\lambda(k-\nu)$$

and use the following theorem from the theory of Fourier integrals[1]: If $f(x)$ is a differentiable function in $[a, \infty]$, if $f''(x_0)$, $x_0 > a$, exists, and if $f(x)/x$ and $f'(x)/x$ are both absolutely integrable in $[a, \infty]$, then, as $R\to\infty$,

$$\int_a^\infty f(x)\frac{\sin R(x-x_0)}{x-x_0}\,dx = \pi f(x_0) + O\left(\frac{1}{R}\right), \quad \mathrm{PV}\int_a^\infty f(x)\frac{\cos R(x-x_0)}{x-x_0}\,dx = O\left(\frac{1}{R}\right). \tag{13.16}$$

Remembering that both r^{-1} and r_1^{-1} are $O(R^{-1})$, one finds

$$\varphi_1(x, y, z) = -4\nu\,\mathrm{e}^{\nu(y+b)}\int_0^1(1-\lambda^2)^{-\frac{1}{2}}\sin R\lambda\nu\,d\lambda + O(R^{-1}).$$

The asymptotic expansion of this integral is well known[2] and we may write

$$\varphi_1(x, y, z) = -2\pi\nu\,\mathrm{e}^{\nu(y+b)}\sqrt{\frac{2}{\pi R\nu}}\sin\left(R\nu-\frac{\pi}{4}\right) + O\left(\frac{1}{R}\right).$$

If we can find a harmonic function φ_2 satisfying 1., 2. and 4. and having the asymptotic behavior

$$\varphi_2(x, y, z) = 2\pi\nu\,\mathrm{e}^{\nu(y+b)}\sqrt{\frac{2}{\pi R\nu}}\cos\left(R\nu-\frac{\pi}{4}\right) + O\left(\frac{1}{R}\right),$$

then

$$\varphi_1\cos\sigma t + \varphi_2\sin\sigma t = -2\pi\nu\,\mathrm{e}^{\nu(y+b)}\sqrt{\frac{2}{\pi R\nu}}\sin\left(R\nu-\sigma t-\frac{\pi}{4}\right) + O\left(\frac{1}{R}\right)$$

will be a solution. The following function fulfils the requirements[3]:

$$\varphi_2(x, y, z) = 2\pi\nu\,\mathrm{e}^{\nu(y+b)}J_0(R\nu).$$

We note in passing that φ_1 has the same asymptotic behavior as

$$-2\pi\nu\,\mathrm{e}^{\nu(y+b)}Y_0(R\nu).$$

The final result is

$$\left.\begin{aligned}\Phi(x, y, z, t) &= \left[\frac{1}{r} + \mathrm{PV}\int_0^\infty\frac{k+\nu}{k-\nu}\,\mathrm{e}^{k(y+b)}J_0(kR)\,dk\right]\cos\sigma t + \\ &\quad + 2\pi\nu\,\mathrm{e}^{\nu(y+b)}J_0(\nu R)\sin\sigma t, \qquad \nu = \sigma^2/g.\end{aligned}\right\} \tag{13.17}$$

Haskind (1954), using a derivation having some similarity to that used below for the two-dimensional case, has found the following form for Φ:

$$\left.\begin{aligned}\Phi(x, y, z, t) = \left[\frac{1}{r} + \frac{1}{r_1} + 2\nu\,\mathrm{e}^{\nu y}\int_\infty^y\frac{\mathrm{e}^{-\nu y}}{r_1}\,dy - 2\pi\nu\,\mathrm{e}^{\nu(y+b)}Y_0(\nu R)\right]\cos\sigma t + \\ + 2\pi\nu\,\mathrm{e}^{\nu(y+b)}J_0(\nu R)\sin\sigma t.\end{aligned}\right\} \tag{13.17'}$$

It is sometimes convenient to use the complex form for the potential, $\varphi\,\mathrm{e}^{-i\sigma t}$, with

$$\varphi(x, y, z) = \frac{1}{r} + \mathrm{PV}\int_0^\infty\frac{k+\nu}{k-\nu}\,\mathrm{e}^{k(y+b)}J_0(kR)\,dk + i\,2\pi\nu\,\mathrm{e}^{\nu(y+b)}J_0(\nu R), \tag{13.17''}$$

---

[1] See, e.g.: S. Bochner, Vorlesungen über Fouriersche Integrale, Leipzig, 1932, ch. I and § 8.

[2] See, e.g., A. Erdélyi: Asymptotic expansions, p. 43. Dover, New York 1956.

[3] See the first Eq. (13.7) and G. N. Watson: Bessel functions, p. 199. Cambridge 1949.

for then $\operatorname{Re}\varphi\, e^{-i\sigma t}$ gives (13.17) and $\operatorname{Im}\varphi\, e^{-i\sigma t}$ the source potential for an outgoing wave with singularity of the form $r^{-1}\sin\sigma t$. Eq. (13.17′) may be written analogously. By deforming the path of integration in a familiar way one may also express $\varphi(x, y, z)$ in the form [cf. HAVELOCK (1942, 1955)]:

$$\left.\begin{aligned}\varphi(x,y,z) = \frac{1}{r} + \frac{1}{r_1} - \frac{4\nu}{\pi}\int_0^\infty [\nu\cos k(y+b) - k\sin k(y+b)]\frac{K_0(kR)}{k^2+\nu^2}\,dk\, - \\ - 2\pi\nu\, e^{\nu(y+b)}\, Y_0(\nu R) + i\,2\pi\nu\, e^{\nu(y+b)}\, J_0(\nu R).\end{aligned}\right\}\quad(13.17''')$$

In the analogous problem for finite depth h one replaces 4. by 4′. $\varphi_y(x, -h, z) = 0$ and proceeds somewhat similarly. However, in order to satisfy 4′. it is convenient to look for a solution in the form

$$\Phi(x,y,z,t) = [r^{-1} + r_2^{-1} + \varphi_0(x,y,z)]\cos\sigma t + \varphi_2(x,y,z)\sin\sigma t,$$

where

$$r_2^2 = (x-a)^2 + (y+2h+b)^2 + (z-c)^2.$$

Eq. (13.11) then becomes

$$\tilde{\varphi}_0 = A_0(k,\vartheta)\cosh k(y+h)$$

and (13.14), now more complicated because of r^{-1} and r_2^{-1}, becomes

$$A_0(k,\vartheta) = \frac{2(k+\nu)\, e^{-kh}\cosh k(b+h)}{k\sinh kh - \nu\cosh kh}\, e^{-ik(a\cos\vartheta + c\sin\vartheta)}.$$

The final formula for the velocity potential is

$$\left.\begin{aligned}&\Phi(x,y,z,t) \\ &= \left[\frac{1}{r} + \frac{1}{r_2} + \mathrm{PV}\int_0^\infty \frac{2(k+\nu)\, e^{-kh}\cosh k(b+h)\cosh k(y+h)}{k\sinh kh - \nu\cosh kh} J_0(kR)\,dk\right]\cos\sigma t\, + \\ &+ \frac{2\pi(m_0+\nu)\, e^{-m_0 h}\sinh m_0 h\cosh m_0(b+h)\cosh m_0(y+h)}{\nu h + \sinh^2 m_0 h} J_0(m_0 R)\sin\sigma t,\end{aligned}\right\}\quad(13.1)$$

where $m_0\tanh m_0 h - \nu = 0$, $\nu = \sigma^2/g$. The form of the last term of (13.18) may be altered by using the identities

$$\frac{e^{-m_0 h}\sinh m_0 h}{\nu h + \sinh^2 m_0 h} = \frac{2e^{-m_0 h}\cosh m_0 h}{2m_0 h + \sinh 2m_0 h} = \frac{m_0 - \nu}{m_0^2 h - \nu^2 h + \nu}.$$

JOHN (1950, p. 95) has derived the following series for Φ, the analogue of (13.17‴)

$$\left.\begin{aligned}\Phi(x,y,z,t) = 2\pi\frac{\nu^2 - m_0^2}{h m_0^2 - h\nu^2 + \nu}\cosh m_0(y+h)\cosh m_0(b+h)\, \times \\ \times [Y_0(m_0 R)\cos\sigma t - J_0(m_0 R)\sin\sigma t]\, + \\ + 4\sum_{k=1}^\infty \frac{m_k^2 + \nu^2}{h m_k^2 + h\nu^2 - \nu}\cos m_k(y+h)\cos m_k(b+h)\, K_0(m_k R)\cos\sigma t,\end{aligned}\right\}\quad(13.19)$$

where m_k, $k > 0$, are the positive real roots of $m\tan mh + \nu = 0$. Either expression may also be given in complex form as in (13.17″).

Potential functions satisfying the condition (13.9), but with $r^{-1}\cos\sigma t$ in 3. replaced by a higher-order singularity have been given by THORNE (1953) and HAVELOCK (1955). In fact, THORNE gives a rather complete census of the possible singular solutions for both two and three dimensions and for finite and infinite depth. Included are series expansions as well as integrals. For infinite depth the general expression which includes (13.17) is

$$\left.\begin{aligned}\Phi(x,y,z,t) = \left[\frac{P_n^m(\cos\Theta)}{r^{n+1}} + \frac{(-1)^m}{(n-m)!}\,\mathrm{PV}\int_0^\infty \frac{k+\nu}{k-\nu}\,k^n\,\mathrm{e}^{k(y+b)}\,J_m(kR)\,dk\right]\times\\ \times\cos m\alpha\cos\sigma t + \frac{(-1)^m}{(n-m)!}\,2\pi\nu^{n+1}\,\mathrm{e}^{\nu(y+b)}\,J_m(\nu R)\cos m\alpha\sin\sigma t,\end{aligned}\right\}\quad(13.20)$$

where $\cos\Theta = (y-b)/r$, $x = R\cos\alpha$, $z = R\sin\alpha$. Here P_n^m are the associated Legendre polynomials defined by

$$P_n^m(\mu) = (1-\mu^2)^{m/2}\,\frac{d^m}{d\mu^m}\,P_n(\mu), \qquad m \leqq n.$$

The asymptotic behavior of (13.20) is given by

$$\Phi(x,y,z,t) = \frac{(-1)^{m+1}}{(n-m)!}\,2\pi\nu^{n+1}\,\mathrm{e}^{\nu(y+b)}\sqrt{\frac{2}{\pi\nu R}}\,\sin\left(\nu R - \sigma t - \frac{2m+1}{4}\,\pi\right) + O\left(\frac{1}{R}\right).$$

It has been pointed out by both HAVELOCK (1955) and MACCAMY (1954) that solutions can be constructed which vanish much faster than this at infinity. Let the function of (13.20) be denoted by Φ_n. Then $\Phi_{n+1} - \nu(n-m+1)^{-1}\,\Phi_n$ is the following function:

$$\left.\begin{aligned}\left[\frac{P_{n+1}^m(\cos\Theta)}{r^{n+2}} - \frac{\nu}{n-m+1}\,\frac{P_n^m(\cos\Theta)}{r^{n+1}} + (-1)^m\,\frac{P_{n+1}^m(-\cos\Theta_1)}{r_1^{n+1}} + \right.\\ \left. + (-1)^m\,\frac{\nu}{n-m+1}\,\frac{P_n^m(-\cos\Theta_1)}{r_1^{n+1}}\right]\cos m\alpha\cos\sigma t,\end{aligned}\right\}\quad(13.21)$$

where $\cos\Theta_1 = (y+b)/r_1$, $r_1^2 = (x-a)^2 + (y+b)^2 + (z-c)^2$. For $y=0$ and large R these solutions are $O(R^{-n-1})$ if m and n are both odd, $O(R^{-n-2})$ if one is even and one odd, and $O(R^{-n-3})$ if both are even. Although they have the form of standing waves, they satisfy the radiation condition because they decrease so rapidly with large R.

In addition to the papers cited above, one can find treatments of the submerged source of pulsating strength in KOCHIN (1940), HAVELOCK (1942), JOHN (1950, p. 92ff.), where a detailed discussion is given for the case of finite depth, HASKIND (1944), and LIU (1952). The definition of the improper integral in (13.15) and following is not always the same in these different treatments. In some cases the variable k is treated as complex and the path of integration deflected around the singularity $k=\nu$ by following a small semi-circle in the lower half of the k-plane. The radiation condition is then automatically satisfied if one writes Φ in the complex form $\varphi\,\mathrm{e}^{-i\sigma t}$, $\varphi = \varphi_1 + i\varphi_2$. Other treatments achieve the same end by introducing a "fictitious viscosity" $i\mu$ which has the effect of replacing the singularity at $k=\nu$ by one at $k=\nu+i\mu$ and thus placing the path of integration below the singularity. In the end one must find the limit of the solution as $\mu\to 0$. The fictitious viscosity has no relation to real viscosity and may be considered a mathematical device to enable one to interpret an improper integral in a suitable way (for the purpose it seems to be infallible).

Source and vortex of pulsating strength in two dimensions. The two-dimensional problem can be formulated analogously to (13.9), and solutions found in a similar manner. The fundamental singularities will now be of the form $\log r\cos\sigma t$, $r^{-n}\cos n\Theta\cos\sigma t$ and $r^{-n}\sin n\Theta\cos\sigma t$, $n=1,2,\ldots$. The results are given in the paper of THORNE (1953) cited earlier. We shall follow a different method here in order to illustrate the use of complex variables to solve such problems.

We shall consider simultaneously a source of strength Q and a vortex of intensity Γ at the point $c = a + ib$, $b < 0$. In the notation used at the end of Sect. 11, we shall be looking for a function $f(z, t)$ analytic in z and of the form

$$\left.\begin{aligned} f(z,t) &= \left[\frac{\Gamma + iQ}{2\pi i}\log(z-c) + f_0(z)\right]\cos\sigma t + f_2(z)\sin\sigma t, \qquad \operatorname{Im} c < 0, \\ &= f_1(z)\cos\sigma t + f_2(z)\sin\sigma t, \end{aligned}\right\} \tag{13.22}$$

where f_0 and f_2 have no singularities in the lower half-plane. In addition, f_1 and f_2 must each satisfy the free-surface condiction (11.7) which we write

$$\operatorname{Im}\{f_k'(x - i0) + i\nu f_k(x - i0)\} = 0, \qquad \nu = \sigma^2/g, \qquad k = 1, 2. \tag{13.23}$$

Condition 4 of (13.9) will be taken in the somewhat stronger form,

$$|f_k'| \leqq M \quad \text{for} \quad |z| \geqq m \quad \text{and} \quad \lim_{y\to-\infty} |f_k'| = 0, \tag{13.24}$$

where m and M are given constants. The radiation condition becomes:

$$\lim_{x\to\pm\infty} \operatorname{Re}\{f_1' \pm \nu f_2\} = 0, \qquad \lim_{x\to\pm\infty} \operatorname{Re}\{f_2' \mp \nu f_1\} = 0. \tag{13.25}$$

Following a method apparently originally due to LEVI-CIVITA (see TONOLO, 1913), but used frequently by KELDYSH (1935), KOCHIN (e.g., 1939), STOKER (1947), LEWY (1946) and others, we introduce the functions

$$A_k(z) = f_k'(z) + i\nu f_k(z). \tag{13.26}$$

Then (13.23) becomes

$$\operatorname{Im} A_k(x - i0) = 0, \qquad k = 1, 2, \tag{13.27}$$

and (13.22) becomes: the two functions

$$A_0(z) = A_1(z) - \frac{\Gamma + iQ}{2\pi i}\,\frac{1}{z-c} - \nu\,\frac{\Gamma + iQ}{2\pi}\log(z-c)$$

and $A_2(z)$ are both regular everywhere in the lower half-plane. A function $A(z)$ with $\operatorname{Im} A(x - i0) = 0$ may be continued into the upper half-plane by defining $A(x+iy) = \overline{A(x-iy)}$, $y > 0$, the bar indicating complex conjugate. Since A_2 is regular in the lower half-plane, the extended function will be regular in the whole plane. In addition, one may derive easily from (13.24) that $|A_2(z)| < C|z| + D$ for sufficiently large $|z|$: then, from the regularity of A_2, such an inequality holds in the whole half-plane and hence in the whole plane after reflection. It then follows from a known generalization of LIOUVILLE's Theorem[1] that $A_2(z) = az + b$, where a and b are constants. It follows from (13.27) that a and b are real. The differential equation

$$f_2'(z) + i\nu f_2(z) = az + b$$

has the solution

$$f_2(z) = C\,e^{-i\nu z} - \frac{ia}{\nu}\,z - \frac{ib}{\nu} + \frac{a}{\nu^2}.$$

The condition $\lim_{y\to-\infty} |f_2'| = 0$ requires $a = 0$. Thus, finally

$$f_2(z) = C_2\,e^{-i\nu z} + i B_2, \qquad B_2 \text{ real}.$$

One may set $B_2 = 0$ without loss of generality. Incidentally, this provides a proof of the theorem of STOKER and KELDYSH mentioned earlier [shortly after Eq. (13.6)].

[1] See C. CARATHÉODORY: Theory of functions of a complex variable, Vol. I, § 168. Chelsea, New York 1954.

The function $A_1(z)$, after extension into the upper half-plane, will consist of four singular terms plus a function regular in the whole complex plane, say $A_3(z)$:

$$A_1(z) = \frac{\Gamma+iQ}{2\pi i}\frac{1}{z-c} + \nu\frac{\Gamma+iQ}{2\pi}\log(z-c) - \frac{\Gamma-iQ}{2\pi i}\frac{1}{z-\bar{c}} + \\ + \nu\frac{\Gamma-iQ}{2\pi}\log(z-\bar{c}) + A_3(z).$$

Since A_1 satisfies (13.27), and the four singular terms taken together also have vanishing imaginary part for $y=0$, the same must hold for A_3. Hence A_3 must have the same form as A_2. Substituting the resulting expression for A_1 in (13.26), one has a differential equation for $f_1(z)$. The solution is

$$f_1(z) = \frac{\Gamma+iQ}{2\pi i}\log(z-c) + \frac{\Gamma-iQ}{2\pi i}\log(z-\bar{c}) - \\ - \frac{\Gamma-iQ}{\pi i}e^{-i\nu z}\int\limits_{\infty}^{z}\frac{e^{i\nu u}}{u-\bar{c}}du + C_1 e^{-i\nu z} + i B_1,$$

where B_1 is real and the path of integration is in the lower half-plane. As in the case f_2, we may set $B_1=0$. C_1 and C_2 must now be chosen to satisfy (13.25). Making use of

$$\int\limits_{-\infty}^{\infty}\frac{e^{i\nu u}}{u-\bar{c}}du = 2\pi i\, e^{i\nu\bar{c}},$$

one can show that

$$f_1' + \nu f_2 = -i\nu C_1 e^{-i\nu z} + \nu C_2 e^{-i\nu z} + O(z^{-1}) \quad \text{as} \quad x\to+\infty,$$
$$f_1' - \nu f_2 = -2i(\Gamma-iQ)e^{-i\nu(z-\bar{c})} - i\nu C_1 e^{-i\nu z} - \nu C_2 e^{-i\nu z} + O(z^{-1}) \quad \text{as} \quad x\to-\infty.$$

This gives

$$C_1 = -(\Gamma-iQ)e^{i\nu\bar{c}}, \qquad C_2 = -i(\Gamma-iQ)e^{i\nu\bar{c}}.$$

One may easily verify that this choice of C_1 and C_2 does produce outgoing waves.

If one makes the change of variable $\nu(u-z) = -k(z-\bar{c})$ in the integral term in f_1 and deforms the resulting path to Ox, one finds

$$-e^{-i\nu z}\int\limits_{\infty}^{z}\frac{e^{i\nu u}}{u-\bar{c}}du = \mathrm{PV}\int\limits_{0}^{\infty}\frac{e^{-ik(z-\bar{c})}}{k-\nu}dk + \pi i\, e^{-i\nu(z-\bar{c})}.$$

Substituting this in the expression for f_1, one finally obtains

$$\left.\begin{aligned} f(z,t) = \Big[&\frac{\Gamma+iQ}{2\pi i}\log(z-c) + \frac{\Gamma-iQ}{2\pi i}\log(z-\bar{c}) + \\ &+ \frac{\Gamma-iQ}{\pi i}\mathrm{PV}\int\limits_{0}^{\infty}\frac{e^{-ik(z-\bar{c})}}{k-\nu}dk\Big]\cos\sigma t - i(\Gamma-iQ)e^{-i\nu(z-\bar{c})}\sin\sigma t. \end{aligned}\right\} \tag{13.28}$$

Singularities of higher order may be found by differentiating (13.28) with respect to z. The expression for $f'(z,t)$ may be put into a somewhat different form by using

$$\frac{\Gamma-iQ}{2\pi i}\frac{1}{z-\bar{c}} = \frac{\Gamma-iQ}{2\pi}\int\limits_{0}^{\infty}e^{-ik(z-c)}dk.$$

Then

$$\left.\begin{aligned} f'(z,t) = &\left[\frac{\Gamma+iQ}{2\pi i}\,\frac{1}{z-c} - \frac{\Gamma-iQ}{2\pi}\,\mathrm{PV}\int_0^\infty \frac{k+\nu}{k-\nu}\,e^{-ik(z-\bar{c})}\,dk\right]\cos\sigma t - \\ &-\nu(\Gamma-iQ)\,e^{-i\nu(z-\bar{c})}\sin\sigma t. \end{aligned}\right\} \quad (13.29)$$

One may continue differentiating, using either form for $f'(z,t)$. Thus, from (13.29)

$$\left.\begin{aligned} f^{(n)}(z,t) = &\left[\frac{\Gamma+iQ}{2\pi i}\,\frac{(-1)^{n-1}(n-1)!}{(z-c)^n} - \right. \\ &\left. -\frac{\Gamma-iQ}{2\pi}(-i)^{n-1}\,\mathrm{PV}\int_0^\infty k^{n-1}\frac{k+\nu}{k-\nu}\,e^{-ik(z-\bar{c})}\,dk\right]\cos\sigma t - \\ &-\nu^n(-i)^{n-1}(\Gamma-iQ)\,e^{-i\nu(z-\bar{c})}\sin\sigma t. \end{aligned}\right\} \quad (13.30)$$

By setting $\Gamma=0$, $z-c=r\,e^{i(\frac{1}{2}\pi-\Theta)}=i r\,e^{-i\Theta}$ (rather than the conventional $r\,e^{i\Theta}$ in order to distinguish easily symmetrical from unsymmetrical solutions) and taking the appropriate real or imaginary part, one finds the following formulas for $\Phi(x,y,t)$:

$$\left.\begin{aligned} \Phi(x,y,t) = &\left[\frac{Q}{2\pi}\log\frac{r}{r_1} - \frac{Q}{\pi}\,\mathrm{PV}\int_0^\infty \frac{e^{k(y+b)}\cos k(x-a)}{k-\nu}\,dk\right]\cos\sigma t - \\ &-Q\,e^{\nu(y+b)}\cos\nu(x-a)\sin\sigma t, \\ \Phi(x,y,t) = &\left[\frac{Q}{2\pi}\,\frac{\cos n\Theta}{r^n} - \right. \\ &\left. -\frac{(-1)^{n-1}}{(n-1)!}\,\frac{Q}{2\pi}\,\mathrm{PV}\int_0^\infty k^{n-1}\frac{k+\nu}{k-\nu}\,e^{k(y+b)}\cos k(x-a)\,dk\right]\cos\sigma t - \\ &-\frac{(-1)^{n-1}}{(n-1)!}\,Q\,\nu^n\,e^{\nu(y+b)}\cos\nu(x-a)\sin\sigma t, \\ \Phi(x,y,t) = &\left[\frac{Q}{2\pi}\,\frac{\sin n\Theta}{r^n} + \right. \\ &\left. +\frac{(-1)^{n-1}}{(n-1)!}\,\frac{Q}{2\pi}\,\mathrm{PV}\int_0^\infty k^{n-1}\frac{k+\nu}{k-\nu}\,e^{k(y+b)}\sin k(x-a)\,dk\right]\cos\sigma t + \\ &+\frac{(-1)^{n-1}}{(n-1)!}\,Q\,\nu^n\,e^{\nu(y+b)}\sin\nu(x-a)\sin\sigma t. \end{aligned}\right\} \quad (13.31)$$

In the formula for the logarithmic singularity r_1 may be eliminated and the coefficient of $\cos\sigma t$ written as [see JOHN (1950, p. 100)]:

$$\frac{Q}{2\pi}\log r + \frac{Q}{2\pi}\,\mathrm{PV}\int_0^\infty\left[\frac{k+\nu}{(k-\nu)k}\,e^{k(y+b)}\cos k(x-a) + \frac{1}{k}\,e^{-k}\right]dk.$$

For water of finite depth the method used above does not work as conveniently because of the difficulty of formulating the boundary condition on the bottom, $\mathrm{Im}\,f'(x-ih)=0$, in terms of the function $A(z)$. However, it can be done, yielding a differential-difference equation for $f(z)$ which can be solved by use of

Laplace transforms[1]. The method used for the three-dimensional problem can also be carried through [see HASKIND (1942b), JOHN (1950), and THORNE (1953)].

It is convenient to separate the vortex from the source. The resulting functions are as follows:

vortex:

$$\left.\begin{aligned} f(z,t) = \Bigg[\frac{\Gamma}{2\pi i}\log(z-c) - \frac{\Gamma}{2\pi i}\log(z-c_2) - \\ -\frac{\Gamma}{\pi}\,\mathrm{PV}\int_0^\infty \frac{k+\nu}{k}\,\frac{e^{-kh}\sinh k(h+b)\sin k(z-a+ih)}{k\sinh kh-\nu\cosh kh}\,dk\Bigg]\cos\sigma t - \\ -\Gamma\frac{\nu+m_0}{m_0}\cdot\frac{e^{-m_0 h}\sinh m_0 h\sinh m_0(h+b)\sin m_0(z-a+ih)}{\nu h+\sinh^2 m_0 h}\sin\sigma t; \end{aligned}\right\} \tag{13.32}$$

source:

$$\left.\begin{aligned} f(z,t) = \Bigg[\frac{Q}{2\pi}\log(z-c) + \frac{Q}{2\pi}\log(z-c_2) - \frac{Q}{\pi}\log i h - \\ -\frac{Q}{\pi}\,\mathrm{PV}\int_0^\infty \left\{\frac{k+\nu}{k}\,\frac{e^{-kh}\cosh k(h+b)\cos k(z-a+ih)}{k\sinh kh-\nu\cosh kh} + \frac{e^{-kh}}{k}\right\}dk\Bigg]\cos\sigma t - \\ -Q\frac{\nu+m_0}{m_0}\,\frac{e^{-m_0 h}\sinh m_0 h\cosh m_0(h+b)\cos m_0(z-a+ih)}{\nu h+\sinh^2 m_0 h}\sin\sigma t. \end{aligned}\right\} \tag{13.33}$$

Here $c_2 = a - i b - 2 i h$. The remark following (13.18) concerning the form of the last term of that formula applies also here. The real part of either of these gives the corresponding potential function.

For the source, the integral representation and the series representation analogous to (13.19) are:

$$\left.\begin{aligned} \Phi(x,y,t) &= \Bigg[\frac{Q}{2\pi}\log\frac{r}{h} + \frac{Q}{2\pi}\log\frac{r_2}{h} - \\ &\quad -\frac{Q}{\pi}\,\mathrm{PV}\int_0^\infty \left\{\frac{k+\nu}{k}\,\frac{e^{-kh}\cosh k(h+b)\cosh k(y+h)\cos k(x-a)}{k\sinh kh-\nu\cosh kh} - \frac{e^{-kh}}{k}\right\}dk\Bigg]\cos\sigma t - \\ &\quad -Q\frac{\nu+m_0}{m_0}\,\frac{e^{-m_0 h}\sinh m_0 h\cosh m_0(h+b)\cosh m_0(y+h)\cos m_0(x-a)}{\nu h+\sinh^2 m_0 h}\sin\sigma t, \\ &= Q\frac{1}{m_0}\,\frac{m_0^2-\nu^2}{h m_0^2 - h\nu^2+\nu}\cosh m_0(y+h)\cosh m_0(b+h)\sin[m_0|x-a|-\sigma t] - \\ &\quad -Q\sum_{k=1}^\infty \frac{1}{m_k}\,\frac{m_k^2+\nu^2}{h m_k^2+h\nu^2-\nu}\cos m_k(y+h)\cos m_k(b+h)\,e^{-m_k|x-a|}\sin\sigma t. \end{aligned}\right\} \tag{13.34}$$

THORNE (1953) gives the potential functions for the higher-order singularities and the function for the logarithmic singularity in a form involving r and r_1 and hence more analogous to the one in (13.31). VOITSENYA (1958) has derived the complex potential for a source-vortex situated in an infinitely deep fluid of density ϱ_1 lying beneath another of density $\varrho_2 < \varrho_1$ and of thickness d.

Source of constant strength in uniform motion: three dimensions. We shall assume the source moving in the direction Ox with constant velocity u_0. Let (x, y, z) be coordinates in a system moving with velocity u_0 in direc-

[1] Cf. S. BOCHNER: Vorlesungen über Fouriersche Integrale, pp. 167—168. Leipzig 1932.

tion Ox and let the source be at (a, b, c), $b<0$. Then, from Sect. 11, we wish to find a function $\varphi(x, y, z)$ satisfying

$$\left.\begin{array}{ll}1. & \Delta\varphi = 0 \quad \text{except at } (a, b, c),\\ 2. & \varphi_{xx}(x, 0, z) + \nu\,\varphi_y(x, 0, z) = 0, \qquad \nu = g/u_0^2,\\ 3. & \varphi(x, y, z) = r^{-1} + \varphi_0(x, y, z),\\ & \qquad \text{where } \varphi_0 \text{ is harmonic in the region } y<0,\\ 4. & \lim\limits_{y\to-\infty} \operatorname{grad}\varphi = 0,\\ 5. & \lim\limits_{x\to+\infty} \operatorname{grad}\varphi = 0.\end{array}\right\} \tag{13.35}$$

For fluid of finite depth h, 4. is replaced by 4'. $\varphi_y(x, -h, z) = a$. Without condition 5, demanding vanishing of the motion far ahead of the source, the solution would not be unique. The profile of the free surface is obtained from $\eta(x, z) = u_0 g^{-1} \varphi_x(x, 0, z)$. Strictly speaking, the solution of (13.35) will represent a sink, i.e. a source of strength -1. However, we shall continue to call such solutions sources.

A solution to this problem may be obtained by methods very similar to those used for the source of pulsating strength. The details will not be repeated, but can be found in HAVELOCK (1932), SRETENSKII (1937), KOCHIN (1937), LUNDE (1951), PETERS and STOKER (1957), TIMMAN and VOSSERS (1955) and elsewhere. The result is

$$\left.\begin{aligned}\varphi(x, y, z) = \frac{1}{r} - \frac{1}{r_1} - \frac{4\nu}{\pi}\int_0^{\frac{1}{2}\pi} d\vartheta\, \mathrm{PV}\int_0^\infty \frac{e^{k(y+b)}\cos[k(x-a)\cos\vartheta]\cos[k(z-c)\sin\vartheta]}{k\cos^2\vartheta - \nu}\, dk -\\ - 4\nu\int_0^{\frac{1}{2}\pi} e^{\nu(y+b)\sec^2\vartheta}\sin[\nu(x-a)\sec\vartheta]\cos[\nu(z-c)\sin\vartheta\sec^2\vartheta]\sec^2\vartheta\, d\vartheta,\end{aligned}\right\} \tag{13.34}$$

where

$$r^2 = (x-a)^2 + (y-b)^2 + (z-c)^2, \quad r_1^2 = (x-a)^2 + (y+b)^2 + (z-c)^2, \quad \nu = g/u_0^2.$$

The potential functions for higher-order singularities are unwieldy and will not be given. The one corresponding to $r^{-n-1} P_n(\cos\Theta)$ can be easily obtained by n-fold differentiation with respect to y, if one remembers that

$$\frac{P_n(\cos\Theta)}{r^{n+1}} = \frac{(-1)^n}{n!}\frac{\partial^n}{\partial y^n}\left(\frac{1}{r}\right).$$

The dipole with axis in the direction Ox is obtained by differentiating (13.36) with respect to x and will be used later.

The velocity potential for a source moving in fluid of finite depth has been calculated by SRETENSKII (1937) and by HASKIND (1945b). The form given below is essentially that given by LUNDE (1951):

$$\left.\begin{aligned}\varphi(x, y, z) = \frac{1}{r} + \frac{1}{r_2} -\\ - \frac{4}{\pi}\int_0^{\frac{1}{2}\pi} d\vartheta\, \mathrm{PV}\int_0^\infty \frac{e^{-kh}\cosh k(y+h)\,[\cosh k(b+h)(k\cos^2\vartheta + \nu) - \nu]}{k\cos^2\vartheta\cosh kh - \nu\sinh kh} \times\\ \times \cos[k(x-a)\cos\vartheta]\cos[k(z-c)\sin\vartheta]\, dk -\\ - 4\int_{\vartheta_0}^{\frac{1}{2}\pi} \frac{e^{-k_0 h}\operatorname{sech} k_0 h\cosh k_0(y+h)\,[\cosh k_0(b+h)(k_0\cos^2\vartheta + \nu) - \nu]}{\cos^2\vartheta - \nu h\operatorname{sech}^2 k_0 h} \times\\ \times \sin[k_0(x-a)\cos\vartheta]\cos[k_0(z-c)\sin\vartheta]\, d\vartheta,\end{aligned}\right\} \tag{13.37}$$

where $k_0 = k_0(\vartheta)$ is the real positive root of

$$k_0 - \nu \sec^2\vartheta \tanh k_0 h = 0, \qquad \vartheta_0 < \vartheta < \tfrac{1}{2}\pi,$$

where $\vartheta_0 = \arccos\sqrt{\nu h}$ if $\nu h = gh/u_0^2 < 1$, $\vartheta_0 = 0$ if $\nu h \geqq 1$. As before, $r^2 = (x-a)^2 + (y-b)^2 + (z-c)^2$ and $r_2^2 = (x-a)^2 + (y+2h+b)^2 + (z-c)^2$. We note that $k_0(\vartheta) < \nu\sec^2\vartheta$, $k_0(\vartheta)\to 0$ as $\vartheta\to\vartheta_0$, $k_0(\vartheta)/\nu\sec^2\vartheta\to 1$ as $\vartheta\to\frac{1}{2}\pi$ and $k_0(\vartheta)\to\nu\sec^2\vartheta$ as $h\to\infty$. In the double integral the principal value is necessary only for $\vartheta_0<\vartheta<\frac{1}{2}\pi$, for the singularity does not occur in the denominator for $0\leqq\vartheta<\vartheta_0$. The part of the double integral with $0\leqq\vartheta<\vartheta_0$ approaches zero as $x\to\pm\infty$, so that no correction is necessary in order to satisfy condition 5. This is the explanation of the lower limit ϑ_0 in the second integral. In this integral the denominator vanishes only at $\vartheta=\vartheta_0$. One may verify that the integral is convergent by noting that

$$k_0'(\vartheta) = \frac{k_0 \sin 2\vartheta}{\cos^2\vartheta - \nu h\,\mathrm{sech}^2 k_0 h}$$

and rewriting it as an integral with respect to k_0. When $h\to\infty$, (13.37) reduces to a form of (13.36) in which r_1 is absorbed into the double integral.

For the stationary pulsating source the asymptotic form of the velocity potential for large R was found in the course of deriving the potential function. For the moving source of constant strength the asymptotic form is more difficult to compute. Since the form of the free surface, $\eta = g u_0^{-1}\varphi_x(x, 0, z)$ is of principal physical interest, we shall discuss the asymptotic form of φ_x instead of φ.

Introduce cylindrical coordinates $x-a=R\cos\alpha$, $z-c=R\sin\alpha$ into the x derivative of (13.36):

$$\left.\begin{aligned}\varphi_x(R,\alpha,y) &= \frac{-R\cos\alpha}{[R^2+(y-b)^2]^{\frac{3}{2}}} + \frac{R\cos\alpha}{[R^2+(y+b)^2]^{\frac{3}{2}}} + \\ &+ \frac{2\nu}{\pi}\int_0^{\frac{1}{2}\pi}\sec\vartheta\,d\vartheta\,\mathrm{PV}\int_0^\infty e^{k(y+b)}\frac{\sin[kR\cos(\vartheta-\alpha)]+\sin[kR\cos(\vartheta+\alpha)]}{k-\nu\sec^2\vartheta}k\,dk - \\ &- 2\nu^2\int_0^{\frac{1}{2}\pi} e^{\nu(y+b)\sec^2\vartheta}\{\cos[\nu R\sec^2\vartheta\cos(\vartheta-\alpha)] + \\ &\qquad + \cos[\nu R\sec^2\vartheta\cos(\vartheta+\alpha)]\}\sec^3\vartheta\,d\vartheta.\end{aligned}\right\} \quad (13.38)$$

For large R the first two terms taken together are $O(R^{-3})$. Apply the theorem (13.16) to the integral with respect to k. This gives, after combining with the second integral,

$$\left.\begin{aligned}\varphi_x(R,\alpha,y) &= 2\nu^2\int_0^{\frac{1}{2}\pi}\sec^3\vartheta\, e^{\nu(y+b)\sec^2\vartheta}\times \\ &\times\{\cos[\nu R\sec^2\vartheta\cos(\vartheta-\alpha)][-1+\mathrm{sgn}\cos(\vartheta-\alpha)] + \\ &+\cos[\nu R\sec^2\vartheta\cos(\vartheta+\alpha)][-1+\mathrm{sgn}\cos(\vartheta+\alpha)]\}\,d\vartheta + O(R^{-1}).\end{aligned}\right\} \quad (13.39)$$

Since φ_x is symmetric in α, we consider only $0\leqq\alpha\leqq\pi$. We have for $0\leqq\alpha\leqq\frac{1}{2}\pi$

$$\varphi_x(R,\alpha,y) = -4\nu^2\int_{\frac{1}{2}\pi-\alpha}^{\frac{1}{2}\pi}\sec^3\vartheta\, e^{\nu(y+b)\sec^2\vartheta}\times \cos[\nu R\sec^2\vartheta\cos(\vartheta+\alpha)]\,d\vartheta + O(R^{-1}),$$

and for $\frac{1}{2}\pi<\alpha\leqq\pi$

$$\left.\begin{aligned}\varphi_x(R,\alpha,y) &= -4\nu^2\int_0^{\frac{1}{2}\pi}\sec^3\vartheta\, e^{\nu(y+b)\sec^2\vartheta}\cos[\nu R\sec^2\vartheta\cos(\vartheta+\alpha)]\,d\vartheta - \\ &- 4\nu^2\int_0^{\alpha-\frac{1}{2}\pi}\sec^3\vartheta\, e^{\nu(y+b)\sec^2\vartheta}\cos[\nu R\sec^2\vartheta\cos(\vartheta-\alpha)]\,d\vartheta + O(R^{-1}).\end{aligned}\right\} \quad (13.40)$$

Consider the two integrals containing $\cos(\vartheta+\alpha)$ and let

$$\lambda = \sec^2\vartheta \cos(\vartheta+\alpha).$$

Then for $0 \leqq \alpha \leqq \frac{1}{2}\pi$ the integral becomes

$$-4\nu^2 \int_0^{-\infty} \frac{2\,e^{\nu(y+b)\sec^2\vartheta}}{\sin(2\vartheta+\alpha) - 3\sin\alpha} \cos\nu R\lambda\, d\lambda.$$

If $\frac{1}{2}\pi < \alpha < \pi$, the lower limit is $\cos\alpha$. In either case one may show that the coefficient of $\cos\nu R\lambda$ is single-valued, continuous, absolutely integrable and monotonically decreasing as a function of λ. By integration by parts one may then establish the following estimates as $R\to\infty$ (cf. S. BOCHNER, Vorlesungen über Fouriersche Integrale, Leipzig, 1932, § 3):

for $0 \leqq \alpha \leqq \frac{1}{2}\pi$

$$O(R^{-2});$$

for $\frac{1}{2}\pi < \alpha < \pi$

$$-4\nu \frac{e^{\nu(y+b)}}{R\sin\alpha} \sin(\nu R\cos\alpha) + O(R^{-2}).$$

If $\alpha = \pi$, the two integrals in (13.40) combine to give

$$\begin{aligned} 8\nu^2 \int_1^{\infty} e^{\nu(y+b)\lambda^2} \frac{\lambda^2}{\sqrt{\lambda^2-1}} \cos\nu R\lambda\, d\lambda \\ = 4\nu^2\sqrt{2\pi}\, \frac{e^{\nu(y+b)}}{\sqrt{\nu R}} \cos\left(\nu R - \frac{3}{4}\pi\right) + O(R^{-1}). \end{aligned}$$

Consider now the remaining integral in (13.40), and let

$$\mu(\vartheta) = \sec^2\vartheta\cos(\vartheta-\alpha).$$

The integral takes the form

$$-8\nu^2 \int_{\cos\alpha}^{0} e^{\nu(y+b)\sec^2\vartheta} \frac{\cos\nu R\mu}{\sin(2\vartheta-\alpha) + 3\sin\alpha}\, d\mu.$$

The denominator now becomes zero when

$$\tan\vartheta = -\tfrac{1}{4}\cot\alpha\left[1 \pm \sqrt{1 - 8\tan^2\alpha}\right], \tag{13.41}$$

an equation which has real roots when $\tan^2\alpha \leqq \frac{1}{8}$, i.e. when

$$180^\circ - 19^\circ 28' \ldots < \alpha < 180^\circ.$$

When $\frac{1}{2}\pi < \alpha < \pi - \arcsin\frac{1}{3} = \alpha_c$, the Fourier-integral estimate used for the other integral may be applied to give

$$4\nu\, e^{\nu(y+b)} \frac{\sin(\nu R\cos\alpha)}{R\sin\alpha} + O(R^{-2}).$$

When $\alpha_2 < \alpha < \pi$, ϑ is a two-branched function of μ and the resulting two integrals each have singularities at one of the limits. Thus the elementary method of analysis used above can no longer be applied. However, a modification of the method above can be carried through[1]; the classical treatment is by the method of stationary phase which is well discussed in STOKER (1957, Chap. 8).

[1] See, e.g., A. ERDÉLYI: Asymptotic expansions, pp. 46—56. Dover, New York 1956.

The estimates already derived are of the same order as the remainder term in (13.39). Analysis of this term produces terms which cancel the terms in $1/R \sin\alpha$ already derived, thus removing an apparent singular behavior near the x-axis

The asymptotic form for the surface $\eta(R,\alpha)$ above a source of strength m (i.e. $-m/r$) is as follows:

for $0 \leqq \alpha < \pi - \arcsin\frac{1}{3} = \alpha_c$

$$\eta(R,\alpha) = O\left((\nu R)^{-2}\right); \tag{13.42}$$

for $\alpha = \alpha_c$

$$\eta(R,\alpha_c) = 4\cdot 3^{\frac{5}{6}}\,\Gamma\left(\frac{1}{3}\right)\frac{m\nu}{u_0}(\nu R)^{-\frac{1}{3}}\,e^{\frac{3}{2}\nu b}\cos\left(\frac{\sqrt{3}}{2}\nu R\right) + O\left((\nu R)^{-\frac{2}{3}}\right);$$

for $\alpha_c < \alpha < \pi$

$$\eta(R,\alpha) = 4\sqrt{2\pi}\,\frac{m\nu}{u_0}\,\frac{(\nu R)^{-\frac{1}{2}}}{[1-9\sin^2\alpha]^{\frac{1}{4}}}\times$$

$$\times\left\{\sec^{\frac{3}{2}}\vartheta_1\,e^{\nu b\sec^2\vartheta_1}\cos\left(\nu R\mu_1 - \tfrac{1}{4}\pi\right) + \sec^{\frac{3}{2}}\vartheta_2\,e^{\nu b\sec^2\vartheta_2}\cos\left(\nu R\mu_2 + \tfrac{1}{4}\pi\right)\right\} + O\left((\nu R)^{-1}\right);$$

for $\alpha = \pi$

$$\eta(R,\pi) = -4\sqrt{2\pi}\,\frac{m\nu}{u_0}(\nu R)^{-\frac{1}{2}}\,e^{\nu b}\cos\left(\nu R - \tfrac{3}{4}\pi\right) + O\left((\nu R)^{-1}\right).$$

Here ϑ_1 and $\vartheta_2 > \vartheta_1$ are the two roots (13.41) and $\mu_1 (<0)$ and $\mu_2 < \mu_1$ the corresponding values of $\sec^2\vartheta\cos(\vartheta - \alpha)$. As $\alpha \to \alpha_c$, $\vartheta_i \to \arctan\frac{1}{2}\sqrt{2}$, $\mu_i \to -\frac{1}{2}\sqrt{3}$; as $\alpha \to \pi$, $\vartheta_1 \to 0$, $\mu_1 \to -1$, $\vartheta_2 \to \frac{1}{2}\pi$, $\mu_2 \to -\infty$. In order to have some idea of the form of the free surface far behind the source, one may graph the curves

$$\nu R\mu_1(\alpha) - \tfrac{1}{4}\pi = -2n\pi, \qquad \nu R\mu_2(\alpha) + \tfrac{1}{4}\pi = -2n\pi, \qquad n > 0,$$

showing the traces of the wave crests in the region

$$\pi - \arcsin\tfrac{1}{3} < \alpha < \pi + \arcsin\tfrac{1}{3}.$$

This gives the well known pattern shown in Fig. 1a. The first equation gives the transverse waves, the second one the diverging waves. The wavelength along $\alpha = \pi$ is $2\pi/\nu$ and along the boundary lines $4\pi\sqrt{3}/3\nu$. The expansion is not suitable in the region near the boundary lines $\alpha = \alpha_c$. As $\alpha \to \alpha_c$, $\alpha > \alpha_c$, the term $[1 - 9\sin^2\alpha]^{\frac{1}{4}} \to 0$ and the amplitudes become infinite. A special investigation of the region near $\alpha = \alpha_c$ is necessary and shows $(\nu R)^{-\frac{1}{3}}$ as leading term; η may be expressed in terms of Airy functions [cf. URSELL (1960)].

Essentially the same pattern is produced by a moving concentrated pressure on the free surface; it was first analysed by KELVIN (1906 = Papers, Vol. IV, pp. 407—413). The asymptotic behavior for moving pressure distributions has been extensively studied [e.g., HOGNER (1923), TETURÔ INUI (1936), PETERS (1949), BARTELS and DOWNING (1955)]. LAMB (1926) has given the asymptotic form of the surface over a moving submerged dipole. The form of the surface near the moving dipole has been investigated by HAVELOCK (1928), who gives traces of the profile on planes $\alpha =$ const for several values of α between $\frac{1}{2}\pi$ and π (the radial lines of Figs. 1b and c) and for $|b\nu| = \frac{1}{2}$ and 4. HAVELOCK's computations were later used by WIGLEY (1930) to produce the contour curves shown in Figs. 1b and c.

A similar analysis may be made for (13.37), a source moving in fluid of finite depth. For a moving pressure distribution the problem has been treated by HAVELOCK (1908) and TETURÔ INUI (1936). The pattern is modified as follows.

If $\nu h > 1$, the pattern is qualitatively like that for $h = \infty$. However, the wedge within which the disturbance is chiefly contained has a wider aperture and as $\nu h \to 1$ the aperture approaches $\frac{1}{2}\pi$ radians on each side of the line of motion.

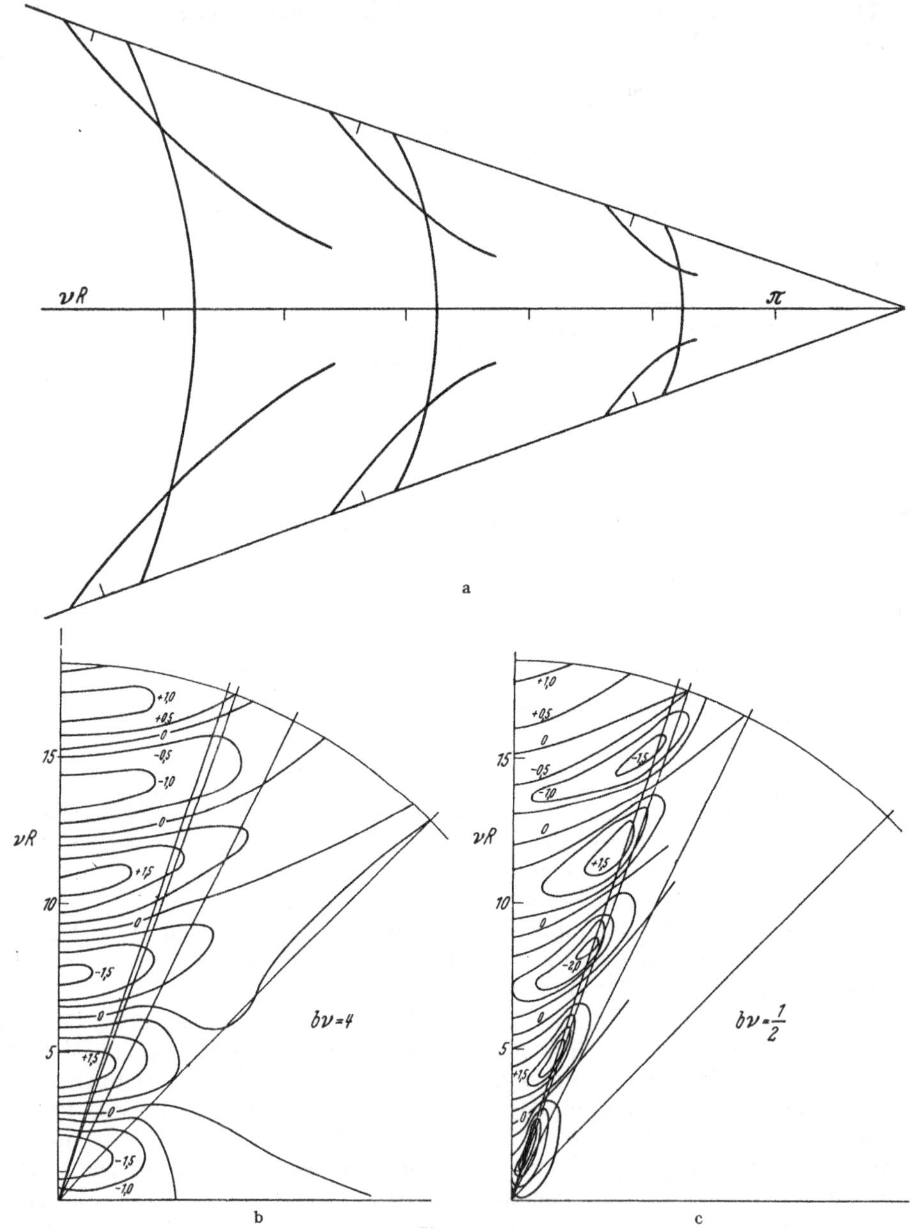

Fig. 1 a—c.

In addition, the wave length of the transverse wave system increases and approaches infinity as $\nu h \to 1$. When $\nu h \geqq 1$, the transverse wave system is missing completely, but diverging waves still occur in a wedge of aperture varying from π to 0 as $\nu h \to 0$. [See also EKMAN (1906), who has considered the free surface over a dipole on a flat bottom.]

Fig. 2 from HAVELOCK (1908) shows the half-angle of the aperture.

KOCHIN (1938c) has gone further in this type of problem. He has derived the potential for a source situated in a fluid of density ϱ_1 and depth h, bounded below by a plane, over which is lying another fluid of density $\varrho_2 < \varrho_1$, extending infinitely far upwards. The lower fluid moves with velocity c_1, the upper with velocity c_2 in the same direction. He also finds the asymptotic behavior of the solution.

Singularities of constant strength in uniform motion: two dimensions. For submerged sources and vortices in two-dimensional motion the complex-variable method used for the pulsating source may again be applied. For infinitely deep fluid, the computation has been carried out in this way by KELDYSH and LAVRENT'EV (1937) and KOCHIN (1937); a detailed exposition is given in the textbook of KOCHIN, KIBEL' and ROZE (1948, Chap. VIII, § 19). HAVELOCK (1927) and SRETENSKII (1938) have treated the problem by different methods. The complex velocity potential for a combined source of strength Q and vortex of intensity Γ at $c = a + i b$, $b < 0$, is given by

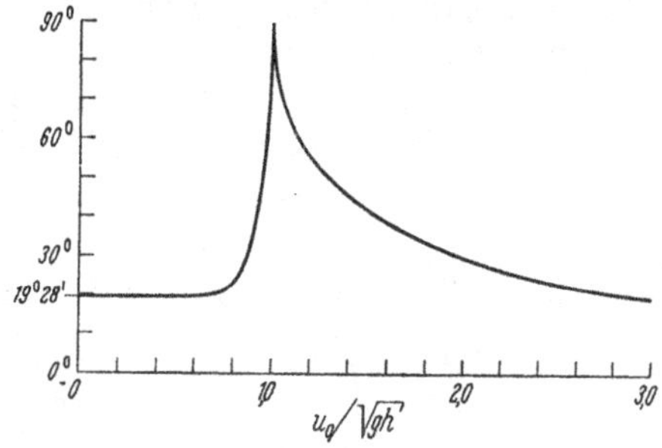

Fig. 2.

$$
\left.
\begin{aligned}
f(z) &= \frac{\Gamma + i Q}{2\pi i} \log(z - c) - \frac{\Gamma - i Q}{2\pi i} \log(z - \bar{c}) + \frac{\Gamma - i Q}{\pi i} e^{-i\nu z} \int_{\infty}^{z} \frac{e^{i\nu u}}{u - \bar{c}}\, du, \\
&= \frac{\Gamma + i Q}{2\pi i} \log(z - c) - \frac{\Gamma - i Q}{2\pi i} \log(z - \bar{c}) - 2(\Gamma - i Q)\, e^{-i\nu(z-\bar{c})} + \\
&\qquad + \frac{\Gamma - i Q}{\pi i} e^{-i\nu z} \int_{-\infty}^{z} \frac{e^{i\nu u}}{u - \bar{}}\, du, \\
&= \frac{\Gamma + i Q}{2\pi i} \log(z - c) - \frac{\Gamma - i Q}{2\pi i} \log(z - \bar{c}) - \frac{\Gamma - i Q}{\pi i} \mathrm{PV} \int_{0}^{\infty} \frac{e^{-ik(z-\bar{c})}}{k - \nu}\, dk - \\
&\qquad - (\Gamma - i Q)\, e^{-i(z-\bar{c})}.
\end{aligned}
\right\} \quad (13.43)
$$

The real velocity potential for, say, a submerged source can be obtained from any of these equations. The last one gives a form analogous to (13.36):

$$
\left.
\begin{aligned}
\varphi(x, y) &= \frac{Q}{2\pi} \log r + \frac{Q}{2\pi} \log r_1 + \\
&\quad + \frac{Q}{\pi} \mathrm{PV} \int_{0}^{\infty} \frac{e^{k(y+b)} \cos k(x - a)}{k - \nu}\, dk + Q\, e^{\nu(y+b)} \sin \nu(x - a).
\end{aligned}
\right\} \quad (13.44)
$$

Higher-order singularities can be obtained by differentiating (13.43). The complex velocity potential for a dipole of moment M and axis in the direction $e^{i\alpha}$ is given by

$$
\left.
\begin{aligned}
f(z) &= -\frac{M}{2\pi} \frac{e^{i\alpha}}{z - c} + \frac{M}{2\pi} \frac{e^{-i\alpha}}{z - \bar{c}} - \frac{i M \nu}{\pi} e^{-i\alpha} e^{-i\nu z} \int_{\infty}^{z} \frac{e^{i\nu u}}{u - \bar{c}}\, du, \\
&= -\frac{M}{2\pi} \frac{e^{i\alpha}}{z - c} + \frac{M}{2\pi} \frac{e^{-i\alpha}}{z - \bar{c}} + \frac{i M \nu}{\pi} e^{-i\alpha}\, \mathrm{PV} \int_{0}^{\infty} \frac{e^{-ik(z-\bar{c})}}{k - \nu}\, dk - \\
&\qquad - M \nu\, e^{-i\alpha} e^{-i\nu(z-\bar{c})}.
\end{aligned}
\right\} \quad (13.45)
$$

If in the last term of the first equation of either (13.43) or (13.45), one makes use of the identity

$$e^{-i\nu z}\int_{\infty}^{z}\frac{e^{i\nu u}}{u-\bar{c}}\,du=\int_{-\infty}^{0}\frac{-e^{-i\nu u}}{z-u-\bar{c}}\,du,$$

it is not difficult to see that this last term is equivalent to a distribution of dipoles on the ray from $\bar{c}$ to $-\infty$ parallel to the x-axis. The moment density and axis can be determined for the three cases, source, vortex and dipole, by comparison of the integrand with the first term of (13.45).

For the case of finite depth the complex velocity potential has been calculated by TIKHONOV (1940) and is also given by HASKIND (1945a) for both source and vortex. We give separately source, vortex and dipole:

source:

$$\left.\begin{aligned}f(z)&=\frac{Q}{2\pi}\log(z-c)+\frac{Q}{2\pi}\log(z-c_2)+\\&\quad+\frac{2Q}{\pi}\,\mathrm{PV}\int_0^\infty\frac{k+\nu}{k}\,e^{-kh}\frac{\cosh k(b+h)}{\nu\sinh kh-k\cosh kh}\sin^2\frac{1}{2}k(z-a+ih)\,dk-\\&\quad-\frac{Q\nu}{k_0}\,\frac{\cosh k_0(b+h)}{\nu h-\cosh^2 k_0 h}\sin k_0(z-a+ih);\end{aligned}\right\}\tag{13.46}$$

vortex:

$$\left.\begin{aligned}f(z)&=\frac{\Gamma}{2\pi i}\log(z-c)-\frac{\Gamma}{2\pi i}\log(z-c_2)-\\&\quad-\frac{\Gamma}{\pi}\,\mathrm{PV}\int_0^\infty\frac{k+\nu}{k}\,e^{-kh}\frac{\sinh k(b+h)}{\nu\sinh kh-k\cosh kh}\sin k(z-a+ih)\,dk+\\&\quad+\frac{\Gamma\nu}{k_0}\,\frac{\sinh k_0(b+h)}{\nu h-\cosh^2 k_0 h}\cos k_0(z-a+ih);\end{aligned}\right\}\tag{13.47}$$

dipole:

$$\left.\begin{aligned}f(z)&=-\frac{M}{2\pi}\,\frac{e^{i\alpha}}{z-c}-\frac{M}{2\pi}\,\frac{e^{-i\alpha}}{z-c_2}-\\&\quad-\frac{M}{2\pi}\,\mathrm{PV}\int_0^\infty(k+\nu)\,e^{-kh}\frac{e^{i\alpha}\sin k(z-c)+e^{-i\alpha}\sin k(z-c_2)}{\nu\sinh kh-k\cosh kh}\,dk+\\&\quad+\frac{\nu M}{2}\,\frac{e^{i\alpha}\cos k_0(z-c)+e^{-i\alpha}\cos k_0(z-c_2)}{\nu h-\cosh^2 k_0 h}.\end{aligned}\right\}\tag{13.48}$$

Here $c_2=a-ib+2ih$ and the last summand in each of (13.46) to (13.48) is to be deleted if $\nu h=gh/c^2\leqq 1$; k_0 is the positive real root of $\nu\sinh kh-k\cosh kh=0$, which exists only if $\nu h>1$.

Asymptotic form of these functions as $x\to-\infty$ is easily seen to be given by double the last term in each expression. When $\nu h<1$, the disturbance is only local, a fact which corresponds to the absence of transverse waves behind the three-dimensional source for $\nu h<1$.

KOCHIN (1937a, b) has derived the complex velocity potential when fluid of density ϱ_2 overlies the fluid of density $\varrho_1>\varrho_2$ containing the singularity. The lower fluid may be of infinite or finite depth; the upper one is taken infinitely deep. Their velocities may be different.

Source of variable strength, starting from rest and following an arbitrary path. Consider now a source whose position and strength at time

$t \geqq 0$ are given by $(a(t), b(t), c(t))$ and $m(t)$, where $b(t) < 0$. Let $m(t) = 0$ for $t < 0$. The conditions to be satisfied by the velocity potential $\Phi(x, y, z, t)$ are

1. $\Delta \Phi = 0, \quad y < 0, \quad (x, y, z) \neq (a(t), b(t), c(t))$,
2. $\Phi_{tt}(x, 0, z, t) + g\,\Phi_y(x, 0, z, t) = 0$,
3. $\Phi(x, y, z, t) = m(t)\, r^{-1} + \Phi_0(x, y, z, t)$, Φ_0 harmonic everywhere in $y < 0$,
4. $\lim\limits_{y \to -\infty} \operatorname{grad} \Phi = 0$ for all x, z and t,
5. $\lim\limits_{R \to \infty} \operatorname{grad} \Phi = 0$ for all t,
6. $\Phi(x, 0, z, 0) = \Phi_t(x, 0, z, 0) = 0$.

Here $r^2 = (x - a(t))^2 + (y - b(t))^2 + (z - c(t))^2$, $R^2 = (x - a(t))^2 + (z - c(t))^2$.

If one assumes a solution in the form

$$\Phi = m\, r^{-1} - m_1^{-1}\, r + \Phi_1$$

where $r_1^2 = (x - a)^2 + (y + b)^2 + (z - c)^2$, then Φ_1 must be harmonic in $y < 0$ and satisfy 4., 5., 6. and

$$\Phi_{1tt}(x, 0, z, t) + g\,\Phi_{1y}(x, 0, z, t) = -2g\, m(t)\, b(t)\, [(x-a)^2 + b^2 + (z-c)^2]^{-\frac{3}{2}}, \quad t \geqq 0.$$

It follows from the conditions that, for $t < 0$, $\Phi_1 = \text{const}$, which we may take as zero. Let $\overline{\Phi}_1$ be the Laplace transform of Φ_1:

$$\overline{\Phi}_1(x, y, z, s) = \int_0^\infty e^{-st}\, \Phi_1(x, y, z, t)\, dt.$$

Then $\overline{\Phi}_1$ is a harmonic function in $y < 0$ satisfying 4. and 5. for each s and also, after making use of 6., the condition

$$s^2 \overline{\Phi}_1(x, 0, z, s) + g\,\overline{\Phi}_{1y}(x, 0, z, s) = -2g \int_0^\infty e^{-st}\, m(t)\, b(t)\, [(x-a)^2 + b^2 + (z-c)^2]^{-\frac{3}{2}}\, dt.$$

Since

$$s^2\, \overline{\Phi}_1(x, y, z, s) + g\, \overline{\Phi}_{1y}(x, y, z, s) + \\ + 2g \int_0^\infty e^{-st}\, m(t)\, (y + b(t))\, [(x-a)^2 + (y+b)^2 + (z-c)^2]^{-\frac{3}{2}}\, dt$$

is a harmonic function in $y < 0$ vanishing on $y = 0$ and at infinity, it is identically zero. Making use of (13.12) differentiated with respect to y and changing the order of integration, one obtains

$$\begin{aligned} s^2\, \overline{\Phi}_1(x, y, z, s) &+ g\, \overline{\Phi}_{1y}(x, y, z, s) \\ &= \frac{g}{\pi} \int_0^\infty k\, dk \int_0^\infty dt\, e^{-st} m(t)\, e^{k(y+b)} \int_{-\pi}^{\pi} d\vartheta\, e^{ik[(x-a)\cos\vartheta + (z-c)\sin\vartheta]} \\ &= 2g \int_0^\infty k\, dk \int_0^\infty dt\, e^{-st}\, m(t)\, e^{k(y+b)}\, J_0\big(k\, [(x-a)^2 + (z-c)^2]^{\frac{1}{2}}\big). \end{aligned}$$

The solution for $\overline{\Phi}_1$ is

$$\overline{\Phi}_1(x, y, z, s) = 2g \int_0^\infty dk\, \frac{k}{s^2 + g k} \int_0^\infty dt\, e^{-st}\, m(t)\, e^{k(y+b)}\, J_0\big(k\, R(t)\big).$$

Making use of the convolution theorem and the fact that $(s^2+gk)^{-1}$ is the transform of $(gk)^{-\frac{1}{2}}\sin(gk)^{\frac{1}{2}}t$, one may find the original function $\Phi_1(x, y, z, t)$:

$$\Phi_1(x, y, z, t) = 2\int_0^\infty dk\,(g\,k)^{\frac{1}{2}}\int_0^t d\tau \sin\left[(g\,k)^{\frac{1}{2}}(t-\tau)\right] m(\tau)\, e^{k(y+b(\tau))} J_0\big(k\,R(\tau)\big).$$

For fixed t one may easily verify, using known properties of the Fourier-Bessel transform[1], that Φ_1 is $o(R^{-\frac{1}{2}})$ and hence that 5. is satisfied. One has then the result

$$\left.\begin{aligned}\Phi(x,y,z,t) &= \frac{m(t)}{r(t)} - \frac{m(t)}{r_1(t)} + 2\int_0^\infty dk\,(g\,k)^{\frac{1}{2}}\int_0^t d\tau\, m(\tau)\sin\left[(g\,k)^{\frac{1}{2}}(t-\tau)\right]\times\\ &\qquad\times e^{k(y+b)} J_0\big(k\,R(\tau)\big)\\ &= \frac{m(t)}{r(t)} - \frac{m(t)}{r_1(t)} + \frac{1}{\pi}\int_{-\pi}^{\pi} d\vartheta\int_0^\infty dk(g\,k)^{\frac{1}{2}}\int_0^t d\tau\, m(\tau)\sin\left[(g\,k)^{\frac{1}{2}}(t-\tau)\right]\times\\ &\qquad\times e^{k[y+b(\tau)+i(x-a(\tau))\cos\vartheta+i(z-c(\tau))\sin\vartheta]}.\end{aligned}\right\}\quad(13.49)$$

By a more refined analysis of the behavior for large R [cf. STOKER (1957, pp. 190 to 191)] one may establish that Φ is $O(R^{-2})$ and Φ_R and Φ_y are $O(R^{-3})$ as $R\to\infty$.

For some time $t>t_0\geqq 0$, one may write Φ in the form

$$\begin{aligned}\Phi(x, y, z, t) &= 2\int_0^\infty dk\,(g\,k)^{\frac{1}{2}}\int_0^{t_0} d\tau\, m(\tau)\sin\left[(g\,k)^{\frac{1}{2}}(t-\tau)\right] e^{k[y+b(\tau)]} J_0\big(k\,R(\tau)\big) +\\ &\quad+\frac{m(t)}{r(t)} - \frac{m(t)}{r_1(t)} + 2\int_0^\infty dk\,(g\,k)^{\frac{1}{2}}\int_0^{t-t_0} d\tau\, m(\tau+t_0)\sin\left[(g\,k)^{\frac{1}{2}}(t-t_0-\tau)\right]\times\\ &\quad\times e^{k[y+b(\tau+t_0)]} J_0\big(k\,R(\tau+t_0)\big) = \Phi_2(x, y, z, t) + \Phi_3(x, y, z, t).\end{aligned}$$

Here the first summand Φ_2 represents the effect at time $t>t_0$ of the action of the source from $t=0$ to $t=t_0$. The remaining terms, Φ_3, are the same as (13.49) with t measured from $t_0\big(m(t)=m(t-t_0+t_0)$, etc.$\big)$, and show the effect at time t of the action of the source from $t=t_0$ to $t=t$. (This is, of course, what one would expect from the linearity of the problem and the fact that the choice of $t=0$ is arbitrary.) When $t=t_0$, Φ_3 reduces to

$$\frac{m(t_0)}{r(t_0)} - \frac{m(t_0)}{r_1(t_0)}.$$

Thus

$$\Phi_3(x, 0, z, t_0) = 0.$$

This fact provides a basis for HAVELOCK's procedure in similar problems, a procedure originating with KELVIN in the treatment of moving pressure distributions. The idea is roughly as follows. Divide the path of the source into small segments of time span Δt. If Δt is small enough, the effect of gravity upon the fluid motion produced by the source during this time interval will be negligible, and one may take the boundary condition at the free surface to be $\Phi=0$. The distortion of the surface resulting from the action of the source during this short interval is found and the future behavior of the distortion computed while taking account of gravity. Summing over all Δt and taking the limit leads to the potential function.

[1] Cf. G.N. WATSON: Bessel functions, § 14.41. Cambridge 1944.

The expression (13.49) has been essentially given by HASKIND (1946b) and BRARD (1948a). Special choices of $m(t)$ and of the motion of the source lead to cases similar to those treated earlier. Thus, if $m(t) = m \cos \sigma t$ and (a, b, c) is fixed, one has the potential function for a stationary source of oscillating strength, starting to oscillate at $t=0$. Carrying out the t integration and taking a limit by using, say, the Fourier Integral Theorem (13.16) allow one to derive (13.17). The radiation condition is automatically satisfied. The velocity potential for finite values of t may be written in the form

$$\left.\begin{aligned} \Phi(x, y, z, t) = {} & \frac{m \cos \sigma t}{r} - \frac{m \cos \sigma t}{r_1} + \\ & + 2m \cos \sigma t \, \mathrm{PV} \int_0^\infty \frac{k}{k - \sigma^2/g} \, e^{k(y+b)} J_0(kR) \, dk - \\ & - 2m \, \mathrm{PV} \int_0^\infty \frac{k}{k - \sigma^2/g} \cos (gk)^{\frac{1}{2}} t \, e^{k(y+b)} J_0(kR) \, dk. \end{aligned}\right\} \tag{13.50}$$

The leading term in the asymptotic expansion of the last summand gives the last summand of (13.17).

If one takes $m(t) = m = $ a constant, $a(t) = a_0 + u_0 t$, $b(t) = b_0$, $c(t) = c_0$, one obtains the velocity potential for a source suddenly brought into existence at $t=0$ and moving with constant velocity in the direction Ox [cf. LUNDE (1951, p. 18)]. A limit as $t \to \infty$ will give (13.36), the proper boundary conditions at infinity being again automatically satisfied. For finite t the velocity potential in a coordinate system moving with velocity u_0 in direction Ox $(\bar{x} = x - u_0 t$, so that $\Phi(x, y, z, t) = \varphi(\bar{x}, y, z, t))$ is given by

$$\left.\begin{aligned} \varphi(\bar{x}, y, z, t) = {} & \frac{m}{r} - \frac{m}{r_1} + \\ & + \frac{m}{\pi} \int_{-\pi}^{\pi} d\vartheta \int_0^\infty dk \, (gk)^{\frac{1}{2}} e^{k[y + b_0 + i\omega(\vartheta)]} \int_0^t d\tau \sin \tau (gk)^{\frac{1}{2}} e^{i k u_0 \tau \cos \vartheta}, \\ & \omega(\vartheta) = (\bar{x} - a_0) \cos \vartheta + (z - c_0) \sin \vartheta. \end{aligned}\right\} \tag{13.51}$$

The two cases just discussed may be combined by choosing $m(t) = m \cos \sigma t$ and $a(t) = a_0 + u_0 t$, $b(t) = b_0$, $c(t) = c_0$. The modification of (13.51) is simple: a factor $\cos \sigma t$ must be put with the first two terms and a factor $\cos \sigma (t - \tau)$ put at the end of the integral. The asymptotic form as $t \to \infty$ can again be found by use of the Fourier Integral Theorem (13.16) or simple modifications. However, if the resulting formula is written out as principal-value integrals plus other terms, the expression is very unwieldy; it may be found in HAVELOCK (1958). Use of complex integrals allows one to compress the formula. Let

$$\varphi(\bar{x}, y, z, t) = m \cos \sigma t \left(\frac{1}{r} - \frac{1}{r_1} \right) + m \, \mathrm{Re} \, e^{-i\sigma t} \varphi_0, \qquad \varphi_0 = \varphi_1 + i \varphi_2.$$

Then

$$\varphi_0 = \frac{2g}{\pi} \int_0^\gamma d\vartheta \int_0^\infty dk \, F(\vartheta, k) + \frac{2g}{\pi} \int_\gamma^{\frac{1}{2}\pi} d\vartheta \int_{L_1} dk \, F(\vartheta, k) + \frac{2g}{\pi} \int_{\frac{1}{2}\pi}^{\pi} d\vartheta \int_{L_2} dk \, F(\vartheta, k),$$

$$F(\vartheta, k) = \frac{k \, e^{k[y + b_0 + i(\bar{x} - a_0) \cos \vartheta]} \cos [k(z - c_0) \sin \vartheta]}{gk - (\sigma + k u_0 \cos \vartheta)^2}, \tag{13.52}$$

where

$$\tau = u_0\,\sigma/g\,,$$

$$\gamma = \begin{cases} 0 & \text{if} \quad \tau < \frac{1}{4} \\ \arccos \frac{1}{4\tau} & \text{if} \quad \tau \geqq \frac{1}{4}\,, \end{cases}$$

$$\sqrt{g\,k_1}\,,\ \sqrt{g\,k_3} = \frac{1-\sqrt{1-4\tau\cos\vartheta}}{2\tau\cos\vartheta}\,\sigma\,,$$

$$\sqrt{g\,k_2}\,,\ -\sqrt{g\,k_4} = \frac{1+\sqrt{1-4\tau\cos\vartheta}}{2\tau\cos\vartheta}\,\sigma\,.$$

This potential has been derived by HASKIND (1946a), BRARD (1948a, b), HANAOKA (1953), SRETENSKII (1954), the last with an unfortunate mistake in sign

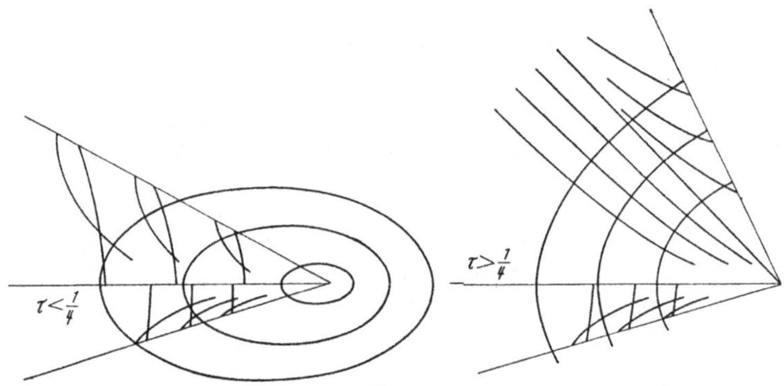

Fig. 3.

in one term, EGGERS (1957), and HAVELOCK (1958). HANAOKA, BRARD, EGGERS, and SRETENSKII have each considered the asymptotic form of the surface for large R. Fig. 3 shows qualitatively (cf. BECKER 1958) the curves of equal phase, say the crests, for the various systems of waves formed. The patterns must be completed by reflection in the x-axis.

Motion of a source on a circular path of radius D may be treated by taking $a = D\cos\sigma t$, $c = D\sin\sigma t$ in (13.49). For constant m this problem has been considered by SRETENSKII (1946a, b, 1957), HAVELOCK (1950), and STOKER (1957).

One may derive a formula analogous to (13.49) when the source moves in the presence of both a horizontal bottom at $y = -h$ and a free surface. The derivation may be carried out along lines similar to those used in deriving (13.49). The resulting velocity potential is [cf. LUNDE (1951, p. 32)]

$$\left.\begin{aligned} \Phi(x, y, z, t) &= \frac{m(t)}{r(t)} + \frac{m(t)}{r_2(t)} - 2m(t)\int\limits_0^\infty e^{-kh}\,\frac{\cosh k\,(h+b(t))}{\cosh k\,h}\,\times \\ &\times \cosh k\,(y+h)\,J_0\big(k\,R(t)\big)\,d\,k + 2\int\limits_0^\infty d\,k\,\sqrt{g\,k}\,\frac{\cosh k\,(y+h)}{\cosh^2 k\,h\,\sqrt{\tanh k\,h}}\,\times \\ &\times \int\limits_0^t d\tau \sin\big[(t-\tau)\sqrt{g\,k\tanh k\,h}\big]\,m(\tau)\cosh k\big(h+b(\tau)\big)\,J_0\big(k\,R(\tau)\big)\,, \end{aligned}\right\} \qquad (13.53)$$

where
$$r_2^2 = (x - a(t))^2 + (y + 2h + b(t))^2 + (z - c(t))^2.$$

Two-dimensional formulas corresponding to (13.49) and (13.53) may also be derived. They are as follows, with the source and vortex separated for finite depth:

infinite depth:

$$\left.\begin{aligned} f(z,t) &= \frac{\Gamma(t) + i\,Q(t)}{2\pi i}\log\bigl(z - c(t)\bigr) + \frac{\Gamma(t) + i\,Q(t)}{2\pi i}\log\bigl(z - \bar{c}(t)\bigr) + \\ &+ \frac{g}{\pi i}\int_0^t [\Gamma(\tau) - i\,Q(\tau)]\,d\tau \int_0^\infty \frac{1}{\sqrt{g\,k}}\,e^{-ik(z-\bar{c}(\tau))}\sin[\sqrt{g\,k}\,(t-\tau)]\,dk; \end{aligned}\right\} \quad (13.54)$$

depth h, source:

$$\left.\begin{aligned} f(z,t) &= \frac{Q(t)}{2\pi}\log\bigl(z - c(t)\bigr) + \frac{Q(t)}{2\pi}\log\bigl(z - \bar{c}(t) + 2i\,h\bigr) + \\ &+ \frac{Q(t)}{\pi}\int_0^\infty \frac{e^{-kh}}{k\cosh k\,h}\cosh k\bigl(b(t) + h\bigr)\cos k\bigl(z - a(t) + i\,h\bigr)\,dk - \\ &- \frac{g}{\pi}\int_0^\infty \frac{\operatorname{sech}^2 k\,h}{\sqrt{g\,k\tanh k\,h}}\,dk\int_0^t Q(\tau)\cosh k\bigl(b(\tau) + h\bigr)\cos k\bigl(z - a(\tau) + i\,h\bigr)\times \\ &\times\sin[\sqrt{g\,k\tanh k\,h}\,(t-\tau)]\,d\tau; \end{aligned}\right\} \quad (13.55)$$

depth h, vortex:

$$\left.\begin{aligned} f(z,t) &= \frac{\Gamma(t)}{2\pi i}\log\bigl(z - c(t)\bigr) - \frac{\Gamma(t)}{2\pi i}\log\bigl(z - \bar{c}(t) + 2i\,h\bigr) + \\ &+ \frac{\Gamma(t)}{\pi}\int_0^\infty \frac{e^{-kh}}{k\cosh k\,h}\sinh k\bigl(b(t) + h\bigr)\sin k\bigl(z - a(t) + i\,h\bigr) - \\ &- \frac{g}{\pi}\int_0^\infty \frac{\operatorname{sech}^2 k\,h}{\sqrt{g\,k\tanh k\,h}}\,dk\int_0^t \Gamma(\tau)\sinh k\bigl(b(\tau) + h\bigr)\sin k\bigl(z - a(\tau) + i\,h\bigr)\times \\ &\times\sin[\sqrt{g\,k\tanh k\,h}\,(t-\tau)]\,dk. \end{aligned}\right\} \quad (13.56)$$

Higher-order singularities may be generated by taking derivatives with respect to z. One may transfer to moving coordinates, etc., just as in the three-dimensional case [see HAVELOCK (1949) for (13.54) in moving coordinates]. The velocity potential for a steadily moving source of pulsing strength in two dimensions has been given by HASKIND (1954, p. 23 ff.), who also gives the asymptotic expressions for large values of $\pm x$. When $\tau < \frac{1}{4}$, there exist one wave far ahead of the moving source propagating in the same direction and three far behind, one propagating in the same direction and two in the opposite direction; when $\tau > \frac{1}{4}$, there exist two waves far behind propagating in the opposite direction. The analysis for finite depth has been given by BECKER (1956).

14. Some simple physical solutions. In this section we consider periodic waves in an ocean of infinite horizontal extent, either infinitely deep or with a horizontal bottom, in canals, and at an interface. The linearizing parameter ε of Sect. 10α may be taken to be the ratio of amplitude to wave length.

α) Standing waves in an infinite ocean. It is appropriate to the physical problem to require that the motion remain bounded everywhere.

Consider first two-dimensional motion. Then, from Sect. 13β, the only solutions of the form $\Phi = \varphi\cos(\sigma t + \tau)$ are given by

$$\Phi(x, y, t) = a\,e^{\nu y}\cos(\nu x + \alpha)\cos(\sigma t + \tau), \quad \nu = \sigma^2/g \qquad (14.1)$$

for infinite depth, and

$$\left.\begin{aligned} \Phi(x, y, t) &= a \cosh m_0 (y + h) \cos(m_0 x + \alpha) \cos(\sigma t + \tau), \\ m_0 \tanh m_0 h - \nu &= 0, \end{aligned}\right\} \tag{14.2}$$

for finite depth.

The corresponding forms of the free surface are given by

$$\eta(x, t) = A \cos(\nu x + \alpha) \sin(\sigma t + \tau)$$

and

$$\eta(x, t) = A \cos(m_0 x + \alpha) \sin(\sigma t + \tau),$$

respectively. These represent standing waves according to our definition in Sect. 7. We recall that $m_0 > \nu$.

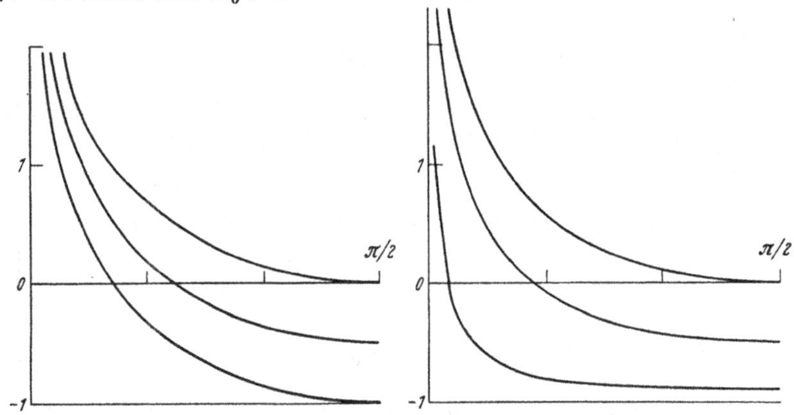

Fig. 4 a and b.

It is of interest to examine the streamlines and the paths of the individual fluid particles. The streamlines of the motion can be easily found from

$$\frac{dy}{dx} = \frac{\Phi_y}{\Phi_x} = \cot(\nu x + \alpha)$$

and

$$\frac{dy}{dx} = \frac{\Phi_y}{\Phi_x} = -\tanh m_0 (y + h) \cot(m_0 x + \alpha),$$

respectively. The streamlines are then

$$e^{\nu(y - y_m)} |\sin(\nu x + \alpha)| = 1$$

and

$$\sinh m_0 (y + h) |\sin(m_0 x + \alpha)| = \sinh m_0 (y_m + h), \quad 0 \geqq y_m \geqq h, \tag{14.3}$$

for infinite and finite depth respectively; here y_m is the lowest point of the streamline. If the fluid is infinitely deep, the streamlines are all congruent. Fig. 4a shows three of them for a quarter wave length and $\alpha = 0$, $\nu = 1$. The vertical line $x = 0$ is also a streamline. If the fluid is of finite depth, the streamlines vary with depth. Fig. 4b shows streamlines corresponding to $y_m = 0, -0.5, -0.9$ for $\alpha = 0$, $h = 1$, $m_0 = 1$. The horizontal line $y = -1$ and the vertical line $x = 0$ are also streamlines.

Since the streamlines are time-independent, they also contain the curves for the trajectories of individual particles. However, the trajectory of an individual particle will be an oscillating motion of small amplitude along a segment of the streamline passing through the point. Thus, in Fig. 4b the particles on the bottom simply oscillate back and forth about an equilibrium position, those directly beneath a crest, i.e. at $x=0$, oscillate vertically, etc. In view of the infinitesimal-wave approximation used in this chapter the streamlines have physical significance for only a small distance above the equilibrium free surface, $y=0$.

In order to investigate, at least approximately, the behavior of the trajectories more fully, we may replace the actual trajectory by its tangent at an average position, say (x_0, y_0), an approximation consistent with the assumptions made in linearizing. Then the equations describing the trajectory become (setting $\alpha=\tau=0$)

$$\frac{dx}{dt} = -a\nu\, e^{\nu y_0} \sin \nu x_0 \cos \sigma t, \qquad \frac{dy}{dt} = a\nu\, e^{\nu y_0} \cos \nu x_0 \cos \sigma t$$

for infinite depth, and

$$\frac{dx}{dt} = -a\, m_0 \cosh m_0 (y_0 + h) \sin m_0 x_0 \cos \sigma t,$$

$$\frac{dy}{dt} = a\, m_0 \sinh m_0 (y_0 + h) \cos m_0 x \cos \sigma t$$

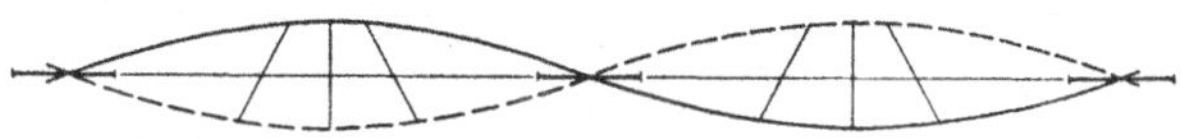

Fig. 5.

for finite depth. The approximate trajectories are then

$$x = x_0 - a\sigma^{-1} \nu\, e^{\nu y_0} \sin \nu x_0 \sin \sigma t, \qquad y = y_0 + a\sigma^{-1} \nu\, e^{\nu y_0} \cos \nu x_0 \sin \sigma t \tag{14.4}$$

for infinite depth, and

$$\left.\begin{aligned} x &= x_0 - a\sigma^{-1} m_0 \cosh m_0 (y_0 + h) \sin m_0 x_0 \sin \sigma t, \\ y &= y_0 + a\sigma^{-1} m_0 \sinh m_0 (y_0 + h) \cos m_0 x_0 \sin \sigma t \end{aligned}\right\} \tag{14.5}$$

for finite depth. For infinite depth, the amplitude of oscillation drops off very rapidly as depth of the equilibrium position increases, the ratio of the amplitude at depth y_0 to the amplitude at the free surface being $e^{\nu y_0}$. The same ratio for the case of finite depth is

$$\frac{\sinh^2 m_0 (y_0 + h) + \sin^2 m_0 x_0}{\sinh^2 m_0 h + \sin^2 m_0 x_0}.$$

Thus, on the bottom, when $y_0 = -h$, the amplitude is zero under the crests and maximum under the nodes. As is evident from the equations of the trajectories, the path lines of particles on the free surface are approximately as in Fig. 5. In order to explain the apparently inconsistent behavior at the nodes one mus. go to a higher approximation than the linearized theory used in this chaptert

Let us now consider three-dimensional solutions. The standing-wave solutions are of the form

$$\Phi(x, y, z, t) = e^{\nu y} \chi(x, y) \cos(\sigma t + \tau) \qquad \text{for finite depth},$$

or

$$\Phi(x, y, z, t) = \cosh m_0 (y + h)\, \chi(x, z) \cos(\sigma t + \tau) \qquad \text{for finite depth},$$

where $\chi(x, z)$ is a solution of

$$\Delta_2 \chi + \nu^2 \chi = 0 \quad \text{or} \quad \Delta_2 \chi + m_0 \chi = 0,$$

respectively, regular everywhere in $y \leqq 0$.

Two particular cases are of especial interest. The first corresponds to separation of variables in rectangular coordinates [see (13.5) and (13.6)]. The solutions are

$$\left.\begin{aligned}\Phi(x, y, z, t) &= a\, e^{\nu y} \cos(k_1 x + \alpha) \cos(k_2 z + \gamma) \cos(\sigma t + \tau),\\ k_1^2 + k_2^2 &= \nu^2 = \sigma^2/g,\end{aligned}\right\} \tag{14.6}$$

for infinite depth, and

$$\left.\begin{aligned}\Phi(x, y, z, t) &= a \cosh m_0(y+h) \cos(k_1 x+\alpha) \cos(k_2 z+\gamma) \cos(\sigma t+\tau),\\ k_1^2 + k_2^2 &= m_0^2, \qquad m_0 \tanh m_0 h - \nu = 0,\end{aligned}\right\} \tag{14.7}$$

for finite depth. The other solutions result from separating variables in polar coordinates [see (13.7) and (13.8)]. They are

$$\Phi(R, \alpha, y, t) = a\, e^{\nu y} J_n(\nu R) \cos(n\alpha + \delta) \cos(\sigma t + \tau), \qquad n = 0, 1, \ldots, \tag{14.8}$$

for infinite depth, and

$$\left.\begin{aligned}\Phi(R, \alpha, y, t) &= a \cosh m_0(y + h)\, J_n(m_0 R) \cos(n\alpha + \delta) \cos(\sigma t + \tau),\\ n &= 0, 1, \ldots,\end{aligned}\right\} \tag{14.9}$$

for finite depth. The form of the free surface may be found immediately from $\eta(x, z, t) = -\Phi_t(x, 0, z, t)/g$. These are all standing waves.

The streamlines and path lines may be found for these two cases with no special difficulty. For the first case for finite depth the streamlines are the intersections of the surfaces

$$\left.\begin{aligned}|\sin k_1 x|^{k_2^2} &= C_1 |\sin k_2 z|^{k_1^2},\\ |\sin k_1 x \sin k_2 z| \sinh m_0(y + h) &= C_2.\end{aligned}\right\} \tag{14.10}$$

The vertical lines, $x = p\pi/k_1$, $z = q\pi/k_2$, passing through the maxima and minima are streamlines. The points on the vertical lines $x = (p + \frac{1}{2})\pi/k_1$, $z = (q + \frac{1}{2})\pi/k_2$ passing through the sattlepoints are all stagnation points. The projection on $y = 0$ of the field of streamlines is indicated qualitatively by Fig. 6. The behavior in a projection on a vertical plane is similar to that for two-dimensional motion.

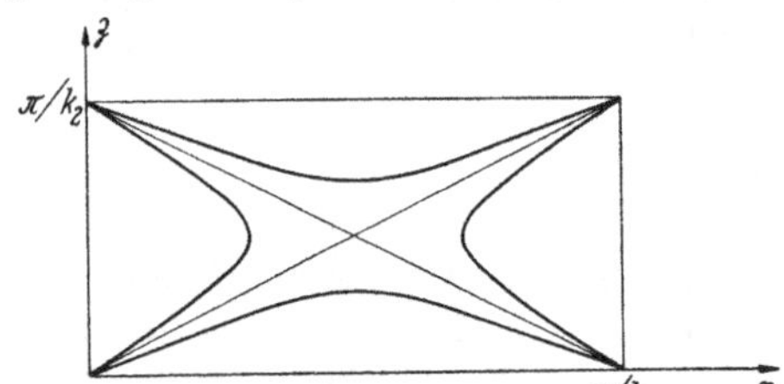

Fig. 6.

In the second case above one may easily visualize the streamlines for the case of pure ring waves, $n = 0$. For finite depht they are given in a plane $\alpha = \text{const}$ by

$$m_0 R\, J_1(m_0 R) \sinh m_0(y+h) = C_1 \tag{14.11}$$

together with the vertical lines at the zeros of $J_1(m_0 R)$. The behavior of the curves is qualitatively similar to that of the two-dimensional case.

In both cases approximations to the path lines can be found as in the two-dimensional case.

β) Progressive waves in an infinite ocean. By taking the proper linear combinations of the standing-wave solutions one may obtain progressive waves. Thus, adding

$$\Phi_1 = a\, e^{\nu y} \cos \nu x \cos \sigma t \quad \text{and} \quad \Phi_2 = a\, e^{\nu y} \sin \nu x \sin \sigma t,$$

one obtains

$$\Phi = a\, e^{\nu y} \cos(\nu x - \sigma t) \tag{14.12}$$

which represents a progressive wave moving to the right with velocity

$$c = \frac{\sigma}{\nu} = \frac{g}{\sigma} = \sqrt{\frac{g}{\nu}} = \sqrt{\frac{g\lambda}{2\pi}}, \tag{14.13}$$

where $\lambda = 2\pi/\nu$ is the wavelength. Subtracting yields a progressive wave moving to the left. If one takes the coefficient in Φ_1 as a_1 and in Φ_2 as a_2, the sum may be written

$$\Phi = \tfrac{1}{2} e^{\nu y} [(a_1 + a_2) \cos(\nu x - \sigma t) + (a_1 - a_2) \cos(\nu x + \sigma t)].$$

This is a superposition of two progressive waves of different amplitudes, one moving to the left and one to the right. If $a_1 = a_2$, a pure progressive wave is obtained; if $a_2 = 0$, one obtains again a standing wave, as a superposition of two progressive waves moving in opposite directions.

For water of finite depth h the corresponding expressions for Φ may be obtained by replacing $e^{\nu y}$ by $\cosh m_0(y+h)$ and ν by m_0. The phase velocity is given by

$$c = \frac{\sigma}{m_0} = \sqrt{\frac{g \tanh m_0 h}{m_0}} = \sqrt{\frac{g\lambda}{2\pi} \tanh \frac{2\pi h}{\lambda}}. \tag{14.14}$$

As $h \to \infty$, the velocity approaches that obtained above for deep water. In fact, if $h/\lambda > 0.2$, the velocity is already within 0.1 of the value for deep water. c is an increasing function of λ, but cannot increase indefinitely as in the case of infinitely deep water, for (14.14) implies

$$c < \sqrt{g h}. \tag{14.15}$$

The streamlines for the progressive wave moving to the right are given by

$$e^{\nu y} |\sin(\nu x - \sigma t)| = C \quad \text{and} \quad \sinh m_0(y+h) |\sin(m_0 x - \sigma t)| = C \tag{14.16}$$

for infinite and finite depth, respectively. At a given instant t these have the same shape relative to a crest as the streamlines for a standing wave. However, since they are time-dependent, the path lines for particles do not lie on the streamlines. The path lines may be found approximately for a particle with equilibrium position (x_0, y_0) from the equations

$$\frac{dx}{dt} = \Phi_x(x_0, y_0, t), \qquad \frac{dy}{dt} = \Phi_y(x_0, y_0, t).$$

This approximation is consistent with the assumptions made in linearizing the boundary condition, as can be seen by assuming a solution in the form

$$x(t) = x_0 + \varepsilon x_1(t) + \cdots, \qquad y(t) = y_0 + \varepsilon y_1(t) + \cdots,$$

where $\varepsilon = a\sigma\nu/2\pi g$ for infinite depth and $\varepsilon = a\sigma m_0/2\pi g$ for finite depth, substituting in the exact path equations, and retaining only first-order terms.

For infinite depth the particle trajectories are given by

$$x = x_0 - a\nu\sigma^{-1} e^{\nu y_0} \cos(\nu x_0 - \sigma t), \quad y = y_0 - a\nu\sigma^{-1} e^{\nu y_0} \sin(\nu x_0 - \sigma t). \tag{14.17}$$

The particles follow circular orbits of radius $a\nu\sigma^{-1} e^{\nu y_0}$ about the equilibrium position (x_0, y_0); at the top of the orbit they are moving in the same direction as the wave. The orbital velocity is $a\nu\, e^{\nu y_0}$, so that the motion dies out quickly as $|y_0|$ increases; for example, at a depth of one wave length the velocity and orbit radius are only $\frac{1}{535}$ the value at the free surface. Although the particles at the crest of a wave are moving in the same direction as the wave, their velocity is not necessarily the same and is, in fact, much smaller in view of the assumed smallness of $\varepsilon = (a\nu/c)(\nu/2\pi)$.

For finite depth the orbits are elliptical with the major axis horizontal:

$$\left.\begin{aligned} x &= x_0 - a\, m_0\, \sigma^{-1} \cosh m_0 (y_0 + h) \cos (m_0\, x - \sigma t)\,, \\ y &= y_0 - a\, m_0\, \sigma^{-1} \sinh m_0 (y_0 + h) \sin (m_0\, x - \sigma t)\,. \end{aligned}\right\} \tag{14.18}$$

The particles again trace the orbit in a clockwise direction except that on the bottom they simply oscillate along a horizontal segment. Fig. 7 from Ruellan and Wallet (1950) shows the path lines for a variety of cases of superposed waves. The topmost picture shows the orbits for a pure progressive wave moving to the right. The bottom picture is a superposition of progressive waves of equal amplitudes moving in opposite directions, i.e. a pure standing wave. The intermediate cases show superpositions with varying ratios of the amplitudes. The intermediate cases are instructive in that not only path lines, but also streamlines are visible.

Since the progressive-wave solutions are steady with respect to a coordinate system moving with the wave, it is clear that we could have obtained a steady-state solution as a small motion superposed upon a uniform flow. If we take a complex velocity potential in the form

$$F(z) = -c\, z + f(z)\,. \tag{14.19}$$

Then [see Eq. (11.6)] f must satisfy

$$\operatorname{Re}\{i\, g\, f + c^2 f'\} = 0 \qquad \text{for } y = 0$$

and either $|f'| \to 0$ as $y \to -\infty$ or $\operatorname{Im} f' = 0$ for $y = -h$. The solution for the first case, infinite depth, is given by

$$f = a\, \mathrm{e}^{-i\nu z} = a\, \mathrm{e}^{\nu y} [\cos \nu\, x - i \sin \nu\, x]\,, \qquad \nu = g/c^2. \tag{14.20}$$

The solution for the finite-depth case is given by

$$\left.\begin{aligned} f &= a \cos m_0 (z + i\, h) \\ &= a\, [\cos m_0\, x \cosh m_0\, (y + h) - i \sin m_0\, x \sinh m_0\, (y + h)]\,, \end{aligned}\right\} \tag{14.21}$$

where m_0 must satisfy

$$c^2\, m_0 - g \tanh m_0\, h = 0\,.$$

The same relation is found in (14.14). A real solution does not exist if $c^2/g\,h > 1$ and in this case there is no wave-like motion consistent with the linearized theory. The streamlines, identical here with the path lines, are obtained from

$$-c\, y + \psi(x, y) = 0\,.$$

One may replace this equation, consistently with the linearization assumptions [cf. (10.18)], by

$$-c\, y + \psi(x, y_0) = 0\,,$$

where y_0 is the mean height of the streamline. Thus, for finite depth, they are given by

$$y = -\frac{a}{c} \sinh m_0\, (y_0 + h) \sin m_0\, x\,, \tag{14.22}$$

an easily constructed family of curves. In the foregoing we have tacitly taken a to be real. However, it may be complex and thus include waves of different phase.

We note finally that (14.8) or (14.9) allow one to construct waves progressing like the spokes of a wheel. However, outwardly progressing waves can be constructed only when the solution involving Y_n is used, and this has a singularity at the origin.

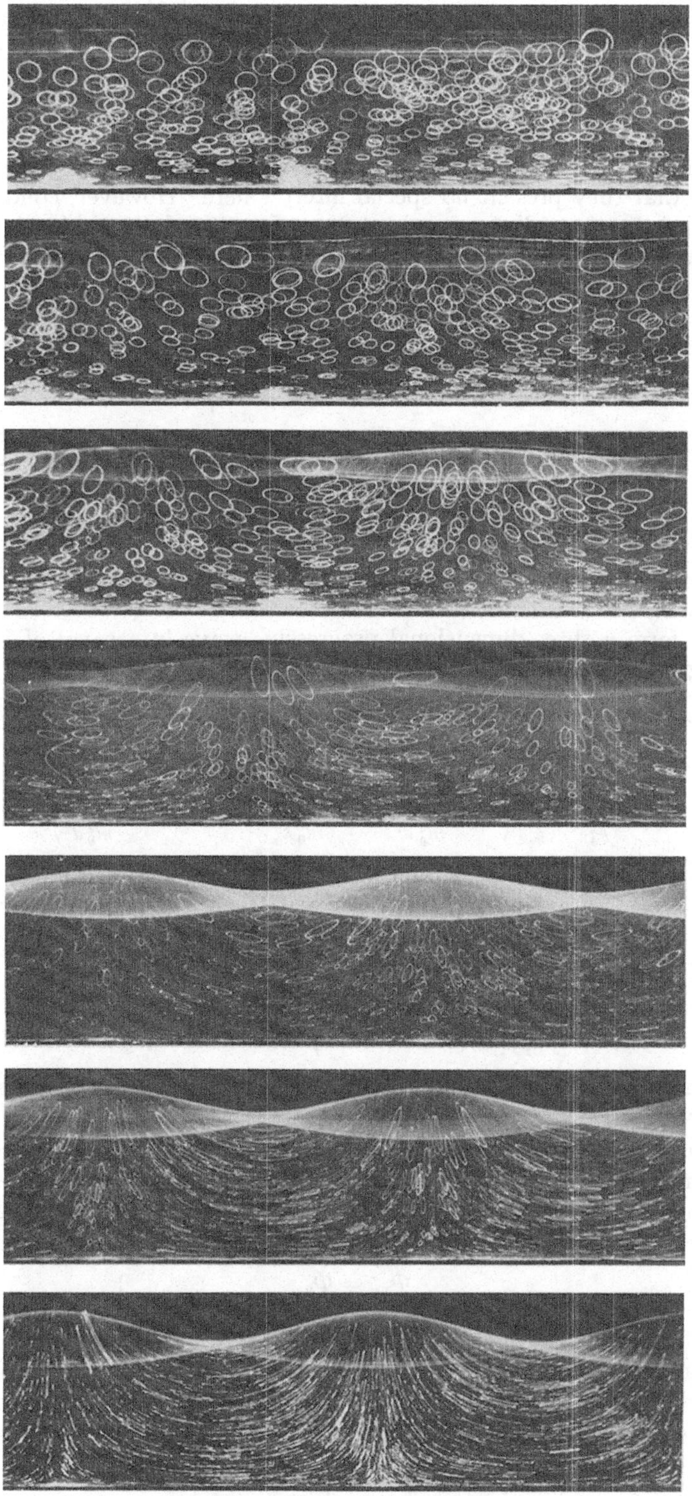

Fig. 7 a—g. Particle trajectories in progressive and standing waves.

γ) Periodic waves in rectangular canals. Let us suppose that the fluid is contained between the planes $z=0$ and $z=d$. Then the velocity potential must satisfy the additional conditions

$$\Phi_z(x, y, 0, t) = \Phi_z(x, y, d, t) = 0. \tag{14.23}$$

This condition is automatically satisfied by the two-dimensional waves discussed in 14α, so that they present no special interest here. However, condition (14.4) does restrict the three-dimensional solutions (14.6) and (14.7), for k_2 must now satisfy (taking $\gamma=0$).

$$k_2 = \frac{n\pi}{d}, \quad n = 1, 2, \ldots.$$

Since $k_1^2 + k_2^2 = \nu^2$ or m_0^2, there can be no solution periodic in x unless

$$n < \frac{\nu d}{\pi} \quad \text{or} \quad n < \frac{m_0 d}{\pi}, \tag{14.24}$$

respectively. Hence, for frequencies below a certain critical frequency σ_1, where

$$\sigma_1 = \sqrt{\frac{\pi g}{d}} \quad \text{or} \quad \sigma_1 = \sqrt{\frac{\pi g}{d} \tanh \frac{\pi h}{d}} \tag{14.25}$$

for infinite or finite depth respectively, there can exist no three-dimensional standing waves in a canal.

Let us form a three-dimensional progressive wave in a canal of finite depth by adding standing-wave solutions:

$$\Phi(x, y, z, t) = a \cosh m_0 (y + h) \cos k_2 z \cos (k_1 x - \sigma t), \qquad k_2 = n\pi/d.$$

The velocity of the progressive wave is given by

$$c^2 = \frac{\sigma^2}{k_1^2} = g h \left(1 - \frac{n^2 \pi^2}{m_0^2 d^2}\right)^{-1} \frac{\tanh m_0 h}{m_0 h} < g h \left(1 - \frac{n^2\pi^2}{m_0^2 d^2}\right)^{-1}. \tag{14.26}$$

As in the case treated above, there can exist no three-dimensional progressive waves unless $\sigma > \sigma_1$. However, if they exist, their velocity is higher than the velocity of two-dimensional waves of the same frequency.

One may define similarly a sequence of critical frequencies $\sigma_1, \sigma_2, \ldots$, where

$$\sigma_k = \sqrt{\frac{k\pi g}{d}} \quad \text{or} \quad \sigma_k = \sqrt{\frac{k\pi g}{d} \tanh \frac{k\pi h}{d}};$$

when $\sigma_k < \sigma < \sigma_{k+1}$, k types of three-dimensional waves are possible with $n = 1, 2, \ldots, k$.

δ) Waves at an interface. Let us now suppose that two fluids are present, one lying over the other. Variables referring to the upper and lower fluids have subscripts 2 and 1 respectively. From (10.7) and (10.8) the linearized boundary conditions for a small disturbance are

$$\left.\begin{aligned} \Phi_{1y} &= \Phi_{2y}, \\ \varrho_1 [\Phi_{1tt} + g \Phi_{1y}] &= \varrho_2 [\Phi_{2tt} + g \Phi_{2y}], \end{aligned}\right\} \tag{14.27}$$

both equations to be satisfied at the equilibrium position of the interface. We shall consider several typical problems.

Let the upper fluid fill the region $y > 0$, and the lower fluid the region $y < 0$. We require of a solution that

$$|\operatorname{grad} \Phi_1| \to 0 \quad \text{as} \quad y \to -\infty \quad \text{and} \quad |\operatorname{grad} \Phi_2| \to 0 \quad \text{as} \quad y \to +\infty.$$

In looking for a standing-wave solution, one may, following Sect. 14α, take

$$\Phi_1 = a_1 e^{my} \varphi(x, z) \cos(\sigma t + \tau), \qquad \Phi_2 = a_2 e^{-my} \varphi(x, z) \cos(\sigma t + \tau),$$

where the relation between a_1 and a_2 and m and σ is to be determined by (14.27), and φ satisfies

$$\Delta_2 \varphi + m^2 \varphi = 0.$$

The first Eq. (14.27) gives immediately that

$$a_1 + a_2 = 0.$$

The second one gives the relation

$$\sigma^2 = \frac{\varrho_1 - \varrho_2}{\varrho_1 + \varrho_2} m g. \tag{14.28}$$

The equation of the interface may be obtained from (10.8):

$$\eta(x, z, t) = \frac{a_1 m}{\sigma} \varphi(x, z) \sin(\sigma t + \tau).$$

Since $a_1 = -a_2$, there is a discontinuity in u (and w if the motion is three-dimensional) as one crosses the interface.

The special choices of $\varphi(x, z)$ made in Sect. 14β may, of course, also be made here. In particular, one may make progressive and standing waves. If one forms two-dimensional progressive waves at the interface, one finds for the velocity

$$c^2 = \frac{\varrho_1 - \varrho_2}{\varrho_1 + \varrho_2} \frac{g}{m}. \tag{14.29}$$

If one assumes the fluids bounded above and below by planes $y = h_2$ and $y = -h_1$, respectively, a similar calculation shows

$$\sigma^2 = \frac{\varrho_1 - \varrho_2}{\varrho_1 \coth m h_1 + \varrho_2 \coth m h_2} g m. \tag{14.30}$$

It is clear from (14.28) and (14.30) that these solutions exist only if $\varrho_2 < \varrho_1$. The case $\varrho_2 > \varrho_1$ will be discussed later.

A more complicated problem of this type is the following [cf. LAMB (1932, § 231), GREENHILL (1887)]. Suppose there is a solid horizontal bottom at $y = -h$, an interface at $y = -d$ and a free surface at $y = 0$. Then, in addition to (14.27) at $y = -d$, Φ_1 and Φ_2 must satisfy

$$\Phi_{2tt} + g \Phi_{2y} = 0 \quad \text{at} \quad y = 0, \qquad \Phi_{1y} = 0 \quad \text{at} \quad y = -h.$$

If one seeks solutions of the form

$$\Phi_2 = (a_2 \cosh m y + b_2 \sinh m y) \varphi(x, z) \cos(\sigma t + \tau),$$
$$\Phi_1 = a_1 \cosh m(y + h) \varphi(x, z) \cos(\sigma t + \tau),$$

substitution in the various boundary conditions yields the following relation between σ and m:

$$\left.\begin{aligned} &\left(\frac{\sigma^2}{g m}\right)^2 [\varrho_1 \coth m d \coth m(h-d) + \varrho_2] - \\ &\quad - \frac{\sigma^2}{g m} \varrho_1 [\coth m d + \coth m(h-d)] + (\varrho_1 - \varrho_2) = 0. \end{aligned}\right\} \tag{14.31}$$

If $\varrho_2 < \varrho_1$, one may establish that there exist two positive solutions for σ^2 for a given m, so that two possible frequencies are possible for a given wave pattern.

If the bottom fluid is taken infinitely deep, one replaces $\coth m(h-d)$ by 1 in (14.31) and the two solutions simplify to

$$\sigma_1^2 = g\,m, \qquad \sigma_2^2 = g\,m\,\frac{\varrho_1-\varrho_2}{\varrho_1\coth m\,d+\varrho_2} < \sigma_1^2. \tag{14.32}$$

The first solution, σ_1, is the same as would be obtained if the two fluids were identical (and there is no discontinuity in u and w at the interface); the second, σ_2, is interpreted below. The inequality $\sigma_2^2<\sigma_1^2$ holds in general, and one may establish

$$\frac{\sigma_2^2}{g\,m} < \{\tanh m\,d,\, \tanh m\,(h-d)\} \leqq \frac{\sigma_1^2}{g\,m} \leqq \min\left\{1,\, \frac{\varrho_1}{\varrho_2}\tanh m\,h\right\}. \tag{14.33}$$

If one computes the ratio of the amplitude of the disturbance at the interface to that at the free surface, one finds, no matter whether h is finite or not,

$$\cosh m\,d - \frac{g\,m}{\sigma^2}\sinh m\,d. \tag{14.34}$$

An examination of the roots of (14.31) shows that the ratio (14.34) is negative for the smaller of the two roots and positive for the larger. Thus, in the solution associated with the smaller root, a maximum of the disturbance at the interface is associated with a minimum of that at the free surface, and vice versa. On the other hand, with the larger root the maxima and minima go together. For the values given in (14.32), the ratio becomes

$$\mathrm{e}^{-m\,d} \quad \text{and} \quad -\frac{\varrho_1}{\varrho_1-\varrho_2}\,\mathrm{e}^{m\,d}, \tag{14.35}$$

respectively. We note that, although the first ratio is <1, the second is in absolute value >1 if $\varrho_2(1+\mathrm{e}^{m\,d})>\varrho_1>\varrho_2$, a condition satisfied if ϱ_1 is only slightly greater than ϱ_2. In fact, the ratio may become very large.

For a given wave length and amplitude of the wave at the free surface one may also compare the amplitudes of the two different modes of motion at the interface. If A_i is the amplitude associated with the frequency σ_i, then for the case $h=\infty$ one finds

$$\left|\frac{A_2}{A_1}\right| = \frac{\varrho_2}{\varrho_1-\varrho_2}\,\frac{1+\tanh m\,d}{1-\tanh m\,d},$$

which may be either less than or greater than 1.

It is of some interest to examine somewhat further the solution associated with the smaller root σ_2 of (14.31). Then, since $a_2/b_2 = g\,m/\sigma^2$, the inequality (14.39) implies that there exists an h_0 with $0<h_0<d$ such that

$$\frac{\sigma_2^2}{g\,m} = \frac{b_2}{a_2} = \tanh m\,h_0 < \tanh m\,d < 1$$

and that

$$\Phi_2 = \sqrt{a_2^2-b_2^2}\cosh m\,(y+h_0)\,\varphi(x,z)\cos(\sigma\,t+\tau).$$

Thus the part of the top fluid between $y=0$ and $y=-h_0$ behaves as if there were a solid boundary at $y=-h_0$; and, of course, the fluid between $y=-h_0$ and $y=-h$ as if it were between solid boundaries. If one has selected solutions for φ which can be combined to form a progressive plane wave, then one may conclude that the velocity $c_2=\sigma_2/m$ associated with this mode of motion has an upper bound:

$$c_2 = \sqrt{\frac{g}{m}\tanh m\,h_0} < \sqrt{g\,d}.$$

In fact, when $h=\infty$, one may verify immediately from (14.32) that

$$c_2=\sqrt{\frac{g}{m}\tanh m\,d\,\frac{\varrho_1-\varrho_2}{\varrho_1+\varrho_2\tanh m\,d}}\leq\sqrt{g\,d\,\frac{\varrho_1-\varrho_2}{\varrho_1}}=c_{2\max}\,.$$

Thus for $h=\infty$ a progressive wave travelling faster than $c_{2\max}$ will consist of only the one mode of motion, i.e. the one associated with σ_1. If $c<c_{2\max}$, there may be two modes of motion excited. This fact is associated with the phenomenon of "dead-water" resistance of ships [see LAMB (1916a), EKMAN (1904), SRETENSKII (1934)].

For superposed fluids one may also find solutions analogous to (14.20) and (14.21). Let us suppose that the first (upper) fluid flows to the left with mean velocity c_2 and the second with mean velocity c_1. We wish to find the possible steady periodic profiles of the interface, assuming as usual that the disturbance is small. The complex velocity potential for each fluid is taken in the form

$$F_1(z)=-c_1\,z+f_1(z)\,,\qquad F_2(z)=-c_2\,z+f_2(z)\,. \tag{14.36}$$

The conditions to be satisfied at the mean common boundary, $y=0$, are:

$$\left.\begin{aligned}c_1^{-1}\operatorname{Im}f_1&=c_2^{-1}\operatorname{Im}f_2\,,\\ \varrho_1\,c_1^{-1}\operatorname{Re}\{i\,g\,f_1+c_1^2\,f_1'\}&=\varrho_2\,c_2^{-1}\operatorname{Re}\{i\,g\,f_2+c_2^2\,f_2'\}\,.\end{aligned}\right\} \tag{14.37}$$

If each fluid extends infinitely far vertically, then

$$f_1=a_1\,e^{-i\,m\,z}\,,\qquad f_2=a_2\,e^{i\,m\,z}$$

give a steady-state solution if

$$\frac{a_1}{c_1}=-\frac{\bar a_2}{c_2}$$

and

$$m=\frac{g\,(\varrho_1-\varrho_2)}{\varrho_1\,c_1^2+\varrho_2\,c_2^2}>0\,, \tag{14.38}$$

where $\bar a_2$ is the complex conjugate of a_2. If the upper fluid is bounded by $y=h_2$ and the lower by $y=-h_1$, then the solution is

$$f_1=a_1\cos m\,(z+i\,h_1)\,,\qquad f_2=a_2\cos m\,(z-i\,h_2)\,,$$

where, letting $a_k=\alpha_k+i\beta_k$, $k=1,2$,

$$\frac{\alpha_1}{c_1}\sinh m\,h_1=-\frac{\alpha_2}{c_2}\sinh m\,h_2\,,\qquad \frac{\beta_1}{c_1}\cosh m\,h_1=\frac{\beta_2}{c_2}\cosh m\,h_2$$

and

$$m=\frac{g\,(\varrho_1-\varrho_2)}{\varrho_1\,c_1^2\coth m\,h_1+\varrho_2\,c_2^2\coth m\,h_2}\,. \tag{14.39}$$

In either case the equation of the interface is given by

$$y=\frac{1}{c_k}\,\psi_k(x,0)\,.$$

SRETENSKII (1952b) has considered a three-dimensional analogue of the above problem in which the direction of flow of one of the fluids makes an angle ϑ with that of the other. Thus, take velocity potentials of the following form:

$$\left.\begin{aligned}\Phi_2(x,y,z)&=-c_2(x\cos\vartheta+z\sin\vartheta)+\varphi_2(x,y,z)\,,\\ \Phi_1(x,y,z)&=-c_1\,x+\varphi_1(x,y,z)\,.\end{aligned}\right\} \tag{14.40}$$

The following are the boundary conditions at the interface $\eta(x, z)$ for small disturbance:

$$\left.\begin{aligned}&\varphi_{2y}(x,0,z)+c_2(\eta_x\cos\vartheta+\eta_z\sin\vartheta)=0, \qquad \varphi_{1y}(x,0,z)+c_1\eta_x=0,\\&g(\varrho_1-\varrho_2)\,\eta=\varrho_1 c_1\varphi_{1x}(x,0,z)-\varrho_2 c_2[\varphi_{2x}(x,0,z)\cos\vartheta+\varphi_{2z}(x,0,z)\sin\vartheta].\end{aligned}\right\}\tag{14.41}$$

For a solution in the form

$$\varphi_1=A_1\,\mathrm{e}^{my}\cos(k_1x+k_2z), \qquad \varphi_2=A_2\,\mathrm{e}^{-my}\cos(k_1x+k_2z), \qquad k_1^2+k_2^2=m^2,$$

the following relations must hold

$$\left.\begin{aligned}&\frac{A_2}{A_1}=-\frac{c_2}{c_1}\,\frac{k_1\cos\vartheta+k_2\sin\vartheta}{k_1}\\ &\text{and}\\ &\varrho_1c_1^2k_1^2+\varrho_2c_2^2(k_1\cos\vartheta+k_2\sin\vartheta)^2=g\,m(\varrho_1-\varrho_2).\end{aligned}\right\}\tag{14.42}$$

These reduce to (14.38) for $\vartheta=0$, $k_1=m_1$ as they should. The equation for the interface is

$$y=-A_1\frac{m}{k_1c_1}\sin(k_1x+k_2z).\tag{14.43}$$

SRETENSKII studies the properties of the solution in more detail.

As a further extension of the preceding cases one may consider a time-dependent disturbance at the interface between two fluids flowing at different velocities. This will be treated in the section on stability of motion.

A natural generalization of the two-fluid system is the n-fluid system [see GREENHILL (1887)] and then the heterogeneous fluid with density given as a series

$$\varrho(x,y,z,t)=\varrho_0(y)+\varepsilon\,\varrho^{(1)}(x,y,z,t)+\varepsilon^2\varrho^{(2)}+\cdots.$$

If one assumes a similar expansion for p and expansions for u, v, w, and η starting with ε, one may derive easily the linearized equations. These, discussion of some periodic solutions, and references to the literature may be found in LAMB (1932, § 235). GROEN (1948) has shown that the period for simple harmonic motion in the linearized problem is a monotonic increasing function of the wave length starting with the minimum $2\pi\sqrt{-\varrho_0(y)/g\,\varrho_0'(y)}$ for $\lambda=+0$. This theorem has been generalized by HEYNA and GROEN (1958) to allow a free upper surface. GROEN (1950) discusses properties of internal waves in an expository way and gives further references to the more recent literature. For some pertinent theorems about waves in heterogeneous fluids see Sect. 32β.

15. Group velocity and the propagation of disturbances and of energy. In the last section we considered periodic waves at a free surface or interface. In this section we wish to consider waves of a given but fairly general initial form and study the way in which they propagate. Although this will entail writing down the solution to a particular initial-value problem, this is of only incidental interest, the chief interest being in the history of the form of the free surface or interface. Initial-value problems as such will be treated in more detail later on. In fact, the remarks below apply equally well to other initial-value problems, for example, an initial distribution of velocity on the surface. What is essential is the resolution of the subsequent motion into a set of waves moving to the right and of ones moving to the left, as in (15.2).

The property of the fluid and its boundaries which is most important for this investigation is the functional relation between the frequency σ and the wave number k. The earlier parts of this chapter have shown that considerable variation is possible in the form of $\sigma(k)$. The two-fluid example with both free surface and interface gave a doubly valued function. A multiply valued function could have been obtained with more layers. However, each branch, or the branch, is a decreasing function of k, approaching zero as $k \to \infty$. When surface tension is taken into account (see Sect. 24), the form of $\sigma(k)$ for large k changes; it then becomes an increasing function, behaving like $k^{\frac{3}{2}}$. If h is large enough, $\sigma(k)$ decreases initially, i.e. for $k<k_m$, reaches a minimum at k_m and then increases; if h is small enough $\sigma(k)$ is everywhere increasing. It will be convenient to extend the definition of $\sigma(k)$ to negative k by setting $\sigma(-k) = -\sigma(k)$.

α) The propagation of an initial elevation. Let us suppose that at time $t=0$ the free surface is given by $y=\eta(x,0)$ and that the fluid is at rest. How does the free surface behave subsequently? One may conveniently think of this as an initial humping up of the fluid near one point, but this is not essential. We shall also suppose that $\eta(x,0)$ is sufficiently restricted to allow a Fourier-integral representation. In part of what follows we shall also assume it to be square integrable, i.e. the total available energy is finite, and on occasion that $x\eta$ is square integrable. Let

$$\left.\begin{aligned}\eta(x,0) &= \int_0^\infty [C(k)\cos kx + S(k)\sin kx]\,dk\\ &= \int_{-\infty}^\infty e^{-ikx}E(k)\,dk = 2\,\mathrm{Re}\int_0^\infty e^{-ikx}E(k)\,dk,\end{aligned}\right\}\tag{15.1}$$

where

$$C(k) = \frac{1}{\pi}\int_{-\infty}^\infty \eta(x,0)\cos kx\,dx, \qquad S(k) = \frac{1}{\pi}\int_{-\infty}^\infty \eta(x,0)\sin kx\,dx,$$

$$E(k) = \frac{1}{2\pi}\int_{-\infty}^\infty \eta(x,0)\,e^{ikx}\,dx = \frac{1}{2}[C(k) + i\,S(k)].$$

We shall call $E(k)$ the *spectrum* of $\eta(x,0)$. Note that $E(-k) = E^*(k)$, the complex conjugate of $E(k)$ (we change notation temporarily in order to avoid conflict with the notation for averages introduced below).

A formal solution for Φ and $\eta(x,t)$ may be written down immediately:

$$\left.\begin{aligned}\Phi(x,y,t) &= -\int_0^\infty \frac{\sigma(k)}{k}\,Y(y)\,[C(k)\cos kx + S(k)\sin kx]\sin\sigma t\,dk\\ &= -\int_{-\infty}^\infty \frac{\sigma(k)}{k}\,Y(y)\,E(k)\,e^{-ikx}\sin\sigma t\,dk\\ &= \frac{1}{2}i\int_{-\infty}^\infty \frac{\sigma(k)}{k}\,Y(y)\,E(k)\,[e^{-i(kx-\sigma t)} - e^{-i(kx+\sigma t)}]\,dk,\\ \eta(x,t) &= \int_0^\infty [C(k)\cos kx + S(k)\sin kx]\cos\sigma t\,dk\\ &= \int_{-\infty}^\infty e^{-ikx}E(k)\cos\sigma t\,dk = \frac{1}{2}\int_{-\infty}^\infty E(k)\,[e^{-i(kx-\sigma t)} + e^{-i(kx+\sigma t)}]\,dk.\end{aligned}\right\}\tag{15.2}$$

Here $Y(y) = \cosh k(y+h)/\sinh kh$ for a single fluid of depth h, $Y(y) = e^{|k|y}\,\mathrm{sgn}\,k$ for infinite depth (the peculiar modification of Y for $h=\infty$ is necessary for

$k<0$). However, more general situations are allowable in which, for example, $\eta(x, t)$ describes an interface. The choice of an expression for Φ has been based upon the kinematic boundary condition $\Phi_y(x, 0, t) = \eta_t(x, t)$ in order not to exclude the possibility of surface tension. For simplicity we also restrict ourselves to single-valued σ's. For more complicated problems, such as the two-fluid problem with both free surface and interface discussed in Sect. 14 δ, the freedom to fix both $\eta_1(x, 0)$ and $\eta_2(x, 0)$ independently requires the determination of two spectra for each surface with relations between them set by (14.34). The remarks below will still apply to motion resulting from each spectrum separately. Finally, we note that a statement concerning specific conditions to be satisfied by $\eta(x, 0)$ for the case of a single free surface may be found in a paper by Kampé de Fériet and Kotik (1953).

It is clear from (15.2) that one may express $\eta(x, t)$ as a sum of two functions, one, say $\eta_R(x, t)$, representing a superposition of waves moving to the right, the other, η_L, waves moving to the left. We consider only η_R since similar remarks apply to η_L with x replaced by $-x$. The spectrum of η_R is given by $\frac{1}{2} E(k)\, e^{i\sigma(k)t}$, so that clearly $\sigma(k)$ plays an important role in the change of shape of η_R. Since each harmonic component in η_R is moving to the right with velocity $\sigma(k)/k$, and since this is not a constant in the cases we have been considering, the different components will move with different velocities and we shall expect η_R to change its shape with time, even though moving as a whole to the right.

In order to get some idea of the overall motion it is reasonable to try to compute an average position of $\eta_R(x, t)$ and find how this moves. One must first decide how to define the average position. One possibility, which, as we shall see presently, is unsatisfactory is to use η_R itself as the weighting function, i.e. to define

$$\bar{x}_R(t) = \int_{-\infty}^{\infty} x\,\eta_R(x, t)\,dx \Big/ \int_{-\infty}^{\infty} \eta_R(x, t)\,dx$$

when this exists. An easy computation shows that

$$\bar{x}_R(t) = \bar{x}_R(0) + \sigma'(0)\,t,$$

i.e. the average motion is, on this definition, independent of the form of $\sigma(k)$ except near $k=0$. For deep-water gravity waves $\sigma'(0) = \infty$; for depth h, $\sigma'(0) = \sqrt{gh}$, the maximum velocity [see Eq. (14.15)]. In conformity with the above one may define the "spread" of the hump to be

$$\int_{-\infty}^{\infty} [x - \bar{x}_R(t)]^2\,\eta_R(x, t)\,dx \Big/ \int_{-\infty}^{\infty} \eta_R(x, t)\,dx.$$

A computation shows that this remains constant in time, when it exists. This definition of average is unsatisfactory, as could have been expected inasmuch as the weighting function can become negative. We note in passing that

$$\int_{-\infty}^{\infty} \eta_R(x, t)\,dx = \int_{-\infty}^{\infty} \eta_R(x, 0)\,dx,$$

an expression of conservation of mass.

Another possible weighting function without this shortcoming, but still allowing ease of computation, is $\eta_R^2(x, t)$. We note first that

$$\int_{-\infty}^{\infty} \eta_R^2(x, t)\,dx = \int_{-\infty}^{\infty} \eta_R^2(x, 0)\,dx = \tfrac{1}{2}\pi \int_{-\infty}^{\infty} E(k)\,E^*(k)\,dk.$$

Let us define two averages, one for functions of x:

$$\bar{f}(t) = \int_{-\infty}^{\infty} f(x)\, \eta_R^2(x, t)\, dx \Big/ \int_{-\infty}^{\infty} \eta_R^2(x, t)\, dx,$$

and one for functions of k:

$$\bar{\Phi} = \int_{-\infty}^{\infty} \Phi(k)\, E(k)\, E^*(k)\, dk \Big/ \int_{-\infty}^{\infty} E(k)\, E^*(k)\, dk.$$

Then, assuming that the various quantities in question exist, one finds, using well known theorems on Fourier transforms[1],

$$\bar{x}_R(t) = \bar{x}_R(0) + \bar{\sigma}' t \tag{15.3}$$

and

$$\left.\begin{aligned}\overline{[x - \bar{x}_R(t)]^2} = \overline{[x - \bar{x}_R(0)]^2} + t\left\{\overline{\sigma'\left[i \log \frac{E^*}{E}\right]'} - \bar{\sigma}'\,\overline{\left[i \log \frac{E^*}{E}\right]'}\right\} + \\ + t^2\left\{\overline{\sigma'^2} - \bar{\sigma}'^2\right\}.\end{aligned}\right\} \tag{15.4}$$

Thus, on this definition the average position of η_R moves to the right with constant velocity $\bar{\sigma}'$ and the hump spreads according to a quadratic law. We note that the coefficient of t^2 is positive except if σ' is a constant, when it vanishes. It may become infinite, and, in fact, does so for infinitely deep water if the gravest modes are present, i.e., if $\int \eta_R\, dx \neq 0$. The coefficient of t vanishes if σ' is constant or if $[i \log E^*/E]'$ is constant; the latter will occur if $\eta(x, 0)$ is either symmetric or antisymmetric about some point x_0, but this does not exhaust all possibilities. The sign of this term does not seem to be determined, so that the spread of the hump may conceivably decrease before starting to increase.

Investigations of the motion of the average position of the hump and of its spread give only a rather crude picture of its behavior. By other methods outlined below one may obtain further insight into the motion.

We begin by applying the analysis of the average motion to that part of η_R resulting from only a narrow band in its spectrum. Let

$$\eta_R(x, t; k_0, \varepsilon) = \mathrm{Re} \int_{k_0-\varepsilon}^{k_0+\varepsilon} \tfrac{1}{2} E(k)\, \mathrm{e}^{-i(kx - \sigma(k)t)}\, dk. \tag{15.5}$$

We shall call this a *wave packet*. The average position satisfies

$$\bar{x}_R(t; k_0, \varepsilon) = \bar{x}_R(0; k_0, \varepsilon) + \bar{\sigma}'(k_0, \varepsilon)\, t,$$

where $\bar{\sigma}'(k, \varepsilon)$ is now the average of $\sigma'(k)$ over the narrow band $[k_0 - \varepsilon, k_0 + \varepsilon]$. The narrower the band, the closer $\bar{\sigma}'(k_0, \varepsilon)$ is to $\sigma'(k_0)$, assuming the latter continuous. As a limiting case we shall say that the wave packet resulting from an infinitesimal band about k_0 moves with velocity $\sigma'(k_0)$. It is customary to call $\sigma'(k)$ the *group velocity*. This is the same as the phase velocity $\sigma(k)/k$ only if $\sigma = a k$. A wave packet will spread with passage of time unless the two velocities are equal, for (15.4) is applicable to the wave packet with the restricted definition of average. As might be expected, the smaller the width of the band, the smaller the coefficient of t^2 and the smaller the rate of growth. However, as we shall see below, the initial spread may be wide for a narrow band.

The wave packet (15.5) may also be investigated by a different method. Let us expand $\sigma(k)$ in the first few terms of a Taylor series about k_0:

$$\sigma(k) = \sigma(k_0) + \sigma'(k_0)\,(k - k_0) + \tfrac{1}{2} \int_{k_0}^{k} \sigma''(\varkappa)\,(k - \varkappa)\, d\varkappa. \tag{15.6}$$

[1] See, e.g., S. Bochner and K. Chandrasekharan: Fourier transforms, Chap. IV, § 2. Princeton 1949.

We may then write

$$\left.\begin{aligned}\eta_R(x,t;k_0,\varepsilon) &= \operatorname{Re}\tfrac{1}{2}\,\mathrm{e}^{-i[k_0x-\sigma(k_0)t]}\left\{\int_{k_0-\varepsilon}^{k_0+\varepsilon} E(k)\,\mathrm{e}^{-i[x-\sigma'(k_0)t](k-k_0)}\,dk\,+\right.\\ &\quad\left.+\int_{k_0-\varepsilon}^{k_0+\varepsilon} E(k)\,\mathrm{e}^{-i[x-\sigma'(k_0)t](k-k_0)}\left[\exp\left(-\tfrac{1}{2}\,i\,t\int_{k_0}^{k}\sigma''(\varkappa)\,(k-\varkappa)\,d\varkappa\right)-1\right]dk\right\}\\ &= \operatorname{Re}\tfrac{1}{2}\,\mathrm{e}^{-i[k_0x-\sigma(k_0)t]}\,M\big(x-\sigma'(k_0)\,t;\,k_0,\varepsilon\big)+R.\end{aligned}\right\} \tag{15.7}$$

Using the inequality $|\mathrm{e}^{iu}-1|\leqq|u|$, one finds

$$|R|\leqq\tfrac{1}{4}\,t\,\varepsilon^3\max_{|k-k_0|<\varepsilon}|E(k)|\cdot\max_{|k-k_0|<\varepsilon}|\sigma''(k)|. \tag{15.8}$$

The remainder can thus be made small by taking ε or t small enough. However, once ε is fixed, R will eventually become large as t increases. Let us suppose, however, that t and ε are small enough so that the first term determines the main features of the motion. The first factor represents a periodic wave of wave number k_0 moving with its phase velocity $\sigma(k_0)/k_0$. The second factor, determining the amplitude of the first, represents a profile being translated to the right with velocity $\sigma'(k_0)$. Thus one may say that the gross outline of the surface is moving to the right with the group velocity. One may see this more clearly if one assumes ε small enough so that we may take $E(k)$ as constant over the band width. Then

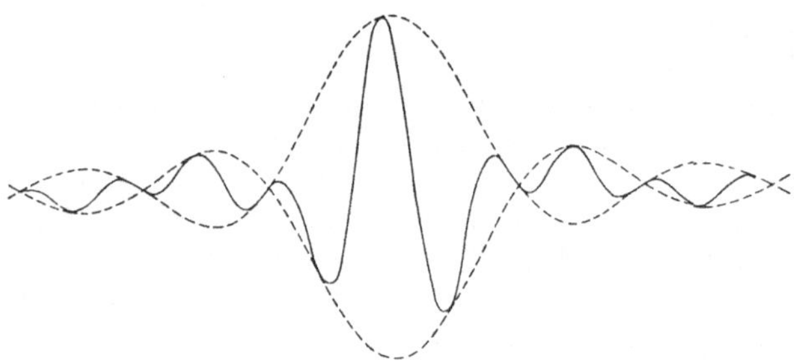

Fig. 8.

$$M\big(x-\sigma'(k_0)\,t;\,k_0,\varepsilon\big)=E(k_0)\,\frac{\sin\big(x-\sigma'(k_0)\,t\big)\,\varepsilon}{x-\sigma'(k_0)\,t},$$

and $\eta_R(x,t;k_0,\varepsilon)$ appears approximately as in Fig. 8. Here the dotted enveloping curves represent $\pm\tfrac{1}{2}M$ and move to the right with velocity $\sigma'(k_0)$, whereas the inscribed solid curves represent the first factor and move to the right with phase velocity $\sigma(k_0)/k_0$. The whole moves as a fixed pattern only if the two velocities are equal. Otherwise, assuming $\sigma'(k_0)<\sigma(k_0)/k_0$, the inscribed curves will progress through the wave packet, gradually disappearing at the right. For a very narrow band the packet will spread wide before its first zero on either side of the maximum.

A disadvantage of this last analysis is that it becomes less and less accurate as t becomes large. However, there exists another approximation to $\eta_R(x,t)$ for large values of t which helps to complete the picture. This ultimate behavior of η_R can to some extent be predicted from the analysis of the average motion of a wave band. If we think of η_R as made up of the contributions from a number of narrow wave bands, we know that each contribution is moving with the average group velocity of the band. Thus after some time we shall expect that these various contributions will have separated from one another, with the bands about the gravest modes, which travel fastest, having progressed the furthest. This prediction will be confirmed.

What is needed for this final approximation is an asymptotic expansion for large t. It is convenient to express η_R in the slightly altered form

$$\eta_R(x,t)=\tfrac{1}{2}\int_{-\infty}^{\infty}E(k)\,\mathrm{e}^{-i\left[k\frac{x}{t}-\sigma(k)\right]t}\,dk \tag{15.9}$$

and to consider it as depending upon the two parameters x/t and t. Then for each value of x/t we shall give an expansion for large values of t. For a derivation of the expansion we refer to STOKER (1957, § 6.8) or ERDÉLYI (1956, § 2.9).

Let the functions $k_r(x/t)$, $r=1, 2, \ldots, n$, be defined by

$$\sigma'(k_r) = x/t; \tag{15.10}$$

i.e. we allow the possibility of several roots. In the situation of interest to us there will be either one or two roots, or none. The asymptotic expression for η_R is then given by

$$\left.\begin{aligned}\eta_R(x,t) &= \mathrm{Re}\sum_r \frac{1}{2} E(k_r)\left[\frac{2\pi}{t\,|\sigma''(k_r)|}\right]^{\frac{1}{2}} \mathrm{e}^{-i[k_r x - \sigma(k_r)t - \frac{1}{4}\pi\,\mathrm{sgn}\,\sigma''(k_r)]} + \\ &\quad + \mathrm{Re}\sum_r \frac{1}{2} E(k_r)\frac{1}{\sqrt{3}}\,\Gamma\!\left(\frac{1}{3}\right)\left[\frac{6}{t\,|\sigma'''(k_r)|}\right]^{\frac{1}{3}} \mathrm{e}^{-i[k_r x - \sigma(k_r)t]} + O(t^{-\frac{2}{3}}),\end{aligned}\right\} \tag{15.11}$$

where the first summation is over all values of r for which $\sigma''(k_r)\neq 0$ and the second over all k_r for which $\sigma''(k_r)=0$ but $\sigma'''(k_r)\neq 0$; further terms would be necessary for values of r for which both vanish but this will not occur in our examples. If some $k_r=0$, then the corresponding term must be multiplied by $\frac{1}{2}$. For a value of x/t for which no solution to (15.10) exists, it is easy to show by a change of variables in (15.9), say $u=k\,x/t-\sigma(k)$, and integration by parts that $\eta_R(x,t)=O(t^{-1})$.

Let us examine in some detail the implications of one term of (15.11), say $r=1$, for the motion of η_R; if several terms are present for a given value of x/t one must superpose the resultant motions.

If x/t is held constant while t increases, then clearly one must set $x=\sigma'(k_1)\,t$, i.e. we are examining η_R from the standpoint of an observer moving with group velocity $\sigma'(k_1)$. Since the coefficient of the harmonic term is $t^{-\frac{1}{2}}$ times a function of k_1, which is being held constant, the gross outline of η_R will appear constant in form, but decreasing in amplitude because of $t^{-\frac{1}{2}}$. However, just as in the analysis of (15.7), there is a harmonic of wave number k_1 moving through the gross outline with phase velocity $\sigma(k_1)/k_1$. The amplitude of the gross outline is proportional to $E(k_1)$, but also depends now upon $\sigma''(k_1)$, in contrast to the situation for small t according to (15.7).

If the value of x/t is such that $\sigma''(k_1)=0$, then one must examine a term from the second summation in (15.11). It is evident that the interpretation is the same except that σ''' occurs in place of σ'' and that the amplitude decreases more slowly because of the $t^{-\frac{1}{3}}$. This situation can happen, for example, in the case of gravity waves in water of depth h for $x=t\sqrt{gh}$. Then $k_1(\sqrt{gh})=0$, $\sigma''(0)=0$, and $\sigma'''(0)=-h^2\sqrt{gh}$. This also occurs for combined gravity-capillary waves when the curve $\sigma'(k)$ has a minimum.

The approximation (15.11) to η_R will obviously be very poor for a value x/t such that $\sigma''(k_r)$ is near to zero for some r unless t is extremely large. It is shown elsewhere[1] how an Airy function may be used to modify the relevant term in the second summand to give a useful asymptotic expansion for k_r near a zero of σ''.

If x/t is fixed at a value for which (15.10) has no solution, then for an observer moving with this velocity the disturbance of the surface is very small, for it has been dying out as t^{-1}. The first term of the expansion may, of course, be com-

[1] H. JEFFREYS and B. JEFFREYS: Methods of mathematical physics, 3rd ed., § 17.09. Cambridge 1956. — See also C. CHESTER, B. FRIEDMAN and F. URSELL: Proc. Cambridge Phil. Soc. 53, 599–611 (1957).

puted as indicated above. This situation will occur for a disturbance in water of depth h if $x/t > \sqrt{gh}$. It will also occur when surface tension is taken into account for $x/t < \sigma'_{\min}$.

The asymptotic expansion (15.11) may also be used in a different fashion. Let us fix our attention upon one value of x and let t increase. Then x/t will decrease and the value $k_1(x/t)$ associated with the point x at a given moment will also change; for pure gravity waves it will increase. The observer stationed at x will then observe waves of continually increasing wave number (decreasing wave length) moving by with phase velocities appropriate to their lengths. The amplitudes at a given instant will depend upon the first factor. The gross outline of the waves will pass the observer at the group velocity appropriate to the wave number present at the moment, and, of course, the amplitude is decreasing as $t^{-\frac{1}{2}}$. In the case of a disturbance on water of depth h, if the observer is initially far from the hump, then even for large enough values of t for the asymptotic expansion to be valid the value of x/t may be greater than $\sqrt{gh}$. Then the observer will see practically no disturbance until the gravest modes begin to reach him. We note again that he must anticipate the arrival of a given wave number by its group velocity, not phase velocity, for it is the former which controls the amplitude. In the case of combined gravity-capillary waves, when t is large enough one will have $x/t < \sigma'_{\min}$ and the disturbance will be negligible.

It is also possible to find an asymptotic expansion for $\eta_R(x, t)$ for x/t fixed and large x. It turns out to be the same as (15.11) with $O(t^{-\frac{3}{2}})$ replaced by $O(x^{-\frac{3}{2}})$. This expansion allows one, so to speak, to take snapshots of the right-hand end of η_R at different instants of time. If we fix t and let x increase, x/t increases also and $k_1(x/t)$ decreases for pure gravity waves. Thus the wave length increases as one moves to the right; the observed amplitude will depend upon the first factor. For gravity waves on water of depth h, if x is large enough, $x/t > \sqrt{gh}$ and the disturbance will be small of order x^{-1}.

Finally, we use the asymptotic expansion to investigate the motion of a particular phase of $\eta_R(x, t)$, say a zero, for large t. Such a point will be determined by

$$\alpha(x, t) \equiv k_1 x - \sigma(k_1)\, t = \text{const},$$

where, as usual, $k_1 = k_1(x/t)$; solving for x gives $x = x(t)$. One may find $\dot{x}(t)$ from

$$\dot{x}(t) = -\frac{\alpha_t}{\alpha_x} = -\frac{-k_1'\frac{x}{t^2} - \sigma(k_1) + \sigma'(k_1)\frac{x}{t^2}\,t}{k_1 + k_1'\frac{x}{t} - \sigma'(k_1)\,k_1'} = \frac{\sigma(k_1)}{k_1}.$$

Thus a particular phase travels with the phase velocity of the harmonic component associated with it at the moment. However, if the group and phase velocities are different, it is then moving at a different velocity from a point just keeping pace with waves of a given wave number. In particular, for gravity waves it is moving faster, hence moves into region of lower wave number and higher velocity and is accelerating. Acomputation of $\ddot{x}$ bears this out:

$$\ddot{x}(t) = -\frac{1}{t\,k_1\,\sigma''}\left[\frac{\sigma}{k_1} - \sigma'\right]^2,$$

for this is always positive for gravity waves. The right-hand side is, of course, a function of x and t. For deep-water gravity waves the function $x(t)$ may easily be found from the earlier equation:

$$x = \frac{\sigma(k_1)}{k_1}\,t - \frac{a}{k_1} = 2\sigma'(k_1)\,t - \frac{a}{k_1} = 2\,\frac{x}{t}\,t - \frac{4a}{g}\,\frac{x^2}{t^2}$$

or

$$x(t) = \frac{g\,t^2}{4a}.$$

Hence $\ddot{x} = g/2a$ and for large t the acceleration is constant. If the depth is finite, the computation is no longer simple, although it is possible to show that $x(t)$ varies from $x(t) = t\sqrt{g\,h}$ for a phase associated with $k=0$ to $x(t) = A\,t^2$ for a phase associated with very large k.

Fig. 9 is taken from a paper of KELVIN's (1907), and shows the computed values of $\eta(x, t)$ for an initial displacement given by

$$\eta(x, 0) = \frac{[1+(1+x^2)^{\frac{1}{2}}]^{\frac{1}{2}}}{2^{\frac{3}{2}}(1+x^2)^{\frac{3}{2}}}\,[2-(1+x^2)^{\frac{1}{2}}]$$

and for $t/\pi^{\frac{1}{2}} = \frac{1}{2}$, 1, $\frac{3}{2}$, 4, 8 (the units have been chosen so that $g=4$). The description of the behavior of $\eta_R(x, t)$ outlined in the preceding paragraphs can be easily

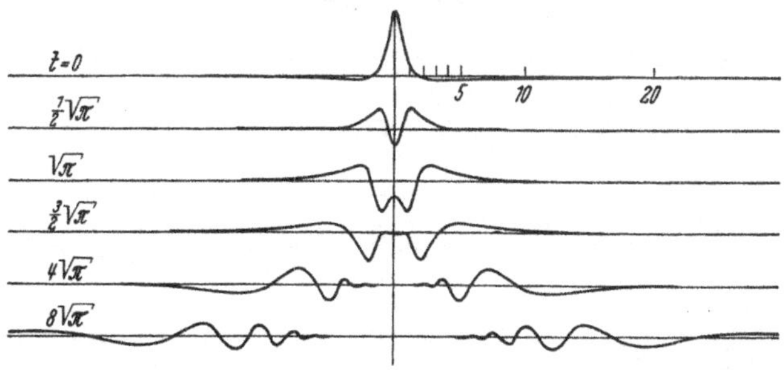

Fig. 9.

verified qualitatively by inspection of the successive snapshots of $\eta_R(x, t)$ GREEN (1909) has shown that if one estimates the wave length at any maximum as double the distance between the two including zeros, then the position is very close to that which would be estimated by using the group velocity (cf. HAVELOCK, 1914, p. 37).

Fig. 10 from a report by J.E. PRINS (1956; also 1958b) shows measured time histories taken at various distances from the center of an initial rectangular hump of length $2L$ and height Q in water of depth h for specific values shown in the figure. In general, the features of the motion described above were well verified by this experimental investigation.

We assemble here the expressions for $\sigma(k)$ and $k\sigma'/\sigma$ for a number of cases of water waves.

1. Deep-water gravity waves:

$$\sigma(k) = \sqrt{g\,k}, \qquad \frac{k\,\sigma'}{\sigma} = \frac{1}{2}.$$

2. Gravity waves at the interface of two fluids, each of infinite vertical extent:

$$\sigma(k) = \sqrt{\frac{\varrho_1 - \varrho_2}{\varrho_1 + \varrho_2}\,g\,k}, \qquad \frac{k\,\sigma'}{\sigma} = \frac{1}{2}.$$

3. Gravity waves in water of depth h:

$$\sigma(k) = \sqrt{g\,k \tanh k\,h}\,, \qquad \frac{k\,\sigma'}{\sigma} = \frac{1}{2}\left[1 + \frac{2k\,h}{\sinh 2k\,h}\right].$$

4. Gravity waves for a layer of thickness d of one fluid over a deep layer of a heavier one:

$$\sigma_1(k) = \sqrt{g\,k}\,, \qquad \frac{k\,\sigma_1'}{\sigma_1} = \frac{1}{2}\,,$$

$$\sigma_2(k) = \sqrt{\frac{\varrho_1 - \varrho_2}{\varrho_1 \coth k\,d + \varrho_2}\,g\,k}\,,$$

$$\frac{k\,\sigma_2'}{\sigma_2} = \frac{1}{2}\left[1 + \frac{2\varrho_1\,k\,d}{\varrho_1 \sinh 2k\,d + 2\varrho_2 \sinh^2 k\,d}\right].$$

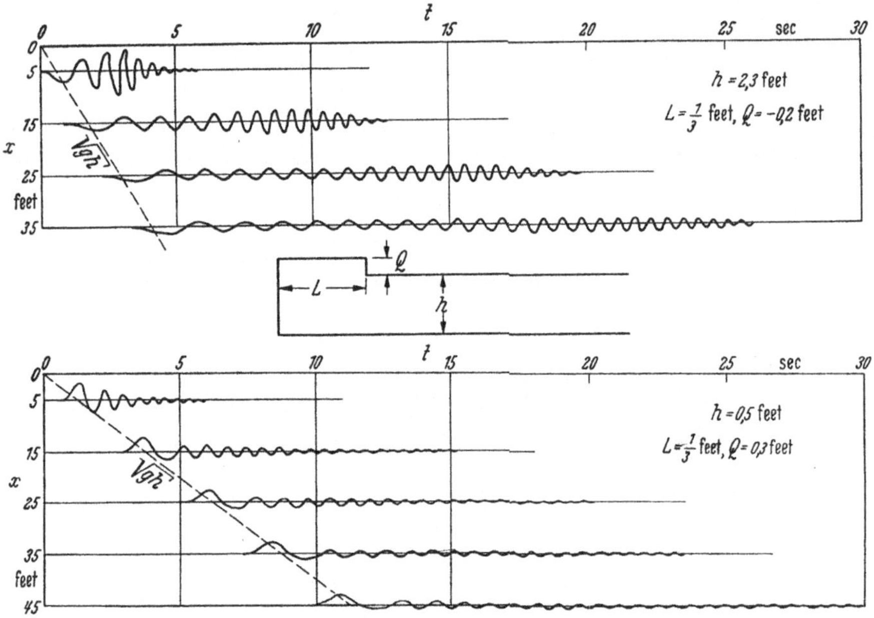

Fig. 10.

5. Waves at a free surface of a deep fluid with both gravity and surface tension acting:

$$\sigma(k) = \sqrt{g\,k + \frac{T\,k^3}{\varrho}}\,, \qquad \frac{k\,\sigma'}{\sigma} = \frac{1}{2}\,\frac{1 + 3\,T\,k^2/\varrho\,g}{1 + T\,k^2/\varrho\,g}\,.$$

6. Waves at a free surface of a fluid of depth h with both gravity and surface tension acting:

$$\sigma(k) = \sqrt{\left(g\,k + \frac{T\,k^3}{\varrho}\right)\tanh k\,h}\,,$$

$$\frac{k\,\sigma'}{\sigma} = \frac{1}{2}\left[1 + \frac{2h\,k}{\sinh 2h\,k} + 2\,\frac{T\,k^2/\varrho\,g}{1 + T\,k^2/\varrho\,g}\right].$$

In cases 1 to 4 σ'' is always negative if $k > 0$. In case 5 it crosses the k-axis at $k = [g\,\varrho\,T^{-1}\,\frac{1}{3}(2\sqrt{3} - 3)]^{\frac{1}{2}}$ and becomes positive. In cases 1 to 4 $\sigma' < \sigma/k$ for $k > 0$. In case 5 $\sigma' < \sigma/k$ for $0 < k < \sqrt{g\,\varrho/T}$; then σ' crosses σ/k at the minimum of the latter and thereafter remains larger. (Note that σ' always passes through a stationary value of σ/k, passing from beneath to above in going through a minimum, and the reverse at a maximum.) We shall not discuss 6 in detail. For $h > h_c = \sqrt{3\,T/2\varrho g}$, σ/k has a minimum for some k_0, $0 < k_0 < \sqrt{\varrho g/T}$ and σ' a minimum to the left of this. For $h \leqq h_c$, σ/k is an increasing function, starting at $\sqrt{g h}$ for $k = 0$, and σ' is also increasing, $\sigma' > \sigma/k$ for $k > 0$, $\sigma'(0) = \sqrt{g h}$. Fig. 11 shows graphs of σ, σ/k and σ' for 1, 3, 5, and 6 (the scales were chosen for convenience).

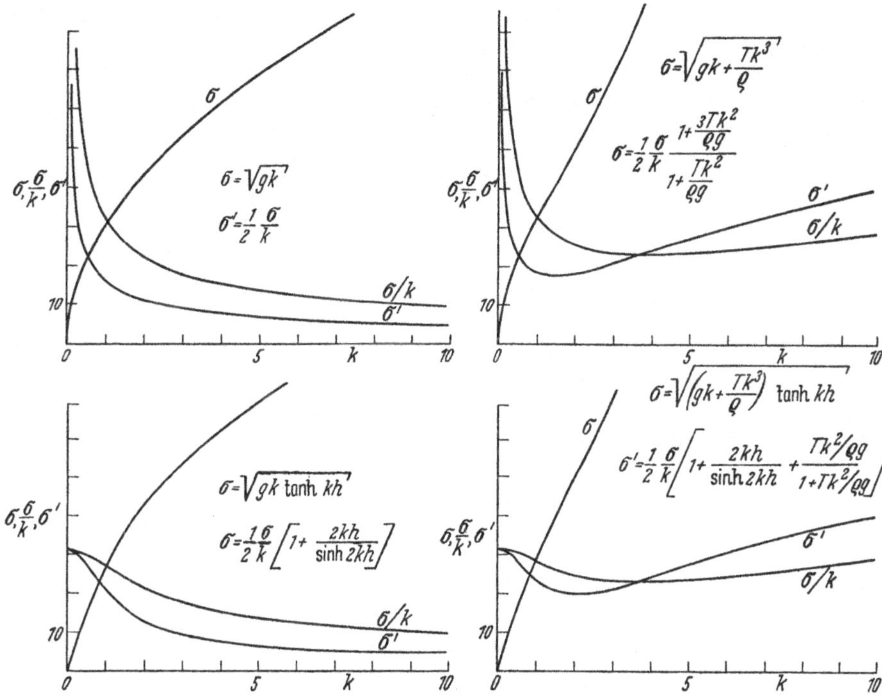

Fig. 11.

One may also take $\lambda = 2\pi/k$ as the independent variable, and then express the phase velocity c and group velocity U as functions of λ. An easy computation shows that

$$\lambda \frac{dc}{d\lambda} = c - U.$$

This equation has a simple interpretation in the geometry of the curve for $c(\lambda)$, as was shown by LAMB (1932, p. 382): For a given value of λ, U is the intercept on the vertical axis of the tangent to the $c(\lambda)$ curve at the point $(\lambda, c(\lambda))$. One value of U may correspond to more than one value of λ, as, for example, in the case of gravity-capillary waves. See HAVELOCK (1914, § 11).

β) The propagation of energy. It seems intuitively clear that as long as the right-moving part of an initial hump keeps its integrity the energy associated with the motion will in some sense move with the hump. We wish to consider in what sense this is true. We limit ourselves in the following discussion to a single fluid of depth h, where h may become infinite. However, surface tension may act upon the free surface.

We first introduce the notion of energy density for a given value of x. It will be convenient to separate potential, kinetic and surface energy. Let

$$\left.\begin{aligned}
\mathscr{V}(x,t) &= \varrho g \int_0^{\eta_R(x,t)} y\,dy = \tfrac{1}{2}\varrho g \eta_R^2(x,t),\\
\mathscr{T}(x,t) &= \tfrac{1}{2}\varrho \int_{-h}^{0} (\Phi_x^2+\Phi_y^2)\,dy = \tfrac{1}{2}\varrho \int_{-h}^{0} (\Phi\,\Phi_x)_x\,dy + \tfrac{1}{2}\varrho\,\Phi\,\Phi_y(x,0,t),\\
\mathscr{S}(x,t) &= \tfrac{1}{2} T \eta_{Rx}^2(x,t)
\end{aligned}\right\} \tag{15.12}$$

be the densities of potential, kinetic and surface energies, respectively, where here Φ is the velocity potential corresponding to η_R.

These functions may now be treated in the same way as η_R was in Sect. 15α. We may ask for the average position of the distributions of the several densities. They are defined by

$$\left.\begin{aligned}
\bar{x}_V(t) &= \int_{-\infty}^{\infty} x\,\mathscr{V}(x,t)\,dx \Big/ \int_{-\infty}^{\infty} \mathscr{V}(x,t)\,dx,\\
\bar{x}_T(t) &= \int_{-\infty}^{\infty} x\,\mathscr{T}(x,t)\,dx \Big/ \int_{-\infty}^{\infty} \mathscr{T}(x,t)\,dx,\\
\bar{x}_S(t) &= \int_{-\infty}^{\infty} x\,\mathscr{S}(x,t)\,dx \Big/ \int_{-\infty}^{\infty} \mathscr{S}(x,t)\,dx,
\end{aligned}\right\} \tag{15.13}$$

respectively. Since all three densities are non-negative, one avoids the difficulty met with in defining the average position of η_R. In fact, it is obvious that the definitions of $\bar{x}_R$ and $\bar{x}_V$ coincide, so that the conclusions concerning $\bar{x}_R$ can be applied immediately to $\bar{x}_V(t)$. In particular,

$$\bar{x}_V(t) = \bar{x}_V(0) + \bar{\sigma}' t. \tag{15.14}$$

Consider now $\bar{x}_T(t)$. First we note that, from GREEN's Theorem,

$$\int_{-\infty}^{\infty} \mathscr{T}(x,t)\,dx = \tfrac{1}{2}\varrho \int_{-\infty}^{\infty} \Phi(x,0,t)\,\Phi_y(x,0,t)\,dx + \\ + \lim_{\substack{x_1\to-\infty\\ x_2\to+\infty}} \tfrac{1}{2}\varrho \int_{-h}^{0} [-\Phi(x_1,y,t)\,\Phi_x(x_1,y,t) + \Phi(x_2,y,t)\,\Phi_x(x_2,y,t]\,dy.$$

From the assumed square-integrability of η_R, the limit vanishes. Use of the identity $x(\Phi_x^2+\Phi_y^2) = (x\Phi)_x\,\Phi_x + (x\Phi)_y\,\Phi_y - \Phi\,\Phi_x$ and GREEN's Theorem gives

$$\int_{-\infty}^{\infty} x\,\mathscr{T}(x,t)\,dx = \tfrac{1}{2}\varrho \int_{-\infty}^{\infty} x\,\Phi(x,0,t)\,\Phi_y(x,0,t)\,dx - \\ - \lim_{\substack{x_1\to-\infty\\ x_2\to+\infty}} \tfrac{1}{4}\varrho \int_{-h}^{0} [\Phi^2(x_2,y,t) - \Phi^2(x_1,y,t)]\,dy + \\ + \lim_{\substack{x_1\to-\infty\\ x_2\to\infty}} \tfrac{1}{2}\varrho \int_{-h}^{0} [-x_1\,\Phi(x_1,y,t)\,\Phi_x(x_1,y,t) + x_2\,\Phi(x_2,y,t)\,\Phi_x(x_2,y,t)]\,dy,$$

where again the last two limits vanish. A similar computation shows

$$\int_{-\infty}^{\infty} x^2\,\mathscr{T}(x,t)\,dx = \tfrac{1}{2}\varrho \int_{-\infty}^{\infty} x^2\,\Phi(x,0,t)\,\Phi_y(x,0,t)\,dx + \tfrac{1}{2}\varrho \int_{-\infty}^{\infty}\int_{-h}^{0} \Phi^2(x,y,t)\,dx\,dy + \\ + \lim_{\substack{x_1\to-\infty\\ x_2\to\infty}} \tfrac{1}{2}\varrho \int_{-h}^{0} [-x_1^2\,\Phi(x_1,y,t)\,\Phi_x(x_1,y,t) + x_2^2\,\Phi(x_2,y,t)\,\Phi_x(x_2,y,t)]\,dy - \\ - \lim_{\substack{x_1\to-\infty\\ x_2\to\infty}} \tfrac{1}{2}\varrho \int_{-h}^{0} [-x_1\,\Phi^2(x_1,y,t) + x_2\,\Phi^2(x_2,y,t)]\,dy.$$

Collecting these results we have

$$\left.\begin{aligned}
\int_{-\infty}^{\infty} \mathcal{T}(x,t)\,dx &= \tfrac{1}{2}\varrho \int_{-\infty}^{\infty} \Phi(x,0,t)\,\Phi_y(x,0,t)\,dx,\\
\int_{-\infty}^{\infty} x\,\mathcal{T}(x,t)\,dx &= \tfrac{1}{2}\varrho \int_{-\infty}^{\infty} x\,\Phi(x,0,t)\,\Phi_y(x,0,t)\,dx,\\
\int_{-\infty}^{\infty} x^2\,\mathcal{T}(x,t)\,dx &= \tfrac{1}{2}\varrho \int_{-\infty}^{\infty} x^2\,\Phi(x,0,t)\,\Phi_y(x,0,t)\,dx +\\
&\quad + \tfrac{1}{2}\varrho \int_{-\infty}^{\infty}\int_{-h}^{0} \Phi^2(x,y,t)\,dx\,dy.
\end{aligned}\right\} \tag{15.15}$$

Since from (15.2),

$$\eta_R(x,t) = \tfrac{1}{2}\int_{-\infty}^{\infty} E(k)\,\mathrm{e}^{-i(kx-\sigma t)}\,dk$$

and

$$\Phi(x,y,t) = \frac{1}{2}\int_{-\infty}^{\infty} i\,\frac{\sigma(k)}{k}\,Y(y)\,E(k)\,\mathrm{e}^{-i(kx-\sigma t)}\,dk, \tag{15.16}$$

one finds easily

$$\Phi(x,0,t) = \frac{1}{2}\int_{-\infty}^{\infty} i\,\frac{\sigma(k)}{k}\coth kh\,E(k)\,\mathrm{e}^{-i(kx-\sigma t)}\,dk,$$

$$\Phi_y(x,0,t) = \tfrac{1}{2}\int_{-\infty}^{\infty} i\,\sigma(k)\,E(k)\,\mathrm{e}^{-i(kx-\sigma t)}\,dk.$$

One may now apply again, as in Sect. 15a, theorems on Fourier transforms to obtain

$$\left.\begin{aligned}
\int_{-\infty}^{\infty} \mathcal{T}(x,t)\,dx &= \frac{1}{4}\pi\varrho\int_{-\infty}^{\infty} E(k)\,E^*(k)\,\frac{\sigma^2}{k}\coth kh\,dk,\\
\int_{-\infty}^{\infty} x\,\mathcal{T}(x,t)\,dx &= \frac{1}{4}\pi\varrho\int_{-\infty}^{\infty} i\,E(k)\,E^{*\prime}(k)\,\frac{\sigma^2}{k}\coth kh\,dk +\\
&\quad + \frac{1}{4}\pi\varrho\,t\int_{-\infty}^{\infty} E(k)\,E^*(k)\,\sigma'(k)\,\frac{\sigma^2}{k}\coth kh\,dk,\\
\int_{-\infty}^{\infty} x^2\,\mathcal{T}(x,t)\,dx &= \int_{-\infty}^{\infty} x^2\,\mathcal{T}(x,0)\,dx +\\
&\quad + \frac{1}{2}\pi\varrho\,t\int_{-\infty}^{\infty} i\,E(k)\,E^{*\prime}(k)\,\sigma'\,\frac{\sigma^2}{k}\coth kh\,dk +\\
&\quad + \frac{1}{4}\pi\varrho\,t^2\int_{-\infty}^{\infty} E(k)\,E^*(k)\,\sigma'^2(k)\,\frac{\sigma^2}{k}\coth kh.
\end{aligned}\right\} \tag{15.17}$$

If one uses the definition introduced earlier for average of a function of k, one now finds

$$\bar{x}_T(t) = \bar{x}_T(0) + t\,\frac{\overline{\sigma'\sigma^2 k^{-1}\coth kh}}{\overline{\sigma^2 k^{-1}\coth kh}} \tag{15.18}$$

and a rather unwieldy expression for $\overline{[x-\bar{x}_T(t)]^2}$, similar in character to (15.4). We note that if we are dealing with pure gravity waves, so that $\sigma^2 = gk\tanh kh$, then formulas (15.17) simplify considerably and become identical with those for V. In this case the potential and kinetic energies are equal and propagate with the same velocities.

We may now carry out similar calculations for $S(x, t)$. The corresponding formulas follow

$$\left.\begin{aligned}
&\int_{-\infty}^{\infty} \mathcal{S}(x,t)\,dx = \frac{1}{4}\pi T \int_{-\infty}^{\infty} k^2 E(k)\, E^*(k)\, dk,\\
&\int_{-\infty}^{\infty} x\,\mathcal{S}(x,t)\,dx = \frac{1}{4}\pi T \int_{-\infty}^{\infty} i\,k^2 E^{*\prime}(k)\, E(k)\, dk + \\
&\qquad\qquad + \frac{1}{4}\pi T t \int_{-\infty}^{\infty} k^2 \sigma'(k)\, E(k)\, E^*(k)\, dk,\\
&\int_{-\infty}^{\infty} x^2\,\mathcal{S}(x,t)\,dx = \int_{-\infty}^{\infty} x^2\,\mathcal{S}(x,0)\,dx + \frac{1}{2}\pi T t \int_{-\infty}^{\infty} k^2 \sigma' E^{*\prime}(k)\, E(k)\, dk + \\
&\qquad\qquad + \frac{1}{4}\pi T t^2 \int_{-\infty}^{\infty} k^2 \sigma'^2 E(k)\, E^*(k)\, dk,
\end{aligned}\right\} \tag{15.19}$$

and

$$\bar{x}_S(t) = \bar{x}_S(0) + t\,\frac{\overline{k^2\sigma'}}{\overline{k^2}} \tag{15.20}$$

and again a formula for $\overline{[x-\bar{x}_S(t)]^2}$ similar in character to (15.4).

One should note that the total potential, kinetic and surface energies associated with $\eta_R(x,t)$ each remain constant in time. If $T \neq 0$, then the mean positions of the three energy densities propagate with different velocities, each velocity being an average, in some sense, of σ'. If one considers a wave packet (15.5), then as the width 2ε of the band of wave numbers approaches zero the velocity of propagation of the individual energy densities will each approach $\sigma'(k_0)$, the group velocity.

Consider now the total energy density,

$$\mathcal{E}(x,t) = \mathcal{V}(x,t) + \mathcal{T}(x,t) + \mathcal{S}(x,t).$$

Making use of the form of $\sigma(k)$,

$$\sigma^2(k) = (g\,k + T\,k^3/\varrho)\tanh k\,h,$$

one finds

$$\left.\begin{aligned}
&\int_{-\infty}^{\infty} \mathcal{E}(x,t)\,dx = \frac{1}{4}\pi\varrho \int_{-\infty}^{\infty} \left[g + \frac{\sigma^2}{k}\coth k\,h + \frac{T}{\varrho}k^2\right] E(k)\, E^*(k)\, dk\\
&\qquad = \frac{1}{2}\pi \int_{-\infty}^{\infty} [g\varrho + T k^2]\, E(k)\, E^*(k)\, dk,\\
&\int_{-\infty}^{\infty} x\,\mathcal{E}(x,t)\,dx = \frac{1}{2}\pi \int_{-\infty}^{\infty} [g\varrho + T k^2]\, i\, E^{*\prime}(k)\, E^*(k)\, dk + \frac{1}{2}\pi t \int_{-\infty}^{\infty} \sigma'(k)\, \times\\
&\qquad\qquad \times [g\varrho + T k^2]\, E\, E^*\, dk,\\
&\int_{-\infty}^{\infty} x^2\mathcal{E}(x,t)\,dx = \int_{-\infty}^{\infty} x^2\,\mathcal{E}(x,0)\,dx + \pi t \int_{-\infty}^{\infty} \sigma'(g\varrho + T k^2]\, i\, E^{*\prime}\, E\, dk + \\
&\qquad\qquad + \frac{1}{2}\pi t^2 \int_{-\infty}^{\infty} \sigma'^2 [g\varrho + T k^2]\, E\, E^*\, dk,
\end{aligned}\right\} \tag{15.21}$$

and

$$\bar{x}_E(t) = \bar{x}_E(0) + t\,\frac{\overline{\sigma'[\varrho g + T k^2]}}{\varrho g + T k^2}\,. \tag{15.22}$$

At any instant t half of the total energy is kinetic energy and the other half is divided between potential and surface energy.

There is another way of considering the energy transported by surface waves which, at first glance, is different from the preceding treatment. Consider a fixed plane $x = \text{const}$. Then from the results in Sect. 8 one may compute the rate at which energy is being transported through this plane, the so-called *energy-flux*. Let us denote it by $\mathcal{F}(x, t)$. After appropriate linearization, formula (8.10) gives

$$\mathcal{F}(x, t) = -\int_{-\infty}^{0} \varrho\, \Phi_t(x, y, t)\, \Phi_x(x, y, t)\, dy - T\,\eta_t(x, t)\,\eta_x(x, t)\,. \tag{15.23}$$

The expression for the flux has an advantage over the expressions for mean positions considered above in that no strong restrictions upon η are required for it to exist. In fact, it can be computed for a single harmonic wave

$$\eta = A \sin(k x - \sigma t)\,. \tag{15.24}$$

With

$$\Phi = -A\,\frac{\sigma}{k}\,\frac{\cosh k(y+h)}{\sinh k h}\cos(k x - \sigma t)\,,$$

one finds by a straightforward calculation

$$\mathcal{F}(x, t) = A^2 T k \sigma \cos^2(k x - \sigma t) + A^2 \varrho\,\frac{\sigma^3}{2k^2}\coth k h\left[1 + \frac{2 k h}{\sinh 2 k h}\right]\sin^2(k x - \sigma t)\,.$$

Averaging over a wavelength (or over a period, it makes no difference which), one finds

$$\left.\begin{aligned} \mathcal{F}_{\text{av}} &= A^2\,\frac{1}{4}\,\frac{\sigma}{k}\left\{2 T k^2 + \sigma^2 \varrho\,\frac{\coth k h}{k}\left[1 + \frac{2 k h}{\sinh 2 k h}\right]\right\}\\ &= \frac{1}{2} A^2 (g \varrho + T k^2)\,\sigma'(k)\,. \end{aligned}\right\} \tag{15.25}$$

Thus the group velocity enters again in connection with energy propagation, even though no "group" is present. The energy density and average energy per wave length for (15.24) are

$$\left.\begin{aligned} \mathcal{E}(x, t) &= A^2\left\{\frac{1}{2}\,\varrho g \sin^2(k x - \sigma t) + \frac{1}{2}\,T k^2 \cos^2(k x - \sigma t) + \right.\\ &\quad \left. + \frac{1}{4}\,\varrho\,\frac{\sigma^2}{k}\coth k h\left[1 - \frac{2 k h}{\sinh 2 k h}\cos 2(k x - \sigma t)\right]\right\},\\ \mathcal{E}_{\text{av}} &= \tfrac{1}{2} A^2 (g \varrho + T k^2)\,. \end{aligned}\right\} \tag{15.26}$$

If one is dealing with a composite wave, averaging over a wave length is possible only if the resulting wave is periodic. However, even without this restriction, one may compute both the average flux and average energy per unit length from

$$\left.\begin{aligned} \mathcal{F}_{\text{av}} &= \lim_{L\to\infty}\frac{1}{2L}\int_{-L}^{L}\mathcal{F}(x, t)\,dx\,,\\ \mathcal{E}_{\text{av}} &= \lim_{L\to\infty}\frac{1}{2L}\int_{-L}^{L}\mathcal{E}(x, t)\,dx\,. \end{aligned}\right\} \tag{15.27}$$

Then if a composite wave propagating to the right is given by

$$\eta(x,t) = \sum_{j=-\infty}^{\infty} E_j \mathrm{e}^{-i(k_j x - \sigma t)} = \sum_{j=1}^{\infty} a_j \cos(k_j x - \sigma_j t) + b_j \sin(k_j x - \sigma_j t), \quad (15.28)$$

with

$$\Phi = \sum_{j=-\infty}^{\infty} i\, E_j \frac{\cosh k_j(y+h)}{\sinh k_j h} \frac{\sigma_j}{k_j} \mathrm{e}^{-i(k_j x - \sigma_j t)},$$

where $E_j = E_{-j}^* = \frac{1}{2}(a_j + b_j)$, $k_{-j} = -k_j$, $\sigma_j = \sigma(k_j) = -\sigma_{-j}$, one finds

$$\left.\begin{aligned} \mathscr{E}_{\mathrm{av}} &= \frac{1}{2}\sum_{-\infty}^{\infty} |E_j|^2 \left[T k_j^2 + \varrho g + \varrho \frac{\sigma_j^2}{k_j} \coth k_j h\right] \\ &= \sum_{-\infty}^{\infty} |E_j|^2 [\varrho g + T k_j^2] = \frac{1}{2}\sum_{1}^{\infty} (a_j^2 + b_j^2)[\varrho g + T k_j^2] \end{aligned}\right\} \quad (15.29)$$

and

$$\left.\begin{aligned} \mathscr{F}_{\mathrm{av}} &= \sum_{-\infty}^{\infty} |E_j|^2 \left\{T k_j \sigma_j + \varrho \frac{\sigma_j^3}{2k_j^2} \coth k_j h \left[1 + \frac{2k_j h}{\sinh 2k_j h}\right]\right\} \\ &= \sum_{-\infty}^{\infty} |E_j|^2 [\varrho g + T k_j^2]\, \sigma_j'. \end{aligned}\right\} \quad (15.30)$$

In order to obtain these relatively simple formulas in which the contributions from the individual harmonics are isolated, it is essential that the averages be taken. Otherwise, for $\mathscr{E}(x,t)$ or $\mathscr{F}(x,t)$ one obtains a complicated double summation, and the role of the group velocity is not apparent.

A similar analysis may be carried through for the right-moving initial hump (15.16). However, an average of either $\mathscr{F}$ or $\mathscr{E}$ computed according to (15.27) would vanish. Instead we take the total flux and total energy, respectively:

$$\mathscr{F}_{\mathrm{total}} = \int_{-\infty}^{\infty} \mathscr{F}(x,t)\, dx, \qquad \mathscr{E}_{\mathrm{total}} = \int_{-\infty}^{\infty} \mathscr{E}(x,t)\, dx. \quad (15.31)$$

The resulting formulas are analogous to (15.29) and (15.30):

$$\left.\begin{aligned} \mathscr{E}_{\mathrm{total}} &= \tfrac{1}{2}\pi \int_{-\infty}^{\infty} [g\varrho + T k^2]\, E(k)\, E^*(k)\, dk, \\ \mathscr{F}_{\mathrm{total}} &= \tfrac{1}{2}\pi \int_{-\infty}^{\infty} \sigma'(k)\, [g\varrho + T k^2]\, E(k)\, E^*(k)\, dk. \end{aligned}\right\} \quad (15.32)$$

If the last result is applied tc ı narrow wave band, such as (15.5), then one finds the limiting relationship

$$\lim_{\varepsilon\to 0} \frac{\mathscr{F}_{\mathrm{total}}}{\mathscr{E}_{\mathrm{total}}} = \sigma'(k_0).$$

In the first method of treating the propagation of energy, i.e. in terms of the motion of the mean position of the energy density, it was not surprising that σ' should appear, for it is a familiar property of Fourier transforms that taking the derivative of the transform is associated with multiplying the function by the variable. Thus, if

$$g(k) = \int_{-\infty}^{\infty} f(x)\, \mathrm{e}^{ikx}\, dx,$$

then

$$g'(k) = \int_{-\infty}^{\infty} i\, x\, f(x)\, \mathrm{e}^{ikx}\, dx.$$

In the cases considered above the transform contained $e^{i\sigma t}$ as a factor, and the derivative contained $\sigma' t$ in one summand. However, the appearance of σ' in the formulas for $\mathcal{F}_{\text{av}}$ or $\mathcal{F}_{\text{total}}$ seems in some ways coincidental: One makes a calculation, and after gathering and manipulating terms discovers that a certain combination of them indeed contains σ'. That this is not really coincidence is indicated by the following theorem for the case (15.21):

$$\frac{\partial}{\partial t}\int_{-\infty}^{\infty} x\,\mathcal{E}(x,t)\,dx = \mathcal{F}_{\text{total}}. \tag{15.33}$$

It may be proved as follows. From the definition of $\mathcal{E}(x,t)$

$$\int_{-\infty}^{\infty} x\,\mathcal{E}(x,t)\,dx = \int_{-\infty}^{\infty} x\left[\tfrac{1}{2}\varrho g\eta^2 + \tfrac{1}{2}T\eta_x^2 + \tfrac{1}{2}\varrho\int_{-h}^{0}(\Phi_x^2+\Phi_y^2)\,dy\right]dx.$$

Hence

$$\frac{\partial}{\partial t}\int_{-\infty}^{\infty} x\,\mathcal{E}(x,t)\,dx = \int_{-\infty}^{\infty} x\left[\varrho g\eta\eta_t + T\eta_x\eta_{xt} + \varrho\int_{-h}^{0}(\Phi_x\Phi_{xt}+\Phi_y\Phi_{yt})\,dy\,dx\right].$$

Integrating the second and third terms by parts and taking account of the assumed behavior of η and Φ at $\pm\infty$, one finds

$$\int_{-\infty}^{\infty} x\left[\varrho g\eta\eta_t - T\eta_{xx}\eta_t + \varrho\int_{-h}^{0}(\Phi_y\Phi_{yt}-\Phi_{xx}\Phi_t)\,dy\right]dx -$$
$$-\int_{-\infty}^{\infty}\left[T\eta_x\eta_t + \varrho\int_{-h}^{0}\Phi_x\Phi_t\,dy\right]dx.$$

Since $\Phi_{xx}+\Phi_{yy}=0$, one may express the third summand in the first integral as

$$\varrho\int_{-h}^{0}(\Phi_y\Phi_{yt}+\Phi_{yy}\Phi_t)\,dy = \varrho\int_{-h}^{0}(\Phi_y\Phi_t)_y\,dy = \varrho\,\Phi_y(x,0,t)\,\Phi_t(x,0,t) = \varrho\,\eta_t\,\Phi_t.$$

Hence the first integral may be written

$$\int_{-\infty}^{\infty} x\,\eta_t\left[\varrho g\eta - T\eta_{xx} + \varrho\,\Phi_t(x,0,t)\right]dx,$$

which vanishes, since the term in brackets is just the dynamical boundary condition at the free surface. The second integral above is just $\mathcal{F}_{\text{total}}$, so that (15.33) is proved.

A similar line of reasoning allows one to establish the following relation between $\mathcal{E}$ and $\mathcal{F}$:

$$\frac{\partial\mathcal{E}(x,t)}{\partial t} = -\frac{\partial\mathcal{F}(x,t)}{\partial x}, \tag{15.34}$$

essentially an expression of the conservation of energy. Eq. (15.33) may also be derived from (15.34) by writing the latter in the form

$$\frac{\partial(x\mathcal{E})}{\partial t} = -\frac{\partial(x\mathcal{F})}{\partial x} + \mathcal{F}$$

and integrating.

Although (15.33) may explain the presence of σ' in the energy flux for a continuous spectrum and finite total energy, one is still left with the apparently paradoxical situation that even for (15.24), when only one frequency is present, σ' enters into the expression for $\mathcal{F}_{\text{av}}$. One would expect the occurrence of σ' only if one were dealing not only with a specific value k but also with neighboring values. There is no useful analogue to (15.33) for the discrete spectrum, because there is no Fourier integral to connect in a natural way the mean position of a hump with σ'. However, if one approximates (15.24) or (15.28) by considering only the segment of η between $-L$ and L and taking $\eta=0$ outside this segment, then one has approximated η by η_L, where the latter has a continuous spectrum and finite energy. For η_L it is reasonable that $\sigma'(k)$ should enter into the energy propagation. The definitions adopted for $\mathcal{F}_{\text{av}}$ and $\mathcal{E}_{\text{av}}$ in (15.27) reflect this approximation of η by η_L and then a passage to the limit in such a way as to keep these quantities finite. Thus it is perhaps not surprising after all that σ' has entered into the computation of $\mathcal{F}_{\text{av}}$, for the method of averaging $\mathcal{F}$ and $\mathcal{E}$ is such that one replaces the discrete spectrum by a continuous one and then takes a limit. A different explanation of this paradoxical situation has been given by RAYLEIGH [*Theory of sound*, Vol. I p. 479]; generally it seems to be overlooked.

One should note that the definitions of velocity of propagation of mean positions of humps and energy distributions for finite total energy and of total or average energy flux all retain meaning even if the boundary condition at the free surface has not been linearized. The comparative simplicity of the formulas when the boundary condition is linearized and the occurrence in them of σ' both result from the special form of the spectrum, namely, $E(k,t)=E(k,0)\,e^{i\sigma(k)t}$, and the applicability of properties of Fourier transforms of convolutions.

For further information one may consult the monograph of HAVELOCK (1914) already cited, papers by BOURGIN (1936), ROSSBY (1945, 1947), ECKART (1948), BROER (1951), and POINCELOT (1953, 1954), JEFFREYS and JEFFREYS, *Methods of mathematical physics* (3rd ed., Cambridge, 1956, pp. 511–518) and standard texts such as LAMB (1932, Sects. 236, 237, 240, 241) and KOCHIN, KIBEL' and ROZE (1948, Chap. 8, Sect. 8).

16. The solution of special boundary problems. In the next several sections we shall be considering a variety of problems, each associated with some special geometrical configuration.

In treating a particular boundary configuration one must first consider whether it is tractable at all by the theory of infinitesimal waves, i.e. whether it is possible to select a perturbation parameter ε satisfying the requirements mentioned in Sect. 10. On this basis, for example, it would appear unreasonable to try to apply infinitesimal-wave theory to the waves generated by a vertical circular cylinder moving with constant velocity, for the slope of free surfaces may be expected to become very large near the front of the cylinder. On the other hand, in certain similar situations, notably the theory of planing surfaces, it is possible to strain the theory to accomodate such a situation. The choice of parameter will be discussed in each individual case. We call attention to the fact that in many cases it is a consequence of the linearization procedure that the boundary condition on a solid boundary is no longer to be satisfied on the physical boundary, but instead on some neighboring surface. The same situation occurred earlier in linearizing the free-surface condition. This should not be considered as a further approximation, but rather as one consistent with the infinitesimal-wave approximation.

The methods for finding a solution to a boundary-value problem, once it has been properly formulated, seem to fall into two or possibly three groups. One method is a combination of separation of variables and expansion of the factors in Fourier-type series or integrals. This requires, of course, a geometric configuration related in a suitable way to the coordinate surfaces of a set of variables which allows separation and a complete set of associated elementary solutions to be used in the expansion. If a Fourier-series expansion is to be used, orthogonality of the elementary solutions is desirable.

If the motion is harmonic in time with frequency σ and if the fluid is of finite depth h, then the functions

$$\{\cosh m_0(y+h), \cos m_i(y+h)\} \tag{16.1}$$

occurring as factors in (13.2) and (13.4), in (13.6), and in (13.8) may be shown easily by direct computation to be orthogonal on the interval $0 \geqq y \geqq -h$. Completeness follows from known criteria[1]. However, both orthogonality and completeness are consequences of the general theory of Sturm-Liouville systems. The result may be used in the following way, for example. Suppose fluid occupies the region

$$x>0, \quad 0>y>-h, \quad 0<z<l,$$

and that the boundary conditions on the walls and bottom are

$$\left.\begin{aligned} \Phi_x(0,y,z,t) &= F(y,z)\cos\sigma t, \\ \Phi_z(0,x,y,t) &= \Phi_z(l,x,y,t) = 0, \\ \Phi_x(x,y,-h,t) &= 0. \end{aligned}\right\} \tag{16.2}$$

Then, by expressing $F(y, z)$ as a double series

$$\left.\begin{aligned} F(y,z) &= \sum a_{0q}\cosh m_0(y+h)\cos\frac{\pi q}{l}z \\ &+ \sum\sum a_{pq}\cos m_p(y+h)\cos\frac{\pi q}{l}z \end{aligned}\right\} \tag{16.3}$$

(with appropriate restrictions upon F), one may construct a solution from the elementary solutions in (13.6). Further conditions relating to boundedness and behavior as $x\to\infty$ are necessary in order to ensure a unique solution, but will not be discussed here. The elementary solutions (13.8) can be used in a similar way for the region exterior to a vertical cylindrical boundary. Still other configurations are possible corresponding to the various coordinate systems allowing separation of $\Delta_2\varphi \pm m\varphi = 0$.

If the fluid is infinitely deep, it is possible to construct a Fourier-integral expansion using the function.

$$\{e^{\nu y}, k\cos ky + \nu\sin ky\}, \quad \nu = \sigma^2/g, \quad 0<k<\infty. \tag{16.4}$$

In fact, HAVELOCK (1929b) has remarked that the usual Fourier-integral representation of a function may be altered to give

$$\left.\begin{aligned} f(y) &= \frac{2}{\pi}\int_0^\infty\int_{-\infty}^0 f(\eta)\,\frac{(k\cos k\,y+\nu\sin k\,y)\,(k\cos k\,\eta+\nu\sin k\,\eta)}{k^2+\nu^2}\,d\eta\,dk \\ &+2\nu\, e^{\nu y}\int_{-\infty}^0 f(\eta)\, e^{\nu\eta}\,d\eta. \end{aligned}\right\} \tag{16.5}$$

If the problem is such that rectangular coordinates may be used conveniently, then (16.5) may be combined with a Fourier-series or Fourier-integral expansion in z and the elementary solutions (13.5) used to construct a solution analogous to (16.3). The necessary expressions in both rectangular and cylindrical coordinates can be found in the cited paper of HAVELOCK.

If the fluid is of bounded horizontal extent and is bounded by vertical surfaces which are constant-coordinate surfaces in one of the coordinate systems

[1] See, e.g., N. LEVINSON: Gap and density theorems. Amer. Math. Soc. Colloq. Publ. No. 27, Chap. I. New York 1940.

allowing separation of $\Delta_2\varphi \pm m\varphi = 0$, the various possible modes of motion of the fluid may be obtained as the solution of an eigenvalue problem of a classical type. If the container is of more general shape, it is more difficult to obtain explicit solutions. The problem will be discussed in Sect. 23.

The orthogonal functions (16.1) were associated with a single value of the frequency σ. It is possible to derive another result concerning orthogonality of solutions associated with different values of σ. Let $\varphi_1(x, y, z)\cos\sigma_1 t$ and $\varphi_2(x, y, z)\cos\sigma_2 t$, $\sigma_1 \neq \sigma_2$, be regular velocity potentials of harmonic oscillations of different frequencies. Furthermore, let any solid boundaries be fixed and, if the fluid is not bounded in extent, we suppose that $|\operatorname{grad}\varphi| = O(R^{-1-\varepsilon})$ as $R^2 = x^2 + z^2 \to \infty$. Consider the fluid contained within a large cylinder Ω_R of radius R and above the plane $y = -R$. The fluid will be bounded partly by free surface F_R, partly by solid boundaries S_R, partly by the horizontal plane B_R and partly by the cylinder Ω_R. Applying GREEN'S theorem to the two potential function, one obtains

$$\left.\begin{aligned} 0 &= \iint\limits_{F_R+S_R+B_R+\Omega_R} (\varphi_1\varphi_{2n} - \varphi_{1n}\varphi_2)\,d\sigma \\ &= \iint\limits_{F_R} (\varphi_1\varphi_{2y} - \varphi_{1y}\varphi_2)\,d\sigma + \iint\limits_{B_R+\Omega_R} (\varphi_1\varphi_{2n} - \varphi_{1n}\varphi_2)\,d\sigma. \end{aligned}\right\} \tag{16.6}$$

As $R\to\infty$, the integral over $\Omega_R + B_R \to 0$, and one has

$$\iint\limits_{F} (\varphi_1\varphi_{2y} - \varphi_{1y}\varphi_2)\,d\sigma = 0. \tag{16.7}$$

From the free-surface condition

$$\varphi_{iy}(x, 0, z) = -\frac{\sigma_i^2}{g}\varphi_i(x, 0, z), \qquad i = 1, 2, \tag{16.8}$$

and (16.7) becomes

$$\frac{\sigma_1^2 - \sigma_2^2}{g}\iint\limits_{F} \varphi_1(x, 0, z)\,\varphi_2(x, 0, z)\,d\sigma = 0, \tag{16.9}$$

or simply

$$\iint\limits_{F} \varphi_1(x, 0, z)\,\varphi_2(x, 0, z)\,d\sigma = 0. \tag{16.10}$$

Hence φ_1 and φ_2 are orthogonal over the free surface of the fluid. This theorem can be used for certain initial-value problems in a manner analogous to that in which the orthogonality of (16.1) can be used for boundary-value problems. This will be done in Sect. 23 α.

A second method for solving special problems is the method of GREEN'S functions or source functions [cf. VOLTERRA (1934)]. In this method one constructs first a potential function of the form

$$\left.\begin{aligned} G(x, y, z; \xi, \eta, \zeta) &= \frac{1}{r} + G_0(x, y, z; \xi, \eta, \zeta), \\ r^2 &= (x-\xi)^2 + (y-\eta)^2 + (z-\zeta)^2, \end{aligned}\right\} \tag{16.11}$$

such that G_0 is regular in $y < 0$ and such that G satisfies the free-surface condition, conditions at infinity appropriate to the problem at hand, and, if the fluid is of finite depth, the boundary condition on the bottom. Such solutions are, of course, just the singular solutions derived in Sect. 13 γ. Next, if there are surfaces S in (or on) the fluid upon which certain further boundary conditions must be satisfied, we attempt to satisfy them by a distribution of the modified sources (16.11) over the surface(s) S:

$$\Phi(x, y, z, t) = \iint\limits_{S} \gamma(\xi, \eta, \zeta, t)\,G(x, y, z; \xi, \eta, \zeta; t)\,d\sigma. \tag{16.12}$$

Here γ is an unknown function which is to be determined from the boundary condition on S. In most problems this boundary condition consists in specifying Φ_n on S. Well known properties of surface distributions of sources then allow one to formulate an integral equation for γ:

$$\left.\begin{aligned}\Phi_n(x, y, z, t) = -2\pi\gamma(x, y, z, t) + \iint_S \gamma(\xi, \eta, \zeta, t)\, G_n(x, y, z; \xi, \eta, \zeta; t)\, d\sigma, \\ (x, y, z) \text{ on } S,\end{aligned}\right\} \quad (16.13)$$

where n is the exterior normal to the surface S (taken here as a closed surface). When it is convenient, one may also use distributions of dipoles.

It is also possible, and sometimes advantageous, to construct solutions satisfying given boundary conditions on a closed surface S by distributing the singular solutions on surfaces, lines or points completely inside S. Examples will occur later.

A third method of approach is to seek first, instead of $\Phi(x, y, z, t)$ or $f(z, t)$ the functions

$$\chi = \Phi_{tt} + g\Phi_y \quad \text{or} \quad F = f_{tt} + i g f'.$$

These functions satisfy a simpler condition on the plane $y=0$:

$$\chi(x, 0, z, t) = 0 \quad \text{or} \quad \operatorname{Re} F(x - i0, t) = 0.$$

If the other boundary conditions are such that they can be formulated simply in terms of χ or F, the new problem may be simpler to solve. After finding χ or F, one must then solve a differential equation in order to obtain the desired solution Φ or f. This procedure is called the "reduction" method by WEINSTEIN (1949). It was apparently first introduced by LEVI-CIVITA and has since been much exploited by CISOTTI, KELDYSH,, KOCHIN, SEDOV, HASKIND, LEWY, STOKER and others. It has already been used in the derivation of (13.28) and will be applied in several other problems[1]. The solution of the reduced problem may, of course, be carried out by one of the two methods already described above, or any other one which is convenient.

The methods outlines above do not exhaust the possible ones for finding analytic solutions. However, they will occur frequently in the next several sections. Several of the special problems treated in the following sections can be solved by each of the three approaches. The choice of a particular one has been made either to illustrate a method or because it happens to be convenient. Techniques for finding numerical solutions will not be discussed.

17. Two-dimensional progressive and standing waves in unbounded regions with fixed boundaries. In this and the following section we shall consider situations in which the region occupied by fluid extends to infinity horizontally, the solid boundaries are fixed, but of more complicated shape than the simple flat bottom considered up to now, and the motion of the fluid at infinity is prescribed, or at least partly so. We shall assume that the velocity is bounded at all interior points of the fluid and also at the infinite limits of the fluid. The motion is taken to be periodic everywhere with period σ. Hence we shall assume (cf. Sect. 11) that

$$\Phi(x, y, t) = \varphi_1(x, y)\cos\sigma t + \varphi_2(x, y)\sin\sigma t = \operatorname{Re}\varphi\, e^{-i\sigma t}.$$

The restriction to standing or progressive waves can be properly applied only at $x = \pm\infty$. Thus, we shall look for solutions which at $x = \infty$ behave like

$$(A\cos m x + B\sin m x)\cos\sigma t$$

---

[1] The method is used also by MUSKHELISHVILI [Singular integral equations, Noordhoff, Groningen, 1953, § 74] to reduce a mixed boundary condition of more complicated type to a simple one.

or

$$A\cos(m x+\sigma t)+B\cos(m x-\sigma t),$$

and similarly at $x=-\infty$ if the fluid extends in that direction. As we shall see below, the coefficients cannot be chosen independently if φ remains bounded everywhere.

The parameter of linearization may be chosen as

$$\varepsilon=\max(A m, B m).$$

If the solution φ is bounded everywhere, then as $\varepsilon\to 0$, $\varphi\to 0$ uniformly. However, if a singularity is allowed, then $\varphi\to 0$ uniformly only in a region excluding a neighborhood about the singularity. One may presume that the solution to the linearized problem loses physical significance within such a neighborhood. [It is assumed by STOKER (1947, p. 5) that singularities at the surface are associated with breaking of the waves.]

We shall discuss below two types of problems: obstacles in an infinite ocean and sloping beaches. For each type a special case will be discussed in some detail.

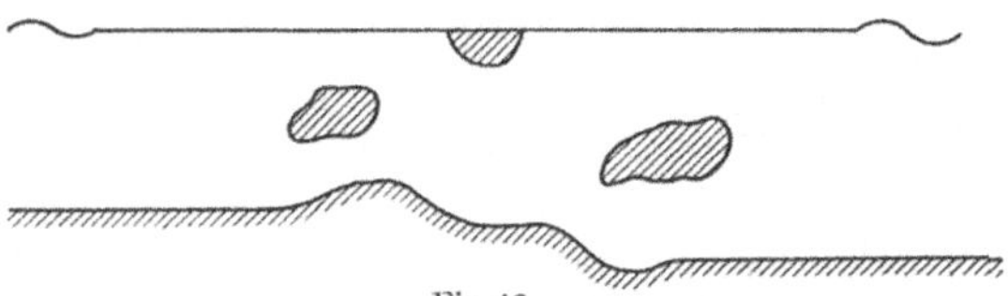

Fig. 12.

α) Obstructions in an infinitely long canal. Consider first the following situation. The fluid extends from $x=-\infty$ to $x=+\infty$; the bottom is given by $y=-h(x)$, where $h(x)=h_1>0$ for $x\geqq x_1$, $h(x)=h_2>0$ for $x\leqq x_2<x_1$; fixed obstacles may be present in the fluid or on the surface (see Fig. 12). The surface at $x=+\infty$ will be assumed to behave like

$$\eta=A_1\cos(m_1 x+\sigma t+\alpha_1)+B_1\cos(m_1 x-\sigma t+\beta_1)$$

and at $x=-\infty$ like

$$\eta=A_2\cos(m_2 x+\sigma t+\alpha_2)+B_2\cos(m_2 x-\sigma t+\beta_2).$$

A proof of the existence of a solution to this problem does not seem to exist for the general case. One would not expect a uniqueness theorem since no statement has been made about singularities or circulation. For infinite depth and a submerged body KOCHIN (1939) has proved the existence for sufficiently large values of m (the situation is slightly different, but the proof carries over). KREISEL (1949) has established the existence of a solution and its uniqueness in two cases. In the first case $h_1=h_2$, only obstacles on the bottom are allowed, Φ is assumed bounded, and a certain constant, defined in terms of the wavelength and the conformal mapping of the fluid region onto the strip $0<y<h_1$, must be less than 1. Included are theorems comparing the values of this constant for different types of obstructions. The second result allows a shallow obstruction in the surface, but requires a flat bottom and sufficiently long waves and again bounded Φ. ROSEAU (1952) has proved existence and uniqueness for no obstructions within the fluid, but for $h_1\neq h_2$; the curve joining the two ends is of a special sort. JOHN (1950, p. 78ff.) has proved uniqueness for a flat bottom and for a body in the free surface with the property that every vertical line intersects either the free surface or the body just once; certain regularity properties of Φ must also be assumed. If the body is convex and intersects the free surface perpendicularly, he is able to prove also existence of a solution.

Existence and uniqueness theorems have also been proved for several special configurations. In most of these cases explicit solutions are given. A vertical-line barrier extending from the free surface to a depth l in an infinite fluid has been considered by DEAN (1945), URSELL (1947) and HASKIND (1948). Both DEAN and URSELL, and also MARNYANSKII (1954), also consider a barrier extending from $-\infty$ to a distance l below the surface. JOHN (1948) has generalized both these problems to the case of a slanting barrier of slope $\pi/2n$, and obtained a more general solution even for the vertical barrier. DEAN (1948) and URSELL (1950) have also considered submerged circular cylinders in an infinitely deep fluid, and URSELL has established a uniqueness theorem for this case. A horizontal obstruction of finite width on the water surface (the "finite-dock problem") has been treated by RUBIN (1954), who proved existence of a solution by a variational method. Other references concerning the dock problem will be given below. BARTHOLOMEUSZ (1958) treats the long-wave approximation for reflection at a step in the bottom.

Reflection and transmission coefficients. If one assumes the existence of a solution to the general problem stated above, one may establish the form of the solution for $x>x_1$ and $x<x_2$ by using the completeness of the functions [cf. (16.1)]

$$\{\cosh m_0 (y+h), \quad \cos m_n (y+h)\}$$

in the interval $-h \leqq y \leqq 0$ (cf. KREISEL 1949, pp. 26—29; JOHN 1948, p. 152). It is

$$\left.\begin{aligned}\Phi(x, y, t) &= [A_i \cos (m_0^{(i)} x + \sigma t + \alpha_i) + B_i \cos (m_0^{(i)} x - \sigma t + \beta_i)] \times \\ &\times \cosh m_0^{(i)} (y+h_i) + \sum_{n=1}^{\infty} (a_{in} \cos\sigma t + b_{in} \sin\sigma t) \exp(-m_n^{(i)} |x|) \cos m_n^{(i)} (y+h_i),\end{aligned}\right\} \tag{17.1}$$

where $i=1, 2$ and $\sigma^2 = g m_0^{(i)} \tanh m_0^{(i)} h_i = -g m_n^{(i)} \tan m_n^{(i)} h_i$.

Let us now apply the formula for dE/dt in Eq. (8.2) to the region of fluid bounded by the planes $x=c_2<x_2$, $x=c_1>x_1$, the bottom and any other obstructions, which we take to be between these two planes. Then, if $\varphi_x^2+\varphi_y^2$ is bounded in the region considered,

$$\frac{dE}{dt} = \int_{-h_1}^{0} \varrho\, \Phi_t\, \Phi_x(c_1, y, t)\, dy - \int_{-h_2}^{0} \varrho\, \Phi_t\, \Phi_x(c_2, y, t)\, dy,$$

since on the "physical" boundaries [cf. Eq. (8.3)] either $p=0$ or $\Phi_n=0$. Anticipating that we are interested only in the asymptotic values for $c_1\to\infty$ and $c_2\to-\infty$, we compute the above expression using only the first term in (17.1) and average over a period $2\pi/\sigma$:

$$\begin{aligned}\left[\frac{dE}{dt}\right]_{\text{av}} &= \pi\, m_0^{(1)} h_1 \left[1 + \frac{\sinh 2 m_0^{(1)} h_1}{2 m_0^{(1)} h_1}\right] [A_1^2 + B_1^2] - \\ &\quad - \pi\, m_0^{(2)} h_2 \left[1 + \frac{\sinh 2 m_0^{(2)} h_2}{2 m_0^{(2)} h_2}\right] [A_2^2 - B_2^2].\end{aligned}$$

Since the average energy in the region is constant,

$$m_0^{(1)} h_1 \left[1 + \frac{\sinh 2 m_0^{(1)} h_1}{2 m_0^{(1)} h_1}\right] [A_1^2 - B_1^2] = m_0^2 h_2 \left[1 + \frac{\sinh 2 m_0^{(2)} h_2}{2 m_0^{(2)} h_2}\right] [A_2^2 - B_2^2]. \tag{17.2}$$

This is, of course, a statement of the conservation of energy. If A_1 is given $\neq 0$ and $B_2=0$, then A_2, B_1 are uniquely determined. For suppose two solutions

Φ and Φ' are possible, both with the same A_1 and $B_2=0$, but one with A_2, B_1, the other with A_2', B_1'. Apply (17.2) to the difference $\Phi-\Phi'$:

$$-m_0^{(1)} h_1\left[1+\frac{\sinh 2m_0^{(1)} h_1}{2m_0^{(1)} h_1}\right](B_1-B_1^1)^2 = m_0^{(2)} h_2\left[1+\frac{\sinh 2m_0^{(2)} h_2}{2m_0^{(2)} h_2}\right](A_2-A_2^1)^2.$$

Each side must be zero since they differ in sign and are equal. Hence $A_2=A_2'$ and $B_1=B_1'$. This does not, of course, imply the uniqueness of Φ itself.

If $h_1=h_2$, then (17.2) simplifies in an obvious way:

$$A_1^2-B_1^2=A_2^2-B_2^2. \tag{17.3}$$

Here h may also be infinite.

Setting $B_2=0$ and fixing A_1 as above corresponds physically to giving the amplitude of an incoming wave far to the right. B_1 is then the amplitude of the reflected wave and A_2 of the transmitted wave. The theorem of the preceding paragraph states that A_1 fixes them uniquely. We define $|B_1/A_1|$ as the *reflection coefficient* R and $|A_2/A_1|$ as the *transmission coefficient* T. They are uniquely determined and $R^2+T^2=1$. Properly one should define both left and right coefficients since the channel is not symmetric. However, the uniqueness theorem implies that both have the same value [see KREISEL (1949) or MEYER (1955)]. One can clearly arrange the phases so that A_1 and A_2 have the same sign. If this is done, $\alpha_2-\alpha_1$ will be the *phase shift* caused by the obstacles.

KREISEL (1949) has proved several general theorems concerning the reflection coefficient if $h_1=h_2$. In particular, if there are no obstacles within the fluid, he determines upper and lower bounds for the reflection coefficient in terms of the conformal mapping $z(\zeta)$ of the infinite strip $0>\eta>-h$ onto the region occupied by fluid, with infinities corresponding. His bounds become closer as the wavelength increases. He gives, for example, asymptotic expressions as $m_0\to 0$ for the reflection coefficient from a horizontal reef of width a and height ε and from a flat plate in the surface of beam b, namely,

$$\frac{\varepsilon}{h}\,\frac{2m_0 h\,|\sin 2m_0 a|}{\sinh 2m_0 h'(1+2m_0 h/\sinh 2m_0 h)}$$

and

$$\frac{m_0 b}{1+2m_0 h/\sinh 2m_0 h}.$$

Other general considerations will be found in BIESEL and LE MÉHAUTÉ (1955).

An interesting special result of DEAN (1947) [see also URSELL (1950)] is that the reflection coefficient from a submerged circular cylinder in infinitely deep water vanishes. The proof may be briefly sketched. Let a be the radius and let the center be at $(0,-b)$, $b>a$. Let the velocity potential be written as a sum of an incoming wave and a diverging wave:

$$\Phi = A\,\nu\,e^{\nu y}\cos(\nu x+\sigma t)+\Phi_0;$$

and suppose that Φ_0 can be expressed as a sum of multipoles (13.31), starting with dipoles:

$$\Phi_0=\sum a_n\,\Phi_n^{(s)}(x,y,t)+b_n\,\Phi_n^{(a)}(x,y,t)+c_n\,\Phi_n^{(s)}\left(x,y,t+\frac{\pi}{2\sigma}\right)+d_n\,\Phi_n^{(a)}\left(x,y,t+\frac{\pi}{2\sigma}\right),$$

where $\Phi_n^{(s)}$ is the potential for the symmetric potential of order n and strength $Q=1$, and $\Phi_n^{(a)}$ that for the antisymmetric one. The boundary condition on the cylinder [using the notation of (13.31)],

$$\left.\frac{\partial \Phi_0}{\partial r}\right|_{r=a} = A\nu e^{-\nu b} e^{\nu a \cos\vartheta} [\{\sin(\nu a \sin\vartheta)\sin\vartheta - \cos(\nu a \cos\vartheta)\cos\vartheta\}\cos\sigma t + \\ + \{\cos(\nu a \sin\vartheta)\sin\vartheta + \sin(\nu a \sin\vartheta)\cos\vartheta\}\sin\sigma t],$$

gives the relation $a_n = -d_n$, $b_n = c_n$. The reflected wave at $+\infty$ from the anti-symmetric functions then just cancels that from the symmetric functions. They reinforce each other at $x = -\infty$. The phase change for $b/a = \frac{5}{4}$, $\sigma^2 a/g = \frac{4}{3}$ was computed numerically by both DEAN and URSELL and for this case was very close to 90°.

As mentioned above, straight-line barriers have been considered by DEAN (1945, 1946), URSELL (1947), HASKIND (1948), JOHN (1948), and LEVINE (1957). The last three authors use the reduction method, whereas the first two use a Fourier-integral method which leads to a singular integral equation. We shall treat this problem by the reduction method. DEAN and JOHN also treat barriers inclined at an angle $\pi/2n$. LEVINE and RODEMICH (1958) solve the vertical-barrier problem by several methods, including the cited ones, and then apply one of them to the problem of waves incident upon two parallel vertical barriers.

Vertical barrier. Let the barrier extend along the y-axis from $y=0$ to $y=-l$ and suppose an incoming wave is given at $x=+\infty$ as

$$\eta = A\cos(\nu x + \sigma t + \alpha), \qquad \sigma^2 = g\nu.$$

We shall look for a velocity potential Φ having the form

$$\Phi = -A\frac{g}{\sigma} e^{\nu y}\sin(\nu x + \sigma t + \alpha) + \varphi_1 \cos\sigma t + \varphi_2 \sin\sigma t$$

and satisfying the following boundary conditions on the free surface and the barrier:

$$\Phi_{tt} + g\Phi_y(x, 0, t), \quad |x| > 0 \quad \text{and} \quad \Phi_x(0, y, t) = 0, \quad 0 > y > -l.$$

As $x \to \pm\infty$, $\varphi_1 \cos\sigma t + \varphi_2 \sin\sigma t$ must represent outgoing waves. In the neighborhood of $(0, -l)$ it will be assumed that

$$\lim\,[x^2 + (y+l)^2]\,(\Phi_x^2 + \Phi_y^2) = 0 \quad \text{as} \quad (x, y) \to (0, -l).$$

In the neighborhood of the intersection of the barrier and the surface (0, 0) as well as in the region of fluid bounded away from the barrier, we shall assume $\Phi_x^2 + \Phi_y^2$ bounded. It should be noted, however, that this assumption excludes a large class of solutions of possible physical interest (cf. JOHN 1948).

If we introduce the stream functions Ψ, ψ_1, and ψ_2 corresponding to Φ, φ and φ_2 and the corresponding complex potentials F, f_1 and f_2, we have

$$F = \left(-\frac{Ag}{\sigma} i e^{-i(\nu z + \alpha)} + f_1\right)\cos\sigma t + \\ + \left(-\frac{Ag}{\sigma} e^{-i(\nu z+\alpha)} + f_2\right)\sin\sigma t = F_1\cos\sigma t + F_2 \sin\sigma t$$

and the boundary conditions

$$\begin{aligned} \operatorname{Re}\{-\nu F_n + i F_n'\} &= 0, \quad y=0, \quad |x|>0, \quad n=1,2, \\ \operatorname{Re} F_n' &= 0, \quad x=0, \quad 0>y>l, \quad n=1,2. \end{aligned}$$

After finding F_1 and F_2 satisfying these conditions, constants occurring in the solutions must be adjusted so that f_1 and f_2 satisfy the radiation conditions:

$$\lim_{x\to\pm\infty}(f_1' \pm \nu f_2) = 0, \qquad \lim_{x\to\pm\infty}(f_2' \mp f_1) = 0.$$

Consider the function

$$G_1 = F_1' + i\nu F_1 = e^{-i\nu z}(e^{i\nu z} F_1)'.$$

Then the boundary conditions imply that G_1 satisfies

$$\operatorname{Im} G_1 = 0 \quad \text{for} \quad y=0, \quad |x|>0,$$
$$\operatorname{Im} G_1' = 0 \quad \text{for} \quad y=0, \quad |x|>0 \quad \text{and} \quad x=0, \quad 0>y>-l.$$

The function G_1 may be extended into the upper half-plane by defining $G_1(x+iy) = \overline{G}_1(x-iy)$ for $y>0$. Since we have assumed $|F'|\leqq B$ for $|z|>b>l$, we may conclude that $|G_1|<B+C|z|$ for $|z|>b$ and expand G_1 in a Laurent series

$$G_1(z) = cz + d + \frac{a_1}{z} + \frac{a_2}{z^2} + \cdots, \qquad |z|>b>l,$$

where all coefficients are real since $\operatorname{Im} G_1(x+i0)=0$. The condition $|F_1'|\to 0$ as $y\to-\infty$ implies $|G_1'|\to 0$ as $y\to-\infty$ and hence $c=0$. We may arbitrarily set $d=0$ by redefinition of Ψ_1. Further, we may show as follows that $a_1=0$. Consider a contour containing the obstruction and lying in the region $|z|>b$. Then

$$\oint G_1(z)\,dz = 2\pi i\,a_1.$$

Let this contour be contracted onto the barrier. Then, from the assumed behavior of F_1 on the barrier, the integral vanishes; hence $a_1=0$. Thus

$$G_1(z) = \frac{a_2}{z^2} + \frac{a_3}{z^3} + \cdots.$$

Let us now exploit the boundary condition for G_1' by mapping the z-plane into a ζ-plane by the mapping

$$z = \sqrt{\zeta^2 - l^2},$$

where the branch of the square root is chosen which makes $z\cong\zeta$ for large ζ. This maps the z-plane cut from $-il$ to il, i.e. along the barrier, onto the z-plane cut from $-l$ to $+l$, with infinities and upper and lower half-planes corresponding. Then $G_1'(z(\zeta)) = H_1(\zeta)$ is analytic in the whole lower half-plane with a singularity only at $\zeta=0$, corresponding to $z=-il$, and $\operatorname{Im} H_1(\xi+i0)=0$. Since $H_1(\zeta)$ must agree with $G_1'(z)$ for large z, $H_1(\zeta)$ must have the form

$$H_1(\zeta) = \frac{b_3}{\zeta^3} + \frac{b_4}{\zeta^4} + \cdots, \qquad b_n \text{ real}.$$

The condition on Φ near the edge of the barrier, implies that $|z+il|\cdot|F_1'|\to 0$ as $z\to -il$, or $|\zeta^4 H_1(\zeta)|\to 0$ as $\zeta\to 0$ and hence that $b_n=0$, $n\geqq 4$. Thus

$$H_1(\zeta) = \frac{C_1}{\zeta^3}, \qquad C_1 \text{ real},$$

or

$$G_1'(z) = \frac{C_1}{(z^2+l^2)^{\frac{3}{2}}}.$$

Integrating, and writing $D_1 = C_1/l^2$,

$$G_1(z) = D_1\frac{z-\sqrt{z^2+l^2}}{\sqrt{z^2+l^2}} = D_1\frac{z}{\sqrt{z^2+l^2}} - D_1,$$

where the constant of integration has been chosen so as to make $G_1(z)$ behave like z^{-2} for large z. Then

$$F_1(z) = E_1 e^{-i\nu z} + D_1 e^{-i\nu z} \int_{i\infty}^{z} \frac{z - \sqrt{z^2+l^2}}{\sqrt{z^2+l^2}} e^{i\nu z} dz,$$

where the path of integration will be taken around the right-hand side of the barrier. The boundary condition $\operatorname{Re} F_1'(0+iy)=0$, $0<|y|<l$, relates E_1 and D_1 as follows. From $F_1(z)$:

$$F_1'(z) = e^{-i\nu z}\left[-i\nu E_1 + D_1 \frac{z}{\sqrt{z^2+l^2}} - i\nu D_1 \int_{i\infty}^{z} \frac{z e^{i\nu z}}{\sqrt{z^2+l^2}} dz\right].$$

Take the path of integration along the y-axis, so that the integral becomes

$$\begin{aligned}\int_{i\infty}^{z} &= -i\int_{l}^{\infty} \frac{y e^{-\nu y}}{\sqrt{y^2-l^2}} dy \mp \int_{l}^{y} \frac{y e^{-\nu y}}{\sqrt{l^2-y^2}} dy \\ &= -i l K_1(\nu l) \mp \int_{l}^{y} \frac{y e^{-\nu y}}{\sqrt{l^2-y^2}} dy, \qquad x = \pm 0.\end{aligned}$$

Hence

$$\operatorname{Re} F_1'(0+iy) = e^{\nu y}\left[+\nu \operatorname{Im} E_1 - \nu l D_1 K_1(\nu l)\right] = 0$$

or

$$\operatorname{Im} E_1 = +D_1 l K_1(\nu l).$$

Let $E_1 = e_1 + i l D K_1(\nu l)$. Then

$$F_1(z) = e^{-i\nu z}\left[e_1 + i l D_1 K_1(\nu l) + D_1 \int_{i\infty}^{z} \frac{z - \sqrt{z^2+l^2}}{\sqrt{z^2+l^2}} e^{i\nu z} dz\right],$$

We now compute the asymptotic expressions for $F_1(z)$ for $x \to \pm\infty$. If the path of integration is taken on a large arc of radius R in the first quadrant and then to z, and if R is allowed to become infinite, it follows from JORDAN'S lemma that the integral may also be written

$$\int_{\infty}^{z} \frac{z - \sqrt{z^2+l^2}}{\sqrt{z^2+l^2}} e^{i\nu z} dz.$$

Clearly,

$$F_1(z) \sim e^{-i\nu z}\left[e_1 + i l D_1 K_1(\nu l)\right] \quad \text{as} \quad x \to +\infty.$$

As $x \to -\infty$,

$$F_1(z) \sim e^{-i\nu z}\left[e_1 + i l D_1 K_1(\nu l) + D_1 \int_{\infty}^{-\infty} \frac{z - \sqrt{z^2+l^2}}{\sqrt{z^2+l^2}} e^{i\nu z} dz\right],$$

where the path of integration passes below the barrier. By completing this path by a large semicircle in the upper half-plane, which gives a zero contribution in the limit, and then contracting the contour about the barrier, one sees that

$$\begin{aligned}\int_{\infty}^{-\infty} \frac{z - \sqrt{z^2+l^2}}{\sqrt{z^2+l^2}} e^{i\nu z} dz &= +2\int_{-l}^{l} \frac{y e^{-\nu y}}{\sqrt{l^2-y^2}} dy \\ &= -2\nu \int_{-l}^{l} e^{-\nu y}\sqrt{l^2-y^2}\, dy = -2\pi l I_1(\nu l).\end{aligned}$$

Hence

$$F_1(z) \sim e^{-i\nu z}\left[e_1 + i\,l\,D_1 K_1(\nu l) - 2\pi l\,D_1 I_1(\nu l)\right] \quad \text{as} \quad x \to -\infty.$$

Similar expressions hold for $F_2(z)$ with constants e_2 and D_2.

For f_1 and f_2 we have the asymptotic expressions:

$$\left.\begin{aligned} f_1(z) &\sim e^{-i\nu z}\left[\frac{A g}{\sigma}\, i\, e^{-i\alpha} + e_1 + i\,l\,K_1(\nu l)\,D_1\right], \\ f_2(z) &\sim e^{-i\nu z}\left[\frac{A g}{\sigma}\, e^{-i\alpha} + e_2 + i\,l\,K_1(\nu l)\,D_2\right] \end{aligned}\right\} \quad \text{as} \quad x \to +\infty,$$

$$\left.\begin{aligned} f_1(z) &\sim e^{-i\nu z}\left[\frac{A g}{\sigma}\, i\, e^{-i\alpha} + e_1 + (i\,K_1 - 2\pi I_1)\,l\,D_1\right], \\ f_2(z) &\sim e^{-i\nu z}\left[\frac{A g}{\sigma}\, e^{-i\alpha} + e_2 + (i\,K_1 - 2\pi I_1)\,l\,D_2\right] \end{aligned}\right\} \quad \text{as} \quad x \to -\infty.$$

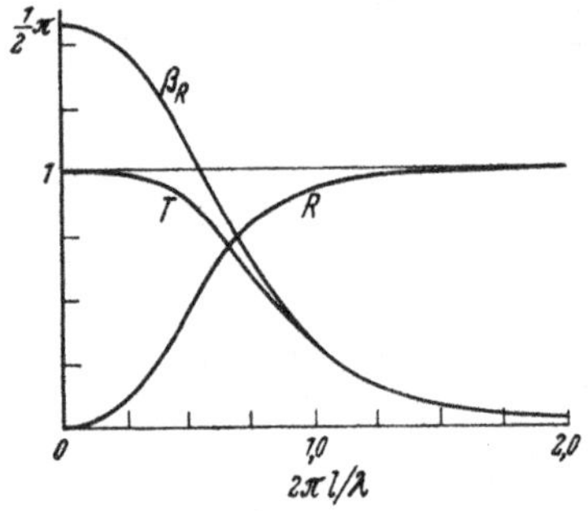

Fig. 13.

The radiation condition gives simultaneous equations for the determination of e_1, e_2, D_1 and D_2. The solution may be written

$$l(D_1 + i\,D_2) = -\frac{A g}{\sigma}\, i\, e^{-i\alpha}\frac{1}{\pi I_1 + i\,K_1},$$

$$e_1 + i\,e_2 = -\frac{A g}{\sigma}\, i\, e^{-i\alpha}\left(1 + \frac{\pi I_1}{\pi I_1 + i\,K_1}\right).$$

Substitution in the expressions for f_1 and f_2, and computation of $F_1 \cos\sigma t + F_2 \sin\sigma t$ give, after a somewhat tedious calculation, the following asymptotic expressions for Φ:

$$\left.\begin{aligned} \Phi &\sim \frac{A g}{\sigma}\, e^{\nu y}\left\{-\sin(\nu x + \sigma t + \alpha) + \frac{\pi I_1}{\sqrt{\pi I_1^2 + K_1^2}}\sin(\nu x - \sigma t - \alpha - \beta_R)\right\}, \\ &\qquad\qquad\qquad\qquad\qquad\qquad\qquad\qquad x \to +\infty, \\ \Phi &\sim \frac{-A g}{\sigma}\, e^{\nu y}\frac{K_1}{\sqrt{\pi^2 I_1^2 + K_1^2}}\sin(\nu x + \sigma t + \alpha + \beta_T), \qquad x \to -\infty, \end{aligned}\right\} \tag{17.4}$$

where $\tan\beta_R = K_1/\pi I_1 = \cot\beta_T$, and $I_1 = I_1(\nu l)$, $K_1 = K_1(\nu l)$. Clearly the reflection and transmission coefficients are

$$R = \frac{\pi I_1}{\sqrt{\pi I_1^2 + K_1^2}}, \qquad T = \frac{K_1}{\sqrt{\pi I_1^2 + K_1^2}}. \tag{17.5}$$

R, T and $\beta_R = \frac{1}{2}\pi - \beta_T$ are shown in Fig. 13 as functions of $2\pi l/\lambda = \nu l$, The reflection coefficient is practically one if $l/\lambda \geqq \frac{1}{4}$.

One may now use the velocity potential to find the behavior of the fluid near the barrier, in particular, the water height and the pressure. The calculations will not be carried through, but may be found in HASKIND (1948). The elevation on either side of the barrier is given by

$$\eta(\pm 0, t) = A\left[\cos(\sigma t + \alpha) \mp \frac{1 + \nu l\, S(\nu l)}{\sqrt{\pi^2 I_1^2 + K_1^2}}\cos(\sigma t + \alpha + \beta_R)\right] \tag{17.6}$$

where

$$S(\nu l) = \frac{\pi}{2\nu l}\left[I_1(\nu l) + L_1(\nu l)\right] = \int_0^1 e^{\nu l y}\sqrt{1 - y^2}\, dy,$$

L_1 being a Struve function of imaginary argument[1]. Let the force and moment about the origin, per unit length of barrier, be denoted by X and M, the former being positive if directed along OX and the latter counterclockwise. Then

$$\left.\begin{aligned} X &= +2\varrho g A l X_0 \cos(\sigma t + \alpha + \beta_R), \\ M &= +2\varrho g A l^2 M_0 \cos(\sigma t + \alpha + \beta_R), \end{aligned}\right\} \tag{17.7}$$

where

$$X_0 = \frac{S}{\sqrt{\pi I_1^2 + K_1^2}}, \qquad M_0 = \frac{1}{\nu l \sqrt{\pi I_1^2 + K_1^2}} \left(S - \frac{\pi}{4}\right).$$

HASKIND also computes the average force and moment per unit length of the barrier. The results are:

$$\left.\begin{aligned} -X_{\text{av}} &= \frac{1}{2}\varrho g A^2 \frac{\pi^2 I_1^2}{\pi^2 I_1^2 + K_1^2} = \frac{1}{2}\varrho g A^2 R^2, \\ -M_{\text{av}} &= \frac{1}{2}\varrho g A^2 l \left[S(-\nu l) - T(-\nu l) - \frac{\pi I_1(\nu l)}{2\nu l}\right] \frac{\pi I_1}{\pi_1^2 I_1^2 + K_1^2}, \end{aligned}\right\} \tag{17.8}$$

where

$$\nu l S(-\nu l) = \tfrac{1}{2}\pi\,[I_1(\nu l) - L_1(\nu l)],$$
$$T(-\nu l) = \tfrac{1}{2}\pi\,[I_0(\nu l) - L_0(\nu l)].$$

Fig. 14 displays all four functions in dimensionless form.

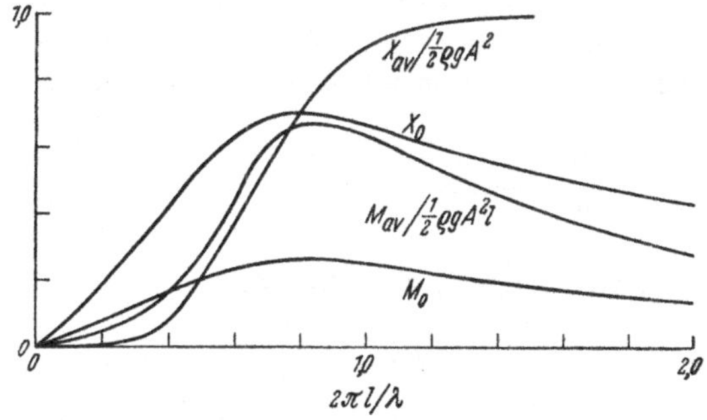

Fig. 14.

The method of integral equations. This method for finding solutions has been frequently used, especially by KOCHIN (1937, 1939, 1940) and his colleagues. One of its advantages is that approximate solutions to the integral

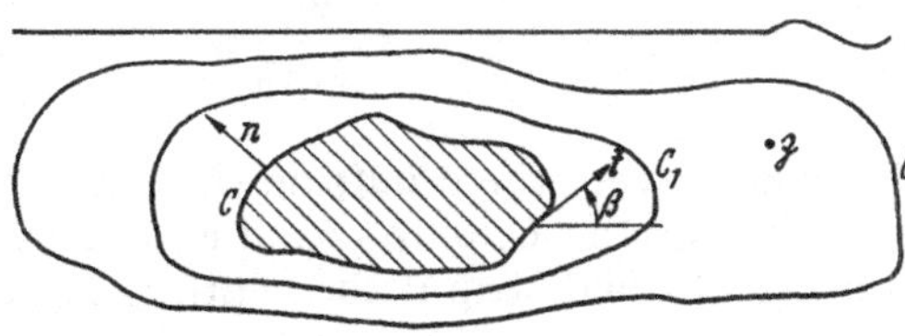

Fig. 15.

equation can frequently be found even when an explicit solution cannot be easily obtained. The following exposition follows approximately KOCHIN (1937) and KELDYSH and LAVRENT'EV (1937).

Consider a submerged obstacle whose contour C is given parametrically by $z = z(s)$ and is oriented counterclockwise. Let $\beta(s)$ be the angle between the tangent vector and the positive x-direction (see Fig. 15). We shall assume that

[1] G. N. WATSON: Bessel functions, p. 329; L_1 is tabulated in J. Math. Phys. **25**, 252—259 (1946).

as $x \to \infty$ the motion approximates to a standing wave:

$$\Phi(x, y, t) \sim A \frac{g}{\sigma} e^{\nu y} \cos(\nu x + \alpha) \cos(\sigma t + \tau), \qquad \nu = \frac{\sigma^2}{g}. \tag{17.9}$$

The other boundary conditions in terms of the complex potential $f(z) = \varphi(x, y) + i\psi(x, y)$ are

$$\left.\begin{aligned} \operatorname{Im}\{f'(x) + i f(x)\} &= 0, \\ \operatorname{Im}\{f'(z(s))\, e^{i\beta(s)}\} &= 0, \\ \lim_{y \to -\infty} |f'| &= 0. \end{aligned}\right\} \tag{17.10}$$

Write $f(z)$ in the form

$$f(z) = f_1(z) + \frac{A g}{\sigma} e^{-i(\nu z + \alpha)} = f_1(z) + a\, e^{-i\nu z}. \tag{17.11}$$

Then $f_1(z)$ must satisfy

$$\left.\begin{aligned} \lim_{x \to \infty} f_1(z) &= 0, \\ \operatorname{Im}\left[f_1'(z(s)) - i a \nu e^{-i\nu z(s)}\right] e^{i\beta(s)} &= 0, \end{aligned}\right\} \tag{17.12}$$

as well as the free surface condition and the condition as $y \to -\infty$.

We shall try to express $f_1(z)$ as a distribution of vortices over the contour C. However, the vortices are chosen so that the conditions on the free surface, at $x = \infty$ and at $y = -\infty$ are satisfied. As is apparent from the derivation of (13.28), the complex velocity potential for such vortices is given by

$$f_v(z; c) = \frac{\Gamma}{2\pi i} \left\{\log(z - c)(z - \bar{c}) - 2 e^{-i\nu z} \int_{\infty}^{z} \frac{e^{i\nu u}}{u - \bar{c}}\, du\right\}. \tag{17.13}$$

We set $\Gamma = 1$ and try to express $f_1(z)$ as follows:

$$f_1(z) = \int_C \gamma(s)\, f_v(z; z(s))\, ds, \tag{17.14}$$

where $\gamma(s)$ must be chosen so that the boundary condition on the body is satisfied.

In order to derive an integral equation for $\gamma(s)$, consider the following expression for $f_1'(z)$, a direct consequence of CAUCHY's integral:

$$f_1'(z) = \frac{1}{2\pi i} \int_{C_1} \frac{f'(\zeta)}{z - \zeta}\, d\zeta - \frac{1}{2\pi i} \int_{C_2} \frac{f'(z)}{z - \zeta}\, d\zeta = g_1(z) + g_2(z).$$

The function $g_1(z)$ is regular everywhere outside C_1 and $g_2(z)$ is regular everywhere inside C_2. One may contract C_1 onto C and extend $g_2(z)$ analytically into the whole lower half-plane (or fluid strip if the depth is finite).

Consider now (for infinite depth; the finite-depth case is analogous) the following function:

$$g(z) = \frac{1}{2\pi i} \int_{C_1} \frac{f_1'(\zeta)}{z - \zeta}\, d\zeta + \frac{1}{2\pi i} \int_{C_2} \overline{f_1'(\zeta)} \left[\frac{1}{z - \bar{\zeta}} - 2 i \nu e^{-i\nu z} \int_{\infty}^{z} \frac{e^{i\nu u}}{u - \bar{\zeta}}\, du\right] d\bar{\zeta}.$$

The first summand is identical with $g_1(z)$ and the second is also regular in the whole half-plane. $g(z)$ satisfies the same boundary conditions as $f_1'(z)$. Hence $f_1'(z) - g(z)$ is regular in the whole lower half-plane, satisfies the free-surface condition and vanishes as $y \to -\infty$ and $x \to +\infty$. The uniqueness argument used in the derivation of (13.28) shows that $f_1'(z)\equiv g(z)$. Thus we have

$$f_1'(z)=\frac{1}{2\pi i}\int\limits_{C_1}\frac{f_1'(\zeta)}{z-\zeta}\,d\zeta-\frac{1}{2\pi i}\int\limits_{C_1}\overline{f_1'(\zeta)}\left[\frac{1}{z-\bar\zeta}-2i\nu\,\mathrm{e}^{-i\nu z}\int\limits_{\infty}^{z}\frac{\mathrm{e}^{i\nu u}}{u-\bar\zeta}\,du\right]d\bar\zeta. \tag{17.15}$$

Now contract C_1 to C. Then

$$[f_1'(\zeta)-i\,a\,\nu\,\mathrm{e}^{-i\nu\zeta}]\,\mathrm{e}^{i\beta}=v_t+i\,v_n=v_t$$

or

$$f_1'(\zeta)=v_t\,\mathrm{e}^{-i\beta}+i\,a\,\nu\,\mathrm{e}^{-i\nu\zeta}. \tag{17.16}$$

If one substitutes above, one finds that the contribution from the second summand in $f_1'(\zeta)$ vanishes and that, since $d\zeta/ds=\mathrm{e}^{i\beta(s)}$,

$$\left.\begin{aligned}f_1'(z)&=\frac{1}{2\pi i}\int\limits_{C}v_t(s)\left[\frac{1}{z-\zeta}-\frac{1}{z-\bar\zeta}+2i\nu\,\mathrm{e}^{-i\nu z}\int\limits_{\infty}^{z}\frac{\mathrm{e}^{i\nu u}}{u-\bar\zeta}\,du\right]ds\\&=\int\limits_{C}v_t(s)\,f_v'(z;z(s))\,ds.\end{aligned}\right\} \tag{17.17}$$

This identifies $\gamma(s)$ as the tangential velocity $v_t(s)$ at a point of the contour.

Let us now consider the effect of letting $z\to z(s')$, a point of the contour C. Then, according to the Theorem of PLEMELJ-SOKHOTSKII,

$$\left.\begin{aligned}\int\limits_{C}\gamma(s)\,f_v'(z;z(s))\,ds=\int\limits_{C}\gamma(s)\,f_v'(z;z(s))\,\mathrm{e}^{-i\beta(s)}\,dz(s)\to\tfrac{1}{2}\gamma(s')\,\mathrm{e}^{-i\beta(s')}+{}\\+\mathrm{PV}\int\limits_{C}\gamma(s)\,f_v'(z(s');z(s))\,ds,\end{aligned}\right\} \tag{17.18}$$

whereas

$$f_1'(z)\to v_t(s')\,\mathrm{e}^{-i\beta(s')}+i\,a\nu\,\mathrm{e}^{-i\nu z(s')}=\gamma(s')\,\mathrm{e}^{-i\beta(s')}+i\,a\nu\,\mathrm{e}^{-i\nu z(s')}.$$

Hence we have the integral equation for $\gamma(s)$:

$$-\tfrac{1}{2}\gamma(s')+\mathrm{PV}\int\limits_{C}\gamma(s)\,f_v'(z(s');z(s))\,\mathrm{e}^{i\beta(s')}\,ds=i\,A\,\sigma\,\mathrm{e}^{-i[\nu z(s')-\beta(s')+\alpha]}. \tag{17.19}$$

This is really two integral equations. The imaginary part gives a singular integral equation of the first kind:

$$\mathrm{PV}\int\limits_{C}\gamma(s)\,K(s',s)\,ds=-2\pi A\,\sigma\,\mathrm{e}^{\nu y(s')}\cos[\nu x(s')-\beta(s')+\alpha]. \tag{17.20}$$

The real part gives a Fredholm equation of the second kind with continuous kernel:

$$-\frac{1}{2}\gamma(s')+\frac{1}{2\pi}\int\limits_{C}\gamma(s)\,L(s',s)\,ds=2\pi A\,\sigma\,\mathrm{e}^{\nu y(s')}\sin[\nu x(s')-\beta(s')+\alpha]. \tag{17.21}$$

Here

$$f_v'(z(s');z(s))\,\mathrm{e}^{i\beta(s')}=\frac{1}{2\pi i}[K(s',s)+iL(s',s)]. \tag{17.22}$$

The kernel $K(s',s)$ is of the form

$$K(s',s)=\frac{1}{s'-s}+C(s',s), \tag{17.23}$$

where $C(s',s)$ is continuous; the first term comes from $\mathrm{e}^{i\beta(s')}/[z(s')-z(s)]$. If the curve C is sufficiently smooth,

$$\lim_{s'\to s}\mathrm{Im}\frac{\mathrm{e}^{i\beta(s')}}{z(s')-z(s)}=\frac{1}{\varrho(s)}, \tag{17.24}$$

where $\varrho(s)$ is the radius of curvature of C at $z(s)$.

If the obstacle consists of a smooth arc, an analogous argument leads to only the singular integral equation above, but with $\gamma(s)$ now identified with the jump in $v_t(s)$ as one goes from the left to the right side of the arc.

There does not seem to be a published proof that a solution to either integral equation exists for all ν. However, KOCHIN (1936, pp. 119—126) shows the existence of a solution for both sufficiently large and sufficiently small values of ν for the equation of the second kind when the body is completely submerged.

By adjusting the phases in (17.9) one may obtain two Φ's which may be added to give an outgoing progessive wave. The behavior as $x \to -\infty$ will then be a superposition of an incoming and an outgoing wave. However, one may also modify the preceding arguments in order to treat the pregressive-wave problem directly. One specifies, say, an incoming wave from the right, writes

$$\Phi(x, y, t) = \frac{A g}{\sigma} e^{\nu y} \cos(\nu x + \sigma t) + \Phi^*(x, y, t), \tag{17.25}$$

where Φ^* must satisfy the radiation condition, and tries to express the corresponding complex potential as a distribution of the vortices (13.28) since they already satisfy the radiation condition. We shall not dwell on the details except to remark that the problem leads to a pair of coupled integral equations since one needs a distribution not only of (13.20) as it stands, but also of the vortices obtained by replacing t by $t - \pi/2\sigma$. This method could have been applied, for example, to the problem of the vertical barrier considered above.

Dock problems. This term is generally applied to water-wave problems in which the obstruction is a horizontal plane of finite or semi-infinite extent, either submerged or lying on the surface. The solution for the semi-infinite dock in infinitely deep water was given by FRIEDRICHS and LEWY (1948), and at about the same time the same problem in water of finite depth was treated by A. HEINS (1948) who also allowed a restricted type of three-dimensional motion. The methods were quite different. Subsequently HEINS (1950) and GREENE and HEINS (1953) extended the treatment to submerged docks in water of finite and infinite depth. As was remarked earlier, RUBIN (1954) has shown the existence of a solution for the finite dock in infinitely deep water. SPARENBERG (1957) has deduced an integral equation of the second kind for this problem.

As an example, consider a submerged dock at depth b and extending from $x = -a$ to $x = a$. The integral equation (17.20) then becomes

$$\mathrm{PV}\int_{-a}^{a} \gamma(\xi)\, K(x, \xi)\, d\xi = -2\pi A \sigma e^{-\nu b} \cos(\nu x + \alpha), \tag{17.26}$$

where $K(x, \xi) = K(x - \xi)$ with

$$\left.\begin{aligned} K(x) &= \mathrm{Re}\left\{\frac{1}{x} - \frac{1}{x - 2ib} + i\,\frac{2\nu}{\pi} e^{-i\nu x} \int_{\infty}^{x} \frac{e^{-i\nu u}}{u - 2ib}\, du\right\} \\ &= \frac{1}{x} - \frac{x}{x^2 + 4b^2} - \frac{2\nu}{\pi} \int_{\infty}^{x} \frac{u \sin\nu(u - x) + 2b \cos\nu(u - x)}{u^2 + 4b^2}\, du. \end{aligned}\right\} \tag{17.27}$$

Without actually establishing the existence of a solution to (17.26), KELDYSH and LAVRENT'EV (1937) in treating the flow about thin hydrofoils (see Sect. 20β) propose an approximate method of solution by expanding $\gamma(x)$ and $K(x)$ in a series in $\tau = a/2b$:

$$\gamma(x) = \gamma_0(x) + \gamma_1(x)\,\tau + \cdots,$$
$$K(x) = \frac{1}{x} + a \sum_n K_n \left(\frac{x}{a}\right)^n \tau^{n+1}$$

and determining the $\gamma_n(x)$ recursively. In the problem treated by them the total vorticity was fixed by the Kutta-Joukowski condition, in the present problem the corresponding condition is still to be determined.

If the dock extends from $-\infty$ to 0, one may modify the earlier arguments so as to apply to an unbounded body and derive the integral equation

$$\mathrm{PV}\int_{-\infty}^{0} \gamma(\xi)\, K(x-\xi)\, d\xi = -2\pi A\,\sigma\, e^{-\nu b} \cos(\nu x + \alpha). \tag{17.28}$$

An integral equation of this form is known as a Wiener-Hopf integral equation and in many cases can be solved by use of Fourier transforms. It does not seem possible to expound the method briefly, so we refer to the paper of GREENE and HEINS (1953) where this problem is treated, but with the kernel expressed differently.

When the semi-infinite dock is on the surface, the dock may be considered as a limiting case of a beach in which the angle between the bottom and the free surface is 180°. Although waves on beaches are discussed in the next section, the methods which allow extension of the angle to 180° are also difficult and will not be considered there. They may be found in STOKER'S *Water waves* (1957, § 5.4).

β) Waves on beaches. Let the fluid at rest be contained in the wedge defined by

$$\tan\gamma \leqq \frac{-y}{x} \leqq 1, \quad x>0, \quad y>0,$$

i.e., the bottom is the plane $x\sin\alpha + y\cos\alpha = 0$. For such a body of fluid one may look for periodic waves which are either standing or progressive. The appropriate mathematical problem for standing wave is to find a velocity potential

$$\Phi(x,y,t) = \varphi(x,y)\cos(\sigma t + \tau) \tag{17.29}$$

satisfying

1. $\Delta\varphi = 0$,
2. $\varphi_y(x,0) - \frac{\sigma^2}{g}\varphi(x,0) = 0$,
3. $\varphi_x \sin\gamma + \varphi_y\cos\gamma = 0$ for $x\sin\gamma + y\cos\gamma = 0$,
4. $\lim\limits_{x^2+y^2\to\infty} \varphi_x^2 + \varphi_y^2 = 0$ for $x\sin\gamma + y\cos\gamma = 0$.

This problem, in both this form and the three-dimensional form to be considered in Sect. 18, has received intensive study in recent years (e.g., MICHE 1944, LEWY 1945, STOKER 1947, FRIEDRICHS 1948, ISAACSON 1948, 1950, WEINSTEIN 1949, PETERS 1950, 1952, ROSEAU 1952, LEHMAN 1954, BRILLOUËT 1957). In particular, the cited work of BRILLOUËT and Chap. 5 of STOKER'S *Water waves* (1957) contain a general exposition of the mathematical theory. We shall restrict the present treatment to simple cases.

KIRCHHOFF (1879) was apparently the first one to treat the two-dimensional case. The problem was taken up again by MACDONALD (1896), POCKLINGTON (1921), and by HANSON (1926), who considered both the two and three-dimensional cases. All these authors restricted the solution to be bounded everywhere. This has the effect of excluding a physically important class of solutions with singularities at the origin. One may see this easily if $\gamma = 90°$, i.e. when there is a vertical cliff. A bounded solution is obviously $\varphi(x,y) = A\, e^{\nu y}\cos\nu x$, $\nu = \sigma^2/g$. This generates a standing wave behaving like $\cos \nu x$ at $x=\infty$. However, if we wish to construct a solution behaving, say, like an incoming wave at infinity we need also a standing-wave solution behaving like $\sin \nu x$ at infinity. No such solution exists which is bounded everywhere. However, as we shall see, it is possible to construct such a solution by allowing a singularity at the origin. If the two standing-wave solutions are used to construct an incoming progressive wave, the consequent loss of energy associated with the singularity is sometimes interpreted physically as representing loss of energy in breaking of the waves, at least when α is sufficiently small for this to happen. There is, of course, no a priori method of selecting the mathematical solution best representing the physical phenomena. The comparison between physical waves and mathematical solutions is discussed briefly in Stoker (1957, pp. 69–77).

Kirchhoff's approach to the solution is interesting historically because of its similarity to the method used later by Peters (1950) and Roseau (1951). His reasoning runs as follows, with a slight change in notation. Let $f(z)=\varphi+i\psi$ be the complex potential. Then

$$2\varphi(x, y) = f(x+iy) + \bar{f}(x-iy),$$
$$2i\psi(x, y) = f(x+iy) - \bar{f}(x-iy).$$

The free-surface condition becomes

$$i\left[f'(x) - \bar{f}'(x)\right] = \nu\left[f(x) + \bar{f}(x)\right], \qquad \nu = \sigma^2/g.$$

But then also

$$i\left[f'(z) - \bar{f}'(z)\right] = \nu\left[f(z) + \bar{f}(z)\right]. \tag{17.30}$$

The bottom must be a streamline. Hence

$$f(r\,e^{-i\gamma}) - \bar{f}(r\,e^{i\gamma}) = \text{const};$$

we may take this constant as 0. From this

$$\bar{f}(z) = f(z\,e^{-i2\gamma}). \tag{17.31}$$

Hence

$$\frac{d}{dz}\left[f(z) - f(z\,e^{-i2\gamma})\right] = -i\nu\left[f(z) + f(z\,e^{-i2\gamma})\right]. \tag{17.32}$$

This differential-difference equation must hold for all z for which $f(z)$ and $f(z\,e^{-i2\gamma})$ are both defined, namely for

$$-\gamma < \arg z < \gamma.$$

Kirchhoff's formal arguments need to be supported in terms of analytic continuation by the reflection principle, but the essential idea is the same as that used more recently (cf., e.g., Lehman, 1954, § 3, or Peters, 1950, § 3).

Kirchhoff proceeds to solve this equation in the special case $\gamma = m\pi/n$, m and n relatively prime integers, by assuming

$$f(z) = \sum_{k=0}^{n-1} A_k \exp(i\lambda\nu z\beta^k), \qquad \beta = e^{-i\frac{2m\pi}{n}}. \tag{17.33}$$

Substitution in (17.32) gives

$$A_k(\beta^k\lambda + 1) = A_{k-1}(\beta^k\lambda - 1), \qquad k = 0, \ldots, n-1, \tag{17.34}$$

with $A_{-1} \equiv A_{n-1}$. Multiplying all equations together and remembering that 1, $\beta, \ldots, \beta^{n-1}$ are all n-th roots of unity, one finds

$$\lambda^n - (-1)^n = \lambda^n - 1,$$

which can hold only if n is even, say $n=2q$ (hence m is odd). With $\lambda=-1=\beta^q$, the above equations determine successively $A_1, \ldots, A_{q-1}$ in terms of A_0, and $A_q=\cdots=A_{n-1}=0$:

$$A_k = i A_{k-1} \cot k\gamma = i^k A_0 \cot\gamma \cot 2\gamma \ldots \cot k\gamma. \tag{17.35}$$

Then

$$f(z) = \sum_{k=0}^{q-1} A_k \exp(-i\nu\beta^k z). \tag{17.36}$$

A_0 is still an arbitrary complex constant. The differential-difference equation is a necessary condition for $f(z)$, but not sufficient to ensure that all boundary conditions are satisfied. If one substitutes the above expression for $f(z)$ in (17.31), one finds after some computation that one must take

$$A_0 = B_0\, e^{-i\pi(q-1)/4}, \tag{17.37}$$

where B_0 is pure imaginary (say $i B_0'$) if both $\frac{1}{2}(m+1)$ and q are even and otherwise is real. With this choice of A_0 one has

$$A_{q-k} = \bar{A}_{k-1}. \tag{17.38}$$

As Kirchhoff points out, the solution is physically acceptable for the problem at hand only if $m=1$; otherwise, φ does not remain bounded as $x\to+\infty$. If $m=1$, then for $y=0$, the dominant term as $x\to\infty$ is given by

$$\left.\begin{aligned} f(x) &\sim B_0 \exp\left(-i\nu x - i\pi\frac{q-1}{4}\right) \\ \text{or}\qquad \varphi(x,0) &\sim B_0 \cos\left(\nu x + \pi\frac{q-1}{4}\pi\right). \end{aligned}\right\} \tag{17.39}$$

Here are several easily computable special cases of (17.36):

$\gamma=90^\circ$ $(m=1, q=1, \beta=-1)$:

$$f(z) = B_0\, e^{-i\nu z} = B_0\, e^{\nu y}(\cos\nu x - i\sin\nu x); \tag{17.40}$$

$\gamma=45^\circ$ $(m=1, q=2, \beta=-i)$:

$$\left.\begin{aligned} f(z) &= B_0\, e^{-i\frac{\pi}{4}}\left[e^{-i\nu z} + i\, e^{-\nu z}\right] \\ &= B_0\left[e^{\nu y}\cos\left(\nu x+\frac{\pi}{4}\right) + e^{-\nu x}\cos\left(\nu y-\frac{\pi}{4}\right)\right] - \\ &\quad - i B_0\left[e^{\nu y}\sin\left(\nu x+\frac{\pi}{4}\right) + e^{-\nu x}\sin\left(\nu y-\frac{\pi}{4}\right)\right]; \end{aligned}\right\} \tag{17.41}$$

$\gamma=30^\circ$ $(m=1, q=3, \beta=\frac{1}{2}(\sqrt{3}-i))$:

$$\begin{aligned} f(z) &= B_0\, e^{-i\frac{\pi}{2}}\left[e^{-i\nu z} + i\sqrt{3}\, e^{-\frac{1}{2}(\sqrt{3}+i)\nu z} - e^{-\frac{1}{2}(\sqrt{3}-i)\nu z}\right] \\ &= B_0\{-e^{\nu y}\sin\nu x - e^{-\frac{1}{2}\nu(x\sqrt{3}+y)}\sin\tfrac{1}{2}\nu(x-y\sqrt{3}) + \\ &\quad + \sqrt{3}\, e^{-\frac{1}{2}\nu(x\sqrt{3}-y)}\cos\tfrac{1}{2}\nu(x+y\sqrt{3})\} + \\ &\quad + i B_0\{-e^{\nu y}\cos\nu x + e^{-\frac{1}{2}\nu(x\sqrt{3}+y)}\cos\tfrac{1}{2}\nu(x-y\sqrt{3}) - \\ &\quad - \sqrt{3}\, e^{-\frac{1}{2}\nu(x\sqrt{3}-y)}\sin\tfrac{1}{2}\nu(x+y\sqrt{3})\}. \end{aligned}$$

Numerical computations for $\varphi(x, y)$ for $\gamma=6^\circ$ $(q=15)$ as well as for the above cases were carried out by STOKER (1947) and are presented graphically in his paper.

KIRCHHOFF's solution is limited to the special choice of angle noted above and furthermore presents only solutions which are bounded at the origin. The solution of the differential-difference equation (17.32) for arbitrary γ, $0<\gamma\leqq\pi$, has been given by both PETERS (1950), ISAACSON (1950), and ROSEAU (1952, Chap. V). All use Laplace transforms. However, the method cannot be expounded briefly and we refer to either the original papers or STOKER's *Water waves* for the details.

The special case $\gamma=\pi/2q$ can be treated fairly simply by the reduction method used in the problem of the vertical barrier.

From (17.32) we have

$$f^{(k+1)}(z)+i\nu f^{(k)}(z)=\beta^{k+1}f^{(k+1)}(\beta z)-i\nu\beta^k f^{(k)}(\beta z),\quad k=0,1,\dots. \tag{17.42}$$

The free surface condition [cf. (11.7)] implies

$$\mathrm{Im}\{f^{(k+1)}(x)+i\nu f^{(k)}(x)\}=0,\quad x>0. \tag{17.43}$$

Hence also

$$\mathrm{Im}\{\beta^{k+1}f^{(k+1)}(x\beta)-i\nu\beta^k f^{(k)}(x\beta)\}=0,\quad x>0.$$

This last equation can also be written

$$\mathrm{Im}\{\beta^{k+1}f^{(k+1)}(z)-i\nu\beta^k f^{(k)}(z)\}=0\quad\text{for}\quad z=r\,e^{-2i\gamma}. \tag{17.44}$$

If the numbers a_k and a_k' are real, (17.43) and (17.44) imply

$$\left.\begin{aligned}&\mathrm{Im}\left\{\sum_{k=0}^{s}a_k\left[f^{(k+1)}(x)+i\nu f^{(k)}(x)\right]\right\}=0,\\&\mathrm{Im}\left\{\sum_{k=0}^{s}a_k'\beta^k\left[\beta f^{(k+1)}(r\,e^{-2i\gamma})-i\nu f^{(k)}(r\,e^{-2i\gamma})\right]\right\}=0.\end{aligned}\right\} \tag{17.45}$$

We wish to find numbers $\{a_k\}$ and $\{a_k'\}$ such that

$$\sum_{k=0}^{s}a_k\left[f^{(k+1)}(z)+i\nu f^{(k)}(z)\right]\equiv\sum_{k=0}^{s}a_k'\beta^k\left[\beta f^{(k+1)}(z)-i\nu f^{(k)}(z)\right]. \tag{17.46}$$

Comparing coefficients of derivatives of the same order, one finds

$$\left.\begin{aligned}a_0&=-a_0',\\a_{k-1}+i\nu a_k&=\beta^k(a_{k-1}'-i\nu a_k'),\quad k=1,\dots,s,\\a_s&=\beta^{s+1}a_s'.\end{aligned}\right\} \tag{17.47}$$

These relations will be satisfied if one takes $s=q-1$ (for $\beta^q=-1$) and

$$\left.\begin{aligned}a_k&=-a_k'=a_{k-1}\frac{1}{i\nu}\frac{\beta^k+1}{\beta^k-1}=a_{k-1}\frac{1}{\nu}\cot k\gamma,\\&=\frac{a_0}{\nu^k}\cot\gamma\cot 2\gamma\dots\cot k\gamma,\quad k=1,\dots,s.\end{aligned}\right\} \tag{17.48}$$

We note that $\nu^{q-k}a_{q-k}=\nu^{k-1}a_{k-1}$. With this choice of the coefficients $\{a_k\}$, define

$$\left.\begin{aligned}g(z)&=\sum_{k=0}^{q-1}a_k\{f^{(k+1)}(z)+i\nu f^{(k)}(z)\}=P\left(\frac{d}{dz}\right)\left(\frac{d}{dz}+i\nu\right)f(z),\\&=-\sum_{k=0}^{q-1}a_k\beta^k\{\beta f^{(k+1)}(z)-i\nu f^{(k)}(z)\}=-P\left(\beta\frac{d}{dz}\right)\left(\beta\frac{d}{dz}-i\nu\right)f(z)\\&=\sum_{k=0}^{q-1}a_k\{f^{(k+1)}(\beta z)-i\nu f^{(k)}(\beta z)\}=P\left(\frac{d}{dz}\right)\left(\frac{d}{dz}-i\nu\right)f(\beta z)\end{aligned}\right\} \tag{17.49}$$

where the last equation follows from (17.42) and where

$$P(\lambda) = \sum_{k=0}^{q-1} a_k \lambda^k. \tag{17.50}$$

From the assumptions originally made concerning $f(z)$ and from the method of selecting the $\{a_k\}$ it follows that $g(z)$ is regular everywhere in the wedge

$$-2\gamma \leqq \vartheta \leqq 0$$

except possibly at the origin, that

$$\operatorname{Im}\{g(z)\} = 0 \quad \text{for} \quad z = x > 0 \quad \text{and} \quad z = r\,e^{-2i\gamma},$$

and finally, from the last of Eqs. (17.24), that

$$g(\beta z) = -g(z).$$

Since $f(z)$ is assumed bounded as $x \to \infty$, this is true also of $g(z)$. These various conditions imply that $g(z)$ must have the form

$$g(z) = \sum_{n=0}^{\infty} \frac{b_n}{z^{(2n+1)q}}, \qquad b_n \text{ real}. \tag{17.51}$$

We have thus shown that $f(z)$ satisfies the differential equation

$$P\left(\frac{d}{dz}\right)\left(\frac{d}{dz} + i\nu\right) f(z) = \sum_{n=0}^{\infty} \frac{b_n}{z^{(2n+1)q}}, \qquad b_n \text{ real}. \tag{17.52}$$

From the definition of $P(\lambda)$ it follows that

$$P(\lambda)(\lambda + i\nu) \equiv P(\beta\lambda)(-\beta\lambda + i\nu).$$

Since the coefficients in $P(\lambda)$ are real, $\bar{\lambda}$ is a root or $P(\lambda) = 0$ if λ is a root. Furthermore, from the identity above also $\beta\lambda$ is a root providing $\beta\lambda \neq i\nu$. Since $\lambda = -i\nu$ is an obvious root of the left hand member, $-i\beta\nu$ is also a root and hence $-i\beta^2\nu$, $-i\beta^3\nu, \ldots$. Since $\beta^q = -1$, no new roots are added by going further than $-i\beta^{q-1}\nu$, and since $i\beta^{-k}\nu = -i\beta^{q-k}\nu$, a complete set of roots of $P(\lambda)(\lambda + i\nu)$ is

$$-i\nu, \quad -i\beta\nu, \quad -i\beta^2\nu, \ldots, \quad -i\beta^{q-1}\nu.$$

Thus the solution of the homogeneous equation can be expressed in the form

$$\sum_{k=0}^{q-1} A_k \exp(-i\nu\beta^k z). \tag{17.53}$$

This is, of course, exactly the form of KIRCHHOFF'S solution of (17.36). Since we have already determined the necessary form of the A_k in order to satisfy the boundary condition on the bottom, we need not pursue further the solution of the homogeneous equation.

The solution of the nonhomogeneous equation is straightforward. However, just as for the homogeneous equation, one must take care to satisfy the boundary condition on the bottom, i.e. $\operatorname{Im}\{e^{-i\gamma} f'(r\,e^{-i\gamma})\} = 0$. The detailed considerations may be found in the several cited papers; BRILLOUËT (1957) treats the matter thoroughly. If one considers (17.52) with the right-hand side replaced by only one of its summands, say $b_n z^{-(2n+1)q}$, then the complete solution can be put in the following form, as shown by BRILLOUËT:

$$f(z) = \sum_{k=0}^{q-1} A_k \exp(-i\nu\beta^k z)\left[c_n + \frac{1}{2}(-1)^{nq+q-1} \frac{b_n}{\sqrt{q}} \int_{\Gamma_k} \frac{e^t\,dt}{t^{(2n+1)q}}\right], \tag{17.54}$$

where c_n is an arbitrary real constant, B_0 of (17.37) has been set equal to 1, and where Γ_k indicates that the integral is to be carried out over each of the paths Γ_k^+ and Γ_k^- shown in Fig. 16. However, one may obtain a variety of other forms for the solution.

An asymptotic expression as $x \to \infty$ and for $y=0$ is given by

$$\left.\begin{aligned} f(x) &\sim \left[c_n + i b_n \frac{(-1)^{nq+q-1}\pi}{(2nq+q-1)!\sqrt{q}}\right] \exp\left(-i\nu x - i\pi \frac{q-1}{4}\right) \\ &\text{or} \\ \varphi(x,0) &\sim c_n \cos\left(\nu x + \pi\frac{q-1}{4}\right) + b_n \frac{(-1)^{nq+q-1}\pi}{(2nq+q-1)!\sqrt{q}} \sin\left(\nu x + \pi \frac{q-1}{4}\right). \end{aligned}\right\} \tag{17.55}$$

In the neighborhood of $z=0$, $f(x)$ behaves like $\log z$ for $n=0$ and like z^{-2nq} for $n>0$.

It is not clear physically what type of singularity at $z=0$ most nearly describes the behavior of real waves. However, most writers have restricted their treatment to the weakest possible singularity, i.e., the logarithmic one.

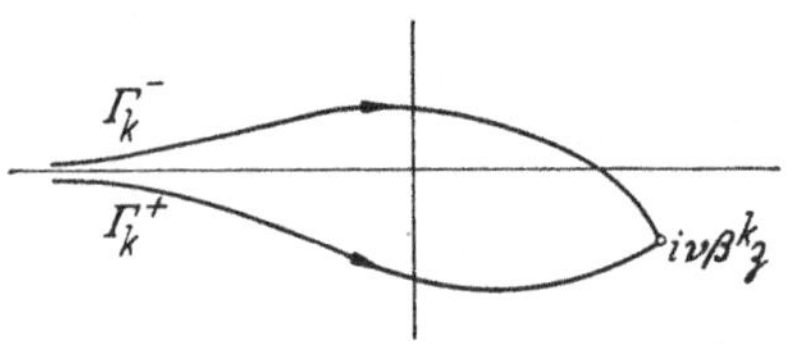

Fig. 16.

From the asymptotic expansion as $x \to \infty$ one sees that it is now possible to construct an incoming progressive wave by proper choice of the constants c_n and b_n. Thus, if we select

$$c_n = a\cos(\sigma t + \tau), \qquad b_n = -(-1)^{nq+q-1}\pi^{-1}(2nq+q-1)!\sqrt{q}\, a \sin(\sigma t + \tau),$$

then the resulting solution will behave like

$$a\cos(\nu x + \sigma t + \tau)$$

as $x \to \infty$ for $y=0$. In connection with (17.55) and the selection of b_n just made it is apparent that the formulas (17.54) and (17.55) will be more directly connected with parameters with a simple physical interpretation if we replace b_n by

$$d_n = b_n \frac{(-1)^{nq+q-1}\pi}{(2nq+q-1)\sqrt{q}}.$$

For $n=0$ companion singular solutions to the regular solutions (17.40) and (17.41) are not difficult to write out:

$\gamma = 90°\ (q=1,\ n=0)$:

$$\varphi(x,y) = d_0 e^{\nu y} \sin\nu x - \frac{d_0}{\pi}\int_0^\infty e^{-\sigma x} \frac{\sigma\cos\sigma y + \nu \sin\sigma y}{\nu^2+\sigma^2}\, d\sigma; \tag{17.56}$$

$\gamma = 45°\ (q=2,\ n=0)$:

$$\left.\begin{aligned} \varphi(x,y) = \frac{d_0}{\pi} e^{\nu y}\Big[&\left(\frac{\pi}{2} + \mathrm{Si}(\nu x)\right)\sin\left(\nu x + \frac{\pi}{4}\right) + \\ &+ \mathrm{Ci}(\nu x)\cos\left(\nu x + \frac{\pi}{4}\right) + \frac{1}{2}\sqrt{\nu}\, e^{-\nu x}\,\mathrm{Ei}(\nu x)\Big]. \end{aligned}\right\} \tag{17.57}$$

Further formulas for $\gamma = 30°$ and $\gamma = 6°$ may be found in BRILLOUËT (1957, p. 93 ff.).

18. Three-dimensional progressive and standing waves in unbounded regions with fixed boundaries. The general remarks at the beginning of Sect. 17 apply here also. Although most of the solvable problems in the present category are such that they can be reduced to two-dimensional ones (however, see the end of Sect. 19β), the methods of complex-function theory are no longer applicable to the same extent. The division of topics is the same as in the last section, namely, diffraction of waves by obstacles and waves on beaches.

α) Diffraction of water waves. In a horizontally unbounded ocean of uniform depth h assume that an incoming wave is specified by

$$\Phi_I(x, y, z, t) = \frac{A g}{\sigma} \cosh m(y+h) \cos(m x + \sigma t + \alpha) \tag{18.1}$$

and that it is scattered by one or more obstacles in the water. We wish to find the velocity potential for the motion of the water in the form

$$\Phi(x, y, z, t) = \Phi_I + \Phi_S, \tag{18.2}$$

where Φ_S is the scattered wave and satisfies the radiation condition if the body is of bounded extent.

As usual, we may write Φ in the form

$$\Phi(x, y, z, t) = \mathrm{Re}\, \varphi(x, y, z)\, \mathrm{e}^{-i\sigma t}, \qquad \varphi = \varphi_1 + i\varphi_2, \tag{18.3}$$

where φ must be a potential function satisfying

$$\left.\begin{aligned} &\varphi_y(x, 0, z) - \nu\varphi(x, 0, z) = 0, \qquad \nu = \sigma^2/g, \\ &\varphi_n = \varphi_{In} + \varphi_{Sn} = 0 \quad \text{on the obstacles,} \\ &\lim_{R\to\infty} \sqrt{R}\left(\frac{\partial\varphi_S}{\partial R} - i\nu\varphi_S\right) = 0, \quad \sqrt{R}\,\varphi_S = O(1) \quad \text{as } R\to\infty. \end{aligned}\right\} \tag{18.4}$$

General obstructions. Consider a single submerged obstacle bounded by the surface S. We shall try to express the scattered wave $\Phi_S = \mathrm{Re}\,\varphi_S\,\mathrm{e}^{-i\sigma t}$ by a distribution of sources over S. However, in order to satisfy the various boundary conditions, we take sources in the complex form (13.18) or, in the case of infinite depth, in the form (13.17″):

$$\varphi_S(x, y, z) = \frac{1}{4\pi} \iint_S \gamma(\xi, \eta, \zeta)\, G(x, y, z; \xi, \eta, \zeta)\, dS, \tag{18.5}$$

where we have written $G = G_1 + i G_2$ for the complex form of (13.18). The boundary condition on the body now becomes

$$\left.\begin{aligned} 0 = \frac{\partial\varphi_I}{\partial n} + \frac{\partial\varphi_S}{\partial n} &= \frac{\partial\varphi_I}{\partial n} - \frac{1}{2}\gamma(x, y, z) + \\ &+ \frac{1}{4\pi}\iint_S \gamma(\xi, \eta, \zeta) \frac{\partial}{\partial n} G(x, y, z; \xi, \eta, \zeta)\, dS \end{aligned}\right\} \tag{18.6}$$

or

$$\gamma(x, y, z) = 2\frac{\partial\varphi_I}{\partial n} + \frac{1}{2\pi}\iint \gamma(\xi, \eta, \zeta) \frac{\partial G}{\partial n}\, dS.$$

Since $\partial\varphi_I/\partial n$ is a known function, this is a Fredholm integral equation of the second kind for $\gamma(x, y, z)$. (We note in passing that if the motion of the surface S had been prescribed to be $\partial\varphi_I/\partial n$, then the same integral equation for γ would have been obtained.)

This equation has been considered by KOCHIN (1940) in the case of infinite depth, and he proves that a solution exists if $\nu = \sigma^2/g$ is large enough and the body is submerged. Iterative procedures for computing γ follow from the theory. HASKIND (1946) has extended the argument to finite depth.

JOHN (1950) has treated both the uniqueness and existence problem in great detail and has shown that a unique solution exists for a body whose surface intersects the free surface perpendicularly and which can be represented as a single-valued function over the area enclosed in the intersection. His result holds for all values of m (or ν if the depth is infinite). He also reduces the existence problem to solution of an integral equation.

Vertical cylinders. When the obstacle or obstacles are vertical cylinders extending from above the free surface to the bottom, it is possible to reduce the problem to one in diffraction of sound waves for which many special solutions are known [see, e.g., HAVELOCK (1940)]. In this case we may separate the y variable in the manner shown in Sect. 13α:

$$\left.\begin{aligned}\varphi(x,y,z) &= \varphi(x,z)\,Y(y)\\ Y(y) &= \cosh m(y+h)\,\varphi(x,z)\end{aligned}\right\} \tag{18.7}$$

where

and

$$\varphi_{xx}+\varphi_{zz}+m^2\varphi=0. \tag{18.8}$$

Here m must be the same as in (18.1) since the frequency is fixed by the incoming wave. $\varphi(x,z)$ must now satisfy (18.8) and the second two conditions of (18.4). This is exactly the same mathematical problem encountered in the diffraction of sound waves by a cylindrical body (in that case the air pressure replaces φ). Thus, any solutions known for sound diffraction by cylinders may be taken over immediately for water-wave diffraction. For example, if the obstacle is a vertical circular post of radius a, the velocity potential of the scattered wave is given by[1]

$$\varphi_S(R,\vartheta,y)=\frac{-A g}{\sigma}\cosh m(y+h)\,\Sigma(-i)^n\,\varepsilon_n\,e^{-i\gamma_n}\sin\gamma_n\cos\vartheta\,H_n^{(1)}(mR), \tag{18.9}$$

where

$$\tan\gamma_n=J_n'(ma)/Y_n'(ma)$$

and

$$\varepsilon_0=1,\qquad \varepsilon_n=2\quad\text{for}\quad n\geqq 1.$$

Various approximations for large and small values of ma are known. The maximum wave amplitude at any point is given by $\frac{\sigma}{g}|\varphi|$.

The diffraction of water waves by a vertical half-plane may also be treated by transferring known solutions due to SOMMERFELD for sound and electromagnetic waves to the present context. This has been done by HASKIND (1948) for normal incidence and by PENNEY and PRICE (1952a) for both normal and oblique incidence. PETERS and STOKER (1954) [see also STOKER (1956) and (1957, pp. 109 to 133)] have also solved this problem by a new and rather easy method, following an investigation of boundary conditions which will ensure uniqueness. The solution has an obvious application in predicting the effect of breakwaters. Let the breakwater be the half-plane $z=0$, $x>0$ and the incoming wave be given by

$$\begin{aligned}\eta &= A\cos(m\,x\cos\alpha+m\,z\sin\alpha+\sigma t),\\ &= A\cos(m\,R\cos(\vartheta-\alpha)+\sigma t),\end{aligned}$$

where α is the angle between $-0x$ and the direction of propagation, measured clockwise. Then the solution given by PETERS and STOKER is

$$\varphi(R,\vartheta,y)=\frac{A g}{\sigma}\cosh m(y+h)\left[J_0(R)+2\sum_1^\infty e^{i n\pi/4}J_{n/2}(R)\cos\frac{n\alpha}{2}\cos\frac{n\vartheta}{2}\right]. \tag{18.10}$$

[1] See P.M. MORSE: Vibration and sound, 2nd ed., pp. 347ff., 449. New York 1948.

The result can also be expressed by means of integrals. In the case of normal incidence these reduce to Fresnel integrals, for which tables exist. Graphical representations of the behavior of the wave amplitudes may be found in PENNEY and PRICE (1952a).

PENNEY and PRICE also apply this analysis to an approximate treatment of diffraction by a breakwater of finite length and through a gap. The results are presumably applicable if the wavelength is small compared to the length of the breakwater or the gap.

Periodic solutions for horizontal cylindrical obstacles. In two physical situations the dependence upon z may be precipitated out, leaving a two-dimensional problem which in many cases can be solved by methods analogous to those used for the two-dimensional problems of Sect. 17.

Let the obstruction be an infinitely long horizontal cylinder parallel to Oz. This might be, for example, a semi-infinite dock or submerged plane barrier, say $y=-b$, $x<0$, a finite horizontal barrier, say $y=-b$, $|x|<a$, a vertical barrier, $x=0$, $-b<Y\leqq 0$, a beach, $y=-x\tan\gamma$, etc. Let an incoming plane wave at infinity propagate at an angle α to the x axis:

$$\eta_I(x, y, z, t) = A\cos\left[m(x\cos\alpha + z\sin\alpha) + \sigma t\right]. \tag{18.11}$$

Although one will not expect the velocity potential Φ to be periodic in x, it seems reasonable to assume that it will be periodic in z. In fact, we shall assume that

$$\Phi(x, y, z, t) = \varphi(x, y)\, e^{-i(mz\sin\alpha+\sigma t)}, \tag{18.12}$$

where $\varphi(x, y)$ must now satisfy, with $k=m\sin\alpha$,

$$\varphi_{xx} + \varphi_{yy} - k^2\varphi = 0 \tag{18.13}$$

and the usual conditions on the free surface and rigid boundaries.

We should have come to the same conclusion if we had assumed an incoming wave at infinity of the form

$$\eta_I(x, y, z, t) = A\cos kz\cos(k_1 x + \sigma t), \quad k^2 + k_1^2 = m^2, \tag{18.14}$$

a so-called short-crested wave (note that we assume $k^2<m^2$). That is, we shall now look for a solution in the form

$$\Phi(x, y, z, t) = \varphi(x, y)\cos kz\, e^{-i\sigma t} \tag{18.15}$$

satisfying Eq. (18.13) and the conditions on the free surface and rigid boundaries. Thus, a solution for one of these cases carries over easily to the other.

The problem is thus reduced to one almost identical with that of Sect. 17, with the exception that the two-dimensional Laplacian is replaced by (18.13). Many of the same methods may be carried over, e.g., the reduction method and the integral-equation method. HASKIND (1953) has considered some general aspects of the problem which will be outlined below, has derived the source solution of (18.13), and has treated the diffraction about a vertical barrier (an analogue of the problem treated in Sect. 17α) and a finite dock, all in infinitely deep water. MACCAMY (1957) has derived a source solution of (18.13) and treated the finite dock problem in water of finite depth. HEINS (1948, 1950, 1953) has given source solutions of (18.13) for finite depth and formulated and solved Wiener-Hopf integral equations for semi-infinite docks and submerged horizontal barriers. GREENE and HEINS (1953) treat the submerged barrier in water of infinite depth. The literature for beaches will be given in Sect. 18β.

Suppose the fluid infinitely deep and let a cross-section of the obstacle have contour C. We wish then to find a solution $\varphi(x, y) = \varphi_1 + i\varphi_2$ of (18.13) such that

$$\left.\begin{aligned}
&\varphi_n = 0 \quad \text{on} \quad C,\\
&\varphi_y(x, 0) - \nu\varphi(x, 0) = 0 \qquad \text{on the free surface},\\
&\varphi \sim \frac{Ag}{\sigma}\, e^{\nu y}\, e^{-ik_1 x} + \frac{B^+ g}{\sigma}\, e^{\nu y}\, e^{ik_1 x} \quad \text{as} \quad x \to +\infty,\\
&\varphi \sim \frac{Ag}{\sigma}\, e^{\nu y}\, e^{-ik_1 x} + \frac{B^- g}{\sigma}\, e^{\nu y} e^{-ik_1 x} \quad \text{as} \quad x \to -\infty,
\end{aligned}\right\} \tag{18.16}$$

where $k_1^2 < \nu^2$. HASKIND (1953) applies the reduction method in the following manner (we follow his presentation closely). Introduce the function $f(x, y)$ by

$$\frac{\partial f}{\partial y} = \frac{\partial \varphi}{\partial y} - \nu\varphi. \tag{18.17}$$

Then f also satisfies (18.13) and

$$f_y(x, 0) = 0 \qquad \text{on the free surface}. \tag{18.18}$$

Consequently, f may be extended into the upper half-plane by defining $f(x, -y) = f(x, y)$ and f now satisfies (18.13) in the whole plane outside the contour C and its mirror image $\bar{C}$. Moreover, $|f| \to 0$ as $x^2 + y^2 \to \infty$. Assuming that f is known, one must now try to reconstruct φ from f in such a way that conditions (18.16) are satisfied. In order to do this, HASKIND differentiates (18.17) with respect to y, subtracts from

$$f_{xx} + f_{yy} - k^2 f = 0,$$

and after some easy manipulation obtains

$$\frac{\partial^2}{\partial x^2}(\varphi - f) + k_1^2(\varphi - f) = -\nu\left(\frac{\partial f}{\partial y} + \nu f\right). \tag{18.19}$$

Treating this as a differential equation for $\varphi - f$, he finds the following solution for φ:

$$\left.\begin{aligned}
\varphi = f - \frac{\nu}{2ik_1}\Big\{&e^{ik_1 x}\int_{\infty}^{x} e^{-ik_1\xi}(f_y + \nu f)\,d\xi -\\
&- e^{-ik_1 x}\int_{-\infty}^{x} e^{ik_1\xi}(f_y + \nu f)\,d\xi\Big\} + \frac{Ag}{\sigma}\, e^{\nu y}\, e^{-ik_1 x},
\end{aligned}\right\} \tag{18.20}$$

the integrals being taken along half-lines parallel to the x-axis and below C. One may verify without great difficulty that φ satisfies (18.13). The asymptotic form of φ as $x \to \pm\infty$ may be written down immediately, and gives

$$\frac{\nu}{2ik_1}\, e^{\mp ik_1 x}\int_{-\infty}^{\infty} e^{\pm ik_1\xi}(f_y + \nu f)\,d\xi + \frac{Ag}{\sigma}\, e^{\nu y}\, e^{-ik_1 x}, \tag{18.21}$$

the path of integration being a line below the body. Consider now the region D bounded externally by this line and a large semicircle containing $C + \bar{C}$ and internally by $C + \bar{C}$. Application of GREEN's Theorem to f and $\chi = \exp(-\nu y + ik_1 x)$ shows that

$$e^{-\nu y}\int_{-\infty}^{\infty} e^{ik_1\xi}(f_y + \nu f)\,d\xi = \int_{C+\bar{C}} (f\chi_n - \chi f_n)\,ds,$$

$$e^{-\nu y}\int_{-\infty}^{\infty} e^{-ik_1\xi}(f_y + \nu f)\,d\xi = \int_{C+\bar{C}} (f\bar{\chi}_n - \bar{\chi} f_n)\,ds.$$

Hence, the asymptotic conditions are satisfied and, moreover,

$$\frac{B^+ g}{\sigma} = \frac{\nu}{2 i k_1} \int\limits_{C+\bar{C}} (f \chi_n - \chi f_n)\, ds, \qquad \frac{B^- g}{\sigma} = \frac{\nu}{2 i k_1} \int\limits_{C+\bar{C}} (f \bar{\chi}_n - \bar{\chi}_n f)\, ds. \tag{18.22}$$

By a similar application of Green's Theorem Haskind shows that one may also write

$$\varphi = f + \nu\, e^{\nu y} \int\limits_{\infty}^{y} f\, e^{-\nu \eta}\, d\eta + \frac{B^{\pm} g}{\sigma} e^{\nu y \mp i k_1 x} + \frac{A g}{\sigma} e^{\nu y - i k_1 x}, \tag{18.23}$$

where the plus sign is used for points to the right of C and the minus sign for points to the left. It is easy to verify directly that φ satisfies (18.13) and (18.17); however, (18.20) allows one to investigate the asymptotic behavior more simply. If φ has no singularities, then (17.3) must also hold here, i.e., $(B^+)^2 + (B^-)^2 + 2 A B^- = 0$.

This result may be used to find the source solutions giving outgoing waves at $\pm\infty$. For Eq. (18.13) the singular solutions for the whole plane are known to be the Bessel functions $K_n(kr)$, where $r^2 = (x-a)^2 + (y-b)^2$. To find the solution corresponding to (13.22), one assumes it may be expressed as

$$G(x, y; a, b) = \varphi_0 + K_0(k r) - K_0(k r_1),$$

with $r_1^2 = (x-a)^2 + (y+b^2)$, where φ_0 has no singularities for $y<0$. Then $f_{0y} = \varphi_{0y} - \nu \varphi_0$ may be extended as a regular solution of (18.13) to the whole plane. Also,

$$f_{0y}(x, 0) = 2 \frac{\partial}{\partial y} K_0(k r_1)\big|_{y=0}.$$

One may then show that this relation holds for all $y \leqq 0$:

$$f_{0y}(x, y) = 2 \frac{\partial}{\partial y} K_0(k r_1), \qquad y \leqq 0,$$

or

$$f_0(x, y) = 2 K_0(k r_1).$$

Substitution in (18.23) with $A=0$ and direct computation of $B^{\pm}$ from (18.22) by taking C as a small circle about the singularity gives

$$G = K_0(k r) + K_0(k r_1) + 2\nu\, e^{\nu y} \int\limits_{\infty}^{-y} e^{-\nu y} K_0(k r_1)\, dy - 2\pi i \frac{\nu}{k_1} e^{\nu(y+b) \mp i k_1 (x-a)}. \tag{18.24}$$

For Haskind's application of this method to the diffraction about a vertical and a horizontal barrier we refer to the original paper. Force and moment are obtained in terms of Mathieu functions. For the horizontal barrier in water of finite depth we refer to MacCamy's paper (1957) where a formula analogous to (18.24) is derived.

β) Waves on beaches. Much of the immediately preceding discussion of diffraction of plane waves approaching at an angle or of short-crested waves approaching normally applies also to this case. One is led to the following boundary-value problem for $\varphi(x, y) = \varphi_1 + i \varphi_2$:

$$\left.\begin{array}{ll}
1. & \varphi_{xx} + \varphi_{yy} - k^2 \varphi = 0, \quad k^2 < \nu^2, \\
2. & \varphi_y(x, 0) - \nu \varphi(x, 0) = 0, \\
3. & \varphi_x \sin\gamma + \varphi_y \cos\gamma = 0 \quad \text{for } y + x \tan\gamma = 0, \\
4. & \varphi \sim \frac{A g}{\sigma} e^{\nu y} e^{-i k_1 x} \quad \text{as} \quad x \to \infty, \quad k = \nu^2 - k^2, \\
5. & \varphi_x^2 + \varphi_y^2 \to 0 \quad \text{as} \quad x^2 + y^2 \to \infty \quad \text{along} \quad y + x \tan\gamma = 0.
\end{array}\right\} \tag{18.25}$$

Many of the authors cited in Sect. 17β considered this problem along with the two-dimensional one. In particular, we refer to HANSON (1926), MICHE (1944), STOKER (1947), WEINSTEIN (1949), ROSEAU (1952), and PETERS (1952). Both PETERS and ROSEAU solve the problem for arbitrary angle γ, $0<\gamma\leqq\pi$ [thus including the semi-infinite dock problem treated differently by HEINS (1948)]. The use of the reduction method limits one here, as in the two-dimensional case, to angles $\gamma=p\pi/2q$. We shall illustrate the procedure briefly for $\gamma=\pi/4$ and $\gamma=\pi/2$, following essentially WEINSTEIN'S (1949) treatment [see also BRILLOUËT (1957, Chaps. I, II)].

Since the boundary condition on the free surface and bottom is the same in the two- and three-dimensional cases, we may make use of the auxiliary function g constructed in (17.49) by using only the real part of the complex potential. Thus, for $\gamma=\pi/4$ one finds from (17.48) that $a_1=a_0/\nu$. Hence, from (17.50)

$$p(\lambda) = a_0(1+\lambda/\nu),$$

and

$$g(z) = \frac{a_0}{\nu}\left(\frac{d}{dz}+\nu\right)\left(\frac{d}{dz}+i\nu\right)(\varphi+i\psi),$$

$$\operatorname{Im} g(z) = \frac{a_0}{\nu}\left(\frac{\partial}{\partial x}+\nu\right)\left(-\frac{\partial}{\partial y}+\nu\right)\varphi.$$

Thus, the boundary conditions 2 and 3 of (18.25) imply that

$$\left.\begin{array}{ll} h(x,y)\equiv\left(\frac{\partial}{\partial x}+\nu\right)\left(\frac{\partial}{\partial y}-\nu\right)\varphi(x,y)=0 & \text{on}\quad y=0,\ x>0 \\ & \text{and}\quad x=0,\ y<0. \end{array}\right\} \tag{18.26}$$

We recall that the definition of $\varphi(x,y)$ has been extended from the original wedge by reflection in the bottom. One must now find a function $h(x,y)$ satisfying equation 1 of (18.25) and the boundary conditions (18.26) and which is regular everywhere in the extended wedge, $0\geqq\vartheta\geqq\frac{1}{2}\pi$, except possibly at the origin, bounded as $x^2+y^2\to\infty$, and symmetric about the line $y=-x$. It is known that the general solution of this problem is given by

$$h(x,y)=\left(\frac{\partial}{\partial x}+\nu\right)\left(\frac{\partial}{\partial y}-\nu\right)\varphi(x,y)=\sum_{n=0}^{\infty}A_n K_{2(2n+1)}(kr)\sin 2(2n+1)\vartheta. \tag{18.27}$$

A similar analysis for waves approaching a vertical cliff ($\gamma=\frac{1}{2}\pi$) leads to

$$h(x,y)\equiv\left(\frac{\partial}{\partial y}-\nu\right)\varphi(x,y)=\sum_{n=0}^{\infty}A_n K_{2n+1}(kr)\sin(2n+1)\vartheta. \tag{18.28}$$

Let us take the weakest possible singularity in each case, i.e., K_1 for the 90° cliff and K_2 for the 45° beach. Consider first the vertical cliff. Taking account of the relation $K_0'(u)=-K_1(u)$, we have

$$\left(\frac{\partial}{\partial y}-\nu\right)\varphi(x,y)=-\frac{A_0}{k}\frac{\partial}{\partial y}K_0(kr).$$

We may then identify $-A_0K_0/k$ with f and from (18.23), with $B^{\pm}=0$, we have

$$\varphi=-\frac{A_0}{k}K_0(kr)-A_0\frac{\nu}{k}e^{\nu y}\int_{\infty}^{y}e^{-\nu\eta}K_0\left(k\sqrt{x^2+y^2}\right)d\eta+\frac{Ag}{\gamma'}e^{\nu y-ik_1x},$$

where A_0 must still be determined so that $\varphi_x(0, y) = 0$, $y<0$. In computing φ_x as $x \to 0$, one must remember that $K_0(u) \sim \ln(2/u)$ as $u \to 0$. Hence, one finds

$$\varphi_x(0, y) = -\frac{A_0}{k} \nu e^{\nu y} \lim_{\varepsilon \to 0} \lim_{x \to 0} \int_{-\varepsilon}^{\varepsilon} \frac{-x}{x^2+y^2} dy - i \frac{A g k_1}{\sigma} e^{\nu y}$$

$$= \frac{A_0}{k} \nu \pi e^{\nu y} - i \frac{A g k_1}{\sigma} e^{\nu y}.$$

Setting this equal to zero, one finds

$$\frac{A_0}{k} = \frac{A g}{\sigma} \cdot \frac{i k_1}{\pi \nu}.$$

Substituting above and separating the real and imaginary parts of $\varphi = \varphi_1 + i \varphi_2$, we obtain an everywhere regular solution φ_1 and a solution φ_2 with a singularity at the origin and 90° out of phase at $x = \infty$:

$$\left.\begin{aligned} \varphi_1(x, y) &= \frac{A g}{\sigma} e^{\nu y} \cos k_1 x, \\ \varphi_2(x, y) &= -\frac{A g}{\sigma} \frac{k_1}{\pi \nu} \left[K_0(k r) + \nu e^{\nu y} \int_{\infty}^{y} e^{-\nu \eta} K_0\left(k \sqrt{x^2+\eta^2}\right) d\eta \right] + \\ &\quad + \frac{A g}{\sigma} e^{\nu y} \sin k_1 x. \end{aligned}\right\} \tag{18.29}$$

The corresponding equation for (18.27) can be written in the form

$$\left(\frac{\partial}{\partial x} + \nu\right)\left(\frac{\partial}{\partial y} - \nu\right) \varphi(x, y) = A_0 K_2(k r) \sin 2\vartheta = \frac{2 A_0}{k^2} \frac{\partial^2}{\partial x \, \partial y} K_0(k r). \tag{18.30}$$

One can find its integration discussed in ROSEAU (1952, Chap. IV). A solution for the next simplest case, $\gamma = 30°$, does not seem to have been published. For $\gamma = 45°$ the regular solution φ_1, and singular solution φ_2 as given by ROSEAU, but corrected according to personal communications from ROSEAU and LEHMAN, are

$$\left.\begin{aligned} \varphi_1 &= A_1 \{e^{\nu y} [k_1 \cos k_1 x - \nu \sin k_1 x] + e^{-\nu x} [k_1 \cos k_1 y + \nu \sin k_1 y]\}, \\ \varphi_2 &= A_2 \{e^{\nu y} [\nu \cos k_1 x + k_1 \sin k_1 x] + e^{-\nu x} [\nu \cos k_1 y - k_1 \sin k_1 y]\} + \\ &\quad + A_2 \frac{\nu^2+k_1^2}{\pi \nu} \Big\{ -K_0\left(k \sqrt{x^2+y^2}\right) + \nu e^{-\nu x} \int_{-\infty}^{x} e^{\nu \xi} K_0\left(k \sqrt{\xi^2+y^2}\right) d\xi + \\ &\quad + \nu e^{\nu y} \int_{y}^{\infty} e^{-\nu \eta} K_0\left(k \sqrt{x^2+\eta^2}\right) d\eta - \nu^2 e^{\nu y} \int_{y}^{\infty} d\eta \, e^{-\nu \eta} \left(e^{-\nu x} \int_{-\infty}^{x} d\xi \, e^{\nu \xi} K_0\left(k \sqrt{\xi^2+\eta^2}\right)\right)\Big\}. \end{aligned}\right\} \tag{18.31}$$

In order to satisfy condition 4 of (18.25) one must take

$$A_1 = \frac{A g}{\sigma} \cdot \frac{k_1 + i \nu}{k_1^2 + \nu^2}, \qquad A_2 = \frac{A g}{\sigma} \cdot \frac{\nu - i k_1}{k_1^2 + \nu^2}, \qquad \varphi = \varphi_1 + \varphi_2.$$

Edge waves. In the investigation of diffraction of waves on horizontal cylindrical obstacles and of waves on beaches, it was specifically assumed that $k^2 < m^2$. This was automatically fulfilled for plane waves approaching at an angle, but needed to be assumed for short-crested waves. For the short-crested waves there also exist standing-wave solutions which can be exhibited in certain cases for $k^2 > m^2$. Such solutions were apparently first noticed by STOKES (1846, p. 7 = 1880, p. 167) in connection with the propagation of waves in a canal of non-rectangular cross-section. Certain peculiarities of these solutions have been pointed out by URSELL (1951, 1952).

Consider the first three conditions of (18.25) for waves on a sloping beach, but with $k^2 > \nu^2$. Then one may verify directly that

$$\varphi(x, y) = \mathrm{e}^{k[y \sin\gamma - x\cos\gamma]}$$

is a solution. This gives a velocity potential for standing waves:

$$\Phi(x, y, z, t) = \mathrm{e}^{k[y \sin\gamma - x\cos\gamma]} \cos(k z + \varepsilon) \cos(\sigma t + \tau), \tag{18.32}$$

where

$$k \sin\gamma = \sigma^2/g.$$

The wave amplitude is bounded at the origin and drops off very quickly as x increases. Clearly, one must have $\gamma < \frac{1}{2}\pi$. URSELL has pointed out other interesting aspects. For a given γ and σ there is only one allowable k, i.e., it is a discrete point of the spectrum. In the case discussed earlier with $k^2 < \nu^2$ all values of k between 0 and ν were allowable. In addition, the total energy per unit length in the z direction is finite for the Stokes edge wave.

From (18.29) one may construct a progressive wave moving in the direction Oz with velocity.

$$c = \frac{g \sin\gamma}{\sigma}.$$

There is evidence that such waves have been observed in nature (cf. MUNK, SNODGRASS and CARRIER 1956; DONN and EWING 1956).

URSELL (1952) has shown that (18.32) is only the first in a sequence of solutions of this nature for a sloping beach. He shows, in fact, that the following velocity potential also satisfies the condition:

$$\left.\begin{aligned}\Phi(x, y, z, t) = \Big\{ &\mathrm{e}^{-k[x\cos\gamma - y\sin\gamma]} + \sum_{m=1}^{n} A_{mn} \big[\mathrm{e}^{-k[x\cos(2m-1)\gamma + y\sin(2m-1)\gamma} + \\ &+ \mathrm{e}^{-k[x\cos(2m+1)\gamma - y\sin(2m+1)\gamma]}\big]\Big\} \cos(k z + \varepsilon)\cos(\sigma t + \tau),\end{aligned}\right\} \tag{18.33}$$

where

$$A_{mn} = (-1)^m \prod_{r=1}^{n} \frac{\tan(n - r + 1)\gamma}{\tan(n + r)\gamma}, \qquad \sigma^2 = g k \sin(2n + 1)\gamma.$$

It follows from the last condition that one must have

$$(2n + 1)\gamma \leqq \frac{\pi}{2} \quad \text{or} \quad n < \frac{\pi}{4\gamma} + \frac{1}{2},$$

where $n = 0$ will be taken to indicate the Stokes edge wave. Thus, for fixed wave number k, the above formula gives one frequency σ if $\frac{1}{2}\pi > \gamma > \frac{1}{6}\pi$, two if $\frac{1}{6}\pi > \gamma > \frac{1}{10}\pi$, etc. An experiment carried out by URSELL confirms the existence of these other modes of motion. The solutions (18.33) for $\gamma = \pi/2(2n+1)$ have also been given by MACDONALD (1896). At these critical angles the solution (18.33) does not vanish as $x \to \infty$. MACDONALD apparently discarded the other solutions as being of little interest, not "being sensible at a distance from the edge". ROSEAU (1958) has recently carried through a systematic investigation of edge waves, including ones with singular behavior at the edge.

KELDYSH (1936) has stated without proof that for $\gamma = 45°$ the Stokes edge wave and the function φ_1 from (18.31) constitute a complete set of bounded solutions in the sense that for any absolutely integrable function $f(x, y)$, $x = 0$, the following Fourier-integral-like theorem holds [cf. formula (16.5)]:

$$f(x,z)=\frac{2}{\pi^2}\int_0^\infty\int_0^\infty\frac{dk\,dk_1}{k^2+2k_1^2}\int_{-\infty}^{\infty}d\zeta\cos(z-\zeta)\int_0^\infty d\xi\, f(\xi,\zeta)\times$$
$$\times\{[k_1 e^{-\nu x}+k_1\cos k_1 x-\nu\sin k_1 x][k_1 e^{-\nu\xi}+k_1\cos k_1\xi-\nu\sin k_1\xi]+$$
$$+2k^2\exp\left(-k(x+\xi)/\sqrt{2}\right)\}.$$

It is possible to construct other types of edge waves. First we rederive the Stokes wave from the third formula in (13.5) with $a=0$. A surface satisfying $\Phi_n=0$ is defined by

$$\frac{dy}{dx}=\frac{\Phi_y}{\Phi_x}=-\frac{\nu}{\sqrt{k^2-\nu^2}},$$

or

$$y=-x\tan\gamma+C,\qquad \tan\gamma=\nu/\sqrt{k^2-\nu^2},$$

where we may set $C=0$ since it does not provide essentially different solutions for the bottom. This is just STOKES' solution.

One may expect to find a different type of solution by using the third equation of (13.6) with $a=0$. Here the corresponding solution is

$$-\log\frac{\sinh m_0(y+h)}{\sinh m_0 h}=\frac{m_0^2}{\sqrt{k^2-m_0^2}}x,\tag{18.34}$$

where again we have dropped an added constant. This describes a bottom which starts as a sloping beach and approaches, as $x\to\infty$, a flat bottom at depth h. The initial slope of the beach is $\sigma^2/g\sqrt{k^2-m_0^2}$. The velocity potential describes edge waves for such a configuration.

One may proceed in the same fashion with the last formulas of (13.5) and (13.6). They turn out to give identical bottoms:

$$\log\frac{\sin m_i(y+h)}{\sin m_i h}=\frac{m_i^2}{\sqrt{k^2+m_i^2}}x.\tag{18.35}$$

This corresponds to edge waves along an overhanging cliff in water of finite depth. The initial backward slope of the cliff is $\sigma^2/g\sqrt{k^2+m_i^2}$.

A particularly interesting sort of edge wave, although the name is now a misnomer since there is no edge, has been discovered by URSELL (1951). He has shown the existence of standing waves of the form

$$\varphi(x,y)\cos kz\cos\sigma t$$

in the neighborhood of a fixed submerged cylinder of radius a if ka is small enough. The waves are symmetric about the vertical plane through the axis of the cylinder and decay exponentially as $|x|$ increases. One can, of course, also construct waves progressing along the cylinder.

URSELL calls such modes of motion "trapping modes" since, if they occur in a canal with sides given by $z=0$ and $z=n\pi/k$, no energy is radiated away, even though there is a path of escape. In fact, the motion is similar in this respect to standing waves in a basin of finite extent. The edge waves considered above also can be used to construct trapping modes.

γ) Waves in canals. The propagation of periodic waves along a canal leads to problems similar to those occurring in the propagation of waves parallel to a beach. Let the canal be parallel to Oz with cross-sectional contour C. We wish to find

$$\Phi(x, y, z, t) = \varphi(x, y)\cos(kz - \sigma t)$$

where $\varphi(x, y)$ satisfies

$$\varphi_{xx} + \varphi_{yy} - k^2\varphi = 0, \tag{18.36}$$

$$\varphi_y(x, 0) - \nu\varphi(x, 0) = 0, \quad \nu = \sigma^2/g,$$

on the free surface,

$$\varphi_n = 0 \quad \text{on } C.$$

It will also be assumed that $\varphi_x^2 + \varphi_y^2$ is bounded.

Clearly the same equations arise in searching for standing-wave solutions in a horizontal cylindrical basin with cross-sectional contour C bounded at either end by vertical walls at a distance l apart. In this case k is restricted to the values $n\pi/l$. For progressive waves solutions with $k=0$ are, of course, of no interest.

The special case when C is a rectangle has already been discussed in Sect. 14γ. The configuration for C which seems to have attracted the next most attention is a triangular one in which the two sides are inclined at the same angle. KELLAND (1844) was apparently the first to consider this problem for infinitesimal waves, limiting his treatment to angles of 45°. The matter was treated systematically by MACDONALD (1894) who states that a solution with the properties of (18.36) exists only for angles $\gamma = 45°$ and $\gamma = 30°$. This does not exclude the possibility of the existence for other angles of a periodic progressive wave with a curved wave front, for these would not be described by the assumed form of Φ.

The solutions for $\gamma = 45°$ can be obtained from the fundamental solutions of (13.6), but it is nearly as easy to find them directly. In the third formula of (13.6) let $a = b = \frac{1}{2}A$, $k^2 = 2m_0^2$. This gives the velocity potential, after forming a progressive wave,

$$\Phi(x, y, z, t) = A\cosh\frac{k}{\sqrt{2}}(y+h)\cosh\frac{k}{\sqrt{2}}x\cos(kz - \sigma t). \tag{18.37}$$

Let the sides of the canal be given by $y = \pm x - h$. Then it is easy to verify that

$$\Phi_n|_{y=x-h} = -\Phi_x + \Phi_y|_{y=x-h} = 0, \quad \Phi_n|_{y=-x-h} = \Phi_x + \Phi_y|_{y=-x-h} = 0,$$

so that the boundary conditions are all satisfied. Since

$$\sigma^2 = g m_0 \tanh m_0 h = \frac{1}{\sqrt{2}} g k \tanh\frac{1}{\sqrt{2}} k h,$$

the wave velocity is given by

$$c^2 = \frac{g}{k\sqrt{2}}\tanh\frac{kh}{\sqrt{2}}. \tag{18.38}$$

If $m_0^2 > m_i^2$ [in the notation of Eq. (13.6)], there will be i further symmetric modes. In (13.6), formula 4, set $a = b = \frac{1}{2}A$ and add this to formula 1 with $a = A$, $b = 0$. This gives

$$\begin{aligned}\Phi(x, y, z, t) = A\big[&\cos m_i(y+h)\cosh\sqrt{k^2+m_i^2}\,x + \\ &+ \cosh m_0(y+h)\cos\sqrt{m_0^2-k^2}\,x\big]\cos(kz-\sigma t).\end{aligned}$$

One may again verify easily that $\Phi_n = 0$ on the two sides of the canal if $k^2 = m_0^2 - m_i^2$. Hence this mode of motion will exist only if $m_0^2 > m_i^2$. For given σ there will be no modes of this sort if h is small enough, for then $m_0^2 < m_1^2$. The number gradually increases as h increases. If h and k are fixed and σ allowed to increase, there will be an infinite sequence $\sigma_1, \sigma_2, \ldots$ for which $k^2 = m_0^2 - m_i^2$ will be satisfied; $\sigma_n^2 h/g \to (n + \frac{3}{4})\pi$ as $n \to \infty$. The situation is easily visualized by plotting on one graph $\tanh mh$, $-\tan mh$ and $(\sigma^2 h/g)/mh$. One may write the potential function in the form

$$\Phi(x, y, z, t) = A\,[\cos m_i(y + h) \cosh m_0 x + \cosh m_0(y + h) \cos m_i x] \cos(kz - \sigma t),$$

where

$$m_0 \tanh m_0 h = \nu, \qquad m_i \tan m_i h = -\nu; \qquad k^2 = m_0^2 - m_i^2. \tag{18.39}$$

The velocity is given by

$$c^2 = \frac{g\, m_0 \tanh m_0 h}{m_0^2 - m_i^2}. \tag{18.40}$$

Asymmetric modes of motion also exist, having first been noticed by GREENHILL (1886). These cannot be deduced from (13.6) but must be found directly. The velocity potential corresponding to (18.37) is

$$\Phi(x, y, z, t) = A \sinh \frac{k}{\sqrt{2}}(y + h) \sin \frac{k}{\sqrt{2}} x \cos(kz - \sigma t) \tag{18.41}$$

The wave velocity is

$$c^2 = \frac{g}{k\sqrt{2}} \coth \frac{kh}{\sqrt{2}}, \tag{18.42}$$

which approaches infinity as $kh \to 0$. In addition to this mode, other asymmetric modes may exist under conditions similar to those required for (18.39). The velocity potential for these modes is

$$\left.\begin{aligned}\Phi(x, y, z, t) = A\,[\sin n_i(y + h) \sinh n_0 x + \sinh n_0(y + h) \sin n_i x] \times \\ \times \cos(kz - \sigma t),\end{aligned}\right\} \tag{18.43}$$

where

$$n_0 \coth n_0 h = \nu, \qquad n_i \cot n_i h = \nu, \qquad k^2 = n_0^2 - n_i^2.$$

The velocity of propagation is given by

$$c^2 = \frac{n_0 \coth n_0 h}{n_0^2 - n_i^2}. \tag{18.44}$$

The solution for the angle $\gamma = 30°$ will not be discussed here. It can be found in LAMB's Hydrodynamics (1932, p. 449) as well as in MACDONALD's paper cited above.

One may construct other possible contours for the canal cross-section by starting from one of the solutions (13.5) or (13.6) and finding surfaces for which $\Phi_n = 0$. Thus, from the third equation of (13.5) form

$$\Phi = A\, e^{\nu y} \sinh x \sqrt{k^2 - \nu^2} \cos(kz - \sigma t).$$

Solution of the differential equation $dy/dx = \Phi_y/\Phi_x$ leads easily to

$$y + h = \frac{\nu}{k^2 - \nu^2} \log \cosh x \sqrt{k^2 - \nu^2}$$

as an equation for the contour of a possible canal. The contour is reasonably shaped but varies with the choice of k. Also, the method is unsatisfactory in that it gives no information about other possible modes of motion.

19. Problems with steadily oscillating boundaries. Such problems include waves resulting from forced oscillation of a submerged body and the waves associated with steady oscillations of a freely floating body in oncoming waves.

In this section we shall assume the fluid of infinite extent. Waves in an oscillating bounded basin will be discussed later. The mathematical treatment has much in common with that of the last two sections, the scattered wave of those sections becoming the forced wave of this one.

α) Forced oscillations. Suppose that the surface of the oscillator in its equilibrium position is represented by $F(x, y, z)=0$. Let us take it, for example, to be oscillating vertically with amplitude ε. Then the equation of the oscillating surface S may be written $F(x, y, z, t)=F(x, y+\varepsilon a \sin\sigma t, z)=0$ where a is some length dimension of the oscillator. This ε will be taken as the expansion parameter in the perturbation procedure. In the perturbation theory of Sect. 10, we were concerned only with the functions $\Phi(x, y, z, t)$ and $\eta(x, y, t)$. However, we must similarly expand F before substituting it into the boundary condition satisfied on the surface of the oscillator, namely,

$$F_x\Phi_x+F_y\Phi_y+F_z\Phi_z+F_t=0 \quad \text{on} \quad F(x, y, z, t)=0. \tag{19.1}$$

The expansion for this case is

$$\left.\begin{aligned} &F(x, y+\varepsilon a\sin\sigma t, z)\\ &\quad=F(x, y, z)+\varepsilon a\sin\sigma t\, F_y(x, y, z)+\tfrac{1}{2}\varepsilon^2 a^2\sin^2\sigma t\, F_{yy}(x, y, z)+\cdots.\end{aligned}\right\} \tag{19.2}$$

We don't wish to restrict ourselves to this one mode of motion for the oscillator, but an examination of the form of this and similar expansions indicates that we may assume in general that the surface of the oscillator can be represented by the series

$$\left.\begin{aligned} F(x, y, z, t) &= F^{(0)}(x, y, z)+\varepsilon\left[F_1^{(1)}(x, y, z)\cos\sigma t+F_2^{(1)}(x, y, z)\sin\sigma t\right]+\\ &\quad+\text{time-periodic terms in higher powers of } \varepsilon=0,\end{aligned}\right\} \tag{19.3}$$

where $F^{(0)}(x, y, z)=0$ is the equilibrium position of the oscillator. We may now assume either that Φ is periodic, i.e.,

$$\Phi(x, y, z, t)=\sum\varphi_{1n}(x, y, z)\cos n\sigma t+\varphi_{2n}(x, y, z)\sin n\sigma t \tag{19.4}$$

or, more simply, that it is simple harmonic,

$$\Phi(x, y, z, t)=\varphi_1(x, y, z)\cos\sigma t+\varphi_2(x, y, z)\sin\sigma t, \tag{19.5}$$

where each function φ_{in} or φ_i is still to be expanded in a perturbation series. The two assumptions are not quite equivalent, even for the first-order theory, but since under certain conditions (19.4) leads to the same first-order equations as (19.5), we shall assume the latter form, together with

$$\eta(x, z, t)=\eta_1(x, z)\cos\sigma t+\eta_2(x, z)\sin\sigma t. \tag{19.6}$$

Substitution of the perturbation series into the exact equations and boundary conditions, as in Sect. 10, then leads to the following first-order equation and boundary conditions:

$$\left.\begin{aligned} &1.\quad \Delta\varphi_k^{(1)}=0, \qquad k=1,2,\\ &2.\quad \varphi_{ky}^{(1)}(x, 0, z)-\frac{\sigma^2}{g}\varphi_k^{(1)}(x, 0, z)=0, \quad k=1,2,\\ &3.\quad \operatorname{grad}F^{(0)}\cdot\operatorname{grad}\varphi_1^{(1)}+\sigma F_2^{(1)}=0 \quad \text{on} \quad F(x, y, z)=0,\\ &4.\quad \operatorname{grad}F^{(0)}\cdot\operatorname{grad}\varphi_2^{(1)}-\sigma F_1^{(1)}=0 \quad \text{on} \quad F(x, y, z)=0.\end{aligned}\right\} \tag{19.7}$$

One should note that it is a natural consequence of the method that the boundary condition on the oscillator is to be satisfied at its equilibrium position.

If we let

$$A_1(x,y,z) = \frac{-\sigma F_2^{(1)}}{|\operatorname{grad} F^{(0)}|}, \quad A_2(x,y,z) = \frac{\sigma F_1^{(1)}}{|\operatorname{grad} F^{(0)}|} \quad \text{for } F^{(0)}(x,y,z) = 0, \tag{19.8}$$

then conditions 3. and 4. of (19.7) may be written

$$\varphi_n^{(1)} = A(x,y,z) \quad \text{on} \quad F^{(0)} = 0, \tag{19.9}$$

where

$$\varphi^{(1)} = \varphi_1^{(1)} + i\,\varphi_2^{(1)} \quad \text{and} \quad A = A_1 + i A_2.$$

We shall henceforth drop the superscripts and consider only the first-order equations. In addition to Eqs. (19.7) the functions φ_i must also satisfy the usual conditions on fixed solid boundaries, $\varphi_{in} = 0$, and, if the fluid is infinitely deep, $|\operatorname{grad}\varphi| \to 0$ as $y \to -\infty$. Finally, one needs a boundary condition to ensure only outgoing waves at infinity. As has been pointed out by URSELL (1951), the foregoing conditions are not always sufficient to guarantee uniqueness of solution.

KOCHIN (1939, 1940) has considered the general mathematical problem in water of infinite depth for both two and three dimensions. HASKIND (1942b, 1944, 1946) has extended KOCHIN's methods to water of finite depth. The frequently-cited paper by JOHN (1950) treats the theoretical aspects of the problem in a thorough manner and includes many of the results of KOCHIN and HASKIND. Special problems have been considered by numerous authors. HAVELOCK (1929b) considers the waves generated by oscillation of a vertical plate extending to the bottom in water of infinite depth for both two and three dimensions, and in water of finite depth for two dimensions; he also considers waves generated by oscillations of a vertical cylinder. MACCAMY (1957) has treated the three-dimensional problem in water of finite depth. KENNARD (1949) has treated the two-dimensional problem as an initial-value problem. URSELL (1948) has considered waves generated by oscillation of a vertical strip with finite depth of immersion in water of infinite depth; the treatment is two dimensional. ALBAS (1958) treats a similar three-dimensional problem in which the motion is periodic along the length of the strip. In a later paper URSELL (1949b) considered the waves generated by the rolling of long cylinders of ship-like cross-section. In addition, URSELL has treated the waves generated by a heaving half-submerged circular cylinder (1949a, 1953c, 1954) and by a pulsing submerged cylinder (1950). HAVELOCK (1955) has treated the wave motion generated by a half-submerged sphere. Certain mathematical aspects of this last problem have been examined in more detail by MACCAMY (1954). Because of its interest in connection with the heaving motion of a ship there exist many papers attempting to compute approximately the force and moment on a heaving shiplike body resulting from wave formation. We mention particularly one by GRIM (1953). In the cited papers by KOCHIN and HASKIND certain special problems are solved approximately; by improving the approximation, LEVINE (1958) has clarified certain anomalous results of KOCHIN for an oscillating horizontal plate. In addition, HASKIND (1942, 1943b) has considered the motion resulting from forced oscillation of a plate, or a system of plates, on the surface. In a more recent paper HASKIND (1953a) has developed a method for finding solutions, and in particular the force and moment on the body, for a wide class of two-dimensional contours of ship-like cross-section. One should also consult a recent expository paper by MARUO (1957). A general survey of methods of generating waves in the laboratory, including some account of theoretical results, may be found in a recent paper by BIESEL and SUQUET (1951, 1952).

This brief summary of papers on forced water waves is by no means complete but lists many of the important papers and indicates the richness of the literature.

As was stated in the introductory remarks, the theory of forced water waves is mathematically almost identical with the diffraction theory. If one interprets the value of $-\partial\Phi_I/\partial n$ on the body as the function describing the motion of the oscillator, then it is clear that the problems are the same. Hence, the general remarks about existence of solutions, uniqueness, and special methods carry over directly and will not be repeated. However, we wish to consider here one further topic in the general theory.

KOCHIN'S H-function. The H-function was apparently first introduced by KOCHIN (1937) in connection with the theory of wave resistance. He later extended it (1939, 1940) to waves generated by oscillating bodies, and it has become a standard device among other Russian workers in this field, especially HASKIND, who has extended its definition to other situations.

Each potential function φ satisfying the boundary conditions has associated with it an H-function which is related to it much in the same way that the Fourier transform of a function is related to the function. One of its chief virtues is that it allows one to give compact formulas for force and moment on an oscillating body (in the present context) as well as certain other quantities. It is also sometimes helpful in suggesting approximate solutions.

Let us suppose that the surface S of a body of bounded extent is oscillating in some manner in fluid of infinite depth and let $\varphi=\varphi_1+i\varphi_2$ be the solution to the potential-theory problem formulated earlier. Let S_1 and S_2 be two closed surfaces lying below $y=0$ with S_2 enclosing S_1 and S_1 enclosing S. Let us denote the source potential introduced in (13.17″) by $G(x, y, z; \xi, \eta, \zeta)$, where (ξ, η, ζ) are the coordinates of the singularity, and let us write it as a contour integral:

$$G(x, y, z; \xi, \eta, \zeta) = \frac{1}{r} + \frac{1}{2\pi}\int_{-\pi}^{\pi} d\vartheta \int_{0\,(L)}^{\infty} dk\, \frac{k+\nu}{k-\nu}\, e^{k(y+\eta-i(x-\xi)\cos\vartheta-i(z-\zeta)\sin\vartheta)}, \tag{19.10}$$

where the path L passes below the singularity at $k=\nu=\sigma^2/g$. [The residue at this point gives exactly the imaginary part of (13.17″).]

Now apply GREEN'S Theorem to the region between S_1 and S_2 (the following argument is very similar to a two-dimensional one used in Sect. 17α in discussing the integral-equation method):

$$\left.\begin{aligned}\varphi(x, y, z) &= -\frac{1}{4\pi}\iint_{S_1}\left[\frac{1}{r}\frac{\partial\varphi}{\partial n} - \varphi\frac{\partial}{\partial n}\left(\frac{1}{r}\right)\right]d\sigma + \frac{1}{4\pi}\iint_{S_2}\left[\frac{1}{r}\frac{\partial\varphi}{\partial n} - \varphi\frac{\partial}{\partial n}\left(\frac{1}{r}\right)\right]d\sigma\\ &= \varphi^{(1)} + \varphi^{(2)}.\end{aligned}\right\} \tag{19.11}$$

Then $\varphi^{(1)}$ may be extended to a function harmonic in the whole space exterior to S_1. $\varphi^{(2)}$ is harmonic in the whole interior of S_2, but since S_2 may be indefinitely enlarged as long as it remains below $y=0$, $\varphi^{(2)}$ may be extended to be harmonic in the whole half-space, $y\leqq 0$. Consider now the function

$$\psi(x, y, z) = -\frac{1}{4\pi}\iint_{S_1}\left[G\frac{\partial\varphi}{\partial n} - \varphi\frac{\partial G}{\partial n}\right]d\sigma. \tag{19.12}$$

ψ satisfies the free-surface condition and the condition at infinity. Moreover, since $G=r^{-1}+$a function harmonic in the lower half-space, $\varphi-\psi$ is harmonic in the lower half-space and satisfies the other boundary conditions. But then $\varphi-\psi\equiv 0$, as follows from a uniqueness theorem proved by KOCHIN (1940, Sect. 1).

Hence, we may write

$$\varphi(x, y, z) = \frac{1}{4\pi} \iint_{S_1} [\varphi(\xi, \eta, \zeta)\, G_n(x, y, z; \xi, \eta, \zeta) - G\, \varphi_n]\, d\sigma. \tag{19.13}$$

Now define

$$\left.\begin{aligned} H(k, \vartheta) &= \iint_{S_1} e^{k[\eta + i\xi\cos\vartheta + i\zeta\sin\vartheta)]} \{\varphi_n(\xi, \eta, \zeta) - \\ &\quad - k\,\varphi[\cos(n, \eta) + i\cos\vartheta\cos(n, \xi) + i\sin\vartheta\cos(n, \zeta)]\}\, d\sigma, \\ &= \iint_{S_1} e^{k[\eta + i\xi\cos\vartheta + i\zeta\sin\vartheta]} \{\varphi_n(\xi, \eta, \zeta) + \\ &\quad + i\cos\vartheta[\varphi_\eta\cos(n, \xi) - \varphi_\xi\cos(n, \eta)] + \\ &\quad + i\sin\vartheta[\varphi_\eta\cos(n, \zeta) - \varphi_\zeta\cos(n, \eta)]\}\, d\sigma. \end{aligned}\right\} \tag{19.14}$$

Then, after some manipulation with (19.13), one can show that

$$\left.\begin{aligned} \varphi(x, y, z) &= \frac{1}{4\pi} \iint_{S_1} \left[\varphi \frac{\partial}{\partial n}\left(\frac{1}{r}\right) - \frac{1}{r}\frac{\partial \varphi}{\partial n}\right] d\sigma - \\ &\quad - \frac{1}{8\pi^2} \int_{-\pi}^{\pi} d\vartheta \int_{0\,(L)}^{\infty} dk\, \frac{k+\nu}{k-\nu}\, e^{k(y - ix\cos\vartheta + iz\sin\vartheta)}\, H(k, \vartheta). \end{aligned}\right\} \tag{19.15}$$

We give a few of KOCHIN's derived formulas. The asymptotic form of the free surface in a direction α is given by

$$\eta(R, \alpha, t) \cong \operatorname{Re}\left[\frac{\sigma}{g}\sqrt{\frac{\nu}{2\pi R}}\, \overline{H}(\nu, \alpha)\, e^{i\left(\nu R - \sigma t - \frac{\pi}{4}\right)}\right] \quad \text{as} \quad R \to \infty. \tag{19.16}$$

The rate at which energy is being carried off by the waves (and hence also the power input) is given by

$$N = \frac{1}{8\pi}\frac{\varrho\,\sigma^3}{g} \int_0^{2\pi} |H(\nu, \vartheta)|^2\, d\vartheta. \tag{19.17}$$

The force components on the oscillating body, averaged over a period, are given by

$$\left.\begin{aligned} X_{\text{av}} &= \frac{\varrho\,\nu^2}{8\pi} \int_{-\pi}^{\pi} |H(\nu, \vartheta)|^2 \cos\vartheta\, d\vartheta, \\ Y_{\text{av}} &= \varrho\, g\, V + \frac{\varrho}{16\pi^2} \int_{-\pi}^{\pi} \mathrm{PV} \int_0^{\infty} k\, \frac{k+\nu}{k-\nu} |H(k, \vartheta)|^2\, dk\, d\vartheta, \\ Z_{\text{av}} &= \frac{\varrho\,\nu^2}{8\pi} \int_{-\pi}^{\pi} |H(\nu, \vartheta)|^2 \sin\vartheta\, d\vartheta. \end{aligned}\right\} \tag{19.18}$$

The formulas can be derived from (8.4), (9.4), and asymptotic expressions for φ.

In formulas (19.14) and (19.15) the surface S_1 over which the integrals are taken may be contracted to S. This sometimes makes it possible to express H directly in terms of known boundary values. If φ can be expressed by means of a source distribution, say

$$\varphi(x, y, z) = \frac{1}{4\pi} \iint_S \gamma(\xi, \eta, \zeta)\, G(x, y, z; \xi, \eta, \zeta)\, d\sigma, \tag{19.19}$$

then one has
$$H(k,\vartheta) = -\iint_S \gamma(\xi,\eta,\zeta)\, e^{k[\eta+i\xi\cos\vartheta+i\zeta\sin\vartheta]}\, d\sigma. \tag{19.20}$$

In order to find approximate answers, Kochin frequently uses the distribution γ which would be proper in an infinite fluid without free surface, substitutes this in (19.20) and then uses the resulting approximation to H in (19.17) and (19.18) above. The procedure may be looked upon as the first two steps in an alternating type of approximation in which one first satisfies the boundary condition on the body, neglecting the free surface, next corrects this so as to satisfy the free-surface condition, but now disturbing the condition on the body, then corrects again to satisfy the condition on the body, etc. This method of approximation has frequently been used by HAVELOCK (e.g., 1929a).

KOCHIN (1939) has also defined the H-function for two-dimensional wave motion excited by an oscillating body. We simply reproduce the formulas. Let, as usual, $f(z,t) = f_1(z)\cos\sigma t + f_2(z)\sin\sigma t$ be the complex potential and let C_1 and C_2 be two contours in the lower half-plane containing C, C_1 inside C_2. Define
$$H_s(k) = \int_{C_1} e^{-ik\zeta} f_s'(\zeta)\, d\zeta, \qquad s = 1, 2. \tag{19.21}$$

Then
$$\left.\begin{aligned} f_s'(z) &= \frac{1}{2\pi i}\int_{C_1} \frac{f_s'(\zeta)}{z-\zeta}\, d\zeta - \frac{1}{2\pi}\int_0^\infty \overline{H}_s(k)\, e^{-ikz}\, dk - \\ &\quad - \frac{\nu}{\pi}\,\mathrm{PV}\int_0^\infty \frac{\overline{H}_s(k)}{k-\nu}\, e^{-ikz}\, dk + (-1)^{s+1} H_{s+1}(\nu)\, e^{-i\nu z}, \end{aligned}\right\} \tag{19.22}$$

where $H_3 \equiv H_1$. This follows immediately from a formula similar to (17.15). For the asymptotic form of the waves one gets
$$\left.\begin{aligned} \eta(x,t) &\cong \mathrm{Re}\,\frac{i\nu}{\sigma}\,[\overline{H}_1(\nu) - i\overline{H}_2(\nu)]\, e^{-i(\nu x - \sigma t)} \quad \text{as} \quad x \to +\infty, \\ \eta(x,t) &\cong \mathrm{Re}\,\frac{-i\nu}{\sigma}\,[\overline{H}_1(\nu) + i\overline{H}_2(\nu)]\, e^{-i(\nu x + \sigma t)} \quad \text{as} \quad x \to -\infty. \end{aligned}\right\} \tag{19.23}$$

The rate of dissipation of energy is
$$N = \tfrac{1}{2}\varrho\sigma\,[|H_1(\nu)|^2 + +|H_2(\nu)|^2]. \tag{19.24}$$

The mean values of the force and moment, averaged over a period, are
$$\left.\begin{aligned} X_{\mathrm{av}} &= \varrho\nu\,\mathrm{Im}\,\{\overline{H}_1(\nu)\, H_2(\nu)\}, \\ Y_{\mathrm{av}} &= \frac{\varrho}{4\pi}\,\mathrm{PV}\int_0^\infty \frac{k+\nu}{k-\nu}\,\{|H_1(k)|^2 + |H_2(k)|^2\}\, dk, \\ M_{\mathrm{av}} &= -\frac{\varrho}{4\pi}\,\mathrm{Im}\left\{\mathrm{PV}\int_0^\infty \frac{k+\nu}{k-\nu}\,[H_1'\overline{H}_1 + H_2'\overline{H}_2]\, dk\right\} + \\ &\quad + \tfrac{1}{2}\nu\varrho\,\mathrm{Im}\,\{H_1'(\nu)\,\overline{H}_2(\nu) - H_2'(\nu)\,\overline{H}_1(\nu)\}. \end{aligned}\right\} \tag{19.25}$$

Roughly the same remarks apply to the use of the two-dimensional formulas as of the three-dimensional ones.

Waves from an oscillator in a wall. In order to illustrate the use of the H-function, we consider the following problem. Let the (y,z)-plane be a rigid wall expect for a certain bounded area S in which there is a membrane oscillating according to a given law

$$x = F(y, z) \sin \sigma t, \qquad (y, z) \text{ in } S. \tag{19.26}$$

The boundary condition which has to be satisfied on the plane $x=0$ is then

$$\varphi_x(0, y, z) = \left\{\begin{matrix} \sigma F(y, z), & (y, z) \text{ in } S \\ 0, & (y, z) \text{ not in } S, \end{matrix}\right\} \tag{19.27}$$

where we still have $\varphi = \varphi_1 + i\varphi_2$.

This boundary condition, as well as the ones at infinity, can be satisfied by distributing "modified" sources (13.17″) or (19.10) over S with density $-\sigma F(y, z)/2\pi$:

$$\varphi(x, y, z) = \frac{-\sigma}{2\pi} \iint\limits_S F(\eta, \zeta)\, G(x, y, z; 0, \eta, \zeta)\, d\eta\, d\zeta. \tag{19.28}$$

In order to compute the H-function, we shall interpret the source distribution as representing a thin body making symmetric pulsations in an infinite fluid. Hence, we may assume that the wall is removed and the membrane replaced by a doubled one. (That the requisite motion is physically impossible doesn't invalidate the considerations; a more realistic model can easily be devised.) In (19.14) we take S_1 to be both sides of the thin body. Then, remembering that

$$\varphi_n(+0, \eta, \zeta) = \varphi_x(0, \eta, \zeta) = \sigma F, \qquad \varphi_n(-0, \eta, \xi) = -\varphi_x(0, \eta, \zeta) = \sigma F,$$

$$\cos(n, \xi) = 1 \quad \text{for } x > 0 \quad \text{and} \quad \cos(n, \xi) = -1 \quad \text{for } x < 0,$$

one finds easily that

$$H(k, \vartheta) = 2\sigma \iint\limits_S F(\eta, \zeta)\, e^{k(\eta + i\zeta \sin\vartheta)}\, d\eta\, d\zeta. \tag{19.29}$$

From (19.17) one then finds immediately, after carrying out the ϑ integration, that the rate of dissipation of energy to one side is given by

$$N = \frac{\varrho \sigma^5}{4\pi g} \iint\limits_S dy\, dz \iint\limits_S d\eta\, d\zeta\, F(y, z)\, F(\eta, \zeta)\, e^{\nu(y+\eta)}\, J_0(\nu(z - \zeta)). \tag{19.30}$$

Expressions for Y_{av} and Z_{av} may also be written down. The result $X_{av}=0$ is not really significant because the integral is over both sides of the thin pulsing body.

The theory for generation of two-dimensional waves in a semi-infinite channel by a vertical wave maker in the end-wall is easily derived in the same way. If the motion of the wave-maker is described by

$$x = F(y) \sin \sigma t, \qquad a \leqq y \leqq b \leqq 0 \tag{19.31}$$

then

$$H_1(k) = \int_a^b e^{k\eta} F(\eta)\, d\eta, \qquad H_2(k) = 0, \tag{19.32}$$

and, for example, the rate of dissipation of energy is given by

$$N = \varrho \sigma^3 \left[\int_a^b e^{\nu y} F(y)\, dy\right]^2. \tag{19.33}$$

The generation of short-crested waves is subject to the limitations described in Sect. 14γ. Suppose, for example, that the water is of depth h, the channel of breadth b, and that the motion of the wave-maker is described by

$$x = F(y) \cos kz \sin \sigma t, \qquad k = n\pi/b, \qquad -h \leqq y \leqq 0. \tag{19.34}$$

Then, since $\cos m_i(y+h)$, $\cosh m_0(y+h$ form a complete set of functions in $-h\leq y\leq 0$, there is no difficulty in representing $F(y)$ by a series of the fundamental solutions (13.6), but if $k^2>m_0^2$, no progressive waves will move down the tank (within the limits of applicability of the linearized theory, of course). The analysis of the filtering effect of the tank on more complicated wave-maker motions can easily be carried through by Fourier analysis.

Waves from an oscillator not in a wall. Let us now suppose that we have a two-dimensional oscillator in infinitely deep water moving according to the law

$$x = F(y)\sin\sigma t, \qquad a<y<b\leq 0, \tag{19.35}$$

but with no wall present. This small change complicates the solution of the problem in a substantial way, the complication being associated with the now possible flow under (and over if $b<0$) the oscillator. In addition, in order to ensure a unique solution some further condition analogous to the Kutta-Joukowski condition in airfoil theory is required; here the last two conditions of (19.36) play this role. Then the boundary conditions to be satisfied on the oscillator by the velocity potential

$$\Phi(x,y,t) = \varphi_1\cos\sigma t + \varphi_2\sin\sigma t$$

are

$$\left.\begin{aligned}\Phi_x(0,y,t) &= \sigma F(y)\cos\sigma t, \quad a<y<b\leq 0,\\ \Phi_y(0,a,t) &= 0,\\ \Phi_y(0,b,t) &= 0 \quad \text{if} \quad b<0,\end{aligned}\right\} \tag{19.36}$$

The problem is clearly closely related to that of diffraction of plane waves by a vertical barrier and could be treated by a modification of the method used in Sect. 17α for that problem. It may also be solved by the integral-equation method discussed in Sect. 17α. A modification of this method has been used by URSELL (1948).

Introduce the complex potential

$$\left.\begin{aligned}\Phi + i\Psi &= \mathrm{Re}_j\{f(z)\,e^{-j\sigma t}\},\\ \text{where}\qquad f(z) &= f_1(z)+jf_2(z) = (\varphi_1+j\varphi_2)+i(\psi_1+j\psi_2).\end{aligned}\right\} \tag{19.37}$$

We try to construct a solution by means of a distribution of vortices of the form (13.28)

$$\left.\begin{aligned}f_v(z;\zeta) &= \frac{1}{2\pi i}\log(z-\zeta)(z-\bar\zeta) + \frac{1}{\pi i}\,\mathrm{PV}\int_0^\infty \frac{e^{-ik(z-\bar\zeta)}}{k-\nu}\,dk - j\,i\,e^{-i\nu(z-\bar\zeta)}\\ &= f_{v1}+jf_{v2}\end{aligned}\right\} \tag{19.38}$$

with intensity

$$\gamma(\eta) = \gamma_1 + j\gamma_2, \qquad a<y<b, \tag{19.39}$$

along the oscillator:

$$f(z) = \int_a^b \gamma(\eta)\,f_v(z;i\eta)\,d\eta. \tag{19.40}$$

An analysis almost identical with that in Sect. 17α leads quickly to the integral equation

$$\mathrm{Re}_i\int_a^b \gamma(\eta)\,f_v'(iy;i\eta)\,d\eta = \sigma F(y) + j\cdot 0, \qquad a<y<b. \tag{19.41}$$

Separating γ_1 and γ_2 and noting that $f'_v(iy; i\eta)$ is real with respect to i, one finds

$$\left.\begin{aligned}\int_a^b [\gamma_1(\eta) f'_{v1}(iy; i\eta) - \gamma_2(\eta) f'_{v2}(iy; i\eta)]\, d\eta = \sigma F(y),\\ \int_a^b [\gamma_1(\eta) f'_{v2}(iy; i\eta) + \gamma_2(\eta) f'_{v1}(iy; i\eta)]\, d\eta = 0.\end{aligned}\right\} \tag{19.42}$$

The equations can be uncoupled by applying the operator $[\partial/\partial y - \nu]$ to each (so that the reduction method enters after all!). Introducing

$$\mu_k = \gamma'_k - \nu \gamma_k, \qquad G(y) = F' - \nu F, \tag{19.43}$$

one finally obtains the pair of equations

$$\left.\begin{aligned}\int_a^b \mu_1(\eta) \frac{d\eta}{y^2 - \eta^2} &= \frac{\gamma_1(b)}{y^2 - b^2} - \pi \sigma \frac{G(y)}{y},\\ \int_a^b \mu_2(\eta) \frac{d\eta}{y^2 - \eta^2} &= \frac{\gamma_2(b)}{y^2 - b^2},\end{aligned}\right\} \tag{19.44}$$

where we have taken advantage of the fact that $\varphi_y(\pm 0, y) = \mp \gamma(y)$ and hence $\gamma(a) = 0$; if $b < 0$ also $\gamma(b) = 0$. The integral equations are easily reduced to a known type occurring in airfoil theory[1] by the transformation

$$r = y^2 - \tfrac{1}{2}(a^2 + b^2), \qquad \varrho = \eta^2 - \tfrac{1}{2}(a^2 + b^2).$$

The solution may be written in the form

$$\left.\begin{aligned}\mu_1(\eta) &= \frac{2\eta}{\pi \sqrt{(a^2 - \eta^2)(\eta^2 - b^2)}} \times\\ &\quad \times \left[\nu \int_a^b \gamma_1(y)\, dy + 2\varepsilon\sigma\, \mathrm{PV} \int_a^b G(y) \sqrt{(a^2 - y^2)(y^2 - b^2)} \frac{dy}{\eta^2 - y^2}\right],\\ \mu_2(\eta) &= \frac{2\eta\nu}{\pi \sqrt{(a^2 - \eta^2)(\eta^2 - b^2)}} \int_a^b \gamma_2(y)\, dy.\end{aligned}\right\} \tag{19.45}$$

It is evident that the solution is not uniquely determined without some statement about the total circulation. Fixing the total circulation is equivalent to fixing $\gamma(b)$, as follows easily from the form of $\mu(\eta)$ and the relation

$$\gamma(\eta) = e^{\nu\eta} \int_a^\eta e^{-\nu s} \mu(s)\, ds. \tag{19.46}$$

It is possible to compute the H-function directly in terms of $\mu(s)$. First, we note that

$$\begin{aligned}H(\lambda) &= \oint_{C_1} e^{-i\lambda\zeta} f'(\zeta)\, d\zeta = \oint_{C_1} e^{-i\lambda\zeta} d\zeta \int_a^b \gamma(y) f'_v(\zeta; iy)\, dy\\ &= \int_a^b \gamma(y)\, dy \oint_{C_1} e^{-i\lambda\zeta} f'_v(\zeta; iy)\, d\zeta = \int_a^b \gamma(y)\, e^{\lambda y}\, dy.\end{aligned}$$

It then follows from (19.46) that

$$H(\lambda) = \frac{e^{\lambda b}}{\lambda + \nu} \gamma(b) - \frac{1}{\lambda + \nu} \int_a^b \mu(y)\, e^{\lambda y}\, dy. \tag{19.47}$$

[1] See, e.g., W. SCHMEIDLER: Integralgleichungen ..., pp. 55–56. Leipzig: Akademische Verlagsgesellschaft 1950, or S. G. MIKHLIN: Integral'nye uravneniya ..., pp. 149–154, Gostekhizdat, Moscow 1949.

One may now apply formulas (19.23) to (19.25) to find the quantities indicated there (note that $\bar{H}=H$).

One notes again that the function $H(\lambda)$ is determined only after $\gamma(b)$ is fixed. Taking $\gamma(b)\neq 0$ is equivalent to having a singularity at the end. If the oscillator is totally submerged, it seems reasonable to set $\gamma(b)=0$, as we assumed in (19.36), for then the vertical velocity is continuous at the end, i.e., $\varphi_y(+0, b)=\varphi_y(-0, b)$, as has already been assumed for the lower end at $y=a$. It is not clear what is the proper assumption if $b=0$, i.e., if the oscillator extends through the surface. In the similar problem of diffraction by a vertical plate, treated by the reduction method in Sect. 17α, the assumption of no singularity at the surface is equivalent to assuming $\gamma(0)=0$. We note that if $\gamma(b)=0$, then it follows from (19.46) and the form of μ_2 in (19.45) that $\mu_2\equiv 0$, and hence that $\gamma_2\equiv 0$. This is not true, of course, for γ_1.

Waves generated by a heaving hemisphere. We describe briefly a procedure used by HAVELOCK (1955) and MACCAMY (1954), and before them also by URSELL (1949a) for an analogous two-dimensional problem. Let a hemisphere of radius a have its center on the free surface in its undisturbed position and let it undergo forced vertical oscillations described by

$$x^2+(y-b_0\sin\sigma t)^2+z^2=a^2. \tag{19.48}$$

Then the boundary condition to be satisfied by $\varphi(x,y,z)=\varphi_1+i\varphi_2$ on the hemisphere is

$$\frac{\partial\varphi_1}{\partial r}=b_0\sigma\frac{y}{a}=b_0\sigma\cos\vartheta,\qquad \frac{\partial\varphi_2}{\partial r}=0\quad\text{on}\quad x^2+y^2+z^2=a^2,\quad y\leqq 0. \tag{19.49}$$

φ must, of course, also satisfy the free-surface condition and the radiation condition, as stated in (19.7).

The method of the above-named authors is to represent φ as a series in which the first term is a source at the center, say (13.17), and the remaining terms represent only local disturbances of the sort shown in (13.21), with $m=0$ since we have radial symmetry. The source term is actually taken in the form (13.17‴). Since the source is at (0, 0, 0), $r=r_1$ in the formulas and certain terms cancel and others double. Let

$$\left.\begin{aligned}\varphi^{(0)}&=\frac{1}{r}-\frac{2\nu}{\pi}\int_0^\infty[\nu\cos k y-k\sin k y]\frac{K_0(kR)}{k^2+\nu^2}\,dk-\\&\quad-\pi\nu e^{\nu y}Y_0(\nu R)+i\pi\nu e^{\nu y}J_0(\nu R),\\ \varphi^{(n)}&=\frac{-\nu}{2n}\frac{P_{2n-1}(\cos\vartheta)}{r^{2n}}+\frac{P_{2n}(\cos\vartheta)}{r^{2n+1}}.\end{aligned}\right\} \tag{19.50}$$

Then the assumed form for φ is

$$\varphi(x,y,z)=\sum_{n=0}^{\infty}a^{n+2}(A_n+iB_n)\,\varphi^{(n)}(x,y,z). \tag{19.51}$$

Substitution in the boundary condition (19.49) leads to an infinite set of linear equations for the coefficients A_n, B_n. Numerical methods may then be used to find any desired number of terms.

Having found φ approximately, one may proceed to compute the vertical hydrodynamic force on the sphere by integrating the pressure $p=-\varrho\,\partial\Phi/\partial t$

over the hemisphere. HAVELOCK carried through an approximate calculation, expressing the result in the form

$$\left.\begin{aligned} Y &= \frac{2}{3}\pi\varrho a^3 b_0 \sigma^2 [k \sin\sigma t - 2h\cos\sigma t] \\ &= -M\cdot k\cdot\frac{d^2 y_0}{dt^2} - M\cdot 2h\sigma\cdot\frac{dy_0}{dt}, \end{aligned}\right\} \tag{19.52}$$

where M is the mass of displaced fluid and y_0 the coordinate of the center. The parameter k is usually called the added-mass coefficient; h is called the damping parameter. Fig. 17 from HAVELOCK's paper shows k and $2h$ as functions of νa. As $\nu a \to \infty$, $2h \to 0$ and $k \to \frac{1}{2}$; $k(0) = 0.828\ldots$. The average rate at which work is being done by the sphere is $\frac{2}{3}\pi\varrho a^3 b_0^2 \sigma^3 h$ and does not involve k.

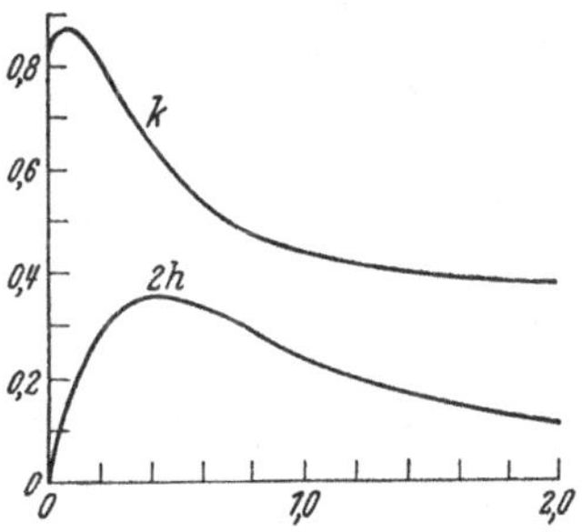

Fig. 17.

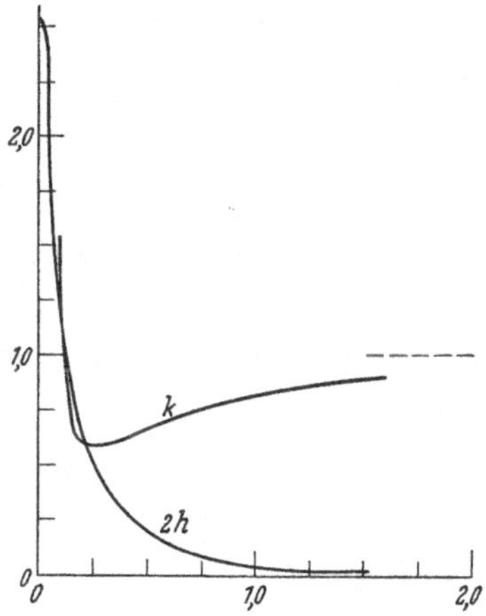

Fig. 18.

It is of interest to compare the same parameters as computed by URSELL (1949a) for a circular cylinder (per unit length). They are shown in Fig. 18. The asymptotic behavior of k is given by

$$\left.\begin{aligned} k(\nu a) &= \frac{8}{\pi^2}\left[\log\frac{1}{\nu a} + \frac{3}{2} - 2\log 2 - \gamma\right] + o(\nu a) \\ &= \frac{8}{\pi^2}\left[\log\frac{1}{\nu a} - 0.46\right] + o(\nu a) \quad \text{as} \quad \nu a \to 0, \\ k(\nu a) &= 1 - \frac{4}{3\pi\nu a} + o\left(\frac{1}{\nu a}\right) \quad \text{as} \quad \nu a \to \infty. \end{aligned}\right\} \tag{19.53}$$

β) Steady oscillations of a freely floating body in waves. Let us suppose that a rigid body is floating in an infinite ocean with prescribed plane waves approaching from a fixed direction, say from $x = +\infty$. If the motion has persisted for some time, we may suppose that the body is moving with a simple periodic motion of the same frequency as the waves. With this assumption the proper formulation of the linearized equations and boundary conditions has been derived by JOHN (1949).

Suppose the body is at rest in still water and let (x_0, y_0, z_0) be the coordinates of its center of gravity. Let $\bar{O}\bar{x}\bar{y}\bar{z}$ be a coordinate system fixed in the body with $\bar{O}$ at the center of gravity and the axes parallel to the space axes $Oxyz$. When the body is displaced, one may describe its position by giving the new position of the center of gravity $(\xi, \eta, \zeta) = (x_0 + \varepsilon x_1, y_0 + \varepsilon y_1, z_0 + \varepsilon z_1)$ and the Eulerian angles $\varepsilon\alpha$, $\varepsilon\beta$, $\varepsilon\gamma$ (we change notation from the customary φ, ϑ, ψ to avoid confusion with our other use of these letters). Thus the choice of ε implies that the amplitude of motion is small compared to some typical body length. The assumption of Sect. 10α that $\Phi = \varepsilon\,\Phi^{(1)} + \cdots$ implies that the amplitude of the prescribed incoming waves is also small compared to this length. The relationship between the two sets of coordinates may be easily found from the usual formulas concerning Eulerian angles to be of the form

$$\left.\begin{aligned}\bar{x} &= x - x_0 - \varepsilon\,[x_1 - \gamma(y - y_0) + \beta(z - z_0)] + \varepsilon^2\,[\cdots] + \cdots,\\ \bar{y} &= y - y_0 - \varepsilon\,[\gamma(x - x_0) + y_1 - \alpha(z - z_0)] + \cdots,\\ \bar{z} &= z - z_0 - \varepsilon\,[-\beta(x - x_0) + \alpha(z - z_0) + z_1] + \cdots.\end{aligned}\right\}\tag{19.54}$$

Let the surface of the body be described by

$$F(\bar{x}, \bar{y}, \bar{z}) = 0 \tag{19.55}$$

in body coordinates. To find the position in space coordinates one must substitute from (19.54) in (19.55). The kinematic boundary condition [see Eq. (19.1)] then becomes

$$\left.\begin{aligned}&\varepsilon\{\operatorname{grad} F(x-x_0, y-y_0, z-z_0)\cdot\operatorname{grad}\Phi^{(1)} + F_x(x-x_0, y-y_0, z-z_0)\times\\ &\quad\times[-\dot{x}_1 - \dot{\gamma}(y-y_0) - \dot{\beta}(z-z_0)] + F_y[-\dot{y}_1 + \dot{\alpha}(z-z_0) - \dot{\gamma}(x-x_0)] +\\ &\quad + F_z[-\dot{z}_1 + \dot{\beta}(x-x_0) - \dot{\alpha}(y-y_0)]\} + \varepsilon^2\{\cdots\} + \cdots = 0.\end{aligned}\right\}\tag{19.56}$$

Letting n_x, n_y, n_z be the components of the unit inward normal vector to the surface at rest, i.e.,

$$F(x - x_0, y - y_0, z - z_0) = 0 \tag{19.57}$$

(we shall call this surface S_0), and $\boldsymbol{q} = (\boldsymbol{r} - \boldsymbol{r}_0)\times\boldsymbol{n}$, i.e.

$$\left.\begin{aligned}&q_x = (y - y_0)\,n_z - (z - z_0)\,n_y, \qquad q_y = (z - z_0)\,n_x - (x - x_0)\,n_z,\\ &q_z = (x - x_0)\,n_y - (y - y_0)\,n_x,\end{aligned}\right\}\tag{19.58}$$

we may express the first-order term in (19.56), after dropping the superscript, in the form

$$\Phi_n = \dot{x}_1 n_x + \dot{y}_1 n_y + \dot{z}_1 n_z + \dot{\alpha}\,q_x + \dot{\beta}\,q_y + \dot{\gamma}\,q_z \quad \text{for} \quad (x, y, z) \text{ on } S_0. \tag{19.59}$$

We call attention to the fact that it follows as a natural consequence of the linearization that the boundary condition is to be satisfied on the surface in its undisturbed position.

In order to state the dynamical conditions on the body we introduce the following notations. Let M be its mass and I_x, I_y, I_z, I_{xx}, I_{xy}, ... its moments and moments and products of inertia about the body axes selected above. Let V be the volume bounded by the plane $y = 0$ and the submerged part of the surface in its rest position, and let I^V, I^V_x, I^V_y, I^V_z, I^V_{xx}, I^V_{xy}, ... be the volume, the moments and the moments and products of inertia of this volume about the body axes in their rest position. Let A be the intersection of the body in its rest position with the surface $y = 0$, and let I^A, I^A_x, I^A_z, I^A_{xx}, I^A_{xz}, I^A_{zz} denote the area, the moments and the moments and products of inertia of A with respect to axes through $(x_0, 0, z_0)$ and parallel to the body axes; e.g.,

$$I^A_{xz} = \iint_A (x - x_0)\,(z - z_0)\,dx\,dz.$$

The exact dynamical equations are

$$\left.\begin{aligned} M\ddot{\xi} &= \iint_S p\cos(n, x)\, d\sigma, \\ M\ddot{\eta} &= \iint_S p\cos(n, y)\, d\sigma - M g, \\ M\ddot{\zeta} &= \iint_S p\cos(n, z)\, d\sigma, \end{aligned}\right\} \tag{19.60}$$

where S is the wetted surface of the body in its (to-be-determined) position at time t and

$$p = -\varrho g y - \varrho \Phi_t - \tfrac{1}{2}\varrho\, |\operatorname{grad} \Phi|^2;$$

and three similar equations for $\ddot{\alpha}, \ddot{\beta}, \ddot{\gamma}$. Substitution of the perturbation series gives for the zero-order terms

$$M = \varrho I^V, \qquad I_x^V = I_y^V = 0, \tag{19.61}$$

i.e., ARCHIMEDES' law and the statement that the center of buoyancy and center of gravity are on the same vertical line. The first-order equations, after dropping superscripts, are

$$\left.\begin{aligned} M\ddot{x}_1 &= -\varrho \iint_{S_0} \Phi_t n_x\, d\sigma, \\ M\ddot{y}_1 &= -\varrho \iint_{S_0} \Phi_t n_y\, d\sigma + \varrho g\left(-I^A y_1 - I_x^A \gamma + I_z^A \alpha\right), \\ M\ddot{z}_1 &= -\varrho \iint_{S_0} \Phi_t n_z\, d\sigma, \\ -(I_{yy}+I_{zz})\ddot{\alpha} + I_{xy}\ddot{\beta} + I_{xz}\ddot{\gamma} &= \varrho \iint_{S_0} \Phi_t q_x\, d\sigma - \varrho g\left[I_z^A y_1 + I_{xz}^A \gamma - I_{zz}^A \alpha - I_y^V \alpha\right], \\ I_{xy}\ddot{\alpha} - (I_{xx}+I_{zz})\ddot{\beta} + I_{yz}\ddot{\gamma} &= \varrho \iint_{S_0} \Phi_t q_y\, d\sigma, \\ I_{xz}\ddot{\alpha} + I_{yz}\ddot{\beta} - (I_{xx}+I_{yy})\ddot{\gamma} &= \varrho \iint_{S_0} \Phi_t q_z\, d\sigma + \varrho g\left[I_x^A y_1 + I_{xx}^A \gamma - I_{xz}^A \alpha + I_y^V \gamma\right]. \end{aligned}\right\} \tag{19.62}$$

We note that the boundary conditions have been derived for general motions of the body and fluid, not just for the simply periodic ones for which they will be used below.

JOHN (1949) has used the equations to investigate the stability of a floating body. We shall not reproduce the results but remark that he shows that the usual condition for stability, namely that the metacenter lie above the center of gravity, derived from purely hydrostatic considerations, is in fact still a sufficient condition for stability when the hydrodynamic equations are considered (within the limitations of the linearized theory).

It is also shown by JOHN that the above equations have a unique solution for an initial-value problem, i.e., if at some instant the position and velocity of body and fluid are prescribed. However, for the problem with which we are concerned in this section, steady simple harmonic motion with a prescribed incoming wave, he proves uniqueness only for sufficiently large values of σ and for bodies such that a vertical line intersects the immersed surface only once (e.g., a floating sphere with its center at or above the free surface).

Knowledge of the motion of a floating body in surface waves is obviously of great importance to ship designers, and, as might be expected, there is a large amount of specialized literature. However, most of this literature may be considered irrelevant to this article for it is based upon the assumption that one may neglect the kinematic boundary condition (19.59) completely and, in the dynamic boundary condition (19.62), that one may take for Φ simply the velocity potential for the oncoming wave, thus neglecting the effect of the diffracted waves and the waves generated by the ship's own motion. This assumption is usually called the Froude-Krylov Hypothesis. W. FROUDE (1861) introduced it in connection with an investigation of ship rolling in waves and A. N. KRYLOV (1896, 1898) investigated its implications rather thoroughly for general motions. In spite of its apparent crudeness this assumption has been useful in elucidating many aspects of ship motions.

In recent years there have appeared a number of papers in which an attempt has been made to take account of the proper boundary conditions, but no attempt will be made to summarize this literature. The most systematic investigation of the matter has been made by JOHN (1949, 1950), HASKIND (1946a), and PETERS and STOKER (1957). The papers by JOHN consider the proper formulation of the linearized problem for a body with no average forward speed and the uniqueness and existence of solutions. Both HASKIND and PETERS and STOKER are primarily concerned with ships having a constant average forward speed. PETERS and STOKER treat carefully the proper formulation of the linearized problem and conclude that HASKIND's fundamental equations are not properly formulated in that some of his terms really belong with the second-order terms and should have been discarded. The objection applies also to part of his results for a stationary ship. The other part will be summarized below.

The motion of a ship in waves when it has a nonzero translational velocity will not be considered in this article. For this theory one should refer to the cited papers, to STOKER's *Water waves* (1957, Chap. 9), or to a recent survey by MARUO (1957). The transient oscillatory motion of a floating body in calm water will be considered later.

Let us return to the problem of steady oscillation of a floating body in oncoming waves. Since we assume steady oscillation, we shall write

$$\left.\begin{aligned} \Phi = \operatorname{Re}\{\varphi\, e^{-i\sigma t}\}, \qquad (x_1, y_1, z_1) &= \operatorname{Re}\{(a_0, b_0, c_0)\, e^{-i\sigma t}\},\\ (\alpha, \beta, \gamma) &= \operatorname{Re}\{(\alpha_0, \beta_0, \gamma_0)\, e^{-i\sigma t}\},\end{aligned}\right\} \tag{19.63}$$

where $\varphi = \varphi_1 + i\,\varphi_2$, $a_0 = a_0' + i\,a_0''$, etc. The unknown function φ and the constants $a_0, \ldots, \gamma_0$ are to be determined from the equations and boundary conditions.

We shall assume that Φ can be expressed as the sum of the velocity potentials of the incoming wave, say

$$\Phi^e = \frac{A\sigma}{g}\, e^{\nu y} \cos(\nu x + \sigma t) \tag{19.64}$$

if the fluid is infinitely deep, a diffracted wave $\Phi^0 = \varphi^0\, e^{-i\sigma t}$ and a forced wave $\Phi_f = \varphi_f\, e^{-i\sigma t}$ resulting from the body's own motion:

$$\Phi = \Phi^e + \Phi^0 + \Phi_f. \tag{19.65}$$

We shall express Φ_f in the following form (following HASKIND):

$$\Phi_f = \operatorname{Re}\{\varphi^1 \dot{x}_1 + \varphi^2 \dot{y}_1 + \varphi^3 \dot{z}_1 + \varphi^4 \dot{\alpha} + \varphi^5 \dot{\beta} + \varphi^6 \dot{\gamma}\}. \tag{19.66}$$

Then the kinematic boundary condition (19.59) implies:

$$\left.\begin{aligned} &\varphi_n^0 = -\varphi_n^e,\\ &\varphi_n^1 = n_x, \quad \varphi_n^2 = n_y, \quad \varphi_n^3 = n_z,\\ &\varphi_n^4 = q_x, \quad \varphi_n^5 = q_y, \quad \varphi_n^6 = q_z,\end{aligned}\right\} \tag{19.67}$$

all to be satisfied on S_0, the rest position of the body. The functions φ^k, $k=0, 1, \ldots, 6$, are to satisfy also the radiation condition and the condition at $y=-\infty$ (or at $y=-h$ for a flat bottom). The dynamical condition (19.62) will be used to determine the amplitudes and phases (i.e., the complex amplitudes), but first we introduce some notation. Let

$$\mu_{jk} + \frac{i}{\sigma} \lambda_{jk} = \varrho \iint\limits_{S_0} \varphi^j \frac{\partial \varphi^k}{\partial n} \, d\sigma. \tag{19.68}$$

The constants μ_{jk} and λ_{jk} depend only upon the geometry of the body. It may be shown by an application of GREEN's Theorem that $\mu_{kj}=\mu_{jk}$ and $\lambda_{kj}=\lambda_{jk}$.

Let us now substitute the expanded expression for Φ into, say, the first of Eqs. (19.62) (the others may be treated similarly), remembering that $n_x=\varphi_n^1$ on S_0:

$$M\ddot{x}_1 = -\varrho \iint\limits_{S_0} (\Phi^e + \Phi^0)_t \, n_x \, d\sigma - \varrho \iint\limits_{S_0} (\varphi^1 \ddot{x}_1 + \cdots + \varphi^6 \ddot{\gamma}) \, \varphi_n^1 \, d\sigma. \tag{19.69}$$

Consider, for example, the second term of the second integral:

$$\varrho \iint\limits_{S_0} \varphi^2 \varphi_n^1 \ddot{y}_1 \, d\sigma = \left(\mu_{21} + \frac{i}{\sigma} \lambda_{21}\right) \ddot{y}_1 = \mu_{21} \ddot{y}_1 + \lambda_{21} \dot{y}_1, \tag{19.70}$$

where we have used the special form of $y_1=b_0 \, \mathrm{e}^{-i\sigma t}$. Thus, (19.69) may be written

$$(M+\mu_{11}) \ddot{x}_1 + \mu_{21} \ddot{y}_1 + \cdots + \mu_{61} \ddot{\gamma} + \lambda_{11} \dot{x}_1 + \lambda_{21} \dot{y}_1 + \cdots + \lambda_{61} \dot{\gamma} = F_{ex} + F_{0x}, \tag{19.71}$$

where $F_{ex}=f_{ex} \, \mathrm{e}^{-i\sigma t}$ and $F_{0x}=F_{0x} \, \mathrm{e}^{-i\sigma t}$ represent the x-components of the forces resulting from the incoming and diffracted waves and are to be computed from the first integral in (19.69). The form of (19.71) explains the names given to the μ_{ij} and λ_{ij}: the μ_{ij} are called *added masses*, the λ_{ij}, *damping coefficients*. If one now writes $x_1, \ldots, \gamma$ in their assumed forms in (19.63) and substitutes in (19.71), one obtains

$$\left.\begin{aligned} &-\sigma^2 (M+\mu_{11}) a_0 - \sigma^2 \mu_{21} b_0 - \cdots - \sigma^2 \mu_{61} \gamma_0 - \\ &\qquad - i\sigma \lambda_{11} a_0 - \cdots - i\sigma \lambda_{61} \gamma_0 = f_{ex} + f_{0x} \end{aligned}\right\} \tag{19.72}$$

and five similar equations. Since the amplitudes $a_0, \ldots, \gamma_0$ are complex, this gives twelve equations to determine the twelve unknown quantities. It is thus clear that, providing these equations can be uniquely solved, the problem of finding the steady oscillatory motion of a freely floating body can be reduced to the solution of several problems of the type studied in Sects. 18 and 19α. From the form of (19.72) and the similar equations, it is clear that the complex amplitudes are all proportional to the amplitude A of the incoming wave as would be expected.

HASKIND has applied the method outlined above to a body symmetric with respect to the (x, y)-plane, e.g., a ship heading into waves. The only possible motions are heaving, pitching and surging. In carrying out some numerical computations he makes a further approximation that the kinematic boundary condition on the body may be satisfied on its plane of symmetry rather than on the surface. Although this approximation is perfectly consistent with the linearized theory in certain contexts, as will be seen in Sect. 21, it is not consistent with the theory as formulated here and must be considered to be a further approximation of some sort.

Freely floating sphere. Computation of the motion of a freely floating sphere with its center at the undisturbed water level can be carried through without an unreasonable amount of numerical work. The procedure for the heaving motion has been carried up to the point of numerical computation by Barakat (1958) [in an earlier investigation by MacCamy (1954) the multipole terms in the potential for the diffracted wave were omitted]. Part of the problem has already been solved in Sect. 19α, i.e., the waves resulting from the forced motion.

Since the phase at infinity must be kept arbitrary, one must replace (19.48) by

$$x^2 + (y - b_0' \cos\sigma t - b_0'' \sin\sigma t)^2 + z^2 = a^2 \tag{19.72}$$

However, the solution of that problem may be taken over with practically no change, for the velocity potential φ^2 in the notation of (19.66) must satisfy

$$\frac{\partial}{\partial r}\varphi^2 = \frac{y}{a} = \cos\vartheta \qquad \text{for } x^2 + y^2 + z^2 = a^2, \qquad y \leqq 0. \tag{19.73}$$

Thus we need only set $b_0\sigma = 1$ in (19.49) and later. In fact, from formula (19.52)

$$\mu_{22} = \tfrac{2}{3}\pi\varrho a^3 \cdot k, \qquad \lambda_{22} = \tfrac{2}{3}\pi\varrho a^3 \cdot 2h. \tag{19.74}$$

Finding the diffracted wave requires finding an outgoing wave satisfying

$$\left.\frac{\partial\varphi^0}{\partial r}\right|_{r=a} = -\frac{Ag}{\sigma}\nu\left[\cos\vartheta - i\sin\vartheta\cos\alpha\right] e^{\nu a\cos\vartheta} e^{-i\nu a\sin\vartheta\cos\alpha}, \tag{19.75}$$

where $x = r\sin\vartheta\cos\alpha$, $y = r\cos\vartheta$, $z = r\sin\vartheta\sin\alpha$. Barakat shows that φ^0 can be found as a series in functions of the form (13.21), with $b=0$ and account taken of certain symmetries, and functions of the form (13.20) with $b=0$ and $m=n$. Let

$$\left.\begin{aligned}
G_{2k}^{2m} &= \left[\frac{P_{2k}^{2m}(\cos\vartheta)}{r^{2k+1}} - \frac{\nu}{2k-2m}\,\frac{P_{2k-1}^{2m}(\cos\vartheta)}{r^{2k}}\right]\cos 2m\alpha,\\
&\qquad\qquad k = 1, 2, \ldots; \qquad m = 0, \ldots, k-1;\\
G_{2k}^{2m-1} &= \left[\frac{P_{2k+1}^{2m-1}(\cos\vartheta)}{r^{2k+2}} - \frac{\nu}{2k-2m+2}\,\frac{P_{2k}^{2m-1}(\cos\vartheta)}{r^{2k+1}}\right]\cos(2m-1)\alpha,\\
&\qquad\qquad k = 1, 2, \ldots; \qquad m = 1, \ldots, k;\\
\Phi_n &= \left[\frac{P_n^n(\cos\vartheta)}{r^{n+1}} + (-1)^n \,\mathrm{PV}\!\int_0^\infty \frac{k+\nu}{k-\nu}\,k^n e^{ky} J_n(kR)\,dk + \right.\\
&\qquad \left. + 2\pi i(-1)^n \nu^{n+1} e^{\nu y} J_n(\nu R)\right]\cos n\alpha, \qquad n = 1, 2, \ldots.
\end{aligned}\right\} \tag{19.76}$$

Then φ^0 may be expressed as follows

$$\varphi^0 = \sum_{k=1}^{\infty}\sum_{m=0}^{k-1} A_{2k}^{2m} a^{2k+2} G_{2k}^{2m} + i\sum_{k=1}^{\infty}\sum_{m=1}^{k} B_{2k}^{2m-1} a^{2k+3} G_{2k}^{2m-1} + \sum_{n=0}^{\infty} C_n \Phi_n, \tag{19.77}$$

where the complex coefficients A_{2k}^{2m}, B_{2k}^{2m-1}, C_n are to be determined from (19.75). No numerical computations seem to be available.

20. Motions which may be treated as steady flows. In this section we shall consider several problems which are time-independent, either by their formulation or by introduction of moving coordinates. The flow associated with a constant discharge rate through a canal is of the first type; the waves associated with a ship which has moved with constant velocity C over a long period is typical of the second.

The boundary conditions at the free surface have been derived in Sects. 10 and 11. For three-dimensional motion the velocity potential must satisfy [see Eq. (11.3)]

$$\varphi_y(x, 0, z) + \frac{c^2}{g}\varphi_{xx}(x, 0, z) = 0; \tag{20.1}$$

the equation of the free surface is

$$y = \eta(x, z) = \frac{c}{g}\varphi_x(x, 0, z). \tag{20.2}$$

In two-dimensional motion, if the complex potential $f = \varphi + i\psi$ is used, then the boundary condition may be written

$$\mathrm{Re}\left\{f''(x + i\,0) + i\frac{g}{c^2}f'(x + i\,0)\right\} = 0. \tag{20.3}$$

If the potential has been taken in the form $F(z) = -cz + f(z)$ with $\Psi = 0$ as the free-surface streamline, then

$$\mathrm{Re}\left\{f'(x + i\,0) + i\frac{g}{c^2}f(x + i\,0)\right\} = 0; \tag{20.3'}$$

the surface is given by

$$y = \eta(x) = \frac{1}{c}\psi(x, 0). \tag{20.4}$$

On obstructions, which are now all fixed, one has as usual

$$\varphi_n = 0 \quad \text{or} \quad \psi = \text{const}. \tag{20.5}$$

Far ahead of, or far upstream of, the obstruction the motion must approach either rest, or a uniform flow, respectively.

The general theory of steady free-surface flow about a submerged obstacle in infinitely deep fluid has been considered by KOCHIN (1937) for both two and three dimensions. HASKIND (1945a, b) has extended KOCHIN's treatment to fluid of constant finite depth. The methods used for waves generated by oscillating bodies carry over with only slight change, so that we shall not consider here the general aspects of the theory but consider instead several special problems.

α) Flow over an uneven bottom. Let us first derive the proper boundary condition on the bottom. We shall assume that the bottom may be represented in the form

$$y = -h + \varepsilon b^{(1)}(x) \tag{20.6}$$

and that the fluid flows from the right with discharge rate $q = ch$. As in the derivation of (10.19) we take

$$F(z) = -cz + \varepsilon f^{(1)}(z) + \varepsilon^2 f^{(2)}(z) + \cdots. \tag{20.7}$$

Then the condition that the bottom be a streamline is

$$-c(-h + \varepsilon b^{(1)} + \cdots) + \varepsilon\psi^{(1)}(x, -h + \varepsilon b^{(1)} + \cdots) + \cdots = ch. \tag{20.8}$$

Expansion in the manner of Sect. 10 and grouping of coefficients leads to the boundary condition for $\psi^{(1)}$:

$$\psi^{(1)}(x, -h) = c\,b^{(1)}. \tag{20.9}$$

We may hereafter write ψ for $\varepsilon\psi^{(1)}$ and b for $\varepsilon b^{(1)}$. We note that the choice of ε indicates that the amplitude of unevenness of the bottom must be small compared with h for the linearized theory to be applicable.

Consider now a bottom of the form [see Lamb (1932, p. 409), Wien (1900, p. 200)]

$$y = -h + b_0 \cos kx. \tag{20.10}$$

We look for a solution in the form

$$f(z) = A \cos kz + B \sin kz, \tag{20.11}$$

where A and B are complex. Substitution in (20.9), with $b^{(1)} = b_0 \cos kx$, shows that A must be pure imaginary, say iA', and B real, and further that

$$A' \cosh kh - B \sinh kh = c b_0. \tag{20.12}$$

Substitution in (20.3), i.e., $\psi_y(x, 0) - g c^{-2} \psi(x, 0) = 0$ yields

$$kB = \frac{g}{c^2} A'. \tag{20.13}$$

One then finds easily that

$$f(z) = \frac{\nu \sin kz + i k \cos kz}{k \cosh kh - \nu \sinh kh} c b_0, \qquad \nu = \frac{g}{c^2}, \tag{20.14}$$

$$\eta(x) = \frac{k b_0}{k \cosh kh - \nu \sinh kh} \cos kx.$$

An interesting consequence is that when $c^2/gh < 1$, i.e., when the flow is subcritical, the crests and troughs just oppose those of the bottom, whereas, if $c^2/gh > 1$, they occur together. If $c^2/gh = 1$, there is no steady flow. Also, when c^2/gh is close to 1, it is clear that the assumption of small perturbations is no longer satisfied.

By use of the Fourier Integral Theorem one may now construct solutions for an arbitrarily shaped bottom, within the limitations of the theorem. For from

$$b(x) = \frac{1}{\pi} \int_0^\infty dk \int_{-\infty}^\infty b(\xi) \cos k(x - \xi) \, d\xi \tag{20.15}$$

one may derive

$$\left. \begin{aligned} f(z) &= \frac{c}{\pi} \, \mathrm{PV} \int_0^\infty dk \int_{-\infty}^\infty b(\xi) \frac{\nu \sin k(z-\xi) + i k \cos k(z-\xi)}{k \cosh kh - \nu \sinh kh} \, d\xi, \\ \eta(x) &= \frac{1}{\pi} \, \mathrm{PV} \int_0^\infty \frac{k}{k \cosh kh - \nu \sinh kh} \, dk \int_{-\infty}^\infty b(\xi) \cos k(x-\xi) \, d\xi. \end{aligned} \right\} \tag{20.16}$$

An examination of the asymptotic properties of these integrals as $x \to +\infty$ shows that they do not vanish if $\nu h = gh/c^2 > 1$. Conditions for the validity of the Fourier Integral Theorem, e.g., that $b(x)$ is of bounded variation and absolutely integrable, indicate that it applies to situations in which the bottom unevenness is somewhat localized. Hence, it is reasonable to require the additional boundary condition

$$\lim_{x \to \infty} \eta(x) = 0.$$

Thus we must amend the solutions (20.10) if $gh/c^2 > 1$ by adding, respectively,

$$\left. \begin{aligned} &\frac{-\nu c}{\cosh^2 k_0 h - \nu h} \int_{-\infty}^\infty b(\xi) \cos k_0 (z - \xi + ih) \, d\xi, \\ &\frac{+\nu \sinh k_0 h}{\cosh^2 k_0 h - \nu h} \int_{-\infty}^\infty b(\xi) \sin k_0 (x - \xi) \, d\xi, \end{aligned} \right\} \tag{20.17}$$

where k_0 is the real solution of $k \cosh kh - \nu \sinh kh = 0$. We note that the other boundary conditions are not spoiled, for the first expression in (20.17) satisfies (20.3) and its imaginary (stream-function) part vanishes for $y = -h$ so that (20.3) is still satisfied.

Thus, if $c^2 > gh$ there is a local disturbance of the fluid in the region of unevenness which eventually smooths out. If $c^2 < gh$ there is also a local disturbance given by (20.16), but as $x \to -\infty$ there remains a disturbance given by twice the expressions in (20.17).

We remark in passing that we might have obtained this solution by distributing along the bottom dipoles of the form (13.48) with $\alpha = 0$ and with moment density $c\,b(x)$.

Various special cases of $b(x)$ may be considered. LAMB (1932, p. 410) replaces the unevenness by a single dipole. WIEN (1900, p. 202) takes $b(x) = \operatorname{arc\,tan} e x$ and in the limit lets $e \to \infty$ in order to find the flow over a small step. However, KOCHIN (1938) has treated this problem by a different method and finds that WIEN has made an error by a factor of two in the downstream waves (he had not satisfied the upstream condition) [see also LAMB (1934)]. The flow about a vertical plate in the bottom may be treated by distributing vortices (13.47) along the plate with the intensity to be determined by solving an integral equation.

One will find an attractive discussion of the subject in four papers of W. THOMSON (Lord KELVIN) (1886, 1887). EKMAN (1906) has applied the same method to three-dimensional flow. First he finds the form of the free surface over a doubly periodic bottom, then applies the double Fourier integral theorem to construct flows over irregular bottoms. He analyzes the asymptotic behavior of the surface for the case of an isolated dipole on the bottom and presents graphs showing the change in wave amplitude for different radial sections. The method of analysis may also be extended to superposed fluids of different densities (see KOCHIN (1937a, b, 1938c), LONG (1953, § 4)].

β) Flow about submerged obstacles. Linearization. The procedure for linearizing may be carried through in at least two ways, leading to somewhat different boundary conditions for the body. Consider a body moving in a fluid. For the time being, in order to achieve somewhat greater generality, we shall not restrict the velocity to be constant. If the dimensions of the body are sufficiently small compared with the depth of submersion, it will not disturb the surface appreciably, and one will expect to be able to use the infinitesimal-wave approximation. However, the same end is obtained if the body approximates to a flat disc moving in its plane, a line segment moving along its line, a piece of a cylindrical surface moving along the cylinder, etc., various combinations being easily visualized. We consider the two situations separately.

Let $F(x, y, z, t) = 0$ describe the surface of a bounded body at time t, and let a be some typical dimension of the body, say its maximum diameter, and let h be the depth of submersion measured to some point $(x_0, -h, z_0)$ of the body. Now, consider the family of flows associated with the family of surfaces

$$F^{(\varepsilon)}(x, y, z, t) = F\left(\frac{x - x_0}{\varepsilon} + x_0, \frac{y + h}{\varepsilon} - h, \frac{z - z_0}{\varepsilon} + z_0, t\right) = 0 \qquad (20.18)$$

where $\varepsilon = a/h$. As $\varepsilon \to 0$ the surface $F^{(\varepsilon)} = 0$ contracts to a point and the fluid approaches a state of rest. Hence, as in Sect. 10α, it is allowable to expand Φ and η in a perturbation series

$$\Phi = \varepsilon \Phi^{(1)} + \varepsilon^2 \Phi^{(2)} + \cdots, \qquad \eta = \varepsilon \eta^{(1)} + \varepsilon^2 \eta^{(2)} + \cdots. \qquad (20.19)$$

The boundary condition to be satisfied on the body, namely,

$$\operatorname{grad} F^{(\varepsilon)} \cdot \operatorname{grad} \Phi + F_t^{(\varepsilon)} = 0, \tag{20.20}$$

becomes

$$\operatorname{grad} F \cdot \operatorname{grad} \Phi^{(1)} + F_t + \varepsilon \operatorname{grad} F \cdot \operatorname{grad} \Phi^{(2)} + \cdots = 0,$$

and $\Phi^{(1)}$ must satisfy

$$\operatorname{grad} F \cdot \operatorname{grad} \Phi^{(1)} + F_t = 0 \quad \text{on} \quad F^{(\varepsilon)} = 0.$$

Thus, one finds that, in this method of linearizing, the boundary condition to be satisfied on the body is the exact one

$$\operatorname{grad} F \cdot \operatorname{grad} \Phi + F_t = 0 \quad \text{on} \quad F(x, y, z, t) = 0. \tag{20.21}$$

The boundary condition satisfied by Φ on the free surface is, of course, the linearized one. The approximation to the exact solution is better, the deeper the relative submergence.

The second method of linearization will be illustrated with the so-called *thin-ship* approximation. Let the equation of a ship hull be given in the form

$$\bar{z} = \pm F(\bar{x}, \bar{y}). \tag{20.22}$$

in coordinates fixed in the ship. Let us write this in the form

$$\bar{z} = \pm \varepsilon F^{(1)}(\bar{x}, \bar{y}) \tag{20.23}$$

where ε is, say, the beam/length. Suppose the ship moves in direction OX with velocity $c(t)$ and consider the family of flows generated by the motion of such bodies for different ε. Let the velocity potential be $\Phi(x, y, z, t; \varepsilon)$. Then, since as $\varepsilon \to 0$ the hull approaches a flat disc S_0, the ship's centerplane section, the motion of the fluid will also approach a state of rest and we may expand

$$\Phi = \varepsilon \Phi^{(1)} + \varepsilon^2 \Phi^{(2)} + \cdots \tag{20.24}$$

and similarly for η. The assumed forms for Φ and η lead immediately as in Sect. 10α to the linearized free-surface boundary condition for $\Phi^{(1)}$. The exact condition on the hull is

$$F_x\Big(x - \int^t c(\tau)\, d\tau, y\Big)\, \Phi_x\Big(x, y, F\Big(x - \int^t c\, d\tau, y\Big), t\Big) + F_y \Phi_y - \Phi_z - c(t)\, F_x = 0. \tag{20.25}$$

After substituting (20.23) and (20.24) in (20.25), one finds that $\Phi^{(1)}$ must satisfy

$$\Phi_z^{(1)}(x, y, \pm 0, t) = \mp c(t)\, F_x^{(1)}\Big(x - \int^t c(\tau)\, d\tau, y\Big), \tag{20.26}$$

$\Phi^{(2)}$ must satisfy

$$\Phi_z^{(2)}(x, y, \pm 0, t) = \pm \left[F_x^{(1)}(x, y)\, \Phi_x^{(1)}(x, y \pm 0, t) + F_y^{(1)} \Phi_y^{(1)} - F^{(1)} \Phi_{zz}^{(1)}\right], \tag{20.27}$$

and $\Phi^{(i)}$ a relation of the form

$$\Phi_z^{(i)}(x, y, \pm 0, t) = \pm C_i\{F^{(1)}, \Phi^{(1)}, \ldots, \Phi^{(i-1)}\}, \tag{20.28}$$

where C_i is a functional of the functions in braces. We note especially that it is a consequence of the linearization that the boundary condition imposed by the presence of the body is to be satisfied on the centerplane section and not on the actual surface. One will expect this linearized theory to be more accurate the smaller ε is, i.e., the smaller the beam-to-length ratio.

It is clear that one may proceed similarly in the situations mentioned earlier. We record the results in several cases for reference.

First consider the *thin-wing approximation* for two-dimensional hydrofoils. In a coordinate system $\bar{O}\bar{x}\bar{y}$ fixed in the hydrofoil let the trailing edge of the hydrofoil be at $(-a, -h)$, and let the upper and lower surfaces be given by

$$\bar{y} = -h + u(\bar{x}) \quad \text{and} \quad \bar{y} = -h + b(\bar{x}), \qquad -a \leq \bar{x} \leq a, \tag{20.29}$$

respectively. Define

$$r(\bar{x}) = \tfrac{1}{2}[u(\bar{x}) + b(\bar{x})], \qquad s(\bar{x}) = \tfrac{1}{2}[u(\bar{x}) - b(\bar{x})], \tag{20.30}$$

so that now the top and bottom are given by

$$\bar{y} = -h + r(\bar{x}) + s(\bar{x}) \quad \text{and} \quad \bar{y} = -h + r(\bar{x}) - s(\bar{x}), \quad -a \leqq \bar{x} \leqq a, \tag{20.31}$$

respectively. The class of profiles in a form analogous to (20.23) is now given by

$$\bar{y} = -h + \varepsilon[r^{(1)}(\bar{x}) \pm s^{(1)}(\bar{x})], \qquad -a \leqq \bar{x} \leqq a. \tag{20.32}$$

It is clear that as $\varepsilon \to 0$, the profiles approach the line segment $\bar{y} = -h$, $0 \leqq \bar{x} \leqq a$, so that the perturbation procedure is allowable. The analysis leads to the linearized boundary condition

$$\left.\begin{aligned} \varphi_y(x, -h \pm 0, t) = -c(t)\, r'\Big(x - \int^t c(\tau)\, d\tau\Big) \mp c(t)\, s'\Big(x - \int^t c(\tau)\, d\tau\Big), \\ -a \leqq x - \int^t c\, d\tau \leqq a. \end{aligned}\right\} \tag{20.33}$$

The *slender-body approximation* is also consistent with the linearized free-surface condition. Let the body be defined by

$$(\bar{y} + h)^2 + \bar{z}^2 - r^2(\bar{x}) = 0, \qquad |\bar{x}| < a, \qquad h > 0, \tag{20.34}$$

in a coordinate system fixed in the body. If one considers the class of bodies defined by $\varepsilon r^{(1)}(\bar{x})$, then the appropriate condition to be satisfied by $\Phi^{(1)}$ is

$$\lim_{\varepsilon \to 0} \Phi_r^{(1)}\Big(x - \int^t c(\tau)\, d\tau, -h + \varepsilon r^{(1)} \cos\vartheta, \varepsilon r^{(1)} \sin\vartheta\Big)\, r^{(1)}\Big(x - \int^t c\, d\tau\Big) = -c r^{(1)} r^{(1)\prime}. \tag{20.35}$$

We note that the same problem may be approached by two linearized theories. For example, in approximating the flow about a hydrofoil, one may either consider it as a relatively deeply submerged body and satisfy the exact conditions on the surface, or consider it as a thin hydrofoil and use the conditions (20.33). Each method will have its own domain of excellence, but it is not proper in the present context to say that the thin-hydrofoil approximation is less exact than the other one, even though this is true in an unbounded fluid.

The H-functions. Kochin's H-function, introduced in Sect. 19α, may also be used effectively for the flows considered in the present section. The definition for three dimensions is identical in appearance with (19.14). For two dimensions (19.21) isreplaced by

$$H(k) = \int_{C_1} \mathrm{e}^{-ik\zeta} f'(\zeta)\, d\zeta. \tag{20.36}$$

However, the formulas for the force on the body are somewhat different. For three dimensions they are

$$\left.\begin{aligned} X &= -\frac{\nu^2 \varrho}{2\pi} \int_{-\frac{1}{2}\pi}^{\frac{1}{2}\pi} |H(\nu \sec^2\vartheta, \vartheta)|^2 \sec^3\vartheta\, d\vartheta, \\ Y &= \varrho g V - \frac{\varrho}{8\pi^2} \int_0^\infty \int_{-\pi}^{\pi} |H(k, \vartheta)|^2 k\, d\vartheta\, dk + \\ &\quad + \frac{\nu^2 \varrho}{4\pi^2} \int_{-\pi}^{\pi} \mathrm{PV} \int_{-\infty}^{1} \left|H\left(\frac{\nu(1-\lambda)}{\cos^2\vartheta}, \vartheta\right)\right|^2 \frac{1-\lambda}{\lambda} \sec^4\vartheta\, d\lambda\, d\vartheta, \\ Z &= -\frac{\nu^2 \varrho}{2\pi} \int_{-\frac{1}{2}\pi}^{\frac{1}{2}\pi} |H(\nu \sec^2\vartheta, \vartheta)|^2 \sec^4\vartheta \sin\vartheta\, d\vartheta, \end{aligned}\right\} \tag{20.37}$$

where V is the displaced volume of fluid and $\nu = g/c^2$. The two-dimensional formulas are

$$\left.\begin{aligned} X &= -\nu\varrho\,|H(\nu)|^2,\\ Y &= \varrho g A + \varrho c\Gamma - \frac{\varrho}{2\pi}\int_0^\infty |H(k)|^2\,dk + \frac{\varrho\nu}{\pi}\,\mathrm{PV}\int_{-\infty}^{1} |H(\nu - k\nu)|^2\,\frac{dk}{k},\\ M &= -g\varrho A x_c - \varrho c\,\mathrm{Re}\{iH'(0)\} - \varrho\,\mathrm{Re}\Big\{\frac{1}{2\pi i}\int_0^\infty H'(k)\,\overline{H(k)}\,dk + {}\\ &\quad + \nu H'(\nu)\,\overline{H(\nu)} + \frac{\nu i}{\pi}\,\mathrm{PV}\int_{-\infty}^{1} H'(\nu - k\nu)\,\overline{H(\nu - k\nu)}\,\frac{dk}{k}\Big\}, \end{aligned}\right\} \tag{20.38}$$

where A is the area of the profile, (x_c, y_c) are the coordinates of its centroid, and Γ is the circulation. The remarks made in Sect. 19α concerning the use of the H-function apply also here.

Submerged circular cylinder. The appropriate linearization for the circular cylinder is the one associated with deep submergence. Hence, one must try to satisfy the exact boundary condition on the cylinder.

This problem has been treated by LAMB (1913; see also 1932, § 247), HAVELOCK (1927, 1929a, 1936), SRETENSKII (1938), who considers also finite depth, KOCHIN (1937) and HASKIND (1945a), who applies KOCHIN's methods for finite depth. COOMBS (1950) considers the flow about a pair of submerged cylinders, and, as a preliminary, also about a single cylinder; numerical computations are carried though for two cases, one with the centers on a horizontal line and one with them on a vertical line. COOMB's method has wider applicability than just to circular cylinders. In all the cited papers, with the exception of HAVELOCK's and COOMBS', the problem is solved by placing at the center of the circle a dipole modified to satisfy the free-surface condition, i.e., (13.45) with $\alpha = 0$ and $M = 2\pi c a^2$, where a is the radius and c the velocity [the c of (13.45) is taken as $-ih$, h the depth of the center]. This provides, of course, only an approximate solution, for in the presence of a free surface a dipole in a stream no longer generates a circle, as is testified to by the fact that the contour actually generated is subject to a moment. HAVELOCK (1927, 1929a) gave second approximations for drag and lift and later (1936b) a complete solution.

The problem may be treated by a combination of MILNE-THOMSON's Circle Theorem (1956, p. 151) and a formula of KOCHIN's. The former states that if $f(z)$ is the complex potential for a flow with its singularities all at a distance greater than a from the origin and with no solid boundaries, then

$$f(z) + \bar{f}\left(\frac{a^2}{z}\right) \tag{20.39}$$

is the complex potential for a flow with the same singularities but now with a circle of radius a and center at the origin situated in the fluid.

KOCHIN (1937) has proved that if $f(z)$ is a single-valued complex potential for a bounded contour C under a free surface, then

$$f(z) = \frac{1}{2\pi i}\int_{C_1} \frac{f(\zeta)}{z-\zeta}\,d\zeta + \frac{1}{2\pi i}\int_{C_1} \overline{f(\zeta)}\left[\frac{1}{z-\bar{\zeta}} - 2i\nu\,e^{-i\nu z}\int_\infty^z \frac{e^{i\nu u}}{u-\bar{\zeta}}\,du\right]d\bar{\zeta}, \tag{20.40}$$

where C_1 is any contour in the lower half-plane containing C. The formula and its proof are almost identical with that given in (17.15). The first integral in

(20.40) represents a function regular everywhere outside C, the second integral a function regular everywhere in the lower half-plane. If one starts with a function $f(z)$ whose only singularities are contained inside C, then the operation

$$f + \mathfrak{N}\{f\}, \tag{20.41}$$

where

$$\left.\begin{aligned} \mathfrak{N}\{f\} &= \frac{1}{2\pi i}\int\limits_{C_1} \overline{f(\zeta)}\left[\frac{1}{z-\bar{\zeta}} - 2i\nu\, e^{-i\nu z}\int\limits_{\infty}^{2} \frac{e^{i\nu u}}{u-\bar{\zeta}}\,du\right] d\bar{\zeta}, \\ &= \frac{1}{2\pi i}\int\limits_{C_1} \overline{f(z)}\left[\frac{1}{z-\bar{\zeta}} - 2\pi\nu\, e^{-i\nu(z-\bar{\zeta})} + 2i\nu\,\mathrm{PV}\int\limits_{0}^{\infty} \frac{e^{-iu(z-\bar{\zeta})}}{u-\nu}\,du\right] d\bar{\zeta} \end{aligned}\right\} \tag{20.42}$$

yields a complex potential function satisfying the free-surface condition and having the same singularities as f in the lower half-plane.

On the other hand, if one starts with a complex potential $f(z)$ whose only singularities are in the upper half-plane, then

$$f + \mathfrak{M}\{f\} \tag{20.43}$$

where

$$\mathfrak{M}\{f\} = \bar{f}\left(\frac{a^2}{z+ih} + ih\right) + ich, \qquad a < h; \tag{20.44}$$

will be a complex potential for a flow about a circle of radius a and center $-ih$ and with the same singularities as f in the upper half-plane, the singularities of $\mathfrak{M}\{f\}$ being all inside the circle.

We start with the flow $f_0(z) = -cz$ representing a uniform flow from the right; the free-surface condition is satisfied for $y=0$ in a trivial manner. Now form the sequence

$$f_0,\ f_1 = \mathfrak{M}\{f_0\},\quad f_2 = \mathfrak{N}\{f\},\ \ldots,\ f_{2n+1} = \mathfrak{M}\{f_{2n}\},\ f_{2n+2} = \mathfrak{N}\{f_{2n+1}\},\ \ldots. \tag{20.45}$$

Then $f_0+f_1, f_2+f_3, f_4+f_5, \ldots$ each represent flows satisfying the boundary condition on the circle; hence, also their sum if the series converges. On the other hand, $f_0, f_1+f_2, f_3+f_4, \ldots$ each represent flows satisfying the free surface condition, and, hence, also the sum if it exists.

Let us now consider the two operators $\mathfrak{M}$ and $\mathfrak{N}$. $\mathfrak{M}$ is always being applied to a function regular and bounded in the lower half-plane. Since $a^2(z+ih)^{-1}+ih$ maps the exterior of the circle on to the the interior, the maximum of $|\mathfrak{M}\{f\}|$ for z in the lower half-plane does not exceed that for $|f|$ within or on the circle. We write this as

$$|\mathfrak{M}\{f\}| \leqq \|f\| = \max_C |f|. \tag{20.46}$$

In particular,

$$\|f_{2n+1}\| \leqq \|f_{2n}\|. \tag{20.47}$$

The operator $\mathfrak{N}$ is always applied here to functions regular everywhere outside and on the boundary of the circle. Hence C_1 may be contracted to C and one can establish the following estimate for z in the lower half-plane:

$$\left.\begin{aligned} |\mathfrak{N}\{f\}| &\leqq a\max_C |f|\left[\frac{1}{|z-\bar{\zeta}|} + 2\pi\nu\, e^{\nu(y+\eta)} + 2\nu\, e^{\nu(y+\eta)}\,|\mathrm{Ei}(\nu|y+\eta|)|\right] \\ &\leqq a\left[\frac{1}{h-a} + 2\pi\nu\, e^{-\nu(h-a)} + 2\nu\, e^{-\nu(h-a)}\,\mathrm{Ei}\left(\nu(h-a)\right)\right]\max_C |f| \\ &\leqq K\|f\|, \end{aligned}\right\} \tag{20.48}$$

where in the second inquality h must be large enough that $\nu(h-a)>0.4$. For fixed values of νa one may select h/a large enough so that K is as small as one wishes, in any case, less than 1. Thus, in particular,

$$\begin{aligned}\|f_{2n}\| &\leqq K\,\|f_{2n-1}\| < \|f_{2n-1}\| \\ &\leqq K^n\,\|f_0\|, \\ \|f_{2n+1}\| &\leqq \|f_{2n}\| \leqq K^n\,\|f_0\|.\end{aligned} \tag{20.50}$$

From this it follows easily that the series

$$f_0+f_1+f_2+\cdots+f_{2n}+f_{2n+1}+\cdots \tag{20.51}$$

with terms defined by (20.45) converges uniformly in the part of the lower half-plane exterior to $|z+ih|<a$.

One may extend the method to flows about more general cylinders by combining the operator $\mathfrak{M}$ with another defined in terms of conformal mapping of the given profile into a circle. The procedure carried through above is a natural generalization of the procedure used by HAVELOCK in his first two papers (1926, 1929a) to find the second approximation. However, in his later paper (1939b) he used a different procedure, one which has also been used by URSELL in analogous problems. This consists in expressing the potential as a sum of multipoles situated at the center and, of course, already modified so as to satisfy the condition on the free surface and as $x\to\infty$. This leads to an infinite set of equations for the coefficients. The method is quite suitable for approximate computation.

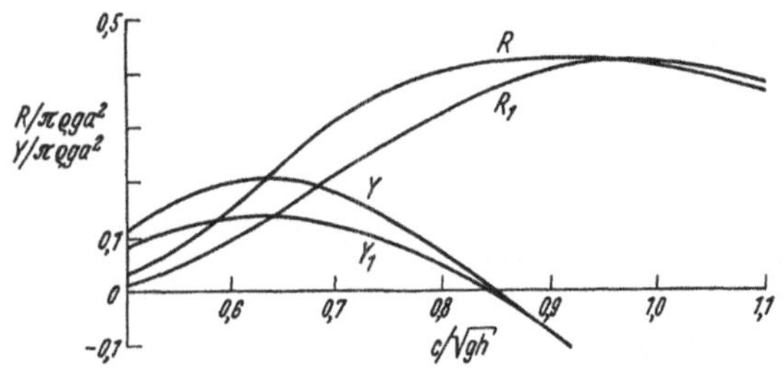

Fig. 19.

After computation of $H(k)$, the formulas (20.38) can be used to find the force and moment. In the computation of H only the terms with odd indices contribute. This leads to a considerable saving in effort. For example, if one had approximated the flow by the first three terms of (20.51) and computed the force by integrating the pressure over the cylinder, the result would be the same as that obtained by using the H-function evaluated for f_1 alone, and without the need of finding f_2. HAVELOCK has frequently made use of this device without specifically introducing the H-function. Fig. 19 from HAVELOCK (1936b) shows $R=-X$ and Y plotted in units of $\pi g\varrho a^2$ with abscissa $1/\sqrt{\nu h}$ for $a/h=\frac{1}{2}$. The curves labelled R_1 and Y_1 give the result when only the first approximation is used, i.e.,

$$\left.\begin{aligned}H(k)&=2\pi c\,a^2 k\,e^{-kh}\\ R_1&=\pi\varrho g a^2\cdot\pi\left(\frac{a}{h}\right)^2\left(\frac{2gh}{c^2}\right)^2 e^{-2gh/c^2},\\ Y_1&=-\pi\varrho g a^2\cdot\left(\frac{a}{h}\right)^2\frac{c^2}{2gh}\left[1+\frac{2gh}{c^2}-\left(\frac{2gh}{c^2}\right)^2-\left(\frac{2gh}{c^2}\right)^3 e^{-2gh/c^2}\,\mathrm{Ei}\left(\frac{2gh}{c^2}\right)\right].\end{aligned}\right\} \tag{20.52}$$

Computation of M gives, on this approximation, the anomalous result

$$M_1=h\left[1-\frac{c^2}{gh}\right]\cdot R.$$

In the formula for Y the terms resulting from buoyancy and circulation are omitted. HAVELOCK (1928b) has investigated the form of the surface over a moving dipole, i.e., over a sphere to the same degree of approximation.

Some submerged three-dimensional bodies. The flow about submerged ellipsoids and bodies of revolution in general has obvious interest in connection with the wave resistance of submarines. As a result there is a considerable amount of both theoretical and experimental work available, and even some tables for computation of wave resistance. Most of the theoretical work does not go beyond the approximation in which one represents the body by the singularity distribution appropriate to an unbounded fluid, but with the potential function for the singularity modified to satisfy the conditions on the free surface and at $x = +\infty$. Thus, to find the flow about a submerged sphere one will in this approximation use a modified dipole with axis in the direction Ox and moment $\frac{1}{2} c a^3$. One should realize, however, that the boundary condition on the body appropriate to deep submergence has not been satisfied. The necessary refinements could be carried through for the sphere in a manner similar to that used for the circular cylinder. Since the sphere (and even more, the circular cylinder) is a poor shape to which to apply perfect-fluid theory, such a computation is of only moderate interest. Both POND (1951, 1952) and HAVELOCK (1952) have considered methods for improving the accuracy with which the boundary condition on bodies of revolution is satisfied. This is particularly important in estimating the moment about the transverse horizontal axis, but, as POND shows, of less importance for the wave resistance.

HAVELOCK (1931a) treated by the approximate method the wave resistance of prolate and oblate spheroids moving both along and at right angles to their axes. Later (1931b) he extended the results to general ellipsoids moving in the direction of the longest axis. WEINBLUM (1936) has considered bodies of revolution using the slender-body approximation, but satisfying it only in the approximate sense described above; his aim was to find forms of minimum wave resistance. WEINBLUM (1951) returned to the problem, taking up in particular numerical computation of the wave resistance for a given shape. Tables and graphs are given to facilitate the computation for certain classes of bodies. Experiments were made by WEINBLUM, AMTSBERG and BOCK (1936) on three forms at several depths. Presumably, more recent experiments exist whose results are not publicly available. A general survey of the theory may be found in BESSHO (1957).

If one has once computed the function $H(k, \vartheta)$ for a source and a dipole, it is usually straightforward to compute it for bodies generated by distributions of sources and dipoles, and hence to compute the force. Let S_1 be a surface containing a single submerged source of strength m at the point (a, b, c), $b < 0$ [i.e., (13.36) multiplied by $-m$]; one finds

$$H(k, \vartheta) = 4\pi m \, e^{k b} \, e^{i k (a \cos\vartheta + c \sin\vartheta)}. \tag{20.53}$$

For a dipole of moment M in the direction Ox one finds

$$H(k, \vartheta) = 4\pi i M k \, e^{k b} \, e^{i k (a \cos\vartheta + c \sin\vartheta)} \cos\vartheta. \tag{20.54}$$

These may now be superposed as necessary for either discrete or continuous distributions. Thus, if we write $G(x, y, z; \xi, \eta, \zeta)$ for the function (13.36) with (a, b, c) replaced by (ξ, η, ζ), and if we can express φ for some problem by

$$\varphi = \iint_S \gamma(\xi, \eta, \zeta) \, G(x, y, z, \xi, \eta, \zeta) \, d\sigma, \tag{20.55}$$

then [cf. Eq. (19.20)]

$$H(k, \vartheta) = -4\pi \iint_S e^{k\eta} \, e^{i k (\xi \cos\vartheta + \zeta \sin\vartheta)} \gamma(\xi, \eta, \zeta) \, d\sigma. \tag{20.56}$$

Prolate spheroid. We give an example of the preceding remarks. A prolate spheroid of major semi-axis a and minor semi-axes b moving with velocity c in the direction of its major axis can be represented in an unbounded fluid by a distribution of dipoles of moment density

$$\mu(\xi) = A\,c\,(a^2 e^2 - \xi^2), \qquad |\xi| < a e, \tag{20.57}$$

where

$$A^{-1} = \frac{4e}{1-e^2} - 2\log\frac{1+e}{1-e}, \qquad e^2 = 1 - \frac{b^2}{a^2},$$

placed along the major axis between the two foci. Hence with the center at depth h one has in this approximation

$$\left.\begin{aligned} H(k\ \vartheta) &= 4\pi i A c k \,\mathrm{e}^{-kh} \cos\vartheta \int_{-ae}^{ae} (a^2 e^2 - \xi^2)\, \mathrm{e}^{ik\xi\cos\vartheta}\, d\xi \\ &= 8\sqrt{2\pi}\, i A c (a e)^{\frac{3}{2}} \frac{\mathrm{e}^{-kh}}{k^{\frac{1}{2}} \cos^{\frac{1}{2}}\vartheta} J_{\frac{3}{2}}(a e k \cos\vartheta). \end{aligned}\right\} \tag{20.58}$$

Substituting in the first formula of (20.37) one finds

$$R = -X = +128\pi \varrho \nu c^2 a^3 e^3 A^2 \int_0^{\frac{1}{2}\pi} \mathrm{e}^{-2\nu h \sec^2\vartheta} \left[J_{\frac{3}{2}}(a e \nu \sec\vartheta)\right]^2 \sec^2\vartheta\, d\vartheta. \tag{20.59}$$

Fig. 20 from HAVELOCK (1931a) shows a graphical presentation of $R/\pi\varrho a b^3$ for spheroids with various ratios of a/b and for $h = 2b$. In comparing the different curves one should keep in mind the selected vertical scale; one based on displaced fluid, i.e. $R/\frac{4}{3}\pi\varrho g a b^2$, would give the comparison a different aspect.

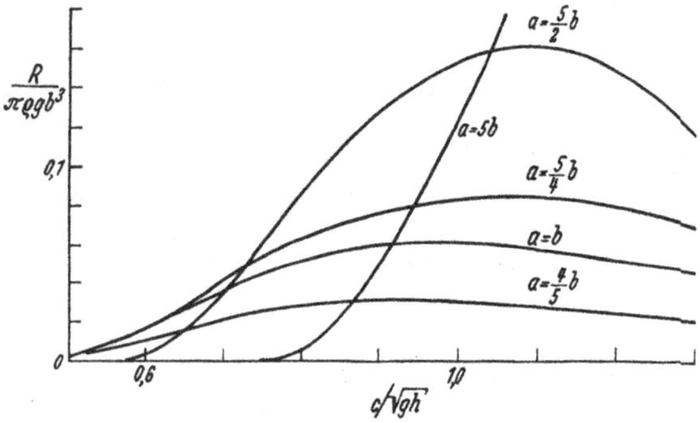

Fig. 20.

As mentioned earlier, it has been shown by both POND and HAVELOCK that this approximate treatment of the boundary condition on the body is inadequate for computation of the moment. Fig. 21 is from POND (1951, 1958) and shows the computed moment about the center for a Rankine ovoid, i.e., for the body generated in an unbounded fluid by a source and sink of equal strengths moving in the direction of their axis. The dashed curves show the result with the approximate computation; the solid curves were computed by a method in which the boundary condition on the body is more closely satisfied. The length l of the body is 10.5 times the maximum diameter $d = 2b$. A positive moment is nose-up.

Slender bodies. It is known that, for bodies of revolution given in the form (20.34) the slender-body boundary condition (20.35) can be satisfied in an infinite fluid by a distribution of sources along the axis of strength density

$$\gamma(x) = \frac{1}{4} c \frac{d}{dx} r^2(x). \tag{20.60}$$

If one assumes that this same distribution of the modified sources (13.36) will satisfy approximately (20.35) then

$$\varphi(x, y, z) = \int_{-a}^{a} \gamma(\xi)\, G(x, y, z; \xi, -h, 0)\, d\xi \tag{20.61}$$

and

$$H(k,\vartheta) = -4\pi\, e^{-kh} \int_{-a}^{a} e^{ik\xi\cos\vartheta}\, \gamma(\xi)\, d\xi. \tag{20.62}$$

From this one finds easily from (20.37)

$$\left.\begin{aligned} R &= -X \\ &= 16\pi\varrho\nu^2 \int_0^{\pi/2} d\vartheta \sec^3\vartheta\, e^{-2\nu h\sec^2\vartheta} \int_{-a}^{a}\int_{-a}^{a} \gamma(\xi)\gamma(\xi')\cos[k(\xi-\xi')\sec\vartheta]\, d\xi\, d\xi' \\ &= 16\pi\varrho\nu^2 \int_0^{\pi/2} \sec^3\vartheta\, e^{-2\nu h\sec^2\vartheta} [P^2+Q^2]\, d\vartheta, \end{aligned}\right\} \tag{20.63}$$

where

$$P(\vartheta) = \int_{-a}^{a} \gamma(\xi)\cos(\nu\xi\sec\vartheta)\, d\xi, \qquad Q(\vartheta) = \int_{-a}^{a} \gamma(\xi)\sin(\nu\xi\sec\vartheta)\, d\xi.$$

As mentioned earlier, WEINBLUM (1951) has published tables allowing one to compute R for γ-s representable as certain polynomials. An earlier paper (1936) considers the minimization of R among certain classes of polynomial γ-s. POND (1952) treats the necessary refinements to (20.61) in order to compute the moment. CUMMINS (1954) finds the additional effect of a train of waves on the surface.

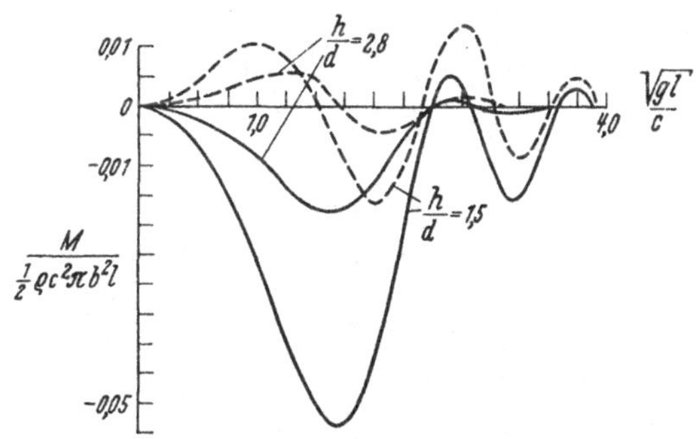

Fig. 21.

Thin ships. Let the equation of the hull be given as in (20.22) by $z=\pm F(x,y)$ in a coordinate system moving with the ship which we take to move with constant velocity c in the direction Ox. If we assume that a steady state has been reached, the boundary condition for the hull appropriate to the thin-ship approximation is, from (20.26) with $\Phi(x,y,z,t) = \varphi(x-ct, y, z)$ and a change to amoving coordinate system,

$$\varphi_z(x,y,\pm 0) = \mp c\, F_x(x,y) \tag{20.64}$$

for $(x,y,0)$ in S_0, the centerplane section of the ship at rest. For $(x,y,0)$ not in S_0 one has $\varphi_z(x,y,\pm 0)=0$ from symmetry considerations. φ must, of course, also satisfy (20.1) and the condition of vanishing motion as $x\to\infty$.

The boundary conditions may be satisfied immediately by distributing sources (13.26) over S_0. If we again denote the potential function in (13.36) by $G(x,y,z;\xi,\eta,\zeta)$, then, for infinitely deep fluid, the solution is

$$\varphi(x,y,z) = \frac{c}{2\pi}\iint_{S_0} G(x,y,z;\xi,\eta,0)\, F_x(\xi,\eta)\, d\sigma. \tag{20.65}$$

This follows easily from known theorems in potential theory[1]. (The part of G regular in $y\leq 0$ does not interfere with the satisfying of (20.64) since the z-derivative of these terms vanishes for $z=0$.)

The quantity of chief interest is the resistance resulting from the waves. This may be computed by using again (20.56) and (20.37) (and remembering to

[1] See, e.g., O. D. KELLOGG: Foundations of potential theory, pp. 160–166. Berlin: Springer 1929.

take account of both halves of the hull), or by direct integration of the pressure over the hull after taking account of linearization, i.e.

$$R = 2\varrho c \iint_{S_0} \varphi_x(x, y, 0)\, F_x(x, y)\, dx\, dy. \tag{20.66}$$

If the latter form is used, only the single-integral term in G gives a non-vanishing contribution. In either case one finds immediately, again for infinitely deep fluid,

$$\left.\begin{aligned} R &= \frac{4g^2\varrho}{\pi c^2} \int_0^{\pi/2} \sec^3\vartheta \left[P^2(\vartheta) + Q^2(\vartheta)\right] d\vartheta, \\ P &= \iint_{S_0} F_x(x, y)\, e^{\nu y \sec^2\vartheta} \cos(\nu x \sec\vartheta)\, dx\, dy, \\ Q &= \iint_{S_0} F_x(x, y)\, e^{\nu y \sec^2\vartheta} \sin(\nu x \sec\vartheta)\, dx\, dy. \end{aligned}\right\} \tag{20.67}$$

The result may be, and has been, put into a variety of different forms by change of variable and order of integration. We give one of them. Let $\lambda = \sec\vartheta$. Then one may verify that

$$\left.\begin{aligned} R &= \frac{4g^2\varrho}{\pi c^2} \iint_{S_0} dx\, dy \iint_{S_0} d\xi\, d\eta\, F_x(x, y)\, F_\xi(\xi, \eta)\, M\big(\nu(x-\xi),\ \nu(y+\eta)\big), \\ M(x, y) &= \int_1^\infty \frac{\lambda^2}{\sqrt{\lambda^2 - 1}}\, e^{\lambda^2 y} \cos\lambda x\, d\lambda. \end{aligned}\right\} \tag{20.68}$$

This expression for R in terms of the hull form and velocity was first given by MICHELL (1898), but derivations by different methods have since been given by many other, e.g. HAVELOCK (1932, 1951), SRETENSKII (1937), KOCHIN (1937), LUNDE (1951), and TIMMAN and VOSSERS (1955). It is usually called "MICHELL's integral".

Because MICHELL's integral gives R directly in terms of the hull geometry it has been intensively investigated by several persons in order to throw light upon the influence of variations of hull form upon wave resistance. Foremost among these investigators has been HAVELOCK, who in a series of notable papers (1923, 1925a, b, 1926a, 1932a, b, 1935) studied the effects of various systematic variations described by the titles of the papers. Much of this work is summarized in HAVELOCK (1926). In addition, there are numerous papers by G.P. WEINBLUM and W.C.S. WIGLEY devoted to comparison of experiment and theory. One can find surveys of much of this and related work, as well as further bibliography, in WIGLEY (1930, 1935, 1949), WEINBLUM (1950), HAVELOCK (1951), LUNDE (1957), and WEHAUSEN (1957), LUNDE's 1951 paper contains derivations of practically all the general theoretical results, including the effect of finite depth, walls, and acceleration. TAKAO INUI (1954) has given an extensive survey of Japanese investigations on wave resistance and related topics, and in a later paper (1957) a complete survey.

In order to allow better exploitation of MICHELL's integral much attention has been given in recent years to its numerical computation. One can find a general discussion in BIRKHOFF and KOTIK (1954), and various special proposals in KABACHINSKII (1947), REINOV (1951), GUILLOTON (1951) and WEINBLUM (1955). The last two papers both contain sets of tables to be used in evaluating MICHELL's integral.

In making a comparison of the theoretically predicted wave resistance with measured wave resistance one must examine critically the experimental method for estimating the wave resistance. The standard method consists in measuring the total resistance, estimating the part of the total resulting from the effects of viscosity, and attributing the difference to wave making. Thus the accuracy of the experimentally estimated wave resistance depends upon the accuracy of the estimated viscous resistance and upon the validity of the assumption that the two may be added. In the case of a very thin body this estimate can be made accurately and, in addition, the physical assumption in the thin-ship linearization is well realized. Fig. 22 from a report by WEINBLUM, KENDRICK and TODD (1952) shows a comparison between estimated and computed values of $R_w/\frac{1}{2}\varrho c^2 S$ for a towed "friction plane" 21 feet long with parabolic ends and 3 foot draft. These experimental data present MICHELL's integral in a most favorable light. For more ship-like forms the separation of viscous from wave-making resistance is more difficult and the compared values seldom show such striking quantitative agreement, although it is still fair in many cases. We call attention to the fact that MICHELL's integral predicts the same wave-making resistance no matter in which direction the ship moves.

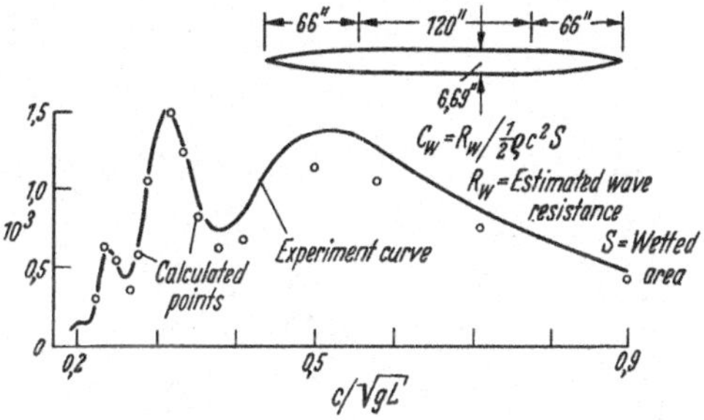

Fig. 22.

So far we have discussed vessels moving in an infinitely deep fluid. However, if for our function G we had taken (13.37) instead of (13.36), the same analysis would have led us to the following expression, first given by SRETENSKII (1937):

$$\left.\begin{aligned} R &= \frac{2\varrho g c}{\pi}\int_{\mu_h}^{\infty}[P^2(\mu)+Q^2(\mu)]\sqrt{\frac{\mu}{\mu-\nu\tanh\mu h}}\,d\mu,\\ P(\mu) &= \iint_{S_0} F_x(x,y)\,\frac{\cosh\mu(y+h)}{\cosh\mu h}\cos\left(x\sqrt{\nu\mu\tanh\mu h}\right)dx\,dy,\\ Q(\mu) &= \iint_{S_0} F_x(x,y)\,\frac{\cosh\mu(y+h)}{\cosh\mu h}\sin\left(x\sqrt{\nu\mu\tanh\mu h}\right)dx\,dy. \end{aligned}\right\} \tag{20.69}$$

Here μ_h is the nonzero solution of $\mu=\nu\tanh\mu h$ if such exists, i.e. if $c^2/gh>1$; otherwise $\mu_h=0$. As $h\to\infty$, $\mu_h\to\nu$ and one obtains one of the forms of MICHELL's integral.

An expression for the wave resistance of a thin ship moving down the center of a rectangular canal was derived independently by SRETENSKII (1936, 1937) and KELDYSH and SEDOV (1937). The result may be found in LUNDE (1951).

One may naturally ask how the wave pattern illustrated in Fig. 1 for a moving source is related to that for a ship. In the thin-ship approximation, the ship is replaced by a distribution of sources on the centerplane, so that each infinitesimal area of the centerplane contributes to such a pattern according to its strength. However, in many large vessels the middle part of the ship is cylindrical, so that $F_x=0$ in this region and only the bow and stern regions contribute a nonvanishing source density. Thus, if one replaces the ship by a single source in the bow region and a single sink in the stern region, the resulting wave pattern will approximate to that of a ship, the approximation being better at higher values of the Froude number $c/\sqrt{gL}$. Depending upon the value of $c/\sqrt{gL}$, the transverse wave systems from the two singularities may either reinforce or partially cancel one another. When they are in phase, a larger amount of energy is being left behind in the wave system and the resistance curve shows a maximum, when they are out of phase a minimum, the so-called "humps and hollows" of the resistance curve; these show clearly in Fig. 22. Replacing the ship by a source and sink is, of course, a gross simplification. However, it serves to explain qualitatively certain aspects of a ship's wave pattern and resistance curve, and, in fact, can be given a certain

Fig. 23. Ship waves.

amount of validity as an approximate computation of MICHELL's integral for sufficiently large $c/\sqrt{gL}$. For very large values of $c/\sqrt{gL}$, the wave length of the transverse waves along the path, $2\pi c^2/g$, becomes much larger than L and one may approximate the ship by a dipole. Many photographs of the wave pattern made by a fast motor boat fall into this class. The photograph reproduced in Fig. 23 shows clearly the diverging waves from the bow and stern, a third set possibly originating at the forward shoulder, and also the transverse waves, which presumably are here nearly in phase.

The angular opening of the wedge containing the wave pattern should be, according to (13.42) and Fig. 1, 38° 56′ in deep water. This is confirmed only approximately in photographs; difficulty in determining boundaries makes precise confirmation difficult. For ships moving in water of finite depth h the angular opening changes as shown in Fig. 2, and for supercritical velocities, i.e. $c^2 > gh$, there are no transverse waves.

JINNAKA (1957) has recently published a brief survey of the theory of ship waves.

Thin hydrofoils. We take the hydrofoil as described in (20.29) and treat the problem two-dimensionally. Assuming constant velocity c and steady motion and taking our coordinate system moving with the hydrofoil, the boundary condition (20.33) becomes

$$\varphi_y(x, -h \pm 0) = -c\,r'(x) \mp c\,s'(x), \qquad -a < x < a. \tag{20.70}$$

This problem has been treated very thoroughly by KELDYSH and LAVRENT'EV (1936). They follow a procedure quite analogous to that used in Sect. 19α to find the waves generated by a vertical oscillator not in a wall. Distribute vortices of intensity $\gamma(x)$ and sources of strength $\sigma(x)$ along the line $-a < x < a$, $y = -h$, but taking them, of course, modified as in (13.43) in order to satisfy the free-surface condition and the conditions at infinity. To start with, we take $\sigma(x) = -2cs'(x)$. It then follows from the theorem of PLEMELJ-SOKHOTSKII [cf. Eq. (17.18)] that

$$\varphi_y(x, -h+0) - \varphi_y(x, -h-0) = -2cs'(x), \tag{20.71}$$

a step toward satisfying (20.70). We now look for a complex potential in the form

$$f(z) = \int_{-a}^{a} [-2c\,s'(\xi)\,f_s(z;\xi - i\,h) + \gamma(\xi)\,f_v(z;\xi - i\,h)]\,d\xi, \tag{20.72}$$

where we have separated the source and vortex potentials in (13.43). The boundary condition (20.70) now yields an integral equation for $\gamma(x)$ in much the same manner that (17.18) was derived:

$$\left.\begin{aligned} \operatorname{Im}\int_{-a}^{a}[-2cs'(\xi)\,f_s'(x-i\,h;\,\xi-i\,h)+\gamma(\xi)\,f_v'(x-i\,h;\,\xi-i\,h)]\,d\xi = c\,r'(x), \\ -a < x < a. \end{aligned}\right\} \tag{20.73}$$

Noting from the third expression in (13.43) that f_s and f_v are functions of the difference $x - \xi$, we define

$$\begin{aligned} 2\pi i\,f_s'(x - i\,h, -i\,h) &= i\left[\frac{1}{x} - \frac{1}{x - 2i\,h}\right] - 2\nu\,\mathrm{e}^{-i\nu x}\int_{\infty}^{x}\frac{\mathrm{e}^{i\nu t}}{t - 2i\,h}\,dt \\ &= H(x) + i\,J(x), \end{aligned}$$

$$\begin{aligned} 2\pi i\,f_v'(x - i\,h, -i\,h) &= \frac{1}{x} + \frac{1}{x - 2i\,h} - 2i\,\nu\,\mathrm{e}^{-i\nu x}\int_{\infty}^{x}\frac{\mathrm{e}^{i\nu t}}{t - 2i\,h}\,dt \\ &= K(x) + i\,I(x). \end{aligned}$$

Here, for example,

$$K(x) = \frac{1}{x} + \frac{x}{x^2 + 4h^2} + 2\nu\int_{\infty}^{x}\frac{2h\cos\nu(t - x) + t\sin\nu(t - x)}{t^2 + 4h^2}\,dt.$$

The integral equation (20.73) can now be written in the following form:

$$\int_{-a}^{a}\gamma(\xi)\,K(x - \xi)\,d\xi = -2\pi\,c\,r'(x) + \int_{-a}^{a}2c\,s'(\xi)\,H(x - \xi)\,d\xi, \tag{20.74}$$

where the right-hand side is a known function.

This integral equation is the hydrofoil analogue of the thin-wing integral equation of airfoil theory:

$$\int_{-a}^{a} \gamma(\xi)\, \frac{1}{x-\xi}\, d\xi = -\, 2\pi c\, r'(x). \tag{20.75}$$

In the latter equation the kernel is simpler and in addition only the function r' describing the camber and the angle of attack occurs on the right side. Since the wing thickness does not enter into the determination of γ in (20.75) it may be neglected, for only γ is needed to find the lift. The situation is clearly different for hydrofoils. Even a symmetric wing with zero angle of attack may have a circulation, and hence lift. This is a consequence, of course, of the presence of the free surface and the associated wave motion.

KOCHIN (1936) has also considered hydrofoils, but from a somewhat different viewpoint. He has essentially used the "deep-submersion" linearization described first in this section. Thus he must satisfy the exact boundary conditions on the wing as well as the Kutta-Joukowski condition. However, one cannot say here, as one could for an infinite fluid, that his method is more exact than that of KELDYSH and LAVRENT'EV. Their approximation is more accurate the thinner the wing, for a given submersion. KOCHIN'S is more accurate the deeper the submersion, for a given wing.

Eq. (20.74) is not sufficient to determine $\gamma(x)$ uniquely. One must still add some further condition. We shall assume a finite velocity at the trailing edge, i.e.

$$\varphi_y(-a-0, -h) = \int_{-a}^{a} [\gamma(\xi)\, K(-a-\xi) - 2c\, s'(\xi)\, H(-a-\xi)]\, d\xi \quad \text{finite}. \tag{20.76}$$

KELDYSH and LAVRENT'EV propose solving the integral equation (without actually proving that a solution exists) by expanding K, H and γ in a power series in $\tau = a/\nu h$ and then determining recursively the coefficients. Let

$$\left.\begin{aligned} K(x) &= \frac{1}{x} + \frac{\tau}{a} \sum_{n=0}^{\infty} K_n(2\nu h)\, \tau^n \left(\frac{x}{a}\right)^n, \\ H(x) &= \frac{\tau}{a} \sum_{n=0}^{\infty} H_n(2\nu h)\, \tau^n \left(\frac{x}{a}\right)^n, \\ \gamma(x) &= \sum_{n=0}^{\infty} \gamma_n(x)\, \tau^n. \end{aligned}\right\} \tag{20.77}$$

Then (20.74) gives the following sequence of integral equations.

$$\left.\begin{aligned} &\int_{-a}^{a} \gamma_0(\xi)\, \frac{d\xi}{x-\xi} = -\, 2\pi c\, r'(x), \\ &\int_{-a}^{a} \gamma_1(\xi)\, \frac{d\xi}{x-\xi} = \frac{1}{a} H_0 \int_{-a}^{a} 2c\, s'(\xi)\, d\xi = 0, \\ &\cdots\cdots\cdots\cdots\cdots\cdots\cdots\cdots \\ &\int_{-a}^{a} \gamma_{n+1}(\xi)\, \frac{d\xi}{x-\xi} = \frac{1}{a^{n+1}} H_n \int_{-a}^{a} 2a\, s'(\xi)\, (x-\xi)^n d\xi - \\ &\qquad - \sum_{k+l=n} \frac{1}{a^{l+1}} K_l \int_{-a}^{a} \gamma(\xi)\, (x-\xi)^l\, d\xi, \\ &\cdots\cdots\cdots\cdots\cdots\cdots\cdots\cdots \end{aligned}\right\} \tag{20.78}$$

This procedure has the obvious advantage of reducing the solution to the airfoil integral equation for which an explicit solution satisfying the trailing-edge condition is known. If we denote temporarily the right-hand sides of the Eqs. (20.78) by $F_n(x)$, respectively, then the general solution is[1]

$$\gamma_n(x) = \frac{1}{\pi^2 \sqrt{a^2 - x^2}} \left[\int_{-a}^{a} \frac{F_n(\xi) \sqrt{a^2 - \xi^2}}{\xi - x} d\xi + \pi \int_{-a}^{a} \gamma_n(\xi)\, d\xi \right], \tag{20.79}$$

where the value of $\int_{-a}^{a} \gamma_n\, d\xi$ is undetermined. In terms of the series expansion, condition (20.76) states that

$$\left.\begin{aligned} \sum_{n=0}^{\infty} \tau^n \int_{-a}^{a} \gamma_n(\xi) \frac{d\xi}{x-\xi} + \sum_{n=1}^{\infty} \tau^n \sum_{k+l=n-1} a^{-l-1} K_l \int_{-a}^{a} \gamma_k(\xi)\, (x-\xi)^l\, d\xi - \\ - \sum_{n=1}^{\infty} a^{-n-1} H_{n-1} \int_{-a}^{a} 2c\, s'(\xi)\, (x-\xi)^{n-1}\, d\xi \end{aligned}\right\} \tag{20.80}$$

must remain finite for $x \to -a$. We assume $s'(-a)$ finite. γ_k may possibly have a singularity of the form $1/\sqrt{a+x}$ near $x = -a$. However, the integral

$$\int_{-a}^{a} \frac{(x-\xi)^l}{\sqrt{a^2 - \xi^2}}\, d\xi = \int_{-\frac{1}{2}\pi}^{\frac{1}{2}\pi} (x - a \sin\alpha)^l\, d\alpha,$$

is a polynomial in x for $l \geqq 0$. Thus the last two summations of (20.80) remain finite at $x = -a$. However, the first summations potentially contributes terms like

$$\int_{-a}^{a} \frac{d\xi}{(x-\xi)\sqrt{a^2 - \xi^2}} = \frac{\pi}{\sqrt{x^2 - a^2}},$$

In order to avoid this singularity at $x = -a$, we select the total circulation $\int_{-a}^{a} \gamma_n\, d\xi$ so that

$$\int_{-a}^{a} \gamma_n(\xi)\, d\xi = -\frac{1}{\pi} \int_{-a}^{a} \frac{F_n(\xi) \sqrt{a^2 - \xi^2}}{\xi + a}\, d\xi. \tag{20.81}$$

Substituting into (20.79), one finds finally

$$\gamma_n(x) = \frac{1}{\pi^2} \sqrt{\frac{a+x}{a-x}} \int_{-a}^{a} F_n(\xi) \sqrt{\frac{a-\xi}{a+\xi}}\, \frac{d\xi}{\xi - x}. \tag{20.82}$$

$\gamma(x)$ itself is given by the sum displayed in (20.77). Although the singularity at the trailing edge has been removed, there is still one at the leading edge; this occurs also in thin-airfoil theory and corresponds roughly to the fact that the conditions of linearization (i.e. of small disturbance) are not satisfied near the leading-edge stagnation point.

Keldysh and Lavrent'ev compute the integrals which will be necessary if $r(x)$ and $s(x)$ are given as polynomials and apply their computational method to a flat-plate and circular-arc airfoil at a small angle of attack.

---

[1] See, e.g., W. Schmeidler: Integralgleichungen ..., pp. 55–56. Leipzig: Akademische Verlagsgesellschaft 1950.

In order to find the force and moment on the wing it is convenient to fall back on the H-function. One finds easily

$$\left.\begin{aligned} H(k) &= e^{-kh}\int_{-a}^{a}[\gamma(\xi) - 2ic\,s'(\xi)]\,e^{-i\lambda s}\,ds,\\ |H(k)|^2 &= e^{-2hk}\int_{-a}^{a}\int_{-a}^{a}\{[\gamma(\xi)\,\gamma(x) + 4c^2 s'(\xi)\,s'(x)]\cos k(\xi - x) + \\ &\quad + 2c[\gamma(\xi)\,\sigma'(x) - \gamma(x)\,\sigma'(\xi)]\sin k(\xi - x)\}\,d\xi\,dx. \end{aligned}\right\} \tag{20.83}$$

Formulas (20.38) allow one to complete the calculation for special cases.

The theory analogous to that described above for fluid of finite depth h_0 has been carried through by TIKHONOV (1940). He has applied the method to calculate the lift and drag coefficients for a flat plate at a small angle of attack.

Rather than reproduce the graphical presentations of KELDYSH and LAVRENT'EV and TIKHONOV for the flat plate, we shall give instead the lift and drag coefficients for a submerged vortex. Here one may give relatively simple analytic expressions, and the qualitative behavior of the curves is similar to that of a flat plate. The formulas for lift L and drag D are as follows:

$$\left.\begin{aligned} D_\infty &= \varrho\nu\Gamma^2 e^{-2\nu h},\\ L_\infty &= \varrho\Gamma c - \frac{\varrho\Gamma^2}{4\pi h} + \frac{\varrho\Gamma^2}{\pi}\nu e^{-2\nu h}\,\mathrm{Ei}(2\nu h),\\ D_{h_0} &= \varrho\nu\Gamma^2\frac{\sinh^2 m_0(h_0 - h)}{\nu h_0 - \cosh^2 m_0 h_0} \quad \text{if} \quad \nu h_0 > 1, \quad = 0 \quad \text{if} \quad \nu h_0 < 1,\\ L_{h_0} &= \varrho\Gamma c - \frac{\varrho\Gamma^2}{4\pi}\frac{1}{h_0 - h} + \frac{\varrho\Gamma^2}{2\pi}\int_0^\infty(\nu + k)\,e^{-kh_0}\frac{\sinh 2k(h_0 - h)}{\nu\sinh kh_0 - k\cosh kh_0}\,dk, \end{aligned}\right\} \tag{20.84}$$

where m_0 is the real root of $m = \nu\tanh mh_0$. For finite depth the expression for D stems from the last term in (13.47). The dimensionless coefficients $C_D = Dh/\varrho\Gamma^2$ and $C_L = (L - \varrho c\Gamma)\,h/\varrho\Gamma^2$ are shown in Fig. 24a for infinite depth as functions of c^2/gh and in Figs. 24b and c as functions of c^2/gh_0 for various values of $\beta = h/h_0$. For infinite depth C_L starts with a value $1/4\pi$ and tends asymptotically to $-1/4\pi$, crossing the axis at $c^2/gh = 2.47$. For finite depth the coefficients have a discontinuity at $c^2/gh_0 = 1$. As $c^2/gh_0 \to 0$, $C_D \to 0$, and as $c^2/h_0 \to 1$, $C_D \to \frac{3}{2}\beta(1-\beta)^2$. For $c^2/gh_0 > 1$, C_L is always negative and increasing with a vertical asymptote at $c^2/gh_0 = 1$ and a horizontal one as $c^2/gh_0 \to \infty$ at $-\beta/4\sin\beta\pi$; these curves start at $\frac{1}{4}\beta\cot\beta\pi$.

Further development of hydrofoil theory has taken place in several directions. HASKIND (1945a) has extended KOCHIN'S "deep-submersion" theory to water of finite depth. However, he does not discuss the steps necessary for fulfilment of the Kutta-Joukowski condition, as does KOCHIN. The lifting-line theory for airfoils of finite span has been extended to hydrofoils by WU (1954), BRESLIN (1957), and HASKIND (1956). PARKIN, PERRY and WU (1956) and LAITONE (1954, 1955) have investigated both theoretically and experimentally the effect of bringing a given hydrofoil so close to the surface that the infinitesimal-wave approximation breaks down completely. There exists also a considerable amount of work on flow about cavitating hydrofoils. However, since the effect of gravity is neglected, this work is not considered in the present article. Experimental data relevant to the theoretical development outlined above are scanty. Reports by BENSON and LAND (1942) and by LAND (1943) give results of an experimental investigation of the effect of depth of submersion. However, the investigations were not designed to test the validity of the theory and do not, for example, include the region of maximum C_D. AUSMAN (1953), in connection with an experimental investigation of the pressure distribution on the upper surface of a hydrofoil, measured the lift coefficient and compared it with that predicted by the thin-hydrofoil theory. The theory failed when gh/c^2 became too small because the associated free surface over the hydrofoil no longer approximated infinitesimal waves, or, in other words, the thin hydrofoil was not thin enough

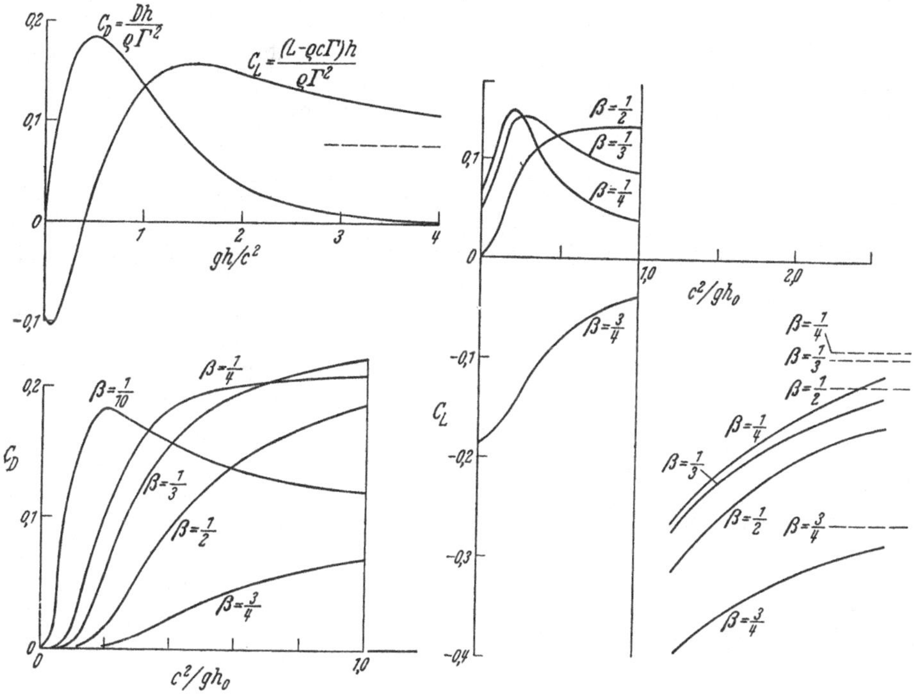

Fig. 24 a—c.

for these values of gh/c^2 for the theory to be applicable. It should also be emphasized that for small values of gh/c^2 the occurrence of cavitation on the upper surface must be taken into account for a complete theory. Recent measurements by NISHIYAMA (1959) show good agreement even for small values of gh/c^2. A comprehensive survey of hydrofoil theory is given in a recent paper by NISHIYAMA (1957).

γ) Planing surfaces. The following discussion is limited to two-dimensional motion, for the theory of three-dimensional planing surfaces for flows with gravity does not appear to have been developed.

For the linearized problem it is natural to consider the planing surface or glider as an approximation to a flat plate moving along the surface of the undisturbed fluid, i.e. the curvature, angle of attack and vertical displacement are all assumed small. In order to formalize the perturbation procedure, let the planing surface be represented by

$$y = h + F(x), \quad |x| \leqq a, \quad F(-a) = 0, \tag{20.85}$$

in coordinates fixed in space, and let the fluid have velocity $-c$ at $x = +\infty$. Thus, we are going to consider the flow to be a perturbation of a uniform flow.

First let us consider briefly in a qualitative fashion the exact solution. There will be a stagnation point A somewhere behind the leading edge and a jet will be thrown out ahead of the glider. We take it to be of thickness b and to make an angle β with OX. If $\Phi=-cx+\varphi(x,y)$ and $\Psi=-cy+\psi(x,y)$ are potential and stream function, respectively, we shall take the free surface ahead of the glider to be given by $\Psi=-bc$ and behind the glider by $\Psi=0$ (see Fig. 25). Then b/a, and AL/a will all be functions of ga/c^2 and gh/c^2. It will be assumed as one of the boundary conditions of the problem, in analogy with the Kutta-Joukowski condition, that the velocity is continuous at the trialing edge. It is obvious that the flow near the leading edge cannot conform to the requirement that it be a small perturbation of a uniform flow. However, we shall give arguments below to indicate that, except in the neighborhood of the leading edge, this effect is of the second order.

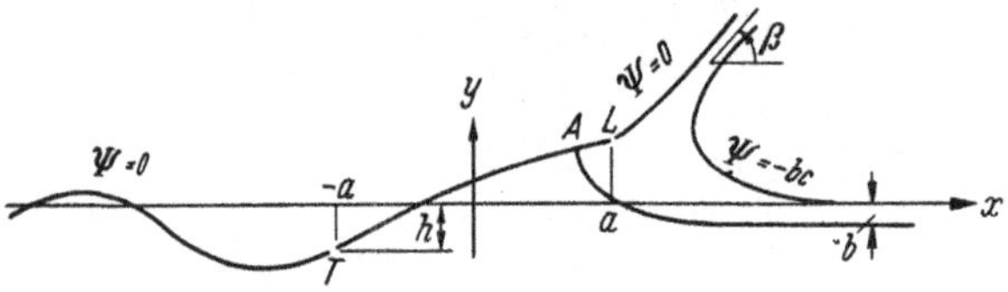

Fig. 25.

In order to get some idea of the relative size of the jet, consider the simpler problem of a flat plate of length l and angle of attack α gliding on a weightless fluid. This problem can be solved exactly [see, e.g., MILNE-THOMSON (1956, § 12.26); A. E. GREEN (1935, 1936)]. The asymptotic expression, for small α, of both the ratios b/l and AL/l is $\frac{\pi}{2}\frac{\alpha^2}{1+\cos\beta}$, i.e. they are both of the second order. We shall suppose that this relation continues to hold when gravity is acting.

We now carry through the perturbation procedure of Sect. 10α [see especially Eq. (10.16)], writing

$$\left.\begin{aligned} \Phi &= -cx+\varepsilon\varphi^{(1)}+\cdots, \quad \Psi=-cy+\varepsilon\psi^{(1)}+\cdots, \quad \eta=\varepsilon\eta^{(1)}+\cdots, \\ h+F(x) &= \varepsilon F^{(1)}(x)+\varepsilon h^{(1)}+\varepsilon^2 h^{(2)}+\cdots, \\ b &= \varepsilon^2 b^{(2)}+\cdots. \end{aligned}\right\} \tag{20.86}$$

Substitution in the exact boundary conditions then yields the following linearized conditions:

$$\left.\begin{aligned} \psi^{(1)}(x,0)-\frac{c^2}{g}\psi_y^{(1)}(x,0) &= 0, \qquad |x|>a, \\ \psi^{(1)}(x,0) &= c\left(h^{(1)}+F^{(1)}(x)\right), \qquad |x|<a. \end{aligned}\right\} \tag{20.87}$$

The free surface is given by

$$\eta^{(1)}(x)=\frac{1}{c}\psi^{(1)}(x,0)=\frac{c}{g}\varphi_x^{(1)}(x,0), \qquad |x|>a. \tag{20.88}$$

We require as usual that the disturbance vanish as $x\to\infty$. One will expect that the behavior near the leading edge will reflect in some manner the inconsistency of the exact solution with the notion of a small perturbation. It will turn out that it will be necessary to allow a singularity at the leading edge. (Almost the same situation exists in the thin-hydrofoil or thin-wing theory since the stagnation point near the leading edge also prevents the flow in that region from being a small perturbation of a uniform flow.) A singularity at the trailing edge, although mathematically possible, has been specifically proscribed. The strength of the singularity at the leading edge and the elevation h of the trailing edge will be determined as functions of ga/c^2 in the course of solving the problem.

The problem as formulated above has been considered, for infinite depth, by SRETENSKII (1933, 1940), SEDOV (1937), KOCHIN (1938), and MARUO (1951). HASKIND has extended SEDOV's analysis to finite depth (1943a), and later (1955) has treated a glider moving on a wavy surface. YU. S. CHAPLYGIN (1940) has apparently carried through a fairly comprehensive numerical analysis for a flat plate making use of SEDOV's method of analysis [see SRETENSKII (1951, p. 83)]. SRETENSKII's papers are expounded in terms of a flat plate, but the method clearly has wider applicability, as he remarks in his first paper. SRETENSKII's 1940 paper gives the results of rather extensive calculations for flat plates. MARUO's paper is conceptually very similar to those of SRETENSKII, but his method is not quite as efficient for computation. However, he also gives computational results and includes a correction to take account of the failure of the linearized theory near the leading edge. More recently the problem has been considered again by authors unaware of the earlier work. SQUIRE (1957) has analyzed a gliding flat plate by a method similar to that used by SRETENSKII and MARUO. Certain integrals involved in this method have been tabulated by MILLER (1957). CUMBERBATCH (1958) has used a method similar to SEDOV's. Both authors add new results to the earlier work.

Both SEDOV and KOCHIN introduce the complex potential $f(z)=\varphi+i\psi$ and thereafter the function $f'+i\nu f$, $\nu=g/c^2$. Although the two methods are not by any means the same, they have much in common with the treatment of hydrofoils given above. Consequently, we shall outline below the method followed by SRETENSKII.

As a preliminary we need a result from Sect. 21 below. Suppose that a pressure distribution $p(x)$,which we take to be absolutely integrable, is given on the free surface. Then the complex velocity potential must satisfy

$$\operatorname{Re}\{f'(x+i0)+i\nu f(x+i0)\}=\frac{1}{\varrho c}\,p(x), \tag{20.89}$$

and the free surface is given by

$$y=\eta(x)=\frac{1}{c}\,\psi(x,0)=\frac{c}{g}\,\varphi_x(x,0)-\frac{1}{\varrho g}\,p(x). \tag{20.90}$$

The function $f(z)$ which satisfies (20.89) and which vanishes as $x\to\infty$ can be written in several forms, of which we select the following [see Eq. (21.38)]:

$$f(z)=\frac{1}{\pi i\varrho c}\int_{-\infty}^{\infty}d\xi\,p(\xi)\,\mathrm{PV}\int_{0}^{\infty}\frac{-\,\mathrm{e}^{-i\lambda(z-\xi)}}{\lambda-\nu}\,d\lambda-\frac{1}{\varrho c}\int_{-\infty}^{\infty}p(\xi)\,\mathrm{e}^{-i\nu(z-\xi)}\,d\xi. \tag{20.91}$$

The free surface is given by

$$\left.\begin{aligned}\eta(x)=&\lim_{y\to-0}\frac{1}{\pi\varrho c^2}\int_{-\infty}^{\infty}d\xi\,p(\xi)\,\mathrm{PV}\int_{0}^{\infty}\frac{\cos\lambda(x-\xi)}{\lambda-\nu}\,\mathrm{e}^{\lambda y}\,d\lambda+\\&+\frac{\mathrm{e}^{\nu y}}{\varrho c}\int_{-\infty}^{\infty}p(\xi)\sin\nu(x-\xi)\,d\xi;\end{aligned}\right\} \tag{20.92}$$

the reason for leaving y explicitly in the formulas will appear below.

When a glider is moving on a free surface, the streamline $y=c^{-1}\psi(x,0)$ will consist partly of free surface, where $p(x)=0$, and partly of the wetted surface of the glider, where $p(x)$ is some unknown function.

Eq. (20.92) may then be written as the following integral equation for this unknown function $p(x)$:

$$\left.\begin{aligned} h+F(x) &= \lim_{y\to -0} \frac{1}{\pi \varrho c^2} \int_{-a}^{a} d\xi\, p(\xi)\, \mathrm{PV}\int_0^\infty \frac{\cos\lambda(x-\xi)}{\lambda-\nu}\, e^{\lambda y}\, d\lambda \,+ \\ &\quad + \frac{e^{\nu y}}{\varrho c} \int_{-\infty}^{\infty} p(\xi) \sin\nu(x-\xi)\, d\xi, \qquad |x|<a. \end{aligned}\right\} \tag{20.93}$$

Once $p(x)$ has been determined, one may substitute back into (20.92) in order to find the form of the free surface for $|x|>a$.

It is possible to work directly with (20.93), and this is the procedure followed by MARUO. However, SRETENSKII differentiates twice with respect to x and adds ν^2 times (20.93). This yields

$$\left.\begin{aligned} F''(x)+\nu^2 F(x)+\nu^2 h &= \lim_{y\to -0} \frac{-1}{\pi \varrho c^2} \int_{-a}^{a} d\xi\, p(\xi) \int_0^\infty (\lambda+\nu)\cos\lambda(x-\xi)\, e^{\lambda y}\, d\lambda \\ &= \lim_{y\to 0} \frac{1}{\pi \varrho c^2} \int_{-a}^{a} p(\xi)\left(\nu+\frac{\partial}{\partial y}\right) \frac{y}{(x-\xi)^2+y^2}\, d\xi \\ &= -\frac{\nu}{\varrho c^2}\, p(x) - \lim_{y\to 0} \frac{1}{\pi \varrho c^2} \frac{\partial}{\partial x} \int_{-a}^{a} p(\xi)\, \frac{x-\xi}{(x-\xi)^2+y^2}\, d\xi \\ &= \frac{-\nu}{\varrho c^2}\, p(x) - \frac{1}{\pi \varrho c^2} \frac{\partial}{\partial x}\, \mathrm{PV}\int_{-a}^{a} \frac{p(\xi)}{x-\xi}\, d\xi. \end{aligned}\right\} \tag{20.94}$$

Although this last equation is a necessary condition for $p(x)$, it obviously cannot determine it uniquely, for the last term of (20.93), assuring vanishing of the disturbance far ahead of the glider, was lost in the formation of (20.94). Thus one still has need for (20.93). Eq. (20.94) is essentially the equation derived by SRETENSKII.

Let us now integrate (20.94) with respect to x from $x=-a$ to x, and denote

$$P(x) = \frac{1}{\varrho c^2} \int_{-a}^{x} p(\xi)\, d\xi.$$

Then Eq. (20.94) becomes

$$F'(x)-F'(-a)+\nu^2 \int_{-a}^{x} F(\xi)\, d\xi + \nu^2 h(x+a) = -\nu\, P(x) - \frac{1}{\pi} \int_{-a}^{a} \frac{P'(\xi)}{x-\xi}\, d\xi, \tag{20.95}$$

where an additive constant has been discarded since h itself is an undetermined constant. Eq. (20.95) is just PRANDTL's integro-differential equation for the circulation about an airfoil of finite span[1]. Thus known methods for solving the airfoil equation can be carried over to the study of this equation. However, the solutions themselves cannot be taken over directly, for different boundary conditions are impcsed: in the airfoil equation the unknown function is the circulation $\Gamma(x)$ and it is usually assumed that $\Gamma(-x)=\Gamma(x)$ and $\Gamma(-a)=\Gamma(a)=0$;

[1] See, e.g., N. I. MUSKHELISHVILI: Singular integral equations, Chap. 17. Groningen: Noordhoff 1953.

in the present problem $P(x)$ is not necessarily symmetric and $P(-a) = P'(-a) = 0$, but $P(a)$ is not restricted except to be finite. The theory of the Prandtl integro-differential equation without the customary additional requirements associated with airplane wings has been developed by L. G. MAGNARADZE [Soobshch. Akad. Nauk Gruzin. SSR **3**, 503—508 (1942)].

The equation can be solved by an extension of GLAUERT's method[1]. This is the method which has been used by both MARUO and SRETENSKII. However, each expands $P' = p$ rather than P in a Fourier series in order to obtain the correct behavior at the two end points. Introduce the new variables ϑ and γ by the equations

$$x = -a\cos\vartheta, \qquad \xi = -a\cos\gamma$$

and assume the following expansion for $p(x)$:

$$\left.\begin{aligned} \frac{1}{\varrho c^2}\,p(x) &= \frac{1}{\varrho c^2}\,p(-a\cos\vartheta) = a_0\tan\frac{1}{2}\vartheta + a_1\sin\vartheta + \cdots + a_n\sin n\vartheta + \cdots \\ &= a_0\sqrt{\frac{a-x}{a+x}} + a_1\sqrt{a^2-x^2} + \cdots. \end{aligned}\right\} \quad (20.96)$$

MARUO substitutes (20.96) into (20.93), SRETENSKII into (20.94). The latter, which seems less laborious, leads to

$$\left.\begin{aligned} &a\left[F''(-a\cos\vartheta) + \nu^2 F(-a\cos\vartheta) + \nu^2 h\right]\sin\vartheta \\ &\quad = -\nu a a_0(1-\cos\vartheta) - \nu a\sum_{n=1}^{\infty} a_n\sin\vartheta\sin n\vartheta - \sum_{n=1}^{\infty} n a_n\sin n\vartheta. \end{aligned}\right\} \quad (20.97)$$

We shall not discuss SRETENSKII 's further steps to determine the coefficients a_n. However, they lead to expressions of the following kinds for the coefficients:

$$\left.\begin{aligned} a_{2n-1} &= A_{2n-1} a\nu^2 h + B_{2n-1} a\nu a_0 + C_{2n-1}, \\ a_{2n} &= B_{2n} a\nu a_0 + C_{2n}, \quad n = 1, 2, \ldots, \end{aligned}\right\} \quad (20.98)$$

where A_n, B_n, C_n are functions of νa. Substitution of the coefficients into (20.93) and into (20.93) differentiated once with respect to x and evaluated at $x = -a$ results in equations of the form

$$\left.\begin{aligned} \nu h &= Q_1 a_0 + R_1\nu h + S_1, \\ F'(-a) &= Q_2 a_0 + R_2\nu h + S_2, \end{aligned}\right\} \quad (20.99)$$

where Q_i, R_i, S_i are functions of νa; these equations may be used to determine νh and a_0 as functions of νa. As long as $a_0 \neq 0$ there will be a singularity at the leading edge.

Once $p(x)$ has been determined approximately, one can compute the lift, drag and moment about, say, the center. To the order of approximation appropriate to the linear theory they are

$$\left.\begin{aligned} L &= \int_{-a}^{a} p(x)\,dx, \\ R &= \int_{-a}^{a} p(x)\,f'(x)\,dx, \\ M &= \int_{-a}^{a} p(x)\,x\,dx. \end{aligned}\right\} \quad (20.100)$$

[1] H. GLAUERT: The elements of airfoil and airscrew theory, Chap. XI. Cambridge 1943.

For the flat-plate glider it is possible to give the following asymptotic expressions for these quantities when $\nu a \to 0$.

$$\left.\begin{aligned} L &= \pi a \varrho c^2 \alpha \left[1 - \nu a \left(\pi + \frac{4}{\pi}\right)\right] + O(\nu^2 a^2 \log \nu a), \qquad R = \alpha L, \\ M &= \frac{1}{2} \pi a^2 \varrho c^2 \alpha \left[1 - \nu a \left(\pi + \frac{8}{3\pi}\right)\right] + O(\nu^4 a^4 \log \nu a). \end{aligned}\right\} \quad (20.101)$$

There were first given by SEDOV, but are also derived in the papers by KOCHIN and SRETENSKII.

Fig. 26 reproduces several of SRETENSKII's computed pressure distributions for a flat plate. The predictions of the linearized theory cannot, of course, be expected to be

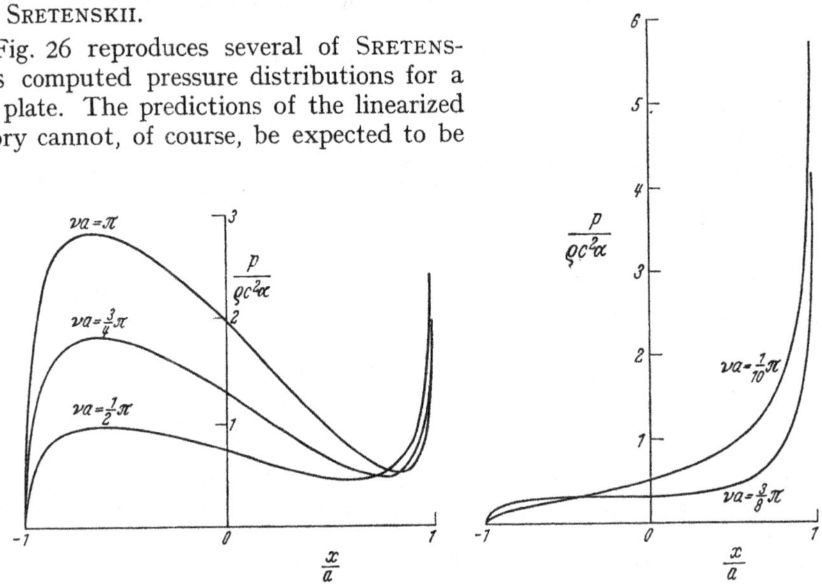

Fig. 26.

accurate near the leading edge. MARUO has corrected his computed points in this region by using the exact theory for a weightless fluid. Both MARUO and SRETENSKII give further computational results which we do not reproduce. MARUO (1959) has also provided experimental confirmation of the predicted pressure distributions.

21. Waves resulting from variable pressure distributions. In the situations considered up to now in this chapter the pressure at the free surface has been taken as constant. We now consider the result of allowing the pressure over the free surface to be a given function of both position and time. Otherwise the fluid is taken to be infinite in horizontal extent and to be either infinitely deep or of uniform depth h. The time variation in pressure will be limited to two cases. In Sect. 21 α a periodically varying pressure is considered; in Sect. 21 β the pressure is taken to move with uniform velocity; Sect. 21 γ gives some references to a combination of these two. In Sect. 22 waves from pressure distributions will be considered again in connection with initial-value problems. Since the methods for finding the velocity potential are similar in most respects to those used in finding the velocity potential for a source, we shall, with one exception, give the results without proof.

Just as in the cases of the stationary source of periodic strength and the moving source of constant strength treated in Sect. 13 γ, we must in the present situation impose boundary conditions at infinity in order to ensure a unique solution. The imposed conditions, namely the radiation condition and the vanishing of the fluid motion far ahead, respectively, are selected as being physically reasonable. However, one may proceed differently, derive formulas analogous to (13.50) and (13.51) and then find the limit as $t \to \infty$. The resulting velocity potentials automatically satisfy the correct boundary conditions at infinity. This method has been used, for example, by G. GREEN (1948) in the two-dimensional problems considered in the following two sections, and also by STOKER (1953, 1954).

The theory of wave generation by pressure distributions has an obvious application in oceanographic problems. However, the theory was apparently first developed in an attempt to explain the wave pattern produced by a ship. We shall not attempt to disentangle the history of the subject. For the material covered in Sect. 21β we call attention to a survey by J. K. LUNDE (1951b) which also contains a useful bibliography.

α) *Pressure distributions periodic in time.* Three dimensions. The boundary conditions have already been given in Sect. 11. If Φ and p are represented by

$$\left.\begin{aligned} \Phi(x, y, z, t) &= \operatorname{Re} \varphi(x, y, z)\, e^{-i\sigma t}, \qquad p(x, y, z t) = p(x, z)\, e^{-i\sigma t}, \\ \varphi &= \varphi_1 + i\varphi_2, \qquad p = p_1 + i p_2, \end{aligned}\right\} \tag{21.1}$$

then the condition on the free surface may be written

$$\varphi_y(x, 0, z) - \frac{\sigma^2}{g} \varphi(x, 0, z) = \frac{i\sigma}{\varrho g} p(x, z), \tag{21.2}$$

and the form of the surface is given by

$$\eta(x, z, t) = \operatorname{Re}\left\{\frac{i\sigma}{g} \varphi(x, 0, z) - \frac{1}{\varrho g} p(x, z)\right\} e^{-i\sigma t}. \tag{21.3}$$

In addition, a radiation condition is assumed at infinity [see fifth Eq. (13.9)] and a condition appropriate to the depth of fluid. We shall also assume $p(x, z)$ to be absolutely integrable.

The velocity potential can be expressed as follows:

$$\left.\begin{aligned} \varphi(x, y, z) = {} & \frac{-i\sigma}{2\pi\varrho g} \int_{-\infty}^{\infty}\int_{\infty}^{\infty} p(\xi, \zeta)\, d\xi\, d\zeta \,\mathrm{PV}\int_{\infty}^{\infty} \frac{k\, e^{ky}}{k-\nu} J_0\left(k\sqrt{(x-\xi)^2 + (z-\zeta)^2}\right) dk + \\ & + \frac{\sigma\nu\, e^{\nu y}}{2\varrho g} \int_{-\infty}^{\infty}\int_{-\infty}^{\infty} p(\xi, \zeta)\, J_0\left(\nu\sqrt{(x-\xi)^2 + (z-\zeta)^2}\right) d\xi\, d\zeta, \qquad \nu = \frac{\sigma^2}{g}; \end{aligned}\right\} \tag{21.3}$$

and in cylindrical coordinates $x = R\cos\alpha$, $z = R\sin\alpha$ in the form

$$\left.\begin{aligned} \varphi(R, \alpha, y) = {} & \frac{-i\sigma}{2\pi\varrho g} \int_0^{2\pi}\int_0^{\infty} p(R', \alpha')\, R'\, dR'\, d\alpha' \,\mathrm{PV}\int_{-\infty}^{\infty} \frac{k\, e^{ky}}{k-\nu} \times \\ & \times J_0\left(k\sqrt{R^2 + R'^2 - 2RR'\cos(\alpha'-\alpha)}\right) dk + \frac{\sigma\nu\, e^{\nu y}}{2\varrho g} \int_0^{2\pi}\int_0^{\infty} p(R', \alpha') \times \\ & \times J_0\left(\nu\sqrt{R^2 + R'^2 - 2RR'\cos(\alpha'-\alpha)}\right) R'\, dR'\, d\alpha'. \end{aligned}\right\} \tag{21.4}$$

The addition theorem for J_0 allows one to write[1]

$$\left.\begin{aligned} J_0\left(k\sqrt{R^2+R'^2-2RR'\cos(\alpha'-\alpha)}\right) &= \sum_{n=0}^{\infty}\varepsilon_n J_n(kR)\,J_n(kR')\cos n(\alpha'-\alpha),\\ \varepsilon_0=1,\quad \varepsilon_n&=2,\quad n\geqq 1.\end{aligned}\right\} \tag{21.5}$$

If p is independent of α, one may derive easily

$$\left.\begin{aligned}\varphi(R,y) = \frac{-i\sigma}{\varrho g}\int_0^\infty p(R')\,R'\,dR'\,\mathrm{PV}\int\frac{k\,e^{ky}}{k-\nu}J_0(kR)\,J_0(kR')\,dk+ \\ +\frac{\pi\sigma\nu e^{\nu y}}{\varrho g}J_0(\nu R)\int_0^\infty p(R')\,J_0(\nu R')\,R'\,dR'.\end{aligned}\right\} \tag{21.6}$$

The asymptotic form for large R of (21.6) is a relatively simpel expression:

$$\varphi(R,y)\sim\frac{\pi\sigma\nu e^{\nu y}}{\varrho g}\sqrt{\frac{2}{\pi\nu R}}\,e^{i\left(\nu R-\frac{\pi}{4}\right)}\int_0^\infty p(R')\,J_0(\nu R')\,R'\,dR'. \tag{21.7}$$

We note in passing that the potential function (21.3) or (21.4) can also be obtained as a distribution of sources on the surface [see HUDIMAC (1953, p. 78)]. This may be easily verified as follows. In (13.17'') let $b=0$. Then, using (13.12), one obtains (substituting ξ,ζ for a, c)

$$2\,\mathrm{PV}\int^\infty\frac{k}{k-\nu}e^{ky}J_0\left(k\sqrt{(x-\xi)^2+(z-\zeta)^2}\right)dk+i\,2\pi\nu\,e^{\nu y}J_0\left(\nu\sqrt{(x-\xi)^2+(z-\zeta)^2}\right).$$

A distribution over the plane $y=0$ of these sources of strength $+i\sigma p(\xi,\zeta)/4\pi\varrho g$ yields (21.3) (we recall that a source of strength m behaves like $-m/r$ near the singularity).

The rate at which the pressure distribution does work upon the fluid can be calculated directly or by using Eq. (8.2). Consider the volume of fluid contained in a large cylinder of radius R_0. Then, from (8.2), after appropriate linearization, the rate of increase of energy of the fluid is given by

$$\begin{aligned}\frac{dE}{dt} = \mathrm{Re}\Bigg\{&\int_0^{2\pi}\int_0^{R_0} p(R,\alpha,t)\,\Phi(R,\alpha,0,t)\,R\,dR\,d\alpha+\\ &+\varrho\int_0^{2\pi}\int_{-\infty}^{0}\Phi_t(R_0,\alpha,y,t)\,\Phi_R(R_0\,\alpha,y,t)\,R_0\,dy\,d\alpha\Bigg\}.\end{aligned}$$

Now substitute (21.1) and take the average over a period, which will clearly be zero. The result may be written:

$$\left.\begin{aligned}0=\left[\frac{dE}{dt}\right]_{\mathrm{av}} = \mathrm{Re}\Big\{&-\tfrac{1}{2}\iint p(R,\alpha)\,\overline{\varphi}_y(R,\alpha,0)\,R\,dR\,d\alpha+\\ &+\tfrac{1}{2}\varrho\sigma\iint i\,\varphi_R(R_0,\alpha,y)\,\overline{\varphi}(R_0,\alpha,y)\,R_0\,dy\,d\alpha\Big\}.\end{aligned}\right\} \tag{21.8}$$

---

[1] See G.N. WATSON: A treatise on the theory of Bessel functions, p. 353. Cambridge 1944.

The first integral gives the average rate W_{av} at which the pressure distribution is working upon the fluid. It must equal minus the second integral. If $p(R,\alpha) = p(R)$, then we may apply (21.7) to obtain a relatively simple expression for the average rate over the whole fluid:

$$W_{av} = \frac{\pi^2 \sigma \nu^2}{\varrho g} \left| \int_0^\infty p(R') J_0(\nu R') R' \, dR' \right|^2. \tag{21.9}$$

To carry through the computation when p is not circularly symmetric is more complicated arithmetically, but can be carried through by use of (21.5).

One can find an investigation of the waves resulting from a doubly modulated pressure distribution over a rectangular domain,

$$p = A\,e^{-i\sigma t} \cos m\,x \cos n\,z, \qquad |x| \leqq a, \qquad |z| \leqq b,$$

in a paper of SRETENSKII (1956).

If the fluid is of uniform depth h, the expression for the velocity potential is

$$\left.\begin{aligned} \varphi(x,y,z) = \frac{-i\sigma}{2\pi\varrho g} \iint_{-\infty}^{\infty} p(\xi,\zeta)\, d\xi\, d\zeta\, \mathrm{PV}\int_0^\infty \frac{k \cosh k(y+h)}{k \sinh k h - \nu \cosh k h} \times \\ \times J_0\left(k\sqrt{(x-\xi)^2+(z-\zeta)^2}\right) + \frac{\sigma}{2\varrho g} \frac{m_0 \cosh m_0 (y+h) \sinh m_0 h}{\nu h + \sinh^2 m_0 h} \times \\ \times \iint_{-\infty}^{\infty} p(\xi,\zeta)\, J_0\left(m_0\sqrt{(x-\xi)^2 + (z-\zeta^2)}\right) d\xi\, d\zeta, \end{aligned}\right\} \tag{21.10}$$

where, as usual, m_0 is the real solution of

$$m_0 \tanh m_0 h - \nu = 0.$$

Other forms of this expression similar to (21..), (21.6) and (21.7) can be found with no difficulty. We give only the analogue of (21.9):

$$W_{av} = \frac{1}{2}\pi^2 \frac{\sigma \nu m_0}{\varrho g} \frac{\sinh 2m_0 h}{\nu h + \sinh^2 m_0 h} \left| \int_0^\infty p(R') J_0(\nu R') R' \, dR' \right|^2. \tag{21.11}$$

The identities following (13.18) may be used to put both (21.10) and (21.11) into other forms.

Two dimensions. The derivation of the velocity potential will be carried through, at least in part, since it illustrates a nice application of the Plemelj-Sokhotskii formulas. Two complex units will be introduced, as described at the end of Sect. 11. That is, we shall write

$$\left.\begin{aligned} \Phi(x,y,t) &= \varphi_1(x,y)\cos\sigma t + \varphi_2(x,y)\sin\sigma t = \mathrm{Re}_j\, \varphi\, e^{-j\sigma t}, \\ p(x,t) &= p_1(x)\cos\sigma t + p_2(x)\sin\sigma t = \mathrm{Re}_j\, p\, e^{-j\sigma t}, \\ \varphi &= \varphi_1 + j\varphi_2, \qquad p = p_1 + j p_2, \end{aligned}\right\} \tag{21.12}$$

and also introduce a stream function $\psi = \psi_1 + j\psi_2$ and a complex potential

$$f(z) = f_1(z) + j f_2(z), \qquad f_k = \varphi_k + i\psi_k, \qquad k = 1, 2. \tag{21.13}$$

Then the boundary condition on the free surface may be formulated as follows

$$\mathrm{Im}_i \{f'(x - i0) + i\nu f(x - i0)\} = -\frac{\sigma}{\varrho g} j\, p(x). \tag{21.14}$$

The definition of $g \equiv f' + i\nu f$ may be extended to the whole complex z-plane by reflection, i.e.

$$g(x+iy) = \overline{g(x-iy)}, \quad y>0.$$

Then the condition (21.14) may be written in the form

$$\mathrm{Im}_i\{f'(x \pm i0) + i\nu f(x \pm i0)\} = \pm \frac{\sigma}{\varrho g} j\, p(x). \tag{21.15}$$

We shall suppose $p(x)$ to be absolutely integrable on the infinite interval. In addition, we shall suppose that either $p(x)$ satisfies a Hölder condition (and is hence continuous) on the whole infinite interval, or else that there are a finite number of segments $(-\infty, b_1)$, $(a_2, b_2), \ldots, (a_r, \infty)$, $b_n < a_{n+1}$, such that $p(x)$ satisfies a Hölder condition on any closed segment interior to one of the above segments, and at an end-point may be expressed in the form

$$p(x) = \frac{q(x)}{(x-c)^\alpha}, \quad 0 \leqq \alpha < 1, \quad c = a_i \text{ or } b_i,$$

where $q(x)$ satisfies a Hölder condition at the end. Here a Hölder condition means that for any pair of points x_1, x_2, $p(x)$ satisfies

$$|p(x_1) - p(x_2)| < A\,|x_1 - x_2|^\mu, \quad \mu > 0.$$

In the first case f' will be assumed to have no singularities in the whole lower half-plane. In the second case the behavior of $f'(z)$ near an end-point c will be restricted so that it must satisfy

$$|f'(z)| < \frac{C}{|z-c|^\alpha}, \quad 0 \leqq \alpha < 1.$$

As usual it will be assumed that $|f'|$ is bounded as $z \to \infty$ and that only outgoing waves are generated.

The solution of this boundary-value problem for the function $g(z) = f' + i\nu f$ is determined, up to an additive real constant which may be discarded here, by[1]

$$f'(z) + i\nu f(z) = j \frac{\sigma}{\pi \varrho g} \int_{-\infty}^{\infty} \frac{p(s)}{z-s}\, ds, \quad y < 0. \tag{21.16}$$

After integrating the differential equation and selecting the solution so as to represent outgoing waves at $x = \pm\infty$, one obtains finally [the derivation is similar to that of (13.28)]

$$f(z) = j \frac{\sigma}{\pi \varrho g} e^{-i\nu z} \int_{-\infty}^{\infty} p(s)\, ds \int_{\infty}^{z} \frac{e^{i\nu\zeta}}{\zeta - s}\, d\zeta + \frac{\sigma}{\varrho g} (1 + ij)\, e^{-i\nu z} \int_{-\infty}^{\infty} p(s)\, e^{i\nu s}\, ds, \tag{21.17}$$

where the path of integration for ζ is taken in the lower half-plane. The asymptotic form of the time-dependent velocity potential is given by

$$\mathrm{Re}_j\, f(z)\, e^{-j\sigma t} \sim \frac{\sigma}{\varrho g} e^{-i(\nu z \mp \sigma t)} \int_{-\infty}^{\infty} e^{i\nu s} [p_1(s) \mp i\, p_2(s)]\, ds \quad \text{as} \quad x \to \pm\infty, \tag{21.18}$$

---

[1] See, e.g., N. I. MUSKHELISHVILI: Singular integral equations, §§ 43, 78. Groningen: Noordhoff 1953.

and the asymptotic form of the free surface by

$$\eta(x,t) \sim \mathrm{Im}_i \left\{ \frac{\sigma^2}{\varrho g^2} e^{-i(\nu x \mp \sigma t)} \int_{-\infty}^{\infty} e^{i\nu s} [p_1 \mp i p_2]\, ds \right\} \quad \text{as} \quad x \to \pm\infty. \tag{21.19}$$

From this last expression one can easily derive the average rate at which the pressure system is transferring energy to the fluid:

$$\left.\begin{aligned} W_{\mathrm{av}} &= \frac{1}{4}\frac{\sigma\nu}{\varrho g}\left\{\left|\int_{-\infty}^{\infty}(p_1 - i p_2)\, e^{i\nu s}\, ds\right|^2 + \left|\int_{-\infty}^{\infty}(p_1 + i p_2)\, e^{i\nu s}\, ds\right|^2\right\} \\ &= \frac{1}{2}\frac{\sigma\nu}{\varrho g}\left\{\left[\int_{-\infty}^{\infty} p_1 \cos\nu s\, ds\right]^2 + \left[\int_{-\infty}^{\infty} p_1 \sin\nu s\, ds\right]^2 + \right. \\ &\quad \left. + \left[\int_{-\infty}^{\infty} p_2 \cos\nu s\, ds\right]^2 + \left[\int_{-\infty}^{\infty} p_2 \sin\nu s\, ds\right]^2\right\}. \end{aligned}\right\} \tag{21.20}$$

The expression for $f(z)$ can be put into several different forms by changing variables and deforming the path of integration appropriately. Thus, if one introduces a new variable λ by $\nu(\zeta - z) = -\lambda(z-s)$ and deforms the resulting path to the x-axis, one obtains

$$f(z) = -j\frac{\sigma}{\pi\varrho g}\int_{-\infty}^{\infty} p(s)\, ds\, \mathrm{PV}\int_{0}^{\infty}\frac{e^{-ik(z-s)}}{k-\nu}\, dk + \frac{\sigma}{\varrho g} e^{-i\nu z}\int_{-\infty}^{\infty} p(s)\, e^{i\nu s}\, ds. \tag{21.21}$$

A different deformation of the path leads to

$$\left.\begin{aligned} f(z) = &-j\frac{\sigma}{\pi\varrho g}\int_{-\infty}^{x} p(s)\, ds\int_{0}^{\infty}\frac{e^{-\mu(z-s)}}{\mu - i\nu}\, d\mu - j\frac{\sigma}{\pi\varrho g}\int_{x}^{\infty} p(s)\, ds\int_{0}^{\infty}\frac{e^{\mu(z-s)}}{\mu + i\nu}\, d\mu + \\ &+ \frac{\sigma}{\varrho g}(1 + ij)\int_{-\infty}^{x} p(s)\, e^{-i\nu(z-s)}\, ds + \frac{\sigma}{\varrho g}(1 - ij)\int_{x}^{\infty} p(s)\, e^{-i\nu(z-s)}\, ds. \end{aligned}\right\} \tag{21.22}$$

For fluid of depth h an expression for the complex velocity potential analogous to (21.10) and (21.21) is

$$\left.\begin{aligned} f(z) = &-j\frac{\sigma}{\pi\varrho g}\int_{-\infty}^{\infty} ds\, p(s)\, \mathrm{PV}\int_{0}^{\infty}\frac{\cos k(z - s + ih)}{k\sinh kh - \nu\cosh kh}\, dk + \\ &+ \frac{\sigma}{\varrho g}\frac{\sinh m_0 h}{\nu h + \sinh^2 m_0 h}\int_{-\infty}^{\infty} p(s)\cos m_0 (z - s + ih)\, ds. \end{aligned}\right\} \tag{21.23}$$

One will find both the two- and the three-dimensional case of a periodic pressure distribution over infinitely deep water discussed in LAMB (1904, pp. 387–393). STOKER (1957, Chap. 4) discusses in considerable detail the two-dimensional problem of waves generated by a periodic uniform pressure applied over a finite interval.

β) Moving pressure distributions. In this section we shall suppose that a fixed pressure distribution is moving with a constant velocity c. Thus the motion may be treated as time-independent in a coordinate system moving with the pressure distribution. The boundary condition at the free surface is given by [see (11.3)]

$$\varphi_y(x,0,z)+\frac{1}{\nu}\varphi_{xx}(x,0,z)=\frac{c}{\varrho g}p_x(x,z),\quad \nu=\frac{g}{c^2}, \tag{21.24}$$

and the form of the free surface by

$$\eta(x,z)=\frac{c}{g}\varphi_x(x,0,z)-\frac{1}{\varrho g}p(x,z). \tag{21.25}$$

In addition, we shall assume vanishing of the fluid motion far ahead, i.e. as $x\to+\infty$, and the usual conditions appropriate to infinite or finite depth. p will be assumed to be absolutely integrable and to vanish for sufficiently large values of x^2+z^2; however, the latter condition can be weakened.

Results will be given without proof since their derivations are similar to those in Sects. 13γ and 21α. The results for two and three dimensions will be separated.

Three dimensions. The expression for the velocity potential for infinite depth of fluid can be given as follows:

$$\left.\begin{aligned}\varphi(x,y,z)=&\frac{1}{\pi^2\varrho c}\iint\limits_{-\infty}^{\infty}d\xi\,d\zeta\,p(\xi,\zeta)\int\limits_0^{\frac{1}{2}\pi}d\vartheta\sec\vartheta\,\mathrm{PV}\int\limits_0^{\infty}dk\frac{k\,e^{ky}}{k-\nu\sec^2\vartheta}\times\\&\times\sin[k(x-\xi)\cos\vartheta]\cos[k(z-\zeta)\sin\vartheta]-\\&-\frac{\nu}{\pi\varrho c}\iint\limits_{-\infty}^{\infty}d\xi\,d\zeta\,p(\xi,\zeta)\int\limits_0^{\frac{1}{2}\pi}d\vartheta\sec^3\vartheta\,e^{-\nu\sec^2\vartheta}\times\\&\times\cos[\nu(x-\xi)\sec\vartheta]\cos[\nu(z-\zeta)\sec^2\vartheta\sin\vartheta].\end{aligned}\right\} \tag{21.26}$$

The rate at which the pressure distribution is transferring energy to the fluid is given by

$$W=-\iint\limits_{-\infty}^{\infty}p(x,z)\,\varphi_y(x,-0,z)\,dx\,dz. \tag{21.27}$$

This may be computed directly from (21.26). The first term gives no contribution since it is an odd function of $x-\xi$ [cf. the evaluation of (20.66)]. The final result may be expressed as follows:

$$\left.\begin{aligned}W=&\frac{\nu^2}{\pi\varrho c}\int\limits_0^{\frac{1}{2}\pi}d\vartheta\sec^5\vartheta\iint\limits_{-\infty}^{\infty}dx\,dz\iint\limits_{-\infty}^{\infty}d\xi\,dz\,p(x,y)\,p(\xi,\zeta)\times\\&\times\cos\left(\nu\sec^2\vartheta\,[(x-\xi)\cos\vartheta+(z-\zeta)\sin\vartheta]\right).\\=&\frac{\nu^2}{\pi\varrho c}\int\limits_0^{\frac{1}{2}\pi}d\vartheta\sec^5\vartheta\,[P^2(\vartheta)+Q^2(\vartheta)],\\P(\vartheta)=&\iint\limits_{-\infty}^{\infty}dx\,dz\,p(x,z)\cos[\nu\sec^2\vartheta\,(x\cos\vartheta+z\sin\vartheta)],\\Q(\vartheta)=&\iint\limits_{-\infty}^{\infty}dx\,dz\,p(x,z)\sin[\nu\sec^2\vartheta\,(x\cos\vartheta+z\sin\vartheta)].\end{aligned}\right\} \tag{21.28}$$

If the pressure distribution is given in cylindrical coordinates, $p=p(R,\alpha)$, $x=R\cos\alpha$, $z=R\sin\alpha$, then one may express, say, $P(\vartheta)$ in the form

$$P(\vartheta)=\int\limits_0^{\infty}dR\int\limits_0^{2\pi}d\alpha\,R\,p(R,\alpha)\cos[\nu R\sec^2\vartheta\cos(\alpha-\vartheta)] \tag{21.29}$$

and a similar formula for $Q(\vartheta)$. If p depends only upon R, then $Q(\vartheta)\equiv 0$ and

$$P(\vartheta)=\int_0^\infty 2\pi R\,p(R)\,J_0(\nu R\sec^2\vartheta)\,dR. \tag{21.30}$$

If the fluid is of depth h, the velocity potential is given by

$$\left.\begin{aligned}\varphi(x,y,z)&=\frac{1}{\pi^2\varrho c}\iint_{-\infty}^{\infty} d\xi\,d\zeta\,p(\xi,\zeta)\int_0^{\frac{1}{2}\pi} d\vartheta\sec\vartheta\times\\&\times \mathrm{PV}\int_0^\infty dk\frac{k\cosh k(y+h)\operatorname{sech} kh}{k-\nu\sec^2\vartheta\tanh kh}\sin[k(x-\xi)\cos\vartheta]\cos[k(z-\zeta)\sin\vartheta]-\\&-\frac{1}{\pi\varrho c}\iint_{-\infty}^{\infty} d\xi\,d\zeta\,p(\xi,\zeta)\int_{\vartheta_0}^{\frac{1}{2}\pi} d\vartheta\sec\vartheta\frac{k_0\cosh k_0(y+h)\operatorname{sech} k_0h}{1-\nu h\sec^2\vartheta\operatorname{sech}^2 k_0h}\times\\&\qquad\times\cos[k_0(x-\xi)\cos\vartheta]\cos[k_0(z-\zeta)\sin\vartheta],\end{aligned}\right\} \tag{21.31}$$

where $k_0=k_0(\vartheta)$ is the positive real root of

$$k-\nu\sec^2\vartheta\tanh kh=0,\qquad \vartheta_0<\vartheta<\tfrac{1}{2}\pi,$$

and

$$\vartheta_0=\begin{cases}\operatorname{Arc\,cos}\sqrt{\nu h} & \text{if } \nu h\equiv\dfrac{gh}{c^2}<1,\\ 0 & \text{if } \nu h>1.\end{cases}$$

The rate of transfer of energy may again be computed from (21.27) and again only the second integral gives a nonvanishing contribution. The result may be expressed in several forms analogous to (21.28) to (21.30):

$$\left.\begin{aligned}W&=\frac{c}{\pi\varrho g}\int_{\vartheta_0}^{\frac{1}{2}\pi} d\vartheta\frac{k_0^3\cos\vartheta}{1-\nu h\sec^2\vartheta\operatorname{sech}^2 k_0h}\iint_{-\infty}^{\infty} dx\,dz\iint_{-\infty}^{\infty} d\xi\,d\zeta\,p(x,z)\,p(\xi,\zeta)\times\\&\qquad\times\cos\left(k_0[(x-\xi)\cos\vartheta+(z-\zeta)\sin\vartheta]\right),\\&=\frac{c}{\pi\varrho g}\int_{\vartheta_0}^{\frac{1}{2}\pi} d\vartheta\frac{k_0^3\cos\vartheta}{1-\nu h\sec^2\vartheta\operatorname{sech}^2 k_0h}\left[P^2(\vartheta)+Q^2(\vartheta)\right],\\P(\vartheta)&=\iint_{-\infty}^{\infty} p(x,z)\cos[k_0(x\cos\vartheta+z\sin\vartheta)]\,dx\,dy,\\Q(\vartheta)&=\iint_{-\infty}^{\infty} p(x,z)\sin[k_0(x\cos\vartheta+z\sin\vartheta)]\,dx\,dy.\end{aligned}\right\} \tag{21.32}$$

In cylindrical coordinates formulas (21.29) and (21.30) carry over to the present situation with ν replaced by k_0.

The asymptotic form of the free surface for either infinite or finite depth is much more complicated to analyze than for the stationary periodically oscillating pressure distribution of the preceding section. Although it is not strictly necessary to do so, it has been customary in this type of analysis to consider the special case of a "concentrated pressure point". To derive the velocity potential for the pressure point consider the pressure distribution defined by

$$p(R)=\left\{\begin{aligned}&\frac{p_0}{\pi R_0^2}, && R\leqq R_0,\\&0, && R>R_0.\end{aligned}\right\} \tag{21.33}$$

Substitute in (21.26) or (21.31) and take the limit as $R_0 \to 0$. Then (21.26) becomes

$$\left.\begin{aligned}\varphi(x, y, z) = \frac{p_0}{\pi^2 \rho c}\int_0^{\frac{1}{2}\pi} d\vartheta \sec\vartheta \,\mathrm{PV}\int_0^{\infty} dk \frac{k\, e^{ky}}{k-\nu\sec^2} \sin(k x\cos\vartheta)\cos(k z\sin\vartheta) - \\ - \frac{\nu p_0}{\pi\varrho c}\int_0^{\frac{1}{2}\pi} d\vartheta \sec^3\vartheta\, e^{y\nu\sec^2\vartheta}\cos(\nu x\sec\vartheta)\cos(\nu z\sec^2\vartheta\sin\vartheta).\end{aligned}\right\} \quad (21.34)$$

The velocity potential for the pressure point in fluid of finite depth is derived similarly. The equation representing the free surface may now be obtained immediately from (21.25):

$$\eta(x, z) = \frac{c}{g}\varphi_x(x, 0, z), \qquad x^2 + z^2 > 0. \quad (21.35)$$

The velocity potential (21.34) is very similar to that of a submerged source in steady motion [see Eq. (13.36)] and the method sketched in Sect. 13 for the derivation of the asymptotic expression (13.42) can be carried over directly to the moving pressure point.

The result, expressed in cylindrical coordinates, is as follows:

$$\left.\begin{aligned}
&\text{for}\quad 0 \leqq \alpha < \pi - \operatorname{Arc\,sin}\tfrac{1}{3} = \alpha_c: \\
&\qquad \eta(R, \alpha) = O(\nu R)^{-2}; \\
&\text{for}\quad \alpha = \alpha_c: \\
&\qquad \eta(R, \alpha_c) = -\frac{p_0 \nu^2}{\pi\varrho g} 2^{-\frac{3}{2}} 3^{\frac{4}{3}} \Gamma\left(\frac{1}{3}\right)(\nu R)^{-\frac{1}{3}} \sin\left(\frac{\sqrt{3}}{2}\nu R\right) + O\left((\nu R)^{-\frac{2}{3}}\right); \\
&\text{for}\quad \alpha_c < \alpha < \pi: \\
&\qquad \eta(R, \alpha) = \frac{p_0\nu^2}{\pi\varrho g}\sqrt{\frac{2\pi}{\nu R}}\frac{1}{[1-9\sin^2\alpha]^{\frac{1}{4}}}\left\{\sec^{\frac{5}{2}}\vartheta_1 \sin\left(\nu R\mu_1 - \frac{\pi}{4}\right) + \right. \\
&\qquad\qquad \left. + \sec^{\frac{5}{2}}\vartheta_2 \sin\left(\nu R\mu_2 + \frac{\pi}{4}\right)\right\} + O\left((\nu R)^{-1}\right); \\
&\text{for}\quad \alpha = \pi: \\
&\qquad \eta(R, \pi) = -\frac{p_0\nu^2}{\pi\varrho g}\sqrt{\frac{2\pi}{\nu R}}\sin\left(\nu R + \frac{\pi}{4}\right).
\end{aligned}\right\} \quad (21.36)$$

The variables $\vartheta_1, \vartheta_2, \mu_1, \mu_2$ are the same ones defined following (13.42), where certain properties are also given. For the pressure point Fig. 1a is not quite accurate as a description of the wave crests in the region $|\pi - \alpha| < \alpha_c$ since the phase in (21.36) has been shifted by $\frac{1}{2}\pi$; the wave crests in Fig. 1a should be moved back a distance $\frac{1}{2}\pi$. Also, in the neighborhood of $\alpha = \alpha_c$ the expressions in (21.36) are inaccurate; in this region η may be expressed in terms of Airy functions [see URSELL (1960)].

The wave pattern resulting from a moving pressure distribution has been the subject of many investigations, starting apparently with KELVIN (1906). His aim was to explain the typical wave pattern found behind a ship. The procedure is quite reasonable as a method for obtaining a qualitative prediction of a ship's wave pattern, since a moving ship has associated with it a pressure distribution around the wetted hull. The obvious disadvantage of the method is that it gives no connection between the geometry of the hull and the wave pattern. For this the "thin ship" approximation of Sect. 20β is better within its range of applicability. The single pressure point can be taken to represent approximately a ship moving at high speed (more accurately, at high Froude number $c/\sqrt{Lg}$, where L is the length), say a fast motor boat.

For detailed investigation of the asymptotic expression one should refer to HOGNER (1923), PETERS (1949), BARTELS and DOWNING (1955) (who do not restrict themselves to a pressure point) and STOKER (1957, Chap. 8). The necessary modifications for finite depth have been made by HAVELOCK (1908) and TETURÔ INUI (1936) and are described qualitatively in the discussion following (13.42). One can find an exposition of the theory of waves generated by moving pressure distributions in a report by LUNDE (1951b). Several papers by HAVELOCK (1909, 1914b, 1919, 1922) take up the wave resistance ($=W/c$) of a pressure distribution. HOGNER (1928) has also considered the wave resistance and gives essentially (21.28).

In the preceding considerations we have assumed that, as $x^2+z^2\to\infty$, $p(x, z)$ approached zero sufficiently quickly so that it might be represented as a Fourier integral. It is also possible to proceed somewhat differently, assume $p(x, y)$ periodic in one or both variables and use a Fourier series representation. This has been done, for example, by VOIT (1957a), who has considered for both infinite and finite depth a moving pressure distribution of the following form:

$$p(x, z) = \begin{cases} P(z)\sum\limits_{n=1}^{\infty} a_n \cos n k x, & |z| < h \\ 0, \quad |z| > h. \end{cases}$$

The waves resulting from a pressure point moving parallel to beaches forming angles of 30 and 45° with the surface have been treated by HANSON (1926); in the same paper he also treats the waves formed by a pressure point moving over a two-layered fluid. A detailed investigation of this last topic is given in a paper of SRETENSKII'S (1934).

Two dimensions. By introducing a stream function $\psi(x, y)$ and a complex potential $f=\varphi+i\psi$, the free surface boundary condition can be put into a form analogous to (21.14), namely,

$$\mathrm{Re}\{f'(x-i0)+i\nu f(x-i0)\} = \frac{1}{\varrho c}\,p(x), \qquad \nu=\frac{g}{c^2}. \tag{21.37}$$

In addition, we assume $|f'|$ bounded as $z\to\infty$ and also $\lim\limits_{x\to\infty}|f'|=0$. We shall assume $p(x)$ subject to the same limitations as in Sect. 21 α.

One may apply the same method of analysis to derive the following forms for the complex velocity potential:

$$\left.\begin{aligned} f(z) &= \frac{e^{-i\nu z}}{\pi i\varrho c}\int\limits_{-\infty}^{\infty} ds\, p(s)\int\limits_{\infty}^{z}\frac{e^{i\nu\zeta}}{\zeta-s}\,d\zeta \\ &= -\frac{1}{\pi i\varrho c}\int\limits_{-\infty}^{\infty} ds\, p(s)\,\mathrm{PV}\int\limits_{0}^{\infty}\frac{e^{-i\lambda(z-s)}}{\lambda-\nu}\,d\lambda-\frac{e^{-i\nu z}}{\varrho c}\int\limits_{-\infty}^{\infty} p(s)\,e^{i\nu s}\,ds \\ &= -\frac{1}{\pi i\varrho c}\int\limits_{-\infty}^{x} ds\, p(s)\int\limits_{0}^{\infty}\frac{e^{-\mu(z-s)}}{\mu-i\nu}\,d\mu-\frac{1}{\pi i\varrho c}\int\limits_{x}^{\infty} ds\, p(s)\int\limits_{0}^{\infty}\frac{e^{\mu(z-s)}}{\mu+i\nu}\,d\mu- \\ &\quad -\frac{2}{\varrho c}\,e^{-i\nu z}\int\limits_{x}^{\infty} p(s)\,e^{i\nu s}\,ds, \end{aligned}\right\} \tag{21.38}$$

where the path of integration for ζ in the first expression is taken in the lower half-plane. The rate at which the pressure distribution transfers energy to the fluid is easily found from formula (21.27) and the second expression for $f(z)$ to be

$$W = \frac{\nu}{\varrho c} \int_{-\infty}^{\infty} \int_{-\infty}^{\infty} p(x)\, p(\xi) \cos \nu (x - \xi)\, dx\, d\xi. \tag{21.39}$$

If the fluid is of depth h, the complex velocity potential may be given in a form analogous to the second formula of (21.37):

$$\left.\begin{aligned} f(z) = {} & \frac{\nu c}{\pi \varrho g} \int_{-\infty}^{\infty} ds\, p(s)\, \mathrm{PV} \int_{-\infty}^{\infty} dk \frac{\sin k (z - s + i h) \operatorname{sech} k h}{k - \nu \tanh k h} - \\ & - \frac{\nu c}{\varrho g} \int_{-\infty}^{\infty} p(s) \frac{\cos k_0 (z - s + i h) \operatorname{sech} k_0 h}{1 - \nu h \operatorname{sech}^2 k_0 h} ds, \end{aligned}\right\} \tag{21.40}$$

where k_0 is the real positive root of

$$k_0 - \nu \tanh k_0 h = 0$$

and exists only if $\nu h = g h/c^2 > 1$; if $\nu h \leqq 1$, the last term in (21.40) must be deleted. The rate at which the pressure is doing work upon the fluid is given by

$$W = \begin{cases} \dfrac{k_0^2 c}{\varrho g} \displaystyle\iint_{-\infty}^{\infty} p(x)\, p(\xi) \frac{\cos k_0 (x - \xi)}{1 - \nu h \operatorname{sech}^2 k_0 h} dx\, d\xi & \text{if } \nu h > 1, \\ 0 \quad \text{if } \nu h < 1. \end{cases} \tag{21.41}$$

The absence of the second term in (21.40) and the vanishing of W when $\nu h < 1$ correspond to the absence of an infinite train of trailing waves. A similar phenomenon occurs behind a moving singularity in two dimensions [cf. (13.46) to (13.48) and the following remarks].

For either (21.37) or (21.39) the form of the free surface can be written down immediately from the formula

$$\eta(x) = \frac{1}{c} \psi(x, 0). \tag{21.41}$$

We shall not carry out the details. The asymptotic form of the surface behind a two-dimensional "pressure point", or also a distributed pressure, is much easier in two than in three dimensions and we again omit a detailed statement. However, the problem has been treated by KELVIN (1905) and is discussed in LAMB (1932, § 242 to 245), both for infinite and finite depth. It is also discussed in the paper of PETERS (1949) already cited in connection with the three-dimensional problem.

Derivations of the complex velocity potential may be found in the papers of SRETENSKII (1934, 1940), SEDOV (1936), KOCHIN (1939) and HASKIND (1943a) already cited in connection with planing surfaces. We refer also to papers of DEAN (1947) and TIMMAN and VOSSERS (1955).

γ) Moving periodic pressure distributions. It is clearly possible to combine the cases considered in Sects. 21α and β and consider the waves resulting from a pressure distribution expressible in the form

$$p(x, z, t) = p_1(x - c t, z) \cos \sigma t + p_2(x - c t, z) \sin \sigma t,$$

where the coordinates are fixed in space. The resulting velocity potential will be analogous to (13.52) for the three-dimensional case, if one is dealing with a "pressure point".

We shall not reproduce the formulas here. However, the analogues of (13.49) and (13.53) for pressure distributions may be found in Eqs. (22.48) and (22.49) or in the cited report of LUNDE (1951b), and from these the required velocity potential may be found. For two-dimensional motion the details, carried out by this procedure, may be found in papers by KAPLAN (1957) and WU (1957).

22. Initial-value problems. In the special problems considered in Sects. 17 to 21, the dependence upon the time has been precipitated out, either by assuming the motion steady in a moving coordinate system or by assuming a harmonic dependence upon the time. In this section we shall consider motions in which the displacement and velocity of the surface are specified at some instant of time, say $t=0$, the motions of any solid boundaries are given for each instant $t \geqq 0$ (except at the very end where freely floating bodies are considered) and the pressure distribution over the free surface is a given function for $t \geqq 0$.

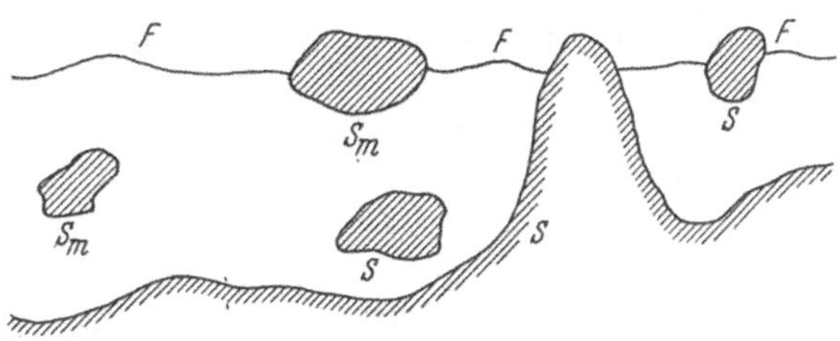

Fig. 27.

It is not usually possible in the most general situations to give explicit solutions for such problems. However, VOLTERRA (1934) has proved a uniqueness theorem and has shown how to reduce the problem to finding an appropriate GREEN's function. His results were later rediscovered and extended to a wider class of problems by FINKELSTEIN (1957) [see also STOKER (1957, Chap. 6)]. However, the use of GREEN's functions for initial-value problems extends back even earlier, at least to the papers of HADAMARD (1910, 1916) and BOULIGAND (1913). These theorems are discussed in Sect. 22α.

One of the classical problems in this category is associated with the names CAUCHY (1827) and POISSON (1815). In this problem the fluid is infinite in horizontal extent, without obstructions, and either infinitely deep or of uniform depth h. At the initial instant $t=0$, the form of the surface and its vertical velocity are given and one seeks the subsequent motion. Such problems have already been discussed at some length in Sect. 15. However, in that section interest centered upon investigation of certain aspects of the subsequent motion rather than upon obtaining the solution. In addition, the treatment of that section was limited to two-dimensional motion, although the methods could have been extended to three-dimensional motion.

The history of this problem, including an exposition of the methods used by various authors, is included in a paper by RISSER (1924, pp. 113—144). Another expository account can be found in VERGNE (1928, Chap. I). The problem is discussed here in Sect. 22β.

In Sect. 22γ several special initial-value problems are discussed.

α) Some general theorems. Let the fluid be bounded by the free surface F, fixed surfaces S and the surfaces of a finite number of bodies of bounded extent undergoing specified motions of small amplitude about equilibrium positions S_m (see Fig. 27). Let the pressure distribution on the free surface F, also be a given function $p(x, z, t)$. Furthermore, at time $t=0$ let the initial displacement and vertical velocity of the free surface be given functions

$$\eta(x, z, 0), \qquad \eta_t = (x, z, 0). \tag{22.1}$$

The boundary conditions to be satisfied by the velocity potential $\Phi(x, y, z, t)$ are [see Eq. (11.1)]

$$\left.\begin{aligned} &\Phi_{tt}(x, 0, z, t) + g\Phi_y(x, 0, z, t) = -\frac{1}{\varrho} p_t(x, z, t) \quad \text{on} \quad F, \\ &\Phi_n = 0 \quad \text{on} \quad S, \\ &\Phi_n = V_n(t) \quad \text{on} \quad S_m, \\ &\Phi_t(x, 0, z, 0) = -g\eta(x, z, 0) - \frac{1}{\varrho} p(x, z, 0) \quad \text{on} \quad F, \\ &\Phi_y(x, 0, z, 0) = \eta_t(x, z, 0) \quad \text{on} \quad F. \end{aligned}\right\} \tag{22.2}$$

Here F means that part of the plane $y=0$ occupied by fluid when everything is at rest. In addition, it will be assumed that, for each t, there is a bound B and a distance r_0 such that $|\Phi|$, $|\Phi_t|$, $|\text{grad}\,\Phi|$ and $|\text{grad}\,\Phi_t|$ are each less than B for $x^2+y^2+z^2>r_0^2$.

Let us now suppose that it is possible to find a source function G of the following nature:

$$G(x, y, z; \xi, \eta, \zeta; t, \tau) = \frac{1}{r} + H(x, y, z; \xi, \eta, \zeta; t, \tau), \tag{22.3}$$

where as usual $r^2 = (x-\xi)^2 + (y-\eta)^2 + (z-\zeta)^2$ and H is harmonic for $y \leqq 0$, and in addition G satisfies

$$\left.\begin{aligned} &G(x, y, z; \xi, \eta, \zeta; t, t) = G_t(x, y, z; \xi, \eta, \zeta; t, t) = 0, \\ &G_n(x, y, z; \xi, \eta, \zeta; t, \tau) = 0 \quad \text{for} \quad (x, y, z) \text{ on } S \text{ and for all } t. \end{aligned}\right\} \tag{22.4}$$

This function has already been constructed in two cases. If there are no fixed boundaries and the fluid is infinitely deep, the function defined in (13.49) satisfies the conditions after slight modifications: replace (a, b, c) by (ξ, η, ζ), set $m(t)=1$, and extend the definition of Φ [of (13.49)] to negative t by $\Phi(x, y, z, -t) = \Phi(x, y, z, t)$. Then we take

$$G(x, y, z; \xi, \zeta; t, \tau) = \Phi(x, y, z; t-\tau) = G(x, y, z; \xi, \eta, \zeta; \tau, t).$$

Similarly, the function defined in (13.53) allows one to construct G when the fixed boundary consists of a horizontal bottom at $y=-h$. For the first G, FINKELSTEIN (1957, Appendix) has shown that G is $O(R^{-2})$ and G_R and G_y are $O(R^{-3})$ as $R\to\infty$, where $R^2=(x-\xi)^2+(z-\zeta)^2$; for the second G, FINKELSTEIN (1957, § 3) has shown that G, G_R and G_y are $o\left(\exp\left(\frac{-\pi}{2h}+\varepsilon\right)R\right)$ as $R\to\infty$ for arbitrary $\varepsilon>0$.

Now apply GREEN's theorem to the functions Φ_t and G and the region of fluid bounded by the surfaces S_m, the fixed boundaries S and a large sphere Ω of radius ϱ and center at the origin, where ϱ is chosen large enough to include all the surfaces S_m. Only parts of F, S and Ω will serve as bounding surfaces, and we shall call these parts F', S' and Ω', respectively. Then

$$\Phi_t(x, y, z, t) = \frac{1}{4\pi} \iint\limits_{F'+S_m+S'+\Omega'} [G(\xi, \eta, \zeta; x, y, z; t, \tau)\,\Phi_{t\nu}(\xi, \eta, \zeta, t) - \Phi_t G_\nu]\,d\sigma, \tag{22.5}$$

where ν is the exterior normal. The right-hand side is actually independent of τ since τ enters only through the function H which is harmonic. The integral over

S' vanishes since both Φ_n and G_n are zero on S. We shall assume that the behavior of G and Φ as $R\to\infty$ is such that the integral over Ω' vanishes as $\varrho\to\infty$. If the fluid is of bounded extent, the situation considered by VOLTERRA (1934), this presents, of course, no difficulty. In the two cases for which G has been given above, it has been shown by FINKELSTEIN that this is true. For finite depth the proof presents no difficulty once the estimates for G are obtained; for infinite depth the analysis is more troublesome and we refer to his paper or to STOKER (1957, pp. 193/194) for proof. After letting $\varrho\to\infty$, one then has

$$\Phi_t(x,y,z,t)=\frac{1}{4\pi}\iint\limits_F [G\Phi_{t\eta}-G_\eta\Phi_t]\,d\sigma+\frac{1}{4\pi}\iint\limits_{S_m}[G\Phi_{t\nu}-G_\nu\Phi_t]\,d\sigma. \tag{22.6}$$

In the integral over F we may replace G_η by $-g^{-1}G_{tt}$ because of the boundary condition at F. Now interchange t and τ and integrate with respect to τ between limits 0 and t. This gives, following an integration by parts,

$$\left.\begin{aligned}
\Phi(x,y,z,t)-\Phi(x,y,z,0)&=\frac{1}{4\pi}\iint\limits_F\left\{\left[G\Phi_\eta+\frac{1}{g}\Phi_tG_t\right]\Big|_0^t-\int\limits_0^t\left[G_t\Phi_\eta+\frac{1}{g}\Phi_{tt}G_t\right]d\tau\right\}+\\
&\quad+\frac{1}{4\pi}\int\limits_0^t d\tau\iint\limits_{S_m}[G\Phi_{tn}-G_n\Phi_t]\,d\sigma\\
&=\frac{1}{4\pi}\iint\limits_F\Big\{G(\xi,0,\zeta,x,y,z;t,t)\,\Phi_y(\xi,0,\zeta,t)+\frac{1}{g}\Phi_t(\ldots,t)\,G_t(\ldots;t,t)-\\
&\quad-G(\ldots;0,t)\,\Phi_y(\ldots,0)-\frac{1}{g}\Phi_t(\ldots,0)\,G_t(\ldots;0,t)+\\
&\quad+\frac{1}{\varrho g}\int\limits_0^t G_t(\ldots;\tau,t)\,p_t(\xi,\zeta,\tau)\,d\tau\Big\}\,d\xi\,d\zeta+I
\end{aligned}\right\} \tag{22.7}$$

where I stands for the last integral. (G_t always represents the derivative with respect to the seventh variable.) Recalling the properties of G in (22.4), one finds

$$\left.\begin{aligned}
\Phi(x,y,z,t)&=\Phi(x,y,z,0)+\frac{1}{4\pi}\iint\limits_F\Big\{-G(\xi,0,\zeta;x,y,z;0,t)\,\eta_t(\xi,\zeta,0)+\\
&\quad+G_t(\xi,0,\zeta,x,y,z;0,t)\left[\eta(\xi,\zeta,0)+\frac{1}{\varrho g}p(\xi,\zeta,0)\right]+\\
&\quad+\frac{1}{\varrho g}\int\limits_0 G(\xi,0,\zeta;x,y,z,\tau,t)\,p_t(\xi,\zeta,\tau)\,d\tau\Big\}\,d\xi\,d\zeta+I\\
&=\Phi(x,y,z,0)+\frac{1}{4\pi}\iint\limits_F\Big\{-G_t(\xi,0,\zeta,x,y,z;0,t)\,\eta_t(\xi,\zeta,0)+\\
&\quad+G_t(\xi,0,\zeta,x,y,z;0,t)\,\eta(\xi,\zeta,0)-\\
&\quad-\frac{1}{\varrho g}\int\limits_0 G_{tt}(\xi,0,\zeta,x,y,z;\tau,t)\,p(\xi,\zeta,\tau)\,d\tau\Big\}\,d\xi\,d\zeta+I,
\end{aligned}\right\} \tag{22.8}$$

where $\Phi(x,y,z,0)$ is determined up to an additive constant as the solution to a Neumann problem, since $\Phi_n(x,y,z,0)$ is given on all boundaries and bounded at infinity. In the integral I we note that $\Phi_{tn}=V_n'(t)$ is known on S_m, but Φ_t is not.

If there are no moving bodies in the fluid, then the integral I is not present and Φ is determined by the initial displacement and velocity of the free surface and the given pressure distribution over it. This is VOLTERRA's result as extended to unbounded fluids by FINKELSTEIN. If surfaces S_m are present, one may still use (22.8) to derive an integral equation in the same way that (16.13) was derived. For as (x, y, z) is made to approach a point (x_0, y_0, z_0) of S_m,

$$\frac{1}{4\pi}\iint_{S_m} G_\nu(\xi,\eta,\zeta,x,y,z;\tau,t)\,\Phi_t(\xi,\eta,\zeta,\tau)\,d\sigma \to \frac{1}{2}\Phi_t(x_0,y_0,z_0,t) + \frac{1}{4\pi}\iint_{S_m} G_\nu(\xi,\eta,\zeta,x_0,y_0,z_0;\tau,t)\,\Phi_t\,d\sigma\,^1.$$

Thus, after carrying out the integration with respect to τ, one has an integral equation for $\Phi(x, y, z, t)$ for each value of $t>0$. This may be used to find the value of Φ, and hence Φ_t, on the surface S_m, providing that the integral equation can be solved. One may then use (22.8) to determine $\Phi(x, y, z, t)$ for all values of (x, y, z) in the fluid. The integral equation has the same appearance as (22.8) except that the first two terms have coefficients $\frac{1}{2}$ and (x, y, z) is understood to be a point of S_m. This further extension of VOLTERRA's analysis is also due to FINKELSTEIN.

Uniqueness of $\Phi(x, y, z, t)$, at least up to an additive constant, may be proved a follows. Let Φ_1 and Φ_2 be two solutions satisfying the boundary conditions. Then $\Phi=\Phi_1-\Phi_2$ satisfies (22.8) with f, F, p and V_n all identically zero, i.e.

$$\Phi(x, y, z, t) = \text{const} - \frac{1}{4\pi}\int_0^t d\tau \iint_{S_m} G_n\,\Phi_t\,d\sigma.$$

If we assume that G_n is $O(R^{-1-\varepsilon})$ as $R\to\infty$, then Φ_t and grad Φ will have the same behavior and the integrals we shall write below may be shown to exist. As has been mentioned above, G_n vanishes much quicker than is required in the cases when the fluid is infinitely deep and when the fixed surface consists of a horizontal bottom; if the fluid is bounded in extent, no such condition is necessary to make the integrals converge.

Consider then, following VOLTERRA,

$$\begin{aligned}\Omega &= \frac{1}{2}\frac{\partial}{\partial t}\iint_F \frac{1}{g}\Phi_t^2\,d\sigma \\ &= \iint_F \frac{1}{g}\Phi_t\Phi_{tt}\,d\sigma = -\iint_F \Phi_t\Phi_y\,d\sigma \\ &= -\iint_{F+S+S_m} \Phi_t\Phi_n\,d\sigma\end{aligned}$$

since Φ_n vanishes on S and S_m. Now apply GREEN's theorem and denote the volume occupied by fluid by T:

$$\Omega = -\iiint_T \operatorname{grad}\Phi_t\cdot\operatorname{grad}\Phi\,d\tau = -\frac{1}{2}\frac{\partial}{\partial t}\iiint_T (\operatorname{grad}\Phi^2)\,d\tau.$$

Hence

$$\frac{\partial}{\partial t}\left\{\iint_F \frac{1}{g}\Phi_t^2\,d\sigma + \iiint_T (\operatorname{grad}\Phi)^2\,d\tau\right\} = 0$$

and

$$\iint_F \frac{1}{g}\Phi_t^2\,d\sigma + \iiint_T (\operatorname{grad}\Phi)^2\,d\tau = \text{const}. \tag{22.9}$$

[1] Cf. O.D. KELLOGG: Foundations of potential theory, p. 167. Berlin: Springer 1929.

Since $\Phi_n = 0$ on F, S and S_m for $t=0$, $\Phi(x, y, z, 0) = C$, a constant; hence grad $\Phi = 0$ for $t=0$. Also $\Phi_t(x, y, z, 0) = 0$. Hence the constant in (22.9) is zero and Φ_t and grad Φ vanish for all t. Thus $\Phi(x, y, z, t) =$ const and the solution of the initial-value problem is determined up to a constant.

β) The Cauchy-Poisson problem. In this classical problem of water-wave theory, the pressure over the free surface is constant, say zero, the fluid is infinitely deep or bounded below by a horizontal bottom, no obstructions are present and the initial displacement and velocity of the free surface are given. The two- and three-dimensional cases will be separated in order to illustrate different methods of approach.

Three dimensions. The velocity potential may be obtained directly from (22.8) after setting $p(x, z, t)$ and I equal to zero. However, the explicit expressions for G and G_t are needed. As was noted in Sect. 22α, these can be written down immediately from (13.49) for infinite depth and (13.53) for depth h. The resulting expressions, after setting $\eta = 0$, are as follows:

infinite depth:

$$\left.\begin{aligned} G(x, y, z; \xi, 0, \zeta; 0, t) &= 2\int_0^\infty [1 - \cos(\sqrt{g k}\, t)]\, e^{k y} J_0(kR)\, dk, \\ G_t(x, y, z; \xi, 0, \zeta; 0, t) &= -2\int_0^\infty \sin(\sqrt{g k}\, t)\, e^{k y} J_0(kR) \sqrt{g k}\, dk; \end{aligned}\right\} \quad (22.10)$$

depth h:

$$\left.\begin{aligned} G(x, y, z; \xi, 0, \zeta; 0, t) &= 2\int_0^\infty [1 - \cos(\sqrt{g k \tanh k h}\, t)] \frac{\cosh k(y+h)}{\sinh k h} J_0(kR), \\ G_t(x, y, z; \xi, 0, \zeta; 0, t) &= -2\int_0^\infty \sqrt{g k \tanh k h}\, \sin(\sqrt{g k \tanh k h}\, t) \frac{\cosh k(y+h)}{\sinh k h} J_0(kR), \end{aligned}\right\} \quad (22.11)$$

where $R^2 = (x-\xi)^2 + (z-\zeta)^2$.

There still remains to find $\Phi(x, y, z, 0)$ where

$$\Phi_y(x, 0, z, 0) = \eta_t(x, z, 0)$$

and

$$\lim_{y \to -\infty} \Phi_y(x, y, z, 0) = 0 \quad \text{or} \quad \Phi_y(x_1 - h, z, 0) = 0.$$

The solution of these two problems is well known:

infinite depth:

$$\left.\begin{aligned} \Phi(x, y, z, 0) &= \frac{1}{2\pi} \iint_{-\infty}^{\infty} \frac{\eta_t(\xi, \zeta, 0)}{[(x-\xi)^2 + y^2 + (z-\zeta)^2]^{\frac{1}{2}}}\, d\xi\, d\zeta \\ &= \frac{1}{2\pi} \iint_{-\infty}^{\infty} \eta(\xi, \zeta, 0)\, d\xi\, d\zeta \int_0^\infty e^{k y} J_0(kR)\, dk; \end{aligned}\right\} \quad (22.12)$$

depth h:

$$\Phi(x, y, z, 0) = \frac{1}{2\pi} \iint_{-\infty}^{\infty} \eta_t(\xi, \zeta, 0)\, d\xi\, d\zeta \int_0^\infty \frac{\cosh k(y+h)}{\sinh k h} J_0(kR)\, dk, \quad (22.13)$$

where R is defined as above.

Substituting the several expressions in (22.8), one obtains the expressions for the velocity potential:

infinite depth:

$$\left.\begin{aligned}\Phi(x,y,z,t) &= \frac{1}{2\pi}\iint\limits_{-\infty}^{\infty}\eta_t(\xi,\zeta,0)\,d\xi\,d\zeta\int\limits_0^{\infty} e^{ky}\cos\sigma t\,J_0(kR)\,dk - \\ &\quad -\frac{1}{2\pi}\iint\limits_{-\infty}^{\infty}\eta(\xi,\zeta,0)\,d\xi\,d\zeta\int\limits_0^{\infty}\sigma\,e^{ky}\sin\sigma t\,J_0(kR)\,dk,\\ \sigma^2 &= gk;\end{aligned}\right\}\qquad(22.14)$$

depth h:

$$\left.\begin{aligned}\Phi(x,y,z,t) &= \frac{1}{2\pi}\iint\limits_{-\infty}^{\infty}\eta_t(\xi,\zeta,0)\,d\xi\,d\zeta\int\limits_0^{\infty}\frac{\cosh k(y+h)}{\sinh kh}\cos\sigma t\,J_0(kR)\,dk - \\ &\quad -\frac{1}{2\pi}\iint\limits_{-\infty}^{\infty}\eta(\xi,\zeta,0)\,d\xi\,d\zeta\int\limits_0^{\infty}\sigma\frac{\cosh k(y+h)}{\sinh kh}\sin\sigma t\,J_0(kR)\,dk,\\ \sigma^2 &= gk\tanh kh.\end{aligned}\right\}\qquad(22.15)$$

The equations describing the free surface are as follows:

infinite-depth:

$$\left.\begin{aligned}\eta(x,z,t) &= \frac{1}{2\pi g}\iint\limits_{-\infty}^{\infty}\eta_t(\xi,\zeta,0)\,d\xi\,d\zeta\int\limits_0^{\infty}\sigma\sin\sigma t\,J_0(kR)\,dk + \\ &\quad +\frac{1}{2\pi g}\iint\limits_{-\infty}^{\infty}\eta(\xi,\zeta,0)\,d\xi\,d\zeta\int\limits_0^{\infty}\sigma^2\cos\sigma t\,J_0(kR)\,dk,\\ \sigma^2 &= gk;\end{aligned}\right\}\qquad(22.16)$$

depth h:

$$\left.\begin{aligned}\eta(x,z,t) &= \frac{1}{2\pi g}\iint\limits_{-\infty}^{\infty}\eta_t(\xi,\zeta,0)\,d\xi\,d\zeta\int\limits_0^{\infty}\sigma\sin\sigma t\coth kh\,J_0(kR)\,dk + \\ &\quad +\frac{1}{2\pi g}\iint\limits_{-\infty}^{\infty}\eta(\xi,\zeta,0)\,d\xi\,d\zeta\int\limits_0^{\infty}\sigma^2\cos\sigma t\coth kh\,J_0(kR)\,dk,\\ \sigma^2 &= gk\tanh kh.\end{aligned}\right\}\qquad(22.17)$$

It has been shown by KOCHIN (1935) that the integrals with respect to k in (22.16) can be evaluated. Consider the integral

$$K = \int\limits_0^{\infty}\sigma^{-1}\sin\sigma t\,J_0(kR)\,dk,\qquad \sigma^2 = gk.\qquad(22.18)$$

Then the first integral with respect to k in (22.16) is $-K_{tt}$ and the second one is $-K_{ttt}$ To evaluate K make first the following change of variables:

$$\varkappa^2 = kR,\qquad \omega^2 = gt^2/4R.\qquad(22.19)$$

Then

$$K = \frac{2}{\sqrt{gR}} \int_0^\infty \sin 2\omega\varkappa\, J_0(\varkappa^2)\, d\varkappa$$

$$= \frac{2}{\sqrt{gR}} \int_0^\infty d\varkappa \int_0^1 \frac{\sin 2\omega\varkappa \cos v\varkappa^2}{\sqrt{1-v^2}}\, dv$$

$$= \frac{2}{\sqrt{gR}} \int_0^\infty d\varkappa \int_0^1 [\sin(2\omega\varkappa + v\varkappa^2) + \sin(2\omega\varkappa - v\varkappa^2]\, \frac{dv}{\sqrt{1-v^2}}.$$

In the first integral let $u = v\varkappa + \omega$, in the second let $u = v\varkappa - \omega$. Then

$$K = \frac{2}{\sqrt{gR}} \int_{+\omega}^\infty du \int_0^1 \sin\left(\frac{u^2-\omega^2}{v}\right) \frac{dv}{v\sqrt{1-v^2}} - \frac{2}{\sqrt{gR}} \int_{-\omega}^\infty du \int_0^1 \sin\left(\frac{u^2-\omega^2}{v}\right) \frac{dv}{v\sqrt{1-v^2}}$$

$$= + \frac{2}{\sqrt{gR}} \int_{-\omega}^{\omega} du \int_1^\infty \frac{\sin(\omega^2 - u^2)\, v'}{\sqrt{v'^2 - 1}}\, dv' \quad (v' = 1/v),$$

$$= \frac{4}{\sqrt{gR}} \frac{\pi}{2} \int_0^\omega J_0(\omega^2 - u^2)\, du,$$

and, after setting $u = \sqrt{2}\omega \sin \frac{1}{2}\vartheta$,

$$K = \frac{\sqrt{2}\pi}{\sqrt{gR}}\, \omega \int_0^{\frac{1}{2}\pi} J_0\left(2\frac{\omega^2}{2}\cos\vartheta\right) \cos\frac{1}{2}\vartheta\, d\vartheta.$$

Finally, from an identity in WATSON'S *Bessel functions* [§ 5.43, Eq. (1)] one finds

$$K(\omega) = \frac{\pi^2}{\sqrt{2gR}}\, \omega\, J_{\frac{1}{4}}\left(\frac{1}{2}\omega^2\right) J_{-\frac{1}{4}}\left(\frac{1}{2}\omega^2\right). \tag{22.20}$$

In order to use the results in (22.16) one needs the first three derivatives with respect to t. Since

$$\frac{\partial}{\partial t} = \frac{1}{2}\sqrt{\frac{g}{R}}\, \frac{\partial}{\partial \omega},$$

the derivatives can be computed by taking derivatives with respect to ω and multiplying by an appropriate factor. After some rather tedious computation one finds

$$\left.\begin{aligned}
\frac{\partial}{\partial t} K(\omega) &= -\frac{1}{2\sqrt{2}}\pi^2 \frac{\omega^2}{R}\left[J_{\frac{1}{4}}\left(\frac{1}{2}\omega^2\right) J_{\frac{3}{4}}\left(\frac{1}{2}\omega^2\right) - J_{-\frac{1}{4}}\left(\frac{1}{2}\omega^2\right) J_{-\frac{3}{4}}\left(\frac{1}{2}\omega^2\right)\right],\\
\frac{\partial^2}{\partial t^2} K(\omega) &= -\frac{1}{2}\pi^2\omega^3 \sqrt{\frac{g}{2R^3}}\left[J_{\frac{1}{4}}\left(\frac{1}{2}\omega^2\right) J_{-\frac{1}{4}}\left(\frac{1}{2}\omega^2\right) + J_{\frac{3}{4}}\left(\frac{1}{2}\omega^2\right) J_{-\frac{3}{4}}\left(\frac{1}{2}\omega^2\right)\right],\\
\frac{\partial^3}{\partial t^3} K(\omega) &= -\frac{\pi^2 g}{2R^2}\omega^2\Big[J_{\frac{1}{4}}\left(\frac{1}{2}\omega^2\right) J_{-\frac{1}{4}}\left(\frac{1}{2}\omega^2\right) -\\
&\quad - \omega^2\left\{J_{\frac{1}{4}}\left(\frac{1}{2}\omega^2\right) J_{\frac{3}{4}}\left(\frac{1}{2}\omega^2\right) - J_{-\frac{1}{4}}\left(\frac{1}{2}\omega^2\right) J_{-\frac{3}{4}}\left(\frac{1}{2}\omega^2\right)\right\}\Big].
\end{aligned}\right\} \tag{22.21}$$

These are KOCHIN'S formulas, but derived somewhat differently from his original paper; still another derivation may be found in KOCHIN, KIBEL and ROZE (1948, Chap. 8, § 21). Similar formulas for (22.17) do not seem to have been discovered.

It should be noted that in the final form of (22.16) the dependence upon t is through the dimensionless variable $\omega^2 = gt^2/4R$. Hence, if one examines the contribution to the surface profile from a given locality, say the neighborhood of (ξ, ζ), then a given phase of this contribution, say a maximum, will be described by $gt^2/4R = \text{const}$; i.e., the phase is moving away from (ξ, ζ) with constant acceleration proportional to g. The amplitude of the contribution is modulated by either $R^{-\frac{3}{2}}$ or R^{-2} according as one is considering the first or second summand in (22.16). KOCHIN'S 1935 paper is of some methodological interest inasmuch as he started his analysis with dimensional considerations. This method will be introduced for the two-dimensional case.

One may obtain without great difficulty series expansions for the k-integrals in (22.14) and (22.16), as was first done by CAUCHY and POISSON. We refer to LAMB'S *Hydrodynamics* (1932, § 255) for the derivation and exact expressions. They can also be derived from the the known expansions for $J_{\frac{1}{4}}$, etc., as can asymptotic expressions for large ω. One may also carry out an analysis of the changing shape of the surface profile following the methods of Sect. 15.

It is evident that one can solve explicitly other similar initial-value problems for which the GREEN'S function can be given. For example, the method of images allows one to give an explicit solution for various cases when vertical walls are present as boundaries. Such cases have been considered by RISSER (1925). The Cauchy-Poisson problem in the presence of a vertical half-plane, $z = 0$, $x > 0$, has been treated by BOIKO (1938), but by more complex methods.

Two dimensions. Rather than repeat the methods used for three-dimensional motion, we shall introduce a method making use of the complex potential and thus special to two-dimensional motion. It is analogous to the method used in deriving (13.28).

Let $f(z, t) = \Phi(x, y, t) + i\Psi(x, y, t)$ be the complex velocity potential. The initial conditions will be taken in the form

$$-\frac{1}{g}\operatorname{Re} f_t(x - i0, 0+) = \eta(x, 0), \quad -\operatorname{Im} f'(x - i0, 0+) = \eta_t(x, 0). \tag{22.22}$$

Let us consider infinite depth first. For $t > 0$ we assume that $f(z, t)$ is regular and $|f'| < M(t)$, $|f_{tt}| < M(t)$ for $y < 0$ and that both f' and f_{tt} approach zero as $y \to -\infty$. Consider now the function

$$G(z, t) = f_{tt}(z, t) + i g f'(z, t). \tag{22.23}$$

From the assumptions about f it follows that, for $t > 0$, $G(z, t)$ is regular for $y < 0$, that $|G| < B(t)$ for $y < 0$ and that $G \to 0$ as $y \to -\infty$. Moreover, it follows from the condition at the free surface, (11.5), that $\operatorname{Re} G(x - i0, t) = 0$. Hence, the definition of G may be extended into the upper half-plane by defining

$$G(x + iy) = -\overline{G(x - iy)}. \tag{22.24}$$

But then since G is regular and bounded in the whole finite z-plane, it follows from LIOUVILLE'S theorem that $G = \text{const}$; the constant must equal zero from the assumed behavior as $y \to -\infty$. Hence the fundamental differential equation for the Cauchy-Poisson problem in two dimensions is

$$f_{tt}(z, t) + i g f'(z, t) = 0, \qquad t > 0, \tag{22.25}$$

an observation usually credited to LEVI-CIVITA (cf. TONOLO, 1913).

Let us now find the analogous equation for finite depth. The function f will be assumed regular in the strip $0>y>-h$. The boundary condition on the bottom is

$$\operatorname{Im} f'(x-ih,t)=0. \tag{22.26}$$

Hence f may be extended analytically into the strip $-2h<y<-h$ by defining

$$f(x-i(y+2h))=\overline{f(x+iy)}, \quad 0>y\geqq -h. \tag{22.27}$$

We may also, as before, extend the function $f_{tt}+igf'$ into the strip $h\geqq y\geqq 0$. The condition $\operatorname{Re}\{f_{tt}+igf'\}=0$ for $y=0$ implies $\operatorname{Re}\{f_{tt}-igf'\}=0$ for $y=-2h$. Hence the function $f_{tt}-igf'$ can be extended by reflection into the strip $-3h\leqq y\leqq -2h$. Now consider the function

$$\left.\begin{aligned} H(z,t) &= [f_{tt}(z+ih,t)+f_{tt}(z-ih,t)]+ig[f'(z+ih,t)-f'(z-ih,t)] \\ &= \{f_{tt}(z+ih,t)+igf'(z+ih,t)\}+\{f_{tt}(z-ih,t)-igf'(z-ih,t)\}. \end{aligned}\right\} \tag{22.28}$$

As a result of the various extended regions of definition, one may verify easily that H is defined for all z in the strip $-2h<y<0$ and is regular there. Moreover, it follows that

$$H(x-ih,t)=0; \tag{22.29}$$

for from (22.27) it follows that the two pairs of summands in the first form of (22.28) are real for $z=x-ih$, whereas from the boundary conditions at $y=0$ and $y=-2h$ it follows that the terms in curly brackets in the second form of (22.28) have zero real parts. Since $H(z,t)$ is regular in the strip $0>y>-2h$ and vanishes on $y=-h$, it must vanish identically in the strip. Hence we have the following differential-difference equation of Cisotti (1918):

$$f_{tt}(z+ih,t)+f_{tt}(z-ih,t)+ig[f'(z+ih,t)-f'(z-ih,t)]=0. \tag{22.30}$$

Let us now turn to the solution of (22.25) with initial conditions (22.22). We shall follow closely an exposition of Sedov's (1948, 1957). However, the idea of the derivation is Kochin's (1935) and, in fact, really goes back to Tonolo (1913). The use of dimensional analysis can be extended to the three-dimensional problem; this was also done by Kochin.

We first remark that the initial-value problem can be solved by solving it for two special cases of (22.22), namely, first with $\eta(x,0)\equiv 0$ and then with $\eta_t(x,0)\equiv 0$. The sum of these two solutions will satisfy (22.22). Next we note that $\eta(x,0)$ has the dimension "length" and $\eta_t(x,0)$ the dimension "velocity", and that the solution f in each of the two initial-value problems will be proportional to some typical parameters associated with $\eta(x,0)$ or $\eta_t(x,0)$, respectively. Let us suppose that a is such a parameter with dimension $L^p T^q$ and that f is proportional to a. Since f has dimension L^2T^{-1} and g has dimension LT^{-2}, the Π theorem of dimensional analysis then states that f can be expressed as follows:

$$f(z,t)=a z^{\alpha} g^{\beta}\chi\left(\frac{igt^2}{4z}\right), \tag{22.31}$$

where

$$\alpha=\tfrac{3}{2}-p-\tfrac{1}{2}q, \quad \beta=\tfrac{1}{2}(q+1). \tag{22.32}$$

(The factor $i/4$ in the argument of χ is chosen for later convenience.) Now substitute (22.31) into (22.25). One finds after some computation that

$$f_{tt}+igf'=iaz^{\alpha-1}g^{\beta+1}[\zeta\chi''(\zeta)+(\tfrac{1}{2}-\zeta)\chi'(\zeta)+\alpha\zeta]=0, \tag{22.33}$$

where $\zeta = i g t^2/4z$. The differential equation obtained by setting the expression in square brackets equal to zero determines χ in terms of confluent hypergeometric functions:

$$\chi(\zeta) = A\, {}_1F_1(-\alpha, \tfrac{1}{2}; \zeta) + B\zeta^{\frac{1}{2}}\, {}_1F_1(\tfrac{1}{2} - \alpha, \tfrac{3}{2}; \zeta). \tag{22.34}$$

From this it follows that

$$\left.\begin{aligned} f(z,t;\alpha) &= a z^\alpha g^\beta\left[A\, {}_1F_1\left(-\alpha, \frac{1}{2}; \frac{i g t^2}{4z}\right) + B\left(\frac{i g t^2}{4z}\right)^{\frac{1}{2}} {}_1F_1\left(\frac{1}{2}, -\alpha, \frac{3}{2}; \frac{i g t^2}{4z}\right)\right] \\ &= A f_1(z,t;\alpha) + B f_2(z,t;\alpha). \end{aligned}\right\} \tag{22.35}$$

Remembering that

$${}_1F_1(\gamma, \delta; 0) = 1, \qquad {}_1F_1'(\gamma, \delta; 0) = \gamma/\delta,$$

one may easily derive the following:

$$\left.\begin{aligned} f(z,0) &= A f_1(z,0) = A\, a g^\beta z^\alpha, \\ f'(z,0) &= A f_1'(z,0) = A\, a \alpha g^\beta z^{\alpha-1}, \\ f_t(z,0) &= B f_{2t}(z,0) = \tfrac{1}{2} B\, a\, i^{\frac{1}{2}} g^{\beta+\frac{1}{2}} z^{\alpha-\frac{1}{2}}. \end{aligned}\right\} \tag{22.36}$$

The solution (22.35) may be further generalized by replacing t by $t - t_0$ and z by $z - x_0$ (i.e., by a different choice of the dimensionless variable ζ). One may then further superimpose these solutions. For the purpose at hand it will be sufficient to retain $t_0 = 0$. Then we may form the solution

$$f(z,t) = \int_{-\infty}^{\infty} A(x_0)\, f_1(z - x_0, t; \alpha_1)\, d x_0 + \int_{-\infty}^{\infty} B(x_0)\, f_2(z - x_0, t; \alpha_2)\, d x_0. \tag{22.37}$$

One finds from (22.36) that

$$\left.\begin{aligned} f(z,0) &= a_1 g^{\beta_1} \int_{-\infty}^{\infty} A(x_0)\, (z - x_0)^{\alpha_1}\, d x_0, \\ f'(z,0) &= a_1 \alpha_1 g^{\beta_1} \int_{-\infty}^{\infty} A(x_0)\, (z - x_0)^{\alpha_1 - 1}\, d x_0, \\ f_t(z,0) &= \tfrac{1}{2} a_2\, i^{\frac{1}{2}} g^{\beta_2 + \frac{1}{2}} \int_{-\infty}^{\infty} B(x_0)\, (z - x_0)^{\alpha_2 - \frac{1}{2}}\, d x_0. \end{aligned}\right\} \tag{22.38}$$

Let us now make some special choices of a, and hence of α and β. As a parameter describing the initial profile of the surface take

$$a_2 = \int_{-\infty}^{\infty} \eta(x,0)\, dx; \tag{22.39}$$

as a parameter describing the initial velocity distribution take

$$a_1 = \int_{-\infty}^{\infty} dx \int_{-\infty}^{x} \eta_t(\xi, 0)\, d\xi. \tag{22.40}$$

Then a_1 has the dimension $L^3 T^{-1}$, corresponding to $\alpha_1 = -1$, $\beta_1 = 0$, and a_2 the dimension L^2, corresponding to $\alpha_2 = -\frac{1}{2}$, $\beta_2 = \frac{1}{2}$. With these choices of α_1 and α_2 in (22.37) we take

$$\left.\begin{aligned} A(x_0) &= \frac{-1}{a_1 \pi} \int_{-\infty}^{x_0} \eta_t(\xi, 0)\, d\xi, \\ B(x_0) &= \frac{-2}{a_2 \pi\, i^{\frac{3}{2}}}\, \eta(x_0, 0). \end{aligned}\right\} \tag{22.41}$$

Then the last two equations of (22.38) become (after an integration by parts in the first one)

$$f'(z,0)=\frac{1}{\pi}\int_{-\infty}^{\infty}\frac{\eta_t(x_0,0)}{x_0-z}\,dx_0,$$

$$f_t(z,0)=\frac{g}{\pi i}\int_{-\infty}^{\infty}\frac{\eta(x_0,0)}{x_0-z}\,dx_0.$$

From the Plemelj-Sokhotskii theorem we have

$$\left.\begin{aligned}f'(x-i\,0,0)&=-i\eta_t(x,0)+\frac{1}{\pi}\,\mathrm{PV}\int_{-\infty}^{\infty}\frac{\eta_t(x_0,0)}{x_0-x}\,dx_0,\\ f_t(x-i\,0,0)&=-g\eta(x,0)+\frac{g}{\pi i}\,\mathrm{PV}\int_{-\infty}^{\infty}\frac{\eta(x_0,0)}{x_0-x}\,dx_0.\end{aligned}\right\}\tag{22.42}$$

Thus the initial conditions (22.22) are satisfied.

There remains to point out that for the special choices of $\alpha_1=-1$ and $\alpha_2=-\frac{1}{2}$ the corresponding confluent hypergeometric functions in (22.35) may be expressed in terms of Fresnel integrals or integrals of these. In fact, if we write (22.37) the form

$$f(z,t)=\int_{-\infty}^{\infty}\Omega_1(z-x_0,t)\int_{-\infty}^{x_0}\eta_t(\xi,0)\,d\xi\,dx_0+\int_{-\infty}^{\infty}\Omega_2(z-x_0,t)\,\eta(x_0,0)\,dx_0,\tag{22.43}$$

then

$$\left.\begin{aligned}\Omega_1(z,t)&=-\frac{2i}{z}\,e^{-i\frac{\pi}{2}\omega^2}\,\omega-\frac{2i}{z}\int_0^{\omega}du\int_0^{u}e^{-i\frac{\pi}{2}(v^2-\omega^2)}\,dv,\\ \Omega_2(z,t)&=2i\sqrt{\frac{2g}{\pi z}}\int_0^{\omega}e^{-i\frac{\pi}{2}(u^2-\omega^2)}\,du,\end{aligned}\right\}\tag{22.44}$$

where

$$\omega^2=\frac{g\,t^2}{2\pi z}.$$

One should also consult the discussion in LAMB's Hydrodynamics (1932, § 238, 239), where graphs are given which display the behavior of the surface profile corresponding to an initial elevation concentrated in the neighborhood of one point, i.e., essentially $-g^{-1}\Omega_{2t}(x-i\,0,t)$, and to a concentrated impulse, i.e., essentially $-g^{-1}\Omega_{1t}(x-i\,0,t)$. However, general aspects of the development of the surface profile have already been discussed in Sect. 15α.

It should be noted that the velocity potential (22.37) represents a much wider class of time-dependent gravity-wave motions than does (22.43). The initial-value problems corresponding to other values of α have been determined by SEDOV (1948) but the discussion will not be repeated here.

A class of solutions of (22.30) analogous to that found by SEDOV for (22.25) does not seem to have been given in the published literature. CISOTTI (1920) expands $f(z,t)$ in a power series in t, thus replacing (22.30) by a recursive set of difference equations. We refer to the original paper for his discussion of this set of equations.

In (15.22) we have already given the velocity potential and surface profile corresponding to a given initial profile; the derivation was based upon a Fourier analysis of the initial profile and the result was valid for either finite or infinite depth. The same procedure may be used for an initial velocity distribution. The combined result for the complex velocity potential and surface profile is given below in such a way as to include the possible presence of surface tension:

infinite depth:

$$\left.\begin{aligned} f(z,t) &= \frac{1}{\pi}\int_0^\infty dk\int_{-\infty}^\infty d\xi\,\frac{1}{k}\left[-\sigma(k)\,\eta(\xi,0)\sin\sigma t+\eta_t(\xi,0)\cos\sigma t\right]e^{ik(z-\xi)},\\ \eta(x,t) &= \frac{1}{\pi}\int_0^\infty dk\int_{-\infty}^\infty d\xi\left[\eta(\xi,0)\cos\sigma t+\frac{\sigma}{g k}\,\eta_t(\xi,0)\sin\sigma t\right]\cos k(x-\xi),\end{aligned}\right\} \tag{22.45}$$

where

$$\sigma^2 = g k + T k^3/\varrho;$$

depth h:

$$\left.\begin{aligned} f(z,t) &= \frac{1}{\pi}\int_0^\infty dk\int_{-\infty}^\infty d\xi\,\frac{1}{k\sinh k h}\times\\ &\quad\times\left[-\sigma(k)\,\eta(\xi,0)\sin\sigma t+\eta_t(\xi,0)\cos\sigma t\right]\cos k(z-\xi+i h),\\ \eta(x,t) &= \frac{1}{\pi}\int_0^\infty dk\int_{-\infty}^\infty d\xi\times\\ &\quad\times\left[\eta(\xi,0)\cos\sigma t+\frac{\sigma(k)}{g k\tanh k h}\,\eta_t(\xi,0)\sin\sigma t\right]\cos k(x-\xi),\end{aligned}\right\} \tag{22.46}$$

where

$$\sigma^2 = (g k + T k^3/\varrho)\tanh k h.$$

If $T=0$, the coefficients of $\eta_t(\xi, 0)$ in the formulas for $\eta(x, t)$ reduce to σ^{-1}. When $T=0$, (22.45) is, of course, another form of (22.43).

It has already been indicated in Sect. 15 that the Cauchy-Poisson problem can also be solved for superposed fluids. SRETENSKII (1955) has investigated a further generalization in which the two fluids are each flowing with constant velocities for $t<0$ and then when $t=0$ a disturbance is suddenly created at the interface.

γ) Some other time-dependent problems. It is possible to solve a number of initial-value problems either by using Eq. (22.8) or by using the time-dependent Green functions (13.49) or (13.53) directly. The special situations treated, below fall roughly into the following four categories: wave motions resulting from a pressure distribution suddenly brought into existence at time $t=0$; waves resulting from a body set into motion at time $t=0$; waves resulting from an underwater explosion or a sudden movement of the bottom (tsunamis); and waves resulting from an initially displaced freely floating body.

Time-dependent pressure distributions. Suppose that the fluid is undisturbed for $t<0$ and that starting with $t=0$ the pressure over the free surface is a given function $p(x, z, t)$. The consequent motion of the fluid may be easily obtained, for this is just the problem formulated in (22.2) if we put $\eta(x, z, 0) = \eta_t(x, z, 0) = 0$. Formula (22.8) then gives the velocity potential in the form

$$\Phi(x,y,z,t) = \frac{-1}{4\pi\varrho g}\iint_F d\xi\,d\zeta\int_0^t G_{tt}(\xi,0,\zeta,x,y,z;\tau,t)\,p(\xi,\zeta,\tau)\,d\tau + I. \tag{22.47}$$

In the two situations for which explicit GREEN'S functions have been given, Eqs. (22.10) and (22.11), we may give explicit solutions for Φ:

infinite depth:

$$\Phi(x,y,z,t) = \frac{-1}{2\pi\varrho}\iint\limits_{-\infty}^{\infty} d\xi\, d\zeta \int\limits_0^t p(\xi,\zeta,\tau)\, d\tau \int\limits_0^\infty \cos\left(\sqrt{g k}(\tau-t)\right) e^{k y} J_0(k R)\, k\, d k; \tag{22.48}$$

depth h:

$$\left.\begin{aligned}\Phi(x,y,z,t) = \frac{-1}{2\pi\varrho}\iint\limits_{-\infty}^{\infty} d\xi\, d\zeta \int\limits_0^t p(\xi,\zeta,\tau)\, d\tau \int\limits_0^\infty \left(\cos\sqrt{g k \tanh k h}\,(\tau-t)\right) \times \\ \times \frac{\cosh k(y+h)}{\cosh k h} J_0(k R)\, k\, d k,\end{aligned}\right\} \tag{22.49}$$

where, as usual, $R^2 = (x-\xi)^2 + (z-\zeta)^2$.

The velocity potential for a moving pressure distribution is obtained from these expressions simply by letting

$$p(\xi,\zeta,\tau) = p_0(\xi - c\tau, \zeta, \tau).$$

If $p_0(\xi - c\tau, \zeta, \tau) = p_0(\xi - c\tau, \zeta)\cos\sigma\tau$ the resulting Φ is the velocity potential for a steadily moving pressure distribution of oscillating strength. LUNDE (1951b) has investigated the special case when $p(\xi,\zeta,\tau) = p\left(\sqrt{(\xi - c\tau)^2 + \zeta^2}\right)$ and has shown that as $t\to\infty$ the expressions (22.48) and (22.49), after a change to moving coordinates, approach asymptotically to the expressions (21.26) or (21.31) properly modified for circular symmetry (the assumed symmetry is not essential). The computation is interesting but will not be carried out here. This procedure for obtaining (21.26) or (21.31) yields the velocity potential without necessitating the extra boundary condition requiring the motion to vanish as $x\to+\infty$.

As was mentioned in connection with the solution of the Cauchy-Poisson problem, the GREEN'S function for some other simple configurations can be found by the method of reflection.

The complex velocity potentials for two-dimensional motion which correspond to (22.48) and (22.49) are as follows:

infinite depth:

$$f(z,t) = \frac{-1}{\pi\varrho}\int\limits_{-\infty}^{\infty} d\xi \int\limits_0^t p(\xi,\tau)\, d\tau \int\limits_0^\infty \cos\left(\sqrt{g k}(\tau-t)\right) e^{-i k(z-\xi)}\, d k; \tag{22.50}$$

depth h:

$$f(z,t) = \frac{-1}{\pi\varrho}\int\limits_{-\infty}^{\infty} d\xi \int\limits_0^t p(\xi,\tau)\, d\tau \int\limits_0^\infty \cos\left(\sqrt{g k \tanh k h}\,(\tau-t)\right) \frac{\cos k(z-\xi+i h)}{\cosh k h}\, d k. \tag{22.51}$$

Certain special cases have been investigated in more detail. STOKER (1953) [see also WURTELE (1955)] has treated the motion resulting when a pressure distribution constant in time for $t>0$ is suddenly applied to a uniformly moving stream of depth h. The velocity potential may be obtained from (22.51) by taking $p(\xi,\tau) = p_0(\xi - c\tau)$ and transferring to moving coordinates. His aim, as was that of LUNDE in the computations described above, was to show that the potential (21.40) can be derived without a special assumption about its behavior as $x\to+\infty$. The same can be carried through with (22.50) to derive (21.38). If one assumes $p(\xi,\tau) = p_1(\xi)\cos\sigma\tau + p_2(\xi)\sin\sigma\tau$, then one may also derive (21.21) or (21.23) from (22.49) or (22.50), respectively, as asymptotic expressions for large t without having to impose a radiation condition. VOIT (1957b) has investigated the surface profile for large t when $p(\xi, \tau) = p(\tau)$ for $\xi < c\tau < cT$, $p(\xi, \tau) = 0$ for $\xi \geqq c\tau$ or for $\tau > T$.

Waves resulting when a body is set into motion. Many of the problems solved in Sects. 17 to 20 by means of source distributions can be formulated as initial-value problems and solved by the same procedure if one uses the appropriate time-dependent GREEN's function. We shall consider briefly several examples, omitting details.

In (19.28) the velocity potential was given for the motion resulting from an oscillator in a wall, described by (19.26). It was assumed there that a steady situation had been reached in which the motion was purely harmonic in the time. Suppose instead that the motion of the oscillator described by (19.26) is to start at $t=0$ and that for $t<0$ the oscillator and fluid are at rest. It is easy to verify that the time-dependent velocity $\Phi(x, y, z, t)$ potential is still given by (19.28) if for the GREEN's function G one uses (13.50) with $m=1$. The last term in (13.50) will give the transient aspects of the motion. For two-dimensional motion the time-dependent wave-maker has been considered by KENNARD (1949), who also gives an estimate of time necessary for the transient terms to die out.

In (20.65) the velocity potential has been given for a "thin" ship moving with constant velocity c; it is assumed there that a steady state has been reached. Let us now suppose the same ship to move with velocity $c(t)$, $t>0$, but that it and the fluid have been at rest for $t<0$. As in (20.64) we take a coordinate system moving with the ship. Then from (20.26) it follows that the velocity potential $\Phi(x, y, z, t)$ must satisfy the boundary condition

$$\Phi_z(x, y, \pm 0, t) = \mp c(t) F_x(x, y).$$

A GREEN's function enabling us to construct Φ can be easily obtained from either (13.49) or (13.53). However, let us take $c(t) = C$, a constant, for $t>0$, i.e. we suppose the ship to attain instantaneously its final velocity. The GREEN's function for this situation has already been written out explicitly in (13.51). Setting there $u_0 = c$, $a_0 = \xi$, $b_0 = \eta$, $c_0 = \zeta$ and calling the resulting function $G(x, y, z, \xi, \eta, \zeta, t)$, the velocity potential for the problem at hand is

$$\Phi(x, y, z, t) = \frac{c}{2\pi} \iint_{S_0} G(x, y, z, \xi, \eta, 0, t) F_x(\xi, \eta)\, d\xi\, d\eta. \tag{22.52}$$

Having found Φ, one may then compute the force upon the ship and obtain formulas analogous to (20.67) or (20.69). The computations for infinite depth was originally made by SRETENSKII (1937); LUNDE (1951a) gives an exposition of this result and extends it to include thin ships moving in an infinite expanse of fluid of depth h and down the center of a canal of width b and depth h. In these computations c is allowed to be an arbitrary function of t. We refer to LUNDE's paper for the results.

HAVELOCK (1948, 1949) has considered the accelerated motion of a submerged horizontal circular cylinder in fluid of infinite depth. The complex velocity potential is expanded in a Laurent series about the center, starting with a dipole. In order to satisfy the other boundary conditions, one makes use of (13.54) to obtain singularities of the proper sort. The boundary condition on the circle then yields an infinite set of equations for determining the coefficients in the Laurent series. After finding as many terms as seems necessary for a suitable approximation, one may compute the force on the cylinder. HAVELOCK has carried this out for the first two singularities [a slight inconsistency in the approximation is corrected in MARUO (1957)] and has made numerical computations for an impulsive start and for a constant acceleration. Consider an impulsive start with instantaneous acceleration to constant speed c, and let the cylinder have radius a and its center be submerged to depth h. Then the two leading terms in the resistance are given by R_0, the steady-state resistance given in Eq. (20.52), and by the transient term

$$R_1 = \frac{1}{2}\pi g \varrho a^4 \nu^2 \left(\frac{\pi}{\nu c t}\right)^{\frac{1}{2}} e^{-\frac{1}{2}\nu h} \sin\left(\frac{1}{4}\nu c t - \frac{\pi}{4}\right), \quad \nu = \frac{g}{c^2}. \tag{22.53}$$

Fig. 28, taken from MARUO (1957), shows $(R_0+R_1)/R_0$ plotted against ct/h for $c/\sqrt{gh}=1$.

An exposition of the theory of accelerated motion of submerged bodies is given by MARUO (1957, Chap. 3). Both two- and three-dimensional problems in fluid of infinite or finite depth are considered. We note that the use of KOCHIN's H-function may be extended with no difficulty to time-dependent motion; this has been done by MARUO and earlier by HASKIND (1946b).

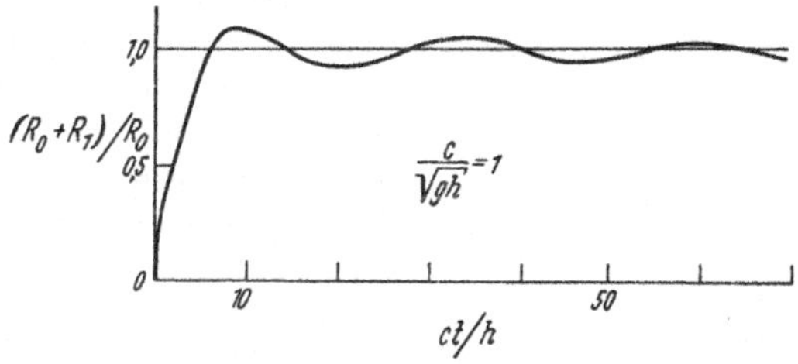

Fig. 28.

An investigation of PALM (1953) also fits into the category of problems under consideration. In considering flow over an uneven bottom in Sect. 20α, it was necessary to impose an upstream boundary condition in order to obtain uniqueness if the velocity is subcritical. In order to avoid this extra condition he formulated an initial-value problem in which the fluid is at rest and the bottom suddenly starts to move. The asymptotic expression for large t in a coordinate system moving with the bottom agrees with the results in Sect. 20α.

Tsunamis and submarine explosions. A tsunami is an ocean wave originating from a sudden upheaval or recession of the ocean floor. If one assumes an ocean of uniform depth h and if the disturbance occurs in a region S of the bottom, one may approximate this situation by the boundary-value problem in which

$$\Phi_y(x,-h,z,t) = \begin{cases} V(x,z,t), & 0<t<T, \ (x,z) \text{ in } S, \\ 0, & \text{otherwise}. \end{cases} \tag{22.54}$$

If the time-interval of the disturbance is short (i.e., if gT^2/h is small), the solution for Φ is given approximately by distributing over S sources of a form easily derived from (13.53). In fact, in (13.53) let $a=\xi$, $b=-h$, $c=\zeta$, and let $2m(t) = 2m(\xi,\zeta,t) = -\Phi_y(\xi,-h,\zeta,t)/2\pi$; denote the resulting function by $\Phi_s(x,y,z,\xi,-h,\zeta,t)$. Then

$$\Phi(x,y,z,t) = \iint_S \Phi_s(x,y,z,\xi,-h,\zeta,t)\,d\xi\,d\zeta \tag{22.55}$$

is the approximate solution. If one assumes $V(x,z,t)=V(x,z)$ for $0<t<T$, then Φ_s takes the following simple form for $t>T$:

$$\left.\begin{aligned} \Phi_s(x,y,z,\xi,-h,\zeta,t) = -\frac{1}{2\pi} V(\xi,\zeta) \int_0^\infty \frac{\cosh k(y+h)\, J_0(kR)}{\sinh kh \cosh kh} \times \\ \times [\cos\sigma(t-T) - \cos\sigma t]\,dk, \end{aligned}\right\} \tag{22.56}$$

where $\sigma^2 = gk \tanh kh$. If the deformation is assumed to take place so quickly that one may let $T \to 0$ while keeping $VT = L(\xi, \zeta)$ constant (i.e., keeping the same total deformation), (22.56) becomes

$$\Phi_s(x, y, z; \xi, -h, \zeta, t) = \frac{-1}{2\pi} L(\xi, \zeta) \int_0^\infty \frac{\cosh k(y+h)\, J_0(kR)}{\sinh kh \cosh kh} \sigma(k) \sin \sigma(k)\, t\, dk, \quad (22.57)$$

and the solution (22.55) is no longer approximate for the formulated problem.

A further approximation may be obtained by assuming the area of disturbance to be so localized that one may assume the whole disturbance to originate at one point, say $(0, -h, 0)$. Then (22.55) becomes simply (22.57) with L replaced by $Q = \iint L\, d\xi\, d\zeta$ and $R^2 = x^2 + z^2$. Although this may be a reasonable approximation to the explosion of a mine on the ocean floor, it is not in general suitable for a tsunami since the diameter of the region of disturbance in the latter may be many times the depth of fluid.

A comparison of (22.55), with (22.57) for Φ_s, with (22.15) shows that one may expect the same qualitative behavior for tsunamis as for waves resulting from an initial deformation of the free surface. In fact, if one makes the substitution (22.19) in the expressions for the surface profiles, they become the following, respectively, for the initially displaced surface and the tsunami:

$$\left.\begin{aligned} \eta(x, z, t) &= \frac{1}{\pi} \iint_{-\infty}^{\infty} \eta(\xi, \zeta, 0) \frac{1}{R^2}\, d\xi\, d\zeta \int_0^\infty \varkappa^3 J_0(\varkappa^2) \cos\left(2\omega\varkappa \sqrt{\tanh \varkappa^2 \frac{h}{R}}\right) d\varkappa, \\ \eta(x, z, t) &= \frac{1}{\pi} \iint_{-\infty}^{\infty} L(\xi, \zeta) \frac{1}{R^2}\, d\xi\, d\zeta \int_0^\infty \varkappa^3 \frac{J_0(\varkappa^2)}{\cosh^2\left(\varkappa^2 \frac{h}{R}\right)} \cos\left(2\omega\varkappa \sqrt{\tanh \varkappa^2 \frac{h}{R}}\right) d\varkappa. \end{aligned}\right\} \quad (22.58)$$

One may study the development of η along the lines worked out in Sect. 15 for two dimensions.

Many of the investigations of tsunamis have been devoted to an examination of the profile for a given type of initial bottom disturbance. The classical papers on tsunamis are by SANO and HASEGAWA (1915) and SYONO (1936). They have recently been investigated by TAKAHASHI (1942, 1945, 1947), ICHIYE (1950), GAZARYAN (1955) and others. Since the shape of the bottom and the configuration of the shore are of obvious importance in a geophysical application of the theory, much recent attention has been given to this aspect of the propagation of tsunamis. GRIGORASH (1957a) has given a brief survey of the literature together with a substantial bibliography.

The waves resulting from an exploding submerged mine may be represented approximately by using the velocity potential for a source whose strength $m(t)$ has the form of a square pulse of duration T. One may then determine Φ from either (13.49) or (13.53). If one assumes T very small and forms the limit as $T \to 0$ while keeping $mT = Q$ constant, one finds easily the following expressions for Φ:

infinite depth:

$$\Phi = 2Q \int_0^\infty e^{k(y+b)} J_0(kR)\, \sigma(k) \sin \sigma t\, dk, \quad \sigma^2 = gk; \quad (22.59)$$

depth h:

$$\Phi = 2Q \int_0^\infty \frac{\cosh k(h+b) \cosh k(y+h)}{\sinh kh \cosh kh} \sigma(k) \sin \sigma t\, dk, \quad \sigma^2 = gk \tanh kh. \quad (22.60)$$

Again one may examine the development of the surface profile by the methods developed in Sect. 15.

Investigations of waves generated by a sudden pulse of the above or similar sort have been made by LAMB (1913, 1922) and TERAZAWA (1915); both took the fluid to be infinitely deep. SRETENSKII (1950, 1949) has made a similar study when the source (two-dimensional) is situated on the bottom of a rectangular basin and within a fluid layer covering a solid sphere. SEZAWA (1929a, b) has included the effect of compressibility of the fluid.

One should recognize that such studies can elucidate only a small part of the phenomena associated with underwater explosions. An investigation of the migration and oscillation of the explosion bubble requires different analytical methods. Furthermore, if the explosion is too violent the linearized boundary condition on the free surface may not be a useful approximation.

Freely floating bodies. The motion of a freely floating body following an initial displacement is of considerable interest and practical importance, but also leads to a difficult mathematical problem. Uniqueness of solution follows from the argument in Sect. 22α. For the sake of perspicuity let us restrict ourselves to motion constrained to be vertical, i.e., heaving motion. Then from (19.59) and (19.62) the boundary conditions to be satisfied on the surface of the body in its equilibrium position, S_0, are

$$\Phi_n(x, y, z, t) = \dot{y}_1(t)\, n_y(x, y, z), \quad (x, y, z) \text{ on } S_0, \tag{22.61}$$

$$M\ddot{y}_1(t) + \varrho g I^A y_1(t) = -\varrho \iint_{S_0} \Phi_t(\xi, \eta, \zeta, t)\, n_y(\xi, \eta, \zeta)\, d\sigma. \tag{22.62}$$

(The notation is explained in Sect. 19β.) In addition Φ must satisfy the free-surface condition

$$\Phi_{tt}(x, 0, z, t) + g\,\Phi_y(x, 0, z, t) = 0 \tag{22.63}$$

and initial conditions, say

$$\Phi_t(x, 0, z, 0) = \Phi_y(x, 0, z, 0) = 0, \tag{22.64}$$

$$\dot{y}_1(0) = \dot{y}_{10}, \qquad y_1(0) = y_{10}. \tag{22.65}$$

As in many previous cases one may reduce the problem to the solution of an integral equation by use of a GREEN'S function. In (13.49) replace (a, b, c) by (ξ, η, ζ) and $m(t)$ by $\gamma(\xi, \eta, \zeta, t)$; denote the resulting function by Φ_s:

$$\left.\begin{aligned} &\Phi_s(x, y, z, \xi, \eta, \zeta, t) = \gamma(\xi, \eta, \zeta, t)\left[\frac{1}{r} - \frac{1}{r_1}\right] + \\ &\quad + 2\int_0^\infty (g k)^{\frac{1}{2}}\, e^{k(y+\eta)} J_0(kR)\, dk \int_0^t \gamma(\xi, \eta, \zeta, \tau) \sin\left[(g k)^{\frac{1}{2}}(t-\tau)\right] d\tau. \end{aligned}\right\} \tag{22.66}$$

We now attempt to express Φ by the integral

$$\Phi(x, y, z, t) = \iint_{S_0} \Phi_s(x, y, z, \xi, \eta, \zeta, t)\, d\sigma, \tag{22.67}$$

for then (22.63) and (22.64) will be satisfied. One should note especially that the relation of Φ to γ is more complicated here than in problems typified by (16.12), for the past history of γ is involved in Φ_s. The conditions (22.61) and (22.62) now become

$$-2\pi\gamma(x, y, z, t) + \iint_{S_0} \Phi_{sn}(x, y, z, \xi, \eta, \zeta, t)\, d\sigma = \dot{y}_1(t)\, n_y(x, y, z), \tag{22.68}$$

$$M\ddot{y}_1(t) + \varrho g I^A y_1(t) = -\varrho \iint_{S_0} d\sigma \iint_{S_0} dS\, \Phi_{st}(x, y, z, \xi, \eta, \zeta, t)\, n_y(\xi, \eta, \zeta), \tag{22.69}$$

where γ also enters into the equations through Φ_s. The two equations form a pair of coupled integro-differential equations for γ and y_1. It is evident that one can probably not hope for an analytic solution even for simple configurations.

SRETENSKII (1937b), for two dimensions, and later HASKIND (1946b) for three dimensions simplified the problem further by assuming the body to be "thin", i.e., if the surface is given by $z = \pm F(x, y)$, by replacing the boundary condition (22.61) by

$$\Phi_z(x, y, \pm 0, t) = \mp \dot{y}_1(t) F_y(x, y) \tag{22.70}$$

and S_0 by the projection of S_0 on the plane $z=0$ [cf. (20.26) and (20.64)]. With this further assumption one can immediately satisfy (22.68) by taking

$$\gamma(x, y, t) = -\frac{1}{2\pi} \dot{y}_1(t) F_y(x, y). \tag{22.71}$$

Eq. (22.69) then becomes an integro-differential equation for $y_1(t)$.

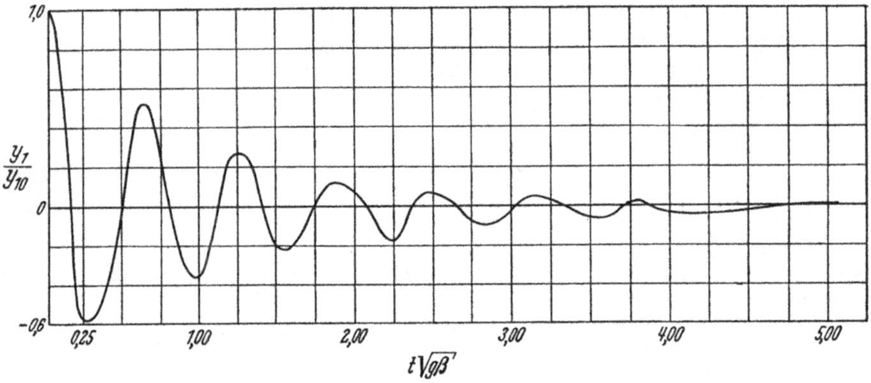

Fig. 29.

The procedure is open to some objection in that the substituted boundary condition (22.70) does not seem to fit into the general perturbation scheme as developed in Sects. 10α, 19α and 20β. It is thus not clear what physical problem really corresponds to the mathematical problem. However, this seems to be the closest anyone has come to reducing the equations to a manageable form. SRETENSKII solved his resulting integro-differential equation numerically for a surface described by $F(y) = l\, e^{\beta|y|}$, where

$$l = \frac{\pi g}{800} = 3.85 \text{ cm}, \qquad \beta = \frac{100}{g} = 0.104 \text{ cm}^{-1}.$$

The resulting graph of y_1/y_{10} is shown in Fig. 29 with a dimensionless abscissa $t\sqrt{g\beta}$. In spite of the questionableness of the formulation of the problem, the graph is instructive in showing the difference between a damped harmonic oscillation and the solution of SRETENSKII's integro-differential equation. Approximate methods of solution to the problem which assume that the fluid motion at any instant is independent of its past history lead to damped harmonic oscillations.

23. Waves in basins of bounded extent. The study of wave motion in a basin presents no special difficulties not already encountered earlier, and has a particular interest because of the many opportunities of observing such waves. Certain general aspects of the problem may be considered as being contained in earlier sections. For example, the general discussion of initial-value problems in Sect. 22α applies to motion in a basin. However, in order to make use of the results, in particular of Eq. (22.8), in constructing a solution, one must have prior knowledge of the time-dependent GREEN's function for the geometric boundary. Although the method of images can be used together with (13.49) or (13.53) to construct the GREEN's function for certain simple configurations, an explicit analytic solution is generally not available.

The time-dependent problem has also been approached in another manner by HADAMARD (1910, 1916), who derived an integro-differental equation for the function $\eta(x, y, t)$ describing the free surface. HADAMARD's short notes have been worked out by BOULIGAND (1912, 1926, 1927) and developed further. Certain of BOULIGAND's investigations indicate that singularities which may occur at the intersection of the plane $y=0$ with the basin walls are a result of linearizing the free-surface boundary condition. For an exact statement one should consult the original papers. There is a brief treatment of HADAMARD's equation in VERGNE (1928, § 10, 14). MOISEEV (1953) has developed a treatment of the time-dependent problem which generalizes somewhat the method used in Sect. 23α.

In Sect. 23α we give some general theorems concerning motions periodic in time, and another solution of the initial-value problem. In Sect. 23β wave motions for several special configurations of the boundary are given. In Sect. 23γ the theory of wave motion in movable basins is considered.

α) *Periodic waves in basins: general theorems.* If the motion is periodic in time, the velocity potential may be found by solving a Fredholm integral equation, obtained after introduction of an appropriate GREEN's function. Assume $\Phi(x, y, z, t) = \varphi(x, y, z)\cos(\sigma t + \tau)$; then φ must satisfy the boundary conditions

$$\left.\begin{aligned} &\varphi_y(x, 0, z) - \nu\varphi(x, 0, z) = 0, \quad (x, z) \text{ in } F, \quad \nu = \sigma^2/g, \\ &\varphi_n = 0, \quad (x, y, z) \text{ on } S, \end{aligned}\right\} \tag{23.1}$$

where F is the part of the plane $y=0$ occupied by the free surface at rest and S is the surface of the basin. Let $G(x, y, z, \xi, \eta, \zeta)$ be the GREEN's function for NEUMANN's problem, i.e.,

$$G = \frac{1}{r} + G_0,$$

where G_0 is regular in the region occupied by fluid and G satisfies the conditions

$$G_n = c \text{ on } S, \quad G_y(x, 0, z, \xi, \eta, \zeta) = c \text{ on } F, \tag{23.2}$$

where c is an arbitrary nonzero constant. In addition, in order to make the definition of φ unique we require

$$\iint_{S+F} \varphi\, d\sigma = 0.$$

It then follows from GREEN's theorem that

$$\varphi(x, y, z) = \frac{1}{4\pi}\iint_F G\varphi_y\, d\xi\, d\zeta = \frac{\nu}{4\pi}\iint_F G\varphi(\xi, 0, \zeta)\, d\xi\, d\zeta. \tag{23.3}$$

If one now lets $y \to 0$, one obtains

$$\varphi(x, 0, z) = \frac{\nu}{4\pi}\iint_F G(x, 0, z, \xi, 0, \zeta)\,\varphi(\xi, 0, \zeta)\, d\xi\, d\zeta, \tag{23.4}$$

a homogeneous Fredholm integral equation for $\varphi(x, 0, z)$. If $\varphi(x, 0, z)$ can be found, then $\varphi(x, y, z)$ is determined by (23.2). From the theory of such integral equations there will exist a sequence $\nu_1, \nu_2, \ldots$ of eigenvalues for which (23.4) will yield solutions $\varphi_1, \varphi_2, \ldots$. The functions φ_i corresponding to different ν_i-s are orthogonal on F, as shown in (16.10). If several ν_i-s have the same value, the corresponding φ_i-s can be orthogonalized. The φ_i also form a complete set on F. Each solution φ_i yields a standing wave in the basin.

It is possible to use these solutions to solve the initial-value problem formulated in (22.2), but with $p=0$. Let $\eta(x, z, 0)$ and $\eta_t(x, z, 0)$ be given. We try to express $\Phi(x, 0, z, t)$ in the following form:

$$\Phi(x, 0, z, t) = \sum_{i=1}^{\infty} a_i \varphi_i(x, 0, z) \cos \sigma_i t + b \varphi_i(x, 0, z) \sin \sigma_i t. \tag{23.5}$$

Then

$$\left.\begin{aligned} -g\,\eta(x, z, 0) &= \Phi_t(x, 0, z, 0) = \sum \sigma_i b_i \varphi_i(x, 0, z), \\ -g\,\eta_t(x, z, 0) &= \Phi_{tt}(x, 0, z, 0) = -\sum \sigma_i^2 a_i \varphi_i(x, 0, z). \end{aligned}\right\} \tag{23.6}$$

Since the φ_i form a complete set of orthogonal functions over F, the coefficients a_i and b_i can be determined in the usual manner. $\Phi(x, y, z, t)$ is then determined by (23.5) and (23.3).

In order to use the integral equation (23.4) one must first find G, the GREEN's function to a Neumann problem for a region having a corner along the curve of intersection of the plane $y=0$ and the basin walls. The difficulty with the corner can be overcome in certain cases. If the basin wall intersects the plane perpendicularly, then the basin plus its reflection in the plane $y=0$ has a boundary without this corner. If $\gamma(x, y, z, \xi, \eta, \zeta)$ is a GREEN's function for the Neumann problem for the extended region, then

$$G(x, y, z, \xi, \eta, \zeta) = \tfrac{1}{2}\left[\gamma(x, y, z, \xi, \eta, \zeta) + \gamma(x, -y, z, \xi, \eta, \zeta)\right] \tag{23.7}$$

is a GREEN's function for the original region. For some other special regions one may construct a GREEN's function by the method of images, even though the intersection with the plane $y=0$ is not perpendicular.

As mentioned above, each φ_i represents a standing wave of frequency σ_i. It may happen, as we shall see presently, that two or more σ_i-s are equal. Let σ be such an eigenvalue and $\varphi^{(1)}$ and $\varphi^{(2)}$ two of the corresponding potential functions. By forming the standing-wave solution.

$$[\lambda_1 \varphi^{(1)} + \lambda_2 \varphi^{(2)}] \cos \sigma t, \qquad \lambda_1 + \lambda_2 = 1, \tag{23.8}$$

one may vary continuously the position of the nodal curves, say. If n independent φ_i correspond to σ, then the possible nodal curves form an $(n-1)$-parameter family of curves in F. With the two solutions $\varphi^{(1)}$ and $\varphi^{(2)}$ one may also form the solution

$$\Phi(x, y, z, t) = \varphi^{(1)}(x, y, z) \cos \sigma t + \varphi^{(2)}(x, y, z) \sin \sigma t. \tag{23.9}$$

The nodal curves will now migrate from those of $\varphi^{(1)}$ to those of $\varphi^{(2)}$, and then on again to those of $\varphi^{(1)}$. If $\varphi^{(1)}$ and $\varphi^{(2)}$ have a common zero at, say, (x_0, z_0), then a nodal curve for Φ will always pass through (x_0, z_0). Near (x_0, z_0) the waves will appear to progress about (x_0, z_0) like spokes moving about a wheel. There may, of course, be several such centers.

β) Some special boundaries. It is possible to give explicit solutions for standing waves for several particular configurations of the basin. The variety of such configurations, however, is rather small. As a preliminary we note that if the basin has a flat bottom at depth h and if the side walls form a vertical cylinder making a section C with $y=0$ then, from Sect. 13α, we have

$$\left.\begin{aligned} \Phi(x,y,z,t) &= \varphi(x,z)\cosh m_0(y+h)\cos(\sigma t+\tau), \\ m_0\tanh m_0 h - \frac{\sigma^2}{g} &= 0, \end{aligned}\right\} \tag{23.10}$$

where $\varphi(x,z)$ is a solution of

$$\varphi_{xx}+\varphi_{zz}+m_0^2\varphi=0 \tag{23.11}$$

satisfying

$$\varphi_n = 0 \text{ on } C. \tag{23.12}$$

The boundary condition (23.12) will limit m_0, and hence σ, to a discrete sequence of eigenvalues

$$m_0^{(1)}, m_0^{(2)}, \ldots; \sigma_1, \sigma_2, \ldots. \tag{23.13}$$

In a coordinate system in which (23.11) can be separated it is usually possible to find the standing waves in basins whose side walls are constant-coordinate surfaces. These statements will be illustrated below for rectangular and cylindrical coordinates.

In connection with the special cases treated below we call attention to papers by Honda and Matsushita (1913) and Sasaki (1914). The authors investigated experimentally in a systematic way the various modes of motion in rectangular, triangular, circular, ring-shaped, circular-sectorial and ring-sectorial basins and compared measured with calculated periods. In most cases the agreement is with 2%. Photographs showing the various modes were obtained by sprinkling the surface with aluminum powder and exposing a photographic plate for about one period. The nodes then show up as dots, the rest as streaks. In connection with a study of the excitations of waves in a port, McKnown (1953) has also investigated experimentally the standing waves in circular and square basins; some striking photographs are included. Apté (1957) has studied further the theory of the excitation of standing waves in a square basin and has also given experimental results. Perhaps the first theoretical investigation was by Rayleigh (1876, pp. 272—279); he compared his predicted frequencies with observations of his own and of Guthrie (1875).

Rectangular basin. Let the basin walls be given by

$$x=0,\quad x=l,\quad z=0,\quad z=b,\quad y=-h.$$

Then from (13.6) one may write down immediately the solution

$$\left.\begin{aligned} &\Phi = A\cosh m_0(y+h)\cos\frac{2\pi}{l}x\cos\frac{p\pi}{b}z\cos(\sigma t+\tau), \\ &m_0^2=\pi^2\left(\frac{q^2}{l^2}+\frac{p^2}{b^2}\right),\quad \frac{\sigma^2}{g}=m_0\tanh m_0 h,\quad p,q=0,1,2,\ldots. \end{aligned}\right\} \tag{23.14}$$

Thus the choice of the integers p and q determines m_0 and then σ. If the basin is square, i.e., $l=b$, then the same values of m_0 and σ may correspond to two different solutions obtained by interchanging p and q, assuming $p\neq q$. However, this may also occur for other rectangular basins if b and l are commensurate.

Circular-cylinder basin. Let the basin have radius a. Then from (13.8) we find the solutions

$$\left.\begin{aligned} &\Phi = A\cosh m_0(y+h)\,J_n(m_0 R)\cos(n\alpha+\delta)\cos(\sigma t+\tau),\quad n=0,1,2,\ldots, \\ &J_n'(m_0 a)=0,\quad \frac{\sigma^2}{g}=m_0\tanh m_0 h. \end{aligned}\right\} \tag{23.15}$$

Thus m_0 must be selected so that $m_0 a$ is one of the zeros of J_n'; this then determines σ. For $n=0$ the wave crests and nodes lie on concentric circles, the number of such nodal circles depending upon which zero of J_n' is used to determine m_0. If $n \geqq 1$, then to the same σ there correspond two independent solutions ($\delta=0$, $\delta=\frac{1}{2}\pi$), and the remarks made in connection with (23.8) and (23.9) apply.

The standing waves in a basin shaped like a sector of a circle may be obtained from (13.8). If α_0 is the angle of the sector, then

$$\Phi = A \cosh m_0(y+h)\, J_{\frac{n\pi}{\alpha_0}}(m_0 R) \cos\frac{n\pi}{\alpha_0}\alpha \cos(\sigma t+\tau), \qquad n=0,1,\ldots,$$

$$J'_{\frac{n\pi}{\alpha_0}}(m_0 a) = 0, \qquad \frac{\sigma^2}{g} = m_0 h \tanh m_0 h.$$

If the basin is ring-shaped, with inner radius b and outer radius a, then from (13.8) one finds [cf. Sano (1913), Campbell (1953)]:

$$\left.\begin{aligned} \Phi &= A \cosh m_0(y+h)\,[Y_n'(m_0 b)\, J_n(m_0 R) - \\ &\qquad - J_n'(m_0 b)\, Y_n(m_0 R)] \cos(n\alpha+\delta)\cos(\sigma t+\tau), \\ &Y_n'(m_0 b)\, J_n'(m_0 a) - J_n'(m_0 b)\, Y_n'(m_0 a) = 0, \\ &\frac{\sigma^2}{g} = m_0 h \tanh m_0 h, \qquad n = 0, 1, \ldots. \end{aligned}\right\} \tag{23.16}$$

Formulas for sectors of a ring may be obtained and are similar to (23.15).

Basins with sloping side-walls. There are very few explicit solutions known when the sides are not vertical. If the basin is a horizontal cylinder bounded at either end by vertical walls at, say, $z=0$ and $z=l$, the theory of progressive waves in canals, developed in Sect. 18γ, can be carried over with only small changes, namely replacement of $\cos(kz-\sigma t)$ by $\cos kz \cos(\sigma t+\tau)$ where now k is restricted to the values $n\pi/l$ and σ correspondingly. Thus (18.39) and (18.43) give the velocity potentials, after the indicated modifications, for various modes of oscillation of a fluid in a basin of triangular section whose sides form an angle of 45° with the horizontal. However, even though these formulas may be used also for the two-dimensional modes, when $k=0$, they do not give the gravest two-dimensional mode except by a limiting process [described, e.g., in Lamb (1932, p. 443)].

The two-dimensional modes of motion in triangular basins whose sides form an angle $\gamma = m\pi/n$ with the horizontal may also be studied by use of the methods introduced in Sect. 17β for standing waves on beaches. Indeed, it is apparent that Kirchhoff (1879) considered his investigation of waves on beaches as a preliminary to the problem at hand. Because his approach is systematic we shall describe it.

In order to use the results of 17γ we take one side as $y=-x\tan\gamma$, i.e., $z = r\,e^{i\gamma}$; let the other side be given by

$$z = 2a - r\,e^{i\gamma}. \tag{23.17}$$

Then the complex potential $f(z)$ must satisfy not only (17.31) and (17.32), but also

$$f(2a - r\,e^{i\gamma}) = \bar{f}(2a - r\,e^{-i\gamma}), \tag{23.18}$$

which, taken with (17.31), implies that

$$f(z) = f\big(z\,e^{-i4\gamma} + 2a\,e^{-i2\gamma}(1-e^{-i2\gamma})\big). \tag{23.19}$$

In order to satisfy (17.31), (17.32) and (23.19) KIRCHHOFF first takes

$$f(z) = B_h z^h + \cdots + B_{h+k} z^{h+k}. \tag{23.20}$$

Substitution in (17.32) yields (with $\beta = e^{-2i\gamma}$ as before)

$$1 - \beta^h = 0, \quad B_{n+1} = \frac{-i\nu}{n+1} \frac{1+\beta^n}{1-\beta^{n+1}} B_n, \quad 1 + \beta^{h+k} = 0. \tag{23.21}$$

Thus, since $\gamma = m\pi/n$, one must have $h = pn$, $p = 0, 1, \ldots$, and $k = \frac{1}{2}n = q$, an integer. If one takes $p = 0$, then (23.20) becomes

$$f(z) = B_0 \left\{ 1 + \sum_{s=1}^{q} (-1)^s z^s \frac{\nu^s}{s!} \frac{e^{is\gamma}}{\cos s\gamma} \cot\gamma \ldots \cot s\gamma \right\}. \tag{23.22}$$

Conditions (17.31) requires B_0 to be real. Condition (23.18) or (23.19) remains to be satisfied. The function $f(z)$ in its assumed form, is apparently overdetermined, and it is possible to show that for $q > 3$ not all conditions can be satisfied. For $q = 2$, $m = 1$ and $q = 3$, $m = 1$, (23.18) can be satisfied. The potential functions are as follows:

$$\left.\begin{aligned} &\gamma = \pi/4: \\ &f(z) = B_0[1 - (1+i)\nu z + \tfrac{1}{2} i\nu^2 z^2] = \tfrac{1}{2} i B_0 (\nu z - 1 + i)^2 \\ &\quad = B_0[(1-\nu x)(1+\nu y) - i\nu(x+y)(1-\nu(x-y))], \quad a = 1/\nu,\ h = 1/\nu; \end{aligned}\right\} \tag{23.23}$$

$$\left.\begin{aligned} &\gamma = \pi/6: \\ &f(z) = B_0[1 - (\sqrt{3}+i)\nu z + \tfrac{1}{2}(1+i\sqrt{3})\nu^2 z^2 - \tfrac{1}{6} i\nu^3 z^3] \\ &\quad = -\tfrac{1}{6} i B_0 [2 + i(\nu z - \sqrt{3} + i)^3] \\ &\quad = -\tfrac{1}{6} B_0 [2 + (\nu y + 1)[(\nu y + 1)^2 - 3(\nu x - \sqrt{3})^2] + \\ &\qquad + i(\nu x + \nu y\sqrt{3})(\nu x - \nu y\sqrt{3} - 2\sqrt{3})(\nu x - \sqrt{3})], \quad a = \sqrt{3}/\nu,\ h = 1/\nu. \end{aligned}\right\} \tag{23.24}$$

Here h is the depth of fluid at the deepest point. The surface profile for $\gamma = 45°$ is a straight line, for $\gamma = 30°$ a parabola.

In order to find the higher modes of oscillation KIRCHHOFF returns to the form (17.33) for $f(z)$. It then follows as before that (17.34) must hold and that n must be even, say $2q$. Now, however, instead of taking $\lambda = 1$ it is left to be determined by (23.19). Substitution of (17.33) into (23.19) gives

$$A_{k+2} = A_k \exp[i\, 2\lambda\nu a \beta^{k+1}(1-\beta)], \quad k = 0, 1, \ldots, n-3. \tag{23.25}$$

Altogether there are then $n - 1 + n - 2 = 2n - 3$ independent equations to determine $A_1, \ldots, A_{n-1}$ and also λ and νa. Again the conditions can be satisfied for $\gamma = \pi/4$ and $\gamma = \pi/6$.

The solutions for $\gamma = \pi/4$ are as follows, where C is an arbitrary real constant:

$$\left.\begin{aligned} &f(z) = C[\cos\lambda\nu(z - a(1-i)) + \cosh\nu\lambda(z - a(1-i))], \\ &\lambda = \coth\lambda\nu a = -\cot\lambda\nu a; \\ &f(z) = C[\cos\lambda\nu(z - a(1-i)) - \cosh\lambda\nu(z - a(1-i))], \\ &\lambda = \tanh\lambda\nu a = \tan\lambda\nu a. \end{aligned}\right\} \tag{23.26}$$

The values of λ and ν can easily be determined graphically. For the first set of solutions the values of $\lambda\nu a$ will be slightly more than $3\pi/4$, $7\pi/4$, ..., for the second set slightly less than $7\pi/4$, $11\pi/4$,... These two sets of solutions correspond, respectively to (18.39) and (18.43) with $k=0$; the eigenvalues λ may be identified with m_i/ν and n_i/ν, respectively. KIRCHHOFF and HANSEMANN (1880) carried out an experimental investigation of the first three antisymmetric modes [Eq. (23.23) gives the first one]; they compare frequencies and positions of maxima and minima. The agreement seems satisfactory, although corrections for surface tension were necessary for the two higher modes.

The solution for $\gamma=30°$ is the following:

$$\left.\begin{aligned} f(z) = C\Big[&\frac{1}{\lambda+1}\,e^{i\lambda\nu[z-a]} + \frac{1}{\lambda-1}\,e^{-i\lambda\nu[z-a]} + \\ &+\frac{1}{\lambda+1}\,e^{i\beta^2\lambda\nu[z-a-\beta^2 a]} + \frac{1}{\lambda-1}\,e^{-i\beta^2\lambda\nu[z-a-\beta^2 a]} + \\ &+\frac{1}{\lambda+1}\,e^{-i\beta\lambda\nu[z-a-\beta a]} + \frac{1}{\lambda-1}\,e^{i\beta\lambda\nu[z-a-\beta a]}\Big], \end{aligned}\right\} \tag{23.27}$$

where C is an arbitrary real constant, $\beta=\frac{1}{2}(1-i\sqrt{3})$, $\beta^2=-\bar{\beta}=-\frac{1}{2}(1+i\sqrt{3})$, and the eigenvalues for λ and ν are determined by the equations

$$\left.\begin{aligned} \frac{\lambda^2-1}{\lambda} = -\sqrt{3}\cot\lambda\nu a, \qquad \frac{\lambda^2+\beta}{\lambda} = -i(1+\beta)\cot\beta\lambda\nu a, \\ \frac{\lambda^2+\bar{\beta}}{\lambda} = +i(1+\bar{\beta})\cot\beta\lambda\nu a. \end{aligned}\right\} \tag{23.28}$$

If λ is a solution of (23.28), then also $-\lambda$, $\bar{\lambda}$, $\beta\lambda$ and $\bar{\beta}\lambda$ are solutions. There exists a real solution which may be found from the equations

$$\cosh\sqrt{3}\,\lambda\nu a = 2\sec\lambda\nu a - \cos\lambda\nu a, \qquad \lambda = \frac{\sinh\sqrt{3}\,\lambda\nu a - \sqrt{3}\sin\lambda\nu a}{\cosh\sqrt{3}\,\lambda\nu a - \cos\lambda\nu a}. \tag{23.29}$$

The other solutions which may be generated from these do not lead to expressions different from (23.27). The first eigenvalue for $\lambda\nu a$ is a trifle to the right of $3\pi/2$. The form of the free surface corresponding to (23.17) is given by

$$\left.\begin{aligned} \eta(x,t) = \frac{\sigma}{g}\,C\Big\{&\frac{-2}{\lambda^2-1}\cos\lambda\nu(x-a) + \\ &+\Big[\frac{1}{\lambda+1}\,e^{\frac{1}{2}\sqrt{3}\lambda\nu x} + \frac{1}{\lambda-1}\,e^{-\frac{1}{2}\sqrt{3}\lambda\nu x}\Big]\cos\frac{1}{2}\lambda\nu(x-2a) + \\ &+\Big[\frac{1}{\lambda+1}\,e^{-\frac{1}{2}\sqrt{3}\lambda\nu(x-2a)} + \frac{1}{\lambda-1}\,e^{\frac{1}{2}\sqrt{3}\lambda\nu(x-2a)}\Big]\cos\frac{1}{2}\lambda\nu x\Big\}\sin(\sigma t+\tau). \end{aligned}\right\} \tag{23.30}$$

Note that (23.24) and (23.27) both give only symmetric modes. MACDONALD (1896) states that antisymmetric modes, if they exist, cannot be represented in the assumed forms (23.20) or (17.33).

VINT (1923) has succeeded in finding an infinite number of modes of motion in an inverted four-sided pyramid, each of whose sides makes a 45° angle with the horizontal. We refer to the original paper for the exact formulas.

Additional solutions have been obtained by inverse methods by SEN (1927) and by STORCHI (1949, 1952). STORCHI's result, although restricted to two-dimensional motion, is neat. Suppose that the form of the free surface is given as $\eta(x,t)=\eta(x)\sin(\sigma t+\tau)=F'(x)\sin(\sigma t+\tau)$, where $F(x)$ is analytic. Then, since $\eta(x)=\sigma g^{-1}\varphi(x,0)$ and $\varphi_y(x,0)=\nu\varphi(x,0)$,

$$f'(x-i0)=\varphi_x(x,0)-i\varphi_y(x,0)=\varphi_x(x,0)-i\nu\varphi(x,0)=\frac{g}{\sigma}\left[F''(x)-i\nu F'(x)\right]$$

and

$$f(x - i0) = \frac{g}{\sigma}[F'(x) - i\nu F(x)] + \text{const}.$$

We may take the constant as zero and write

$$f(x + iy) = \frac{g}{\sigma}[F'(x + iy) - i\nu F(x + iy)], \tag{23.31}$$

where $F(z)$ is the analytic function determined by $F(x)$. From this we have

$$\left.\begin{aligned}\varphi(x, y) &= \frac{g}{2\sigma}\{F'(x + iy) + F'(x - iy) - i\nu[F(x + iy) + F(x - iy)]\},\\ \psi(x, y) &= \frac{-g}{2\sigma}\{i[F'(x+iy) + F'(x-iy)] - \nu[F(x+iy) + F(x-iy)]\}.\end{aligned}\right\} \tag{23.32}$$

Any streamline, defined by ψ = real const, can now be taken as determining a possible basin shape corresponding to the assumed standing-wave profile. STORCHI applies the procedure to several special choices of F. An obvious disadvantage of this method, as well as of SEN'S, is that only one mode of motion is obtained for a resulting basin shape.

γ) Waves in movable basins. In several preceding sections, especially 19 and 22γ, we considered the wave motion occurring in the presence of an oscillating body when the fluid is exterior to the body. One may attempt analogous problems when the fluid is situated inside the body. Such problems occur in many situations of practical interest, for example, the sloshing of oil in a partly filled compartment of a tanker and the sloshing of fuel in an airplane or rocket. In each of these cases interest centers upon the dynamics of the whole system as well as upon the effect upon the walls of the container. A further interest in such problems arises from the interpretation of the experiments on standing waves, referred to earlier, carried out by HONDA and MATSUSHITA (1913), SASAKI (1914), and KIRCHHOFF and HANSEMANN (1880). The results were intended for comparison with theoretical prediction of standing waves in fixed basins. The waves were actually generated by oscillating the basin and finding the frequencies at which resonance appeared to occur.

We shall not consider the most general motions of the basin consistent with linearization of the free surface conditions, but shall limit ourselves here to a particular problem with small horizontal oscillations. In Sect. 26α small vertical oscillations of the basin will be considered. The general problem of motion of a body containing fluid with a free surface has been treated by MOISEEV (1953) and NARIMANOV (1956, 1957). However, both are primarily concerned with small oscillations. KREIN and MOISEEV (1957) have also considered certain mathematical aspects of this problem. OKHOTSIMSKII (1957) and RABINOVICH (1957) have considered the special case when the fluid is situated in a vertical, or almost vertical cylinder; NARIMANOV also gives special attention to this case. (Publication of the work of these three authors was apparently delayed; it is stated that, for the most part, it was carried out independently of and prior to MOISEEV'S papers.) A particular problem, the one discussed below, was treated by SRETENSKII (1951) and later by MOISEEV (1952a, b, 1953). Two later papers by MOISEEV (1954a, b) apply the theory to engineering problems, especially ships. Waves resulting from a special type of forced oscillation of a rectangular tank have been studied by BINNIE (1941) and TAMIYA (1958). A problem somewhat related to those of this section is the motion of a freely floating body in a fixed bounded basin (there is now no dissipation of energy as in the problem treated at the end of Sect. 22γ). This problem has been dealt with by PERZHNYANKO (1956) and MOISEEV (1958).

Waves in a basin with elastic restoring force. Consider the configuration shown in Fig. 30. The coordinate system OXY is fixed, the system $\overline{O}\overline{X}\overline{Y}$ moves with the carriage. Let $x_0(t)=O\overline{O}$, $u_0=\dot{x}_0$. The bottom of the fluid is at $\bar{y}=-h$, the side walls at $\bar{x}=\pm a$. The motion will be taken as two-dimensional. Denote the mass of the carriage, per unit width, by m_c, that of the fluid by m_f and the total by $m=m_c+m_f$. Let the spring constant be mk^2. We suppose as usual that the motion may be described by a velocity potential $\Phi(x, y, t)$. Following the notation at the end of Sect. 2, let $\bar{\bar{\Phi}}(\bar{x}, \bar{y}, t)$ describe the motion relative to the basin, i.e.

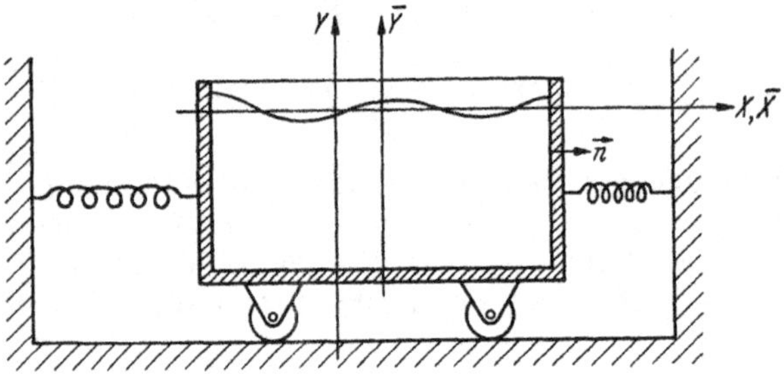

Fig. 30.

$$\left.\begin{aligned}&\Phi(x, y, t)\\ &=\bar{\bar{\Phi}}(\bar{x},\bar{y}, t)+u_0\bar{x}.\end{aligned}\right\}\quad(23.33)$$

We shall assume that x_0 and u_0 are both small, and of the same order as $\bar{\bar{\Phi}}$, i.e., in the notation of Sect. 10α, we assume expansions of the form

$$\left.\begin{aligned}&x_0=\varepsilon x_0^{(1)}, \qquad u_0=\varepsilon u_0^{(1)},\\ &\bar{\bar{\Phi}}=\varepsilon\bar{\bar{\Phi}}^{(1)}+\varepsilon^2\bar{\bar{\Phi}}^{(2)}+\cdots.\end{aligned}\right\}\quad(23.34)$$

We omit the formal details of substitution of the perturbation series in the exact boundary conditions. They lead to the following linearized boundary conditions for $\bar{\bar{\Phi}}$:

$$\left.\begin{aligned}&\bar{\bar{\Phi}}_{tt}(\bar{x}, 0, t)+g\bar{\bar{\Phi}}_{\bar{y}}(\bar{x}_y, 0, t)+\ddot{u}_0\bar{x}=0,\\ &\bar{\bar{\Phi}}_{\bar{x}}(\pm a,\bar{y}, t)=0,\\ &\bar{\bar{\Phi}}_{\bar{y}}(\bar{x}, -h, t)=0.\end{aligned}\right\}\quad(23.35)$$

The pressure, after discarding higher-order terms, is given by

$$p=-\varrho\Phi_t=-\varrho[\bar{\bar{\Phi}}_t+\dot{u}_0\bar{x}].\quad(23.36)$$

The motion of the carriage is determined by the equation

$$m_c\ddot{x}_0=\int p\cos(n,\bar{x})\,ds-mk^2x_0,\quad(23.37)$$

where the integral is taken over the wetted surface when the system is at rest. Substitution of (23.36) gives

$$\left.\begin{aligned}m\ddot{x}_0&=-\varrho\int\bar{\bar{\Phi}}_t\,ds-mk^2x_0\\ &=-\varrho\int_{-h}^{0}\int_{-a}^{a}\bar{\bar{\Phi}}_{tx}\,dx\,dy-mk^2x_0.\end{aligned}\right\}\quad(23.38)$$

[Eq. (23.38) is also a direct consequence of conservation of momentum.]

The velocity potential $\bar{\bar{\Phi}}$ and the displacement x_0 must be determined together from Eqs. (23.35) and (23.38) and either initial conditions or the further assumption that the motion is harmonic in t.

As a preliminary we shall first suppose that the basin motion, i.e. x_0, is given, so that only (23.35) need be satisfied. One may try separation of variables and express $\bar{\bar{\Phi}}$ in the form

$$\bar{\bar{\Phi}} = \sum T_n(t)\, X_n(\bar{x})\, Y_n(\bar{y}). \tag{23.39}$$

LAPLACE's equation and the last two condition of (23.35) are satisfied by

$$\left.\begin{aligned} X_{2n} Y_{2n} &= \cos\frac{2n}{2a}\pi\bar{x}\cosh\frac{2n\pi}{2a}(\bar{y}+h),\\ X_{2n+1} Y_{2n+1} &= \sin\frac{2n+1}{2a}\pi\bar{x}\cosh\frac{2n+1}{2a}\pi(\bar{y}+h). \end{aligned}\right\} \tag{23.40}$$

In order to find the corresponding T_n, expand x in a Fourier series:

$$x = \sum_{n=0}^{\infty}(-1)^n \frac{8a}{(2n+1)^2\pi^2}\sin\frac{2n+1}{2a}\pi x \tag{23.41}$$

and substitute (23.39) and (23.41) in the first condition of (23.35):

$$\left.\begin{aligned} &\sum_{n=0}^{\infty}\left[\ddot{T}_{2n}\cosh\frac{2n}{2a}\pi h + T_{2n}\, g\,\frac{2n\pi}{2a}\sinh\frac{2n}{2a}\pi h\right]\cos\frac{2n}{2a}\pi x +\\ &\quad+\sum_{n=0}^{\infty}\left[\ddot{T}_{2n+1}\cosh\frac{2n+1}{2a}\pi h + T_{2n+1}\, g\,\frac{2n+1}{2a}\pi\sinh\frac{2n+1}{2a}\pi h +\right.\\ &\quad\left.+\dddot{x}_0(-1)^n\frac{2a}{(2n+1)^2\pi^2}\right]\sin\frac{2n+1}{2a}\pi x = 0. \end{aligned}\right\} \tag{23.42}$$

Let us set

$$\sigma_n^2 = g\pi\frac{n}{2a}\tanh\frac{n}{2a}\pi h,\quad b_{2n+1} = -(-1)^n\frac{2a}{(2n+1)^2\pi^2}\operatorname{sech}\frac{2n+1}{2a}\pi h. \tag{23.43}$$

Then Eq. (23.42) yields the infinite set of differential equations

$$\left.\begin{aligned} &\ddot{T}_{2n} + \sigma_{2n}^2 T_{2n} = 0,\\ &\ddot{T}_{2n+1} + \sigma_{2n+1}^2 T_{2n+1} = b_{2n+1}\dddot{x}_0. \end{aligned}\right\} \tag{23.44}$$

The solution of the first set, $T_{2n} = A_{2n}\cos(\sigma_{2n}t + \tau_{2n})$, is independent of the motion of the basin and yields the symmetric modes of oscillation in a fixed basin. The solution to the second set may also be found by elementary methods, but will not be given here. However, we note that, if x_0 is harmonic, it confirms that resonance occurs at the frequencies of the asymmetric modes for a fixed basin.

We now turn to the joint solution of (23.35) and (23.38). Substitute (23.39) into (23.38). Then, after evaluating the integral, one finds

$$m\ddot{x}_0 + \frac{4a\varrho}{\pi}\sum_{n=0}^{\infty}\frac{(-1)^n}{2n+1}\sinh\frac{2n+1}{2a}\pi h\,\dot{T}_{2n+1} + m k^2 x_0 = 0. \tag{23.45}$$

The Eqs. (23.44) and (23.45) taken together may now be used to determine x_0 and the T_n. If we formulate an initial-value problem by requiring, say,

$$x_0(0) = c_0,\quad \dot{x}_0(0) = 0,\quad \bar{\bar{\Phi}}_{\bar{y}}(\bar{x},\bar{y},0) = 0,\quad \bar{\bar{\Phi}}_t(\bar{x},\bar{y},0) = 0, \tag{23.46}$$

then the T_{2n} are all zero and the T_{2n+1} and x_0 must be determined from the differential equations. As usual, one looks for a solution in the form

$$x_0 = c\,e^{-i\omega t}, \qquad T_{2n+1} = d_{2n+1}\,e^{-i\omega t}, \tag{23.47}$$

where both c and d_{2n+1} may, of course, be complex. Substitution in (23.44) and (23.45), followed by elimination of d_{2n+1}, yields the following equation for determining ω:

$$\omega^2 - k^2 = \frac{32 a^2 \varrho}{\pi^3 m}\,\omega^4 \sum_{n=0}^{\infty} \frac{1}{(2n+1)^3} \tanh \frac{2n+1}{2a}\,\pi h\,\frac{1}{\omega^2 - \sigma_{2n+1}^2}. \tag{23.48}$$

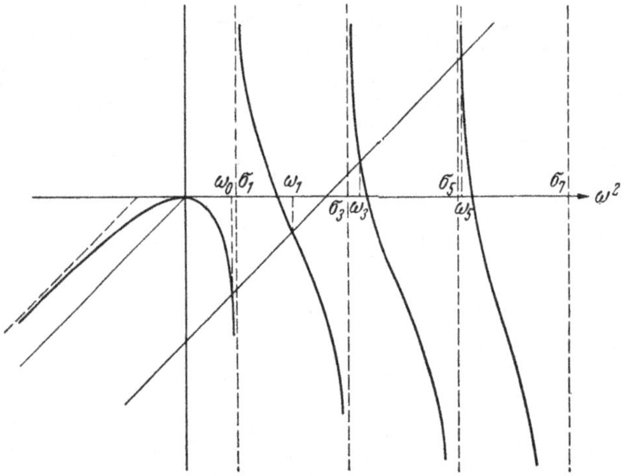

Fig. 31.

One may find the solutions graphically by plotting each side of the equation as function of ω^2. Fig. 31 gives a qualitative idea of the distribution of solutions $\omega_0, \omega_1, \ldots$. As $n \to \infty$, $\omega_{2n+1}^2 - \sigma_{2n+1}^2 \to 0$; this fact, which can be proved analytically and which seems clear from Fig. 31, would not have been so evident if we had divided (23.48) by ω^4 before plotting. A point of importance is that there is

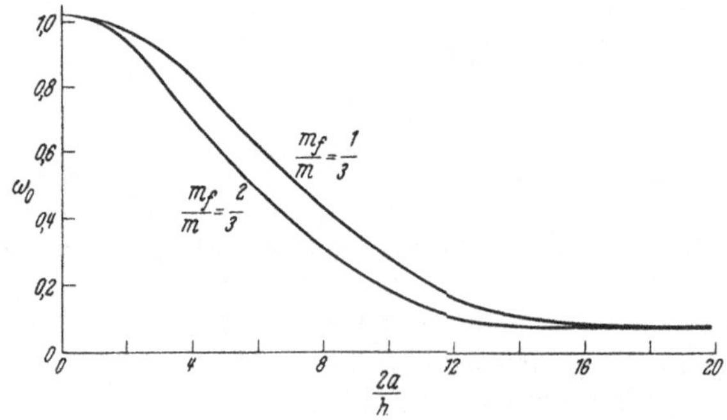

Fig. 32.

no intersection for $\omega^2 < 0$; as a result the motion is stable. This may be proved as follows. Since $x^{-1} \tanh x \leq 1$, the right hand side of (23.48), for $\omega^2 < 0$, is greater than or equal to

$$\frac{32 a^2 \varrho}{\pi^3 m}\,\omega^2 \sum_{n=0}^{\infty} \frac{1}{(2n+1)^3}\,\frac{2n+1}{2a}\,\pi h = \frac{2 a h \varrho}{m}\,\omega^2 = \frac{m_f}{m}\,\omega^2 > \omega^2. \tag{23.49}$$

Hence the line $\omega^2 - k^2$ lies below the left-hand branch of the curve for $k^2 \geqq 0$. The eigenvalues ω_i depend upon the parameters k^2, $2a/h$ and $2\varrho a h/m = m_f/m$. Fig. 32 from MOISEEV (1953) shows the dependence of the fundamental mode ω_0 upon $2a/h$ for two values of m_f/m and $k^2 = 1$.

The general solution for x_0 and T_{2n+1} is

$$x_0(t) = \mathrm{Re} \sum_{s=0}^{\infty} c_s \, \mathrm{e}^{-i\omega_s t}, \qquad T_{2n+1}(t) = \mathrm{Re} \sum_{s=0}^{\infty} d_{2n+1,s} \, \mathrm{e}^{-i\omega_s t}. \tag{23.50}$$

The solution of the initial-value problem formulated in (23.46) will not be completed. It involves solution of infinite sets of linear equations. Approximate solutions can be obtained by considering only a finite number of equations and variables.

The general theory of stability of such systems is discussed in MOISEEV's 1953 paper. In an earlier papers (1952b) he studies the special case of a basin containing fluid and serving as the bob of a pendulum. If the suspension is by a parallelogram linkage, so that the container moves parallel to itself, the motion is always stable; if the suspension is by a rod rigidly attached to the container, the motion may be, under certain circumstances, unstable.

The last cited paper by MOISEEV describes briefly the results of an experiment with a pendulum; the measured and computed fundamental frequencies for the two systems of suspension agreed with 0.1%.

24. Gravity waves in the presence of surface tension. Apparently the first one to investigate the theory of waves in a fluid acted upon by both gravity and surface tension was KELVIN (1871a, b). However, many of the essential features had been discovered earlier through observation by RUSSELL (1844) and others; references may be found in KELVIN's papers. A good account of the classical researches of KELVIN and RAYLEIGH may be found in LAMB (1932, § 265 to 272). Also, Chap. XX of RAYLEIGH's *Theory of Sound* (Cambridge 1929; Dover, N. Y., 1945) contains an exposition of many of his own fundamental researches on surface-tension phenomena, including waves.

The chief mathematical complication added by the action of surface tension is a somewhat more elaborate dynamical boundary condition at an interface or free surface. The difference of primary physical interest lies in the existence of a minimum wave velocity and of two wave lengths with the same velocity. Many of the special problems considered in preceding sections can also be solved when surface tension is acting. However, there has been little motivation for carrying through such analyses for wave motion associated with solid boundaries, since it has been recognized that the additional forces would be small. A further difficulty also appears when the solid boundary pierces the surface, for an additional boundary condition is required at the intersection. As a result, most of the investigations have dealt with waves analogous to those considered in Sects. 14α, β and δ, 15, and 22β. In fact, the complex velocity potential for the Cauchy-Poisson initial-value problem including the effect of surface tension, has already been given in Eqs. (22.45) and (22.46). A topic of particular geophysical interest, the stability of an interface, will be dealt with in Sect. 26. Waves on the surface of a viscous fluid, including surface tension, are considered in Sect. 25.

Boundary conditions. The linearized conditions to be satisfied at the interface of two fluids have already been given in Eqs. (10.7) and (10.8) (we recall that subscript 1 refers to the lower fluid). If one eliminates η from these two equations and makes use of the fact that LAPLACE's equation is satisfied on either side of the boundary, one has the following condition:

$$\Delta\Phi_1 = 0 \quad \text{for} \quad y < 0, \qquad \Delta\Phi_2 = 0 \quad \text{for} \quad y > 0. \tag{24.1}$$

$$\eta_t(x, z) = \Phi_{1y}(x, 0, z, t) = \Phi_{2y}(x, 0, z, t), \tag{24.2}$$

$$\varrho_1\left[\Phi_{1tt}(x, 0, z, t) + g\,\Phi_{1y} + \frac{T}{\varrho_1}\Phi_{1yyy}\right] = \varrho_2\left[\Phi_{2tt} + g\,\Phi_{2y}\right]. \tag{24.3}$$

If the upper fluid is absent, one sets ϱ_2 and Φ_2 equal to zero and may, of course, drop the subscript.

If the motion is two-dimensional one may introduce a stream function Ψ and a complex potential $F(z, t) = \Phi + i\Psi$ and express (24.2) and (24.3) as follows:

$$\eta_t(x) = \operatorname{Im} F_1'(x - i\,0) = \operatorname{Im} F_2'(x + i\,0), \tag{24.4}$$

$$\operatorname{Re}\varrho_1\left\{F_{1tt}(x - i\,0) + i\,g\,F_1' - i\frac{T}{\varrho_1}F_1'''\right\} = \operatorname{Re}\varrho_2\left\{F_{2tt}(x + i\,0) + i\,g\,F_2'\right\}. \tag{24.5}$$

If the upper fluid is absent and if the lower fluid is infinitely deep, one may extend the reasoning which led up to LEVI-CIVITA's differential equation (22.25) to derive the following one which must be satisfied for all z:

$$F_{tt}(z, t) + i\,g\,F' - i\frac{T}{\varrho}F''' = 0. \tag{24.6}$$

Furthermore, if the fluid is of constant depth h, CISOTTI's equation (22.30) may also be extended to include the effect of surface tension:

$$\left.\begin{aligned} &F_{tt}(z + i\,h, t) + F_{tt}(z - i\,h) + i\,g\left[F'(z + i\,h) - F'(z - i\,h)\right] - \\ &\qquad - i\frac{T}{\varrho}\left[F'''(z + i\,h) - F'''(z - i\,h)\right] = 0 \quad \text{for} \quad -2h < y < 0. \end{aligned}\right\} \tag{24.7}$$

Elementary solutions. Let us suppose first that only one fluid is present, and in addition that

$$\Phi(x, y, z, t) = \varphi(x, y, z)\cos(\sigma t + \tau).$$

Then φ must be a potential function satisfying

$$-\sigma^2\varphi(x, 0, z) + g\,\varphi_y + \frac{T}{\varrho}\varphi_{yyy} = 0. \tag{24.8}$$

Just as in Sect. 13α, we may separate out the y-variable and obtain the following expressions:

infinite depth:

$$\varphi(x, y, z) = A\,e^{my}\varphi(x, y) \tag{24.9}$$

where

$$\Delta_2\varphi + m^2\varphi = 0$$

and

$$\sigma^2 = g\,m + \frac{T}{\varrho}m^3;$$

depth h:

$$\varphi(x, y, z) = A\cosh m_0(y + h)\,\varphi(x, z), \tag{24.10}$$

where

$$\Delta_2\varphi + m_0^2\varphi = 0$$

and

$$\sigma^2 = \left(g\,m_0 + \frac{T}{\varrho}m_0^3\right)\tanh m_0 h.$$

One may also with no difficulty construct solutions analogous to (13.3) and (13.4), namely

$$\varphi(x, y, z) = A\left[m\left(1 - \frac{T}{\varrho g} m^2\right)\cos m y + \frac{\sigma^2}{g}\sin m y\right]\varphi(x, z) \tag{24.11}$$

and

$$\varphi(x, y, z) = A\cos m_i(y+h)\,\varphi(x, z) \tag{24.12}$$

for infinite and finite depth, respectively, where m_i in (24.12) must satisfy

$$\sigma^2 = \left(-g\,m_i + \frac{T}{\varrho}m_i^3\right)\tan m_i h$$

and $\varphi(x, z)$ must be a solution of

$$\Delta_2\varphi - m^2\varphi = 0.$$

Unfortunately, the set of function

$$\{\cosh m_0(y+h),\ \cos m_i(y+h)\}$$

is no longer orthogonal in general, so that the convenience of general solutions like (16.3) is lost.

It is not necessary to repeat the computations of Sect. 13 since they remain unaltered. Essentially the only change is in the relation between the frequency σ and the wave number m. Here the fact of predominant physical interest is that for small values of m the relation is controlled chiefly by the gravitational constant g and for large values of m by T/ϱ.

If one forms two-dimensional progressive waves by superposing the standing-wave solutions obtained from (24.9) and (24.10), a further significant physical fact appears: the wave velocity now has a minimum for some value of $m>0$, except for very shallow depth. These facts are displayed graphically in Fig. 11 and further information is given in the associated discussion (the curves were computed for water at 20° C and $h=\infty$ or 1 cm). Formulas for the position of the minimum and various associated values are given for infinite depth in the following table; the numerical values are for water at 20° C ($T=72.8$ dynes/cm, $\varrho=0.998$ gm/cm³):

$$\left.\begin{aligned} m_m &= \sqrt{\varrho g/T} = 3.66\ \text{cm}^{-1}, \\ \lambda_m &= 2\pi\sqrt{T/\varrho g} = 1.71\ \text{cm}, \\ c_m &= \sqrt[4]{4g\,T/\varrho} = 23.1\ \text{cm/sec}, \\ \sigma_m &= \sqrt[4]{4\varrho g^3/T} = 84.8\ \text{radians/sec} = 13.5\ \text{cycles/sec}. \end{aligned}\right\} \tag{24.13}$$

When $h \leqq \sqrt{3T/2\varrho g}$ there is no longer a minimum value of c for $m>0$; in this case c increases monotonically with m. The critical depth for water is about 0.33 cm. Except in this latter case every value of c has associated with it waves of two different lengths, each of which travels with velocity c. KELVIN suggested that the shorter waves, whose behavior is controlled chiefly by surface tension be called "ripples". The suggestion has been followed for the most part (French: "rideaux"; German: "Rippeln" or "Kräuselwellen"; Russian: "ryabi"), but they are frequently also called "capillary waves" in contrast with "gravity waves".

The relation between σ and m was subjected to a rather thorough experimental investigation by MATTHIESSEN (1889). He made measurements with water, mercury, alcohol, ethyl ether and carbon disulfide with frequencies ranging from 8.4 to roughly 2000 cycles per second. Agreement between theory and measurement is generally within 5% with the greatest discrepancies occurring near the minima. RAYLEIGH (1890) and MICHIE SMITH (1890) were apparently the first to use the theoretical relation as a means of experimental determination of T, and it has become one of the standard experimental procedures. For more recent developments and further references see BROWN (1936) and TYLER (1941).

Solutions for standing or progressive interfacial waves, analogous to those considered in Sect. 14δ, can be found by application of the same methods. Since the analysis is similar we give only the relation between σ and m. If the two fluids fill the whole space, with their interface at $y=0$, then

$$\sigma^2 = \frac{\varrho_1 - \varrho_2}{\varrho_1 + \varrho_2} g\,m + \frac{T}{\varrho_1 + \varrho_2} m^3. \tag{24.14}$$

If the lower fluid is of depth d_1, the upper of depth d_2 and the interface at $y=0$, then

$$\sigma^2 = \frac{(\varrho_1 - \varrho_2)\, m\, g + T\, m^3}{\varrho_1 \coth m\, d_1 + \varrho_2 \coth m\, d_2}. \tag{24.15}$$

In both (24.14) and (24.15) $\sigma^2 > 0$ if

$$\varrho_2 < \varrho_1 + \frac{T\, m^2}{g}; \tag{24.16}$$

thus the motion may be stable even when the lower fluid is less dense than the upper one. This is not true in the absence of surface tension, as inspection of (14.29) and (14.30) shows.

The analogue of the next example of Sect. 14δ is somewhat more complex, for two surface tensions are necessary. Let T be the surface tension at the free surface $y=0$, and T_{12} that at the interface $y=-d_1$; let the rigid bottom be at $y=-h=-d_1-d_2$. Then the relation analogous to (14.31) is

$$\left.\begin{aligned}
&\left(\frac{\sigma^2}{g\,m}\right)^2 [\varrho_1 \coth m\, d_1 \coth m\, d_2 + \varrho_2] - \\
&\quad - \frac{\sigma^2}{g\,m}\varrho_1 \left[\left(1 + \frac{T+T_{12}}{\varrho_1 g} m^2\right) \coth m\, d_2 + \varrho_1\left(1 + \frac{T}{\varrho_2 g} m^2\right) \coth m\, d_1\right] + \\
&\quad + \left(1 + \frac{T}{\varrho_2 g} m^2\right)\left[\varrho_1\left(1 + \frac{T+T_{12}}{\varrho_1 g} m^2\right) - \varrho_2\left(1 + \frac{T}{\varrho_2 g} m^2\right)\right] = 0.
\end{aligned}\right\} \tag{24.17}$$

The assumption of $d_1 = \infty$ no longer results in any notable simplification of the equation. However, one may show that the solutions $\sigma^2/g\,m$ are always real, and that they are positive if

$$\varrho_2 < \varrho_1 + \frac{T_{12}}{g} m^2. \tag{24.18}$$

This is the same condition for stability as was found in (24.16) (and is still necessary as well as sufficient). Much of the rest of the pure gravity-wave analysis of Sect. 14δ may be carried through. Thus, if σ_1 is the larger and σ_2 the smaller root of (24.17) for a given m, then one may establish the inequality

$$\left.\begin{aligned}
&0 < \frac{\sigma_2^2}{g\,m} < \left(1 + \frac{T}{\varrho_2 g} m^2\right) \tanh m\, d_2 < \frac{\sigma_1^2}{g\,m} < \\
&\quad < \left(1 + \frac{T}{\varrho_2 g} m^2\right)\left(1 + \frac{T+T_{12}}{\varrho_1 g} m^2\right) \min\left\{1, \frac{\varrho_1}{\varrho_2} \tanh m\, h\right\}.
\end{aligned}\right\} \tag{24.19}$$

If η and η_{12} are the profiles of the free surface and interface, respectively, then one finds, analogously to (14.34),

$$\frac{\eta_{12}}{\eta} = \cosh m\, d_2 - \frac{g\, m}{\sigma_2}\left(1 + \frac{T}{\varrho_2\, g}\, m^2\right) \sinh m\, d_2; \tag{24.20}$$

again, it follows from (24.18) that this ratio is positive for the larger and negative for the smaller of the two roots of (24.17). The discussion of the nature of the motion associated with the root σ_2 may be taken directly from Sect. 14δ; however, the upper bound for the velocity c_2 of a progressive wave of wave number m is now given by

$$c_2^2 = \frac{\sigma_2^2}{m^2} = \frac{\sigma_2^2}{g\, m} \cdot \frac{g}{m} < \frac{g}{m}\left(1 + \frac{T}{\varrho_2\, g}\, m^2\right) \tanh m\, d_2 < g\, d_2 \left(1 + \frac{T}{\varrho_2\, g}\, m^2\right). \tag{24.21}$$

Let us turn next to the situation in which the two fluids are moving and look for possible steady motions. Assume each fluid to move to the left with mean velocity c_i and let $F_i(z)$, $i = 1, 2$, be the complex velocity potentials. We again look for solutions in the form [cf. Eq. (14.36)]

$$F_i(z) = -\, c_i\, z + f_i(z)\,, \qquad i = 1, 2, \tag{24.22}$$

where f_i is assumed small with respect to $c_i z$. Then the linearized boundary conditions corresponding to (14.37) are

$$\left.\begin{aligned} \eta(x) &= \frac{1}{c_1}\, \mathrm{Im}\, f_1(x - i\,0) = \frac{1}{c_2}\, \mathrm{Im}\, f_2(x + i\,0)\,,\\ \frac{\varrho_2}{c_2}\, \mathrm{Re}\{i g f_2(x + i 0) + c_2^2 f_1'(x + i 0)\} &= \frac{\varrho_1}{c_1}\, \mathrm{Re}\Big\{i g f_1(x - i 0) + c_1^2 f_1'(x - i 0) -\\ &\qquad - i\, \frac{T}{\varrho_1}\, f_1''(x - i\,0)\Big\}. \end{aligned}\right\} \tag{24.23}$$

If we look for a steady motion of the form

$$f_1 = a_1\, e^{-i m z}, \qquad f_2 = a_2\, e^{i m z}, \tag{24.24}$$

then substitution in (24.23) yields

$$\frac{a_1}{c_1} = -\frac{\bar a_2}{c_2}$$

and

$$m(\varrho_1 c_1^2 + \varrho_2 c_2^2) = (\varrho_1 - \varrho_2)\, g + T\, m^2. \tag{24.25}$$

The last equation will not have a real solution for m, assuming $\varrho_1 > \varrho_2$, unless

$$4 g(\varrho_1 - \varrho_2)\, T \leqq (\varrho_1 c_1^2 + \varrho_2 c_2^2)^2. \tag{24.26}$$

There are then two solutions of the form (24.24). The effect of surface tension may be seen more clearly if one graphs each side of (24.25) and finds the intersections, if any. It will be shown in Sect. 26 that this type of motion is unstable if $|c_1 - c_2|$ becomes too large.

Singular solutions. The methods used in Sect. 13γ for finding source-type solutions can generally be extended to take account of surface tension. Aside from the algebraic complications the chief difficulties are associated with selecting the proper boundary conditions at infinity. For a stationary source of pulsating strength one may still impose a radiation condition as in (13.9) and obtain the correct solution. However, for the steadily moving source of constant strength the proper choice is no longer clear. Although it is possible to fall back upon arguments based upon considerations of group velocity, they are not completely convincing, so that it seems safer to formulate first an initial-value problem which can yield either of the two cases mentioned above as a limit when $t\to\infty$. First we give the velocity potential for a source of variable strength $m(t)$, $t\geqq 0$, moving on an arbitrary path $(a(t), b(t), c(t))$. The potential function Φ must satisfy the same conditions given on p. 491 except that 2 is now replaced by

$$\Phi_{tt}(x,0,z,t)+g\,\Phi_y(x,0,z,t)+\frac{T}{\varrho}\,\Phi_{yyy}(x,0,z,t)=0. \tag{24.27}$$

There is no special difficulty involved in finding Φ. For infinite depth, it is as follows:

$$\left.\begin{aligned}\Phi(x,y,z,t)&=\frac{m(t)}{r(t)}-\frac{m(t)}{r_1(t)}+\\&+2\int_0^\infty dk\sqrt{gk+T'k^3}\int_0^t d\tau\, m(\tau)\sin\left[(t-\tau)\sqrt{gk+T'k^3}\right]e^{k[y+b(\tau)]}J_0\big(kR(\tau)\big),\end{aligned}\right\} \tag{24.28}$$

where we have written T' for T/ϱ. One may similarly find the function analogous to (13.53) by replacing gk by $gk+T'k^3$. Knowledge of these functions allows one now to repeat, at least in part, the considerations of Sects. 22α and 22β.

For a stationary source at (a, b, c) with strength $m\cos\sigma t$, the velocity potential may be easily derived from (24.28). It is as follows:

$$\left.\begin{aligned}\Phi(x,y,z,t)&=\left[\frac{1}{r}+\frac{1}{r_1}+2\sigma^2\int_0^\infty\frac{1}{T'k^3+gk-\sigma^2}\,e^{k(y+b)}\,J_0(kR)\,dk\right]m\cos\sigma t+\\&+2\pi m\,\frac{\sigma^2}{g+3T'k_0^2}\,e^{k_0(y+k)}\,J_0(k_0R)\sin\sigma t,\end{aligned}\right\} \tag{24.29}$$

where k_0 is the real solution of $\sigma^2=gk+T'k^3$. If the fluid is of depth h, then

$$\left.\begin{aligned}&\Phi(x,y,z,t)\\&=\left[\frac{1}{r}+\frac{1}{r_2}+2\int_0^\infty\frac{T'k^3+gk+\sigma^2}{T'k^3+gk-\sigma^2\coth kh}\cdot\frac{e^{kh}\cosh k(b+h)\cosh k(y+h)}{\sinh kh}\,J_0(kR)\,dk\right]\times\\&\times m\cos\sigma t+2\pi m\,\frac{T'k_0^3+gk_0+\sigma^2}{\sigma^2h+(3T'k_0^2+g)\sinh^2k_0h}\times\\&\times e^{k_0h}\sinh k_0h\cosh k_0(b+h)\cosh k_0(y+h)\cdot J_0(k_0R)\sin\sigma t,\end{aligned}\right\} \tag{24.30}$$

where k_0 is the real root of

$$T'k^3+gk-\sigma^2\coth k\cdot h=0.$$

The velocity potential for a source moving in the direction Ox with constant velocity u_0 may also be obtained from (24.38) by a suitable limiting procedure, although the computation is somewhat more tedious. In a coordinate system moving with the source it is as follows for $h=\infty$:

$$\left.\begin{aligned}\varphi(x,y,z)&=\frac{m}{r}-\frac{m}{r_1}+\frac{4m}{\pi}\int_0^{\frac{1}{2}\pi}d\vartheta\,\mathrm{PV}\int_0^\infty dk\,\frac{g+T'k^2}{g+T'k^2-ku_0^2\cos^2\vartheta}\times\\&\times e^{k(y+b)}\cos\left[k(x-a)\cos\vartheta\right]\cos\left[k(z-c)\sin\vartheta\right]+\\&+4m\sum_{i=1,2}(-1)^{i-1}\int_0^{\vartheta_0}d\vartheta\,k_i(\vartheta)\,\frac{T'k_i^2+g}{T'k_i^2-g}\times\\&\times e^{k_i(y+b)}\sin\left[k_i(x-a)\cos\vartheta\right]\cos\left[k_i(z-c)\sin\vartheta\right],\end{aligned}\right\} \tag{24.31}$$

where

$$\vartheta_0 = \begin{cases} \arc\cos (4g\,T')^{\frac{1}{4}}/u_0 & \text{if} \quad 4g\,T' \leqq u_0^4 \\ 0 & \text{if} \quad 4g\,T' \geqq u_0^4 \end{cases}$$

and

$$\left.\begin{matrix} k_1(\vartheta) \\ k_2(\vartheta) \end{matrix}\right\} = \frac{u_0^2 \cos^2\vartheta \pm \sqrt{u_0^4 \cos^4\vartheta - 4g\,T'}}{2\,T'}.$$

One may easily show that

$$\nu \sec^2\vartheta < k_1(\vartheta) \leqq \sqrt{g/T'} \leqq k_2(\vartheta) \leqq u_0^2 \cos^2\vartheta/T'.$$

As $T'\to 0$ it is then evident that the integral involving k_2 vanishes and that (24.31) reduces to (13.36).

One may carry out an asymptotic investigation of (24.31), or of φ_x, along the lines of (13.38) and following. However, the analysis is considerably more complicated. The behavior of the wave pattern is roughly as follows. For $u_0^4/4g\,T' \leqq 1$, $\varphi_x(R, \alpha, y)$ is $O(R^{-1})$ for all α, and the disturbance is chiefly local. There is a constant $c>1$ such that when $1<u_0^4/4g\,T'<c$ the wave pattern is a superposition of two sets of waves corresponding to the two roots k_1 and k_2. Those corresponding to k_2 are capillary waves which precede the source and bend around it so that their crests eventually make an angle $\frac{1}{2}\pi+\vartheta_0$ with the x-axis. The gravity waves corresponding to k_1 behave similarly except that they follow the source and are longer. If $u_0^4/4g\,T'>c$, a second angle, say ϑ_1, appears, where $\vartheta_1<\vartheta_0$. There are now three sets of waves. Those associated with k_2 behave as described above. The gravity waves, however, consist of both transverse waves spanning the angle between $\pm(\frac{1}{2}\pi+\vartheta_1)$ and diverging waves which now lie in the wedge bounded by $\frac{1}{2}\pi+\vartheta_1$ and $\frac{1}{2}\pi+\vartheta_0$ and its reflection. One will find a sketch in LAMB's *Hydrodynamics* (1932, p. 470) which was computed for the similar problem of a moving pressure point, the so-called "fishline problem". A precise value for the constant c does not seem to be known. The free surface η may be computed from

$$\eta(x, z) = \frac{u_0}{g}\left[\varphi_x(x, 0, z) + \frac{T}{\varrho\, u_0^2}\,\varphi_{xy}(x, 0, z)\right].$$

In spite of the general complexity of the asymptotic analysis of (24.31), it is relatively easy to find the asymptotic form of η directly ahead ($\alpha=0°$) and directly behind ($\alpha=180°$):

$\alpha=0$:

$$\left.\begin{aligned} \eta(x, z) = -\,8m\,\frac{u_0}{g}\,k_2^2\left[1+\frac{T'}{u_0^2}\,k_2\right]\sqrt{\frac{\pi}{2k_2 R}}\;\frac{T' k_2^2 + g}{[(T' k_2^2 - g)(3\,T' k_2^2+g)]^{\frac{1}{2}}}\times \\ \times\, e^{k_2 b}\cos\left(k_2 R - \tfrac{1}{4}\pi\right) + O(R^{-1}); \end{aligned}\right\} \tag{24.32}$$

$\alpha=180°$:

$$\left.\begin{aligned} \eta(x, z) = 8m\,\frac{u_0}{g}\,k_1^2\left[1+\frac{T'}{u_0^2}\,k_1\right]\sqrt{\frac{\pi}{2k_1 R}}\;\frac{T' k_1^2 + g}{[(g - T' k_1^2)(3\,T' k_1^2+g)]^{\frac{1}{2}}}\times \\ \times\, e^{k_1 b}\cos\left(k_1 R - \tfrac{3}{4}\pi\right) + O(R^{-1}); \end{aligned}\right\} \tag{24.33}$$

here

$$k_1 = k_1(0) = \frac{u_0^2}{2T'}\left[1-\sqrt{1-4T' g/u_0^4}\right],$$

$$k_2 = k_2(0) = \frac{u_0^2}{2T'}\left[1+\sqrt{1-4T' g/u_0^4}\right]$$

and we assume $u_0^2 > 4T'g$. One may see rather clearly the effect upon k_1 and k_2 of varying T' and u_0 by finding them as the intersection of the graphs of $T'k^2+g$ and $u_0^2 k$.

There is no special difficulty in finding source solutions for two-dimensional motion, and the asymptotic behavior is of course easier to determine. The related problem of a moving concentrated pressure is treated in LAMB (1932, §§ 270,1). For this problem a paper by DEPRIMA and WU (1957) is particularly instructive, for they obtain the solution by first formulating the initial-value problem and then finding the limit as $t \to \infty$. In addition, they analyze the form of the surface for large but finite values of t.

25. Waves in a viscous fluid. If one abandons the assumption of a perfect fluid with irrotational motion, one loses at the same time many convenient and powerful mathematical tools from potential theory and the theory of functions of a complex variable. However, the simplifications introduced by the infinitesimal-wave approximation are sufficient to allow obtaining a number of solutions of interest, most of which have been known for many years. However, discovery of errors in early work has resulted in several recent papers. Furthermore, in connection with the theory of stability of interfaces the subject has again attracted attention; this work will be summarized in Sect. 26. One will find general expositions of many of the fundamental results in LAMB (1932, §§ 348 to 351), and LEVICH (1952, pp. 467—497). LONGUET-HIGGINS (1953b) gives a valuable discussion of the perturbation procedure and carries through certain second-order computations.

Subject to the limitations of the approximation one can find solutions for periodic standing waves in fluid of both infinite and finite depth with a free surface, at the interface of two different fluids in which either may have a fixed horizontal plane as its other boundary, and at the interface and free surface when two different fluids are superposed, the upper one having a free surface. In all cases the presence of surface tension may be admitted. By making use of such solutions together with Fourier analysis one can find the solution to the Cauchy-Poisson initial-value problem [cf. SRETENSKII (1941)].

In general, in the investigation of standing waves one is particularly interested in two things, the effect of viscosity upon the relation between wave-length and frequency, and the rate of decay of amplitude. As an alternative to examining the rate of decay, one may instead assume that a space-and time-periodic pressure has been applied to the free surface and determine the rate of transfer of energy necessary to maintain a steady oscillation.

One may still, as for perfect fluids, combine standing-wave solutions which are out of phase in order to form progressive waves. In a coordinate system moving with the waves the wave system will be stationary but the motion will not be steady for, as a result of viscosity, it will decay unless a periodic pressure distribution is moving with the waves and doing work upon the fluid. Fourier analysis may be used to obtain the fluid motion resulting from an arbitrary moving pressure distribution. Indeed, one need not restrict oneself to a pressure distribution but may also include a distribution of shearing stress at the free surface. If a pressure and shearing distribution of localized extent is moving over the fluid the dissipation of wave energy in viscosity will show up in a diminution of amplitude, as one moves away from the pressure area, which is more rapid than for a perfect fluid. Such problems have been investigated by SRETENSKII (1941, 1957) and by WU and MESSICK (1958). The latter include the effect of surface tension and make a particularly thorough study of the behavior of the solution; they restrict themselves to two-dimensional motion. One should keep in mind that if the fluid is of finite depth it is no longer equivalent to formulate a problem in which the pressure distribution is fixed and the fluid moves with a constant mean velocity.

Instead of attempting to construct a steady progressive-wave solution by means of a moving pressure distribution, one may instead assume that the progressive waves have been somehow initiated and then study their rate of decay with distance from the wave-maker. (This is, of course, closely related to finding the decay with time of an initially given progressive wave.) Studies of this nature have been made by BIESEL (1949) and CARRY (1956), who investigated especially the effect of the bottom, by URSELL (1952), who investigated the effect of side walls for infinite depth, and by HUNT (1952), who combined the two. Dissipation with distance when no walls are present has been treated by DMITRIEV (1953) in connection with the theory of the wave-maker. A point of physical interest in these studies is the relative contribution to dissipation of shearing motion near the surface, near the bottom, near the walls, and within the fluid. CASE and PARKINSON (1957) have studied the damping of standing waves in a circular cylinder of finite depth, making use of the linearized equations of this section; their experimental data seem to confirm the theoretical predictions when the cylinder walls are sufficiently polished. KEULEGAN (1959) has made further measurements with rectangular basins; he finds a striking difference between fluids which wet the container walls and those which do not, but confirms the theory for large enough containers.

The fluid motion resulting from a submerged stationary source of pulsing strength has been derived by DMITRIEV (1953) for two-dimensional motion and infinite depth. SRETENSKII (1957) has carried through the calculations for steady motion of a source in three dimensions. Unfortunately, the source function is not now as useful a tool for constructing solutions to special boundary-value problems as it is for perfect fluids. In particular, one can no longer satisfy the proper boundary condition on a steadily moving body by means of distributions of sources and sinks, as was possible in Sect. 20β. On the other hand, distributions of pulsating sources may still be used to satisfy the linearized boundary conditions on certain types of stationary oscillating bodies. Thus, if the motion is such that the linearized boundary condition specifies the velocity normal to the surface together with zero tangential velocity, then a source distribution may prove useful. For example, the wave-maker problems formulated in (19.26) and (19.31) may be treated in this fashion; DMITRIEV (1953) has done this.

A fundamental assumption of the preceding remarks is that the motion is laminar. Such an assumption seems to be in harmony with the asumption of small motions which is made in deriving the equations of the present section. However, the possible occurrence of turbulent motion in progressive waves has been reported by DMITRIEV and BONCHKOVSKAYA (1953) who found experimental evidence for it near the surface, where the vorticity was highest. The photographs in Fig. 7 do not seem to show any evidence of it, but this may result from special circumstances of the experiments. BOWDEN (1950) has essayed a theory based on VON KÁRMÁN'S similarity hypothesis; further references are given there. In the case of steady free-surface flow in a channel the importance of turbulence in modifying the mean-velocity profile is almost obvious. However, investigations have been confined to the necessary modifications of the shallow-water approximation and will be discussed elsewhere.

Finally, we note that much of the theory given below for a constant surface tension T can, in fact, be extended to a more general surface condition. This is indicated in Lamb (1932, §§ 351) and carried out by Dorrestein (1951) in some detail for infinite depth. He includes compressibility of the surface film, hysteresis and a "surface viscosity". An earlier investigation of the effect of generalized surface conditions is due to Wieghardt (1943).

α) Linearized equations and simple solutions. The linearized equations and boundary conditions have already been derived in Sect. 10. For a stratified fluid with interface at $y=0$ the zeroth-order equations are given in (10.2), the first-order in (10.3). For a single fluid with free surface they are given in (10.4). It is customary and convenient to combine the zeroth- and first-order equations. Thus, if in (10.4) we let $p=p^{(0)}+\varepsilon p^{(1)}$ and $\boldsymbol{v}=\varepsilon\boldsymbol{v}^{(1)}$, then the equations become

$$\left.\begin{aligned} &u_x+v_y+w_y=0 \\ &\boldsymbol{v}_t=-\frac{1}{\varrho}\operatorname{grad}(p+\varrho g y)+\nu\Delta\boldsymbol{v}, \\ &u_y+v_x=w_y+v_z=0 && \text{for}\quad y=0, \\ &p-\varrho g\eta-2\mu v_y=-T(\eta_{xx}+\eta_{zz})+\bar{p} && \text{for}\quad y=0, \\ &\eta_t(x,z,t)=v(x,0,z,t). \end{aligned}\right\} \tag{25.1}$$

One may clearly combine (10.2) and (10.3) in the same way. In order to obtain the proper equations in a coordinate system moving to the right with velocity u_0, one need only replace $\partial/\partial t$ by $\partial/\partial t-u_0\,\partial/\partial x$.

The standard procedure for solving the equations is to represent the motion as a potential flow plus a rotational flow and to determine the pressure from the potential part. Thus, let

$$\boldsymbol{v}=\boldsymbol{v}^{(p)}+\boldsymbol{v}^{(r)} \tag{25.2}$$

where

$$\boldsymbol{v}^{(p)}=\operatorname{grad}\Phi \tag{25.3}$$

and let

$$p=-\varrho\Phi_t-\varrho g y. \tag{25.4}$$

It then follows from the second equation in (25.1) that $\boldsymbol{v}^{(r)}$ must satisfy

$$\frac{\partial}{\partial t}\boldsymbol{v}^{(r)}=\nu\Delta\boldsymbol{v}^{(r)}. \tag{25.5}$$

The relation between $\boldsymbol{v}^{(p)}$ and $\boldsymbol{v}^{(r)}$ is established through the boundary conditions. In the several examples treated below the motion is two-dimensional. However, there is no difficulty in principle and not much additional algebraic complexity in solving the analogous three-dimensional problems. The essential simplification in two dimensions is that the components of $\boldsymbol{v}^{(r)}$ may be expressed, as a consequence of the continuity equation, in terms of a single function Ψ:

$$u^{(r)}=\frac{\partial\Psi}{\partial y},\qquad v^{(r)}=-\frac{\partial\Psi}{\partial x}. \tag{25.6}$$

It then follows easily from (25.5) that

$$\frac{\partial\Psi}{\partial t}=\nu\Delta\Psi. \tag{25.7}$$

Standing waves-infinite depth. We shall try to find a solution to the equations which has a profile of the form

$$\eta(x,t)=A(t)\cos(mx+\alpha). \tag{25.8}$$

If such a solution exists, the nature of $A(t)$ will, of course, be of especial interest. We take Φ and Ψ of the form

$$\Phi = F(y, t)\cos(m x + \alpha), \qquad \Psi = G(y, t)\sin(m x + \alpha). \tag{25.9}$$

Eq. (25.7) then implies that

$$\Psi = (c\, e^{l y} + d\, e^{-l y})\, e^{\omega t}\sin(m x + \alpha), \tag{25.10}$$

wheren

$$l^2 = m^2 + \frac{\omega}{\nu}. \tag{25.11}$$

Neither l nor ω need be real. The form of Φ is further determined by $\Delta\Phi = 0$ and its relation to Ψ through the third boundary condition in (25.1). It must be

$$\Phi = (a\, e^{m y} + b\, e^{-m y})\, e^{\omega t}\cos(m x + \alpha). \tag{25.12}$$

If, as usual, we require the motion to remain bounded as $y \to -\infty$, we must take $b = 0$. If l has a non-vanishing real part, which we assume for the present, we may without loss of generality take it to be positive. Hence one must have $d = 0$. It follows from the third condition of (25.1) that

$$a = c\,\frac{l^2 + m^2}{2 m^2}. \tag{25.13}$$

Substitution in the formula for η_t and integration with respect to t yield

$$\eta = c\,\frac{1}{2\nu m}\, e^{\omega t}\cos(m x + \alpha) = A_0\, e^{\omega t}\cos(m x + \alpha). \tag{25.14}$$

Finally, one must substitute into the dynamical boundary condition in (25.1). There p is computed from (25.4) with $y = 0$. For future use we retain the external pressure distribution $\bar{p}$, which we take in the form

$$\bar{p} = p_0\, e^{\omega t}\cos(m x + \alpha), \tag{25.15}$$

where p_0 may be complex. The boundary condition yields an equation relating l and m:

$$\nu^2 (l^2 + m^2)^2 - 4\nu^2 m^3 l + g m + T' m^3 = -m\,\frac{p_0}{\varrho}\,\frac{2 m \nu}{c} = -m\,\frac{p_0}{\varrho A_0}, \tag{25.16}$$

or, by making use of (25.11), an equation relating ω and m:

$$(\omega + 2 m^2 \nu)^2 - 4\nu^2 m^3 \sqrt{m^2 + \frac{\omega}{\nu}} + g m + T' m^3 = -m\,\frac{p_0}{\varrho A_0}. \tag{25.17}$$

Consider first Eq. (25.16) with $p_0 = 0$ and let

$$z = \frac{l}{m}, \qquad K = \frac{g m + T' m^3}{\nu^2 m^4}. \tag{25.18}$$

Then (25.16) takes the dimensionless form

$$(z^2 + 1)^2 - 4 z + K = 0.$$

An examination of this equation shows that two of its roots are always complex with negative real parts. These roots are discarded since the corresponding motion would not die out as $y \to -\infty$; in fact, we explicitly assumed earlier that l has a positive real part. [Note that if we had made the other possible assumption, i.e., that l had a negative real part, the resulting equation corresponding to (25.16) would have had roots with positive real part, again to be discarded.]

The other two roots have positive real part. Whether or not there is an imaginary part depends upon the value of K. There is a critical value $K_c \approx 0.581$ such that if $K < K_c$ the two allowable solutions are both real. If $K > K_c$, the solutions are complex conjugates. Let the two complex roots of positive real part be denoted by $l_1 \pm i l_2$. Then one may establish that $l_1/m > 0.683$. When the two admissible roots are real, both of them lie between 0 and m.

One may find the values of ω associated with the two admissible roots from (25.11). If they are both real ($K < K_c$), then $\omega = -\nu(m^2 - l^2) < 0$. In this case the motion is critically damped and the initial configuration of the surface gradually subsides. This occurs for a given m if ν is large enough. On the other hand, no matter how small ν is, it also occurs when m is large enough, i.e., for very small wavelength. If the two admissible roots are complex ($K > K_c$), then

$$\left.\begin{aligned} \omega &= -\nu m^2\left(1 - \frac{l_1^2}{m^2} + \frac{l_2^2}{m^2} \pm 2i\frac{l_1 l_2}{m^2}\right) \\ \text{and}\qquad & \\ e^{\omega t} &= 2e^{-\nu m^2\left(1 - \frac{l_1^2}{m^2} + \frac{l_2^2}{m^2}\right)t} \cos 2\frac{l_1 l_2}{m^2} t. \end{aligned}\right\} \tag{25.19}$$

One may establish that $1 - l_1^2/m^2 + l_2^2/m^2 > 0.534$, so that this motion consists of damped standing-wave oscillations. The larger m is, the more quickly it is damped.

Because of the relative complexity of Eqs. (25.16) and (25.17), it is convenient and leads to more perspicuous results to find the relation between ω and m in the two limiting cases of small and large viscosity. First consider the case of small viscosity. If in (25.17) one lets $\nu \to 0$, one regains the relation $\omega^2 = -gm - T'm^3$ of (24.9); let $\sigma_0^2 = gm + T'm^3$. However, if one retains all terms of the first power in ν, (25.17) becomes

$$\omega^2 + 4\nu m^2\omega + gm + T'm^3 = 0, \tag{25.20}$$

which has roots

$$-2m^2\nu \pm \sqrt{4m^4\nu^2 - gm - T'm^3} \approx -2m^2\nu \pm i\sigma_0 \tag{25.21}$$

if $4m^4\nu^2 \ll gm + T'm^3$. Hence the surface profile can be described by

$$\eta = A_0 e^{-2m^2\nu t}\cos(\sigma_0 t + \tau)\cos(mx + \alpha). \tag{25.22}$$

To this order of approximation, the frequency σ_0 is related to m as in a perfect fluid, but the amplitude is gradually damped. To have some idea of the orders of magnitude involved in the damping, one should consult the table on p. 645 where the row τ_0 gives computations relevant to this.

In order to find the behavior for large ν, divide equation (25.17) by $4m^4\nu^2$ and expand the term $[1 + \omega/m^2\nu]^{\frac{1}{2}}$ in a series. If one retains only terms in ν^{-1} and ν^{-2}, the resulting equation leads to

$$3\omega^2 + 4m^2\nu\omega + 2(gm + T'\omega^3) = 0. \tag{25.23}$$

The two solutions, both real and negative, are approximately, if $4m^4\nu^2 \gg gm + T'm^3$,

$$\omega_1 = -\frac{gm + T'm^3}{2m^2\nu}, \qquad \omega_2 = -\frac{4}{3}m^2\nu. \tag{25.24}$$

Here $|\omega_1| < |\omega_2|$ and hence ω_1 is the more important root inasmuch as it represents a slower damping of the motion. As is pointed out by LAMB (1932, p. 628), the root ω_1 corresponds to a value of l only slightly less than m, so that the motion is nearly irrotational. It should also be noted that by different methods of analyzing (25.17) for large ν one may obtain somewhat different coefficients for ω_2.

In the preceding analysis it was assumed explicitly that l had a non-vanishing real part. If l is pure imaginary, $l=il'$, another family of solutions exists. It is now convenient two write Φ and Ψ in the forms

$$\left.\begin{aligned}\Phi &= a\, e^{my}\, e^{\omega t} \cos(m x+\alpha),\\ \Psi &= (c\cos l' y + d \sin l' y)\, e^{\omega t} \sin(m x+\alpha),\end{aligned}\right\} \tag{25.25}$$

where

$$\omega = -\nu(l'^2+m^2) < 0. \tag{25.26}$$

The motion is thus a purely subsiding one. The boundary conditions determine the following relations between a, c, and d:

$$a = c\frac{m^2-l'^2}{2m^2}, \qquad d = c\frac{1}{4\nu^2 m^3 l}\left[\nu^2(m^2-l'^2)^2+g m+T' m^3\right]. \tag{25.27}$$

All real values of l' are now admissible. The surface profile is given by

$$\eta = c\frac{1}{2m\nu}\, e^{\omega t}\cos(m x+\alpha). \tag{25.28}$$

The two sets of solutions may now be used to investigate the development of an initial disturbance [cf. SRETENSKII (1941)].

Forced standing waves. We may apply Eq. (25.16) or (25.17) to answer the following question. Suppose that m is given. Can we determine p_0 in such a way that a steady standing wave

$$\eta = A_0\, e^{-i\sigma t}\cos(m x+\alpha) \tag{25.29}$$

of prescribed frequency σ is maintained? From (25.17) p_0 is then determined by

$$-m\frac{p_0}{\varrho A_0} = (2m^2\nu - i\sigma)^2 - 4\nu^2 m^3\sqrt{m^2 - i\frac{\sigma}{\nu}} + g m + T' m^3. \tag{25.30}$$

If, for small viscosity, one discards terms higher than the first in ν, one obtains

$$p_0 = 4 i\sigma\mu m A_0 - \sigma^2 + g m + T' m^3. \tag{25.31}$$

If we take $\sigma^2 = gm + T'm^3$, the frequency obtained from perfect-fluid theory, the necessary pressure distribution becomes

$$\bar p = 4\sigma\mu m A_0 i\, e^{-i\sigma t}\cos(m x+\alpha). \tag{25.32}$$

Thus the pressure must lead the surface displacement by a quarter of a period.

Standing waves-finite depth. If the fluid is of depth h, the analysis is similar to that above, but yields expressions of much greater complexity. The functions Φ and Ψ may be shown to have the forms

$$\left.\begin{aligned}\Phi &= \frac{1}{m}\left[d l \cosh m(y+h) + c m \sinh m(y+h)\right] e^{\omega t}\cos(m x+\alpha),\\ \Psi &= \left[c\cosh m(y+h) + d \sinh m(y+h)\right] e^{\omega t}\sin(m x+\alpha),\end{aligned}\right\} \tag{25.33}$$

where again

$$\omega = \nu(l^2 - m^2). \tag{25.34}$$

Let

$$L = \cosh l h, \qquad L' = \sinh l h, \qquad M = \cosh m h, \qquad M' = \sinh m h.$$

Then c and d are related by

$$2m(c m M + d l M') - (l^2+m^2)(c L + d L') = 0. \tag{25.35}$$

The relation between l and m corresponding to (25.16) becomes

$$\left.\begin{aligned}\nu^2(l^2+m^2)^2\,\frac{(l^2+m^2)\,(l\,L\,M-m\,L'\,M')-2m^2\,l}{(l^2+m^2)\,(l\,L\,M'-m\,L'\,M)}-\\ -4\nu^2 m^3 l\,\frac{2m\,(m\,ML-l\,M'\,L')-(l^2+m^2)}{2m\,(m\,M\,L'-l\,M'\,L)}+g\,m+T'\,m^3=-\,m\,\frac{p_0}{\varrho\,A_0}\end{aligned}\right\}\qquad(25.36)$$

and the surface profile is

$$\eta=\frac{1}{2\nu\,m}\,(c\,L+d\,L')\,e^{\omega t}\cos(m\,x+\alpha)=A_0\,e^{\omega t}\cos(m\,x+\alpha).\qquad(25.37)$$

The formulas become more perspicuous in the case of small viscosity and no external pressure and exhibit the importance of the presence of the bottom. If in (25.36) one sets $p_0=0$ and retains only terms of order ν^0, $\nu^{\frac{1}{2}}$ and ν, the following equation results:

$$\left.\begin{aligned}\omega^3-m\sqrt{\nu}\tanh m\,h\,\omega^{\frac{5}{2}}+\tfrac{9}{2}\,m^2\nu\,\omega^2+(g\,m+T'\,m^3)\tanh m\,h\,\omega-\\ -(g\,m+T'\,m^3)\,m\sqrt{\nu}\,\omega^{\frac{1}{2}}+\tfrac{1}{2}\,(g\,m+T'\,m^3)\,m^2\nu\tanh m\,h=0.\end{aligned}\right\}\qquad(25.38)$$

One may solve this equation by expanding ω in powers of $\nu^{\frac{1}{2}}$,

$$\omega=\omega_0+\omega_1\sqrt{\nu}+\omega_2\nu+\cdots,$$

substituting in (25.38) and keeping only terms in ν^0, $\nu^{\frac{1}{2}}$ and ν. The term independent of ν yields $\omega_0=\pm i\sigma_0$, where σ_0 is given in (24.10) and is the frequency for an inviscid fluid. To the order of accuracy consistent with (25.38), one finds

$$\omega=\pm i\sigma_0-(1\pm i)\,\frac{1}{2}\,m\sqrt{2\sigma_0\nu}\,\operatorname{cosech}2m\,h-2m^2\nu\,\frac{\cosh 4m\,h+\cosh 2m\,h-1}{\cosh 4m\,h-1}.\qquad(25.39)$$

The first two terms were given by HOUGH (1897). The correct expression (25.39) was first given by BIESEL (1949); HOUGH had given $-2m^2\nu$ for the last term but the apparently made an error in calculation, for (25.39) was derived independently of BIESEL's work and has also been checked by CARRY (1956) [BASSET's analysis (1888, p. 314) overlooks the terms in $\nu^{\frac{1}{2}}$].

The formula (25.39) should be compared with (25.21), the corresponding formula for infinite depth. There the effect of viscosity enters only with the first power of ν. The dissipation of energy in the body of the fluid is evidently of less importance than in the vicinity of the bottom. When two fluids are superposed, a similar phenomenon occurs in the neighborhood of the interface [cf. (25.44)].

Standing waves-stratified fluids. Consider now the situation in which a fluid typified by ϱ_1 and μ_1 fills the space $y<0$ and another typified by $\varrho_2<\varrho_1$ and μ_2 the space $y>0$. The equations to be satisfied in the two fluids and at their interface are given in (10.3). The method of solution is analogous to that used for a single fluid. However, separate functions Φ_1, Ψ_1, and Φ_2, Ψ_2 are needed for the lower and upper fluids. For a standing-wave solution they may be taken in the form

$$\left.\begin{aligned}\Phi_1&=a_1\,e^{\omega t}\,e^{m y}\cos(m\,x+\alpha), &\Psi_1&=b_1\,e^{\omega t}\,e^{l_1 y}\sin(m\,x+\alpha),\\ \Phi_2&=a_2\,e^{\omega t}\,e^{-m y}\cos(m\,x+\alpha), &\Psi_2&=b_2\,e^{\omega t}\,e^{-l_2 y}\sin(m\,x+\alpha),\end{aligned}\right\}\qquad(25.40)$$

where we assume both l_1 and l_2 to have positive real parts. ω, l_1, l_2 and m are related by the equation

$$\omega=\nu_1(l_1^2-m^2)=\nu_2(l_2^2-m^2).\qquad(25.41)$$

Substitution of (25.40) in the various boundary conditions at $y=0$ gives four homogeneous equations relating a_1, a_2, b_1, and b_2. The determinant of the coefficients set equal to zero yields another relation between ω_1, l_1, l_2 and m:

$$\left.\begin{aligned}&[(\varrho_1+\varrho_2)\,\omega^2+(\varrho_1-\varrho_2)\,g\,m+T\,m^3]\,[\mu_1\,m+\mu_2\,l_2+\mu_2\,m+\mu_1\,l_1]+\\&\quad+4\omega\,m\,(\mu_1\,m+\mu_2\,l_2)\,(\mu_2\,m+\mu_1\,l_1)=0.\end{aligned}\right\}\tag{25.42}$$

In the limiting case of small viscosity, (25.42) gives

$$\omega^2+\frac{4m}{\varrho_1+\varrho_2}\,\frac{\sqrt{\varrho_1\,\varrho_2\,\mu_1\,\mu_2}}{\sqrt{\mu_1\,\varrho_1}+\sqrt{\mu_2\,\varrho_2}}\,\omega^{\frac{3}{2}}+\frac{(\varrho_1-\varrho_2)\,g\,m+T\,m^3}{\varrho_1+\varrho_2}=0.\tag{25.43}$$

This has the approximate solutions, when the coefficient of $\omega^{\frac{3}{2}}$ is small relative to the last term,

$$\omega=\pm i\,\sigma_0-\frac{1\pm i}{\sqrt{2}}\sqrt{\sigma_0}\,\frac{2m}{\varrho_1+\varrho_2}\cdot\frac{\sqrt{\varrho_1\,\varrho_2\,\mu_1\,\mu_2}}{\sqrt{\varrho_1\,\mu_1}+\sqrt{\varrho_2\,\mu_2}}-\frac{2m^2}{\varrho_1\,\varrho_2}\cdot\frac{\varrho_1\,\mu_1^{\frac{3}{2}}+\varrho_2\,\mu_2^{\frac{3}{2}}}{\left(\sqrt{\varrho_1\,\mu_1}+\sqrt{\varrho_2\,\mu_2}\right)^2}\tag{25.44}$$

where σ_0 is the perfect-fluid frequency given in Eq. (24.14). This solution was first given by HARRISON (1908). The most significant physical fact about (25.44) when compared with (25.21) is that, to the order of approximation considered, the latter shows a rate of decay proportional to $m^2\nu$ and no influence of viscosity on the frequency, whereas (25.44) shows a rate of decay and an alteration of the frequency proportional to $m\sqrt{\nu}$ (in a dimensional sense). The greater importance of viscosity for stratified fluids may be ascribed to the different boundary condition at the interface. HARRISON computed the wave velocity and modulus of decay (time required for the amplitude to decrease by a factor e^{-1}) for an air-water interface at 17° C ($\varrho_1=1$, $\varrho_2=0.00129$, $\nu_1=0.0109$, $\nu_2=0.139$, $T=74$ in c.g.s. units). In the following table reproduced from HARRISON's paper v_0, v_c

Wavelength (cm)	1	10	100	1000
v_0 (cm/sec)	12.48	39.46	124.79	394.62
v_c	24.90	40.05	124.81	394.62
v	24.89	40.04	124.81	394.62
τ_0	1.ˢ162	1ᵐ 56.ˢ2	3ʰ 12ᵐ 39.ˢ4	321ʰ 5ᵐ 40ˢ
τ	1.ˢ125	1ᵐ 34.ˢ1	1ʰ 21ᵐ 40.ˢ6	36ʰ 50ᵐ 36ˢ
τ_c	1.ˢ106	1ᵐ 34.ˢ0	1ʰ 21ᵐ 40.ˢ3	36ʰ 50ᵐ 34ˢ

and v are the wave velocities neglecting, respectively, both surface tension and viscosity, viscosity, and neither; τ_0, τ, τ_c are the moduli of decay taking account of the water viscosity only, a water-air interface without surface tension and a water-air interface with surface tension. A striking aspect is the apparent importance of the air-water interface in damping long waves and almost total lack of influence on wave velocity [the latter fact is obvious from (25.44)].

For very large viscosities the results are analogous to those for a single fluid. The two values of ω analogous to those in (25.24) are

$$\omega_1=-\frac{(\varrho_1-\varrho_2)\,g\,m+T\,m^3}{\varrho_1+\varrho_2}\,\frac{1}{2m^2}\,\frac{\varrho_1+\varrho_2}{\mu_1+\mu_2},\qquad\omega_2=-m^2\,\frac{\mu_1+\mu_2}{\varrho_1+\varrho_2}.\tag{25.45}$$

The analysis of the roots of (25.42) for general values of ν_1 and ν_2 is difficult. However, it has been carried through by CHANDRASEKHAR (1955, especially pp. 170—173) for the special situation $\nu_1=\nu_2$ and $T=0$. In this case $l_1=l_2$.

The behavior is similar to that described for (25.17) except that the critical value K_c separating a steadily decaying motion from an oscillatory decaying one is now a function of $(\varrho_1-\varrho_2)/(\varrho_1+\varrho_2)$. This value (actually a different one since he chooses a different parameter) is tabulated for a variety of density combinations. Further analysis of (25.17) may be found in a paper by HIDE (1955) and TCHEN (1956b).

KUSAKOV (1944) has carried through an analysis similar to HARRISON's when the upper fluid is of depth h_2, the lower of depth h_1. However, the results do not seem to be consistent with HARRISON'S (or those above) when h_1 and h_2 become large. This problem has also been considered by HIDE (1955), but with an approximation that neglects the viscous boundary conditions on the walls. HARRISON, in the same paper, has treated also the problem when the upper fluid is of finite depth and with a free surface. We shall not reproduce the results except to remark that his computations show that a thin layer of fluid of slightly different density exerts a very marked influence on the damping. The effect of a variable surface tension upon wave motion is investigated briefly in LAMB (1932, § 351) and at some length in LEVICH (1952, pp. 477—490).

Pulsing stationary source. DMITRIEV (1953) has derived the form of the functions Φ and Ψ and the surface profile in the presence of a submerged source of pulsating intensity $-Q\cos\sigma t$. We shall give here only his expression for the surface profile and an asymptotic expression for large distances from the source. Let the source be located at $(0, -h_0)$ and let

$$h=h_0\sqrt{\frac{\sigma}{\nu}},\qquad \bar{x}=x\sqrt{\frac{\sigma}{\nu}},\qquad \bar{y}=y\sqrt{\frac{\sigma}{\nu}},\qquad \varepsilon=\frac{\sigma^2}{g}\sqrt{\frac{\nu}{\sigma}}.$$

The surface profile is then represented by

$$\left.\begin{aligned}\eta&=\operatorname{Re}\frac{Q\,e^{i\sigma t}}{\pi}\,\frac{\sigma}{g}\int_0^\infty\frac{1-2i\chi^2}{4\varepsilon\left[i\chi^3\left(\chi-(i+\chi^2)^{\frac{1}{2}}\right)-\chi^2\right]+i(\chi-\varepsilon)}\,e^{-h\chi}\cos\bar{x}\chi\,d\chi\\&=Q\frac{\sigma}{g}\,(1+100\,\varepsilon^4)^{\frac{1}{2}}\,e^{-h\varepsilon-4\varepsilon^3\xi\bar{x}}\cos(\sigma t-\varepsilon\bar{x}+4\varepsilon^3 h-\arctan 10\,\varepsilon^2)+\cdots.\end{aligned}\right\}\quad(25.46)$$

26. Stability of free surfaces and interfaces. In this section we wish to examine the circumstances under which a small disturbance of a free surface or of an interface between two fluids will increase in magnitude with time. The energy for this increase may come either from available potential energy, e.g. if the lower fluid is lighter than the upper one, available kinetic energy in the case of flowing fluids, from forced motion of solid boundaries, or possibly some other source such as a given pressure distribution over a free surface. Surface tension and viscosity may be expected to have a stabilizing effect, so that special interest attaches to the study of their influence. We shall use the nature of the energy source as a convenient one for separating classes of problems, even though not every situation falls clearly into one of them.

Since the boundary conditions and equations which we shall use for the mathematical analysis have been linearized, following the assumption that the disturbances are small, one cannot expect the predictions of the theory to be valid quantitatively much beyond the initiation of an unstable motion. However, a great advantage in the use of linearized theory is that an arbitrary initial disturbance can be analyzed into Fourier components and the behavior of individual components examined separately.

α) Interface between stationary superposed fluids. Following our earlier notation, let us identify quantities referring to the lower fluid by the subscript 1 and to the upper fluid by 2. Let a sinusoidal disturbance of wave number m exist at the interface. Consider first the case of perfect fluids with no surface tension. Then, if both fluids are infinitely deep, the relation (14.28) must hold. If $\varrho_1 > \varrho_2$, the standing-wave solution of Sect. 14δ obtains. However, if $\varrho_1 < \varrho_2$, then $\sigma^2 < 0$ and σ is imaginary. Let $\omega^2 = -\sigma^2$, i.e.

$$\omega^2 = \frac{\varrho_2 - \varrho_1}{\varrho_2 + \varrho_1} g\, m. \tag{26.1}$$

Then one must replace $\cos(\sigma t + \tau)$ in the Φ_i of that section by, say, $\sinh \omega t$. The profile of the free surface is then, according to (10.8), given by

$$\eta = A \sin m\, x \cosh \omega\, t. \tag{26.2}$$

The amplitude of the initial corrugations of the surface evidently increases very rapidly with time, and the solution is a valid approximation for only a limited time interval. The nature of the disturbance need not have been restricted to $\sin m\, x$; any function $\varphi(x, z)$ satisfying $\Delta\varphi + m^2\varphi = 0$ would have yielded the same behavior. Eq. (26.1) still holds if the two fluids are bounded below and above, respectively, by $y = -h_1$ and $y = h_2$ except that ω is given by

$$\omega^2 = \frac{\varrho_2 - \varrho_1}{\varrho_2 \coth m\, h_2 + \varrho_1 \coth m\, h_1} g\, m < \frac{\varrho_2 - \varrho_1}{\varrho_2 + \varrho_1} g\, m. \tag{26.3}$$

The surface is still unstable, but the rate of growth of the amplitude is slower.

Effect of surface tension. Let us now suppose that surface tension acts at the interface. Then the relation between σ and m given in (24.14) or (24.15) must hold, and a standing-wave solution is possible even if $\varrho_2 > \varrho_1$, provided that (24.16) holds, i.e.

$$\varrho_2 < \varrho_1 + \frac{T\, m^2}{g}. \tag{26.4}$$

Thus the interface is stable under small disturbances of sufficiently small wave length. If the unequality in (26.4) is reversed and we again set $\omega^2 = -\sigma^2$, then (26.2) holds once more and the solution is unstable. However, the value of ω^2 is less than that when $T = 0$, so that the rate of growth of the disturbance is retarded. It is also clear from the form of the relationship between ω^2 and m that there is a wave number for which ω^2, that is the rate of growth of the disturbance, is a maximum. If both fluids are of infinite depth this mode of maximum instability occurs when

$$m^2 = (\varrho_2 - \varrho_1)\, g/3\, T. \tag{26.5}$$

The effect of finite depth of the fluids is to displace the position of the maximum to higher values of m (smaller wavelengths) but a precise calculation requires solving a transcendental equation.

Effect of viscosity. The influence of viscosity in stabilizing interfacial disturbances has been the subject of a number of recent papers, in particular Bellman and Pennington (1954), Chandrasekhar (1955), Hide (1955) and Tchen (1956). The relevant equation relating ω and m is now (25.42). Because of the high degree of this equation it is not easy to give a complete discussion of its admissible roots. However, it is easy to establish that if

$$(\varrho_1 - \varrho_2)\, g + T\, m^2 < 0, \tag{26.6}$$

then (25.42) has a positive real root ω_0 satisfying

$$0 < \omega_0 < \sqrt{(\varrho_2 - \varrho_1)\, g\, m - T\, m^3}. \tag{26.7}$$

Thus the presence of viscosity does not alter the conditions for instability, as the presence of surface tension did, but it does have a stabilizing effect in that the rate of growth of a disturbance is slower.

In order to show the existence of a positive root under condition (26.6), one can write (25.42) in the form

$$(\varrho_1+\varrho_2)\,\omega^2+(\varrho_1-\varrho_2)\,g\,m+T\,m^3=-\,4\omega\,m\,\frac{(\mu_1 m+\mu_2 l_2)\,(\mu_2 m+\mu_1 l_1)}{\mu_1 m+\mu_2 l_2+\mu_2 m+\mu_1 l_1} \tag{26.8}$$

and sketch as functions of ω the curves represented by the two sides of the equation (remembering that l_1 and l_2 are functions of ω). The statement above then follows easily from the fact that both curves are continuous and the one represented by the right-hand function starts at the origin like

$$-\,2\,m^2\,(\mu_1+\mu_2)\,\omega$$

and goes to $-\infty$ in the fourth quadrant, behaving as $\omega\to\infty$ like

$$-\,4\,\omega^{\frac{3}{2}}\,m\,\frac{\sqrt{\varrho_1\,\varrho_2\,\mu_1\,\mu_2}}{\sqrt{\varrho_1\,\mu_1}+\sqrt{\varrho_2\,\mu_2}}\,.$$

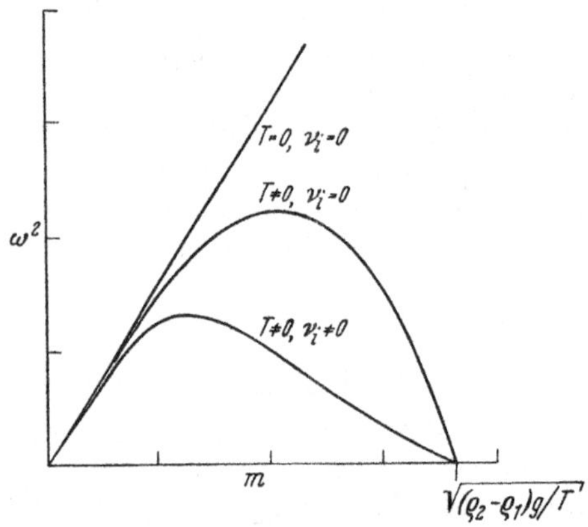

Fig. 33.

A more elaborate discussion of the roots is given by BELLMAN and PENNINGTON (1954).

The behavior of ω_0 as a function of m in the interval defined by (26.7) and in particular the mode of maximum instability has been investigated by the authors cited earlier. CHANDRASEKHAR has computed the curves $\omega_0(m)$ for $\nu_1=\nu_2$, $T=0$ and a number of values of $(\varrho_2-\varrho_1)/(\varrho_2+\varrho_1)$. HIDE has recomputed these by an approximate method and then applied the method further to a fluid of finite depth with a continuous density variation $\varrho_0\,e^{\beta y}$. TCHEN has devised a different method of approximate computation and includes the effect of surface tension. Fig. 33, which is chiefly qualitative, shows the variation of ω^2 as a function of m in the interval of instability.

Accelerated fluid. If the whole system of fluid is being accelerated in the y-direction by a constant amount $\dot v_0=g_1$, then the relative motion in a moving coordinate system is the same as if the system were at rest and g had been replaced by $g+g_1$, as is immediately evident from Eq. (2.15). With this change the reasoning of the preceding paragraphs still applies. This fact was pointed out by G.I. TAYLOR (1950) who, on the basis of it, formulated the following rule (neglecting the influence of surface tension): If the fluids are being accelerated in a direction from the more to the less dense fluid, the interface is stable; in the converse case it is unstable. Experiments carried out by LEWIS (1950) for large accelerations, about 50 g, confirm TAYLOR's observation and the predicted initial rate of growth. TAYLOR's paper gave rise to a number of others treating various aspects of the instability of accelerated interfaces. In addition to those cited in the last paragraph, we mention INGRAHAM (1954), PLESSET (1954), BIRKHOFF (1954), KELLER and KOLODNER (1954), and LAYZER (1955) but shall not summarize the contents. The effect of an imposed acceleration oscillating in magnitude will be discussed in Sect. 26γ.

β) Interface between moving fluids. Consider the situation in which the fluid occupying the region $y<0$ $(y>0)$ is moving to the left with velocity $-c_1$ $(-c_2)$, and suppose that a small disturbance exists near the interface. If we suppose that the fluid is perfect and the motion in each fluid irrotational, then we may describe it by the velocity potentials

$$\Phi_i(x, y, z, t) = -c_i x + \phi_i(x, y, z, t). \tag{26.9}$$

We shall assume $c_1 \neq c_2$.

The kinematic boundary condition at the interface may be written, after linearization appropriate to the assumption of a small disturbance, in the form:

$$\eta_t(x, y, t) = c_1 \eta_x + \phi_{1y}(x, 0, z, t) = c_2 \eta_x + \phi_{2y}(x, 0, z, t). \tag{26.10}$$

The dynamical boundary condition (3.9) yields the following generalization of (10.8):

$$\varrho_1(\phi_{1t} - c_1 \phi_{1x}) - \varrho_2(\phi_{2t} - c_2 \phi_{2x}) + (\varrho_1 - \varrho_2) g \eta = T(\eta_{xx} + \eta_{zz}) \quad \text{for} \quad y = 0. \tag{26.11}$$

If η is eliminated between (26.10) and (26.11), one finds

$$\left.\begin{aligned} \varrho_1(\phi_{1tx} - c_1 \phi_{1xx}) - \varrho_2(\phi_{2tx} - c_2 \phi_{2xx}) + \frac{\varrho_1 - \varrho_2}{c_1 - c_2} g(\phi_{2y} - \phi_{1y}) + \\ + \frac{1}{c_1 - c_2} T(\phi_{2yyy} - \phi_{1yyy}) = 0. \end{aligned}\right\} \tag{26.12}$$

Let us now restrict our attention to two-dimensional motion of fluids bounded above by $y = h_2$ and below by $y = -h_1$, and let the initial displacement be $\eta(x, 0)$. Then from (15.2) we know that the subsequent motion may be resolved into harmonic progressive waves moving to the right and left. It will be sufficient for our purpose to examine a single component of the spectrum. Hence, we look for a solution in the form

$$\left.\begin{aligned} \phi_1 &= a_1 \cosh m(y + h_1) \, e^{i(mx - \sigma t)}, \\ \phi_2 &= a_2 \cosh m(y - h_2) \, e^{i(mx - \sigma t)}. \end{aligned}\right\} \tag{26.13}$$

It follows from (26.10) that $(c_1 - c_2)\eta_x = -\phi_{1y} + \phi_{2y}$. Hence

$$\eta = \frac{-i}{c_1 - c_2} [a_1 \sinh m h_1 + a_2 \sinh m h_2] \, e^{i(mx - \sigma t)}. \tag{26.14}$$

It then follows from (26.10) that

$$\frac{a_1 \sinh m h_1 + a_2 \sinh m h_2}{c_1 - c_2} = \frac{a_1 m}{\sigma + c_1 m} \sinh m h_1 = -\frac{a_2 m}{\sigma + c_2 m} \sinh m h_2. \tag{26.15}$$

Substitution of (26.13) in (26.12) and use of (26.15) yield the following relation between σ and m:

$$\varrho_1(\sigma + c_1 m)^2 \coth m h_1 + \varrho_2(\sigma + c_2 m)^2 \coth m h_2 - (\varrho_1 - \varrho_2) g m - T m^3 = 0. \tag{26.16}$$

The solution may be expressed as follows:

$$\left.\begin{aligned} \frac{\sigma}{m} = -\frac{c_1 \varrho_1 \coth m h_1 + c_2 \varrho_2 \coth m h_2}{\varrho_1 \coth m h_1 + \varrho_2 \coth m h_2} \pm \\ \pm \sqrt{\frac{(\varrho_1 - \varrho_2)\frac{g}{m} + T m}{\varrho_1 \coth m h_1 + \varrho_2 \coth m h_2} - (c_1 - c_2)^2 \frac{\varrho_1 \varrho_2 \coth m h_1 \coth m h_2}{(\varrho_1 \coth m h_1 + \varrho_2 \coth m h_2)^2}}. \end{aligned}\right\} \tag{26.16}$$

It is evident from the form of the term under the radical that σ cannot be real unless

$$(\varrho_1 - \varrho_2)\frac{g}{m} + T m > (c_1 - c_2)^2 \frac{\varrho_1 \varrho_2 \coth m h_1 \coth m h_2}{\varrho_1 \coth m h_1 + \varrho_2 \coth m h_2}. \tag{26.17}$$

It is thus evident that there are no real solutions unless the left-hand side is positive and that there may even then exist an interval of wave numbers for which the disturbance is unstable (if both g and T are zero, such a velocity discontinuity is always unstable). If one assumes $\varrho_1 > \varrho_2$, the minimum value of the left-hand side is

$$2\sqrt{(\varrho_1-\varrho_2)\,g\,T} \tag{26.18}$$

and occurs for $m^2=(\varrho_1-\varrho_2)\,g/T$. Since

$$\frac{\varrho_1\,\varrho_2 \coth m\,h_1 \coth m\,h_2}{\varrho_1 \coth m\,h_1 + \varrho_2 \coth m\,h_2} > \frac{\varrho_1\,\varrho_2}{\varrho_1+\varrho_2}, \tag{26.19}$$

the disturbance will be unstable for some wave numbers whenever

$$(c_1-c_2)^2 > 2\,\frac{\varrho_1+\varrho_2}{\varrho_1-\varrho_2}\sqrt{(\varrho_1-\varrho_2)\,g\,T}. \tag{26.20}$$

One may conclude from (26.19) that the horizontal walls have a destabilizing effect in the sense that wave numbers which are stable for infinitely deep fluids may become unstable modes in the presence of walls. For an air-water interface the right side of (26.20) is about (646 cm/sec)^2. The corresponding wavelength is 1.71 cm; if the water is at rest ($c_1=0$), then the wave velocity is 0.84 cm/sec in the direction of the wind.

Let us suppose that c_1 and c_2 are both positive, i.e. that both fluids really do flow to the left. Then it follows from (26.16) that, if the roots are real, one of them is always negative and thus, from (26.13), represents a wave moving along the interface in the direction of the stream. The other will propagate upstream if

$$(\varrho_1-\varrho_2)\,\frac{g}{m} + T\,m > \varrho_1\,c_1^2 \coth m\,h_1 + \varrho_2\,c_2^2 \coth m\,h_2, \tag{26.21}$$

otherwise also downstream.

An investigation along the above lines of the stability of an interface between flowing fluids was first given by KELVIN (1871). Similar treatments with additional information may be found in many texts, especially LAMB (1932, §§ 232, 268) and RAYLEIGH'S *Theory of Sound* (Cambridge 1929, § 365). KELVIN'S intention was to try to predict the minimum wind velocity which will cause a small disturbance on smooth water to increase in amplitude, and to find the unstable wave lengths. The predicted minimum velocity, roughly 650 cm/sec, is much higher than the observed minimum which is about 100 cm/sec. An evident objection to the analysis above is that viscosity of both air and water has been neglected. Since this alters in an essential way the behavior of the fluids near the interface, it is not surprising that the prediction is not accurate. One should not expect confirmation except in circumstances in which it is possible to show that the effect of viscosity is confined to a neighborhood of the interface small with respect to the minimum wave lengths considered. The subject of wind generation of waves is still in an unsettled state. One will find summaries of the present status in the article by H. U. ROLL in Vol. XLVIII of this Encyclopedia, especially pp. 703—717, and also in a critical exposition by URSELL (1956). A summary of some of the work in the USSR on wave generation is included in SHULEIKIN (1956).

The inclusion of viscosity in the analysis above leads to a somewhat more difficult development than in the case of standing waves. An exposition of the present achievements in this theory will be omitted; they consist chiefly of papers by WUEST (1949) and LOCK (1951, 1954).

γ) Vertically oscillated basins. Let S denote the wetted surface of a basin and F the water surface when the basin is at rest. We shall suppose that the basin is being oscillated in the y-direction according to some given law, which may be specified by giving $v_0(t)$, the velocity of a point of the basin. It will be most convenient to describe the motion of the fluid in coordinates fixed in the basin; these will be denoted by x, y, z. We shall assume the oscillations and the resulting motion to be of small amplitude so that we may linearize the equations and boundary conditions.

If Φ is the velocity potential for the motion relative to the basin and η the profile of the surface, both in coordinates fixed in the basin, then it follows easily from (2.17) that the only necessary change is to replace g by $g+\dot{v}_0$ in the boundary conditions at the free surface. They become:

$$\eta_t(x, z, t) = \Phi_y(x, 0, z, t), \tag{26.22}$$

$$(g+\dot{v}_0)\,\eta + \Phi_t(x, 0, z, t) = T'(\eta_{xx}+\eta_{zz}), \qquad T' = T/\varrho. \tag{26.23}$$

On the basin walls one must have

$$\Phi_n = 0 \quad \text{on} \quad S. \tag{26.24}$$

We wish, as usual, to investigate the character of the motion of the fluid.

The problem formulated above is clearly related to the problem considered in Sect. 23γ. However, the resulting motions are quite different. Rayleigh (1883) appears to have made the first theoretical investigation of this problem. More recently it has been studied by Moiseev (1953, 1954), Benjamin and Ursell (1954), Schultz-Grunow (1955) and Bolotin (1956). Moiseev's analysis is the most general in that the only restriction upon the basin shape is that it should allow construction of a Green's function for the Neumann problem; surface tension is not taken into account. Benjamin and Ursell restrict themselves to basins in the form of a vertical cylinder with horizontal bottom, but include the effect of surface tension. However, at the intersection with the walls they assume a 90° angle of contact with the free surface. This is in contradiction with the observed behavior of fluids but simplifies the mathematical treatment. In spite of this shortcoming it seems desirable to include the effect of surface tension, and this will be done below. Bolotin's paper considers a modification for viscous damping. The treatment below follows closely that of Benjamin and Ursell.

Let the basin be of depth h, let C denote the intersection of the walls with the plane $y=0$, and let $\boldsymbol{n}$ be a normal to the wall at a point of C. Then, from (26.22) and (26.24) it follows that $\eta_{tn}=\Phi_{yn}=0$, or $\eta_n=$const at each point of C; we take this constant to be zero, thus assuming a 90° contact angle with the wall. It then follows from (26.23) that $(\eta_{xx}+\eta_{zz})_n=0$.

Let $\varphi_k(x, y, z)$ be a set of functions harmonic in the region bounded by the basin and the plane $y=0$ and satisfying (26.24), and such that $\varphi_k(x, 0, z)$ form a complete set of orthonormal functions in the area of the (x, z)-plane bounded by C. Then $\Phi(x, 0, z, t)$, $\eta(x, z, t)$ and $\eta_{xx}+\eta_{zz}$ can each be expanded in series in $\varphi_k(x, 0, z)$. The expansion of $\Phi(x, 0, z, t)$ determines $\Phi(x, y, z, t)$ as as eries in $\varphi_k(x, y, z)$. In the case at hand, when the basin is a vertical cylinder, one may separate variables as in Sect. 12α and construct a set φ_k in the form

$$\varphi_k(x, y, z) = \frac{\cosh m_k(y+h)\,\varphi_k(x, z)}{\cosh m_k h}, \tag{26.25}$$

where

$$\left(\frac{\partial^2}{\partial x^2} + \frac{\partial^2}{\partial z^2}\right) \varphi_k(x, z) + m_k^2 \varphi_k(x, z) = 0. \tag{26.26}$$

The eigenvalues m_k^2 are determined by the boundary condition on the contour C, namely $(\partial/\partial n)\ \varphi_k = 0$.

Let the expansion for η be written in the form

$$\eta(x, z, t) = \sum_{n=1}^{\infty} a_k(t)\ \varphi_k(x, z). \tag{26.27}$$

Then, by differentiating (26.27) and using (26.26) one gets

$$\eta_{xx} + \eta_{zz} = -\sum a_k(t)\ m_k^2 \varphi_k(x, z). \tag{26.28}$$

If

$$\Phi(x, 0, z, t) = \sum b_k(t)\ \varphi_k(x, z),$$

then

$$\Phi_y(x, y, z, t) = \sum b_k(t)\ m_k \frac{\sinh m_k(y+h)}{\cosh m_k h} \varphi_k(x, z)$$

and, from (26.22),

$$b_k(t)\ m_k \tanh m_k h = \dot{a}_k(t).$$

Hence

$$\Phi(x, y, z, t) = \sum \dot{a}_k(t) \frac{\cosh m_k(y+h)}{m_k \sinh m_k h} \varphi_k(x, z). \tag{26.29}$$

Now substitute (26.27) to (26.29) in the remaining boundary condition (26.23):

$$\sum \left[(g + \dot{v}_0)\ a_k + T' m_k^2 a_k + \frac{1}{m_k} \ddot{a}_k \coth m_k h\right] \varphi_k = 0.$$

Since the φ_k are orthogonal, we may set each coefficient of φ_k equal to zero. With the special choice

$$\dot{v}_0 = c \cos \sigma t \tag{26.30}$$

the following set of differential equations determine the a_k:

$$\ddot{a}_k(t) + [(g m_k + T' m_k^3) \tanh m_k h + c m_k \tanh m_k h \cos \sigma t]\ a_k(t) = 0. \tag{26.31}$$

If we set

$$\left.\begin{aligned} \tau &= \frac{1}{2} \sigma t, \quad p_k = \frac{4}{\sigma^2} (g m_k + T' m_k^3) \tanh m_k h = 4 \frac{\sigma_k^2}{\sigma^2}, \\ q_k &= -\frac{2}{\sigma^2} c m_k \tanh m_k h, \end{aligned}\right\} \tag{26.32}$$

where σ_k is the frequency of free oscillations in the mode m_k when the basin is fixed, then (26.31) takes one of the standard forms for the Mathieu equation:

$$\frac{d^2}{d\tau^2} a_k + [p_k - 2 q_k \cos 2\tau]\ a_k = 0. \tag{26.33}$$

Of particular interest in the present context is the behavior of the solutions a_k as τ, or t, becomes large. It is known from the theory of differential equations with periodic coefficients that a pair of fundamental solutions can be given in the form

$$e^{\mu\tau} Q(\tau), \quad e^{-\mu\tau} Q(-\tau), \tag{26.34}$$

where Q is of period π, unless $i\mu$ is an integer. In the latter case there exists a periodic solution, of period π if $i\mu$ is even and of period 2π if odd, and another independent nonperiodic solution. The coefficient μ will be a function of the parameters p_k, q_k and it is particularly pertinent to the present investigation to know for what regions in the (p, q)-plane μ has a nonzero real part. These regions have been investigated for other purposes and may be found, for example, in N. W. McLachlan's *Theory and application of Mathieu functions* (Oxford, 1947, pp. 40, 41). In Fig. 34, reproduced from Benjamin and Ursell, the shaded regions represent the unstable regions of the (p, q)-plane where μ has a nonzero real part. In the unshaded regions μ is pure imaginary (but not an integer) and the two solutions (26.34) are bounded for all τ. The boundaries between regions correspond to the periodic solutions occurring when $i\mu$ is an integer. In the unstable regions the periodicity behavior of the solutions is of two types. In the second, fourth, ... regions μ is real and the solutions (26.34) are functions of period π multiplied by exponentials. In the first, third, ... regions $\mu = \mu_1 + i$, μ_1 real, and the solutions (26.34) now become functions of period 2π multiplied by exponentials. In terms of t the two sets of regions correspond, respectively, to frequencies σ and $\frac{1}{2}\sigma$.

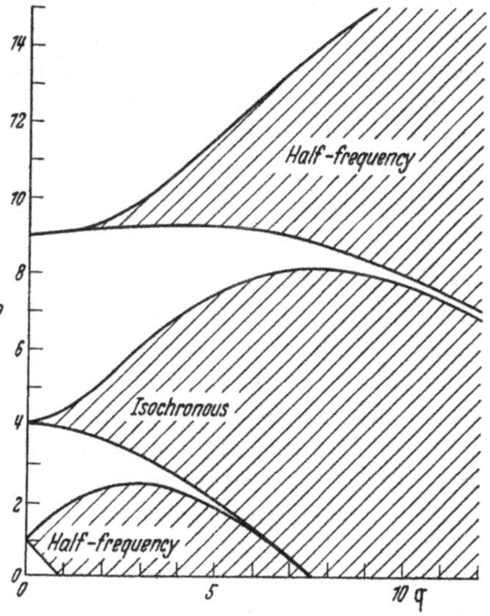

Fig. 34.

For a given mode of oscillation m_k one must compute p_k and q_k and plot (p_k, q_k) on the stability chart in order to find out whether the mode is stable or not. It seems likely, and, in fact, has been proved by Moiseev (1954, p. 44), that for any given values of σ and c some of the possible modes will be unstable. However, the analysis above has neglected the damping effect of viscosity and it may be supposed that the only unstable modes which actually occur are those associated with the smaller values of m_k. In any case, as has been emphasized earlier, the analysis is only suitable for describing the initial stages of the motion.

If the half-frequency of oscillation $\frac{1}{2}\sigma$ is equal, or nearly so, to one of the frequencies σ_k for free oscillation of the fluid, or to a subharmonic of σ_k, i.e. $\frac{1}{2}\sigma = \sigma_k/n$, then $p_k = 1$, or n^2, and it is evident from Fig 34 that (p_k, q_k) will be in an unstable region. If $\frac{1}{2}\sigma = \sigma_k$, (p_k, q_k) will lie in the lowest region and standing waves with half the frequency of the basin will be generated. If $\sigma = \sigma_k$, (p_k, q_k) will lie in the second region and the generated standing waves will have the same frequency as the basin. Thus the mode σ_k can be excited by oscillating the basin with frequency either σ_k or $2\sigma_k$. It is pointed out by Benjamin and Ursell that an apparent discrepancy between experimental observations of Faraday and Rayleigh and of Matthiessen can be explained by the above remarks.

Benjamin and Ursell made an experimental investigation with a circular cylinder in order to determine by experiment the boundaries of the lowest region of instability. The measurements provide a surprisingly good confirmation within certain limitations.

27. Higher-order theory of infinitesimal waves. It is implicit in the theory of infinitesimal waves developed in the preceding sections of this chapter that the approximation given by first-order theory to the solution of a particular problem, assuming that one exists, can be improved by including further terms in the perturbation series. The solution of the resulting boundary-value problems, at least in the simplest cases, can be carried through in a manner similar to that of the first-order theory, although the computations become more and more tedious the higher the order of approximation. Nevertheless, in view of the interest of the results, the computations have been carried through by a number of persons and by a variety of methods.

STOKES (1849) was apparently the first to make the calculation for progressive waves; in fact, the method used below in Sect. 27α is not essentially different from STOKES' first method. Later, in connection with the publication of his collected papers, STOKES (1880) added a supplement describing a different procedure. RAYLEIGH turned to the problem several times (1876, 1911, 1915, 1917) and introduced still another method of approximation. It should be noted, however, that both STOKES' second method and RAYLEIGH's method are limited to two-dimensional irrotational progressive waves. RAYLEIGH (1915) seems to be the first to have given an adequate treatment of the higher-order theory of standing waves. In addition to these classical papers there have been many others extending or improving the earlier theory; some of these will be noted below.

In all such computations, and indeed in the numerous first-order computations carried out in the earlier sections of this chapter, there is the tacit assumption that there exists an "exact solution" which is being approximated and which can be approached more and more closely by pursuing the selected method of approximation. Unfortunately, it is seldom that one is able to prove the existence of an exact solution or of convergence of the method of approximation, and, in fact, BURNSIDE (1916) cast doubt upon the usefulness of the Stokes-Rayleigh type of approximation of periodic progressive waves of permanent type. BURNSIDE's objection was later met by NEKRASOV's (1921, 1922, 1951), LEVI-CIVITA's (1925) and STRUIK's (1926) proofs of the existence of such waves for both infinite and finite depth. However, the existence of a standing wave satisfying the exact boundary conditions has not been demonstrated as yet. The same is true of the more complicated problems considered in earlier sections. However, this mathematical shortcoming is possibly of no more importance than the neglect in many problems of relevant physical parameters such as viscosity.

One should bear in mind that the higher-order infinitesimal waves considered below are not the only higher-order approximations. The solitary and cnoidal waves of the next chapter bear a similar relation to the first-order shallow-water theory. In addition, in the last chapter another method of approximating exact waves, due to HAVELOCK (1919a), will be described.

α) *Periodic progressive waves.* In the following we shall be seeking a wave which moves without change of form, i.e. a progressive wave in the sense of Sect. 7. Hence we shall expect to be able to represent Φ and η in the form

$$\Phi(x, y, z, t) = \varphi(x - ct, y, z), \qquad y = \eta(x - ct, z), \tag{27.1}$$

where c is the velocity of the wave. It will be convenient to represent the motion in a moving coordinate system, say $\bar{x} = x - ct$. However, we shall henceforth drop the bar over the x. The boundary conditions at the free surface are then the following:

$$\eta_x(x, z)\, \varphi_x(x, \eta(x, z)\, z) - \varphi_y + \eta_z \varphi_z - c\varphi_x = 0, \tag{27.2}$$

$$-c\varphi_x(x, \eta(x, z), z) + \tfrac{1}{2}(\operatorname{grad}\varphi)^2 + g\eta - T'(R_1^{-1} + R_2^{-1}) = 0, \tag{27.3}$$

where $R_1^{-1}+R_2^{-1}$ is given by (3.5′) and, as usual, $T'=T/\varrho$. Surface tension is being taken into account both for the intrinsic interest of the results and because of an interesting phenomenon which occurs in the higher-order approximations. We shall suppose that the wave length $\lambda=2\pi/m$ of the wave system has been given, so that c is still an unknown of the problem.

Let us now, as in Sect. 10α, assume that φ, η and c may all be expanded in a perturbation series in some parameter ε:

$$\left.\begin{aligned} \varphi &= \varepsilon\varphi^{(1)}+\varepsilon^2\varphi^{(2)}+\cdots,\\ \eta &= \varepsilon\eta^{(1)}+\varepsilon^2\eta^{(2)}+\cdots,\\ c &= c_0+\varepsilon c_1+\varepsilon^2 c_2+\cdots. \end{aligned}\right\} \tag{27.4}$$

After substituting in (27.2) and (27.3) and collecting terms in the manner of Sect. 10α, one obtains the following boundary conditions which must be satisfied successively by $\varphi^{(1)}, \eta^{(1)}, c_0$; $\varphi^{(2)}, \eta^{(2)}, c_1$; $\varphi^{(3)}, \eta^{(3)}, c_2$:

$$c_0\eta_x^{(1)}+\varphi_y^{(1)}=0, \qquad g\eta^{(1)}-c_0\varphi_x^{(1)}-T'(\eta_{xx}^{(1)}+\eta_{zz}^{(1)})=0; \tag{27.5}$$

$$\left.\begin{aligned} &c_0\eta_x^{(2)}+\varphi_y^{(2)}=\varphi_x^{(1)}\eta_x^{(1)}+\varphi_z^{(1)}\eta_z^{(1)}-\eta^{(1)}\varphi_{yy}^{(1)}-c_1\eta_x^{(1)},\\ &g\eta^{(2)}-c_0\varphi_x^{(2)}-T'(\eta_{xx}^{(2)}+\eta_{zz}^{(2)})=c_1\varphi_x^{(1)}-\tfrac{1}{2}(\operatorname{grad}\varphi^{(1)})^2+c_0\eta^{(1)}\varphi_{xy}^{(1)}; \end{aligned}\right\} \tag{27.6}$$

$$\left.\begin{aligned} c_0\eta_x^{(3)}+\varphi_y^{(3)}={}&\varphi_x^{(2)}\eta_x^{(1)}+\varphi_z^{(2)}\eta_x^{(1)}-\varphi_{yy}^{(2)}\eta^{(1)}-c_2\eta_x^{(1)}+\varphi_x^{(1)}\eta_x^{(2)}+\varphi_z^{(1)}\eta_z^{(2)}-\\ &-\varphi_{yy}^{(1)}\eta^{(2)}-c_1\eta_x^{(2)}+\eta^{(1)}[\varphi_{xy}^{(1)}\eta_x^{(1)}+\varphi_{zy}^{(1)}\eta_z^{(1)}]-\tfrac{1}{2}\varphi_{yyy}^{(1)}\eta^{(1)2},\\ g\eta^{(3)}-c_0\varphi_x^{(3)}-T'(\eta_{xx}^{(3)}+\eta_{zz}^{(3)})={}&c_2\varphi_x^{(1)}+c_1\varphi_x^{(2)}+c_1\varphi_{xy}^{(1)}\eta^{(1)}+c_0\varphi_{xy}^{(2)}\eta^{(1)}+\\ &+c_0\varphi_{xy}^{(1)}\eta^{(2)}+\tfrac{1}{2}c_0\varphi_{xyy}^{(1)}\eta^{(1)2}+\operatorname{grad}\varphi^{(1)}\cdot\operatorname{grad}\varphi^{(2)}+\\ &+\eta^{(1)}\operatorname{grad}\varphi^{(1)}\cdot\operatorname{grad}\varphi_y^{(1)}-T'[\eta_{xx}^{(1)}\eta_z^{(1)2}+\eta_{zz}^{(1)}\eta_x^{(1)2}-\\ &-2\eta_{xz}^{(1)}\eta_x^{(1)}\eta_z^{(1)}-\tfrac{3}{2}(\eta_{xx}^{(1)}+\eta_{zz}^{(1)})(\eta_x^{(1)2}+\eta_z^{(1)2})], \end{aligned}\right\} \tag{27.7}$$

where all conditions are to be satisfied on the plane $y=0$. It is possible, of course, to carry the approximations further, but three steps are ample to illustrate the procedures. The solution will be carried through in outline through the third order for infinite depth and through the second order for finite depth. As an expansion parameter we may take $\varepsilon=Am$, where A is a length determining the amplitude of the waves. The motion will be restricted to be two-dimensional.

Infinite depth. The solutions of (27.5) are already known from (13.5). We take them in the following form

$$\varphi^{(1)}=\frac{c_0}{m}e^{my}\sin mx, \qquad \eta^{(1)}=\frac{1}{m}\cos mx, \qquad c_0^2 m=g+m^2T'. \tag{27.8}$$

After substitution in (27.9), one finds

$$\left.\begin{aligned} &\varphi_y^{(2)}+c_0\eta_x^{(2)}=c_1\sin mx-c_0\sin 2mx,\\ &c_0\varphi_x^{(2)}-g\eta^{(2)}+T'\eta_{xx}^{(2)}=-c_1c_0\cos mx-\tfrac{1}{2}c_0^2\cos 2mx. \end{aligned}\right\} \tag{27.9}$$

Elimination of $\eta^{(2)}$ yields

$$c_0^2\varphi_{xx}^{(2)}+g\varphi_y^{(2)}-T'\varphi_{yxx}^{(2)}=2c_1c_0^2 m\sin mx-3c_0m^2T'\sin 2mx \tag{27.10}$$

as the boundary condition to be satisfied by $\varphi^{(2)}$. If $c_1\neq 0$, one cannot find a periodic potential function satisfying (27.10). Hence we set

$$c_1=0, \tag{27.11}$$

A solution of LAPLACE'S equation satisfying (27.10) with $c_1=0$ and vanishing as $y\to-\infty$ is easily found to be

$$\varphi^{(2)}=\frac{3}{2}\frac{c_0}{m}\frac{m^2T'}{g-2m^2T'}e^{2my}\sin 2mx, \tag{27.12}$$

providing $m^2\neq g/2T'$. The corresponding expression for $\eta^{(2)}$ is

$$\eta^{(2)}=\frac{1}{2}\frac{1}{m}\frac{g+m^2T'}{g-2m^2T'}\cos 2mx. \tag{27.13}$$

One could, of course, add terms of the form given in (27.8) but with arbitrary multipliers. However, such solutions are discarded since we wish to allow only first-order terms of this form.

Two striking facts show up in (27.12) and (27.13): First, if surface tension is neglected, $\varphi^{(2)}$ vanishes and $\varphi^{(1)}$ gives the velocity potential correctly to at least the second order. The second fact is the zero in the denominator in both $\varphi^{(2)}$ and $\eta^{(2)}$, which shows that $\varphi^{(2)}$ and $\lambda^{(2)}$ become unbounded as m approaches $\sqrt{g/2T'}$. One may argue, of course, that this simply shows that validity of the perturbation method is limited to smaller and smaller values of Am the closer one comes to $\sqrt{g/2T'}$. However, it seems also to be an indication that near $m=\sqrt{g/2T'}$ the mode represented by $\varphi^{(2)}$ is of the same order of magnitude as that represented by $\varphi^{(1)}$. That this is indeed the case is clear from an examination of the equation determining $\varphi^{(1)}$ and $\varphi^{(2)}$ when $m=\sqrt{g/2T'}$. In fact, $\varphi^{(2)}$ was not determined by (27.10) for this value of m and, furthermore, (27.8) does not give the complete solution of (27.5). The solution with which we must start in this case is

$$\varphi^{(1)}=\frac{c_0}{m}\left[e^{my}\sin mx+a\,e^{2my}\sin 2mx+b\,e^{2my}\cos 2mx\right], \tag{27.14}$$

where a and b are as yet undetermined constants. Thus these two modes of motion are of the same order for $m=\sqrt{g/2T'}$. One may now substitute (27.14) and the corresponding $\eta^{(1)}$ into (27.9). By reasoning similar to that used earlier in setting $c_1=0$, we now find

$$a=\pm\tfrac{1}{2},\qquad b=0,\qquad c_1=\pm\tfrac{1}{4}c_0. \tag{27.15}$$

There are thus two possible first-order modes depending upon the sign of a. $\varphi^{(2)}$ is now a sum of terms with modes $\sin 3mx$ and $\sin 4mx$, but will not be given here. The wave profile, including modes through $\cos 2mx$, may be written as follows:

$$\eta=A\left[\cos mx+\frac{1}{2}Am\frac{g+m^2T'}{g-2m^2T'}\cos 2mx\right],\qquad m\neq\sqrt{\frac{g}{2T'}}, \tag{27.16}$$

$$\eta=A\left[\cos mx\pm\frac{1}{2}\cos 2mx\right],\qquad m=\sqrt{\frac{g}{2T'}}. \tag{27.17}$$

The two signs in the second solution correspond roughly to the change of sign occurring in the first when k passes through $\sqrt{g/2T'}$. Comparison of the two cases also gives an indication of the limitations upon Am necessary in the first solution, namely,

$$|Am|<\left|\frac{g-2m^2T'}{g+m^2T'}\right|. \tag{27.18}$$

A reversal of curvature at the center of the wave trough for $m<\sqrt{g/2T'}$, or of the crest for $m>\sqrt{g/2T'}$, will occur when

$$|Am|>\frac{1}{2}\left|\frac{g-2m^2T'}{g+m^2T'}\right|. \tag{27.19}$$

The existence of the singularity in the expressions for $\eta^{(2)}$ and $\varphi^{(2)}$ was first noticed by HARRISON (1909). WILTON (1915) examined the matter more carefully, found the solutions (27.17) and, in fact, carried all approximations further. Some of WILTON'S computed profiles are shown in Fig. 35. Although WILTON casts doubt upon the existence of the solution (27.17) with $+\frac{1}{2}$, such profiles seen to have been observed by KAMESVARA RAV (1920). However, the matter apparently still awaits a thorough experimental investigation, as do also similar higher modes mentioned below.

Let us now turn to the next order, assuming $m \neq \sqrt{g/2T'}$. Substitution of (27.8) and (27.11) to (27.13) into (27.7) and elimination of $\eta^{(3)}$ yield the following boundary condition to be satisfied by $\varphi^{(3)}$ on $y=0$:

$$\left.\begin{aligned} c_0^2\varphi_{xx}^{(3)}+g\varphi_y^3-T'\varphi_{yxx}^{(3)}&=c_0^2 m\left[2c_2-\frac{1}{2}c_0\frac{2g-m^2T'}{g-2m^2T'}+\frac{3}{8}c_0\frac{m^2T'}{g+m^2T'}\right]\sin mx+\\ &+\frac{9}{8}c_0^3 m\left[\frac{4m^2T'}{g-2m^2T'}-\frac{m^2T'}{g+m^2T'}\right]\sin 3mx.\end{aligned}\right\}\qquad(27.20)$$

Again in order to avoid an unbounded solution we must set the coefficient of $\sin mx$ equal to zero. This yields a value for c_2:

$$c_2=\frac{1}{2}c_0\left[1+\frac{\frac{3}{2}m^2T'}{g-2m^2T'}-\frac{3}{8}\frac{m^2T'}{g+m^2T'}\right].\qquad(27.21)$$

One may now find a potential function satisfying (27.20) and vanishing as $y\to-\infty$. The solutions for $\varphi^{(3)}$ and $\eta^{(3)}$ are as follows:

$$\varphi^{(3)}=-\frac{9}{16}\frac{c_0}{m}\frac{m^2T'(g+2m^2T')}{(g-2m^2T')(g-3m^2T')}e^{3my}\sin 3mx;\qquad(27.22)$$

$$\left.\begin{aligned}\eta^{(3)}&=\frac{1}{m}\left[\frac{1}{8}+\frac{3}{16}\frac{m^2T'(5g+2m^2T')}{(g+m^2T')(g-2m^2T')}\right]\cos mx+\\ &+\frac{3}{16}\frac{1}{m}\frac{2g^2-gT'm^2-30(m^2T')^2}{(g-2m^2T')(g-3m^2T')}\cos 3mx,\end{aligned}\right\}\qquad(27.23)$$

for $m\neq\sqrt{g/2T'}$, $\sqrt{g/3T'}$. From (27.22) one sees again that $\varphi^{(3)}$ would vanish if surface tension were neglected. Although we shall not carry through the computation, this does not happen for $\varphi^{(4)}$. It is also evident that another singularity has appeared at $m=\sqrt{g/3T'}$. In fact, when one examines the reason for the appearance of the singularities, it is evident that a mode of the form $\cos nmx$ will always show a singularity at $m=\sqrt{g/nT'}$. In each such case the reason is the same as in the situation discussed earlier with $n=2$: for $m=\sqrt{g/nT'}\equiv m_n$ the proper first-order solution is of the form

$$\varphi^{(1)}=\frac{c_0}{m}\left[e^{my}\sin mx+a_n e^{nmy}\sin mx\right],$$

with a_n to be determined subsequently (according to WILTON only a_2 is not unique). Thus (27.8) should be qualified by $m^2\neq g/nT'$. One should note that, although m_n is getting small (and hence λ_n large) as n increases, the wave number of the second first-order mode is $\sqrt{ng/T'}$. Hence, on the basis of the results in Sect. 25, one will expect this mode to be quickly damped for large values of n. However, one may presume the first few to be observable. We remark that these special associated pairs of first-order waves always straddle the wave number for minimum c_0, namely m_1.

The wave profile, velocity potential and wave velocity are now given by

$$\left.\begin{aligned} \eta &= A m \eta^{(1)} + A^2 m^2 \eta^{(2)} + A^3 m^3 \eta^{(3)} + \cdots, \\ \varphi &= A m \varphi^{(1)} + A^2 m^2 \varphi^{(2)} + A^3 m^3 \varphi^{(3)} + \cdots, \\ c &= c_0 + A m c_1 + A^2 m^2 c_2 + \cdots. \end{aligned}\right\} \tag{27.24}$$

To the third order the profile for pure gravity waves ($T'=0$) is represented by the following function:

$$\left.\begin{aligned} \eta &= A\{[1 + \tfrac{1}{8} A^2 m^2] \cos m x + \tfrac{1}{2} A m \cos 2 m x + \tfrac{3}{8} A^2 m^2 \cos 3 m x + \cdots\} \\ &= A'\{\cos m x + \tfrac{1}{2} A' m \cos 2 m x + \tfrac{3}{8} A'^2 m^2 \cos 3 m x + \cdots\}, \end{aligned}\right\} \tag{27.25}$$

where $A' = A[1 + \frac{1}{8} A^2 m^2]$; the velocity becomes

$$c = \sqrt{\frac{g}{m}} \left(1 + \frac{1}{2} A^2 m^2 + \cdots\right). \tag{27.26}$$

The velocity potential to the third order is

$$\varphi = A \sqrt{\frac{g}{m}}\, e^{m y} \sin m x. \tag{27.27}$$

If one sets $g=0$, then the wave profile for pure capillary waves becomes

$$\eta = A\{[1 - \tfrac{1}{16} A^2 m^2] \cos m x - \tfrac{1}{4} A m \cos 2 m x - \tfrac{15}{16} A^2 m^2 \cos 3 m x + \cdots\} \tag{27.28}$$

and the velocity

$$c = \sqrt{T' m}\, [1 - \tfrac{1}{16} A^2 m^2 + \cdots]. \tag{27.29}$$

For pure gravity waves the approximations were carried to the fifth order by STOKES, RAYLEIGH (1917) and others.

It is of interest to compare the profiles represented in (27.25) and (27.28). The effect of including higher-order terms in pure gravity waves is to sharpen and raise the crests and to broaden and raise the troughs. For pure capillary waves the effect is just the reverse. For combined gravity-capillary waves the increasing importance of the second-order term near $m = \sqrt{g/2T'}$ will first show up as a reversal of curvature at the middle of the flattened part of the wave; formula (27.19) gives the condition for the first occurrence. In Fig. 35 are shown a pure gravity wave as computed by WILTON (1914) for $Am = 0.86$ (here A is the amplitude), and five gravity-capillary waves, the last two corresponding to the solutions (27.17), also computed by WILTON (1915) for a value of $T/\varrho g = 0.075$. It should be remarked that the value of $Am = 0.86$ is much larger than any for which it is possible to prove convergence of the perturbation series and is, in fact, very close to the value of Am for the highest possible irrotational wave of permanent type (see Sect. 33α), namely 0.891.

Finite depth. When a solid bottom is present at $y = -h$, the only necessary modification of the preceding analysis is substitution of the boundary condition $\varphi_y^{(i)}(x, -h) = 0$ for $\varphi_y^{(i)} \to 0$ as $y \to -\infty$. This increases the computational labor by a substantial amount, but otherwise introduces no difficulties. However, we call attention to the remarks on the definition of wave velocity in Sect. 7; the velocity c below is the one defined there also as c.

The wave profile, velocity potential and wave velocity, including the effect of surface tension, are as follows, to the second order:

$$\eta = A\left\{\cos m + \frac{1}{2}\,A\,m\,\frac{(2+\cosh 2m\,h)\,\operatorname{cosech} 2m\,h}{\tanh^2 m\,h - 3\,T'\,m^2\,(g+T'\,m^2)^{-1}}\cos 2m\,x\right\}, \tag{27.30}$$

$$\left.\begin{aligned}\varphi &= A\,c_0\left\{\frac{\cosh m\,(y+h)}{\sinh m\,h}\sin m\,x + \right.\\ &\left.+\frac{3}{4}\,A\,m\,\frac{(g+3\,T'\,m^2)\coth m\,h-(g+T'\,m^2)\tanh m\,h}{(g+T'\,m^2)\tanh^2 m\,h-3\,T'\,m^2}\,\frac{\cosh 2m\,(y+h)}{\sinh 2m\,h}\sin 2m\,x\right\},\end{aligned}\right\} \tag{27.31}$$

$$c^2 = c_0^2 = \left(\frac{g}{m} + T'\,m\right)\tanh m\,h. \tag{27.32}$$

The velocity is the same as in the first-order theory; this occurred also for infinite depth. In contrast to the case of infinite depth, the term $\varphi^{(2)}$ does not vanish

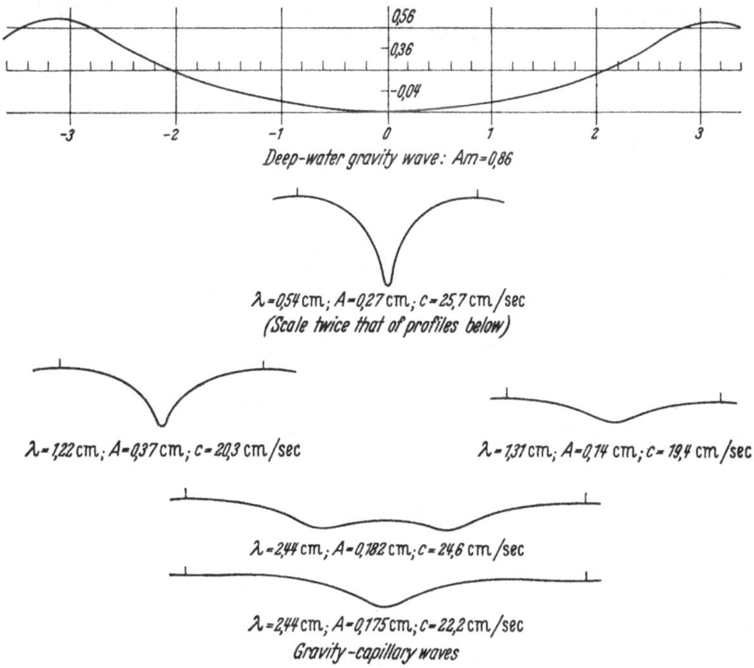

Fig. 35.

when $T'=0$. The singularity in the coefficient of $\cos 2m\,x$ still persists provided that $h>\sqrt{3T'/g}$. The earlier discussion of this phenomenon is still relevant, and a detailed one will be omitted here. However, even if surface tension is neglected in (27.30), the second-order term may still become large for small values of $m\,h$, as has been emphasized by MICHE (1944). If one again takes as an indication of increasing predominance of the second-order term a reversal of curvature at the bottom of the trough, one finds that this occurs for

$$A\,m > \frac{1}{2}\,\frac{\tanh^2 m\,h\sinh 2m\,h}{2+\cosh 2m\,h}, \tag{27.33}$$

or approximately

$$A\,m > \tfrac{1}{3}\tanh m\,h\sinh^2 m\,h$$

as given by MICHE. The occurrence of this secondary crest when mh is small has frequently been observed. It has been investigated experimentally by MORISON and CROOKE (1953) and by HORIKAWA and WIEGEL (1959).

The wave profile and velocity computations were carried by STOKES to the third order, and by DE (1955) to the fifth order, for pure gravity waves in fluid of finite depth. The following expressions are taken from a report by SKJELBREIA (1959):

$$\left.\begin{aligned}\eta = A\Big\{\cos m\,x &+ \frac{1}{4}A\,m\,\frac{\cosh m\,h\,(2+\cosh 2m\,h)}{\sinh^3 m\,h}\cos 2m\,x + \\ &+ \frac{3}{64}A^2 m^2\,\frac{8\cosh^6 m\,h+1}{\sinh^6 m\,h}\cos 3m\,x + \cdots\Big\},\\ c^2 = \frac{g}{m}\tanh m\,h\Big[1 &+ A^2 m^2\,\frac{8+\cosh 4m\,h}{8\sinh^4 m\,h} + \cdots\Big].\end{aligned}\right\}\tag{27.34}$$

SKJELBREIA has provided comprehensive tables allowing easy computation of η, φ and many other quantities of interest, all to the third order.

Particle orbits. A particularly interesting phenomenon occurs when higher-order approximations are used in the computation of the paths of individual particles. The equations which the coordinates of a particle must satisfy are

$$\frac{dx}{dt} = \varphi_x(x - c\,t,\,y), \qquad \frac{dy}{dt} = \varphi_y(x - c\,t,\,y). \tag{27.35}$$

Since φ depends upon the parameter ε, the solutions x and y also will. We assume then that x and y may be expanded into series of the form

$$x(t) = x_0 + \varepsilon\,x_1(t) + \cdots, \qquad y(t) = y_0 + \varepsilon\,y_1(t) + \cdots, \tag{27.36}$$

substitute them into (27.35) together with the appropriate expansion of φ in powers of ε, and then equate the several powers of ε separately. This results in a sequence of equations of which the first two are as follows

$$\frac{dx_1}{dt} = \varphi_x^{(1)}(x_0 - c_0\,t,\,y_0), \qquad \frac{dy_1}{dt} = \varphi_y^{(1)}(x_0 - c_0\,t,\,y_0); \tag{27.37}$$

$$\left.\begin{aligned}\frac{dx_2}{dt} &= x_1(t)\,\varphi_{xx}^{(1)}(x_0 - c_0\,t,\,y_0) + y_1\,\varphi_{xy}^{(1)} + \varphi_x^{(2)},\\ \frac{dy_2}{dt} &= x_1(t)\,\varphi_{xy}^{(1)}(x_0 - c_0\,t,\,y_0) + y_1\,\varphi_{yy}^{(1)} + \varphi_y^{(2)}.\end{aligned}\right\}\tag{27.38}$$

The first set, (27.37), was already solved in (14.17) and (14.18) and to the first order of approximation gave circular or elliptical orbits. The solution for higher orders is facilitated by neglecting surface tension and assuming $h = \infty$, for then $\varphi^{(2)}$ and $\varphi^{(3)}$ both vanish. From (27.8) one finds easily the orbit to the second order:

$$\left.\begin{aligned}x(t) &= x_0 - A\,e^{m y_0}\sin m(x_0 - c_0 t) + A^2 m^2 c_0\,e^{2m y_0}\,t,\\ y(t) &= y_0 + A\,e^{m y_0}\cos m(x_0 - c_0 t).\end{aligned}\right\}\tag{27.39}$$

The circular orbits of first-order theory are now modified by a general drift in the direction of wave motion. The total amount of fluid transported per unit time (and width) is $\frac{1}{2}A^2 m c_0$. As the formula shows, this additional flow is concentrated chiefly near the surface.

When the depth is finite, or when surface tension is taken into account, the orbits become more complicated. Let

$$K = \frac{(g + 3\,T' m^2)\coth m\,h - (g + T' m^2)\tanh m\,h}{(g + T' m^2)\tanh^2 m\,h - 3m^2\,T'}\,. \tag{27.40}$$

The particle orbits, accurate to the second order, are as follows:

$$\left.\begin{aligned} x(t) &= x_0 - A\,\frac{\cosh m\,(y_0+h)}{\sinh m\,h}\sin m\,(x_0 - c_0\,t) + \frac{1}{2}\,A^2 m^2\,c_0\,t\,\frac{\cosh 2m\,(y_0+h)}{\sinh^2 m\,h} + \\ &\quad + \frac{1}{4}\,A^2 m\left[\operatorname{cosech}^2 m\,h - 3K\,\frac{\cosh 2m\,(y_0+h)}{\sinh 2m\,h}\right]\sin 2m\,(x_0 - c_0\,t)\,,\\ y(t) &= y_0 + A\,\frac{\sinh m\,(y_0+h)}{\sinh m\,h}\cos m\,(x_0 - c_0\,t) + \\ &\quad + \frac{3}{4}\,A^2 m\,K\,\frac{\sinh 2m\,(y_0+h)}{\sinh 2m\,h}\cos 2m\,(x_0 - c_0\,t)\,. \end{aligned}\right\} \tag{27.41}$$

The mass-transport term in $x(t)$ is still present, and in fact, persists to the very bottom. The elliptical orbits of the first-order theory are now modified not only by the forward drift at all levels, but also by another superposed cyclic motion of twice the frequency. The effect of this is to make the orbits approximately epitrochoidal (neglecting for a moment the drift) with a small hump at the bottom which in extreme cases can become a cusp or a loop. This behavior has, in fact, been observed by MORISON and CROOKE (1953). For capillary waves the situation is reversed and a dimple appears at the top.

The existence of mass transport will be reconsidered in the last chapter, where it will be demonstrated that it is a general consequence of irrotational motion when the exact boundary conditions are satisfied. The theoretically predicted monotonically decreasing forward drift with increasing depth is not confirmed experimentally for small values of $m h$, say $m h < 2$. Instead, with respect to a coordinate system moving with the mean velocity of the fluid, there is an observed forward flow near the bottom and top and a backward flow in the middle portions. It is not surprising that the perfect-fluid model does not give a good prediction for small $m h$, for the high shear rate near the bottom indicates that viscosity should not be neglected. LONGUET-HIGGINS (1953 b) has, in fact, devoted a long monograph to development of the higher-order theory of waves in a viscous fluid and finds theoretical drift curves agreeing qualitatively with observed ones. We shall not carry through the details here and refer to LONGUET-HIGGINS' paper.

Wave energy. One of the striking facts about progressive first-order pure gravity waves is that the kinetic and potential energy per wave length are equal (see Sect. 15β). This equal division of energy no longer holds when higher-order terms are taken into account. It is particularly easy to show this for $h = \infty$, for then we may use (27.25) and (27.27). The average potential energy in a wave-length is

$$\mathscr{V}_{\text{av}} = \frac{m}{2\pi}\int_0^{2\pi/m} dx \int_0^{\eta} \varrho\, g\, y\, dy = \frac{m}{2\pi}\int_0^{2\pi/m} \frac{1}{2}\,\varrho\, g\,\eta^2\, dx = \frac{1}{4}\,\varrho\, g\, A^2\left[1 + \frac{1}{2}\,A^2 m^2\right]. \tag{27.42}$$

The average kinetic energy is

$$\left.\begin{aligned} \mathscr{T}_{\text{av}} &= \frac{m}{2\pi}\int_0^{2\pi/m} dx \int_0^{\infty} \frac{1}{2}\,\varrho\,(\varphi_x^2 + \varphi_y^2)\, dy = \frac{m}{2\pi}\int_0^{2\pi/m} \frac{1}{4}\,\varrho\, A^2 c_0^2\, m\, e^{2m\eta}\, dx \\ &= \frac{1}{4}\,\varrho\, A^2 g\,[1 + A^2 m^2]\,. \end{aligned}\right\} \tag{27.43}$$

Composite waves. Previously in this section we have been discussing a wave of permanent type whose prototype is the first-order progressive wave of the form $\eta = A\cos m(x-ct)$. It is natural to inquire into the behavior of higher-order waves whose first-order prototype is composite, say

$$\eta = A_1 \cos m_1 (x - c_1 t) + A_2 \cos m_2 (x - c_2 t). \tag{27.44}$$

To find the corresponding second-order terms one may use Eqs. (10.11) and (10.12); the computations are tedious but not difficult. The third order would introduce modifications of both c_1 and c_2 and lead to a much longer computation. As might be expected in analogy with the theory of sound, the second-order terms introduce waves of wave numbers $m_1 - m_2$ and $m_1 + m_2$, as well as $2m_1$ and $2m_2$. The velocity potential to the second order is given by

$$\left.\begin{aligned}\Phi = {} & A_1 c_1 e^{m_1 y} \sin m_1 (x - c_1 t) + A_2 c_2 e^{m_2 y} \sin m_2 (x - c_2 t) + \\ & + 2A_1 A_2 \frac{m_1 m_2 (c_1 - c_2) g}{g(m_1 - m_2) - (m_1 c_1 - m_2 c_2)^2} e^{(m_1 - m_2) y} \sin\left[(m_1 - m_2) x - (m_1 c_1 - m_2 c_2) t\right].\end{aligned}\right\} \tag{27.41}$$

The profile is then computed from BERNOULLI'S law

$$\eta = -\frac{1}{g}\left[\Phi_t(x, \eta, t) + \frac{1}{2}(\Phi_x^2 + \Phi_y^2)\right]$$

with retention of only terms of first or second order [cf. Eqs. (10.9) and (10.11)]. We omit the rather long expression.

BIESEL (1952) has derived formulas for a composite wave with a finite number of components and for h finite. He computes a number of quantities of interest. However, the formulas are very long and will not be reproduced here.

Three-dimensional waves. By using the full three-dimensional equations as given in (27.5) to (27.7) one may develop a higher-order theory of doubly modulated waves analogous to those considered in Sect. 14γ by first-order theory. This has been done by FUCHS (1952) and SRETENSKII (1954) to whose papers we refer for the resulting motion.

Further references. Development of systematic methods of computation of higher-order approximations has recently attracted the attention of several persons. Among these are SRETENSKII (1952), BORGMAN and CHAPPELEAR (1957), DAUBERT (1957, 1958) in a series of notes, JOLAS (1958) and NORMANDIN (1957). SRETENSKII (1953, 1955) has investigated the higher-order theory of wave motion resulting from a moving pressure distribution and waves in a circular canal.

β) Standing waves. As will be evident below, the formulation of a higher-order theory of standing waves is somewhat clumsier than that for progressive waves of permanent type. Part of the difficulty stems from the fact that one necessarily must deal with one more variable, namely t. The type of motion we are seeking will be represented by a profile $\eta(x, t)$ and a velocity potential $\Phi(x, y, t)$ periodic in both x and t:

$$\eta(x + r\lambda,\ t + s\tau) = \eta(x, t), \quad \Phi(x + r\lambda, y, t + s\tau) = \Phi(x, y, t). \tag{27.46}$$

If we fix the wave length $\lambda = 2\pi/m$, then the period $\tau = 2\pi/\sigma$ will have to be determined as one of the unknowns of the problem. In addition, we wish to have the first-order standing wave $\eta = A\cos mx \cos\sigma t$ of Sect. 14α serve as a prototype and first-order solution of the more general problem. As a further condition, we shall suppose the motion to be symmetric with respect to a vertical line through a crest.

RAYLEIGH (1915) was apparently the first to consider this problem. It was later attacked in an entirely different way, using Lagrangian coordinates, by SEKERZH-ZENKOVICH (1947, 1951 a, b, 1952), who treated both two- and three-dimensional waves for infinite depth, two-dimensional waves for finite depth, and composite waves for infinite depth. PENNEY and PRICE (1952), following approximately RAYLEIGH's method, carried the approximation for two-dimensional motion and $h=\infty$ to the fifth order, and to the second order for h finite and for doubly modulated standing waves. The method used below is a modification of theirs. The two-dimensional problem has recently been studied in a series of notes by CHABERT D'HIÈRES (1957, 1958). CARRY (1953) has carried to the second-order the superposition of two standing waves of the same wave length but 90° out of phase and of differing first-order amplitudes. INGRAHAM (1954) has carried to the second order the stability analysis of superposed two-fluid systems discussed at the beginning of Sect. 26α.

Since η and Φ are periodic in both x and t, we may expand each in a double Fourier series. However, it is also necessary to bring into the form of the series some indications of orders of magnitudes of the components, and in such a way that the first-order term is of the desired sort. We assume the following expansions for an infinitely deep fluid:

$$\left.\begin{aligned}\sigma &= \sigma_0+\varepsilon\,\sigma_1+\varepsilon^2\sigma_2+\cdots,\\ \eta(x,t) &= \sum_{r=1}^{\infty}\varepsilon^r\eta^{(r)}=\sum_{r=1}^{\infty}\varepsilon^r\sum_{p,q=0}^{r}\left[a_{pq}^{(r)}\cos q\,\sigma t+b_{pq}^{(r)}\sin q\,\sigma\,t\right]\cos p\,m\,x,\\ \Phi(x,y,t) &= \sum_{r=1}^{\infty}\varepsilon^r\Phi^{(r)}=\sum_{r=1}^{\infty}\varepsilon^r\sum_{p,q=1}^{r}\left[c_{pq}^{(r)}\cos q\,\sigma\,t+d_{pq}^{(r)}\sin q\,\sigma t\right]e^{p m y}\cos p\,m\,x.\end{aligned}\right\}\tag{27.47}$$

We may immediately set $d_{p0}^{(r)}=0$, $b_{p0}^{(r)}=0$ and with no loss of generality also $c_{00}^{(r)}=0$. Since the mean water level has been fixed at $y=0$, we must also have $a_{00}^{(r)}=0$. We shall again take $\varepsilon=A\,m$, where A is the amplitude of the first-order term.

Substitution of (27.47) into the exact kinematic and dynamic boundary conditions,

$$\left.\begin{aligned}&\Phi_x(x,\eta,t)\,\eta_x-\Phi_y+\eta_t=0\\ &\Phi_t+\tfrac{1}{2}(\Phi_x^2+\Phi_y^2)+g\,\eta-T'(R_1^{-1}+R_2^{-1})=0,\end{aligned}\right\}\tag{27.48}$$

results, as in Sects. 10α and 27α, in a series of equations for successive determination of the coefficients $a_{pq}^{(r)},\ldots,d_{pq}^{(r)}$ and $\sigma_0,\sigma_1,\ldots$. Because of the assumed form of the solution, the equations are now always linear equations between the coefficients. The boundary conditions for $\Phi^{(1)}$ and $\eta^{(1)}$, namely,

$$\left.\begin{aligned}&\Phi_y^{(1)}-\eta_t^{(1)}=0,\\ &\Phi_t^{(1)}+g\,\eta^{(1)}-T'\,\eta_{xx}^{(1)}=0,\end{aligned}\right\}\tag{27.49}$$

yield

$$\left.\begin{aligned}&-\sigma_0\,a_{01}^{(1)}\sin\sigma_0 t+\sigma_0\,b_{01}^{(1)}\cos\sigma_0 t=0,\\ &-\sigma_0\,a_{11}^{(1)}\sin\sigma_0 t+\sigma_0\,b_{11}^{(1)}\cos\sigma_0 t-k\left[c_{10}^{(1)}+c_{11}^{(1)}\cos\sigma_0 t+d_{11}^{(1)}\sin\sigma_0 t\right]=0;\end{aligned}\right\}\tag{27.50}$$

$$\left.\begin{aligned}&g\left[a_{01}^{(1)}\cos\sigma_0 t+b_{01}^{(1)}\sin\sigma_0 t\right]+\left[-\sigma_0\,c_{01}^{(1)}\sin\sigma_0 t+\sigma_0\,d_{01}^{(1)}\cos\sigma_0 t\right]=0,\\ &(g+m^2T')\left[a_{10}^{(1)}+a_{11}^{(1)}\cos\sigma_0 t+b_{11}^{(1)}\sin\sigma_0 t\right]+\\ &\qquad+\left[-\sigma_0\,c_{11}^{(1)}\sin\sigma_0 t+\sigma_0\,d_{11}^{(1)}\cos\sigma_0 t\right]=0.\end{aligned}\right\}\tag{27.51}$$

From these follow immediately

$$a_{01}^{(1)} = b_{01}^{(1)} = c_{10}^{(1)} = a_{10}^{(1)} = 0, \quad d_{11}^{(1)} = -\frac{\sigma_0}{m} a_{11}^{(1)}, \quad c_{11}^{(1)} = -\frac{\sigma_0}{m} b_{11}^{(1)}, \tag{27.52}$$

and

$$\sigma_0^2 = g m + m^3 T'. \tag{27.53}$$

We shall in addition fix the phase by making the arbitrary choice

$$a_{11}^{(1)} = \frac{1}{m}, \quad b_{11}^{(1)} = 0, \tag{27.54}$$

so that

$$\eta^{(1)} = \frac{1}{m} \cos m x \cos \sigma_0 t, \quad \Phi^{(1)} = -\frac{c_0}{m^2} \cos m x \sin \sigma_0 t. \tag{27.55}$$

This is a rather clumsy way to derive a first-order solution which was found much more directly earlier in Sect. 14α. However, it provides a caricature of the procedure necessary at each new stage of approximation. Since the higher-order approximations lead to extremely tedious calculations, they will be completely omitted and only the results given.

The profile and velocity potential through the second order are given by

$$\left.\begin{aligned} \eta &= A \cos \sigma_0 t \cos m x + \frac{1}{4} A^2 m \frac{g + m^2 T'}{g + 4m^2 T'} \cos 2m x + \\ &\quad + \frac{1}{4} A^2 m \frac{g + m^2 T'}{g - 2m^2 T'} \cos 2\sigma_0 t \cos 2m x, \\ \Phi &= -A \frac{\sigma_0}{m} \sin \sigma_0 t \, e^{m y} \cos m x + \frac{1}{4} A^2 \sigma_0 \sin 2\sigma_0 t - \\ &\quad - \frac{1}{4} A^2 \sigma_0 \frac{3m^2 T'}{g - 2m^2 T'} \sin 2\sigma_0 t \, e^{2m y} \cos 2m y, \end{aligned}\right\} \tag{27.56}$$

for $m^2 \neq g/2T'$; here $\sigma_1 = 0$. If $m^2 = g/2T'$, the situation is similar to that discussed in Sect. 27α following (27.13). For this value of m we must start with a first-order solution of the form:

$$\left.\begin{aligned} \Phi^{(1)} &= -\frac{\sigma_0}{m^2} [\sin \sigma_0 t \, e^{m y} \cos m x + (b_1 \sin 2\sigma_0 t - b_2 \cos 2\sigma_0 t) e^{2m y} \cos 2m x], \\ \eta^{(1)} &= \frac{1}{m} [\cos \sigma_0 t \cos m x + (b_1 \cos 2\sigma_0 t + b_2 \sin 2\sigma_0 t) \cos 2m x]. \end{aligned}\right\} \tag{27.57}$$

The values of b_1, b_2 and σ_1 are now determined by the second-order equations and are

$$b_1 = \pm \tfrac{1}{2} \quad b_2 = 0, \quad \sigma_1 = \pm \tfrac{1}{8} \sigma_0. \tag{27.58}$$

Thus the first-order profile for $m^2 = g/2T'$ is

$$\eta = A \cos \sigma_0 t \cos m x \pm \tfrac{1}{2} A \cos 2\sigma_0 t \cos 2m x. \tag{27.59}$$

The amplitude relation between the two first-order modes is the same as for progressive waves of this length.

The expression for the third-order standing wave is very clumsy if T' is retained. Also, as might be expected from analogy with the progressive wave, another apparent singularity appears for $m^2 = g/3T'$. If one sets $T' = 0$, the expressions for η and Φ become much simpler and are as follows:

$$\left.\begin{aligned}
\eta &= A\cos\sigma t\cos m x + \frac{1}{4}A^2 m\cos 2m x + \frac{1}{4}A^2 m\cos 2\sigma t\cos 2m x + \\
&\quad + \frac{1}{32}A^3 m^2\,[-2\cos 3\sigma t\cos m x + 9\cos\sigma t\cos 3m x + \\
&\qquad\qquad + 3\cos 3\sigma t\cos 3m x;\\
\Phi &= -\frac{\sigma A}{m}\sin\sigma t\, e^{m y}\cos m x + \frac{1}{4}\sigma A^2\sin 2\sigma t + \\
&\quad + \frac{5}{32}\sigma m A^3\sin 3\sigma t\, e^{m y}\cos m x + \frac{3}{16}\sigma m A^3\sin 3\sigma t\, e^{3m y}\cos 3m x;\\
\sigma &= \sqrt{g m}\left(1 - \frac{1}{8}A^2 m^2\right).
\end{aligned}\right\}\qquad(27.60)$$

As has been mentioned earlier, the approximation has been carried to the fifth order by PENNEY and PRICE (1952). However, it is not necessary to carry the approximation so far in order to see some important features of the motion, namely the sharpening of the crests and flattening of the troughs, the absence of any nodal points and the decrease of frequency with amplitude. One interesting feature does require carrying the approximation to at least the fourth order: this is the absence of any time during a period when the surface is completely flat. In connection with an experimental test of a predicted standing wave of greatest amplitude-length ratio by PENNEY and PRICE, G.I. TAYLOR (1953) has also provided an experimental verification of the correctness of the theory in an extreme case.

Orbits. The method of computation of orbits including higher-order terms is the same as that outlined at the end of Sect. 27α and we omit a detailed exposition. For infinite depth and $T' = 0$ the orbits to the second order are given by

$$\left.\begin{aligned}
x &= x_0 - A\, e^{m y_0}\sin m x_0\cos\sigma_0 t,\\
y &= y_0 + A\, e^{m y_0}\cos m x_0\cos\sigma_0 t + \tfrac{1}{4}A^2 m\, e^{2m y_0}\cos 2\sigma_0 t.
\end{aligned}\right\}\qquad(27.61)$$

The effect of the last term in y is easily seen to be a small wiggle superposed on the first-order straight-line trajectories discussed in Sect. 14α, except directly beneath the crests where the trajectory is still vertical but with the midpoint somewhat above the equilibrium position.

Pressure distribution. A particularly interesting consequence of keeping second-order terms appears in the behavior of the pressure distribution. From (27.56) and BERNOULLI's theorem one finds for the average pressure over a wave length at depth y

$$\left.\begin{aligned}
\overline{p - p_0} &= \frac{1}{\lambda}\int_0^\lambda (p - p_0)\,dx = -\varrho g y - \frac{1}{4}\varrho A^2\sigma_0^2 e^{2m y} + \\
&\quad + \frac{1}{4}\varrho A^2\sigma_0^2 e^{2m y}\cos 2\sigma_0 t - \frac{1}{2}\varrho A^2\sigma_0^2\cos 2\sigma_0 t.
\end{aligned}\right\}\qquad(27.62)$$

The terms with e^{2my} as a factor drop off quickly. However, the last term is independent of y and at all depths yields a fluctuation about the hydrostatic pressure with double the frequency of the standing waves. The existence of this depth-independent fluctuation, deriving from the term Φ_t in BERNOULLI's theorem and the purely time-dependent term in Φ, was pointed out by MICHE (1944, p. 73). The matter has been investigated more intensively by LONGUET-HIGGINS (1950) who has extended the theory to include a more general wave motion and compressibility of the fluid. He has further applied the theory to give a plausible explanation of recorded microseisms. KIERSTEAD (1952) has extended LONGUET-HIGGINS' analysis to include two-fluid systems. COOPER and LONGUET-HIGGINS (1951) have carried out laboratory experiments showing excellent agreement with the predicted pressure distribution for both progressive and standing waves.

Finite depth. Computations of the surface profile, particle orbits and other quantities for finite depth have been carried to the third order by SEKERZH-ZENKOVICH (1951) and CARRY and CHABERT D'HIÈRES (1957). We reproduce here the results only to the second order (for pure gravity waves):

$$\left.\begin{aligned}
\eta &= A\cos\sigma t\cos m x + \frac{1}{8}A^2 m\tanh m h\times\\
&\quad\times\left[1+\coth^2 m h-\coth^2 m h\,(3\coth^2 m h-1)\cos 2\sigma t\right]\cos 2m x;\\
\Phi &= -A\frac{\sigma}{m}\,\frac{\cosh m(y+h)}{\sinh m h}\sin\sigma t\cos m x+\frac{1}{16}A^2\sigma(3+\coth^2 m h)\sin 2\sigma t+\\
&\quad+\frac{3}{8}A^2\sigma\frac{\coth m h}{\sinh^2 m h}\,\frac{\cosh 2m(y+h)}{\sinh 2m h}\sin 2\sigma t\cos 2m x,\\
\sigma^2 &= \sigma_0^2 = g\,m\tanh m h,\qquad \sigma_1=0.
\end{aligned}\right\}\qquad(27.63)$$

The pressure averaged over a wave length [cf. (27.62)] is

$$\left.\begin{aligned}
\overline{p-p_0} &= -\varrho g y+\frac{1}{8}\,\frac{A^2\sigma^2}{\sinh^2 m h}\left[1-\cosh 2m(y+h)-\right.\\
&\quad\left.-(2\cosh 2m h-\cosh 2m(y+h)-1)\cos 2\sigma t\right].
\end{aligned}\right\}\qquad(27.64)$$

On the bottom, $y=-h$, one finds

$$\overline{p-p_0}=\varrho g h-\tfrac{1}{2}\varrho A^2\sigma^2\cos 2\sigma t.\qquad(27.65)$$

We note that here also, as in the case of progressive waves, the importance of the second-order terms in η and Φ increases as $mh\to 0$.

γ) Waves in a viscous fluid. The Eqs. (10.2) to (10.4), used in Sect. 25 in developing the first-order theory of waves in a viscous fluid, may be considered as the first in a sequence for the determination of higher-order approximations. Although the formulation of the equations appears to be straight forward, if laborious, the higher-order theory does not seem to have attracted many investigators. HARRISON (1909) made a second-order investigation of progressive waves and LONGUET-HIGGINS (1953) has recently made an elaborate study of both progressive and standing waves in an attempt to explain certain observed features of mass transport velocities. We shall not attempt to summarize either paper. However, the following results, taken from HARRISON, may be of interest. For the wave profile to the second order he gives the following expression when ν is small [cf. Eq. (25.22)]:

$$\left.\begin{aligned}
\eta &= A\,e^{-2\nu m^2 t}\cos(m x-\sigma_0 t)+\\
&\quad+A^2 e^{-4\nu m^2 t}\left[\frac{1}{2}m\cos 2(m x-\sigma_0 t)-m^2\left(\frac{\nu^2}{4g m}\right)^{\frac{1}{4}}\sin 2(m x-\sigma_0 t)\right],
\end{aligned}\right\}\qquad(27.66)$$

where $\sigma_0^2=gm$. The effect of viscosity, besides damping, is to make the leading side of the crest steeper than the trailing side. According to HARRISON the average horizontal velocity of a particle, again for small ν, is

$$\left.\begin{aligned}
&A^2\sigma_0 m\,e^{2my-4\nu m^2 t}-A^2 m^2\sqrt{\tfrac{1}{2}\sigma_0\nu}\times\\
&\qquad\times\left[(4\cos l_2 y+\sin l_2 y)\,e^{(m+l_1)y}+\sin 2m y\right]e^{-4\nu m^2 t}+\\
&\qquad+A^2 m^3\nu\left[4\,e^{(m+l_1)y}\sin l_2 y+3\,e^{2l_1 y}\right]e^{-4\nu m^2 t},
\end{aligned}\right\}\qquad(27.67)$$

where, as in (25.19), $l = l_1 + i l_2$ and $\nu(l^2 - m^2) = \omega \approx -2\nu m^2 + i\sigma_0$. This formula should be compared with $A^2 m^2 c_0 \, e^{2my}$ computed from (27.39), to which it reduces when $\nu = 0$.

E. Shallow-water waves

This chapter will deal with special solutions based on the shallow-water approximation, following the method of FRIEDRICHS (1948) as presented in subsection 10β. The shallow-water approximation for the waves over a rigid bottom yields a set of nonlinear equations [cf. (10.32)] even in the first approximation. If these equations are then linearized, they result in a hyperbolic-type equation which reduces to the simple wave equation for a flat horizontal bottom. Consequently, the solutions resulting from the shallow-water approximation are completely different in character from those resulting from the infinitesimal-wave approximation of subsection 10α and Chap. D, which resulted in linear equations and linear boundary conditions. That is, the shallow-water approximation leads to nonlinear hyperbolic-type equations, whereas the infinitesimal-wave approximation leads to a set of linear equations satisfying the boundary conditions and having each successive approximation to the velocity potential satisfy the simplest elliptic equation, namely the Laplace equation.

After the first-order shallow-water approximation (10.32) has been applied to several problems, the method of FRIEDRICHS (1948) and KELLER (1948) will be extended to obtain the second and third approximations of the shallow-water theory and thereby present, for the first time, the exact second approximation to the cnoidal wave of KORTEWEG and DE VRIES (1895), and the solitary wave of BOUSSINESQ (1871), and RAYLEIGH (1876). These higher-order approximations lead directly to relations predicting the maximum heights of cnoidal waves and solitary waves.

28. The fundamental equations for the first approximation. The shallow-water expansion method introduced by FRIEDRICHS (1948) is discussed in Sect. 10. For this application the expansion parameter ε was selected so that the first approximation would be identical to the nonlinear equations of the classical shallow-water theory, which is based on the assumption of hydrostatic pressure variation throughout and neglect of the variation of the horizontal velocity components with depth, so that the complicated boundary-value problem is greatly simplified to the following nonlinear equations:

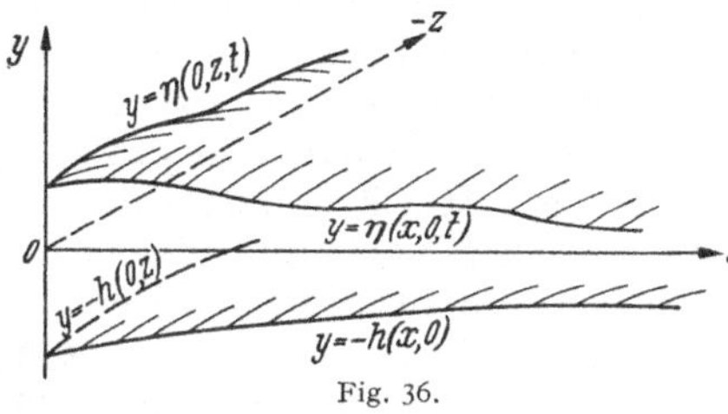

Fig. 36.

$$
\left.
\begin{aligned}
u_t + u\,u_x + w\,u_z &= -g\,\eta_x,\\
w_t + u\,w_x + w\,w_z &= -g\,\eta_z,\\
\eta_t + [u(\eta + h)]_x + [w(\eta + h)]_z &= 0
\end{aligned}
\right\}
\tag{28.1}
$$

[see LAMB (1932, p. 254) or STOKER (1957, p. 23)]. The coordinates and notation are shown in Fig. 36.

The set of nonlinear equations (28.1) is identical to (10.32) and is the first approximation in FRIEDRICHS' (1948) shallow-water expansion method as discussed in Sect. 10; this provides some mathematical justification for these classical equations. It is evident that the higher-order approximations following (10.23) and (10.33) also require that ε be sufficiently small; consequently, as will be shown, this expansion method is applicable if the product of water depth and surface curvature is small. Therefore, in some cases, this shallow-water theory is applicable to extremely large water depths as long as the wave length is sufficiently long, the most common application being to tidal waves, that is, the oceanic tides produced by the gravitational action of the sun and the moon [see, e.g., Lamb (1932) or Defant's article in Vol. XLVIII of this Encyclopedia].

The mathematical justification for this shallow-water expansion method, at least for special cases, lies in the existence proof of Friedrichs and Hyers (1954) for the solitary wave, and the existence proof of Littman (1957) for the more general cnoidal waves. Both of these proofs demonstrate that this expansion method converges to the exact solutions for these particular problems.

The nonlinear first approximation given by (28.1) is considerably simplified if the rigid bottom surface $h(x, z)$ is flat and horizontal, as may be seen by letting $h=$const so that (28.1) may be written as

$$\left.\begin{aligned} &u_t + u u_x + w u_z = -g(\eta + h)_x, \\ &w_t + u w_x + w w_z = -g(\eta + h)_z, \\ &(\eta + h)_t + [u(\eta + h)]_x + [w(\eta + h)]_z = 0. \end{aligned}\right\} \tag{28.2}$$

This is identical to the well known two-dimensional gas-dynamics equation [see, e.g., Lamb (1932)] if we write

$$\left.\begin{aligned} &\varrho(x, z, t) = [\eta(x, z, t) + h], \\ &\frac{\gamma p}{\varrho^2} = \frac{c^2}{\varrho} = \frac{c^2}{\eta + h} = g = \text{const}. \end{aligned}\right\} \tag{28.3}$$

Since the isentropic gas relationship is $p=\text{const}\times\varrho^{\gamma}$, the first-order nonlinear shallow-water approximation for a flat horizontal bottom is identical to the isentropic two-dimensional gas flow having a specific heat ratio of $\gamma=2$. This is the basis of the so called hydraulic analogy which has been used for many experimental investigations [see, e.g., Stoker (1957)].

It must be noted, however, that this hydraulic analogy is only valid for a flat horizontal bottom, as may be seen by comparing (28.1) and (28.2), and even more important, it is valid only as a first approximation even for the nonlinear case. It will be shown in Sect. 31 that the second approximation to shallow-water theory yields finite-amplitude waves (the solitary wave or cnoidal waves) which can be propagated without a change in shape or form, a fact which completely invalidates the hydraulic analogy to compressible gas flow since (28.2), or the gas dynamics equation, predicts that any finite disturbance quickly forms a finite discontinuity, e.g. [see, e.g., Lamb (1932), pp. 278, 481)].

In Sect. 29, immediately following, it will be shown that even for the linearized first approximation the hydraulic analogy to compressible gas flow is limited to a flat horizontal bottom.

29. The linearized shallow-water theory. The first approximation to shallow-water theory can now be linearized by two different methods, each suitable for various problems. We shall assume that $u_z=w_x$, so that a velocity potential $\Phi(x, z, t)$ exists. The first method is more appropriate for investigating steady water flow in canals or rivers and consists of the following approximations for carrying out the linearization:

$$u(x, z) = U + \varphi_x \approx U, \quad w(x, z) = \varphi_z \ll U, \tag{29.1}$$

$$\eta(x, z) \ll h(x, z), \tag{29.2}$$

so that (28.1) is linearized to

$$\left(1 - \frac{U^2}{c^2}\right) \varphi_{xx} + \varphi_{zz} + \frac{(U + \varphi_x)\, h_x}{h} + \frac{\varphi_z h_z}{h} = 0, \tag{29.3}$$

$$c^2 = g\, h(x, z). \tag{29.4}$$

In agreement with the previous discussion, (29.3) corresponds to the linearized gas dynamics equation only if the bottom is flat and horizontal, i.e. if h is constant.

The second method of linearization corresponds to the classical tidal-wave theory, or long-wave theory [see, e.g., LAMB (1932, p. 254) or Eqs. (10.36)] and can be obtained by writing

$$u(x, z, t) = \Phi_x \ll 1, \qquad w(x, z, t) = \Phi_z \ll 1, \tag{29.5}$$

$$\eta(x, z, t) \ll h(x, z), \tag{29.6}$$

so that (28.1) is linearized to

$$(\Phi_{xx} + \Phi_{zz}) + \left(\frac{h_x}{h} \Phi_x + \frac{h_z}{h} \Phi_z\right) = \frac{1}{g\, h} \Phi_{tt}. \tag{29.7}$$

Again, as before, (29.7) corresponds to the linearized gas dynamic case, or the simple acoustic wave-propagation equation, only if the bottom is flat and horizontal. In this case the general solution of (29.7) for one-dimensional flow is the well known d'Alembert solution of the simple wave equation,

$$\Phi(x, t) = F(x - c\,t) + f(x + c\,t) \qquad c = \sqrt{g\,h} = \text{const}, \tag{29.8}$$

which is used to study long-wave-length oscillations in canals when the water is either at rest or moving with a velocity $U \ll c$. The limitation to small perturbations and constant h for one-dimensional flow allows (28.1) to be linearized to

$$\left.\begin{aligned} \Phi_{xt} &= u_t = -g\,\eta_x, \\ \eta_{xx} &= \frac{1}{gh}\,\eta_{tt} = \frac{1}{c^2}\,\eta_{tt}; \end{aligned}\right\} \tag{29.9}$$

various applications of this, including the canal theory of tides, are given in LAMB (1932, pp. 254—273) and DEFANT (1957).

For the case of a canal having a non-rectangular but constant cross-section, we may generalize (29.9) by defining the mean depth h as the undisturbed cross-sectional area S divided by the width b of the canal at the undisturbed free water surface [see LAMB (1932), p. 256]. When the canal has a variable depth $h(x)$ and the disturbance may be considered one-dimensional, then (29.7) may be written in terms of the varying cross-sectional area $S(x)$ for constant width b as follows

$$\frac{1}{g\,h} \Phi_{tt} = \frac{1}{h} (h\,\Phi_x)_x = \frac{1}{S} (S\,\Phi_x)_x, \qquad S(x) = b\,h(x), \qquad b = \text{const}. \tag{29.10}$$

Then from (29.9) we obtain

$$\frac{1}{S} (S\,\eta_x)_x = \frac{1}{g\,h} \eta_{tt}, \tag{29.11}$$

which is the same as the expression derived by GREEN (1838) for a canal that is varying in both width b and depth h so that

$$S(x) = h(x)\, b(x).$$

However, the exact linearized first order approximation is (29.7), and the form of this equation indicates that large values of $b'(x)$ would invalidate the one-dimensional assumption, especially if h_z is relatively large. This is also indicated by LAMB (1932, p. 274). However, (29.7) provides the rigorous proof that (29.10) is applicable to one-dimensional, long-wavelength, small-amplitude disturbances in a canal of rectangular cross-section having a constant width and a varying depth.

If we now limit our analysis to long wave lengths having a simple harmonic oscillation of frequency $\sigma/2\pi$, so that we may write

$$\eta(x,t) = \eta(x)\sin(\sigma t+\tau), \qquad \Phi(x,t) = \varphi(x)\cos(\sigma t+\tau),$$

Eqs. (29.10) and (29.11) reduce to

$$\frac{1}{S}(S\varphi_x)_x + \frac{\sigma^2}{g\,h}\varphi = 0, \qquad \frac{1}{S}(S\eta_x)_x + \frac{\sigma^2}{g\,h}\eta = 0. \tag{29.12}$$

If we solve these equations in order to determine the harmonic oscillations in long canals with various special choices of varying cross-section, boundary conditions at the ends of the canal or finiteness conditions may further limit the allowable values of the frequency to a sequence of eigenvalues or fundamental frequencies, $\sigma_1, \sigma_2, \ldots$. Associated with each σ_i there is an η_i and Φ_i. The general solution of the Eqs. (29.12) is then a superposition of these characteristic solutions,

$$\eta(x,t) = \sum A_n\eta_n(x)\sin(\sigma_n t+\tau_n), \qquad \Phi(x,t) = \sum A_n\varphi_n(x)\cos(\sigma_n t+\tau_n),$$

where A_n and τ_n are arbitrary. Emphasis, however, is usually upon finding the fundamental mode σ_0, η_0 and the first few higher modes. We consider two special problems in order to illustrate the procedure. Other more complex situations are analyzed in LAMB (1932, p. 275 ff.) or in DEFANT (1957).

Let the canal be of rectangular cross-section with $h=h_0$, $b=\beta x$. We shall suppose it to be bounded at the ends by vertical walls at $x=x_1>0$, $x=x_2>x_1$. The Eq. (29.12) for φ now becomes

$$\varphi_{xx} + \frac{1}{x}\varphi_x + \frac{\sigma^2}{g\,h}\varphi = 0,$$

BESSEL's equation of order zero. The general solution is of the form

$$c\,J_0(\sigma x/c) + D\,Y_0(\sigma x/c), \qquad c^2 = g\,h.$$

The boundary conditions at the ends, $\phi'(x_1)=\phi'(x_2)=0$, can be satisfied only if

$$J_1(\sigma x_1/c)\,Y_1(\sigma x_2/c) - J_1(\sigma x_2/c)\,Y_1(\sigma x_1/c) = 0.$$

This equation determines the eigenvalues $\sigma_1, \sigma_2, \ldots$. The various modes of motion are then of the form

$$\left.\begin{aligned}\Phi_n = A_n\left[Y_1(\sigma_n x_2/c)\,J_0(\sigma_n x/c) - J_1(\sigma_n x_2/c)\,Y_0(\sigma_n x/c)\right]\cos(\sigma_n t+\tau_n),\\ n=1,2,\ldots\end{aligned}\right\} \tag{29.13}$$

If $x_1=0$, the solution Y_0 must be excluded because of its singularity at the origin and the eigenvalues are determined simply from $J_1(\sigma_n x_2/c)=0$, $n=1,2,\ldots$.

A solvable case in which h is variable is the canal of rectangular cross-section with $b=b_0$ and

$$h(x) = h_0\left(1-\frac{x^2}{L^2}\right).$$

Eq. (29.12) now becomes

$$\left(1-\frac{x^2}{L^2}\right)\varphi_{xx}-2\frac{x}{L^2}\varphi_x+\frac{\sigma^2}{g\,h_0}\varphi=0, \tag{29.14}$$

the equation for the spherical harmonics $P_\nu(x/L)$, $Q_\nu(x/L)$ with $(\sigma L)^2/g h_0=\nu(\nu+1)$. The condition that the solution should be finite on $|x|\leqq L$ requires one to discard Q_ν and further restricts ν to integers, thus determining the fundamental frequencies:

$$\sigma_n^2=\frac{g\,h_0}{L^2}\,n\,(n+1).$$

The fundamental solutions are then formed with Legendre polynomials:

$$\Phi_n=A_n\,P_n(x/L)\cos(\sigma_n t+\tau_n). \tag{29.15}$$

Motions of the type considered above may be identified with the long period oscillations called seiches which occur in certain lakes or canals throughout the world. Many applications are presented by CHRYSTAL (1905, 1906) and the periods observed in several lochs and lakes seem to correspond to those calculated by the linear shallow-water theory. The linear shallow-water equation (29.7) should be very suitable for the study of seiches because of their long period and relatively small amplitude. Usually the complete Eq. (29.7) must be solved numerically by the method of finite differences because the contour of the body of water is quite irregular and the depth variation is important.

When the motion cannot be considered one-dimensional, one must use the complete two-dimensional equations (29.7). If the motion is harmonic with frequency $\sigma/2\pi$, so that

$$\eta(x,z,t)=\eta(x,z)\sin(\sigma t+\tau),\qquad \Phi(x,z,t)=\varphi(x,z)\cos(\sigma t+\tau),$$

then the right-hand side of (29.7) is replaced by $-(\sigma^2/g h)\Phi$. However, just as in the one-dimensional case, the allowable values of σ may be restricted by the boundary conditions or finiteness conditions to a sequence of eigenvalues σ_1, $\sigma_2,\ldots$ with associated functions $\eta_1,\eta_2,\ldots,\Phi_1,\Phi_2,\ldots$. The general solution is again a superposition. We illustrate with several typical examples, but refer again to LAMB (1932, p. 282ff.) or DEFANT (1957) for a more comprehensive treatment.

Consider first a rectangular basin of constant depth h bounded by $x=0$, $x=x_0$, $z=0$, $z=z_0$. Then (29.7) becomes

$$\Phi_{xx}+\Phi_{zz}+\frac{\sigma^2}{g\,h}\Phi=0$$

and the boundary conditions are

$$\Phi_x(0,z,t)=\Phi_x(x_0,z,t)=\Phi_z(x,0,t)=\Phi_z(x,z_0,t)=0.$$

It is easy to verify that the fundamental solutions are

$$\Phi_{mn}=A_{mn}\cos\frac{m\pi x}{x_0}\cos\frac{n\pi z}{z_0}\cos(\sigma_{mn}t+\tau_{mn}) \tag{29.16}$$

where the eigenvalues σ_{mn} are given by

$$\sigma_{mn}=\pi\sqrt{g\,h}\sqrt{\left(\frac{m}{x_0}\right)^2+\left(\frac{n}{z_0}\right)^2}.$$

The result should be compared with (23.14) which reduces to this when $m_0 h$ is small enough so that $\tanh m_0 h \cong m_0 h$.

As another example consider a basin of circular planform of radius a and depth h. In polar coordinates, $x = r\cos\vartheta$, $y = r\sin\vartheta$, Eq. (29.7) becomes

$$\Phi_{,rr} + \frac{1}{r}\Phi_{,r} + \frac{1}{r^2}\Phi_{\vartheta\vartheta} + \frac{\sigma^2}{g h}\Phi = 0$$

and Φ must satisfy $\Phi_{,r}(a, \vartheta, t) = 0$. The fundamental solutions are easily found by separation of variables to be

$$\Phi_{mn} = A_{mn} J_n(\sigma_{mn} r/c)\cos(n\vartheta + \delta_{mn})\cos(\sigma_{mn} t + \tau_{mn}), \qquad c^2 = g h, \tag{29.17}$$

where the fundamental frequencies σ_{mn} are the roots of

$$J_n'(\sigma_{mn} a/c) = 0, \qquad m = 1, 2, \ldots.$$

The solution (23.15) again reduces to this if $\tanh m_0 h \cong m_0 h$.

If the planform is ring-shaped with the rings having radii a and $b<a$, then one needs also the solution Y_n in order to satisfy the boundary condition on $r = b$. (The singularity of Y_n at the origin obviously causes no difficulty, for it is not in the fluid). The fundamental solutions now become

$$\left.\begin{aligned}\Phi_{mn} = A_{mn}\left[Y_n'\left(\frac{\sigma_{mn}}{c}b\right)J_n\left(\frac{\sigma_{mn}}{c}r\right) - J_n'\left(\frac{\sigma_{mn}}{c}b\right)Y_n\left(\frac{\sigma_{mn}}{c}r\right)\right]\times\\ \times\cos(n\vartheta + \delta_{mn})\cos(\sigma_{mn} t + \tau_{mn}),\end{aligned}\right\} \tag{29.18}$$

where the fundamental frequencies σ_{mn} are determined from the equation

$$J_n'\left(\frac{\sigma a}{c}\right)Y_n'\left(\frac{\sigma b}{c}\right) - J_n'\left(\frac{\sigma b}{c}\right)Y_n'\left(\frac{\sigma a}{c}\right) = 0. \tag{29.19}$$

As before, the solution (23.16) reduces to this one for small $m_0 h$.

As a final example of two-dimensional seiches we consider the long-period simple harmonic oscillation in a shallow circular basin with depth variation depending only on r. Then, in polar coordinates (29.7) becomes

$$\Phi_{,rr} + \frac{1}{r}\Phi_{,r} + \frac{1}{r^2}\Phi_{,\vartheta\vartheta} + \frac{h_r}{h}\Phi_{,r} + \frac{\sigma^2}{g h}\Phi = 0. \tag{29.20}$$

If the depth variation is parabolic,

$$h(r) = h_0\left(1 - \frac{r^2}{a^2}\right),$$

LAMB (1932, p. 291) has shown that the fundamental solutions are given by

$$\Phi_{mn} = A_{mn}\left(\frac{r}{a}\right)^m\cos(m\vartheta + \delta_{mn})\,F\left(\alpha, \beta, \gamma; \frac{r^2}{a^2}\right)\cos(\sigma_{mn} t + \tau_{mn}), \tag{29.21}$$

where F is the hypergeometric series

$$F\left(\alpha, \beta, \gamma; \frac{r}{a}\right) = 1 + \frac{\alpha\beta}{1\cdot\gamma}\left(\frac{r}{a}\right)^2 + \frac{\alpha(\alpha+1)\beta(\beta+1)}{1\cdot 2\cdot\gamma(\gamma+1)}\left(\frac{r}{a}\right)^4 + \cdots,$$

and

$$\alpha = m + n, \qquad \beta = 1 - n, \qquad \gamma = m + 1.$$

The fundamental frequencies σ_{mn} are determined from

$$\frac{\sigma_{mn}^2 a^2}{g h_0} = 2m(2n-1) + 4n(n-1).$$

Both m and n must be integers in the above formulas. They simplify in an obvious fashion for the symmetric mode $m=0$.

It has been pointed out above in connection with several of the examples that the results obtained by analyzing the problem by means of the infinitesimal-wave approximation reduce to those obtained by the linearized shallow-water approximation if $mh=2\pi h/\lambda$ is small enough so that $\tanh mh \cong mh$. One should note that this holds also for the velocity of propagation of a periodic wave:

$$c=\sqrt{\frac{g\lambda}{2\pi}\tanh\frac{2\pi h}{\lambda}}=\sqrt{gh}\left[1-\frac{1}{6}\left(\frac{2\pi h}{\lambda}\right)^2+\cdots\right]=\sqrt{gh}\left[1+O\left(\frac{h}{\lambda}\right)^2\right]. \tag{29.22}$$

The remainder $O(h/\lambda)^2$ confirms the appropriateness of the term "long-wave approximation" sometimes applied to the shallow-water theory.

This exact agreement of the linearized results in the limiting case is encouraging justification for both the shallow-water approximation and the infinitesimal-wave approximation since they originate not only from different physical considerations, but also by entirely different types of mathematical approximation, as discussed in Sect. 10. The shallow-water approximation leads to hyperbolic type nonlinear equations, while the infinitesimal-wave approximation deals with linear elliptic equations. Stoker (1947, p. 32) gives a detailed comparison of the two linearized approximations for the case of wave motion over a flat bottom at a 6° slope.

α) Linearized shallow-water theory applied to two-dimensional steady flow. The first method of linearizing the shallow-water theory, as given by (29.3), is applicable to the determination of the variation in water depth for the steady flow in a shallow open channel or river. However, in practically all cases (29.3) must be solved numerically, so that it does not entail a prohibitive amount of extra labor to solve directly the more exact original nonlinear first-order equations (28.1) using the methods discussed in the next section (30) on nonlinear first-order theory. As a matter of fact, for supercritical flow, defined by $U>\sqrt{gh}$, the method of characteristics is very easy to use in the numerical solution for a nearly horizontal open channel having a flat bottom and varying width, as shown in Sect. 30. The subcritical case, having a flow velocity everywhere less than $\sqrt{gh}$, can be satisfactorily approximated by the one-dimensional hydraulic theory which assumes that the velocity at each cross-section $S(x)$ is independent of y and z. This method would yield, of course, a constant depth over a given cross-section and would therefore not be satisfactory for predicting the rise in water level about an island, or a jetty, or a pile in a swiftly moving relatively wide stream. For this particular application the linearized form of (29.3) is very useful, especially for subcritical flow, i.e. for $U^2/gh<1$.

We now consider the application of (29.3) to the problem of determining the water depth variation about a two-dimensional cylinder that is perpendicular to the bottom and has a narrow cross-section parallel to the flow as shown in Fig. 37. If the bottom is approximately flat and horizontal everywhere near the vertical cylinder, then we may consider h as constant and, providing that $U^2/gh<1$, write (29.3) as

$$\left.\begin{aligned}\beta^2\varphi_{xx}+\varphi_{zz}=0 \quad\text{or}\quad \frac{\partial^2\varphi}{\partial x^2}+\frac{\partial^2\varphi}{\partial(\beta z)^2}&=0,\\ \beta^2=1-F^2=1-\frac{U^2}{gh}&=\text{const}>0.\end{aligned}\right\} \tag{29.23}$$

The fundamental solution of (29.23), in view of (29.1), for two-dimensional profiles which may be considered symmetrical about the z-axis as shown in Fig. 37, is

$$\left.\begin{aligned} \varphi(x,z) &= +\frac{1}{2\pi}\int_0^L f(\xi)\ln\sqrt{(x-\xi)^2+(\beta z)^2}\,d\xi,\\ u &= U+\varphi_x = U+\frac{1}{2\pi}\int_0^L \frac{(x-\xi)\,f(\xi)\,d\xi}{(x-\xi)^2+(\beta z)^2},\\ w &= \varphi_z = \frac{\beta^2 z}{2\pi}\int_0^L \frac{f(\xi)\,d\xi}{(x-\xi)^2+(\beta z)^2}. \end{aligned}\right\} \tag{29.24}$$

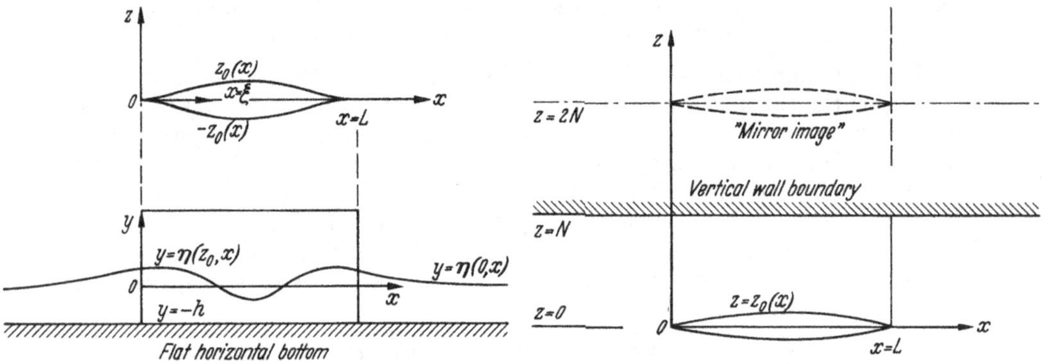

Fig. 37. Shallow-water flow about a two-dimensional symmetrical cylinder perpendicular to the flat horizontal bottom.

The boundary condition for the two-dimensional shape (see Fig. 37) is

$$\frac{dz_0}{dx} = \frac{w}{u} = \frac{\varphi_z(x,z_0)}{U+\varphi_x(x,z_0)} = \frac{\varphi_z}{U}\left[1+O\left(\frac{\varphi_x}{U}\right)\right], \tag{29.25}$$

where the same linearization procedure has been applied to the boundary condition as was used in deriving (29.3). Therefore (29.24) may also be similarly linearized by writing

$$U\frac{dz_0}{dx} = \varphi_z(x,z_0) = \frac{\beta}{2\pi}\int_0^L \frac{f(\xi)\left(\frac{d\xi}{\beta z_0}\right)}{1+\left(\frac{x-\xi}{\beta z_0}\right)^2}.$$

Hence if we let $\frac{(x-\xi)}{\beta z_0}=p$, then for $z_0 \geqq 0$,

$$U\frac{dz_0}{dx} = \frac{\beta}{2\pi}\int_{-\frac{L-x}{\beta z_0}}^{\frac{x}{\beta z_0}} \frac{f(x-\beta z_0 p)\,dp}{1+p^2} = \frac{1}{2}\beta f(x) + O(z_0^2).$$

Therefore

$$f(\xi) = \frac{2U}{\beta} z_0'(\xi) + O(z_0^2), \tag{29.26}$$

so that the linearized form of (29.24) is

$$\left.\begin{aligned} \frac{u(x,z)}{U} &= 1+\frac{1}{\pi\beta}\int_0^L \frac{(x-\xi)\,z_0'(\xi)\,d\xi}{(x-\xi)^2+(\beta z)^2} = 1+\frac{\varphi_x}{U},\\ \frac{w(x,z)}{U} &= \frac{z\beta}{\pi}\int_0^L \frac{z_0'(\xi)\,d\xi}{(x-\xi)^2+(\beta z)^2} = \frac{\varphi_z}{U}. \end{aligned}\right\} \tag{29.27}$$

On the actual surface of the two-dimensional profile (29.27) may be further linearized to

$$\left.\begin{aligned} \frac{u(x,z_0)}{U} &= 1+\frac{1}{\pi\beta}\left[\lim_{\varepsilon\to 0}\int\limits_0^{x-\varepsilon}+\int\limits_{x+\varepsilon}^{L}\frac{z_0'(\xi)\,d\xi}{x-\xi}\right] = 1+\frac{1}{\pi\beta}\,\mathrm{PV}\int\limits_0^L\frac{z_0'(\xi)}{x-\xi}\,d\xi,\\ \frac{w(x,z_0)}{U} &= z_0'(x). \end{aligned}\right\} \tag{29.28}$$

On the other hand, for large values of z we may write

$$\frac{u(x,z)}{U}\approx 1+\frac{1}{\pi\beta^3 z^2}\int\limits_0^L(x-\xi)\,z_0'(\xi)\,d\xi,\qquad \frac{w(x,z)}{U}\approx\frac{1}{\pi\beta z}\int\limits_0^L z_0'(\xi)\,d\xi. \tag{29.29}$$

The change $\eta(x,z)$ in the original constant water depth h can then be determined by the linearized relations corresponding to (29.1) and (29.2) as

$$\frac{\eta(x,z)}{h}+O\left(\frac{\eta}{h}\right)^2 = -F^2\frac{\varphi_x}{U}+O\left(\frac{\varphi_x^2+\varphi_z^2}{U^2}\right),\quad F^2=\frac{U^2}{g\,h}, \tag{29.30}$$

where for any (x, z) we obtain φ_x and φ_z from (29.27). For example, on the surface of the two-dimensional profile ($z=z_0$), (29.30) reduces to

$$\frac{\eta(x,z_0)}{h} = -\frac{U^2}{g\,h}\,\frac{1}{\pi\beta}\left[\mathrm{PV}\int\limits_0^L\frac{z_0'(\xi)\,d\xi}{x-\xi}+O(z_0^2)\right] \tag{29.31}$$

where φ_x and φ_z are both of $O(z_0')$.

These relations are, of course, completely restricted to flows that are everywhere subcritical since (29.23) shows that the Froude number ($F=U/\sqrt{g h}$) must be everywhere less than unity to keep $\beta>0$. The effect of increasing Froude number is to increase φ_x, and therefore decrease η, since β decreases. It is seen that this effect increases as z increases, the greatest effect being on $\varphi_x \sim 1/\beta^3$ in the limiting case of very large values of z as shown in (29.29). This relation, or preferably (29.27), could be used to predict the additional change in $\eta(x, z)$ due to a finite stream width by using the increment of φ_x from one mirror image to represent the first approximation to the channel boundary wall as indicated in Fig. 37. For slender cylinders in a narrow channel the "one-dimensional" approximation of Sect. 30γ is generally used, this allows an approximation for frictional head loss which becomes relatively more important as the channel width decreases.

For supercritical flow ($F=U/\sqrt{g h}>1$), (29.23) must be written as

$$\left.\begin{aligned} &B^2\varphi_{xx}-\varphi_{zz}=0\quad\text{or}\quad\frac{\partial^2\varphi}{\partial x^2}=\frac{\partial^2\varphi}{\partial(B z)^2},\\ &B^2=F^2-1=\frac{U^2}{g\,h}-1=\text{const}>0. \end{aligned}\right\} \tag{29.32}$$

Now, however, (29.32) cannot provide a satisfactory approximation of the change in water depth at some distance from the two-dimensional profile since its general solution is

$$\varphi(x,z)=G(x-B z)+g(x+B z), \tag{29.33}$$

which predicts no change, even upon approaching infinity, along the lines of constant slope $dz/dx=\pm 1/B=\pm[F^2-1]^{-\frac{1}{2}}$. Consequently the nonlinear method of characteristics, which will be described in Sect. 30, must be used in predicting the depth variation at any finite distance from the profile. Although the method of characteristics will directly and easily give the velocity distribution or depth variation on the profile itself, we will also derive the variation on the profile surface according to the linearized theory. The result will be of crucial importance in evaluating the validity of the nonlinear first-order shallow-water theory (28.1), since any great discrepancy between the linearized result and the nonlinear results from (28.1) would indicate that the perturbations involved are sufficiently large that the second-order shallow-water theory of Sect. 31 must be introduced.

The linearized solution of (29.32) for any sharp-nosed slender two-dimensional profile, as in Fig. 37, is obtained from the general solution (29.33) and the following linearized boundary condition:

$$z_0'(x) = \frac{\varphi_z(x, z_0)}{U} = -\frac{B}{U} G'(x - B z_0); \qquad z = z_0 > 0.$$

Therefore $G'(x - Bz_0) = -Uz_0'(x)/B$, so that on the profile surface, $z = z_0(x)$,

$$\left.\begin{aligned} u(x, z_0) &= U + \varphi_x = U + G'(x - B z_0) = U\left[1 - \frac{z_0'(x)}{B}\right], \\ w(x, z_0) &= U z_0'(x). \end{aligned}\right\} \tag{29.34}$$

Then the variation in water depth on the profile surface is given by (29.30) as

$$\frac{\eta(x, z_0)}{h} = F^2\left[\frac{z_0'(x)}{B} + O(z_0'^2)\right] \tag{29.35}$$

for flow that is everywhere supercritical, i.e. $B^2 = F^2 - 1 > 0$.

It should be noted that (29.23) and (29.32) are identical to the linearized potential equations for two-dimensional steady subsonic flow and supersonic flow, respectively, if we simply replace the Froude number ($F = U/\sqrt{gh}$) by the Mach number ($M = U/c$) [see (28.3)]. This is in complete accord with the statement that the hydraulic analogy is valid for the flow over a flat horizontal bottom (i.e., the flow is equivalent to the two-dimensional isentropic flow of a fictitous perfect gas having a specific heat ratio $\gamma = 2$). Consequently, Eqs. (29.24) through (29.29) are identical to these for subsonic flow about slender two-dimensional profiles in free air or in a wind tunnel of rectangular cross-section as derived by LAITONE (1946). These equations confirm the known result that the linearized equations are independent of the value of the specific heat ratio γ. Similarly, Eq. (29.34) is identical to the well-known linearized two-dimensional supersonic-flow solution if we let $F^2 - 1 \equiv B^2 = M^2 - 1 > 0$.

Although these linearized results are very satisfactory for slender sharp-nosed profiles, they only apply for Froude numbers that are not too near unity, that is they are not applicable to flows near the critical velocity $U = \sqrt{gh} = c$, equivalent to sonic flow. For these cases we must return to the nonlinear equation (28.1), as discussed in Sect. 30.

30. Nonlinear shallow-water theory. This section will primarily discuss methods for obtaining solutions of the nonlinear equations (28.1) which provide the first-order approximation of the shallow-water theory. The special cases to be considered are the one-dimensional unsteady flow and the two-dimensional steady flow in open channels. This will provide a basis for discussing the one-dimensional assumption of open-channel flow. Finally the hydraulic jumps, and their relation to the first-order shallow-water theory, will be discussed.

α) One-dimensional non-steady, first-order, shallow-water theory. By assuming one-dimensional flow in the x direction only, the nonlinear equations (28.1) reduce to

$$\left.\begin{aligned} u_t + u\,u_x + g(\eta + h)_x &= g\,h_x, \\ (\eta + h)_t + [u(\eta + h)]_x &= h_t = 0. \end{aligned}\right\} \tag{30.1}$$

Again it should be noted that these are equivalent to the gas dynamic equations, upon introducing (28.3), only if the bottom is flat and horizontal, i.e. $h_x = 0$.

Now, if we let

$$\left.\begin{aligned} c^2(x,t) &= g[\eta(x,t) + h(x)], \\ 2c\,c_x &= g(\eta+h)_x, \qquad 2c\,c_t = g(\eta+h)_t, \end{aligned}\right\} \tag{30.2}$$

and give the initial conditions as $du/d\alpha$ and $dc/d\alpha$ along a curve in the (x, t)-plane defined by $x(\alpha)$, $t(\alpha)$, then we may write (30.1) as

$$\left.\begin{aligned} u\,u_x + u_t + 2c\,c_x + 0 &= g\,h_x, \\ c\,u_x + 0 + 2u\,c_x + 2c_t &= 0, \\ x_\alpha u_x + t_\alpha u_t + 0 + 0 &= \frac{du}{d\alpha}, \\ 0 + 0 + x_\alpha c_x + t_\alpha c_t &= \frac{dc}{d\alpha}, \end{aligned}\right\} \tag{30.3}$$

This set of four equations can be solved uniquely for u_x, u_t, c_x, c_t in terms of u, c, h_x and the initial conditions as long as the determinant of the coefficients in (30.3) does not vanish. This condition is violated along the characteristic curves $x(\alpha)$, $t(\alpha)$ defined by

$$\begin{vmatrix} u & 1 & 2c & 0 \\ c & 0 & 2u & 2 \\ x_\alpha & t_\alpha & 0 & 0 \\ 0 & 0 & x_\alpha & t_\alpha \end{vmatrix} = 0, \tag{30.4}$$

which may be easily expanded by the minors of the bottom row to give

$$x_\alpha^2 - 2u\,x_\alpha t_\alpha + (u^2 - c^2)\,t_\alpha^2 = [x_\alpha - (u-c)\,t_\alpha]\,[x_\alpha - (u+c)\,t_\alpha] = 0.$$

Therefore the characteristic curves, C_+ and C_-, are defined by

$$\frac{x_\alpha}{t_\alpha} = \left(\frac{dx}{dt}\right)_{C_\pm} = u(x,t) \pm c(x,t). \tag{30.5}$$

Since h_x is given, and appears only on the right-hand side of the first equation in (30.3), therefore the characteristic curves as defined in (30.5) are identical to those in the gas-dynamics case [see, e.g., COURANT and FRIEDRICHS (1948)]. However, the Riemann invariants, or quantities that can be constant along a characteristic curve, now depend upon the bottom slope, as may be seen by adding the two equations in (30.1) after introducing (30.2) so as to obtain

$$(u+2c)_t + (u+c)\,(u+2c)_x = \left[\frac{\partial}{\partial t} + (u+c)\frac{\partial}{\partial x}\right][u(x,t) + 2c(x,t)] = g\,h_x. \tag{30.6}$$

These give the same Riemann invariants as in the isentropic one-dimensional unsteady gas flow with a specific heat ratio $\gamma = 2$ only if $h_x = 0$ [see, e.g.,

COURANT and FRIEDRICHS (1948, p. 87)]. No simple Riemann invariant involving only u and c is possible if h_x varies with x; however, if h_x is constant, so that $g h_x = m = \text{const}$, then (30.6) may be written

$$\left[\frac{\partial}{\partial t} + (u + c)\frac{\partial}{\partial x}\right][u + 2c - m t] = 0. \tag{30.7}$$

Similarly, by subtracting the two equations in (30.1), we obtain

$$\left[\frac{\partial}{\partial t} + (u - c)\frac{\partial}{\partial x}\right][u - 2c - m t] = 0. \tag{30.8}$$

Consequently, the basic statements relating the characteristic curves and Riemann invariants of Eq. (30.1) with $g h_x = m = \text{const}$ may be summarized as follows:

$$\left.\begin{aligned} u + 2c - m t = R(x, t) &= \text{const along a curve } C_+ \\ &\text{defined by } \frac{dx}{dt} = u + c; \\ u - 2c - m t = -S(x, t) &= \text{const along a curve } C_- \\ &\text{defined by } \frac{dx}{dt} = u - c. \end{aligned}\right\} \tag{30.9}$$

Fig. 38 shows typical sets of curves in the (x, t)-plane. The above equations show that in any given region in the (x, t)-plane there are three basic types of solutions, namely:

(1) the constant steady state in which u and c remain constant everywhere in the region, so that all characteristics form straight lines;

(2) the general flow in which neither R nor S is constant in a finite region;

(3) the special case of a simple wave over a flat horizontal bottom ($m = 0$) wherein a constant steady-state region is separated from a varying region by a straight characteristic line along which either R or S is constant.

The first type of solution obviously has R and S constant throughout the region only if the bottom is flat and horizontal ($m = 0$). The second type of solution is complicated and can best be obtained by the method of finite differences [see, e.g., STOKER (1957, pp. 293—300)]. The third type of solution will now be discussed since it has considerable physical significance for many problems concerning the propagation of a disturbance into water that is originally at constant depth and constant velocity, and extends an unlimited distance for $x > 0$.

When a disturbance moves into still water at constant depth over a flat horizontal bottom ($m = 0$), then it is obvious that $(dx/dt)_0 = c(\infty)$ is the characteristic, now a straight line, which must continually separate the steady-state region from the disturbance region in the (x, t)-plane, as indicated in Fig. 38. This characteristic curve must be a straight line since there is a constant steady state always ahead of it so that $(dx/dt)_0 = \text{const}$ and therefore $x_0 = c(\infty)\, t$. Also, either R or S must be constant along the characteristic, and since R_0 corresponds to C_+^0 or $(dx/dt)_0 = c(\infty) > 0$, as in Fig. 38, therefore $R_0 = 2c(\infty) = \text{const}$. This type of simple wave, having $(dx/dt)_0 = c(\infty) > 0$ and $R_0 = 2c(\infty) = \text{const.}$, is called a forward-facing wave since the particle paths enter from the side with greater values of x, as in Fig. 38. The value of R varies as one passes from one to another C_+ characteristic inside the region of the disturbance since u and c both vary due to the disturbance and none of the C_+ characteristic lines can ever intersect C_+^0. However, every C_- characteristic intersects C_+^0, as shown in Fig. 38, and since S remains constant on any given C_- characteristic curve, therefore S is everywhere constant since every C_- characteristic must have the same value $S(x, t) = R_0 = 2c(\infty) = \text{const}$ on C_+^0.

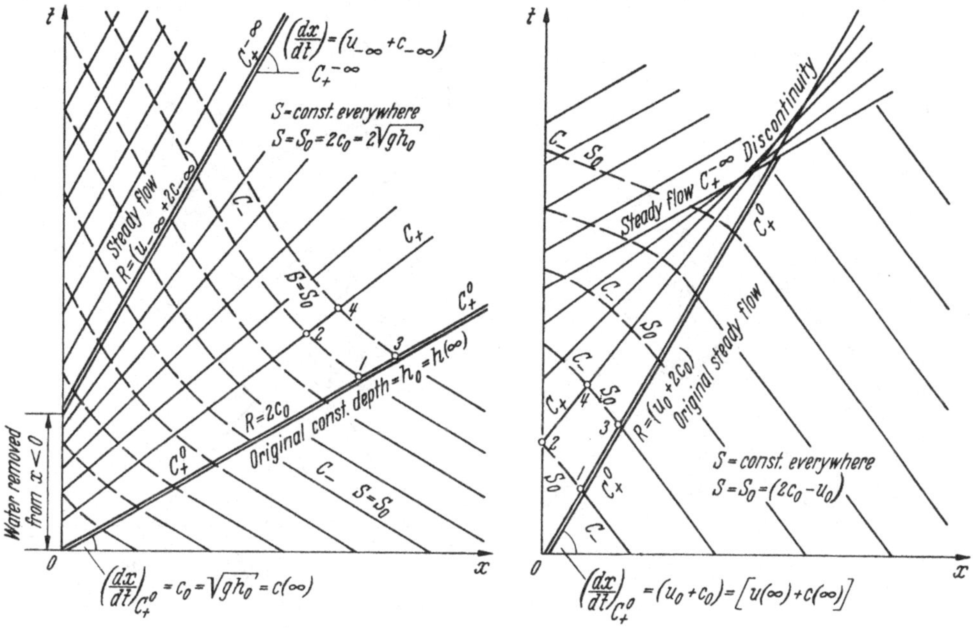

Fig. 38. Simple forward-facing waves, $S = \text{const.}$

The same considerations are true even if the water of constant depth into which the disturbance is being propagated is flowing with a constant velocity $u(\infty) < c(\infty)$. The only change is that now the following are constant:

$$\left(\frac{dx}{dt}\right)_0 = u(\infty) + c(\infty) > 0, \qquad R_0 = 2c(\infty) + u(\infty)$$

on C_+^0 only, while on all C_-,

$$2c(x, t) - u(x, t) = S(x, t) = 2c(\infty) - u(\infty) = \text{const.}$$

Similarly all R in the disturbance region vary as

$$R(x, t) = 2c(x, t) + u(x, t),$$

as indicated in Fig. 38 for the simple forward-facing (C_+^0) wave. As shown in Fig. 39 a simple backward-facing (C_-^0) occurs if $R = \text{const}$ and $S = 2c - u$. These waves are called simple waves because all the characteristics of the family for which the Riemann invariant takes on a different constant for each line form straight lines. For example, referring to Fig. 38, the forward-facing waves $(dx/dt > 0)$ have $S(x, t)$ constant everywhere and $R(x, t)$ varying so that the C_+ characteristics form straight lines. On the other hand, in Fig. 39 the backward-facing wave $(dx/dt < 0)$ has $R(x, t)$ constant everywhere and $S(x, t)$ varying, so that now only the C_- characteristics form straight lines. The characteristics of one family only must form straight lines in a simple wave because only one of the Riemann invariants (S or R) is constant in the entire region of the disturbance. For example, in the case of the forward-facing simple wave in Fig. 38, we have S constant in the region of the disturbance. Therefore from (30.9) and Fig. 38 we may write,

$$-S_1 = -S_2 = -S_3 = -S_4 = u_1 - 2c_1 = u_2 - 2c_2 = u_3 - 2c_3 = u_4 - 2c_4 = \text{const},$$
$$R_1 = R_3 = u_1 + 2c_1 = u_3 + 2c_3 \neq R_2 = R_4 = u_2 + 2c_2 = u_4 + 2c_4.$$

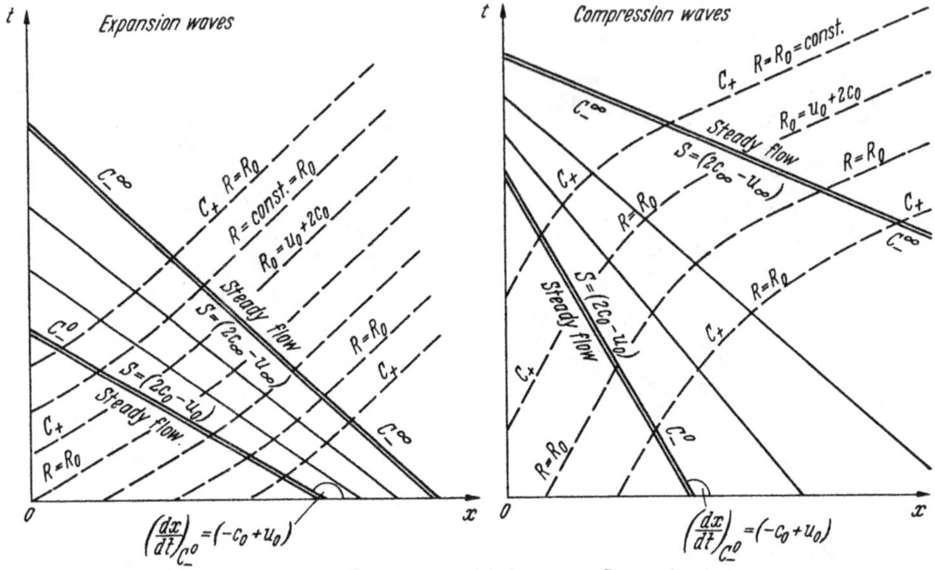

Fig. 39. Simple backward-facing waves, $R=\text{const.}$

Consequently, $u_1 = u_3$, $c_1 = c_3$, $u_2 = u_4$, $c_2 = c_4$, and $u_1 \neq u_2$, $c_1 \neq c_2$, $u_3 \neq u_4$, $c_3 \neq c_4$, so that

$$\left(\frac{dx}{dt}\right)_{C_-} = u_1 - c_1 \neq u_2 - c_2 \neq \text{const}, \qquad (C_- \text{ curved}),$$

$$\left(\frac{dx}{dt}\right)_{C_+^0} = u_1 + c_1 = u_3 + c_3 = \text{const}, \qquad (C_+^0 \text{ straight}),$$

$$\left(\frac{dx}{dt}\right)_{C_+} = u_2 + c_2 = u_4 + c_4 = \text{const}, \qquad (C_+ \text{ straight}).$$

It is important to note that these simple waves can exist only over a flat horizontal bottom, i.e. when $m=0$.

We have now shown how the method of characteristics for one-dimensional unsteady flow has resulted in the concept of the simple wave which quickly gives a numerical evaluation of the propagation of a one-dimensional disturbance into water of constant depth moving at constant speed. The solution of this problem in the (x, t)-plane can be obtained by direct application of (30.9). For example, the usual case of a forward-facing wave having S everywhere constant, and straight C_+ characteristic lines, as shown in Fig. 38, has the slope of the C_+ straight lines determined directly by the time history of the disturbance at $x=0$, and Eqs. (30.2) and (30.9) which show that

$$\left(\frac{dx}{dt}\right)_{C_+} = u(0,t) + c(0,t) = \text{const} = u(0,t) + \sqrt{g[h+\eta(0,t)]}. \qquad (30.10)$$

Along any given C_+ straight line having this constant slope

$$R(x,t)_{C_+} = u(0,t) + 2c(0,t) = \text{const} = u(0,t) + 2\sqrt{g[h+\eta(0,t)]}. \qquad (30.11)$$

Consequently, the values of u and c are determined in the entire region shown in Fig. 38 by the given values on the t-axis. The curved C_- characteristics need not be calculated, since the desired numerical solution is independent of them. Their existence, however, can lead to a simplification in the numerical calculation of (30.10) since, in the case shown in Fig. 38, each curved C_- characteristic extends from the C_+^0 characteristic to the t-axis, and on each and every C_- characteristic, $-S=u(\infty)-2c(\infty)=\text{const}$. Therefore, at every point on the t-axis that can be reached by a C_- characteristic we must have

$$S=2c(0,t)-u(0,t)=2c(\infty)-u(\infty)=2\sqrt{gh}-u(\infty)=\text{const}. \tag{30.12}$$

Of course the C_- characteristics can continue from C_+^0 to the t-axis only if $(dx/dt)_{C_-}=u-c<0$, or $u<c$, so that in this case (30.10) may be simplified to

$$\left.\begin{aligned}\left(\frac{dx}{dt}\right)_{C_+}&=u(0,t)+c(0,t)=\text{const}\\&=\tfrac{3}{2}u(0,t)-\tfrac{1}{2}[u(\infty)-2c(\infty)]=\tfrac{3}{2}u(0,t)-\tfrac{1}{2}u(\infty)+\sqrt{gh}\\&=3c(0,t)+[u(\infty)-2c(\infty)]=3\sqrt{g[h+\eta(0,t)]}+u(\infty)-2\sqrt{gh}.\end{aligned}\right\} \tag{30.13}$$

Consequently, the problem is solved in the region so defined if either $u(0,t)$ or $c(0,t)$ is alone given. The surface elevation is given by (30.2) as

$$h+\eta(x,t)=\frac{c^2(x,t)}{g},\qquad h=\frac{c^2(\infty)}{g}=\text{const} \tag{30.14}$$

in every case of disturbance propagations into a constant water depth over a flat horizontal bottom ($m=0$).

Many other physical problems can be simulated by giving the data along a prescribed curve in the (x,t)-plane for $x\leqq 0$; e.g., see STOKER (1957) where the disturbance created by the breaking of a dam, and the effect of moving a vertical end plate in a tank of still water of rectangular cross-section, $u(\infty)=0$, are considered. Since the bottom is flat and horizontal ($m=0$), all of the equations following (30.9) are equivalent to the gas-dynamics equations with a specific heat ratio $\gamma=2$. Consequently, the problems solved in COURANT and FRIEDRICHS (1948) for channels of finite length which produce wave reflections at either end are also applicable. In this hydraulic analogy to compressible flow it is important to remember that (30.13) is only applicable to subcritical flow, which is equivalent to subsonic gas flow, since we must have $(dx/dt)_{C_-}=u-c<0$, or $u(\infty)<c(\infty)=\sqrt{gh}$. When the flow is supercritical, so that $u(\infty)>c(\infty)=\sqrt{gh}$, corresponding to supersonic gas flow, then the slopes of both the C_+ and C_- characteristics are positive. Consequently the two families can meet in a cusp, and the C_- characteristics cannot intersect both the t-axis and the undisturbed steady supercritical state that lies at, and to the right of, C_+^0. Therefore, in order to apply (30.13) for supercritical flow, the region of the constant value of S, as given by (30.12), must be very carefully defined.

Another limitation on all the preceding equations is indicated for the compression wave depicted in Figs. 38 and 39. This limitation is defined by the envelope of the straight characteristic lines that *must* always form for a compression wave in this first-order theory, as will be proven later. This envelope of the straight characteristic lines corresponds to a discontinuity that can be interpreted as a discontinuity in η, or the breaking of the wave crest. This leads to the hydraulic jump or surge that will be discussed later. The gas dynamic case has the envelope of the straight characteristic lines interpreted as a steady-state shock wave [see, e.g., COURANT and FRIEDRICHS (1948, pp. 110–181)].

β) Two-dimensional, steady, supercritical flow by the first-order shallow-water theory. We will now investigate the characteristic curves of the nonlinear equations of the first-order shallow-water theory for the case of steady two-dimensional flow. We will find that real characteristic curves, which are a great aid to numerical calculations, exist only in the regions wherein the flow is everywhere supercritical.

If we consider the steady two-dimensional flow over a flat horizontal bottom, then we may write (28.2) as

$$\left.\begin{aligned}
&u\,u_x + w\,u_z = -g(\eta+h_0)_x = -(c^2)_x,\\
&u\,w_x + w\,w_z = -g(\eta+h_0)_z = -(c^2)_z,\\
&[u(\eta+h_0)]_x + [w(\eta+h_0)]_z = 0 \quad\text{or}\quad (u\,c^2)_x + (w\,c^2)_z = 0,\\
&u=\varphi_x,\quad w=\varphi_z,\quad u_z = w_x = \varphi_{xz}.
\end{aligned}\right\}\tag{30.15}$$

By multiplying the first equation by $u=\varphi_x$ and the second by $w=\varphi_z$, and adding, we obtain

$$\varphi_x^2\varphi_{xx} + 2\varphi_x\varphi_z\varphi_{xz} + \varphi_z^2\varphi_{zz} = -[\varphi_x(c^2)_x + \varphi_z(c^2)_z] = (\varphi_{xx}+\varphi_{zz})\,c^2. \tag{30.16}$$

Therefore

$$\left(\frac{\varphi_x^2}{c^2}-1\right)\varphi_{xx} + 2\frac{\varphi_x\varphi_z}{c^2}\varphi_{xz} + \left(\frac{\varphi_z^2}{c^2}-1\right)\varphi_{zz} = 0 \tag{30.17}$$

or

$$\left(1-\frac{u^2}{c^2}\right)\varphi_{xx} - 2\frac{u\,w}{c^2}\varphi_{xz} + \left(1-\frac{w^2}{c^2}\right)\varphi_{zz} = 0 \tag{30.18}$$

where $c^2(x,z)=g[h_0+\eta(x,z)]$ and h_0 now is the still water depth found whenever $(u^2+w^2)=0=\eta$. Note that (30.18) immediately linearizes to (29.3), so that the numerical differences between the solutions of (29.3) and (30.18) will provide an estimate of whether or not the second-order shallow-water theory, as discussed in Sect. 31, must be introduced.

The characteristic curves of (30.18) may be found in a manner similar to that used for (30.3) by finding the curve $[x(\alpha), z(\alpha)]$ in the (x,z)-plane along which prescribed values of φ_x and φ_z cannot determine φ_{xx}, φ_{xz} and φ_{zz}. Therefore we write

$$\left.\begin{aligned}
&\left(1-\frac{u^2}{c^2}\right)\varphi_{xx} + \left(-2\frac{u\,w}{c^2}\right)\varphi_{xz} + \left(1-\frac{w^2}{c^2}\right)\varphi_{zz} = 0,\\
&x_\alpha\varphi_{xx} + z_\alpha\varphi_{xz} + 0 = \frac{d\varphi_x}{d\alpha},\\
&0 + x_\alpha\varphi_{xz} + z_\alpha\varphi_{zz} = \frac{d\varphi_z}{d\alpha},
\end{aligned}\right\}\tag{30.19}$$

which may not have a solution if the determinant of the coefficient is zero, that is if

$$\begin{vmatrix} 1-\dfrac{u^2}{c^2} & -\dfrac{2u\,w}{c^2} & 1-\dfrac{w^2}{c^2}\\ x_\alpha & z_\alpha & 0\\ 0 & x_\alpha & z_\alpha \end{vmatrix} = 0, \tag{30.20}$$

or

$$\left.\begin{aligned}
&\left(1-\frac{u^2}{c^2}\right)z_\alpha^2 + 2\frac{u\,w}{c^2}z_\alpha x_\alpha + \left(1-\frac{w^2}{c^2}\right)x_\alpha^2 = 0,\\
&\frac{z_\alpha}{x_\alpha} = \left(\frac{dz}{dx}\right)_{C_\pm} = \frac{\dfrac{u\,w}{c^2} \pm \sqrt{\dfrac{u^2+w^2}{c^2}-1}}{\dfrac{u^2}{c^2}-1}
\end{aligned}\right\}\tag{30.21}$$

which therefore gives the slopes of the two families (C_+ and C_-) of characteristic curves. Now, however, entirely unlike the previous one-dimensional unsteady flow solution, the characteristic curves exist only for supercritical flow, i.e., for $u^2+w^2>c^2=g(h_0+\eta)$. The fact that the characteristic curves are real for supercritical flow means that in this case the nonlinear equation (30.17) is hyperbolic. However, for subcritical flow, since the characteristic curves are then complex functions, it is of elliptic type [see, e.g., COURANT and FRIEDRICHS (1948, pp. 40—55) or PREISWERK (1938)].

We can obtain a solution for the behavior of the quantity

$$F(x,z)=\sqrt{\frac{u^2+w^2}{c^2}}\geqq 1 \tag{30.22}$$

(which defines the Froude number of the supercritical flow) along a characteristic curve by transforming (30.18) into the (u, w)-plane, called the hodograph plane, through the use of the Legendre contact transformation which is given by [see, e.g., COURANT and FRIEDRICHS (1948, p. 249) or PREISWERK (1938)]

$$\begin{aligned}\chi &= (x\,\varphi_x+z\,\varphi_z-\varphi)=(x\,u+z\,w-\varphi),\\ d\chi &= (x\,du+u\,dx+z\,dw+w\,dz-d\varphi)=(x\,du+z\,dw).\end{aligned}$$

Hence

$$\begin{aligned}x&=\chi_n,\quad z=\chi_w,\\ dx&=x_u\,du+x_w\,d_w=\chi_{uu}\,du+\chi_{uw}\,dw,\\ dy&=z_u\,du+z_w\,dw=\chi_{uw}\,du+\chi_{ww}\,dw.\end{aligned}$$

Solving for du and dw, we obtain

$$\begin{aligned}du&=N^{-1}(\chi_{ww}\,dx-\chi_{uw}\,dz)=d\varphi_x=\varphi_{xx}\,dx+\varphi_{xz}\,dz,\\ dw&=N^{-1}(-\chi_{uw}\,dx+\chi_{uu}\,dz)=d\varphi_z=\varphi_{xz}\,dx+\varphi_{zz}\,dz,\end{aligned}$$

where

$$N=\begin{vmatrix}\chi_{uu} & \chi_{uw}\\ \chi_{uw} & \chi_{ww}\end{vmatrix}\neq 0,$$

so that

$$\varphi_{xx}=\frac{\chi_{ww}}{N},\qquad \varphi_{xz}=-\frac{\chi_{uw}}{N},\qquad \varphi_{zz}=\frac{\chi_{uu}}{N}.$$

The nonlinear equation (30.17) in the physical (x, z)-plane is transformed into the following linear equation in the hodograph (u, w)-plane:

$$\left(\frac{w^2}{c^2}-1\right)\chi_{uu}-2\frac{u\,w}{c^2}\chi_{uw}+\left(\frac{u^2}{c^2}-1\right)\chi_{ww}=0. \tag{30.23}$$

The same procedure used in (30.19) through (30.21), or a simple comparison of (30.17), (30.21), and (30.23), shows that the characteristic curves of (30.23) in the hodograph (u, w)-plane are defined by

$$\frac{w_\alpha}{u_\alpha}=\left(\frac{dw}{du}\right)_{\Gamma_\pm}=\frac{-\dfrac{u\,w}{c^2}\pm\sqrt{\dfrac{u^2+w^2}{c^2}-1}}{\dfrac{w^2}{c^2}-1}. \tag{30.24}$$

The characteristic curve Γ_- in the hodograph (u, w)-plane is orthogonal to the characteristic curve C_+ in the physical (x, z)-plane if we superimpose the two planes so that the velocity vectors coincide. This may be easily shown by rotating the axes for (30.21) and (30.24) so that $w=0$ (see Fig. 40); then the equations for the slopes of the characteristic curves C_+ and Γ_- simplify to

$$\left(\frac{dz}{dx}\right)_{C_+} = \frac{1}{\sqrt{\frac{u^2}{c^2}-1}} = \frac{1}{\sqrt{F^2-1}} = -\frac{1}{\left(\frac{dw}{du}\right)_{\Gamma_-}}. \tag{30.25}$$

Similarly, Γ_+ is orthogonal to C_- when the planes are superimposed so that the velocity vectors are coincident (see Fig. 40).

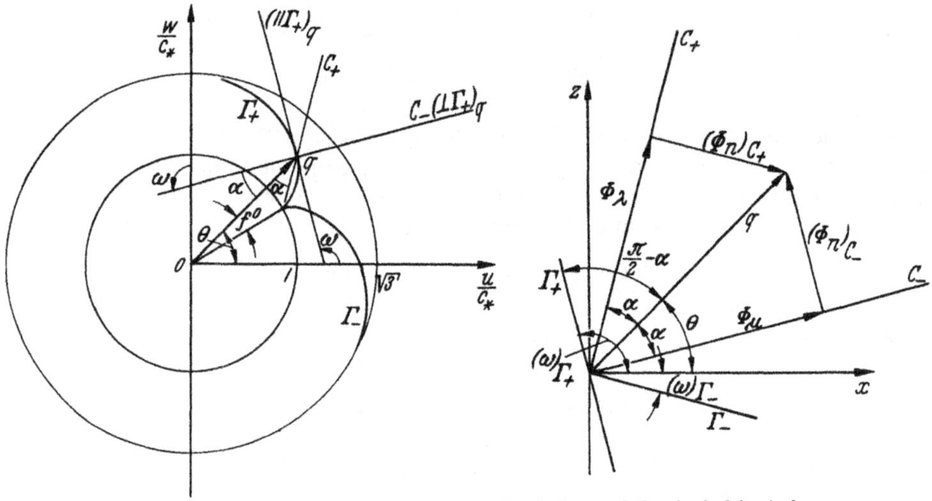

Fig. 40. Characteristic directions in the hodograph (u, w)-plane and the physical (x, z)-plane.

Eqs. (30.24) and (30.25) show that along any characteristic curve there exists a simple solution which is independent of the boundary conditions of a particular problem, for we can directly integrate (30.25) with axes rotated so that $w=0$ and hence $dw = u\,d\vartheta$:

$$\left(\frac{dw}{du}\right)_{\Gamma_-} = \left(\frac{u\,d\vartheta}{du}\right)_{\Gamma_-} = -\sqrt{\frac{u^2}{c^2}-1} = -\sqrt{F^2-1},$$

We integrate[1] (30.25) as follows:

$$\left.\begin{aligned} \int_{\Gamma_-} d\vartheta &= -\int_{\Gamma_-} \sqrt{\frac{u^2}{c^2}-1}\,\frac{du}{u} = -\int_{\Gamma_-} \frac{\sqrt{F^2-1}}{1+\frac{1}{2}F^2}\,\frac{dF}{F}, \\ |\varDelta\vartheta| &\equiv \sqrt{3}\tan^{-1}\sqrt{\tfrac{1}{3}(F^2-1)} - \tan^{-1}\sqrt{F^2-1} = f(F). \end{aligned}\right\} \tag{30.26}$$

Consequently (30.26) provides a general solution, independent of the boundary conditions in the physical plane, for any two-dimensional potential flow that possesses the property of having simple waves in the given region, so that the end of the velocity vector follows Γ_- in the hodograph plane. The numerical values from (30.26) are indicated in Fig. 40 and are tabulated in Table 1 on page 688 [taken from Preiswerk (1938)].

The useful relation between c and (u, w) that was used to integrate (30.26) and calculate Table 1 is obtained by multiplying the first equation in (30.15)

[1] See (30.27) and (30.29) which show that with $w=0$

$$\frac{du}{u} = -\frac{d(c^2)}{u^2} = +\frac{c_0^2}{u^2}\left(1+\frac{1}{2}F^2\right)^{-2} F\,dF = +\frac{dF}{F(1+\frac{1}{2}F^2)}.$$

by dx and the second equation by dz, and then adding them. One obtains successively

$$\left.\begin{aligned} &u(u_x\,dx + u_z\,dz) + w(w_x\,dx + w_z\,dz) = -\left[(c^2)_x\,dx + (c^2)_z\,dz\right],\\ &u\,du + w\,dw = -\,d(c^2) = -\,g\,d\eta,\\ &\tfrac{1}{2}(u^2+w^2) + c^2 = \text{const} = \tfrac{1}{2}(u^2+w^2) + g(h_0+\eta). \end{aligned}\right\} \quad (30.27)$$

Therefore

$$\left.\begin{aligned} \tfrac{1}{2}(u^2+w^2) + c^2 &= \tfrac{1}{2}(u^2+w^2) + g(h_0+\eta)\\ &= g\,h_0 = \tfrac{1}{2}(u^2+w^2)_{\max} = \tfrac{3}{2}c_*^2 = \tfrac{1}{2}q^2 + c^2 = \tfrac{1}{2}q_{\max}^2, \end{aligned}\right\} \quad (30.28)$$

where (see Fig. 41) h_0 is the still water depth (or stagnation total head depth) that corresponds to $u_0^2+w_0^2=0=\eta_0$, $(u^2+w^2)_{\max}$ is the limiting resultant velocity

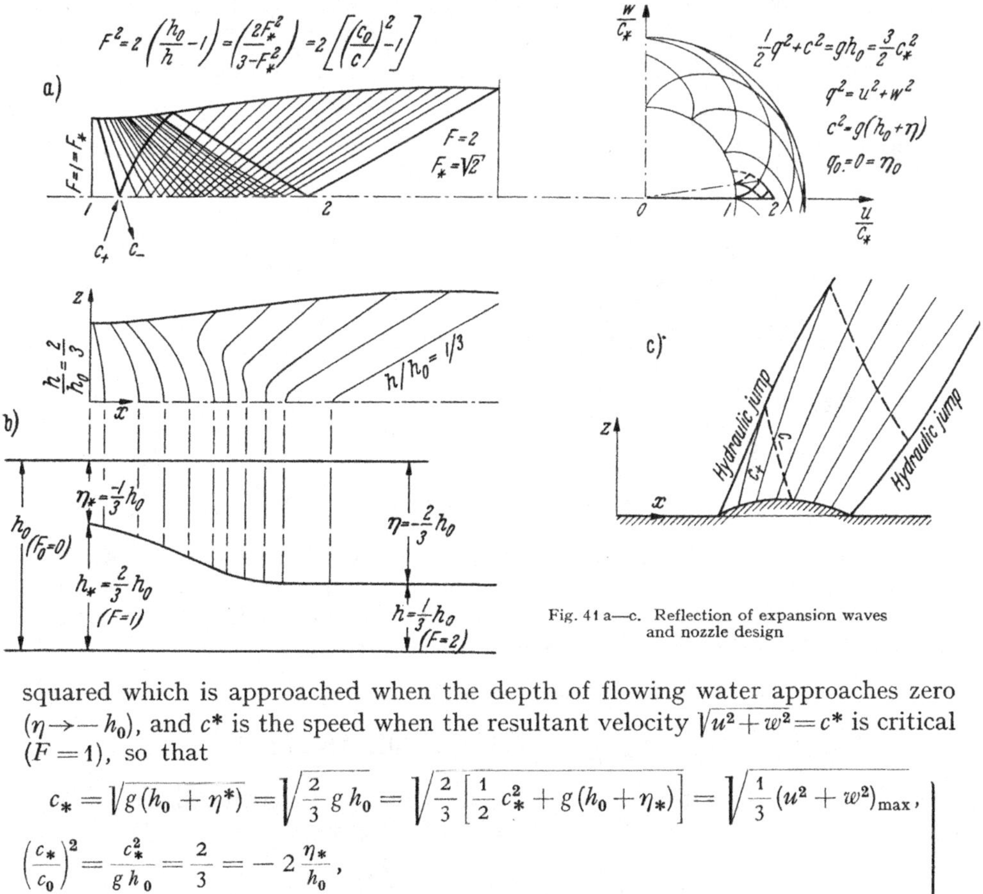

Fig. 41 a—c. Reflection of expansion waves and nozzle design

squared which is approached when the depth of flowing water approaches zero ($\eta\to-h_0$), and c^* is the speed when the resultant velocity $\sqrt{u^2+w^2}=c^*$ is critical ($F=1$), so that

$$\left.\begin{aligned} &c_* = \sqrt{g(h_0+\eta^*)} = \sqrt{\frac{2}{3}g\,h_0} = \sqrt{\frac{2}{3}\left[\frac{1}{2}c_*^2 + g(h_0+\eta_*)\right]} = \sqrt{\frac{1}{3}(u^2+w^2)_{\max}},\\ &\left(\frac{c_*}{c_0}\right)^2 = \frac{c_*^2}{g\,h_0} = \frac{2}{3} = -2\frac{\eta_*}{h_0},\\ &\frac{\eta_*}{h_0} = -\frac{1}{3}, \qquad \frac{h_0+\eta_*}{h_0} = \frac{2}{3}, \qquad \frac{(u^2+w^2)_{\max}}{c_*^2} = 3,\\ &\left(\frac{c_0}{c}\right)^2 = \frac{g\,h_0}{c^2} = 1 + \frac{u^2+w^2}{2c^2} = 1 + \frac{1}{2}F^2 = \left(\frac{c_0}{c_*}\right)^2\left(\frac{c_*}{c}\right)^2 = \frac{3}{2}\left(\frac{c_*}{c}\right)^2 = \frac{h_0}{h},\\ &F^2 = \frac{u^2+w^2}{c^2} = \left(\frac{c_*}{c}\right)^2 F_*^2 = \frac{2F_*^2}{3-F_*^2},\\ &F_*^2 = \frac{u^2+w^2}{c_*^2} = \left(\frac{c}{c_*}\right)^2 F^2 = \frac{3F^2}{2+F^2}. \end{aligned}\right\} \quad (30.29)$$

It is very useful to note that Eqs. (30.26) through (30.29) may all be obtained directly from the two-dimensional isentropic gas-flow equations by simply letting the specific heat ratio $\gamma = 2$ and $F \equiv M$, $F_* \equiv M_*$, as had been previously shown by PREISWERK (1938) [see also COURANT and FRIEDRICHS (1948)].

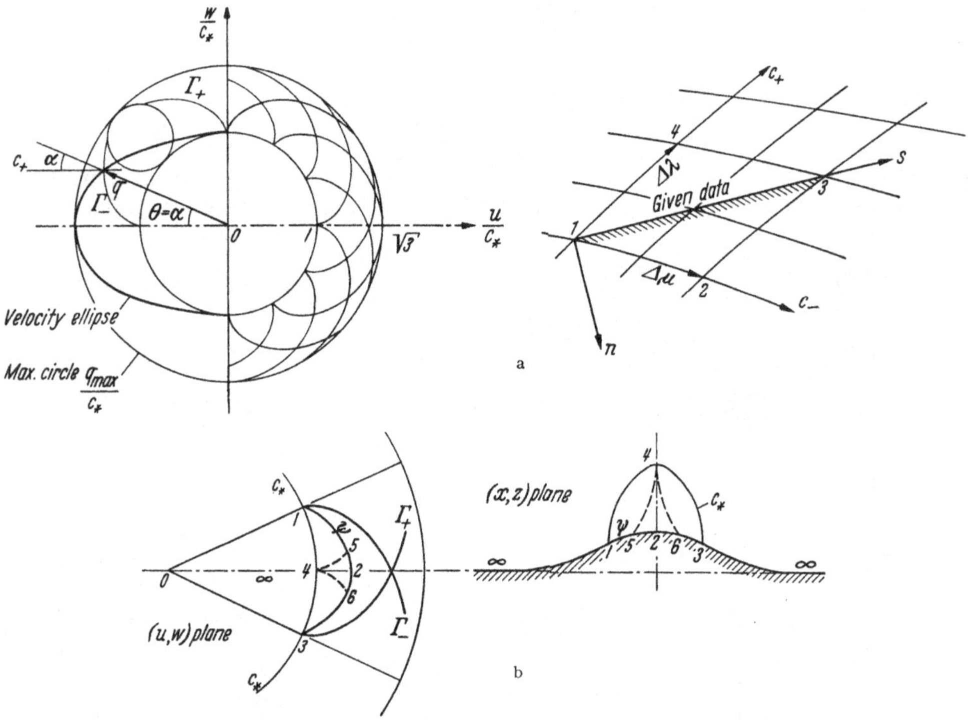

Fig. 42 a and b. Hodograph (u, w)-plane characteristic epicycloids.

The Riemann invariants for the characteristic curves (C_+, C_-) will now be determined. First we can show that the velocity component normal to the characteristic curves is always the local velocity of the shallow-water wave propagation, $c(x, z)$. We do this by writing (30.21) as

$$\left.\begin{aligned} (u\,dz - w\,dx)^2_{C_+} &= c^2\left[(dx)^2 + (dz)^2\right]_{C_+} = c^2 (d\lambda)^2_{C_+}, \\ c^2 &= \left(u\frac{dz}{d\lambda} - w\frac{dx}{d\lambda}\right)^2_{C_+} = (\varphi_x x_n + \varphi_z z_n)^2_{C_+} = (\varphi_n)^2_{C_+}, \end{aligned}\right\} \tag{30.30}$$

since the relation between the normal direction (n), and the tangential direction (λ) along the characteristic curve (C_+) is given by (see Fig. 40)

$$\left(\frac{dx}{dn}\right)_{C_+} = \left(\frac{dz}{d\lambda}\right)_{C_+}, \quad \left(\frac{dz}{dn}\right)_{C_+} = -\left(\frac{dx}{d\lambda}\right)_{C_+}.$$

Similarly, if μ is the tangential direction along C_-,

$$c^2 = \left(u\frac{dz}{d\mu} - w\frac{dx}{d\mu}\right)^2_{C_-} = (\varphi_x x_n + \varphi_z z_n)^2_{C_-} = (\varphi_n)^2_{C_-}.$$

Also, from Fig. 40 and (30.21)

$$\tan\alpha = \left(\frac{dz}{dx}\right)_{C_\pm,\,\vartheta=0} = \pm\frac{1}{\sqrt{F^2-1}}, \qquad \sin\alpha = \frac{\pm 1}{F} = \pm\frac{c}{q} \tag{30.31}$$

where q is the resultant velocity magnitude. Hence

$$\left.\begin{aligned} &q^2 = (u^2 + w^2) = (\varphi_\lambda^2 + c^2) = (\varphi_\mu^2 + c^2), \quad \vartheta = \tan^{-1}\frac{w}{u}, \\ &u = q\cos\vartheta, \quad w = q\sin\vartheta, \\ &\varphi_n = c = q\sin\alpha, \quad \varphi_\lambda = q\cos\alpha = \varphi_\mu. \end{aligned}\right\} \tag{30.32}$$

Substituting (30.31) and (30.32) into (30.21) and (30.24) we obtain

$$\left(\frac{dz}{dx}\right)_{C_\pm} = \frac{\dfrac{\cos\vartheta\sin\vartheta}{\sin^2\alpha} \pm \dfrac{1}{\tan\alpha}}{\dfrac{\cos^2\vartheta}{\sin^2\alpha} - 1} = \tan(\vartheta \pm \alpha), \tag{30.33}$$

$$\left(\frac{dw}{du}\right)_{\Gamma_\mp} = \frac{-\dfrac{\cos\vartheta\sin\vartheta}{\sin^2\alpha} \pm \dfrac{1}{\tan\alpha}}{\dfrac{\sin^2\vartheta}{\sin^2\alpha} - 1} = -\cot(\vartheta \pm \alpha). \tag{30.34}$$

Therefore, as proven before in (30.25),

$$\left(\frac{dz}{dx}\right)_{C_+}\left(\frac{dw}{du}\right)_{\Gamma_-} = -1 = \left(\frac{dz}{dx}\right)_{C_-}\left(\frac{dw}{du}\right)_{\Gamma_+}, \tag{30.35}$$

that is, as shown in Fig. 40, the C_+ characteristic curves in the physical (x, z) plane are at every corresponding point orthogonal to the Γ_- characteristic curves in the hodograph (u, w) plane. All these results are the same as in the gas-dynamics case where the $C_\pm$ characteristic curves are referred to as the Mach lines since, as shown by (30.30), the normal velocity component is always the local speed of sound.

Now, as shown in Fig. 40,

$$\begin{aligned} &\left(\frac{dw}{du}\right)_{\Gamma_-} = \tan\omega_{\Gamma_-}; \quad \omega_{\Gamma_-} = \vartheta + \alpha - \frac{1}{2}\pi, \\ &\left(\frac{dw}{du}\right)_{\Gamma_+} = \tan\omega_{\Gamma_+}; \quad \omega_{\Gamma_+} = \vartheta - \alpha + \frac{1}{2}\pi; \end{aligned}$$

therefore (30.33) may be written as

$$\begin{aligned} &\left(\frac{dz}{dx}\right)_{C_+} = \tan(\vartheta + \alpha) = \tan\left(\omega_{\Gamma_-} + \frac{1}{2}\pi\right), \\ &\left(\frac{dz}{dx}\right)_{C_-} = \tan(\vartheta - \alpha) = \tan\left(\omega_{\Gamma_+} - \frac{1}{2}\pi\right). \end{aligned}$$

Consequently the Riemann invariants are given by

$$R = \vartheta + \alpha - \omega_{\Gamma_-} - \tfrac{1}{2}\pi, \qquad S = \vartheta - \alpha - \omega_{\Gamma_+} + \tfrac{1}{2}\pi.$$

These may be simplified by calculating

$$\omega_{\Gamma_-} = \arctan\left(\frac{dw}{du}\right)_{\Gamma_-} = -\sqrt{3}\,\mathrm{arc\,cot}\left[\sqrt{\frac{3}{F^2-1}}\right] = -\sqrt{3}\arctan\left[\sqrt{\frac{F^2-1}{3}}\right]$$

from (30.24) and Fig. 40 since

$$\frac{1}{\sqrt{F^2-1}} = |\tan\alpha| = \sqrt{\frac{1}{3}}\left|\cot\left(\frac{\omega}{\sqrt{3}}\right)\right|$$

[or see Courant and Friedrichs (1948, p. 266)].

Table 1.

f (deg.)	$1+\frac{\eta}{h_0}$	F_*	F	K	f (deg.)	$1+\frac{\eta}{h_0}$	F_*	F	K
0	2/3	1.000	1.000	∞	26	0.234	1.516	2.56	−0.160
1	0.624	1.062	1.098	2.68	27	0.223	1.527	2.64	−0.177
2	0.598	1.101	1.160	2.07	28	0.212	1.538	2.73	−0.196
3	0.576	1.129	1.214	1.40	29	0.201	1.549	2.82	−0.216
4	0.555	1.156	1.267	1.014	30	0.190	1.559	2.92	−0.234
5	0.535	1.182	1.319	0.758	31	0.180	1.569	3.02	−0.252
6	0.516	1.207	1.371	0.590	32	0.170	1.579	3.13	−0.271
7	0.498	1.229	1.422	0.476	33	0.160	1.588	3.24	−0.291
8	0.481	1.249	1.470	0.394	34	0.151	1.597	3.36	−0.313
9	0.464	1.269	1.520	0.318	35	0.141	1.605	3.49	−0.336
10	0.448	1.288	1.570	0.263	36	0.132	1.613	3.63	−0.36
11	0.432	1.306	1.622	0.215	37	0.123	1.621	3.78	−0.38
12	0.417	1.323	1.674	0.170	38	0.115	1.629	3.93	−0.40
13	0.402	1.340	1.727	0.133	39	0.107	1.637	4.01	−0.43
14	0.387	1.356	1.781	0.103	40	0.099	1.644	4.26	−0.46
15	0.373	1.372	1.835	0.072	41	0.092	1.651	4.44	−0.49
16	0.359	1.387	1.89	0.046	42	0.085	1.657	4.63	−0.52
17	0.345	1.402	1.95	0.020	43	0.078	1.663	4.85	−0.54
18	0.331	1.416	2.01	−0.004	44	0.072	1.669	5.08	−0.58
19	0.318	1.430	2.07	−0.028	45	0.066	1.675	5.33	−0.62
20	0.305	1.444	2.13	−0.050	46	0.060	1.681	5.62	−0.66
21	0.292	1.457	2.20	−0.071	47	0.054	1.686	5.95	−0.70
22	0.280	1.470	2.27	−0.089	48	0.048	1.691	6.30	−0.75
23	0.268	1.482	2.34	−0.108	49	0.043	1.696	6.68	−0.81
24	0.256	1.494	2.41	−0.126	50	0.038	1.700	7.11	−0.86
25	0.245	1.505	2.48	−0.143	65°53′	0	$\sqrt{3}$	∞	$-\infty$

Therefore

$$R = \vartheta + \arctan\frac{1}{\sqrt{F^2-1}} - \frac{\pi}{2} + \sqrt{3}\arctan\sqrt{\frac{F^2-1}{3}}$$

$$= \vartheta + \sqrt{3}\arctan\sqrt{\frac{F^2-1}{3}} - \arctan\sqrt{F^2-1} = \vartheta + f(F),$$

$$S = \vartheta - \arctan\frac{1}{\sqrt{F^2-1}} + \frac{\pi}{2} - \sqrt{3}\arctan\sqrt{\frac{F^2-1}{3}}$$

$$= \vartheta - \sqrt{3}\arctan\sqrt{\frac{F^2-1}{3}} + \arctan\sqrt{F^2-1} = \vartheta - f(F),$$

where $f(F)$ is given by (30.26) and Table 1. Consequently the Riemann invariants are very simply expressed for the characteristic curves in the physical plane as

$$\vartheta - f(F) = \text{const on } C_+, \qquad \vartheta + f(F) = \text{const on } C_-. \tag{30.36}$$

The function $f(F)$, which was derived from the fact that the end point of the velocity vector follows a characteristic in the hodograph plane in (30.26), is seen to have important physical significance, and directly provides the Riemann invariants for the steady two-dimensional potential flow. In gas dynamics $f(F) \equiv f(M)$ is referred to as the Prandtl-Meyer expansion function and, in the form in which it is given in Table 1, it corresponds to the supersonic free expansion about a sharp corner as shown in Fig. 43 for the centered simple wave with a specific heat ratio $\gamma = 2$. Since this Prandtl-Meyer function is so important, let us re-derive it on another basis that will further illustrate its physical significance. From the fact that f forms the Riemann invariant it is evident that u and w cannot be independent of one another on any such simple characteristic.

Consequently, if we write the original potential equation (30.18) in the physical plane as

$$\left(\frac{u^2}{c^2}-1\right)u_x+2\frac{u\,w}{c^2}u_y+\left(\frac{w^2}{c^2}-1\right)w_y=0$$

and introduce $w=w(u)$ so that

$$w_y=u_y\,w'(u),\quad u_x\,w'(u)=w_x=\varphi_{xz}=u_y,$$

we obtain

$$\left(\frac{u^2}{c^2}-1\right)\frac{u_y}{w'}+2\frac{u\,w}{c^2}u_y+\left(\frac{w^2}{c^2}-1\right)u_y\,w'=0$$

or

$$\frac{dw}{du}=w'(u)=\frac{\dfrac{-u\,w}{c^2}\pm\sqrt{\dfrac{u^2+w^2}{c^2}-1}}{\dfrac{w^2}{c^2}-1}. \tag{30.24'}$$

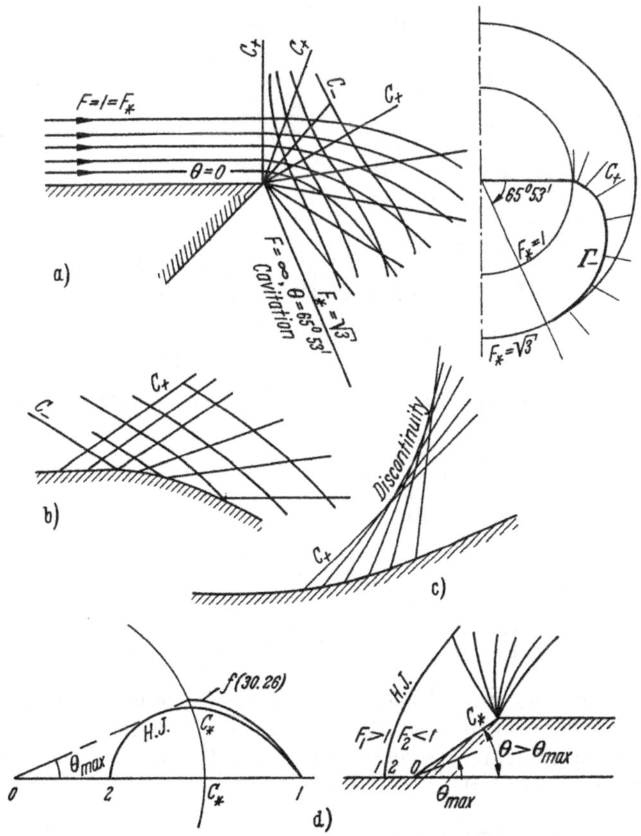

Fig. 43a–d. Simple waves and the formation of hydraulic jumps. (a) Complete centered simple expansion wave. (b) Simple(C_+) expansion waves. (c) Simple (C_+) compression waves forming a discontinuous (hydraulic jump) increase in water depth. (d) Detached hydraulic jump.

This derivation gives exactly the same result as in (30.24) and verifies the fact that discontinuities can occur in the first derivatives normal to a characteristic curve. If we introduce (30.32) into (30.24) we obtain the equivalent of (30.26)

$$\frac{1}{q}\frac{dq}{d\vartheta}=\tan\alpha=\frac{1}{\sqrt{F^2-1}} \tag{30.37}$$

which again has $f(F)$ as the general integral because (30.28) shows that

$$\frac{d\vartheta}{\sqrt{F^2-1}} = \frac{dq}{q} = \frac{d(q/c_*)}{(q/c_*)} = \frac{dF_*}{F_*} = \frac{1}{1+\frac{1}{2}F^2}\,\frac{dF}{F}\,. \tag{30.38}$$

However, neither of these methods gives the direct proof that $f(F)$ provides the Riemann invariant. This fact may be proved directly by the following derivation which utilizes the velocity component φ_λ along the C_+ characteristic, and $\varphi_n = c$, from (30.30), normal to C_+ as shown in Fig. 40. Hence

$$\left.\begin{aligned} \varphi_\lambda &= q\cos\alpha, \qquad \varphi_n = c = q\sin\alpha,\\ d\varphi_\lambda &= \cos\alpha\,dq - q\sin\alpha\,d\alpha = c\,(d\vartheta - d\alpha),\\ \lambda_z &= 1/\sin(\vartheta+\alpha), \qquad \mu_z = 1/\sin(\vartheta-\alpha) \end{aligned}\right\} \tag{30.39}$$

since, from (30.37),

$$dq = \frac{\sin\alpha}{\cos\alpha}\,q\,d\vartheta\,.$$

Then from (30.28) and (30.32) we have

$$\left.\begin{aligned} \tfrac{1}{2}q^2 + c^2 &= \tfrac{1}{2}(\varphi_\lambda^2 + \varphi_n^2) + c^2 = \tfrac{1}{2}\varphi_\lambda^2 + \tfrac{3}{2}c^2 = \tfrac{1}{2}q_{\max}^2 = \tfrac{3}{2}c_*^2\,,\\ c^2 &= \tfrac{1}{3}(q_{\max}^2 - \varphi_\lambda^2) = c_*^2 - \tfrac{1}{3}\varphi_\lambda^2\,, \end{aligned}\right\} \tag{30.40}$$

so that (30.39) may be written

$$\left.\begin{aligned} \vartheta - \alpha + \text{const} &= \int \frac{d(\varphi_\lambda)}{c} = \int \frac{d(\varphi_\lambda)}{\sqrt{\frac{1}{3}(q_{\max}^2 - \varphi_\lambda^2)}} = \sqrt{3}\int \frac{d(\varphi_\lambda/q_{\max})}{\sqrt{1-(\varphi_\lambda/q_{\max})^2}}\\ &= \sqrt{3}\arcsin\left(\frac{\varphi_\lambda}{q_{\max}}\right) = \sqrt{3}\arcsin\left(\frac{\varphi_\lambda}{\sqrt{3c^2+\varphi_\lambda^2}}\right) = \sqrt{3}\arctan\left(\frac{\varphi_\lambda}{c\sqrt{3}}\right) \end{aligned}\right\} \tag{30.41}$$

This may be finally written in terms of (F) alone by noting from (30.39) that

$$\frac{\varphi_\lambda}{c} = \frac{\varphi_\lambda}{\varphi_n} = \frac{q\cos\alpha}{q\sin\alpha} = \frac{1}{\tan\alpha} = \sqrt{F^2-1}\,.$$

Consequently (30.41) reduces to

$$\left.\begin{aligned} &\vartheta(F) - \sqrt{3}\arctan\left(\sqrt{\frac{1}{3}(F^2-1)}\right) - \arctan\frac{1}{\sqrt{F^2-1}} + \text{const}\\ &= \vartheta(F) - \sqrt{3}\arctan\left(\sqrt{\frac{1}{3}(F^2-1)}\right) + \arctan\sqrt{F^2-1} = \vartheta - f(F) = \text{const}, \end{aligned}\right\} \tag{30.42}$$

where $f(F)$ is the same Prandtl-Meyer function as given in (30.26) and Table 1. Therefore we have proven that the Riemann invariants are given by (30.36) and (30.26). In addition to the relation between f and F in (30.26) it is sometimes convenient to use one of the following:

$$\left.\begin{aligned} f(\alpha) &= \sqrt{3}\,\mathrm{arc\,cot}\,(\sqrt{3}\tan\alpha) + \alpha - \frac{1}{2}\pi\\ &= f(F_*) = \sqrt{3}\arctan\sqrt{\frac{F_*^2-1}{3-F_*^2}} - \arctan\sqrt{\frac{F_*^2-1}{1-\frac{1}{3}F_*^2}} = \frac{\lambda+\mu}{2} \end{aligned}\right\} \tag{30.26'}$$

It now follows that a numerical solution can be obtained for the general problem in which both families of characteristics represent curved non-simple waves by carrying on a simultaneous finite-difference solution in the physical (x, z)-plane with (30.33), and in the hodograph (u, w)-plane by (30.26), (30.34), and (30.36). Almost any initial- or boundary-value data can be handled in this manner as long as the curve on which the data are given is not coincident with a characteristic curve. The solution cannot be obtained in the neighborhood of any portion of the boundary-value curve that happens to be tangent to any characteristic curve, because, as proven by (30.20), the solution is indeterminate for boundary-value data given on a characteristic. It is easily seen by this finite-difference method that the data along a smooth non-characteristic curve can only determine the solution inside the quadrilateral formed by the characteristic curves passing through its end points (Fig. 42) [see, e.g., PREISWERK (1938) or COURANT and FRIEDRICHS (1948)]. This well-known behavior of hyperbolic-type partial differential equations is most directly demonstrated by writing them in their normal or canonical form by transforming the coordinates to curvilinear axes which are the characteristic curves themselves. For example, PREISWERK (1938) transforms the equivalent of (30.23) onto the curvilinear characteristic-coordinate (λ, μ) system to obtain

$$\left.\begin{aligned} f(F_*) &= \tfrac{1}{2}(\lambda+\mu), \qquad \vartheta = \tfrac{1}{2}(\lambda-\mu), \\ \chi_{\lambda\mu} &= -K(\lambda,\mu)\,(\chi_\lambda+\chi_\mu), \\ K(\lambda,\mu) &= \frac{F_*^2(1-\frac{1}{2}F_*^2)}{\sqrt{3}\,(3-F_*^2)^{\frac{1}{2}}\,(F_*^2-1)^{\frac{3}{2}}}. \end{aligned}\right\} \tag{30.43}$$

This normal or canonical form is so useful in carrying out the finite-difference method of solution that the values of K have also been included in Table 1. It can be used in the following type of approximation, as indicated in Fig. 42 where (1, 3) are known values and (2, 4) are to be calculated,

$$\left.\begin{aligned} &\chi_\lambda = \frac{\chi_4-\chi_1}{\lambda_4-\lambda_1}, \qquad \chi_\mu = \frac{\chi_2-\chi_1}{\mu_2-\mu_1}, \\ &\left(-\frac{K_1+K_3}{2}\right)(\chi_\lambda+\chi_\mu) = \chi_{\lambda\mu} = \frac{(\chi_3+\chi_1)-(\chi_4+\chi_2)}{(\lambda_4-\lambda_1)\,(\mu_2-\mu_1)}. \end{aligned}\right\} \tag{30.43'}$$

Consequently, if the data were given on only one characteristic curve the method would fail since the values must be known on *both* characteristics, or on the non-characteristic curve s in Fig. 42, so that one can also write

$$\begin{aligned} \chi_s &= \chi_\lambda\,\lambda_s + \chi_\mu\,\mu_s = g(s), \\ \chi_n &= \chi_\lambda\,\lambda_n + \chi_\mu\,\mu_n = G(s). \end{aligned}$$

The numerical method of solution by finite differences following (30.43') is known as the "lattice-point method" and replaces the original partial differential equation (30.43) by a set of linear algebraic equations. The other commonly used semi-graphical method of solving hyperbolic partial differential equations is called the network or "mesh method" and can be illustrated by writing (30.43) in the form

$$\begin{aligned} \Delta\chi_\lambda &= -K(\chi_\lambda+\chi_\mu)\,\Delta\mu, \\ \Delta\chi_\mu &= -K(\chi_\lambda+\chi_\mu)\,\Delta\lambda. \end{aligned} \tag{30.43''}$$

The average value at the center of each mesh formed by the characteristic network is used for the trial and error numerical calculation of each Δ increment. The increments are drawn tangent to the characteristic curves as indicated in Fig. 42. The simultaneous semi-graphical solution must be carried out in the physical plane as shown in Fig. 42 by using (30.33) and writing (30.39) and (30.41) in finite difference form.

As a further aid to numerical and graphical solutions it is useful to plot $f(F_*)$ from (30.26′) or Table 1 on the hodograph $(u/c_*, w/c_*)$-plane as shown in Fig. 42. The single curve defined by Table 1 may be drawn and then rotated by equal increments of $\Delta\vartheta$, or the construction may be accomplished entirely by graphical means as indicated in Fig. 42 by rotating the small circle upon the inner unit circle representing critical flow, while the outer maximum circle has a radius of $\sqrt{3}$ representing $q_{\max}/c_*$ from (30.29). This geometrical construction yields $f(F_*)$ since it is an epicycloid, as proven by PREISWERK (1938), or COURANT and FRIEDRICHS (1948, p. 262). All simple waves must follow the characteristic epicycloid in the hodograph plane because simple waves are defined by (30.37) which has been proven to have $f(F_*)$ as its integral. It can be shown that all streamlines corresponding to non-simple waves must lie within the corresponding characteristic epicycloids as indicated in Fig. 42, since the streamline must have

$$\left|\frac{1}{F_*}\left(\frac{dF_*}{d\vartheta}\right)_{\psi\,\mathrm{const}}\right| \leq \tan\alpha \tag{30.44}$$

unless a finite discontinuity corresponding to a hydraulic jump (or shock wave in a gas) is formed.

Another useful aid in the hodograph graphical construction is the velocity ellipse, which is also drawn in Fig. 42. Wherever the velocity vector q touches the curve of the ellipse, it will be found that the major axis of the ellipse is in the direction of the tangent to the corresponding characteristic (either C_+ or C_-) in the physical plane because, as a consequence of (30.29) and (30.32), if we assume that $\vartheta = \alpha$ then

$$(w/c_*)^2 = F_*^2 \sin^2\alpha = F_*^2/F^2 = \tfrac{1}{2}(3 - F_*^2),$$

$$(u/c_*)^2 = F_*^2(1 - \sin^2\alpha) = \tfrac{3}{2}(F_*^2 - 1),$$

$$\tfrac{2}{3}(u/c_*)^2 + 1 = F_*^2 = 3 - 2(w/c_*)^2,$$

or

$$\tfrac{1}{3}(u/c_*)^2 + (w/c_*)^2 = 1 = [\tfrac{1}{3}(\varphi_\mu/c_*)^2 + (c/c_*)^2]_{C_-,\vartheta=0} = [\tfrac{1}{3}(\varphi_\lambda/c_*)^2 + (c/c^*)^2]_{C_+,\vartheta=0}. \tag{30.45}$$

This gives the velocity ellipse shown in Fig. 42 with a major axis of $\sqrt{3}$ and a minor axis of unity. The major axis is always at the Mach angle α with respect to the velocity vector q because we find from (30.28) and (30.32) that when $\alpha = \vartheta$

$$(w/c_*)^2 = (\varphi_n/c_*)^2 = (c/c_*)^2 = 1 - \tfrac{1}{3}(u/c_*)^2.$$

As in the previous case of unsteady one-dimensional flow over a flat bottom we can obtain very simple solutions for the case of simple waves. In this case there is an analogy between the (t, x) diagram and the (x, z) diagram [see, e.g., COURANT and FRIEDRICHS (1948)]. As before, the simple wave corresponds to having the characteristics in the (x, z)-plane of one family become straight lines, as in the examples shown in Fig. 43, so that $(q, \vartheta, \alpha, \eta)$ are all constant on the straight line $dz/dx = \mathrm{const}$ in the physical plane. Therefore any given straight characteristic line has all of its properties determined by $f(F_*)$ from (30.26) and each of the straight lines in the physical plane maps onto a single point of the same single characteristic epicycloid in the hodograph plane. The characteristics of the other family remain curved in the physical plane and map in a unique continuous manner upon the corresponding characteristic epicycloid arcs in the hodograph plane. As before, in a simple wave these curved characteristics are not required for a numerical solution.

Common examples of simple-wave problems are shown in Fig. 43, and they always occur whenever a region of constant uniform properties adjoins a region having any variation in its properties, the two regions always being joined by a straight-line physical characteristic ($dz/dx = \text{const}$) as long as no finite discontinuities, corresponding to hydraulic jumps or shock waves, have been formed. These finite discontinuities correspond to an envelope of the straight characteristic lines that must form whenever the boundary-surface curves towards the oncoming flow, resulting in a flow compression or decrease of velocity and increase in water depth as indicated in Fig. 43. The solution is no longer single valued at, or downstream of the envelope so this region must be replaced by a hydraulic jump having a finite discontinuity.

If the local flow velocity and water depth are required only on the curved boundary itself, then neither family of characteristics has to be determined (except as a precaution to verify that no finite discontinuities have formed near the boundary due to flow compression). The solution on the curved boundary itself is given directly from Table 1 by simply measuring $f(F_*)$ as the value corresponding to (see Fig. 43)

$$f[F_*(\vartheta)] = f[F_{*\infty}] \pm \vartheta. \tag{30.46}$$

If this expression becomes zero it signifies that the supercritical flow has been compressed to critical speed and a detached hydraulic jump can occur as in Fig. 43.

Whenever disturbance waves enter along both families, either due to another boundary or by reflection from a hydraulic jump, as in Fig. 41, then the mixed region contains non-simple waves, and only a numerical solution, similar to the ones discussed in conjunction with (30.43), can yield the exact solution. However, an approximate solution for the particular cases shown in Fig. 41 can be obtained by approximating the curved characteristics in the non-simple region by means of simple-wave straight characteristic lines. The geometrical construction assumes that the curved boundary wall of the nozzle can be replaced by a series of straight chord lines which each have the same magnitude of $\Delta\vartheta$ at every corner, as depicted in Fig. 44. At each expansion corner it is assumed that the centered simple wave (corresponding to a portion of the complete Prandtl-Meyer expansion, f) can be approximated by a single physical characteristic that is the average of the actual expansion fan of characteristics. This is the (dz/dx) straight line that is normal to the midpoint of the $\Delta\vartheta$ epicycloid arc representing the expansion-angle change at this corner, as shown in Fig. 44. Similarly, the compression corner that turns into the flow is represented by the single compression simple wave that is normal to the midpoint of the $\Delta\vartheta$ epicycloid arc representing the compression angle change at this corner. It will be shown that the angle of this single average compression wave is actually the correct limiting value for a weak hydraulic jump. The geometrical construction is carried out in the manner indicated in Fig. 44. Whenever a streamline crosses one of these finite amplitude construction characteristics the flow is assumed to bend through the $\Delta\vartheta$ associated with the finite corner bend which supposedly produced this single finite wave. The corresponding construction in the hodograph plane transfers to the epicycloid arc that is normal to the single finite wave in the physical plane as shown in Fig. 44.

Also shown in Fig. 44 are the geometric constructions required for the reflection of these simple finite waves in the physical plane from either solid boundaries, or from free boundaries with constant water depth. In the reflection from a solid boundary the original boundary slope is again attained by the velocity vector after passing through the reflected wave which has the same strength for flow deflection as the original oncoming finite simple wave. In the hodograph plane the streamline has gone from one family of epicycloids to the other, ending at the same value of ϑ. The completed solution for the flow inside a channel

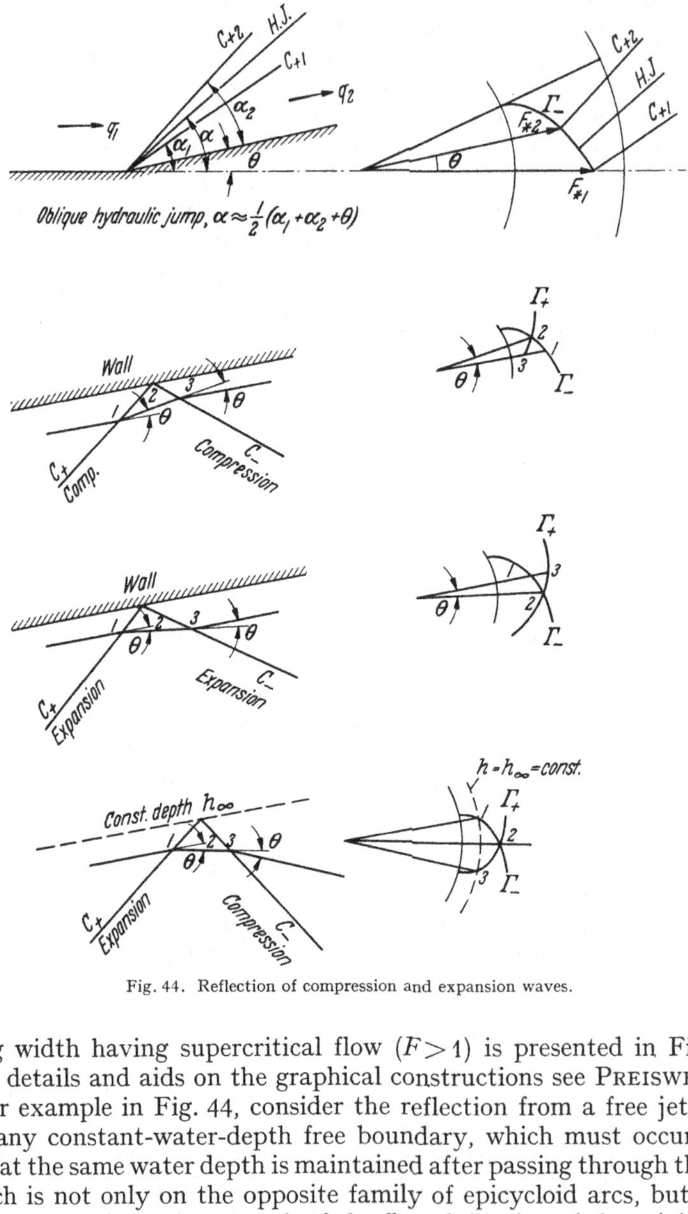

Fig. 44. Reflection of compression and expansion waves.

of varying width having supercritical flow ($F>1$) is presented in Fig. 41. For additional details and aids on the graphical constructions see PREISWERK (1938). As another example in Fig. 44, consider the reflection from a free jet, hydraulic jump, or any constant-water-depth free boundary, which must occur in such a manner that the same water depth is maintained after passing through the reflected wave which is not only on the opposite family of epicycloid arcs, but now must have the negative algebraic strength of the flow deflection of the original oncoming wave; consequently, the value of $\Delta\vartheta$ is exactly doubled after passing through the reflected wave. That is to say, unlike the ordinary reflection from a solid boundary, the reflection from a constant-depth free boundary results in the opposite type of wave, an expansion wave becoming a compression wave and vice versa.

In conclusion it must be noted that this two-dimensional steady-flow analysis is only valid for a flat horizontal bottom, as was already shown by (29.3) for the linearized equations. If the bottom slope varies, then the Riemann invariants do not exist, simple waves do not occur, and the numerical solution is much more complicated. However, there is an even more important criterion that must be satisfied before *any* of the solutions given so far can be applicable. This is the necessary requirement that all the perturbation quantities involved $(u-U, w, \eta)$ must be sufficiently small so that it is not necessary to introduce the second-order terms from Sect. 31. A satisfactory evaluation of this criterion, at least for F not too near unity, can be obtained by comparing the solutions of the non-linear equation (30.17) with the linearized equation (29.23) or (29.32). As is well-known in gas dynamics, and is apparent by inspection, (29.23) and (29.32) are not satisfactory for F approaching unity since additional terms must then be retained. For example, on the boundary profile itself, (29.3) for a flat horizontal bottom must include the additional term $3F^2(\varphi_x/U)\varphi_{xx}$, which corresponds to the "transonic approximation" of the gas dynamic equation (with a specific heat ratio $\gamma=2$) in the limit as F approaches unity. However, for the solution of the steady flow everywhere about a two-dimensional profile it may be necessary to use

$$(1-F^2)\,\varphi_{xx}+\varphi_{zz}=F^2\left[3\,\frac{\varphi_x}{U}\,\varphi_{xx}+2\,\frac{\varphi_z}{U}\,\varphi_{xz}\right] \tag{30.47}$$

since (29.29) indicates that far from the profile $w/U=\varphi_z/U\sim 1/z$, whereas $(u-U)/U=\varphi_x/U\sim 1/z^2$. In any case any radical increase in the order of magnitude of any perturbation term immediately indicates that the second-order terms discussed in Sect. 31 must be introduced, since the non-linear equation (28.1) and all the preceding results are based only on the first-order terms of the shallow-water theory.

γ) One-dimensional, steady, open-channel hydraulics and the hydraulic jump. The reltations given in Eqs. (30.27) (30.28) and (39.29), and shown in Fig. 41, can be used in what is commonly known as the steady "one-dimensional" hydraulics of open-channel flow. Here we assume that even though the channel width $b(x)$ is varying, still the values of $q(x)$ and $\eta(x)$ do not depend upon z and therefore do not vary on any given cross-section. In conjunction with the steady "one-dimensional" concept it is necessary that $w\approx 0\approx\vartheta$. Consequently the basic equations to be used for a flat horizontal bottom are given by $q(x)=u(x)$ in (30.27), (30.28) and (30.29), and, in addition, by the "one-dimensional" continuity equation

$$b(x)\,h(x)\,u(x)=A(x)\,u(x)=Q=\left(\frac{\text{meters}^3}{\text{sec}}\right)=\text{const}, \tag{30.48}$$

where, from Fig. 41, $h(x)=h_0+\eta(x)=A(x)/b(x)$.

The validity of the "one-dimensional" assumption can be considerably in error if $b'(x)$ is large since it is obvious that in this case w or ϑ cannot be small. However, the "one-dimensional" approximation gives surprisingly good numerical values, even in supercritical flow *if* the channel is well designed as in Fig. 41 so as to maintain the flow as uniform as possible. However in supercritical flow the velocity over any cross-section remains uniform only near the design Froude number (F). Preiswerk (1938) gives the calculated and measured water depths in a Laval-type nozzle (the same one duplicated in Fig. 41) at various supercritical Froude numbers ($F>1$). His results indicate that "one-dimensional" hydraulics gives a satisfactory approximation, having an error probably less than 10 %, even for critical or supercritical flow. This method should be especially useful for subcritical flow since the more exact numerical solution is now very difficult to obtain because the simple method of characteristics is no longer applicable.

The most useful, and obviously the most accurate, application of "one-dimensional" hydraulics is to the constant-width rectangular-cross-section open-channel flow. In this application the friction effect of the vertical channel walls generally has a greater effect on the variation of $q(x, y, z)$ than would any of the more exact terms of the complete first-order shallow-water equations (28.1) which have been derived on the assumption of negligible viscosity effects. Consequently the

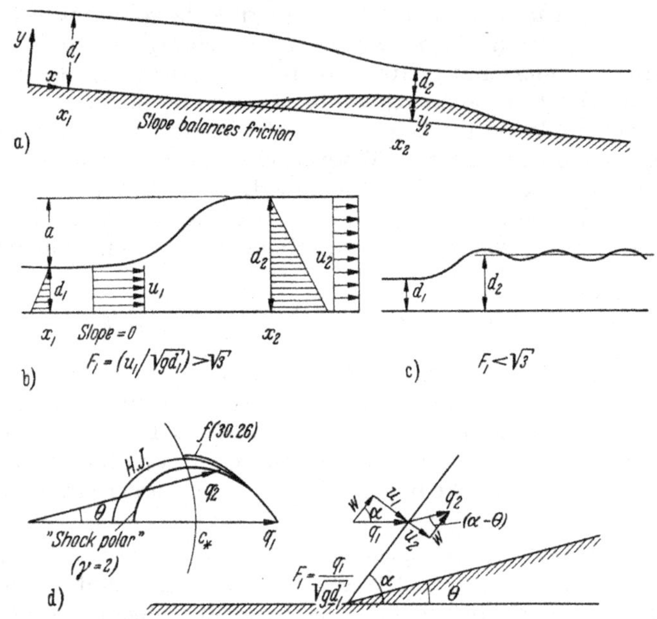

Fig. 45a–d. (a) One-dimensional flow over a sloping bottom, $|dy/dx| \ll 1$ so centrifugal force negligible. (b) Normal hydraulic jump. (c) Undulating hydraulic jump. (d) Oblique hydraulic jump with $w = w_1 = w_2$.

"one-dimensional" assumption that $q = u(x)$ provides a satisfactory approximation for the constant width (b), rectangular-cross-section, vertical-wall channel having $A(x) = b\,d(x)$. Even more important, this open-channel flow analysis may be further generalized, with but little additional difficulty, to apply to a bottom slope varying also with x. The "one-dimensional" continuity equation (30.48) then becomes

$$u(x)\,d(x) = \frac{Q}{b} = \text{const}, \tag{30.49}$$

where $d(x)$ is measured vertically from the varying bottom as shown in Fig. 45. The generalization of the Bernoulli equation (30.27) to include extraneous head losses (h_L), other than those due to friction, and local variations in the bottom contour $y(x)$, as shown in Fig. 45, may be written as

$$\left.\begin{aligned} (\text{specific energy}) &= \left(\frac{\text{kg meters}}{\text{kg}}\right) = (\text{meters}) = d(x) + \frac{u^2(x)}{2g} + y(x) + h_L(x) \\ &= d(x) + \frac{(Q/b)^2}{2g\,d^2(x)} + y(x) + h_L(x) = \text{const}. \end{aligned}\right\} \tag{30.50}$$

This assumes that in steady flow the work of gravity, through the known average slope of the flow, is wholly spent in overcoming the frictional resistance.

Another relation, that is necessary for calculating the sudden additional head loss h_L in hydraulic jumps or other discontinuous flow phenomenon, is given by the impulse-momentum relation [see KELVIN (1886), RAYLEIGH (1914), or BAKHMETEFF (1932)],

$$\left.\begin{aligned}(\text{specific momentum}) &= \left(\frac{\text{meters}^3}{\sec^2}\right) = \frac{1}{2} g\, d^2(x) + d(x)\, u^2(x)\\ &= \frac{1}{2} g\, d^2(x) + \frac{(Q/b)^2}{d(x)}\end{aligned}\right\} \tag{30.51}$$

which is constant across the hydraulic jump over a flat horizontal bottom as shown in Fig. 45.

Eq. (30.50) with zero additional head loss ($h_L = 0$), gives the "one-dimensional" solution for the open-channel flow that has no finite discontinuities in the flow itself, and either has the hydraulic frictional resistance exactly balanced by the given average slope for steady flow (so that if $y = 0$ the surface slope is parallel to the bottom), or the hydraulic frictional resistance can be approximated by the Chézy formula for the case of varying open-channel flow [see BAKHMETEFF (1932) or STOKER (1957)]. A useful concept for nearly all solutions is the definition of the critical depth d_*, which corresponds to our previous definition of critical flow velocity in (30.28), that is, with $w \approx 0 \approx \vartheta$ we assume that

$$\left.\begin{aligned}u_* &= c_* = \sqrt{\frac{2}{3} g\, h_0} = \frac{u_{\max}}{\sqrt{3}}, \qquad F_* = 1 = F,\\ d_* &= \frac{2}{3} h_0 = \frac{c_*^2}{g} = \left[\frac{(Q/b)^2}{g}\right]^{\frac{1}{3}}.\end{aligned}\right\} \tag{30.52}$$

The last relation for d_* can be obtained either directly from (30.27), or by substituting the expression $u_* = \sqrt{\frac{2}{3} g h_0}$ into (30.50) with y and h_L both zero. Also from (30.29) we have

$$\left.\begin{aligned}\frac{d}{h_0} &= \frac{2}{2+F^2}, \qquad F_*^2 = F^2\left(\frac{c}{c_*}\right)^2 = \frac{3F^2}{2+F^2},\\ F^2 &= \frac{u^2}{g\,d} = \frac{(Q/b)^2}{g\,d^3} = \left(\frac{d_*}{d}\right)^3.\end{aligned}\right\} \tag{30.53}$$

As an example, if we apply Eqs. (30.50) and (30.52) to determine the flow relations between stations (1) and (2) in Fig. 45a we obtain, since h_L is generally negligible for a smooth variation in y_2,

$$\frac{d_2}{d_*} + \frac{1}{2}\left(\frac{d_*}{d_2}\right)^2 = \frac{d_1}{d_*} + \frac{1}{2}\left(\frac{d_*}{d_1}\right)^2 - \frac{y_2}{d_*}$$

as a satisfactory approximation for "one-dimensional" hydraulics, at least as long as y_2/d_* is sufficiently small. It is interesting to note that here is another resemblance to gas-dynamics behaviour since

$$\frac{y_2 + d_2}{d_*} < \frac{d_1}{d_*} > 1 \qquad \text{for } F_1 < 1,$$

$$\frac{y_2 + d_2}{d_*} > \frac{d_1}{d_*} < 1 \qquad \text{for } F_1 > 1.$$

As another example, if we consider the hydraulic jump shown in Fig. 45b, now we find that a solution can only be obtained by using the impulse-momentum relation (30.51), thereby proving that the discontinuous change occurring in a hydraulic jump must result in a head loss. If the bottom slope is negligible, as indicated in Fig. 45b, then the impulse-momentum relation (30.51) may be written, with $Q/b = u_1 d_1 = u_2 d_2$, in the following manner, first given by RAYLEIGH (1914):

$$\left.\begin{aligned} \frac{1}{2} g\, d_1^2 + d_1 u_1^2 &= \frac{1}{2} g\, d_2^2 + \frac{(Q/b)^2}{d_2} = \frac{1}{2} g\, d_2^2 + \frac{(u_1 d_1)^2}{d_2}, \\ F_1^2 = \frac{(Q/b)^2}{g\, d_1^3} &= \frac{u_1^2}{g\, d_1} = \frac{1}{2}\,\frac{\left(\frac{d_2}{d_1}\right)^2 - 1}{1 - \frac{d_1}{d_2}} = \frac{1}{2}\,\frac{d_2}{d_1}\left(1 + \frac{d_2}{d_1}\right), \end{aligned}\right\} \tag{30.54}$$

or, if we let the actual rise in water level be $a = d_2 - d_1$,

$$F_1 = \frac{u_1}{\sqrt{g\, d_1}} = \left[1 + \frac{3}{2}\,\frac{a}{d_1} + \frac{1}{2}\left(\frac{a}{d_1}\right)^2\right]^{\frac{1}{2}}. \tag{30.55}$$

where

$$1 + \frac{a}{d_1} = \frac{d_2}{d_1} = \frac{1}{2}\left[\sqrt{1 + 8\,\frac{(Q/b)^2}{g\, d_1^3}} - 1\right] = \frac{1}{2}\left[\sqrt{1 + 8 F_1^2} - 1\right]. \tag{30.56}$$

Similarly, (30.54) can also be solved for

$$F_2^2 = \frac{(Q/b)^2}{g\, d_2^3} = \frac{u_2^2}{g\, d_2} = \frac{1}{2}\,\frac{d_1}{d_2}\left(1 + \frac{d_1}{d_2}\right) = F_1^2 \left(\frac{d_1}{d_2}\right)^3. \tag{30.57}$$

Eqs. (30.54) and (30.57) may be multiplied together to yield

$$u_1 u_2 = \frac{1}{2} g\,(d_1 + d_2) = c_{1*}^2 \left[\frac{3}{2 + F_1^2}\right]\left[\frac{1 + \sqrt{1 + 8 F_1^2}}{4}\right] < c_{1*}^2 > c_{2*}^2. \tag{30.58}$$

The last inequality in (30.58) is obtained from (30.50) and (30.52) by noting that in any finite hydraulic jump the head loss must also be finite, so that $h_L = h_{0_1} - h_{0_2} > 0$ and (30.50) must be written as

$$\left.\begin{aligned} d_1^2 + \frac{u_1^2}{2g} &= d_2^2 + \frac{u_1^2}{2g}\left(\frac{d_1}{d_2}\right)^2 + h_{0_1} - h_{0_2}, \\ h_L = h_{0_1} - h_{0_2} &= \frac{1}{2} F_1^2\, \frac{(1 - d_1/d_2)^3}{1 + d_2/d_1}\, d_2 \\ &= \frac{1}{4} F_1^2 \left[\sqrt{1 + 8 F_1^2} - 1\right] \frac{(1 - d_1/d_2)^3}{1 + d_2/d_1}\, d_1. \end{aligned}\right\} \tag{30.59}$$

Thus (30.52) and (30.59) give the total head ratio, and therefore the critical speed ratio, as

$$\left.\begin{aligned} \frac{h_{02}}{h_{01}} &= \left(\frac{c_{2*}}{c_{1*}}\right)^2 = 1 - \frac{h_L}{h_{01}} = \frac{d_{2*}}{d_{1*}} \\ &= \frac{(\sqrt{1 + 8 F_1^2} - 1)^3 + 4 F_1^2}{(\sqrt{1 + 8 F_1^2} - 1)^2\,(2 + F_1^2)} < 1. \end{aligned}\right\} \tag{30.60}$$

Consequently there is no direct analogy between finite hydraulic jumps and gas-dynamic shock waves, as was pointed out by PREISWERK (1938), since in gas dynamics the well-known Prandtl relation for normal shock waves gives $u_1 u_2 = c_*^2$, and c_* is constant through the shock wave [see, e.g., COURANT and FRIEDRICHS (1948, p. 146)]. The equations are similar only for the limiting case as the hydraulic jump vanishes so that $F_1 = F_2 = 1$, $d_2 = d_1$, and $h_{0_2} = h_{0_1}$. However this limiting process corresponds to the isentropic potential-flow case where there is an analogy for small perturbations over a flat horizontal bottom, as previously discussed. Also, as indicated by (30.59) the head loss and variation in c_* could be neglected until the third-order terms become important, so that for F_1 near unity the first- and second-order terms of the hydraulic-jump relations correspond to the gas-dynamic shock-wave relations having a specific heat ratio $\gamma=2$. However, this is identical to the known fact that weak shock waves may be considered isentropic to the third order of approximation; consequently the hydraulic analogy to compressible gas dynamics exists only for small perturbations in potential flow.

There is no direct analogy between the finite hydraulic jump and the gas-dynamic shock wave because the hydraulic jump has a head loss that must be included in the specific-energy equation (30.50). This head loss results in a loss of kinetic energy that is no longer available as flow energy since it is converted into an insignificant temperature rise in the water itself. In the gas dynamics energy equation the entropy increase through a shock wave of course corresponds to a loss of kinetic energy, but this is converted, through the increase of the temperature of the gas, into an adiabatic enthalpy increase that maintains constant flow energy through the shock wave [see, e.g., COURANT and FRIEDRICHS (1948, p. 125)]. The most unusual effect of this loss in flow energy (or h_L) in the hydraulic jump is revealed in (30.58) which shows that the flow velocity downstream of a hydraulic jump is always less than in the corresponding gas dynamics case, which maintains c_* constant so that $u_1 u_2 = c_*^2$. For example, in the gas-dynamic case when $F_1 \to \infty$, then $u_1/c_* \to \sqrt{3}$ (for $\gamma=2$), and therefore $u_2/c_* \to 1/\sqrt{3}$. However, in a hydraulic jump (30.58) shows that $u_2/c_* \to 0$ when $F_1 \to \infty$ (or $u_1/c_* \to \sqrt{3}$).

The experimental investigations by BAKHMETEFF (1932) have shown that the hydraulic jumps in a horizontal rectangular channel are in excellent agreement with the predictions of the "one-dimensional" hydraulic equations (30.54) through (30.60). BAKHMETEFF found that depth increases as high as $10 d_1$ were in excellent agreement with (30.56). However, he found that for Froude numbers of the oncoming flow less than $\sqrt{3}$ (i.e., $F_1 < \sqrt{3}$) the profile of the normal hydraulic jump developed undulations, and the relative length of transition became indeterminate because the undulating surface made the region of parallel flow increasingly remote from the start of the wave front, as indicated in Fig. 45c. It is interesting to note that $F_1 = \sqrt{3}$ corresponds to the maximum absolute elevation that a hydraulic jump can reach with a given h_0 (although there is no limit to d_2/d_1), since (30.53) and (30.56) may be combined to give

$$\frac{d_2}{h_{0_1}} = \frac{1}{2+F_1^2}\left(\sqrt{1+8F_1^2}-1\right) \leqq \frac{4}{5}$$

which attains its maximum elevation of $\frac{4}{5} h_{0_1}$ above the channel bottom only for $F_1 = \sqrt{3}$; at this condition we have

$$\frac{d_2}{d_1} = 2, \quad F_2^2 = \frac{3}{8}, \quad \frac{h_L}{d_1} = \frac{1}{8}, \quad \frac{h_{0_2}}{h_{0_1}} = \frac{19}{20}, \quad u_1 u_2 = \frac{9}{10} c_{1*}^2 .$$

This shows that for all the undulating hydraulic jumps ($F_1 < \sqrt{3}$) the change in total head is less than 5%; consequently these jumps can be approximated by the isentropic, potential-flow relations. This is of great aid in calculating the slant or oblique hydraulic jumps, as shown in Fig. 45d, since the characteristic epicycloid values of $f(F_*)$, as given in Table 1, may be used in the manner indicated in (30.46) to approximate the change in F_*, F, or η upon turning through an angle ϑ by means of an oblique hydraulic jump whenever $d_2/d_1 \to 1$. The comparison between the value given by $f(F_1)$ in Table 1 for compression to $F_2=1$, is compared with the exact values for the corresponding oblique hydraulic jumps in Fig. 46. It is seen that, although the gas-dynamic shock wave is not a satisfactory approximation for $F_1 > \sqrt{3}$, still the isentropic potential relation $f(F_1)$ provides an excellent approximation for much greater values of F_1 since the criterion for oblique hydraulic jumps is that the flow component normal to the discontinuity satisfy $F_1 \sin\alpha < \sqrt{3}$.

The exact relations for the oblique hydraulic jump are given by PREISWERK (1938) and can be obtained by simply adding the same velocity component ($w = w_1 = w_2$) tangent to both faces of the hydraulic jump as shown in Fig. 45d.

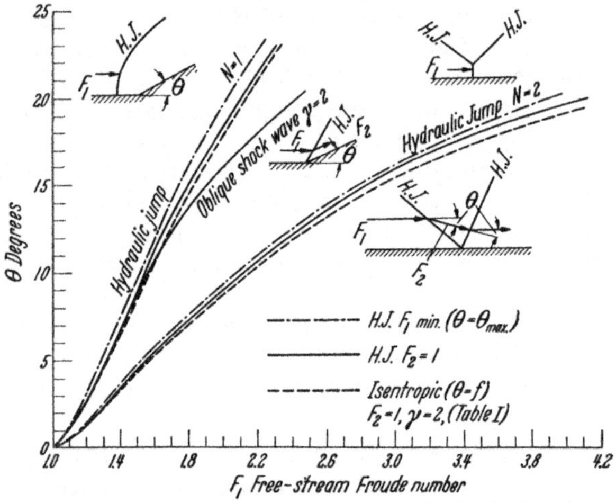

Fig. 46. Maximum flow deflection ($n = 1$), and reflection ($n = 2$).

This results in the following equations (which reduce to the preceding ones for a normal hydraulic jump by simply letting $\vartheta \to 0$ and $\alpha \to \pi/2$):

$$\left.\begin{aligned}
F_1^2 &= \frac{1}{2\sin^2\alpha}\,\frac{\tan\alpha}{\tan(\alpha-\vartheta)}\left[1+\frac{\tan\alpha}{\tan(\alpha-\vartheta)}\right] = \left[\frac{u_1/\sqrt{g\,d_1}}{\sin\alpha}\right]^2,\\
F_2^2 &= F_1^2\left(\frac{d_1}{d_2}\right)^3\left[\frac{\sin\alpha}{\sin(\alpha-\vartheta)}\right]^2 = \left[\frac{u_2/\sqrt{g\,d_2}}{\sin(\alpha-\vartheta)}\right]^2,\\
\frac{d_2}{d_1} &= \frac{\tan\alpha}{\tan(\alpha-\vartheta)} = \frac{1}{2}\left[\sqrt{1+8F_1^2\sin^2\alpha}-1\right] = \frac{u_1}{u_2},\\
\tan\vartheta &= \frac{\tan\alpha\left[\sqrt{1+8F_1^2\sin^2\alpha}-3\right]}{2\tan^2\alpha-1+\sqrt{1+8F_1^2\sin^2\alpha}},\\
\sin\alpha &= \frac{1}{F_1}\sqrt{\frac{1}{2}\left(\frac{d_2}{d_1}\right)\left(1+\frac{d_2}{d_1}\right)} = \frac{u_1/\sqrt{g\,d_1}}{F_1},\\
\sin(\alpha-\vartheta) &= \frac{1}{F_2}\sqrt{\frac{1}{2}\left(\frac{d_1}{d_2}\right)\left(1+\frac{d_1}{d_2}\right)} = \frac{u_2/\sqrt{g\,d_2}}{F_2}.
\end{aligned}\right\} \tag{30.61}$$

The last two equations in (30.61) clearly show how the oblique hydraulic jump approaches the same value as given by the isentropic, potential relation $f(F_1)$ at *any* value of F_1 as long as $d_2/d_1 \to 1$, since they reduce to the isentropic, potential characteristic curve given by (30.31) whenever $\vartheta \to 0$ and $d_2/d_1 \to 1$. As a matter of fact, as previously mentioned, (30.61) shows that the oblique hydraulic jump angle (α) can be approximated as in Fig. 44 by

$$\alpha = \frac{1}{2}\left(\arcsin F_1^{-1} + \arcsin F_2^{-1} + \vartheta\right) + O\left(\frac{d_2}{d_1}-1\right)^2 + O(\vartheta^3), \tag{30.62}$$

that is, by taking α as defined by the average line between the two characteristics in the physical plane, or the line normal to the midpoint of the corresponding characteristic epicycloid arc in the hodograph plane (see Fig. 44). The close approximation of the characteristic epicycloid as given by f from (30.26) and Table 1, to the oblique hydraulic-jump relations (30.61) is strikingly illustrated when they are both plotted in the hodograph plane as in Fig. 45d [see also PREISWERK (1938)]. The trace of the endpoint of the velocity vector for the oblique shock wave of gas dynamics, generally called the "shock polar", is also shown in Fig. 45d for a specific heat ratio $\gamma=2$.

As long as f from (30.26) is in close agreement with the oblique hydraulic jump relations (30.61), then the problems involving the interaction and reflection of hydraulic jumps can be closely approximated by the same procedure as detailed previously for the characteristic epicycloids involving compression waves (see Figs. 43 and 44). Whenever the required flow deflection ϑ is greater than that provided by the epicycloid passing through F_1, as shown in Fig. 43, then subcritical flow follows the curved or normal hydraulic jump as indicated by $N=1$ in Fig. 46. Similarly, $N=2$ defines the maximum flow reflection angle (ϑ) that can occur without ending in subcritical flow with a curved or normal hydraulic jump. In both cases two curves are shown for the oblique hydraulic jump: one shows the turning angle ϑ that will make the flow critical ($F_2=1$), and the other one is the maximum possible turning angle $\vartheta_{\max}$ for any oblique hydraulic jump at the given value of F_1. The latter always produces subcritical flow ($F_2<1$) as indicated in Fig. 43.

All of the preceding results primarily hold for hydraulic jumps in rectangular cross-section channels with a nearly horizontal bottom. BAKHMETEFF (1932) shows experimentally the various effects of steepening bottom slopes. He also generalizes (30.51) so that it will apply to any constant cross-section shape. However, it must be noted that our Eq. (29.3) shows conclusively that (30.51) which completely neglects the w velocity component, cannot be applicable to channel walls that are not nearly vertical. Sloping sides on a channel would increase the vertical velocity gradients, make a normal hydraulic jump impossible, and induce unsteady vortex motions.

It must also be noted that all of the preceding results are valid only for relatively small bottom slopes, as indicated by the direct comparison of (30.50) and (30.51) with (28.1) and (29.3). When the flow is rapidly varying because of large changes in the bottom slope, then the change in surface profile curvature is so pronounced that the pressure variation can no longer be considered as hydrostatic. For example, over the spillway of a dam the centrifugal force due to the streamline curvature can actually exceed the hydrostatic pressure, thereby leading to a pressure less than atmospheric resulting in flow separation or violent oscillations. At present spillway design is based on semi-empirical methods or model tests since no satisfactory mathematical analysis is available.

31. Higher-order theories and the solitary and cnoidal waves. It will now be shown that many of the preceding methods and results based on the shallow-water approximation are valid only if the local variations in water depth are not too large, and the average or undisturbed water depth is sufficiently small. The first requirement implies that the solutions of the first-order nonlinear shallow-water equations (28.1) do not greatly differ (at least for Froude numbers not near unity) from the linearized solutions given by (29.3) or (29.7). The second requirement essentially demands that the depth h be much less than the effective wavelength λ in any application, say $\frac{h}{\lambda} < \frac{1}{10}$, in order to reduce the effects associated with the infinitesimal-wave approximation.

As already discussed, the infinitesimal-wave approximation predicts that the fluid particle motion varies with the distance below the free surface, and also that the propagation velocity depends upon the wavelength, as shown in Sect. 15. There it was proved that the velocity defined by

$$c = \sqrt{\frac{g\lambda}{2\pi} \tanh\left(\frac{2\pi h}{\lambda}\right)} = \sqrt{g h}\left[1 - \frac{1}{6}\left(\frac{2\pi h}{\lambda}\right)^2 + \cdots\right] \tag{31.1}$$

can only be considered a phase velocity while the actual rate of propagation of energy is associated with the group velocity defined by

$$c - \lambda \frac{dc}{d\lambda} = \frac{1}{2} c\left(1 + \frac{4\pi h/\lambda}{\sinh 4\pi h/\lambda}\right) = c\left[1 - \frac{1}{3}\left(\frac{2\pi h}{\lambda}\right)^2 + \cdots\right] \tag{31.2}$$

[see also Lamb (1932, p. 381)]. Any such variation will directly interfere with the applicability of the shallow-water results. However, as long as $\frac{h}{\lambda} < \frac{1}{10}$ it is seen that the phase velocity and the group velocity are both satisfactory approximations to the shallow-water first-order result that $c = \sqrt{g h}$ and is independent of the effective wave length.

This means that if small-scale model tests are used to simulate results appropriate to the shallow-water theory, then the undisturbed water depth should be less than $\frac{1}{10}$ the principal model dimensions. Consequently, if models less than 10 cm in effective dimensions are used, the depth of test water should be less than 1 cm, so the capillary ripples produced by surface tension must be considered. As shown in Sects. 15 and 24 the effect of the surface tension T is to increase the phase velocity for the short wavelength capillary ripples so that (31.1) is replaced by

$$c = \sqrt{\left(\frac{g\lambda}{2\pi} + \frac{2\pi T}{\lambda\varrho}\right) \tanh \frac{2\pi h}{\lambda}}. \tag{31.1'}$$

For ordinary water (at 20° C, $T = 72.8$ dynes/cm, $\varrho = 0.998$ gm/cm^3) this gives the interesting result that both the phase velocity and the group velocity are closely approximated by $\sqrt{g h}$ for all $\lambda > 2$ cm if $h \approx \frac{1}{2}$ cm. However, in any small-model tests the surface wave patterns formed by the capillary ripples must be ignored since they are short-wavelength surface waves that are never in accord with the long-wavelength shallow-water theory.

Except for the section on hydraulic jumps the preceding shallow-water results have all been based entirely on (28.1), the first approximation to shallow-water theory, and this will now be shown to be limited to relatively small wave amplitudes even though the complete nonlinear equation (28.1) be used, and even though the bottom surface be flat and horizontal. The second approximation to shallow-water theory will be shown to immediately yield particular solutions corresponding to continuous permanent wave profiles of finite amplitude that can be propagated without a change in form or shape if viscosity effects are neglected. These permanent, finite-amplitude wave forms are the cnoidal waves discovered by Korteweg and de Vries (1895) which reduce, in the limiting case of essentially infinite wavelength, to the solitary wave of Russell (1837, 1844) which was first analyzed theoretically by Boussinesq (1871, 1872) and Rayleigh (1876).

The second approximation to shallow-water theory will show that the limitation of the nonlinear first approximation to relatively small amplitudes is primarily due to the fact that the variation in the vertical velocity cannot be neglected as the wave amplitude is increased. This of course invalidates even the rectangular channel hydraulic analogy to compressible gas flow, since, as previously discussed, the principal assumption of the hydraulic analogy is that the vertical acceleration be negligible.

The third approximation to shallow-water theory will then be presented to obtain new relations which will predict the limiting heights of the continuous finite-amplitude steady-state wave forms and give, for the first time, the complete second approximation to the cnoidal and solitary waves. It will be found that the pressure is no longer hydrostatic, thereby violating the remaining principal assumption of the hydraulic analogy and the ordinary classical shallow-water theory.

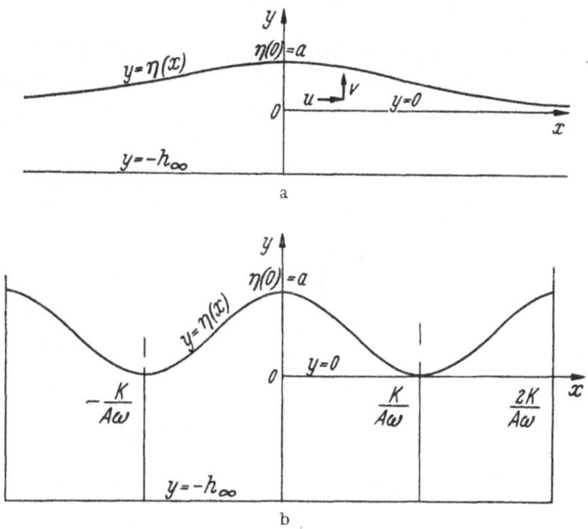

Fig. 47a and b. (a) Solitary wave over a flat horizontal bottom. (b) Cnoidal wave $\eta(x) = a\,\mathrm{cn}^2(A\,\omega\,x, k)$.

α) *The first and second approximations to the cnoidal and solitary waves.* We will now extend the perturbation method of FRIEDRICHS (1948), which was used to derive the nonlinear first-order approximation (28.1) to shallow-water theory, to obtain the second and higher orders of approximations for the special case of the steady-state propagation of a wave independent of z and t over a flat horizontal bottom described by $y=-h_\infty=\text{const}$ as in Fig. 47.

First we will show that the only steady-state finite-amplitude solution of the first-order equation (28.2) is $y^{(0)}=\eta_0=\text{const}$ and $u^{(0)}=u_0=\text{const}$. This is most easily proved by substituting the solution of the zeroth-order terms in (10.24) for steady water flow over a flat horizontal bottom, namely

$$u^{(0)}=u^{(0)}(x), \qquad v^{(0)}=0, \qquad p^{(0)}=0, \qquad \eta^{(0)}=\eta^{(0)}(x), \tag{31.3}$$

into the first-order terms in (10.27) to obtain

$$u^{(0)}=u_0=\text{const}, \qquad v^{(1)}=0, \qquad p_y^{(0)}=-\varrho\,g, \qquad \eta^{(0)}=\eta_0=\text{const} \tag{31.4}$$

since $\eta_x^{(0)}=0=p_x^{(0)}$. Consequently the only finite-amplitude first-order steady-state solution must have $\eta_x^{(0)}=0$, which would permit only the hydraulic jump as a solution since $\eta_x^{(0)}=0$ and $u^{(0)}=\text{const}$ on each side of the finite discontinuity defining the hydraulic jump. This is in agreement with the well-known fact that the gas-dynamics equation or (28.2), predicts that any finite amplitude disturbance must form a finite discontinuity which is a shock wave, or hydraulic jump [see, e.g., LAMB (1932, pp. 278, 481)]. However, the second-order approximation of shallow-water theory (10.33) does yield a permanent finite-amplitude, steady-state wave profile that does not form a discontinuity. These are called the cnoidal waves, discovered by KORTEWEG and DE VRIES (1895), and the solitary waves of RUSSELL (1837, 1844), BOUSSINESQ (1871, 1872), and RAYLEIGH (1876). In order to obtain the higher-order approximations and limiting heights of these waves, it is more convenient to use exactly the same non-dimensional variables introduced by FRIEDRICHS (1948), and also used by KELLER (1948), namely,

$$\left.\begin{aligned}
&\varepsilon=\omega^2 h^2,\quad \alpha=\omega x,\quad \beta=y/h,\quad H=h_\infty/h,\\
&u(\alpha,\beta)=u(x,y)/\sqrt{g h},\qquad v(\alpha,\beta)=\frac{v(x,y)}{\sqrt{g h}}\,\omega h,\\
&Y(\alpha)=\eta(x)/h,\qquad \pi(\alpha,\beta)=p(x,y)/\varrho g h,\\
&Y(\alpha)=Y^{(0)}+\varepsilon Y^{(1)}+\varepsilon^2 Y^{(2)}+\cdots,\\
&\eta(x)=h Y^{(0)}+\omega^2 h^3 Y^{(1)}+\omega^4 h^5 Y^{(2)}+\cdots,
\end{aligned}\right\}\tag{31.5}$$

the only difference in notation being that x and y are now defined as in Fig. 47; consequently, the flat horizontal bottom is given by $y=-h_\infty$ or $\beta=-h_\infty/h=-H$, and the expansion parameter $\varepsilon=\omega^2 h^2$ is used as defined in (10.23), with (31.5) replacing (10.21).

Introducing the transformation defined by (31.5) into (31.4) and into the corresponding equivalent of (10.33), we obtain

$$\left.\begin{aligned}
&v^{(0)}=0=v^{(1)}\\
&u^{(0)}(\alpha,\beta)=u_0=\text{const},\qquad Y^{(0)}(\alpha)=Y_0=\text{const}=\eta_0/h,\\
&\pi_\beta^{(0)}=-1,\quad \pi_\beta^{(1)}=0,\quad u_\beta^{(1)}=v_\alpha^{(1)}=0,\quad u_\alpha^{(1)}=-v_\beta^{(2)},\\
&u_0 u_\alpha^{(1)}+\pi_\alpha^{(1)}=0,\qquad v^{(2)}(Y_0)=u_0 Y_\alpha^{(1)},\\
&v^{(2)}(-H)=0,\qquad \pi^{(1)}(Y_0)=-Y^{(1)}\pi_\beta^{(0)}=Y^{(1)}(\alpha).
\end{aligned}\right\}\tag{31.6}$$

These expressions may be integrated to obtain

$$\left.\begin{aligned}
&u^{(1)}(\alpha,\beta)=f(\alpha)=u^{(1)}(\alpha),\\
&v^{(2)}(\alpha,\beta)=v^{(2)}(\alpha,\beta)-v^{(2)}(\alpha,-H)\\
&\qquad=-\int_{-H}^{\beta} f_\alpha\,d\beta=-(\beta+H)\,f_\alpha,\\
&\pi^{(1)}(\alpha,\beta)=-\int u_0 f_\alpha\,d\alpha=-(u_0 f+C)=\pi^{(1)}(\alpha)=Y^{(1)}(\alpha),\\
&Y_\alpha^{(1)}=\pi_\alpha^{(1)}=-u_0 f_\alpha=\frac{v_2(Y_0)}{u_0}=-\left(\frac{Y_0+H}{u_0}\right)f_\alpha.
\end{aligned}\right\}\tag{31.7}$$

The identities in the last equation show that the solution for constant u_0 is restricted to the unique value defined by

$$u(x,y)=u_0\sqrt{g h}=\sqrt{g h(Y_0+H)}=\sqrt{g(\eta_0+h_\infty)}=\text{const}\tag{31.8}$$

which corresponds to the infinitesimal-wave propagation velocity (28.3) and shows that the steady-state solution will be in the neighborhood of the critical speed defined by a Froude number of unity. However, $u^{(1)}=f(\alpha)$ now provides a finite-amplitude steady-state solution that does not form a discontinuity; consequently, the behavior of the second-order shallow-water theory is mathematically completely different from the first-order (28.2) shallow-water theory or the gas-dynamics equations. The pressure variation is still hydrostatic, since $\pi^{(1)}$ does not depend upon β, and only $v^{(2)}$ has a direct dependence upon $\beta\,(=y/h)$.

Now in order to continue the solution and determine $f(\alpha)$ we must introduce some ε^3 terms. By following the same procedure as used in collecting the ε^2 terms for (10.33) we obtain for the particular case of steady flow over a flat horizontal bottom the following additional terms that are required for completing the second order solution:

$$\left.\begin{aligned}
&u_\beta^{(2)} = v_\alpha^{(2)}, \quad u_\alpha^{(2)} = -v_\beta^{(3)},\\
&u_0 u_\alpha^{(2)} + u^{(1)} u_\alpha^{(1)} + \pi_\alpha^{(2)} = 0, \quad u_0 v_\alpha^{(2)} + \pi_\beta^{(2)} = 0,\\
&v^{(3)}(Y_0) = u_0 Y_\alpha^{(2)} + u^{(1)} Y_\alpha^{(1)} - v_\beta^{(2)} Y^{(1)}, \quad v^{(3)}(-H) = 0,\\
&\pi^{(2)}(Y_0) = - Y^{(2)} \pi_\beta^{(0)} = Y^{(2)}.
\end{aligned}\right\} \tag{31.9}$$

These expressions were first given by KELLER (1948) and they may be directly integrated to give the following:

$$\left.\begin{aligned}
u^{(2)} &= \int v_\alpha^{(2)} d\beta = - f_{\alpha\alpha} \int (\beta + H)\, d\beta\\
&= -\tfrac{1}{2}(\beta^2 + 2H\beta) f_{\alpha\alpha} + R(\alpha) = u^{(2)}(\alpha,\beta)\\
\pi^{(2)}(\alpha,\beta) &= Y^{(2)}(\alpha) - [\tfrac{1}{2}(Y_0^2 - \beta^2) + H(Y_0 - \beta)]\, u_0 f_{\alpha\alpha},\\
v^{(3)}(\alpha,\beta) &= v^{(3)}(\alpha,\beta) - v^{(3)}(\alpha,-H) = \int_{-H}^{\beta} - u_\alpha^{(2)}\, d\beta\\
&= [\tfrac{1}{6}(\beta^3 + H^3) + \tfrac{1}{2} H(\beta^2 - H^2)] f_{\alpha\alpha\alpha} - (\beta + H) R_\alpha,\\
v^{(3)}(Y_0) &= u_0 Y_\alpha^{(2)} - u_0 f f_\alpha - (u_0 f + C) f_\alpha\\
&= \tfrac{1}{6}(Y_0^3 + 3H Y_0^2 - 2H^3) f_{\alpha\alpha\alpha} - (Y_0 + H) R_\alpha.
\end{aligned}\right\} \tag{31.10}$$

The last equation for $v^{(3)}$ gives the following expression for the ε^2 term in the surface profile

$$u_0 Y^{(2)}(\alpha) = u_0 f^2 + C f + \tfrac{1}{6}(Y_0^3 + 3H Y_0^2 - 2H^3) f_{\alpha\alpha} - (Y_0 + H) R + \text{const}, \tag{31.11}$$

while a similar expression may be obtained directly from $\pi^{(2)}(Y_0)$ by equating its relation in (31.9) and (31.10) so as to obtain

$$\left.\begin{aligned}
u_0 Y^{(2)}(\alpha) &= u_0 \pi^{(2)}(Y_0) = - u_0 [u_0 u^{(2)} + \tfrac{1}{2} u^{(1)2}]_{\beta = Y_0}\\
&= \tfrac{1}{2} u_0^2 (Y_0^2 + 2H Y_0) f_{\alpha\alpha} - u_0^2 R(\alpha) - \tfrac{1}{2} u_0 f^2 + \text{const}.
\end{aligned}\right\} \tag{31.12}$$

Since (31.11) and (31.12) must be identical, we may equate them and find that $f(\alpha)$ must satisfy the ordinary differential equation

$$f_{\alpha\alpha} - \frac{9}{2u_0^5} f^2 - \frac{3C}{u_0^6} f + C_0 = 0 \tag{31.13}$$

after having introduced (31.8) to eliminate Y_0. Eq. (31.13) may be integrated to

$$\tfrac{1}{3} u_0^6 f_\alpha^2 - u_0 f^3 - C f^2 + \tfrac{2}{3} u_0^6 C_0 f = \text{const}. \tag{31.14}$$

Upon noting from (31.7) that $f(\alpha) = u^{(1)}(\alpha)$ and

$$\frac{u_0}{C} f = -\left[1 + \frac{Y^{(1)}(\alpha)}{C}\right],$$

it is evident that (31.13) and (31.14) are the same equations as obtained by BOUSSINESQ (1871, 1872), RAYLEIGH (1876), KORTEWEG and DE VRIES (1895), LAVRENT'EV (1943), and KELLER (1948). The physical significance of each term in (31.14) was first pointed out by BENJAMIN and LIGHTHILL (1954), who derived (31.14) in an entirely different manner, starting with the same series expansion of the stream function as was introduced by RAYLEIGH (1876). BENJAMIN and LIGHTHILL (1954) use the continuity equation (30.49), the specific-energy equation (30.50), and the specific-momentum equation (30.51) to derive the equivalent of (31.14), and then they give a very useful discussion of the mathematical and physical behavior of its solutions.

The appropriate solution of (31.13) for the boundary conditions shown in Fig. 47 is given by the square of the Jacobian elliptic function "cn" having the modulus $0<k\leqq 1$ and the real period

$$4K(k) = 4\int_0^{\pi/2} \frac{d\vartheta}{\sqrt{1-k^2\sin^2\vartheta}} > 2\pi .$$

Substituting

$$f(\alpha) = -B\,\mathrm{cn}^2(A\,\alpha, k) \tag{31.15}$$

into (31.13) we find that (31.15) is a solution if, and *only* if, $0<k\leqq 1$ and

$$B = \frac{4}{3}u_0^5 A^2 k^2 = \frac{C}{u_0}\,\frac{k^2}{2k^2-1} = \frac{C_0}{2A^2(1-k^2)}\,. \tag{31.16}$$

Substituting (31.15) into (31.7) and (31.5) we obtain

$$\eta(x) = \eta_0 + \omega^2 h^3 B\,u_0\left[\mathrm{cn}^2(A\,\omega\,x, k) - \frac{2k^2-1}{k^2}\right] + O(\varepsilon^2)\,.$$

The boundary conditions in Fig. 47 then yield

$$\left.\begin{aligned}
\eta(0) &= \eta_0 + \omega^2 h^3 B\,u_0\,\frac{1-k^2}{k^2} = a\,,\\
\eta\left(\frac{K}{A\,\omega}\right) &= \eta_0 - \omega^2 h^3 B\,u_0\,\frac{2k^2-1}{k^2} = 0\,,\\
a &= \omega^2 h^3 B\,u_0\,, \qquad \eta_0 = a\,\frac{2k^2-1}{k^2}\,,\\
\eta(x) &= a\,\mathrm{cn}^2(A\,\omega\,x, k)\,.
\end{aligned}\right\} \tag{31.17}$$

Then upon introducing (31.8) and (31.16) into (31.17) we obtain

$$\left.\begin{aligned}
A\,\omega\,x &= \frac{x}{h_\infty}\sqrt{\frac{3}{4k^2}\,\frac{a/h_\infty}{(1+\eta_0/h_\infty)^3}} = \frac{x}{h_\infty}\sqrt{\frac{3}{4k^2}\,\frac{a/h_\infty}{[1+(a/h_\infty)(2k^2-1)/k^2]^3}}\\
&= \frac{x}{h_\infty}\sqrt{\frac{3}{4k^2}\,\frac{a}{h_\infty}}\left\{1-\frac{3}{2}\,\frac{a}{h_\infty}\,\frac{2k^2-1}{k^2}+\cdots\right\}\\
&= \frac{x}{h_\infty}\sqrt{\frac{3}{4k^2}\,\frac{a}{h_\infty}} + O\left[\frac{a}{h_\infty}\,\frac{2k^2-1}{k^2}\right]^{\frac{3}{2}}
\end{aligned}\right\} \tag{31.18}$$

as the exact second-order shallow-water theory solution for the first approximation to the cnoidal waves of Korteweg and de Vries (1895). The remaining terms in (31.6) and (31.7) may similarly be solved to give

$$\left.\begin{aligned}
\frac{p(x,y)}{\varrho g h_\infty} &= \frac{\eta(x)-y}{h_\infty} + O\left[\frac{a}{h_\infty}\,\frac{2k^2-1}{k^2}\right]^2,\\
\frac{u(x)}{\sqrt{g\,h_\infty}} &= 1 + \left(1-\frac{1}{2k^2}\right)\frac{a}{h_\infty} - \frac{\eta(x)}{h_\infty} + O\left[\frac{a}{h_\infty}\,\frac{2k^2-1}{k^2}\right]^2,\\
\frac{v(x,y)}{\sqrt{g\,h_\infty}} &= -\left(1+\frac{y}{h_\infty}\right)\sqrt{\frac{3}{k^2}\left(\frac{a}{h_\infty}\right)^3}\,\mathrm{cn}(A\,\omega\,x,k)\,\mathrm{sn}(A\,\omega\,x,k)\,\mathrm{dn}(A\,\omega\,x,k) + {}\\
&\quad + O\left[\frac{a}{h_\infty}\,\frac{2k^2-1}{k^2}\right]^{\frac{5}{2}}\\
&= +\left(1+\frac{y}{h_\infty}\right)\frac{d\eta(x)}{dx} + O\left[\frac{a}{h_\infty}\,\frac{2k^2-1}{k^2}\right]^{\frac{5}{2}},
\end{aligned}\right\} \tag{31.19}$$

where cn, sn, dn are the Jacobian elliptic functions with the argument $A\omega x$ defined by (31.18). It must be noted that $0<k\leqq 1$; k can never become identically zero for two reasons. First, because for $k=0$

$$\mathrm{cn}^2(A\,\omega\,x,0)=\cos^2(A\,\omega\,x)$$

is not a solution of (31.13) or (31.14), and second, because the asymptotic expansions given in (31.18) and (31.19) are only valid as $a^2/k^2\to 0$.

The limiting case of $k=1$ corresponds to an essentially infinite wavelength since $K(k^2)\to\infty$ as $k^2\to 1$, and the cnoidal-wave solutions reduce to

$$\left.\begin{aligned}
\frac{\eta(x)}{h_\infty}&=\frac{a}{h_\infty}\,\mathrm{sech}^2\left(\frac{x}{h_\infty}\sqrt{\frac{3}{4}\frac{a}{h_\infty}}\right)+O\left(\frac{a}{h_\infty}\right)^2,\\
\frac{p(x,y)}{\varrho\,g\,h_\infty}&=\frac{\eta(x)-y}{h_\infty}+O\left(\frac{a}{h_\infty}\right)^2,\\
\frac{u(x)}{\sqrt{g\,h_\infty}}&=1+\frac{1}{2}\frac{a}{h_\infty}-\frac{\eta(x)}{h_\infty}+O\left(\frac{a}{h_\infty}\right)^2,\\
\frac{v(x,y)}{\sqrt{g\,h_\infty}}&=\left(1+\frac{y}{h_\infty}\right)\frac{d\eta(x)}{dx}+O\left(\frac{a}{h_\infty}\right)^{\frac{5}{2}},
\end{aligned}\right\}\qquad(31.20)$$

which provides the exact first approximation to the solitary wave.

All of these solutions for the cnoidal wave and the solitary wave are in exact agreement with the expressions first given by KORTEWEG and DE VRIES (1895, pp. 430—431) if one neglects the terms of $O(a/h_\infty)^2$. It will now be proved that the terms of $O(a/h_\infty)^2$ must be neglected in these first approximations because the second approximations introduce additional terms having this order of magnitude.

We can continue to the next order of approximation by collecting the remaining terms corresponding to ε^3, and adding some of the ε^4 terms that are necessary in order to complete the solution

$$\left.\begin{aligned}
&\pi^{(3)}(Y_0)=Y^{(3)}-Y^{(1)}\,\pi_\beta^{(2)}(Y_0),\\
&u_\beta^{(3)}=v_\alpha^{(3)},\qquad u_\alpha^{(3)}=-\,v_\beta^{(4)},\\
&u_0\,u_\alpha^{(3)}+u^{(1)}\,u_\alpha^{(2)}+u^{(2)}\,u_\alpha^{(1)}+\pi_\alpha^{(3)}+v^{(2)}\,u_\beta^{(2)}=0,\\
&u_0\,v_\alpha^{(3)}+u^{(1)}\,v_\alpha^{(2)}+\pi_\beta^{(3)}+v^{(2)}\,v_\beta^{(2)}=0,\\
&v^{(4)}(Y_0)=u_0\,Y_\alpha^{(3)}+u^{(1)}\,Y_\alpha^{(2)}+u^{(2)}\,Y_\alpha^{(1)}-v_\beta^{(2)}\,Y^{(2)}-v_\beta^{(3)}\,Y^{(1)},\\
&v^{(4)}(-H)=0.
\end{aligned}\right\}\qquad(31.21)$$

Now we can combine the expression for $v^{(3)}$ in (31.10) with that in (31.21) to write

$$\left.\begin{aligned}
u^{(3)}(\alpha,\beta)=\int v_\alpha^{(3)}\,d\beta&=\tfrac{1}{24}(\beta^4+4H\beta^3-8H^3\beta)\,f_{\alpha\alpha\alpha\alpha}-\\
&-\tfrac{1}{2}(\beta^2+2H\beta)\,R_{\alpha\alpha}+S(\alpha).
\end{aligned}\right\}\qquad(31.22)$$

Then the expression for $v^{(4)}$ in (31.21) yields

$$\left.\begin{aligned}
v^{(4)}(\alpha,\beta)&=-\int_{-H}^{\beta}u_\alpha^{(3)}(\alpha,\beta)\,d\beta=-\,[(\beta+H)\,S_\alpha+\\
&+\tfrac{1}{120}(\beta^5+5H\beta^4-20H^3\beta^2+16H^5)\,f_{\alpha\alpha\alpha\alpha\alpha}-\tfrac{1}{6}(\beta^3+3H\beta^2-2H^3)\,R_{\alpha\alpha\alpha}].
\end{aligned}\right\}\qquad(31.23)$$

The boundary condition defined by the expression for $v^{(4)}$ in (31.21) thereby gives one relation for $Y^{(3)}$ that may be written as

$$\left.\begin{aligned} u_0 Y_\alpha^{(3)}(\alpha) &= v^{(4)}(Y_0) - [u^{(1)} Y_\alpha^{(2)} - v_\beta^{(2)} Y^{(2)}] - [u^{(2)} Y_\alpha^{(1)} - v_\beta^{(3)} Y^{(1)}] \\ &= v^4(Y_0) - [u^{(1)} Y^{(2)}]_\alpha - [u^{(2)} Y^{(1)}]_\alpha \end{aligned}\right\} \tag{31.24}$$

which may be directly integrated, upon substituting (31.23) for $v^{(4)}$, as

$$\left.\begin{aligned} u_0 Y_\alpha^{(3)} &= \int v^{(4)}(Y_0)\, d\alpha - u^{(1)}(Y_0)\, Y^{(2)} - u^{(2)}(Y_0)\, Y^{(1)} \\ &= \text{const} - \Big\{\frac{1}{120}(Y_0^5 + 5H Y_0^4 - 20H^3 Y_0^2 + 16H^5) f_{\alpha\alpha\alpha\alpha} + \\ &\quad - \frac{1}{6}(Y_0^3 + 3H Y_0^2 - 2H^3) R_{\alpha\alpha} + (Y_0 + H)\, S(\alpha) + \\ &\quad + \frac{f}{u_0}\Big[\frac{1}{2} u_0^2 (Y_0^2 + 2H Y_0) f_{\alpha\alpha} - u_0^2 R - \frac{1}{2} u_0 f^2\Big] + \\ &\quad + \Big[\frac{1}{2}(Y_0^2 + 2H Y_0) f_{\alpha\alpha} - R(\alpha)\Big][u_0 f + C]\Big\}. \end{aligned}\right\} \tag{31.25}$$

Another relation for $Y^{(3)}$ may also be obtained from the other boundary condition defined by the expression for $\pi^{(3)}$ in (31.21), namely

$$Y^{(3)}(\alpha) = \pi^{(3)}(Y_0) + Y^{(1)} \pi_\beta^{(2)}(Y_0), \tag{31.26}$$

where $\pi^{(3)}$ itself may be obtained by integrating the expressions for $\pi_\alpha^{(3)}$ and $\pi_\beta^{(3)}$ in (31.21) to obtain

$$\pi^{(3)}(\alpha, \beta) = - u_0 u^{(3)} - u^{(1)} u^{(2)} - \tfrac{1}{2}[v^{(2)}]^2 + \text{const}. \tag{31.27}$$

Then substituting $\pi^{(3)}$ from (31.27), $\pi_\beta^{(2)}$ from (31.10), $u^{(1)}=f$, $u^{(2)}$ from (31.10), $u^{(3)}$ from (31.22), $y^{(1)} = -(u_0 f + C)$ and $Y^{(2)}$ from (31.12) into (31.26) we obtain another relation for $Y^{(3)}$, namely,

$$\left.\begin{aligned} u_0 Y^{(3)}(\alpha) &= -\{\tfrac{1}{24} u_0^2 (Y_0^4 + 4H Y_0^3 - 8H^3 Y_0) f_{\alpha\alpha\alpha\alpha} - \tfrac{1}{2} u_0^2 (Y_0^2 + 2H Y_0) R_{\alpha\alpha} + \\ &\quad + u_0^2 S - \tfrac{1}{2} u_0 (Y_0^2 + 2H Y_0) f f_\alpha + u_0 f R + \tfrac{1}{2} u_0^5 (f_\alpha)^2 + \\ &\quad + (u_0 f + C)\, u_0^4 f_{\alpha\alpha} + \text{const}\}. \end{aligned}\right\} \tag{31.28}$$

These two expressions for $Y^{(3)}$, (31.25) and (31.28), must be identically equal; therefore, since u_0 is defined by (31.8), we find that the unknown function R must satisfy the ordinary differential equation

$$\left.\begin{aligned} &\frac{1}{3} u_0^5 R_{\alpha\alpha} - \Big(\frac{C}{u_0} + 3f\Big) R + \text{const} \\ &\quad = \frac{1}{30} u_0^5 (u_0^4 - 5H^2) f_{\alpha\alpha\alpha\alpha} - \frac{1}{2}(u_0^4 - 3H^2) f f_{\alpha\alpha} + \\ &\quad + \frac{C}{2u_0}(u_0^4 + H^2) f_{\alpha\alpha} + \frac{1}{2} u_0^4 (f_\alpha)^2 + \frac{1}{2u_0} f^3, \end{aligned}\right\} \tag{31.29}$$

the other unknown function $S(\alpha)$ having been eliminated since $u_0^2 = Y_0 + H$.

When $f(\alpha)$ is given by (31.15), then the solution of (31.29) is

$$\left.\begin{aligned} R(\alpha) = \frac{C^2}{u_0^3}\Big\{&\Big(\frac{k^2}{2k^2-1}\Big)^2\Big(1-\frac{9}{4}\frac{H^2}{u_0^4}\Big)\,\mathrm{cn}^4(A\,\alpha,k) + \\ &+\frac{k^2}{2k^2-1}\Big(1+\frac{3}{2}\frac{H^2}{u_0^4}\Big)\mathrm{cn}^2(A\,\alpha,k) - \frac{3}{10}\frac{k^2(1-k^2)}{(2k^2-1)^2}\Big(1-\frac{5}{2}\frac{H^2}{u_0^4}\Big)-\frac{3}{5}\Big\}\end{aligned}\right\}\quad(31.30)$$

and (31.11) or (31.12) give the ε^2 term of the wave profile as

$$\left.\begin{aligned} Y^{(2)}(\alpha) = \Big(\frac{C}{u_0}\Big)^2\Big\{&\frac{3}{4}\Big(\frac{k^2}{2k^2-1}\Big)^2\mathrm{cn}^4(A\,\alpha,k) - \\ &-\frac{5}{2}\frac{k^2}{2k^2-1}\mathrm{cn}^2(A\,\alpha,k)+\frac{12-57k^2+57k^4}{20(2k^2-1)^2}\Big\}.\end{aligned}\right\}\quad(31.31)$$

Consequently, the second approximation to the cnoidal-wave profile is obtained from the preceding and (31.5) as

$$\left.\begin{aligned} \eta(x) &= \eta_0 + \omega^2 h^3\, Y^{(1)} + \omega^4 h^5\, Y^{(2)} + O(\varepsilon^3) \\ &= \eta_0 - \eta_1\Big[1-\frac{k^2}{2k^2-1}\mathrm{cn}^2(A\,\omega\,x,k)\Big] + \\ &+\frac{\eta_1^2}{\eta_0+h_\infty}\Big[\frac{3}{4}\Big(\frac{k^2}{2k^2-1}\Big)^2\mathrm{cn}^4(A\,\omega\,x,k)-\frac{5}{2}\frac{k^2}{2k^2-1}\mathrm{cn}^2(A\,\omega\,x,k) + \\ &+\frac{12-57k^2+57k^4}{20(2k^2-1)^2}\Big],\end{aligned}\right\}\quad(31.32)$$

where

$$\begin{aligned}\eta_1 = C\,\omega^2 h^3 &= (A\,\omega)^2\,\tfrac{4}{3}\,(2k^2-1)\,(h\,u_0^2)^3 \\ &= (A\,\omega)^2\,\tfrac{4}{3}\,(2k^2-1)\,(\eta_0+h_\infty)^3.\end{aligned}$$

Then the boundary conditions shown in Fig. 37c yield the relations:

$$\left.\begin{aligned} \eta(0) = a &= \eta_0 - \eta_1\frac{k^2-1}{2k^2-1}+\frac{\eta_1^2}{\eta_0+h_\infty}\,\frac{12-7k^2-28k^4}{20(2k^2-1)^2}, \\ \eta\Big(\frac{K}{A\,\omega}\Big) = 0 &= \eta_0-\eta_1+\frac{\eta_1^2}{\eta_0+h_\infty}\Big[\frac{12-57k^2+57k^4}{20(2k^2-1)^2}\Big],\end{aligned}\right\}\quad(31.33)$$

which may be solved to give the second approximation

$$\left.\begin{aligned} \frac{\eta_0}{h_\infty} &= \Big(\frac{a}{h_\infty}\Big)\frac{2k^2-1}{k^2}+\Big(\frac{a}{h_\infty}\Big)^2\frac{38-128k^2+113k^4}{20k^4}+O\Big(\frac{a}{h_\infty}\Big)^3, \\ \frac{\eta_1}{h_\infty} &= \frac{2k^2-1}{k^2}\Big(\frac{a}{h_\infty}\Big)\Big[1+\Big(\frac{a}{h_\infty}\Big)\Big(\frac{85k^2-50}{20k^2}\Big)\Big]+O\Big(\frac{a}{h_\infty}\Big)^3, \\ \frac{\eta(x)}{h_\infty} &= \Big(\frac{a}{h_\infty}\Big)\mathrm{cn}^2(A\,\omega\,x,k)-\frac{3}{4}\Big(\frac{a}{h_\infty}\Big)^2\mathrm{cn}^2(A\,\omega\,x,k)\times \\ &\times[1-\mathrm{cn}^2(A\,\omega\,x,k)]+O\Big(\frac{a}{h_\infty}\Big)^3,\end{aligned}\right\}\quad(31.34)$$

where now

$$\left.\begin{aligned} A\,\omega\,x &= \frac{x}{h_\infty}\sqrt{\frac{3}{4k^2}\Big(\frac{a}{h_\infty}\Big)}\Big(1+\frac{\eta_0}{h_\infty}\Big)^{-\frac{3}{2}}\Big[1+\Big(\frac{a}{h_\infty}\Big)\frac{85k^2-50}{40k^2}\Big] \\ &= \frac{x}{h_\infty}\sqrt{\frac{3}{4k^2}\Big(\frac{a}{h_\infty}\Big)}\Big[1-\Big(\frac{a}{h_\infty}\Big)\frac{7k^2-2}{8k^2}\Big]+O\Big[\Big(\frac{a}{h_\infty}\Big)\frac{2k^2-1}{k^2}\Big]^{\frac{5}{2}}.\end{aligned}\right\}\quad(31.35)$$

The remaining ε^2 terms from (31.10) may then be combined with the ε terms from (31.7), by means of (31.5), to give

$$\left.\begin{aligned}
\frac{p(x,y)}{\varrho g h_\infty} &= \frac{\eta(x)-y}{h_\infty} - \left(\frac{a}{h_\infty}\right)^2 \frac{3}{4k^2}\left(2\frac{y}{h_\infty}+\frac{y^2}{h_\infty^2}\right)[1-k^2+ \\
&\quad +2(2k^2-1)\,\mathrm{cn}^2(A\omega x,k)-3k^2\,\mathrm{cn}^4(A\omega x,k)]+O\left[\left(\frac{a}{h_\infty}\right)\frac{2k^2-1}{k^2}\right]^3, \\
\frac{u(x,y)}{\sqrt{g h_\infty}} &= 1+\left(\frac{a}{h_\infty}\right)\left(1-\frac{1}{2k^2}\right)-\left(\frac{a}{h_\infty}\right)^2\frac{21k^4-6k^2-9}{40k^4}+ \\
&\quad -\left(\frac{a}{h_\infty}\right)\left[1-\left(\frac{a}{h_\infty}\right)\frac{7k^2-2}{4k^2}-\left(\frac{a}{h_\infty}\right)\frac{3}{2}\left(2-\frac{1}{k^2}\right)\left(2\frac{y}{h_\infty}+\frac{y^2}{h_\infty^2}\right)\right]\mathrm{cn}^2(A\omega x,k)- \\
&\quad -\left(\frac{a}{h_\infty}\right)^2\left[\frac{5}{4}+\frac{9}{4}\left(2\frac{y}{h_\infty}+\frac{y^2}{h_\infty^2}\right)\right]\mathrm{cn}^4(A\omega x,k)+ \\
&\quad +\left(\frac{a}{h_\infty}\right)^2\frac{3}{4}\left(\frac{1}{k^2}-1\right)\left(2\frac{y}{h_\infty}+\frac{y^2}{h_\infty^2}\right)+O\left[\left(\frac{a}{h_\infty}\right)\frac{2k^2-1}{k^2}\right]^3, \\
\frac{v(x,y)}{\sqrt{g h_\infty}} &= -\sqrt{\frac{3}{k^2}\left(\frac{a}{h_\infty}\right)^3}\left(1+\frac{y}{h_\infty}\right)\mathrm{cn}(A\omega x,k)\,\mathrm{sn}(A\omega x,k)\,\mathrm{dn}(A\omega x,k)\times \\
&\quad \times\left\{1-\left(\frac{a}{h_\infty}\right)\left(\frac{5k^2+2}{8k^2}\right)-\left(\frac{a}{h_\infty}\right)\left(1-\frac{1}{2k^2}\right)\left(2\frac{y}{h_\infty}+\frac{y^2}{h_\infty^2}\right)-\right. \\
&\quad \left.-\frac{1}{2}\left(\frac{a}{h_\infty}\right)\left(1-6\frac{y}{h_\infty}-3\frac{y^2}{h_\infty^2}\right)\mathrm{cn}^2(A\omega x,k)\right\}+O\left[\left(\frac{a}{h_\infty}\right)\left(\frac{2k^2-1}{k^2}\right)\right]^{\frac{7}{2}}.
\end{aligned}\right\} \quad (31.36)$$

For the solitary wave we have $k=1$ and essentially infinite wavelength, so that (31.36) reduces to

$$\left.\begin{aligned}
\frac{\eta(x)}{h_\infty} &= \frac{a}{h_\infty}\,\mathrm{sech}^2(A\omega x)-\frac{3}{4}\left(\frac{a}{h_\infty}\right)^2\mathrm{sech}^2(A\omega x)\times \\
&\quad \times[1-\mathrm{sech}^2(A\omega x)]+O\left(\frac{a}{h_\infty}\right)^3, \\
A\omega x &= \frac{x}{h_\infty}\sqrt{\frac{3}{4}\left(\frac{a}{h_\infty}\right)}\left\{1-\frac{5}{8}\left(\frac{a}{h_\infty}\right)\right\}+O\left(\frac{a}{h_\infty}\right)^{\frac{5}{2}}, \\
\frac{p}{\varrho g h_\infty} &= \frac{\eta(x)-y}{h_\infty}-\left(\frac{a}{h_\infty}\right)^2\frac{3}{4}\left(2\frac{y}{h_\infty}+\frac{y^2}{h_\infty^2}\right)\times \\
&\quad \times[2\,\mathrm{sech}^2(A\omega x)-3\,\mathrm{sech}^4(A\omega x)]+O\left(\frac{a}{h_\infty}\right)^3, \\
\frac{u(x,y)}{\sqrt{g h_\infty}} &= 1+\frac{1}{2}\left(\frac{a}{h_\infty}\right)-\frac{3}{20}\left(\frac{a}{h_\infty}\right)^2-\frac{\eta(x)}{h_\infty}+ \\
&\quad +\frac{1}{2}\left(\frac{a}{h_\infty}\right)^2\left[1+6\left(\frac{y}{h_\infty}\right)+3\left(\frac{y}{h_\infty}\right)^2\right]\mathrm{sech}^2(A\omega x)+ \\
&\quad -\frac{1}{2}\left(\frac{a}{h_\infty}\right)^2\left[1+9\left(\frac{y}{h_\infty}\right)+\frac{9}{2}\left(\frac{y}{h_\infty}\right)^2\right]\mathrm{sech}^4(A\omega x)+O\left(\frac{a}{h_\infty}\right)^3, \\
\frac{v(x,y)}{\sqrt{g h_\infty}} &= -\sqrt{3}\left(\frac{a}{h_\infty}\right)^{\frac{3}{2}}\left(1+\frac{y}{h_\infty}\right)\mathrm{sech}^2(A\omega x)\tanh(A\omega x)\times \\
&\quad \times\left\{1-\frac{7}{8}\left(\frac{a}{h_\infty}\right)-\left(\frac{a}{h_\infty}\right)\left(\frac{y}{h_\infty}+\frac{1}{2}\frac{y^2}{h_\infty^2}\right)-\frac{1}{2}\left(\frac{a}{h_\infty}\right)\times\right. \\
&\quad \left.\times\left(1-6\frac{y}{h_\infty}-3\frac{y^2}{h_\infty^2}\right)\mathrm{sech}^2(A\omega x)\right\}+O\left(\frac{a}{h_\infty}\right)^{\frac{7}{2}}.
\end{aligned}\right\} \quad (31.37)$$

The celerity or propagation velocity c of a solitary wave is defined by (31.37) as the constant uniform motion attained as $x\to\infty$,

$$\frac{c}{\sqrt{g h_\infty}} = \frac{u(\infty)}{\sqrt{g h_\infty}} = 1+\frac{1}{2}\left(\frac{a}{h_\infty}\right)-\frac{3}{20}\left(\frac{a}{h_\infty}\right)^2+O\left(\frac{a}{h_\infty}\right)^3. \quad (31.38)$$

In Fig. 48, Eq. (31.38) is shown to be in better agreement with recent experimental data than is the commonly used Boussinesq (1871)-Rayleigh (1876) propagation velocity given by

$$\frac{c}{\sqrt{g\,h_\infty}} \approx \sqrt{1+\left(\frac{a}{h_\infty}\right)} \approx 1+\frac{1}{2}\left(\frac{a}{h_\infty}\right)-\frac{1}{8}\left(\frac{a}{h_\infty}\right)^2+\cdots.$$

The past success of the Boussinesq-Rayleigh equation, as opposed to the propagation velocities derived by McCowan (1891), as indicated in Fig. 48, is easily explained when one notices the close numerical agreement of the coefficients of the Boussinesq-Rayleigh equation with the exact second approximation given by (31.38).

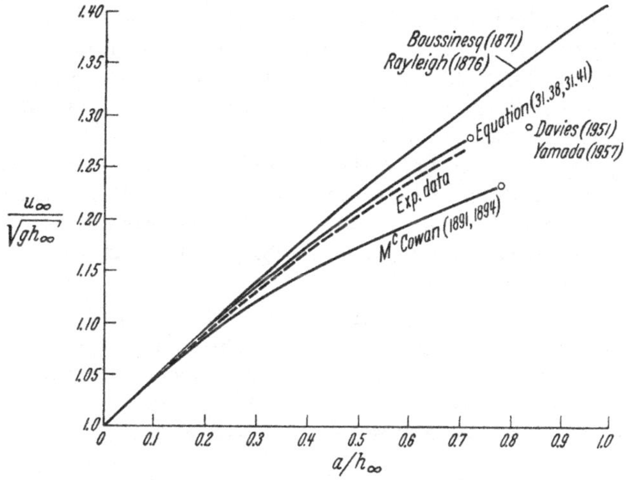

Fig. 48. Propagation velocity of solitary waves.

A comparison of the second approximations with the first approximations to the cnoidal waves proves conclusively that only the proper order of a/h_∞ must be retained for each order of approximation. For example, a comparison of (31.18) with (31.35) shows that a completely erroneous second approximation would be obtained by trying to extend the first approximation to include an additional a/h_∞ term. The reason for this is evident upon comparing the first and second approximations for η_0 in (31.17) and (31.34). Each successive approximation directly affects all the coefficients of the corresponding a/h_∞ terms. Fig. 49 shows the effect of the second approximation on a solitary wave.

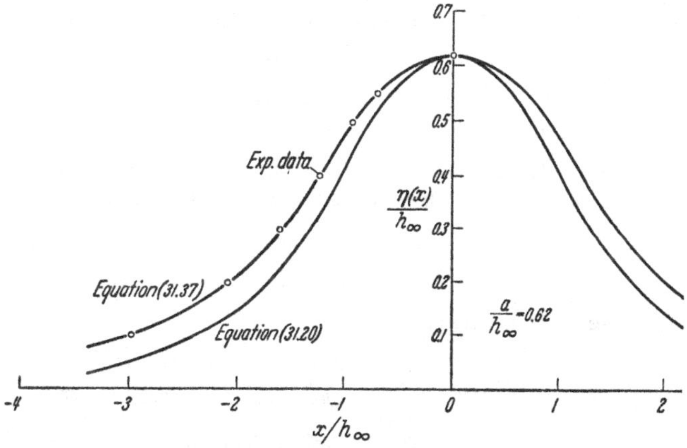

Fig. 49. Comparison of first (31.17) and second (31.37) approximation to the solitary-wave profile, $\eta(x)$.

Of course it must be remembered that the expansion method of Friedrichs (1948), which was used to obtain all the preceding results, is applicable only to shallow water, or long-wavelength wave propagations. However, this is precisely the nature of the solitary wave, especially if the amplitude a/h_∞ is relatively small, since its wavelength, as a member of the family of cnoidal waves, is essentially infinite since $K(k^2)\to\infty$ when $k\to 1$. Also, Friedrichs and Hyers (1954) proved that this expansion method does yield an existence proof for the solitary wave, and thereby demonstrate that it will at least provide asymptotic descriptions of the exact solution for the solitary-wave problem. The corresponding existence proof for cnoidal waves (in the neighborhood of the critical speed defined by a Froude number of unity) was given by LITTMAN (1957). Again this justified the Friedrichs expansion method, at least as an asymptotic type of series development. An additional discussion of these existence proofs is given in Sect. 35.

β) The limiting height and velocity of propagation of cnoidal and solitary waves. It is interesting to note that with the second approximation the pressure is still hydrostatic for $y \approx 0$, but is no longer hydrostatic as the bottom $(y = -h_\infty)$ is approached. Similarly, the variation of the horizontal velocity component with depth below the surface becomes important in the second approximation only upon approaching the flat horizontal bottom. However, the finite vertical velocity component is now seen to be the principal variation from the basic assumptions of first-order shallow-water theory. The first approximation given in (31.19) gives a monotonic variation in $v(y)$ that is obviously necessary from physical considerations in order to satisfy the continuity equation. However, this monotonic variation in $v(y)$ is of the higher order $(a/h_\infty)^{\frac{3}{2}}$ so that it can be neglected in the first-order equations (28.2) as long as the resulting local variations in η are sufficiently small.

The second approximation to the vertical velocity component, as given in (31.36), now shows that the variation of $v(y)$ will no longer be monotonic as a/h_∞ increases. This leads one to suspect that there is a limiting value to a/h_∞ for cnoidal and solitary waves. For example, (31.36) shows that in the neighborhood of the wave crest, where $x \approx 0$ and

$$\mathrm{cn}^2(A\,\omega\,x) \approx 1 - (A\,\omega\,x)^2 = 1 - \frac{3}{4k^2}\left(\frac{a}{h_\infty}\right)\left(\frac{x}{h_\infty}\right)^2 + O\left(\frac{a}{h_\infty}\right)^2,$$

$v(y)$ actually has a reversal in its direction if a/h_∞ exceeds the value given by

$$\left(\frac{a}{h_\infty}\right)_{\max} = \frac{8k^2}{9k^2+2} \tag{31.39}$$

for any value of $y \geqq 0$.

This limiting value can be substantiated, at least in the limit as $k \to 1$, by noting that (31.33) has a real solution for η_1 only if

$$\frac{2k^2-1}{k^2}\,\frac{a}{h_\infty} < \frac{\eta_0}{h_\infty} \leqq \frac{5(2k^2-1)^2}{7-37k^2(1-k^2)},$$

leading to a limiting value of

$$\left(\frac{a}{h_\infty}\right)_{\max} < \frac{5k^2(2k^2-1)}{7-37k^2(1-k^2)}. \tag{31.40}$$

The most interesting application of these results is to the solitary wave, defined by $k=1$, in which case we find from (31.38), (31.39) and (31.40) that the limiting heights and the corresponding total velocity at infinity are given by

$$\left(\frac{a}{h_\infty}\right)_{\max} = \frac{8}{11} = 0.7273 > \frac{5}{7} = 0.7143, \qquad \left[\frac{u(\infty)}{\sqrt{g\,h}}\right]_{\max} = 1.284 > 1.281. \tag{31.41}$$

Either of these limiting heights would be satisfactory for a solitary wave since recent experimental investigations by IPPEN and KULIN (1955), DAILY and STEPHAN (1952), and PERROUD (1957) have shown that under properly controlled conditions most solitary waves have $a/h_\infty < 0.7$, the maximum recorded value being 0.72. Not only are the limiting values given by (31.39) or (31.40) in excellent agreement with recent experimental data, but they are consistent with the order of approximation involved. The value $\frac{8}{11}$ is derived from the vertical velocity variation given to the order $(a/h_\infty)^{\frac{5}{2}}$ by (31.36), while the value $\frac{5}{7}$ corresponds to the terms governed by ε^2 or $(a/h_\infty)^2$ in (31.32).

Many attempts have been made to determine the limiting height of a solitary wave. However, nearly all of the theoretical calculations have been based on STOKES' (1880, p. 227) relation which assumes that for the limiting heights of any wave the wave crest must form a sharp peak or corner having an enclosed angle of 120° in order to reduce the relative local velocity to zero at the crest itself [see, e.g., Sect. 33 or LAMB (1932, p. 418)]. This 120° enclosed angle at the wave crest was assumed by MCCOWAN (1894), STOKES (1905), GWYTHER (1900), DAVIES (1952), PACKHAM (1952), GOODY and DAVIES (1957) and YAMADA (1957). Several of these values are compared in Fig. 48 with experimental data, and with the theoretical values given by (31.38) and (31.41). It is seen that none of these limiting heights for solitary waves are in as good an agreement with the experimental data as is (31.41). A reasonable explanation of the failure of the 120° sharp crest wave to provide a satisfactory limiting height for a solitary wave may be obtained by noting that KORTEWEG and DE VRIES (1895) proved that any finite-amplitude profile that did not correspond to (31.17) or (31.20) would not be steady with respect to time. Consequently, (31.37) defines the only possible steady-state solitary wave, and when $a/h_\infty > \frac{8}{11}$ the vertical velocity variation reverses its direction near the crest. This probably leads to an unsteady wave crest that breaks unsymmetrically.

Eqs. (31.38) or (32.52) show that the solitary wave occurs only in supercritical flow since the Froude number corresponding to the propagation velocity is always greater than unity. Its velocity of propagation is always less than that of the corresponding hydraulic jump of the same height as may be seen by comparing (30.55) with (31.38), after expanding it in powers of $a/d_1 = a/h_\infty$:

$$F_1 = \frac{u_1}{\sqrt{g\,h_\infty}} = 1 + \frac{3}{4}\left(\frac{a}{h_\infty}\right) - \frac{1}{32}\left(\frac{a}{h_\infty}\right)^2 + O\left(\frac{a}{h_\infty}\right)^3. \tag{30.55'}$$

However, the cnoidal wave can occur in subcritical as well as in supercritical flow, and as shown by BENJAMIN and LIGHTHILL (1954), the undulating flow in the subcritical region behind a hydraulic jump produced at all Froude numbers less than $\sqrt{3}$ may well be represented by these cnoidal waves. The fact that cnoidal waves can form in subcritical flow is easily shown, even in the first approximation, by writing the horizontal velocity component from (31.19) as

$$\left.\begin{aligned} \frac{u(0)}{\sqrt{g\,h_\infty}} &= 1 - \frac{1}{2k^2}\left(\frac{a}{h_\infty}\right) < 1 \qquad \text{for all } k \leqq 1, \\ \frac{u\left(\frac{K}{A\,\omega}\right)}{\sqrt{g\,h_\infty}} &= 1 + \left(1 - \frac{1}{2k^2}\right)\left(\frac{a}{h_\infty}\right) < 1 \qquad \text{for } k^2 < \frac{1}{2}. \end{aligned}\right\} \tag{31.42}$$

Therefore (31.19) shows that any definition of the wave propagation velocity would be subcritical when $k^2 < \frac{1}{2}$. STOKES (1847) (see Sect. 7) has given two logical definitions of the celerity or propagation velocity of permanent periodic wave forms, and each one would define a critical celerity corresponding to a different value of k, varying as $\frac{1}{2} \leqq k^2 < 1$, the solitary wave ($k = 1$) being always supercritical for a finite amplitude. However, the existence proof for cnoidal waves by LITTMAN (1957) is only valid for average velocities (defined as the velocity of the vertical plane that would have zero average flux across it) that are near critical. An interesting physical and mathematical explanation of these flow restrictions is given by BENJAMIN and LIGHTHILL (1954). The main consideration, as shown in Fig. 56 on page 754 and Sect. 35, is that the finite-amplitude periodic waves corresponding to $k^2<0.9$ may be better described by using infinitesimal wave theory. This becomes necessary because the wavelength of the cnoidal waves decreases rapidly with k^2 when k is near unity. Fig. 56 indicates that not only must the wavelength be large compared to the water depth, in order to satisfy the shallow-water expansion method, but also the amplitude of the cnoidal wave must become extremely small for values of $k^2<0.9$, or for subcritical flow.

KORTEWEG and DE VRIES (1895) have also shown how negative cnoidal or solitary waves can be formed when the water is very shallow and surface tension T is considered. Their correct first approximation may be written as

$$\left.\begin{aligned} \frac{\eta(x)}{h_\infty} &= \pm\left(\frac{a}{h_\infty}\right)\mathrm{cn}^2(A\,\omega\,x,k),\\ A\,\omega\,x &= \frac{x}{h_\infty}\sqrt{\frac{3}{4k^2}\,\frac{a/h_\infty}{|1-3T/\varrho g h_\infty^2|}}, \end{aligned}\right\}\tag{31.43}$$

where the negative algebraic value is assigned to the surface profile whenever

$$h_\infty < \sqrt{\frac{3T}{\varrho g}} \approx \frac{1}{2}\,\mathrm{cm} \qquad \text{for water.}\tag{31.44}$$

These negative waves have a very small amplitude and a very large wavelength, but can create a surprising particle motion. It is interesting to notice that the depth of $h_\infty=\frac{1}{2}$ cm, which, if it could be maintained, would eliminate both solitary and cnoidal waves, is the same depth found from (31.1′) and (31.2) to give nearly the same value of $\sqrt{g h_\infty}$ for both the propagation velocity and the group velocity of infinitesimal waves (also see Sect. 15). Consequently the depth of $\frac{1}{2}$ cm seems to be the optimum for ordinary water ($T=72.8$ dynes/cm) whenever one uses small models to simulate results appropriate to the first-order shallow-water theory of (28.2), since this particular depth minimizes the effect of group velocity and variation with wavelength for the infinitesimal waves, and minimizes the second-order effect due to the existence of finite-amplitude cnoidal or solitary waves. However, the variation of η must remain sufficiently small since a finite increase in η above $h_\infty=\frac{1}{2}$ cm could still produce cnoidal or solitary waves. Also, the short-wavelength or capillary ripples that will form must be neglected in these model tests.

F. Exact solutions.

The word "exact" in this context is generally understood to mean solutions in which there has been no approximation in the equations or boundary conditions. However, this usage of the word does not exclude neglect of viscosity and, in fact, since positive results have been obtained only for perfect fluids, the treatment below will be restricted to them. Indeed, the present results in the theory of exact solutions are restricted, with few exceptions, to a very special class of motions, namely, those which can be represented as steady two-dimensional flows.

In Sect. 32 some general theorems will be established. In Sect. 33 waves of maximum amplitude-to-length ratio are discussed; because the methods are intimately related, we have also included in this section a discussion of HAVELOCK's method of approximating periodic waves. Sect. 34 treats methods of obtaining explicit exact solutions and of various ones which have been obtained. In Sect. 35, the last, existence theorems are discussed, but only in a descriptive way, for proofs are highly technical and lengthy.

32. Some general theorems. This section will be devoted to several theorems of a rather general nature concerning the motion of a fluid with free surface in a gravitational field. The theorems in subsection 32α are mostly of a kinematical nature and are associated with the phenomenon of mass transport already discussed in subsection 27α. The last part of this section is devoted to several theorems on energy and momentum. In subsection 32β some theorems concerning waves in heterogeneous fluids will be established. In subsection 32γ several different ways of formulating the problem of motion with a free surface will be described.

α) *Kinematical theorems—mass transport—energy integrals.* The first theorem, due to M. S. LONGUET-HIGGINS (1953), is independent of the presence of a free surface or of the nature of the time dependence. Let $f(z) = \Phi + i\Psi$ describe a space-periodic motion, i.e. $f(z + n\lambda) = f(z)$. The definition of φ will be normalized so that

$$\int_0^\lambda \Phi(x, y, t)\, dx = 0. \tag{32.1}$$

Note that if this condition holds for one value of y, it holds for all since

$$\frac{\partial}{\partial y}\int_0^\lambda \Phi\, dx = \int_0^\lambda \Phi_y\, dx = -\int_0^\lambda \Psi_x\, dx = -\Psi(\lambda, y, t) + \Psi(0, y, t) = 0.$$

In Eq. (2.10′) we shall write

$$\frac{p}{\varrho} = \frac{p_0}{\varrho} - g y + \frac{p_d}{\varrho}, \qquad \frac{p_d}{\varrho} = A(t) - \Phi_t - \frac{1}{2}(u^2 + v^2). \tag{32.2}$$

In the following we define an average by

$$\overline{F}(y, t) = \frac{1}{\lambda}\int_0^\lambda F(x, y, t)\, dx. \tag{32.3}$$

Theorem. In a non-uniform space-periodic motion $\overline{u^2}$, $\overline{v^2}$, $-\overline{p_d}$ each decrease with increasing depth, provided either $\Phi_y(x, -h, t) = 0$ or $\lim_{y\to-\infty} \Phi_y = 0$.

This may be proved as follows. Consider first $q^2 = u^2 + v^2$. Then

$$\left.\begin{aligned}
\frac{\partial}{\partial y}\overline{q^2} &= \frac{\partial}{\partial y}\frac{1}{\lambda}\int_0^\lambda (\Phi_x^2 + \Phi_y^2)\, dx = \frac{2}{\lambda}\int_0^\lambda (\Phi_x \Phi_{xy} + \Phi_y \Phi_{yy})\, dx \\
&= \frac{2}{\lambda}\int_0^\lambda [(\Phi_y \Phi_x)_x - 2\Phi_y \Phi_{xx}]\, dx \\
&= \frac{2}{\lambda}[\Phi_y \Phi_x]_0^\lambda - \frac{4}{\lambda}\int_0^\lambda \Phi_y \Phi_{xx}\, dx \\
&= -\frac{4}{\lambda}\int_0^\lambda \Phi_y \Phi_{xx}\, dx.
\end{aligned}\right\} \tag{32.4}$$

By a similar computation it follows that

$$\frac{\partial^2}{\partial y^2}\overline{q^2} = \frac{4}{\lambda}\int_0^\lambda (\Phi_{xx}^2 + \Phi_{xy}^2)\,dx > 0, \tag{32.5}$$

since we have assumed that Φ_x is not constant. It is evident from (32.4) that, if the fluid is bounded below by $y=-h$, then

$$\frac{\partial}{\partial y}\overline{q^2}(-h, t) = 0; \tag{32.6}$$

if it is infinitely deep, it is an assumed boundary condition that $\Phi_y \to 0$ as $y\to-\infty$ and hence

$$\frac{\partial}{\partial y}\overline{q^2} \to 0 \quad \text{as} \quad y\to-\infty. \tag{32.7}$$

In either case it then follows from (32.5) that $\partial q^2/\partial y$ is an increasing function of y and hence

$$\frac{\partial}{\partial y}\overline{q^2} \geqq 0, \tag{32.8}$$

with equality occurring only for $y=-h$. In fact, even more can be concluded, for (32.5) is like $\overline{q^2}$ itself with Φ replaced by $2\Phi_x$. Hence, by repeating the above reasoning one may establish that

$$\frac{\partial^{2n}}{\partial y^{2n}}\overline{q^2} > 0, \qquad \frac{\partial^{2n-1}}{\partial y^{2n-1}}\overline{q^2} \geqq 0, \qquad n = 1, 2, \ldots. \tag{32.9}$$

Next consider $\overline{u^2} - \overline{v^2}$. A similar computation shows that

$$\frac{\partial}{\partial y}(\overline{u^2} - \overline{v^2}) = \frac{2}{\lambda}\int_0^\lambda (\Phi_x \Phi_y)_x\,dx = \frac{2}{\lambda}[\Phi_x \Phi_y]_0^\lambda = 0.$$

Hence

$$\overline{u^2} - \overline{v^2} = C(t) = \overline{u^2}|_{y=-h \text{ or } -\infty}. \tag{32.10}$$

It follows from (32.8) that

$$\overline{u^2} = \tfrac{1}{2}[\overline{q^2} + C], \qquad \overline{v^2} = \tfrac{1}{2}[\overline{q^2} - C], \qquad -\overline{p_d} = \tfrac{1}{2}\overline{q^2} - A(t) \tag{32.11}$$

are each increasing functions of y, i.e., they decrease with increasing depth. For infinite depth LONGUET-HIGGINS shows further that, if axes are chosen such that $u=0$ at $y=-\infty$, then the quantites $|u|$, $|v|$ and $|p_d|$ all decrease exponentially to zero. He had shown earlier (1950) for exact waves (we shall not carry through the proof) that

$$\overline{p_d} = \frac{1}{2}\frac{\partial^2}{\partial t^2}\overline{\eta^2} - \overline{v^2}. \tag{32.12}$$

Hence it follows that

$$p_d|_{y=-h \text{ or } -\infty} = \frac{1}{2}\frac{\partial^2}{\partial t^2}\overline{\eta^2}. \tag{32.13}$$

For purely progressive waves this quantity vanishes, but we recall that for standing waves we found earlier a constant pressure fluctuation of double the wave frequency [see (27.62) and (27.65)] if second-order terms were retained.

Mass transport. In Sect. 27 [see (27.39) and (27.41)] it was shown that a forward drift, called "mass transport", occurred in progressive waves if second-order terms were taken into account. It was shown by RAYLEIGH (1876) in a proof valid only for infinitely deep fluid that mass transport must always occur. The proof is independent of the dynamical free-surface condition. LEVI-CIVITA (1912) and later URSELL (1953) developed methods of analysis to include both finite depth and nonperiodic waves; essentially URSELL's analysis has also been given by NEKRASOV (1951) for infinite depth. The analysis given below is due to LONGUET-HIGGINS (1953) and is similar to that used in the preceding theorem. We note that STARR (1945) has also given an instructive and perspicuous derivation of RAYLEIGH's theorem for infinite depth.

Take the wave as moving to the left with velocity c (in the sense of Sect. 7) and impose a uniform velocity c in the opposite direction, so that the motion is reduced to a steady one, generally in the positive x-direction in the sense that $u > \varepsilon > 0$. We may then write the complex potential in the form

$$f(z) = \Phi + i\Psi = c z + \varphi + i\psi, \tag{32.14}$$

where $\operatorname{Re} f' > \varepsilon > 0$ and

$$\Phi(x + n\lambda, y) = n c \lambda + \Phi(x, y), \qquad \Psi(x + n\lambda, y) = \Psi(x, y). \tag{32.15}$$

We take $\Phi = 0$ at a crest and assume $\Psi = 0$ as the free-surface streamline and $\Psi = -Q$ as the bottom streamline if the depth is finite. One may invert the relation $f = f(z)$ and obtain $z = z(f)$. Then, since $q^2 \neq 0$,

$$z'(f) = \frac{1}{f'(z)} = \frac{\Phi_x + i\Phi_y}{\Phi_x^2 + \Phi_y^2} = \frac{1}{q^2}(u + i v) = x_\Phi + i y_\Phi. \tag{32.16}$$

Denote by $T(\Psi)$ the time required for a given particle to progress one wavelength along a streamline $\Psi = \text{const}$. In the original wave motion, the time elapsed between the passage of two successive crests over a given point is λ/c. If $T > \lambda/c$, the particle is being transported with the wave and it will be reasonable to call

$$U(\Psi) = c - \frac{\lambda}{T(\Psi)} \tag{32.17}$$

the mass transport in the direction of wave motion. The following theorem is true.

Theorem. Both T and U decrease with increasing depth, and, with the assumed definition of c, $U > 0$.

The theorem may be proved by the following computation:

$$T(\Psi) = \int_0^{s(\lambda)} \frac{1}{q}\, ds = \int_0^{c\lambda} \frac{1}{q} \frac{\partial s}{\partial \Phi}\, d\Phi = \int_0^{c\lambda} (x_\Phi^2 + y_\Phi^2)\, d\Phi = \int_0^{c\lambda} (x_\Phi^2 + x_\Psi^2)\, d\Phi, \tag{32.18}$$

$$T'(\Psi) = 4 \int_0^{c\lambda} x_\Phi\, x_{\Phi\Psi}\, d\Phi, \tag{32.19}$$

$$T''(\Psi) = 4 \int_0^{c\lambda} (x_{\Psi\Phi}^2 + x_{\Psi\Psi}^2)\, d\Phi. \tag{32.20}$$

The details of the computation are almost identical with those used in deriving (32.4) and (32.5). Since

$$x_\Psi = -y_\Phi = -\frac{1}{q^2}\Phi_y = 0 \quad \text{on} \quad \Psi = -Q,$$

it follows from (32.19) that $T'(-Q) = 0$. Then, since $T''(\Psi) > 0$ unless the flow is uniform, it follows that

$$T'(\Psi) \geqq 0, \tag{32.21}$$

with equality holding only for $\Psi = -Q$. As in the earlier theorem, the computations can be extended to yield

$$T^{(2n)}(\Psi) > 0, \qquad T^{(2n-1)}(\Psi) \geqq 0. \tag{32.22}$$

It now follows immediately from (32.17) that

$$U'(\Psi) \geqq 0, \tag{32.23}$$

with the equality holding only for the bottom streamline. If the fluid is infinitely deep, then $U' > 0$ for all Ψ. To complete the proof we must show that $U > 0$. If the fluid is infinitely deep, it is evident that

$$\lim_{\psi \to -\infty} T(\Psi) = \frac{\lambda}{c}. \tag{32.24}$$

Hence $\lim U = 0$ as $\Psi \to -\infty$ and the conclusion follows from $U' > 0$. If the depth is finite, we compute

$$T(-Q) = \int_0^{c\lambda} x_\Phi^2 \, d\Phi > \frac{1}{c\lambda} \left[\int_0^{c\lambda} x_\Phi \, d\Phi \right]^2 = \frac{1}{c\lambda} \lambda^2 = \frac{\lambda}{c}; \tag{32.25}$$

here use has been made of the Schwarz-Bunyakovskii inequality. (We have written $>$ rather than $\geqq$, for the equal sign will hold only in the trivial case of a uniform flow.) It now follows that $U(-Q) > 0$ and hence that

$$U(\Psi) > 0 \tag{32.26}$$

since $U' \geqq 0$. This completes the proof of the theorem.

The method of analysis can be extended to prove an analogous theorem for nonperiodic steady motions which approach uniform flows as $x \to \pm\infty$, in particular, to the solitary wave.

Momentum and energy integrals. We close this section with several momentum and energy integrals, most of which have been found by Levi-Civita (1912, 1921), Starr (1947a, b, 1948) and Starr and Platzman (1948).

Let us again take the wave as moving to the left without change of form and impose an opposite velocity c which brings the profile to rest (or, equivalently, consider the motion relative to a coordinate system moving with the wave). Let the velocity potential be as in (32.14). Consider the area bounded by two streamlines $\Psi = \Psi_1$ and $\Psi = \Psi_2$, say $y = \eta_1(x)$ and $y = \eta_2(x)$ and two vertical lines a wavelength λ apart. To this area apply the theorem

$$\iint (\Phi_x^2 + \Phi_y^2) \, d\sigma = \oint \Phi \Phi_n \, ds. \tag{32.27}$$

This yields

$$\iint [(c+u)^2 + v^2] \, d\sigma = \int_{\eta_1(x_0+\lambda)}^{\eta_2(x_0+\lambda)} \Phi(x_0+\lambda, y) \, \Phi_x \, dy - \int_{\eta_1(x_0)}^{\eta_2(x_0)} \Phi(x_0, y) \, \Phi_x \, dy \tag{32.28}$$

since $\Phi_n = 0$ on the streamlines. Moreover, since $\Phi(x+\lambda, y) = c\lambda + \Phi(x, y)$, $\Phi_x(x+\lambda, y) = \Phi_x(x, y) = c + u$ and $\eta_i(x+\lambda) = \eta_i(x)$, the right-hand side of (32.28) may be written as

$$c^2\lambda[\eta_2(x_0) - \eta_1(x_0)] + c\lambda \int_{\eta_1(x_0)}^{\eta_2(x_0)} \varphi_x(x_0, y) \, dy = c^2\lambda[\eta_2(x) - \eta_1(x)] + c\lambda \int_{\eta_1(x)}^{\eta_2(x)} u \, dy. \tag{32.29}$$

Expanding $(c+u)^2$ and rearranging give

$$\iint (u^2 + v^2) \, d\sigma + 2c \iint u \, d\sigma + c^2 \iint d\sigma = c^2\lambda[\eta_2(x) - \eta_1(x)] + c\lambda \int_{\eta_1(x)}^{\eta_2(x)} u \, dy. \tag{32.30}$$

If one now applies the operator $\lambda^{-1}\int_0^\lambda \ldots dx$ to (32.30), one obtains

$$\iint (u^2+v^2)\,d\sigma + c\iint u\,d\sigma = 0 \tag{32.31}$$

or, after multiplying by $\frac{1}{2}\varrho$ and rearranging,

$$\iint \tfrac{1}{2}\varrho(u^2+v^2)\,d\sigma = \tfrac{1}{2}c\iint -\varrho\,u\,d\sigma, \tag{32.32}$$

i.e., the kinetic energy per wavelength between two streamlines equals $\frac{1}{2}c$ times the momentum in the direction of the wave (here to the left).

Next let us write the integral (2.10′) in the form

$$\tfrac{1}{2}\varrho\left[(c+u)^2+v^2\right]+\varrho g y + p = \tfrac{1}{2}\varrho c_1^2, \tag{32.33}$$

the form of the constant having been chosen for later convenience. Write the terms $p+\varrho g y$ as follows:

$$\left.\begin{aligned} p+\varrho g y &= \frac{\partial}{\partial y}\left[y(p+\varrho g y)\right] - y\frac{\partial}{\partial y}(p+\varrho g y)\\ &= \frac{\partial}{\partial y}\left[y(p+\varrho g y)\right] + y\frac{D}{Dt}v\\ &= \frac{\partial}{\partial y}\left[y(p+\varrho g y)\right] - v^2 + \frac{D}{Dt}(y v).\end{aligned}\right\} \tag{32.34}$$

Here we have used the second equation of (2.6). We may now write (32.33) as follows

$$\frac{1}{2}\varrho(u^2-v^2)+\varrho c u + \frac{\partial}{\partial y}\left[y(p+\varrho g y)\right] + \frac{D}{Dt}(y v) = \frac{1}{2}\varrho(c_1^2-c^2). \tag{32.35}$$

Next let us integrate Eq. (32.35) over the same area as is described in the preceding paragraph. First consider $D(yv)/Dt$. Since the motion is steady in the selected coordinate system,

$$\frac{D}{Dt}(y v) = (u+c)\frac{\partial(yv)}{\partial x} + v\frac{\partial(y v)}{\partial y} = \frac{\partial}{\partial x}(u+c)\,y v + \frac{\partial}{\partial y}\,y v^2,$$

where the last equality follows from the continuity equation. Hence

$$\iint \frac{D}{Dt}(y v)\,d\sigma = \oint y v(u+c,\,v)\cdot\boldsymbol{n}\,ds = \oint y v\,\Phi_n\,ds = 0 \tag{32.36}$$

since $\Phi_n=0$ on the streamline boundaries and the integrals over the vertical boundaries cancel from periodicity. The integrated equation then becomes

$$\left.\begin{aligned} \iint \tfrac{1}{2}\varrho(u^2-v^2)\,d\sigma + c\iint \varrho u\,d\sigma + \int_0^\lambda \{\eta_2(x)\left[p(x,\eta_2)+\varrho g\eta_2\right] -\\ -\eta_1(x)\left[p(x,\eta_1)+\varrho g\eta_1\right]\}\,dx = \tfrac{1}{2}\varrho(c_1^2-c^2)\iint d\sigma.\end{aligned}\right\} \tag{32.37}$$

If one eliminates the second integral by means of (32.32), one obtains

$$\left.\begin{aligned} \iint \tfrac{1}{2}\varrho u^2\,d\sigma + 3\iint \tfrac{1}{2}\varrho v^2\,d\sigma - \int_0^\lambda \{\eta_2\left[p(x,\eta_2)+\varrho g\eta_2\right] -\\ -\eta_1\left[p(x,\eta_1)+\varrho g\eta_1\right]\}\,dx = \tfrac{1}{2}\varrho(c^2-c_1^2)\iint d\sigma.\end{aligned}\right\} \tag{32.38}$$

Eq. (32.38) has a simpler aspect if the two streamlines are taken as the free surface $\eta(x)$ and the bottom $y=-h$. Then $p(x,\eta(x))=0$ and the third integral becomes

$$\int_0^\lambda \varrho g\eta_2^2(x)\,dx + h\int_0^\lambda \left[p(x,-h)-\varrho g h\right]dx.$$

Moreover,

$$\int_0^\lambda [p(x,-h) - \varrho g h]\, dx = 0 \tag{32.39}$$

if the x-axis is taken at the mean water level. This follows from the following sequence of equations, similar to those used in (32.36):

$$\left.\begin{aligned}\int_0^\lambda [p(x,-h) - \varrho g h]\, dx &= \iint \frac{\partial}{\partial y}[p(x,y) + \varrho g y]\, d\sigma - \\ &- \iint \left[(u+c)\frac{\partial}{\partial x} v + v \frac{\partial}{\partial y} v\right] d\sigma = -\iint \left[\frac{\partial}{\partial x} v(u+c) + \frac{\partial}{\partial y} v^2\right] d\sigma \\ &= \oint v(u+c, v)\cdot \boldsymbol{n}\, d\sigma = -\oint v\, \Phi_n\, d\sigma = 0.\end{aligned}\right\} \tag{32.40}$$

Eq. (32.39) now allows us to give a simple physical interpretation of the constant c_1 in (32.33). For if (32.33) is integrated along $y=-h$, and account is taken of (32.39), one finds

$$\frac{1}{\lambda}\int_0^\lambda (u+c)^2\, dx = c_1^2 \geqq c^2, \tag{32.41}$$

i.e., c_1^2 is the mean square velocity of fluid along the bottom. The inequality follows easily from

$$\int_0^\lambda u(x,-h)\, dx = \int_0^\lambda \varphi_x(x,-h)\, dx = \varphi(\lambda,-h) - \varphi(0,-h) = 0. \tag{32.42}$$

If the fluid is infinitely deep, $u \to 0$ as $y \to -\infty$, and (32.41) reduces to

$$c^2 = c_1^2. \tag{32.43}$$

If, following (15.27), we let $\mathcal{T}_{\text{av}}, \mathcal{T}_{x\,\text{av}}, \mathcal{T}_{y\,\text{av}}, \mathcal{V}_{\text{av}}, \mathcal{M}_{\text{av}}$ denote the average kinetic energy, the contributions to this due to the x and y velocity components, the potential energy, and the momentum in the direction of wave motion, respectively, then (32.32) and (32.88) may be expressed as follows:

$$2\mathcal{T}_{\text{av}} = c\,\mathcal{M}_{\text{av}}, \qquad \mathcal{T}_{x\,\text{av}} + 3\,\mathcal{T}_{y\,\text{av}} = 2\,\mathcal{V}_{\text{av}} - \tfrac{1}{2}\varrho\,(c_1^2 - c^2)\,h, \tag{32.44}$$

where the last term of the second equation is zero for $h=\infty$. The first equation is essentially due to LEVI-CIVITA (1912, 1921), the second to STARR (1947b).

We note another simple consequence of (32.41), due to LEVI-CIVITA (1924). Let us integrate (32.33) along the free surface for a wavelength and divide by $\frac{1}{2}\varrho\lambda$. Then, remembering our choice of x-axis as the mean water level, we find

$$\frac{1}{\lambda}\int_0^\lambda [(c+u)^2 + v^2]\, dx = c_1^2. \tag{32.45}$$

On the other hand, if we compute the velocity at the intersection of the mean water level and the profile, we also find

$$(c+u)^2 + v^2 \big|_{y=0} = c_1^2. \tag{32.46}$$

Hence the absolute value of the velocity at the mean water level equals the root-mean-square velocity along the surface profile or along the bottom, or, indeed, along any streamline, for in the reasoning in (32.40) we could have substituted any streamline $y=\eta_1(x)$ for $y=-h$ and obtained

$$\int_0^\lambda [p(x, \eta_1(x)) - \varrho g \eta_1(x)]\, dx = 0. \tag{32.47}$$

STARR and PLATZMAN (1948) have used the relations above to derive some general relations concerning the flow of energy in a periodic wave. We recall that the average flux of energy in the direction of wave motion is given by [cf. Sect. 8 and Eqs. (15.23) and (15.27)]

$$\mathscr{F}_{\text{av}} = \int_0^\lambda dx \int_{-h}^{\eta(x)} \varrho\, c\, \varphi_x^2(x, y)\, dy = 2\mathscr{T}_{x\,\text{av}}. \tag{32.48}$$

It follows from the second formula in (32.44) that

$$2\mathscr{T}_{x\,\text{av}} = 3\mathscr{T}_{\text{av}} - 2\mathscr{V}_{\text{av}} + \tfrac{1}{2}\varrho(c_1^2 - c^2)\, h. \tag{32.49}$$

Hence, with $\mathscr{E}_{\text{av}} = \mathscr{T}_{\text{av}} + \mathscr{V}_{\text{av}}$, we obtain from (32.48)

$$\frac{\mathscr{F}_{\text{av}}}{\mathscr{E}_{\text{av}}} = \frac{1}{2} + \frac{5}{2}\,\frac{\mathscr{T}_{\text{av}} - \mathscr{V}_{\text{av}} + \frac{1}{5}\varrho(c_1^2 - c^2)\, h}{\mathscr{E}_{\text{av}}}. \tag{32.50}$$

This should be compared with the result derived in Sect. 15 for infinitesimal waves with neglect of surface tension [cf. (15.25) and (15.26)], namely, $\mathscr{F}_{\text{av}} = \frac{1}{2}\mathscr{E}_{\text{av}}$. Eq. (32.50) is consistent with this, for to the order of approximation involved, $\mathscr{T}_{\text{av}} = \mathscr{V}_{\text{av}}$ and $c_1^2 = c^2$. However, for waves of finite height it was shown in Sect. 27α [cf. Eqs. (27.42), (27.43)] that to the second order of approximation $\mathscr{T}_{\text{av}} > \mathscr{V}_{\text{av}}$. PLATZMAN (1947) has verified that this remains true when 4th-order terms are kept.

Several of the above results have analogues for steady motion of nonperiodic waves, provided that $\eta(x)\to 0$ as $x\to\pm\infty$ in such a way that $\int_{-\infty}^{\infty}\eta\, dx$ is finite. Under such circumstances $c_1^2 = c^2$ and the following results may be established [the notation is that of (15.31) with obvious extensions]:

$$\left.\begin{aligned}
&\mathscr{M}_{\text{total}} = c\int_{-\infty}^{\infty}\eta\, dx = c\,\mathscr{A}_{\text{total}},\\
&\mathscr{T}_{x\,\text{total}} - \mathscr{T}_{y\,\text{total}} = \mathscr{V}_{\text{total}},\\
&\mathscr{T}_{x\,\text{total}} - \mathscr{T}_{y\,\text{total}} + 2\mathscr{V}_{\text{total}} + (g\,h - c^2)\,\mathscr{A}_{\text{total}} = 0.
\end{aligned}\right\} \tag{32.51}$$

For details of the proof one may refer to STARR (1947b). From the last two equations follows

$$c^2 = g\,h + 3\mathscr{V}_{\text{total}}/\mathscr{A}_{\text{total}} > g\,h. \tag{32.52}$$

We note that the second equation of (32.51) is a special case of a more general one applying to any steady motion:

$$\mathscr{T}_x(x) - \mathscr{T}_y(x) - \mathscr{V}(x) = \text{const} \tag{32.53}$$

where the constant is zero under the conditions of (32.51). The proof is analogous to that of (8.6). Here

$$\mathscr{T}_x(x) = \int_{-h}^{\eta(x)} \tfrac{1}{2}\varrho\, u^2(x, y)\, dy, \qquad \text{etc.}$$

β) Waves in heterogeneous fluids. The first two theorems proved below are also true for homogeneous fluids and were first proved for this case. The last theorems deal specifically with heterogeneous fluids. In the extended form they are all due to DUBREIL-JACOTIN (1932).

A flow will be called barotropic if both the pressure and density are constant along streamlines. We first derive the energy integral for such flows. The Eqs. (2.6) may be written in the following form in two dimensions:

$$-\frac{1}{\varrho}\frac{\partial p}{\partial x} = \frac{\partial E}{\partial x} - v\zeta + \frac{\partial u}{\partial t}, \qquad -\frac{1}{\varrho}\frac{\partial p}{\partial y} = \frac{\partial E}{\partial y} + u\zeta + \frac{\partial v}{\partial t}, \tag{32.54}$$

where

$$E = g y + \frac{1}{2}(u^2 + v^2), \qquad \zeta = \frac{\partial v}{\partial x} - \frac{\partial u}{\partial y}.$$

Since p is assumed constant on a streamline, $u p_x + v p_y = 0$; it follows from (32.27) and the definition of E that

$$0 = u\frac{\partial E}{\partial x} + v\frac{\partial E}{\partial y} + \frac{1}{2}\frac{\partial}{\partial t} q^2 = g v + \frac{1}{2}\frac{D}{Dt} q^2 = \frac{D}{Dt} E. \tag{32.55}$$

In particular, if the flow is steady, E is also constant along a streamline. For steady flow it is a consequence of the incompressibility condition that ϱ is also constant along a stream-line.

The following theorem was proved by BURNSIDE (1915) for a homogeneous fluid. He gives two proofs, of which the second can be carried over to the present more general situation with no change. It will perhaps give more substance to the theorem if we remark that GERSTNER'S wave (see subsection 34β), which is not irrotational, satisfies the other conditions of the theorem.

Theorem. The only steady two-dimensional irrotational motion of a fluid subject to gravity for which all streamlines are also lines of constant pressure is a uniform flow.

Let the streamlines be given by $\psi(x, y) = \text{const}$. Since, from the remark following (32.55), $E = \text{const}$ along a streamline, we may write

$$\tfrac{1}{2}(\psi_x^2 + \psi_y^2) + g y = E(\psi). \tag{32.56}$$

[BURNSIDE shows that one may generalize (32.56) by replacing $g y$ by a function $g(y)$.] Since the motion is irrotational, $\Delta\psi = 0$ and hence also

$$\Delta \log(\psi_x^2 + \psi_y^2) = 0.$$

But then

$$\Delta \log[E(\psi) - g y] = 0,$$

which yields after some computation

$$2 y E'(\psi)\,\psi_y = 2(E - g y)\,[E'^2 - (E - g y) E''] + g^2. \tag{32.57}$$

We write this in the form

$$\psi_y(x, y) = G(\psi, y). \tag{32.58}$$

It then follows from (32.56) and (32.58) that

$$\psi_{xx} = E' - G G_\psi, \qquad \psi_{yy} = G_\psi \psi_y + G_y$$

or

$$E'(\psi) + G_y(\psi, y) = 0.$$

But then

$$\psi_y = -\, y\, E'(\psi) + \text{const}$$

and ψ is a function of y only. Hence, since $\Delta\psi = \psi_{yy} = 0$, ψ_y is a constant and the flow is uniform.

The next theorem was first proved by LEVI-CIVITA (1925) for homogeneous fluids. FENCHEL (1931) showed that his hypotheses could be weakened and DUBREIL-JACOTIN (1932) extended FENCHEL's proof to heterogeneous fluids. The gist of the theorem is that if the surface profile moves without change of form, then the whole velocity field is steady in a coordinate system moving with the surface. The theorem will be formulated in the moving coordinate system.

Theorem. Let a possibly heterogeneous fluid, bounded below by a horizontal plane $y = -h$, be flowing irrotationally in the x-direction with discharge rate $Q(t)$ and with a fixed surface profile $y = \eta(x)$. If η and u satisfy the conditions

$$-h < b_1 < \eta < b_2, \qquad u > \varepsilon > 0, \tag{32.59}$$

then the velocity potential $f(z)$ is independent of t.

First we derive a boundary condition at the free surface. From the condition of constant pressure and the assumption that the surface profile is an invariant streamline it follows that

$$u\frac{\partial p}{\partial x} + v\frac{\partial p}{\partial y} = 0 \qquad \text{on } \psi = 0.$$

It then follows as in (32.55) that

$$\frac{DE}{Dt} = g\,v + \frac{Dq^2}{Dt} = 0 \qquad \text{on } \psi = 0. \tag{32.60}$$

However, this conclusion holds now only on this one streamline.

The complex potential $f(z,t) = \varphi + i\psi$ maps the region of the z-plane occupied by fluid onto the strip $-Q(t) \leqq \psi \leqq 0$, where the free surface corresponds to $\psi = 0$, the bottom to $\psi = -Q$ and $x = \pm\infty$ to $\varphi = \pm\infty$. Let $F(z) = \Phi + i\Psi$ be the mapping, unique up to an additive real constant, of the fluid region onto the strip $-1 \leqq \Psi \leqq 0$ with $x = \pm\infty$ corresponding to $\Phi = \pm\infty$. Then

$$f(z,t) = Q(t)\,F(z) \tag{32.61}$$

evidently satisfies the requirements for $f(z,t)$ and, in fact, is determined uniquely, up to the added constant in F, by $Q(t)$ and $\eta(x)$. Now substitute $\varphi(x,y,t) = Q(t)\,\Phi(x,y)$ into (32.60):

$$g\,Q\,\Phi_y + Q\,Q'[\Phi_x^2 + \Phi_y^2] + Q^2[\Phi_{xx}\,\Phi_x^2 + 2\Phi_{xy}\,\Phi_x\,\Phi_y + \Phi_{yy}\,\Phi_y^2] = 0, \tag{32.62}$$

which we may write in the form

$$Q' + AQ + B = 0 \qquad \text{on } \Psi = 0 \tag{32.63}$$

where A and B are independent of t. Division by $\Phi_x^2 + \Phi_y^2$ is possible since (32.59) implies that this does not vanish. Note also that

$$B = g\frac{\Phi_y}{\Phi_x^2 + \Phi_y^2} = g\,y_\Phi\big|_{\Psi=0} = g\frac{d}{d\Phi}\eta, \tag{32.64}$$

and that both A and B may be considered as functions of Φ. Consider two cases: (a) $A = \text{const}$, (b) $A \neq \text{const}$. (a) In this case, since B is independent of t and

$Q' + AQ$ is independent of Φ, it follows from (32.63) that both equal constants. It now follows from (32.64) that, unless this constant is zero, the profile $\eta(x)$ will be unbounded and the first part of (32.59) will be contradicted. Hence, in case (a) $\eta = \text{const}$ and the mapping F must be of the form $F = az + b$, a and b real. It then follows that $A = 0$ and hence $Q' = 0$, i.e. the flow is uniform. (b) Let A_1, A_2 be two different values of A, $A_1 \neq A_2$. Write Eq. (32.63) for each value and subtract. This yields

$$Q = -\frac{B_1 - B_2}{A_1 - A_2}. \tag{32.65}$$

But then Q is evidently independent of t. Hence also $f(z, t) = QF(z)$ is also independent of t. This completes the proof.

The next theorem, due to DUBREIL-JACOTIN (1932), specifically requires that the fluid be heterogeneous.

Theorem. Suppose the motion of an incompressible heterogeneous fluid to be irrotational, the free surface to move without change of form, and that, in a coordinate system moving with the surface, conditions (32.59) are satisfied. Then not only is the velocity field steady, but also E, p and ϱ are constant along the streamlines.

It follows from the preceding theorem that the velocity is steady, hence that $E = E(x, y)$. However, we may still conceivably have $p = p(x, y, t)$, $\varrho = \varrho(x, y, t)$. The Eqs. (32.54) may now be written in the form

$$-\frac{1}{\varrho}\frac{\partial p}{\partial x} = \frac{\partial E}{\partial x}, \qquad -\frac{1}{\varrho}\frac{\partial p}{\partial y} = \frac{\partial E}{\partial y}. \tag{32.66}$$

Elimination of first ϱ, then p between these two equations yields

$$\frac{\partial(p, E)}{\partial(x, y)} = 0, \qquad \frac{\partial(\varrho, E)}{\partial(x, y)} = 0.$$

We assume that the corresponding functional relations may be solved and write

$$p = p(E, t), \qquad \varrho = \varrho(E, t), \tag{32.67}$$

where, from (32.66)

$$\varrho = -\frac{\partial p}{\partial E}. \tag{32.68}$$

From the equation expressing incompressibility, namely,

$$\frac{D\varrho}{Dt} = \frac{\partial \varrho}{\partial t} + u\frac{\partial \varrho}{\partial x} + v\frac{\partial \varrho}{\partial y} = 0,$$

follows

$$\frac{\partial \varrho}{\partial t} + \frac{\partial \varrho}{\partial E}\left(u\frac{\partial E}{\partial x} + v\frac{\partial E}{\partial y}\right) = \frac{\partial \varrho}{\partial t} + \frac{\partial(E, \psi)}{\partial(x, y)}\frac{\partial \varrho}{\partial E} = 0. \tag{32.69}$$

We shall assume $\partial\varrho/\partial E \neq 0$ everywhere, and may thus write

$$\frac{\partial(E, \psi)}{\partial(x, y)} = -\frac{\varrho_t(E, t)}{\varrho_E(E, t)}. \tag{32.70}$$

Since the left-hand side is independent of t, it follows from the form of the right-hand side that we may set both sides equal to $k(E)$, i.e.

$$\frac{\partial \varrho}{\partial t} + k(E)\frac{\partial \varrho}{\partial E} = 0. \tag{32.71}$$

Let us suppose $k(E) \not\equiv 0$, e.g. $k(E_1) \neq 0$. Then ϱ must be a function of the form

$$\varrho = \varrho\left(t - \int_{E_1}^{E} \frac{dE}{k(E)}\right) \tag{32.72}$$

in some neighborhood of E_1. If $k(E)$ vanishes for some values of E, let E_0 be the first zero larger than E_1. Then from (32.71) and (32.72)

$$\varrho_t(E, t) = \varrho'\left(t - \int_{E_1}^{E} \frac{dE}{k(E)}\right) \to 0 \quad \text{as } E \to E_0. \tag{32.73}$$

But (32.73) can hold for all t only if $\varrho' = 0$, i.e. if $\varrho = \text{const}$, which is contrary to the assumed heterogeneity. Moreover, at least one such zero of k exists, for we already know from (32.60) that E is constant along the free surface, so that in steady motion the Jacobian in (32.70) vanishes for $\psi = 0$. Hence $k(E) = 0$ for the corresponding value of E. We must conclude that $k(E) \equiv 0$. This implies, from (32.70) that $E = E(\psi)$ and $\varrho = \varrho(E)$. From (32.68) and the condition $p_t = 0$ on the free surface, it follows that also $p = p(E)$. Hence p, ϱ and E are all constant on streamlines.

The last in this complex of theorems is also due to DUBREIL-JACOTIN (1932).

Theorem. There cannot exist irrotational waves in a heterogeneous fluid such that the profile is propagated without change of form.

This follows immediately from the first and last theorems proved above, and is, of course, subject to condition (32.59). This striking result is all the more so in view of the fact that GERSTNER'S wave (subsection 34β) does provide a steadily propagating wave, even in a heterogeneous fluid. The theorem also casts some doubt upon the significance of the linearized theory of irrotational wave motion in a heterogeneous fluid as developed, for example, in LAMB (1932, § 235). Such a wave evidently cannot be considered as a first approximation to an exact steady solution.

γ) Some transformations of the boundary-value problem. By means of introduction of new variables or other devices, it is possible to formulate the boundary-value problem for exact solutions in a variety of ways. Several such formulations will be considered in subsection 34α on inverse methods. Here we give a few which seem to be of general interest.

Inversion of $f(z)$. One elementary but important transformation has already been introduced in subsection 32α in the discussion of mass transport. This is the inversion of the velocity potential $f(z)$ when $|f'|$ vanishes nowhere within the fluid, and treatment of f as the independent variable. This has the advantage that under certain circumstances the domain of definition of the independent variable can be given exactly; when z is the independent variable, the domain of definition is one of the unknowns of the problem. For example, if the motion is reducible to a steady flow with discharge rate Q, one may take the surface profile to correspond to $\psi = 0$ and the bottom streamline to correspond to $\psi = -Q$. Hence the domain of definition of $z(f)$ is the strip $0 \geqq \psi \geqq -Q$; if the fluid is infinitely deep, the domain is the half-plane $\psi \leqq 0$. Whenever f can be taken as the independent variable, then one can also express $w = f'$ as a function of f. It has been established independently by GERBER (1951) and LEWY (1952a) that the equation describing the free surface, $z = z(\varphi)$, is an analytic function of φ at all points for which $w \neq 0$.

STOKES' "second method". In the introduction to Sect. 27 it was mentioned that STOKES (1880), in a supplement to an earlier paper in his collected works, developed a method for approximating exact periodic waves which is different from the straightforward generalization of infinitesimal-wave theory expounded in that section. This method is based upon use of f as the independent variable and expansion of z as a Fourier series in f:

$$c\,z = f + i\,\frac{c\,\lambda}{2\pi}\sum_{n=0}^{\infty} a_n\,e^{-i n 2\pi f/c\lambda} \tag{32.74}$$

or

$$c\,z = f + i\,\frac{c\,\lambda}{2\pi}\,a_0 + \frac{c\,\lambda}{2\pi}\sum_{n=1}^{\infty} a_n \sin n\,\frac{2\pi}{c\,\lambda}\,(f + i\,Q) \tag{32.75}$$

for infinite and finite depth respectively; the a_n may be taken to be real. Here ψ is taken as in the preceding paragraph. The coefficients a_n are to be determined from the condition that the pressure be constant on the surface, i.e. from

$$q^2 + 2g\,y = C \qquad \text{for } \psi = 0. \tag{32.76}$$

If the mean water level is taken at $y=0$ and the fluid is infinitely deep, then $C=c^2$; we shall consider only this case here. Then Eq. (32.76) may be expressed as

$$(c^2 - 2g\,y)\,|z'|^2 = 1. \tag{32.77}$$

Substitution of (32.74) in (32.77) yields

$$\left.\begin{aligned}&\left(1 - \frac{g\,\lambda}{\pi\,c^2}\sum_{n=0}^{\infty} a_n \cos\frac{2\pi\,n\,\varphi}{c\,\lambda}\right)\times\\ &\times\left(1 + 2\sum_{n=1}^{\infty} n\,a_n \cos\frac{2\pi\,n\,\varphi}{c\,\lambda} + \sum_{n,m=1}^{\infty} n\,m\,a_n\,a_m \cos(n-m)\,\frac{2\pi\,\varphi}{c\,\lambda}\right) = 1.\end{aligned}\right\} \tag{32.78}$$

After multiplying the two factors and reducing the cosine products to cosines of sums and differences, the resulting expression may be put into the form

$$\sum_{n=0}^{\infty}\left(\frac{g\,\lambda}{\pi\,c^2}\,b_n + c_n\right)\cos\frac{2\pi\,n\,\varphi}{c\,\lambda} = 0, \tag{32.79}$$

where the b_n's and c_n's are forms of the third degree in the a_n's. The coefficients of the individual cosine terms must then be equated to zero. This results in an infinite sequence of equations, each involving all the a_n's and $g\lambda/\pi c^2$. In order to proceed further, one must devise some method for approximate determination of the a_n's. STOKES' procedure was to assume that each a_n could be expanded in a power series in some parameter, the initial term in the series having the power n. This allows one to carry through a step-by-step improvement in the approximation of the a_n's by including successively higher powers of the parameter. We shall not pursue the matter further, but remark that the most systematic arrangement of such computations seems to have been devised by SRETENSKII (1952).

LEVI-CIVITA's differential-difference equation. The following theorem, due to LEVI-CIVITA (1907), reduces determination of $w(f)$ for steady flow over a horizontal bottom to solution of a differential-difference equation.

Theorem. The complex velocity $w=u-iv$ of an irrotational gravity flow with constant discharge rate Q and with $u\geqq\varepsilon>0$ must satisfy the differential-difference equation

$$\frac{d}{df}\left[w(f+iQ)\,w(f-iQ)\right]-ig\left[\frac{1}{w(f+iQ)}-\frac{1}{w(f-iQ)}\right]=0. \tag{32.80}$$

Conversely, any function $w(f)$ satisfying (32.80) which is regular in the strip $-2Q\leqq\operatorname{Im}f\leqq 0$, finite at ∞, real on $\operatorname{Im}f=-Q$ and has $u>\varepsilon>0$ represents such a flow.

In order to derive (32.80), we note first that the functions $w(f)$ and $z(f)+ih$ both have vanishing real parts for $\psi=-Q$ and consequently can be extended by reflection to the strip $-Q\geqq\psi\geqq-2Q$:

$$w(\bar f-2iQ)=\overline{w(f)},\qquad z(\bar f-2iQ)+ih=\overline{z(f)}-ih. \tag{32.81}$$

The free-surface condition may be expressed by the equation

$$\frac{\partial}{\partial\varphi}w\overline{w}+2g\frac{\partial}{\partial\varphi}y=0\qquad\text{for }\psi=0, \tag{32.82}$$

or, by making use of the extended definitions of w and z, by

$$\frac{\partial}{\partial\varphi}\{w(\varphi)\,w(\varphi-2iQ)-ig\,[z(\varphi)-z(\varphi-2iQ)]\}=0. \tag{32.83}$$

Consider the function

$$H(f)=w(f+iQ)\,w(f-iQ)-ig[z(f+iQ)-z(f-iQ)]. \tag{32.84}$$

Evidently, H is defined and is regular on the line $\psi=-iQ$ and thus in some neighborhood of this line. From (32.83) it follows that $H'(\varphi-iQ)=0$, hence that $H'(f)\equiv 0$ in its region of definition. Eq. (32.80) follows from the fact that $z'(f\pm iQ)=1/w(f\pm iQ)$. For proof of the converse we refer to LEVI-CIVITA's paper. LEVI-CIVITA also gives a special form of (32.80) appropriate to a space-periodic flow. CISOTTI (1919) generalized the preceding theorem to include a variable discharge rate. The Eq. (32.80) may be considered to contain the Eq. (22.30), when in that equation $f(z,t)=f(z-ct)$, in the sense that linearization of (32.80) by assuming

$$w=c(1+\varepsilon w_1+\cdots)$$

yields (22.30).

RUDZKI's transformation. The following transformation was apparently first introduced by RUDZKI (1898). It has later been used by many others in the investigation of exact water waves. The validity of the reformulated boundary condition is not limited to periodic waves. However, it is assumed that a coordinate system has been selected with respect to which the flow is steady. It is again assumed that $u>\varepsilon>0$. Let ϑ be the angle between the velocity vector $\overline{w}=u+iv$ and the positive x-axis. Then one may write

$$w=u-iv=q\,e^{-i\vartheta}=c\,e^{-i\omega} \tag{32.85}$$

where

$$\omega=\vartheta+i\tau,\qquad q=c\,e^{\tau}. \tag{32.86}$$

Here c is some typical velocity, say the wave velocity as defined in Sect. 7. We consider ω as a function of f and let $\psi=0$ correspond to the free surface. The free-surface condition may then be expressed by

$$g\frac{\partial y}{\partial\varphi}+q\frac{\partial q}{\partial\varphi}=0\quad\text{for }\psi=0. \tag{32.87}$$

But [see (32.16)]

$$\frac{\partial y}{\partial \varphi} = \frac{1}{q^2}\frac{\partial \varphi}{\partial y} = \frac{1}{q}\sin\vartheta$$

and, from (32.86),

$$\frac{\partial q}{\partial \varphi} = c\,\mathrm{e}^{\tau}\frac{\partial \tau}{\partial \varphi} = q\frac{\partial \tau}{\partial \varphi}.$$

Hence (32.87) becomes

$$\frac{\partial \tau}{\partial \varphi} = -g\frac{1}{q^3}\sin\vartheta = -\frac{g}{c^3}\mathrm{e}^{-3\tau}\sin\vartheta \quad \text{for} \quad \psi = 0, \tag{32.88}$$

or, since $\partial\tau/\partial\varphi = -\partial\vartheta/\partial\psi$ from the Cauchy-Riemann equations,

$$\frac{\partial \vartheta}{\partial \psi} = \frac{g}{c^3}\mathrm{e}^{-3\tau}\sin\vartheta \qquad \text{for } \psi = 0. \tag{32.89}$$

It one can find a function $\omega(f)$ regular in the strip $0 \geqq \psi \geqq -Q$, with $|\vartheta| < \frac{1}{2}\pi - \varepsilon'$, and with its real and imaginary parts satisfying (32.88) or (32.89) on $\psi = 0$, one may then construct from it a free-surface flow with gravity. Of course, further conditions must be imposed at $\psi = -Q$ or as $\psi \to -\infty$.

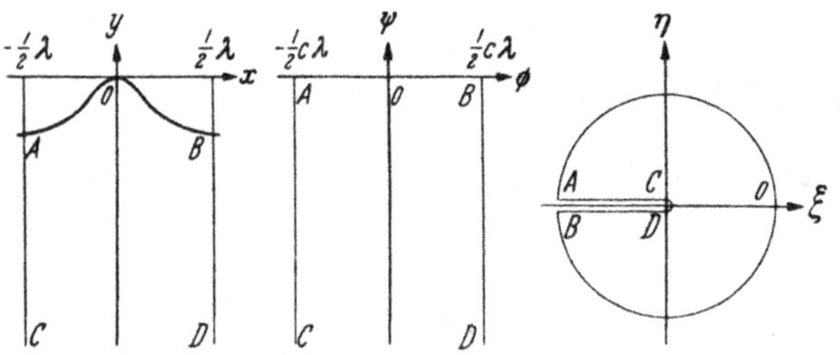

Fig. 50.

NEKRASOV'S transformation. The following transformation is due to NEKRASOV (1921, 1951). It will be assumed that the surface is periodic with period λ, symmetric about a crest and that the fluid is infinitely deep and $\lim_{y\to-\infty} w = c$. Let the origin in the z-plane be taken at a crest, $\psi = 0$ be the free surface, and assume $u > \varepsilon > 0$. In addition to the z- and f-planes, we introduce a ζ-plane,

$$\zeta = \xi + i\eta = \varrho\,\mathrm{e}^{i\gamma}, \tag{32.90}$$

related to the f-plane through

$$\zeta = \mathrm{e}^{-\frac{2\pi i}{\lambda c}f}. \tag{32.91}$$

With a cut along the negative ξ-axis there is a one-to-one correspondence between the various domains $CAOBD$ shown in Fig. 50.

The relation between the z- and ζ-planes will be determined by

$$\frac{dz}{d\zeta} = -\frac{\lambda}{2\pi i}\frac{h(\zeta)}{\zeta}, \quad h(\zeta) = 1 + a_1\zeta + a_2\zeta^2 + \cdots, \quad a_k \text{ real}, \tag{32.92}$$

where $h(\zeta)$ is regular in the disc and is related to w by

$$w = \frac{df}{d\zeta}\frac{d\zeta}{dz} = \frac{c}{h(\zeta)}. \tag{32.93}$$

The form of h shown in (32.92) follows from the assumed properties of the motion. Since $\varrho = 1$ on the free surface, the condition of constant pressure may be expressed by

$$2g\frac{\partial y}{\partial \gamma} + \frac{\partial q^2}{\partial \gamma} = 0 \quad \text{for} \quad \varrho = 1. \tag{32.94}$$

But

$$\left.\frac{\partial y}{\partial \gamma}\right|_{\varrho=1} = \operatorname{Im} \left.\frac{dz}{d\zeta}\frac{d\zeta}{d\gamma}\right|_{\varrho=1} = \operatorname{Im} \left.\frac{-\lambda}{2\pi i}\frac{h(\zeta)}{\zeta} i\zeta\right|_{\varrho=1} = \frac{-\lambda}{2\pi} \operatorname{Im} h(e^{i\gamma}). \tag{32.95}$$

It then follows from (32.93) and (32.95) that

$$\frac{d}{d\gamma}\frac{1}{h(e^{i\gamma})\,h(e^{-i\gamma})} = \frac{\lambda g}{\pi c^2} \operatorname{Im} h(e^{i\gamma}). \tag{32.96}$$

In this formulation of the problem one seeks a function $h(\zeta)$, regular in the disc $|\zeta| \leqq 1$, real on the real axis, $h(0)=1$, and satisfying (32.96). From such a function one can easily construct a periodic gravity flow with free surface.

NEKRASOV's integral equation. NEKRASOV also considers the function ω of (32.92), but as a function of ζ. Let us start from (32.88) and compute

$$\frac{\partial \tau}{\partial \gamma} = \frac{\partial \tau}{\partial \varphi}\frac{\partial \varphi}{\partial \gamma} = -\frac{g}{c^3} e^{-3\tau} \sin\vartheta \cdot \frac{-\lambda c}{2\pi} = \frac{g\lambda}{2\pi c^2} e^{-3\tau} \sin\vartheta \quad \text{for} \quad \varrho = 1. \tag{32.97}$$

One may formally integrate this equation and obtain

$$e^{3\tau} = \frac{3}{2\pi}\frac{g\lambda}{c^2\mu}\left[1 + \mu\int_0^\gamma \sin\vartheta(\alpha)\, d\alpha\right], \tag{32.98}$$

where $1/\mu$ is the integration constant; μ is related to the velocity at the crest, $q_0 = \tau(1) = c/h(1)$, by

$$\mu = \frac{3}{2\pi}\frac{g\lambda c}{q_0^3} > 0. \tag{32.99}$$

Substitution of (32.98) into (32.99) yields the following equation for the relation between τ and ϑ on the boundary:

$$\frac{d\tau(\gamma)}{d\gamma} = \frac{1}{3}\frac{\mu\sin\vartheta(\gamma)}{1 + \mu\int_0^\gamma \sin\vartheta(\alpha)\, d\alpha} \tag{32.100}$$

[it follows from (32.98) that the denominator does not vanish]. It is known from the theory of functions of a complex variable (see, e.g., CARATHÉODORY, Funktionentheorie, Bd. 1, § 147—149, Birkhäuser, Basel, 1950) that, if a function is regular within and on a closed Jordan curve, it is determined up to an additive constant by giving either its real or imaginary part on the boundary. In particular, in the case at hand we may express the value of ϑ on the boundary $|\zeta| = 1$ in terms of τ on the boundary:

$$\vartheta(\gamma) = \text{const} - \frac{1}{2\pi} \operatorname{PV} \int_0^{2\pi} \tau(\beta) \cot\frac{1}{2}(\gamma - \beta)\, d\beta, \tag{32.101}$$

where the constant $= i\vartheta|_{\zeta=0} = 0$. An integration by parts gives

$$\vartheta(\gamma) = -\frac{1}{\pi}\int_0^{2\pi} \frac{d\tau}{d\beta} \log\left|\sin\frac{1}{2}(\gamma-\beta)\right| d\beta. \tag{32.102}$$

From the assumed symmetry about a crest follows $\tau'(-\beta) = -\tau'(\beta)$, so that (32.102) may be expressed as follows:

$$\vartheta(\gamma) = \frac{1}{2\pi}\int_0^{2\pi} \frac{d\tau}{d\beta} \log\left|\frac{\sin\frac{1}{2}(\gamma+\beta)}{\sin\frac{1}{2}(\gamma-\beta)}\right| d\beta. \tag{32.103}$$

Substitution of (32.100) into (32.103) yields NEKRASOV's nonlinear integral equation for $\vartheta(\gamma)$:

$$\vartheta(\gamma) = \frac{1}{\sigma\pi} \int_0^{2\pi} \frac{\mu \sin\vartheta(\beta)}{1+\mu \int_0^\beta \sin\vartheta(\alpha)\, d\alpha} \log \left| \frac{\sin\frac{1}{2}(\gamma+\beta)}{\sin\frac{1}{2}(\gamma-\beta)} \right| d\beta. \tag{32.104}$$

If one can find ϑ satisfying (32.104), one can then reconstruct $\omega(\zeta)$ and hence the whole flow.

NEKRASOV (1928, 1951) carried through a similar analysis when the depth is finite. We shall only sketch it. In Fig. 50 suppose that $y = -h_1$ represents the bottom (h_1 is not the mean depth) and $\psi = -Q$ the corresponding streamline. In the ζ-plane this maps into a circle of radius $\varrho_0 < 1$, where

$$\varrho_0 = e^{-\frac{2\pi Q}{\lambda c}}. \tag{32.105}$$

In (32.92) $h(\zeta)$ becomes a Laurent series. The integral equation for $\vartheta(\gamma)$ remains the same in form as (32.104), but the kernel function $\log|\ldots|$ is now replaced by

$$\sum_{n=1}^{\infty} \frac{2}{n} \tanh \frac{2\pi Q}{\lambda c} \sin n\gamma \sin n\beta. \tag{32.106}$$

MOISEEV (1957b) has further generalized NEKRASOV's equation so as to allow a wavy bottom.

The solution $\vartheta(\gamma)$ of (32.104) will, of course, depend upon the parameter μ, except for the trivial solution $\vartheta \equiv 0$ corresponding to a uniform flow. It is possible to show that not all μ's are allowable. Let

$$M = \max |\vartheta(\gamma)|. \tag{32.107}$$

It then follows from (32.102) that

$$0 \leqq |\vartheta(\gamma)| < \frac{1}{6\pi} \frac{\mu \sin M}{1-\pi\mu \sin M} \int_0^{2\pi} -\log\left|\sin\frac{1}{2}(\gamma-\beta)\right| d\beta < \frac{1}{3} \frac{\mu \sin M}{1-\pi\mu\sin M}, \tag{32.108}$$

hence that

$$0 \leqq M < \frac{1}{3} \frac{\mu \sin M}{1-\pi\mu\sin M}. \tag{32.109}$$

From this follows

$$\frac{1}{\pi \sin M} > \mu > \frac{1}{\pi \sin M + \frac{1}{3} M^{-1} \sin M} > \frac{3}{1+3\pi}. \tag{32.110}$$

VILLAT's integral equation. Even though we shall not consider its contents in any detail, it would be improper not to mention an important paper of VILLAT (1915). VILLAT wished to find the steady motion of a fluid in a canal of given bottom profile and also with a given top profile over the part of the fluid upstream of some point. Downstream of this point the top profile is one of constant pressure. The boundary condition on the free surface, (32.89), is modified by introduction of new variables, and a pair of integral equations, one of them nonlinear, is derived. The method is also applicable if the upstream "cover" is absent and, in fact, becomes a little simpler. The chief use made of the procedure by VILLAT is as an inverse method in which the free surface is given and the corresponding bottom and cover determined.

33. Waves of maximum amplitude. In the higher-order theory of infinitesimal waves one of the important effects of including higher-order terms was to make the profile more peaked at the crests and flatter in the troughs. The effect was the same for either steady progressive waves or standing waves. Since the peakedness increased with increase of the amplitude-to-wavelength ratio, it seems reasonable to conjecture that there is some bound to this ratio and that, if a wave of maximum amplitude-to-length ratio exists, it will be characterized by a corner or a cusp at the crest, at least if capillarity is neglected. It has never been proved that such waves exist. However, if one assumes their existence, it is possible to prove some necessary properties. This will be done below.

Following an earlier erroneous investigation of RANKINE (1865), STOKES (1880, p. 225) showed that, if a corner occurs in steady motion, the angle included between the tangents must be 120°. MICHELL (1893) assumed that a periodic highest progressive wave exists and showed how to compute the coefficients of an associated series, but without proving convergence. HAVELOCK (1919) made MICHELL's procedure the basis of a general method of approximation to periodic progressive waves, again with no proof of convergence. MICHELL'S wave was later investigated by a different procedure by NEKRASOV (1920). However, NEKRASOV did not carry his computations to the same degree of accuracy as MICHELL and HAVELOCK, so that the numerical results are discrepant. More recently YAMADA (1957) rediscovered NEKRASOV's method and carried through the calculations with the necessary accuracy; the results are now in substantial agreement with those of HAVELOCK and MICHELL.

PENNEY and PRICE (1952b), in their work on standing waves of finite amplitude, include an analysis intended to show that, if there exists a standing wave of maximum amplitude with a corner at the crest, then the angle must be 90°. G.I. TAYLOR (1953) has questioned the validity of the proof, and it appears, in fact, to be untenable. On the other hand, in the same paper TAYLOR reports the results of experiments which appear to confirm PENNY and PRICE'S prediction. In view of the present unsatisfactory state of the theory, it will not be further discussed here.

STOKES' theorem. We prove first STOKES' theorem on the angle at a corner in steady flow. Let the corner be at the origin $z=0$, the free surface be the streamline $\psi=0$, and $\varphi=0$ at the corner. Since $z=0$ is assumed to be a corner, it must also be a stagnation point and the constant-pressure condition on the surface may be taken in the form

$$q^2+2g\eta(x)=0. \tag{33.1}$$

In the mapping from the z- to the f-plane the point $z=0$ must be a branch point, so that in the neighborhood of $z=0$ the complex velocity potential will take the form

$$f=A\,z^n. \tag{33.2}$$

If $\alpha_+<0$ is the angle between the right-hand tangent to the corner and OX, then near $z=0$ Eq. (33.1) can be written

$$|A|^2 n^2 r^{2n-2}+2g\,r\sin\alpha_+=0.$$

This can hold for all small r only if

$$n=\tfrac{3}{2}. \tag{33.3}$$

It also follows that, if α_- is the angle between the left-hand tangent and OX, then $\sin\alpha_-=\sin\alpha_+$ and $\alpha_-=-180°-\alpha_+$ so that the surface is symmetrical about

OX near the corner. If $\psi \leqq 0$ corresponds to the region occupied by fluid and if the branch of $z = r\,e^{i\alpha}$ with $-\frac{3}{2}\pi < \alpha < \frac{1}{2}\pi$ is taken, then the complex velocity potential has the following form near $z=0$:

$$\left.\begin{aligned} f(z) &= -\tfrac{2}{3}\sqrt{g}\,(-i\,z)^{\frac{3}{2}} \\ &= -\tfrac{2}{3}\sqrt{g}\,r^{\frac{3}{2}}\left[\cos\tfrac{3}{2}(\alpha - \tfrac{1}{2}\pi) + i\sin\tfrac{3}{2}(\alpha - \tfrac{1}{2}\pi)\right]. \end{aligned}\right\} \tag{33.4}$$

The streamline $\psi=0$ has a corner at $z=0$ with included angle 120°. In this case the flow is to the right. The inversion of (33.4) gives

$$z(f) = \left[\frac{3}{2\sqrt{g}}\right]^{\frac{2}{3}} e^{-i\pi/6} f^{\frac{2}{3}} \tag{33.5}$$

for f near 0.

α) Periodic wave of maximum height. Let us suppose that a periodic progressive wave of maximum amplitude-length ratio exists. We may take this as a steady flow with complex velocity potential $f(z) = \varphi + i\psi$ and with

$$\lim_{y\to-\infty} f'(z) = c. \tag{33.6}$$

Let the origin of the z-plane be at one of the crests, the surface profile correspond to $\psi=0$, and the origin of the f-plane to that of the z-plane. Then the free surface condition may be taken in the form (33.1).

MICHELL'S method. First we give MICHELL'S procedure for finding $f'(z)$. As we have done earlier, we shall write

$$f'(z) = q\,e^{-i\vartheta}, \qquad z'(f) = \frac{1}{q}\,e^{i\vartheta}. \tag{33.7}$$

From the assumed periodicity and symmetry, ϑ is an odd periodic function of φ with period $c\lambda$ for $\psi=0$. From (33.7) follows

$$\frac{d}{df}\log z'(f) = -\frac{\partial}{\partial\varphi}\log q + i\,\frac{\partial\vartheta}{\partial\varphi}. \tag{33.8}$$

For $\psi=0$, $\partial\vartheta/\partial\varphi$ is an even periodic function of φ with removable singularities at the crests; we expand it in a Fourier series:

$$\frac{\partial\vartheta}{\partial\varphi} = \frac{\pi}{c\lambda}\left[a_0 + a_1\cos\frac{2\pi\varphi}{c\lambda} + a_2\cos\frac{4\pi\varphi}{c\lambda} + \cdots\right]. \tag{33.9}$$

The a_k are real. Substitute (33.9) into (33.8) and rewrite it in the following way:

$$\left[\frac{d}{df}\log z'(f) - i\,\frac{\pi}{c\lambda}\sum_{n=0}^{\infty} a_n\,e^{-i2n\pi f/c\lambda}\right]_{\psi=0} = -\frac{\partial}{\partial\varphi}\log q - \frac{\pi}{c\lambda}\sum_{n=1}^{\infty} a_n \sin\frac{2n\pi\varphi}{c\lambda}. \tag{33.10}$$

Now consider the function

$$Z(f) = \frac{d}{df}\log z'(f) - i\,\frac{\pi}{c\lambda}\sum_{n=0}^{\infty} a_n\,e^{-i2n\pi f/c\lambda}. \tag{33.11}$$

$Z(f)$ is defined in the whole lower half-plane, is regular for $\psi<0$, and, as $\psi\to-\infty$, $Z(f)\to -i\pi a_0/c\lambda$. Moreover, from (33.10) Z is also real on the real axis and hence may be extended by reflection to the upper half-plane. Z is then a function with only singularities on the real axis at the points $\varphi = nc\lambda$ associated with the crests. The form of the singularity may be determined from (33.5). In fact, near $f=0$

$$\frac{d}{df}\log z' = -\frac{1}{3}\,\frac{1}{f}. \tag{33.12}$$

Hence $Z(f)$ has singularities of the form

$$-\frac{1}{3}\frac{1}{f-nc\lambda}, \qquad n=0, \pm 1, \pm 2, \ldots, \tag{33.13}$$

along the real axis, and only these, so that it must have the form

$$Z(f) = -\frac{\pi}{3c\lambda}\cot\frac{\pi f}{c\lambda} + b, \qquad b = \text{const.} \tag{33.14}$$

From (33.14) $Z \to b - i\pi/3c\lambda$ as $\psi \to -\infty$. Then from (33.11)

$$b = -i\frac{\pi}{c\lambda}\left(a_0 - \frac{1}{3}\right).$$

Since Z must be real for real f, it follows that

$$a_0 = \tfrac{1}{3} \tag{33.15}$$

and

$$Z(f) = -\frac{\pi}{3c\lambda}\cot\frac{\pi f}{c\lambda}. \tag{33.16}$$

It now follows from the definition of $Z(f)$ that

$$\frac{d}{df}\log z'(f) = -\frac{\pi}{3c\lambda}\cot\frac{\pi f}{c\lambda} + i\frac{\pi}{3c\lambda} + i\frac{\pi}{c\lambda}\sum_{n=1}^{\infty} a_n\, e^{-i2n\pi f/c\lambda}, \tag{33.17}$$

which yields, after integration, inversion of the logarithm and use of (33.6) to evaluate a multiplicative constant,

$$\left.\begin{aligned} z'(f) &= \frac{1}{c\sqrt[3]{2}}\, e^{\frac{1}{3}i\pi f/c\lambda}\left(i\sin\frac{\pi f}{c\lambda}\right)^{-\frac{1}{3}}\prod_{n=1}^{\infty}\exp\left(\frac{-c\lambda}{2n\pi}\, a_n\, e^{-i2n\pi f/c\lambda}\right)\\ &= \frac{1}{c\sqrt[3]{2}}\, e^{\frac{1}{3}i\pi f/c\lambda}\left(i\sin\frac{\pi f}{c\lambda}\right)^{-\frac{1}{3}}\sum_{n=0}^{\infty} b_n\, e^{-i2n\pi f/c\lambda}, \end{aligned}\right\} \tag{33.18}$$

where $b_0 = 1$ and the b_n are real. The branch of the root must be chosen so that its argument lies between $\pm\frac{1}{2}\pi$ for $\psi = 0$. From (33.18) one finds immediately also

$$f'(z) = c\sqrt[3]{2}\, e^{-\frac{1}{3}i\pi f/c\lambda}\left(i\sin\frac{\pi f}{c\lambda}\right)^{\frac{1}{3}}\sum_{n=0}^{\infty} c_n\, e^{-i2n\pi f/c\lambda}, \qquad c_0 = 1,\ c_n \text{ real.} \tag{33.19}$$

Aside from the first one, the coefficients in (33.18) or (33.19) are still to be determined. The constant-pressure condition for the surface profile is still available for this purpose, for we have made use of the Eq. (33.4) or (33.5) only through the value of the exponent. The value of the gravitation constant has not entered into (33.18) or (33.19). In fact, a comparison of (33.5) after differentiation and (33.18) in the neighborhood of $f = 0$ yields immediately

$$\frac{c^2}{g\lambda} = \frac{3}{4\pi}\left[1 + b_1 + b_2 + \cdots\right]^3 = \frac{3}{4\pi}\left[1 + c_1 + c_2 + \cdots\right]^{-3}, \tag{33.20}$$

so that, once the b_n or c_n are determined, the relation between wavelength and velocity may be found. This method could presumably be pursued to obtain a sequence of further equations for determination of the b_n. However, MICHELL proceeds somewhat differently. If we differentiate (33.1) with respect to φ, we may write the free surface condition as follows [cf. (32.54) and following]:

$$\left.\begin{aligned} &\frac{\partial}{\partial \varphi} q^2 = -\frac{g}{q^2}\frac{\partial \varphi}{\partial y} \qquad \text{for } \psi = 0 \\ \text{or} \qquad & \\ &\frac{\partial}{\partial \varphi} |f'|^4 = 4g \operatorname{Im} f' \qquad \text{for } \psi = 0. \end{aligned}\right\} \tag{33.21}$$

Substitution of (33.19) in (33.21) yields an equation of the following form

$$\left.\begin{aligned} &\frac{4}{3}\pi 2^{\frac{4}{3}} \frac{c^3}{\lambda} \sin^{\frac{4}{3}} \frac{\pi\varphi}{c\lambda}\left\{A_1 \cos\frac{\pi\varphi}{c\lambda} + A_3 \cos\frac{3\pi\varphi}{c\lambda} + \cdots\right\} \\ &\quad = \frac{4}{3} g c \sin^{\frac{1}{3}} \frac{\pi\varphi}{c\lambda}\left\{B_1 \cos\frac{\pi\varphi}{c\lambda} + B_3 \cos\frac{3\pi\varphi}{c\lambda} + \cdots\right\}, \end{aligned}\right\} \tag{33.22}$$

where the B_n's depend linearly upon the c_n's, and the A_n's depend upon them in a more complicated manner. The derivation of (33.22), especially of the right-hand part, and of the particular dependence of the A_n's and B_n's upon the c_n's is rather tedious and we refer to either MICHELL's original paper or preferably to HAVELOCK's more general and systematic treatment. Equating coefficients of the individual cosine terms leads to a set of equations relating $c^2/g\lambda$, $c_1, c_2, \ldots$. The values as computed by HAVELOCK, which we assume to be somewhat more accurate than MICHELL's own, are as follows:

$$\frac{g\lambda}{c^2} = 0.833 \cdot 2\pi, \quad c_1 = 0.0414, \quad c_2 = 0.0114, \quad c_3 = 0.0042, \quad c_4 = 0.0014. \tag{33.23}$$

The value for $g\lambda/c^2$ should be compared with that for infinitesimal waves, namely 2π. Substitution of $\frac{1}{2}c\lambda$ for f in (33.19) yields the velocity at a trough:

$$u = c\sqrt[6]{2}\,[1 - c_1 + c_2 - c_3 + \cdots] \approx 1.219c. \tag{33.24}$$

From $q^2 + 2g\eta = 0$ one may now find η for the trough and hence the amplitude-wavelength ratio:

$$\left|\frac{\eta}{\lambda}\right| = \frac{1}{\sqrt[3]{2}}\frac{c^2}{g\lambda}[1 - c_1 + c_2 - \cdots]^2 \approx 0.1418. \tag{33.25}$$

H. JEFFREYS (1951) has recently reexamined the basis of the Michell-Havelock method of approximation and concludes that an apparent discrepancy between the values in (33.23) and Eq. (33.20) does not really indicate a numerical error in the computations.

We note in passing that MICHELL also gave the form of $f'(z)$ analogous to (33.19) which must hold if a highest wave with corner exists in a fluid of finite depth.

Method of NEKRASOV and YAMADA. This method makes use of the ζ-plane introduced in (32.57) and related to the f-plane by (32.58). We may again make use of Fig. 50 but must keep in mind that in the z-plane there is now a corner at 0 with an included angle of 120°. Hence (32.59), the equation relating the z- and ζ-planes, must be replaced by

$$\frac{dz}{d\zeta} = -\frac{\lambda}{2\pi i}\frac{h(\zeta)}{\zeta(1-\zeta)^{\frac{1}{3}}}, \qquad h(\zeta) = 1 + a_1\zeta + a_2\zeta^2 + \cdots \tag{33.26}$$

and (32.60) by

$$w = \frac{df}{dz} = c\frac{(1-\zeta)^{\frac{1}{3}}}{h(\zeta)}. \tag{33.27}$$

The coefficients a_n are now to be determined by the constant-pressure condition on the free surface taken in the form (32.94). From

$$q^2|_{\varrho=1} = c^2 \frac{[(1-e^{i\gamma})(1-e^{-i\gamma})]^{\frac{1}{3}}}{h(e^{i\gamma})\,h(e^{-i\gamma})} = c^2 \frac{(2\sin\frac{1}{2}\gamma)^{\frac{2}{3}}}{h(e^{i\gamma})\,h(e^{-i\gamma})} \tag{33.28}$$

and

$$\frac{dz}{d\gamma}\bigg|_{\varrho=1} = -\frac{\lambda}{2\pi}\,\frac{h(e^{i\gamma})}{(1-e^{i\gamma})^{\frac{1}{3}}} = -\frac{\lambda}{2\pi}\left(2\sin\frac{1}{2}\gamma\right)^{-\frac{1}{3}} e^{-i(\gamma-\pi)/6}\,h(e^{i\gamma}) \tag{33.29}$$

one obtains as the equation analogous to (32.96)

$$\frac{d}{d\gamma}\,\frac{(2\sin\frac{1}{2}\gamma)^{\frac{2}{3}}}{h(e^{i\gamma})\,h(e^{-i\gamma})} = \frac{g\lambda}{\pi c^2}\left(2\sin\frac{1}{2}\gamma\right)^{-\frac{1}{3}} \operatorname{Im}\{e^{-i(\gamma-\pi)/6}\,h(e^{i\gamma})\}. \tag{33.30}$$

This yields a set of equations for determination of $g\lambda/c^2$, a_1, a_2, ... The actual computation appears to be as tedious as that of MICHELL's method and, in fact, NEKRASOV's (1920) computations do not seem to have yielded as accurate results as MICHELL's. However, as mentioned earlier, YAMADA (1957) has set up a systematic computation procedure and has obtained results in substantial agreement with those of MICHELL and HAVELOCK. Once $g\lambda/c^2$ and the a_n have been determined, the surface profile can be found in parametric form by integrating (33.29) with respect to γ from 0 to γ. Fig. 51, reproduced from YAMADA's cited paper, shows the form of the profile.

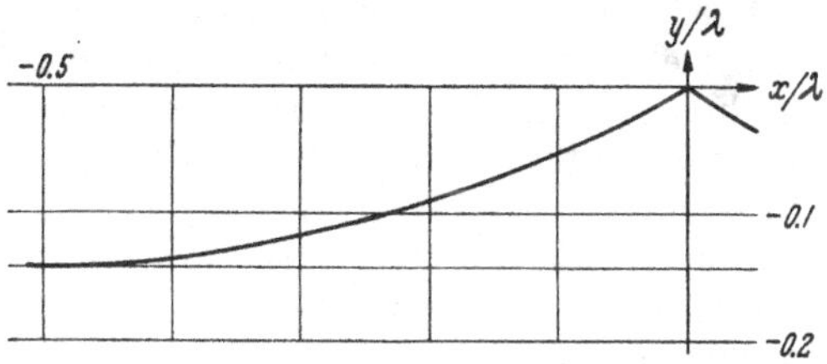

Fig. 51.

β) Havelock's approximation for gravity waves. In a paper already cited several times above HAVELOCK (1919) extended MICHELL's method of construction of periodic waves of maximum amplitude, outlined in the preceding section, to one for construction of periodic waves of any allowable amplitude-length ratio. Up to the present, no one has proved the series involved to converge. However, as HAVELOCK points out, the method has attractive theoretical features: the parameter describing the family of waves occurs in the form $e^{-\beta}$ where β varies from 0, corresponding to the highest wave, to ∞, corresponding to infinitesimal waves.

The method starts out exactly like MICHELL's up to Eq. (33.19) except that it is not assumed that $\psi=0$ corresponds to the free surface. We recall that in MICHELL's analysis the constant-pressure condition did not enter completely until after (33.19), in particular, in (33.21). HAVELOCK assumes instead that this condition is to be satisfied on some other streamline, $\psi=-\alpha$, which will then be taken to correspond to the free surface. The condition may still be written in the form (33.21) provided that one replaces $\psi=0$ by $\psi=-\alpha$. For $\psi=-\alpha$ one may write

$$f' = c\sqrt[3]{2}\,e^{-\frac{1}{3}i\pi(\varphi-i\alpha)/c\lambda}\left(i\sin\pi(\varphi-i\alpha)/c\lambda\right)^{\frac{1}{3}} \sum_{n=0}^{\infty}\gamma_n\,e^{-i2n\pi\varphi/c\lambda}, \tag{33.31}$$

where $\gamma_n = c_n\,e^{-2n\pi\alpha/c\lambda}$, the c_n being the same as those in (33.19). HAVELOCK shows that one may express $\partial|f'|^4/\partial\varphi$ in the following form

$$\left.\begin{aligned} \frac{\partial}{\partial\varphi}|f'|^4 &= \frac{4}{3}\pi\,2^{\frac{4}{3}}\frac{c^3}{\lambda}\sin\frac{\pi\varphi}{c\lambda}\left[\sinh^2\frac{\pi\alpha}{c\lambda}+\sin^2\frac{\pi\varphi}{c\lambda}\right]^{-\frac{1}{3}}\times\\ &\quad\times\left[A_1\cos\frac{\pi\varphi}{c\lambda}+A_3\cos\frac{3\pi\varphi}{c\lambda}+\cdots\right]\end{aligned}\right\} \tag{33.32}$$

and Im f' in the form

$$\left.\begin{aligned}\operatorname{Im} f' = \frac{1}{3}\, c\, e^{-4\pi\alpha/3c\lambda} \sin\frac{\pi\varphi}{c\lambda}\left[\sinh^2\frac{\pi\alpha}{c\lambda} + \sin^2\frac{\pi\varphi}{c\lambda}\right]^{-\frac{1}{3}}\times\\ \times\left[B_1\cos\frac{\pi\varphi}{c\lambda} + B_3\cos\frac{3\pi\varphi}{c\lambda} + \cdots\right].\end{aligned}\right\} \tag{33.33}$$

Here the A_n's are rather complicated expressions in the γ_n's but also involve $\cosh \pi\alpha/c\lambda$ linearly; the B_n's are linear expressions in the γ_n with coefficients which are functions of $e^{-2\pi\alpha/c\lambda}$. HAVELOCK finds complete expressions for the B_n's; for A_1, A_3, A_5, A_7 he finds the dependence upon the first few γ_n's. One must refer to the original for details, especially for the scheme for approximate solution for the γ_n's.

When $\alpha = 0$ the above analysis is precisely that for the highest wave. The numerical results of HAVELOCK's computations for this case were given in the last section. He also computes $g\lambda/c^2$, γ_1, γ_2 (also γ_3 for the first) for two further cases: $e^{-2\pi\alpha/c\lambda} = 0.75$ and 0.3. The agreement with results computed by other methods, either those of subsection 27α or similar ones, is very close. However, to establish the validity of the method, one must prove convergence of the series $\Sigma|\gamma_n|$.

The relation of this method of approximation to STOKES' "second method" (see subsection 32γ) is also clarified by HAVELOCK. For this we refer to the original paper.

34. Explicit solutions. Although it is not in general possible to give an explicit exact solution to a particular problem of interest, it is possible to give various classes of exact solutions and then to determine the associated solid boundaries. This is sometimes referred to as an "inverse method". Several such methods for constructing exact solutions will be discussed in subsection 34α. In addition, there is one periodic wave in infinitely deep fluid which satisfies the boundary conditions exactly, the Gerstner wave. This will be discussed in subsection 34β. In subsection 34γ we shall discuss briefly what may be called pseudo-exact solutions due to DAVIES and PACKHAM. In these the exact boundary condition is replaced by a closely related one which allows exact solution. They derive their interest from the fact that they contain in one family waves ranging from the smallest amplitude-length ratio up to a counterpart of the Michell wave. Furthermore, the procedure also can be used for pseudo solitary and cnoidal waves. Subsection 34δ will be devoted to an exact solution for pure capillary waves recently discovered by CRAPPER (1957).

α) *Inverse methods.* SAUTREAUX's method. Possibly the earliest method capable of generating a wide class of steady irrotational solutions is due to C. SAUTREAUX (1893, 1894, 1901). It has been rediscovered several times subsequently, e.g., by BLASIUS (1910), WILTON (1913), RICHARDSON (1920) and LEWY (1952). F. AIMOND (1929) has given a very comprehensive treatment of the method and of various related ones. The method may be easily generalized to include an arbitrary impressed pressure distribution on the free surface (see the papers of RICHARDSON or AIMOND).

Let $z = x + iy$, $f = \varphi + i\psi$, and take f as the independent variable. The free surface will be represented by $\psi = 0$. We further assume $q^2 > \varepsilon > 0$. In the constant-pressure condition on the surface, $\frac{1}{2}q^2 + g\eta = \text{const}$, it will be convenient to take the position of the x-axis so that the constant is zero, and hence $\eta \leq 0$. This condition may then be expressed in terms of $z(f)$ as follows [cf. (32.56)]:

$$z'(\varphi)\,\overline{z'(\varphi)}\left[z(\varphi) - \overline{z(\varphi)}\right] = -ig. \tag{34.1}$$

Define

$$\mu(f) = \tfrac{1}{2} i \left[z(f) - \overline{z(f)}\right]. \tag{34.2}$$

Then $-\mu(\varphi) = y(\varphi)$, the y-coordinate of the free surface. Hence, from (34.1)

$$\mu(\varphi) = \frac{1}{2g} \frac{1}{z'(\varphi)\,\overline{z'(\varphi)}}, \tag{34.3}$$

From (34.2) and (34.3) one may now derive

$$2\,[g\,\mu(\varphi)]^{-1} - 4\mu'^2(\varphi) = \left[z'(\varphi) + \overline{z'(\varphi)}\right]^2. \tag{34.4}$$

Elimination of $\overline{z'}$ between (34.2) and (34.4) yields

$$z'(\varphi) = -i\,\mu'(\varphi) + \sqrt{(2g\,\mu)^{-1} - \mu'^2}, \tag{34.5}$$

where

$$\mu(\varphi) > 0, \quad 2g\,\mu(\varphi)\,\mu'^2(\varphi) \leqq 1. \tag{34.6}$$

But then, since z' is an analytic function of f, at least near $\psi = 0$,

$$z'(f) = -i\,\mu'(f) + \sqrt{(2g\,\mu)^{-1} - \mu'^2} \tag{34.7}$$

and

$$z(f) = -i\,\mu(f) + \int \sqrt{(2g\,\mu)^{-1} - \mu'^2}\,df. \tag{34.8}$$

One may now reverse the procedure, select an arbitrary analytic function $\mu(f)$ satisfying (34.6) and construct the function $z(f)$ by means of (34.8). The resulting function will describe a flow for which $z(\varphi)$ is the free surface. If (34.6) is satisfied only for some range of φ, then for the remaining range one must treat the streamline $\psi = 0$ as a solid boundary.

One can use the preceding result to construct a flow if the form of the free surface is given. Let the surface be given in the form $x = \xi(y)$ in a neighborhood of some point of the surface. Since $y(\varphi) = -\mu(\varphi)$ on the surface, we may define $\sigma(\mu) = \xi'(y) = x'(\varphi)/y'(\varphi)$; σ is an analytic function of μ for real μ as follows from the theorem of LEWY and GERBER cited near the beginning of subsection 32γ. Hence, from (34.7),

$$\sigma(\mu) = -\left[(2g\,\mu)^{-1} - \mu'^2\right]^{\frac{1}{2}}/\mu'(\varphi). \tag{34.9}$$

Solving for $1/\mu'$, one finds

$$\frac{d\varphi}{d\mu} = \sqrt{2g\,\mu\,(1+\sigma^2)}. \tag{34.10}$$

Since μ is also an analytic function of φ, the same relation holds for $df/d\mu$ when μ is complex, and consequently

$$f = \int \sqrt{2g\,\mu\,(1+\sigma^2(\mu))}\,d\mu. \tag{34.11}$$

It follows similarly from (34.8) and (34.9), first for real μ, then for complex μ that

$$z = -i\,\mu - \int \sigma(\mu)\,d\mu. \tag{34.12}$$

Eqs. (34.11) and (34.12) thus provide a relation between f and z determined by the form of $\sigma(\mu)$ for real μ.

RUDZKI's method. RUDZKI (1898) has given a different formula for deriving exact solutions. The derivation and statement of the formula below differ somewhat from RUDZKI's, but the result is equivalent.

Let $z=z(f)$ and write

$$z'=\frac{1}{q}\,\mathrm{e}^{i\vartheta},\qquad q=q(\varphi,\psi),\qquad \vartheta=\vartheta(\varphi,\psi). \tag{34.13}$$

The free-surface condition may be expressed as follows, from (32.61) and the equation preceding it,

$$q^2\frac{\partial q}{\partial\varphi}=-g\sin\vartheta\quad\text{for}\quad\psi=0. \tag{34.14}$$

Hence

$$q=[-3g\int\sin\vartheta(\varphi,0)\,d\varphi]^{\frac{1}{3}}\quad\text{for}\quad\psi=0, \tag{34.15}$$

where the branch of the cube root is taken which is real for real numbers. Combining (34.15) with (34.13) gives

$$z'(\varphi)=\mathrm{e}^{i\vartheta(\varphi,0)}\,[-3g\int\sin\vartheta(\varphi,0)\,d\varphi]^{-\frac{1}{3}}. \tag{34.16}$$

This relation must then hold also for $\psi\neq 0$, i.e.

$$z'(f)=\mathrm{e}^{i\vartheta(f,0)}\,[-3g\int\sin\vartheta(f,0)\,df]^{-\frac{1}{3}}. \tag{34.17}$$

As in Sautreaux's method, we may now reverse the above procedure, take $\vartheta(f)$ as an arbitrary analytic function of f such that ϑ is real for f real, and construct $z'(f)$ from (34.17).

Richardson's method. From (34.17) one can derive immediately a formula due to Richardson (1920) for constructing exact solutions. Let $G'(f)=-\sin\vartheta(f)$. Then $\mathrm{e}^{i\vartheta}=\sqrt{1-G'^2}-iG'$ and (34.17) becomes

$$z'(f)=[3g\,G(f)]^{-\frac{1}{3}}\left[-i\,G'(f)+\sqrt{1-G'^2}\right]. \tag{34.18}$$

Again, inversely, if $G(f)$ is any analytic function such that, for real f, G', $\sqrt{1-G'^2}$ and G are real, (34.18) gives a corresponding exact free-surface flow.

Examples. The largest collections of specific flows constructed by one of the preceding methods are in the paper of Richardson (1920) and a report of Vitousek (1954). Several examples are given below.

1. In (34.17) let $\vartheta(f)=\text{const}=\alpha<0$. Further, take the constant of integration as zero even though this results in a singularity in z' on the surface. One finds easily

$$f=\tfrac{2}{3}\sqrt{-2g\sin\alpha}\,(z\,\mathrm{e}^{-i\alpha})^{\frac{3}{2}}. \tag{34.19}$$

The free surface will consist of only the ray $z=r\,\mathrm{e}^{i\alpha}$ unless $\alpha=\pi/6$. However, the ray $z=r\,\mathrm{e}^{i(\alpha-\frac{2}{3}\pi)}$ is also a streamline, but not one along which the pressure is constant unless $\alpha=-\pi/6$. Hence it must be taken as a solid boundary in general. The special case $\alpha=-\pi/6$ is just the flow (33.4) considered earlier and has a corner. One may, of course, take any other streamline $\psi=\psi_0<0$ as another solid boundary representing a bottom. The pressure remains positive everywhere only if $-\pi/6<\alpha<0$. This special family of flows was discussed by Weingarten (1904).

2. If in (34.8) one takes $\mu(f)=f/c$ or in (34.18) takes $G(f)=\frac{2}{3}\sqrt{2g/c^3}\,f^{\frac{3}{2}}$, where c is some fixed velocity, one finds

$$c\,z'(f)=-i+\sqrt{(2g\,f/c^3)^{-1}-1}. \tag{34.20}$$

This yields a flow of the sort shown in Fig. 52c, taken from Richardson. The internal solid boundary represents some streamline $\psi=\psi_0<0$. The free surface corresponds to the segment $\psi=0$, $0<\varphi<c^3/2g$ in the f-plane.

3. Let c be some fixed velocity and let

$$G(f)=\frac{3c^3}{g}\left[B+\tanh\left(\alpha\frac{g}{3c^3}f\right)\right],\qquad B>1,\ \alpha<1,$$

in (34.18). Then

$$c\,z'(f)=\left[B+\tanh\left(\alpha\frac{g}{3c^3}f\right)\right]^{-\frac{1}{3}}\left\{-i\,\alpha\,\mathrm{sech}^2\left(\alpha\frac{g}{3c^3}f\right)+\sqrt{1-\alpha^2\,\mathrm{sech}^4\left(\alpha\frac{g}{3c^3}f\right)}\right\}. \quad (34.21)$$

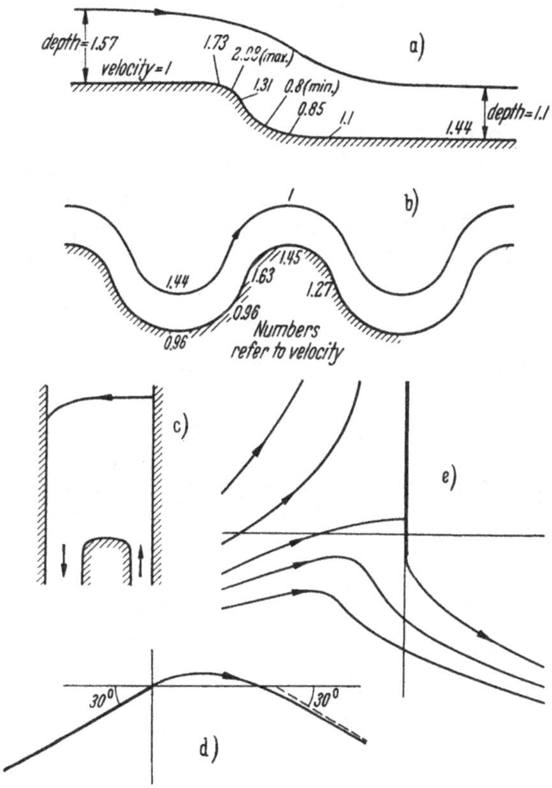

Fig. 52 a—d.

Here $\psi=0$ corresponds to the free surface. The choice of the bottom streamline is restricted by the necessity of avoiding having the singularity at $B=\tanh\frac{1}{3}\alpha g c^{-3} f$ within the fluid. Fig. 52a, also from RICHARDSON, shows a flow computed from (34.21) for $B=2$, $\alpha=\frac{1}{2}$ and $c=1$.

4. Flow over a corrugated bottom has been investigated by both RICHARDSON and RUDZKI by essentially the same method. Following RICHARDSON, we let

$$G(f)=\frac{3c^3}{g}\left[B-\cos\alpha\frac{g}{3c^3}f\right],\qquad B>1,\ \alpha<1.$$

Then

$$c\,z'(f)=\left[B-\cos\alpha\frac{g}{3c^3}f\right]^{-\frac{1}{3}}\left\{-i\,\alpha\sin\alpha\frac{g}{3c^3}f+\sqrt{1-\alpha^2\cos^2\alpha\frac{g}{3c^3}f}\right\}. \quad (34.22)$$

Fig. 52b shows a flow computed from this formula for $B=2$, $\alpha=0.9$.

5. Flows similar to flows over a weir, under a sluice gate, through an opening, etc. have been considered by a number of the cited authors. SAUTREAUX (1901)

applied his formula (34.8) with $\mu = (c^2/2g)\, e^{-2gf/c^3}$ to obtain a number of different flows of this nature. Fig. 52d shows one of them. LAUCK (1925) has also constructed such flows. RICHARDSON obtained a flow through an opening by selecting

$$G(f) = \frac{3c^3}{g}\left[B - e^{gf/3c^3}\right].$$

Possibly the simplest such flow, studied by both BLASIUS and VITOUSEK, is obtained by taking $\mu = \sqrt{cf/g}$ in (34.8); this yields

$$\frac{g}{c^2}\, z = -i\sqrt{\frac{gf}{c^3}} + \frac{1}{3}\left[2\sqrt{\frac{gf}{c^3}} - 1\right]^{\frac{3}{2}}. \tag{34.23}$$

The flow is shown in Fig. 52e.

FRITZ JOHN's method. FRITZ JOHN (1953) has devised a method for constructing exact irrotational two-dimensional free-surface flows which may be time-dependent. Let $F(z, t) = \Phi + i\Psi$ denote the complex velocity potential. The flow of particles on the free surface, $y = \eta(x, t)$, will also be described in a Lagrangian system:

$$z = e(\alpha, t), \tag{34.24}$$

where α is a real number associated with a particular particle. Then

$$\frac{dz}{dt} = \frac{\partial e}{\partial t} = F'\big(x + i\eta(x, t), t\big), \tag{34.25}$$

where F' denotes the partial derivative with respect to z. The equations of motion (2.7), reduced to two dimensions and to motion along the free surface, give

$$\frac{\partial^2 x}{\partial t^2}\frac{\partial x}{\partial \alpha} + \left(g + \frac{\partial^2 y}{\partial t^2}\right)\frac{\partial y}{\partial \alpha} = -\frac{1}{\varrho}\frac{\partial p}{\partial \alpha}. \tag{34.26}$$

Since $p = \text{const}$ on the surface, $\partial p/\partial\alpha = 0$ and (34.26) states that $\partial^2 z/\partial t^2 + ig$ is perpendicular to $\partial z/\partial\alpha$, or that

$$e_{tt} + ig = i\, r(\alpha, t)\, e_\alpha, \tag{34.27}$$

where $r(\alpha, t)$ is a real function. Thus $e(\alpha, t)$ must satisfy the parabolic partial differential equation (34.27).

If $e(\alpha, t)$ is a solution of (34.27) for some $r(\alpha, t)$ which is real for real α and if e and e_t are analytic functions of α and real for real α, then one may construct the velocity potential $F(z, t)$ for a free-boundary flow as follows. Actually, we shall construct F as a function of α and t, i.e. we shall construct a function G related to F by $G(\alpha, t) = F\big(e(\alpha, t), t\big)$. For real α it follows from (34.25) that

$$G_\alpha = F'\frac{\partial z}{\partial \alpha} = \overline{e_t(\alpha, t)}\, e_\alpha(\alpha, t) = \overline{e_t(\bar\alpha, t)}\, e_\alpha(\alpha, t). \tag{34.28}$$

One may now use the last expression in (34.28) to extend analytically G_α, and hence G from real to complex α's. By inverting $z = e(\alpha, t)$, one may now construct $F(z, t)$ [invertibility follows from (2.4) which implies $e_\alpha \bar e_\alpha = 1$].

It is possible to prescribe $\eta(x, t)$ and then construct the associated function $r(\alpha, t)$. For it follows from (34.26) with $y = \eta\big(x(\alpha, t), t\big)$ that

$$\frac{\partial^2 x}{\partial t^2} + \eta_x\left[\eta_x\frac{\partial^2 x}{\partial t^2} + \eta_{xx}\left(\frac{\partial x}{\partial t}\right)^2 + 2\eta_{xt}\frac{\partial x}{\partial t} + \eta_{tt} + g\right] = 0. \tag{34.29}$$

Any set of solutions $x(\alpha, t)$ depending upon a parameter α yields a function $e(\alpha, t)$ defined by

$$e(\alpha, t) = x(\alpha, t) + i\eta\big(x(\alpha, t), t\big). \tag{34.30}$$

The function $r(\alpha, t)$ for real α is given by

$$r(\alpha, t) = \frac{e_{tt} + i g}{i e_\alpha} = -\frac{x_{tt}}{\eta_x x_\alpha}, \tag{34.31}$$

where (34.29) has been used in obtaining the last expression.

We shall suppose now that the motion is steady and make the following special choice of Lagrangian parameter α. Select some fixed point z_0 of the surface $y = \eta(x)$ and for any particle on the surface let $-\alpha$ be the time at which the particle was at z_0. Since the motion is steady, all particles take the same amount of time to travel from z_0 to any given point z and hence

$$e(\alpha, t) = e(0, t + \alpha) \equiv e(t + \alpha). \tag{34.32}$$

It then follows from (34.27) that also

$$r(\alpha, t) = r(\alpha + t). \tag{34.33}$$

Hence (34.22) becomes an ordinary differential equation in a single variable, say $\tau = t + \alpha$:

$$e''(\tau) - i r(\tau)\, e'(\tau) + i g = 0. \tag{34.34}$$

It follows next from (34.28) that $G(\alpha, t) = G(\alpha + t)$ and thus, if $e(\tau)$ is an analytic solution of (34.34), real for real τ,

$$G'(\tau) = \overline{e'(\bar\tau)}\, e'(\tau). \tag{34.35}$$

In this case each choice of a function $r(\tau)$, real for real τ, results in a function $e(\tau)$ as a solution of (34.34), and then in a function $G(\tau)$ obtained by a quadrature of (34.35). One may invert $z = e(\alpha + t)$ and find F as a function of z as in the last paragraph or else regard

$$z = e(\tau), \qquad F = G(\tau) \tag{34.36}$$

as parametric equations with complex parameter τ.

Several examples are considered by JOHN, two of which are time-dependent. A simple and interesting steady flow is obtained by taking $r(\tau) = \nu$, a constant, in (34.34). Then (34.34) and (34.35), after setting the constants of integration equal to zero, yield

$$z = \frac{g}{\nu}\tau + A\,e^{i\nu\tau}, \qquad F = \left(\frac{g^2}{\nu^2} + \nu^2 A^2\right)\tau - 2\frac{g}{\nu} A \cos\nu\tau. \tag{34.38}$$

The free surface, obtained by taking τ real in the first formula, is a trochoidal curve without self-intersections if $A < g/\nu^2$; the wavelength is $\lambda = 2\pi g/\nu^2$ and the amplitude is A. If $A < g/\nu^2$, then $|dF/dz| > 0$ and $A/\lambda < 1/2\pi$. However, dF/dz can become infinite if $dz/d\tau = 0$. Such points occur at

$$z = \left(n + \frac{1}{4}\right)\lambda + i\frac{\lambda}{2\pi}\left(1 - \log\frac{\lambda}{2\pi A}\right). \tag{34.39}$$

In order to avoid having them within the fluid, the bottom must be chosen as a streamline which passes above or through these points. Fig. 53, taken from JOHN's paper, shows several profiles and the associated streamlines through the branch points (34.39) computed for various values of the constant A when

$\lambda=2\pi$ (this is equivalent to graphing $2\pi z/\lambda$ for various values of $2\pi A/\lambda$). The surface profile and bottom come closer together as $A\to 1$ and draw further apart as $A\to 0$. For $A=0.9$ they are so close that they cannot be conveniently separated in the figure; in such cases one may, of course have reservations about the applicability of the perfect-fluid model.

The surface profile in this example is exactly the same as in the Gerstner wave treated in the next section. However, the Gerstner wave is defined for infinite depth and is not irrotational. The flow described above may also be obtained by Sautreaux's method. Vitousek (1954) has studied it by this procedure.

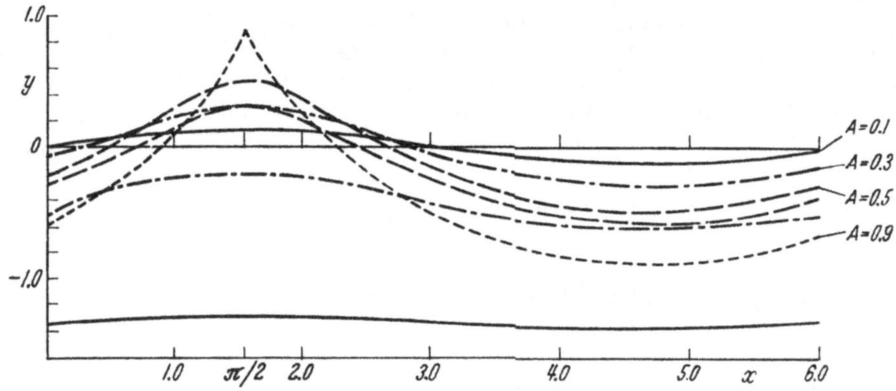

Fig. 53.

Methods of Villat and Poncin. At the end of Sect. 32γ brief mention was made of a pair of integral equations derived by Villat (1915) for the determination of flows over some given bottom profile and with the top profile also given upstream of some point. The method seems to be chiefly useful as an inverse method in which the free surface is given and the other profiles sought. Villat has worked out one case, but not in complete detail, where the top cover is missing and the bottom has a declivity.

Poncin (1932) has further elaborated Villat's method in the direction of starting with the fixed profile and finding the free profile. Actually, he does not really achieve this. Instead, he is able to construct a flow for a fixed profile of the same general behavior as the given one, but not identical with it. The method is applied to a number of interesting special cases, including flow over wavy bottoms and over bottoms with a declivity. The solutions are generally for large values of the velocity. The method and results do not lend themselves to a brief summary.

β) Gerstner's wave. Gerstner's wave (1802) is apparently the first flow to have been discovered which satisfies exactly the condition of constant pressure on the surface, and is, in fact, one of the earliest papers on water-wave theory. It was subsequently rediscovered by Rankine (1863). As will be shown below, the motion is not irrotational. This fact itself would not be enough to rule it out as a mathematical model for real periodic waves. However, the direction of the vorticity is such that it is difficult to conceive of a scheme whereby such a wave could be generated in nature.

The motion is most easily described in Lagrangian coordinates. Each particle is associated with a pair of parameters (a, b), $b\leq 0$. However, (a, b) does not represent a particular position of the particle at some time t_0, but instead a mean position. Hence, instead of (2.3) and (2.4) we need require instead only that the determinant D of those formulas be independent of t. The motion is described by the equations

$$x = a + A\,e^{mb}\sin(m\,a + \sigma\,t)\,, \qquad y = b - A\,e^{mb}\cos(m\,a + \sigma\,t)\,. \tag{34.40}$$

If $b=0$ is taken as the free surface, the motion evidently represents a wave moving to the left with velocity $c=\sigma/m$, while the particles themselves describe in a counter-clockwise direction circular paths about the points (a, b) associated with the particles. The surface $b=0$ is a trochoid and, in fact, each curve $b=\text{const}<0$ is also a trochoid. In order that there shall be no self-intersections, one must have

$$A \leqq \frac{1}{m}\,. \tag{34.41}$$

In order to verify that the motion is kinematically possible, it is necessary to show, as noted above, only that the Jacobian $\partial(x, y)/\partial(a, b)$ is independent of t. An easy computation shows

$$\frac{\partial(x, y)}{\partial(a, b)} = 1 - m^2 A^2 e^{-2mb}\,, \tag{34.42}$$

so that the continuity condition is satisfied. Next one must show that the pressure is constant along the free surface. We shall, in fact, show more, namely, that it is constant along any line $b=\text{const}<0$, provided $\sigma^2=g\,m$. To see this, introduce the Eq. (34.40) into the first of Eqs. (2.7). A straightforward computation yields

$$A(g\,m - \sigma^2)\,e^{mb}\sin(m\,a + \sigma\,t) = -\frac{1}{\varrho}\,\frac{\partial p}{\partial a}\,. \tag{34.43}$$

If the pressure is constant along the surface, then $\partial p/\partial a=0$. This can only hold if

$$\sigma^2 = g\,m\,. \tag{34.44}$$

However, if $\sigma^2=g\,m$, then $\partial p/\partial a=0$ for all b, so that each curve $b=\text{const}$ is an isobaric curve. Although we shall verify this fact directly, it now follows immediately from BURNSIDE's theorem in subsection 32β that the motion cannot be irrotational. A direct computation of the vorticity vector is facilitated by noting that

$$\left.\begin{aligned} u &= \frac{\partial x}{\partial t} = A\,\sigma\,e^{mb}\cos(m\,a + \sigma\,t) = -\,\sigma(y - b)\,,\\ v &= \frac{\partial y}{\partial t} = A\,\sigma\,e^{mb}\sin(m\,a + \sigma\,t) = \sigma(x - a)\,. \end{aligned}\right\} \tag{34.45}$$

Hence

$$\frac{\partial v}{\partial x} - \frac{\partial u}{\partial y} = \sigma\left(2 - \frac{\partial a}{\partial x} - \frac{\partial b}{\partial y}\right). \tag{34.46}$$

The right-hand side of (34.46) may be computed from (34.40) by application of the rules of inversion for partial derivatives. The final result is

$$\frac{\partial v}{\partial x} - \frac{\partial u}{\partial y} = -\frac{2\sigma\,m^2 A^2 e^{2mb}}{1 - m^2 A^2 e^{2mb}}\,; \tag{34.47}$$

the negative sign indicates that the vorticity is directed oppositely to the orbital motion of the particles. The relatively strong vorticity of Gerstner waves when $m\,A$ is not quite small has been pointed out by TRUESDELL (1953), who measures it with a dimensionless vorticity number (see Sect. 27 of SERRIN's article in Vol. VIII/1).

We shall omit a discussion of the construction of the curves $b = \text{const}$, streamlines in a coordinate system moving with the waves; it may be found in Lamb (1932, § 251), Milne-Thomson (1956, § 14.81) and in Kochin, Kibel and Roze (1948, Chap. 8, § 16) together with reproductions of Gerstner's original curves. It is, however, of interest to note that there is, according to (34.41), a "highest" wave of ratio $2A/\lambda = 1/\pi = 0.318$, a figure which may be compared with the value 0.142 for Michell's wave. The highest Gerstner wave has a cusp at the crests, a further indication that the motion cannot be irrotational.

The pressure distribution may be found by substituting (34.40) in the second equation of (2.7), using (34.44), and integrating. The result is

$$p = p_0 - \varrho g b - \tfrac{1}{2}\varrho \sigma^2 A (1 - e^{2mb}). \tag{34.48}$$

A computation of the potential and kinetic energies over a wave length yields

$$T = V = \frac{1}{4}\lambda g \varrho A^2 \left[1 - \frac{2\pi^2}{\lambda^2} R^2\right]. \tag{34.49}$$

Finally, we note that nowhere in the preceding analysis have we made use of homogeneity of the fluid, i.e. the Gerstner wave also represents an exact solution for an arbitrary heterogeneous fluid (with ϱ constant along streamlines). Moreover, Dubreil-Jacotin (1935) has shown that the Gerstner wave is unique in this respect.

Gerstner's wave is defined only for infinite depth. One may ask if a similar wave exists for finite depth. "Similar", in this context, will be taken to mean a periodic wave which satisfies exactly the constant-pressure condition on the free surface and for which the particle orbits are closed. Dubreil-Jacotin (1934) has proved the existence of such a wave and showed that it is unique when the period is fixed. However, this motion cannot be given explicitly except in the case of infinite depth. Methods of approximate computation of the wave have been given by Kravtchenko and Daubert (1957) and Gouyon (1958).

γ) Pseudo-exact solutions. Although the solutions of this section are not really solutions satisfying the exact boundary conditions formulated earlier, they are exact solutions to a closely related problem, also with a nonlinear boundary condition. Their interest derives from the fact that it is possible to encompass within one explicit formula waves of all amplitudes up to a highest wave analogous to Michell's wave. The procedure also allows explicit construction of solitary and cnoidal waves. It is possibly a misnomer to call these solutions pseudo-exact, for one may also interpret them as the first term in a certain series solution of the correctly posed problem. In this sense they are analogous to Havelock's approximation procedure described in subsection 33β. The work to be described has appeared in a series of papers by Davies (1951, 1952), Packham (1952) and Goody and Davies (1957).

The motion will be described in terms of the variables introduced in (32.85), $\omega = \vartheta + i\tau$. The alteration in the boundary condition consists in replacing (32.89) by

$$\frac{\partial \vartheta}{\partial \psi} = l \frac{g}{c^3} e^{-3\tau} \sin 3\vartheta \quad \text{for } \psi = 0, \tag{34.50}$$

where l is some fixed constant chosen so that $l \sin 3\vartheta$ is a good approximation to $\sin \vartheta$. If one wishes to consider (34.50) as the first term in a series approximation to (32.89), one may expand $\sin \vartheta$ in a series in $\sin 3\vartheta$ and express (32.89) as

$$\frac{\partial \vartheta}{\partial \psi} = \frac{g}{c^3} e^{3\tau} \left[\frac{1}{3} \sin 3\vartheta + \frac{4}{81} \sin^3 3\vartheta + \cdots\right]. \tag{34.51}$$

In this case (34.50) with $l=\frac{1}{3}$ represents the approximation obtained by keeping only the first term of (34.51). However, we shall not pursue the approximation procedure and refer to DAVIES (1951) for further information. It will be convenient to reformulate the boundary condition (34.50) as follows:

$$\operatorname{Im}\left\{\frac{d\omega}{df}+l\frac{g}{c^3}\,\mathrm{e}^{i3\omega}\right\}=0 \quad \text{for } \psi=0, \tag{34.52}$$

or, after introducing the new variable $\chi(f)=\mathrm{e}^{-i3\omega}=\omega^3/c^3$, as

$$\operatorname{Im}\left\{\frac{1}{\chi}\left(i\frac{d\chi}{df}+3l\frac{g}{c^3}\right)\right\}=0 \quad \text{for } \psi=0. \tag{34.53}$$

In order to proceed further, we must further specify the nature of the wave motion. Let us suppose the motion to be periodic with wavelength λ and take the fluid to be infinitely deep. We again introduce the ζ-plane of (32.91) and take coordinates as in Fig. 50. The expression in curly brackets in (34.53) is a regular analytic function of ζ inside the unit disc of the ζ-plane with vanishing imaginary part on the boundary, hence is a constant. Since, for $\zeta=0$ (i.e. as $\psi\to-\infty$), $\chi=1$ and $d\chi/df=0$, the constant must be $3lg/c^3$. Thus χ must satisfy the differential equation

$$i\frac{d\chi}{df}-3l\frac{g}{c^3}\chi=-3l\frac{g}{c^3}. \tag{34.54}$$

The solution is easily found to be

$$\chi=1+A\,\mathrm{e}^{-i3lgf/c^3}. \tag{34.55}$$

Referring to Fig. 50, we see that, if $f=0$, $\chi=q_0^3/c^3$, where q_0 is the absolute velocity at a crest. Hence

$$A=\frac{q_0^3}{c^3}-1. \tag{34.56}$$

Since ϑ must also vanish at $\varphi=\pm\frac{1}{2}c\lambda$, i.e. the left-hand side of (34.55) must be real, we must also have $(3lg/c^3)\,\frac{1}{2}c\lambda=\pi$, or

$$c^2=3lg\,\lambda/2\pi. \tag{34.57}$$

Note that if $l=\frac{1}{3}$, the relation between c^2 and λ is the same as in the infinitesimal-wave theory. The solution (34.55) may now be put into the following form:

$$\chi=\frac{w^3}{c^3}=1-\left(1-\frac{q_0^3}{c^3}\right)\mathrm{e}^{-i2\pi f/c\lambda}, \tag{34.58}$$

where $0\leqq q_0\leqq c$. If $q_0=c$, then $w=c$ and the flow is uniform. If $q_0=0$, then

$$w^3=c^3\left[1-\mathrm{e}^{-i2\pi f/c\lambda}\right], \tag{34.59}$$

and near $f=0, \pm c\lambda, \pm 2c\lambda, \ldots$ there is a corner in the wave profile with the two tangents making the same angle 120° as in STOKES' theorem [near $f=0$, (33.5) gives $w^3=i\,\frac{3}{2}gf$, (34.58) gives $w^3=i\,3lgf$]. Hence this wave corresponds to MICHELL's highest periodic wave. The ratio of amplitude to length of this wave may be computed from the following expression for the trough:

$$\frac{1}{2}\lambda-i\,a=\frac{1}{c}\int_0^{\frac{1}{2}c\lambda}\left[1-\mathrm{e}^{-i2\pi\varphi/c\lambda}\right]^{-\frac{1}{3}}d\varphi.$$

By expanding in a series and integrating term by term, one finds

$$\frac{a}{\lambda}=0.127. \tag{34.60}$$

We recall that the value for MICHELL'S wave was 0.142.

If the depth of fluid is finite, one must add the additional boundary condition, $\mathrm{Im}\{\chi\}=0$ for $\psi=-Q$, as well as for $\varphi=0$ and $\pm\frac{1}{2}\lambda c$ if the motion is to be periodic. The determination of χ now becomes too difficult to carry through briefly. However, an explicit solution is still possible and has been worked out by DAVIES (1952) and further investigated by GOODY and DAVIES (1957). Similarly, a "solitary wave" can be explicitly constructed which satisfies the boundary conditions $\mathrm{Im}\{\chi\}=0$ for $\psi=-Q$ and for $\varphi=0$, $0>\psi\geqq-Q$ and $\chi\to1$ as $\varphi\to\pm\infty$. This has been done by PACKHAM (1952). Either of these problems leads to the following differential-difference equation for $\chi(f)$:

$$\frac{1}{\chi(f+iQ)}\left[\chi'(f+iQ)-3l\frac{g}{c^3}i\right]+\frac{1}{\chi(f-iQ)}\left[\chi'(f-iQ)+3l\frac{g}{c^3}i\right]=0; \tag{34.61}$$

it may be established in a manner similar to that used in deriving (22.30) or (32.80).

δ) Pure capillary waves. The first investigation of periodic progressive capillary waves satisfying the exact boundary condition is apparently due to N.A. SLËZKIN (1937). He formulated the boundary-value problem in the same manner as will be done below, reduced it to solution of a nonlinear integral equation analogous to NEKRASOV'S and proved existence and uniqueness of a solution. However, he apparently did not observe that an explicit solution was possible for infinite depth of fluid. This was discovered by CRAPPER (1957), following a different and, in fact, more elementary method.

We shall consider the motion as a steady one in which the fluid moves to the right with velocity c as $y\to-\infty$. The existence of a complex velocity potential $f(z)=\varphi+i\psi$ will be assumed and the free surface $y=\eta(x)$ will be taken to correspond to the streamline $\psi=0$ as usual. It will also be convenient to make use of the variable $\omega=\vartheta+i\tau$ introduced in (32.85). If p_0 is atmospheric pressure, then from BERNOULLI'S integral

$$p+\tfrac{1}{2}\varrho q^2=p_0+\tfrac{1}{2}\varrho c^2 \tag{34.62}$$

(we recall that gravity is being neglected). The dynamical condition at the free surface [see (3.8) and (3.9)] is

$$p-p_0=\frac{T}{R}=T\frac{\eta''}{[1+\eta'^2]^{\frac{3}{2}}}. \tag{34.63}$$

Before combining (34.62) and (34.63), we recall that the curvature of a streamline at any of its points is given by $d\vartheta/ds$ where s is arc length along the streamline. Hence, we may combine (34.62) and (34.63) to obtain the following boundary condition

$$\frac{1}{2}\varrho(c^2-q^2)=T\frac{d\vartheta}{ds}=T\frac{\partial\vartheta}{\partial\varphi}\frac{d\varphi}{ds}=Tq\frac{\partial\vartheta}{\partial\varphi}\quad\text{for}\quad\psi=0, \tag{34.64}$$

or

$$\frac{\varrho c}{2T}\left(\frac{c}{q}-\frac{q}{c}\right)=\frac{\partial\vartheta}{\partial\varphi}\quad\text{for}\quad\psi=0. \tag{34.65}$$

Since $q = c e^{\tau}$ and since $\partial\vartheta/\partial\varphi = \partial\tau/\partial\psi$ from the Cauchy-Riemann equations, (34.65) may be written

$$\frac{\partial\tau}{\partial\psi} = \frac{\varrho c}{2T}\left(e^{-\tau} - e^{\tau}\right) = -\frac{\varrho c}{T}\sinh\tau \quad \text{for} \quad \psi = 0. \tag{34.66}$$

The problem is now to find a function $\omega(f)$ analytic for $\psi \leqq 0$, such that $\omega \to 0$ as $\psi \to -\infty$ and such that the imaginary part τ satisfies (34.66). However, since the boundary condition (34.66) involves only τ, unlike its analogue (32.89) for pure gravity waves, it is possible to solve first for the harmonic function $\tau(\varphi, \psi)$ and then to find ϑ later.

We assume that a solution can be found which satisfies

$$\frac{\partial\tau}{\partial\psi} = -h(\psi)\sinh\tau, \qquad h(0) = \frac{\varrho c}{T}, \tag{34.67}$$

and proceed to verify the assumption. Integrating (34.67), we obtain

$$\log\tanh\tfrac{1}{2}\tau = -H(\psi) + G(\varphi), \tag{34.68}$$

where $H'(\psi) = h(\psi)$ and $G(\varphi)$ is an arbitrary function, or

$$\tau = \log\frac{e^{H} + e^{G}}{e^{H} - e^{G}} = \log\frac{X(\psi) + Y(\varphi)}{X(\psi) - Y(\varphi)}. \tag{34.69}$$

Since τ is a harmonic function of φ and ψ, LAPLACE's equation must be satisfied by (34.69). This yields an equation to be satisfied by X and Y in which the two variables can be separated. We shall not repeat the detailed analysis, which is typical of that occurring in separation-of-variables problems. The final result is that X and Y must satisfy

$$\left.\begin{aligned} X'^2 &= a_1 + a_2 X^2 + a_3 X^4, \\ Y'^2 &= -a_1 - a_2 Y^2 - a_3 Y^4, \end{aligned}\right\} \tag{34.70}$$

where a_1, a_2, a_3 are arbitrary constants. CRAPPER states that the full equations may be used to construct a solution for fluid of finite depth, but that it is sufficient to set $a_3 = 0$ for infinite depth (in view of SLËZKIN's result, this is presumably also necessary). Since τ is real, we shall also take X and Y to be real. If one does set $a_3 = 0$ and assumes $a_1 < 0$, $a_2 > 0$, the following give real solutions of (34.70):

$$X(\psi) = \sqrt{\frac{-a_1}{a_2}}\cosh\left(\sqrt{a_2}\,\psi + E\right), \quad Y(\varphi) = \sqrt{\frac{-a_1}{a_2}}\cos\left(\sqrt{a_2}\,\varphi + F\right), \tag{34.71}$$

where E and F are real constants. A glance at (34.69) shows that τ is independent of the choice of a_1. It will be convenient to let $a_2 = m^2/c^2$, where $m > 0$. One may determine E from (34.67), for

$$\frac{\varrho c}{T} = H'(0) = \frac{d}{d\psi}\log X\big|_{\psi=0} = \frac{m}{c}\tanh E$$

or

$$e^{2E} = \frac{m T/\varrho c^2 + 1}{m T/\varrho c^2 - 1} = B^{-2}. \tag{34.72}$$

Since E is to be real, we must evidently have

$$\frac{m T}{\varrho c^2} \geqq 1. \tag{34.73}$$

Since F adds only a real constant to φ we may select it at our convenience; we take $F=0$. Substitution of (34.71) into (34.69) gives

$$\left.\begin{aligned}\tau &= \log\frac{\cosh(m\psi/c+E)+\cos(m\varphi/c)}{\cosh(m\psi/c+E)-\cos(m\varphi/c)}\\ &= \log\frac{\cos(i\,m\psi/c+i\,E)+\cos(m\varphi/c)}{\cos(i\,m\psi/c+i\,E)-\cos(m\varphi/c)}\\ &= \log\left[\cot\frac{1}{2}(m\,f/c+i\,E)\cot\frac{1}{2}(m\,\bar{f}/c-i\,E)\right].\end{aligned}\right\} \tag{34.74}$$

The analytic function ω, which has τ as imaginary part and which approaches zero as $\psi\to-\infty$, is given by

$$\omega = i\log\left[-\cot^2\tfrac{1}{2}(m\,f/c+i\,E)\right]. \tag{34.75}$$

We then have

$$\frac{df}{dz} = c\,e^{-i\omega} = -c\cot^2\frac{1}{2}(m\,f/c+i\,E) = c\coth^2\frac{1}{2}(i\,m\,f/c-E). \tag{34.76}$$

From this one may solve for z in terms of f:

$$\left.\begin{aligned}c\,z &= f-\frac{2c}{m}\tan\frac{1}{2}\left(\frac{m\,f}{c}+i\,E\right)+\text{const}\\ &= f-i\frac{4c}{m}\frac{1}{1+e^{(i\,m\,f/c-E)}}+\text{const}\\ &= f-i\frac{4c}{m}\frac{1}{1+B\,e^{i\,m\,f/c}}+i\frac{4c}{m},\end{aligned}\right\} \tag{34.77}$$

where the constant has been chosen so as to make cz reduce to f when $B=0$. It is evident that

$$z\left(f+\frac{2\pi c}{m}\right) = z(f)+\frac{2\pi}{m},$$

so that the streamlines are periodic in the x-direction with wavelength $\lambda=2\pi/m$.

The surface streamline is obtained by setting $\psi=0$. After separation of real and imaginary parts in (34.77) the equation for the surface becomes:

$$\left.\begin{aligned}x &= \frac{\varphi}{c}-\frac{4}{m}\,\frac{B\sin m\varphi/c}{1+B^2+2B\cos m\varphi/c},\\ y &= \frac{4}{m}-\frac{4}{m}\,\frac{1+B\cos m\varphi/c}{1+B^2+2B\cos m\varphi/c},\end{aligned}\right\} \tag{34.78}$$

with φ serving as a parameter. There is a crest when $\varphi=0$ and a trough when $\varphi=\pi c/m$. The difference in the values of y yields the following expression for the ratio of total amplitude to wavelength:

$$\frac{a}{\lambda} = \frac{4}{\pi}\,\frac{B}{1-B^2}. \tag{34.79}$$

Eq. (34.72) then provides a relation between A/λ and $m\,T/\varrho c^2$, which we may write, for example, as

$$\left.\begin{aligned}c &= \sqrt{\frac{T\,m}{\varrho}}\left(1+\frac{1}{16}a^2m^2\right)^{-\frac{1}{4}}\\ &= \sqrt{\frac{T\,m}{\varrho}}\left(1-\frac{1}{64}a^2m^2+\cdots\right).\end{aligned}\right\} \tag{34.80}$$

If this formula is compared with (27.29), it should be kept in mind that a is here the total amplitude and that in (27.29) A is a length associated with the half amplitude. The formulas are consistent.

As a/λ increases, the surface profile becomes steeper and steeper near the troughs until the two sides finally touch. This occurs for $a/\lambda = 0.730$. A wave of these proportions may be considered as a "highest" capillary wave, an analogue of MICHELL'S wave, although the nature of the limitation is different. Fig. 54, reproduced from CRAPPER'S paper, shows the profile of this wave together with other streamlines. It is a consequence of the form of the dependence in (34.77) that the other streamlines in Fig. 54 may also serve as surface profiles for different values of a/λ, i.e. for different values of B. It is not surprising, of course, that the profiles are similar to the middle three of Fig. 35.

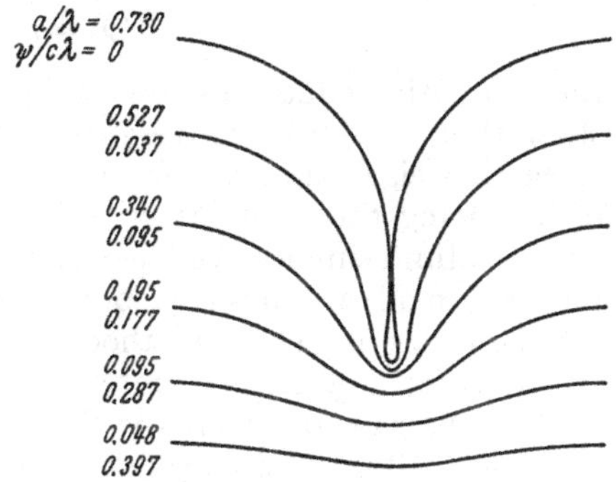

Fig. 54.

35. Existence theorems. In the various applications of the approximate theories of Chapt. D and E it is tacitly assumed that there is an exact solution which is being approximated. Without knowledge of conditions for existence and uniqueness of a solution to a particular problem, the status of an approximate solution is somewhat anomalous and one must rely upon comparison with experimental results for conviction concerning the correctness of the solution. However, such comparison is not a satisfactory criterion, for in the original formulation of a problem one will usually have already made a decision about the mathematical model of a fluid which will be used. Thus, if one has assumed a perfect fluid (as we usually have) and then made a further mathematical approximation in solving the problem at hand, the validity of this approximate solution must first be established before comparison of the predicted results with experimental measurements can be used as a criterion of the applicability of the fluid model. Without this additional knowledge, the comparison of approximate solutions with experimental results must be considered in some sense to be second best, even though valuable evidence may be provided by good agreement in a wide variety of situations.

Unfortunately, existence and uniqueness proofs in exact water-wave theory have generally been difficult to establish, and have usually been obtained for only rather restricted, although physically important, situations. Many of them are very recent and some rely upon methods of functional or topological analysis which cannot be briefly expounded. Although some proofs are so constructed that approximation methods are inherent in them, others are only able to assert the existence of a solution with no indication of how to obtain it approximately. Proofs are still lacking for many relatively simple but important problems, for example, MICHELL'S highest wave and standing water waves.

No attempt will be made to give an exposition of the mathematical methods which have been used in establishing the various existing theorems. Instead only a discursive account will be given of the nature and limitations of the known theorems.

α) Irrotational waves—infinite depth. Proof of the existence of periodic waves of permanent type in infinitely deep water was first given by NEKRASOV (1921,

1922) in a journal of very restricted distribution. Shortly thereafter LEVI-CIVITA (1925) gave another proof along quite independent lines. Further proofs were given by NEUMANN (1929) and by LICHTENSTEIN (1931), these being more closely related to NEKRASOV'S. A new treatment of LEVI-CIVITA'S proof, due to LITTMAN and NIRENBERG, is contained in STOKER'S *Water waves* (1957, § 12.2). Also, NEKRASOV (1951) has recently published his researches in a more accessible form.

NEKRASOV'S method requires proving that there exists a solution $\vartheta(\gamma)$ to his nonlinear integral equation (32.104). His procedure, in brief, is to assume an expansion of $\vartheta(\gamma)$ in powers of the parameter $\mu'=\mu-3>0$,

$$\vartheta(\gamma)=\mu'\vartheta_1+\mu'^2\vartheta_2+\cdots, \tag{35.1}$$

then to derive equations relating each ϑ_k to ones of lower index, and finally to show that the series converges. $\mu=3$ is chosen as a starting point because it is the first eigenvalue of the "linearized" equation (32.104), i.e. the one obtained by replacing the quotient containing $\sin\vartheta$ by simply $\mu\vartheta(\beta)$. This corresponds to the infinitesimal-wave theory. Proof of convergence requires that μ' be sufficiently small, and positive, but no estimate of radius of convergence is obtained. On the other hand, the method does allow computation of explicit approximate formulas for quantities of interest.

LEVI-CIVITA also works with the variable ω, treating it as a function of the variable ζ introduced in (32.90). Hence his formulation of the problem is essentially the same as NEKRASOV'S, i.e. he is seeking a function $\omega(\zeta)$, regular in the disc $|\zeta|<1$, vanishing at $\zeta=0$ and satisfying (32.97) on $|\zeta|=1$ and some further condition assuring that $|w/c-1|<\beta<1$. His procedure for finding such a function is to expand both ω and $k\equiv 1-g\lambda/2\pi c^2$ in a power series in a parameter $\mu>0$:

$$\omega=\sum_{n=1}^{\infty}\omega_n(\zeta)\,\mu^n,\qquad k=\sum_{n=1}^{\infty}k_n\mu^n, \tag{35.2}$$

where the functions $\omega_n(\zeta)$ and the constants k_n are to be determined by the boundary conditions. The first terms, $\omega_1=-i\zeta$, $k_1=0$, correspond to infinitesimal waves, so that the parameter μ is essentially the amplitude/wavelength of this approximation. LEVI-CIVITA establishes the convergence of the series (35.2) for sufficiently small values of μ. No estimate of a radius of convergence is given, but HUNT (1953) has stated that an examination and refinement of LEVI-CIVITA'S inequalities show that convergence is established for amplitude-wavelength ratios up to $\frac{1}{98}$. The procedure lends itself to explicit computation of higher-order computations, and, in fact, he carries them out through $n=5$. LEVI-CIVITA further derives the interesting theorem that irrotational waves of permanent type must be symmetric about vertical lines through crest and trough. NEKRASOV assumed this at the outset.

NEUMANN and LICHTENSTEIN (the latter's approach is simpler) derive a coupled pair of nonlinear integral equations and put them into a form such that SCHMIDT'S theory of nonlinear integral equations is applicable. Iterative methods of solution can be used to obtain approximate formulas.

β) Irrotational waves—horizontal bottom. When the fluid is infinitely deep and the motion periodic, the only independent dimensionless parameter besides the amplitude-wavelength ratio is $c^2/g\lambda$. When the fluid is bounded below by a horizontal plane at mean depth h, then a new parameter, say c^2/gh, must enter into any solution. However, other independent sets of parameters may be used, and, in particular, different choices of a perturbation parameter have led earlier to different approximate solutions for finite depth. Thus, in Sects. 14β and 27 one finds the first and higher approximations for periodic waves of permanent type when A/λ is taken as a perturbation parameter, whereas in Sect. 31 one finds approximations to two further types of waves of permanent type, one of them periodic, corresponding to a different choice of parameter and a different method of approximation. In each of these cases there arises the question as to whether there exist waves of permanent type satisfying the exact boundary conditions for which these waves may be considered approximations. In each case the answer is affirmative.

Waves of small amplitude. The first proof of the existence of periodic progressive waves in fluid of finite depth is due to STRUIK (1926). His method of analysis is similar to LEVI-CIVITA's for infinite depth. Existence of the desired wave is established for each value of $c^2/gh<1$ and for each sufficiently small value of A/λ, where the bound on A/λ depends upon c^2/gh. HUNT (1953) has recently corrected some errors in the proof which did not invalidate it but which resulted in incorrect approximate formulas.

NEKRASOV (1928, 1951) was also able to show that his integral equation for ϑ, as modified for finite depth [see (32.104) and (32.106)], had a solution, thus providing an independent proof. As was the case for infinite depth, NEKRASOV assumes that the waves are symmetric about verticals through trough and crest; STRUIK proves this. KRASNOSELSKII (1956) has recently applied topological methods of analysis to NEKRASOV's equation and established not only existence of solutions for μ in the neighborhood of the eigenvalues of the linearized equation, but also their uniqueness and continuous dependence upon μ.

Solitary and cnoidal waves. LAVRENT'EV (1943, 1947) was the first one to establish the existence of cnoidal and solitary waves. Cnoidal waves are not mentioned by him by name, but, in fact, their existence for sufficiently large wavelength is established along with that of the solitary wave, the latter being obtained as a limiting case. The detailed exposition of the results (1947) is unfortunately both difficult of access and difficult to read, and relies upon earlier theorems of the author. Although the perturbation parameter appears at first glance to be taken as $\varepsilon^2=-1+gh^3/Q^2$, which for the solitary wave would be in contradiction with (32.52), the quantity h is not really mean depth but a related quantity which varies with the wavelength of the approximating periodic wave. FRIEDRICHS and HYERS (1954) by a completely different procedure have established the existence of the solitary wave. Their perturbation parameter is essentially $\varepsilon^2=1-gh^3/Q^2$ [actually it is $a^2=-\frac{1}{3}\log(gh^3/Q^2)$]. The point of departure is again the boundary condition (32.89) for the function ω. However, an integral equation is formulated, then altered by a change of variable $\hat{\varphi}=a\varphi$, $\hat{\psi}=\psi$. The different rates of stretching correspond to those of subsection 10β. (Something like this also occurs in LAVRENT'EV's proof, but is disguised in his theorems on conformal mapping of narrow strip-like regions.) An iterative procedure is used to prove existence of a solution for sufficiently small values of ε^2.

LITTMAN (1957) has used a method somewhat similar to that of FRIEDRICHS and HYERS to establish the existence of cnoidal waves satisfying the exact boundary conditions. However, as a parameter he has used essentially h/λ, where h is the mean depth and λ the wavelength. It is demonstrated that solutions exist for values of c^2/gh which are both greater and less than 1. Fig. 55, modified slightly from one in LITTMAN's paper, shows in a qualitative fashion the relation between the dimensionless parameters. The dotted lines enclose values of the parameters, again in a purely qualitative way, for which solutions have been demonstrated to exist. (Here k is the modulus of the elliptic functions

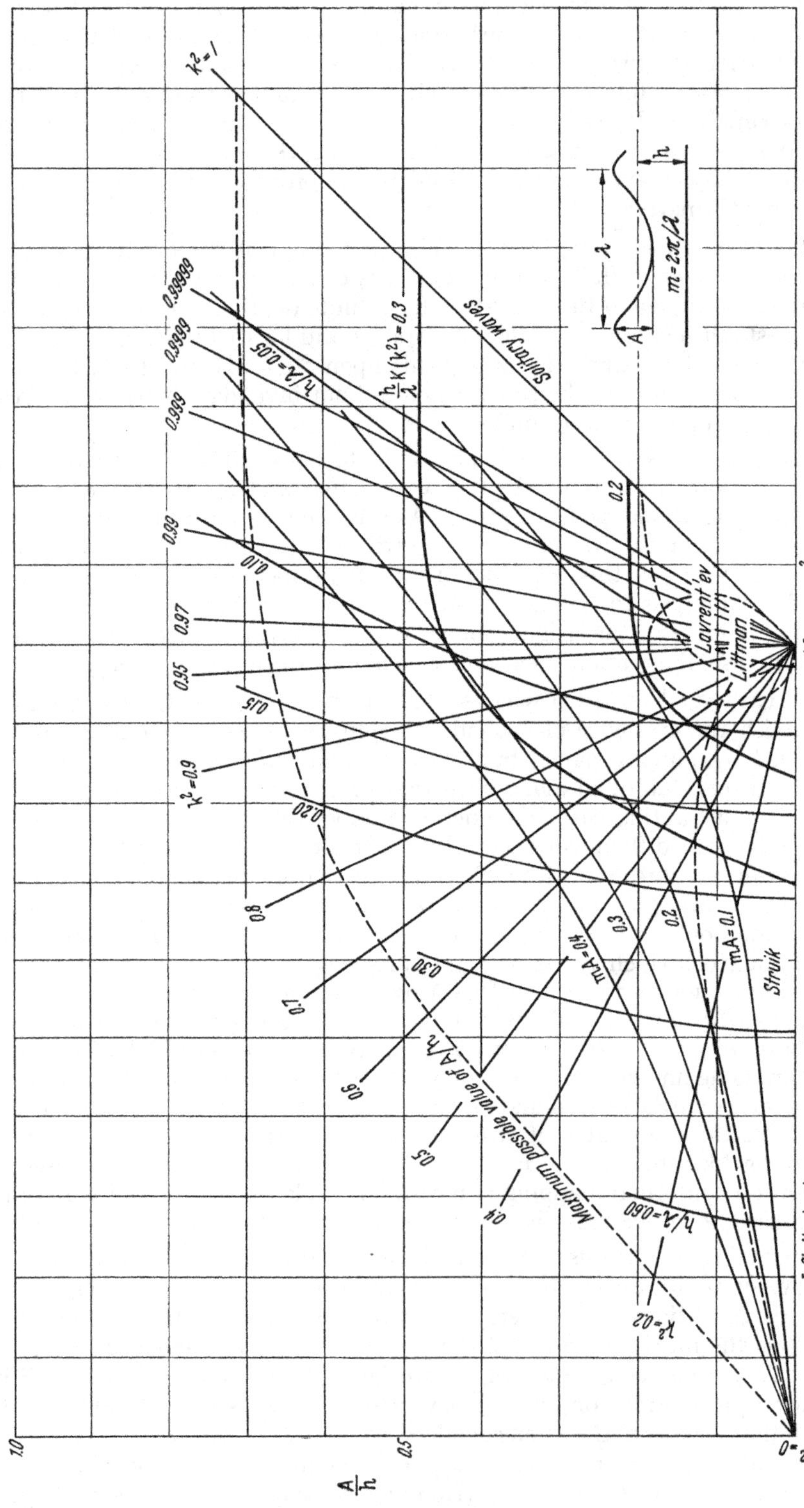
Solitary waves
Infinitesimal waves
Maximum possible value of A/h
Lavrent'ev
Littman
Struik
m=2π/λ

Fig. 55.

and K is the complete elliptic integral of the first kind. k serves as a parameter in certain approximate formulas.)

Fig. 55 was prepared by first computing the curves shown by means of both the cnoidal-wave theory and the theory of higher-order infinitesimal waves as developed in subsection 27α. [SKJELBREIA'S tables (1959) facilitated the computation for the latter method.] Curves were then faired by eye in such a way as to pass smoothly from one set to the other. Hence, although they are claimed to be only qualitative, they have in fact a quantitative basis. An additional error has been introduced by taking the curves $k^2 = \text{const}$ as straight lines; they should, in fact, show some curvature as the radial distance from $c^2/gh = 1$, $A/h = 0$ increases. This additional complication of the computation did not seem necessary for the purpose at hand.

Although it is not strictly relevant to the material of the present section, it seems of interest to display the two sets of curves mentioned above, for they indicate in a rough way the ranges of validity of the two fundamental methods of approximation and show how they fit together. They are shown in Fig. 56. One expects the curves based on the infinitesimal-wave theory to be accurate near the horizontal axis, $A/h = 0$, those based on cnoidal-wave theory to be accurate near $c^2/gh = 1$, and the two to agree where these two regions overlap. The curves confirm this expected behavior. Computations based on the second-order cnoidal-wave theory of Eqs. (31.37) may be expected to produce better agreement over a wider range.

γ) Irrotational waves—other configurations. Flow over a wavy bottom. In connection with the study of inverse methods in subsection 34α an explicit example of a steady flow over a wave-shaped bottom was exhibited. However, there the surface profile was given and the bottom profile calculated. The direct problem, in which the bottom profile and other flow data are given, has also been considered by several persons. LAVRENT'EV (1943) announced theorems concerning this problem, but did not include them in his later (1947) exposition. GERBER (1955) has given a comprehensive treatment of the "supercritical" case and has announced further results for the "subcritical" case (1956). Let the bottom profile S be periodic and symmetric about vertical lines through the maxima and minima; let $\vartheta(s)$ be its intrinsic equation where s is arc length measured from a maximum and ϑ is the angle between the tangent and the x-direction. Let Q be the discharge rate for the fluid, and let q_0 be the velocity at a crest. In the first paper he considers flows in which the slope of the surface has the same sign as that of the bottom (we recall the two possible flows occurring in the linearized theory of subsection 20α). GERBER shows that there exists at least one solution of this type provided the following inequalities are satisfied in the interval between a maximum and the first minimum to the right:

$$\left.\begin{aligned} &\frac{gQ}{q_0^3} + \max|\vartheta| \leqq \pi - \varepsilon_1, \\ &-\frac{1}{2}\pi + \varepsilon_2 \leqq \vartheta(s) \leqq 0, \end{aligned}\right\} \tag{35.3}$$

where ε_1 and ε_2 are arbitrary small but positive quantities. If certain other inequalities, further limiting gQ/q_0^3, are satisfied, he is also able to prove uniqueness provided $\vartheta(s) \not\equiv 0$. In the second paper he announces that there exists at least one solution such that the profile has slope of opposite sign to that of the bottom if

$$\frac{gQ}{q_0^3} > (1+\varepsilon)\frac{\pi^3}{2} \tag{35.4}$$

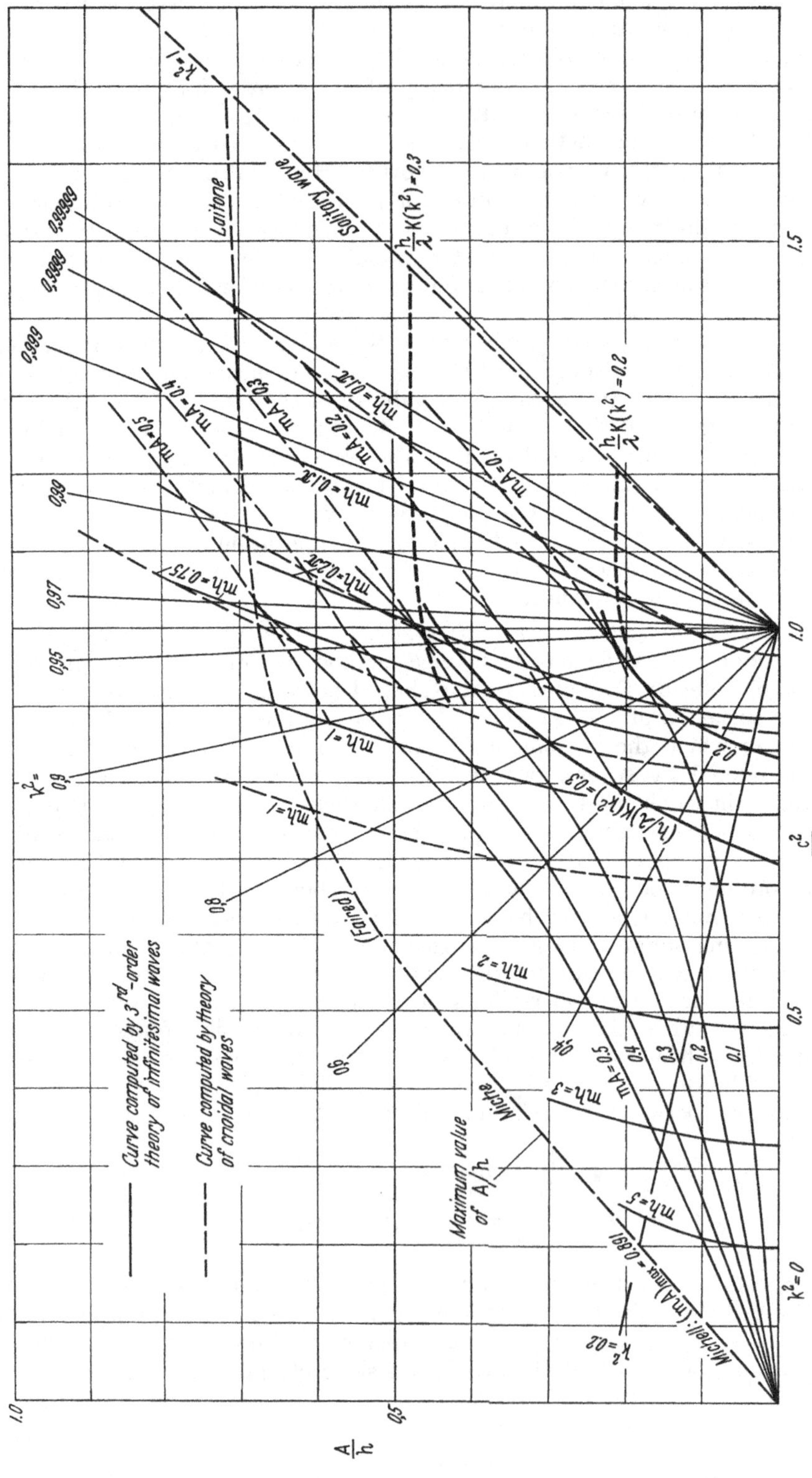
Curve computed by 3rd-order theory of infinitesimal waves
Curve computed by theory of cnoidal waves
Solitary wave
Laitone
(Faired)
Michie
Maximum value of A/h
Michell: (mA)max = 0.881
A/h
c²/gh
k² = 0

Fig. 56.

and $Q/L_0 q_0$ and $q_0 \Delta/Q$ are small enough; here L_0 is the arclength from a maximum to a minimum of S and Δ is the vertical distance. GERBER's methods are topological (Schauder-Leray theory) and do not yield effective methods of approximation.

MOISEEV (1957) has also considered this problem. By a modification of the method used to derive NEKRASOV's integral equation (32.104), he derives a pair of nonlinear integral equations to which the Lyapunov-Schmidt method is applicable. Let c be the average velocity defined by (7.5) for an allowable value of y (thus φ increases by $c\lambda$ over a wavelength), and let Q be the discharge rate. Then MOISEEV finds that there exists a sequence of velocities $c_1 > c_2 > \cdots > 0$ associated with the eigenvalues of a certain linear operator, such that, if $c \neq c_n$, there exists a unique flow provided the slope of the bottom is sufficiently small. Also, if $c > c_1$ or $c_{2n+1} < c < c_{2n}$ then the solution is such that the slopes of bottom and surface are of the same sign; if $c_{2n} < c < c_{2n+1}$, the slopes are of opposite sign.

Flow over a bottom with a declivity. Let the flow be from left to right and suppose the bottom profile to be asymptotic to horizontal lines as $x \to \pm\infty$, the one on the right being lower than that on the left. The discharge rate Q and velocity c at $x = -\infty$ should then be sufficient to determine the flow. The existence of a steady flow under these circumstances has been investigated by HAIMOVICI (1935) and GERBER (1955). The former derives a pair of nonlinear integral equations, similar to NEKRASOV's, relating ϑ and τ of (32.86). An iterative method is used to prove the existence of a solution. GERBER makes use again of the Schauder-Leray theory. The theorems established by each are very similar, but GERBER's is sharper. Let the bottom be given intrinsically by $\vartheta(s)$, measured from some fixed point. Then a solution exists if

$$\left.\begin{aligned} &\frac{g\,Q}{c^3} < 1, \qquad \max|\vartheta(s)| + \frac{g\,Q}{c^3} < 1, \\ &|\vartheta(s)| \leqq A\,\mathrm{e}^{-\alpha|s|}, \qquad A, \alpha > 0. \end{aligned}\right\} \tag{35.5}$$

The last condition assures a rapid approach to the horizontal asymptotes. The case of subcritical flow does not appear to have been treated in the published literature.

Motion past a submerged vortex. TER-KRIKOROV (1958) has recently investigated steady flow past a submerged vortex of intensity Γ in a channel of depth h when the exact boundary conditions on the free surface are retained. If c is the velocity far upstream of the vortex, he proves existence and uniqueness of the flow provided that $c^2/gh > 1$ and Γ/ch is sufficiently small.

Interfacial waves. In subsection 14δ we considered the linearized theory of waves at an interface between two perfect fluids of different densities, bounded above and below by horizontal planes. The question naturally arises as to whether one can establish the existence of such waves when the exact boundary conditions at the interface are observed. KOCHIN (1927) extended the methods of LEVI-CIVITA and STRUIK to this problem and established the existence of (necessarily symmetric) interfacial waves of finite amplitude.

δ) *Rotational waves.* The explicit construction in subsection 34β of a periodic wave of permanent type which is rotational and the demonstrated existence of irrotational waves of this type which are of finite, if small, amplitude raises the question as to whether each of these waves is a special case of a more general type. This question has been discussed in a notable paper by DUBREIL-JACOTIN (1934) with results which include and generalize those of LEVI-CIVITA and STRUIK. We give a only a bare indication of the results.

Let us suppose that a coordinate system has been chosen so that we may treat the wave motion as a steady flow to the right. Although we do not assume the motion to be irrotational, there will still exist a stream function $\psi(x, y)$ by virtue of the continuity equation. The vorticity of the flow will be given by $-\Delta\psi$, and since by a classical theorem the vorticity is constant along a streamline, the following equation must be satisfied by ψ:

$$\Delta\psi = f(\psi), \tag{35.6}$$

where $f(\psi)$ is some unspecified function. The condition on the free surface $\psi = 0$ may still be derived from the special Bernoulli theorem [see Eq. (2.10'')]

$$g\,\eta(x) + \tfrac{1}{2}\left[\psi_x^2 + \psi_y^2\right] = \text{const}. \tag{35.7}$$

For irrotational waves the function $f \equiv 0$; for GERSTNER'S wave it is given by (34.47) after setting $b = \psi$, $\sigma = cm$. The question which DUBREIL-JACOTIN asked is whether a wave of finite amplitude exists for any distribution of vorticity $f(\psi)$. In order to encompass both of the known finite waves into her results, she limits f to functions of the following sort:

$$f(\psi) = -\mu\frac{Q}{h}\,m^2\,e^{2mh\psi/Q}\,F\!\left(e^{mh\psi/Q}\right), \qquad -Q \leqq \psi \leqq 0, \tag{35.8}$$

where Q is discharge rate, h the mean depth, and the function $F(\varrho)$ is bounded and satisfies a Hölder condition in ϱ; μ is a small parameter. If the depth is infinite, one must replace Q/h by c, the velocity at $\psi = -\infty$ (it is assumed that $\psi_y \to c$ as $\psi \to -\infty$).

DUBREIL-JACOTIN'S theorem is as follows. For any $m = 2\pi/\lambda$, h and $f(\psi)$ satisfying (35.8) there exists a $\delta > 0$ such that for $\mu < \delta$ there exists a unique corresponding progressive wave of permanent type with vorticity distribution $f(\psi)$. The waves are also shown to be symmetric about vertical lines through crest or trough. She also demonstrates that among this class of waves for finite depth there is a unique analogue of the Gerstner wave, in the sense that the trajectories of individual particles are all closed. This wave has recently been investigated by KRAVTCHENKO and DAUBERT (1957). The development of means of calculating rotational waves has been the subject of a recent investigation by GOUYON (1958).

ε) Waves in heterogeneous fluids—internal waves. It has been shown in subsection 32β that irrotational waves of permanent type are not possible in a heterogeneous fluid, but that GERSTNER'S rotational wave still provides a solution for infinite depth. DUBREIL-JACOTIN (1935) has shown that this is the only periodic wave of permanent type in infinitely deep fluid having this property. In a later paper (1937) she returned to this topic and made use of the methods developed by her for rotational waves to investigate the existence theory for the two problems described below. The first problem, a natural generalization of one investigated by KOCHIN and mentioned at the end of subsection 35γ, is the existence of periodic internal waves of permanent type in a heterogeneous fluid bounded both above and below by horizontal planes. In the second problem the upper surface is free.

The two problems may be formulated as follows. First we recall that in a steady flow of a heterogeneous fluid the density must be constant along streamlines. Hence, if ψ is the stream function, we may write $\varrho = \varrho(\psi)$. The equation analogous to (35.6) is now somewhat more complicated. It may be derived from (32.54) as follows. Apply the operators

$$\frac{\varrho'}{\varrho}\psi_y - \frac{\partial}{\partial y}, \qquad \frac{\varrho'}{\varrho}\psi_x - \frac{\partial}{\partial x}$$

to the two equations of (32.54), respectively, and subtract. This yields

$$\frac{\varrho'}{\varrho}\frac{\partial(E,\psi)}{\partial(x,y)} - \frac{\partial(\zeta,\psi)}{\partial(x,y)} = 0.$$

Since $\varrho = \varrho(\psi)$,

$$\frac{\varrho'}{\varrho}\frac{\partial(E,\psi)}{\partial(x,y)} = \frac{\partial(\varrho' E/\varrho,\psi)}{\partial(x,y)}$$

and hence

$$\frac{\partial(\varrho' E/\varrho - \zeta,\psi)}{\partial(x,y)} = 0$$

or, integrating and substituting $\zeta = -\Delta\psi$,

$$\Delta\psi + \frac{\varrho'}{\varrho}\left[g\,y + \frac{1}{2}(\psi_x^2 + \psi_y^2)\right] = f_1(\psi) \tag{35.9}$$

where $f_1(\psi)$ is an arbitrary function. This is the equation which ψ must satisfy. If $\psi = 0$ is the top streamline and $\psi = -Q$ the bottom streamline, then the boundary conditions are,

$$\psi = 0 \quad \text{for} \quad y = 0, \qquad \psi = -Q \quad \text{for} \quad y = -h \tag{35.10}$$

for the first problem, and

$$\left.\begin{aligned} \psi_x^2 + \psi_y^2 + 2g\,y = \text{const} \quad &\text{for} \quad \psi = 0,\\ \psi = -Q \quad &\text{for} \quad y = -h \end{aligned}\right\} \tag{35.11}$$

for the second problem [cf. (32.60)]. The function $\varrho(\psi)$ cannot be considered as an arbitrary given function in the same sense that $f_1(\psi)$ is arbitrary; it must be related to the density distribution when the fluid is at rest. DUBREIL-JACOTIN assumes that $\varrho(\psi)$ is the same as the density at the mean level of the streamline ψ when the fluid is at rest.

In order to obtain results analogous to those of subsection 35 δ, certain restrictions are placed upon the function $f_1(\psi)$ and the density distribution. Both problems are then reducible to integro-differential equations. In general there is no nontrivial solution. However, under certain conditions there are an infinite number of values of the parameter $\lambda g/2\pi c^2$ in the neighborhood of which there exist nontrivial symmetric waves of finite (but small) amplitude.

ζ) Waves with surface tension. It has already been mentioned in subsection 34 δ that SLËZKIN (1935 b, 1937) had derived an integral equation for the motion of pure capillary waves and had proved both existence and uniqueness of solution under certain circumstances. The explicit solution for this problem derived by CRAPPER supersedes in a sense these earlier results.

SEKERZH-ZENKOVICH (1956) has formulated the exact boundary-value problem for combined gravity and capillary waves in terms of the function ω of (32.86) and announced that a proof of existence for sufficiently small amplitude-to-wavelength ratio can be carried out by LEVI-CIVITA's method for pure gravity waves.

G. Bibliography.

The following bibliography is not by any means a complete one for the subject. For the most part it consists of papers to which reference is made in the text. In a few cases a paper has been included because its subject has been neglected in the text, although this can hardly be considered a remedy. In general, the bibliography is weakest in its coverage of papers describing experimental work; many more such exist than are listed here.

The forms of reference are more or less standard and will not be explained. Titles of papers in Russian have been translated into English; these are indicated by an (R) following the title. Otherwise titles are given in the original language. Published translations into another language are occasionally listed after the original. However, no attempt has been made to search for them. Russian names have been transliterated by a system natural to English, i.e. the cyrillic letters which might be transliterated in a phonetic alphabet as "č, š, ž" are here "ch, sh, zh", respectively. However, when Russians write in a language which uses a roman alphabet, they customarily transliterate their names so that the pronunciation is approximately correct according to the rules of the language in which they are writing. Such variants have been included in brackets behind the transliteration used here.

AIMOND, F.: Recherches d'hydrodynamique en vue de la détermination du mouvement de l'eau sur un barrage-déversoir. Ann. Fac. Sci. Univ. Toulouse (3) **21**, 139—221 (1929).

AIRY, G.B.: Tides and waves. Encyclopaedia Metropolitana, Vol. 5, pp. 241*—396* (+6 plates in Vol. 26). London 1845.

ALBLAS, J.B.: On the generation of water waves by a vibrating strip. Appl. Sci. Res. A **7** 224—236 (1958).

APTÉ, ACHYUT S.: Recherches théoriques et expérimentales sur les mouvements des liquides pesants avec surface libre. Publ. Sci. Tech. du Ministère de l'Air, No. 333, Paris 1957, ii + ii + 115 pp.

AUERBACH, F.: Wellenbewegung der Flüssigkeiten. In Handbuch der physikalischen und technischen Mechanik, Bd. V, S. 300—365. Leipzig: Johann Ambrosius Barth 1931.

AUSMAN, J.S.: Pressure limitation on the upper surface of a hydrofoil. Diss. Univ. of California, Berkeley, 1953, 36 pp.

BAKHMETEFF, B.A.: Hydraulics of open channels. 329 pp. New York: McGraw-Hill 1932.

BARAKAT, RICHARD: Diffraction of water waves by a fixed sphere in the long-wave approximation. Unpublished manuscript 1958.

— Linearized solution of the vertical motion of a floating sphere in a regular seaway. Unpublished manuscript 1958.

BARTELS, R.C.F., and A.C. DOWNING jr.: On surface waves generated by travelling disturbances with circular symmetry. Proc. 2nd U.S. Nat. Congr. Appl. Mech., Ann Arbor, 1954, pp. 607—615. Amer. Soc. Mech. Engrs., New York 1955.

BARTHOLOMEUSZ, E.F.: The reflection of long waves at a step. Proc. Cambridge Phil. Soc. **54**, 106—118 (1958).

BASSET, A.B.: A treatise on hydrodynamics, Vol. II. xvi + 328 pp. Cambridge: Deighton, Bell & Co. 1888.

BECKER, E.: Die pulsierende Quelle unter der freien Oberfläche eines Stromes endlicher Tiefe. Ing.-Arch. **24**, 69—76 (1956).

— Das Wellenbild einer unter der Oberfläche eines Stromes schwerer Flüssigkeit pulsierender Quelle. Z. angew. Math. Mech. **38**, 391—399 (1958).

BELLMAN, R., and R.H. PENNINGTON: Effects of surface tension and viscosity on Taylor instability. Quart. Appl. Math. **12**, 151—162 (1954).

BELYAKOVA, V.K.: Oscillation of a flat plate under a free surface, taking account of small quantities of the second order. (R) Prikl. Mat. Mekh. **15**, 504—510 (1951).

BENJAMIN, T.B.: On the flow in channels when rigid obstacles are placed in the stream. J. Fluid Mech. **1**, 227—248 (1956).

—, and M.J. LIGHTHILL: On cnoidal waves and bores. Proc. Roy. Soc. Lond., Ser. A **224**, 448—460 (1954).

—, and F. URSELL: The stability of the plane free surface of a liquid in vertical periodic motion. Proc. Roy. Soc., Lond., Ser. A **225**, 505—515 (1954).

BENSON, J.M., and N.S. LAND: An investigation of hydrofoils in the NACA tank. I. Effect of dihedral and depth of submersion. NACA Wartime Report L-**758** (1942), 12 pp. + 21 figs. (1946).

BESSHO, MASATOSHI: On the wave resistance theory of a submerged body. The Society of Naval Architects of Japan, 60th anniversary series, Vol. 2, pp. 135—172. Tokyo 1957.

Biesel, F.: Calcul de l'amortissement d'une houle dans un liquide visqueux de profondeur finie. Houille bl. **4**, 630—634 (1949).
— Remarques sur la célérité de la houle irrotationnelle exacte au troisième ordre. Houille bl. **6**, 414—416 (1951).
— Equations générales au second ordre de la houle irrégulière. Houille bl. **7**, 372—376 (1952) (1 insert).
—, et B. LeMéhauté: Etude théorique de la réflexion de la houle sur certains obstacles. Houille bl. **10** 130—138 (1955).
—, et F. Suquet: Les appareils générateurs de houle en laboratoire. Houille bl. **6**, 147—165, 475—496, 723—737 (1951); **7**, 779—801 (1952).
— — Laboratory wave-generating apparatus. Houille bl. **9**, (1954), Suppl. to No. 5, 27 pp.
Binnie, A.M.: Waves in an open oscillating tank. Engineering **151**, 224—226 (1941).
Birkhoff, Garrett: Note on Taylor instability. Quart. Appl. Math. **12**, 306—309 (1954).
—, and Jack Kotik: Theory of the wave resistance of ships. II. The calculation of Michell's integral. Trans. Soc. Naval Arch. Marine Engrs. **62**, 372—385 (1954); discussion, 385 to 396.
— — Some transformations of Michell's integral. Publ. Nat. Tech. Univ. Athens, No. **10**, 26 pp. (1954).
Blasius, H.: Funktionentheoretische Methoden in der Hydrodynamik. Z. Math. Phys. **58**, 90—110 (1910).
— Mitteilung zu meiner Abhandlung über: Funktionentheoretische Methoden in der Hydrodynamik. Z. Math. Phys. **59**, 43—44 (1911).
Boĭko, L.A.: Diffraction of waves on the surface of a heavy incompressible fluid. (R) Uchenye Zapiski Moskov. Gos. Univ. **24**, 34—60 (1938).
Bolotin, V.V.: On the motion of a fluid in an oscillating container. (R) Prikl. Mat. Mekh. **20**, 293—294 (1956).
Bondi, H.: Waves on the surface of a compressible liquid. Proc. Cambridge Phil. Soc. **43**, 75—95 (1947).
Borgman, L.E., and J.E. Chappelear: The use of the Stokes-Struik approximation for waves of finite height. Proc. 6th Conf. Coastal Engrg., Florida, 1957, pp. 252—280. Council on Wave Research, Univ. of Calif., Richmond, Calif., 1958.
Borodin, V.A., and Yu. F. Dityakin: Unstable capillary waves on the surface of separation of two viscous fluids. (R) Prikl. Mat. Mekh. **13**, 267—276 (1949). Translated in NACA Tech. Memo. **1281**, 9 pp. (1951).
Bouasse, H.: Houle, rides, seiches et marées. xxii + 516 pp. Paris: Librairie Delagrave 1924.
Bouligand, G.: Sur les équations des petits mouvements de surface des fluides parfaits. Bull. Soc. Math. France **40**, 149—180 (1912).
— Sur les singularités à la paroi dans le problème des ondes liquides. Bull. Sci. Math. (2) **50**, 89—96, 106—112 (1926).
— Sur la continuité et les approximations en dynamique des liquides. J. École polytech. (2) **26**, 1—38 (1927).
Bourgin, D.G.: Interference calculations and wave groups. Phil. Mag. (7) **21**, 1033—1056 (1936).
Boussinesq, J.: Théorie de l'intumescence liquide appelée onde solitaire ou de translation, se propageant dans un canal rectangulaire. C. R. Acad. Sci., Paris **72**, 755—759 (1871).
— Théorie des ondes et des remous qui se propagent le long d'un canal rectangulaire horizontal, en communiquant au liquide contenu dans ce canal des vitesses sensiblement pareilles de la surface au fond. J. Math. pures appl. (2) **17**, 55—108 (1872).
— Théorie des ondes liquides périodiques. Mém. présentés par divers Savants à l'Acad. Sci. Inst. France **20**, 509—615 (1872) (1 plate).
— Essai sur la théorie des eaux courantes. Mém. présentés par divers Savants à l'Acad. Sci. Inst. France **23**, 380—398 (1877).
— Sur une importante simplification de la théorie des ondes que produisent à la surface d'un liquide l'émersion d'un solide ou l'impulsion d'un coup de vent. Ann. Sci. Ecole norm. Sup. (3) **27**, 9—42 (1910).
Bowden, K.F.: The effect of eddy viscosity on ocean waves. Phil. Mag. (7) **41**, 907—917 (1950).
Brard, R.: Introduction à l'étude théorique du tangage en marche. Bull. Assoc. Tech. Mar. Aéro. **47**, 455—471 (1948), discussion, 472—479.
— Vagues engendrées par une source pulsatoire en mouvement horizontal rectiligne uniforme. Application au tangage en marche. C. R. Acad. Sci., Paris **226**, 2124—2125 (1948).
Breslin, J.P.: Application of ship-wave theory to the hydrofoil of finite span. J. Ship Res. **1**, No. 1, 27—35, 55 (1957).
Brillouët, G.: Etude de quelques problèmes sur les ondes liquides de gravité. Publ. Sci. Tech. du Ministère de l'Air, No. 329, Paris, 1957, ii + ii + v + 145 pp.

BROER, L. J. F.: On the propagation of energy in linear conservative waves. Appl. Sci. Res. A 2, 329—344 (1951).

BROWN, R.C.: The ripple method of measuring surface tension. Proc. Phys. Soc. Lond. **48**, 312—322 (1936).

— A method of measuring the amplitude and damping of ripples. Proc. Phys. Soc. Lond. **48**, 323—328 (1936).

BURNSIDE, W.: On the small wave-motions of a heterogeneous fluid under gravity. Proc. Lond. Math. Soc. (1) **20**, 392—397 (1889).

— On the modification of a train of waves as it advances into shallow water. Proc. Lond. Math. Soc. (2) **14**, 131—133 (1915).

— On the steady motion of fluid under gravity. Messenger of Math. **45**, 43—46 (1915/16).

— On periodic irrotational waves at the surface of deep water. Proc. Lond. Math. Soc. (2) **15**, 26—30 (1916).

CAMPBELL, I. J.: Wave motion in an annular tank. Phil. Mag. (7) **44**, 845—853 (1953) (1 plate).

CARRY, C.: Clapotis partiel. Houille bl. **8**, 482—494 (1953).

— Calcul de l'amortissement d'une houle dans un liquide visqueux en profondeur finie. Houille bl. **11**, 75—79 (1956).

—, et G. CHABERT D'HIÈRES: Compléments sur le calcul approché au troisième ordre d'un clapotis parfait monochromatique. C. R. Acad. Sci., Paris **245**, 1377—1379 (1957).

CASE, K.M., and W.C. PARKINSON: Damping of surface waves in an incompressible liquid. J. Fluid Mech. **2**, 172—184 (1957).

CAUCHY, A.-L.: Théorie de la propagation des ondes à la surface d'un fluide pesant d'une profondeur indéfinie. Mém. présentés par divers Savants à l'Acad. Roy. Sci. Inst. France **1**, 3—123 (1827); notes, 124—316.

CHABERT D'HIÈRES, G.: Sur les équations approchées du clapotis parfait monochromatique. C. R. Acad. Sci., Paris **244**, 2474—2476 (1957).

— Calcul approché du troisième ordre d'un clapotis parfait monochromatique. C. R. Acad. Sci., Paris **244**, 2573—2575 (1957).

Sur l'existence d'un potentiel des vitesses pour un clapotis parfait. C. R. Acad. Sci., Paris **246**, 1803—1806 (1958).

CHANDRASEKHAR, S.: The character of the equilibrium of an incompressible heavy viscous fluid of variable density. Proc. Cambridge Phil. Soc. **51**, 162—178 (1955).

CHAPLYGIN, YU. S. [TCHAPLYGUINE, J. S.]: Le glissement d'une lame plane d'une envergure infinie sur la surface d'un fluide pesant. C. R. (Doklady) Acad. Sci. URSS. (N.S.) **26**, 746—750 (1940).

— — Gliding of a flat plate of infinite span on the surface of a heavy fluid. (R) Trudy Tsentral. Aèro-Gidrodinam. Inst., vyp. **508**, 3—45 (1940).

— — Gliding on a fluid of finite depth. (Comparison of the nonlinear and linear theory.) (R) Prikl. Mat. Mekh. **5**, 223—252 (1941).

CHRYSTAL, G.: Some further results in the mathematical theory of seiches. Proc. Roy. Soc. Edinburgh **27**, 637—647 (1905).

— On the hydrodynamical theory of seiches, with bibliographical sketch. Trans. Roy. Soc. Edinburgh **41**, 599—649 (1906).

CISOTTI, U.: Sulle onde simplice di tipo permanente e rotazionale. Ist. Lombardo Sci. Lett. Rend. (2) **46**, 917—925 (1913) = Nuovo Cim. (6) **7**, 251—259 (1914).

— Nuovi tipi di onde periodiche permanenti e rotazionali. Atti Accad. Lincei, Rend. Cl. Sci. Fis. Mat. Nat. (5) **23**, 2° sem., 556—561 (1915); **24**, 1° sem., 129—133 (1915).

— Equazione caratteristica dei piccoli moti ondosi in un canale di qualunque profondità. I, II. Atti Accad. Lincei, Rend. Cl. Sci. Fis. Mat. Nat. (5) **27**, 2° sem., 255—259, 312—316 (1918).

— Sul moto variabile nei canali a fondo orizzontale. Atti Accad. Lincei, Rend. Cl. Sci. Fis. Mat. Nat. (5) **28**, 1° sem., 196—199 (1919).

— Sull'integrazione dell'equazione caratteristica dei piccoli moti ondosi in un canale di qualunque profondità. I. II. Atti Accad. Lincei, Rend. Cl. Sci. Fis. Mat. Nat. (5) **20**, 1° sem., 131—133, 175—180 (1920).

COOMBS, A.: The translation of two bodies under the free surface of a heavy fluid. Proc. Cambridge Phil. Soc. **46**, 453—468 (1950).

COOPER, R. I. B., and M. S. LONGUET-HIGGINS: An experimental study of the pressure variations in standing water waves. Proc. Roy. Soc. Lond., Ser. A **206**, 424—435 (1951).

CORNISH, VAUGHAN: Waves of the sea and other sea waves. 374 pp. London: T. Fisher Unwin 1910.

— Ocean waves and kindred geophysical phenomena. With additional notes by HAROLD JEFFREYS. xv + 164 pp. Cambridge 1934.

COURANT, R., and K.O. FRIEDRICHS: Supersonic flow and shock waves. xvi + 464 pp. New York: Interscience Publishers 1948.

Crapper, G.D.: An exact solution for progressive capillary waves of arbitrary amplitude. J. Fluid Mech. **2**, 532—540 (1957).

Crudeli, U.: Sulle onde progressive, di tipo permanente, oscillatorie (seconda approssimazione) I. II. Atti Accad. Lincei, Rend. Cl. Sci. Fis. Mat. Nat. (5) **28**, 2° sem., 174—178 (1919); **29**, 2° sem., 265—269 (1920).

Cumberbatch, E.: Two-dimensional planing at high Froude number. J. Fluid Mech. **4**, 466—478 (1958).

Cummins, W.E.: Hydrodynamic forces and moments acting on a slender body of revolution moving under a regular train of waves. The David W. Taylor Model Basin, Washington, D.C., Rep. **910**, vi + 33 pp. (1954).

Daily, J.W., and S.C. Stephan jr.: The solitary wave. Proc. 3rd Conf. Coastal Engrg., Cambridge, Mass., 1952, pp. 13—30.

Daubert, A.: Sur les équations approchées des ondes permanentes et périodiques de gravité. C. R. Acad. Sci., Paris **244**, 2472—2474 (1957).

— Calcul approché au troisième ordre d'une houle de gravité. C. R. Acad. Sci., Paris **244**, 2575—2577 (1957).

— Sur une méthode de calcul approché au troisième ordre d'une houle complexe. C. R. Acad. Sci., Paris **245**, 1878—1880 (1957).

— Calcul approché au troisième ordre d'une houle complexe. C. R. Acad. Sci., Paris **245**, 2006—2009 (1957).

— Sur la propagation d'une houle complexe. C. R. Acad. Sci., Paris **246**, 888—890 (1958).

Davies, T.V.: The theory of symmetrical gravity waves of finite amplitude. I. Proc. Roy. Soc. Lond. Ser. A **208**, 475—486 (1951).

— Gravity waves of finite amplitude. III. Steady, symmetrical, periodic waves in a channel of finite depth. Quart. Appl. Math. **10**, 57—67 (1952).

De, S.C.: Contributions to the theory of Stokes waves. Proc. Cambridge Phil. Soc. **51**, 713—736 (1955).

Dean, W.R.: On the reflection of surface waves by a submerged plane barrier. Proc. Cambridge Phil. Soc. **41**, 231—238 (1945).

— On some cases of the reflexion of surface waves by an inclined plane barrier. Proc. Cambridge Phil. Soc. **42**, 24—28 (1946).

— Note on waves on the surface of running water. Proc. Cambridge Phil. Soc. **43**, 96—99 (1947).

— On the reflection of surface waves by a submerged circular cylinder. Proc. Cambridge Phil. Soc. **44**, 483—491 (1948).

Defant, A.: Flutwellen und Gezeiten des Wassers. In Handbuch der Physik, Bd. XLVIII, pp. 846—927. Berlin-Göttingen-Heidelberg: Springer 1957.

De Prima, C.R., and T.Y. Wu: On the theory of surface waves in water generated by moving disturbances. Engineering Division, California Institute of Technology, Pasadena, Calif., Rep. No. **21—23**, 40 pp. (1957).

Dmitriev, A.A.: Waves on the surface of a viscous fluid which are generated by a pulsating source. (R) Izv. Akad. Nauk SSSR., Ser. Geofiz. **1953**, 335—345.

—, and T.V. Bonchkovskaya: On turbulence in a wave. (R) Dokl. Akad. Nauk SSSR. (N.S.) **91**, 31—33 (1953) (1 plate).

Dobroklonskiĭ, S.V.: Extinction of capillary-gravity waves on the surface of clean water. (R) Trudy Morsk. Gidrofiz. Inst. **6**, 43—57 (1955).

—, and V.A. Tyumeneva: On extinction of capillary-gravity waves on the surface of water covered by films of certain surface-active substances in its dependence upon frequency. (R) Izv. Akad. Nauk SSSR., Ser. Geograf. Geofiz. **14**, 425—439 (1950).

Donn, W.L., and M. Ewing: Stokes' edge waves in Lake Michigan. Science **124**, 1238—1242 (1956).

Dorrestein, R.: General linearized theory of the effect of surface films on water ripples. I, II. Proc. Kon. Nederl. Akad. Wetensch., Ser. B **54**, 260—272, 350—356 (1951).

Dubreil-Jacotin, M.-L.: Sur les ondes de type permanent dans les liquides hétérogènes. Atti Accad. Lincei, Rend. Cl. Sci. Fis. Mat. Nat. (6) **15**, 814—819 (1932).

— Sur la détermination rigoureuse des ondes permanentes périodiques d'ampleur finie. J. Math. pures appl. (9) **13**, 217—291 (1934).

— Complément à une Note antérieure sur les ondes de type permanent dans les liquides hétérogènes. Atti Accad. Lincei, Rend. Cl. Sci. Fis. Mat. Nat. (6) **21**, 344—346 (1935).

— Sur la discussion des équations de ramification relatives à certains problèmes d'ondes. Application aux ondes dues aux inégalités du fond. Bull. Soc. Math. France **64**, 1—24 (1936).

— Sur les théorèmes d'existence relatifs aux ondes permanentes périodiques à deux dimensions dans les liquides hétérogènes. J. Math. pures appl. (9) **16**, 43—67 (1937).

ECKART, CARL: The approximate solution of one-dimensional wave equations. Rev. Mod. Phys. **20**, 399—417 (1948).
— The ray-particle analogy. J. Marine Res. **9**, 139—144 (1950).
— Surface waves on water of variable depth. Lecture notes, Fall semester, 1950/51. Wave Report No. 100; SIO Ref. 51—12. La Jolla, Calif., August, 1951, vi + 99 pp. (mimeographed) + 8 plates.
EGGERS, KLAUS: Über das Wellenbild einer pulsierenden Störung in Translation. Schiff u. Hafen **9**, 874—878 (1957).
EKMAN, V.W.: On dead water. The Norwegian North Polar Expedition, 1893—1896. Scientific Results, Vol. 5, No. 15, viii + 152 pp. + 17 plates. Christiania 1906.
— On stationary waves in running water. Ark. Mat. Astronom. Fys. **3**, No. 2, 30 pp. (1906).
ÈPSHTEĬN, L.A. [EPSTEIN]: Quelques nouvelles données expérimentales sur le phénomène du glissement. C. R. (Doklady) Acad. Sci. URSS. (N.S.) **26**, 742—745 (1940).
— Some new experimental material on gliding of flat plates. (R) Trudy Tsentral. Aèro-Gidrodinam. Inst., vyp. **508**, 46—75 (1940).
FAVRE, HENRY: Etude théorique et expérimentale des ondes de translations dans les canaux découverts. 215 pp. + 7 plates + 4 folded pages. Paris: Dunod 1935.
FENCHEL, W.: Sulle onde di canale di tipo permanente. Atti Accad. Lincei, Rend. Cl. Sci. Fis. Mat. Nat. (6) **13**, 740—743 (1931).
FERRARI, MARIA: Flusso di energia e velocità di gruppo. Atti Accad. Lincei, Rend. Cl. Sci. Fis. Mat. Nat. (5) **22**, 1° sem., 761—766 (1913).
FINKELSTEIN, A.B.: The initial value problem for transient water waves. Comm. Pure Appl. Math. **10**, 511—522 (1957).
FRIEDRICHS, K.O.: Water waves on a shallow sloping beach. Comm. Appl. Math. **1**, 109—134 (1948).
— On the derivation of the shallow water theory. Comm. Appl. Math. **1**, 81—85 (1948).
—, and D. H. HYERS: The existence of solitary waves. Comm. Pure Appl. Math. **7**, 517—550 (1954).
—, and HANS LEWY: The dock problem. Comm. Appl. Math. **1**, 135—148 (1948).
FROUDE, W.: On the rolling of ships. Trans. Inst. Naval Arch. **2**, 180—227 (1861) (plates XVIII, XIX); discussion, 227—229; Appendices, **3**, 45—62 (1862) (plates II, III) = The papers of WILLIAM FROUDE, The Institution of Naval Architects, London, 1955, pp. 40—65, 65—76.
FUCHS, R.A.: On the theory of short-crested oscillatory waves. Gravity Waves, pp. 187—200. National Bureau of Standards Circular 521. Washington, D. C.: U.S. Government Printing Office 1952.
GAZARYAN, YU. L.: On surface waves in the ocean caused by submarine earthquakes. (R) Akust. Zh. **1**, 203—217 (1955).
GEPPERT, H.: Die permanenten Wellen in ringförmigen Kanälen. Math. Ann. **101**, 424—445 (1929).
GERBER, ROBERT: Sur une condition de prolongement analytique des fonctions harmoniques. C. R. Acad. Sci., Paris **233**, 1560—1562 (1951).
— Sur les solutions exactes des équations du mouvement avec surface libre d'un liquide pesant. J. Math. pures appl. (9) **34**, 185—299 (1955).
— Sur une classe de solutions des équations du mouvement avec surface libre d'un liquide pesant. C. R. Acad. Sci., Paris **242**, 1260—1262 (1956).
GERSTNER, FRANZ J. v.: Theorie der Wellen sammt einer abgeleiteten Theorie der Deichprofile. Abh. böhm. Ges. Wiss. (3) **1** (1804), Abt. 1, Stück 1 = Ann. Physik **32**, 412—445 (1809) (Tafel III).
— Théorie des vagues, suivie d'un essai sur la théorie des profils des diques. (Traduit et annoté par M. DE SAINT-VENANT.) Ann. Ponts et Chaussées (6) **13**, 31—86 (1887) (1 plate).
GOODY, A. J., and T.V. DAVIES: The theory of symmetrical gravity waves of finite amplitude. IV. Steady, symmetrical, periodic waves in a channel of finite depth. Quart. J. Mech. Appl. Math. **10**, 1—12 (1957).
GOTUSSO, GUIDO: Una proprietà delle onde sulla superficie di un liquido. Boll. Un. Mat. Ital. (3) **8**, 36—40 (1953).
GOUYON, RÉNÉ: Contribution à la théorie des houles. Thèse, Université de Toulouse, 1958, ii + 73 pp.
GREEN, A.E.: The gliding of a plate on a stream of finite depth. I, II. Proc. Cambridge Phil. Soc. **31**, 589—603 (1935); **32**, 67—85 (1936).
— GEORGE (1793—1841): On the motion of waves in a variable canal of small depth and width. Mathematical papers, pp. 223—230. London: Macmillan 1871 = Trans. Cambridge Phil. Soc. **6**, 457—462 (1838).
— Note on the motion of waves in canals. Mathematical papers, pp. 271—280. London: Macmillan 1871 = Trans. Cambridge Phil. Soc. **7**, 87—95 (1842).

GREEN, GEORGE: Waves in a dispersive medium, resulting from a limited initial disturbance. Proc. Roy. Soc. Edinburgh **30**, 242—253 (1909).
— On group velocity and the propagation of waves in dispersive media. Proc. Roy. Soc. Edinburgh **29**, 445—470 (1909).
— Waves in deep water due to a concentrated surface pressure. Phil. Mag. (7) **39**, 738—743 (1948).
GREENE, T.R., and A.E. HEINS: Water waves over a channel of infinite depth. Quart. Appl. Math. **11**, 201—214 (1953).
GREENHILL, A.G.: Wave motion in hydrodynamics. Amer. J. Math. **9**, 62—112 (1887).
— Skating on thin ice. Phil. Mag. (6) **31**, 1—22 (1916).
GREENSPAN, H.P.: The generation of edge waves by moving pressure distributions. J. Fluid Mech. **1**, 574—592 (1956).
GRIGORASH, Z.K.: Survey of works concerning the problem of tsunami waves. (R) Trudy Morsk. Gidrofiz. Inst. **10**, 73—81 (1957).
— Waves on the surface of a fluid caused by a small sudden under-water impulse. (R) Trudy Morsk. Gidrofiz. Inst. **11**, 18—27 (1957).
GRIM, OTTO: Berechnung der durch Schwingungen eines Schiffskörpers erzeugten hydrodynamischen Kräfte. Jb. schiffbautech. Ges. **47**, 277—296 (1953); Erörterung, 296—299.
GRÖBNER, W.: Oberflächenwellen von Flüssigkeiten. Ann. Scuola Norm. Super. Pisa (3) **5**, 175—191 (1951).
GROEN, P.: Two fundamental theorems on gravity waves in inhomogeneous incompressible fluids. Physica, Haag **14**, 294—300 (1948).
— Contribution to the theory of internal waves. Kon. Nederl. Meteorol. Inst. Mededel. Verh. Ser. B Deel. II, No. 11, 23pp. (1948).
— Zwaartegolven in en op de zee. Nederl. Tijdschr. Natuurk. **16**, 250—262 (1950).
GUILLOTON, R.: Potential theory of wave resistance of ships with tables for its calculation. Trans. Soc. Naval Arch. Marine Engrs. **59**, 86—123 (1951); discussion, 123—128.
GUTHRIE, F.: On stationary liquid waves. Phil. Mag. (4) **50**, 290—302, 377—388 (1875).
GWYTHER, R.F.: The classes of progressive long waves. Phil. Mag. (5) **50**, 213—216 (1900).
— An appendix to the paper on the classes of progressive long waves. Phil. Mag. (5) **50**, 308—312 (1900).
— The general motion of long waves, with an examination of the direct reflexion of the solitary wave. Phil. Mag. (5) **50**, 349—352 (1900).
HADAMARD, J.: Sur les ondes liquides. C. R. Acad. Sci., Paris **150**, 609—611, 772—774 (1910).
— Sur les ondes liquides. Atti Accad. Lincei, Rend. Cl. Sci. Fis. Mat. Nat. (5) **25**, 1° sem., 716—719 (1916).
HAIMOVICI, M.: Sur l'écoulement des liquides pesants dans un plan vertical. Ann. Sci. Univ. Jassy **21**, 182—231 (1935).
HANAOKA, TATSURO: On the velocity potential in MICHELL's system and the configuration of the wave-ridges due to a moving ship. J. Zôsen Kyôkai **93**, 1—10 (1953).
— Theoretical investigation concerning ship motion in regular waves. Proceedings, Symposium on the behaviour of ships in a seaway, Wageningen, 1957, pp. 266—285.
HANSON, E.T.: The theory of ship waves. Proc. Roy. Soc. Lond., Ser. A **111**, 491—529 (1926).
HARRISON, W.J.: The influence of viscosity on the oscillations of superposed fluids. Proc. Lond. Math. Soc. (2) **6**, 396—405 (1908).
— The influence of viscosity and capillarity on waves of finite amplitude. Proc. Lond. Math. Soc. (2) **7**, 107—121 (1909).
HASKIND, M.D. (KHASKIND): The plane problem of oscillation of a plate on the surface of a heavy fluid. (R) Izv. Akad. Nauk SSSR., Otd. Tekhn. Nauk **1942**, No. 7—8, 75—94.
— The plane problem of steady oscillation of a wing under the surface of a heavy fluid of finite depth. (R) Izv. Akad. Nauk SSSR., Otd. Tekhn. Nauk **1942**, No. 11—12, 66—86.
— The plane problem of gliding on the surface of a fluid of finite depth. (R) Izv. Akad. Nauk SSSR., Otd. Tekhn. Nauk **1943**, No. 1—2, 67—90.
— Oscillations of a system of plates on the surface of a heavy liquid. (R) Prikl. Mat. Mekh. **7**, 421—430 (1943).
— The oscillation of a body immersed in heavy fluid. (R) Prikl. Mat. Mekh. **8**, 287—300 (1944).
— Translation of bodies under the free surface of a heavy fluid of finite depth. (R) Prikl. Mat. Mekh. **9**, 67—78 (1945). Translated in NACA Tech. Memo. **1345**, 20 pp. (1952).
— Wave resistance of a solid in motion through a fluid of finite depth. (R) Prikl. Mat. Mekh. **9**, 257—264 (1945).
— The hydrodynamic theory of ship oscillations in rolling and pitching. (R) Prikl. Mat. Mekh. **10**, 33—66 (1946). Translated in Soc. Naval Arch. Marine Engrs., Tech. & Res. Bull. No. **1—12**, 3—43 (1953).

Haskind, M.D. (Khaskind): The oscillation of a ship in still water. (R) Izv. Akad. Nauk SSSR. Otd. Tekhn. Nauk **1946**, 23—34. Translated in Soc. Naval Arch. Marine Engrs., Tech. & Res. Bull. No. **1—12**, 45—60 (1953).
— Waves arising from oscillation of bodies in shallow water. (R) Prikl. Mat. Mekh. **10**, 475 to 480 (1946).
— The pressure of waves on a barrier. Inzhen. Sb. **4**, No. 2, 147—160 (1948).
— Oscillations of a floating contour on the surface of a heavy liquid. (R) Prikl. Mat. Mekh. **17**, 165—178 (1953).
— The diffraction of waves about a moving cylindrical vessel. (R) Prikl. Mat. Mekh. **17**, 431—442 (1953).
— On wave motions of a heavy fluid. (R) Prikl. Mat. Mekh. **18**, 15—26 (1954).
— Unsteady gliding on an undulating surface of a heavy fluid. (R) Prikl. Mat. Mekh. **19**, 331—342 (1955).
— Three-dimensional flow about slender bodies. (R) Prikl. Mat. Mekh. **20**, 203—210 (1956).
— Theory of the resistance of ships moving in waves. (R) Izv. Akad. Nauk SSSR., Otd. Tekhn. Nauk, Mekh. Mashinostr. **1959**, No. 2, 46—56.
— Radiation and diffraction of surface waves by a flat plate floating vertically. (R) Prikl. Mat. Mekh. **23**, 546—556 (1959).
Havelock, T.H.: The propagation of groups of waves in dispersive media, with application to waves on water produced by a travelling disturbance. Proc. Roy. Soc. Lond., Ser. A **81**, 398—430 (1908).
— The wave making resistance of ships: a theoretical and practical analysis. Proc. Roy. Soc. Lond., Ser. A **82**, 276—300 (1909).
— The wave-making resistance of ships: a study of certain series of model experiments. Proc. Roy. Soc. Lond., Ser. A **84**, 197—208 (1910).
— On the instantaneous propagation of disturbance in a dispersive medium. Phil. Mag. (6) **19**, 160—168 (1910).
— The propagation of disturbances in dispersive media. Cambridge Tracts in Mathematics and Mathematical Physics, No. 17. viii + 87 pp. Cambridge: Cambridge University Press 1914.
— Ship resistance: the wave making properties of certain travelling pressure disturbances. Proc. Roy. Soc. Lond., Ser. A **89**, 489—499 (1914).
— The initial wave resistance of a moving surface pressure. Proc. Roy. Soc. Lond., Ser. A **93**, 240—253 (1916).
— Some cases of wave motion due to a submerged obstacle. Proc. Roy. Soc. Lond., Ser. A **93**, 520—532 (1917).
— Periodic irrotational waves of finite height. Proc. Roy. Soc. Lond., Ser. A **95**, 38—51 (1919).
— Wave resistance: some cases of three-dimensional fluid motion. Proc. Roy. Soc. Lond., Ser. A **95**, 354—365 (1919).
— The effect of shallow water on wave resistance. Proc. Roy. Soc. Lond., Ser. A **100**, 499 to 505 (1922).
— Studies in wave resistance: influence of the form of the water-plane section of the ship. Proc. Roy. Soc. Lond., Ser. A **103**, 571—585 (1923).
— Studies in wave resistance: the effect of parallel middle body. Proc. Roy. Soc. Lond., Ser. A **108**, 77—92 (1925).
— Wave resistance: the effect of varying draught. Proc. Roy. Soc. Lond., Ser. A **108**, 582 to 591 (1925).
— Wave resistance: some cases of unsymmetrical forms. Proc. Roy. Soc. Lond., Ser. A **110**, 233—241 (1926).
— Some aspects of the theory of ship waves and wave resistance. Trans. North-East Coast Inst. Engrs. and Shipbuilders **42**, 71—83 (1926).
— The method of images in some problems of surface waves. Proc. Roy. Soc. Lond., Ser. A **115**, 268—280 (1927).
— Wave resistance. Proc. Roy. Soc. Lond., Ser. A **118**, 24—33 (1928).
— The wave pattern of a doublet in a stream. Proc. Roy. Soc. Lond., Ser. A **121**, 515—523 (1928).
— The vertical force on a cylinder submerged in a uniform stream. Proc. Roy. Soc. Lond., Ser. A **122**, 387—393 (1929).
— Forced surface-waves on water. Phil. Mag. (7) **8**, 569—576 (1929).
— The wave resistance of a spheroid. Proc. Roy. Soc. Lond., Ser. A **131**, 275—285 (1931).
— The wave resistance of an ellipsoid. Proc. Roy. Soc. Lond., Ser. A **132**, 480—486 (1931).
— Ship waves: the calculation of wave profiles. Proc. Roy. Soc. Lond., Ser. A **135**, 1—13 (1932).

HAVELOCK, T. H: Ship waves: their variation with certain systematic changes of form. Proc. Roy. Soc. Lond., Ser. A **136**, 465—471 (1932).
— The theory of wave resistance. Proc. Roy. Soc. Lond., Ser. A **138**, 339—348 (1932).
— Wave patterns and wave resistance. Trans. Inst. Naval Arch. **76**, 430—442 (1934); discussion, 442—446.
— Ship waves: the relative efficiency of bow and stern. Proc. Roy. Soc. Lond., Ser. A **149**, 417—426 (1935).
— Wave resistance: the mutual action of two bodies. Proc. Roy. Soc. Lond., Ser. A **155**, 460—472 (1936).
— The forces on a circular cylinder submerged in a uniform stream. Proc. Roy. Soc. Lond., Ser. A **157**, 526—534 (1936).
— The pressure of water waves upon a fixed obstacle. Proc. Roy. Soc. Lond., Ser. A **175**, 409—421 (1940).
— The damping of the heaving and pitching motion of a ship. Phil. Mag. (7) **33**, 666—673 (1942).
— The wave resistance of a cylinder started from rest. Quart. J. Mech. Appl. Math. **2**, 325—334 (1949).
— The resistance of a submerged cylinder in accelerated motion. Quart. J. Mech. Appl. Math. **2**, 419—427 (1949).
— The forces on a submerged spheroid moving in a circular path. Proc. Roy. Soc. Lond., Ser. A **201**, 297—305 (1950).
— Wave resistance theory and its application to ship problems. Trans. Soc. Nav. Arch. Marine Engrs. **59**, 13—24 (1951).
— The moment on a submerged solid of revolution moving horizontally. Quart. J. Mech. Appl. Math. **5**, 129—136 (1952).
— Waves due to a floating sphere making periodic heaving oscillations. Proc. Roy. Soc. Lond., Ser. A **231**, 1—7 (1955).
— The effect of speed of advance upon the damping of heave and pitch. Trans. Inst. Naval Arch. **100**, 131—135 (1958).
HEINS, A.E.: Water waves over a channel of finite depth with a dock. Amer. J. Math. **70**, 730—748 (1948).
— Water waves over a channel of finite depth with a submerged plane barrier. Canad. J. Math. **2**, 210—222 (1950).
— On gravity waves. Proc. Symp. Appl. Math., Vol. 4, pp. 75—86. New York: McGraw-Hill 1953.
HEYNA, B., and P. GROEN: On short-period internal gravity waves. Physica, Haag **24**, 383—389 (1958).
HIDE, RAYMOND: The character of the equilibrium of an incompressible heavy viscous fluid of variable density: an approximate theory. Proc. Cambridge Phil. Soc. **51**, 179—201 (1955).
HOGNER, EINAR: Contributions to the theory of ship waves. Ark. Mat. Astr. Fys. **17**, No. 12, 1—50 (1923).
— On the theory of ship wave resistance. Ark. Mat. Astr. Fys. A **21**, No. 7, 11 pp. (1928).
HONDA, KÔTARÔ, and TOKUJIRÔ MATSUSHITA: An investigation of the oscillations of tank water. Sci. Rep. Tôhoku Imp. Univ., Ser. I **2**, 131—148 (1913) (10 plates).
HORIKAWA, KIYOSHI, and R.L. WIEGEL: Secondary wave crest formation. Wave Research Laboratory, Univ. of California, Berkeley, Report, Series 89, Issue 4 (1959), 23 pp.
HOUGH, S.S.: On the influence of viscosity on waves and currents. Proc. Lond. Math. Soc. **28**, 264—288 (1897).
HUDIMAC, A.A.: The motion of a body in a fluid with a free surface and irregular solid boundaries. Diss. Univ. of California, Berkeley, 1958, iii + 114 pp. (Inst. Engrg. Res. Tech. Rep., Ser. 82, Issue 5).
HUNT, J.N.: Amortissement par viscosité de la houle sur un fond incliné dans un canal de largeur finie. Viscous damping of waves over an inclined bed in a channel of finite width. Houille bl. **7**, 836—842 (1952).
— A note on gravity waves of finite amplitude. Quart. J. Mech. Appl. Math. **6**, 336—343 (1953).
— Gravity waves in flowing water. Proc. Roy. Soc. Lond., Ser. A **231**, 496—504 (1955).
— On the solitary wave of finite amplitude. Houille bl. **10**, 197—203 (1955).
ICHIYE, TAKASHI: On the theory of tsunami. Oceanogr. Mag. **2**, 83—100 (1950).
— Some remarks on the non-linear theory of shallow water waves on a sloping beach. Oceanogr. Mag. **4**, 159—166 (1953).
INGRAHAM, R.L.: Taylor instability of the interface between superposed fluids—solution by successive approximations. Proc. Phys. Soc. Lond., Sect. B **67**, 748—752 (1954).

INUI, TAKAO: Japanese developments on the theory of wave-making and wave resistance. 7th Internat. Conf. on Ship Hydrodynamics, Oslo, 1954, 60 pp.; discussion, pp. 61—70.
— Wave-making resistance of ships travelling on a shallow water. Proc. 6th Japan Nat. Congr. Appl. Mech., 1956, pp. 357—360. Science Council of Japan, Tokyo, 1957.
— Study on wave-making resistance of ships. The Society of Naval Architects of Japan, 60th anniversary series, Vol. 2, pp. 173—355. Tokyo 1957.
— TETURÔ: On deformation, wave patterns and resonance phenomenon of water surface due to a moving disturbance. I, II. Proc. Phys.-Math. Soc. Japan (3) **18**, 60—98, 99—113 (1936).
IPPEN, A.T., and GERSHON KULIN: Shoaling and breaking characteristics of the solitary wave. Mass. Inst. Tech., Hydrodynamics Lab., Tech. Rep. No. **15** (1955), vii + 56 pp.
ISAACSON, E.: Waves against an overhanging cliff. Comm. Appl. Math. **1**, 201—209 (1948).
— Water waves over a sloping bottom. Comm. Pure Appl. Math. **3**, 11—31 (1950).
IWASA, YOSHIAKI: Analytical considerations on cnoidal and solitary waves. Mem. Fac. Engrg. Kyoto Univ. **17**, 264—276 (1955).
JACOB, CAÏUS: Sur le problème d'unicité locale concernant l'écoulement des liquides pesants. C. R. Acad. Sci., Paris **198**, 539—541 (1934).
JEFFREYS, HAROLD: On the highest gravity waves on deep water. Quart. J. Mech. Appl. Math. **4**, 385—387 (1951).
JINNAKA, TATSUO: Wave patterns and ship waves. The Society of Naval Architects of Japan, 60th anniversary series, Vol. 2, pp. 83—94. Tokyo 1957.
JOHN, FRITZ: Waves in the presence of an inclined barrier. Comm. Appl. Math. **1**, 149—200 (1948).
— On the motion of floating bodies. I, II. Comm. Pure Appl. Math. **2**, 13—57 (1949); **3**, 45—101 (1950).
— Two-dimensional potential flows with a free boundary. Comm. Pure Appl. Math. **6**, 497—503 (1953).
JOLAS, PIERRE: Sur le calcul approché de la houle irrotationnelle. C. R. Acad. Sci., Paris **246**, 1659—1661 (1958).
JONES, D. S.: The eigenvalues of $\nabla^2 u + \lambda u = 0$ when the boundary conditions are given on semi-infinite domains. Proc. Cambridge Phil. Soc. **49**, 668—684 (1953).
KABACHINSKIĬ, N.N.: On the calculation of the wave profile arising from the motion of a ship. (R) Dokl. Akad. Nauk SSSR. (N. S.) **58**, 1301—1304 (1947).
KAMESVARA RAV, J.C.: On ripples of finite amplitude. Proc. Indian Assoc. Cultivation of Sci. **6**, 175—193 (1920—21) (1 plate).
KAMPÉ DE FÉRIET, J.: Les rides, les vagues et la houle. Rev. Questions Sci. **102**, 181—229 (1932).
—, and J. KOTIK: Surface waves of finite energy. J. Rational Mech. Anal. **2**, 577—585 (1953).
KAPLAN, PAUL: The waves generated by the forward motion of oscillatory pressure distributions. Proc. 5th Midwest. Conf. Fluid Mech., Ann Arbor, Mich., 1957, pp. 316 to 329.
KELDYSH, M.V.: Remarks on certain motions of a heavy fluid. (R) Tsentral. Aero-Gidrodinam. Inst., Tekhn. Zametki No. **52**, 5—9 (1935).
— On the reflexion of waves on the surface of a heavy fluid. (R) Trudy Konferentsii po Teorii Volnovogo Soprotivleniya, Moscow, 1936, pp. 140—142. Tsentral. Aero-Gidrodinam. Inst., Moscow, 1937.
—, and M.A. LAVRENT'EV: On the motion of a wing under the surface of a heavy fluid. (R) Trudy Konferentsii po Teorii Volnovogo Soprotivleniya, Moscow, 1936, pp. 31—64.
—, and L.I. SEDOV: The theory of wave resistance in a channel of finite depth. (R) Trudy Konferentsii po Teorii Volnovogo Soprotivleniya, Moscow, 1936, pp. 143—152.
KELLAND, P.: On the theory of waves. I, II. Trans. Roy. Soc. Edinburgh **14**, 497—545 (1840); **15**, 101—144 (1844).
KELLER, JOSEPH B.: The solitary wave and periodic waves in shallow water. Comm. Appl. Math. **1**, 323—339 (1948).
— Surface waves on water of non-uniform depth. J. Fluid Mech. **4**, 607—614 (1958).
—, and EDWARD GOLDSTEIN: Water wave reflection due to surface tension and floating ice. Trans. Amer. Geophys. Un. **34**, 43—48 (1953).
—, and IGNACE KOLODNER: Instability of liquid surfaces and the formation of drops. J. Appl. Phys. **25**, 918—921 (1954).
—, and MORTIMER WEITZ: Reflection and transmission coefficients for waves entering or leaving an icefield. Comm. Pure Appl. Math. **6**, 415—417 (1953).
KELVIN, Lord: See THOMSON, W.
KENNARD, E.H.: Generation of surface waves by a moving partition. Quart. Appl. Math. **7**, 303—312 (1949).

KEULEGAN, G. H.: Wave motion. (Chap. XI of Engineering hydraulics, edit. by H. ROUSE.) New York: John Wiley & Sons 1950.
— Energy dissipation in standing waves in rectangular basins. J. Fluid Mech. **6**, 33—50 (1959).
—, and G. W. PATTERSON: Mathematical theory of irrotational translation waves. J. Res. Nat. Bur. Stand. **24**, 47—101 (1940).
KHAŠKIND, M. D.: See HASKIND.
KHRISTIANOVICH, S. A.: Unsteady motion in canals and rivers. (R) [Nekotorye novye voprosy mekhaniki sploshnoĭ sredy (Some recent questions in the mechanics of a continuous medium), pp. 13—154.] Moscow-Leningrad: Izdatel'stvo Akad. Nauk SSSR 1938.
KIBEL', I. A. [KIEBEL]: On some two-dimensional motions of a heavy compressible fluid. (R) Prikl. Mat. Mekh. (1) **1**, 51—55 (1933).
— Sur le mouvement permanent, dans un fluide compressible dont les surfaces isobariques sont des plans horizontaux. Assoc. Franc. Advanc. Sci., C. R. 57e Session, Chambéry, 1933, pp. 197—201.
KIERSTEAD, H. A.: Bottom pressure fluctuations due to standing waves in a deep, two-layer ocean. Trans. Amer. Geophys. Un. **33**, 390—396 (1952).
KIRCHHOFF, G.: Über stehende Schwingungen einer schweren Flüssigkeit. Mber. Akad. Wiss. Berlin **1879**, 395—410 (1880) = Ann. Phys. Chem., N.F. **10** (**246**), 34—46 (1880) = Gesammelte Abhandlungen, pp. 428—442. Leipzig: Johann Ambrosius Barth 1882.
— u. G. HANSEMANN: Versuche über stehende Schwingungen des Wassers. Ann. Physik u. Chem., N.F. **10** (**246**), 337—347 (1880).
KISHI, TSUTOMU: On the highest progressive wave in shallow water. Proc. 4th Japan Nat. Congr. Appl. Mech., 1954, pp. 241—244. Science Council of Japan, Tokyo, 1955.
KOCHIN, N. E. [KOTCHINE, KOTCHIN, KOTSCHIN]: Détermination rigoureuse des ondes permanentes d'ampleur finie à la surface de séparation de deux liquides de profondeur finie. Math. Ann. **98**, 582—615 (1927) = Sobranie sochinenii, T. II. pp. 43—75, Akad. Nauk SSSR., 1949.
— On the theory of Cauchy-Poisson waves. (R) Trudy Mat. Inst. im. Steklov. **9**, 167—187 (1935) = Sobranie sochinenii, T. II, pp. 86—104, Akad. Nauk SSSR., Moscow-Leningrad, 1949.
— On the wave resistance and lift of bodies submerged in a fluid. (R) Trudy Konferentsii po Teorii Volnvogo Soprotivleniya, Moscow, 1936, pp. 65—134, Tsentral. Aèro-Gidrodinam. Inst., Moscow, 1937 = Sobranie sochinenii, T. II, pp. 105—182, Akad. Nauk SSSR., 1949. Translated in Soc. Naval Arch. Marine Engrs. Tech. & Res. Bull. No. **1—8** (1951).
— Über den Einfluß des Bodenprofils auf die Wellen an der Grenzfläche von zwei Flüssigkeiten verschiedener Dichte. Izv. Akad. Nauk SSSR., Ser. Geograf. Geofiz. **1937**, 357 to 381 = Sobranie sochinenii, T. I, pp. 448—465, Akad. Nauk SSSR., Moscow-Leningrad, 1949.
— On the influence of the profile of the earth upon waves at the surface of separation of two fluids of different density. (R) Trudy Glav. Geofiz. Observ. **1937**, No. 14, 19—30 = Sobraine sochinenii, T. I, pp. 466—477, Akad. Nauk SSSR., Moscow-Leningrad, 1949.
— The plane problem of gliding of a slightly curved contour on the surface of a heavy incompressible fluid. (R) Trudy Tsentral. Aèro-Gidrodinam. Inst., vyp. **356**, 1—24 (1938) = Sobranie sochinenii, T. II, pp. 215—239, Akad. Nauk SSSR., 1949.
— On the motion of a heavy fluid in a canal with a step in the bottom. (R) Dokl. Akad. Nauk SSSR. (N.S.) **19**, 599—601 (1938) = Sobranie sochinenii, T. II, pp. 240—243, Akad. Nauk SSSR., 1949.
— The three-dimensional problem of waves at the surface of separation of two fluids of different density caused by unevenness of the bottom. (R) Trudy Glav. Geofiz. Observ. **1938**, No. 8, 3—30 = Sobranie sochinenii, T. I, pp. 478—508, Akad. Nauk SSSR., Moscow-Leningrad, 1949.
— The two-dimensional problem of steady oscillations of bodies under the free surface of a heavy incompressible fluid. (R) Izv. Akad. Nauk SSSR., Otdel. Tekhn. Nauk **1939**, No. 4, 37—62 = Sobranie sochinenii, T. II, pp. 244—276, Akad. Nauk SSSR., 1949. Translated in Soc. Naval Arch. Marine Engrs. Tech. & Res. Bull. No. **1—9** (1952).
— The theory of waves generated by oscillations of a body under the free surface of a heavy incompressible fluid. (R) Uchenye Zapiski Moskov. Gos. Univ. **46**, 85—106 (1940) = Sobranie sochinenii, T. II, pp. 277—304, Akad. Nauk SSSR., 1949. Translated in Soc. Naval Arch. Marine Engrs. Tech. & Res. Bull. No. **1—10** (1952).
KOCHIN, N. E., I. A. KIBEL' and N. V. ROZE: Teoreticheskaya gidromekhanika. Chast' I. [Theoretical hydromechanics. Part I.] 4th ed. 535 pp. Moscow-Leningrad: Gosudarstvennoe Izdatel'stvo Tekhniko-Teoreticheskoĭ Literatury 1948,

KOTSCHIN, N. J., I. A. KIBEL u. N. W. ROSE: Theoretische Hydromechanik, Bd. I. xii + 508 pp. Berlin: Akademie-Verlag 1954.

KOCHINA, I. N.: On waves on the surface of separation of two fluids flowing at an angle to each other. (R) Prikl. Mat. Mekh. **19**, 628—634 (1955).

KORTEWEG, D. J., and G. DE VRIES: On the change of form of long waves advancing in a rectangular canal and on a new type of long stationary waves. Phil. Mag. (5) **39**, 422—443 (1895).

KOUSSAKOV: See KUSAKOV.

KRASNOSEL'SKIĬ, M. A.: On NEKRASOV's equation from the theory of waves on the surface of a heavy fluid. (R) Dokl. Akad. Nauk SSSR. **109**, 456—459 (1956).

KRAVTCHENKO, J., et A. DAUBERT: La houle à trajectoires fermées en profondeur finie. Houille bl. **12**, 408—429 (1957).

KREĬN, S. G., and N. N. MOISEEV: On oscillations of a rigid body, containing a fluid with a free surface. (R) Prikl. Mat. Mekh. **21**, 169—174 (1957).

KREISEL, G.: Surface waves. Quart. Appl. Math. **7**, 21—44 (1949).

KRILOFF, A.: A new theory of the pitching motion of ships on waves, and of the stresses produced by this motion. Trans. Inst. Naval Arch. **37**, 326—359 (1896); discussion, 360—368.

— A general theory of the oscillations of a ship on waves. Trans. Inst. Naval Arch. **40**, 135—190 (1898); discussion, 190—196.

KRYLOV, A. N.: See KRILOFF.

KUSAKOV, M. [KOUSSAKOV]: Capillary-gravitational waves at the interface between two viscous liquids of finite depth. Acta physicochim. **19**, 286—294 (1944) = Zh. eksp. teoret. Fiz. **14**, 232—239 (1944).

LAITONE, E. V.: The subsonic and supersonic flow fields about slender bodies. Sixth Internat. Congr. Appl. Mech., Paris, 1946. (Results summarized in J. Aeronaut. Sci. **17**, 250—251 (1950).

— Limiting pressure on hydrofoils at small submergence depths. J. Appl. Phys. **25**, 623—626 (1954).

— Comments on limiting pressure on hydrofoils. J. Appl. Phys. **26**, 1519—1520 (1955).

LAMB, HORACE: On deep water waves. Proc. Lond. Math. Soc. (2) **2**, 371—400 (1904).

— On some cases of wave motion on deep water. Ann. di Mat. **21**, 237—250 (1913).

— On waves due to a traveling disturbance, with an application to waves in superposed fluids. Phil. Mag. (6) **31**, 386—399 (1916).

— On wave-patterns due to a travelling disturbance. Phil. Mag. (6) **31**, 539—548 (1916).

— On water waves due to disturbance beneath the surface. Proc. Lond. Math. Soc. (2) **21**, 359—372 (1922).

— On wave resistance. Proc. Roy. Soc. Lond., Ser. A **111**, 14—25 (1926).

— Hydrodynamics, 6th ed. xv + 738 pp. Cambridge: Cambridge University Press 1932 = New York: Dover Publications, 1945.

— The effect of a step in the bed of a stream. J. Lond. Math. Soc. **9**, 308—315 (1934).

LAND, N. S.: Characteristics of an NACA 66 S-209 section hydrofoil at several depths. NACA Wartime Report L-**757** (1943), 3 pp. + 7 figs. (1946).

LANDAU, L. D., and E. M. LIFSHITS: Mekhanika sploshnykh sred. [Mechanics of continuous media.] 2nd ed. 788 pp. Moscow: Gosudarstvennoe Izdatel'stvo Tekhniko-Teoreticheskoĭ Literatury 1953.

LAUCK, AUGUST: Der Überfall über ein Wehr. Z. angew. Math. Mech. **5**, 1—16 (1925).

LAVRENT'EV, M. A. [LAVRENTIEFF]: A contribution to the theory of long waves. C. R. (Doklady) Acad. Sci. URSS. (N.S.) **41**, 275—277 (1943). Reproduced in Amer. Math. Soc. Transl. No. **102**, 51—53 (1954).

— On the theory of long waves. [Ukrainian.] Akad. Nauk Ukrain. RSR., Zbirnik Prac' Inst. Mat. **1946**, No. 8, 13—69 (1947). Translated in Amer. Math. Soc. Transl. No. **102**, 3—50 (1954).

LAYZER, DAVID: On the instability of superposed fluids in a gravitational field. Astrophys. J. **122**, 1—12 (1955).

LEHMAN, R. S.: Developments in the neighborhood of the beach of surface waves over an inclined bottom. Comm. Pure Appl. Math. **7**, 393—439 (1954).

LEVICH, V. G.: Fiziko-khimicheskaya gidrodinamika. [Physical-chemical hydrodynamics.] 538 pp. Moscow: Izdat. Akad. Nauk SSSR 1952.

LEVI-CIVITA, T.: Sulle onde progressive di tipo permanente. Atti Accad. Lincei, Rend. Cl. Sci. Fis. Mat. Nat. (5) **16**, 2° sem., 770—790 (1907).

— Sulle onde di canale. Atti Accad. Lincei, Rend. Cl. Sci. Fis. Mat. Nat. (5) **21**, 1° sem., 3—14 (1912).

— Qüestions de mecànica clàssica i relativista; conferències donades el gener de 1921. viii + 151 pp. Barcelona: Institut d'Estudis Catalans 1922.

— Fragen der klassischen und relativistischen Mechanik. vi + 110 pp. Berlin: Springer 1924

LEVI-CIVITA, T.: Questioni di meccanica classica e relativista. 185 pp. Bologna: N. Zanichelli 1924.
— Über die Transportgeschwindigkeit in einer stationären Wellenbewegung. Vorträge aus dem Gebiete der Hydro- und Aerodynamik (Innsbruck 1922), pp. 85—96. Berlin: Springer 1924.
— La détermination rigoureuse des ondes permanentes d'ampleur finie. Proc. 1st Internat. Congr. Appl. Mech., Delft, 1924, pp. 129—145.
— Détermination rigoureuse des ondes permanentes d'ampleur finie. Math. Ann. **93**, 264 to 314 (1925).
— Nozione adimensionale di vortice e sua applicazione alle onde trocoidali di Gerstner. Pont. Acad. Sci. Acta **4**, 23—30 (1940).
LEVINE, HAROLD: Scattering of surface waves on an infinitely deep fluid. Proceedings, Symposium on the behaviour of ships in a seaway, Wageningen, 1957, pp. 712—716.
— Wave generation by oscillation of submerged bodies. Applied Mathematics and Statistics Laboratory, Stanford Univ., Calif., Tech. Rep. No. **79**, 38 pp. (1958).
—, and EUGENE RODEMICH: Scattering of surface waves on an ideal fluid. Applied Mathematics and Statistics Laboratory, Stanford Univ., Calif., Tech. Rep. No. **78**, 64 pp. (1958).
LEWY, HANS: Water waves on sloping beaches. Bull. Amer. Math. Soc. **52**, 737—775 (1946).
— A note on harmonic functions and a hydrodynamical application. Proc. Amer. Math. Soc. **3**, 111—113 (1952).
— On steady free surface flow in a gravity field. Comm. Pure Appl. Math. **5**, 413—414 (1952).
LICHTENSTEIN, L.: Vorlesungen über einige Klassen nichtlinearer Integralgleichungen und Integro-differentialgleichungen. x + 164 pp. Berlin: Springer 1931.
LIGHTHILL, M. J.: River waves. Symposium on naval hydrodynamics, Washington, D.C. 1956, pp. 17—40; discussion, pp. 40—44. Washington, D.C.: National Academy of Sciences-National Research Council 1957.
—, and G. B. WHITHAM: On kinematic waves. I. Flood movement in long rivers. Proc. Roy. Soc. Lond., Ser. A **229**, 281—316 (1955).
LITTMAN, WALTER: On the existence of periodic waves near critical speed. Comm. Pure Appl. Math. **10**, 241—269 (1957).
LIU, HSIEN CHIH: Über die Entstehung von Ringwellen an einer Flüssigkeitsoberfläche durch unter dieser gelegene, kugelige periodische Quellensysteme. Z. angew. Math. Mech. **32**, 211—226 (1952).
LOCK, R. C.: The velocity distribution in the laminar boundary layer between parallel streams. Quart. J. Mech. Appl. Math. **4**, 42—63 (1951).
— Hydrodynamic stability of the flow in the laminar boundary layer between parallel streams. Proc. Cambridge Phil. Soc. **50**, 105—124 (1954).
LONG, R. R.: Some aspects of the flow of stratified fluids. I. A theoretical investigation. II. Experiments with a two-fluid system. III. Continuous density gradients. Tellus **5**, 42—58 (1953); **6**, 97—115 (1954); **7**, 341—357 (1955).
— Long waves in a two-fluid system. J. Meteorol. **13**, 70—74 (1956).
LONGUET-HIGGINS, M. S.: A theory of the origin of microseisms. Phil. Trans. Roy. Soc. Lond., Ser. A **243**, 1—35 (1950).
— On the decrease of velocity with depth in an irrotational water wave. Proc. Cambridge Phil. Soc. **49**, 552—560 (1953).
— Mass transport in water waves. Phil. Trans. Roy. Soc. Lond., Ser. A **245**, 535—581 (1953).
LOVE, A. E. H., P. APPELL, H. BEGHIN et H. VILLAT: Développements concernant l'hydrodynamique. 5. Mouvements ondulatoires des fluides incompressibles. Encyclopédie des sciences mathématiques pures et appliquées, Tome IV, Vol. 5, pp. 160—180. Paris: Gauthier-Villars; Leipzig: Teubner 1914.
LOWELL, SHERMAN CABOT: The propagation of waves in shallow water. Comm. Pure Appl. Math. **2**, 275—291 (1949).
LUNDE, J. K.: On the linearized theory of wave resistance for displacement ships in steady and accelerated motion. Trans. Soc. Naval Arch. Marine Engrs. **59**, 24—76 (1951); discussion, 76—85.
— On the linearized theory of wave resistance for a pressure distribution moving at constant speed of advance on the surface of deep or shallow water. Skipsmodelltanken, Norges Tekniske Høgskole, Trondheim, Medd. No. **8**, 48 pp. (1951).
— The linearized theory of wave resistance and its application to ship-shaped bodies in motion on the surface of a deep, previously undisturbed fluid. Soc. Naval Arch. Marine Engrs. Tech. & Res. Bull. No. **1—8**, 70 pp. (1957).
MACCAMY, R. C.: Approximate solution to the problem of a freely floating cylinder in surface waves. Wave Research Laboratory, University of California, Berkeley, Report, Series 61, Issue 3 (1953), 9 + 2 + 2 pp.

MACCAMY, R. C.: The motion of a floating sphere in surface waves. Wave Research Laboratory, University of California, Berkeley, Report, Series 61, Issue 4 (1954), 20 pp.
— Une solution par potentiel de sources pour l'équation des houles à courtes crêtes. A source solution for short-crested waves. Houille bl. **12**, 367—378, 379—389 (1957).
MCCOWAN, J.: On the solitary wave. Phil. Mag. (5) **32**, 45—58 (1891).
— On the highest wave of permanent type. Phil. Mag. (5) **38**, 351—358 (1894).
MCKNOWN, J. S.: Sur l'entretien des oscillations des eaux portuaires sous l'action de la haute mer. Publ. Sci. Tech. du Ministère de l'Air, No. 278, Paris, 1953, ii + ii + 45 pp.
MACDONALD, H.M.: Waves in canals. Proc. Lond. Math. Soc. **25**, 101—111 (1894).
— Waves in canals and on a sloping bank. Proc. Lond. Math. Soc. **27**, 622—632 (1896).
MAGNARADZE, L. G.: On a new integral equation of the theory of airplane wings. (R) Soobshch. Akad. Nauk Gruzin. SSR. **3**, 503—508 (1942).
MARNYANSKIĬ, I.A.: The diffraction of waves about a submerged vertical plate. (R) Prikl. Mat. Mekh. **18**, 233—238 (1954).
MARTIN, J.C., and W. J. MOYCE: An experimental study of the collapse of liquid columns on a rigid horizontal plane. Phil. Trans. Roy. Soc. Lond., Ser. A **244**, 312—324 (1952).
— An experimental study of the collapse of fluid columns on a rigid horizontal plane, in a medium of lower, but comparable density. Phil. Trans. Roy. Soc. Lond., Ser. A **244**, 325—334 (1952) (2 plates).
MARUO, HAJIME: Two-dimensional theory of the hydroplane. Proc. 1st Japan Nat. Congr. Appl. Mech., 1951, pp. 409—415. Science Council of Japan, Tokyo, 1952.
— Hydrodynamic researches of the hydroplane. I, II, III. [Japanese.] J. Zôsen Kiôkai **91**, 9—16 (1956); **92**, 57—63 (1957); **105**, 23—26 (1959).
— Modern developments of the theory of wave-making resistance in the non-uniform motion. The Society of Naval Architects of Japan, 60th anniversary series, Vol. 2, pp. 1—82. Tokyo, 1957.
MATTHIESSEN, L.: Experimentelle Untersuchungen über das Thomson'sche Gesetz der Wellenbewegung auf Flüssigkeiten unter der Wirkung der Schwere und Cohäsion. Ann. Physik u. Chemie, N.F. **38** (**274**), 118—130 (1889) (Tafel I, Fig. 9).
MEYER, R.: Symétrie du coefficient (complexe) de transmission des houles à travers un obstacle quelconque. Houille bl. **10**, 139—140 (1955).
MICHE, ROBERT: Mouvements ondulatoires de la mer en profondeur constante ou décroissante. Forme limite de la houle lors de son déferlement. Application aux digues maritimes. Ann. Ponts Chaussées **114**, 25—78, 131—164, 270—292, 369—406 (1944).
— Propriétés des trains d'ondes océaniques et de laboratoire. Evolution et amortissement compte tenu des résistances passives. (Comité central d'océanographie et d'étude des côtes, No. 135.) xii + 447 pp. Paris: Ministère de la Défense Nationale, Imprimerie nationale 1954.
MICHELL, J.H.: The highest waves in water. Phil. Mag. (5) **36**, 430—437 (1893).
— The wave resistance of a ship. Phil. Mag. (5) **45**, 106—123 (1898).
MICHIE SMITH, C.: The determination of surface-tension by measurement of ripples. Proc. Roy. Soc. Edinburgh **17** (1890), 115—121 (1891).
MILLER, G.F.: On certain integrals occurring in a hydrodynamical problem. Proc. Roy. Soc. Lond., Ser. A **243** (1958), 65—77 (1957).
MILNE-THOMSON, L.M.: Theoretical hydrodynamics, 3rd ed. xxviii + 632 pp. New York: Macmillan Company 1956.
MIYAZAKI, MASAMORI: Mathematical studies of surf and breakers—nonlinear wave theory in shallow water of constant slope. I, II. Oceanogr. Mag. **3**, 65—74 (1951); **4**, 13—21 (1952).
MOISEEV, N.N.: The problem of small oscillations of an open vessel with a fluid under the action of an elastic force. (R) Ukrain. Mat. Zh. **4**, 168—173 (1952).
— On two pendulums filled with liquid. (R) Prikl. Mat. Mekh. **16**, 671—678 (1952).
— The problem of the motion of a rigid body containing a liquid mass having a free surface. (R) Mat. Sbornik, N.S. **32** (**74**), 61—96 (1953).
— Dynamics of a ship having a liquid load. (R) Izv. Akad. Nauk SSSR., Otd. Tekhn. Nauk **1954**, No. 7, 27—45.
— Some questions of the theory of oscillation of vessels with a fluid. (R) Inzhen. Sb. **19**, 167—170 (1954).
— On a problem of the theory of waves on the surface of a bounded volume of fluid. (R) Prikl. Mat. Mekh. **19**, 343—347 (1955).
— On the flow of a heavy fluid over a wavy bottom. (R) Prikl. Mat. Mekh. **21**, 15—20 (1957).
— On the non-uniqueness of the possible forms of steady flows of a heavy fluid for Froude numbers close to one. (R) Prikl. Mat. Mekh. **21**, 860—864 (1957).
— On the oscillations of a body floating in a bounded volume of fluid. (R) Moskov. Fiz.-Tekh. Inst. Issled. Mekh. Prikl. Mat. **1**, 145—166 (1958).

MORISON, J.R., and R.C. CROOKE: The mechanics of deep-water, shallow-water and breaking waves. Beach Erosion Board, Corps of Engrs., U.S. Army, Tech. Memo. No. **40**, 14 pp. (1953).

MUNK, W., F. SNODGRASS and G. CARRIER: Edge waves on the continental shelf. Science **123**, 127—132 (1956).

NARIMANOV, G.S.: On the motion of a body, a hollow of which is partly filled with fluid. (R) Prikl. Mat. Mekh. **20**, 21—38 (1956).

— On the motion of a vessel partly filled with fluid; effect of nonsmallness of the motion of the latter. (R) Prikl. Mat. Mekh. **21**, 513—524 (1957).

NEKRASOV, A.I.: On STOKES' wave. (R) Izv. Ivanovo-Voznesensk. Politekhn. Inst. **1920**, 81—89.

— On waves of permanent type. I, II. (R) Izv. Ivanovo-Voznesensk. Politekhn. Inst. **3**, 52—65 (1921); **6**, 155—171 (1922).

— On waves of permanent type on the surface of a heavy fluid. (R) Trudy Vseross. S'ezda Matematikov, Moskva, 1928, pp. 258—262.

— Tochnaya teoriya voln ustanovivshegosya vida na poverkhnosti tyazheloĭ zhidkosti. [The exact theory of waves of permanent type on the surface of a heavy fluid.] 94 pp. Moscow: Izdatel'stvo Akad. Nauk SSSR. 1951.

NEUMANN, FRANZ: Beitrag zu dem Problem der permanenten wirbelfreien Flüssigkeitsbewegung in Kanälen. Diss. Univ. Leipzig, 1930, 1 + 40 pp.

NISHIYAMA, TETSUO: Study on submerged hydrofoils. The Society of Naval Architects of Japan, 60th anniversary series, Vol. 2, pp. 95—134. Tokyo 1957.

— Experimental investigation of the effect of submergence depth upon the hydrofoil section characteristics. [Japanese.] J. Zôsen Kiôkai **105**, 7—21 (1959).

NORMANDIN, M.: Sur la théorie du second ordre des phénomènes parasites dans un canal à houle. C. R. Acad. Sci., Paris **245**, 1880—1882 (1957).

OKHOTSIMSKIĬ, D.E.: On the theory of motion of a body with hollows partly filled with fluid. (R) Prikl. Mat. Mekh. **20**, 3—20 (1956).

PACKHAM, B.A.: The theory of symmetrical gravity waves of finite amplitude. II. The solitary wave. Proc. Roy. Soc. Lond., Ser. A **213**, 238—249 (1952).

PALATINI, A.P.: Sulla influenza del fondo nella propagazione delle onde dovute a perturbazione locali. I, II. Rend. Circ. Mat. Palermo **39**, 362—384 (1915); **40**, 169—184 (1915).

PALM, ENOK: On the formation of surface waves in a fluid flowing over a corrugated bed and on the development of mountain waves. Astrophys. Norveg. **5**, 61—130 (1953).

PARKIN, R. BLAINE, BYRNE PERRY and T. YAO-TSU WU: Pressure distribution on a hydrofoil near the water surface. J. Appl. Phys. **27**, 232—240 (1956).

PENNEY, W.G., and A.T. PRICE: The diffraction theory of sea waves and the shelter afforded by breakwaters. Phil. Trans. Roy. Soc. Lond., Ser. A **244**, 236—253 (1952).

— Finite periodic stationary gravity waves in a perfect liquid. Phil. Trans. Roy. Soc. Lond., Ser. A **244**, 254—284 (1952).

—, and C.K. THORNHILL: The dispersion, under gravity, of a column of fluid supported on a rigid horizontal plane. Phil. Trans. Roy. Soc. Lond., Ser. A **244**, 285—311 (1952).

PERROUD, P.H.: The solitary wave reflection along a straight vertical wall at oblique incidence. Diss. Univ. of California, Berkeley, 1957 (Inst. Engrg. Res. Tech. Rep., Series 99, Issue 3 (1957), vi + 93 pp.).

PERZHNYANKO, E.A.: On vertical oscillations of a body floating on the surface of a fluid between two parallel walls, and on the waves arising from this. (R) Prikl. Mat. Mekh. **20**, 362—372 (1956).

PETERS, A.S.: A new treatment of the ship wave problem. Comm. Pure Appl. Math. **2**, 123—148 (1949).

— The effect of a floating mat on water waves. Comm. Pure Appl. Math. **3**, 319—354 (1950).

— Water waves over sloping beaches and the solution of a mixed boundary value problem for $\Delta^2\phi - k^2\phi = 0$ in a sector. Comm. Pure Appl. Math. **5**, 87—108 (1952).

—, and J.J. STOKER: A uniqueness theorem and a new solution for SOMMERFELD's and other diffraction problems. Comm. Pure Appl. Math. **7**, 565—585 (1954).

— The motion of a ship, as a floating rigid body, in a seaway. Comm. Pure Appl. Math. **10**, 399—490 (1957).

PIDDUCK, F.B.: On the propagation of a disturbance in a fluid under gravity. Proc. Roy. Soc. Lond., Ser. A **83**, 347—356 (1910).

— The wave problem of CAUCHY and POISSON for finite depth and slightly compressible fluid. Proc. Roy. Soc. Lond., Ser. A **86**, 396—405 (1912).

PLATZMAN, G.W.: The partition of energy in periodic irrotational waves on the surface of deep water. J. Marine Res. **6**, 194—202 (1947).

PLESSET, M.S.: On the stability of fluid flows with spherical symmetry. J. Appl. Phys. **25**, 96—98 (1954).

POCKLINGTON, H.C.: Standing waves parallel to a plane beach. Proc. Cambridge Phil. Soc. **20**, 308—310 (1921).

POINCELOT, P.: Application de la notion de vitesse de groupe à la propagation de la houle en eau profonde. Ann. Géophys. **9**, 158—160 (1953).

— Remarque sur la notion de vitesse de groupe; applications diverses. J. Math. pures appl. (9) **33**, 329—364 (1954).

POISSON, S.D.: Mémoire sur la théorie des ondes. Mém. Acad. Roy. Sci. Inst. France (2) **1** (1816), 71—186 (1818).

PONCIN, HENRI: Sur le mouvement d'un fluide pesant dans un plan vertical. Publ. Sci. Tech. du Ministère de l'Air, No. 16, Paris, 1932, iii + 102 pp.

POND, H.L.: The moment acting on a Rankine ovoid moving under a free surface. The David W. Taylor Model Basin, Washington, D.C., Rep. No. **795**, x + 17 pp. (1951).

— The pitching moment acting on a body of revolution moving under a free surface. The David W. Taylor Model Basin, Washington, D.C., Rep. No. **819**, viii + 13 pp. (1952).

— The moment acting on a Rankine ovoid moving under a free surface. J. Ship Res. **2**, No. 4, 1—9 (1959).

PREISWERK, ERNST: Anwendung gasdynamischer Methoden auf Wasserströmungen mit freier Oberfläche. Mitt. Inst. Aerodynamik, Eidgen. Techn. Hochsch., Zürich, No. **7**, 130 pp. (1938).

— Application of gas dynamics to water flows with free surface. Nat. Adv. Comm. Aeron. Tech. Memo. Nos. **934**, **935**, 75, 70 pp. (1940).

PRINS, J.E.: Characteristics of waves generated by a local surface disturbance. Wave Research Laboratory, Inst. Engrg. Res., Univ. of California, Berkeley, Series 99, Issue 1 (1956), 99 pp.

— Water waves due to a local disturbance. Proc. 6th Conf. Coastal Engrg., Florida, 1957, pp. 147—162. Council on Wave Research, Univ. of California, Richmond, Calif., 1958.

— Characteristics of waves generated by a local disturbance. Trans. Amer. Geophys. Un. **39**, 865—874 (1958).

RABINOVICH, B.I.: On the equations of perturbed motion of a rigid body with a cylindrical hollow partly filled with fluid. (R) Prikl. Mat. Mekh. **20**, 39—50 (1956).

RANKINE, W.J.M.: On the exact form of waves near the surface of deep water, Phil. Trans. Roy. Soc. Lond. **153**, 127—138 (1863).

— Summary of the properties of certain stream-lines. Phil. Mag. (4) **28**, 282—288 (1864).

— Supplement to a paper on stream-lines. Phil. Mag. (4) **29**, 25—28 (1865).

RAYLEIGH, Lord: See STRUTT, J.W.

REĬNOV, M.N.: On the computation of the velocity potential of the motion of a fluid caused by the translation of an immersed body. (R) Dokl. Akad. Nauk SSSR. **77**, 201—204 (1951).

RICHARDSON, A.R.: Stationary waves in water. Phil. Mag. (6) **40**, 97—110 (1920).

RISSER, R.: Essai sur la théorie des ondes par émersion. J. École polytech. (2) **24**, 113—218 (1924); **25**, 1—41 (1925).

ROLL, H.U.: Oberflächen-Wellen des Meeres. In Handbuch der Physik, Bd. XLVIII, S. 671 bis 733. Berlin-Göttingen-Heidelberg: Springer 1957.

ROSEAU, MAURICE: Contribution à la théorie des ondes liquides de gravité en profondeur variable. Publ. Sci. Tech. du Ministère de l'Air, No. 275, Paris, 1952, ii + 91 pp.

— Short waves parallel to the shore over a sloping beach. Comm. Pure Appl. Math. **11**, 433—493 (1958).

ROSSBY, C.-G.: On the propagation of frequencies and energy in certain types of oceanic and atmospheric waves. J. Meteorol. **2**, 187—204 (1945).

— Notes on the distribution of energy and frequency in surface waves. J. Marine Res. **6**, 93—103 (1947).

RUBIN, HANAN: The dock of finite extent. Comm. Pure Appl. Math. **7**, 317—344 (1954).

RUDZKI, M.P.: Über eine Klasse hydrodynamischer Probleme mit besonderen Grenzbedingungen. Math. Ann. **50**, 269—281 (1898).

RUELLAN, F., et A. WALLET: Trajectoires internes dans un clapotis partiel. Houille bl. **5**, 483—489 (1950).

RUSSELL, JOHN SCOTT: Report of the Committee on Waves. Report of the seventh meeting of the British Association for the Advancement of Science, Liverpool, 1837, pp. 417—496 (5 plates). London: John Murray 1838.

— Report on waves. Report of the fourteenth meeting of the British Association for the Advancement of Science, York, 1844, pp. 311—390, plates XLVII—LVII. London: John Murray 1845.

—, R.C.H., and D.H. MACMILLAN: Waves and tides. 348 pp. (15 plates). London: HUTCHINSON's Scientific and Technical Publications 1952.

SAINT-VENANT, B. DE: Des diverses manières de poser les équations du mouvement varié des eaux courantes. Ann. Ponts et Chaussées (6) **13**, 148—228 (1887).

—, et A. A. FLAMANT: De la houle et du clapotis. Ann. Ponts et Chaussées (6) **15**, 705—809 (1888).

SANDGREN, L.: A vibration problem. Medd. Lunds Univ. Mat. Sem. **13**, 1—84 (1955).

SANO, KEIZÔ: On the seiches of Lake Tôya. Proc. Tôkyô Math.-Phys. Soc. (2) **7**, 17—22 (1913).

—, and KEN HASEGAWA: On the wave produced by the sudden depression of a small portion of the bottom of a sea of uniform depth. Proc. Tôkyô Math.-Phys. Soc. (2) **8**, 187—199 (1915—16).

SASAKI, SHIGEKI: On the oscillations of water in circular-sectorial and ring-sectorial vessels. Sci. Rep. Tôhoku Imp. Univ., Ser. I **3**, 257—270 (1914) (5 plates).

SAUTREAUX, C.: Sur une question d'hydrodynamique. Ann. Sci. École norm. Sup. (3) **10**, S. 95—S. 182 (1893).

— Sur une question d'hydrodynamique. Ann. Enseign. Sup. Grenoble **6**, 1—17 (1894).

— Mouvement d'un liquide parfait soumis à la pesanteur. Détermination des lignes de courant. J. Math. pures appl. (5) **7**, 125—159 (1901).

SCHÖNFELD, J. C.: Propagation of tides and similar waves. 232 pp. (12 plates) 's-Gravenhage: Ministerie van Verkeer en Waterstaat. Staatsdrukkerij- en Uitgeverijbedrijf 1951.

SCHOOLEY, ALLEN H.: Profiles of wind-created water waves in the capillary-gravity transition region. J. Marine Res. **16**, 100—108 (1958).

SCHULTZ-GRUNOW, F.: Stabilität einer Flüssigkeitsoberfläche bei periodisch ausgeschalteter Erdschwere. Forsch. Ing.-Wes., Aufg. A **21**, No. 1, 27—28 (1955).

SEDOV, L. I.: The plane problem of gliding on the surface of a heavy fluid. (R) Trudy Konferentsii po Teorii Volnovogo Soprotivleniya, Moscow, 1936, pp. 7—30. Tsentral. Aèro-Gidrodinam. Inst., Moscow, 1937.

— On the theory of waves on the surface of an incompressible fluid. (R) Vestnik Moskov. Gos. Univ. **3**, No. 11, 71—77 (1948).

— Ploskie zadachi gidrodinamiki i aèrodinamiki. (Two-dimensional problems of hydrodynamics and aerodynamics.) 443 pp. Moscow: Gos. Izdat. Tekhn.-Teoret. Lit. 1950.

— Metody podobiya i razmernosti v mekhanike. 4th ed. 375 pp. Moscow: Gos. Izdat. Tekhn.-Teoret. Lit. 1957. Translation: Similarity and dimensional methods in mechanics. xvi + 363 pp. New York-London: Academic Press 1959.

SEKERZH-ZEN'KOVICH, YA. I.: On the theory of standing waves of finite amplitude on the surface of a heavy fluid. (R) Dokl. Akad. Nauk SSSR. (N.S.) **58**, 551—553 (1947).

— On the theory of standing waves of finite amplitude on the surface of a heavy fluid of finite depth. (R) Izv. Akad. Nauk SSSR., Ser. Geograf. Geofiz. **15**, 57—73 (1951).

— Composite standing waves of finite amplitude on the surface of a heavy fluid of infinite extent. (R) Izv. Akad. Nauk SSSR., Ser. Geofiz. **1951**, No. 5, 68—83.

— On the three-dimensional problem of standing waves of finite amplitude on the surface of a heavy liquid. (R) Dokl. Akad. Nauk SSSR. (N.S.) **86**, 35—38 (1952).

— On the theory of steady capillary-gravitational waves of finite amplitude. (R) Dokl. Akad. Nauk SSSR. **109**, 913—915 (1956).

SEN, B. M.: Waves in canals and basins. Proc. Lond. Math. Soc. (2) **26**, 363—376 (1927).

SEZAWA, KATSUTADA: Formation of deep-water waves due to subaqueous shocks. Bull. Earthquake Res. Inst. Tokyo Univ. **6**, 19—46 (1929).

— Formation of shallow-water waves due to subaqueous shocks. Bull. Earthquake Res. Inst. Tokyo Univ. **7**, 15—40 (1929).

SHAPIRO, A., and L. S. SIMPSON: The effect of a broken icefield on water waves. Trans. Amer. Geophys. Un. **34**, 36—42 (1953).

SHAROVA, I. F.: Investigation of particular solutions of the problem of waves on the surface of a viscous fluid. (R) Trudy Morsk. Gidrofiz. Inst. **8**, 33—43 (1956).

SHULEĬKIN, V. V.: Fizika morya. (Physics of the sea.) 3rd ed. 989 pp. Moscow: Izdatel'stvo Akad. Nauk SSSR 1953.

— Teoriya morskikh voln. (Theory of ocean waves.) Trudy Morsk. Gidrofiz. Inst. **9**, 143 pp. (1956).

SKJELBREIA, LARS: Gravity waves. STOKES' third-order approximation. Tables of functions. iii + 337 pp. Richmond, Calif.: Council on Wave Research 1959.

SLËZKIN, N. A. [SLIOSKINE]: Sur la question du mouvement plan des fluides pesants. C. R. Acad. Sci., Paris **201**, 644—647 (1935).

— Sur les ondes capillaires permanentes. C. R. Acad. Sci., Paris **201**, 707—709 (1935).

— On steady capillary waves. (R) Moskov. Gos. Univ. Uch. Zap. **7**, 71—102 (1937).

SOURIAU, J. M.: Une méthode générale de linéarisation des problèmes physiques. Actes du Colloque internationale de Mécanique, Poitiers, 1950, T. IV, pp. 251—268. Publ. Sci. Tech. du Ministère de l'Air, No. 261, Paris, 1952.

SPARENBERG, J.A.: The finite dock. Proceedings, Symposium on the behaviour of ships in a seaway, Wageningen, 1957, pp. 717—728.
SQUIRE, H.B.: The motion of a simple wedge along the water surface. Proc. Roy. Soc. Lond., Ser. A **243** (1958), 48—64 (1957).
SRETENSKIĬ, L.N.: On the motion of a glider on deep water. [English.] Izv. Akad. Nauk SSSR., Otdel. Mat. Estest. Nauk **1933**, 817—835.
— On waves at the surface of separation of two fluids with application to the phenomenon of "dead water". (R) Zh. Geofiz. **4**, 332—370 (1934).
— Teoriya volnovykh dvizhenii zhidkosti. (The theory of wave motions of a fluid.) 303 pp. Moscow-Leningrad: Ob"edinennoe Nauchno-Tekhnicheskoe Izdatel'stvo 1936.
— On the wave-making resistance of a ship moving along in a canal. Phil. Mag. (7) **22**, 1005—1013 (1936).
— A theoretical investigation on wave resistance. (R) Trudy Tsentral. Aèro-Gidrodinam. Inst., vyp. **319**, 56 pp. (1937).
— On damping of the vertical oscillations of the center of gravity of floating bodies. (R) Trudy Tsentral. Aèro-Gidrodinam. Inst., vyp. **330**, 12 pp. (1937).
— Motion of a cylinder under the surface of a heavy fluid. (R) Trudy Tsentral. Aèro-Gidrodinam. Inst., vyp. **346**, 27 pp. (1938). Translated in NACA Tech. Memo. **1335**, 37 pp. (1952).
— On the theory of wave resistance. (R) Trudy Tsentral. Aèro-Gidrodinam. Inst., vyp. **348**, 28 pp. (1939).
— On the theory of the glider. (R) Izv. Akad. Nauk SSSR., Otdel. Tekhn. Nauk **1940**, No. 7, 3—26.
— Concerning waves on the surface of a viscous fluid. (R) Trudy Tsentral. Aèro-Gidrodinam. Inst. No. **541**, 1—34 (1941).
— On the forces acting on a sphere moving on a circular path under the surface of a fluid. (R) Dokl. Akad. Nauk SSSR (N.S.) **54**, 777—778 (1946).
— On the waves produced by a ship moving in a circular path. (R) Izv. Akad. Nauk SSSR. Otd. Tekhn. Nauk **1946**, 13—22 (1 plate).
— On the waves generated by an underwater source situated under the surface of a sphere. (R) Izv. Akad. Nauk SSSR., Ser. Geograf. Geofiz. **13**, 473—496 (1949).
— The plane problem of propagation of waves in a basin, excited by an underwater source. (R) Izv. Akad. Nauk SSSR., Otd. Tekhn. Nauk **1950**, 321—332.
— Refraction and reflection of plane waves in liquids at a transition from one depth to another. (R) Izv. Akad. Nauk SSSR., Otd. Tekhn. Nauk **1950**, 1601—1614.
— Waves. (R) Mekhanika v SSSR za tridtsat' let, 1917—1947, pp. 279—299. Moscow-Leningrad: Gosudarstv. Izdat. Tekh.-Teoret. Lit. 1950.
— Survey of works on the theory of waves during the time from 1917 to 1949. (R) Moskov. Gos. Univ. Uch. Zap. **152**, Mekh. **3**, 76—98 (1951).
— The oscillation of a fluid in a movable basin. (R) Izv. Akad. Nauk SSSR., Otd. Tekhn. Nauk **1951**, 1483—1494.
— On a method of determination of waves of finite amplitude. (R) Izv. Akad. Nauk SSSR., Otd. Tekhn. Nauk **1952**, 688—698.
— On waves on the surface of separation of two liquids flowing at an angle to each other. (R) Izv. Akad. Nauk SSSR., Otd. Tekhn. Nauk **1952**, 1782—1787.
— Waves of finite amplitude arising from a periodically distributed pressure. (R) Izv. Akad. Nauk SSSR., Otd. Tekhn. Nauk **1953**, 505—511.
— The spatial problem of steady waves of finite amplitude. (R) Vestnik Moskov. Univ. **9**, No. 5, 3—12 (1954).
— The motion of a vibrator under the surface of a fluid. (R) Trudy Moskov. Mat. Obshch. **3**, 3—14 (1954).
— Propagation of waves of finite amplitude in a circular canal. (R) Trudy Morsk. Gidrofiz. Inst. Akad. Nauk SSSR. **6**, 3—9 (1955).
— The Cauchy-Poisson problem for the surface of separation of two flowing fluids. (R) Izv. Akad. Nauk SSSR., Ser. Geofiz. **1955**, 505—513.
— On the outward radiation of waves from a region subjected to an external pressure. (R) Prikl. Mat. Mekh. **20**, 349—361 (1956); remark, **21**, 595—596 (1957).
— Sur la résistance due aux vagues d'un fluide visqueux. Proceedings, Symposium on the behaviour of ships in a seaway, Wageningen, 1957, pp. 729—733.
— Computation of the tangential forces of wave resistance of a sphere moving on a circular path. (R) Trudy Morsk. Gidrofiz. Inst. **11**, 3—17 (1957).
STARR, V.P.: Water transport of surface waves. J. Meteorol. **2**, 129—131 (1945).
— Note on individual pressure changes in surface waves. J. Meteorol. **3**, 23—24 (1946).
— A momentum integral for surface waves in deep water. J. Marine Res. **6**, 126—135 (1947).

STARR, V. P.: Momentum and energy integrals for gravity waves of finite height. J. Marine Res. **6**, 175—193 (1947).
— Estimates of water transport produced by wave action. J. Marine Res. **7**, 1—9 (1948).
—, and G. W. PLATZMAN: The transmission of energy by gravity waves of finite height. J. Marine Res. **7**, 229—238 (1948).
STOKER, J. J.: Surface waves in water of variable depth. Quart. Appl. Math. **5**, 1—54 (1947).
— The formation of breakers and bores. The theory of nonlinear wave propagation in shallow water and open channels. With an appendix by K. O. FRIEDRICHS: On the derivation of the shallow water theory. Comm. Appl. Math. **1**, 1—80, 86—87 (1948); appendix, 81—85.
— Unsteady waves on a running stream. Comm. Pure Appl. Math. **6**, 471—481 (1953).
— Some remarks on radiation conditions. Proceedings of Symposia in Applied Mathematics, Vol. V, pp. 97—102. New York: McGraw-Hill 1954.
— On radiation conditions. Comm. Pure Appl. Math. **9**, 577—595 (1956).
— Water waves. The mathematical theory with applications. xxviii + 567 pp. New York: Interscience Publishers 1957.
STOKES, G. G.: Report on recent researches in hydrodynamics. Report of the sixteenth meeting of the British Association for the Advancement of Science, Southampton, 1846, pp. 1—20. London: John Murray 1847. = Mathematical and Physical Papers, Vol. 1, pp. 157—187, Cambridge 1880.
— On the theory of oscillatory waves. Trans. Cambridge Phil. Soc. **8**, 441—455 (1849) = Mathematical and Physical Papers, Vol. 1, pp. 197—229, Cambridge, 1880.
— Supplement to a paper on the theory of oscillatory waves. Mathematical and Physical Papers, Vol. 1, pp. 314—326, Cambridge, 1880.
— On the highest wave of uniform propagation. Proc. Cambridge Phil. Soc. **4**, 361—365 (1883) = Mathematical and Physical Papers, Vol. 5, pp. 140—145, Cambridge, 1905.
— Note on the theory of the solitary wave. Phil. Mag. (5) **31**, 314—316 (1891) = Mathematical and Physical Papers, Vol. 5, pp. 160—162, Cambridge, 1905.
— On the maximum wave of uniform propagation. Being a second supplement to a paper on the theory of oscillatory waves. Mathematical and Physical Papers, Vol. 5, pp. 146 to 159 (1880), Cambridge, 1905.
STORCHI, E.: Legame fra la forma del pelo libero e quella del recipiente nelle oscillazioni di un liquido. Ist. Lombardo Sci. Lett. Rend. Cl. Sci. Mat. Nat. (3) **13** (**82**), 95—112 (1949).
— Piccole oscillazioni dell'acqua contenuta da pareti piane. Atti Accad. Naz. Lincei, Rend. Cl. Sci. Fis. Mat. Nat. (8) **12**, 544—552 (1952).
STRUIK, D. J.: Détermination rigoureuse des ondes irrotationnelles permanentes dans un canal à profondeur finie. Math. Ann. **95**, 595—634 (1926).
STRUTT, JOHN WILLIAM (Baron RAYLEIGH): On waves. Phil. Mag. (5) **1**, 257—279 (1876).
— On progressive waves. Proc. Lond. Math. Soc. (1) **9**, 21—26 (1877).
— On maintained vibrations. Phil. Mag. (5) **15**, 229—235 (1883).
— On the crispations of fluid resting upon a vibrating support. Phil. Mag. (5) **16**, 50—58 (1883).
— The form of standing waves on running water. Proc. Lond. Math. Soc. **15**, 69—78 (1883 to 84).
— On the tension of water surfaces, clean and contaminated, investigated by the method of ripples. Phil. Mag. (5) **30**, 386—400 (1890).
— On the instantaneous propagation of disturbance in a dispersive medium, exemplified by waves on water deep and shallow. Phil. Mag. (6) **18**, 1—6 (1909).
— Hydrodynamical notes. Phil. Mag. (6) **21**, 177—195 (1911).
— On the theory of long waves and bores. Proc. Roy. Soc. Lond., Ser. A **90**, 324—328 (1914).
— Deep water waves, progressive or stationary, to the third order of approximation. Proc. Roy. Soc. Lond., Ser. A **91**, 345—353 (1915).
— On periodic irrotational waves at the surface of deep water. Phil. Mag. (6) **33**, 381—389 (1917).
SYÔNO, S.: On the waves caused by a sudden deformation of a finite portion of the bottom of a sea of uniform depth. Geophys. Mag., Tokyo **10**, 21—41 (1936).
TAIT, P. G.: Note on ripples in a viscous liquid. Proc. Roy. Soc. Edinburgh **17** (1890), 110 to 115 (1891).
TAKAHASHI, RYÛTARÔ: On seismic sea waves caused by deformations of the sea bottom. I, II, III. [Japanese.] Bull. Earthquake Res. Inst., Tokyo **20**, 375—400 (1942); **23**, 23—25 (1945); **25**, 5—8 (1947).
TAMIYA, SHIN: On the dynamical effect of free water surface. [Japanese.] J. Zôsen Kyôkai **103**, 59—67 (1958).
TAYLOR, G. I.: An experimental study of standing waves. Proc. Roy. Soc. Lond., Ser. A **218**, 44—59 (1953).
TCHAPLYGUINE: See CHAPLYGIN.

TCHEN, CHAN-MOU: Interfacial waves in viscoelastic media. J. Appl. Phys. **27**, 431—434 (1956).
— Stability of oscillations of superposed fluids. J. Appl. Phys. **27**, 760—767 (1956).
— Approximate theory on the stability of interfacial waves between two streams. J. Appl. Phys. **27**, 1533—1536 (1956).
TERAZAWA, K.: On deep-sea water waves caused by a local disturbance on or beneath the surface. Proc. Roy. Soc. Lond., Ser. A **92** (1916), 57—81 (1915).
TER-KRIKOROV, A. M.: Exact solution of the problem of motion of a vortex under the surface of a fluid. (R) Izv. Akad. Nauk SSSR., Ser. Mat. **22**, 177—200 (1958).
THOMSON, WILLIAM (Baron KELVIN): Influence of wind and capillarity on waves in water supposed frictionless. Phil. Mag. (4) **42**, 368—377 (1871) = Mathematical and Physical Papers, Vol. 4, pp. 76—85, Cambridge, 1910.
— Ripples and waves. Nature, Lond. **5**, 1—3 (1871) = Mathematical and Physical Papers, Vol. 4, pp. 86—92, Cambridge, 1910.
— On stationary waves in flowing water. I, II, III, IV. Phil. Mag. (5) **22**, 353—357, 445 to 452, 515—530 (1886); **23**, 52—57 (1887) = Mathematical and Physical Papers, Vol. 4, pp. 270—302, Cambridge, 1910.
— On the waves produced by a single impulse in water of any depth, or in a dispersive medium. Phil. Mag. (5) **23**, 252—255 (1887) = Mathematical and Physical Papers, Vol. 4, pp. 303 to 306, Cambridge, 1910.
— On deep-water two-dimensional waves produced by any given initiating disturbance. Phil. Mag. (6) **7**, 609—620 (1904) = Mathematical and Physical Papers, Vol. 4, pp. 338 to 350, Cambridge, 1910.
— On the front and rear of a free procession of waves in deep water. Phil. Mag. (6) **8**, 454 to 470 (1904) = Mathematical and Physical Papers, Vol. 4, pp. 351—367, Cambridge, 1910.
— Deep water ship-waves. Phil. Mag. (6) **9**, 733—757 (1905) = Mathematical and Physical Papers, Vol. 4, pp. 368—393, Cambridge, 1910.
— Deep sea ship-waves. Phil. Mag. (6) **11**, 1—25 (1906) = Mathematical and Physical Papers, Vol. 4, pp. 394—418, Cambridge, 1910.
— Initiation of deep-sea waves of three classes: (1) from a single displacement; (2) from a group of equal and similar displacements; (3) by a periodically varying surface-pressure. Phil. Mag. (6) **13**, 1—36 (1907) = Mathematical and Physical Papers, Vol. 4, pp. 419—456, Cambridge, 1910.
THORADE, H.: Probleme der Wasserwellen. (Probleme der kosmischen Physik, XIII u. XIV.) viii + 219 pp. + 16 plates + 4 charts. Hamburg: H. Grand 1931.
THORNE, R. C.: Multipole expansions in the theory of surface waves. Proc. Cambridge Phil. Soc. **49**, 707—716 (1953).
TIKHONOV, A. I.: The two-dimensional problem of motion of a wing under the surface of a heavy fluid. (R) Izv. Akad. Nauk SSSR., Otd. Tekhn. Nauk **1940**, No. 4, 57—78.
TIMMAN, R., and G. VOSSERS: On the MICHELL's expression for the velocity potential of the flow around a ship. Internat. Shipbldg. Progr. **2**, No. 6, 96—102 (1955).
TOMOTIKA, S.: On the instability of a cylindrical thread of a viscous liquid surrounded by another viscous fluid. Proc. Roy. Soc. Lond., Ser. A **150**, 322—337 (1935).
TONOLO, A.: Nuova risoluzione del problema delle onde di Poisson-Cauchy. Atti Ist. Veneto Sci. Lett. Arti (8) **73**, parte 2a, 545—571 (1913—14).
TRUESDELL, C.: Two measures of vorticity. J. Rational Mech. Anal. **2**, 173—217 (1953).
TYLER, E.: Measurement of surface tension by the ripple method. Phil. Mag. (7) **31**, 209—221 (1941).
UNOKI, SANAE, and MASITO NAKANO: On the Cauchy-Poisson waves caused by the eruption of a submarine volcano. I, II. Oceanogr. Mag. **4**, 119—141 (1953); **5**, 1—13 (1953).
URSELL, F.: The effect of a fixed vertical barrier on surface waves in deep water. Proc. Cambridge Phil. Soc. **43**, 374—382 (1947).
— On the waves due to the rolling of a ship. Quart. J. Mech. Appl. Math. **1**, 246—252 (1948).
— On the heaving motion of a circular cylinder on the surface of a fluid. Quart. J. Mech. Appl. Math. **2**, 218—231 (1949).
— On the rolling motion of cylinders in the surface of a fluid. Quart. J. Mech. Appl. Math. **2**, 335—353 (1949).
— Surface waves on deep water in the presence of a submerged circular cylinder. I, II. Proc. Cambridge Phil. Soc. **46**, 141—152, 153—158 (1950).
— Trapping modes in the theory of surface waves. Proc. Cambridge Phil. Soc. **47**, 347—358 (1951).
— Edge waves on a sloping beach. Proc. Roy. Soc. Lond., Ser. A **214**, 79—97 (1952).
— Mass transport in gravity waves. Proc. Cambridge Phil. Soc. **49**, 145—150 (1953).
— The long-wave paradox in the theory of gravity waves. Proc. Cambridge Phil. Soc. **49**, 685—694 (1953).

URSELL, F.: Short surface waves due to an oscillating immersed body. Proc. Roy. Soc. Lond., Ser. A**220**, 90—103 (1953).
— Water waves generated by oscilating bodies. Quart. J. Mech. Appl. Math. **7**, 427—437 (1954).
— Wave generation by wind. Surveys in mechanics, pp. 216—249. Cambridge 1956.
— On KELVIN's ship-wave pattern. J. Fluid Mech. (1960).
—, R. G. DEAN and Y. S. YU: Forced small-amplitude waves: a comparison of theory and experiment. J. Fluid Mech. **7**, 33—52 (1959).
VENTURELLI, LUCIA: Oscillazioni di piccola ampiezza in un fluido compressibile pesante. Rend. Sem. Mat. Univ. Padova **5**, 1—23 (1934).
VERGNE, H.: Ondes liquides de gravité. Mémorial des Sciences Mathématiques, fasc. 34. 58 pp. Paris: Gauthier-Villars 1928.
VILLAT, HENRI: Sur l'écoulement des fluides pesants. Ann. Sci. Ecole norm. Sup. (3) **32**, 177—214 (1915).
— Quelques récents progrès des théories hydrodynamiques. Bull. Sci. Math. (2) **42**, 43—60, 72—92 (1918).
VINT, J.: Surface waves on limited sheets of water. Proc. Lond. Math. Soc. (2) **22** (1924), 1—14 (1923).
VITOUSEK, M. J.: Some flows in a gravity field satisfying the exact free surface condition. Applied Mathematics and Statistics Laboratory, Stanford Univ., Calif., Tech. Rep. No. **25**, 87 pp. (1954).
VOĬT, S. S. (WOIT): Waves on the surface of a fluid arising from a moving periodic system of pressures. (R) Prikl. Mat. Mekh. **21**, 21—26 (1957).
— Waves on the surface of a fluid arising from a variable system of pressures. (R) Trudy Morsk. Gidrofiz. Inst. **10**, 3—9 (1957).
— Passage of plane waves through a shallow-water zone. (R) Trudy Morsk. Gidrofiz. Inst. **15**, 33—42 (1959).
— Waves on the surface of separation of two fluids arising from a moving periodic system of pressures. (R) Trudy Morsk. Gidrofiz. Inst. **17**, 33—40 (1959).
VOĬTSENYA, V. S.: The two-dimensional problem of the oscillations of a body under the surface of separation of two fluids. (R) Prikl. Mat. Mekh. **22**, 789—803 (1958).
VOLTERRA, VITO: Sur la théorie des ondes liquides et la méthode de GREEN. J. Math. pures appl. (9) **13**, 1—18 (1934).
WEBER, ERNST HEINRICH, u. WILHELM WEBER: Wellenlehre auf Experimente gegründet oder über die Wellen tropfbarer Flüssigkeiten mit Anwendung auf die Schall- und Lichtwellen. xxviii + 574 pp. + 2 folded inserts + 18 plates. Leipzig 1825.
WEHAUSEN, J. V.: Wave resistance of thin ships. Symposium on naval hydrodynamics, Washington, D.C., 1956, pp. 109—133; discussion, 133—137. Washington, D.C.: National Academy of Sciences—National Research Council 1957.
WEINBLUM, G. P.: Rotationskörper geringsten Wellenwiderstandes. Ing.-Arch. **7**, 104—117 (1936).
— Analysis of wave resistance. The David W. Taylor Model Basin, Washington, D.C., Rep. **710**, 102 pp. (1950).
— The wave resistance of bodies of revolution. Appendix II by J. BLUM. The David W. Taylor Model Basin, Washington, D.C., Rep. **758**, 58 pp. (1951).
— A systematic evaluation of MICHELL's integral. The David W. Taylor Model Basin, Washington, D.C., Rep. **886**, 59 pp. (1955).
— Schiffe geringsten Widerstandes. Jb. schiffbautech. Ges. **51**, 175—204 (1957); Erörterung, 205—214.
—, H. AMTSBERG u. W. BOCK: Versuche über den Wellenwiderstand getauchter Rotationskörper. Schiffbau, Schiffahrt, Hafenbau **37**, 411—419 (1936). Translated in David Taylor Model Basin Rep. T-**234**, 22 pp. (1950).
—, J. J. KENDRICK and M. A. TODD: Investigation of wave effects produced by a thin body—TMB Model 4125. The David W. Taylor Model Basin, Washington, D.C., Rep. **840**, 14 pp. (1952).
WEINGARTEN, J.: Ein einfaches Beispiel einer stationären und rotationslosen Bewegung einer tropfbaren schweren Flüssigkeit mit freier Begrenzung. Verh. 3. Internat. Math.-Kongr., Heidelberg, 1904, pp. 409—413.
WEINSTEIN, A.: Sur la vitesse de propagation de l'onde solitaire. Atti Accad. Lincei, Rend. Cl. Sci. Fiz. Mat. Nat. (6) **3**, 463—468 (1926).
— Sur la vitesse de propagation de l'onde solitaire. Proc. 2nd Internat. Congr. Appl. Mech., Zürich, 1926, Vol. 1, pp. 445—448.
— Sur un problème aux limites dans une bande indéfinie. C. R. Acad. Sci., Paris **184**, 497 to 499 (1927).
— On surface waves. Canad. J. Math. **1**, 271—278 (1949).

WEITZ, MORTIMER and JOSEPH B. KELLER: Reflection of water waves from floating ice in water of finite depth. Comm. Pure Appl. Math. **3**, 305—318 (1950).

WENDT, H.: Das Problem der Jungfernquelle. Z. angew. Math. Mech. **32**, 338—358 (1952).

WIEGHARDT, K.: Über Kapillarwellen mit Oberflächenzähigkeit. Physik. Z. **44**, 101—108 (1943).

WIEN, W.: Lehrbuch der Hydrodynamik. xiv + 319 pp. Leipzig: S. Hirzel 1900.

WIGLEY, W.C.S.: Ship wave resistance. Proc. 3rd Internat. Congr. Appl. Mech., Stockholm, 1930, Vol. 1, pp. 58—73.

— Ship wave-resistance. Progress since 1930. Trans. Inst. Nav. Arch. **77**, 223—236 (1935) (plates XXVI, XXVII); discussion, 237—244.

— L'état actuel des calculs de résistance de vagues. Bull. Assoc. Tech. Mar. Aéro. **48**, 513 to 545 (1949).

WILTON, J.R.: On the highest wave in deep water. Phil. Mag. (6) **26**, 1053—1058 (1913).

— On deep water waves. Phil. Mag. (6) **27**, 385—394 (1914).

— On ripples. Phil. Mag. (6) **29**, 688—700 (1915).

WU, T. YAO-TSU: A theory for hydrofoils of finite span. J. Math. Phys. **33**, 207—248 (1954).

— Water waves generated by the translatory and oscillatory surface disturbance. Engineering Division, California Institute of Technology, Pasadena, Calif., Rep. No. **85-3**, 36 pp. (1957).

—, and R.E. MESSICK: Viscous effect on surface waves generated by steady disturbances. Engineering Division, California Institute of Technology, Pasadena, Calif., Rep. No. **85-8**, 31 pp. (1958).

WUEST, WALTER: Beitrag zur Entstehung von Wasserwellen durch Wind. Z. angew. Math. Mech. **29**, 239—252 (1949).

WURTELE, M.G.: The transient development of a lee wave. J. Marine Res. **14**, 1—12 (1955).

YAMADA, HIKOJI: Highest waves of permanent type on the surface of deep water. Rep. Res. Inst. Appl. Mech., Kyushu Univ. **5**, No. 18, 37—52 (1957).

— On the highest solitary wave. Rep. Res. Inst. Appl. Mech., Kyushu Univ. **5**, No. 18, 53—67 (1957).

— On approximate expressions of solitary wave. Rep. Res. Inst. Appl. Mech. **6**, No. 21, 35 to 47 (1958).

Sachverzeichnis.

(Deutsch-Englisch.)

Bei gleicher Schreibweise in beiden Sprachen sind die Stichwörter nur einmal aufgeführt.

Subject Index.

(English-German.)

Where English and German spelling of a word is identical the German version is omitted.

Table des matières

pour la contribution écrite en français:

Henri Cabannes: Théorie des ondes de choc.

Made in the USA
Monee, IL
16 September 2026

63296740R00453